AF598550

The Geochemistry and Geophysics of the Antarctic Mantle

Geological Society books refereeing procedures

This volume is published under an agreement between the Scientific Committee on Antarctic Research and the Geological Society of London and arises from the SCAR Expert Group on Antarctic Volcanism (AntVolc) and the scientific research programme Solid Earth Response and influence on Cryospheric Evolution (SERCE) and INStabilities and Thresholds in ANTarctica (INSTANT).

GSL is the publisher of choice for books related to SCAR's geoscience activities, and SCAR receives a fee for all books published under this agreement.

Books published under this agreement are subject to the Society's standard rigorous proposal and manuscript review procedures.

Submitting a book proposal
More information about submitting a proposal and producing a book for the Society can be found at https://www.geolsoc.org.uk/proposals

It is recommended that reference to all or part of this book should be made in one of the following ways:

Martin, A. P. and van der Wal, W. (eds) 2023. *The Geochemistry and Geophysics of the Antarctic Mantle*. Geological Society, London, Memoirs, **56**, https://doi.org/10.1144/M56

Wiens, D. A., Shen, W. and Lloyd, A. J. 2023. The seismic structure of the Antarctic upper mantle. *Geological Society, London, Memoirs*, **56**, 195–212, https://doi.org/10.1144/M56-2020-18

Geological Society Memoir No. 56

The Geochemistry and Geophysics of the Antarctic Mantle

Edited by

A. P. Martin
GNS Science, New Zealand

and

W. van der Wal
Delft University of Technology, the Netherlands

2023
Published by
The Geological Society
London

The Geological Society of London

The Geological Society of London is a not-for-profit organisation, and a registered charity (no. 210161). Our aims are to improve knowledge and understanding of the Earth, to promote Earth science education and awareness, and to promote professional excellence and ethical standards in the work of Earth scientists, for the public good. Founded in 1807, we are the oldest geological society in the world. Today, we are a world-leading communicator of Earth science – through scholarly publishing, library and information services, cutting-edge scientific conferences, education activities and outreach to the general public. We also provide impartial scientific information and evidence to support policy-making and public debate about the challenges facing humanity. For more about the Society, please go to https://www.geolsoc.org.uk/

The Geological Society Publishing House (Bath, UK) produces the Society's international journals and books, and acts as European distributor for selected publications of the American Association of Petroleum Geologists (AAPG), the Geological Society of America (GSA), the Society for Sedimentary Geology (SEPM) and the Geologists' Association (GA). GSL Fellows may purchase these societies' publications at a discount. The Society's online bookshop is at https://www.geolsoc.org.uk/bookshop

To find out about joining the Society and benefiting from substantial discounts on publications of GSL and other Societies go to https://www.geolsoc.org.uk/membership or contact the Fellowship Department at: The Geological Society, Burlington House, Piccadilly, London W1J 0BG: Tel. +44 (0)20 7434 9944; Fax +44 (0)20 7439 8975; E-mail: enquiries@geolsoc.org.uk

For information about the Society's meetings, go to https://www.geolsoc.org.uk/events. To find out more about the Society's Corporate Patrons Scheme visit https://www.geolsoc.org.uk/patrons

Proposing a book
If you are interested in proposing a book then please visit: https://www.geolsoc.org.uk/proposals

Published by The Geological Society from:
The Geological Society Publishing House, Unit 7, Brassmill Enterprise Centre, Brassmill Lane, Bath BA1 3JN, UK

The Lyell Collection: www.lyellcollection.org
Online bookshop: www.geolsoc.org.uk/bookshop
Orders: Tel. +44 (0)1225 445046, Fax +44 (0)1225 442836

British Library Cataloguing in Publication Data

A catalogue record for this book is available from the British Library.
ISBN 978-1-78620-467-7
ISSN 0435-4052

Distributors

For details of international agents and distributors see:
www.geolsoc.org.uk/agentsdistributors

Typeset by Nova Techset Private Limited, Bengaluru & Chennai, India
Printed and bound by CPI Group (UK) Ltd, Croydon CR0 4YY, UK

Contents

Acknowledgements

This volume was inspired through discussions with Terry Wilson and their efforts to bring together people with a wide range of specialities to better understand the Antarctic mantle. The scientific committee on Antarctic research (SCAR) open science conferences were an excellent forum for developing the cross-disciplinary ideas necessary for this volume. There were numerous conversations that helped build a fascination with the Antarctic mantle and contributed to this volume, including with Alan Cooper, Jörg Ebbing, Fausto Ferracioli, Erik Ivins, Matt King, Bruce Marsh, Richard Price, John Smellie and Pippa Whitehouse.

We are indebted to all the authors who have generously contributed to this Memoir, and to the following reviewers, as well as to the anonymous reviewers, who freely donated their time:

Alan Aitken
Joaquin Bastias
Paterno Castillo
Massimo Coltorti
Alan Cooper
Fausto Ferraccioli
Stephen Foley
John Gamble
Monica Handler
Mark Hoggard
Audrey Huerta
Tom Jordan
Matt King
Phillip Leat
Yucheng Lin
Wesley LeMasurier
Sisir Mondal
Richard Moscati
Károly Németh
Kurt Panter
Massimo Pompilio
Anya Reading
Sergio Rocchi
Kate Selway
Norman Sleep
John Smellie
Andreas Stracke
Pippa Whitehouse
Douglas Wiens
Douglas Wilson
Gerhard Wörner

We thank the entire Geological Society Memoir editorial team (Tamzin Anderson, Jo Armstrong, Caroline Astley, David Boyt, Karen Coldwell, Philip Leat, Samuel Lickiss, Bethan Philips, Danielle Tremeer) for their tireless efforts to guide this through to publication, and we thank Randell Stephenson for advice on editorial decisions. This volume is a contribution to the aims and objectives of the SCAR Expert Group on Antarctic Volcanism (AntVolc: https://www.scar.org/science/antvolc/home/) and the scientific research programme Solid Earth Response and influence on Cryospheric Evolution (SERCE: https://www.scar.org/science/serce/serce/) and INStabilities and Thresholds in ANTarctica (INSTANT: https://www.scar.org/science/instant/home/).

Adam P. Martin (a.martin@gns.cri.nz)
Wouter van der Wal (w.vanderwal@tudelft.nl)

An introduction to the geochemistry and geophysics of the Antarctic mantle

Adam P. Martin[1]*, Wouter van der Wal[2] and Bas de Boer[3]

[1]GNS Science, Private Bag 1930, Dunedin, New Zealand

[2]Faculty of Aerospace Engineering, Delft University of Technology, Kluyverweg 1, 2629 HS Delft, The Netherlands

[3]Earth and Climate Cluster, Faculty of Science, Vrije Universiteit Amsterdam, Amsterdam, the Netherlands

APM, 0000-0002-4676-8344

*Correspondence: a.martin@gns.cri.nz

Abstract: The Antarctic mantle, bounded between the core and the Mohorovičić discontinuity, is one of the most difficult targets of study on Earth because of ice cover and rare outcrops. A multidisciplinary approach is adopted in this volume, using petrology, geochemistry, remote-sensed data and geodesy to characterize the Antarctic mantle. This characterization has application to rates of glacial isostatic adjustment, heat flow, sea-level rise and tectonics. It places the Antarctic mantle domain in a global framework on a scale not attempted before. In this chapter we review the historical development of mantle studies in Antarctica, outline current research directions, introduce the volume chapters and provide a summary and outlook.

Preamble

The Antarctic mantle plays a key role in understanding the driving forces of global plate tectonics (Turcotte and Oxburgh 1967). Mantle attached to plates generates pull forces that account for around 50% of the driving force on plate tectonics (Conrad Clinton and Lithgow-Bertelloni 2002; Bercovici 2003). The Antarctic Plate borders several major plates, and traces of all the past supercontinents are contained in the Antarctic subsurface (Harley *et al.* 2013). The mantle controls large-scale uplift and subsidence of the surface topography and the lithospheric heat flow of Antarctica, thereby influencing the evolution of the Antarctic Ice Sheet via change in surface elevation and melting (e.g. Shapiro and Ritzwoller 2004; Whitehouse *et al.* 2012). Thus, the Antarctica mantle plays a central role in questions of large scientific and societal relevance; however, our understanding of it remains poor due to the inaccessibility of the Antarctic continent. A better understanding of the Antarctic mantle requires novel integration of multiple research fields.

This Memoir is the first dedicated to understanding the mantle properties of the whole Antarctic continent. It combines petrographical, geochemical, remote-sensed and geophysical techniques. The first eight chapters of this volume cover studies from mantle xenoliths and igneous rocks, this is followed by eight chapters covering geophysical studies and the final chapter is a summary. The volume aims to characterize the Antarctic mantle and to discuss its role in the dynamics that shape the Antarctic surface and ice sheets. The field of Earth sciences requires multidisciplinary efforts, and the difficulties of remoteness, weather and ice cover make this especially true when addressing the most important scientific questions in Antarctica. This volume has adopted a multidisciplinary approach in combining chapters from different fields of Earth sciences, namely geochemistry and geophysics, and it is hoped that this perspective will highlight how combining different measurement types allows better characterization of the Antarctic mantle and its role in geodynamic processes.

The top of the mantle is defined by a pronounced change in P-wave velocities from 6.7–7.2 km s^{-1} in the lower crust to 7.6–8.6 km s^{-1} in the mantle, this is known as the Mohorovičić discontinuity (Moho: e.g. Meissner 1986; Grad *et al.* 2009) (Fig. 1). In Antarctica, the onshore Moho depth varies between approximately 20 and 60 km (Pappa *et al.* 2019*a*). The Moho can also be associated with a chemical, thermal, mineralogical and density change from material rich in silica and magnesia (simatic) in the lower crust to peridotitic and pyroxenitic material in the upper mantle (Herzberg *et al.* 1983; Griffin and O'Reilly 1987). The mantle is bounded by the Moho and the core (Fig. 1). It occurs in several layers marked by boundaries, including the lithosphere–asthenosphere boundary, which is a rheological, thermal or seismic boundary separating a relatively rigid (lithospheric) layer and a weak asthenosphere (Karato *et al.* 2008; Eaton *et al.* 2009; Fischer *et al.* 2010). The depth of the lithosphere–asthenosphere transition is influenced by temperature, water content, chemical composition, extent of partial melt and grain size (Kawakatsu *et al.* 2009; Green *et al.* 2010; O'Reilly and Griffin 2010). Deeper boundaries in the mantle have clearer seismic expressions. Seismically defined discontinuities occur at depths of 410 and 660 km in the mantle (Adams 1971; Revenaugh and Jordan 1991; Shearer and Flanagan 1999), and are associated with phase changes of olivine to wadsleyite (Katsura and Ito 1989) and of ringwoodite to perovskite and magnesiowustite (Ito and Takahashi 1989), respectively. The 660 km discontinuity marks the boundary between the upper and lower mantle (e.g. Kaufmann and Lambeck 2000; Paulson *et al.* 2007; Čížková *et al.* 2012).

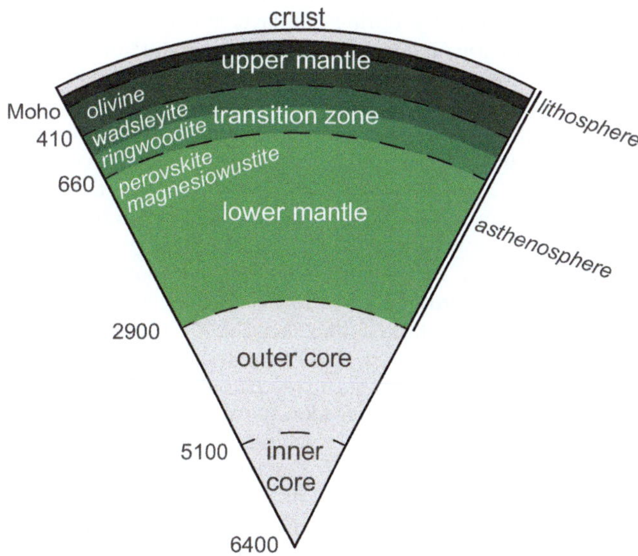

Fig. 1. Schematic cross-section through the Earth showing the crust, mantle and core, and highlighting significant boundaries and phase changes within the mantle (depth in km).

From: Martin, A. P. and van der Wal, W. (eds) 2023. *The Geochemistry and Geophysics of the Antarctic Mantle*. Geological Society, London, Memoirs, **56**, 1–16,
First published online 14 November 2022, https://doi.org/10.1144/M56-2022-21

This volume brings together studies on mantle xenoliths included in igneous rocks, studies on primitive igneous rocks that can inform us about mantle properties, studies that interpret geophysical measurements and studies on geodynamic processes. Mantle xenoliths record information about the lithospheric mantle, whilst primitive igneous rocks include information about the asthenospheric upper mantle. Geophysical techniques include seismology, gravity and magnetic measurements. Together, they can provide information on structure from the uppermost mantle down to the core–mantle boundary. The depth sensitivity of each measurement type varies. Geodynamic processes include mantle convection, tectonics, glacial isostatic adjustment (GIA) and post-seismic deformation that are controlled by rock properties recorded in geochemical studies and with geophysical measurements. Finally, surface changes driven by underlying mantle convection, tectonics and GIA can be detected by geodetic techniques.

This chapter provides a brief historical perspective of Antarctic mantle research. We present a review of the most important Earth sciences questions that drive the study of the Antarctica mantle. An overview of the chapters in this volume is provided here, along with a summary and outlook.

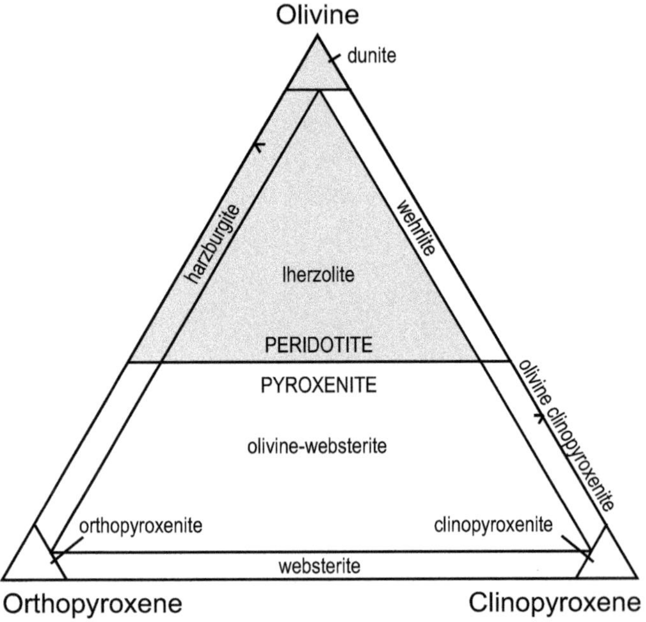

Fig. 2. Olivine–pyroxene ternary diagram. Source: Le Maitre *et al.* (2002).

Background

Mantle geochemistry

Peridotite and pyroxenite xenoliths are fragments of the mantle entrained during volcanic eruptions. They are often an important component in the primitive alkalic rocks commonly found in Antarctica (Smellie *et al.* 2021). It is important to study mantle xenoliths as some information is only determined from physical samples of the mantle such as the mineralogy, the variety of rock types and the interrelationships between rock types (Nixon 1987*a*; Shervais 1988). Mantle xenoliths are critical in understanding the mantle as often they are the only direct sample of mantle available for many areas (Pearson *et al.* 2003). Because of the speed at which mantle xenoliths are brought to the surface they freeze in the mineralogical and chemical signatures of their depth of origin, in contrast to tectonically emplaced slices of the mantle that typically equilibrate extensively during emplacement (Pearson *et al.* 2003). Information on the pressure and temperature of the lithospheric mantle can be derived via thermobarometry studies. Thermobarometry is the experimental study of coexisting minerals that can be utilized to refine the pressure and temperature that mantle xenoliths last crystallized (e.g. Wood and Banno 1973; Macgregor 1974). The results may define a ‘palaeogeotherm’, which is a pressure–temperature curve through the lithosphere (Howells and O’Hara 1978; Berg *et al.* 1989). These curves are similar in shape to modern-day geotherm curves, providing confidence in their interpretation (Condie 2022). Mantle xenoliths can also be studied for the lattice-preferred orientation (LPO) of olivine and pyroxene. These LPO studies can provide evidence of past deformation events, estimates of the propagation speed of seismic waves in the lithospheric mantle and the degree of mantle anisotropy (Mainprice and Silver 1993; Tommasi *et al.* 1999; Kovács *et al.* 2018). Thus, the study of mantle xenoliths shows the composition of the Earth’s lithosphere, how the lithosphere varies from place to place and how the lithosphere evolved (Nixon 1987*a*). The study of mantle xenoliths is vital in building accurate geochemical and physical models of the Earth’s lithosphere (Kovács *et al.* 2018).

Peridotite and pyroxenite xenoliths can be defined based on their modal proportion of olivine, clinopyroxene and orthopyroxene (Fig. 2). Common rock names for peridotite are lherzolite (olivine + clinopyroxene + orthopyroxene), harzburgite (olivine + orthopyroxene) and dunite (olivine). Common rock names for pyroxenite are wehrlite (olivine + clinopyroxene), clinopyroxenite (clinopyroxene), websterite (clinopyroxene + orthopyroxene) and orthopyroxenite (orthopyroxene). An idealized upper-mantle assemblage (Fig. 3) can be drawn from high-pressure experimental data, mantle xenolith compositions and seismic velocity measurements. Idealized geotherms for a continental shield, ocean plate and upwelling mantle beneath an ocean ridge are also shown (Fig. 3). The aluminosilicate assemblage of the lherzolite rock type (olivine + clinopyroxene + orthopyroxene) changes with depth and temperature (Fig. 3) from garnet (deepest) to spinel to plagioclase (shallowest). All three aluminosilicate lherzolite phases are recorded in Antarctica. Spinel lherzolite is the most common. Garnet–spinel lherzolite is reported from East Antarctica and plagioclase-bearing spinel lherzolite is reported from the West Antarctic Rift System and elsewhere (Fig. 4).

Peridotite xenoliths were collected from Antarctica and described during the very earliest Antarctic scientific expeditions (Prior 1902, 1907; Thomson 1916). Around this time, workers on Indian and African specimens first began hypothesizing that peridotite xenoliths were associated with a layer beneath the crust: i.e. the mantle (Fermor 1913; Wagner 1914). This concept was not applied to Antarctic peridotite xenoliths until the mid-twentieth century. For example, Talbot *et al.* (1963) found that their study of peridotite and dunite xenoliths from Kerguelen Island was inadequate to distinguish between a cognate or mantle origin but Forbes (1963), using strain indicators and preferred indicatrix orientations, preferred a non-cognate, mantle origin for the peridotite and dunite xenolith samples they studied from south Victoria Land. This interpretation has since been universally adopted.

Since the 1960s, the chemistry, isotopes and physical properties of Antarctic mantle xenoliths have been characterized to understand the petrogenesis, uplift and thermal regime of the mantle and suprajacent igneous rocks (e.g. Gamble *et al.* 1988; Zipfel and Wörner 1992; Warner and Wasilewski 1995; Wörner and Zipfel 1996; Handler *et al.* 2003; Coltorti *et al.* 2004; Foley *et al.* 2006; Perinelli *et al.* 2012; Bonadiman *et al.* 2014; Martin *et al.* 2015; Day *et al.* 2019). Other notable

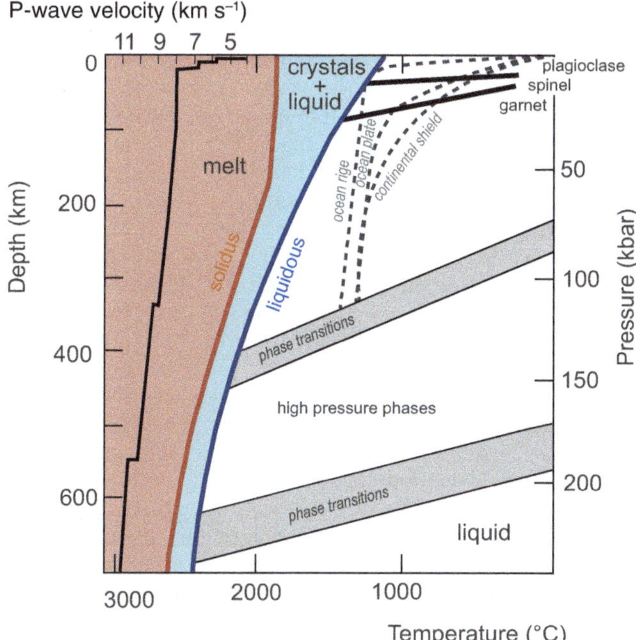

Fig. 3. Schematic phase diagram for the upper mantle with typical geotherms plotted for comparison, after Wyllie (1981). Average P-wave velocities are also shown for reference (Dziewonski and Anderson 1981; Kennett *et al.* 1995).

developments in understanding the Antarctic mantle include proposing a mantle plume beneath Marie Byrd Land (Kyle *et al.* 1994; Weaver *et al.* 1994; LeMasurier and Landis 1996; Storey *et al.* 1999; Lloyd *et al.* 2015) based on crustal doming and the distribution of volcanic rocks and thermal anomalies. The existence of a mantle plume beneath south Victoria Land (Kyle *et al.* 1992; Phillips *et al.* 2018) is much less certain (see **Martin *et al.* 2021*a***, *b* for a discussion).

Studies of primitive igneous rocks in Antarctica have also been used to interpret chemical and isotopic heterogeneity in the Antarctic mantle (Sun and Hanson 1975; Hart *et al.* 1995, 1997; Wörner 1999; Panter *et al.* 2006; Sims and Hart 2006; Cooper *et al.* 2007; Nardini *et al.* 2009; Martin *et al.* 2013), and mantle metasomatism (Sheraton and Black 1982; Futa and Le Masurier 1983; Sheraton 1983). Isotopes of Pb, Sr and Nd were measured in primitive igneous rocks on the Antarctic Peninsula to show similarities in their Sr- and Nd-isotope composition to a HIMU (high μ: enriched in ^{206}Pb and ^{208}Pb) mantle source, and similarities in their Pb-isotope values and some incompatible trace element ratios to a mid-ocean ridge basalt (MORB) mantle source (Hole *et al.* 1993). Volcanic rocks in north Victoria Land (Hart and Kyle 1993) and Marie Byrd Land (Hart *et al.* 1997) showed Sr–Nd–Pb compositions consistent with a HIMU component being involved. Early studies from north Victoria Land and Marie Byrd Land have also proposed an EMII (Enriched Mantle II: low ^{143}Nd/^{144}Nd, high ^{87}Sr/^{86}Sr, and high ^{207}Pb/^{206}Pb and ^{208}Pb/^{204}Pb at a given value of ^{206}Pb/^{204}Pb) or subduction-impregnated subcontinental lithosphere component in the mantle source (Hart and Kyle 1993; Hart *et al.* 1995). These regional variabilities in mantle source have also been measured in Sr–Nd–Pb isotopes and trace element ratios of mantle xenoliths (Wysoczanski 1993; Martin *et al.* 2015), and Martin *et al.* (2015) proposed an EMI (Enriched Mantle I: low ^{143}Nd/^{144}Nd, low ^{87}Sr/^{86}Sr, and high ^{207}Pb/^{206}Pb and ^{208}Pb/^{204}Pb at a given value of ^{206}Pb/^{204}Pb) component in the south Victoria Land mantle. A carbonate-like fluid or melt has also been proposed in the mantle source of volcanic rocks and mantle xenoliths in Antarctica (Foley *et al.* 2006; Kogarko *et al.* 2007; Aviado *et al.* 2015; Martin *et al.* 2015). This brief summary demonstrates that a chemical and isotopic heterogeneity of the Antarctic mantle has been suggested for several decades.

Despite numerous publications on mantle xenoliths from specific localities, to date, there has been no continent-wide review of Antarctic mantle xenoliths, leading to Antarctica being a missing key piece of global compilations of mantle xenolith data (e.g. Ross *et al.* 1954; Jagoutz *et al.* 1979; Nixon 1987*a*; Meisel *et al.* 2001; Pearson *et al.* 2003, who included two Antarctic samples of plagioclase-bearing peridotite in their compilation). The most comprehensive work so far has been the inclusion of some Antarctic regions in a major reference work on mantle xenoliths (Nixon 1987*a*). In this, Nixon (1987*b*) provided the mineral chemistry of one East Antarctic xenolith and the whole-rock chemistry of four other East Antarctic xenoliths; and in Kyle *et al.* (1987), a thorough review of mantle xenolith occurrences in Victoria Land was provided.

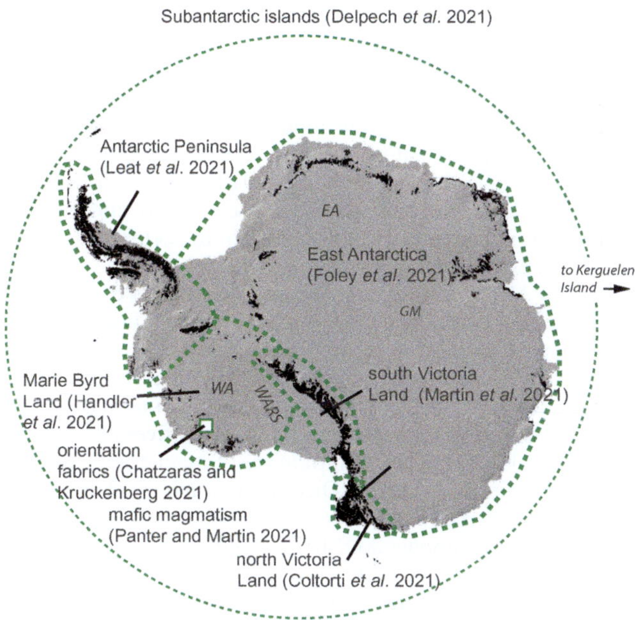

Fig. 4. Diagram showing the location in Antarctica of geochemistry chapters and locations discussed in the text. EA, East Antarctica; GM, Gamburtsev Mountains; WA, West Antarctica; WARS, West Antarctic Rift System. Black polygons indicate exposed bedrock.

Mantle geophysics

Systematic surveying of Antarctica began with the International Geophysical Year (1957–58). Seismic transects (around 13 000 km worth) across East Antarctica and West Antarctica led to the realization that Antarctica formed a single continent (Evison *et al.* 1959; Siegert *et al.* 2018), with thicker crust in East Antarctica compared to West Antarctica (Evison *et al.* 1960). Since that time, the suite of seismic measurements has been expanding, with active seismic measurements first being collected during US and Soviet expeditions (Kaminuma 1979).

In these early geophysical expeditions, a signal from the mantle remained elusive in the seismic measurements. At first, the absence of reflections from mantle depths was thought to illustrate that the Antarctic mantle was fairly homogeneous (Adams 1968). Later, this was shown to be a result of limited earthquake travel paths. The addition of waves from nuclear explosions showed that there was, in fact, considerable inhomogeneity (Adams 1971). The second problem for observing Antarctic Earth structure was what was at the time believed to be an absence of large Antarctic earthquakes compared to other continents (Evison 1967). At this time, seismic measurements remained restricted to a few permanent stations at mainly coastal regions, and models tailored to Antarctica could only be developed from global seismic datasets using occasional earthquake signals that travelled through the Antarctic mantle (Ritzwoller *et al.* 2001). Large improvements in understanding occurred with the initiation of projects from 2000 onwards, and a large array of seismographs deployed across the interior of the continent as part of the Fourth International Polar Year (2007–08) demonstrated for the first time that intraplate earthquakes occur in Antarctica with the same frequency as for other continents (Lough *et al.* 2018).

Gravity measurements started at the end of the 1950s (Harada *et al.* 1958; Bull 1960). However, only when simultaneous ice-thickness measurements were included with the measurements could there be inferences made on the lithospheric structure. A robust feature found from gravity data is a large and sudden increase in lithospheric thickness from West to East Antarctica, at the location of the Transantarctic Mountains (Behrendt *et al.* 1966). It was also hypothesized that asthenospheric mantle effects could be seen in gravity data, notably post-glacial rebound in the Ross Sea (Bennett 1964) and a negative, long-wavelength signal thought to be caused by mantle convection (Kaula 1972).

When using seismic and gravity data to reveal the Earth structure in polar regions, it should be noted that both sets of data need to be corrected to remove the effect caused by the thickness of ice. Therefore, a crucial advance in Antarctic seismic and gravity studies was the airborne campaign undertaken to measure the thickness of, and structure within, ice using radar echo sounding (Drewry 1982). Radar techniques are still to be applied continent-wide in Antarctica and in recent digital elevation maps radar data gaps were filled with gravity data (Fretwell *et al.* 2013).

Interpretations made from seismic and gravity measurements in terms of mantle structure can also be supported by tectonic evolution studies. Early on, the role of the Antarctic Plate in the global picture of tectonics was recognized: for example, from the Ferrar Large Igneous Province signalling the Mesozoic break-up of Gondwana (Dietz and Holden 1970; Elliot 1992). Understanding Antarctica's place in the plate tectonics puzzle also helps when studying parallels between Antarctica and other regions such as Australia and India, and seismic studies showed that the East Antarctic mantle was similar to other, adjacent Precambrian shields (Knopoff and Vane 1978).

The signal of mantle dynamics such as GIA can be recorded in palaeosea-level data. In Antarctic, initially only a handful of palaeosea-level curves were available, which could be interpreted as rebound following the partial demise of the Antarctic Ice Sheet (Clark and Lingle 1979). Quantifying the interpretation requires estimates for the thickness of the Antarctic Ice Sheet through time and numerical models for GIA and sea-level change, which became available in the 1970s. For a long time, the volume of the past Antarctic Ice Sheet could only be derived by subtracting the known contributions from other, non-Antarctic land masses from the global mean sea-level change since the last glacial maximum (LGM: e.g. Lambeck *et al.* 2000). A gradual increase in the number of ice-extent and sea-level data, and the collection of geodetic data of elevation change, improved this situation.

The 1990s saw the first geodetic measurements collected and compiled under the coordination of the Scientific Committee on Antarctic Research (SCAR: Manning 2001). Initially, measurements by GPS receivers placed on bedrock had to be repeated in separate campaigns. At the turn of the century, however, permanent global navigation satellite system (GNSS) stations could be established. This greatly increased the spatial coverage of measurements of elevation change and showed for the first time ongoing changes across the continent that could not be solely explained by ice melt combined with GIA since the LGM (e.g. Thomas *et al.* 2011).

Studies of mantle convection and associated tectonics in Antarctica via geophysical studies is yet to receive as much as attention as GIA studies (although see Austermann *et al.* 2015 and Faccenna *et al.* 2008 for recent exceptions). This is mainly because topography and the geoid have a less direct relationship than that found between GPS measurement of uplift rate and GIA. Despite this, Antarctica has an important part to play in global studies (e.g. Steinberger *et al.* 2012).

A consequence of mantle convection and plate tectonics is post-seismic deformation: the slow readjustment of the mantle after an earthquake. The stress release cannot be predicted from convection or plate tectonic models alone, therefore post-seismic deformation is a separate topic. For studying post-seismic deformation, data at the right location and at the right time have only been available recently, and the study of post-seismic deformation in Antarctica is in its infancy (King and Santamaría-Gómez 2016).

The cost of logistics and extensive ice cover means that satellite missions play an outsized role in the polar regions relative to other areas globally. Measurement conditions for satellite are favourable for high latitudes because of the geometry of satellite orbits. As the satellite moves through its orbit the Earth rotates with smaller ground velocity near the poles, resulting in smaller ground-track spacing and higher resolutions. One limitation is a data gap near the Antarctic pole. Examples of missions that contribute to solid Earth studies now follow.

The satellite gravity missions GRACE (Gravity Recovery and Climate Experiment) and GOCE (Gravity Field and Steady-State Ocean Circulation Explorer) provided increased resolution of the gravity field, to below 100 km, and contributed to the understanding of the Antarctic lithospheric structure (e.g. Pappa *et al.* 2019*b*). The remarkable accuracy of the satellite data means that measurement accuracy hardly plays a role in gravity interpretations. It is so sensitive that mass change of the Antarctic Ice Sheet could be observed in GRACE data (Velicogna and Wahr 2006), which is for a large part due to GIA (King *et al.* 2012). In the absence of direct measurements, forward models are the only method to quantify GIA. This spurred new and updated GIA models for Antarctica (Whitehouse *et al.* 2012; Ivins *et al.* 2013; Argus *et al.* 2014). There has been a concerted effort to fill the satellite polar gap and other regions of importance with airborne gravity data (Forsberg *et al.* 2011, 2018).

The SWARM satellite delivered measurements of the magnetic field that could link lithospheric structures from Antarctica to other continents (e.g. Ebbing *et al.* 2021). To study the mantle, several corrections to the SWARM-derived magnetic measurements are necessary such as for the core and crustal magnetic field and ionospheric effects (see Civet *et al.* 2015 for details). Recently, the compilation of magnetic data from marine and airborne data (ADMAP) was augmented by 2.0 million line-kilometres of new airborne and marine magnetic anomaly data to create ADMAP2 (Golynsky *et al.* 2018). The SWARM satellite data have been used to extend the ADMAP2 dataset, filling residual data gaps (Kim *et al.* 2022).

Radar remote sensing from satellite does not observe the solid Earth directly but is complementary to the interpretation of other geophysical data. The technique transmits microwave pulses and detects the backscattered echo from the surface. Satellite radar measurements were initially designed for

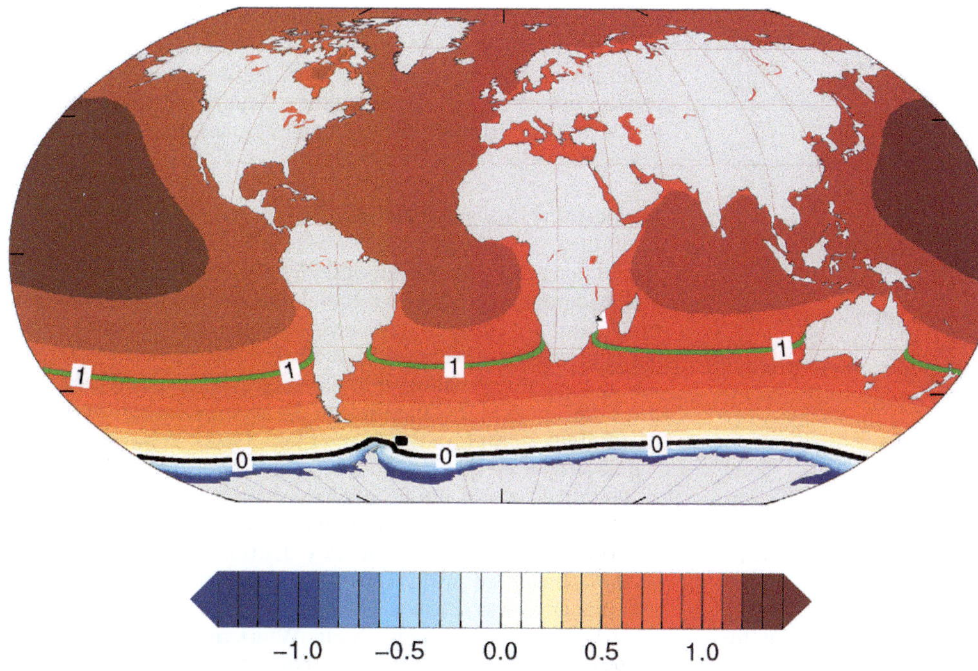

Fig. 5. Sea-level fingerprint of the melting of the Antarctic Ice Sheet normalized relative to 1 mm of global mean sea-level equivalent contribution. Blue is sea level lowering, the black line is no change, red is sea-level rising and the green line (at *c.* 45° S) is the exact global mean. The melt is applied uniformly over the ice sheet, with an elastic Earth, assuming that melting occurs over a short period of time. Calculated using SELEN4 (Spada and Melini 2019), including harmonic degrees 0–128 and rotational effects.

measurements over the sea. Errors over ice are large relative to the open ocean because of the slope of the ice surface (Rémy *et al.* 1989). Despite these challenges, satellite radar accuracy was sufficient to obtain measurements of the average elevation of the ice surface, which, combined with airborne and terrestrial radar measurements of ice thickness, helps in the interpretation of seismic and gravity data. Over time, the errors associated with radar techniques were decreased through improved data-processing approaches such as waveform retracking, which derives more parameters from the returned pulse (Rémy and Parouty 2009). The increased accuracy of ice-elevation observations allowed the detection of changes in elevation of the Antarctic Ice Sheet itself (Davis and Ferguson 2004).

Overview of the impact of the Antarctic mantle on ice sheets and sea level

Ice-sheet melting leads to inhomogeneous sea-level rise globally because of changes in the Earth's gravity field resulting from the ice-sheet melt (Fig. 5). As a result, a large part of the northern hemisphere will experience sea-level increases higher than the mean due to the melting of the Antarctic Ice Sheet (i.e. north of the green line on Fig. 5). The role Antarctica will play in sea-level rise, however, is not well known partly because of uncertainty in characterizing the Antarctic mantle necessary for GIA modelling, which plays a key role in monitoring and predicting the loss in mass of the Antarctica Ice Sheet. The reason for the importance of GIA in monitoring ice loss is that GIA model predictions are required to correct gravity measurements as discussed above, which is vital for measuring current ice-mass loss. Two reasons for the importance of GIA in predicting ice loss are outlined below.

The effect of the mantle on ice loss is through GIA-induced bedrock uplift and accompanying grounding-line migration (e.g. Gomez *et al.* 2010). The West Antarctic Ice Sheet (WAIS) is a marine ice sheet, with the underlying bed in large parts below current sea level by 1000 m or more (Fig. 6), with the bed often deepening upstream. These factors make the WAIS vulnerable to warming ocean temperatures through a process called marine ice sheet instability (MISI: e.g. Pattyn 2018). The ice discharge across the grounding line increases with increasing ice thickness. Therefore, when the ice sheet retreats, this can lead to increased dynamic thinning of the ice sheet. A retreat moves the grounding line into areas that may have the bedrock surface at a lower elevation and covered by a greater thickness of ice relative to the original grounding line, which again leads to increased ice discharge in a process that can become self-repeating (Fig. 7). A retreating grounding line slowed down by GIA-induced bedrock uplift (Gomez *et al.* 2010) can also affect sea level by pushing out seawater during uplift (Gomez *et al.* 2010).

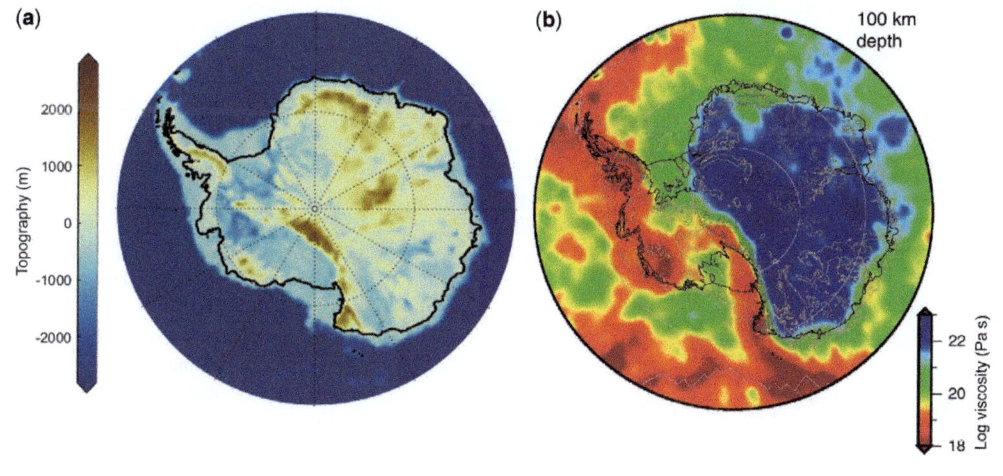

Fig. 6. Topography and viscosity diagrams of Antarctica. (**a**) Bedrock elevation relative to the present sea level of Antarctica (Bedmap2: Fretwell *et al.* 2013). (**b**) Spatial variations in the estimated upper-mantle viscosity underneath Antarctica at a depth of 100 km (fig. 4a from Whitehouse *et al.* 2019). Copyright for both figures: http://creativecommons.org/licenses/by/4.0/

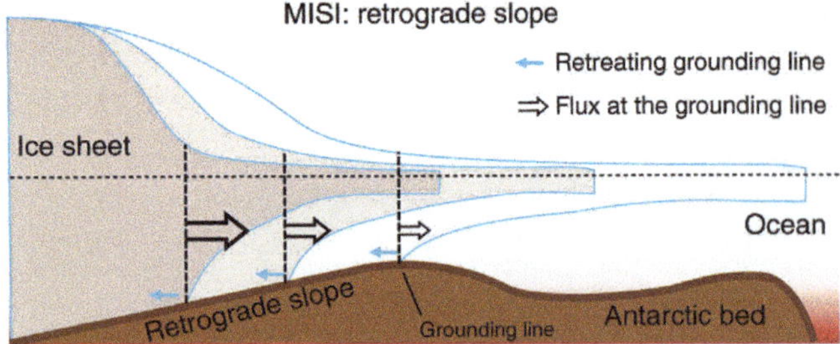

Fig. 7. Schematic representation of marine ice-sheet instability (MISI). Generally, ice discharge increases with increasing ice thickness at the grounding line. The grounding line is the transition from grounded to floating ice. When the bedrock is sloping back towards the interior, this may lead to the unstable retreat of the ice sheet because ice flux increases as the grounding line retreats (MISI). Source: Pattyn (2018, fig. 1a). Copyright: http://creativecommons.org/licenses/by/4.0/

Secondly, heat from the mantle largely controls large-scale variations in geothermal heat flow, which exerts a strong influence on ice-sheet changes (e.g. Sutter *et al.* 2019). Bedrock heating underneath the Antarctic Ice Sheet is important for feeding basal hydrology through the basal melting of ice. Basal sliding of the ice is likely to be enhanced when the ice sheet is temperate at the bottom, leading to a more rapid response in a retreating phase of the ice sheet. It is therefore an important factor to consider in future ice-sheet stability studies. Several geothermal heat-flow models exist (e.g. Shapiro and Ritzwoller 2004; Fox Maule *et al.* 2005; An *et al.* 2015; Martos *et al.* 2017) that show differences between West and East Antarctica related to the crustal structure, tectonic history and heat from the mantle (Fig. 8). However, current uncertainties and differences between models are large (Burton-Johnson *et al.* 2020*a*, *b*; Lösing *et al.* 2020).

Antarctic mantle: open questions

Accurately and precisely characterizing the Antarctic mantle is vital in understanding how the solid Earth will respond to climate-change effects and plate-tectonic forcing. This has been recognized as a key area of research for Antarctica (e.g. Kennicutt *et al.* 2015) as it affects ice-sheet change, sea-level contribution, tectonics and mantle dynamics. Here we establish current research directions and open questions that can be informed by mantle studies. This volume should go part way to addressing several of these questions.

Topography, composition and mantle structure

Antarctica hosts extreme topography: the Gamburtsev Subglacial Mountains are the world's largest intraplate mountain range, and are likely to have been the focal point for the initiation of the East Antarctic Ice Sheet (Paxman *et al.* 2016). Along the western shoulder of the West Antarctic Rift System, the Transantarctic Mountains are the world's largest example of rift-flank uplift (ten Brink and Stern 1992). **What supports Antarctica's largest mountain ranges?**

The thickness of the lithosphere plays a role in explaining current topography because it determines how much topography can be supported isostatically. Estimates of the thickness of the lithosphere in the West Antarctic Rift System suggest that it is abnormally low and similar estimates of East Antarctica show it is abnormally high. **What mantle parameters make the West Antarctic Rift System abnormally depressed and East Antarctica abnormally elevated?** For computational expediency, traditional models of GIA in Antarctica have used a radially varying 1D approximation of the Earth, leading to errors in the estimates of the present-day GIA signal through failure to capture known variations in the Earth structure across Antarctica (Nield *et al.* 2018). **How does lithospheric thickness vary across Antarctica?**

Seismic velocity maps show clear differences between East and West Antarctica. To a first order, seismic velocity anomalies can be interpreted as thermal anomalies (Cammarano *et al.* 2003), which is consistent with the geological history. However, the boundary between the two domains it not well imaged. **How are high-velocity East Antarctica and low-velocity West Antarctica connected?** A mantle plume beneath Marie Byrd Land has long been postulated to explain the regional doming, patterns of volcanism and elevated heat flow (e.g. Kyle *et al.* 1994; Seroussi *et al.* 2017). A mantle plume could be consistent with seismic models but other interpretations of seismic data are possible because of the limited resolution at depth of this method (Hansen *et al.* 2014).

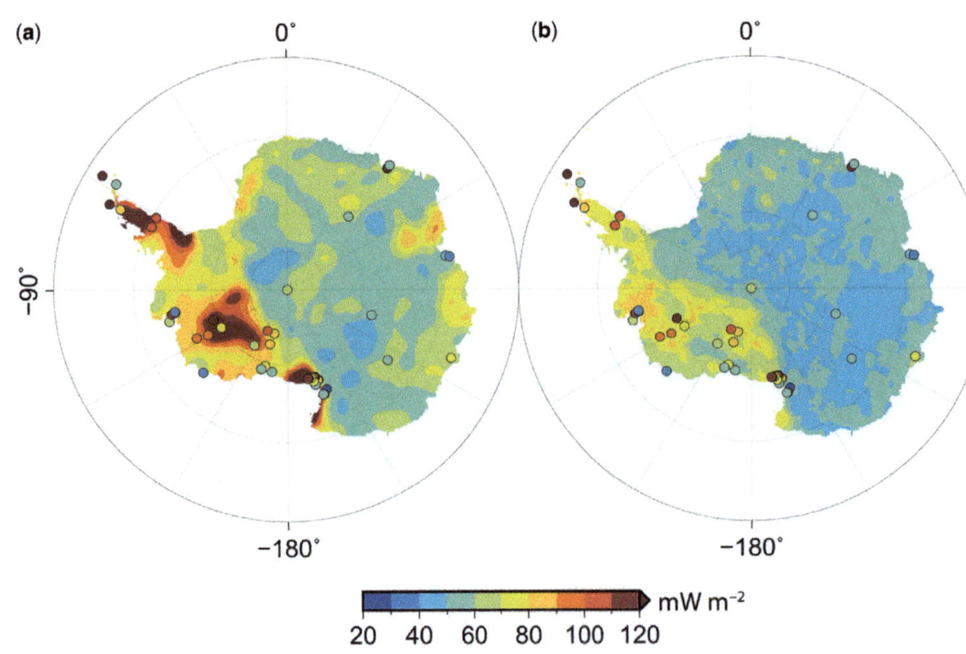

Fig. 8. Geothermal heat flow from (**a**) Martos *et al.* (2017) and (**b**) Shen *et al.* (2020). Heat-flow measurements are represented by circles. Source: Lösing and Ebbing (2021, fig. 8). Copyright: http://creativecommons.org/licenses/by/4.0/

Therefore, other lines of geochemical and geophysical evidence have to be used to arrive at a conclusion on the presence of a plume. **Is a plume beneath Marie Byrd Land required by observations? Are there other plumes beneath Antarctica?**

Geochemistry and geophysical studies have long recognized inhomogeneity in the Antarctic mantle: for example, between East and West Antarctica, and also within regions at the scale of kilometres to thousands of kilometres (Hart *et al.* 1995; Martin *et al.* 2014). The recent inclusion of 3D variability in Earth models has improved GIA signal estimations (van der Wal *et al.* 2015), however more work remains. **How many distinct mantle domains are there in Antarctica? At what scale do mantle domains need to be defined to be useful in modelling?**

A number of mantle reservoirs (HIMU, EMI and EMII) have been identified in Antarctica. **What mantle reservoirs (chemical and isotopic) exist in the Antarctic mantle? Do they vary at a continental-scale? Is there a HIMU-reservoir *sensu stricto*?** This leads onto another question once the Antarctic mantle is defined: **How does the Antarctic mantle domain compare with adjacent mantle domains (e.g. South America, Australia, Zealandia and Africa) at a global scale?**

Change in topography and mantle structure

Models for elastic rebound and Antarctica GIA models do not match well with the estimated velocities made from extensive GPS campaigns (Bevis *et al.* 2009; Thomas *et al.* 2011; Groh *et al.* 2012). In part, this is because ice thickness varies on century- or decadal-long timescales not currently considered in GIA models (Nield *et al.* 2014; Barletta *et al.* 2018). Horizontal GPS velocities are dominated by plate rotation, but neotectonic deformation and GIA also play a role (King *et al.* 2015; Vardić *et al.* 2022). Even though the GIA contribution is small relative to other parameters, its sensitivity is such that even the sign is dependent on the mantle viscosity (Hermans *et al.* 2018). In some regions, post-seismic deformation can be detected (King and Santamaría-Gómez 2016) but the following over-arching question remains: **What mantle processes contribute to observed present-day vertical and horizontal motion, and at which temporal and spatial scales?**

Convection, GIA and post-seismic deformation are a function of mantle viscosity. Some modelling includes 3D viscosity variations (Kaufmann *et al.* 2005; King and Santamaría-Gómez 2016; Gomez *et al.* 2018; Whitehouse 2018). In most cases, spatial variations in viscosity are scaled from seismic velocity anomalies but this is a highly uncertain procedure (e.g. Barnhoorn *et al.* 2011; O'Donnell *et al.* 2017). This raises the question: **How well can viscosity be derived from seismic models and geochemical data, and how can the viscosity information from geochemistry be incorporated into modelling?**

Viscosity is a macroscopic parameter that aggregates material properties such as water content, grain size, temperature, pressure and composition (e.g. Hansen *et al.* 2016). Water content can be detected in xenoliths and with magnetotelluric measurements, and also indirectly in seismic evidence (Emry *et al.* 2015). Xenoliths provide grain size, water content and pressure–temperature variations. Xenoliths and magnetotelluric measurements represent information that is currently underused in geodynamics models and should be incorporated where present. To aim for better viscosity models we need to know: **How do mantle material properties vary laterally across Antarctica and with depth?**

Another important cross-disciplinary consideration is at what scale do continental-sized geophysical, remote-sensed or modelling studies intersect with regional- to local-sized petrological or geochemical studies? This is a fundamental problem of scales, from the macroscopic to the continent. **Do we need to make geophysical studies finer grained or petrological studies coarser grained?**

Integration across multiple temporal scales is also important. Mantle processes operate on a timescale from decades to millions of years. New models allow integration of GIA and post-seismic models (Nield *et al.* 2022). However, mantle-deformation experiments and geodynamic models demonstrate that both short- and long-term viscosity determines post-seismic deformation (e.g. Pollitz and Thatcher 2010), and viscosity changes due to stress that changes through time (e.g. van der Wal *et al.* 2013). In general, viscosity can be seen as dependent on the frequency of loading (Lau *et al.* 2021). **What viscosity is representative for which mantle process in Antarctica and how can they be integrated?**

Influence on ice

The strong feedback between solid Earth and ice through surface heat flow enhances basal sliding, leading to a more rapid response in a retreating phase of the ice sheet. Surface heat flow reflects crustal composition but is also driven by the subadjacent mantle. The effects of geothermal heat flow on ice dynamics has been a long-standing issue for ice-sheet models and estimates vary greatly (e.g. Burton-Johnson *et al.* 2020*a*, *b*). **What is the contribution of the mantle to surface heat flow and, hence, ice-sheet basal melting?** A second feedback between solid Earth and ice affects the grounding line. **What is the effect of GIA on the position of the ice-sheet grounding line in Antarctica?**

The Antarctic mantle also plays a significant role in estimates of the mass balance of the ice sheet as observations need to be corrected for GIA. The GIA signal must first be subtracted from the GRACE data before calculating the true spatial distribution and magnitude of current ice-mass change in Antarctica (e.g. Velicogna and Wahr 2006; Thomas *et al.* 2011). The GIA models do not always agree with each other or with empirical estimates (Martín-Español *et al.* 2016). In addition, new regional (Barletta *et al.* 2018) and 3D GIA models are being developed (Powell *et al.* 2020; Blank *et al.* 2021). Improved constraints on the Antarctica mantle can help to answer the question: **What is the contribution of GIA to the observed mass change and corresponding sea-level change?**

Volume overview

The chapters by **Coltorti *et al.* (2021)**, **Delpech *et al.* (2021)**, **Foley *et al.* (2021)**, **Handler *et al.* (2021)**, **Leat *et al.* (2021)** and **Martin *et al.* (2021*a*)** on mantle xenoliths in volcanic rocks review occurrences from select geographical areas that together cover all of Antarctica (Fig. 4). Important locations covered by **Chatzaras and Kruckenberg (2021)** and **Panter and Martin (2021)** focused on specific topics are also marked in Figure 4. The chapters by **Bredow *et al.* (2021)**, **Ivins *et al.* (2021)**, **Pappa and Ebbing (2021)**, **Paxman (2021)**, **Scheinert *et al.* (2021)**, **van der Wal *et al.* (2022)**, **Wannamaker *et al.* (2021)** and **Wiens *et al.* (2021)** on geophysics, and the overview chapter by **Martin (2021)**, are more continental scale in nature.

Mantle xenoliths of East Antarctica and the mantle properties inferred from igneous rocks of East Antarctica are reviewed by **Foley *et al.* (2021)**. These volcanic rocks have been erupted through rifts and associated cratons of East Antarctica in a setting unique to that of West Antarctica. Mantle xenoliths are only known from three locations and all come from Indo-Antarctica-affiliated lithosphere. The xenoliths are spinel bearing or spinel–garnet bearing, and contain geochemical and mineralogical evidence for multiple enrichment events that can often be linked to activity along major translithospheric structures. The authors note that local modification of the mantle by melt infiltration may influence large-scale images derived from geophysical data.

South Victoria Land as the first locality of mantle xenolith study in Antarctica has more than a 100 year record of scientific data, which **Martin *et al.* (2021*a*)** collate and interpret into a new petrogenetic model. The region has Paleoproterozoic stabilization ages with some evidence for younger regions of mantle. The mantle has been affected by fluids related to oceanic crust subduction at *c.* 0.5 Ga and older events, as well as a thermal and chemical disturbance associated with Cenozoic volcanism. A unique feature is the iron enrichment beneath Ross Island observed in the whole-rock chemistry of primitive magmas, peridotite and pyroxenite xenoliths, and the chemistry of mantle olivine. This iron enrichment is potentially related to iron-rich crust or the exchange of iron across the core–mantle boundary. The peridotite mantle of the entire region has been cross-cut by multiple episodes of pyroxenite veining.

North Victoria Land contains numerous active, dormant and extinct Cenozoic volcanoes with abundant mantle xenolith cargoes. **Coltorti *et al.* (2021)** synthesize the abundant published and unpublished data to provide a petrogenetic overview of the mantle from this region. A palaeogeotherm is calculated, oxygen fugacity of the lithospheric mantle is reported relative to the fayalite–magnetite–quartz buffer and the water content of nominally anhydrous minerals is reported. A long history from old depletion to recent refertilization and metasomatism is discussed, including a refertilization episode related to the Jurassic Ferrar Large Igneous Province emplacement and metasomatism related to Cenozoic magmatism.

Marie Byrd Land has numerous Cenozoic volcanoes with entrained mantle xenoliths, the occurrences and petrogenetic implications of which are reviewed by **Handler *et al.* (2021)**. They show mantle compositions that have experienced 10–25% melt depletion and which have been variably metasomatized and refertilized. The timing of events show a Proterozoic lithospheric mantle and evidence for Ordovician events related to the complex tectonic history of the region, consistent with a protracted history of Paleozoic convergence and subduction.

In the Antarctic Peninsula, there are multiple generations of mantle xenolith-bearing igneous rocks erupted under differing tectonic settings. **Leat *et al.* (2021)** review the mantle xenolith occurrences and contextualize them according to their tectonic setting and age. Xenoliths are restricted to post-subduction volcanic rocks emplaced in forearc or back-arc positions. The xenoliths are spinel bearing, and may be residues from a mid-ocean ridge setting from the upper mantle, and were subsequently accreted to form harzburgitic, oceanic lithosphere.

Studies of mantle xenoliths on the subantarctic islands have greatly informed our understanding of the mantle but are included too infrequently in discussions on the Antarctic mantle. **Delpech *et al.* (2021)** review the information from mantle xenoliths afforded by these island windows into the mantle, as well as discussing serpentinized peridotite dredged from fracture zones and suprasubduction zone peridotite from the Sandwich Plate. They show that the subcontinental and oceanic lithospheric mantle contains ancient fragments that underwent depletion significantly earlier than the formation of the overlying crust, and there are frequent signs of subsequent metasomatic enrichment.

Other focused chapters have been invited. **Panter and Martin (2021)** provide an overview of the West Antarctic mantle deduced from mafic magmatism, and show that mantle composition varies by tectonic setting and time. The mantle-source changes coincide with changes in crustal structure and composition that are also visible in tomographic models and seismic data. Marie Byrd Land magmatism results from plume activity, with chemical and isotopic evidence for a component of subduction-modified mantle that is distinct from the mantle in Victoria Land. Relative to Victoria Land, Marie Byrd Land mantle has a weaker focal zone (FOZO) component and more of the high-μ (HIMU) component.

Chatzaras and Kruckenberg (2021) review crystallographic preferred orientation studies on Antarctic mantle xenoliths to inform seismic anisotropy studies. They report the petrology, microstructure and seismic properties of the mantle beneath Marie Byrd Land. There is evidence for diachronous reactive melt percolation and refertilization, which is likely to have modified seismic velocities and anisotropy. They show that seismic properties are mainly controlled by the strength of the olivine crystallographic texture, and suggest that variations in the mantle structure are related to 3D deformation during the region's prolonged tectonic history.

The chapters in the geophysics section of the volume focus on remote measurements and their interpretation, as well as on geodynamic processes. Measurement campaigns and surveys are described as appropriate to acknowledge the effort required to obtain the data, to highlight which parts of Antarctica remain understudied and to draw awareness to uncertainties stemming from the practical limitations of working in the testing Antarctic environment.

Magnetotelluric measurements are collected along select profiles in Antarctica to study lithosphere and mantle temperature, water content, and to make compositional inferences. **Wannamaker *et al.* (2021)** reanalyse measurements from three campaigns. Near the South Pole, their study shows low resistivity at Moho depths consistent with elevated thermal data. The Ellsworth–Whitmore block study shows near-cratonic geothermal conditions, as well as lithological variability. Moderate mantle hydration incompatible with extensive melting is indicated in the Ross Sea study.

Wiens *et al.* (2021) describe the seismic models that have been developed for parts, or the entirety, of Antarctica. The chapter reviews different sources (distant earthquakes, active seismic and ambient noise), different waves (S-waves and P-waves) and different techniques (receiver functions, tomography and full-waveform tomography), and their sensitivity to mantle structure and Moho depth. Large contrasts in seismic wave speed are seen across Antarctica, with slow velocities below large parts of West Antarctica and thick lithosphere below East Antarctica.

Pappa and Ebbing (2021) review gravity and magnetic field measurements, as well as surface heat flow in Antarctica. This chapter discusses integration with petrology to derive upper-mantle temperature and to reduce ambiguity due to unknown lithospheric structure. First-order results agree with seismic models and provide extra compositional information. Magnetic anomalies reflect different crustal terranes but can also be inverted for the Curie depth, from which surface heat flow can be derived. Uncertainties in surface heat-flow estimates are reviewed.

Paxman (2021) explains the importance of palaeotopography as a boundary condition for ice-sheet modelling. Paxman reviews how structural inheritance played a role in shaping early Antarctic ice sheets, and the role of the lithosphere and the mantle in shaping the topography. Lithospheric cooling contributed to subsidence, which made the WAIS increasingly sensitive to MISI, and dynamic topography provides support to palaeotopography in Antarctica (e.g. the support of the Transantarctic Mountains) together with other processes.

On relatively long timescales, viscosity controls mantle convection and plumes. **Bredow *et al.* (2021)** review simulations of mantle convection. They discuss simulations that give the directions of Antarctic mantle flow and the locations of possible mantle plumes. These simulations show that the elevated heat flow supports a mantle plume in Marie Byrd Land. Simulations of dynamic topography show it to be negative in East Antarctica, contrary to what is expected from the high topography.

Mantle rheology is utilized to model dynamic processes. Mantle rheology is the science that describes the deformation and flow of Earth materials. **Ivins *et al.* (2021)** describe how seismic tomography, material parameters and empirical constants can be combined to obtain viscosity maps. They use an S-wave velocity model (Lloyd *et al.* 2020) to estimate viscosity at different depths in the upper mantle using three approaches, and discuss the resulting differences.

Geodetic measurements can detect the response to past ice-sheet-thickness changes in terms of crustal motion or gravity change. **Scheinert *et al.* (2021)** explain the processing procedure and the importance of correcting for the effect of current ice-sheet changes. They review regional studies in the Amundsen Sea sector and the Antarctic Peninsula, where fast relaxation is detected. This chapter discusses measurements of ice-height and ice-mass changes, and discusses how they can be combined to improve estimates of ice melt and GIA.

Viscosity maps as input for GIA modelling are reviewed in **van der Wal *et al.* (2022)**. The history of GIA research, GIA observations, modelling techniques, and interactions between GIA and the solid Earth are discussed. An important development is the realization that low viscosity in West Antarctica results in faster response times of the mantle, and higher sensitivity to recent ice-thickness changes. This also implies a key role for GIA in modelling ice-thickness changes of marine-terminating glaciers that are sensitive to elevation change. In addition, this chapter discusses the post-seismic deformation detected in GNSS observations, which also require low viscosity in the upper mantle.

Martin (2021) collates a review of the composition and chemistry of mantle xenoliths from Antarctica using the continent-wide dataset made possible by this Memoir and places this into a global context. The Antarctic mantle is heterogeneous at all scales, which opens up a number of research questions and has implications for geophysical models of the lithospheric mantle.

Summary and outlook

Both the historical overview and the chapters in this volume show how concerted efforts such as the International Polar Year have increased our knowledge of the Antarctic mantle. Technological developments are also an important driver for Antarctic Earth science, from lower detection limits in geochemical studies (**Handler *et al.* 2021**), novel isotopic studies (**Martin 2021**) and micro-fabric studies (**Chatzaras and Kruckenberg 2021**), and better stand-alone GNSS instruments (**Scheinert *et al.* 2021**) to more precise magnetotelluric measurements (**Wannamaker *et al.* 2021**) and satellite measurements (**Pappa and Ebbing 2021**). This in turn drove the developments of numerical models for mantle composition (**Wiens *et al.* 2021**), rheology (**Ivins *et al.* 2021**) or mantle-dynamics studies that can integrate those measurements (**van der Wal *et al.* 2022**). With measurements still limited in most of Antarctica, studying the Antarctic mantle is a truly multidisciplinary effort.

In this Memoir, several open questions are tackled, such as how does the mantle support mountain ranges (**Paxman 2021**) and are there multiple Antarctic mantle plumes (**Bredow *et al.* (2021)**; **Martin *et al.* 2021*a*, *b***)? Definitively answering questions like these will require a combination of several models (lithosphere model, convection and composition).

Large contrasts between West Antarctica and East Antarctica signal changes in temperature and, to a lesser extent, changes in composition that are seen in mantle xenolith studies and in seismic wave speed. Quantifying mantle parameters is necessary for their use in geodynamic models but is subject to considerable uncertainty (**Martin 2021**; **Wiens *et al.* 2021**). The uncertainty in viscosity derived from a single seismic velocity model is discussed in **Ivins *et al.* (2021)**. Other uncertainties that could be usefully refined include uncertainty from seismic models and uncertainties in mantle flow laws that relate deformation to applied stress. Currently, the exact mantle properties everywhere in Antarctica are imprecisely known because of poor mantle xenolith coverage and the coarseness of geophysical observations in some regions. However, it is encouraging that forward modelling of viscosity can recover the low viscosities derived from geodetic measurements in the Amundsen region and the Antarctic Peninsula (**Ivins *et al.* 2021**). Combinations of mantle chemistry from rocks and xenoliths, seismic data and gravity based on petrophysics will also be useful in refining Antarctic mantle properties for more accurate modelling of, for example, GIA and bedrock heat flow (e.g. Fullea *et al.* 2021). The suggested improvements listed in this paragraph, once undertaken, will also improve estimates of flexural rigidity that are crucial for palaeotopography studies (e.g. Paxman *et al.* 2019).

Viscosity is timescale dependent and varies by orders of magnitude from post-seismic deformation to GIA to mantle convection. It is still an open question as to whether stress-dependent viscosity and transient viscosity need to be included in geodynamic models (**Ivins *et al.* 2021**). In the latter case, is considering short-term and long-term viscosity sufficient or does viscosity need to be considered as a function of loading frequency (Lau *et al.* 2021)? A consistent description of viscosity must consider deformation at the microscopic scale from observations on mantle xenoliths (**Chatzaras and Kruckenberg 2021**) and laboratory experiments (e.g. Hansen *et al.* 2016). Further improvements should focus on integrating information from mantle xenoliths (**Martin 2021**) and information from depths where wadsleyite and ringwoodite dominate the mantle composition (Fig. 1) and where the scaling is especially unconstrained.

GIA and post-seismic deformation can be simulated with one numerical model such as that of Nield *et al.* (2022), which means that the same viscosity structure can be tested against observations of post-seismic deformation and GIA in areas where both occur. Even if multiple models are used, at least using the same viscosity description derived from geochemistry and geochemical studies would be beneficial so that observational constraints from different regions and processes are available for the same unknown rheological parameters. Horizontal velocities are underused for GIA, and are sensitive to tectonics, convection and post-seismic deformation. An improved palaeosea-level database would be another very beneficial constraint to the GIA models as such data best capture the relaxation through time.

The best approach for predicting mantle heat flow is an open question actively being considered (e.g. Burton-Johnson *et al.* 2020*a*, *b*). Different approaches yield estimates of different magnitudes at different spatial resolution. Better estimates can be produced by integration of data from boreholes, rock outcrops and mantle xenoliths with geophysical data including seismic, gravity and magnetic (Burton-Johnson *et al.* 2020*a*). Using new techniques such as machine learning can help this integration and assist in quantify uncertainties (e.g. Lösing and Ebbing 2021).

Geodynamic models have been developed for simulating surface elevation at timescales from hours to millions of years due to, for example, tectonics, GIA, sediment erosion or loading, thermal subsidence and mantle-dynamics topography (**Bredow *et al.* 2021**; **Paxman 2021**). An example of subsidence and uplift due to GIA can be seen in Figure 9. Although Figure 9 shows relative sea-level changes, these largely correspond to solid Earth deformation rates as the changes in the equipotential surface that determines the sea level are typically an order of magnitude smaller.

The bedrock topography determines how ice can flow towards the ocean (e.g. Paxman *et al.* 2019); topography changes affect the position on the grounding line. **van der Wal *et al.* (2022)** focused on feedbacks in the last glacial cycle; this influence can be quantified in similar processes across timescales.

Dynamic topography calculated with numerical models of mantle convection (**Bredow *et al.* 2021**) are more frequently being used when interpreting past observations of relative sea level such as during the last interglacial (Austermann *et al.* 2017) and the Pliocene, which is a period that is studied extensively as an analogue to future warming. These dynamic topography studies can also be used in coupled ice–landscape evolution models. This would open up the possibility of better connecting Antarctic ice evolution and palaeotopography to measured sediment deposits located outside Antarctica (cf. Pollard and DeConto 2020).

Under future projections of climate warming, the Antarctic Ice Sheet is likely to contribute several metres of sea-level rise by 2300 (DeConto *et al.* 2021). Therefore, precise estimates of Antarctic ice loss are necessary. Over the next centuries, the Antarctic Ice Sheet will interact with the ocean and the atmosphere but also with the solid Earth. Recent studies have already shown that the fast response of the Earth underneath the WAIS is important to determine its precise response to future warming (Gomez *et al.* 2015; Konrad *et al.* 2015; Larour *et al.* 2019; Kachuck *et al.* 2020; Powell *et al.* 2021) but models with realistic mantle properties and better-constrained viscosity are necessary to quantify the effect.

This volume represents the current state of knowledge on the Antarctic mantle. It should be useful to new students entering the field, as well as to active researchers in Antarctic and global mantle studies.

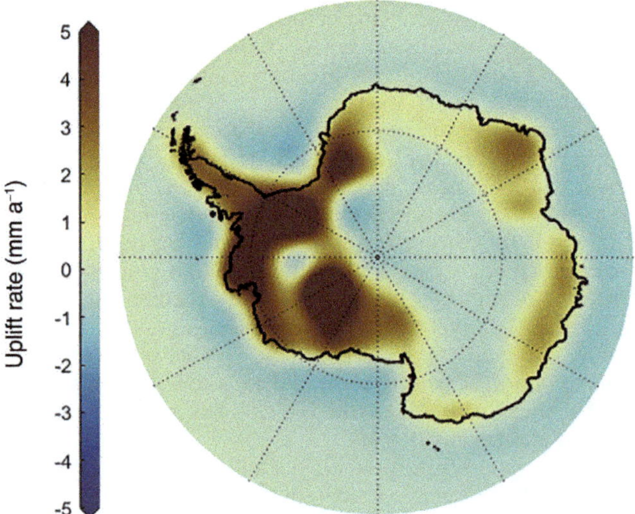

Fig. 9. The uplift rate (in mm a^{-1}) from glacial isostatic adjustment due to the past melting of ice sheets over Antarctica. Past ice-sheet input is the ICE-6G D global model (Peltier *et al.* 2018) and viscosity profile is the radially varying 1D viscosity model VM5a. Source: data were obtained from https://www.atmosp.physics.utoronto.ca/~peltier/data.php

Acknowledgements Thanks go to Pippa Whitehouse and Tom A. Jordan who reviewed this chapter, and to Phillip T. Leat for editorial handling.

Competing interest The authors declare that they have no known competing financial interests or personal relationships that could have appeared to influence the work reported in this paper.

Author contributions **APM**: conceptualization (equal), data curation (equal), investigation (equal), writing – original draft (equal); **WVDW**: conceptualization (equal), investigation (equal), writing – original draft (equal), writing – review & editing (equal); **BDB**: visualization (supporting), writing – original draft (supporting).

Funding This contribution was in part supported by contract ANT1801.

Data availability Data sharing is not applicable to this article as no datasets were generated or analysed during the current study.

References

Adams, R.D. 1968. Early reflections of P′ P′ as an indication of upper mantle structure. *Bulletin of the Seismological Society of America*, **58**, 1933–1947.

Adams, R.D. 1971. Reflections from discontinuities beneath antarctica. *Bulletin of the Seismological Society of America*, **61**, 1441–1451, https://doi.org/10.1785/BSSA0610051441

An, M., Wiens, D.A. *et al.* 2015. Temperature, lithosphere–asthenosphere boundary, and heat flux beneath the Antarctic plate inferred from seismic velocities. *Journal of Geophysical Research: Solid Earth*, **120**, 8720–8742, https://doi.org/10.1002/2015JB011917

Argus, D.F., Peltier, W., Drummond, R. and Moore, A.W. 2014. The Antarctica component of postglacial rebound model ICE-6G_C (VM5a) based on GPS positioning, exposure age dating of ice thicknesses, and relative sea level histories. *Geophysical Journal International*, **198**, 537–563, https://doi.org/10.1093/gji/ggu140

Austermann, J., Pollard, D. *et al.* 2015. The impact of dynamic topography change on Antarctic ice sheet stability during the mid-Pliocene warm period. *Geology*, **43**, 927–930, https://doi.org/10.1130/G36988.1

Austermann, J., Mitrovica, J.X., Huybers, P. and Rovere, A. 2017. Detection of a dynamic topography signal in last interglacial sea-level records. *Science Advances*, **3**, e1700457, https://doi.org/10.1126/sciadv.1700457

Aviado, K.B., Rilling-Hall, S., Bryce, J.G. and Mukasa, S.B. 2015. Submarine and subaerial lavas in the West Antarctic Rift System: Temporal record of shifting magma source components from the lithosphere and asthenosphere. *Geochemistry, Geophysics, Geosystems*, **16**, 4344–4361, https://doi.org/10.1002/2015GC006076

Barletta, V.R., Bevis, M. *et al.* 2018. Observed rapid bedrock uplift in Amundsen Sea Embayment promotes ice-sheet stability. *Science*, **360**, 1335–1339, https://doi.org/10.1126/science.aao1447

Barnhoorn, A., van der Wal, W., Vermeersen, B.L. and Drury, M.R. 2011. Lateral, radial, and temporal variations in upper mantle viscosity and rheology under Scandinavia. *Geochemistry, Geophysics, Geosystems*, **12**, Q01007, https://doi.org/10.1029/2010GC003290

Behrendt, J.C., Meister, L. and Henderson, J.R. 1966. Airborne geophysical study in the Pensacola Mountains of Antarctica.

Science, **153**, 1373–1376, https://doi.org/10.1126/science.153.3742.1373

Bennett, H.F. 1964. *A Gravity and Magnetic Survey of the Ross Ice Shelf Area, Antarctica*. Research Report 64-3. University of Wisconsin, Madison, WI.

Bercovici, D. 2003. The generation of plate tectonics from mantle convection. *Earth and Planetary Science Letters*, **205**, 107–121, https://doi.org/10.1016/S0012-821X(02)01009-9

Berg, J.H., Moscati, R.J. and Herz, D.L. 1989. A petrologic geotherm from a continental rift in Antarctica. *Earth and Planetary Science Letters*, **93**, 98–108, https://doi.org/10.1016/0012-821X(89)90187-8

Bevis, M., Kendrick, E. *et al.* 2009. Geodetic measurements of vertical crustal velocity in West Antarctica and the implications for ice mass balance. *Geochemistry, Geophysics, Geosystems*, **10**, Q10005, https://doi.org/10.1029/2009GC002642

Blank, B., Barletta, V., Hu, H., Pappa, F. and van der Wal, W. 2021. Effect of lateral and stress-dependent viscosity variations on GIA induced uplift rates in the Amundsen Sea Embayment. *Geochemistry, Geophysics, Geosystems*, **22**, e2021GC009807, https://doi.org/10.1029/2021GC009807

Bonadiman, C., Nazzareni, S., Coltorti, M., Comodi, P., Giuli, G. and Faccini, B. 2014. Crystal chemistry of amphiboles: implications for oxygen fugacity and water activity in lithospheric mantle beneath Victoria Land, Antarctica. *Contributions to Mineralogy and Petrology*, **167**, 1–17, https://doi.org/10.1007/s00410-014-0984-8

Bredow, E., Steinberger, B., Gassmöller, R. and Dannberg, J. 2021. Mantle convection and possible mantle plumes beneath Antarctica – insights from geodynamic models and implications for topography. *Geological Society, London, Memoirs*, **56**, https://doi.org/10.1144/M56-2020-2

Bull, C. 1960. Gravity observations in the Wright Valley Area, Victoria Land, Antarctica. *New Zealand Journal of Geology and Geophysics*, **3**, 543–552, https://doi.org/10.1080/00288306.1960.10420142

Burton-Johnson, A., Dziadek, R. and Martin, C. 2020*a*. Geothermal heat flow in Antarctica: current and future directions. *The Cryosphere*, **14**, 3843–3873, https://doi.org/10.5194/tc-14-3843-2020

Burton-Johnson, A., Dziadek, R. *et al.* 2020*b*. *Antarctic Geothermal Heat Flow: Future Research Directions*. SCAR-SERCE White Paper. Scientific Committee on Antarctic Research (SCAR), Cambridge, UK, https://scar.org/scar-library/search/science-4/research-programmes/serce/5454-scar-serce-white-paper-on-antarctic-geothermal-heat-flow/

Cammarano, F., Goes, S., Vacher, P. and Giardini, D. 2003. Inferring upper-mantle temperatures from seismic velocities. *Physics of the Earth and Planetary Interiors*, **138**, 197–222, https://doi.org/10.1016/S0031-9201(03)00156-0

Chatzaras, V. and Kruckenberg, S.C. 2021. Effects of melt-percolation, refertilization and deformation on upper mantle seismic anisotropy: constraints from peridotite xenoliths, Marie Byrd Land, West Antarctica. *Geological Society, London, Memoirs*, **56**, https://doi.org/10.1144/M56-2020-16

Civet, F., Thébault, E., Verhoeven, O., Langlais, B. and Saturnino, D. 2015. Electrical conductivity of the Earth's mantle from the first Swarm magnetic field measurements. *Geophysical Research Letters*, **42**, 3338–3346, https://doi.org/10.1002/2015GL063397

Čížková, H., van den Berg, A.P., Spakman, W. and Matyska, C. 2012. The viscosity of Earth's lower mantle inferred from sinking speed of subducted lithosphere. *Physics of the Earth and Planetary Interiors*, **200–201**, 56–62, https://doi.org/10.1016/j.pepi.2012.02.010

Clark, J.A. and Lingle, C.S. 1979. Predicted relative sea-level changes (18,000 years BP to present) caused by late-glacial retreat of the Antarctic ice sheet. *Quaternary Research*, **11**, 279–298, https://doi.org/10.1016/0033-5894(79)90076-0

Coltorti, M., Beccaluva, L. *et al.* 2004. Amphibole genesis via metasomatic reaction with clinopyroxene in mantle xenoliths from Victoria Land, Antarctica. *Lithos*, **75**, 115–139, https://doi.org/10.1016/j.lithos.2003.12.021

Coltorti, M., Bonadiman, C., Casetta, F., Faccini, B., Giacomoni, P.P., Pelorosso, B. and Perinelli, C. 2021. Nature and evolution of the northern Victoria Land lithospheric mantle (Antarctica) as revealed by ultramafic xenoliths. *Geological Society, London, Memoirs*, **56**, https://doi.org/10.1144/M56-2020-11

Condie, K.C. 2022. The mantle. *In*: Condie, K.C. (ed.) *Earth as an Evolving Planetary System*. 4th edn. Academic Press, London, 81–125.

Conrad Clinton, P. and Lithgow-Bertelloni, C. 2002. How mantle slabs drive plate tectonics. *Science*, **298**, 207–209, https://doi.org/10.1126/science.1074161

Cooper, A.F., Adam, L.J., Coulter, R.F., Eby, G.N. and McIntosh, W.C. 2007. Geology, geochronology and geochemistry of a basanitic volcano, White Island, Ross Sea, Antarctica. *Journal of Volcanology and Geothermal Research*, **165**, 189–216, https://doi.org/10.1016/j.jvolgeores.2007.06.003

Davis, C. and Ferguson, A. 2004. Elevation change of the Antarctic ice sheet, 1995–2000, from ERS-2 satellite radar altimetry. *IEEE Transactions on Geoscience and Remote Sensing*, **42**, 2437–2445, https://doi.org/10.1109/TGRS.2004.836789

Day, J.M.D., Harvey, R.P. and Hilton, D.R. 2019. Melt-modified lithosphere beneath Ross Island and its role in the tectono-magmatic evolution of the West Antarctic Rift System. *Chemical Geology*, **518**, 45–54, https://doi.org/10.1016/j.chemgeo.2019.04.012

DeConto, R.M., Pollard, D. *et al.* 2021. The Paris Climate Agreement and future sea-level rise from Antarctica. *Nature*, **593**, 83–89, https://doi.org/10.1038/s41586-021-03427-0

Delpech, G., Scott, J.M. *et al.* 2021. The subantarctic lithospheric mantle. *Geological Society, London, Memoirs*, **56**, https://doi.org/10.1144/M56-2020-13

Dietz, R.S. and Holden, J.C. 1970. Reconstruction of Pangaea: Breakup and dispersion of continents, Permian to Present. *Journal of Geophysical Research*, **75**, 4939–4956, https://doi.org/10.1029/JB075i026p04939

Drewry, D. 1982. Ice flow, bedrock, and geothermal studies from radio-echo sounding inland of McMurdo Sound, Antarctica. *Antarctic Geoscience*, **1**, 977–983.

Dziewonski, A.M. and Anderson, D.L. 1981. Preliminary reference Earth model. *Physics of the Earth and Planetary Interiors*, **25**, 297–356, https://doi.org/10.1016/0031-9201(81)90046-7

Eaton, D.W., Darbyshire, F., Evans, R.L., Gruetter, H., Jones, A.G. and Yuan, X. 2009. The elusive lithosphere–asthenosphere boundary (LAB) beneath cratons. *Lithos*, **109**, 1–22, https://doi.org/10.1016/j.lithos.2008.05.009

Ebbing, J., Dilixiati, Y., Haas, P., Ferraccioli, F. and Scheiber-Enslin, S. 2021. East Antarctica magnetically linked to its ancient neighbours in Gondwana. *Scientific Reports*, **11**, 5513, https://doi.org/10.1038/s41598-021-84834-1

Elliot, D.H. 1992. Jurassic magmatism and tectonism associated with Gondwanaland break-up: an Antarctic perspective. *Geological Society, London, Special Publications*, **68**, 165–184, https://doi.org/10.1144/GSL.SP.1992.068.01.11

Emry, E., Nyblade, A.A. *et al.* 2015. The mantle transition zone beneath West Antarctica: Seismic evidence for hydration and thermal upwellings. *Geochemistry, Geophysics, Geosystems*, **16**, 40–58, https://doi.org/10.1002/2014GC005588

Evison, F.E. 1967. Note on the aseismicity of Antarctica. *New Zealand Journal of Geology and Geophysics*, **10**, 479–483, https://doi.org/10.1080/00288306.1967.10426752

Evison, F.F., Ingham, C.E. and Orr, R.H. 1959. Thickness of the Earth's crust in Antarctica. *Nature*, **183**, 306–308, https://doi.org/10.1038/183306a0

Evison, F.F., Ingham, C.E., Orr, R.H. and Le Fort, J.H. 1960. Thickness of the Earth's crust in Antarctica and the surrounding oceans. *Geophysical Journal International*, **3**, 289–306, https://doi.org/10.1111/j.1365-246X.1960.tb01704.x

Faccenna, C., Rossetti, F., Becker, T.W., Danesi, S. and Morelli, A. 2008. Recent extension driven by mantle upwelling beneath the Admiralty Mountains (East Antarctica). *Tectonics*, **27**, TC4015, https://doi.org/10.1029/2007TC002197

Fermor, L.L. 1913. Preliminary note on garnet as a geological barometer and on an infra-plutonic zone in the Earth's crust. *Geological Survey of India Records*, **43**, 41–47.

Fischer, K.M., Ford, H.A., Abt, D.L. and Rychert, C.A. 2010. The lithosphere–asthenosphere boundary. *Annual Review of Earth and Planetary Sciences*, **38**, 551–575, https://doi.org/10.1146/annurev-earth-040809-152438

Foley, S.F., Andronikov, A.V., Jacob, D.E. and Melzer, S. 2006. Evidence from Antarctic mantle peridotite xenoliths for changes in mineralogy, geochemistry and geothermal gradients beneath a developing rift. *Geochimica et Cosmochimica Acta*, **70**, 3096–3120, https://doi.org/10.1016/j.gca.2006.03.010

Foley, S.F., Andronikov, A.V., Halpin, J.A., Daczko, N.R. and Jacob, D.E. 2021. Mantle rocks in East Antarctica. *Geological Society, London, Memoirs*, **56**, https://doi.org/10.1144/M56-2020-8

Forbes, R.B. 1963. Ultrabasic inclusion from the basalts of the Hut Point area, Ross Island, Antarctica. *Bulletin Volcanologique*, **26**, 13–21, https://doi.org/10.1007/BF02597270

Forsberg, R., Olesen, A.V., Yildiz, H. and Tscherning, C.C. 2011. Polar gravity fields from GOCE and airborne gravity. *In*: Ouwehand, L. (ed.) *Proceedings of the 4th International GOCE User Workshop, Munich, Germany, 31 March – 1 April 2011 (ESA SP-696, July 2011)*. European Space Agency, Noordwijk, The Netherlands, https://earth.esa.int/eogateway/documents/20142/37627/Polar_Gravity_Fields_GOCE_Airborne_Gravity_R.Forsberg.pdf

Forsberg, R., Oleson, A.V. *et al.* 2018. Exploring the Recovery Lakes region and interior Dronning Maud Land, East Antarctica, with airborne gravity, magnetic and radar measurements. *Geological Society, London, Special Publications*, **461**, 23–34, https://doi.org/10.1144/SP461.17

Fox Maule, C., Purucker, M.E., Olsen, N. and Mosegaard, K. 2005. Heat flux anomalies in Antarctica revealed by satellite magnetic data. *Science*, **309**, 464–467, https://doi.org/10.1126/science.1106888

Fretwell, P. *et al.* 2013. Bedmap2: improved ice bed, surface and thickness datasets for Antarctica. *The Cryosphere*, **7**, 375–393, https://doi.org/10.5194/tc-7-375-2013

Fullea, J., Lebedev, S., Martinec, Z. and Celli, N.L. 2021. WINTERC-G: mapping the upper mantle thermochemical heterogeneity from coupled geophysical–petrological inversion of seismic waveforms, heat flow, surface elevation and gravity satellite data. *Geophysical Journal International*, **226**, 146–191, https://doi.org/10.1093/gji/ggab094

Futa, K. and Le Masurier, W.E. 1983. Nd and Sr isotopic studies on Cenozoic mafic lavas from West Antarctica: Another source for continental alkali basalts. *Contributions to Mineralogy and Petrology*, **83**, 38–44, https://doi.org/10.1007/BF00373077

Gamble, J.A., McGibbon, F., Kyle, P.R., Menzies, M. and Kirsch, I. 1988. Metasomatised xenoliths from Foster Crater, Antarctica: Implications for lithospheric structure and process beneath the Transantarctic Mountain front. *Journal of Petrology*, Special Volume, **1**, 109–138, https://doi.org/10.1093/petrology/Special_Volume.1.109

Golynsky, A.V., Ferraccioli, F. *et al.* 2018. New Magnetic Anomaly Map of the Antarctic. *Geophysical Research Letters*, **45**, 6437–6449, https://doi.org/10.1029/2018GL078153

Gomez, N., Mitrovica, J.X., Huybers, P. and Clark, P.U. 2010. Sea level as a stabilizing factor for marine-ice-sheet grounding lines. *Nature Geoscience*, **3**, 850–853, https://doi.org/10.1038/ngeo1012

Gomez, N., Pollard, D. and Holland, D. 2015. Sea-level feedback lowers projections of future Antarctic Ice-Sheet mass loss. *Nature Communications*, **6**, 8798, https://doi.org/10.1038/ncomms9798

Gomez, N., Latychev, K. and Pollard, D. 2018. A coupled ice sheet–sea level model incorporating 3D Earth structure: variations in Antarctica during the last deglacial retreat. *Journal of Climate*, **31**, 4041–4054, https://doi.org/10.1175/JCLI-D-17-0352.1

Grad, M., Tiira, T. and Group, E.W. 2009. The Moho depth map of the European Plate. *Geophysical Journal International*, **176**, 279–292, https://doi.org/10.1111/j.1365-246X.2008.03919.x

Green, D.H., Hibberson, W.O., Kovács, I. and Rosenthal, A. 2010. Water and its influence on the lithosphere–asthenosphere boundary. *Nature*, **467**, 448–451, https://doi.org/10.1038/nature09369

Griffin, W.L. and O'Reilly, S.Y. 1987. The composition of the lower crust and the nature of the continental Moho-xenolith evidence. *In*: Nixon, P.H. (ed.) *Mantle Xenoliths*. Wiley- Interscience, New York, 413–430.

Groh, A., Ewert, H. *et al.* 2012. An investigation of glacial isostatic adjustment over the Amundsen Sea sector, West Antarctica. *Global and Planetary Change*, **98**, 45–53, https://doi.org/10.1016/j.gloplacha.2012.08.001

Handler, M.R., Wysoczanski, R.J. and Gamble, J.A. 2003. Proterozoic lithosphere in Marie Byrd Land, West Antarctica: Re–Os systematics of spinel peridotite xenoliths. *Chemical Geology*, **196**, 131–145, https://doi.org/10.1016/S0009-2541(02)00410-2

Handler, M.R., Wysoczanski, R.J. and Gamble, J.A. 2021. Marie Byrd Land lithospheric mantle: a review of the xenolith record. *Geological Society, London, Memoirs*, **56**, https://doi.org/10.1144/M56-2020-17

Hansen, L.N., Warren, J.M., Zimmerman, M.E. and Kohlstedt, D.L. 2016. Viscous anisotropy of textured olivine aggregates, Part 1: Measurement of the magnitude and evolution of anisotropy. *Earth and Planetary Science Letters*, **445**, 92–103, https://doi.org/10.1016/j.epsl.2016.04.008

Hansen, S.E., Graw, J.H. *et al.* 2014. Imaging the Antarctic mantle using adaptively parameterized P-wave tomography: Evidence for heterogeneous structure beneath West Antarctica. *Earth and Planetary Science Letters*, **408**, 66–78, https://doi.org/10.1016/j.epsl.2014.09.043

Harada, Y., Suzuki, H. and Ohashi, S. 1958. Determination of the differences in gravity values at Chiba (Japan), Singapore, Cape Town and Antarctica. *Journal of the Geodetic Society of Japan*, **5**, 53–60.

Harley, S.L., Fitzsimons, I.C. and Zhao, Y. 2013. Antarctica and supercontinent evolution: historical perspectives, recent advances and unresolved issues. *Geological Society, London, Special Publications*, **383**, 1–34, https://doi.org/10.1144/SP383.9

Hart, S.R. and Kyle, P.R. 1993. Geochemistry of McMurdo Volcanics Group rocks. *Antarctic Journal of the United States*, **28**, 14–16.

Hart, S.R., Blusztajn, J. and Craddock, C. 1995. Cenozoic volcanism in Antarctica: Jones Mountains and Peter I Island. *Geochimica et Cosmochimica Acta*, **59**, 3379–3388, https://doi.org/10.1016/0016-7037(95)00212-I

Hart, S.R., Blusztajn, J., LeMasurier, W.E. and Rex, D.C. 1997. Hobbs Coast Cenozoic volcanism: implications for the West Antarctic rift system. *Chemical Geology*, **139**, 223–248, https://doi.org/10.1016/S0009-2541(97)00037-5

Hermans, T., van der Wal, W. and Broerse, T. 2018. Reversal of the direction of horizontal velocities induced by GIA as a function of mantle viscosity. *Geophysical Research Letters*, **45**, 9597–9604, https://doi.org/10.1029/2018GL078533

Herzberg, C.T., Fyfe, W.S. and Carr, M.J. 1983. Density constraints on the formation of the continental Moho and crust. *Contributions to Mineralogy and Petrology*, **84**, 1–5, https://doi.org/10.1007/BF01132324

Hole, M.J., Kempton, P.D. and Millar, I.L. 1993. Trace-element and isotopic characteristics of small-degree melts of the asthenosphere: Evidence from the alkalic basalts of the Antarctic Peninsula. *Chemical Geology*, **109**, 51–68, https://doi.org/10.1016/0009-2541(93)90061-M

Howells, S. and O'Hara, M.J. 1978. Low solubility of alumina in enstatite and uncertainties in estimated palaeogeotherms. *Philosophical Transactions of the Royal Society A: Mathematical, Physical and Engineering Sciences*, **288**, 471–486, https://doi.org/10.1098/rsta.1978.0029

Ito, E. and Takahashi, E. 1989. Postspinel transformations in the system Mg_2SiO_4–Fe_2SiO_4 and some geophysical implications. *Journal of Geophysical Research: Solid Earth*, **94**, 10 637–10 646, https://doi.org/10.1029/JB094iB08p10637

Ivins, E.R., James, T.S., Wahr, J.O., Schrama, E.J., Landerer, F.W. and Simon, K.M. 2013. Antarctic contribution to sea level rise observed by GRACE with improved GIA correction. *Journal of Geophysical Research: Solid Earth*, **118**, 3126–3141, https://doi.org/10.1002/jgrb.50208

Ivins, E.R., van der Wal, W., Wiens, D.A., Lloyd, A.J. and Caron, L. 2021. Antarctic upper mantle rheology. *Geological Society, London, Memoirs*, **56**, https://doi.org/10.1144/M56-2020-19

Jagoutz, E., Palme, H. *et al.* 1979. The abundances of major, minor and trace elements in the Earth's mantle as derived from primitive ultramafic nodules. *In*: *Proceedings of the Tenth Lunar and Planetary Science Conference, Houston, Texas, March 19–23, 1979*. Pergamon Press, Oxford, UK, 2031–2050.

Kachuck, S.B., Martin, D.F., Bassis, J.N. and Price, S.F. 2020. Rapid viscoelastic deformation slows marine ice sheet instability at Pine Island Glacier. *Geophysical Research Letters*, **47**, e2019GL086446, https://doi.org/10.1029/2019GL086446

Kaminuma, K. 1979. Seismological bulletin of Syowa Station, Antarctica, 1977. JARE data reports. *Seismology*, **12**, 1–39.

Karato, S.-I., Jung, H., Katayama, I. and Skemer, P. 2008. Geodynamic significance of seismic anisotropy of the upper mantle: New insights from laboratory studies. *Annual Review of Earth and Planetary Sciences*, **36**, 59–95, https://doi.org/10.1146/annurev.earth.36.031207.124120

Katsura, T. and Ito, E. 1989. The system Mg_2SiO_4–Fe_2SiO_4 at high pressures and temperatures: Precise determination of stabilities of olivine, modified spinel, and spinel. *Journal of Geophysical Research: Solid Earth*, **94**, 15 663–15 670, https://doi.org/10.1029/JB094iB11p15663

Kaufmann, G. and Lambeck, K. 2000. Mantle dynamics, postglacial rebound and the radial viscosity profile. *Physics of the Earth and Planetary Interiors*, **121**, 301–324, https://doi.org/10.1016/S0031-9201(00)00174-6

Kaufmann, G., Wu, P. and Ivins, E.R. 2005. Lateral viscosity variations beneath Antarctica and their implications on regional rebound motions and seismotectonics. *Journal of Geodynamics*, **39**, 165–181, https://doi.org/10.1016/j.jog.2004.08.009

Kaula, W.M. 1972. Global gravity and tectonics. *In*: Robertson, E.C. (ed.) *The Nature of the Solid Earth*. McGraw-Hill, New York, 385–405.

Kawakatsu, H. *et al.* 2009. Seismic evidence for sharp lithosphere-asthenosphere boundaries of oceanic plates. *Science*, **324**, 499–502, https://doi.org/10.1126/science.1169499

Kennett, B.L.N., Engdahl, E.R. and Buland, R. 1995. Constraints on seismic velocities in the Earth from traveltimes. *Geophysical Journal International*, **122**, 108–124, https://doi.org/10.1111/j.1365-246X.1995.tb03540.x

Kennicutt, M.C.II, Chown, S.L. *et al.* 2015. A roadmap for Antarctic and Southern Ocean science for the next two decades and beyond. *Antarctic Science*, **27**, 3–18, https://doi.org/10.1017/S0954102014000674

Kim, H.R., Golynsky, A.V., Golynsky, D.A., Yu, H., von Frese, R.R.B. and Hong, J.K. 2022. New magnetic anomaly constraints on the Antarctic crust. *Journal of Geophysical Research: Solid Earth*, **127**, e2021JB023329, https://doi.org/10.1029/2021JB023329

King, M.A. and Santamaría-Gómez, A. 2016. Ongoing deformation of Antarctica following recent Great Earthquakes. *Geophysical Research Letters*, **43**, 1918–1927, https://doi.org/10.1002/2016GL067773

King, M.A., Bingham, R.J., Moore, P., Whitehouse, P.L., Bentley, M. and Milne, G.A. 2012. Lower satellite-gravimetry estimates of Antarctic sea-level contribution. *Nature*, **491**, 586–589, https://doi.org/10.1038/nature11621

King, M.A., Whitehouse, P.L. and van der Wal, W. 2015. Incomplete separability of Antarctic plate rotation from glacial isostatic adjustment deformation within geodetic observations. *Geophysical Journal International*, **204**, 324–330, https://doi.org/10.1093/gji/ggv461

Knopoff, L. and Vane, G. 1978. Age of East Antarctica from surface wave dispersion. *Pure and Applied Geophysics*, **117**, 806–815, https://doi.org/10.1007/BF00879981

Kogarko, L.N., Kurat, G. and Ntaflos, T. 2007. Henrymeyerite in the metasomatized upper mantle of Eastern Antarctica. *The Canadian Mineralogist*, **45**, 497–501, https://doi.org/10.2113/gscanmin.45.3.497

Konrad, H., Sasgen, I., Pollard, D. and Klemann, V. 2015. Potential of the solid-Earth response for limiting long-term West Antarctic Ice Sheet retreat in a warming climate. *Earth and Planetary Science Letters*, **432**, 254–264, https://doi.org/10.1016/j.epsl.2015.10.008

Kovács, I., Patkó, L., Falus, G., Aradi, L.E., Szanyi, G., Gráczer, Z. and Szabó, C. 2018. Upper mantle xenoliths as sources of geophysical information: the Perşani Mts. area as a case study. *Acta Geodaetica et Geophysica*, **53**, 415–438, https://doi.org/10.1007/s40328-018-0231-2

Kyle, P.R., Wright, A.C. and Kirsch, I. 1987. Ultramafic xenoliths in the late Cenozoic McMurdo Volcanic Group, western Ross Sea embayment, Antarctica. *In*: Nixon, P.H. (ed.) *Mantle Xenoliths*. Wiley- Interscience, New York, 287–293.

Kyle, P.R., Moore, J.A. and Thirlwall, M.F. 1992. Petrologic evolution of anorthoclase phonolite lavas at Mount Erebus, Ross Island, Antarctica. *Journal of Petrology*, **33**, 849–875, https://doi.org/10.1093/petrology/33.4.849

Kyle, P.R., Pankhurst, R., Mukasa, S., Panter, K., Smellie, J. and McIntosh, W. 1994. *Sr, Nd and Pb Isotopic Variations in the Marie Byrd Plume, West Antarctica*. US Geological Survey Circular, **1107**.

Lambeck, K., Yokoyama, Y., Johnston, P. and Purcell, A. 2000. Global ice volumes at the Last Glacial Maximum and early Lateglacial. *Earth and Planetary Science Letters*, **181**, 513–527, https://doi.org/10.1016/S0012-821X(00)00223-5

Larour, E., Seroussi, H., Adhikari, S., Ivins, E., Caron, L., Morlighem, M. and Schlegel, N. 2019. Slowdown in Antarctic mass loss from solid Earth and sea-level feedbacks. *Science*, **364**, eaav7908, https://doi.org/10.1126/science.aav7908

Lau, H.C.P., Austermann, J., Holtzman, B.K., Havlin, C., Lloyd, A.J., Book, C. and Hopper, E. 2021. Frequency dependent mantle viscoelasticity via the complex viscosity: Cases from Antarctica. *Journal of Geophysical Research: Solid Earth*, **126**, e2021JB022622, https://doi.org/10.1029/2021JB022622

Leat, P.T., Ross, A.J. and Gibson, S.A. 2021. Ultramafic mantle xenoliths in the Late Cenozoic volcanic rocks of the Antarctic Peninsula and Jones Mountains, West Antarctica. *Geological Society, London, Memoirs*, **56**, https://doi.org/10.1144/M56-2019-44

Le Maitre, R.W., Streckeisen, A., Zanettin, B., Le Bas, M., Bonin, B. and Bateman, P. (eds) 2002. *Igneous Rocks: A Classification and Glossary of Terms. Recommendations of the International Union of Geological Sciences Subcommission of the Systematics of Igneous Rocks*. 2nd edn. Cambridge University Press, Cambridge, UK.

LeMasurier, W.E. and Landis, C.A. 1996. Mantle-plume activity recorded by low-relief erosion surfaces in West Antarctica and New Zealand. *Geological Society of America Bulletin*, **108**, 1450–1466, https://doi.org/10.1130/0016-7606(1996)108<1450:MPARBL>2.3.CO;2

Lloyd, A.J., Wiens, D.A. *et al.* 2015. A seismic transect across West Antarctica: Evidence for mantle thermal anomalies beneath the Bentley Subglacial Trench and the Marie Byrd Land Dome. *Journal of Geophysical Research: Solid Earth*, **120**, 8439–8460, https://doi.org/10.1002/2015JB012455

Lloyd, A.J., Wiens, D.A. *et al.* 2020. Seismic structure of the Antarctic upper mantle imaged with adjoint tomography. *Journal of Geophysical Research: Solid Earth*, **125**, https://doi.org/10.1029/2019JB017823

Lösing, M. and Ebbing, J. 2021. Predicting geothermal heat flow in Antarctica with a machine learning approach. *Journal of Geophysical Research: Solid Earth*, **126**, e2020JB021499, https://doi.org/10.1029/2020JB021499

Lösing, M., Ebbing, J. and Szwillus, W. 2020. Geothermal heat flux in Antarctica: Assessing models and observations by Bayesian inversion. *Frontiers in Earth Science*, **8**, 105, https://doi.org/10.3389/feart.2020.00105

Lough, A.C., Wiens, D.A. and Nyblade, A. 2018. Reactivation of ancient Antarctic rift zones by intraplate seismicity. *Nature Geoscience*, **11**, 515–519, https://doi.org/10.1038/s41561-018-0140-6

Macgregor, I.D. 1974. The system $MgO–Al_2O_3–SiO_2$: Solubility of Al_2O_3 in enstatite for spinel and garnet peridotite compositions. *American Mineralogist*, **59**, 110–119.

Mainprice, D. and Silver, P.G. 1993. Interpretation of SKS-waves using samples from the subcontinental lithosphere. *Physics of the Earth and Planetary Interiors*, **78**, 257–280, https://doi.org/10.1016/0031-9201(93)90160-B

Manning, J. 2001. *The SCAR Geodetic Infrastructure of Antarctica.* Report from the Second SCAR Antarctic Geodesy Symposium, Warsaw, July 1999. SCAR Report Number 20. Scientific Committee on Antarctic Research (SCAR), Cambridge, UK, 22–30.

Martin, A.P. 2021. A review of the composition and chemistry of peridotite mantle xenoliths in volcanic rocks from Antarctica and their relevance to petrological and geophysical models for the lithospheric mantle. *Geological Society, London, Memoirs*, **56**, https://doi.org/10.1144/M56-2021-26

Martin, A.P., Cooper, A.F. and Price, R.C. 2013. Petrogenesis of Cenozoic, alkalic volcanic lineages at Mount Morning, West Antarctica and their entrained lithospheric mantle xenoliths: Lithospheric v. asthenospheric mantle sources. *Geochimica et Cosmochimica Acta*, **122**, 127–152, https://doi.org/10.1016/j.gca.2013.08.025

Martin, A.P., Price, R.C. and Cooper, A.F. 2014. Constraints on the composition, source and petrogenesis of plagioclase-bearing mantle peridotite. *Earth-Science Reviews*, **138**, 89–101, https://doi.org/10.1016/j.earscirev.2014.08.006

Martin, A.P., Price, R.C., Cooper, A.F. and McCammon, C.A. 2015. Petrogenesis of the rifted southern Victoria Land lithospheric mantle, Antarctica, Inferred from petrography, geochemistry, thermobarometry and oxybarometry of peridotite and pyroxenite xenoliths from the Mount Morning eruptive centre. *Journal of Petrology*, **56**, 193–226, https://doi.org/10.1093/petrology/egu075

Martin, A.P., Cooper, A.F., Price, R.C., Doherty, C.L. and Gamble, J.A. 2021*a*. A review of mantle xenoliths in volcanic rocks from southern Victoria Land, Antarctica. *Geological Society, London, Memoirs*, **56**, https://doi.org/10.1144/M56-2019-42

Martin, A.P., Cooper, A.F., Price, R.C., Kyle, P.R. and Gamble, J.A. 2021*b*. Erebus Volcanic Province: petrology. *Geological Society, London, Memoirs*, **55**, 447–489, https://doi.org/10.1144/M55-2018-80

Martín-Español, A., King, M.A., Zammit-Mangion, A., Andrews, S.B., Moore, P. and Bamber, J.L. 2016. An assessment of forward and inverse GIA solutions for Antarctica. *Journal of Geophysical Research: Solid Earth*, **121**, 6947–6965, https://doi.org/10.1002/2016JB013154

Martos, Y.M., Catalán, M., Jordan, T.A., Golynsky, A., Golynsky, D., Eagles, G. and Vaughan, D.G. 2017. Heat flux distribution of Antarctica unveiled. *Geophysical Research Letters*, **44**, 11,417–11,426, https://doi.org/10.1002/2017GL075609

Meisel, T., Walker, R.J., Irving, A.J. and Lorand, J.-P. 2001. Osmium isotopic compositions of mantle xenoliths: a global perspective. *Geochimica et Cosmochimica Acta*, **65**, 1311–1323, https://doi.org/10.1016/S0016-7037(00)00566-4

Meissner, R. 1986. *The Continental Crust: A Geophysical Approach.* International Geophysics Series, **34**. Academic Press, Orlando, FL.

Nardini, I., Armienti, P., Rocchi, S., Dallai, L. and Harrison, D. 2009. Sr–Nd–Pb–He–O isotope and geochemical constraints on the genesis of Cenozoic magmas from the West Antarctic Rift. *Journal of Petrology*, **50**, 1359–1375, https://doi.org/10.1093/petrology/egn082

Nield, G.A., Barletta, V.R. *et al.* 2014. Rapid bedrock uplift in the Antarctic Peninsula explained by viscoelastic response to recent ice unloading. *Earth and Planetary Science Letters*, **397**, 32–41, https://doi.org/10.1016/j.epsl.2014.04.019

Nield, G.A., Whitehouse, P.L., van der Wal, W., Blank, B., O'Donnell, J.P. and Stuart, G.W. 2018. The impact of lateral variations in lithospheric thickness on glacial isostatic adjustment in West Antarctica. *Geophysical Journal International*, **214**, 811–824, https://doi.org/10.1093/gji/ggy158

Nield, G.A., King, M.A., Steffen, R. and Blank, B. 2022. A global, spherical finite-element model for post-seismic deformation using Abaqus. *Geoscientific Model Development*, **15**, 2489–2503, https://doi.org/10.5194/gmd-15-2489-2022

Nixon, P.H. 1987*a*. *Mantle Xenoliths.* Wiley-Interscience, New York.

Nixon, P.H. 1987*b*. Indian–Australian and Antartic plates – Introduction. *In*: Nixon, P.H. (ed.) *Mantle Xenoliths.* Wiley- Interscience, New York, 241–248.

O'Donnell, J.P., Selway, K. *et al.* 2017. The uppermost mantle seismic velocity and viscosity structure of central West Antarctica. *Earth and Planetary Science Letters*, **472**, 38–49, https://doi.org/10.1016/j.epsl.2017.05.016

O'Reilly, S.Y. and Griffin, W.L. 2010. The continental lithosphere–asthenosphere boundary: Can we sample it? *Lithos*, **120**, 1–13, https://doi.org/10.1016/j.lithos.2010.03.016

Panter, K.S. and Martin, A.P. 2021. West Antarctic mantle deduced from mafic magmatism. *Geological Society, London, Memoirs*, **56**, https://doi.org/10.1144/M56-2021-10

Panter, K.S., Blusztajn, J., Hart, S.R., Kyle, P.R., Esser, R. and McIntosh, W.C. 2006. The origin of HIMU in the SW Pacific: Evidence from intraplate volcanism in southern New Zealand and subantarctic islands. *Journal of Petrology*, **47**, 1673–1704, https://doi.org/10.1093/petrology/egl024

Pappa, F. and Ebbing, J. 2021. Gravity, magnetics and geothermal heat flow of the Antarctic lithospheric crust and mantle. *Geological Society, London, Memoirs*, **56**, https://doi.org/10.1144/M56-2020-5

Pappa, F., Ebbing, J. and Ferraccioli, F. 2019*a*. Moho depths of Antarctica: Comparison of seismic, gravity, and isostatic results. *Geochemistry, Geophysics, Geosystems*, **20**, 1629–1645, https://doi.org/10.1029/2018GC008111

Pappa, F., Ebbing, J., Ferraccioli, F. and van der Wal, W. 2019*b*. Modeling satellite gravity gradient data to derive density, temperature, and viscosity structure of the Antarctic lithosphere. *Journal of Geophysical Research: Solid Earth*, **124**, 12 053–12 076, https://doi.org/10.1029/2019JB017997

Pattyn, F. 2018. The paradigm shift in Antarctic ice sheet modelling. *Nature Communications*, **9**, 2728, https://doi.org/10.1038/s41467-018-05003-z

Paulson, A., Zhong, S. and Wahr, J. 2007. Limitations on the inversion for mantle viscosity from postglacial rebound. *Geophysical Journal International*, **168**, 1195–1209, https://doi.org/10.1111/j.1365-246X.2006.03222.x

Paxman, G.J.G. 2021. Antarctic palaeotopography. *Geological Society, London, Memoirs*, **56**, https://doi.org/10.1144/M56-2020-7

Paxman, G.J.G., Watts, A.B., Ferraccioli, F., Jordan, T.A., Bell, R.E., Jamieson, S.S.R. and Finn, C.A. 2016. Erosion-driven uplift in the Gamburtsev Subglacial Mountains of East Antarctica. *Earth and Planetary Science Letters*, **452**, 1–14, https://doi.org/10.1016/j.epsl.2016.07.040

Paxman, G.J.G., Jamieson, S.S.R., Hochmuth, K., Gohl, K., Bentley, M.J., Leitchenkov, G. and Ferraccioli, F. 2019. Reconstructions of Antarctic topography since the Eocene–Oligocene boundary. *Palaeogeography, Palaeoclimatology, Palaeoecology*, **535**, 109346, https://doi.org/10.1016/j.palaeo.2019.109346

Pearson, D., Canil, D. and Shirey, S. 2003. Mantle samples included in volcanic rocks: xenoliths and diamonds. *In*: Carlson, R.W. (ed.) *The Mantle and Core.* Treatise on Geochemistry, **2**, 171–275, https://doi.org/10.1016/B0-08-043751-6/02005-3

Peltier, W.R., Argus, D.F. and Drummond, R. 2018. Comment on 'An assessment of the *ICE-6G_C (VM5a)* glacial isostatic adjustment model' by Purcell *et al. Journal of Geophysical Research. Solid Earth*, **123**, 2019–2018, https://doi.org/10.1002/2016JB013844

Perinelli, C., Andreozzi, G., Conte, A., Oberti, R. and Armienti, P. 2012. Redox state of subcontinental lithospheric mantle and relationships with metasomatism: insights from spinel peridotites from northern Victoria Land (Antarctica). *Contributions to Mineralogy and Petrology*, **164**, 1053–1067, https://doi.org/10.1007/s00410-012-0788-7

Phillips, E.H. , Sims, K.W.W. *et al.* 2018. The nature and evolution of mantle upwelling at Ross Island, Antarctica, with implications for the source of HIMU lavas. *Earth and Planetary Science Letters*, **498**, 38–53, https://doi.org/10.1016/j.epsl.2018.05.049

Pollard, D. and DeConto, R.M. 2020. Continuous simulations over the last 40 million years with a coupled Antarctic ice sheet–sediment model. *Palaeogeography, Palaeoclimatology, Palaeoecology*, **537**, 109374, https://doi.org/10.1016/j.palaeo.2019.109374

Pollitz, F.F. and Thatcher, W. 2010. On the resolution of shallow mantle viscosity structure using postearthquake relaxation data: Application to the 1999 Hector Mine, California, earthquake. *Journal of Geophysical Research: Solid Earth*, **115**, B10412, https://doi.org/10.1029/2010JB007405

Powell, E., Gomez, N., Hay, C., Latychev, K. and Mitrovica, J. 2020. Viscous effects in the solid Earth response to modern Antarctic ice mass flux: Implications for geodetic studies of WAIS stability in a warming world. *Journal of Climate*, **33**, 443–459, https://doi.org/10.1175/JCLI-D-19-0479.1

Powell, E.M., Pan, L., Hoggard, M.J., Latychev, K., Gomez, N., Austermann, J. and Mitrovica, J.X. 2021. The impact of 3-D Earth structure on far-field sea level following interglacial West Antarctic Ice Sheet collapse. *Quaternary Science Reviews*, **273**, 107256, https://doi.org/10.1016/j.quascirev.2021.107256

Prior, G.T. 1902. Report on the rock specimens collected by the Southern Cross Antarctic Expedition. *In*: *Report on Southern Cross Expedition*. British Museum, London, 321–332.

Prior, G.T. 1907. Report on the rock specimens collected during the 'Discovery' Antarctic Expedition, 1901–1904. *Natural History*, **1**, 101–160.

Rémy, F. and Parouty, S. 2009. Antarctic ice sheet and radar altimetry: A review. *Remote Sensing*, **1**, 1212–1239, https://doi.org/10.3390/rs1041212

Rémy, F., Mazzega, P., Houry, S., Brossier, C. and Minster, J. 1989. Mapping of the topography of continental ice by inversion of satellite-altimeter data. *Journal of Glaciology*, **35**, 98–107, https://doi.org/10.3189/002214389793701419

Revenaugh, J. and Jordan, T.H. 1991. Mantle layering from ScS reverberations: 2. The transition zone. *Journal of Geophysical Research: Solid Earth*, **96**, 19 763–19 780, https://doi.org/10.1029/91JB01486

Ritzwoller, M.H., Shapiro, N.M., Levshin, A.L. and Leahy, G.M. 2001. Crustal and upper mantle structure beneath Antarctica and surrounding oceans. *Journal of Geophysical Research*, **106**, 30 645–30 670, https://doi.org/10.1029/2001JB000179

Ross, C.S., Foster, M.D. and Myers, A.T. 1954. Origin of dunites and of olivine-rich inclusions in basaltic rocks1. *American Mineralogist*, **39**, 693–737.

Scheinert, M., Engels, O., Schrama, E.J.O., van der Wal, W. and Horwath, M. 2021. Geodetic observations for constraining mantle processes in Antarctica. *Geological Society, London, Memoirs*, **56**, https://doi.org/10.1144/M56-2021-22

Seroussi, H., Ivins, E.R., Wiens, D.A. and Bondzio, J. 2017. Influence of a West Antarctic mantle plume on ice sheet basal conditions. *Journal of Geophysical Research: Solid Earth*, **122**, 7127–7155, https://doi.org/10.1002/2017JB014423

Shapiro, N.M. and Ritzwoller, M.H. 2004. Inferring surface heat flux distributions guided by a global seismic model: particular application to Antarctica. *Earth and Planetary Science Letters*, **223**, 213–224, https://doi.org/10.1016/j.epsl.2004.04.011

Shearer, P.M. and Flanagan, M.P. 1999. Seismic velocity and density jumps across the 410- and 660-kilometer discontinuities. *Science*, **285**, 1545–1548, https://doi.org/10.1126/science.285.5433.1545

Shen, W., Wiens, D.A., Lloyd, A.J. and Nyblade, A.A. 2020. A geothermal heat flux map of Antarctica empirically constrained by seismic structure. *Geophysical Research Letters*, **47**, e2020GL086955, https://doi.org/10.1029/2020GL086955

Sheraton, J.W. 1983. Geochemistry of mafic igneous rocks of the northern Prince Charles Mountains, Antarctica. *Journal of the Geological Society of Australia*, **30**, 295–304, https://doi.org/10.1080/00167618308729257

Sheraton, J.W. and Black, L.P. 1982. Geochemistry and geochronology of Proterozoic tholeiite dykes of East Antarctica: evidence for mantle metasomatism. *Contributions to Mineralogy and Petrology*, **78**, 305–317, https://doi.org/10.1007/BF00398925

Shervais, J.W. 1988. A review of Volcanic Rocks: *Mantle Xenoliths*. Peter H. Nixon, Ed. Wiley- Interscience, New York, 1987. xviii, 844 pp. *Science*, **241**, 366–367, https://doi.org/10.1126/science.241.4863.366

Siegert, M.J., Jamieson, S.S.R. and White, D. 2018. Exploration of subsurface Antarctica: uncovering past changes and modern processes. *Geological Society, London, Special Publications*, **461**, 1–6, https://doi.org/10.1144/SP461.15

Sims, K.W.W. and Hart, S.R. 2006. Comparison of Th, Sr, Nd and Pb isotopes in oceanic basalts: Implications for mantle heterogeneity and magma genesis. *Earth and Planetary Science Letters*, **245**, 743–761, https://doi.org/10.1016/j.epsl.2006.02.030

Smellie, J.L., Panter, K.S. and Geyer, A. (eds) 2021. *Volcanism in Antarctica: 200 Million Years of Subduction, Rifting and Continental Break-up*. Geological Society, London, Memoirs, **55**, https://doi.org/10.1144/M55

Spada, G. and Melini, D. 2019. SELEN4 (SELEN version 4.0): a Fortran program for solving the gravitationally and topographically self-consistent sea-level equation in glacial isostatic adjustment modeling. *Geoscientific Model Development*, **12**, 5055–5075, https://doi.org/10.5194/gmd-12-5055-2019

Steinberger, B., Torsvik, T.H. and Becker, T.W. 2012. Subduction to the lower mantle – a comparison between geodynamic and tomographic models. *Solid Earth*, **3**, 415–432, https://doi.org/10.5194/se-3-415-2012

Storey, B.C., Leat, P.T., Weaver, S.D., Pankhurst, R.J., Bradshaw, J.D. and Kelley, S. 1999. Mantle plumes and Antarctica–New Zealand rifting: evidence from mid-Cretaceous mafic dykes. *Journal of the Geological Society, London*, **156**, 659–671, https://doi.org/10.1144/gsjgs.156.4.0659

Sun, S.S. and Hanson, G.N. 1975. Origin of Ross Island basanitoids and limitations upon the heterogeneity of mantle sources for alkali basalts and nephelinites. *Contributions to Mineralogy and Petrology*, **52**, 77–106, https://doi.org/10.1007/BF00395006

Sutter, J., Fischer, H., Grosfeld, K., Karlsson, N.B., Kleiner, T., Van Liefferinge, B. and Eisen, O. 2019. Modelling the Antarctic Ice Sheet across the mid-Pleistocene transition–implications for Oldest Ice. *The Cryosphere*, **13**, 2023–2041, https://doi.org/10.5194/tc-13-2023-2019

Talbot, J.L., Hobbs, B.E., Wilshire, H.G. and Sweatman, T.R. 1963. Xenoliths and xenocrysts from lavas of the Kerguelen Archipelago. *American Mineralogist*, **48**, 159–179.

ten Brink, U. and Stern, T. 1992. Rift flank uplifts and Hinterland Basins: Comparison of the Transantarctic Mountains with the Great Escarpment of southern Africa. *Journal of Geophysical Research: Solid Earth*, **97**, 569–585, https://doi.org/10.1029/91JB02231

Thomas, I.D., King, M.A. *et al.* 2011. Widespread low rates of Antarctic glacial isostatic adjustment revealed by GPS observations. *Geophysical Research Letters*, **38**, L22302, https://doi.org/10.1029/2011GL049277

Thomson, J.A. 1916. *Report on the Inclusions of the Volcanic Rocks of the Ross Archipelago (with Appendix by F. Cohen)*. Report of the British Antarctic Expedition, 1907–1909.

Tommasi, A., Tikoff, B. and Vauchez, A. 1999. Upper mantle tectonics: three-dimensional deformation, olivine crystallographic fabrics and seismic properties. *Earth and Planetary Science Letters*, **168**, 173–186, https://doi.org/10.1016/S0012-821X(99)00046-1

Turcotte, D.L. and Oxburgh, E.R. 1967. Finite amplitude convective cells and continental drift. *Journal of Fluid Mechanics*, **28**, 29–42, https://doi.org/10.1017/S0022112067001880

van der Wal, W., Barnhoorn, A., Stocchi, P., Gradmann, S., Wu, P., Drury, M. and Vermeersen, B. 2013. Glacial isostatic adjustment model with composite 3-D Earth rheology for Fennoscandia. *Geophysical Journal International*, **194**, 61–77, https://doi.org/10.1093/gji/ggt099

van der Wal, W., Whitehouse, P.L. and Schrama, E.J.O. 2015. Effect of GIA models with 3D composite mantle viscosity on GRACE mass balance estimates for Antarctica. *Earth and Planetary Science Letters*, **414**, 134–143, https://doi.org/10.1016/j.epsl.2015.01.001

van der Wal, W., Barletta, V., Nield, G. and van Calcar, C. 2022. Glacial isostatic adjustment and post-seismic deformation in Antarctica. *Geological Society, London, Memoirs*, **56**, https://doi.org/10.1144/M56-2022-13

Vardić, K., Clarke, P.J. and Whitehouse, P.L. 2022. A GNSS velocity field for crustal deformation studies: The influence of glacial isostatic adjustment on plate motion models. *Geophysical Journal International*, **231**, 426–458, https://doi.org/10.1093/gji/ggac047

Velicogna, I. and Wahr, J. 2006. Measurements of time-variable gravity show mass loss in Antarctica. *Science*, **311**, 1754, https://doi.org/10.1126/science.1123785

Wagner, P.A. 1914. *The Diamond Fields of Southern Africa*. Transvaal Leader, Johannesburg, South Africa.

Wannamaker, P.E., Stodt, J.A., Hill, G.J., Maris, V. and Kordy, M.A. 2021. Thermal regime and state of hydration of the Antarctic upper mantle from regional-scale electrical properties. *Geological Society, London, Memoirs*, **56**, https://doi.org/10.1144/M56-2020-4

Warner, R.D. and Wasilewski, P.J. 1995. Magnetic petrology of lower crust and upper mantle xenoliths from McMurdo Sound, Antarctica. *Tectonophysics*, **249**, 69–92, https://doi.org/10.1016/0040-1951(95)00014-E

Weaver, S.D., Storey, B.C., Pankhurst, R.J., Mukasa, S.B., DiVenere, V.J. and Bradshaw, J.D. 1994. Antarctica–New Zealand rifting and Marie Byrd Land lithospheric magmatism linked to ridge subduction and mantle plume activity. *Geology*, **22**, 811–814, https://doi.org/10.1130/0091-7613(1994)022<0811:ANZRAM>2.3.CO;2

Whitehouse, P.L. 2018. Glacial isostatic adjustment modelling: historical perspectives, recent advances, and future directions. *Earth Surface Dynamics*, **6**, 401–429, https://doi.org/10.5194/esurf-6-401-2018

Whitehouse, P.L., Bentley, M.J., Milne, G.A., King, M.A. and Thomas, I.D. 2012. A new glacial isostatic adjustment model for Antarctica: calibrated and tested using observations of relative sea-level change and present-day uplift rates. *Geophysical Journal International*, **190**, 1464–1482, https://doi.org/10.1111/j.1365-246X.2012.05557.x

Whitehouse, P.L., Gomez, N., King, M.A. and Wiens, D.A. 2019. Solid Earth change and the evolution of the Antarctic Ice Sheet. *Nature Communications*, **10**, 503, https://doi.org/10.1038/s41467-018-08068-y

Wiens, D.A., Shen, W. and Lloyd, A.J. 2021. The seismic structure of the Antarctic upper mantle. *Geological Society, London, Memoirs*, **56**, https://doi.org/10.1144/M56-2020-18

Wood, B.J. and Banno, S. 1973. Garnet–orthopyroxene and orthopyroxene–clinopyroxene relationships in simple and complex systems. *Contributions to Mineralogy and Petrology*, **42**, 109–124, https://doi.org/10.1007/BF00371501

Wörner, G. 1999. Lithospheric dynamics and mantle sources of alkaline magmatism of the Cenozoic West Antarctic Rift System. *Global and Planetary Change*, **23**, 61–77, https://doi.org/10.1016/S0921-8181(99)00051-X

Wörner, G. and Zipfel, J. 1996. A mantle P–T path for the Ross Sea Rift margin (Antarctica) derived from Ca-in-olivine zonation patterns in peridotites xenoliths of the Plio-Pleistocene Mt. Melbourne Volcanic Field. *Geologisches Jahrbuch B*, **89**, 157–167.

Wyllie, P.J. 1981. Plate tectonics and magma genesis. *Geologische Rundschau*, **70**, 128–153, https://doi.org/10.1007/BF0176 4318

Wysoczanski, R.J. 1993. *Lithospheric xenoliths from the Marie Byrd Land volcanic province, West Antarctica*. PhD thesis, Victoria University of Wellington, Wellington, New Zealand.

Zipfel, J. and Wörner, G. 1992. Four- and five-phase peridotites from a continental rift system: Evidence for upper mantle uplift and cooling at the Ross Sea margin (Antarctica). *Contributions to Mineralogy and Petrology*, **111**, 24–36, https://doi.org/10.1007/BF00296575

Mantle rocks in East Antarctica

Stephen F. Foley[1]*, Alexandre V. Andronikov[2], Jacqueline A. Halpin[3], Nathan R. Daczko[1] and Dorrit E. Jacob[4]

[1]Department of Earth and Environmental Sciences and CCFS, Macquarie University, North Ryde, New South Wales 2109, Australia

[2]Department of Environmental Geochemistry and Biogeochemistry, Czech Geological Survey, Geologická 6, 15200 Praha 5, Czech Republic

[3]Institute for Marine and Antarctic Studies, University of Tasmania, Hobart, Tasmania 7001, Australia

[4]Research School of Earth Sciences, Australian National University, Canberra, ACT 2600, Australia

SFF, 0000-0001-7510-0223; JAH, 0000-0002-4992-8681; NRD, 0000-0002-3737-3818; DEJ, 0000-0003-4744-6627

*Correspondence: stephen.foley@mq.edu.au

Abstract: Only three localities of mantle xenoliths are known from all of East Antarctica, occurring at the Jetty Peninsula (Lambert–Amery Rift), Vestfold Hills and Gaussberg volcano. The latter two are spinel-facies peridotites, whereas the Jetty Peninsula rocks also include garnet-spinel lherzolites; all come from Indo-Antarctica. The mantle xenoliths of Jetty Peninsula and Vestfold Hills contain abundant geochemical and mineralogical evidence for multiple enrichment events that are attributed to infiltration of melts and their fluid products. Many of these episodes are spatially related to precursory activity along major trans-lithospheric structures that eventually led to the separation of India from Antarctica. Mantle rocks also occur at Schirmacher Oasis (Dronning Maud Land) and Haskard Highlands (Shackleton Ranges) as blocks tectonically emplaced in high-grade crustal rocks. These show varying degrees of alteration due to reaction with silicic crustal rocks or hydrous fluids: none correspond to unchanged mantle compositions. Geophysical surveys are our only information on the mantle lithosphere beneath the inland ice, and these can be used to infer the locations of thicker lithosphere probably related to cratons by southward extrapolation of coastal geological correlations. Intense local modification of the mantle lithosphere by melt infiltration and fluid movements may influence the large-scale images derived from geophysical data, and may be incorrectly interpreted as homogeneous compositions.

Known occurrences of mantle rocks are exceptionally rare in East Antarctica, which is almost certainly due to the paucity of outcrop relative to that in the Transantarctic Mountains and Antarctic Peninsula. Mantle xenoliths in volcanic rocks have been reported at only three localities, and there are three further occurrences of lenses of ultramafic rocks believed to be derived from the mantle that are included in high-grade metamorphic rocks of the continental crust. One of these, consisting of two lenses of serpentinite at Jetty Peninsula, has not been described in the literature.

This contribution summarizes mantle rock occurrences in East Antarctica (Fig. 1). Despite the rarity of occurrences, they represent examples from a variety of geological and tectonic settings, including cratons, orogenic belts, and rifts. We concentrate on the information that can be extracted from mantle xenoliths as the least-modified mantle samples, but include information on other rocks of probable mantle origin that are hosted in continental crust, as well as indications about the composition, physical state, and history of the mantle that can be gleaned from the compositions of mantle-derived igneous rocks and from correlations of crustal blocks with formerly neighbouring continents.

Summary of localities of mantle rocks in eastern Antarctica

There are five known localities in eastern Antarctica where rocks of definite or suspected mantle origin occur (shown in Fig. 1 and listed in Table 1): (1) spinel- and garnet-spinel peridotite, and rare garnet-free pyroxenite xenoliths in Mesozoic alkaline–ultramafic rocks at the Jetty Peninsula, in the northern Prince Charles Mountains (Andronikov 1990; 1992; Foley *et al.* 2006; Solovova *et al.* 2015); (2) xenocrysts and xenoliths of spinel lherzolite in dykes of Proterozoic ultramafic lamprophyres in the Vestfold Hills (Delor and Rock 1991; Andronikov *et al.* 1994); (3) a single spinel lherzolite xenolith described from the Quaternary Gaussberg volcano in Kaiser Wilhelm II Land (Sheraton and Cundari 1980); (4) spinel peridotite blocks variously altered by reaction with silicic pegmatites in gneisses metamorphosed at granulite to amphibolite facies at 575–530 Ma at Schirmacher Oasis, Dronning Maud Land (Markl *et al.* 2003); and (5) lenses of peridotite and garnet pyroxenite in Pan-African supracrustal gneisses in the Haskard Highlands, Shackleton Range, Coats Land (Schmaedicke *et al.* 2015).

The mantle peridotite xenoliths from East Antarctica are made up of olivine (Ol), orthopyroxene (Opx), clinopyroxene (Cpx), spinel (Sp) and/or garnet (Grt), whereas pyroxenites are represented by websterite (Opx+Cpx+Sp). The deepest-derived mantle rocks are found in a single intrusion, the so-called Southern intrusion of the Jetty Peninsula; these are lherzolite xenoliths that contain both spinel and garnet, and show equilibration pressures of up to 2.4 GPa, equivalent to depths of 75–80 km (Foley *et al.* 2006; Beliatsky and Andronikov 2009). All other mantle xenoliths and orogenic peridotites are garnet-free. Garnet occurs in pyroxenites, also from 2.5–2.3 GPa at the Haskard Highlands, but these are probably metamorphosed cumulates of picritic melts (Schmaedicke *et al.* 2015).

Petrological and geochemical characteristics of eastern Antarctic mantle rocks

Jetty Peninsula, northern Prince Charles Mountains

The Jetty Peninsula area, on the western margin of the Lambert–Amery Rift (Fig. 2), contains the most variable and best-studied locality of mantle xenoliths in East Antarctica. They are hosted by alkaline–ultramafic rocks that have been described as alkaline picrites and sometimes alnöitic ultramafic lamprophyres (Andronikov 1990; Andronikov and Egorov 1993; Foley *et al.* 2002) whose compositions and petrology are consistent with an origin by melting of the mantle at depths of 120–150 km (Foley *et al.* 2002).

From: Martin, A. P. and van der Wal, W. (eds) 2023. *The Geochemistry and Geophysics of the Antarctic Mantle*. Geological Society, London, Memoirs, **56**, 17–32,
First published online 3 June 2021, https://doi.org/10.1144/M56-2020-8

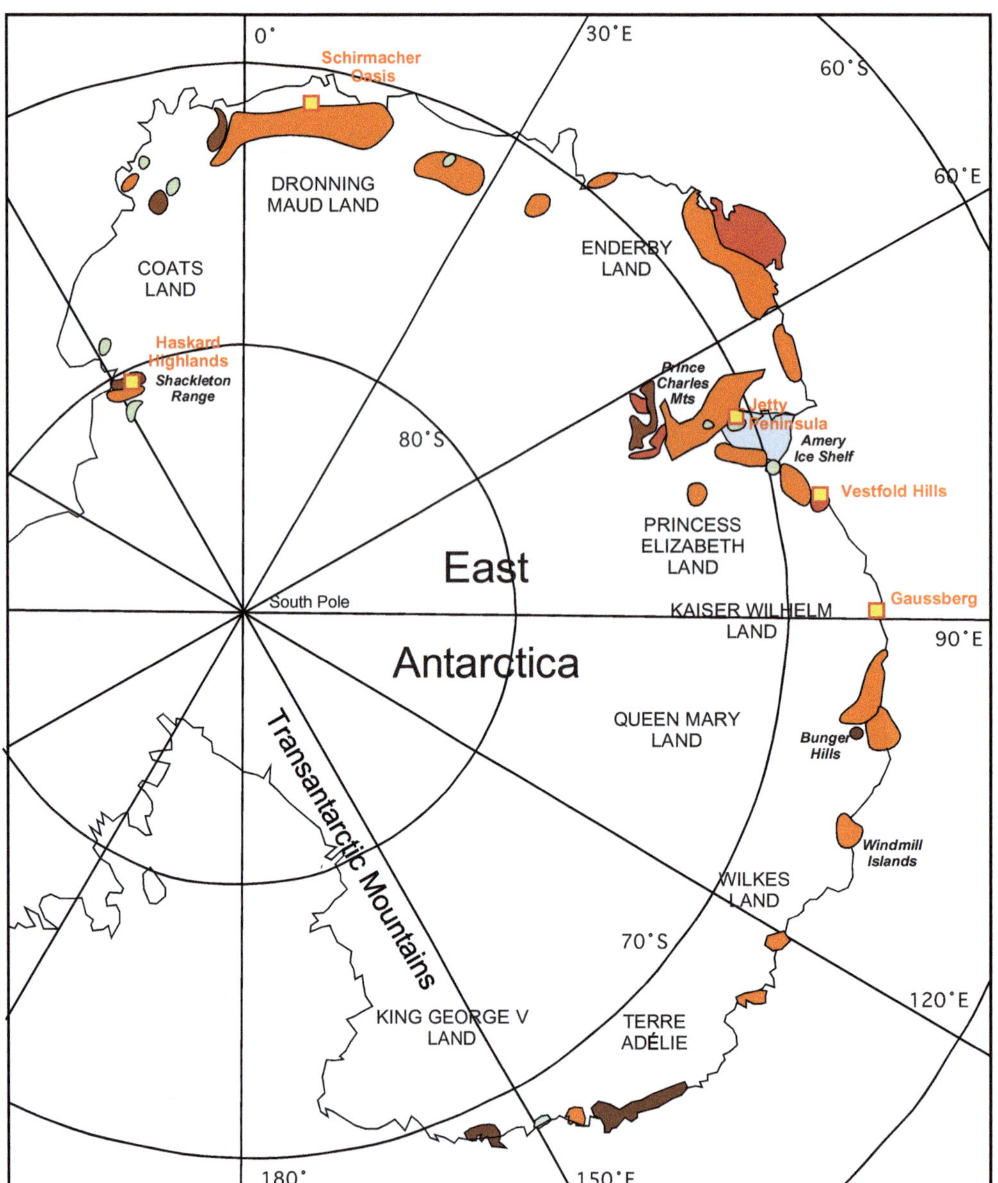

Fig. 1. Mantle rock locations (yellow squares) in eastern Antarctica, with generalized main areas of rocks coloured according to age: red, Archean; dark brown, Paleoproterozoic; orange, Late Proterozoic; green, Phanerozoic. Modified after James and Tingey (1983). Rock outcrop areas are exaggerated in this map: a current map of outcrop geology for Antarctica (GeoMAP) is available online at https://www.scar.org/science/geopmap/resources

The xenoliths may comprise up to 30 vol% of the erupted rocks and are up to 50 cm across (Beliatsky and Andronikov 2009; Solovova *et al.* 2015), and so are some of the largest known xenoliths in host rocks that are not kimberlites. Three major groups of peridotite xenoliths were recognized by Foley *et al.* (2006): clinopyroxene-poor spinel lherzolites, clinopyroxene-rich spinel lherzolites and spinel-garnet lherzolites. The first group appears to be little-modified by later mantle metasomatism, possibly because the xenoliths were torn from shallow levels of the mantle that were not reached by low-degree melts from below (Foley *et al.* 2006). The Cpx-rich and garnet-bearing peridotites have experienced multiple episodes of infiltration by silicate and carbonate melts, which have introduced a variety of metasomatic features, including textural features such as recrystallization and partial melting of clinopyroxene and spinel margins, and mineralogical features such as carbonates, sulfides and rare alkali titanate, apatite and humite minerals. These have been the subject of several recent studies aimed at characterizing the melts and fluids involved in the metasomatism (Kogarko *et al.* 2007; Buikin *et al.* 2014; Solovova *et al.* 2015).

The Cpx-poor spinel lherzolites are medium to coarse-grained (mostly 1–3 mm) inequigranular rocks that have little evidence for infiltration of melts along grain boundaries (Fig. 3a, b) with the exception of rare carbonates and sulfides. Mineral modes average 80% Ol, 12% Opx, 5% Cpx and 3% Sp, which are typical for depleted upper mantle peridotite compositions except for the higher spinel modes. This is reflected in whole-rock compositions, which show slightly higher Mg# (100Mg/(Mg+Fe)) of 90.2 to 91.2 but distinctly lower CaO and Al_2O_3 than the other two groups (Fig. 4). The history of both significant depletion and later metasomatic enrichment are reflected in chondrite-normalized rare earth element (REE) diagrams for the Cpx-poor lherzolites, which are characteristically U-shaped with normalized heavy REE (HREE) concentrations of ≤ 1, mid REE (MREE) ≥ 0.1, and light REE (LREE) up to 5 (Fig. 5).

The Cpx-rich spinel lherzolites comprise crystals of similar or slightly smaller size that are more thoroughly cracked and altered by more than one episode of melt infiltration. They contain higher modes of Cpx (average 71% Ol, 14% Opx, 12% Cpx and 2.5% Sp), which, together with most spinel grains, characteristically exhibit recrystallized edges that may appear to consume the entirety of smaller crystals but only the margins of larger crystals (Fig. 3c, d). Close-ups indicate that Cpx in the rims are rich in melt and fluid inclusions and have grown from the grain margins, presumably catalysed by melt movement along the grain boundaries (Fig. 3e). Resorbed spinel rims frequently contain silicic glass (Andronikov 1997). This group includes considerably less depleted

Table 1. *Localities of mantle rocks in Eastern Antarctica*

Region	Location	Rock types	Description	Host rock	Setting	References
Jetty Peninsula	70°33'S, 68° 47'E	Spinel lherzolites; spinel-ganret lherzolites; harzburgites, dunites, rare websterites	xenoliths	Alkaline picrite and rare ultramafic lamprophyres (alnoite)	Rift	Andronikov (1990, 1992); Foley *et al.* (2006); Kogarko *et al.* (2007); Beliatsky and Andronikov (2009); Buikin *et al.* (2014); Solovova *et al.* (2015)
Vestfold Hills	68°34S, 78°E	spinel peridotites, plagioclase lherzolites, wehrlites, disaggregated dunites	xenoliths	Ultramafic lamprophyres	rift margin?	Andronikov *et al.* (1994); Andronikov and Mikhalsky (1997); Delor and Rock (1991)
Schirmacher Oasis	71°S, 11° 23'E	Peridotite blocks	Spinel lherzolite reacted with crustal pegmatite	Granulite facies metamorphic rocks	collisional orogen	Markl *et al.* (2003)
Gaussberg	66°48'S, 89° 11'E	Spinel lherzolite	Xenolith	Lamproite	continental margin	Sheraton and Cundari (1980)
Haskard Highlands, Shackleton Range	80°26' S, 29° 50'W	Spinel peridotite; olivine-garnet pyroxenite; hornblendite; amphibolite	Lenses in high-grade gneisses		collisional orogen	Schmaedicke *et al.* (2015)

whole-rock compositions (Fig. 4; Mg# 86.8–90.9; CaO 2.0–3.3 wt%; Al_2O_3 2.2–3.0 wt%), but the greater range reflects re-enrichment of previously depleted rocks. Chondrite-normalized REE patterns for the Cpx-rich spinel lherzolites are less depleted than the Cpx-poor spinel lherzolites. Equally, metasomatic enrichment is generally less pronounced (Fig. 5).

The spinel-garnet lherzolites at Jetty Peninsula are the only occurrence of mantle peridotite with both spinel and garnet in Antarctica, and one of few in the world. Original garnet modes (average 3.7%) are higher than spinel (0.4%) but most of the garnet is completely replaced by kelyphite (Fig. 3e) which is interpreted as a result of garnet breakdown under spinel facies conditions (Beliatsky and Andronikov 2009). Apart from the presence of garnet, this group is similar to the Cpx-rich spinel lherzolites, which is consistent with an origin at slightly higher pressures, but having experienced similar metasomatic

Table 2. *Representative whole-rock analyses of Jetty Peninsula xenoliths*

Sample #	Rock type	SiO_2	TiO_2	Al_2O_3	Cr_2O_3	FeO	MnO	MgO	CaO	Na_2O	K_2O	P_2O_5	LOI	Total	Mg #	T°C	Ref
U-3	Cpx-poor spinel lherzolite	42.0	0.04	1.20	0.45	8.41	0.12	44.2	0.90	0.10	n.d.	0.02	1.60	99.26	91.2	1058	1
D-13	Cpx-poor spinel lherzolite	41.6	0.10	0.60	0.43	8.20	0.13	43.6	0.90	0.10	n.d.	0.04	5.25	100.95	90.5	1026	2
D-N2	Cpx-poor spinel lherzolite	41.2	0.02	1.17	0.66	8.00	0.17	45.3	1.03	0.05	0.01	0.01	1.52	99.31	90.7	1057	1
XLT-5	Cpx-rich spinel lherzolite	41.3	0.09	3.02	0.19	7.44	0.12	39.5	3.34	0.22	0.01	n.d.	3.64	99.05	89.6	834	1
U-4/9-2	Cpx-rich spinel lherzolite	43.9	0.04	1.99	0.31	6.94	0.11	40.1	1.96	0.09	0.03	0.04	3.41	99.12	87.7	1093	1
U-9	Spinel-garnet lherzolite	43.0	0.13	3.40	0.34	8.40	0.13	38.3	2.80	0.30	n.d.	0.02	3.90	100.72	89.0	1050	2
U-16	Spinel-garnet lherzolite	43.4	0.07	4.10	0.24	7.57	0.15	38.0	3.35	0.19	0.04	0.01	1.75	99.51	89.9	1130	1
DK-8	Spinel-garnet lherzolite	41.7	0.08	2.80	0.41	7.60	0.13	38.9	2.30	0.10	n.d.	0.03	5.50	99.55	90.1	1124	2
D-N4	Spinel-garnet lherzolite	43.2	0.16	3.00	0.25	8.14	0.15	38.8	2.89	0.19	n.d.	n.d.	3.15	100.09	89.0	1125	1

Ref 1 = Foley *et al.* (2006); Ref 2 = Beliatsky and Andronikov (2009).
T°C, temperature by the Ca-in-Opx thermometer of Brey and Köhler (1990).

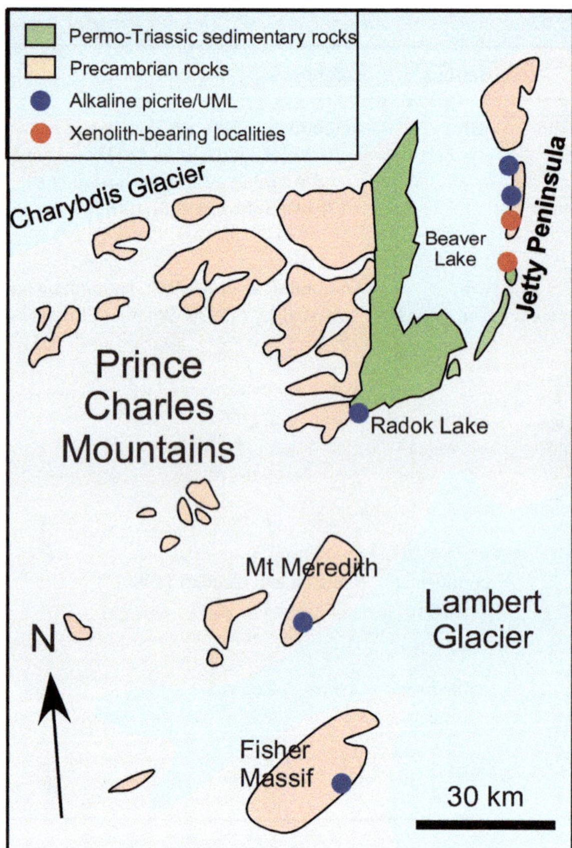

Fig. 2. Rock outcrops in the northern Prince Charles Mountains region, East Antarctica, showing xenolith-bearing localities of the Jetty Peninsula (red dots) and other occurrences of ultramafic alkaline rocks (blue dots).

overprints by melts derived from below (Foley *et al.* 2006). Their whole-rock compositions are similar to the Cpx-rich spinel lherzolites (Fig. 4; Mg# 89.0–91.2; CaO 2.3–3.4 wt%; Al_2O_3 2.3–4.1wt%). REE patterns for the Sp-Grt lherzolites are similar to those of the Cpx-rich lherzolites, although the latter display stronger metasomatic enrichment (Fig. 5).

In-situ mineral analyses for major elements by electron microprobe (Table 2) and trace elements by laser-ICP-MS are consistent with, and permit further decoding of, the depletion and enrichment history outlined above. Olivines in the Cpx-poor spinel lherzolites are uniform in composition (Mg# 90.7–91.4), as are orthopyroxenes, which have low Al_2O_3 contents typical of spinel lherzolite xenoliths worldwide (Table 3). These rocks have just one generation of Cpx, whereas the Cpx-rich spinel lherzolites have two: cores range in composition from similar to those in the Cpx-poor rocks to more Fe-rich compositions. Recrystallized Cpx rims in the Cpx-rich rocks have low Na_2O (average <0.6wt%) and Al_2O_3 contents as well as slightly higher Mg# (average 91.5), which may reflect partial melting and melt removal along grain boundaries (Foley *et al.* 2006). Garnets in the spinel-garnet lherzolites are poorly preserved, but pyrope-rich (Mg#82.1–86.8). CaO contents are mostly around 5% but rare sub-calcic cores (≈1.3 wt%) appear reminiscent of cratonic garnet that are frequently used as diamond indicator minerals. However, Cr_2O_3 contents are only 1.1–2.0 wt%, much lower than in cratonic garnets sampled by kimberlites (>4 wt%; Jacob *et al.* 1997). Spinel tells a similar story, with Cr# (100Cr/(Cr+Al)) highest in the Cpx-poor spinel lherzolites (26–35; average 30.3) and lowest in the spinel-garnet rocks (average 17.8). The lower Cr# in the Cpx-rich rocks results from the stronger metasomatic overprint and from the competition for Cr between garnet and spinel in the spinel-garnet lherzolites with a high modal garnet/spinel ratio. Spinel grains in the latter rocks are similar to those in spinel-garnet peridotite xenoliths on other continents, such as Pali Aike (Patagonia; Kempton *et al.* 1999), Nushan (China; Xu *et al.* 1998) and Vitim (Siberia; Ionov *et al.* 1993; Glaser *et al.* 1999).

The Jetty Peninsula xenoliths contain evidence for several stages of enrichment through melts and fluids, whereby the latter may be released by late-stage separation from melts (Beliatsky and Andronikov 2009; Buikin *et al.* 2014). Silicate melts are witnessed directly by glasses along grain boundaries, and, in particular, along resorbed rims of Cpx and Sp: they range from sodic basaltic melts with SiO_2 contents around 50 wt% to K-rich phonolitic melts (54–57 wt% SiO_2, 6 wt% K_2O). These are distinct from host-rock melts and from the breakdown products of garnet (kelyphite). Carbonate melts were implied in earlier publications but were clearly implicated by Kogarko *et al.* (2007), who identified calcite, dolomite, apatite and magnesite in a spinel lherzolite. Products of carbonate melt infiltration, including sulfide, are the only metasomatic features in the spinel lherzolites, implying that the carbonate melts reached closer to the surface than any silicate metasomatism.

In combination with isotopic data on rocks and minerals (Andronikov and Beliatsky 1995; Beliatsky and Andronikov 2009), a long history can be reconstructed for the mineralogical and geochemical features in the Jetty Peninsula xenoliths, much of which may be related to rifting that eventually led to the separation of India and Australia from Antarctica during the breakup of Gondwana (Foley *et al.* 2006).

The first stage was partial melting that moved the rocks along the depletion trend towards higher Mg# than primitive mantle. This may be reflected in Sm–Nd ages for the xenoliths of 2.6–2.4 Ga (Mukasa and Andronikov 1999; Sushchevskaya *et al.* 2018) that may correspond to Rb–Sr ages for crustal granitic gneisses further south (2.8–2.7 Ga; Tingey 1991). The crust of the northern Prince Charles Mountains is younger than this (1.99–1.74 Ma from zircon Hf and Nd whole-rock model ages; Liu *et al.* 2017), but this late Archean event may be a deep lithospheric expression of the cratonization that affected the Ruker Province in the southern Prince Charles Mountains. The second event was garnet formation, which is believed to be a crustal thickening event related to the protracted (>1170–900 Ma) accretionary Rayner Orogeny that records peak high-grade metamorphism and magmatism mainly between 945 and 915 Ma (Liu *et al.* 2017). Clinopyroxenes bear evidence for two further events: the third was the main enrichment event that reintroduced abundant Cpx in the Cpx-rich and spinel-garnet lherzolites, and the fourth led to the later formation of recrystallized rims on Cpx and Sp (Fig. 2c–e). The main enrichment (third event) post-dated the CaO-rich garnet population and the survival of vestigial CaO-poor garnets that must have grown in harzburgites before enrichment. Sm–Nd dating of clinopyroxenes has narrowed the timing of the third event to 390–375 Ma (Andronikov and Beliatsky 1995), which corresponds to the early stages of the Lambert–Amery Rift (cf. Leitchenkov *et al.* 2018). The low Na and Al and increased Mg# of the recrystallized Cpx rims can be explained by melting along grain boundaries in a heating event that presumably corresponds to further widening of the rift.

The silicate glasses and mineralogical evidence for carbonate melts document later melt movements beneath the evolving rift, indicating long-lived, episodic development of the rift (Foley *et al.* 2006). The timing of silicate melt infiltration must have been shortly before eruption of the host magmas at 117–110 Ma (Walker and Mond 1971; Laiba *et al.* 1987) in order to preserve glasses. Some of the glass compositions are fairly similar to K-rich trachybasalts that erupted either later (40 Ma; Andronikov and Beliatsky 1996; Andronikov

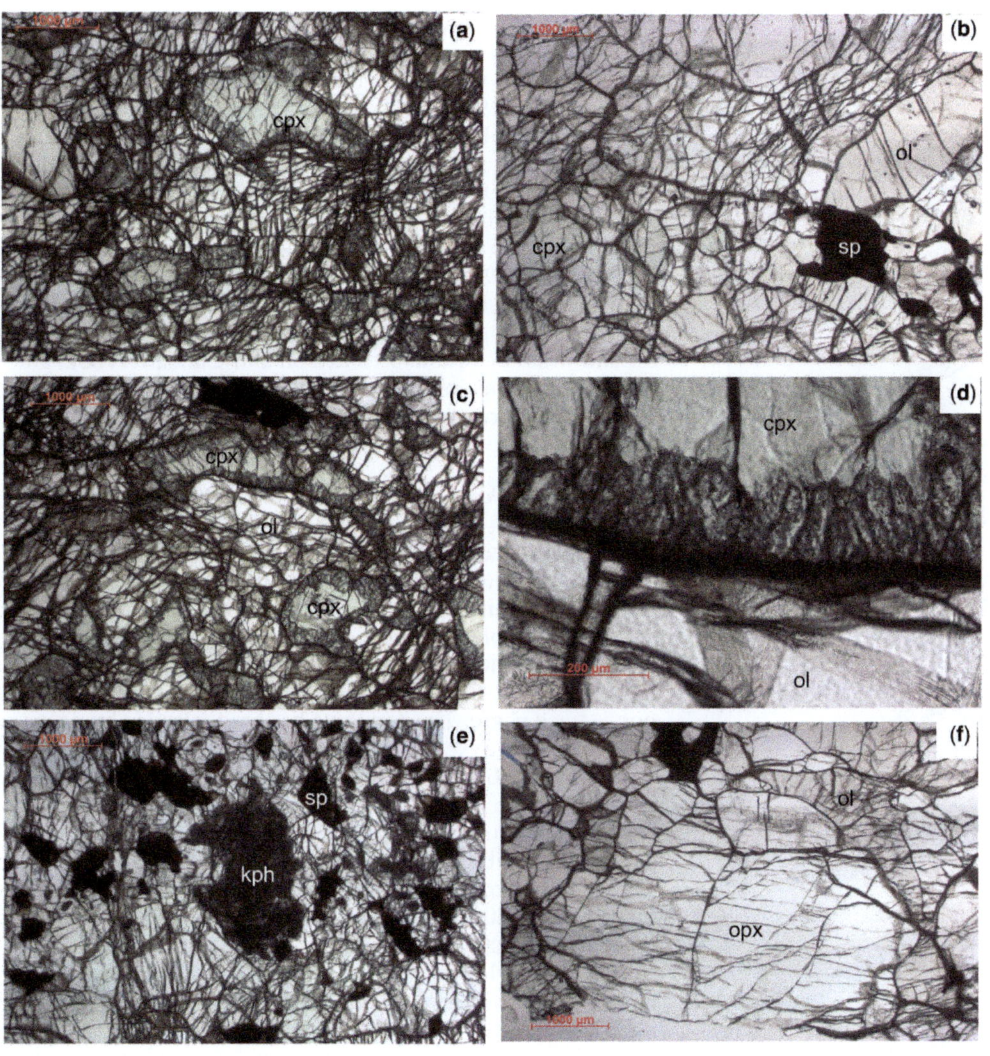

Fig. 3. Thin section photos of Jetty Peninsula peridotite xenoliths: (**a**) Cpx-rich spinel lherzolite; (**b**) Cpx-poor spinel lherzolite; (**c**) recrystallization of Cpx (green) due to melt infiltration, large grains have recrystallized rims whereas small grains are entirely recrystallized; (**d**) close-up of recrystallized Cpx rim showing inward growth from the grain edge. Note fluid-inclusion-rich sub-domains in olivine (bottom); (**e**) spinel-rich region in spinel-garnet lherzolite with completely kelyphitized garnet in the centre; (**f**) fresh Cpx-poor spinel lherzolite showing restricted fluid inclusion trails (lower right). Scale bars are all 1 mm except (d) (0.2 mm). Mineral abbreviations: ol, olivine; cpx, clinopyroxene; opx, orthopyroxene; sp, spinel; kph, kelyphite after garnet.

et al. 1997) or earlier (320 Ma; Leitchenkov *et al.* 2018) in the nearby Manning Massif, indicating that melts of this composition occurred beneath the rift. The age of carbonate melt infiltration is obscure but may be a mantle expression of the Mesozoic alkaline–ultramafic magmatism. The enrichment events here differ from those in North Victoria Land, which were affected by voluminous silica-saturated magmas associated with the Ferrar large igneous province (Coltorti *et al.* 2021), which also resulted in numerous cumulate xenoliths that are not known in the Lambert–Amery rift area.

Vestfold Hills, Princess Elizabeth Land

Situated 450 km to the NE of Jetty Peninsula, ultramafic xenoliths of several types occur in Proterozoic ultramafic lamprophyre dykes in the Vestfold Hills. These include mantle spinel peridotites that have experienced various enrichment episodes attributed to fluids, silicate melts and carbonatite melts (Seitz 1991; Andronikov and Mikhalsky 1997), plus plagioclase-bearing lherzolites, iron-rich wehrlites (Andronikov *et al.* 1994), pyroxenites (Andronikov and Mikhalsky 1997) and disaggregated dunites (Delor and Rock 1991). The enrichment episodes introduced hydrous phases and apatite, increased titanium contents of minerals and the modal abundance of Cpx, and so appear reminiscent of the enrichment types seen at Jetty Peninsula. However, the lamprophyres in which they occur are much older than the Jetty Peninsula intrusive rocks, probably 1.4–1.3 Ga (Kuehner 1987), so that the metasomatic events must be distinct. These are much rarer than abundant dolerite dykes that occur throughout the Vestfold Hills (Sheraton *et al.* 1987) that are both older (*c.* 2.4 Ga) and younger (1.38–1.25 Ga) than the lamprophyres (Seitz 1991; Hoek and Seitz 1995).

Harzburgites and lherzolites show varying levels of depletion but include Cr-rich spinels (>52 wt% Cr_2O_3; Andronikov *et al.* 1994) and olivines with Mg# up to 93 (Table 4). Other ultramafic rock types include much lower Mg#: for example, plagioclase lherzolites have Mg# of 82–73 in olivine and 80–76 in Opx, and olivines in websterites are as low as Mg# 74 (Table 4). This led Andronikov *et al.* (1994) and Andronikov and Mikhalsky (1997) to suspect a lower crustal origin for the most Fe-rich samples, but this is problematic given temperature estimates of 1050–1190°C for the websterites. Fe-rich olivine and pyroxene have since been attributed to interaction with Fe-rich, fractionated melts percolating through the mantle beneath the Cameroon Line in western Africa (Njombie Wagsong *et al.* 2018; Temdjim *et al.* 2020); a similar origin within the upper mantle may also apply for the Vestfold Hills xenoliths. The websterites often contain hydrous minerals in veinlets, which is a similar mode of enrichment to that in the peridotites, further supporting an origin in the upper mantle.

The interaction with infiltrating silicate melts is indicated by increasing TiO_2 contents in Cpx and spinels. Phlogopite is often found as inclusions in primary minerals, and its compositions are Si-rich at high Mg#, which is typical for phlogopite in garnet peridotites (Table 5; Arai 1984; Andronikov and Mikhalsky 1997). The action of carbonatite melts is indicated by the replacement of Opx by Cpx, as well as the

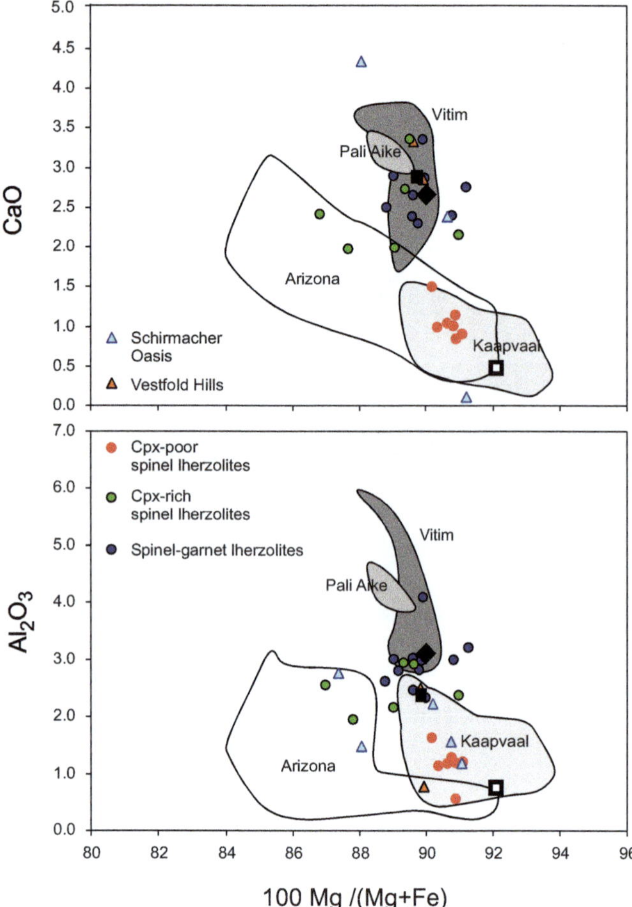

Fig. 4. Plot of whole-rock peridotite compositions for Antarctic mantle rocks against a background of compositions for garnet peridotites from Kaapvaal (Nixon *et al.* 1981; Boyd and Mertzmann 1987) and Arizona (Ehrenberg 1982), and spinel-garnet peridotites from Vitim, Siberia (Ionov *et al.* 1993; Glaser *et al.* 1999) and Pali Aike, Patagonia (Kempton *et al.* 1999). Black symbols are median compositions for spinel lherzolites from continental intraplate (squares) and rift (diamonds) and garnet peridotites from cratons (open square) (Pearson *et al.* 2003).

occurrence of apatite and Na-bearing carbonates (Andronikov and Mikhalsky 1997).

The metasomatic enrichment episodes have led to inter-mineral disequilibrium, resulting in disparate results from two-mineral thermometers on the order of 300°C for the Vestfold Hills rocks (Andronikov and Mikhalsky 1997).

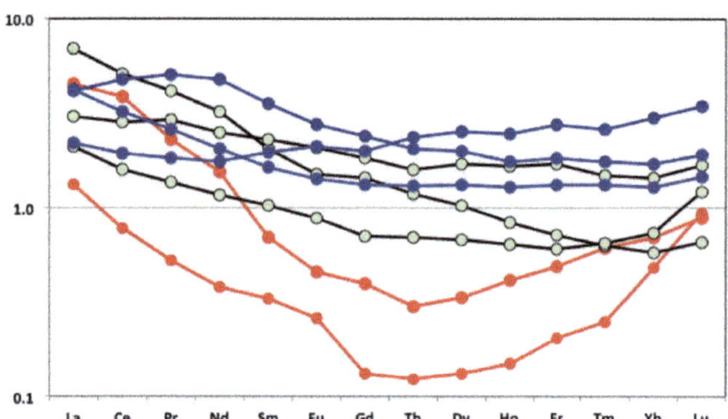

Fig. 5. Rare earth element patterns (chondrite-normalized) for Jetty Peninsula xenoliths. Symbols as in Figure 4: red circles, Cpx-poor spinel lherzolites; green circles, Cpx-rich spinel lherzolites; blue circles, spinel-garnet lherzolites.

Gaussberg volcano, Kaiser Wilhelm Land

The Gaussberg volcano on the coast at 90°E is a Quaternary (Tingey *et al.* 1983) glassy leucitite or leucite lamproite that has been the subject of a number of publications because of its unusual, exceptionally K_2O-rich composition (Sheraton and Cundari 1980; Murphy *et al.* 2002: Foley and Jenner 2004) but only one mantle xenolith has been reported (Sheraton and Cundari 1980). This is a mildly depleted spinel lherzolite with Mg# 91 in olivine and 92 in Opx. Cpx is chrome diopside with 2.1 wt% Cr_2O_3, a typical to high value for spinel lherzolites, and spinel has intermediate composition with Cr# 36 and Mg# 67 (Table 5; Sheraton and Cundari 1980). The xenolith contains interstitial glass with high K_2O (>9 wt%), indicating an origin by infiltration of the host melt.

Schirmacher Oasis, Dronning Maud Land

Rare, metre-sized lenses of harzburgite to spinel lherzolite occur as tectonically emplaced fragments in high-grade pelitic gneisses at Schirmacher Oasis in Dronning Maud Land. The mineral assemblage of 60% olivine and 27% Opx is typical for mantle peridotite, whereas the mode of spinel is exceptionally high at 12–13% (Markl *et al.* 2003). Clinopyroxene is only a trace phase. The peridotite has been intruded by a tonalitic pegmatite dyke of lower crustal origin, causing reaction between the SiO_2-rich melt and the peridotite, which resulted in intermediate zones consisting of biotite, clinoamphibole and zircon; cummingtonite, biotite intergrown with rutile and clinoamphibole; and cummingtonite, biotite and chromite. The peridotite zone is said to represent the original mantle rock (Markl *et al.* 2003), but the chemical characteristics are not those of unreacted mantle peridotite. Olivine and Opx have low average Mg# at 83, although NiO contents in olivine are mantle-like (Table 5; 0.34–0.45 wt%), and spinels also have low Mg# (33) but retain high Cr# (40). Oxygen isotopes show disequilibrium between spinel and olivine in the peridotite, indicating incomplete reaction with a crustal derived fluid at *c.* 650°C. The reaction resulted in diffusion of Mg and Cr from the peridotite towards the vein and Fe and Al in the opposite direction, resulting in lowering of Mg# of all minerals and Cr# of peridotite. It is unlikely that any part of the peridotite reflects the original mantle composition.

Haskard Highlands, Shackleton Range, Coats Land

The Haskard Highlands locality is similar to the Schirmacher Oasis occurrence in that it consists of ultramafic lenses included in high-grade supracrustal gneisses (Schmaedicke *et al.* 2015). Spinel peridotite is a subordinate rock type together with amphibolite, whereas olivine-bearing garnet pyroxenites dominate. A harzburgite with Mg# 91 (Table 4) is thought to be the only rock with close to its original mantle composition, but peridotites are all altered and hydrated, with loss on ignition values of 3–8 wt%. Most peridotites analysed have SiO_2 contents too high (46–50wt%) for unaltered mantle rocks and must have high pyroxene modes. The pyroxenites are interpreted as cumulates from picritic melts metamorphosed at 23–25 kbar pressure (75–80 km) during Pan-African eclogite-facies metamorphism. Serpentinization has severely affected the original mineral grain sizes and textural relationships.

Evidence for conditions in the mantle

Only the first three localities (Jetty Peninsula, Vestfold Hills and Gaussberg) can be used to gauge conditions in the mantle,

Table 3. *Representative mineral analyses of Jetty Peninsula xenoliths*

Rock type	Sample #	Mineral	SiO_2	TiO_2	Al_2O_3	Cr_2O_3	FeO	MnO	MgO	NiO	CaO	Na_2O	K_2O	Total	Mg#	Cr#
Cpx-poor spinel lherzolite	U-3	Olivine	41.00	0.008	0.042	0.049	8.59	0.134	50.10	0.37	0.096	0.010		100.40	91.2	
Cpx-poor spinel lherzolite	D-N2	Olivine	40.26	0.035	0.038	0.072	8.70	0.168	49.13	0.40	0.086	0.015		98.90	91.0	
Cpx-rich spinel lherzolite	XLT-5	Olivine	41.45	0.024	0.009		10.04	0.127	49.12	0.38	0.03	0.013		101.16	89.7	
Cpx-rich spinel lherzolite	U-4/9-2	Olivine	41.32	0.022	0.024	0.053	10.51	0.120	48.57	0.38	0.09	0.001		101.09	89.2	
Spinel-garnet lherzolite	U-16	Olivine	40.69	0.035	0.043	0.053	9.47	0.158	47.76	0.31	0.14	0.015		98.68	90.0	
Spinel-garnet lherzolite	D-N4	Olivine	40.76	0.014	0.064	0.022	10.15	0.141	49.05	0.37	0.10	0.017		100.69	89.6	
Cpx-poor spinel lherzolite	U-3	Orthopyroxene	55.46	0.023	3.83	0.748	5.40	0.154	33.59	0.104	0.96	0.071		100.34	91.7	
Cpx-poor spinel lherzolite	D-N2	Orthopyroxene	54.72	0.020	3.51	0.672	5.38	0.194	33.14	0.136	0.94	0.076		98.79	91.7	
Cpx-rich spinel lherzolite	XLT-5	Orthopyroxene	56.24	0.084	3.86	0.252	6.97	0.149	33.47	0.066	0.34	0.035	0.008	101.47	89.5	
Cpx-rich spinel lherzolite	U-4/9-2	Orthopyroxene	54.86	0.191	4.62	0.715	6.59	0.099	31.92	0.150	1.09	0.151	0.003	100.38	89.6	
Spinel-garnet lherzolite	U-16	Orthopyroxene	54.33	0.076	5.05	0.707	5.94	0.171	32.21	0.106	1.18	0.143		99.92	90.6	
Spinel-garnet lherzolite	D-N4	Orthopyroxene	54.72	0.203	5.02	0.690	6.29	0.166	32.24	0.117	1.15	0.178		100.77	90.1	
Cpx-poor spinel lherzolite	U-3	Clinopyroxene	52.95	0.061	4.24	1.15	2.77	0.102	17.40	0.073	20.46	0.80		100.01	91.8	
Cpx-poor spinel lherzolite	D-N2	Clinopyroxene	52.25	0.040	3.80	1.14	2.50	0.090	17.24	0.063	21.58	0.76		99.46	92.5	
Cpx-rich spinel lherzolite	XLT-5	Clinopyroxene	52.77	0.634	6.83	0.72	2.21	0.063	14.22	0.016	22.14	1.96	0.010	101.57	92.0	
Cpx-rich spinel lherzolite	U-4/9-2	Clinopyroxene	52.13	0.595	5.93	1.29	3.68	0.093	16.29	0.015	19.22	1.46		100.70	88.7	
Spinel-garnet lherzolite	U-16	Clinopyroxene	51.66	0.387	4.52	1.18	3.29	0.108	16.88	0.047	20.80	0.78		99.66	90.1	
Spinel-garnet lherzolite	D-N4	Clinopyroxene	51.40	0.608	6.85	1.19	3.73	0.177	16.63	0.056	18.54	1.22		100.41	88.8	
Cpx-poor spinel lherzolite	U-3	Spinel		0.08	39.2	27.9	12.59	0.14	18.2	0.269				98.35	72.0	32.2
Cpx-poor spinel lherzolite	D-N2	Spinel		0.06	37.6	29.8	12.29	0.18	17.4	0.297				97.60	71.6	34.6
Cpx-rich spinel lherzolite	XLT-5	Spinel	0.02	0.04	59.6	9.1	11.67	0.13	19.8	0.361				100.71	75.2	9.2
Cpx-rich spinel lherzolite	U-4/9-2	Spinel	0.09	0.63	42.4	23.2	14.73	0.22	18.2	0.376				99.85	68.8	26.7
Spinel-garnet lherzolite	U-16	Spinel		0.14	49.2	17.1	11.50	0.18	19.3	0.305				97.71	74.9	18.8
Spinel-garnet lherzolite	D-N4	Spinel		0.44	50.4	15,3	12.31	0.21	19.0	0.356				98.00	73.3	16.8
Spinel-garnet lherzolite	U-16	Garnet	42.04	0.130	22.32	1.91	6.83	0.367	20.54		5.57	0.013		99.72	84.3	
Spinel-garnet lherzolite	D-N4	Garnet	41.78	0.135	21.61	1.50	7.73	0.405	21.28		5.54	0.015		99.98	83.1	

All analyses from Foley *et al.* (2006).

Table 4. *Representative whole-rock analyses of eastern Antarctic mantle rocks*

Locality	Sample #	Rock type	SiO_2	TiO_2	Al_2O_3	Cr_2O_3	FeO	MnO	MgO	CaO	Na_2O	K_2O	P_2O_5	NiO	LOI	Total	Mg #	Ref
Vestfold Hills		Spinel lherzolite	44.77	0.23	0.72	1.02	7.64	0.17	37.62	3.36	0.61	0.01			3.72	99.87	89.8	1
Vestfold Hills		Spinel lherzolite	43.81	0.23	2.53	0.54	7.44	0.18	37.63	2.86	0.35	0.01			4.30	99.88	90.0	1
Vestfold Hills		Spinel lherzolite (Fe-rich)	41.61	0.29	2.91	0.14	15.26	0.19	27.38	4.31	0.54	0.09			7.20	99.92	76.2	1
Vestfold Hills		Harzburgite	43.99	0.03	1.00	0.91	7.65	0.09	40.41	0.85	0.53	0.01			4.50	99.97	90.4	1
Vestfold Hills		Wehrlite	42.49	0.16	0.77	0.62	13.59	0.33	31.77	7.03	0.01	0.01			3.20	99.98	80.7	1
Schirmacher Oasis		Spinel lherzolite	40.03	0.05	4.26	3.97	16.09	0.13	35.48	0.08	0	0		0.25		100.34	79.7	2
Haskard Highlands	247	Spinel peridotite	43.13	0.02	1.16		6.65	0.1	38.01	0.19	0.06	0.35		0.29	8.55	98.51	91.1	3
Haskard Highlands	249	Spinel peridotite	46.30	0.04	1.56		6.38	0.09	35.01	2.33	0.14	0.31		0.24	6.62	99.02	90.7	3
Haskard Highlands	257	Spinel peridotite	47.62	0.06	1.37		7.70	0.18	31.87	4.35	0.17	0.03	0.01	0.18	5.22	98.76	88.1	3
Haskard Highlands	259m	Spinel peridotite	49.90	0.15	2.82		6.15	0.11	24.22	9.92	0.4	0.09	0.01	0.03	2.97	96.77	87.5	3
Haskard Highlands	271	Spinel peridotite	46.50	0.14	2.05		5.59	0.11	29.69	7.67	0.27	0.1	0.01	0.2	6.56	98.89	90.4	3
Haskard Highlands	219	Pyroxenite	47.15	1.82	6.35		11.37	0.16	19.75	9.44	0.66	0.19	0.17		0.98	98.04	75.6	3
Haskard Highlands	228	Pyroxenite	44.57	1.48	8.59		12.67	0.22	18.91	7.71	0.97	0.11	0.13		1.8	97.16	72.7	3
Haskard Highlands	278	Pyroxenite	46.43	1.27	9.72		14.34	0.22	18.8	5.72	0.58	0.12	0.15		0.35	97.70	70.0	3
Haskard Highlands	284	Amphibolite	44.68	1.15	14.0		9.70	0.18	11.17	12.78	1.72	1.08	0.11		2.44	99.01	67.2	3
Haskard Highlands	560	Amphibolite	49.79	0.99	15.5		11.66	0.17	6.7	10.15	2.43	0.61	0.08		0.82	98.90	50.6	3

References: 1 = Andronikov and Mikhalsky (1997); 2 = Markl *et al.* (2003); 3 = Schmaedicke *et al.* (2015).

because neither the Schirmacher Oasis nor the Haskard Highlands peridotites retain their original compositions. The Jetty Peninsula peridotites yield several pieces of information on mantle processes over considerable geological time during the episodic development of a large rift. The position of the Vestfold Hills is marginal to this rift, but since the xenoliths were sampled by lamprophyres at *c.* 1.4–1.3 Ga, they should document processes that occurred long before rifting developed and possibly during the Rayner Orogeny. However, controversy remains as to the timing of juxtaposition of the Vestfold Hills block with adjacent terranes (e.g. Mikhalsky *et al.* 2019). Only the Gaussberg spinel peridotite represents a piece of pristine upper mantle far removed from this area, but a single xenolith yields only limited information.

Mineral assemblages and pressure–temperature estimates indicate that all samples come from the uppermost 80 km or so of the mantle and probably from the mantle lithosphere. The deepest sample occurs at Jetty Peninsula and yields a pressure of 2.4 GPa (Foley *et al.* 2006). Several of the deepest samples are garnet-bearing but also contain spinel and so come from close to the spinel/garnet transition depth at 2–2.4 GPa. Pressures for the spinel lherzolites are notoriously difficult to pinpoint but lie between 1 and 2 GPa based on their mineral assemblages. Geothermometers yield partly conflicting temperatures due to partial equilibration amongst mineral pairs: this is particularly because of the later introduction and recrystallization of Cpx, which compromises the application of two-pyroxene thermometers. Nevertheless, evidence for two geothermal gradients at different times beneath the rift at Jetty Peninsula could be extracted: most garnet-spinel lherzolites yield temperatures of 1050–1130°C for pressures of 2–2.4 GPa, whereas spinel lherzolites yield cooler temperatures of 1005–1105°C consistent with their higher position in the lithosphere (Table 2; Foley *et al.* 2006; Beliatsky and Andronikov 2009). However, occasional temperatures of 835–850°C are believed to be real, representing remnants of former conditions at the same depth, but earlier in the development of the rift, rather than originating from shallower depths on the geotherm. Spinel lherzolites from the Vestfold Hills give similar temperatures, whereas websterites appear to form two populations: those assigned to a mantle origin also show low temperatures (807–899°C), whereas higher temperature websterites (1000–1131°C) were thought to be of lower crustal origin (Andronikov and Mikhalsky 1997). It should be noted here that the Jetty Peninsula xenoliths reflect mantle conditions in the Mesozoic, whereas the xenoliths from the Vestfold Hills reflect mantle conditions in the Proterozoic.

Oxygen fugacities were calculated for the Jetty Peninsula xenoliths (Andronikov and Sheraton 1996; Foley *et al.* 2006) using the oxygen barometer of Ballhaus *et al.* (1990) and the single-mineral Ca-in Opx thermometer (Brey and Köhler 1990) to by-pass issues related to Opx–Cpx disequilibrium. Results lie mostly 2.6 to 0.1 log units below the fayalite-quartz-magnetite (FMQ) buffer, whereby the least metasomatic Cpx-poor spinel lherzolites give the lowest values and the Cpx-rich rocks are restricted almost universally to >FMQ −1.0. This indicates increased oxidation state as rifting progressed, which is often found to result from metasomatic effects caused by the infiltration of melts (Woodland *et al.* 1996; Yaxley *et al.* 2017).

Minerals in the Jetty Peninsula xenoliths have been analysed for their trace element contents (Foley *et al.* 2006). We will not dwell on these here, other than to note that the spinel peridotites that yield the anomalous lower temperatures also exhibit distinct trace element contents in their minerals, namely lower Ti, Cr, V and Zn in olivine and Opx, and lower Cu, Ni, Ga and Li in Opx, which are also signals of earlier conditions beneath the rift that were subsequently overprinted by metasomatic events in most rocks.

Table 5. *Representative mineral analyses from eastern Antarctic mantle rocks*

Locality	Rock type	Mineral	SiO_2	TiO_2	Al_2O_3	Cr_2O_3	Fe_2O_3	FeO	MgO	CaO	Na_2O	K_2O	P_2O_5	MnO	NiO	LOI	Total	Mg#	Cr#	Ref.
Vestfold Hills	Harzburgite	Olivine	40.52	0.02	0.03	0.08		7.84	50.32	0.09	0.03			0.13	0.35		99.41	91.96		1
Vestfold Hills	Lherzolite	Olivine	40.60	0.05	0.02	0.06		7.99	50.45	0.08	0.02			0.14	0.38		99.79	91.84		1
Vestfold Hills	Websterite	Olivine	39.31	0.02	0.01	0.02		10.93	47.46	0.02	0.01			0.20	0.41		98.39	88.56		1
Vestfold Hills	Wehrlite	Olivine	40.02	0.05	0.01	0.05		12.16	47.41	0.04	0.01			0.20	0.40		100.35	87.42		1
Vestfold Hills	Harzburgite	Orthopyroxene	55.47	0.39	1.96	0.52		6.64	32.84	1.09	0.20			0.15	0.10		99.36	89.81		1
Vestfold Hills	Lherzolite	Orthopyroxene	56.55	0.24	0.95	0.36		6.88	33.54	0.79	0.15			0.12	0.07		99.65	89.68		1
Vestfold Hills	Websterite	Orthopyroxene	52.62	0.46	3.21	0.21		7.28	31.87	1.11	0.17			0.09	0.09		97.11	88.64		1
Vestfold Hills	Lherzolite	Clinopyroxene	52.40	1.05	1.84	1.16		3.99	17.13	21.03	0.86			0.11	0.03		99.60	88.44		1
Vestfold Hills	Websterite	Clinopyroxene	53.31	0.12	1.58	1.06		2.77	16.65	23.12	0.65			0.16	0.05		99.47	91.46		1
Vestfold Hills	Wehrlite	Clinopyroxene	52.31	0.99	2.06	1.69		2.97	16.36	21.42	1.23			0.15	0.03		99.21	90.76		1
Vestfold Hills	Harzburgite	Spinel		0.16	22.31	42.43	3.77	18.86	9.26					0.39			97.18	46.67	56.07	1
Vestfold Hills	Lherzolite	Spinel		2.29	10.54	51.13	2.64	24.05	6.96					0.37	0.08		98.06	34.03	76.50	1
Vestfold Hills	Websterite	Spinel		0.12	7.99	54.41	3.97	25.92	3.73					0.64	0.14		96.92	20.41	82.05	1
Vestfold Hills	Wehrlite	Spinel		5.97	11.62	40.18	4.80	31.37	4.56					0.47	0.12		99.09	20.58	69.88	1
Vestfold Hills	Harzburgite	Phlogopite	42.14	0.85	11.50	0.15		2.52	26.54	0.08	0.96	7.46	0.02	0.01	0.16		92.39	94.94		1
Vestfold Hills	Harzburgite	Amphibole	53.20	0.65	1.67	0.44		3.32	23.20	12.86	1.01	0.19		0.12	0.08		96.74	92.57		1
Vestfold Hills	Wehrlite	Amphibole	55.31	0.67	1.14	0.36		3.60	21.82	8.16	5.23	0.32	0.01	0.14	0.04		96.80	91.53		1
Gaussberg	Spinel lherzolite	Olivine	40.40					8.30	50.10					0.14	0.47		99.41	91.50		2
Gaussberg	Spinel lherzolite	Orthopyroxene	55.80	0.10	2.80	0.73		4.90	33.50	0.93	0.15			0.14	0.12		99.17	92.40		2
Gaussberg	Spinel lherzolite	Clinopyroxene	52.00	0.55	5.00	2.10		1.70	14.10	21.30	2.40			0.08	0.10		99.33	93.70		2
Gaussberg	Spinel lherzolite	Spinel		0.14	37.50	31.60		13.50	15.70					0.30	0.17		98.91	67.50	36.12	2
Schirmacher Oasis	Spinel lherzolite	Olivine	40.57					15.73	43.43	0.01				0.13	0.38		100.25	83.10		3
Schirmacher Oasis	Spinel lherzolite	Orthopyroxene	56.66	0.05	1.12			10.84	30.57	0.27				0.16	0.07		99.74	83.40		3
Schirmacher Oasis	Spinel lherzolite	Spinel	0.00	0.29	31.13	30.63		28.88	7.84					0.08			98.85	32.60	39.77	3

References: 1 = Andronikov and Mikhalsky (1997); 2 = Sheraton and Cundari (1980); 3 = Schmaedicke *et al.* (2015).

The original contents of water and other volatiles in the mantle rocks are poorly constrained due to the lack of direct measurements. Only the presence of hydrous and carbonate phases can be used as a guide, but these appear restricted to later metasomatic events. Only the occurrence of phlogopite inclusions in primary minerals in the Vestfold Hills rocks (Andronikov and Mikhalsky 1997) are exceptions that would present exceptionally hydrous conditions at shallow mantle levels. Later metasomatic enrichment introduced water with silicate melts, and CO_2, H_2O, sulfur and nitrogen in carbonate melts and fluids (Kogarko *et al.* 2007; Buikin *et al.* 2014; Solovova *et al.* 2015). However, these probably reflect sub-rift processes and not volatile contents that are characteristic of widespread segments of the East Antarctic upper mantle.

Indications from igneous rocks for conditions and melting events in the mantle

Indirect evidence for the makeup of the upper mantle can be gleaned from the types and distribution of primitive magmatic rocks, some of which require ancient, depleted lithosphere, specific types of mantle metasomatism in reduced or oxidized conditions, or the action of a mantle plume.

The Lambert–Amery Rift and its precursors

Magmatic activity in and around the Lambert–Amery Rift, and its precursor, the Paleozoic Beaver Lake rift has been episodic between the Silurian and Cenozoic. The xenolith-bearing intrusive rocks are Cretaceous, and are related to the breakup of Gondwana, which in this area corresponds to the separation of India from Antarctica (Fig. 6). The earliest activity is an alkali mela-syenite at Mt Bayliss in the southernmost Prince Charles Mountains (430–414 Ma from K–Ar on K-richterite and K-arfvedsonite; Sheraton and England 1980). Its exceptionally high K_2O (9.4 wt%) and K_2O/Na_2O (4.5) and the presence of K-arfvedsonite to K-richterite denote lamproitic affinities. Magmatic activity resumed with alkali basalts, dolerites and lamprophyres in the northern Prince Charles Mountains dated at *c.* 323±31 Ma by U–Pb SHRIMP ages on apatite (Leitchenkov *et al.* 2018), which may represent a surface expression of the mantle event dated at 390–375 Ma by clinopyroxene Sm–Nd ages from the xenoliths (Andronikov and Beliatsky 1995). The main episode of alkaline picrite to ultramafic lamprophyre emplacement followed between 117 and 110 Ma (Walker and Mond 1971; Laiba *et al.* 1987). The youngest activity known is potassic trachybasalts at 40 Ma in the Manning Massif (Andronikov and Beliatsky 1996). All these occurrences are spatially related to a long-lived continental rift (Fig. 6) that is believed to correspond to the failed third arm of the India–Antarctica separation.

The compositions of the parental magmas to these igneous rocks can be explained by a two-stage petrogenesis: firstly migration and resolidification of melts within the mantle form dykes of various ultramafic rocks amongst the peridotite that dominates the mantle. Later re-melting of these dykes contributions to the chemical composition of the magmas that cannot be derived from peridotite. At several localities, Mg-rich melts variously referred to as ultramafic lamprophyres, kimberlites and alkali picrites occur (Andronikov and Egorov 1993; Egorov *et al.* 1993; Andronikov and Foley 2001; Foley *et al.* 2002; Belyatsky *et al.* 2008; Yaxley *et al.* 2013). This association of potassic alkaline ultramafic magmas, particularly those rich in CO_2 as are the Jetty Peninsula rocks (2–6 wt%; Foley *et al.* 2002) is characteristic of areas where a major continental rift abuts against or runs through a craton: other examples include Labrador, where the North Atlantic craton is split (Tappe *et al.* 2006, 2007) and the western branch of the East African rift in Uganda–Congo (Rosenthal *et al.* 2009). This is explained by melting at depths of 120–150 km, including the reactivation of carbonate- and mica-bearing dykes, which is evidenced by abundant radiogenic isotope signatures in Labrador and Uganda (Tappe *et al.* 2006; Rosenthal *et al.* 2009). Notably for the Jetty Peninsula ultramafic lamprophyres, the Sr–Nd isotope signatures of the lamprophyres aged 117–110 Ma can be explained by radiogenic growth in carbonate+mica-bearing dykes for which the isotopic clock started by magmatism in the mantle at the time of the previous volcanic episode at *c.* 323 Ma (Andronikov and Foley 2001). Sm–Nd isochrons and enriched mantle model ages of Jetty Peninsula peridotite xenolith whole-rocks and Cpx and Opx separates have been used to suggest other melt migration events within the mantle beneath this area occurred at around 1.0–0.9 Ga and 0.6 Ga (Andronikov and Beliatsky 1995; Andronikov *et al.* 1998). The Sr and Nd isotope signatures of the 40 Ma alkaline trachybasalts at Manning Massif are consistent with a source consisting of mica-rich dykes in peridotite (Andronikov *et al.* 1998).

The age of the oldest crustal zircons in the Lambert Amery Rift region increases from early Proterozoic in the northern Prince Charles Mountains (Liu *et al.* 2017) to 3.5–3.2 Ga in the south (the Ruker terrane; Boger *et al.* 2008). This may imply the presence of thicker, cratonic lithosphere in the south (Figs 2, 6); this would be consistent with the description of some of the southern intrusive rocks as kimberlite (Yaxley *et al.* 2013), because kimberlites are characteristic for areas of thick cratonic lithosphere (Giuliani and Pearson 2019). It is also compatible with geophysical evidence for a significant boundary at depth between the northern and southern Prince Charles Mountains (Reading 2006).

Xenoliths in the older ultramafic lamprophyres of the Vestfold Hills (*c.* 1.3Ga) also carry evidence for impregnation of the source by carbonate melts (Andronikov and Mikhalsky 1997), which is consistent with the stabilization of the lithosphere at least as early as 2.4 Ga (Hoek and Seitz 1995).

Melting conditions for lamproitic magmas

Gaussberg is an archetypical lamproite, whereas lamproitic affinities have been assigned to geographically widely dispersed igneous rocks in East Antarctica, including Mount Bayliss, minettes at Schirmacher Oasis (Hoch and Tobschall 1998), Kjakbeinet and Sør Rondane in Dronning Maud Land (Romu and Luttinen 2007: Owada *et al.* 2008) and Priestly Peak in Enderby Land (Sheraton and England 1980). The high Mg# and low MgO contents ($<\approx 8$ wt%) of Gaussberg are compatible with melting of mantle in the spinel lherzolite stability field (<75 km depth), but this mantle must contain considerable proportions of other rocks rich in mica and incompatible elements in order to explain the extreme K_2O contents (11–12 wt%) of the rock. Furthermore, the production of perpotassic melts ($K_2O>Al_2O_3$) is favoured in fluorine-rich conditions at low oxygen fugacities in which carbon exists only in reduced form (i.e. no CO_2; Foley 1989). This is consistent with the compositions of phenocryst spinels, which show evidence of oxidation during ascent, as observed in other lamproite localities (Foley 1985). The same source components are probably relevant for the other rocks with lamproitic affinity, but in more diluted form, because melting proceeded to involve more of the surrounding peridotite.

Radiogenic isotopes favour a sedimentary origin, albeit Archean in age, for the enrichment in potassium and other incompatible elements at Gaussberg (Murphy *et al.* 2002). Subducted crust has also been invoked for the 455 Ma Schirmacher Oasis minettes (Hoch *et al.* 2001), possibly involving pelagic

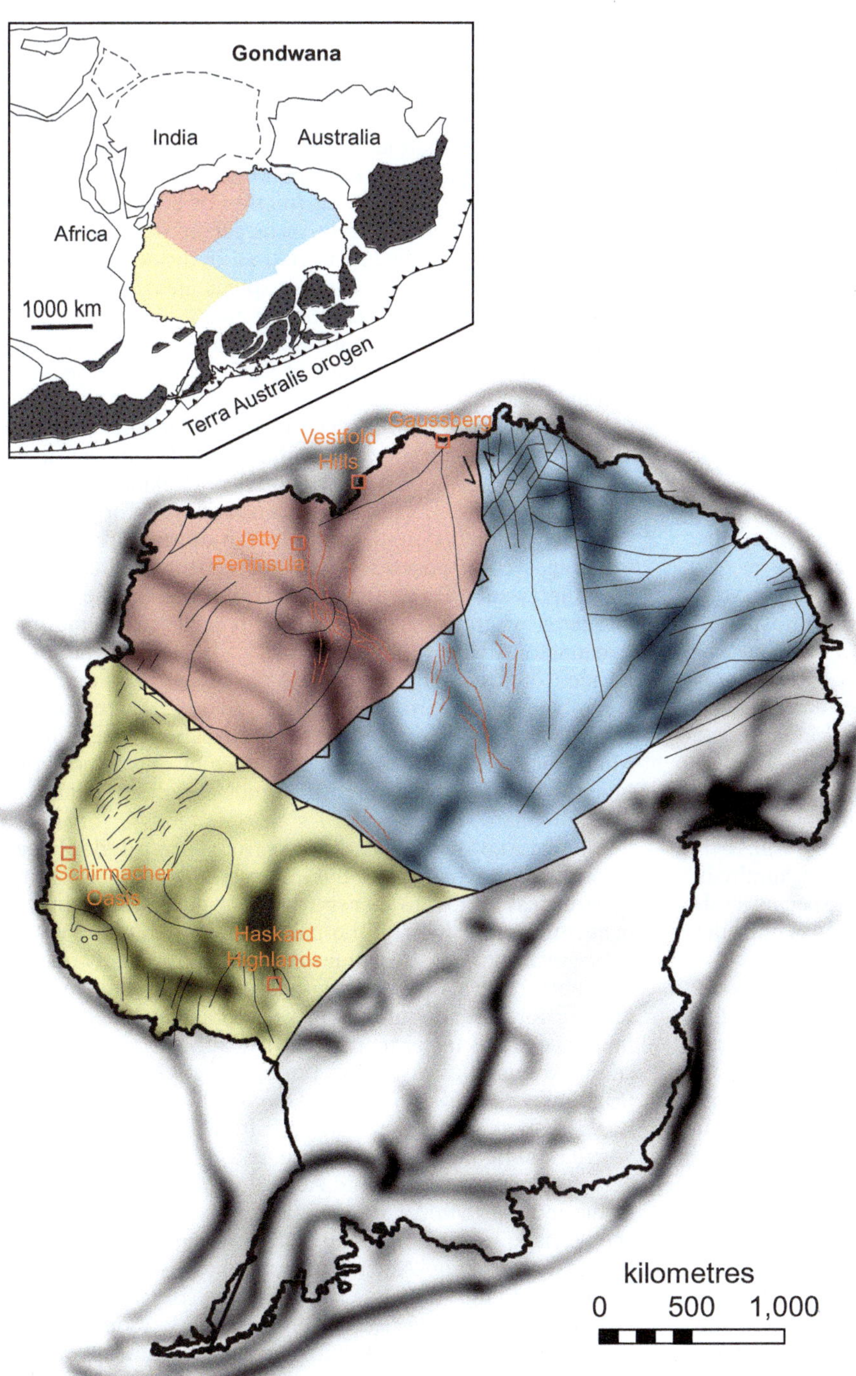

Fig. 6. Correlations of crustal provinces in eastern Gondwana with East Antarctica compiled from coastal rock outcrops, analysis of offshore sedimentary provenance, and subglacial geophysical surveys. Correlations are colour-coded between Antarctica and neighbouring Africa (yellow), India (red) and Australia (blue), with positions in Gondwana shown in the inset. These are first-order tectonic domains for use in plate reconstruction models; each domain comprises numerous cratons and orogenic belts of related affinity prior to Gondwana amalgamation. Lines decorated with triangles indicate inferred subduction/collision zones. Diffuse grey-to-black lines are probable deep lithospheric boundaries derived from a combination of seismic tomography at 150 km depth, gravity data, and topography (Stål *et al.* 2019); darker shading indicates a higher likelihood of a domain boundary; thin black lines are major lineaments interpreted from magnetics (Aitken *et al.* 2014; Jacobs *et al.* 2017); red lines are inferred rift structures (Ferraccioli *et al.* 2011). Positions of the mantle rock localities are shown in red squares.

sediment (Hoch and Tobschall 1998). The Schirmacher ouachitites (damtjernites in modern terminology; Tappe *et al.* 2005) are unrelated to the minettes (Hoch 1999) and indicate a deep hydrous source similar to the Jetty Peninsula ultramafic lamprophyres, but probably with less carbonate.

Flood basalts and plume activity in Dronning Maud Land

Continental flood basalts (CFB) of Jurassic age are widespread in Dronning Maud Land, where they represent the Antarctic extension of the Karoo large igneous province (Luttinen and Furnes 2000). These include many MgO-rich primitive members (ferropicrites, picrobasalts and meimechites; Heinonen *et al.* 2010), some of which have isotopic signatures and olivine phenocryst compositions that indicate derivation from a source enriched in pyroxenites (Heinonen *et al.* 2010, 2013), presumably as dykes. The CFB activity spread eastward over more than 120 km, pre-empting the breakup of Gondwana (Fig. 6; Luttinen and Furnes 2000; Sushchevskaya *et al.* 2009). Radiogenic and oxygen isotopes and trace element characteristics of the Dronning Maud Land CFB permit recognition of distinct lithospheric blocks in the source, which are variably enriched by subduction modification of the Gondwanan mantle (Luttinen and Furnes 2000; Heinonen *et al.* 2018). Karoo magmatism in southern Africa also includes alkali-rich picrites that are explained by melting of mixed asthenosphere and enriched lithosphere sources (Ellam and Cox 1989).

Regional tectonics and potential mantle lithosphere blocks

Although our knowledge of the mantle lithosphere beneath East Antarctica is hampered by the scarcity of xenoliths and the paucity of outcrop, variations in its composition can be inferred from geological correlations of crustal rocks that were its neighbours before the breakup of Gondwana.

Figure 6 illustrates the schematic positions of East Antarctica, Australia, India and South Africa about 180 million years ago, and geological correlations between Antarctica and Africa (yellow), India (red) and Australia (blue). The positions of the mantle rock localities are marked as red squares; these all fall in the Afro-Antarctic and Indo-Antarctic segments. East Antarctica (coloured in Fig. 6) is geophysically distinguished from the Ross Orogen and West Antarctica by high shear wave speeds at 75–150 km depth (Lloyd *et al.* 2019) and lithosphere thicknesses of 150–260 km for 80% of East Antarctica (Pappa *et al.* 2019) from LitMod3D modelling, which combines data from seismological, gravity, heat flow, isostasy and petrological parameters (Fullea *et al.* 2009). The first-order tectonic domains shown in Figure 6 are based mainly on our interpretation of available outcrop geology and magnetic data (Golynsky *et al.* 2018) updating the East Antarctic cratonic elements used in recent plate reconstruction models (Merdith *et al.* 2017). They are overlain on an 'intersect' likelihood map of subglacial deep lithospheric boundaries from Stål *et al.* (2019) based on a combination of seismic tomography at 150 km depth, gravity and topography. Also illustrated are major magnetic lineaments (black lines; Aitken *et al.* 2014; Jacobs *et al.* 2017) and inferred rift structures (red lines; Ferraccioli *et al.* 2011). The inferred polarity of subduction zones are taken from Mulder *et al.* (2019) and Merdith *et al.* (2017). These three coloured domains do not imply uniformity of the lithosphere in each of the areas; on the contrary, on a scale of tens to hundreds of kilometres, significant variations are to be expected both within the deep lithosphere (Stål *et al.* 2019) and within the crust (e.g. Mulder *et al.* 2019).

The isotope characteristics of Mesozoic magmatism associated with the Karoo-Maud plume in the west-central part of the Afro-Antarctic terrane (yellow in Fig. 6) indicate the sampling of several distinct domains, some of which involve subduction-modified mantle lithosphere that may be related to the suture with the Indo-Antarctic domain (red in Fig. 6: Luttinen and Furnes 2000; Heinonen *et al.* 2013, 2018). Within the Afro-Antarctic domain, a suture is recognized by a 7 km offset in Moho depth at *c.* 10°W (Bayer *et al.* 2009), and a small cratonic area with lithosphere thicker than 200 km (Haeger *et al.* 2019) lies in roughly the same area as the widespread potassic magmatism at around 0–25°E (Owada *et al.* 2008). This magmatism may be related to the orangeites of the Kaapvaal craton. Both of the tectonically emplaced mantle rock localities at Schirmacher Oasis and in the Shackleton Ranges are close to suture zones or lithosphere thickness offsets; their emplacement may be related to these.

The most abundant mantle xenoliths (Jetty Peninsula and Vestfold Hills) occur in the central part of the Indo-Antarctic terrane (red in Fig. 6). These have experienced several episodes of enrichment by melt infiltration, which document mantle processes throughout the gradual, punctuated development of the rift that eventually split India from Antarctica. Geochemical and mineralogical evidence in the Proterozoic Vestfold Hills xenoliths shows a broadly similar enrichment history in Archean or Proterozoic times, indicating that small-scale modifications of the mantle lithosphere are the norm rather than the exception and are predicted to apply to all areas that are overprinted by rift structures (Fig. 6). The collective effects of a large number of small-scale enrichments show up as 'undepleted lithosphere' in geophysical compositional models (Haeger *et al.* 2019).

Post-cratonic tectonic events mean that current lithosphere thickness may not correspond to the distribution of Archean crustal rocks: Pb isotopic compositions of feldspar grains from crustal rocks indicate that the Napier Complex of Enderby Land and Vestfold Hills crust were probably both related to Indian cratonic lithosphere at 2.5 Ga (Flowerdew *et al.* 2013), whereas the lithosphere is now thinner in both the Napier and Vestfold regions than is typical for cratons, and this is also true for lithosphere throughout the Rayner Complex due to the partial destruction of lithosphere during Phanerozoic orogenic activity (Haeger *et al.* 2019; Pappa *et al.* 2019). The Archean Ruker crust in the southern Prince Charles Mountains (Fig. 2) may spread inland beneath the ice over much of central–East Antarctica.

The single spinel lherzolite xenolith at Gaussberg volcano tells us little about the history of the lithosphere in Wilhelm II Land, whereas crustal rocks and xenocrysts sampled by the volcano are more fruitful. These include zircon xenocrysts with the same ages of *c.* 980, 500 and 320 Ma seen in the northern Prince Charles Mountains (Mikhalsky *et al.* 2015), where these ages have been interpreted to represent disturbance events in the mantle lithosphere. The first of these ages corresponds to the Rayner Orogeny, the second characterizes the Prydz Belt and likely related to the Kuunga Orogeny, whereas the latter can be related to alkaline magmatic activity in the Lambert–Amery Rift region. These similarities underscore the Indo-Antarctic affinities of the Gaussberg region (red in Fig. 6). Nd model ages of 2.5–2.3 Ga for crustal xenoliths at Gaussberg (Collerson and McCulloch 1983) indicate the existence of crust, and probably mantle lithosphere, at least as old as earliest Proterozoic beneath Gaussberg.

There are no known mantle rocks from the Australo-Antarctic terrane (blue on Fig. 6), but information from crustal rocks nevertheless helps to constrain the nature of the mantle beneath. The position of the palaeo-plate boundary between Indo-Antarctica and Australo-Antarctica is clearly east of Gaussberg (see above): it is thought to be orientated approximately north–south intersecting the coast along the Mirny fault at *c.* 94°E (Daczko *et al.* 2018) and to link inland to a structure running east–west through the Gamburtsev area (Fig. 6; Ferraccioli *et al.* 2011). Dominantly strike-slip motion is interpreted along the Mirny fault (Daczko *et al.* 2018), whereas high-angle convergence and a subduction margin between Indo-Antarctica and Australo-Antarctica is predicted in the interior (Mulder *et al.* 2019). This subduction margin should have led to fluid-modification of the lithosphere. Further east in Australo-Antarctica, coastal rocks of Terre Adelie are related to the Archean Gawler Craton (Fig. 6 inset; Boger 2011), which continue southward as the Mawson Craton in Antarctica. This may link with the thickest lithosphere area beneath the Gamburtsev Subglacial Mountains (80–90°E and >80°S; Pappa *et al.* 2019).

Beneath thick ice, our knowledge is limited to large-scale geophysical patterns, which will be affected by the collective effects of a long history of small-scale melt and fluid infiltration processes as seen wherever we have xenolith evidence. This should be remembered when considering compositional models that implicitly assume the lithosphere to comprise homogeneous peridotite. The mantle lithosphere varies locally to regionally on a scale that cannot be easily sensed by current geophysical models and includes minerals and rock types other than peridotite that result from numerous melt and fluid additions.

Acknowledgements We are grateful to Massimo Coltorti, an anomymous reviewer and editor Adam Martin for comments that have improved the manuscript.

Author contributions SFF: conceptualization (lead), data curation (lead), resources (equal), writing – original draft (lead);

AVA: data curation (supporting), writing – original draft (supporting), writing – review & editing (supporting); **JAH**: writing – review & editing (supporting); **NRD**: writing – review & editing (supporting); **DEJ**: formal analysis (equal), writing – review & editing (supporting).

Funding Funding for this research came from the Australian Research Council (award FL180100134) to Stephen Foley.

Data availability Data sharing is not applicable to this article as no datasets were generated or analysed during the current study.

References

Aitken, A.R.A., Young, D.A. *et al.* 2014. The subglacial geology of Wilkes Land, East Antarctica. *Geophysical Research Letters*, **41**, L059405, https://doi.org/10.1002/2014GL059405

Andronikov, A.V. 1990. Spinel-garnet lherzolite nodules from alkaline–ultrabasic rocks of Jetty Peninsula (East Antarctica). *Antarctic Science*, **2**, 321–330, https://doi.org/10.1017/S0954102090000451

Andronikov, A.V. 1992. The first study of upper mantle inclusions from the Prince Charles Mountains, East Antarctica. *In*: Yoshida, Y., Kaminuma, K. and Shiraishi, K. (eds) *Recent Progress in Antarctic Earth Science*. Terrapub, Tokyo, 163–171.

Andronikov, A.V. 1997. Varying glass compositions in deep-seated inclusions from alkaline–ultamafic rocks of Jetty Peninsula (East Antarctica): Evidence for mantle metasomatism. *In*: Ricci, C.A. (ed.) *The Antarctic Region: Geological Evolution and Processes*. Terra Antarctica, Siena, 901–910.

Andronikov, A.V. and Beliatsky, B.V. 1995. Implication of Sm–Nd isotopic systematics to the events recorded in the mantle-derived xenoliths from the Jetty Peninsula, East Antarctica. *Terra Antarctica*, **2**, 103–110.

Andronikov, A.V. and Beliatsky, B.V. 1996. Manning Massif alkaline trachybasalts (MAT): some characteristics of enriched lithospheric mantle source. Proceedings of 7th Freiberg Isotopic Colloquium, Freiberg, 9–18.

Andronikov, A.V. and Egorov, L.S. 1993. Mesozoic alkaline–ultramafic magmatism of Jetty Peninsula. *In*: Findlay, R.H., Unrug, R., Banks, M.R. and Veevers, J.J. (eds) *Gondwana Eight: Assembly, Evolution and Dispersal*. Balkema, Rotterdam, 547–557.

Andronikov, A.V. and Foley, S.F. 2001. Trace element and Nd–Sr isotopic composition of ultramafic lamprophyres from the East Antarctic Beaver Lake area. *Chemical Geology*, **175**, 291–305, https://doi.org/10.1016/S0009-2541(00)00296-5

Andronikov, A.V. and Mikhalsky, E.V. 1997. Deep-seated xenoliths from Proterozoic ultramafic lamprophyre dykes of the Vestfold Hills (East Antarctica): evidence for the existence of metasomatized lithosphere. *In*: Ricci, C.A. (ed.) *The Antarctic Region: Geological Evolution and Processes*. University of Siena, 911–921.

Andronikov, A.V. and Sheraton, J.W. 1996. The redox state of lithospheric upper mantle beneath the East Antarctic Shield. *Terra Antartica*, **3**, 39–48.

Andronikov, A.V., Mikhalsky, E.V. and Belyatsky, B.V. 1994. Abyssal xenoliths from the lamprophyres of the Vestfold Hills, East Antarctica. *Petrology*, **2**, 288–296.

Andronikov, A.V., Foley, S.F. and Melzer, S. 1997. Trace element composition of the Manning Massif alkaline volcanics (MMAV), Northern Prince Charles Mountains, East Antarctica. *Terra Antarctica*, **4**, 29–32.

Andronikov, A.V., Foley, S.F. and Beliatsky, B.V. 1998. Sm–Nd and Rb–Sr isotopic systematics of the East Antarctic Manning Massif alkaline trachybasalts and the development of the mantle beneath the Lambert–Amery Rift. *Mineralogy and Petrology*, **63**, 243–261, https://doi.org/10.1007/BF01164153

Arai, S. 1984. Pressure-temperature dependent compositional variation of phlogopitic micas in upper mantle peridotites. *Contributions to Mineralogy and Petrology*, **87**, 260–264, https://doi.org/10.1007/BF00373059

Ballhaus, C., Berry, R.F. and Green, D.H. 1990. Oxygen fugacity controls in the Earth's upper mantle. *Nature*, **348**, 437–440, https://doi.org/10.1038/348437a0

Bayer, B., Geissler, W.H., Eckstaller, A. and Jokat, W. 2009. Seismic imaging of the crust beneath Dronning Maud Land, East Antarctica. *Geophysical Journal International*, **178**, 860–876, https://doi.org/10.1111/j.1365-246X.2009.04196.x

Beliatsky, B.V. and Andronikov, A.V. 2009. Deep-seated lherzolite inclusions from alkaline–ultramafic rocks in Jetty Peninsula: the mineral and chemical composition, P–T conditions and Sr–Nd isotope systematic. *In*: *Scientific results of Russian geological and geophysical researches in Antarctica 2*, St Petersburg, VNIIOkeangeologia, 89–109 [in Russian].

Belyatsky, B.V., Antonov, A.V., Rodionov, N.V., Laiba, A.A. and Sergeev, S.A. 2008. Age and composition of carbonatite kimberlite dykes in the Prince Charles Mountains, East Antarctica. 9th International Kimberlite Conference, Extended Abstract 9IKC–00272.

Boger, S.D. 2011. Antarctica – before and after Gondwana. *Gondwana Research*, **19**, 335–371, https://doi.org/10.1016/j.gr.2010.09.003

Boger, S.D., Maas, R. and Fanning, C.M. 2008. Isotopic and geochemical constraints on the age and origin of granitoids from the central Mawson Escarpment, southern Prince Charles Mountains, East Antarctica. *Contributions to Mineralogy and Petrology*, **155**, 379–400, https://doi.org/10.1007/s00410-007-0249-x

Boyd, F.R. and Mertzmann, S. 1987. Comparison and structure of the Kaapvaal lithosphere, southern Africa. *Geochemical Society Special Publication*, **1**, 13–24.

Brey, G.P. and Köhler, T. 1990.Geothermobarometry in four-phase lherzolites II. New thermobarometers, and practical assessment of existing thermobarometers. *Journal of Petrology*, **31**, 1353–1378, https://doi.org/10.1093/petrology/31.6.1353

Buikin, A.I., Solovova, I.P., Verchovsky, A.B., Kogarko, L.N. and Averin, A.A. 2014. PVT parameters of fluid inclusions and the C, O, N, and Ar isotopic composition in a garnet lherzolite xenolith from the Oasis Jetty, East Antarctica. *Geochemistry International*, **52**, 805–821, https://doi.org/10.1134/S0016702914100036

Collerson, K.D. and McCulloch, M.T. 1983. Nd and Sr isotope geochemistry of leucite-bearing lavas from Gaussberg, East Antarctica. *In*: Oliver, R.L., James, P.R. and Jago, J.B. (eds) *Antarctic Earth Science*. Australian Academy of Science, Canberra, 676–680.

Coltorti, M., Bonadiman, C., Casetta, F., Faccini, B., Giacomoni, P.P., Pelorosso, B. and Perinelli, C. 2021. Nature and evolution of the northern Victoria Land lithospheric mantle (Antarctica) as revealed by ultramafic xenoliths. *Geological Society, London, Memoirs*, **56**, https://doi.org/10.1144/M56-2020-11

Daczko, N.R., Halpin, J.A., Fitzsimons, I.C.W. and Whittaker, J.M. 2018. A cryptic Gondwana-forming orogeny located in Antarctica. *Scientific Reports*, **8**, 8371, https://doi.org/10.1038/s41598-018-26530-1

Delor, C.P. and Rock, N.M.S. 1991. Alkaline–ultramafic lamprophyre dykes from the Vestfold Hills, Princess Elizabeth Land (East Antarctica): primitive magmas of deep mantle origin. *Antarctic Science*, **3**, 419–432, https://doi.org/10.1017/S0954102091000512

Egorov, L.S., Melnik, A.Y. and Ukhanov, A.V. 1993. First finding of a kimberlite dyke containing syngenetic calcite carbonatite schlieren in Antarctica. *Reports of the USSR Academy of Sciences*, **328**, 230–233.

Ehrenberg, S.C. 1982. Petrogenesis of garnet lherzolite and megacrystalline nodules from the Thumb, Navajo Volcanic Field. *Journal of Petrology*, **23**, 507–547, https://doi.org/10.1093/petrology/23.4.507

Ellam, R.M. and Cox, K.G. 1989. A Proterozoic source for Karoo magmatism: evidence from the Nuanetsi picrites. *Earth and Planetary Science Letters*, **92**, 207–218, https://doi.org/10.1016/0012-821X(89)90047-2

Ferraccioli, F., Finn, C.A., Jordan, T.A., Bell, R.E., Anderson, L.M. and Damaske, D. 2011. East Antarctic rifting triggers uplift of the Gamburtsev Mountains. *Nature*, **479**, 388–392, https://doi.org/10.1038/nature10566

Flowerdew, M.J., Tyrrell, S., Boger, S.D., Fitzsimons, I.C.W., Harley, S.L., Mikhalsky, E.V. and Vaughan, A.P.M. 2013. Pb isotopic domains from the Indian Ocean sector of Antarctica: implications for past Antarctica-India connections. *Geological Society, London, Special Publications*, **383**, 59–72, https://doi.org/10.1144/SP383.3

Foley, S.F. 1985. The oxidation state of lamproitic magmas. *Tschermaks Mineralogische und Petrographische Mitteilungen*, **34**, 217–238, https://doi.org/10.1007/BF01082963

Foley, S.F. 1989. Experimental constraints on phlogopite chemistry in lamproites: 1. the effect of water activity and oxygen fugacity. *European Journal of Mineralogy*, **1**, 411–426, https://doi.org/10.1127/ejm/1/3/0411

Foley, S.F. and Jenner, G.A. 2004. Trace element partitioning in lamproitic magmas – the Gaussberg olivine leucitite. *Lithos*, **75**, 19–38, https://doi.org/10.1016/j.lithos.2003.12.020

Foley, S.F., Andronikov, A.V. and Melzer, S. 2002. Petroogy, geochemistry and mineral chemistry of ultramafic lamprophyres from the Jetty Peninsula area of the Lambert–Amery Rift, Eastern Antarctica. *Mineralogy and Petrology*, **74**, 361–384, https://doi.org/10.1007/s007100200011

Foley, S.F., Andronikov, A.V., Jacob, D.E. and Melzer, S. 2006. Evidence from Antarctic mantle peridotite xenoliths for changes in mineralogy, geochemistry and geothermal gradients beneath a developing rift. *Geochimica et Cosmochimica Acta*, **70**, 3096–3120, https://doi.org/10.1016/j.gca.2006.03.010

Fullea, J., Afonso, J.C., Connolly, J.A.D., Fernandez, N., Garcia-Castellanos, D. and Zeyen, H. 2009. LitMod3D: an interactive 3D software to model the thermal, compositional, density, rheological and seismological structure of the lithosphere and sub-lithospheric upper mantle. *Geochemistry, Geophysics, Geosystems*, **10**, Q08019, https://doi.org/10.1029/2009GC002391

Giuliani, A. and Pearson, D.G. 2019. Kimberlites: from deep Earth to diamond mines. *Elements*, **15**, 377–380, https://doi.org/10.2138/gselements.15.6.377

Glaser, S.M., Foley, S.F. and Günther, D. 1999. Trace element compositions of minerals in garnet and spinel peridotite xenoliths from the Vitim volcanic field, Transbaikalia, eastern Siberia. *Lithos*, **48**, 263–285, https://doi.org/10.1016/S0024-4937(99)00032-8

Golynsky, A.V., Ferraccioli, F. *et al.* 2018. New magnetic anomaly map of the Antarctic. *Geophysical Research Letters*, **45**, 6437–6449, https://doi.org/10.1029/2018GL078153

Haeger, C., Kaban, M.K., Tesauro, M., Petrunin, A.G. and Mooney, W.D. 2019. 3-D density, thermal, and compositional model of the Antarctic lithosphere and implications for its evolution. *Geochemistry, Geophysics, Geosystems*, **20**, 688–707, https://doi.org/10.1029/2018GC008033

Heinonen, J.S., Carlson, R.W. and Luttinen, A.V. 2010. Isotopic (Sr, Nd, Pb and Os) composition of highly magnesian dikes of Vestfjella, western Dronning Maud Land, Antarctica: a key to the origins of the Jurassic Karoo large igneous province? *Chemical Geology*, **277**, 227–244, https://doi.org/10.1016/j.chemgeo.2010.08.004

Heinonen, J.S., Luttinen, A.V., Riley, T.R. and Michallik, R.M. 2013. Mixed pyroxenite-peridotite sources for mafic and ultramafic dikes from the Antarctic segment of the Karoo continental flood basalt province. *Lithos*, **177**, 366–380, https://doi.org/10.1016/j.lithos.2013.05.015

Heinonen, J.S., Luttinen, A.V., Whitehouse, M.J. 2018. Enrichment of O-18 in the mantle sources of the Antarctic portion of the Karoo large igneous province. *Contributions to Mineralogy and Petrology*, **173**, https://doi.org/10.1007/s00410-018-1447-4.

Hoch, M. 1999. Geochemistry and petrology of ultramafic lamprophyres from Schirmacher Oasis, East Antarctica. *Mineralogy and Petrology*, **65**, 51–67, https://doi.org/10.1007/BF01161576

Hoch, M. and Tobschall, H.J. 1998. Minettes from Schirmacher Oasis, East Antarctica – indicators of an enriched mantle source. *Antarctic Science*, **10**, 476–486, https://doi.org/10.1017/S0954102098000571

Hoch, M., Rehkämper, M. and Tobschall, H.J. 2001. Sr, Nd, Pb and O isotopes of minettes from Schirmacher Oasis, East Antarctica: a case of mantle metasomatism involving subducted continental material. *Journal of Petrology*, **42**, 1387–1400, https://doi.org/10.1093/petrology/42.7.1387

Hoek, J.D. and Seitz, H.M. 1995. Continental mafic dyke swarms as tectonic indicators: an example from the Vestfold Hills, East Antarctica. *Precambrian Research*, **75**, 121–139 https://doi.org/10.1016/0301-9268(95)80002-Y

Ionov, D.A., Ashchepkov, I.V., Stosch, H.G., Witt-Eickschen, G. and Seck, H.A. 1993. Garnet peridotite xenoliths from the Vitim volcanic field, Baikal region: the nature of garnet-spinel-peridotite transition zone in the continental mantle. *Journal of Petrology*, **34**, 1141–1175, https://doi.org/10.1093/petrology/34.6.1141

Jacob, D.E., Jagoutz, E. and Sobolev, N.V. 1997. Neodymium and strontium isotopic measurements on single subcalcic garnet grains from Yakutian kimberlites. *Neues Jahrbuch für Mineralogie, Abhandlungen*, **172**, 357–379, https://doi.org/10.1127/njma/172/1998/357

Jacobs, J., Opås, B. *et al.* 2017. Cryptic sub-ice geology revealed by a U-Pb zircon study of glacial till in Dronning Maud Land, East Antarctica. *Precambrian Research*, **294**, 1–14, https://doi.org/10.1016/j.precamres.2017.03.012

James, P.R. and Tingey, R.J. 1983. The Precambiran geological evolution of the East Antarctic metamorphic shield – a review. *In*: Oliver, R.L., James, P.R. and Jago, J.B. (eds) *Antarctic Earth Science*. Australian Academy of Science, 5–10.

Kempton, P.D., Lopez-Escobar, L., Hawkesworth, C.J., Paerson, D.G., Wright, D.W. and Ware, A.J. 1999. Spinel±garnet lherzolite xenoliths from Pali Aike. Part 1: petrography, mineral chemistry and geothermobarometry. *In*: Dawson, J.B., Gurney, J.J., Gurney, J.L. and Pascoe, M.D. (eds) Proceedings of the 7th International Kimberlite Conference, Red Roof, Cape Town, 403–414.

Kogarko, L.N., Kurat, G. and Ntaflos, T. 2007. Henrymeyerite in the metasomatized upper mantle of Eastern Antarctica. *Canadian Mineralogist*, **45**, 497–501, https://doi.org/10.2113/gscanmin.45.3.497

Kuehner, S.M. 1987. Mafic dykes of the East Antarctic Shield: a note on the Vestfold Hills and Mawson Coast occurrences. *Geological Association of Canada Special Paper*, **34**, 428–429.

Laiba, A.A., Andronikov, A.V., Egorov, L.S. and Fedorov, L.V. 1987. Stock-like and dyke bodies af alkaline–ultrabasic composition in Jetty Peninsula (Prince Charles Mountains, East Antarctica). *In*: Grikurov, G.E. and Ivanov, L.L. (eds) *Geologic and Geophysic Investigations in Antarctica*. Leningrad, Sevmorgeologia, 35–47 [in Russian].

Leitchenkov, G.L., Belyatsky, B.V., Kaminsky, V.D. 2018. The age of rift-related basalts in East Antarctica. *Doklady Akademii Nauk*, **478**, 11–14, https://doi.org/10.1134/S1028334X18010051.

Liu, X., Zhao, Y., Chen, H. and Song, B. 2017. New zircon U–Pb and Hf–Nd isotopic constraints on the timing of magmatism, sedimentation and metamorphism in the northern Prince Charles Mountains, East Antarctica. *Precambrian Research*, **299**, 15–33 https://doi.org/10.1016/j.precamres.2017.07.012

Lloyd, A.J., Wiens, D.A. *et al.* 2019. Seismic structure of the Antarctic upper mantle based on adjoint tomography. *Journal of Geophysical Research Solid Earth*, https://doi.org/10.1029/2019JB017823

Luttinen, A.V. and Furnes, H. 2000. Flood basalts of Vestfjella: Jurassic magmatism across and Archaean-Proterozoic lithospheric boundary in Dronning Maud Land, Antarctica. *Journal of Petrology*, **41**, 1271–1305, https://doi.org/10.1093/petrology/41.8.1271

Markl, G., Abart, R., Vennemann, T. and Sommer, H. 2003. Mid-crustal metasomatic reaction veins in a spinel peridotite. *Journal of Petrology*, **44**, 1097–1120, https://doi.org/10.1093/petrology/44.6.1097

Merdith, A.S., Collins, A.S. *et al.* 2017. A full-plate global reconstruction of the Neoproterozoic. *Gondwana Research*, **50**, 84–134, https://doi.org/10.1016/j.gr.2017.04.001

Mikhalsky, E.V., Belyatsky, B.V. *et al.* 2015. The geological composition of the hidden Wilhelm II Land in east Antarctica: SHRIMP zircon, Nd isotopic and geochemical studies with implications for Proterozoic supercontinent reconstructions. *Precambrian Research*, **258**, 171–185, https://doi.org/10.1016/j.precamres.2014.12.011

Mikhalsky, E.V., Alexeev, N.L., Kamenev, I.A., Egorov, M.S. and Kunakkuzin, E.L. 2019. Mafic dykes in the Rauer Islands and Vestfold Hills (East Antarctica): a chemical and Nd isotopic comparison. *Precambrian Research*, **329**, 273–293, https://doi.org/10.1016/j.precamres.2018.11.014

Mukasa, S.B. and Andronikov, A.V. 1999. Trace element and Sm–Nd and Rb–Sr isotopic evidence for the incipient melting of clinopyroxene in lherzolite xenoliths of the Jetty Peninsula area, East Antarctica. *EOS, Transactions of the American Geophysical Union*, **80**, S379.

Mulder, J.A., Halpin, J.A., Daczko, N.R., Orth, K., Meffre, S., Thompson, J.M. and Morrisey, L.J. 2019. A multiproxy provenance approach to uncovering the assembly of East Gondwana in Antarctica. *Geology*, **47**, 645–649, https://doi.org/10.1130/G45952.1

Murphy, D.T., Collerson, K.D. and Kamber, B.S. 2002. Lamproites from Gaussberg, Antarctica: possible transition zone melts of Archean subducted sediments. *Journal of Petrology*, **43**, 981–1001, https://doi.org/10.1093/petrology/43.6.981

Nixon, P.H., Rogers, N.W., Gibson, I.L. and Grey, A. 1981. *Annual Review of Earth and Planetary Sciences*, **9**, 285–309, https://doi.org/10.1146/annurev.ea.09.050181.001441

Njombie Wagsong, M.P., Temdjim, R., Foley, S.F. 2018. Petrology of spinel lherzolite xenoliths from Youkou volcano, Adamawa Massif, Cameroon Volcanic Line: mineralogical and geochemical fingerprints of sub-rift mantle processes. *Contributions to Mineralogy and Petrology*, **173**, https://doi.org/10.1007/s00410-018-1438-5.

Owada, M., Baba, S., Osanai, Y. and Kagami, H. 2008. Geochemistry of post-kinematic mafic dykes from central to eastern Dronning Maud Land, East Antarctica: evidence for a Pan-African suture in Dronning Maud Land. *Geological Society, London, Special Publications*, **308**, 235–252, https://doi.org/10.1144/SP308.12

Pappa, F., Ebbing, J., Ferraccioli, F. and van der Wal, W. 2019. Modeling satellite gravity gradient data to derive density, temperature, and viscosity structure of the Antarctic lithosphere. *Journal of Geophysical Research Solid Earth*, **124**, https://doi.org/10.1029/2019JB017997

Pearson, D.G., Canil, D. and Shirey, S.B. 2003. Mantle samples included in volcanic rocks: xenoliths and diamonds. *Treatise on Geochemistry*, **2**, 171–275, https://doi.org/10.1016/B0-08-043751-6/02005-3

Reading, A.M. 2006. The seismic structure of Precambrian and early Proterozoic terranes in the Lambert Glacier region, East Antarctica. *Earth and Planetary Science Letters*, **244**, 44–57, https://doi.org/10.1016/j.epsl.2006.01.031

Romu, I. and Luttinen, A. 2007. Lamproite-hosted xenoliths of Vestfjella: implications for lithospheric architecture in western Dronning Maud Land, Antarctica. *Goldschmidt Conference Abstracts*, **A849**.

Rosenthal, A., Foley, S.F., Pearson, D.G., Nowell, G.M. and Tappe, S. 2009 Magmatic evolution at the propagating tip of a continental rift – a geochemical study of primitive alkaline volcanic rocks of the western branch of the East African Rift. *Earth and Planetary Science Letters*, **284**, 236–248, https://doi.org/10.1016/j.epsl.2009.04.036

Schmaedicke, E., Will, T.M. and Mezger, K. 2015. Garnet pyroxenite from the Shackleton Range, Antarctica: intrusion of plume-derived picritic melts in the continental lithosphere during Rodinia breakup? *Lithos*, **238**, 185–206 https://doi.org/10.1016/j.lithos.2015.09.016

Seitz, H.-M. 1991. *Geochemistry and petrogenesis of high-Mg tholeiites and lamprophyres in the Vestfold Hills, Antarctica*. PhD thesis, University of Tasmania, Hobart, 303

Sheraton, J.W. and Cundari, A. 1980. Leucitites from Gaussberg, Antarctica. *Contributions to Mineralogy and Petrology*, **71**, 417–427, https://doi.org/10.1007/BF00374713

Sheraton, J.W. and England, R.N. 1980. Highly potassic mafic dykes from Antarctica. *Journal of the Geological Society of Australia*, **27**, 129–135, https://doi.org/10.1080/00167618008729128

Sheraton, J.W., Thomson, J.W. and Collerson, K.D. 1987. Mafic dyke swarms of Antarctica. *Geological Association of Canada Special Paper*, **34**, 419–432.

Solovova, I.P., Kogarko, L.N. and Averin, A.A. 2015. Conditions of sulphide formation in the metasomatized mantle beneath East Antarctica. *Petrology*, **23**, 519–542, https://doi.org/10.1134/S0869591115060053

Stål, T., Reading, A.M., Halpin, J.A. and Whittakaer, J.M. 2019. A multivariate approach for mapping lithospheric domain boundaries in East Antarctica. *Geophysical Research Letters*, **46**, 10,414–10,416, https://doi.org/10.1029/2019GL083453

Sushchevskaya, N.M., Belyatsky, B.V., Leitchenkov, G.L. and Laiba, A.A. 2009. Evolution of the Karoo-Maud plume in Antarctica and its influence on the magmatism of the early stages of Indian Ocean opening. *Geochemistry International*, **47**, 1–17, https://doi.org/10.1134/S0016702909010017

Sushchevskaya, N.M., Belyatsky, B.V., Tkacheva, D.A., Leitchenkov, G.L., Kuzmin, D.V. and Zhilkina, A.V. 2018. Early Cretaceous alkaline magmatism of East Antarctica: peculiarities, conditions of formation, and relationship with the Kerguelen plume. *Geochemistry International*, **56**, 1051–1070, https://doi.org/10.1134/S0016702918110071

Tappe, S., Foley, S.F., Jenner, G.A. and Kjarsgaard, B.A. 2005. Integrating ultramafic lamprophyres into the IUGS classification of igneous rocks: rationale and implications. *Journal of Petrology*, **46**, 1893–1900, https://doi.org/10.1093/petrology/egi039

Tappe, S., Foley, S.F. *et al.* 2006. Genesis of ultramafic lamprophyres and carbonatites at Aillik Bay, Labrador: a consequence of incipient lithospheric thinning beneath the North Atlantic craton. *Journal of Petrology*, **47**, 1261–1315, https://doi.org/10.1093/petrology/egl008

Tappe, S., Foley, S.F., Stracke, A., Romer, R.L., Heaman, L.M., Kjarsgaard, B.A. and Joyce, N. 2007. Craton reactivation on the Labrador Sea margins: $^{40}Ar/^{39}Ar$ age and Sr–Nd–Hf–Pb isotope constraints from alkaline and carbonatite intrusives. *Earth and Planetary Science Letters*, **256**, 433–454, https://doi.org/10.1016/j.epsl.2007.01.036

Temdjim, R., Njombie, M.P.W. and Foley, S.F. 2020. Upper-mantle heterogeneity beneath the Ngao Bilta volcano, Adamawa Massif (Cameroon volcanic line, West-central Africa): petrography, mineral chemistry and fingerprints of multiple possible origins of mantle xenoliths. *Geoscience Frontiers*, **11**, 665–677, https://doi.org/10.1016/j.gsf.2019.08.002

Tingey, R.J. 1991. The regional geology of Archean and Proterozoic rocks in Antarctica. *In*: Tingey, R.J. (ed.) *The Geology of Antarctica*. Clarendon Press, Oxford, 1–73.

Tingey, R.J., McDougall, I. and Gleadow, A.J.W. 1983. The age and mode of formation of Gaussberg, Antarctica. *Journal of the Geological Society of Australia*, **30**, 241–246, https://doi.org/10.1080/00167618308729251

Walker, K.R. and Mond, A. 1971. Mica lamprophyre (alnoite) from Radok Lake, Pricne Charles Mountains, *Antarctica. Bureau of Mineral Resources Geology and Geophysics Record*, **108**

Woodland, A.B., Kornprobst, J., McPherson, E., Bodinier, J.L. and Menzies, M.A. 1996. Metasomatic interactions in the lithospheric mantle: petrologic evidence from the Lherz Massif, French Pyrenees. *Chemical Geology*, **134**, 83–112, https://doi.org/10.1016/S0009-2541(96)00082-4

Xu, X., O'Reilly, S.Y., Griffin, W.L., Zhou, X. and Huang, X. 1998. The nature of the Cenozoic lithosphere at Nushan, eastern China. *In*: Flower, M.F.J. (ed.) *Mantle Dynamics and plate interaction in East Asia*. American Geophysical Union, Washington, DC, 167–195.

Yaxley, G.M., Kamanetsky, V.S., Nichols, G.T., Maas, R., Belousova, E., Rosenthal, A. and Norman, M. 2013. The discovery of kimberlites in Antarctica extends the vast Gondwanan Cretaceous province. *Nature Communications*, **4**, 1–7, https://doi.org/10.1038/ncomms3921.

Yaxley, G.M., Berry, A.J., Rosenthal, A., Woodland, A.B. and Paterson, D. 2017. Redox preconditioning deep cratonic lithosphere for kimberlite genesis – evidence from the central Slave Craton. *Scientific Reports*, **7**, 30, https://doi.org/10.1038/s41598-017-00049-3

A review of mantle xenoliths in volcanic rocks from southern Victoria Land, Antarctica

A. P. Martin[1]*, A. F. Cooper[2], R. C. Price[3], C. L. Doherty[4] and J. A. Gamble[5,6]

[1]GNS Science, Private Bag 1930, Dunedin, New Zealand

[2]Department of Geology, University of Otago, PO Box 56, Dunedin, New Zealand

[3]Science and Engineering, University of Waikato, Hamilton, New Zealand

[4]Environmental and Occupation Health Sciences Institute, Rutgers University, Piscataway, NJ 08854, USA

[5]School of Geography, Environment and Earth Sciences, Victoria University of Wellington, Wellington, New Zealand

[6]School of Biological, Earth and Environmental Science, University College Cork, Cork, Ireland

APM, 0000-0002-4676-8344; CLD, 0000-0001-8831-6330

*Correspondence: a.martin@gns.cri.nz

Abstract: Mantle xenoliths from southern Victoria Land have been collected and extensively studied for over a century. In this chapter, chemical and petrological data are, for the first time, comprehensively collated, and petrogenetic models for the regional mantle are reviewed and assessed. The most common lithologies are spinel lherzolite and harzburgite; plagioclase lherzolite also occurs, and pyroxenite xenoliths found across the province comprise <20% of all mantle xenoliths. The lithospheric mantle in the region has Paleoproterozoic stabilization ages, although pockets of younger mantle may exist. This peridotite mantle comprises a HIMU (high $^{238}U/^{204}Pb$ = high μ)-component *sensu stricto*, has been variably carbonated and has undergone multiple melt-depletion events. Regional variations in a sedimentary (EMI: Enriched Mantle I) component to the west, and iron-rich components to the east, reflect a complex history of refertilization and metasomatism. The sources of these fluids are likely to have been oceanic crust subducted during *c.* 0.5 Ga and older events. Peridotites have been cross-cut by pyroxenite veins, probably in multiple episodes, with the geochemistry of some samples reflecting the involvement of an upper continental crust (EMII: Enriched Mantle II) component. Future research directions should apply advanced isotopic, noble gas and volatile techniques to better understand the upper mantle below this dynamic rifting environment.

Supplementary material: A compilation of model parameters used in this study and published mineral and whole-rock chemistry of mantle xenoliths of southern Victoria Land is available at https://doi.org/10.6084/m9.figshare.c.5407266

In many alkalic volcanic provinces fragments of the mantle are entrained as xenoliths in the ascending melts. Such xenoliths record conditions in the upper mantle relevant to understanding igneous petrogenesis, glacial isostatic adjustment and regional tectonics; they also place boundary conditions on remotely and geophysically sensed data. At many locations across a continental rift shoulder in southern Victoria Land, Antarctica, Cenozoic alkalic volcanic rocks are hosts to mantle xenoliths (Fig. 1). This region is of significant geological and petrological interest because: (a) mantle xenoliths and their host volcanic rocks have been intensively studied here for more than a century and, consequently, an extensive petrological and geochemical database can be compiled; (b) the region is the locus of active volcanism; (c) it is the site of continental rifting; (d) it is at the tectonic boundary between the West Antarctic Rift System and the Transantarctic Mountains; and (e) it is accessible from historical and modern Antarctic bases, which means it is an area where many petrological, geophysical and remotely sensed studies have been, and continue to be, focused.

Mantle xenoliths were among some of the earliest geological specimens collected from Antarctica. Prior (1902, p. 328) described:

> A large mass of basalt from Franklin Island is remarkable for the number and large size of the olivine-enstatite nodules … the olivine-nodules are as large as a fist … and consist of a fairly coarse-grained aggregate of pale greenish olivine, yellow enstatite showing polysynthetic twinning, brilliant green chrome-diopside, and pisolite.

The first sketches of microscopic textures of mantle xenoliths from the region were published by Prior (1907), and the first chapter on mantle xenoliths by Thomson (1916) also contained the first published thin-section photomicrographs. Many early works on mantle xenoliths were descriptive (e.g. Smith 1954; Cole and Ewart 1968; McIver and Gevers 1970; Skinner *et al.* 1976). Forbes (1963) used mineral strain fabrics in mantle xenoliths to argue for a non-cognate origin (i.e. not genetically related to the host magma) but the first extensive overview of mantle xenoliths in the region was provided by Kyle *et al.* (1987). This chapter builds on that seminal work. Since that time, reviews of specific mantle xenolith occurrences have been published for Foster Crater (Gamble and Kyle 1987; Gamble *et al.* 1988), White Island (Cooper *et al.* 2007; Martin *et al.* 2014*a*), Mount Morning (Martin *et al.* 2015*b*), Pipecleaner Glacier (Martin *et al.* 2014*a*), and Turtle Rock and Hut Point Peninsula (Niida 1988, 1990; Warner and Wasilewski 1995; Day *et al.* 2019).

In this chapter, data for mantle xenoliths from this region, collected over more than a century, are collated and interpreted. This mantle setting is discussed and the implications for mantle petrogenesis are reviewed. The chapter will not directly consider mantle conditions inferred from volcanic rocks as this has been the subject of recent reviews (Martin *et al.* 2021; Panter and Martin 2021; Smellie and Martin 2021).

Geological setting

The West Antarctic Rift System splits Antarctica, with the Transantarctic Mountains marking the rift's western shoulder (Fig. 1a) (Tessensohn and Wörner 1991; Martin *et al.* 2015*a*; Tinto 2019). The Transantarctic Mountains comprise Precambrian–Cambrian metasedimentary (sedimentary rock that has been metamorphosed) lithologies cross-cut by mainly Cambrian–Ordovician granitic rocks of the Granite Harbour Intrusive Complex, which also includes the ultramafic lamprophyres of the Vanda Dykes (Gunn and Warren 1962; Wu and Berg 1992; Stump 1995; Cox *et al.* 2012). These older rocks are overlain by Devonian–Triassic Beacon

From: Martin, A. P. and van der Wal, W. (eds) 2023. *The Geochemistry and Geophysics of the Antarctic Mantle*.
Geological Society, London, Memoirs, **56**, 33–55,
First published online 16 July 2021, https://doi.org/10.1144/M56-2019-42

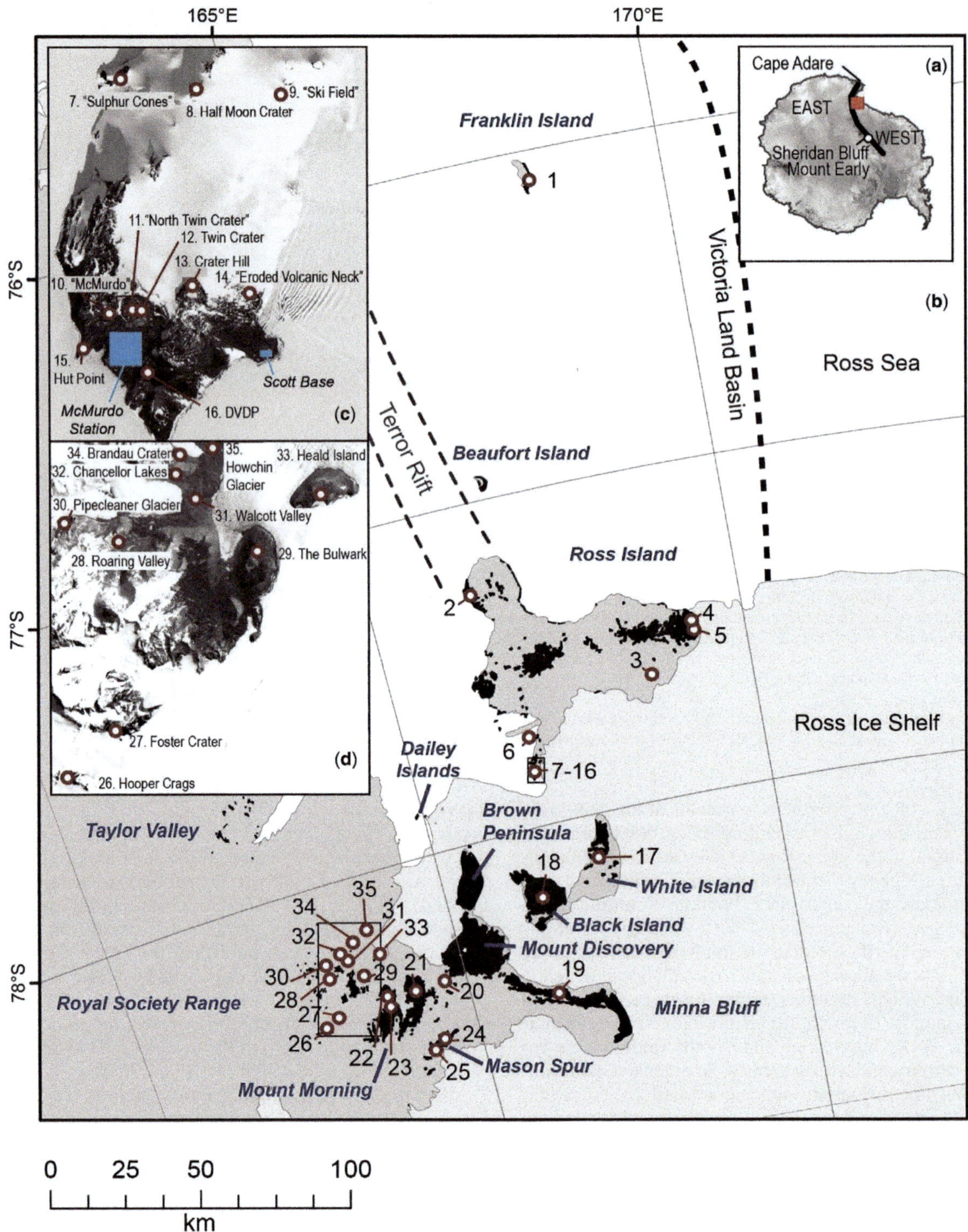

Fig. 1. A locality diagram for mantle xenoliths of southern Victoria Land. (**a**) The continent of Antarctica is divided geographically into East Antarctica and West Antarctica by the Transantarctic Mountains (thick black line). The location of (b) is indicated by the red rectangle. McMurdo Volcanic Group rocks occur between Cape Adare in the north and Mount Early/Mount Sheridan in the south. (**b**) Southern Victoria Land with alkalic igneous rocks of the Erebus Volcanic Province shown in black. The key mantle xenolith localities are shown (red circles) with numbers equating to line entries in Table 2. Major eruptive centre names are shown in blue. Details for localities 7–16 and localities 26–35 are shown in (**c**) and (**d**). DVDP, Dry Valleys Drilling Project.

Supergroup sedimentary rocks (Barrett 1981), which are in turn intruded by the Jurassic Ferrar Dolerite (Burgess *et al.* 2015). Extension in the West Antarctic Rift System commenced in the Jurassic, with the major crustal-rifting phase occurring with the Late Cretaceous Gondwana break-up (Siddoway 2008). A further extensional phase occurred in the Cenozoic along the western rift shoulder associated with significant crustal heating events until *c.* 100 Ma (Molzahn *et al.* 1999), coeval with Transantarctic Mountain uplift (Fitzgerald 2002). A third pulse of Paleogene extension was focused in the Victoria Land Basin (Fig. 1b), an asymmetrical graben along the western edge of the rift (Fitzgerald 2002; Fielding *et al.* 2006). Within the Victoria Land Basin, Neogene rifting has occurred primarily within the Terror Rift (Fig. 1b) (Fielding *et al.* 2006; Martin and Cooper 2010), with some evidence for Quaternary activity (Hall *et al.*

2007). The Moho depth varies across the region from 40 km beneath the Transantarctic Mountains, shallowing eastwards to 18 km across the rift shoulder under the Ross Embayment (e.g. McGinnis *et al.* 1985; Wiens *et al.* 2021).

Cenozoic alkalic volcanic rocks and hypabyssal intrusions, known collectively as the McMurdo Volcanic Group, have been emplaced along the western shoulder of the rift and in the foothills of the Transantarctic Mountains (Harrington 1958). The McMurdo Volcanic Group extends *c.* 2000 km between offshore Cape Adare in the north to Sheridan Bluff and Mount Early in the south (Fig. 1a) (Kyle 1990). It has been subdivided into three provinces, with the southernmost, and the focus of this work, being the Erebus Volcanic Province in southern Victoria Land (Martin *et al.* 2021). This province is further divided into five volcanic fields, which are from north to south, Terror Rift, Ross Island, Mount Discovery, Mount Morning and Southern Local Suite (Fig. 2) (Smellie and Martin 2021). The discussion of data in this chapter is subdivided based on in which of the five volcanic fields the mantle xenoliths were brought to the surface. The volcanic rocks in the Erebus Volcanic Province were the earliest to be recognized as hosting mantle xenoliths (Prior 1902).

Methods

A compilation of published mineral and whole-rock chemistry of mantle xenoliths of southern Victoria Land is provided in the Supplementary material (ESM1). The method of analysis and the original reference is given for every single analysis in the Supplementary material, to which the reader is referred for specific analysis parameters. For plotting, only data with appropriate totals (i.e. 100 ± 1 wt%) were used. Data with totals <99 or >101 wt% are still included in the Supplementary material for completeness. The details of the standard melting models used in the figures in this chapter and discussed in the text are also included in the Supplementary material.

For new analyses in this study, the following methods apply. Whole-rock and electron microprobe analyses were undertaken in the Analytical Facility of Victoria University of Wellington (VUW), New Zealand. Major oxides and selected trace elements were determined using an Automated Philips PW-2002 X-ray fluorescence spectrometer with a Rhodium end-window tube. The USGS and Japan samples were used as monitoring standards. Full details of the analytical methods are given in Palmer (1990).

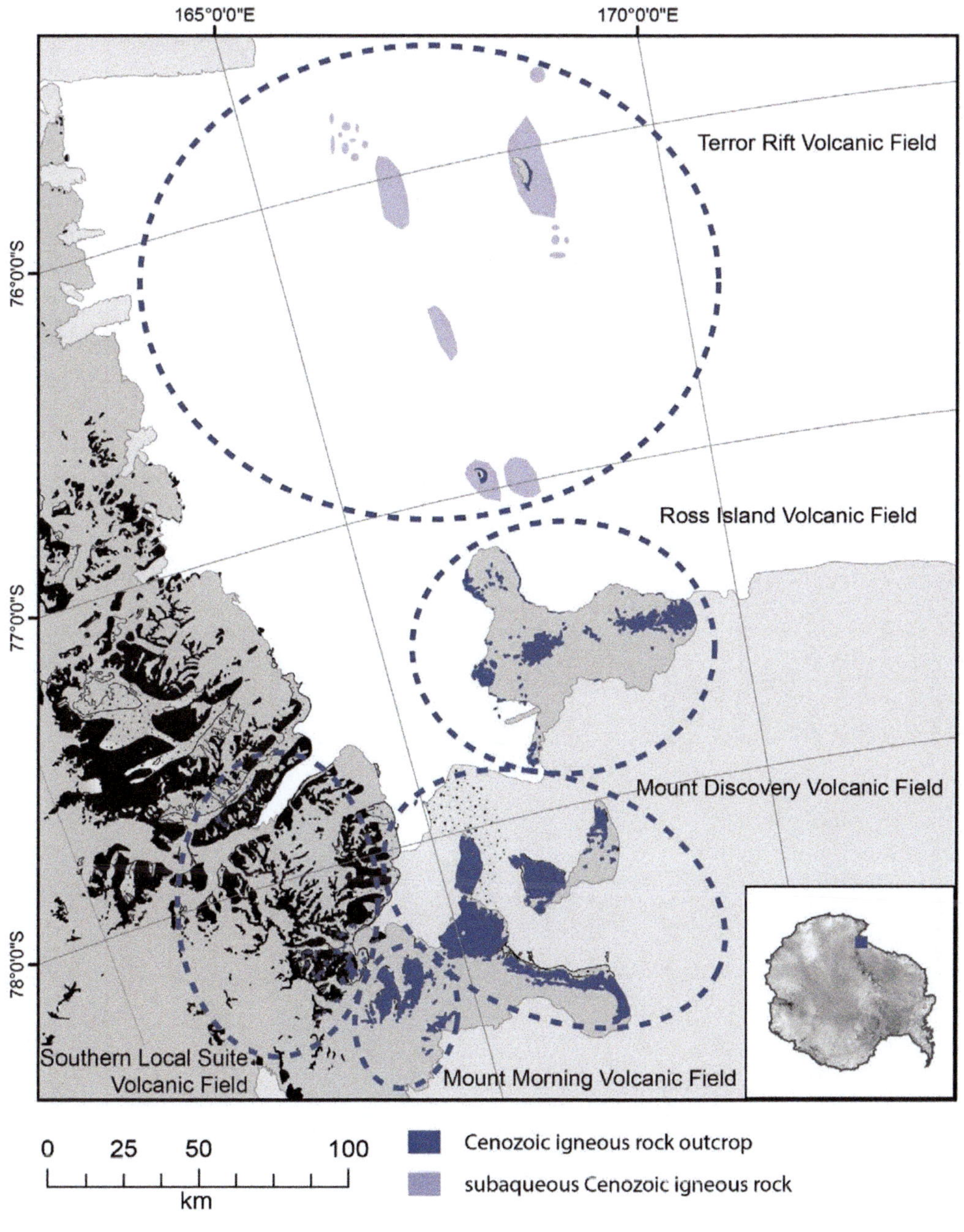

Fig. 2. A diagram showing the location of the five volcanic fields that make up the Erebus Volcanic Province (Martin *et al.* 2021; Smellie and Martin 2021).

Table 1. *Southern Victoria Land locations where mantle xenoliths have been recorded, including relevant references and rock types observed*

Volcanic Field	Location		#	Reference(s)			Xenolith types
				1	2	3	
Terror Rift	Franklin Island		1	19	7	This study	Harzburgite, dunite and wehrlite
	Beaufort Island						Reconnaissance only. Further examination may show xenoliths
	Submarine volcanic rocks						No xenoliths reported to date
Ross Island	Mount Bird	Cape Bird	2	3			Clinopyroxenite
	Mount Terror	Conical Hill	3	7	This study		Harzburgite
		'Coastal cliffs'	4	4	8	22	Dunite, harzburgite, lherzolite and wehrlite
		The Knoll	5	19	7	This study	Lherzolite and harzburgite
	Mount Erebus	Summit area					Area examined but no xenoliths reported
		Dellbridge Islands					Area examined but no xenoliths reported
	Hut Point Peninsula	Turtle Rocks	6	19	7	This study	Dunite, wehrlite and clinopyroxenite
		'Sulphur Cones'	7	21	7	This study	Dunite, harzburgite and lherzolite
		Half Moon crater	8	22	12		Harzburgite, lherzolite, wehrlite and clinopyroxenite
		'Ski Field'	9	21	12		Dunite
		'McMurdo'	10	19	22	6	Harzburgite and clinopyroxenite
		'North Twin Crater'	11	21	12		Dunite
		Twin Crater	12	22	12		Peridotite
		Crater Hill	13	8	21	12	Dunite and peridotite
		'Eroded Volcanic Neck'	14	9	12		Dunite, harzburgite and wehrlite
		Hut Point	15	19	21	12	Dunite and clinopyroxenite
		DVDP	16	23	12	6	Dunite
Mount Discovery	White Island	Mount Hayward and Camp Crater	17	5	14	7	Dunite, harzburgite, lherzolite, plagioclase lherzolite, wehrlite, websterite, clinopyroxenite and pyroxene
	Black Island		18	3	12		Peridotite, clinopyroxenite and wehrlite
	Minna Bluff		19	18			Peridotite
	Mount Discovery	Discovery Saddle	20	12			Dunite and wehrlite
	Brown Peninsula			11			Area examined but no xenoliths reported
	Dailey Islands						Reconnaissance only. Further examination may show xenoliths
Mount Morning	Mount Morning	Hurricane Ridge	21	13	15		Dunite, harzburgite, lherzolite, wehrlite and clinopyroxenite
		Lake Morning	22	12	13	15	Dunite, harzburgite, lherzolite, wehrlite and clinopyroxenite
		Riviera Ridge	23	13	15	7	Dunite, harzburgite, lherzolite, plagioclase lherzolite, wehrlite and clinopyroxenite
	Mason Spur	Windscoop Bluff	24	12	13	15	Dunite, harzburgite, lherzolite, wehrlite and clinopyroxenite
		Anniversary Bluff	25	12	13	15	Dunite, harzburgite, lherzolite, wehrlite and clinopyroxenite
Southern Local Suite	Royal Society Range	Hooper Crags	26	20	12		Dunite
		Foster Crater	27	24	10	7	Dunite, clinopyroxenite, wehrlite, websterite and glimmerite
		Roaring Valley	28	25	12	17	Dunite, peridotite and clinopyroxenite
		The Bulwark	29	2	12	17	Harzburgite, lherzolite and plagioclase lherzolite
		Pipecleaner Glacier	30	25	14	7	Lherzolite, plagioclase lherzolite and wehrlite
		Walcott Valley	31	25	12		Peridotite
		Chancellor Lakes	32	26	12		Peridotite and clinopyroxenite
		Heald Island	33	25	12		Peridotite
		Brandau Crater	34	16	22	26	Dunite, harzburgite, lherzolite, wehrlite, websterite and clinopyroxenite
		Howchin Glacier	35	12			Lherzolite and wehrlite
	Taylor Valley			1			Area examined but no xenoliths reported

The numbers (column labelled #) correspond to locations in Figure 1.

References: 1, Allibone *et al.* (1991); 2, Armstrong (1978); 3, Cole and Ewart (1968); 4, Cole *et al.* (1971); 5, Cooper *et al.* (2007); 6, Day *et al.* (2019); 7, Doherty (2016); 8, Ferrar (1907); 9, Forbes and Banno (1966); 10, Gamble *et al.* (1988); 11, Kyle *et al.* (1979); 12, Kyle *et al.* (1987); 13, Martin (2009); 14, Martin *et al.* (2014*a*); 15, Martin *et al.* (2015*b*); 16, McIver and Gevers (1970); 17, Moscati (1989); 18, Panter *et al.* (2011); 19, Prior (1907); 20, Skinner *et al.* (1976); 21, Smith (1954); 22, Stuckless and Ericksen (1976); 23, Treves and Kyle (1973); 24, Wright (1979*a*); 25, Wright (1979*b*); 26, Wright (1979*c*).

Electron microprobe analyses were undertaken on epoxy-mounted, handpicked mineral grains, polished and carbon coated, and using the VUW Analytical Facility JEOL JXA-733 Superprobe electron microprobe microanalyser. Mineral analyses were made at a 15 kV accelerating voltage and a beam current of 1.2×10^{-8} A using synthetic oxides and natural minerals as standards. Analytical precision is approximately ±1% for elements with an abundance of >10%, and ±5–10% for elements present in lower abundance.

Using the same epoxy mineral mounts, trace elements were determined by ArF excimer laser ablation (Lambda Physik LPX 120i) coupled to a Fisons PlasmaQuad inductively coupled plasma mass spectrometer in the Research School of Earth Sciences, Australian National University. Analytical procedures followed methods outlined in Eggins *et al.* (1997). NIST-610 glass and BHVO-1 were used as monitoring standards.

Results

Location of mantle xenoliths and other mantle rocks

The locations of all reported occurrences of mantle xenoliths in southern Victoria Land are shown in Figure 1, with the major references provided and the prominent rock types summarized in Table 1. The locations where only crustal xenoliths have been found are reported in Kyle *et al.* (1987) and this information is not replicated here. Additional information about localities where xenoliths have been searched for, and not found, is also included in Table 1. For example, no mantle xenoliths have been observed on Beaufort Island based on field exploration of the southern portion of the island (R.J. Moscati pers. comm.), and no mantle-derived xenoliths have been reported in the Taylor Valley (Allibone *et al.* 1991). Other areas, such as the Dailey Islands or northern Beaufort Island, have only been explored at a reconnaissance level and further study may reveal new mantle xenolith occurrences.

Sheridan Bluff and Mount Early (Fig. 1a), to the south of Erebus Volcanic Province, comprise McMurdo Volcanic Group rocks that have recently been mapped in detail and found not to contain mantle xenoliths or mantle-derived megacrysts (Panter 2021). The most primitive portions of the Ferrar Dolerite (e.g. Bédard *et al.* 2007) and the lamprophyric Vanda Dykes (Wu and Berg 1992) are beyond the scope of this chapter.

Petrography

Mantle xenoliths in the region are typically less than 10 cm in maximum dimension; however, much larger specimens (up to 1 m) have been documented (Fig. 3a). Peridotite and pyroxenite type xenoliths are commonly found together (Fig. 3b). The petrography of several mantle xenolith localities is summarized in Table 2, and typical hand-specimen and thin-section photographs are shown in Figure 4. Garnet is not present in mantle xenoliths of southern Victoria Land, although mantle clinopyroxene grains high in Na, relative to other mantle clinopyroxene grains from the same Mount Morning locality, have been interpreted as originating in the garnet stability field (Martin *et al.* 2015*b*). Plagioclase has been identified in some peridotite xenoliths. Amphibole is conspicuously absent from the Mount Morning Volcanic Field relative to all other southern Victoria Land mantle xenolith localities. Only a single phlogopite-bearing sample, from Mount Morning, has been reported. Glass is reported in several xenolith types from Foster Crater but is uncommon elsewhere in the province. Rare, texturally equilibrated carbonate and FeNi-sulfide grains have been described in Mount Morning peridotite xenoliths and carbonate is present in a pyroxenite xenolith from Foster Crater.

Fig. 3. Examples of mantle xenoliths in the field. (**a**) A large peridotite xenolith from Franklin Island. The hammer is *c.* 1 m long. Photograph: R.J. Moscati. (**b**) Mantle xenoliths in outcrop at Mount Morning, including a spinel lherzolite xenolith (beneath the hand lens), a spinel harzburgite xenolith with a strong lineation fabric (bottom right), a clinopyroxene megacryst (upper right, black) and a wehrlite xenolith (middle). The hand lens is *c.* 6 cm long. Photograph: A.F. Cooper.

The different peridotite rock types reflect different mantle processes. In a study of Mount Morning mantle xenoliths, Martin *et al.* (2015*b*) found that spinel lherzolite made up 53% of the peridotite xenoliths collected, spinel harzburgite 35%, plagioclase-bearing spinel lherzolite 7% and spinel dunite 5%. These abundances are applicable across the wider province, where spinel lherzolite is the most common xenolith type, followed closely by spinel harzburgite (Kyle *et al.* 1987) (Tables 1 and 2). Spinel dunite is uncommon relative to lherzolite or harzburgite but has been found at every significant xenolith locality (Table 2). Plagioclase-bearing spinel lherzolite (plagioclase lherzolite hereafter) is rare, and only reported at White Island, Mount Morning and several Southern Local Suite locations, including Roaring Valley, Pipecleaner Glacier, The Bulwark and Foster Crater (Gamble *et al.* 1988; Moscati 1989; Cooper *et al.* 2007; Martin *et al.* 2014*a*).

Porphyroclastic describes a heterogranular metamorphic texture characterized by volumetrically significant amounts of both porphyroclasts and neoblasts. It reflects low-strain plastic deformation and dynamic recrystallization relative to other mantle textures (Harte 1977). In the Erebus Volcanic Province, the evidence to date suggests that peridotite textures are typically porphyroclastic but become more tabular granuloblastic (i.e. strained) in the Mount Morning Volcanic Field than elsewhere. This could suggest that the most intense locale

Table 2. *Typical petrography of mantle xenoliths from southern Victoria Land*

VF	Location	Rock type				Mineralogy								Tex	Size
		D	H	L	Other	Ol	Cpx	Opx	Spl	Pl	Amp	Phl	Other		
TR	Franklin Island	Y	Y	Y									Apatite		
RI	Mount Bird			Y											
					Peridotite								Ol		
	Mount Terror	Y	Y	Y											
	Hut Point Peninsula	Y	Y	Y											
					Wehrlite										
					Megacrysts								Ol + Cpx		
MD	White Island	Y	Y	Y									anorthoclase megacrysts	Pr	15–25 cm
					Ol clinopyroxenite										Rare
					Ols websterite										Vein 8 mm wide
					Wehrlite										Band 1.3 cm wide
					Pyroxene										
					Kaersutite										
	Black Island		Y	Y											
					Clinopyroxenite										
					Wehrlite										
	Minna Bluff		Y	Y											
					Kaersutite										
	Mount Discovery	Y													
					Wehrlite										
MM	Mount Morning	Y	Y	Y									Cb	TG	≤50 cm
					Orthopyroxenite									EG	≤30 cm
					Ol websterite									EG	≤30 cm
					Websterite									EG	≤30 cm
					Wehrlite									Eq	≤30 cm
					Clinopyroxenite									CP	≤30 cm
					Clinopyroxene										5 cm
					Clinopyroxenite										
	Mason Spur	Y	Y	Y	Megacrysts									TG	≤15 cm

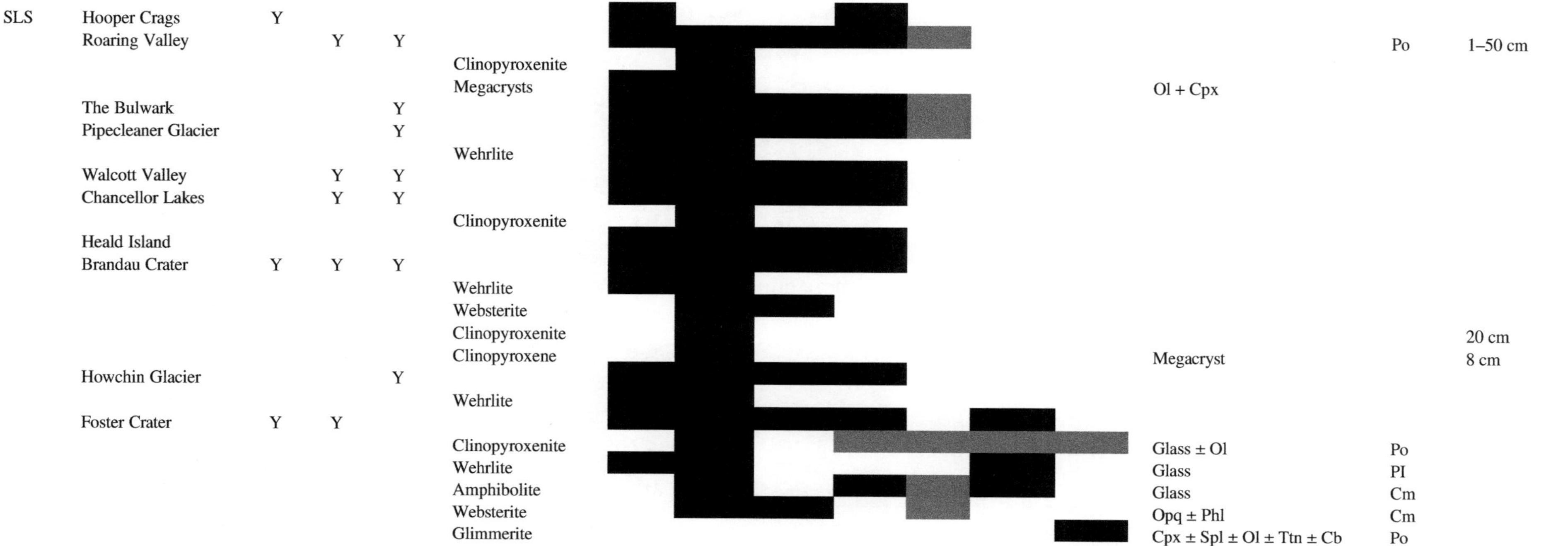

SLS	Hooper Crags	Y						
	Roaring Valley		Y	Y			Po	1–50 cm
					Clinopyroxenite			
					Megacrysts	Ol + Cpx		
	The Bulwark			Y				
	Pipecleaner Glacier			Y				
					Wehrlite			
	Walcott Valley		Y	Y				
	Chancellor Lakes		Y	Y				
					Clinopyroxenite			
	Heald Island							
	Brandau Crater	Y	Y	Y				
					Wehrlite			
					Websterite			
					Clinopyroxenite			20 cm
					Clinopyroxene	Megacryst		8 cm
	Howchin Glacier			Y				
					Wehrlite			
	Foster Crater	Y	Y					
					Clinopyroxenite	Glass ± Ol	Po	
					Wehrlite	Glass	PI	
					Amphibolite	Glass	Cm	
					Websterite	Opq ± Phl	Cm	
					Glimmerite	Cpx ± Spl ± Ol ± Ttn ± Cb	Po	

Abbreviations: D, dunite; H, harzburgite; L, lherzolite; Ol, olivine; Cpx, clinopyroxene; Opx, orthopyroxene; Spl, spinel; Pl, plagioclase; Amp, amphibole; Phl, phlogopite; Ttn, titanite; Cb, carbonate. Black squares indicate dominant mineralogy; grey squares indicate minor mineralogy that is variably present. VF, Volcanic Field; TR, Terror Rift; RI, Ross Island; MD, Mount Discovery; MM, Mount Morning; SLS, Southern Local Suite. Tex, Texture; Pr, protogranular; TG, tabular granuloblastic; EG, equant granuloblastic; Eq, equigranular; CP, coarsely porphyritic; Po, porphyroclastic; PI, poikilitic igneous; Cm, cumulus.

of rift activity is located close to the Transantarctic Mountains beneath the Mount Morning Volcanic Field (e.g. Martin *et al.* 2015*a*). This hypothesis remains to be tested by more comprehensive fabric studies, including quantitative microstructural crystallographic preferred orientation work.

Porphyroclasts of olivine and orthopyroxene are common, the latter frequently showing kink banding or exsolution lamellae of clinopyroxene and containing trails of fluid inclusions. Clinopyroxene porphyroclasts are rare and smaller (*c.* 2 mm) than olivine, orthopyroxene or spinel porphyroclasts (≤4 mm). Orthopyroxene porphyroclasts are in places observed with exsolution lamellae of secondary clinopyroxene. In some cases, spinel oikocrysts poikiloblastically enclose chadocrysts of clinopyroxene. Neoblasts of olivine, orthopyroxene and clinopyroxene are common (*c.* 0.5 mm in maximum dimension). Clinopyroxene neoblasts occur either in textural equilibrium with adjacent grains or as late-stage growth along grain boundaries. Feldspar, when present, occurs only as neoblasts and is always found with spinel. Amphibole occurs as a common accessory mineral. Grains of carbonate, sulfide and apatite are rare and occur as neoblasts.

Pyroxenite xenoliths have been recorded in each volcanic field but in all fields their abundance is low relative to the number of peridotite xenoliths, with one estimate putting the proportions at 80 : 20 peridotite : pyroxenite (Martin *et al.* 2015*b*). A full range of mineralogical variants is present, including wehrlite, olivine clinopyroxenite and clinopyroxenite of the Al-augite series, and orthopyroxenite, websterite and olivine websterite of the Cr-diopside series. A rare rock type, glimmerite (phlogopite pyroxenite), is reported from Foster Crater (Gamble *et al.* 1988). Pyroxenite Cr-diopside series textures are equant or granuloblastic and pyroxenite Al-augite series textures are typically equigranular or coarsely porphyritic.

Mineral major element chemistry

Olivine. The mantle peridotite olivine forsterite content varies between 81.8 and 91.9%. On a plot of Ni/Fe v. Fo content (Fig. 5), olivine compositions from the Ross Island Volcanic Field mostly have lower forsterite contents, at a given Ni/Fe, relative to olivine grains from other volcanic fields in the province (Fig. 5a). This relationship is also true for olivine grains in pyroxenite xenoliths from the Erebus Volcanic Province (Fig. 5b).

Pyroxene. The pyroxene wt% of Al_2O_3 varies inversely with MgO content in pyroxene from both peridotite (Fig. 6a and b) and pyroxenite (Fig. 6c and d) mantle xenoliths. From the available data, there is no systematic variation in pyroxene chemistry between different volcanic fields. Clinopyroxene compositions are predominantly diopside with some augite; orthopyroxene is almost exclusively enstatite (see the Supplementary material).

Fig. 4. Typical petrographical features of mantle xenoliths from southern Victoria Land. The left-hand panels in each pair of columns are hand-specimen photographs with the maximum dimension of each hand specimen provided. The accompanying thin-section photographs (right-hand panels in each pair of columns) are typical fields of view (4 mm across), taken under crossed-polarized light. The textures are: dunite, coarse; harzburgite, tabular granuloblastic; lherzolite, porphyroclastic.

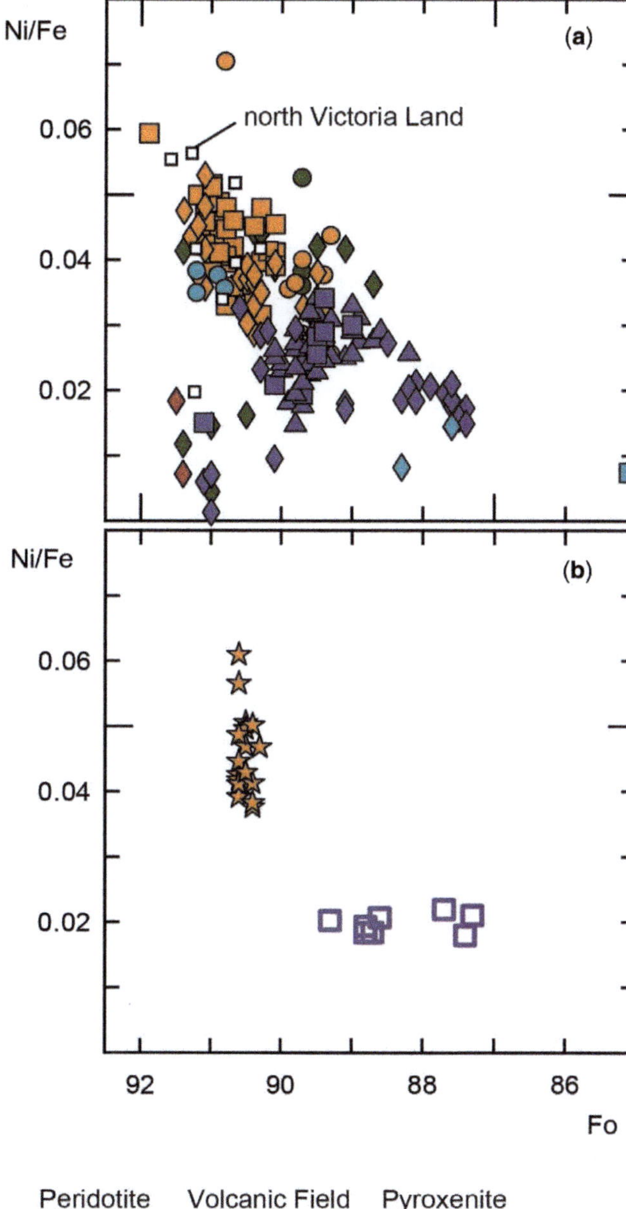

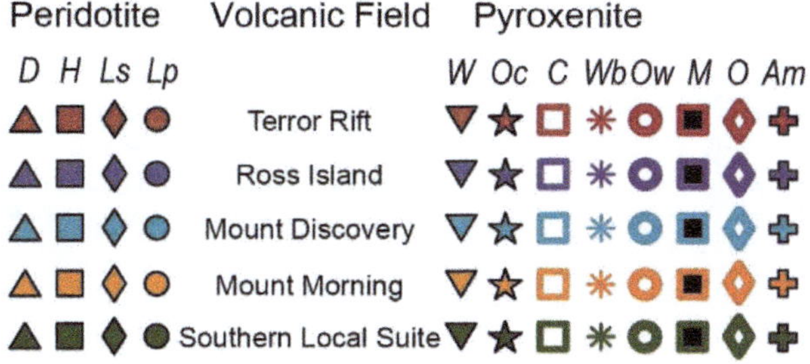

Fig. 5. Olivine mineral chemistry, Ni/Fe v. forsterite (Fo) content for (**a**) peridotite xenoliths and (**b**) pyroxenite xenoliths. The data plotting with the lowest Ni/Fe and Fo content are interpreted as reflecting refertilization. Legend abbreviations are: D, spinel dunite; H, spinel harzburgite; Ls, spinel lherzolite; Lp, plagioclase lherzolite; W, wehrlite; Oc, olivine clinopyroxenite; C, clinopyroxenite; Wb, websterite; Ow, olivine websterite; M, mixed xenolith with multiple rock types; O, orthopyroxenite; Am, amphibolite.

Spinel. Spinel compositional data are available for mantle xenoliths from the Southern Local Suite, Mount Morning and Mount Discovery volcanic fields (Fig. 7a). On a Cr# [(100 × Cr/(Cr + Al)] v. Mg# [100 × Mg/(Mg + Fe^{2+})] diagram (Fig. 7a), all bar one spinel composition from peridotite xenoliths plot within the field typical of continental rifts. Four spinel data from pyroxenite xenoliths plot outside the field typical of continental rifts, with the remaining majority plotting within the field. The data from Mount Morning Volcanic Field covers a wide range of Mg# and Cr# values (Fig. 7a). The spinel data from the Mount Discovery Volcanic Field plot in a relatively tight group and can be clearly distinguished from the Southern Local Suite Volcanic Field data (Fig. 7a).

Feldspars. Feldspars in peridotite xenoliths from the Mount Morning Volcanic Field are andesine in composition (Fig. 7b: anorthite $(An)_{38.4–41.1}$), those from the Mount Discovery Volcanic Field are labradorite and bytownite ($An_{62.6–78.9}$), and those from the Southern Local Suite Volcanic Field range in composition from andesine through labradorite to bytownite ($An_{31.9–72.3}$). Feldspars in pyroxenite xenoliths from locations in the Mount Morning Volcanic Field are andesine to the low end of labradorite (Fig. 7b: $An_{37.1–50.9}$).

Whole-rock major element chemistry

On a whole-rock Al/Si v. Mg/Si plot (Fig. 8), Erebus Volcanic Province mantle peridotite data are inversely correlated. This suggests variable depletion and (re-) enrichment relative to primitive mantle. The data overlap with melting modelled at 2 GPa or even greater depths (Fig. 8). The patterns on the element ratio plot (Fig. 8) are also seen on peridotite and pyroxenite whole-rock wt% MgO v. Al_2O_3 plots, where abundances are inversely correlated (Fig. 9a and b). This is related to increased melting or higher degrees of refertilization (Fig. 9a), as seen in Figure 8.

On a wt% MgO v. Na_2O abundance plot (Fig. 9c), some whole-rock peridotite xenolith compositions overlap with batch and fractional melt model trajectories. Other data for several samples, however, plot away from these model paths, which may be an indication of major element refertilization (Fig. 9c). This is also seen when whole-rock La/Sm is plotted against Na/Sm (Fig. 9d). The pyroxenite data show a weak, inverse relationship between wt% MgO and Na_2O abundances (Fig. 9e).

The FeO contents of peridotite whole-rock samples have been recalculated from the original analyses assuming all iron is Fe^{2+} on an anhydrous basis (see the Supplementary material). The majority of the peridotite samples have FeO <9 wt%, and, on a plot of FeO content v. MgO abundance (Fig. 9f), they lie on the melting grid of Walter (2003). Some samples from the province have FeO >9 wt% and these scatter away from the melting grid, which may indicate major element refertilization (Fig. 9f). This pattern can also be seen when whole-rock La/Sm v. Fe/Sm is plotted (Fig. 9g). On the wt% FeO v. MgO abundance diagram, pyroxenite whole-rock compositions are distributed in a series of arrays, each indicating a particular Mg# value (Fig. 9h). The two amphibolite samples reflect an Mg# of *c.* 60, wehrlite and olivine clinopyroxenite samples have Mg# of *c.* 80, whereas many clinopyroxenite, websterite and orthopyroxenite compositions have Mg# of *c.* 90 (Fig. 9h).

Mineral and whole-rock trace element chemistry and isotopic data

Trace element chemistry is available for clinopyroxene grains from mantle peridotite xenoliths and for whole-rock samples of mantle peridotite xenoliths from each of the five volcanic fields. On primitive-mantle-normalized extended element plots (Fig. 10), clinopyroxene patterns are depleted in Ti and Pb relative to elements of similar compatibility. In general,

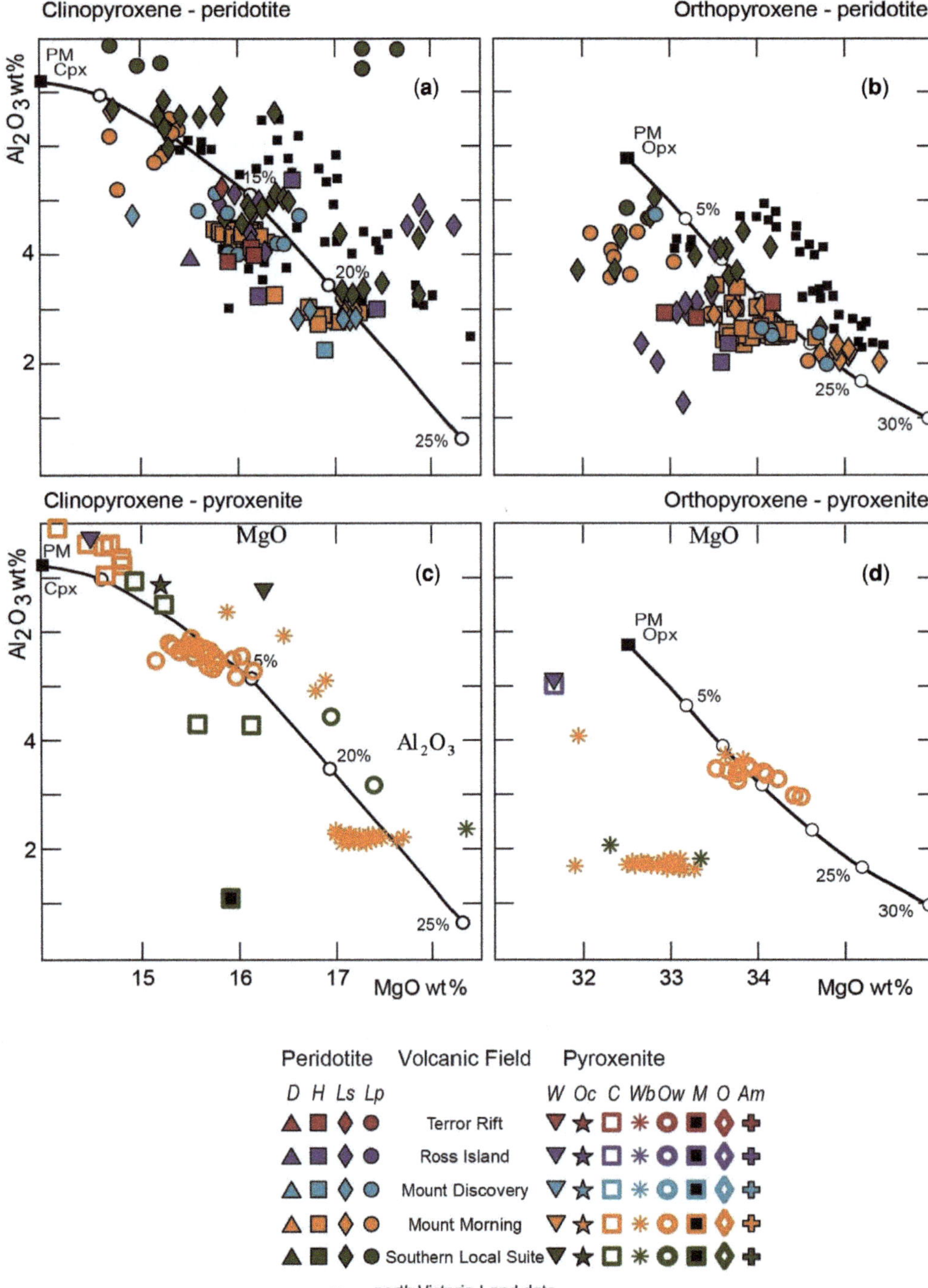

Fig. 6. Pyroxene major element mineral chemistry plotted on wt% MgO–Al_2O_3 diagrams for (**a**) clinopyroxene from peridotite xenoliths, (**b**) orthopyroxene from peridotite xenoliths, (**c**) clinopyroxene from pyroxenite xenoliths and (**d**) orthopyroxene from pyroxenite xenoliths. Primitive mantle (PM) values for clinopyroxene (Cpx) and orthopyroxene (Opx) are from McDonough and Sun (1995), and the theoretical residual trends (solid lines labelled with % melt extraction) follow Upton *et al.* (2011) using equations from Workman and Hart (2005). The legend abbreviations are as in Figure 5. North Victoria Land data (black squares) after Pelorosso *et al.* (2016).

large ion lithophile elements and high field strength elements (HFSEs) are enriched, relative to depleted mid-ocean ridge mantle (Fig. 10). For the Mount Discovery Volcanic Field, whole-rock trace element chemistry is available for only one plagioclase lherzolite sample, whereas analyses of spinel harzburgite have been reported for the other four volcanic fields (see the Supplementary material). On a primitive-mantle-normalized extended element plot, the whole-rock pattern for the Mount Discovery Volcanic Field sample shows Pb enrichment relative to Ce and Nd (Fig. 10). Normalized whole-rock trace element patterns for samples from all other volcanic fields are characterized by depletions in Pb and Ti, relative to elements of similar compatibility; the patterns are comparable to those of clinopyroxene grains (Fig. 10).

Isotopic data for mantle xenoliths from southern Victoria Land are limited. Strontium, Nd and Pb isotope data are available for samples from Mount Morning (Martin *et al.* 2015*b*: *n* = 6), Sr and Nd isotope data for southern Victoria Land are presented in a figure in McGibbon (1991), and $^{87}Sr/^{86}Sr$ data from one sample from the Southern Local Suite and five samples from Ross Island Volcanic Field are reported by Stuckless and Ericksen (1976). The isotopic data for peridotite mantle xenoliths plot close to the HIMU (High-μ: enriched in ^{206}Pb and ^{208}Pb, and relatively depleted in $^{87}Sr/^{86}Sr$) field in $^{87}Sr/^{86}Sr$ v. $^{143}Nd/^{144}Nd$ space (not shown) but radiogenic Pb is never as high as in end-member HIMU (Martin *et al.* 2015*b*). Pyroxenite mantle xenolith isotopic data can be modelled as a mixture between subcontinental lithospheric mantle (SCLM) and enriched mantle I (EMI: low $^{143}Nd/^{144}Nd$, low $^{87}Sr/^{86}Sr$, and high $^{207}Pb/^{206}Pb$ and $^{208}Pb/^{204}Pb$ at a given value of $^{206}Pb/^{204}Pb$) or enriched mantle II (EMII: low $^{143}Nd/^{144}Nd$, high $^{87}Sr/^{86}Sr$, and high $^{207}Pb/^{206}Pb$ and $^{208}Pb/^{204}Pb$ at a given value of $^{206}Pb/^{204}Pb$). The mixing models replicate to a high degree

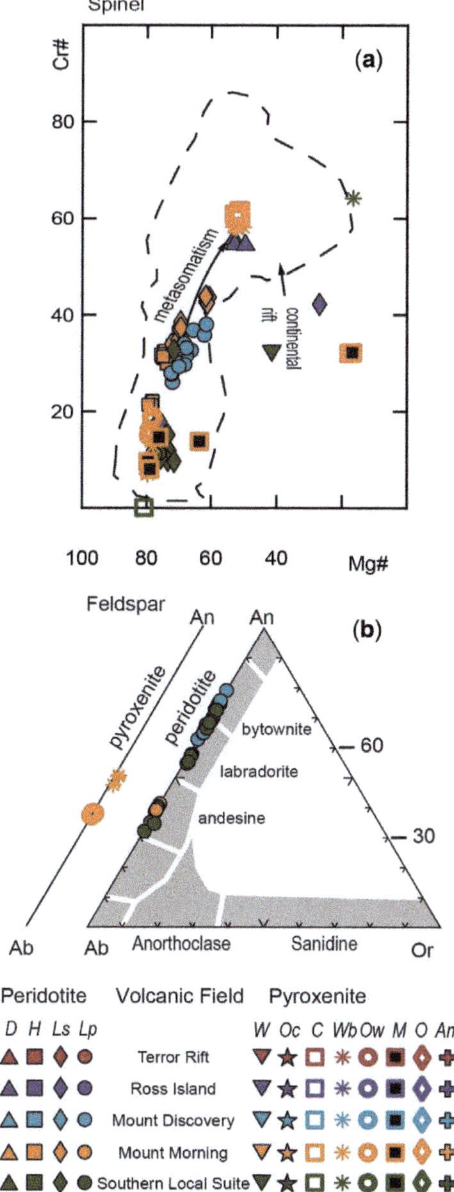

Fig. 7. Spinel and feldspar mineral chemistry for southern Victoria Land mantle xenoliths. (**a**) Spinel Mg# ($100 \times Mg/(Mg + Fe^{2+})$) v. spinel Cr# (($100 \times Cr/(Cr + Al)$). The field of spinel facies peridotite xenoliths from continental rift settings (dashed line) is from Martin *et al.* (2014*b*). Peridotite and pyroxenite data are shown here on the one figure. (**b**) Feldspar mineral chemistry in peridotite and pyroxenite xenoliths. An, anorthite; Ab, albite; Or, orthoclase. The legend abbreviations are as in Figure 5.

the isotopic compositions of mantle xenoliths in the Erebus Volcanic Province (Martin *et al.* 2015*b*).

Discussion

Geothermobarometry

Using data for lower-crustal xenoliths, Berg *et al.* (1989) calculated a geotherm for southern Victoria Land. A petrological geotherm was additionally calculated using mineral compositions from mantle peridotite xenoliths from Mount Morning (Martin *et al.* 2015*b*) and elsewhere in the province (Moscati 1989). Martin *et al.* (2014*a*) recorded that the highest mantle temperatures were from those samples in southern Victoria Land most recently affected by refertilization, which they attributed to entrapment of synrift melts in the lithosphere (Fig. 11). The superimposition of hot peridotite bodies on cold uppermost mantle has also been postulated for northern Victoria Land (Wörner and Zipfel 1996). This can be seen in Figure 11 (hollow diamonds) where plagioclase lherzolites from Martin *et al.* (2014*a*) with pressures and temperatures plotting in the plagioclase mantle stability field are linked (thin dotted line) to relict minerals recording original pressures and temperatures that equilibrated in the spinel mantle stability field. This also explains the plagioclase lherzolite sample (labelled as a relict grain in Fig. 11) plotting in the spinel-bearing mantle stability field; see Martin *et al.* (2014*a*) for further discussion. In this study, geothermobarometry calculated for selected mantle xenolith samples (see the Supplementary material) from each of the volcanic fields in southern Victoria Land is summarized in Figure 11. The calculations were made using the temperature-independent barometer of Putirka (2008) and a thermometer using equilibrated clinopyroxene grains with Mg# >75 (Putirka 2008); these same equations were used by Martin *et al.* (2015*b*) allowing direct comparison. The geothermobarometry calculations used in Figure 11 are highlighted in the Supplementary material. The geothermometer of Brey and Kohler (1990) is also shown in the Supplementary material for comparison but was not used in plotting Figure 11. Most pressure and temperature estimates obtained on mantle xenoliths from each of the volcanic fields suggest that the geotherm reported by Berg *et al.* (1989) is correct and is consistent with previously published pressure–temperature estimates obtained from xenoliths (Fig. 11) (Moscati 1989). Terror Rift Volcanic Field mantle xenolith samples have some of the highest pressures (11.0 and 12.6 kbar) and an average temperature of *c.* 1000°C. The Ross Island Volcanic Field and Mount Discovery Volcanic Field xenolith samples chosen for the study give lower pressures (4.0–6.0 kbar) relative to the Terror Rift Volcanic Field samples, although a more extensive dataset may reveal extended ranges. The pressure–temperature data calculated here indicate that the southern Victoria Land geotherm is relevant to all volcanic fields in the province.

Mantle water content and oxygen fugacity

Water content has not yet been directly estimated for southern Victoria Land mantle samples, so other lines of evidence must be considered. Iacovino *et al.* (2016) performed experiments on volcanic rocks from the Ross Island Volcanic Field to show that there was *c.* 1–2 wt% H_2O in magmas represented by the eruptives thought to be connected to their mantle source via a deep mafic plumbing system. Studying olivine-hosted melt inclusions, Gaetani *et al.* (2019) suggested that the maximum water content was between 2.12 and 1.84 wt% for the Ross Island Volcanic Field and the Mount Morning Volcanic Field. This would be consistent with a volatile content of >1 wt% H_2O calculated for amphibole-bearing northern Victoria Land mantle (Giacomoni *et al.* 2020; Coltorti *et al.* 2021, this volume). Although there are also relatively dry amphibole-free samples in north Victoria Land that only have <0.4 wt% H_2O (Perinelli *et al.* 2006). This suggests that, in general, the Victoria Land mantle is somewhat hydrous relative to, for example, the Antarctic Peninsula where the water content is *c.* 80–130 ppm (Gibson *et al.* 2020).

The mantle peridotite whole-rock data plotted onto a parts per million (ppm) Yb v. V diagram overlap with the fields for oxygen fugacity relative to the fayalite–magnetite–quartz (FMQ) buffer calculated by Parkinson and Pearce (1998) between FMQ −0.5 and FMQ +1 (Fig. 12a). Specific localities

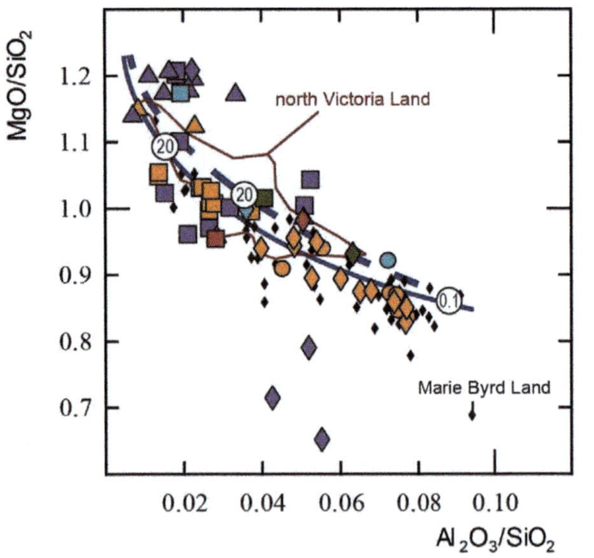

Fig. 8. A whole-rock Al_2O_3/SiO_2 v. MgO/SiO_2 plot showing southern Victoria Land data relative to northern Victoria Land (Coltorti *et al.* 2021, this volume) and Marie Byrd Land data (Handler *et al.* 2021). Melt-extraction curves of Herzberg (2004) are also shown for 2 GPa melt depletion (dashed line) and 1 GPa (solid line) with indications of expected residual composition after 20% melt extraction. The legend abbreviations are as in Figure 5.

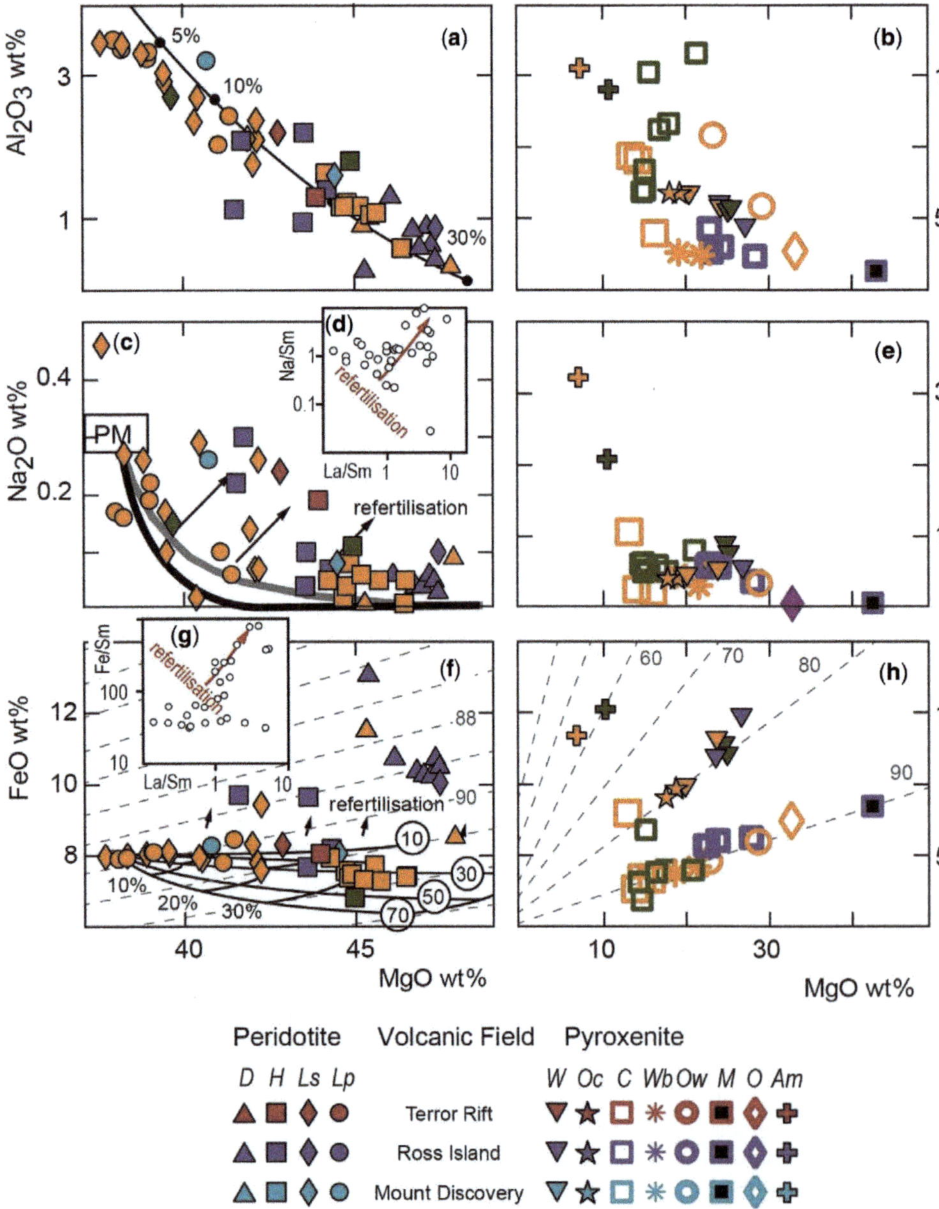

Fig. 9. Whole-rock wt% Al_2O_3, Na_2O and FeO abundances plotted against MgO content for mantle xenolith samples from southern Victoria Land. (**a**), (**c**) and (**f**) show data for peridotite, and (**b**), (**e**) and (**h**) show data for pyroxenites. Whole-rock peridotite xenolith data for (**d**) La/Sm v. Na/Sm or (**g**) Fe/Sm are also shown for comparison. All iron was calculated as Fe^{2+} assuming anhydrous conditions. The melting residues in (a) and (c) were calculated assuming polybaric near-fractional melting between 25 and 15 kbar (black curved lines) or isobaric batch melting between 20 and 10 kbar (grey line) after Niu (1997). Primitive mantle (PM) values are from McDonough and Sun (1995). Whole-rock Mg# are shown as dashed lines for comparison in (e) and (f). (e) shows the melting grid of Walter (2003) as a function of pressure (kbar) and batch melt extraction (%). Trends expected for refertilization are indicated by arrows. The legend abbreviations are as in Figure 5.

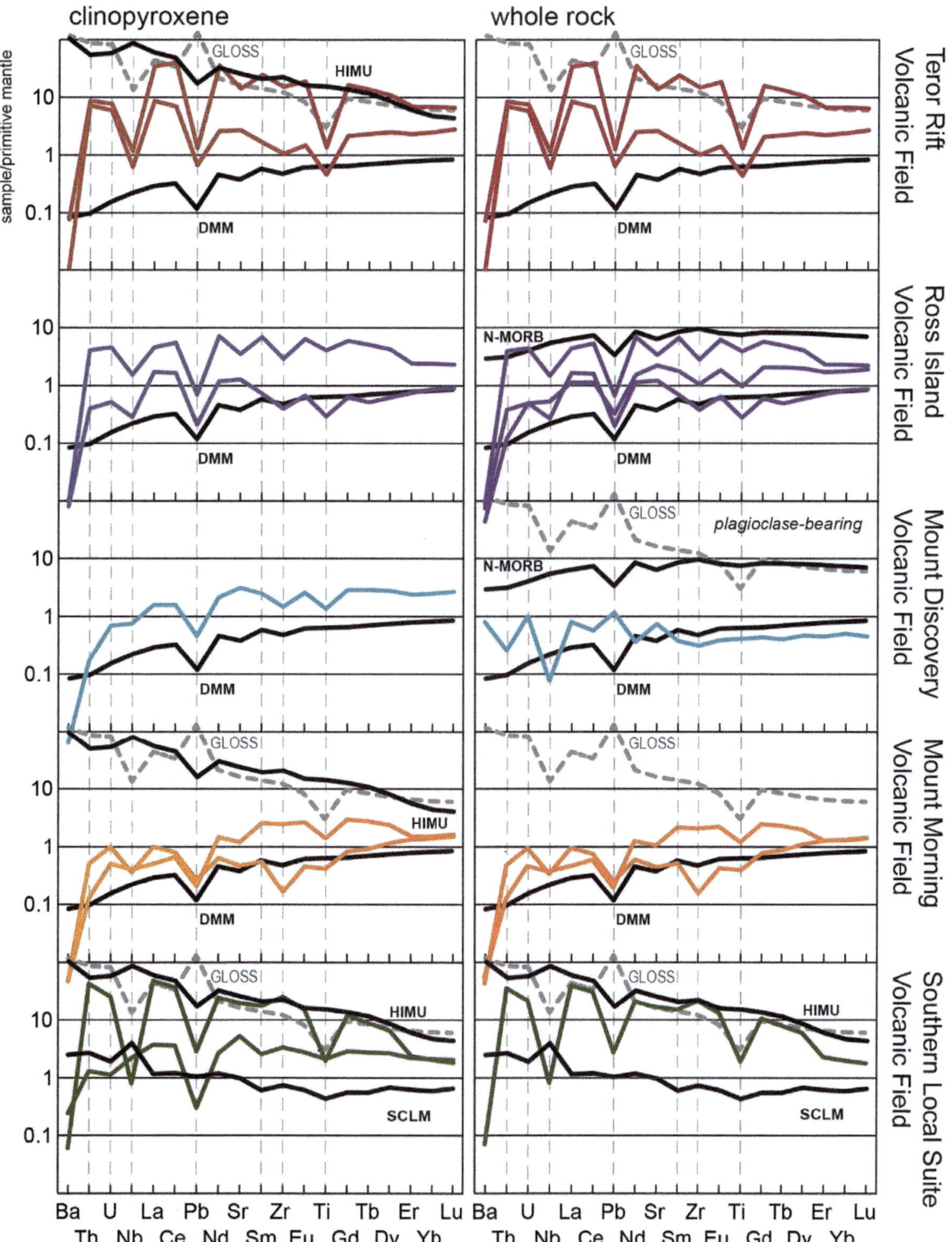

Fig. 10. Primitive-mantle-normalized (McDonough and Sun 1995) extended element plots for clinopyroxene (left column) and whole-rock (right column) data. From left to right, the elements are arranged in order of decreasing incompatibility with respect to peridotitic residue. Note that data from plagioclase lherzolite were used for the Mount Discovery Volcanic Field and spinel harzburgite for other volcanic centres, reflecting the availability of data. The data represent individual analyses from the Supplementary material and are not amalgamated in the totals. Plotted for comparison are typical values for subcontinental lithospheric mantle (SCLM: McDonough 1990), depleted mid-ocean ridge basalt mantle (DMM: Workman and Hart 2005), normal mid-ocean ridge basalt (N-MORB: Gale *et al.* 2013), a high-μ pattern (sample M-11) from Mangaia, Austral Islands (HIMU: Woodhead 1996) and average global subducted sediment (GLOSS: Plank and Langmuir 1998).

have been studied further to refine oxygen fugacity estimates. At Mount Morning, oxygen fugacity calculated from Mössbauer spectroscopy measurements on spinel and pyroxene in mantle xenoliths showed a median value of $-0.6\Delta \log f_{O_2}$ relative to FMQ (Martin *et al.* 2015*b*). This is reduced relative to FMQ 1.3–1.8 $\Delta \log f_{O_2}$ calculated from experiments for the deep plumbing system of primitive magma from the Ross Island Volcanic Field (Iacovino *et al.* 2016). Globally, $\Delta \log f_{O_2}$ in rift settings varies between FMQ −4.1 and FMQ +0.5, with a median rift setting value of $-0.9 \pm 0.1\Delta \log f_{O_2}$

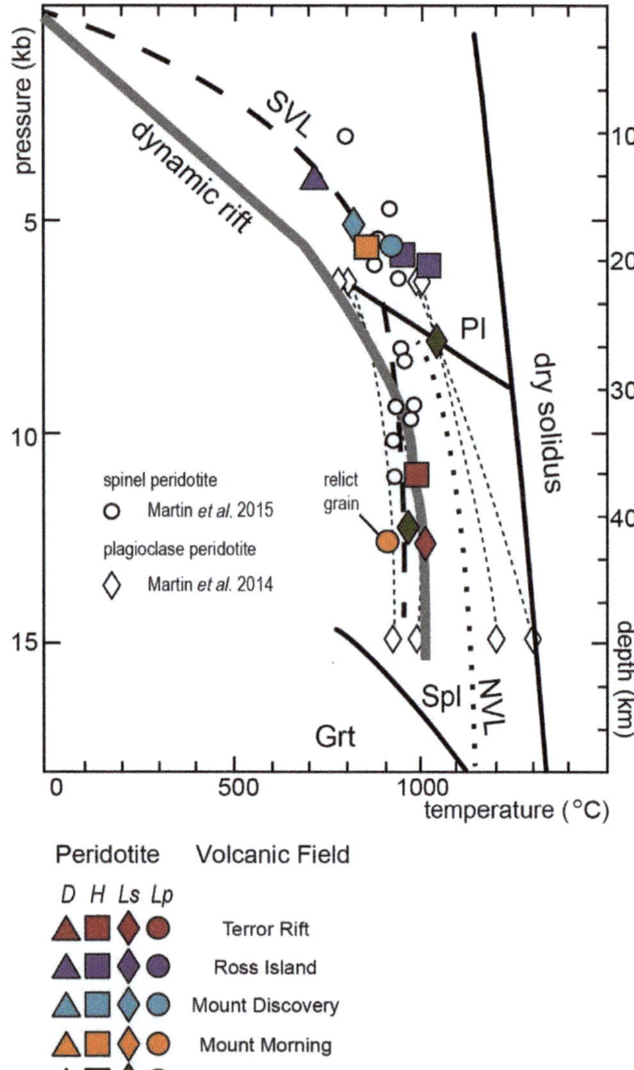

Fig. 11. A temperature–pressure plot showing values calculated using pyroxene compositions in mantle xenoliths (see the Supplementary material) from southern Victoria Land (Putirka 2008). Plotted for comparison are spinel peridotite xenolith data from Mount Morning (hollow circles: Martin *et al.* 2015*b*), and plagioclase lherzolite data from Pipecleaner Glacier, Mount Morning and White Island (hollow diamonds: Martin *et al.* 2014*a*). The mantle fields for plagioclase (Pl), spinel (Spl) and garnet (Grt) peridotite, and the dry solidus follow Borghini *et al.* (2011). Geotherms for southern Victoria Land (SVL: Berg *et al.* 1989), northern Victoria Land (NVL: Perinelli *et al.* 2006) and an idealized dynamic rift (Chapman 1986) are shown for comparison. The legend abbreviations are as in Figure 5.

(Frost and McCammon 2008; Martin *et al.* 2015*a*, *b*). In north Victoria Land, $\Delta \log f_{O_2}$ varies between −1.5 and −0.2 in anhydrous xenoliths, and between −2.52 and −1.32 in amphibole-bearing (hydrous) xenoliths (Perinelli *et al.* 2012; Bonadiman *et al.* 2014; Coltorti *et al.* 2021, this volume). This shows that the f_{O_2} values for the Erebus Volcanic Province overlap with the values typically expected within rift settings in Antarctica, and globally (Martin *et al.* 2015*a*, *b*).

Origin of peridotite xenoliths

Melt depletion. Negative trends on wt% MgO v. Al_2O_3 abundance plots for clinopyroxene (Fig. 6a), orthopyroxene (Fig. 6b) and whole-rock data (Figs 8 and 9a) are commonly interpreted to reflect melt depletion in mantle xenoliths (e.g. Johnson *et al.* 1990; Niu 2004). Southern Victoria Land mantle clinopyroxene data plot adjacent to an equivalent melt trend and indicate up to 25% melting (Fig. 6a). The Mount Morning and Southern Local Suite volcanic fields have the most clinopyroxene data, and these indicate a wide range of melt-depletion values (1–25%). There are fewer clinopyroxene data available for the three other volcanic fields, and these indicate a more restricted range of melt depletion of between *c.* 10 and 25%. The plagioclase lherzolite data from Mount Morning and White Island are interpreted to show less melt depletion than is the case for spinel lherzolite or harzburgite xenoliths from the same localities. These comparisons also apply to the Mount Morning orthopyroxene in peridotite data (Fig. 6b).

Application of a model melting grid to wt% FeO v. MgO abundance data for whole-rock samples indicates that southern Victoria Land peridotites could be residues from the equivalent of 10–35% melting (Fig. 9f). This agrees with a simple melt model applied to a wt% MgO v. Al_2O_3 abundance plot, which indicates up to 30% melt depletion (Fig. 9a). On both plots, spinel harzburgite and spinel dunite xenoliths can be argued to have undergone more melt depletion than spinel lherzolite or plagioclase lherzolite xenoliths. Similar melt-depletion trends are derived when trace element data are compared with calculated models. For example, <25% melting is suggested for a model applied to Yb v. V abundance data (Fig. 10a), 1–>25% melting is calculated for trends on a Yb (ppm) v. Dy/Yb diagram (Fig. 10b) or <5–*c.* 20% melting as shown on a clinopyroxene-mantle-normalized Yb v. Y plot (Fig. 10c).

Melting was modelled in both the spinel and garnet stability fields of the mantle (Fig. 10b), assuming anhydrous conditions. All the southern Victoria Land compositions overlap with the spinel stability field model. Melting pressures modelled for peridotite whole-rock wt% MgO v. FeO abundance plot (Fig. 9f) indicate final melting pressures of 10–30 kbar, equivalent to the spinel facies or spinel–garnet facies of the mantle. This is similar to the 2 GPa or greater pressures estimated from melting models in Figure 8. In summary, the mantle peridotite xenoliths of southern Victoria Land may have undergone up to *c.* 25% melt depletion in the spinel mantle facies.

Evidence for refertilization. On a wt% MgO v. Na_2O abundance plot, several whole-rock mantle xenoliths from southern Victoria Land have compositions that can be explained by fractional or batch melt models (Fig. 9c). However, other samples have elevated wt% Na_2O at a given MgO concentration, relative to the melt models (Fig. 9c), or elevated Na/Sm at a given La/Sm (Fig. 9d), consistent with major element refertilization. This pattern is also observed on a whole-rock wt% MgO v. FeO abundance plot, with one sample from the Southern Local Suite, two samples from Mount Morning and several samples from the Ross Island Volcanic Field with elevated wt % FeO relative to calculated melting grids (Fig. 9f). This can also be seen on a La/Sm v. Fe/Sm plot (Fig. 9g). Re-enrichment of Na and Fe in peridotite xenolith samples is decoupled and is not interpreted to form part of the same event. Similar FeO enrichments documented in abyssal peridotites have been attributed to olivine addition by melt migration through previously depleted mantle residues (e.g. Niu 1997; Rampone *et al.* 2008), which would be consistent with the olivine-rich spinel dunite assemblages of samples from the Ross Island Volcanic Field recording the most elevated FeO abundances (Fig. 9f). Another striking feature of the Ross Island Volcanic Field mantle xenoliths is that olivine grains have lower Ni/Fe and lower forsterite contents than

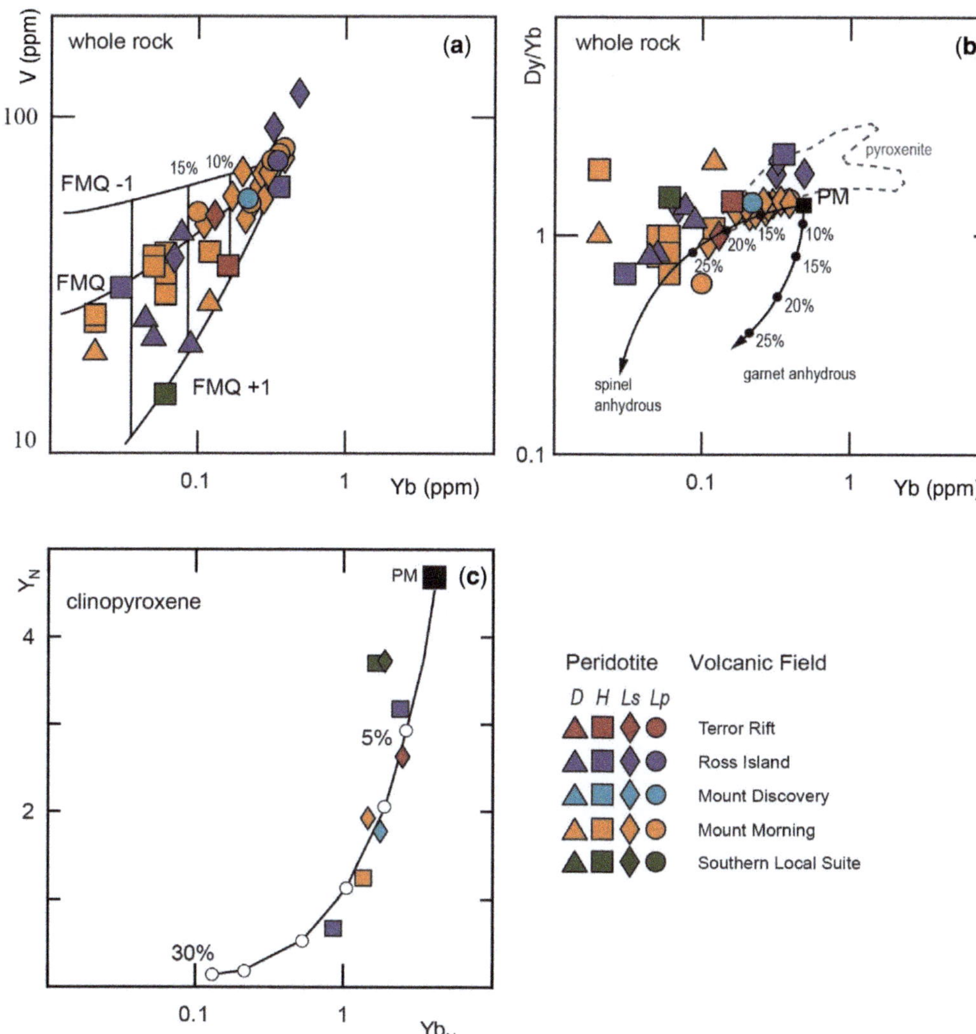

Fig. 12. Whole-rock vanadium (V) abundance (**a**) and Dy/Yb ratio (**b**) plotted against Yb content for southern Victoria Land mantle xenoliths. Clinopyroxene Yb v. Y normalized to primitive mantle (McDonough and Sun 1995) are also shown (**c**). The model in (a) shows predicted melt extraction paths (degree of melting in %) relative to the fayalite–magnetite–quartz (FMQ) oxygen fugacity buffer (Parkinson and Pearce 1998). In (b) data are compared with a model for fractional melting of primitive mantle (PM: McDonough and Sun 1995) using the melting modes and partition coefficients of Johnson *et al.* (1990) for anhydrous garnet and spinel facies conditions. The percentage melt fraction data are shown. The pyroxenite field (dashed line) is from this study (see the Supplementary material). In (c) the melt model shows between <5% and *c.* 20% melt extraction. See the Supplementary material for modelling details. The legend abbreviations are as in Figure 5.

olivine grains from other volcanic fields in the province (Fig. 5). This is the case for olivine in both peridotite and pyroxenite xenoliths (Fig. 5a and b). Another notable feature is the high FeO content (11.19 wt% average) of the Ross Island Volcanic Field spinel lherzolite samples relative to spinel dunite xenoliths (10.35 wt% FeO average). The Mg# is lower in spinel lherzolites (88.5 v. 89.4). These features distinguish mantle xenoliths of the Ross Island Volcanic Field from those of other volcanic fields in the province (Fig. 5).

The primitive-mantle-normalized REE patterns of mantle clinopyroxene and whole-rock peridotite xenoliths from southern Victoria Land are shown in Figure 13a and b. Total REE abundances are elevated relative to those of primitive mantle. Plotted for comparison are modelled melt-depletion curves (Fig. 13b, black solid lines). The mid-REE (MREE) and heavy REE (HREE) patterns of the peridotite xenoliths match reasonably well with the melt-extraction models; however, the light REE (LREE) values are elevated above the modelled curves (Fig. 13b), which can be explained by LREE refertilization (e.g. Müntener *et al.* 2004; Martin *et al.* 2015*b*).

Sources of refertilization. A unique volcanic lineage from the Ross Island Volcanic Field is the enriched iron series (Kyle *et al.* 1992), which is distinguished by elevated whole-rock FeO abundances relative to other volcanic rock series in the Erebus Volcanic Province (Martin *et al.* 2021). In the Ross Island Volcanic Field, elevated FeO abundances occur in rocks representing primitive magmas, in peridotite and pyroxenite xenoliths, and in mantle olivine. High FeO concentrations in the mantle have been attributed to the recycling of iron-rich crust (Kellogg *et al.* 1999; Sobolev *et al.* 2007), exchange of iron across the core–mantle boundary (Garnero 2000; Humayun *et al.* 2004) or primordial sources (Rubie *et al.* 2004). Martin *et al.* (2021) have hypothesized that an eclogitic component is more significant in the mantle beneath the Ross Island Volcanic Field, than is the case in the fields to the west. This would be consistent with the recycling of iron-rich crust. Alternatively, some workers have suggested that a mantle plume is present beneath the Ross Island Volcanic Field (Kyle *et al.* 1992; Phillips *et al.* 2018), which would be more consistent with the core–mantle boundary model for excess FeO. We propose that both the source of the enriched-iron series magmas and mantle xenoliths with elevated FeO have been refertilized by similar melts. Further study of these samples could hold the key to differentiating between a recycled crust and mantle plume origin.

Martin *et al.* (2015*b*) have proposed up to 5 wt% alkalic melt refertilization in peridotite xenoliths from Mount Morning. The effect on REE behaviour of adding alkalic melt to a depleted mantle source is modelled in Figure 13b (black dashed line). The REE data for southern Victoria Land whole-rock mantle xenoliths can be explained by models in which depleted mantle is refertilized by the addition of ≤5 wt% alkalic melt. Martin *et al.* (2014*a*) have also shown that the chemistry of clinopyroxene in plagioclase lherzolite xenoliths from the Mount Discovery, Mount Morning and Southern Local Suite volcanic fields is consistent with up to 6 wt% addition to the mantle of a normal mid-ocean ridge basalt melt.

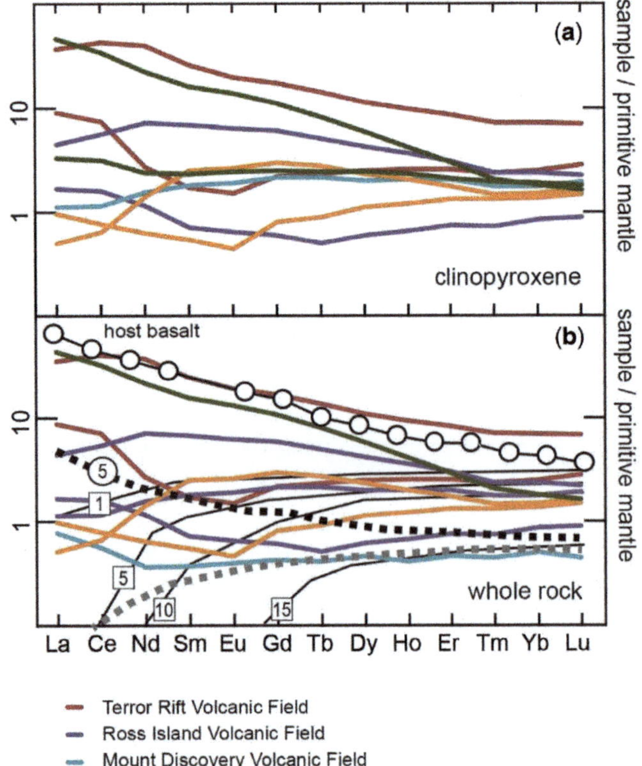

Fig. 13. Primitive-mantle-normalized REE data (McDonough and Sun 1995) for: (**a**) clinopyroxene and (**b**) xenolith whole-rock data. Samples are the same as used in Figure 8. Idealized melt-extraction models are shown (thin black lines) for spinel facies peridotite calculated using the melting modes from Kinzler (1997) and the element distribution coefficients from Kelemen *et al.* (1995) and Liu *et al.* (2012). Melt fractions (%) are shown in the boxes. A mixing model was also calculated between an initial peridotite composition (grey dashed line: sample OU78521) and host basalt (sample OU78540: Martin *et al.* 2013) using the mineral-melt coefficients of Ionov *et al.* (2002). The 5 wt% melt proportion model is shown (circled number on the black dashed line).

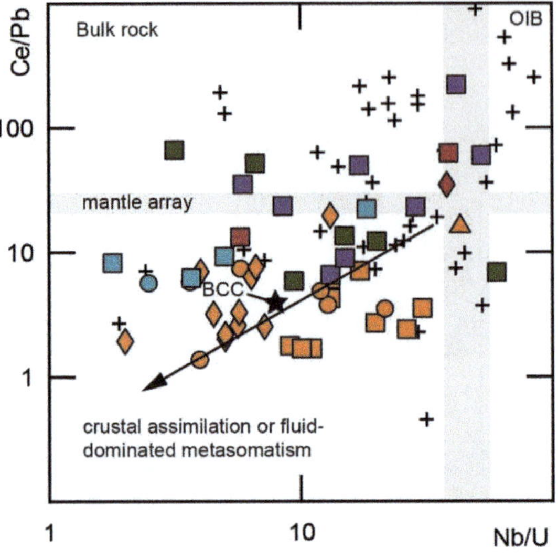

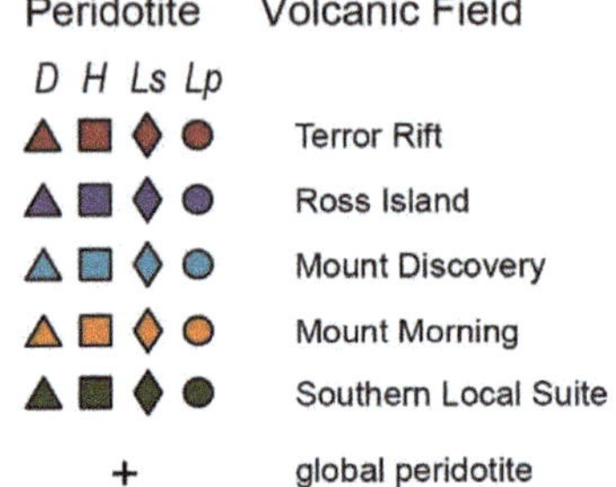

Fig. 14. Niobium/U v. Pb/Ce diagrams for peridotite mantle xenoliths. Mantle array values of Nb/U (47 ± 10) and Ce/Pb (25 ± 5) are from Hofmann *et al.* (1986). Global peridotite data are from Pearson *et al.* (2003). OIB, ocean island basalt; BCC, bulk continental crust. The legend abbreviations are as in Figure 5.

Metasomatism. Evidence for metasomatism in the mantle beneath southern Victoria Land includes the presence in peridotite xenoliths of grains of amphibole, phlogopite, carbonate and sulfide. Spinel mineral chemistry is also consistent with metasomatism in some mantle peridotites (Fig. 7a); as has been seen also in spinel grains in peridotite xenoliths from northern Victoria Land (Perinelli *et al.* 2008). Primitive-mantle-normalized extended element patterns of clinopyroxene and most xenolith whole-rock samples have negative Nb and Ti anomalies (Fig. 10) that are difficult to reconcile solely with melt depletion and refertilization processes. An average primitive-mantle-normalized trace element pattern for subducted sediment calculated by Plank and Langmuir (1998) is characterized by depletions in Ti and Nb relative to mafic melt, and this Ti and Nb depletion is similar to those observed in southern Victoria Land peridotite specimens (Fig. 10). Lead enrichment is also seen in the subducted sediment average (Plank and Langmuir 1998) but this feature is rare in the normalized trace element patterns of southern Victoria Land peridotites. However, it is seen in the Mount Discovery Volcanic Field whole-rock plagioclase lherzolite data (Fig. 10) and has been reported in whole-rock mantle xenolith data from Mount Morning Volcanic Field (Martin *et al.* 2015*a*). To further examine the role of a sedimentary component, the southern Victoria Land whole-rock data are plotted on a Nb/U v. Ce/Pb discrimination diagram (Fig. 14). A common interpretation of trends on this type of diagram is that whole-rock Ce/Pb <9 reflects a sedimentary component (e.g. White 2010). Mount Morning Volcanic Field data have the lowest Ce/Pb values in the province (i.e. less than or equal to the mantle array: Fig. 14), with many values equal to or less than bulk continental crust (Fig. 14). The trace element ratios (Fig. 14) have elsewhere been interpreted as reflecting fluid (with a sedimentary component) metasomatism and this is most obvious in data for the Mount Morning Volcanic Field (Fig. 14). A hint regarding the nature of this sedimentary component can be obtained from $^{87}Sr/^{86}Sr$ and $^{143}Nd/^{144}Nd$ isotope data (Fig. 15), which indicate that one of the Mount Morning spinel harzburgite xenolith compositions is on a mixing line with the EMI isotope end member. Mixing of *c.* 80% HIMU and *c.* 20% EMI isotopic compositions can generate the isotopic composition of this harzburgite specimen. End-member EM1 is commonly equated with lower continental crust, ancient pelagic sediment or delaminated subcontinental lithosphere (Hofmann 2004). Xenolith data from northern Victoria Land also plot along this modelled mixing line (e.g. Melchiorre *et al.* 2011; Perinelli *et al.* 2011) (Fig. 15).

Rare carbonate neoblasts, weak Zr/Sm depletion relative to average subducted sediment (Fig. 10), and strong U/Th fractionation in the whole-rock and clinopyroxene data (Fig. 10) for the Southern Local Suite, Mount Morning and Mount Discovery volcanic fields can be interpreted as being related to carbonatite metasomatism (e.g. Yaxley *et al.* 1991; Pfänder *et al.* 2012; Ackerman *et al.* 2013). Martin *et al.* (2015*b*) calculated that some aspects of the volcanic and peridotite xenolith whole-rock trace element behaviour can be explained by

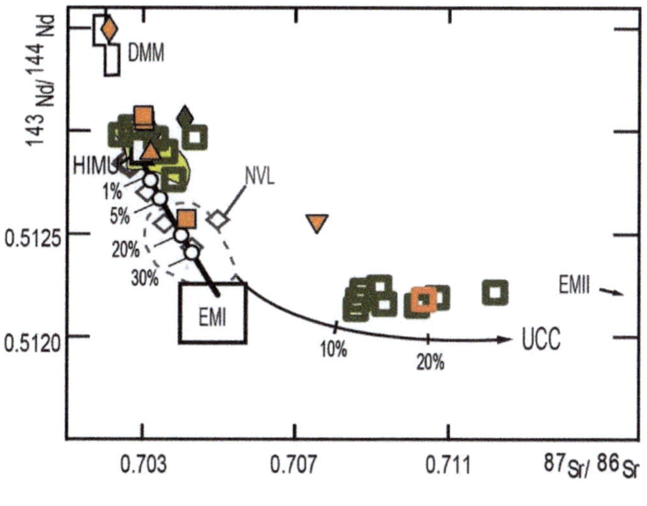

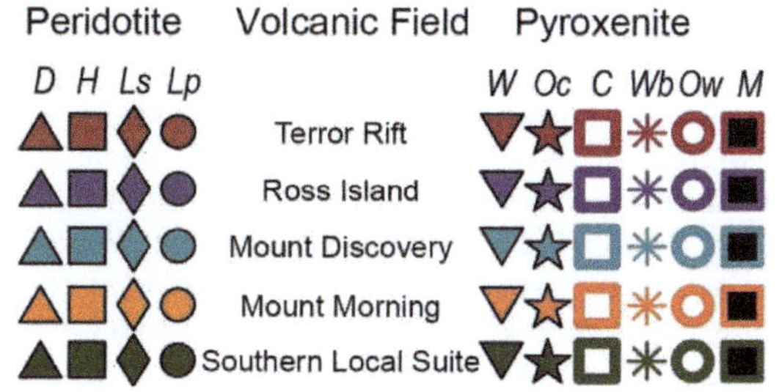

Fig. 15. A ^{143}Nd/^{144}Nd v. ^{87}Sr/^{86}Sr diagram for mantle xenoliths from southern Victoria Land. The Southern Local Suite data are after McGibbon (1991). North Victoria Land (NVL) data are after Melchiorre *et al.* (2011) and Perinelli *et al.* (2011). DMM, depleted mid-ocean ridge mantle; HIMU, high-μ; EMI, enriched mantle I; EMII, enriched mantle II; UCC, upper continental crust. The yellow field represents volcanic rock data (Martin *et al.* 2013) from the diffuse alkaline magmatic province (DAMP: Finn *et al.* 2005); and the dashed field represents typical subcontinental lithospheric mantle (SCLM) values (Jourdan *et al.* 2007). Two melt models have been calculated and plotted for comparison. Melt fractions (%) for both models are shown. The mixing curve (thick black line) between EMI and HIMU uses the equations of Faure (1986) and the compositions of Rolland *et al.* (2009), using EMI: (^{87}Sr/^{86}Sr) = 0.705, 513 ppm Sr; (^{142}Nd/^{144}Nd) = 0.5122, 33 ppm Nd; HIMU: (^{87}Sr/^{86}Sr) = 0.703, 120 ppm Sr; (^{142}Nd/^{144}Nd) = 0.51285, 6.5 ppm Nd. The mixing curve between the SCLM and the UCC (thin black line) uses the values and equations in Jourdan *et al.* (2007), including SCLM: (^{87}Sr/^{86}Sr) = 0.7043, 50 ppm Sr; (^{142}Nd/^{144}Nd) = 0.51244, 2 ppm Nd; UCC: (^{87}Sr/^{86}Sr) = 0.73847, 159 ppm Sr; (^{142}Nd/^{144}Nd) = 0.511800, 26 ppm Nd. The legend abbreviations are as in Figure 5.

the addition of ≤0.1 wt% carbonatite to the Mount Morning mantle. Clinopyroxene and whole-rock data from the Ross Island and Terror Rift volcanic fields show little U/Th fractionation (Fig. 10) but studies of primitive volcanic rocks and mantle xenoliths at Franklin Island have suggested a possible carbonatite component in their source (Aviado *et al.* 2015; Doherty 2016).

Source characteristics and timing. On primitive-mantle-normalized extended element plots, trace element patterns for southern Victoria Land peridotite samples are comparable with those of SCLM or depleted mid-ocean ridge mantle (Fig. 10). Spinel harzburgite and spinel dunite samples have Sr and Nd isotopic compositions similar to those of the HIMU mantle end member (Fig. 15), although the Pb isotope composition (whole-rock ^{206}Pb/^{204}Pb 18.21–19.69) (see the Supplementary material) is less radiogenic than HIMU *sensu stricto*. The peridotite isotopic compositions are comparable to those of primitive volcanic rocks from Victoria Land and the SW Pacific (diffuse alkaline magmatic province (DAMP): Finn *et al.* 2005) (in Fig. 15, the area defined by the yellow field). There is an ongoing debate in the literature as to whether isotopic and chemical source characteristics in Cenozoic DAMP volcanic rocks come exclusively from the asthenosphere or reflect a lithospheric contribution (Lanyon *et al.* 1993; Panter *et al.* 2006; Timm *et al.* 2009; Martin *et al.* 2015*a*, *b*, 2021). One of the spinel lherzolite samples in this study is isotopically more similar to depleted mid-ocean ridge mantle (Fig. 15), and mantle source characteristics for southern Victoria Land have been explained in terms of mixing between depleted mid-ocean ridge mantle and HIMU (Cooper *et al.* 2007; Sims *et al.* 2008; Day *et al.* 2019). It is our favoured interpretation that there is a HIMU component in the lithospheric mantle that has mixed with a depleted mid-ocean ridge mantle component. We cannot discount the presence of a HIMU component also in the asthenosphere. To this depleted HIMU peridotite mantle, various components were added at various times, including sedimentary and carbonatite fractions; there is also evidence for refertilization by various other melts including normal mid-ocean ridge basalt and alkalic compositions, and, beneath the Ross Island Volcanic Field, an iron-rich melt.

A two-stage model-lead age for volcanic rocks from Ross Island Volcanic Field yields a 1500 Ma isochron, which was interpreted by Sun and Hanson (1975) as the time at which chemical heterogeneity developed in the mantle source. This age agrees with Paleoproterozoic aluminachron stabilization ages determined for the southern Victoria Land mantle (Doherty *et al.* 2012, 2013) but contrasts with young (<250 Ma) ages interpreted for Hut Point Peninsula mantle xenoliths from Re–Os isotopes (Day *et al.* 2019). Martin *et al.* (2021) favoured an ancient (>0.5 Ga) age for *in situ* development of HIMU in the mantle source of Erebus Volcanic Province rocks. It is our favoured interpretation that the mantle stabilized during the Paleoproterozoic, allowing *in situ* development of the HIMU isotopic signature (see van der Meer *et al.* 2017 for an example from Zealandia). Melts were added, possibly at <250 Ma beneath Hut Point Peninsula, and alkalic melts have been added since the onset of magmatism in the region at *c.* 25 Ma (Martin *et al.* 2010).

Origin of pyroxenite xenoliths

Pyroxenite xenoliths are reported from all volcanic fields in the province. Detailed pyroxenite xenolith studies have been undertaken at Foster Crater (Gamble and Kyle 1987; Gamble *et al.* 1988; McGibbon 1991) and Mount Morning (Martin *et al.* 2015*b*). McGibbon (1991) used whole-rock Rb–Sr isochrons to yield a 439.2 ± 14.5 Ma formation age on a clinopyroxenite xenolith from Foster Crater. Gamble *et al.* (1988) identified a number of processes that they believed had affected the pyroxenite assemblage at Foster Crater, including potassium metasomatism, dynamic recrystallization, melt generation and infiltration and oxidation. A carbonate grain has also been described in one clinopyroxenite from Foster Crater (Table 2). Martin *et al.* (2015*b*) concluded that Mount Morning pyroxenite xenoliths have formed from melts derived from, or modified by, fluids derived from subducted, eclogitic oceanic crust. North Victoria Land cumulates and pyroxenite xenoliths are discussed in detail by, for example, Wörner *et al.* (1993) and Perinelli *et al.* (2011).

The pyroxenite clinopyroxene data used in this study can be interpreted to suggest that the Al-augite series xenoliths (websterite and olivine websterite) are less fertile than the Cr-diopside series xenoliths (Fig. 6c). The spinel data obtained from pyroxenite xenoliths are consistent with a degree of metasomatism (Fig. 7a). Olivine grains from xenoliths of the Ross Island Volcanic Field have lower Ni/Fe abundances and Fo contents than is observed in olivine in pyroxenite or peridotite xenoliths from elsewhere in the region (Fig. 5). Isotopic data for several Southern Local Suite Volcanic Field xenoliths overlap with the field for the DAMP, isotopic compositions of spinel harzburgite and spinel dunite from southern Victoria Land, and with the HIMU mantle field (Fig. 15). Several other Southern Local Suite Volcanic Field and Mount Morning Volcanic Field samples have high $^{87}Sr/^{86}Sr$ and low $^{143}Nd/^{144}Nd$ relative to other xenoliths (Fig. 15), and these have isotopic compositions on a mixing line with the EMII mantle end member. A mixing model between SCLM and upper continental crust indicates that the high $^{87}Sr/^{86}Sr$ values may be explained by up to 20% mixing with a crustal component, and this is our preferred hypothesis.

The clinopyroxenite xenoliths with HIMU isotopic compositions are likely to be ancient (>0.5 Ga) to allow *in situ* growth of the HIMU component in a manner similar to that envisioned for the peridotite xenoliths. At least some of the southern Victoria Land pyroxenite xenoliths formed in the Ordovician–Silurian (439.2 ± 14.5 Ma: McGibbon 1991). A sedimentary component (EMII), likely to have been derived from the upper continental crust, has influenced the chemistry of some pyroxenite xenoliths. We favour a hypothesis whereby pyroxenite xenoliths formed from fluids derived from (or modified by) a subducting slab, possibly with an eclogitic component. The variability in trace element and isotope chemistry may indicate that modification occurred more than once.

Petrogenesis and regional variability

The cross-section drawn in Figure 16 for south Victoria Land and cross-sections drawn for north Victoria Land across the rift shoulder, for example by Wörner (1999), share a high degree of similarity that suggests commonality in processes along significant stretches of the West Antarctic Rift System. The lithospheric mantle beneath southern Victoria Land stabilized at some time in the Paleoproterozoic but since that time it has undergone a complex history of multiple melt depletions, refertilization and metasomatic events (Fig. 16). There may also be pockets of younger (<250 Ma) mantle beneath Hut Point Peninsula. In some mantle xenoliths, up to 25% melt extraction has occurred in the spinel mantle facies. The HIMU component in the southern Victoria Land mantle must be ancient (>0.5 Ga) to allow for *in situ* growth. All volcanic fields have undergone some degree of alkalic melt refertilization, probably since at least 25 Ma, and this may be contributing to the high heat flow in the province (Wörner and Zipfel 1996; Martin *et al.* 2014*a*). The plagioclase lherzolite samples show evidence of mixing of depleted mantle, with up to 6 wt% of a normal mid-ocean ridge basalt component. In addition, the mantle beneath the Ross Island Volcanic Field has been refertilized by an iron-rich melt that may also be responsible for the Hut Point Peninsula enriched iron series volcanic lineage, emplaced at *c.* 1.3 Ma (Kyle 1981; Smellie and Martin 2021). We attribute this iron-rich mantle source to metasomatism by a fluid from (or modified by) a subducted slab. We suggest the regional variation in iron is real as it is recorded in mantle peridotite xenoliths, pyroxenite xenoliths and primitive magma compositions in the Ross Island Volcanic Field. A carbonatite mantle component is inferred to be present in the mantle beneath all volcanic fields but a sedimentary component, which can be modelled as lower continental crust (EMI), is more significant in the Mount Morning Volcanic Field mantle. It cannot be discounted that the lower continental crust component variability is attributed to selective sampling or temporal variation. Water and oxygen fugacity may be up to *c.* 2 wt% and $-0.6\Delta\log f_{O_2}$.

Cross-cutting the southern Victoria Land peridotite mantle are pyroxenite veins, which were probably emplaced at multiple times: one set has to be ancient (>0.5 Ga), again to allow for *in situ* growth of the HIMU isotopic signature, and another occurred during the Ordovician–Silurian. The pyroxenite samples have undergone up to 25% melt depletion, and at Foster Crater potassium metasomatism has also occurred. The data for olivine from pyroxenite xenoliths in the Ross Island Volcanic Field record the effects of an iron-rich fluid, which is also inferred to have affected the composition of peridotites from the same locality. A sedimentary component can be identified in mantle samples from the Southern Local Suite Volcanic Field and Mount Morning Volcanic Field, and these compositions can be modelled by the mixing of depleted mantle with up to 20% of an upper continental crust (EMII) component.

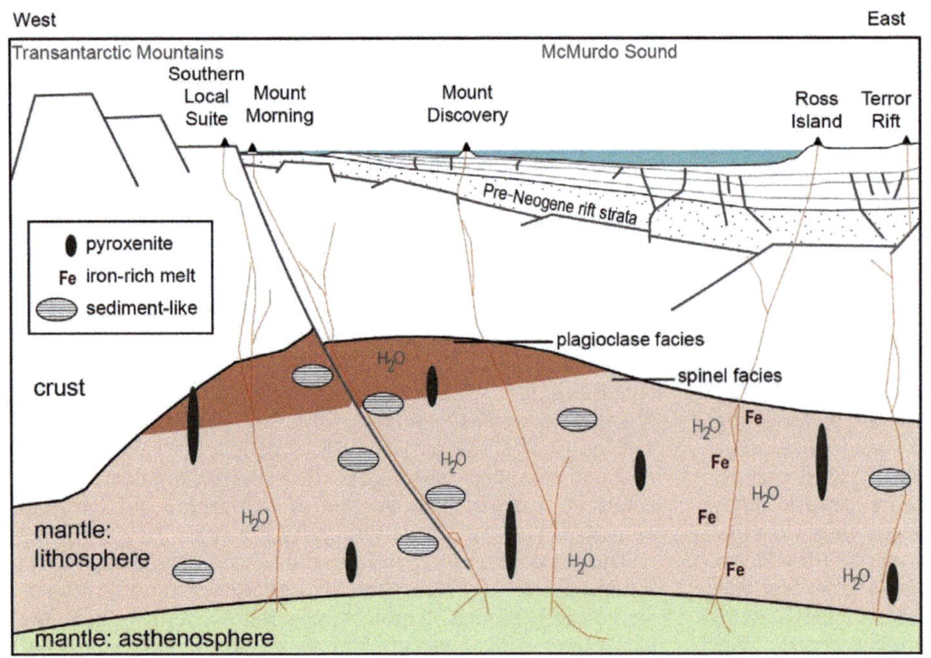

Fig. 16. A schematic interpretive diagram, not to scale, showing structures and composition of the southern Victoria Land mantle. The proposed lithospheric mantle has evolved through the mixing of depleted mantle and HIMU with varying degrees of carbonatite metasomatism and multiple melt extraction events. It is cross-cut by pyroxenite bodies, and the west is enriched in a sedimentary component relative to the east that has an iron-rich melt. Plagioclase facies mantle is restricted to the Mount Discovery Volcanic Field and further west. The crustal section of the cross-section follows seismic data interpretations made in Naish *et al.* (2007).

Summary and conclusions

Mantle xenoliths from southern Victoria Land have been collected and analysed for over 100 years. Key objectives of this chapter have been to provide a synthesis of the petrography of all xenolith samples collected and analysed over the decades (Table 2), and to collate and interpret all the chemical data available (see the Supplementary material). Spinel lherzolite and spinel harzburgite are the most common xenolith types, followed by spinel dunite. A rare type of plagioclase lherzolite is found in several localities. A wide range of pyroxenite xenolith types occur in each volcanic field. The peridotite and pyroxenite xenolith samples indicate that the geochemistry and petrogenetic evolution of the mantle beneath southern Victoria Land is complex and dynamic. The lithospheric mantle stabilized in the Paleoproterozoic but subsequently these depleted peridotitic compositions have been modified by a sequence of further melt depletion and metasomatic (melt and/or fluid) enrichment events. An ancient (>0.5 Ga) depleted, HIMU *sensu stricto* mantle has undergone carbonatite metasomatism and multiple melt-depletion events. The Ross Island Volcanic Field mantle has an iron-rich component relative to other areas in the province, and there may be a component of young (<250 Ma) mantle beneath Hut Point Peninsula. This southern Victoria Land mantle is cross-cut by pyroxenite bodies that were probably emplaced at multiple times and the compositions of some pyroxenite xenoliths can be modelled to have been affected by an upper continental crust (EMII) component.

Despite such a long history of research, there is a limited amount of published age constraints, radiogenic isotope data and volatile data on the southern Victoria Land mantle samples, and consequently it is difficult to rigorously constrain the petrogenetic history. There is a complete absence of H_2O measurements on mantle xenoliths, despite the importance of this parameter to mantle viscosity calculations and melting models. Some localities (e.g. Mount Discovery, northern Beaufort Island and the Dailey Islands) have not been mapped in sufficient detail to exclude the possibility that mantle xenoliths may be present. Furthermore, in other areas (e.g. Minna Bluff) the mantle xenoliths collected have not been analysed. There are many localities yet to be visited, and many more mantle xenoliths to be documented and analysed. Thus, mantle xenolith studies will continue to be an important research area in southern Victoria Land for decades to come.

Acknowledgements We thank R.J. Moscati (United States Geological Survey) for providing material and discussion. We wish to thank G. Wörner and R.J. Mocscati for thorough reviews, and W. van der Wal for editorial handling. Antarctica New Zealand and the United States Antarctic Program are thanked for their logistical support, and J.A. Gamble thanks the VUW Antarctic Research Centre for logistic support over many seasons.

Author contributions **APM**: data curation (lead), formal analysis (equal), writing – original draft (lead); **AFC**: formal analysis (equal), investigation (equal), writing – original draft (supporting); **RCP**: conceptualization (supporting), writing – original draft (supporting); **CLD**: formal analysis (supporting), writing – original draft (supporting); **JAG**: data curation (supporting), investigation (supporting).

Funding This research received no specific grant from any funding agency in the public, commercial, or not-for-profit sectors.

Data availability All data generated or analysed during this study are included in this published article (and its supplementary information files).

Correction notice Figures 1 and 4 have increased in size, the caption to Figure 8 has been amended, Tables 1 and 2 have improved layout and some typographical errors have been corrected. The publisher apologizes for these items that were previously missed.

References

Ackerman, L., ŠpaČek, P., Magna, T., Ulrych, J., Svojtka, M., Hegner, E. and Balogh, K. 2013. Alkaline and carbonate-rich melt metasomatism and melting of subcontinental lithospheric mantle: evidence from mantle xenoliths, NE bavaria, bohemian massif. *Journal of Petrology*, **54**, 2597–2633, https://doi.org/10.1093/petrology/egt059

Allibone, A.H., Forsyth, P.J., Sewell, R.J., Turnbull, I.M. and Bradshaw, M.A. 1991. *Geology of the Thudergut Area, southern Victoria Land, Antarctica. 1:50 000*. New Zealand Geological Survey Miscellaneous Geological Map, **21**. Department of Scientific and Industrial Research, Wellington, New Zealand.

Armstrong, R.L. 1978. K–Ar dating: late Cenozoic McMurdo Volcanic Group and dry valley glacial history, Victoria Land, Antarctica. *New Zealand Journal of Geology and Geophysics*, **21**, 685–698, https://doi.org/10.1080/00288306.1978.10425199

Aviado, K.B., Rilling-Hall, S., Bryce, J.G. and Mukasa, S.B. 2015. Submarine and subaerial lavas in the West Antarctic Rift System: Temporal record of shifting magma source components from the lithosphere and asthenosphere. *Geochemistry, Geophysics, Geosystems*, **16**, 4344–4361, https://doi.org/10.1002/2015GC006076

Barrett, P.J. 1981. History of the Ross Sea region during the deposition of the Beacon Supergroup 400–180 million years ago. *Journal of the Royal Society of New Zealand*, **11**, 447–458, https://doi.org/10.1080/03036758.1981.10423334

Bédard, J.H.J., Marsh, B.D., Hersum, T.G., Naslund, H.R. and Mukasa, S.B. 2007. Large-scale mechanical redistribution of orthopyroxene and plagioclase in the basement sill, Ferrar Dolerites, McMurdo Dry Valleys, Antarctica: petrological, mineral–chemical and field evidence for channelized movement of crystals and melt. *Journal of Petrology*, **48**, 2289–2326, https://doi.org/10.1093/petrology/egm060

Berg, J.H., Moscati, R.J. and Herz, D.L. 1989. A petrologic geotherm from a continental rift in Antarctica. *Earth and Planetary Science Letters*, **93**, 98–108, https://doi.org/10.1016/0012-821X(89)90187-8

Bonadiman, C., Nazzareni, S., Coltorti, M., Comodi, P., Giuli, G. and Faccini, B. 2014. Crystal chemistry of amphiboles: implications for oxygen fugacity and water activity in lithospheric mantle beneath Victoria Land, Antarctica. *Contributions to Mineralogy and Petrology*, **167**, 1–17, https://doi.org/10.1007/s00410-014-0984-8

Borghini, G., Fumagalli, P. and Rampone, E. 2011. The geobarometric significance of plagioclase in mantle peridotites: a link between nature and experiments. *Lithos*, **126**, 42–53, https://doi.org/10.1016/j.lithos.2011.05.012

Brey, G. P. and Kohler, T. 1990. Geothermobarometry in four-phase lherzolites II. New thermobarometers, and practical assessment of existing thermobarometers. *Journal of Petrology*, **31**, 1353–1378.

Burgess, S.D., Bowring, S.A., Fleming, T.H. and Elliot, D.H. 2015. High-precision geochronology links the Ferrar large igneous province with early-Jurassic ocean anoxia and biotic crisis. *Earth and Planetary Science Letters*, **415**, 90–99, https://doi.org/10.1016/j.epsl.2015.01.037

Chapman, D.S. 1986. Thermal gradients in the continental crust. *Geological Society, London, Special Publications*, **24**, 63–70, https://doi.org/10.1144/GSL.SP.1986.024.01.07

Cole, J.W. and Ewart, A. 1968. Contributions to the volcanic geology of the Black Island, Brown Peninsula, and Cape Bird areas, McMurdo sound, Antarctica. *New Zealand Journal of Geology and Geophysics*, **11**, 793–828, https://doi.org/10.1080/00288306.1968.10420754

Cole, J.W., Kyle, P.R. and Neall, V.E. 1971. Contributions to quaternary geology of Cape Crozier, White Island and Hut Point Peninsula, McMurdo Sound region, Antarctica. *New Zealand Journal of Geology and Geophysics*, **14**, 528–546, https://doi.org/10.1080/00288306.1971.10421946

Cooper, A.F., Adam, L.J., Coulter, R.F., Eby, G.N. and McIntosh, W.C. 2007. Geology, geochronology and geochemistry of a basanitic volcano, White Island, Ross Sea, Antarctica. *Journal of Volcanology and Geothermal Research*, **165**, 189–216, https://doi.org/10.1016/j.jvolgeores.2007.06.003

Coltorti, M., Bonadiman, C., Casetta, F., Faccini, B., Giacomni, P.P., Pelorosso, B. and Perinelli, C. 2021. Nature and evolution of the northern Victoria Land lithospheric mantle (Antarctica) as revealed by ultramafic xenoliths. *Geological Society, London, Memoirs*, **56**, https://doi.org/10.1144/M56-2020-11

Cox, S.C., Turnbull, I.M., Isaac, M.J., Townsend, D.B. and Smith Lyttle, B. 2012. *Geology of Southern Victoria Land Antarctica. 1:250 000*. Institute of Geological and Nuclear Sciences Geological Map, **22**. GNS Science, Lower Hutt, New Zealand.

Day, J.M.D., Harvey, R.P. and Hilton, D.R. 2019. Melt-modified lithosphere beneath Ross Island and its role in the tectono-magmatic evolution of the West Antarctic Rift System. *Chemical Geology*, **518**, 45–54, https://doi.org/10.1016/j.chemgeo.2019.04.012

Doherty, C.L. 2016. *Multi-stage Evolution of the Lithospheric Mantle in the West Antarctic Rift System – A Mantle Xenolith Study*. PhD thesis, Columbia University, New York, USA.

Doherty, C., Class, C. *et al.* 2012. Constraining the dynamic response of subcontinental lithospheric mantle to rifting using Re-Os model ages in the Western Ross Sea, Antarctica. Abstract presented at the 2012 Fall Meeting, 3–7 December 2012, San Francisco, California, USA.

Doherty, C., Class, C. *et al.* 2013. Re–Os systematics of the lithospheric mantle beneath the Western Ross Sea area, Antarctica: depletion ages and dynamic response during rifting. Abstract T13A-2516 presented at the 2013 Fall Meeting, 9–13 December 2013, San Francisco, California, USA.

Eggins, S.M., Woodhead, J.D. *et al.* 1997. A simple method for the precise analysis of 40 or more trace elements in geological materials by ICPMS using enriched isotope internal standardization. *Chemical Geology*, **134**, 311–326, https://doi.org/10.1016/S0009-2541(96)00100-3

Faure, G. 1986. *Principles of Isotope Geology*. 2nd edn. Wiley, New York.

Ferrar, H.T. 1907. Report on the field-geology of the region explored during the 'Discovery' Antarctic expedition, 1901–1904. *Natural History*, **1**, 1–100.

Fielding, C.R., Henrys, S.A. and Wilson, T.J. 2006. Rift history of the western Victoria Land Basin: a new perspective based on integration of cores with seismic reflection data. *In*: Futterer, D.K., Damaske, D., Kleinschmidt, G., Miller, H. and Tessensohn, F. (eds) *Antarctica: Contributions to Global Earth Sciences*. Springer, Berlin, 307–316.

Finn, C.A., Müller, R.D. and Panter, K.S. 2005. A Cenozoic diffuse alkaline magmatic province (DAMP) in the southwest Pacific without rift or plume origin. *Geochemistry, Geophysics, Geosystems*, **6**, Q02005, https://doi.org/10.1029/2004gc000723

Fitzgerald, P. 2002. Tectonics and landscape evolution of the Antarctic plate since the breakup of Gondwana, with an emphasis on the West Antarctic Rift System and the Transantarctic Mountains. *Royal Society of New Zealand Bulletin*, **35**, 453–469.

Forbes, R.B. 1963. Ultrabasic inclusion from the basalts of the Hut Point area, Ross Island, Antarctica. *Bulletin Volcanologique*, **26**, 13–21, https://doi.org/10.1007/BF02597270

Forbes, R.B. and Banno, S. 1966. Nickel–iron content of peridotite inclusion and cognate olivine from an alkali-olivine basalt. *American Mineralogist*, **51**, 130–140.

Frost, D.J. and McCammon, C.A. 2008. The redox state of the Earth's mantle. *Annual Review of Earth and Planetary Sciences*, **36**, 389–420, https://doi.org/10.1146/annurev.earth.36.031207.124322

Gaetani, G., Pamukcu, A., Wallace, P., Sims, K.W.W., Le Roux, V. and Klein, F. 2019. Magma storage and ascent beneath the Erebus Volcanic Province. ISAES 2019: XIII International Symposium on Antarctic Earth Sciences, 22–26 July 2019, Songdo Convensia, Incheon, South Korea.

Gale, A., Dalton, C.A., Langmuir, C.H., Su, Y. and Schilling, J.-G. 2013. The mean composition of ocean ridge basalts. *Geochemistry, Geophysics, Geosystems*, **14**, 489–518, https://doi.org/10.1029/2012GC004334

Gamble, J.A. and Kyle, P.R. 1987. The origins of glass and amphibole in spinel-wehrlite xenoliths from Foster Crater, McMurdo Volcanic Group, Antarctica. *Journal of Petrology*, **28**, 755–779, https://doi.org/10.1093/petrology/28.5.755

Gamble, J.A., McGibbon, F., Kyle, P.R., Menzies, M. and Kirsch, I. 1988. Metasomatised xenoliths from Foster Crater, Antarctica: Implications for lithospheric structure and process beneath the transantarctic mountain front. *Journal of Petrology*, Special Lithosphere Issue, **1**, 109–138, https://doi.org/10.1093/petrology/Special_Volume.1.109

Garnero, E.J. 2000. Heterogeneity of the lowermost mantle. *Annual Review of Earth and Planetary Sciences*, **28**, 509–537, https://doi.org/10.1146/annurev.earth.28.1.509

Giacomoni, P. P., Bonadiman, C., Casetta, F., Faccini, B., Ferlito, C., Ottolini, L., Zanetti, A. and Coltorti, M. 2020. Long-term storage of subduction-related volatiles in Northern Victoria Land lithospheric mantle: Insight from olivine-hosted melt inclusions from McMurdo basic lavas (Antarctica). *Lithos*, 378–379, https://doi.org/10.1016/j.lithos.2020.105826

Gibson, S.A., Rooks, E.E., Day, J.A., Petrone, C.M. and Leat, P.T. 2020. The role of sub-continental mantle as both 'sink' and 'source' in deep Earth volatile cycles. *Geochimica et Cosmochimica Acta*, **275**, 140–162, https://doi.org/10.1016/j.gca.2020.02.018

Gunn, B.M. and Warren, G. 1962. Geology of Victoria Land between the Mawson and Mullock Glaciers, Antarctica. *Bulletin of the New Zealand Geological Survey*, **71**, 157.

Handler, M., Wysoczansky, R.J. and Gamble, J.A. 2021. Marie Byrd Land lithospheric mantle: A review of the xenolith record. *Geological Society, London, Memoirs*, **56**, https://doi.org/10.1144/M56-2020-17

Hall, J., Wilson, T., Henrys, S., Cooper, A. and Raymond, C. 2007. Structure of the central Terror Rift, western Ross Sea, Antarctica. *United States Geological Survey Open-File Report*, **2007-1047**, Short Research Paper 108.

Harrington, H.J. 1958. Nomenclature of rock units in the Ross Sea Region, Antarctica. *Nature*, **182**, 290, https://doi.org/10.1038/182290a0

Harte, B. 1977. Rock nomenclature with particular relation to deformation and recrystallization textures in olivine-bearing xenoliths. *Journal of Geology*, **85**, 279–288, https://doi.org/10.1086/628299

Herzberg, C. 2004. Geodynamic information in peridotite petrology. *Journal of Petrology*, **45**, 2507–2530, https://doi.org/10.1093/petrology/egh039

Hofmann, A.W. 2004. Sampling mantle heterogeneity through oceanic basalts; isotopes and trace elements. *In*: Carlson, R.W., Holland, H.D. and Turekian, K.K. (eds) *The Mantle and Core, Volume 2*. Elsevier, Oxford, UK, 61–102.

Hofmann, A.W., Jochum, K.P., Seufert, M. and White, W.M. 1986. Nb and Pb in oceanic basalts: new constraints on mantle evolution. *Earth and Planetary Science Letters*, **79**, 33–45, https://doi.org/10.1016/0012-821X(86)90038-5

Humayun, M., Qin, L. and Norman, M.D. 2004. Geochemical evidence for excess iron in the mantle beneath Hawaii. *Science*, **306**, 91, https://doi.org/10.1126/science.1101050

Iacovino, K., Oppenheimer, C., Scaillet, B. and Kyle, P. 2016. Storage and evolution of mafic and intermediate alkaline magmas beneath Ross Island, Antarctica. *Journal of Petrology*, **57**, 93–118, https://doi.org/10.1093/petrology/egv083

Ionov, D.A., Bodinier, J.-L., Mukasa, S.B. and Zanetti, A. 2002. Mechanisms and Sources of Mantle Metasomatism: Major and trace element compositions of peridotite xenoliths from Spitsbergen in the context of numerical modelling. *Journal of Petrology*, **43**, 2219–2259, https://doi.org/10.1093/petrology/43.12.2219

Johnson, K.T.M., Dick, H.J.B. and Shimizu, N. 1990. Melting in the oceanic upper mantle: An ion microprobe study of diopsides in abyssal peridotites. *Journal of Geophysical Research: Solid Earth*, **95**, 2661–2678, https://doi.org/10.1029/JB095iB03p02661

Jourdan, F., Bertrand, H., Schärer, U., Blichert-Toft, J., Féraud, G. and Kampunzu, A.B. 2007. Major and trace element and Sr, Nd, Hf, and Pb isotope compositions of the Karoo Large Igneous Province, Botswana–Zimbabwe: Lithosphere vs mantle plume contribution. *Journal of Petrology*, **48**, 1043–1077, https://doi.org/10.1093/petrology/egm010

Kelemen, P.B., Shimizu, N. and Salters, V.J.M. 1995. Extraction of mid-ocean-ridge basalt from the upwelling mantle by focused flow of melt in dunite channels. *Nature*, **375**, 747–753, https://doi.org/10.1038/375747a0

Kellogg, L.H., Hager, B.H. and Van Der Hilst, R.D. 1999. Compositional stratification in the deep mantle. *Science*, **283**, 1881–1884, https://doi.org/10.1126/science.283.5409.1881

Kinzler, R.J. 1997. Melting of mantle peridotite at pressures approaching the spinel to garnet transition: Application to mid-ocean ridge basalt petrogenesis. *Journal of Geophysical Research: Solid Earth*, **102**, 853–874, https://doi.org/10.1029/96JB00988

Kyle, P.R. 1981. Mineralogy and geochemistry of a basanite to phonolite sequence at Hut Point Peninsula, Antarctica, based on core from Dry Valley Drilling Project Drillholes 1, 2 and 3. *Journal of Petrology*, **22**, 451–500, https://doi.org/10.1093/petrology/22.4.451

Kyle, P.R. 1990. McMurdo Volcanic Group – western Ross Embayment: Introduction. *American Geophysical Union Antarctic Research Series* **48**, 18–25.

Kyle, P.R., Adams, J. and Rankin, P.C. 1979. Geology and petrology of the McMurdo Volcanic Group at Rainbow Ridge, Brown Peninsula, Antarctica. *Geological Society of America Bulletin*, **90**, 676, https://doi.org/10.1130/0016-7606(1979)90<676:GAPOTM>2.0.CO;2

Kyle, P.R., Wright, A.C. and Kirsch, I. 1987. Ultramafic xenoliths in the late Cenozoic McMurdo Volcanic Group, western Ross Sea embayment, Antarctica. *In*: Nixon, P.H. (ed.) *Mantle Xenoliths*. John Wiley & Sons Ltd, Chichester, UK, 287–293.

Kyle, P.R., Moore, J.A. and Thirlwall, M.F. 1992. Petrologic evolution of anorthoclase phonolite lavas at Mount Erebus, Ross Island, Antarctica. *Journal of Petrology*, **33**, 849–875, https://doi.org/10.1093/petrology/33.4.849

Lanyon, R., Varne, R. and Crawford, A.J. 1993. Tasmanian Tertiary basalts, the Balleny plume, and opening of the Tasman Sea (southwest Pacific Ocean). *Geology* , **21**, 555–558, https://doi.org/10.1130/0091-7613(1993)021<0555:TTBTBP>2.3.CO;2

Liu, J., Carlson, R.W., Rudnick, R.L., Walker, R.J., Gao, S. and Wu, F. 2012. Comparative Sr–Nd–Hf–Os–Pb isotope systematics of xenolithic peridotites from Yangyuan, North China Craton: Additional evidence for a Paleoproterozoic age. *Chemical Geology*, **332**, 1–14.

Martin, A.P. 2009. *Mount Morning, Antarctica: Geochemistry, Geochronology, Petrology, Volcanology, and Oxygen Fugacity of the Rifted Antarctic Lithosphere*. PhD thesis, University of Otago, Dunedin, New Zealand.

Martin, A.P. and Cooper, A.F. 2010. Post 3.9 Ma fault activity within the West Antarctic rift system: onshore evidence from Gandalf Ridge, Mount Morning eruptive centre, southern Victoria Land, Antarctica. *Antarctic Science*, **22**, 513–521, https://doi.org/10.1017/S095410201000026X

Martin, A.P., Cooper, A.F. and Dunlap, W.J. 2010. Geochronology of Mount Morning, Antarctica: two-phase evolution of a long-lived trachyte–basanite–phonolite eruptive center. *Bulletin of Volcanology*, **72**, 357–371, https://doi.org/10.1007/s00445-009-0319-1

Martin, A.P., Cooper, A.F. and Price, R.C. 2013. Petrogenesis of Cenozoic, alkalic volcanic lineages at Mount Morning, West Antarctica and their entrained lithospheric mantle xenoliths: lithospheric v. asthenospheric mantle sources. *Geochimica et Cosmochimica Acta*, **122**, 127–152, https://doi.org/10.1016/j.gca.2013.08.025

Martin, A.P., Cooper, A.F. and Price, R.C. 2014*a*. Increased mantle heat flow with on-going rifting of the West Antarctic rift system inferred from characterisation of plagioclase peridotite in the shallow Antarctic mantle. *Lithos*, **190–191**, 173–190, https://doi.org/10.1016/j.lithos.2013.12.012

Martin, A.P., Price, R.C. and Cooper, A.F. 2014*b*. Constraints on the composition, source and petrogenesis of plagioclase-bearing mantle peridotite. *Earth-Science Reviews*, **138**, 89–101, https://doi.org/10.1016/j.earscirev.2014.08.006

Martin, A.P., Cooper, A.F., Price, R.C., Turnbull, R.E. and Roberts, N.M.W. 2015*a*. The petrology, geochronology and significance of Granite Harbour Intrusive Complex xenoliths and outcrop sampled in western McMurdo Sound, Southern Victoria Land, Antarctica. *New Zealand Journal of Geology and Geophysics*, **58**, 33–51, https://doi.org/10.1080/00288306.2014.982660

Martin, A.P., Price, R.C., Cooper, A.F. and McCammon, C.A. 2015*b*. Petrogenesis of the rifted Southern Victoria Land lithospheric mantle, Antarctica, Inferred from petrography, geochemistry, thermobarometry and oxybarometry of peridotite and pyroxenite xenoliths from the Mount Morning Eruptive Centre. *Journal of Petrology*, **56**, 193–226, https://doi.org/10.1093/petrology/egu075

Martin, A.P., Cooper, A.F., Price, R.C., Kyle, P.R. and Gamble, J.A. 2021. Erebus Volcanic Province: petrology. *Geological Society, London, Memoirs*, **55**, https://doi.org/10.1144/M55-2018-80

McDonough, W.F. 1990. Constraints on the composition of the continental lithospheric mantle. *Earth and Planetary Science Letters*, **101**, 1–18, https://doi.org/10.1016/0012-821X(90)90119-I

McDonough, W.F. and Sun, S.-s. 1995. The composition of the Earth. *Chemical Geology*, **120**, 223–253, https://doi.org/10.1016/0009-2541(94)00140-4

McGibbon, F.M. 1991. Geochemistry and petrology of ultramafic xenoliths of the Erebus Volcanic Province. *In*: Thomson, M.R.A., Crame, J.A. and Thomson, J.W. (eds) *Geological Evolution of Antarctica – Proceedings of the 5th International Symposium on Antarctic Earth Sciences*. Cambridge University Press, Cambridge, UK, 317–321.

McGinnis, L.D., Bowen, R.H., Erickson, J.M., Alfred, B.J. and Kreamer, J.L. 1985. East–West Antarctic boundary in McMurdo Sound. *Tectonophysics*, **114**, 341–356, https://doi.org/10.1016/0040-1951(85)90020-4

McIver, J.R. and Gevers, T.W. 1970. Volcanic vents below the Royal Society Range, Central Victoria Land, Antarctica. *Transactions of the Geological Society of South Africa*, **73**, 65–88.

Melchiorre, M., Coltorti, M., Bonadiman, C., Faccini, B., O'Reilly, S.Y. and Pearson, N.J. 2011. The role of eclogite in the rift-related metasomatism and Cenozoic magmatism of Northern Victoria Land, Antarctica. *Lithos*, **124**, 319–330, https://doi.org/10.1016/j.lithos.2010.11.012

Molzahn, M., Wörner, G., Henjes-Kunst, F. and Rocholl, A. 1999. Constraints on the Cretaceous thermal event in the Transantarctic Mountains from alteration processes in Ferrar flood basalts. *Global and Planetary Change*, **23**, 45–60, https://doi.org/10.1016/S0921-8181(99)00050-8

Moscati, R.J. 1989. *Petrology and Thermobarometry of Pyroxene Granulite and Spinel Lherzolite Xenoliths from a Modern Continental Rift, Royal Society Range, McMurdo Sound Region, Antarctica*. MSc thesis, Northern Illinois University, DeKalb, Illinois, USA.

Müntener, O., Pettke, T., Desmurs, L., Meier, M. and Schaltegger, U. 2004. Refertilization of mantle peridotite in embryonic ocean basins: trace element and Nd isotopic evidence and implications for crust–mantle relationships. *Earth and Planetary Science Letters*, **221**, 293–308, https://doi.org/10.1016/S0012-821X(04)00073-1

Naish, T., Powell, R., Levy, R. and Team, T.A.-M.S. 2007. Background to the ANDRILL McMurdo Ice Shelf Project (Antarctica) and Initial Science Volume. *Terra Antarctica*, **14**, 121–130.

Niida, K. 1988. Metasomatic veins and minerals in mantle-derived xenoliths, Antarctica. *In*: *Proceedings of the NIPR Symposium on Antarctic Geoscience*. National Institute of Polar Research, Tokyo, 67–69.

Niida, K. 1990. Glass in mantle-derived peridotite xenoliths from the McMurdo Volcanic Group, Antarctica. *In*: *Proceedings of the NIPR Symposium on Antarctic Geosciences*. National Institute of Polar Research, Tokyo, 172–180.

Niu, Y. 1997. Mantle melting and melt extraction processes beneath ocean ridges: evidence from abyssal peridotites. *Journal of Petrology*, **38**, 1047–1074, https://doi.org/10.1093/petroj/38.8.1047

Niu, Y. 2004. Bulk-rock major and trace element compositions of abyssal peridotites: Implications for mantle melting, melt extraction and post-melting processes beneath mid-ocean ridges. *Journal of Petrology*, **45**, 2423–2458, https://doi.org/10.1093/petrology/egh068

Palmer, K. 1990. *XRF analyses of Granitoids and Associated Rocks, St John's Range, South Victoria Land, Antarctica*. Research School of Earth Science, Geology Board of Studies Publication, **5**/Analytical Facility Contribution, **13**/Antarctic Data Series, **15**.

Panter, K.S. 2021. Mount Early and Sheridan Bluff: petrology. *Geological Society, London, Memoirs*, **55**, https://doi.org/10.1144/M55-2019-2

Panter, K.S. and Martin, A. 2021. West Antarctic mantle deduced from mafic magmatism. *Geological Society, London, Memoirs*, **56**, https://doi.org/10.1144/M56-2021-10

Panter, K.S., Blusztajn, J., Hart, S.R., Kyle, P.R., Esser, R. and McIntosh, W.C. 2006. The origin of HIMU in the SW Pacific: evidence from intraplate volcanism in southern New Zealand and Subantarctic Islands. *Journal of Petrology*, **47**, 1673–1704, https://doi.org/10.1093/petrology/egl024

Panter, K., Dunbar, N.W., Scanlan, M., Wilch, T., Fargo, A. and McIntosh, W. 2011. Petrogenesis of alkaline magmas at Minna Bluff, Antarctica: evidence for multi-stage differentiation and complex mixing processes. Abstract V31F-2590 presented at the 2011 Fall Meeting, 5–9 December 2012, San Francisco, California, USA.

Parkinson, I.J. and Pearce, J.A. 1998. Peridotites from the Izu–Bonin–Mariana Forearc (ODP Leg 125): Evidence for mantle melting and melt–mantle interaction in a supra-subduction zone setting. *Journal of Petrology*, **39**, 1577–1618, https://doi.org/10.1093/petroj/39.9.1577

Pearson, D.G., Canil, D. and Shirey, S.B. 2003. Mantle samples included in volcanic rocks: xenoliths and diamonds. *Treatise on Geochemistry*, **2**, 568.

Pelorosso, B., Bonadiman, C., Coltorti, M., Faccini, B., Melchiorre, M., Ntaflos, T. and Gregoire, M. 2016. Pervasive, tholeiitic refertilisation and heterogeneous metasomatism in Northern Victoria Land lithospheric mantle (Antarctica). *Lithos*, **248**, 493–505, https://doi.org/10.1016/j.lithos.2016.01.032

Perinelli, C., Armienti, P. and Dallai, L. 2006. Geochemical and O-isotope constraints on the evolution of lithospheric mantle in the Ross Sea rift area (Antarctica). *Contributions to Mineralogy and Petrology*, **151**, 245–266, https://doi.org/10.1007/s00410-006-0065-8

Perinelli, C., Orlando, A., Conte, A.M., Armienti, P., Borrini, D., Faccini, B. and Misiti, V. 2008. Metasomatism induced by alkaline magma in the upper mantle of northern Victoria Land (Antarctica): an experimental approach. *Geological Society, London, Special Publications*, **293**, 279–302, https://doi.org/10.1144/SP293.13

Perinelli, C., Armienti, P. and Dallai, L. 2011. Thermal evolution of the lithosphere in a rift environment as inferred from the geochemistry of mantle cumulates, northern Victoria Land, Antarctica. *Journal of Petrology*, **52**, 665–690, https://doi.org/10.1093/petrology/egq099

Perinelli, C., Andreozzi, G., Conte, A., Oberti, R. and Armienti, P. 2012. Redox state of subcontinental lithospheric mantle and relationships with metasomatism: insights from spinel peridotites from northern Victoria Land (Antarctica). *Contributions to Mineralogy and Petrology*, **164**, 1053–1067, https://doi.org/10.1007/s00410-012-0788-7

Pfänder, J.A., Jung, S., Münker, C., Stracke, A. and Mezger, K. 2012. A possible high Nb/Ta reservoir in the continental lithospheric mantle and consequences on the global Nb budget – Evidence from continental basalts from Central Germany. *Geochimica et Cosmochimica Acta*, **77**, 232–251, https://doi.org/10.1016/j.gca.2011.11.017

Phillips, E.H., Sims, K.W.W. *et al.* 2018. The nature and evolution of mantle upwelling at Ross Island, Antarctica, with implications for the source of HIMU lavas. *Earth and Planetary Science Letters*, **498**, 38–53, https://doi.org/10.1016/j.epsl.2018.05.049

Plank, T. and Langmuir, C.H. 1998. The chemical composition of subducting sediment and its consequence for the crust and mantle. *Chemical Geology*, **145**, 325–394, https://doi.org/10.1016/S0009-2541(97)00150-2

Prior, G.T. 1902. *Report on the Rock Specimens Collected by the Southern Cross Antarctic Expedition*. British Museum.

Prior, G.T. 1907. Report on the rock specimens collected during the 'Discovery' Antarctic Expedition, 1901–1904. *Natural History*, **1**, 101–160.

Putirka, K.D. 2008. Thermometers and barometers for volcanic systems. *Reviews in Mineralogy and Geochemistry*, **69**, 61–120, https://doi.org/10.2138/rmg.2008.69.3

Rampone, E., Piccardo, G.B. and Hofmann, A.W. 2008. Multi-stage melt–rock interaction in the Mt. Maggiore (Corsica, France) ophiolitic peridotites: microstructural and geochemical evidence. *Contributions to Mineralogy and Petrology*, **156**, 453–475, https://doi.org/10.1007/s00410-008-0296-y

Rolland, Y., Galoyan, G., Bosch, D., Sosson, M., Corsini, M., Fornari, M. and Verati, C. 2009. Jurassic back-arc and Cretaceous hot-spot series in the Armenian ophiolites – implications for the obduction process. *Lithos*, **112**, 163–187, https://doi.org/10.1016/j.lithos.2009.02.006

Rubie, D.C., Gessmann, C.K. and Frost, D.J. 2004. Partitioning of oxygen during core formation on the Earth and Mars. *Nature*, **429**, 58–61, https://doi.org/10.1038/nature02473

Siddoway, C. 2008. Tectonics of the West Antarctic rift system: new light on the history and dynamics of distributed intracontinental extension (invited paper). *In*: Cooper, I.K., Barrett, P., Stagg, H., Storey, B., Stump, E. and Wise, W. (eds) *Antarctica: A Keystone in a Changing World*. National Academy of Sciences, Washington, DC, 91–114.

Sims, K.W.W., Blichert-Toft, J. *et al.* 2008. A Sr, Nd, Hf, and Pb isotope perspective on the genesis and long-term evolution of alkaline magmas from Erebus volcano, Antarctica. *Journal of Volcanology and Geothermal Research*, **177**, 606–618, https://doi.org/10.1016/j.jvolgeores.2007.08.006

Skinner, D.N.B., Waterhouse, B.C., Brehaut, G.M. and Sullivan, K. 1976. *New Zealand Geological Survey Antarctic Expedition 1975–76, Skelton–Koettlitz Glaciers, Report DS58*. New Zealand Geological Survey.

Smellie, J.L. and Martin, A.P. 2021. Erebus Volcanic Province: volcanology. *Geological Society, London, Memoirs*, **55**, https://doi.org/10.1144/M55-2018-62

Smith, W.C. 1954. *The Volcanic Rocks of the Ross Archipelago, British Terra Nova Expedition, 1910*. Natural History Report.

Sobolev, A.V., Hofmann, A.W. *et al.* 2007. The amount of recycled crust in sources of mantle-derived melts. *Science*, **316**, 412–417, https://doi.org/10.1126/science. 1138113

Stuckless, J.S. and Ericksen, R.L. 1976. Strontium isotopic geochemistry of the volcanic rocks and associated megacrysts and inclusions from Ross Island and vicinity, Antarctica. *Contributions to Mineralogy and Petrology*, **58**, 111–126, https://doi.org/10.1007/BF00382180

Stump, E. 1995. *The Ross Orogen of the Transantarctic Mountains*. Cambridge University Press, New York.

Sun, S.S. and Hanson, G.N. 1975. Origin of Ross Island basanitoids and limitations upon the heterogeneity of mantle sources for alkali basalts and nephelinites. *Contributions to Mineralogy and Petrology*, **52**, 77–106, https://doi.org/10.1007/BF00395006

Tessensohn, F. and Wörner, G. 1991. The Ross Sea rift system, Antarctica: Structure, evolution, and analogs. *In*: Thompson, M.R.A., Crame, J.A. and Thompson, J.W. (eds) *Geological Evolution of Antarctica*. Cambridge University Press, New York, 273–278.

Thomson, J.A. 1916. *Report on the Inclusions of the Volcanic Rocks of the Ross Archipelago (with Appendix by F. Cohen)*. Report of the British Antarctic Expedition 1907–1909.

Timm, C., Hoernle, K., Van Den Bogaard, P., Bindeman, I. and Weaver, S. 2009. Geochemical evolution of intraplate volcanism at Banks Peninsula, New Zealand: interaction between asthenospheric and lithospheric melts. *Journal of Petrology*, **50**, 989–1023, https://doi.org/10.1093/petrology/egp029

Tinto, K.J. 2019. Petrogenetic models for the evolution of alkalic magmas in the Erebus volcanic province, Antarctica. Abstract A248 presented at ISAES 2019: XIII International Symposium on Antarctic Earth Sciences, 22–26 July 2019, Incheon, South Korea.

Treves, S.B. and Kyle, P.R. 1973. Geology of DVDP 1 and 2, Hut Point Peninsula. Ross Island, Antarctica. *Dry Valley Drilling Project (Northern Illinois University) Bulletin*, **2**, 11–82.

Upton, B.G.J., Downes, H., Kirstein, L.A., Bonadiman, C., Hill, P.G. and Ntaflos, T. 2011. The lithospheric mantle and lower crust–mantle relationships under Scotland: a xenolithic perspective. *Journal of the Geological Society, London*, **168**, 873–886, https://doi.org/10.1144/0016-76492009-172

van der Meer, Q.H.A., Waight, T.E., Scott, J.M. and Münker, C. 2017. Variable sources for Cretaceous to recent HIMU and HIMU-like intraplate magmatism in New Zealand. *Earth and Planetary Science Letters*, **469**, 27–41, https://doi.org/10.1016/j.epsl.2017.03.037

Walter, M.J. 2003. Melt extraction and compositional variability in mantle lithosphere. *In*: Holland, H.D. and Turekian, K.K. (eds) *The Mantle and Core*. Treatise on Geochemistry, **2**. Pergamon, Oxford, UK, 363–394.

Warner, R.D. and Wasilewski, P.J. 1995. Magnetic petrology of lower crust and upper mantle xenoliths from McMurdo Sound, Antarctica. *Tectonophysics*, **249**, 69–92, https://doi.org/10.1016/0040-1951(95)00014-E

White, W.M. 2010. Oceanic island basalts and mantle plumes: the geochemical perspective. *Annual Review of Earth and Planetary Sciences*, **38**, 133–160, https://doi.org/10.1146/annurev-earth-040809-152450

Wiens, D.A., Shen, W. and Lloyd, A. 2021. The seismic structure of the Antarctic Upper Mantle. *Geological Society, London, Memoirs*, **56**, https://doi.org/10.1144/M56-2020-18

Woodhead, J.D. 1996. Extreme HIMU in an oceanic setting: the geochemistry of Mangaia Island (Polynesia), and temporal evolution of the Cook–Austral hotspot. *Journal of Volcanology and Geothermal Research*, **72**, 1–19, https://doi.org/10.1016/0377-0273(96)00002-9

Workman, R.K. and Hart, S.R. 2005. Major and trace element composition of the depleted MORB mantle (DMM). *Earth and Planetary Science Letters*, **231**, 53–72, https://doi.org/10.1016/j.epsl.2004.12.005

Wörner, G. 1999. Lithospheric dynamics and mantle sources of alkaline magmatism of the Cenozoic West Antarctic Rift System. *Global and Planetary Change*, **23**, 61–77, https://doi.org/10.1016/S0921-8181(99)00051-X

Wörner, G. and Zipfel, J. 1996. A mantle P–T path for the Ross Sea Rift margin (Antarctica) derived from Ca-in-olivine zonation patterns in peridotites xenoliths of the Plio-Pleistocene Mt.Melbourne Volcanic Field. *Geologisches Jahrbuch B*, **B89**, 157–167.

Wörner, G., Fricke, A. and Burke, E. 1993. Fluid inclusion studies on lower crustal gabbroic xenoliths from the Mt. Melbourne Volcanic Field (Antarctica): evidence for the post-crystallization uplift history during Cenozoic Ross Sea Rifting. *European Journal of Mineralogy*, **5**, 775–785, https://doi.org/10.1127/ejm/5/4/0775

Wright, A.C. 1979*a*. *McMurdo Volcanics at Foster Crater, Southern Foothills of the Royal Society Range, Central Victoria Land, Antarctica*. New Zealand Geological Survey Internal Report.

Wright, A.C. 1979*b*. *McMurdo Volcanics Northwest of Koettlitz Glacier, Antarctica*. New Zealand Geological Survey Internal Report.

Wright, A.C. 1979*c*. *McMurdo Volcanics on Chancellor Ridge, Southern Foothills of the Royal Society Range, Central Victoria Land, Antarctica*. New Zealand Geological Survey Internal Report.

Wu, B. and Berg, J.H. 1992. Early Paleozoic lamprophyre dikes of southern Victoria Land: geology, petrology and geochemistry *In*: Yoshida, Y., Kaminuma, K. and Shiraishi, K. (eds) *Recent Progress in Antarctic Earth Science*. Terra Scientific, Tokyo, 257–264.

Yaxley, G.M., Crawford, A.J. and Green, D.H. 1991. Evidence for carbonatite metasomatism in spinel peridotite xenoliths from western Victoria, Australia. *Earth and Planetary Science Letters*, **107**, 305–317, https://doi.org/10.1016/0012-821X(91)90078-V

Nature and evolution of the northern Victoria Land lithospheric mantle (Antarctica) as revealed by ultramafic xenoliths

Massimo Coltorti[1,2], Costanza Bonadiman[1], Federico Casetta[1]*, Barbara Faccini[1], Pier Paolo Giacomoni[1], Beatrice Pelorosso[1] and Cristina Perinelli[3]

[1]Department of Physics and Earth Sciences, University of Ferrara, Via Saragat 1, 44122 Ferrara, Italy

[2]Istituto Nazionale di Geofisica e Vulcanologia (INGV), Sezione di Palermo, Via Ugo La Malfa 153, 90146 Palermo, Italy

[3]Department of Earth Sciences, Sapienza University of Rome, Piazzale Aldo Moro 5, 00185 Rome, Italy

MC, 0000-0002-7481-8097; FC, 0000-0002-2935-7420; BF, 0000-0002-3546-7159; BP, 0000-0002-3557-4069; CP, 0000-0002-2308-461X

*Correspondence: cstfrc@unife.it

Abstract: A review of northern Victoria Land ultramafic xenoliths, collected and studied over more than 30 years, was carried out. More than 200 samples were gathered and characterized in a coherent and comparative manner, both for mantle-derived and cumulate xenoliths. Almost 2000 analyses of major elements and more than 300 analyses of trace elements of *in situ* and separated olivine, pyroxenes, amphibole, spinel and glass were taken into consideration. Particular attention was devoted to mantle lithologies in order to emphasize the composition and the evolution of this portion of the subcontinental lithosphere. The three main localities in northern Victoria Land where mantle xenoliths were found (i.e. Mount Melbourne (Baker Rocks), Greene Point and Handler Ridge), over a >200 km distance, were described and compared with ultramafic xenoliths in three other localities (Harrow Peaks, Browning Pass and Mount Overlord) that are mainly cumulate in nature. Altogether, these data enabled us to reconstruct a long evolutionary history, from old depletion to most recent refertilization and metasomatic events, for this large sector of the northern Victoria Land subcontinental lithospheric mantle.

Supplementary material: The petrographical features, representative whole-rock and mineral compositional data, together with the modelled equilibration temperature, pressure and oxygen fugacity of northern Victoria Land ultramafic xenoliths, as well as the degree of partial melting recorded by mantle xenoliths based on major and trace element compositions of olivine, pyroxenes and spinel are available at https://doi.org/10.6084/m9.figshare.c.5303103

The West Antarctic Rift System (WARS: Tessensohn and Wörner 1991) is one of the most developed continental rift systems on Earth, extending for about 3000 km in length from the Ross Sea (Queen Maud Mountains–northern Victoria Land) to the Weddell Sea (Whitmore Mountains–Horlick Mountains) and from 750 to 1000 km in width. It is bordered to the east by the Transantarctic Mountains, representing the uplifted roots of the Early Paleozoic Ross Orogen (*c.* 550–480 Ma). The uplift of the Transantarctic Mountains and the concomitant extensional tectonics of the WARS started during the Late Cretaceous (Fitzgerald and Stump 1997; Chand *et al.* 2001), while magmatic activity initiated during the Eocene. Plutons and dyke swarms of the Meander Intrusive Group were emplaced between 48 and 23 Ma (Müller *et al.* 1991; Tonarini *et al.* 1997; Rocchi *et al.* 2002), predating the McMurdo Volcanic Group (Harrington 1958) which comprises primary and differentiated lavas, as well as pyroclastic deposits younger than 19–15 Ma (Kyle and Muncy 1989; Müller *et al.* 1991; Nardini *et al.* 2009). In Cape Roberts drill core, the only McMurdo Volcanic Group rock found to overlap with the Meander Intrusive Group was dated at 25.7 Ma (McIntosh 2000).

A lively debate has developed in the last few years regarding the origin of this extensional system. The first hypothesis invoked a mantle plume centred beneath Marie Byrd Land (LeMasurier and Landis 1996; Hart *et al.* 1997; Wörner 1999), which was, however, progressively counterbalanced by the large amount of time (>30 myr) elapsed between extension and magmatism (Panter *et al.* 2018), as well as by the homogeneously low $^{3}He/^{4}He$ ratio (7.1 ± 0.4 Ra: Nardini *et al.* 2009; Correale *et al.* 2019 and references therein). Rocchi *et al.* (2005) proposed decompression melting due to the reactivation of translithospheric faults impinging on metasomatized domains created during the Cretaceous amagmatic extension of the rift. Panter *et al.* (2018) added to this model the hypothesis that these enriched domains originated in the sublithospheric mantle during the Ross subduction, which lasted until Cretaceous time in some parts of Marie Byrd Land, furnishing carbonate-rich recycled slab material. This material reacted with the lithospheric mantle, generating portions enriched in amphibole and/or phlogopite, which are the indispensable ingredients for the production of the McMurdo Volcanic Group alkaline magmas.

Irrespective of the origin of the WARS, its basic magmas and pyroclastics often brought ultramafic xenoliths of mantle or cumulate origin to the surface. These provide an extremely useful tool to investigate the nature and the evolution of the lithospheric mantle beneath the rift, as well as its chemical and thermobarometric features.

In northern Victoria Land (Fig. 1) there are numerous localities where ultramafic xenoliths entrained in Cenozoic basic lavas or pyroclastics of the McMurdo Volcanic Group can be found (e.g. Gamble *et al.* 1988; Hornig *et al.* 1991; Zipfel and Wörner 1992; Wörner *et al.* 1993). Around Mount Melbourne, Beccaluva *et al.* (1991) reported xenoliths from Willow Nunatak, Willow Nunatak Beach, Edmondson Point, Baker Rocks and the beach near Baker Rocks (BR) (Fig. 1). This latter locality is the most famous and xenolith-rich in the Mount Melbourne area, being also the closest to the Italian Mario Zucchelli Station. For this reason, although some xenoliths occur all around the Mount Melbourne area (Willow Nunatak, the beach nearby and Edmondson Point; see Beccaluva *et al.* 1991), most of the descriptions reported in this chapter refer to Baker Rocks. Perinelli *et al.* (2006) studied a group of ultramafic xenoliths from Baker Rocks and Greene Point (GP) (Fig. 1), while more detailed studies of BR and GP xenoliths were conducted by Coltorti *et al.* (2004), Melchiorre *et al.* (2011) and Pelorosso *et al.* (2016). Handler Ridge (HR)

From: Martin, A. P. and van der Wal, W. (eds) 2023. *The Geochemistry and Geophysics of the Antarctic Mantle*.
Geological Society, London, Memoirs, **56**, 57–82,
First published online 28 April 2021, https://doi.org/10.1144/M56-2020-11

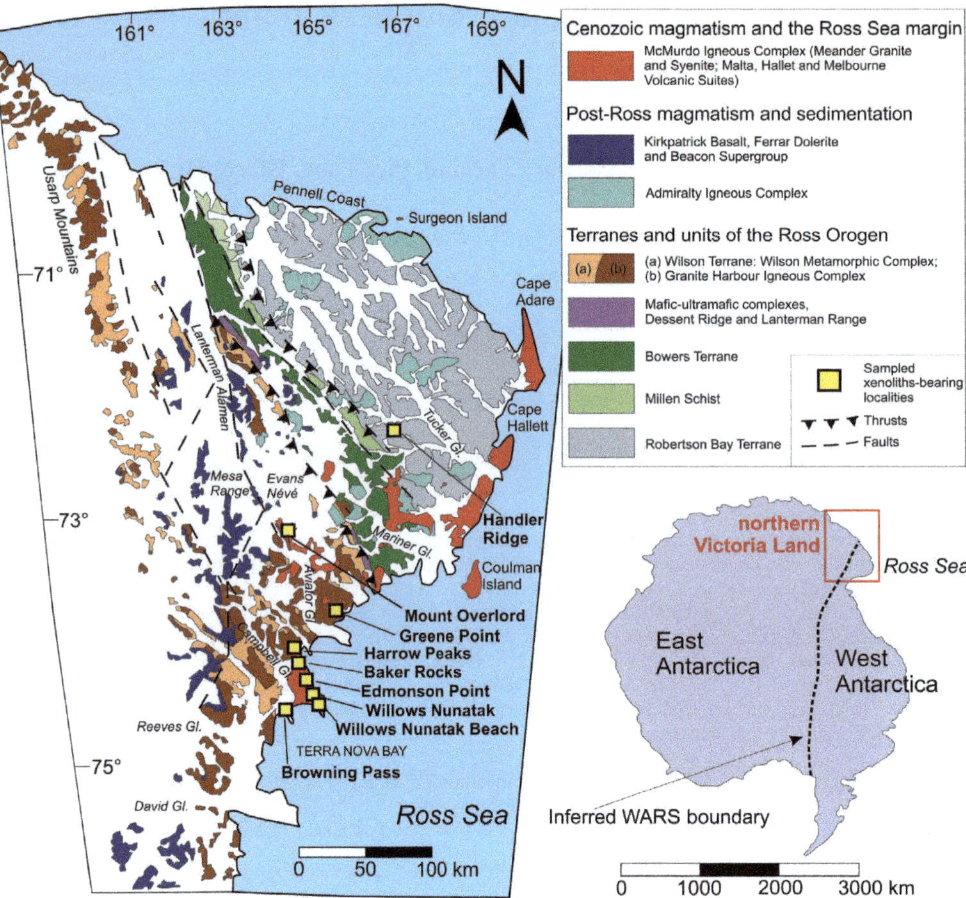

Fig. 1. Simplified geological map of northern Victoria Land, modified from Estrada *et al.* (2016), with permission from Elsevier ©2016, and Pelorosso *et al.* (2019). Squares indicate the locations of the sampled xenoliths: Handler Ridge (Pelorosso *et al.* 2017), Mount Overlord (Perinelli *et al.* 2017), Greene Point (Melchiorre *et al.* 2011; Pelorosso *et al.* 2016), Harrow Peaks (Pelorosso *et al.* 2019), Baker Rocks (Coltorti *et al.* 2004; Perinelli *et al.* 1998, 2006), Edmondson Point, Willow Nunatak and Willow Nunatak Beach (Beccaluva *et al.* 1991), and Browning Pass (Perinelli *et al.* 2011). On the lower right, the sketch map of Antarctica shows the location of northern Victoria Land and the inferred West Antarctic Rift System (WARS) boundary (Panter *et al.* 2018). Gl., glacier.

is the northernmost locality where ultramafic nodules have been found (Pelorosso *et al.* 2017). The most recently studied mantle xenolith locality is Harrow Peaks (HP: Pelorosso *et al.* 2019). Taking into account that this Memoir aims to describe and interpret the petrological features of the northern Victoria Land subcontinental lithospheric mantle (SCLM), attention here will be focused mainly on those localities where mantle xenoliths occur, although some of the cumulate rocks could have been formed at the crust–mantle boundary. Therefore, the three localities that will be considered in the following are BR (74° 12′ 36.07″ S, 164° 49′ 17.81″ E), GP (73° 48′ 33.68″ S, 165° 58′ 58.18″ E) and HR (72° 25′ 17.58″ S, 167° 28′ 42.04″ E), encompassing a mantle length of >200 km (Fig. 1). The ultramafic cumulate rocks from other localities, such as Mount Overlord (OVP) and Browning Pass (BRP: Fig. 1), throughout the text are considered mainly for comparison with mantle lithotypes (see Perinelli *et al.* 2011, 2017 for further details).

Most of the mantle xenoliths from BR, GP and HR are spinel-bearing peridotites varying from harzburgites to lherzolites with subordinate wehrlites. BR is the only locality where amphibole was found both as a disseminated phase and in veins in the mantle xenoliths. GP tends to have a larger proportion of clinopyroxene-poor lherzolites and harzburgites. One occurrence of phlogopite is found in an outcrop near Willow Nunatak. Mantle xenoliths are variably metasomatized with several types of secondary pyrometamorphic textures (Mercier and Nicolas 1975; Pike and Schwarzman 1977), as spongy clinopyroxene, reaction rims around orthopyroxene, and glassy patches intimately intermixed with secondary crystals of olivine, clinopyroxene and spinel. According to Perinelli *et al.* (2006), some GP xenoliths record a partial melting episode that occurred in equilibrium with garnet, while, following Pelorosso *et al.* (2016), this garnet signature can be related to a tholeiitic refertilization event caused by interaction with magmas of the Ferrar dolerites, some of which were generated in garnet peridotite (Riley *et al.* 2005). This hypothesis was reinforced recently through the petrological study of orthopyroxene-rich lherzolites, harzburgites and one orthopyroxenite from Harrow Peaks (HP). Three composite xenoliths made up of a dunitic matrix cross-cut by amphibole–clinopyroxene or orthopyroxene veins are also described. The relatively low forsterite content in olivine in some harzburgites and dunites, and the presence of orthopyroxenites as both primary lithotypes and veins in the peridotite matrix, as well as the barometric estimates, led Pelorosso *et al.* (2019) to consider these xenoliths as cumulate products of the Ferrar Large Igneous Province. Another orthopyroxenite was reported by Martin *et al.* (2015) from southern Victoria Land. It is very likely that the influence of the Ferrar Jurassic event also affected the lithosphere beneath BR, as suggested by the few Re–Os Jurassic minimum ages yielded by sulfide single grains (Melchiorre *et al.* 2011). Furthermore, the presence of two olivine-websterites among the BR xenolith collection testifies to the occurrence of silica-saturated magmas, which are different to the Cenozoic McMurdo alkaline volcanics. The xenoliths from the other localities, namely BRP and OVP, are in fact mainly wehrlites, olivine-clinopyroxenites, clinopyroxenites (sometimes amphibole- and phlogopite-bearing) and hornblendites. The great majority of these rocks are the result of cumulate processes from the alkaline magmas of the Cenozoic McMurdo volcanics. Orthopyroxenes are typically absent from these parageneses (Perinelli *et al.* 2011, 2017).

The aim of this chapter is to gather information on mineralogical and geochemical features of the northern Victoria Land mantle that has developed over almost 30 years of study. In this respect, it presents an opportunity to homogenize the different approaches adopted by the various authors in order to provide a unique vision of the petrological features of the northern Victoria Land lithospheric mantle. Information on grain size, textural features both primary and secondary (metasomatic/refertilized), and mineral phases major and trace element composition will be presented to constrain its chemical and mineralogical features, the oxy-thermobarometric conditions, and the nature of the processes that metasomatized/refertilized these mantle domains. This allows us to speculate on possible links between old melting episode(s) (Ross or pre-Ross), the Jurassic Ferrar large magmatic event and Cenozoic magmatism related to the formation of the WARS.

The evolution of the WARS lithospheric mantle

The geological evolution of the WARS lithospheric mantle, since its aggregation in the Gondwana supercontinent, was characterized by three main geodynamic events:

1. The Paleozoic subduction of oceanic lithosphere along the palaeo-Pacific margin of Gondwana (Kleinschmidt *et al.* 1987), intimately linked to the Ross Orogeny. This subduction continued, perhaps episodically, throughout the Mesozoic, and progressively ceased between 110 and 94 Ma from the west to the east of Marie Byrd Land (Mukasa and Dalziel 2000).
2. The Cambrian–Ordovician Ross Orogeny that formed the basement of the Kukri Peneplain, which was subsequently affected by extensive tholeiitic magmatism of the Ferrar (dolerite) Large Igneous Province in the Jurassic, marking the onset of the break-up of the Gondwana supercontinent (Elliot and Fleming 2018).
3. The Late Cretaceous amagmatic rifting phase that marked the onset of the WARS and which was accompanied by extensive alkaline magmatism from Eocene times onwards. It began with small plutons, plugs and dykes of the Meander intrusive, followed by the effusive and explosive activity of the stratovolcanoes and scoria cones of the McMurdo Volcanic Group (Tessensohn and Wörner 1991; Rocchi *et al.* 2002).

Each event contributed to the modification of both the chemical and modal composition of the lithospheric and sublithospheric mantle from which the WARS magmas were ultimately generated.

The occurrence of a strongly metasomatized mantle source, characterized by the variable abundances of hydrous phases (amphibole and/or phlogopite, as the main host for alkalis and volatile elements), plays a fundamental role in the genesis of silica-undersaturated magmas in both intraplate and continental rift settings. Recent studies (Aviado *et al.* 2015; Hudgins *et al.* 2015) highlighted the presence of multiple recycled lithologies in the mantle source region, including pyroxenite and volatile-rich, amphibole-bearing metasomatized peridotite. These results suggest that the mantle source of magmatic activity along the WARS is both subduction-modified and alkaline-metasomatized. The former could represent a relict of the 500–100 Ma Ross subduction, and the latter is the consequence of the onset of the current rifting and magmatism.

The importance of the Ferrar event has also been underlined recently by Pelorosso *et al.* (2016), who suggested that the northern Victoria Land lithospheric mantle was refertilized by percolation of tholeiitic melts during the Jurassic Ferrar magmatic event, whose effusive (Kirkpatrick Basalt Group) and intrusive (Ferrar Dolerite) products crop out from northern Victoria Land to the Pensacola Mountains. Chemically, two types of Ferrar rocks are distinguished: the first type comprises almost the bulk of the Ferrar province and shows a systematic range of geochemical compositions that point to fractional crystallization accompanied by 5% crustal assimilation; and the second type (1% by volume), although sharing the tholeiitic affinity of the majority of Ferrar dolerites, has a lower Mg# and higher Ti–Zr contents (Elliot and Fleming 2018). Preliminary high-precision zircon geochronology allows a restricted emplacement duration (<0.4 myr) at around 180 Ma to be inferred for most of the Ferrar province (Elliot and Fleming 2018).

However, the effects of the refertilization of mantle lithologies by infiltration and/or percolation of tholeiitic melts has never been experimentally investigated. A lively debate is underway on the possibility that basaltic melts interacting with a depleted harzburgite could generate a pyroxenite (Ionov *et al.* 2018). This is in contrast with what was put forward for the Lherz massif by Le Roux *et al.* (2009), who reversed the common view that harzburgite was the residuum after melt extraction of a lherzolite, proposing instead that the latter was the refertilization product of the former.

The influence of the Ross subduction could, instead, be found in the link between old subduction and recent rifting geodynamic settings. Cox (1978) advanced the idea that continental flood basalts and continental rift magmatism could be related to volatiles reintroduced to the mantle by subduction. More recently, elemental and isotopic analyses of volatiles in melt inclusions have supported this hypothesis (Stefano *et al.* 2011; Cabato *et al.* 2015; Wang *et al.* 2016; Ivanov *et al.* 2018). The investigation of melt inclusions (MI) in phenocrysts from lavas of the Cenozoic East African and Rio Grande rifts highlighted the presence of hydrated/carbonated sublithospheric domains in their mantle source, for which the volatile contents resulted from old subduction events.

Broadley *et al.* (2016) studied WARS ultramafic rocks both of cumulate and mantle origin, and found that the SCLM was enriched in halogens, most probably derived from dewatering of the Paleozoic–Mesozoic subducted slab. Similar theories were proposed by Panter *et al.* (2018), who explained the WARS magma composition by a metasomatic event(s) related to Ross subduction, which introduced recycled slab material into the asthenosphere, which in turn melted as it impinged on the overlying lithosphere. Amphibole overgrowing clinopyroxene, however, points to a more recent metasomatic episode (Coltorti *et al.* 2004). Neither of these models consider the effect of the Ferrar magmatic event, whose refertilization effects have been described in GP (Pelorosso *et al.* 2016) and most probably also in BR (Melchiorre *et al.* 2011), and whose cumulate counterpart can be found in HP (Pelorosso *et al.* 2019).

Olivine-hosted melt inclusions in Cenozoic basanitic and basaltic lavas (Giacomoni *et al.* 2020) reveal that primary melts were characterized by FeO, TiO_2, CaO and K_2O enrichment, as well as by high amounts of CO_2, F and Cl. This suggests the presence of heterogeneous mantle sources, recording a long history of melt/fluid–rock interactions dating back to Ross subduction. Thus, while the return of volatiles and solid material to the asthenosphere by subduction is reasonable and generally accepted, the mechanism by which these materials return to the lithosphere, activate extensional events and produce magmatism is still far from being fully understood. This is particularly true of the WARS, where Ferrar magmatism occurred between Ross subduction and WARS opening and magmatism. The subduction remnants in fact should have found a way to survive this event, which largely affected both the asthenospheric and the lithospheric mantle reservoirs.

Petrography and classification

Modal proportions of the studied xenoliths (Fig. 2; see also Supplementary Table 1) were estimated by point counting with more than 2000 points for each thin section.

Greene Point samples (Fig. 2a) are all anhydrous lherzolites (20) and harzburgites (18). The Mount Melbourne collection is the most heterogeneous (Fig. 2b), including nodules from different sampling points around the main volcanic edifice (Fig. 1): Willow Nunatak, Willow Nunatak Beach, Edmondson Point, Baker Rocks and the beach near Baker Rocks. This last is the most productive locality, where 16 anhydrous (10 lherzolites, five harzburgites and one olivine-websterite), five amphibole-bearing (four lherzolites and one olivine-websterite) and four composite samples (two lherzolites and one harzburgite cut by amphibole veins, and one lherzolite attached to an amphibole-bearing olivine-clinopyroxenite portion) were collected. At Willow Nunatak Beach, 11 anhydrous samples (five lherzolites and six harzburgites) and two amphibole–phlogopite-bearing lherzolites were recovered, while only six anhydrous samples were found at Willow Nunatak (two lherzolites and one olivine-websterite) and Edmondson Point (two harzburgites and one wehrlite).

Handler Ridge ultramafic xenoliths include 10 lherzolites and two wehrlites, all devoid of hydrous phases (Fig. 2c).

At Harrow Peaks, the lithological variety is also high, with five anhydrous (two lherzolites, one harzburgite, one olivine-websterite and one orthopyroxenite), two amphibole-bearing (one lherzolite and one dunite) and three composite samples (one dunite cut by an amphibolitic vein, one amphibole-bearing dunite cut by a clinopyroxenitic vein and one amphibole-bearing harzburgite cut by an orthopyroxenitic vein) found at this locality.

At Mount Overlord, a large number of cumulate xenoliths were found, spanning anhydrous wehrlites (nine) and olivine-clinopyroxenites (four) to amphibole-bearing wehrlites (seven), olivine-clinopyroxenites (six) and clinopyroxenites (two). Two hornblendites and one composite lherzolite/olivine-clinopyroxenites sample are also included (Fig. 2d).

Similar lithologies occur at Browning Pass, with the collection composed of five anhydrous wehrlites, eight anhydrous olivine-clinopyroxenites, one anhydrous clinopyroxenites

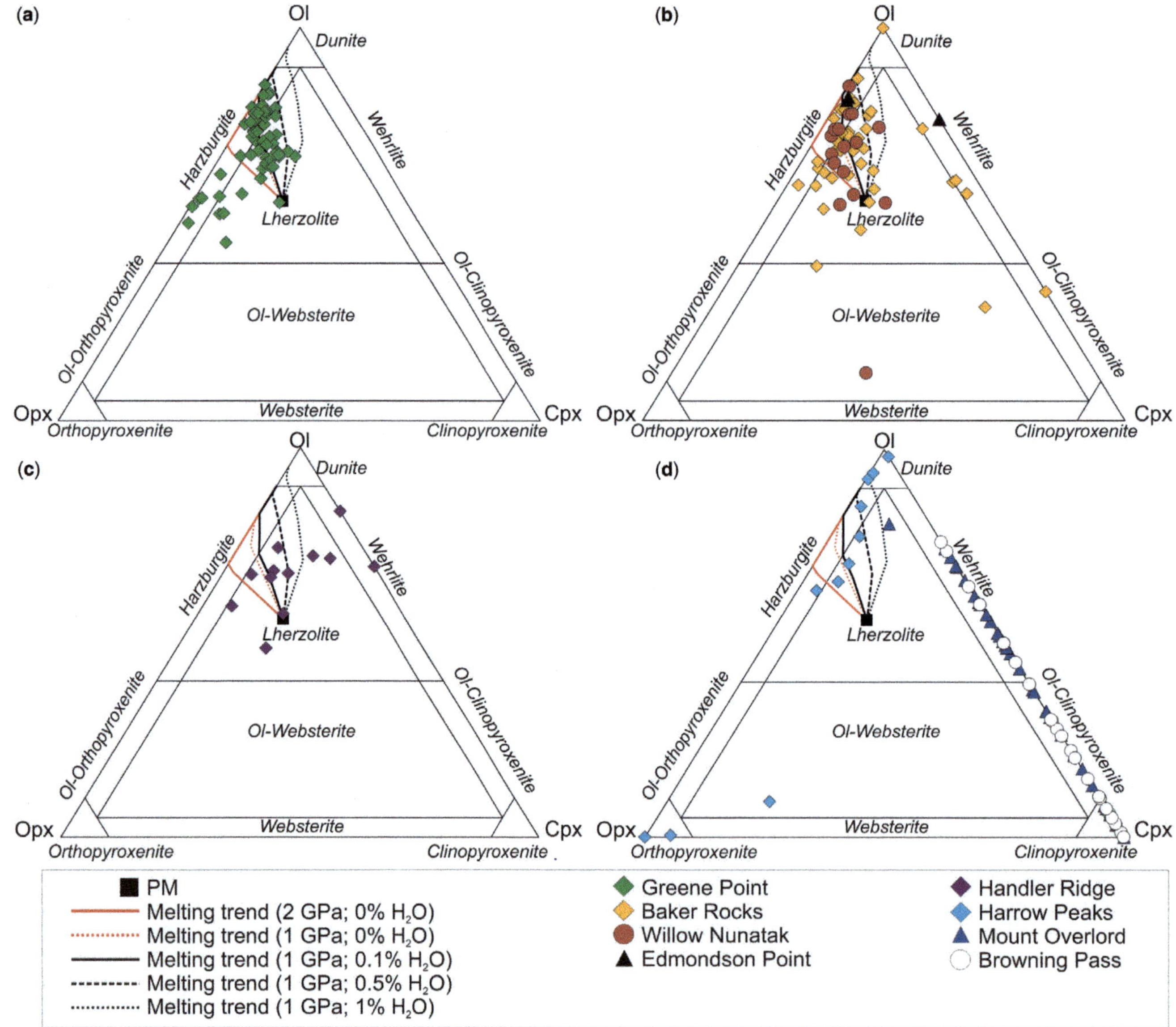

Fig. 2. Ternary diagrams showing the composition of ultramafic xenoliths: (**a**) Greene Point; (**b**) Mount Melbourne (Baker Rocks, Willow Nunatak (including Willow Nunatak Beach) and Edmondson Point); (**c**) Handler Ridge; and (**d**) Harrow Peaks, Mount Overlord and Browning Pass. The primitive mantle (PM) composition is from Johnson *et al.* (1990). Melting trends at 1 GPa and 0–1 wt% H_2O, and at 2 GPa and 0 wt% H_2O (from Niu 1997 and Bénard *et al.* 2018) are also shown for comparison. Cpx, clinopyroxene; Ol, olivine; Opx, orthopyroxene.

and five amphibole-bearing clinopyroxenites. In addition to the ultramafic rocks, amphibole megacrysts were found in the volcanic deposits of Willow Nunatak Beach and Mount Overlord.

Greene Point mantle xenoliths

Greene Point harzburgites and lherzolites are all characterized by a coarse-grained protogranular texture according to the classification of Mercier and Nicolas (1975), with large orthopyroxene and olivine grains (Fig. 3a; see also fig. 2a–g in Pelorosso *et al.* 2016). The maximum olivine size is 1.2 cm. Orthopyroxene can reach 1 cm in size and is associated with strongly kink-banded olivine. Clinopyroxene is generally smaller than orthopyroxene (average 0.5 mm, maximum 3 mm) but it may reach 1 cm in one case (GP9). The largest clinopyroxene grains sometimes show exsolution lamellae of orthopyroxene (fig. 2h in Pelorosso *et al.* 2016). Spinel is the only aluminous phase and has vermicular or, less commonly, lobate shapes, and is always associated with pyroxenes (Fig. 3a; see also fig. 2a–g in Pelorosso *et al.* 2016). The primary paragenesis shows no evidence of phase destabilization in most of the samples, although some xenoliths have spongy clinopyroxenes and/or clinopyroxenes with spongy rims (see fig. 2e, f in Pelorosso *et al.* 2016). However, a 'pyrometamorphic' texture (Mercier and Nicolas 1975; Pike and Schwarzman 1977; Coltorti *et al.* 1999; Beccaluva *et al.* 2001) can be observed in almost all nodules at and close to the contact with the host basalt. This includes partial or total destabilization of orthopyroxene and clinopyroxene that recrystallize into tiny, aligned secondary clinopyroxene immersed in a glassy matrix. When the phase is not completely destabilized, the above-described texture grades into a cloudy, spongy surface that constitutes a transition towards the clear part of the grain.

Contacts between xenoliths and host basalts are characterized by large corrosion gulfs or sharp cuts, the latter only involving olivine. It has to be noted that the host basalt is rich in vesicles, often reaching more than 50% of the volume. Basalt infiltration in the form of veinlets running along grain boundaries is rare. Iddingsitization along olivine fractures and rims, as well as oxidation of glass in the reaction zones, can also be observed.

Mount Melbourne ultramafic xenoliths

The Mount Melbourne suite includes xenoliths from Willow Nunatak, Willow Nunatak Beach, Edmondson Point, Baker Rocks and the beach near Baker Rocks (i.e. BR). The great majority of the Mount Melbourne ultramafic xenoliths are typical mantle lithologies, while a minor amount is made up of cumulates.

Mantle rocks include both anhydrous and amphibole (± phlogopite)-bearing samples whose textures fit the categories porphyroclastic, protogranular (both generation I and generation II) and mosaic equigranular described by Mercier and Nicolas (1975) (Fig. 3b, c). In addition, a group whose texture is transitional between porphyroclastic and protogranular (i.e. with different textural domains in the same sample) has also been recognized. Porphyroclastic, porphyroclastic–protogranular and protogranular xenoliths have grain sizes varying from coarse (i.e. maximum porphyroclast size *c.* 9 mm) to medium grained (i.e. maximum porphyroclast size *c.* 4 mm). Some composite samples consisting of a mixture of mantle lithologies and cumulates occur; they are described separately from the 'pure' mantle rocks and cumulates (Fig. 3d, e).

A common trait to all Mount Melbourne samples is the richness in fluid inclusions, mainly present as trails running along the entire width of grains and/or concentrated along the boundaries. In some cases, the fluid inclusion trails have a high concentration and run parallel to each other, resembling veins.

Porphyroclastic mantle samples. The prevalent textural type of mantle rock in the Mount Melbourne area is porphyroclastic (Fig. 3b). Olivine porphyroclasts (maximum sizes 2.3–9 mm) are characterized by mild to strong kink banding that is totally absent in neoblasts (maximum sizes 0.4–1 mm). Orthopyroxene is the largest phase in some samples; its porphyroclasts (maximum sizes 3.8–8.0 mm) show exsolution lamellae and, sometimes, regular cleavage. In some cases, orthopyroxene can be found also as neoblasts (maximum sizes 0.4–1 mm) that are always devoid of exsolution lamellae. Clinopyroxene crystals are always smaller than olivine and orthopyroxene, and are larger in lherzolites (maximum sizes 1.3–4 mm) than in harzburgites (maximum sizes 0.8–1.5 mm). They may have exsolution lamellae but this feature is not ubiquitous, both between the samples and within the same sample. Moreover, in some samples clinopyroxene has a cloudy surface and/or spongy rims. Spinel is the only aluminous phase and is brown to black in colour; its shape varies between lobate and 'holly leaf', and it is generally associated spatially with the pyroxenes. In many of the porphyroclastic samples from Willow Nunatak Beach, the porphyroclasts show signs of deformation and are characterized by straight, sharp borders. Orthopyroxene sometimes forms peculiar indentation rims with olivine neoblasts.

Where present, amphibole occurs as micrometric rims around spinel (Type D1 texture: see fig. 1a, c in Coltorti *et al.* 2004) and/or as complete veins (Type D4 texture: Coltorti *et al.* 2004) cutting or bordering the samples. In two cases, the process of amphibole vein formation can be observed as 'caught in act'. Amphiboles start growing at the expense of clinopyroxenes when they come into contact with a melt, which is still visible as residual glass around the clinopyroxene–amphibole pairs (see fig. 1b, d & e in Coltorti *et al.* 2004). Amphiboles grow aligned to the direction of melt percolation until a complete vein is formed, with amphibole grains growing into contact with each other (Fig. 3c). In areas of incomplete vein formation, abundant glass, together with secondary clinopyroxenes and spinels, surround the amphibole grains.

Host basalt infiltration is a minor issue. In some samples, it is seen as dark, straight glass veins that cut the peridotite matrix; if the vein narrows, the infiltration becomes a dark film intruding between the mineral grains. Orthopyroxene and clinopyroxene react with the host basalt by developing a spongy rim towards the vein, while olivine does not destabilize.

Porphyroclastic–protogranular mantle samples. Many samples are characterized by a coarse-grained texture showing transitional features between porphyroclastic and protogranular. Olivine always reaches the largest dimensions (up to 1 cm) and has clear kink banding. Orthopyroxene is smaller (maximum size 3.8–5.5 mm) but not as small as clinopyroxene (maximum size 1–1.5 mm); only the largest grains of both pyroxenes have exsolution lamellae. In the harzburgites, clinopyroxene occurs mainly as an 'appendix' to orthopyroxene and around spinel. In a few samples, clinopyroxene has a cloudy surface that may also be combined with spongy rims. As for the porphyroclastic group, spinel is the only aluminium phase and generally has a 'holly-leaf' shape; but, in some cases, it forms complex symplectites with orthopyroxene.

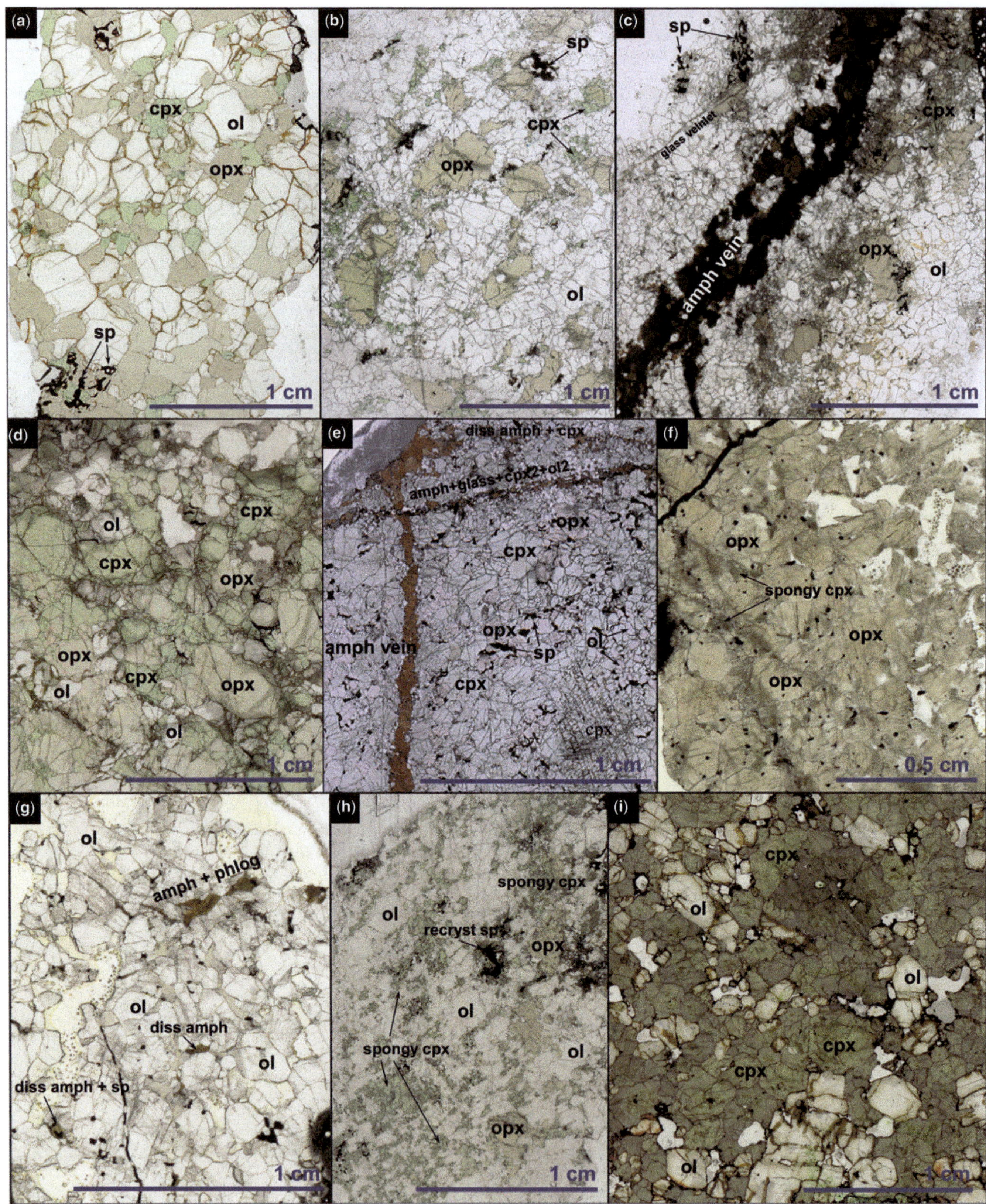

Fig. 3. High-resolution thin-section scans of northern Victoria Land ultramafic xenoliths, showing their main textural types (see the text for further explanation). (**a**) Coarse-grained, anhydrous protogranular lherzolite from Greene Point. (**b**) Anhydrous porphyroclastic lherzolite from Baker Rocks. (**c**) Composite xenolith BR55 from Baker Rocks: anhydrous coarse-grained porphyroclastic lherzolite intruded by a braided hornblendite vein. (**d**) Anhydrous olivine-websterite 265A from Willow Nunatak. (**e**) Composite olivine-websterite 154G from Baker Rocks: anhydrous cumulate cut by later amphibole (± glass and secondary phases) veins; disseminated amphiboles can also be observed in the proximity of the vein crossing point. (**f**) Mesocumulate-textured orthopyroxenite HP163 from Harrow Peaks. Brown coarse-grained orthopyroxene crystals poikilitically enclose euhedral oxides, while spongy clinopyroxenes are the intercumulus phase. (**g**) Mosaic equigranular dunite HP124 from Harrow Peaks. Brown amphibole and phlogopite occur as disseminated grains between olivines or around spinels. (**h**) Medium-grained protogranular lherzolite from Handler Ridge. Primary clinopyroxenes have spongy rims or appear completely spongy and are sometimes associated with glassy patches that contain tiny, idiomorphic secondary phases. Spinels in metasomatic areas recrystallize into clusters of small, idiomorphic grains. (**i**) Coarse-grained olivine-clinopyroxenite from Mount Overlord, with large green clinopyroxene crystals poikilitically enclosing euhedral to subhedral olivine grains. Amph, amphibole; Cpx, clinopyroxene; Ol, olivine; Opx, orthopyroxene; Sp, spinel.

Only two samples with the most pronounced porphyroclastic textures contain hydrous minerals. They are highly deformed xenoliths in which olivine neoblasts are easily recognizable by their small size, lack of kink banding, and their association in clusters around the larger grains of olivine and pyroxenes. Disseminated amphibole (textures D1 and D2: Coltorti *et al.* 2004) is widespread throughout the section and always grows at the expense of clinopyroxene, sometimes with the contribution of spinel. Phlogopite is also present, as small- to medium-sized (up to 2.2 mm) crystals associated with spinel clusters.

Glass veins are observed in a few samples. Secondary idiomorphic and/or subidiomorphic clinopyroxenes, olivines, spinels and rhönites are immersed in glass along these veins and/or around spongy pyroxene rims. They are sometimes clearly related to host basalt infiltration, and associated with acicular plagioclase.

Protogranular and generation-II protogranular mantle samples. Protogranular and generation-II protogranular xenoliths (Mercier and Nicolas 1975) have a similar texture, the difference between the two mainly residing in the spinel shape and its location within the peridotite matrix. Olivine has the largest dimensions (maximum size 2.5–9.0 mm), and it is characterized by curvilinear boundaries, equilibrium triple junctions and kink banding. Orthopyroxene and clinopyroxene are smaller than olivine (maximum sizes 1.3–5.5 and 0.3–4.3 mm, respectively); both generally show cleavage and/or exsolution lamellae. Spinel is the only aluminium phase. It has a lobate shape in the protogranular samples, where it is always associated with the pyroxenes, while it may have idiomorphic or rounded shape in generation-II protogranular rocks. In this latter case, spinel is found between and/or within olivine crystals.

Amphibole is present in two xenoliths. In one case, a glassy phase is also observed throughout the whole sample forming thin veins along or cutting olivine crystals, with no reactions. However, it forms patches when in contact with orthopyroxene and/or clinopyroxene. The pyroxenes have reacted with the melt and may develop spongy rims; the clinopyroxene is generally partially and/or totally transformed into amphibole (Type D1 and D2 textures: see fig. 1a–e in Coltorti *et al.* 2004). Amphibole can be observed in textural continuity with clinopyroxene and/or as crystals immersed in a glassy matrix together with other secondary, idiomorphic phases (clinopyroxene, olivine and spinel).

In the other case, amphiboles forming the veins have an adcumulate texture. In some areas, a messy association of glass and tiny secondary phases fill the veins in place of amphibole. These phases are idiomorphic to subidiomorphic, and include clinopyroxene, olivine, apatite, rhönite and needle-shaped plagioclase. Type D1 texture can be found around spinel close to the veins.

As for the other groups, the host basalt has infiltrated some samples forming dark glass veins rich in needle-shaped plagioclases, occasionally forming glass pools with tiny secondary phases (clinopyroxene and spinel) at the expense of pyroxenes.

Mosaic equigranular group. This texture is rare among Mount Melbourne mantle rocks, including only one sample, an anhydrous medium-grained harzburgite, whose texture is similar to mosaic equigranular (Mercier and Nicolas 1975). Olivine (up to 3 mm) does not show kink banding and grains are generally arranged in equilibrium triple junctions. Orthopyroxene (up to 2 mm) is characterized by very fine exsolution lamellae. The few clinopyroxenes are very small (up to 0.4 mm) and have cloudy surfaces; they can be found associated with orthopyroxene and, to a minor extent, with spinels that have square to roundish shapes and can be found between olivines.

Composite samples. Sample BR53 comprises an anhydrous coarse-grained porphyroclastic harzburgite pervaded by a network of hornblenditic veins with a maximum thickness of 9 mm. In the harzburgitic part, olivine porphyroclasts (up to 1.2 cm) are characterized by a mild kink banding that is not present in the neoblasts (up to 0.8 mm). The neoblast grains often display equilibrium triple junctions and are clustered around the amphibole vein network. Orthopyroxene porphyroclasts (up to 5 mm) are generally clustered together, mostly without exsolution lamellae. Some orthopyroxene neoblasts (up to 0.4 mm) are also present close to the veins. The few clinopyroxenes are small (up to 0.7 mm), and are associated with orthopyroxene and spinel. Spinel is lobate or vermicular in shape and is sometimes surrounded by a thin rim of clinopyroxene. The hornblenditic part has adcumulate texture, made up of large brown-orange amphibole (up to 3 mm), often characterized by remarkable cleavage. The contact between the amphiboles and the harzburgite minerals is sharp. Two veinlets (probably coming from the host basalt) cut both the harzburgite and the hornblendites, establishing a clear temporal relationship. The veinlets cross each other at an angle of about 50° and are composed of glass plus tiny secondary phases (clinopyroxene and phlogopite). Where these veinlets meet minerals, clinopyroxene and amphibole have reacted to develop a spongy rim, while olivine is unaffected.

Sample BR55 is an anhydrous coarse-grained porphyroclastic lherzolite intruded by a network of hornblendite veins with a maximum thickness of 3.5 mm (Fig. 3c). In the lherzolitic part, olivine porphyroclasts (up to 4 mm) are characterized by a mild kink banding that is not present in the neoblasts (up to 1 mm), which often display equilibrium triple junctions. Orthopyroxene porphyroclasts (up to 5 mm) are generally clean, without exsolution lamellae, and some orthopyroxene neoblasts (up to 0.3 mm) are also present. Clinopyroxene (up to 1.5 mm) displays exsolution lamellae more frequently than orthopyroxene, although this is not a ubiquitous feature and has no apparent relation with distance from the amphibole veins. Spinel shows vermicular to holly-leaf shapes, especially if associated with pyroxenes. The hornblenditic part is similar to that of BR53.

Sample BR57 is made up of an anhydrous coarse-grained porphyroclastic lherzolite and an amphibole-bearing olivine-clinopyroxenite. In the lherzolitic part, olivine porphyroclasts (up to 5 mm) are characterized by a mild kink banding that is not present in the neoblasts (up to 1 mm). The neoblasts often display equilibrium triple junctions and are more abundant at the contact with the olivine-clinopyroxenite. Orthopyroxene neoblasts (up to 0.5 mm) are also present. Orthopyroxene porphyroclasts are smaller than olivine (up to 1.4 mm) and generally clean, without exsolution lamellae. Clinopyroxene (up to 1.5 mm) is found in clusters, and, contrary to orthopyroxene, it shows cleavage and exsolution lamellae. Spinel is black and present in small elongated grains. The olivine-clinopyroxenite has an adcumulate texture composed of large clinopyroxene and amphibole crystals (up to 5 mm), as well as by a few olivine crystals. A thin glass veinlet cuts the peridotite matrix without causing major phase destabilization, whereas it destabilizes clinopyroxene and amphibole in the olivine-clinopyroxenite.

Samples BR214 and BR219 are made up of a lherzolite/harzburgite that adheres to a poikilitic wehrlite. The contact between the two lithotypes is quite planar but it is characterized by an area about 1–2 mm wide in which the boundaries of both parts display reaction features and pale, yellow glass. The peridotite parts have a mosaic texture with grains often meeting at equilibrium triple junctions. All silicate phases have a similar size range (0.5–1 mm), while spinel is anhedral and, in general, interstitial between pyroxenes. The wehrlite portions have a cumulate texture, and consist of large poikilitic crystals of clinopyroxene (up to 3.5 mm) enclosing subhedral to anhedral olivine and orthopyroxene.

Sample BR218 consists of an anhydrous coarse-grained lherzolite cut by a 2 mm-wide amphibole vein. The contact is mostly sharp but blurs into a reaction zone located *c.* 1 cm from the edge of the xenolith. In the vein, amphibole occurs as small, brown-orange-coloured euhedral grains (≤0.1 mm in size), often with cleavage. Occasionally, amphiboles are surrounded by a pale yellow glass with microcrystals of clinopyroxene + olivine ± plagioclase. The lherzolitic portion of the xenolith has a porphyroclastic–mosaic type texture. Olivine porphyroclasts (up to 5 mm in diameter) are characterized by mild kink banding that is not present in the neoblasts (up to 1 mm). Orthopyroxene porphyroclasts (up to 4.5 mm) occasionally show thin exsolution lamellae mainly concentrated in the inner part of the grain. Some orthopyroxene neoblasts (up to 0.2 mm) are also present. Clinopyroxenes (≤1.5 mm) are generally clean, without exsolution lamellae. Spinel (up to 2 mm in size) is anhedral, frequently located in the interstices between olivine and orthopyroxene.

Cumulate samples. Sample 265A is an olivine-websterite with a coarse-grained mesocumulitic texture, although partially hindered by high-pressure recrystallization (Fig. 3d). Large (up to 6 mm) orthopyroxene and clinopyroxene co-crystallized as cumulus phases, while olivine (up to 1.5 mm) and spinel occupy the intercumulus areas. No cleavage or exsolution lamellae are present in pyroxenes, and olivine is not kink banded.

Sample 154G is a coarse-grained olivine-websterite intruded by two cross-cutting orthogonal amphibole veins (Fig. 3e). The original texture of the olivine-websterite was probably mesocumulitic but it partially recrystallized and shifted towards a mosaic equigranular texture. Cumulus phases were probably olivine, spinel and orthopyroxene, with clinopyroxene as the intercumulus phase. Olivine (up to 1 mm) is kink banded, and only the largest clinopyroxene crystals (5 mm) have wide exsolution lamellae. Orthopyroxene reaches a maximum size of 1.5 mm. Moreover, the pyroxenes are rich in fluid inclusions. One of the amphibole veins, composed of tabular orange-brown amphibole grains, cuts through the olivine-websteritic matrix; in places the amphibole crystals show small gulfs filled with glass and tiny secondary clinopyroxene and spinel. The second vein has irregular rims, with many amphibole crystals growing in textural continuity with matrix clinopyroxene. The vein is also rich in glass, hosting small secondary clinopyroxene and needle-shaped plagioclase crystals. Rare amphibole grains can be found disseminated within the olivine-websteritic matrix, especially in the proximity of the veins.

Sample 113A is an anhydrous olivine-websterite with a medium-grained mosaic equigranular texture. Orthopyroxene (up to 1.5 mm) is the most abundant phase, and has evident cleavage and exsolution lamellae. Clinopyroxene is slightly smaller (up to 1.3 mm) and is characterized by exsolution lamellae. Olivine is smaller (up to 0.8 mm) than pyroxenes and has no kink banding. Spinel has rounded or slightly elongated shape.

Sample 231 is a wehrlite with a medium-grained porphyroclastic texture. Olivine is strongly kink banded, and is the only phase occurring both as porphyroclasts (up to 4.3 mm) and as neoblasts (<0.5 mm). Clinopyroxene is completely spongy, making the distinction between single grains very difficult. Corrosion gulfs with the host basalt are common, as the sample is small.

The phase assemblage of samples BR215 and BR217 consists of olivine, clinopyroxene, minor amounts of orthopyroxene (<5 vol%) and rare plagioclase (≤1 vol%). The texture is mesocumulitic, in which large oikocrysts of clinopyroxene (up to 4 mm) enclose rounded or subhedral olivine (≤1.2 mm) and anhedral orthopyroxene (≤0.5 mm). Plagioclase occurs as an intercumulus phase. Pale yellow glass crystallizing fine-grained olivine, clinopyroxene and plagioclase occurs in between olivine crystals or in veins.

Handler Ridge ultramafic xenoliths

Protogranular mantle samples. All mantle rock samples, generally small in size (see fig. 2 in Pelorosso *et al.* 2017), are medium- to coarse-grained anhydrous protogranular lherzolites (see fig. 3a–e & h in Pelorosso *et al.* 2017), with highly variable amounts of clinopyroxene. The largest olivine and orthopyroxene grains reach 5 mm in size and are characterized by curvilinear boundaries; clinopyroxene is always smaller (up to 2 mm) than orthopyroxene. Kink banding is common in olivine. Spinel is dark, has lobate shapes, and forms aggregates of smaller, elongated grains generally associated with orthopyroxene and/or clinopyroxene. This xenolith group is texturally well equilibrated, except for two samples in which metasomatic reactions are common (Fig. 3h). Primary clinopyroxene has spongy rims or appears completely spongy, associated with glassy patches containing tiny idiomorphic secondary clinopyroxenes, spinels and olivines. In metasomatic reaction zones, probably independent of host basalt infiltration, spinel occurs as small square grains.

Fluid inclusions are quite common, especially as trails or clusters inside clinopyroxene grains, whereas host basalt infiltration is rare in these xenoliths and visible only as thin, dark glass veins cutting the peridotite matrix.

Cumulate samples. Among Handler Ridge ultramafic xenoliths, two wehrlites with a mesocumulitic texture were found. The cumulus phase is medium-sized olivine (up to 3 mm), sometimes subidiomorphic. Where no intercumulus phases are present, olivine assumes a rather equigranular structure with equilibrium triple junctions (see fig. 3f, g in Pelorosso *et al.* 2017). No kink banding is seen. Clinopyroxene and spinel are intercumulus phases, with spinel often enclosed in clinopyroxene. This latter often has spongy rims or is entirely spongy (see fig. 3f, g in Pelorosso *et al.* 2017), grading towards rims with tiny idiomorphic oxides. Fluid and melt inclusions are common, often arranged in trails crossing the main phases.

Harrow Peaks ultramafic xenoliths

Cumulate samples. The ultramafic cumulates include an orthopyroxenite, three lherzolites, a harzburgite, a dunite and an olivine-websterite.

Sample HP163 is a unicum among all ultramafic xenoliths of northern Victoria Land. It is a medium- to coarse-grained, clinopyroxene-bearing orthopyroxenite with a mesocumulate texture (Fig. 3f). Brownish orthopyroxene (up to 4 mm) is the cumulus phase, with a polygonal shape and equilibrium triple junctions (see fig. 3f in Pelorosso *et al.* 2019), whereas clinopyroxene and olivine constitute the intercumulus assemblage. Orthopyroxene has a remarkable cleavage and is rich in fluid inclusions. Clinopyroxene has a sieve-like appearance due to the presence of submicrometric–micrometric oxide and fluid inclusions.

Most of the xenoliths have mosaic equigranular textures, with grain sizes ranging from 0.8 to 2 mm. Olivine grains, devoid of kink banding, are polygonal, with boundaries converging at 120°. Orthopyroxene and clinopyroxene have no exsolution lamellae but show remarkable cleavage, especially in lherzolites and olivine-websterites. Spinel is either lobate or subidiomorphic, occurring at the junctions between olivines. Amphibole is large (up to 1 mm) only in the dunite (Fig. 3g), where it occurs as tabular disseminated grains between olivines or around spinel (see fig. 3b, d in Pelorosso *et al.* 2019); in one of the lherzolites and in the harzburgite it was found as micrometric traces in reaction areas around clinopyroxene. Pyrometamorphic textures (i.e. glassy patches and sieved pyroxenes or with spongy rims) are rare in this group of ultramafic xenoliths, and consist of thin glassy menisci between minerals (including disseminated amphibole) or small pools (<100 µm) containing microlites of olivine, spinel and clinopyroxene. Host basalt infiltration is also rare.

Two lherzolites have medium-grained protogranular texture, where the silicate phases generally have curvilinear boundaries (see fig. 3e in Pelorosso *et al.* 2019), and spinel has a lobate shape and is associated with orthopyroxene and clinopyroxene. Olivine (up to 2.8 mm) shows mild kink banding, while orthopyroxene (up to 2.4 mm) and clinopyroxene (up to 0.8 mm) have no exsolution lamellae. Where present, amphibole was found only as micrometric spots developed as a clinopyroxene reaction.

Composite samples. Sample HP121 is a pyroxene–amphibole-bearing dunite cut by a hornblendite vein. The dunite has a mosaic equigranular texture, with polygonal, not kink-banded, olivine grains (1–2.5 mm in size). Orthopyroxene has grain size similar to olivine, while clinopyroxene is smaller and completely spongy. Amphibole occurs as tabular crystals (up to 0.5 mm) between olivine and around spinel. The hornblendite vein has an adcumulate texture with tabular brown amphibole (up to 1.25 mm) in sharp contact with olivines of the host dunite.

Sample HP144 is a mosaic equigranular-textured harzburgite, with average olivine and orthopyroxene grain sizes of 2.3 mm. Olivine shows mild kink banding; orthopyroxene has no exsolution lamellae and often has spongy rims, made up of submicrometric clinopyroxene crystals. Clinopyroxene (up to 0.8 mm) is completely spongy. Amphibole occurs around idiomorphic spinel, or as small tabular crystals between olivines. A large (up to 2 mm) adcumulate amphibole-bearing orthopyroxenite vein cuts the host harzburgite. Orthopyroxenes in the vein are fibrous with spongy rims towards the harzburgite that are mainly composed of secondary clinopyroxene.

Mount Overlord ultramafic xenoliths

Cumulate samples. This group comprises subrounded xenoliths, 5–10 cm in size; contacts between the xenoliths and host alkali basalts are sharp or irregular (Perinelli *et al.* 2017). The main minerals are olivine, clinopyroxene, amphibole and spinel; rare but negligible amounts of plagioclase also occur. They have a variable modal composition that corresponds to anhydrous and amphibole-bearing wehrlites, olivine-clinopyroxenites, clinopyroxenites and hornblendites with mesocumulate to adcumulate textures (Fig. 3i; see also fig. 2a–c in Perinelli *et al.* 2017). In some samples, recrystallization and re-equilibration in the solid state has replaced the primary igneous texture with a mosaic texture (see fig. 2d in Perinelli *et al.* 2017).

In wehrlites, olivine-clinopyroxenites and clinopyroxenites, olivine (up to 4.5 mm) is the first cumulus phase and is often euhedral to subhedral (see fig. 2e in Perinelli *et al.* 2017), fractured and devoid of kink banding. Clinopyroxene represents the main intercumulus phase occurring frequently as large (up to 1 cm) poikilitic crystals that enclose subhedral to anhedral olivine. These clinopyroxenes sometimes show local disequilibrium textures (i.e. a 'spongy', cloudy or opaque appearance) or orthopyroxene exsolution lamellae (up to 2 mm in thickness). Spinel commonly forms euhedral grains located either among the cumulus phases or included within olivine or clinopyroxene (see fig. 2b, c & e in Perinelli *et al.* 2017). Amphibole mainly appears as intercumulus poikilitic grains that surround olivine, clinopyroxene and oxide minerals; otherwise, they occur as interstitial crystals replacing clinopyroxene, often associated with patches of brown glass (see fig. 2f, g in Perinelli *et al.* 2017). Textural evidence indicates that these glass patches, also observed among clinopyroxene and olivine, are due to the infiltration of the host lava. Microcrystals of clinopyroxene + olivine ± spinel ± plagioclase ± rhönite often occur in the glass pockets.

Large poikilitic amphiboles (up to 4 cm in size) characterize the hornblendites. Enclosed in these crystals are subhedral to anhedral clinopyroxenes occasionally affected by spongy texture, along with small amounts of olivine and spinel (see fig. 2h in Perinelli *et al.* 2017). Amphibole megacrysts show well-defined edges without corrosions zones; suggesting equilibrium with the host lava.

Composite sample. The composite, amphibole-bearing lherzolite/olivine-clinopyroxenite sample 189A represents a unicum in the Overlord nodule collection. It is a mixture of lherzolite and olivine-clinopyroxenite; the lherzolitic part has porphyroclastic texture, while the olivine-clinopyroxenitic part has a mesocumulitic–cumulitic texture that is partly recrystallized. The contact between the two lithologies is marked by bands of tiny recrystallized olivine (up to 0.1 mm), indicating that the rock underwent strain.

In the lherzolite, olivine porphyroclasts have grains up to 3.8 mm that are characterized by strong kink banding. The neoblasts are very small (up to 0.3 mm). Clinopyroxene may form large crystals (up to 2.8 mm), is rich in fluid and spinel inclusions, and shows diffuse amphibolitization. Orthopyroxenes are smaller than clinopyroxene (up to 1 mm) and have evident cleavage; they may also be found as blebs within clinopyroxene. Spinel is almost completely recrystallized into a cluster of small rounded individuals with lobate shapes.

The olivine-clinopyroxenite is formed by clinopyroxene grains (up to 4 mm) arranged with equilibrium triple junctions and largely amphibolitized. They have exsolution lamellae and abundant fluid inclusions.

The contact between the xenolith and the host basalt is marked by the growth of secondary clinopyroxene, olivine and amphibole (lherzolites) or by the corrosion of orthopyroxene crystals, together with the growth of plagioclase and clinopyroxene (olivine-clinopyroxenite).

Browning Pass ultramafic xenoliths

Browning Pass ultramafic xenoliths are mostly mesocumulate–adcumulate wehrlites, olivine-clinopyroxenites and clinopyroxenites, 5–20 cm in diameter (Perinelli *et al.* 2011). Clinopyroxene and olivine are the predominant minerals, associated with variable amounts of spinel. Coarse olivine has grain size varying from 1 to 4 mm, while clinopyroxene occasionally reaches 10 mm in size, sometimes poikilitically enclosing olivine. Both coarse and poikilitic clinopyroxene sometimes develops secondary features, represented by narrow 'spongy' or cloudy rims occasionally accompanied by amphibole replacement (see fig. 2b in Perinelli *et al.* 2011). Spinel usually appears as interstitial, dark brown or opaque anhedral grains, except in BRP19 where it is associated with olivine as inclusions. Rims of fine-grained ilmenite needles often rim 'primary' spinel, and very scarce accessory plagioclase occurs as an interstitial phase. Diffuse infiltration of host basalt results in dark brown or pale yellow intergranular glass patches containing microcrystalline assemblages of clinopyroxene + olivine ± spinel ± plagioclase ± rhönite (see fig. 2c, d in Perinelli *et al.* 2011). The occurrence of vesicles in these glasses indicates exsolution of a gas phase.

Mineral chemistry of mantle xenoliths

In the following section, a summary of the main compositional features of mantle xenoliths from GP, BR and HR is reported. Further details are given in Supplementary Tables 1, 2 and 3, and in the references therein.

Major elements

Olivine. In GP xenoliths, the forsterite (Fo) content of primary olivine (Ol1) ranges between 90.6 and 92.7 (harzburgites have Fo contents between 91.9 and 92.3: Fig. 4). NiO content ranges between 0.33 and 0.43 wt%, while CaO is always below 0.15 wt%, irrespective of lithology. A few Ol1 rims have CaO values up to 0.27 wt%. Secondary olivine (Ol2) tends to have systematically lower Fo and NiO contents (down to 87.6 and 0.16, respectively), and higher CaO (up to 0.27 wt%) contents.

Willow Nunatak Beach lherzolite is characterized by Fo_{91} olivine, whereas olivine in amphibole-bearing lherzolite 278C has a Fo content between 85.7 and 86.7 (Fig. 4). NiO contents vary from 0.30 to 0.37 wt% in the amphibole-bearing lherzolite, and are 0.41 wt% in the lherzolite. CaO contents are always below 0.07 wt%.

In BR xenoliths, the Fo content of Ol1 in lherzolites varies from 88.0 to 91.9, almost encompassing the whole range of harzburgites and amphibole-bearing lherzolites. The formers have a Fo content of 90.5–91.7, while the latter span the interval between 87.7 and 91.9 (Fig. 4). Both lherzolites and harzburgites are shifted towards a lower Fo content with respect to GP. The NiO content is between 0.28 and 0.54 wt %, while the CaO content is always below 0.13 wt%, irrespective of lithology. A few higher CaO values (up to 0.20 wt%) are found in Ol1 near amphibole veins and/or in glassy patches. Ol2 systematically shows higher CaO values (up to 0.35 wt%), with lower Fo content (88.1) near or within amphibole veins or glassy patches. Several Ol2 with CaO contents between 0.15 and 0.25 wt% present a very high Fo content (up to 92.7), and tend to have NiO down to 0.06 wt%, irrespective of lithology.

In HR lherzolite xenoliths, the Fo content of Ol1 has the lowest value (down to 87.5) and goes up to 91.0, comparable with lherzolites and amphibole-bearing lherzolites from BR (Fig. 4). The NiO content varies from 0.28 to 0.46 wt%, and the CaO content is below 0.20 wt%.

Orthopyroxene. Orthopyroxene in GP lherzolites has a MgO content between 32.5 and 34.4 wt%, and FeO^T ranging from 5.0 to 5.8 wt%. The two lherzolites GP9 and GP13 have the lowest MgO and the highest FeO^T contents. Orthopyroxenes in harzburgites reach slightly higher MgO contents (up to 35.4 wt%: Fig. 5), while FeO^T shows a restricted range (5.0–5.4 wt%). The range of Mg# ((MgO/MgO + FeO^T) mol%) in lherzolites (91.0–92.5) encompasses that in the harzburgites (92.0–92.6). The Al_2O_3 content in orthopyroxene of GP lherzolites and harzburgites cover a similar range, comprising between 2.3 and 4.7 wt% (Fig. 5). Irrespective of lithology, three groups can be distinguished in terms of Al_2O_3 and Mg#. The one with the lowest Mg# consists of lherzolites GP9 and GP13. A second group comprises the large majority of the samples and is typified by intermediate Mg# values, and is represented by lherzolites GP25 and GP84, and harzburgites GP23, GP78 and GP81. A third group has the highest Mg# values, and is made up of lherzolites GP28 and GP30, and harzburgite CD305. Cr_2O_3 is directly correlated with MgO (or Mg#), with its content spanning the range 0.4–1.0 wt%.

Orthopyroxene in Willow Nunatak Beach amphibole-bearing lherzolite has MgO contents between 33.0 and 33.9 wt%, with FeO^T ranging between 6.9 and 7.6 wt% (Mg# = 88.7–89.7); the Al_2O_3 contents range from 1.5 to

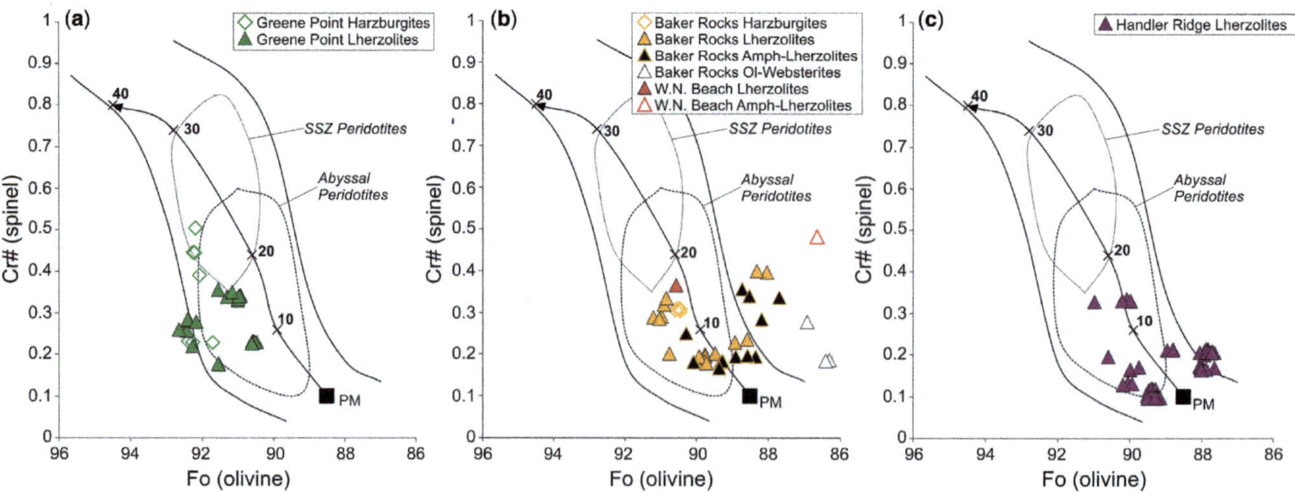

Fig. 4. Olivine–spinel mantle array (OSMA) diagram (Arai 1994) showing the composition of olivine and spinel in the studied mantle xenoliths. (**a**) Greene Point harzburgites and lherzolites. (**b**) Baker Rocks harzburgites, lherzolites, amphibole-bearing lherzolites and olivine-websterites; Willow Nunatak (W.N.) Beach lherzolites and amphibole-bearing lherzolites. (**c**) Handler Ridge lherzolites. Primitive mantle (PM) composition and melting trend are from Arai (1994). The compositional fields of the Supra-Subduction Zone (SSZ: Pearce *et al.* 2000) and Abyssal Peridotites (Dick and Bullen 1984) are also shown for comparison. Amph, amphibole; Ol, olivine.

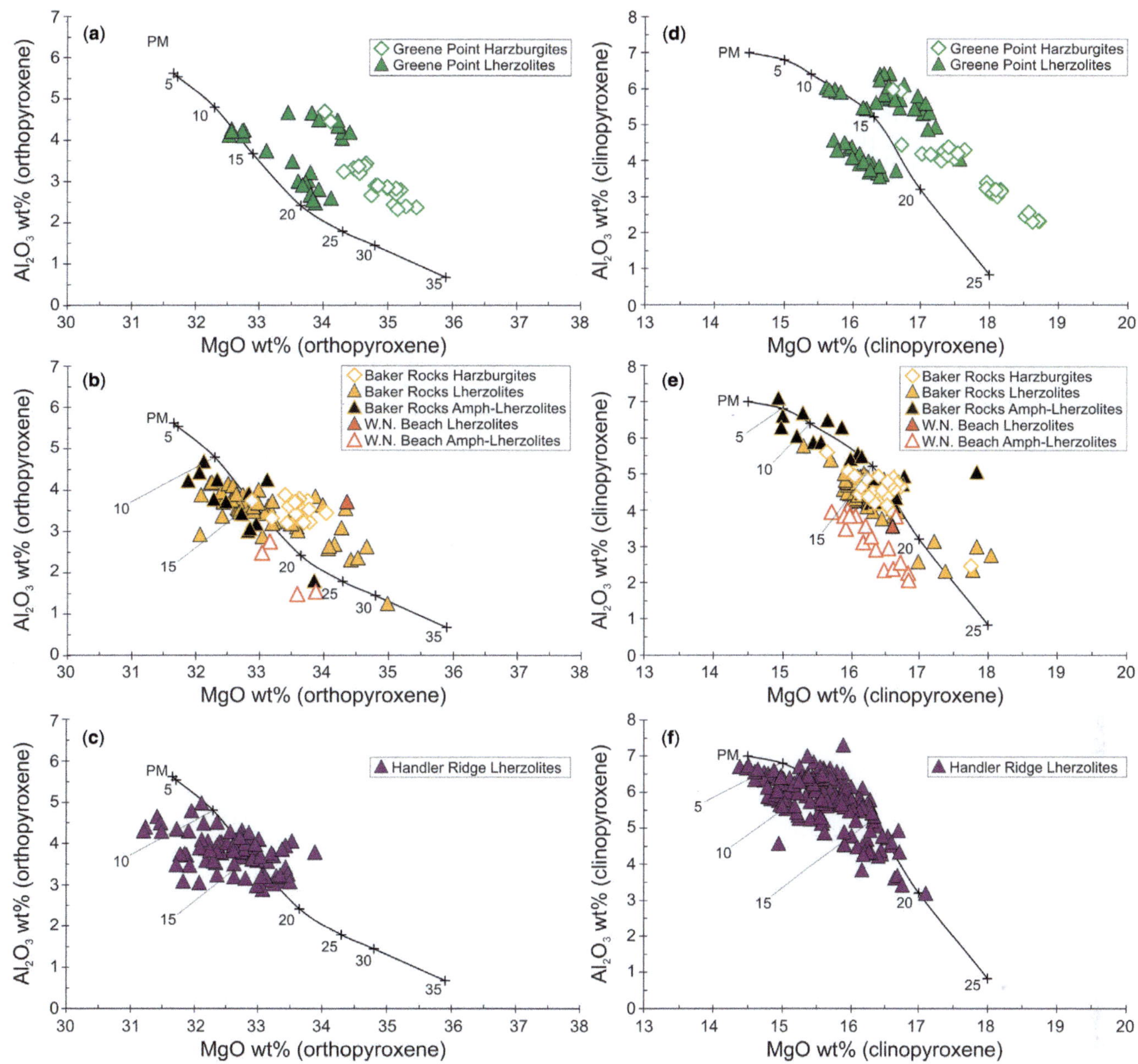

Fig. 5. Al_2O_3 (wt%) v. MgO (wt%) in orthopyroxene and clinopyroxene in the northern Victoria Land mantle xenoliths. (**a**) Orthopyroxene from Greene Point harzburgites and lherzolites. (**b**) Orthopyroxene from Baker Rocks harzburgites, lherzolites, amphibole-bearing lherzolites and olivine-websterites; Willow Nunatak (W.N.) Beach lherzolites and amphibole-bearing lherzolites. (**c**) Orthopyroxene from Handler Ridge lherzolites. (**d**) Clinopyroxene from Greene Point harzburgites and lherzolites. (**e**) Clinopyroxene from Baker Rocks harzburgites, lherzolites, amphibole-bearing lherzolites and olivine-websterites; Willow Nunatak Beach lherzolites and amphibole-bearing lherzolites. (**f**) Clinopyroxene from Handler Ridge lherzolites. The melting curves reported in (a)–(f) refer to the melting model of Bonadiman and Coltorti (2011) and Upton *et al.* (2011), developed from a starting primitive mantle (PM) composition (Sun and McDonough 1989). Numbers along the curves correspond to the partial melting percentages. Amph, amphibole.

2.8 wt% (Fig. 5). Orthopyroxene in lherzolite sample has 3.7 wt% Al_2O_3, 34.4 wt% MgO and 5.5 wt% FeO, corresponding to a Mg# of 91.7.

Orthopyroxene in BR lherzolites has MgO contents between 32.1 and 34.7 wt%, almost encompassing the whole range of harzburgites (32.9–34.0 wt%) and amphibole-bearing lherzolites (31.9–33.1 wt%; Fig. 5). Orthopyroxene in harzburgites shows the lowest FeO^T contents, between 5.0 and 6.1 wt%. In lherzolites and amphibole-bearing lherzolites, the FeO^T content varies from 5.3 to 7.0 and from 6.1 to 7.5 wt%, respectively. As a consequence, the Mg# for orthopyroxene in harzburgites has the highest values (90.8–92.3), while the Mg# of orthopyroxene in lherzolites and amphibole-bearing lherzolites ranges from 89.1 to 91.9 and from 88.4 to 90.7, respectively. The Al_2O_3 content in orthopyroxene from BR lherzolites is between 2.3 and 4.2 wt%, overlapping the interval of orthopyroxene in harzburgites (3.2–3.9 wt%). The Al_2O_3 content in orthopyroxene from amphibole-bearing lherzolites varies from 3.0 to 4.7 wt% (Fig. 5). Irrespective of lithology, three groups can be distinguished on the basis of the Mg#. The first group has the lowest Mg#, and includes the amphibole-bearing lherzolites 154B and 154H, and the lherzolites BR4 and BR36. A second group has intermediate Mg# and includes the large majority of the samples (lherzolites 154E, BR1, BR29 and BR34, and harzburgite BR42). The third group has the highest Mg# and includes only harzburgite BR43. The Cr_2O_3 content is directly correlated with MgO (or Mg#), varying between 0.2 and 0.8 wt%.

The MgO and FeO^T content in orthopyroxene from HR xenoliths ranges from 31.2 to 33.9 (Fig. 5) and from 5.9 to

7.5 wt%, respectively, resulting in an Mg# range of 88.3–90.9. The Al_2O_3 content ranges between 2.9 and 5.0 wt% (Fig. 5). In this locality, two groups may be distinguished based on the Al_2O_3 v. Mg# relationships: one group has a lower Mg# (lherzolites HR2, HR10 and HR10A) and the other has a higher Mg# (lherzolites HR1, HR3, HR4, HR5, HR6 and HR9). The Cr_2O_3 varies from 0.2 to 0.8 wt%, being mostly inversely correlated with MgO or Mg#.

Clinopyroxene. The MgO content of primary clinopyroxene (Cpx1) from GP lherzolites ranges between 15.7 and 17.6 wt%, while in harzburgites it reaches higher values (16.7–18.7 wt%: Fig. 5). The FeO^T content lies in a very restricted range, between 2.1 and 2.6 wt%, irrespective of lithology. As a result, the Mg# varies from 92.4 to 93.9 and from 91.5 to 93.4 in harzburgites and lherzolites, respectively. Clinopyroxene from harzburgites has the lowest Al_2O_3 content (2.3–4.4 wt%, except two spots at 5.9 wt%: Fig. 5). The Al_2O_3 content in clinopyroxene from lherzolites ranges from 3.6 wt% in sample GP9 up to 6.4 wt% in sample GP25. As for orthopyroxene, clinopyroxene from lherzolites GP9 and GP13 presents the lowest MgO content at a comparable Al_2O_3 content (Fig. 5). The Cr_2O_3 content varies between 0.9 and 1.9 wt%, irrespective of lithology. Secondary clinopyroxene (Cpx2) is characterized by a marked decrease in Mg# (down to 88.2) and Al_2O_3 content (down to 0.4 wt%), accompanied by a concomitant increase in SiO_2 content, especially in crystals growing around orthopyroxene in harzburgites.

Cpx1 from Willow Nunatak Beach amphibole-bearing lherzolite has Al_2O_3 contents between 2.1 and 4.0 wt%, MgO contents of 16.8–15.7 wt% (Fig. 5), and FeO^T content ranging between 2.4 and 3.0 wt% (Mg# = 90.5–92.4) (Fig. 6).

Among all the BR xenoliths, Cpx1 from lherzolites 154E has the highest MgO content (18.0 wt%). The lower limit for the MgO content in lherzolites is 15.3 wt%, while Cpx1 in amphibole-bearing lherzolites has even lower values (down to 14.9 wt%). Cpx1 in harzburgites falls in the range 15.6–17.8 wt% (Fig. 5). The FeO^T content varies from 1.6 to 3.8 wt%, with Cpx1 from amphibole-bearing lherzolites presenting the highest values and Cpx1 from the lherzolite 154E the lowest. As a result, the Mg# of Cpx1 from harzburgites, lherzolites and amphibole-bearing lherzolites varies from 91.5 to 93.7, from 89.6 to 95.2 and from 89.4 to 92.6, respectively. The Al_2O_3 content in Cpx1 from lherzolites is between 2.3 and 5.8 wt%, whereas a narrower range characterizes Cpx1 from harzburgites (2.5–5.6 wt%). Cpx1 in amphibole-bearing lherzolites is characterized by the highest Al_2O_3 content (up to 7.1 wt%: Fig. 5). The Cr_2O_3 content varies between 0.3 and 2.1 wt%, irrespective of lithology. Cpx2 in BR xenoliths was described in detail by Coltorti *et al.* (2004), who subdivided them into:

(a) Cpx2 transforming into amphibole (Cpx-A), characterized by a large increase in Al_2O_3 content (up to almost 12 wt%) at constant Mg#; and
(b) Cpx2 growing around orthopyroxene (Cpx2-O), with a very high Mg# (up to 95.2) and a Al_2O_3 content as low as 1.39 wt%.

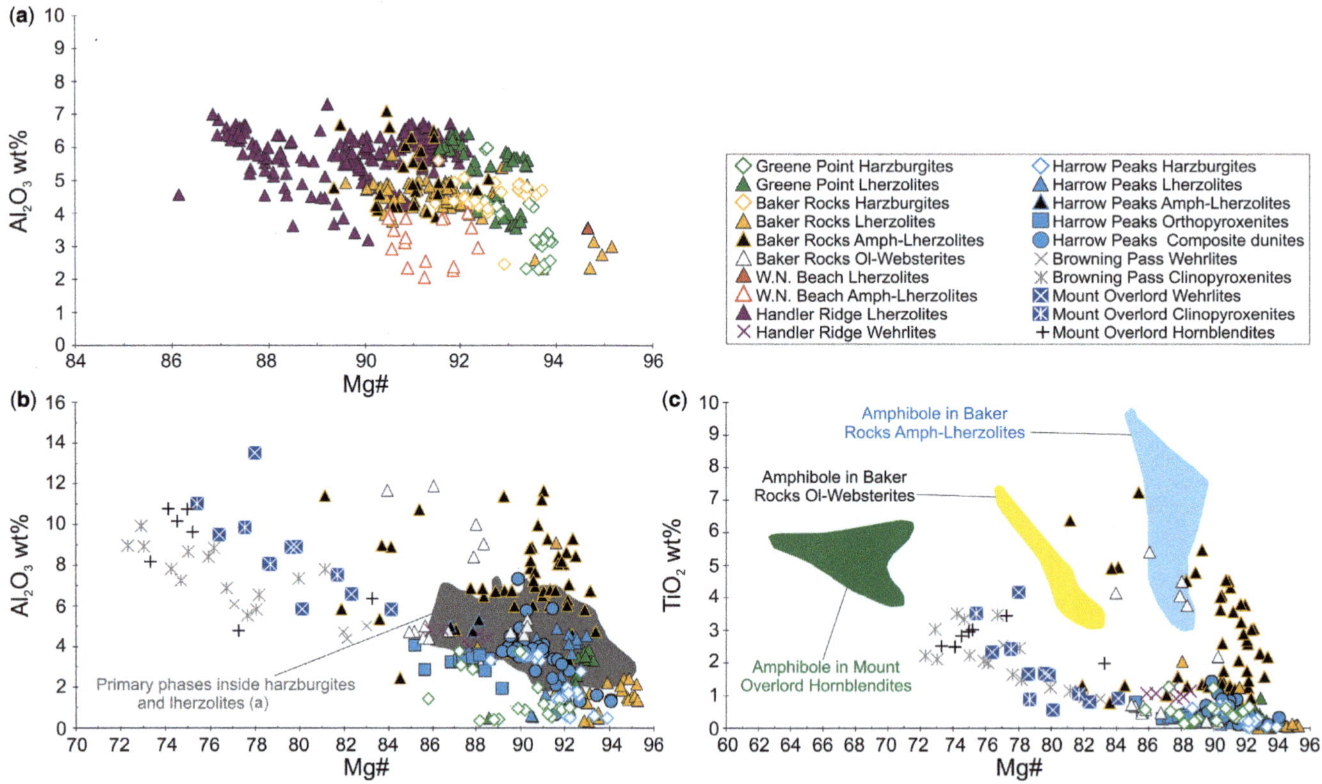

Fig. 6. Composition of clinopyroxene in the northern Victoria Land mantle and cumulate xenoliths. (**a**) Al_2O_3 (wt%) v. Mg# of primary clinopyroxenes (Cpx1) in Greene Point, Baker Rocks, Willow Nunatak (W.N.) Beach and Handler Ridge harzburgites, lherzolites and amphibole-bearing lherzolites. (**b**) Al_2O_3 (wt%) v. Mg# of secondary clinopyroxenes (Cpx2) in Greene Point, Baker Rocks, Willow Nunatak Beach and Handler Ridge harzburgites, lherzolites and amphibole-bearing lherzolites plotted together with clinopyroxene in Baker Rocks olivine-websterite, Handler Ridge wehrlites and cumulate rocks from Harrow Peaks, Browning Pass and Mount Overlord. (**c**) TiO_2 (wt%) v. Mg# of secondary clinopyroxenes (Cpx2) in Greene Point, Baker Rocks, Willow Nunatak Beach and Handler Ridge harzburgites, lherzolites and amphibole-bearing lherzolites plotted together with clinopyroxene in Baker Rocks olivine-websterite, Handler Ridge wehrlites and cumulate rocks from Harrow Peaks, Browning Pass and Mount Overlord. In (c), compositional fields showing the TiO_2 (wt%) v. Mg# distribution in amphibole from Baker Rocks amphibole lherzolites, olivine-websterites and Mount Overlord hornblendites are also reported. Amph, amphibole; Ol, olivine.

Both Cpx1 and Cpx2, when close to the host basalt, are characterized by a remarkable increase in FeO^T (i.e. lower Mg#) and TiO_2.

The MgO content in clinopyroxenes from HR, where only lherzolites are found, have the lowest values, ranging between 14.4 and 17.1 wt% (Fig. 5), while the FeO^T content is in the range 2.2–4.2 wt%. As a result, the Mg# in this clinopyroxene varies from 86.1 to 92.2 (Fig. 6). The Al_2O_3 content is between 3.2 and 7.3 wt%, this variation is almost recognized within a single sample (Fig. 5; Supplementary Table 2). The Cr_2O_3 content is between 0.6 and 1.5 wt%.

Amphibole. Amphibole in mantle xenoliths is documented only in Willow Nunatak Beach and BR samples (see Coltorti *et al.* 2004). In Willow Nunatak Beach lherzolite it occurs as disseminated grains characterized by TiO_2 and Cr_2O_3 contents ranging from 0.4 to 1.3 and from 1.8 to 2.4 wt%, respectively. Its Mg# varies between 87.3 and 90.2.

Disseminated amphibole in BR lherzolites has Mg# ranging from 86.5 to 89.6, at TiO_2 and Cr_2O_3 contents of 2.8–5.1 and 0.7–1.6 wt%, respectively. Amphibole in veins of BR lherzolites has larger TiO_2 and Cr_2O_3 content variations (3.2–8.9 and 0–2.4 wt% respectively) at comparable Mg# (86.1–90.0) with respect to disseminated grains.

Trace element contents in clinopyroxene

The great majority of Cpx1 from GP xenoliths exhibit a 'humped' chondrite-normalized rare earth element (REE) pattern, with Yb_N and La_N varying from 2 to 8 and from 1 to 9, respectively, and Gd_N up to 10 (Fig. 7a). In chondrite-normalized trace element patterns, they have remarkable negative Zr/Hf and Ti, and positive Th–U anomalies (see fig. 5 in Pelorosso *et al.* 2016). Within this group, a subgroup with similar Yb_N values but enriched in light REE (LREE) up to 20 × Ch, thus giving rise to positive fractionated patterns, can be distinguished.

A second group of clinopyroxenes can be identified in GP. This group is characterized by 'classic' LREE-depleted patterns with $(La/Yb)_N = 0.2–0.3$ at Yb_N of around 7 (Fig. 7b). Two samples have spoon-shaped patterns, with Yb_N varying between 2 and 4: to the right of Gd or Tb, their trend is comparable to that of the other clinopyroxenes but it appears enriched in LREE (Fig. 7d).

Among xenolith-bearing localities, BR has by far the largest compositional variety for Cpx1, which would be reflected in the widest heterogeneity in the northern Victoria Land SCLM. In chondrite-normalized REE patterns, Cpx1 in harzburgites vary from moderately depleted ($(La/Yb)_N = 0.3–0.4$) to slightly enriched ($(La/Yb)_N = 2.1–4.6$) at Yb_N between 5 and 9 (Fig. 7b–d). They are characterized by remarkable negative anomalies in both Ti and Zr (see fig. 3 in Coltorti *et al.* 2004). Cpx1 in lherzolites are comparable to those of clinopyroxene from harzburgites, except for one sample with very low Yb_N (2–3) and enriched REE patterns ($La/Yb_N = 3.8–7.7$). The depleted patterns are, however, characterized by a noticeable enrichment in LREE, giving the classical spoon-shaped patterns (Fig. 7d). In this case, the medium REE (MREE) v. heavy REE (HREE) ratio is more appropriate to describe the original, pre-metasomatic patterns. Cpx1 in harzburgites have a $(Sm/Yb)_N$ of around 0.9, while $(Sm/Yb)_N$ in Cpx1

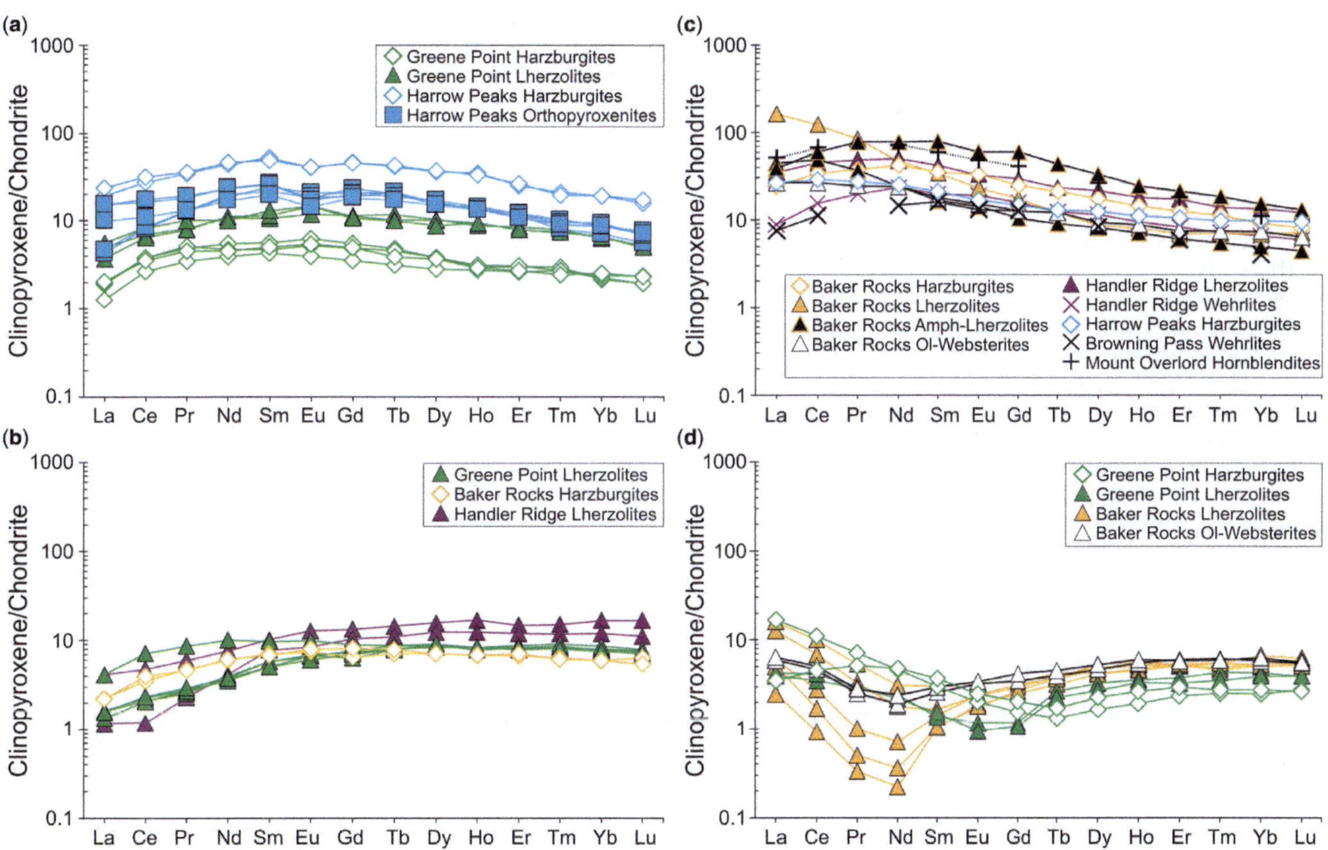

Fig. 7. Chondrite-normalized (Sun and McDonough 1989) REE patterns of clinopyroxene in the studied mantle and cumulate xenoliths. (**a**) Clinopyroxene with a hump-shaped pattern in Greene Point harzburgites and lherzolites, and Harrow Peaks harzburgites and orthopyroxenites. (**b**) Clinopyroxene with a LREE-depleted pattern in Greene Point lherzolites, Baker Rocks harzburgites and Handler Ridge lherzolites. (**c**) Clinopyroxene with a LREE-enriched 'alkaline' pattern in Baker Rocks harzburgites, lherzolites, amphibole lherzolites and olivine-websterites, Handler Ridge lherzolites and wehrlites, Harrow Peaks harzburgites, Browning Pass wehrlites, and Mount Overlord hornblendites. (**d**) Clinopyroxene with a spoon-shaped pattern in Greene Point harzburgites and lherzolites, and Baker Rocks lherolites and olivine-websterites. Amph, amphibole; Ol, olivine.

in lherzolites goes down to 0.2, being indicative of a steeper depletion trend. They are still characterized by remarkable Ti and Zr negative anomalies (see fig. 3 in Coltorti *et al.* 2004). Cpx1 in amphibole-bearing lherzolites present enriched patterns, with $(La/Yb)_N$ varying between 2.7 and 9.1 at Yb_N of around 7. They show spoon-shaped patterns similar to that of Cpx1 in lherzolites but with the LREE and MREE normalized values always above unit.

Cpx1 from HR xenoliths has chondrite-normalized REE patterns comparable to those of the LREE-enriched pattern found at BR (Fig. 7c) or those of the LREE-depleted patterns documented at GP and BR (Fig. 7b). No 'hump' or U-shaped patterns were found.

Trace element contents of amphibole

Both disseminated and vein amphibole in BR lherzolites are characterized by similar patterns in chondrite-normalized incompatible element diagrams (see Coltorti *et al.* 2004). They have high Ba and Nb–Ta positive spikes, variable Zr–Hf negative anomalies and LREE enrichment ($(La/Yb)_N$ = 3.0 and 7.2).

Mineral chemistry of cumulate xenoliths

In the following section, a summary of the main compositional features of cumulate xenoliths from BR, HR, HP, OVP and BRP is reported. Notwithstanding their modal lherzolitic composition, a few xenoliths from BR (i.e. samples BR55 and BR56) have chemical features clearly pointing towards a magmatic origin, and, thus, were considered as cumulate rocks, together with the amphibole-bearing olivine-websteritic sample 154G (see below). For further details, please refer to Supplementary Tables 1, 2 and 3, and the references therein.

Major elements

Olivine. Olivine in BR cumulate lherzolites (samples BR55 and BR56) has a forsterite (Fo) content between 81.6 and 83.4, at NiO and CaO contents ranging from 0.25 to 0.33 wt % and from 0.05 to 0.1 wt%, respectively. Olivine in amphibole-bearing olivine-websterite 154G has a Fo content of 86.3–88.9, NiO content between 0.24 and 0.42 wt%, and CaO ranging from 0.05 to 0.30 wt%. A comparison between olivine in sample 154G and olivine in mantle xenoliths (see the previous section) is reported in Figure 4b.

Olivine in HR wehrlites has Fo contents of 84.6–85.4, at NiO contents between 0.26 and 0.36 wt%. The CaO content ranges from 0.12 to 0.18 wt%.

Fo contents in olivine from HP harzburgites, lherzolites and dunites ranges from 91.4 to 86.2, while it is as low as 80.5 in orthopyroxenites. The NiO content is always between 0.23 and 0.47 wt%. The lowest CaO contents, <0.09 and <0.27 wt%, characterize olivine in peridotites and orthopyroxenite, respectively. One olivine rim is an exception: this has a CaO content of up to 0.65 wt%.

The Fo content in olivine from OVP varies from 87.7 to 77.4 in wehrlites and olivine-clinopyroxenites, and from 82.2 to 74.7 in hornblendites. The NiO content is 0.17–0.30 wt% in wehrlites and olivine-clinopyroxenites, and 0–0.08 wt% in hornblendites. The larger CaO range characterizes olivine in hornblendites (0.11–0.51 wt%), whereas olivine in wehrlites and olivine-clinopyroxenites is between 0.02 and 0.49 wt%.

Olivine from BRP xenoliths has a Fo content varying from 83.6 to 72.4, while the NiO content is between 0.08 and 0.43 wt%. Olivine in clinopyroxenites covers the entire range of CaO content, ranging from 0.10 to 0.47 wt%.

Orthopyroxene. Orthopyroxene in BR cumulate lherzolites (samples BR55 and BR56) has MgO and FeO^T contents of 30.4–30.6 and 10.0–11.1 wt%, respectively, resulting in an Mg# range of 83.0–84.6. Orthopyroxene in BR amphibole-bearing olivine-websterite (sample 154G) has a MgO content ranging from 29.7 to 32.4 wt% and a FeO^T content from 7.8 up to 11.4 wt%, resulting in an Mg# range of 82.3–88.1. The Al_2O_3 content in orthopyroxene from the BR cumulate lherzolites is about 3.6–3.9 wt%, whereas in BR amphibole-bearing olivine-websterite it varies between 7.8 and 11.4 wt%.

Orthopyroxene in HP peridotites and orthopyroxenites has a MgO content of 31.7–33.7 and 27.9–30.9 wt%, respectively. Contemporaneously, the FeO^T content varies between 6.9 and 8.2 wt% in the peridotites, and between 10.1 and 11.3 wt% in the orthopyroxenite, and, as a result, the Mg# ranges from 87.5 to 89.5 in the peridotites and from 82.5 to 84.7 in the orthopyroxenite. The Al_2O_3 content in orthopyroxene from peridotites is between 0.3 and 3.3 wt%, while they have a narrower range in the orthopyroxenite (1.6–2.8 wt%).

Clinopyroxene. Clinopyroxene in BR cumulate lherzolites (samples BR55 and BR56) has a MgO content of 15.1–15.5 wt% and FeO^T content of 5.4–6.0 wt%, resulting in a Mg# of 81.9–83.6. Larger MgO variations (12.4–16.4 wt%) at lower FeO^T contents (3.1–5.0 wt%) characterize clinopyroxene in amphibole-bearing olivine-websterite 154G, resulting in broader Mg# ranges (83.9–90.3: Fig. 6). The Al_2O_3 content of clinopyroxene in BR cumulate lherzolites is almost constant (5.4–5.8 wt%), whereas it is quite variable in olivine-websterite 154G (4.4–11.9 wt%: Fig. 6). Cr_2O_3 varies between 0.4 and 1.4 wt% in BR olivine-websterite, and is around 0.8–0.9 in cumulate lherzolites.

Clinopyroxene in HR wehrlites has a MgO content between 14.8 and 15.5 wt% at a FeO^T content of 3.5–4.3 wt%. The resulting Mg# range from 85.9 to 88.6, whereas the Al_2O_3 content is between 4.0 and 4.9 wt% (Fig. 6). Cr_2O_3 varies from 1.1 to 1.5 wt%.

Clinopyroxene from HP cumulate peridotites has a MgO content between 14.3 and 19.9 wt%, encompassing the range of orthopyroxenites (15.5–17.7 wt%). The FeO^T content of clinopyroxene from orthopyroxenites is between 3.5 and 5.3 wt%, while it ranges from 1.9 to 4.1 wt% in peridotite rocks. As a result, the Mg# of clinopyroxene in peridotites varies from 86.8 to 94.1, while in the orthopyroxenites it goes down to 85.2 (Fig. 6). Al_2O_3 in clinopyroxene from HP lherzolites, harzburgites, dunites and orthopyroxenites is extremely variable, with values as low as 0.5 wt% and as high as 7.3 wt% (Fig. 6). Cr_2O_3 varies between 0.2 and 2.2 wt%, with clinopyroxene in the orthopyroxenite having the highest values.

Clinopyroxenes from BRP wehrlites, olivine-clinopyroxenites and clinopyroxenites have a MgO content between 11.3 and 16.1 wt%. The FeO^T content varies from 5.7 to 7.9 wt%, resulting in an Mg# range of 72.3–83.0 (Fig. 6). The Al_2O_3 content is between 4.4 and 9.9 wt% (Fig. 6), whereas Cr_2O_3 varies between 0 and 1.0 wt%.

Clinopyroxene from OVP wehrlites, olivine-clinopyroxenites, clinopyroxenites and hornblendites has a MgO content between 11.9 and 15.2 wt% at a FeO^T content ranging from 5.1 to 7.7 wt%. As a result, the Mg# varies from 73.3 to 84.1, with hornblendites showing the lowest values (Fig. 6). The Al_2O_3 content varies from 5.8 to 13.5 wt% (Fig. 6), and the Cr_2O_3 content between 0.2 and 1.3 wt%, with the highest values in the wehrlites.

Amphibole. Amphibole occurs in cumulate rocks from BR, HP, BRP and OVP, and is characterized by broad compositional variations.

Amphibole in the BR amphibole-bearing olivine-websterite (sample 154G) occurs in veins. It has a low Mg# (76.9–82.8) at TiO_2 contents ranging between 3.2 and 7.2 wt%, and fairly constant Cr_2O_3 contents (0.5–1.1 wt%).

Amphibole in HP composite dunites is comparable to that of mantle xenoliths, having a high Mg# (87.2–90.6) at TiO_2 and Cr_2O_3 contents of 1.7–4.2 and 1.2–1.9 wt%, respectively. The only amphibole reported from the HP lherzolite (sample HP164) has a lower Mg# (82.9), and both TiO_2 and Cr_2O_3 contents of about 1.3 wt%.

Amphibole in OVP olivine-clinopyroxenites and clinopyroxenites has a Mg# of 69.3–75.9, a TiO_2 content of 3.5–5.3 wt% and a Cr_2O_3 content of about 0.5 wt%. Amphibole from OVP wehrlites is slightly more magnesian (Mg# 74.7–76.7), and has TiO_2 and Cr_2O_3 contents of 2.6–3.5 and *c.* 0.2 wt%, respectively. OVP hornblendites consist of amphibole with low Mg# (63.3–71.3), and TiO_2 contents between 3.9 and 6.2 wt%.

Amphibole in BRP clinopyroxenites is present in small amounts: it has a TiO_2 content of 5.0–5.5 wt%, Cr_2O_3 content of 0.2–0.6 wt% and Mg# of 69.4–71.7.

Trace element contents of clinopyroxene

Particular attention has to be devoted to clinopyroxene from amphibole-bearing olivine-websterite 154G and on the composition of Cpx-A (see Coltorti *et al.* 2004). It has a broad compositional variability, covering the whole spectrum between primary clinopyroxene in lherzolites and amphibole-bearing lherzolites. At Yb_N varying between 5 and 8, La_N varies from 6 to 26. The pattern of Cpx-A is noticeably different, the MREE and HREE are substantially enriched, with a hump between Ce and Eu (Fig. 7c).

Clinopyroxene in HR wehrlites has generally LREE-enriched, convex-upward 'alkaline' patterns (Fig. 7c). It is characterized by $(La/Yb)_N$ ratios between 1.4 and 3.8, at Yb_N ranging from 2.6 up to 11.1.

Clinopyroxene from both HP harzburgites, lherzolites and orthopyroxenites compare well with those found at GP or BR, being characterized by flat to slightly LREE enriched ($(La/Yb)_N$ = 1.4–3.5) and occasionally to convex-upward patterns ($(La/Yb)_N$ = 0.5–1.7) (Fig. 7a, b). Clinopyroxene in HP composite dunites is typified by slightly to markedly LREE-enriched patterns, as testified by the higher $(La/Yb)_N$ ratios (up to 21).

Similarly, the REE patterns of clinopyroxene in OVP and BRP wehrlites, olivine-clinopyroxenites, clinopyroxenites and hornblendites perfectly match the convex-upward shape of clinopyroxenes in Figure 7b (cf. fig. 8 in Perinelli *et al.* 2017 for OVP and fig. 6 in Perinelli *et al.* 2011 for BRP). This is testified by the moderate $(La/Yb)_N$ ratios (1.6–2.3 for BRP wehrlites, olivine-clinopyroxenites and clinopyroxenites; 2.7–4.7 for OVP wehrlites and hornblendites) concomitant with the high constant $(Gd/Yb)_N$ values (2.6–3.8).

Trace element contents of amphibole

Amphibole in veins in the BR olivine-websterite sample 154G has high Ba and Nb–Ta positive spikes, variable Zr–Hf negative anomalies, and a slight enrichment in LREE ($(La/Yb)_N$ = 3.1–4.8).

Amphibole in HP lherzolite and composite dunites is generally characterized by systematic Ba, Nb–Ta and LREE enrichments in chondrite-normalized incompatible element patterns ($(La/Yb)_N$ up to 5.8). An exception is manifest in some crystals in composite dunites, which have relative depletions in Nb–Ta and LREE (see Pelorosso *et al.* 2019).

Amphibole in OVP wehrlites and hornblendites are characterized by moderate to marked Ba and Nb positive anomalies, as well as by LREE enrichment in chondrite-normalized incompatible element patterns (see Perinelli *et al.* 2017). Their $(La/Yb)_N$ ratios vary from 3.3 to 5.1.

Whole-rock chemistry and H_2O content of the northern Victoria Land lithospheric mantle

Whole-rock major/trace elements

The major element composition of northern Victoria Land mantle xenoliths (see Supplementary Table 4) fall within the field of worldwide spinel-bearing peridotites (Perinelli *et al.* 2006), being characterized by a MgO content between 40.1 and 48.6 wt%. The CaO and Al_2O_3 contents are inversely correlated with MgO content, falling in the range of 0.33–2.77 and 0.45–3.00 wt%, respectively. Their Mg# is generally between 88.4 and 91.8 but some composite samples are typified by a Mg# as low as 85.6, suggesting the occurrence of secondary Fe-enrichment processes (Perinelli *et al.* 2006). Whole-rock trace element patterns show a moderate enrichment in the most incompatible components, as well as a concave-upward REE distribution (see fig. 3 in Perinelli *et al.* 2006).

Cumulate xenoliths are clearly distinguishable from mantle rocks in terms of major elements, having CaO, TiO_2 and Al_2O_3 contents up to 16.0, 1.8 and 7.1 wt%, respectively, as well as a MgO content down to 16.0 wt% (see Supplementary Table 4). Consistently, the Mg# of cumulitic rocks lies between 72.0 and 84.4. As expected, the increase in Na_2O and the decrease in compatible trace elements (such as Ni and Co) that characterize cumulate xenoliths is a function of the increasing modal proportions of clinopyroxene (Perinelli *et al.* 2017). Accordingly, trace element patterns of cumulates present a slight enrichment in the most incompatible components, as well as in general Zr and Ti negative anomalies (see fig. 8 of Perinelli *et al.* 2017).

Bulk H_2O content of the SCLM

Measurement of the OH content in nominally anhydrous minerals (NAMs), as well as in amphibole crystals, when present, enabled Bonadiman *et al.* (2009, 2014) to speculate on the retention of volatiles in the northern Victoria Land SCLM. The high H_2O partition coefficient between pyroxenes and olivine in peridotite systems ($^{Opx-Ol}K_{dH_2O}$ = 10–40; $^{Cpx-Ol}K_{dH_2O}$ = 10–80: Hauri *et al.* 2004; Grant *et al.* 2007), coupled with the fast diffusion of H in olivine (Ingrin and Blanchard 2006) and its most probable loss during ascent precluded olivine from retaining significant OH components and, thus, from contributing markedly to the OH storage in these SCLM portions. According to Bonadiman *et al.* (2009), the GP amphibole-free peridotitic xenoliths represent a SCLM portion with an extremely low water content (2–8 ppm), almost completely stored in clinopyroxene (5–16 ppm) and orthopyroxene (9–16 ppm). Preliminary data on HR peridotites show similar results, with bulk H_2O estimates ranging between 7 and 28 ppm (unpublished data). The common presence of amphibole in BR xenoliths testifies to the predominant role played by H_2O in the lithospheric mantle beneath this locality. In BR mantle-derived rocks, amphibole has a H_2O content between 0.84 and 1.42 wt% (Bonadiman *et al.*

2014), whereas orthopyroxene and clinopyroxene are able to retain 39–166 and 82–399 ppm H_2O, respectively (Bonadiman *et al.* 2009). The lithological variability of BR xenoliths can thus be translated in an extremely variable bulk water content within this portion of the Antarctic SCLM, starting from the lowest end represented by amphibole-free peridotites with 6–128 ppm H_2O to peridotites with disseminated amphibole with 354–1120 ppm H_2O, strictly related to the amphibole modal abundance (up to 8%). Amphibole-dominated veins in composite samples, although volumetrically less representative, can retain up to 1.42 wt% H_2O.

Geothermal state and redox conditions

The geothermal state and redox conditions are used here to map lithospheric mantle domains of northern Victoria Land. Calculated f_{O_2} in a rock can preserve the signature of its past history, although the most recent 'event(s)' (depletion or enrichment) can reset this parameter (e.g. Woodland *et al.* 2006). Regional variations in oxidation state can, however, be used to assess the extent of metasomatic interactions and delineate regions of the mantle that have experienced different (or similar) geochemical histories.

For this review, we have recalculated the formula units for the entire major element mineral dataset gathered over more than 20 years of Antarctic research to determine the equilibrium temperature and pressure, and to assess the oxidation state of spinel peridotite xenoliths over a distance of more than 200 km in northern Victoria Land. This was done in order to apply a uniform procedure to the processing of the mineral chemical data to obtain coherent results between the various populations.

Application of the thermometers and thermobarometers to suites of naturally occurring samples indicates that while reactions governing the Cr and Al solubility in orthopyroxene can provide useful estimates of the 'original' mantle temperatures and pressures, comparable reactions for clinopyroxene yield estimates that are variably dependent on the transport of the samples (MacGregor 2015). Similarly, temperature and pressure estimates based on reactions governing Mg and Fe exchange between silicates and spinel may be sensitive to the late transport stage of the samples (e.g. Brey and Köhler 1990; Frost 1991; Woodland *et al.* 1996; Mallmann and O'Neill 2013). To minimize these effects, temperatures (and pressures) of equilibration were determined using a variety of published thermobarometers involving different mineral equilibria (see also Litasov and Ohtani 2009). The application of these combinations of thermobarometers to natural samples indicates internal equilibrium (agreement within the mutual errors) or disequilibrium, allowing a critical evaluation of the resulting temperature and pressure values.

Based on their experiments, Brey and Köhler (1990) showed that combinations of their versions of barometers (Al-in-orthopyroxene and Ca-in-olivine) with various thermometers (two-pyroxenes, Ca-in-orthopyroxene, and Fe–Mg exchange between olivine and spinel) gave internally consistent results. However, pressure is not well constrained for spinel peridotites due to the absence of a strongly pressure-dependent reaction (Köhler and Brey 1990; Wood *et al.* 1990; Uenver-Thiele *et al.* 2014; MacGregor 2015; Davis *et al.* 2017). In turn, the temperature estimates from the individual thermometers are not very sensitive to pressure. For our dataset (both peridotites and pyroxenites), the two-pyroxene thermometers (Wells 1977; Brey and Köhler 1990; Taylor 1998) and the olivine–spinel thermometers (Fabriès 1979; O'Neill and Wall 1987; Ballhaus *et al.* 1991; Jianping *et al.* 1995) show temperature–pressure variations of <2 and <5°C GPa^{-1}, respectively.

Temperatures and pressures for the northern Victoria Land peridotites were calculated using the 'PTEXL3' spreadsheet (http://www.mineralogie.uni-frankfurt.de/petrologie-geochemie/mitarbeiter/brey/downloads/index.html and are reported in Supplementary Table 2. Here, the most reasonable pressure values for each sample are derived by simultaneous iterative calculations, resolving temperature by using a large number of thermometers

Combinations of temperature–pressure pairs using all the two-pyroxene ($T_{Opx–Cpx}$) formulations listed in the PTEXL3 were calculated iteratively, while temperatures from the Fe–Mg exchange between olivine and spinel ($T_{Ol–Sp}$) were obtained using a default pressure of 1.5 GPa, which is roughly the centre of the pressure range for spinel stability. Supplementary Table 2 reports only the calculated pressures that gave petrologically reasonable results. From the compilation of temperature values we extracted the two-pyroxene thermometer data (T_{BKN}) of Brey and Köhler (1990) and the obtained values are listed separately (Supplementary Table 2). This thermometer is insensitive to the presence of Fe^{3+} in clinopyroxene and orthopyroxene (e.g. Matjuschkin *et al.* 2014), and can therefore be used for comparison with other datasets in this Memoir. The temperatures of orthopyroxene-free assemblages were calculated with clinopyroxene (T_{Cpx}) thermometers (Mercier 1980; Nimis and Taylor 2000), notwithstanding that the enstatite–diopside exchange in this phase is sensitive to late-stage perturbations. The temperatures are based on compositions of primary mineral analyses from the interiors of grains, which were homogeneous within a given grain. Differences in temperature calculated from the singular thermometer formulations in each sample are generally in the range of the formula's errors (Lazarov *et al.* 2009). Such differences have little to no impact on the results of our oxygen barometric computations (see below).

Thermobarometry of mantle xenoliths

Greene Point. In this xenolith population, temperatures obtained by the Brey and Köhler (1990) formulation (average T_{BKN}: 1061 ± 64°C) are significantly higher with respect to the estimates obtained by all the other formulations used in the iterative calculation (average $T_{Opx–Cpx}$: 1000 ± 79°C). The olivine–spinel thermometer records temperatures systematically lower by at least *c.* 100°C (up to 180°C) with respect to the $T_{Opx–Cpx}$. Greene Point olivine–spinel pairs record temperatures in the range of 829 ± 7–944 ± 37°C, with an average of 890 ± 61°C. Despite all of the methods applied indicating no appreciable difference in temperatures between harzburgites and lherzolites, the results indicate that the sole $T_{Opx–Cpx}$ estimates respond to the slow responsive solid–solid reactions in the GP mantle segment. The corresponding pressure predicted by multiple regression falls in the range 0.8–1.6 GPa, with most of the samples lying between 1.0 and 1.3 GPa (Fig. 8; Supplementary Table 2).

Mount Melbourne. Just one lherzolite from Willow Nunatak Beach was processed for thermobarometric purposes. A convergence of the temperatures calculated using different mineral equilibria (average $T_{Opx–Cpx}$ = 951°C; T_{BKN} = 933°C; $T_{Ol–Sp}$ = 999°C) is observed. In this sample, the temperatures obtained with the olivine–spinel thermometry were not consistently lower than those obtained using orthopyroxene–clinopyroxene thermometry; this is probably because the olivine and spinel Fe–Mg exchange responds to the rapidly changing conditions during the last-stage processes at mantle depth

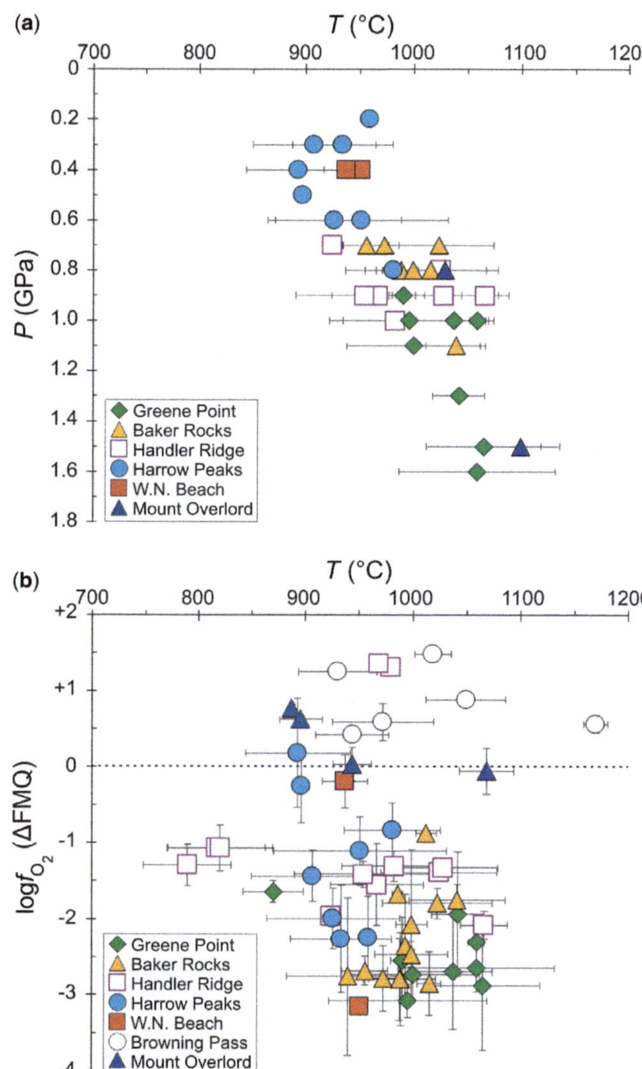

Fig. 8. (**a**) Temperature (T: °C)–pressure (P: GPa) and (**b**) temperature (T: °C)–oxygen fugacity ($\log f_{O_2}$: ΔFMQ) diagrams showing the equilibrium conditions recorded by the studied mantle and cumulate xenoliths. The reported pressures and temperatures were determined by the iteratively solved thermobarometric equation available in the 'PTEXL3' spreadsheet (http://www.mineralogie.uni-frankfurt.de/petrologie-geochemie/mitarbeiter/brey/downloads/index.html). For mantle xenoliths, temperature was determined by means of the orthopyroxene–clinopyroxene ($T_{Opx-Cpx}$) equations, whereas for cumulate samples iterative calculations were limited to single-pyroxene thermometers (T_{Cpx}). In (**b**), the equilibration temperature reported for the Mount Overlord xenoliths was calculated by means of the olivine–spinel thermometer (T_{Ol-Sp}) due to the lack of T–f_{O_2} pairs when T_{Cpx} is considered. Oxygen fugacity was calculated for all the samples by means of the Wood (1990) formulation. For each sample, average values are reported and error bars indicate the standard deviation of the results. W.N. Beach, Willow Nunatak Beach.

(metasomatism) or the transit stage to the surface of the xenolith. The amphibole-bearing lherzolite from Willow Nunatak Beach records temperatures of 844 ± 58 (T_{BKN}), 937 ± 21 ($T_{Opx-Cpx}$) and 1061 ± 34°C (T_{Ol-Sp}). Pressures around 0.4 GPa were obtained from both Willow Nunatak Beaches samples.

Amphibole occurs as disseminated in the peridotite matrix in a limited number of samples of the BR xenolith population, being generally <4%, except for lherzolite 154C (7.5%). Therefore, in principle, it has little to no impact on the results of thermobarometric computation. However, when amphibole occurs, clinopyroxene and spinel show large chemical zonation, making application of the single thermometer problematic. Therefore, the simultaneous application of various thermometers reduces the uncertainties. $T_{Opx-Cpx}$ ranges from 988 ± 33 to 1041 ± 44°C, whereas T_{Ol-Sp} varies from 873 ± 89 to 970 ± 121°C (Fig. 8; Supplementary Table 2). These temperatures are at the lower end of the 900–1100°C range for the equilibration of amphibole-bearing lherzolites and harzburgites given by Perinelli *et al.* (2012) and Bonadiman *et al.* (2014). The anhydrous BR samples (both lherzolites and harzburgites) have, on average, higher temperatures ($T_{Opx-Cpx}$ from 940 ± 57 to 1023 ± 51°C; T_{Ol-Sp} from 750 ± 21 to 1015 ± 22°C). All samples record pressures between 0.7 and 0.8 GPa, except for one harzburgitic xenolith (P = 1. 1 GPa).

Handler Ridge. Most of the Handler Ridge xenoliths are lherzolites dominated by chemically homogeneous minerals, suggesting that thermodynamic equilibrium was achieved at the scale of a hand sample. An overlap is observed in the standard deviations associated with the average temperatures calculated with clinopyroxene–orthopyroxene and olivine–spinel thermometers. Temperatures vary from 905 ± 45 to 1067 ± 26°C (T_{BKN}), from 817 ± 46 to 1028 ± 51°C ($T_{Opx-Cpx}$) and from 894 ± 29 to 951 ± 83°C (T_{Ol-Sp}). This population gives pressure estimates between 0.7 and 1.0 GPa (Fig. 8; Supplementary Table 2), in agreement with the experimental and theoretical calculations for a shallow geotherm (Greenfield *et al.* 2013; MacGregor 2015).

Thermobarometry of cumulate xenoliths

For cumulate peridotites and olivine-websterites from BR and HP, the temperature–pressure conditions of equilibrium were calculated using the same methods applied for mantle rocks (see above). For cumulate clinopyroxenites and wehrlites from Handler Ridge, Mount Overlord and Browning Pass (Perinelli *et al.* 2011), thermobarometric calculations are often difficult to perform, as not all the mineral parageneses are compatible with many established thermobarometers. In these cases, iterative calculations of temperature and pressure limited to single-pyroxene thermometers (T_{Cpx}) and barometers (Mercier 1980; Nimis and Ulmer 1998; Taylor 1998; Nimis and Taylor 2000) were performed to unveil temperature–pressure estimates with respect to the bulk system. For comparison, T_{Ol-Sp} was also calculated in these samples and is likely to yield information on their dynamic evolution.

Mount Melbourne. Cumulate lherzolites from BR record equilibrium temperatures of 966 ± 50 (T_{BKN}), 1013 ± 10 ($T_{Opx-Cpx}$) and about 888°C (T_{Ol-Sp}). Consistently, the amphibole-bearing olivine-websterite (sample 154G) yields comparable values (T_{BKN} of 961 ± 44°C; $T_{Opx-Cpx}$ of 1041 ± 44°C; T_{Ol-Sp} of 873 ± 89°C). No reliable pressure estimates can be obtained from these samples.

Handler Ridge. Equilibrium temperatures for the two wehrlitic xenoliths, calculated by means of the single-pyroxene thermometer, range from 968 ± 8 to 980 ± 8°C. Lower values (from 720 ± 12 to 781 ± 13°C) were yielded by the T_{Ol-Sp} method. As for BR cumulate rocks, no reliable pressure estimates were obtained.

Harrow Peaks. The textural and geochemical features of HP xenoliths are unusual in the northern Victoria Land lithosphere. The predominant dunitic matrix (olivine + spinel) of the composite xenoliths, the lherzolites and harzburgites lithotypes, and the occurrence of orthopyroxenites (absent in all the other populations: see Supplementary Table 1) led to the interpretation that HP xenoliths are deep crustal–mantle cumulates. Such cumulates are likely to have crystallized from a primitive high-Mg melt (close to silica saturation), with primitive olivine crystallizing at a temperature of *c.* 1300°C in a pressure range of *c.* 0.9–1.3 GPa (Pelorosso *et al.* 2019). The fractionation process was modelled to stop at an early stage of evolution, with a predominant assemblage of olivine (spinel) + orthopyroxene (clinopyroxene). Amphibole and most clinopyroxenes in veins are in complete geochemical disequilibrium with the anhydrous matrix, and are interpreted to be products of late alkaline metasomatism related to the host magma of the xenoliths (Gentili *et al.* 2015; Pelorosso *et al.* 2019).

The calculated $T_{Ol–Sp}$ in pyroxene-bearing dunites converge in a narrow range, from 870 ± 54 to 969 ± 29°C (Fig. 8; Supplementary Table 2), being not far from temperatures calculated using the single-clinopyroxene computation (T_{Cpx} from 897 ± 7 to 981 ± 44°C). These data suggest that the mineral assemblages were quite well equilibrated. However, lherzolites and harzburgites have slightly higher olivine–spinel temperatures ($T_{Ol–Sp}$ from 927 ± 44 to 1150 ± 73°C), clearly in disequilibrium with the clinopyroxene (T_{Cpx} from 907 ± 57 to 951 ± 80°C). Olivine–spinel in orthopyroxenite HP163 record temperatures in a range between 790 ± 76 ($T_{Ol–Sp}$) and 893 ± 49°C, suggesting the thermal closure of the fractionation path. Barometric estimates for HP indicate that the fractionation paths evolves in a pressure range from the crustal–mantle boundary (*P* = 0.6–0.8 GPa: Perinelli *et al.* 2011) up to the pure crustal domains (*P* = 0.2–0.5 GPa) (Fig. 8; Supplementary Table 2).

Mount Overlord. This xenolith group includes cumulates with anhydrous assemblages (clinopyroxene ± olivine; amphiboles <5%), and cumulates with parageneses dominated by amphiboles (>50%) with minor, but relevant, contents of clinopyroxene (20–25%: Supplementary Table 1). They were interpreted as products of the crystal fractionation of a hydrous basanitic magma (Perinelli *et al.* 2017). The absence of Cr-spinels prevents the use of the conventional olivine–spinel thermometers in all samples. Some xenoliths have clinopyroxene with appropriate Al–Cr (Na) compositions to calculate the T_{Cpx} (Brey and Köhler 1990; MacGregor 2015). The recorded temperatures vary from 991 ± 59 to 1106 ± 59°C (T_{Cpx}) and from 718 ± 59 to 1224 ± 97°C ($T_{Ol–Sp}$). The few pressure estimates (Mercier 1980; Nimis and Ulmer 1998; Taylor 1998; Nimis and Taylor 2000) suggest that clinopyroxene (and amphibole) compositions point to a fractionation process that operated over a range of high pressures (0.8–1.5 GPa), in agreement with what was already stated by Perinelli *et al.* (2017) (Fig. 8; Supplementary Table 2).

Browning Pass. The clinopyroxene-rich cumulates collected at BRP range in mineralogy from wehrlites to clinopyroxenites. They represent an early stage of the fractional crystallization process of an alkaline basic melt of the McMurdo Volcanic Group (Perinelli *et al.* 2011). Clinopyroxenes belong to both the Cr-diopside (wehrlites) and the Al-augite series (olivine-clinopyroxenites and clinopyroxenites), and discriminate between two thermal regimes. Wehrlites record T_{Cpx} between 1117 ± 43 and 1170 ± 11°C, whereas clinopyroxenites and olivine-clinopyroxenites yield temperatures of 930 ± 36–1049 ± 37 (T_{Cpx}) and 746 ± 9–954 ± 20°C ($T_{Ol–Sp}$), with temperature decreasing with the olivine modal contents. No reasonable values for pressure were obtained from these rocks (Fig. 8; Supplementary Table 2). We therefore refer to the results of Perinelli *et al.* (2011), which indicate a clinopyroxene crystallization pressure in the range 0.6–0.8 GPa.

Oxygen barometry of mantle xenoliths

We determined the northern Victoria Land upper-mantle f_{O_2} directly from peridotites and pyroxenites containing the olivine + orthopyroxene + spinel assemblage, following the reaction: $6Fe_2SiO_4$(olivine) + $2O_2$ = $3Fe_2Si_2O_6$(orthopyroxene) + $2Fe_3O_4$ (spinel) and the pressure–temperature conditions as estimated in each sample (Supplementary Table 2). Several studies have parameterized f_{O_2} based on this equilibrium (O'Neill and Wall 1987; Mattioli and Wood 1988; Wood 1991): we used the formulation of Wood (1990) and applied the Wood and Virgo (1989) correction for the spinel $Fe^{3+}/\Sigma Fe$ ratios determined from microprobe measurements. In orthopyroxene-free assemblages (i.e. most of the clinopyroxenites from Browning Pass and Mount Overlord), the olivine–spinel oxybarometer was applied (Ballhaus *et al.* 1991) using a single clinopyroxene thermometer (Mercier 1980; Taylor 1998) as temperature input. To minimize the effects of uncertainties, mainly due to the equilibration temperature and Fe^{3+} determination in spinel, the f_{O_2} estimates are referred to the fayalite–magnetite–quartz (FMQ) buffer (calibration of Frost 1991) and are reported as log f_{O_2} (ΔFMQ; see Wood 1991). We estimated that temperature and pressure uncertainties of 100°C and 0.3 GPa result in variations in calculated f_{O_2} of only 0.15 and 0.09 log units, respectively. Conservatively, the uncertainties from pressure, temperature and Fe^{3+} in spinel determined by microprobe are estimated to be about *c.* 0.5 log units (Woodland *et al.* 1992, 2006).

Greene Point. GP samples yield f_{O_2} ranging from −3.08 ± 0.22 to −1.64 ± 0.14 ΔFMQ. These values are slightly lower than those calculated previously by Pelorosso *et al.* (2016) by means of the oxybarometer and olivine–spinel thermometer of Ballhaus *et al.* (1991). The differences are attributable to the different calibrations (Woodland *et al.* 1992, 2006) but, as the olivine–spinel equilibrium is more sensitive to late-stage magmatic perturbation, we can also speculate that the highest values obtained previously may record this late event (transport or metasomatism). It is interesting to note that among the northern Victoria Land mantle xenoliths, the GP group records the highest pressure and relatively reduced conditions (Fig. 8), lying at the lowest end of the range for non-cratonic peridotites, close to the cratonic peridotitic field (Foley 2011).

Mount Melbourne. The lherzolitic and amphibole-bearing lherzolitic xenoliths from Willow Nunatak Beach record f_{O_2} values from −3.16 to −0.19 ± 0.35 ΔFMQ. Lherzolite yields values comparable to the reduced conditions recorded by GP mantle lithotypes, whereas amphibole-bearing lherzolite grades towards the most oxidized conditions recorded by cumulate rocks (see below and Fig. 8).

All BR xenoliths have metasomatic overprints, both amphibole-bearing and amphibole-free samples. The calculated f_{O_2} values of BR amphibole-bearing xenoliths vary from −2.80 ± 0.54 to −1.75 ± 0.21 ΔFMQ, whereas those of BR amphibole-free rocks range from −2.85 ± 0.42 to −1.79 ± 0.20 ΔFMQ (Fig. 8). The range of values is similar, confirming that the presence of amphibole (≤8%) does not affect the olivine–orthopyroxene–spinel oxygen equilibrium (Perinelli *et al.* 2012; Bonadiman *et al.* 2014). These f_{O_2} values reveal that amphiboles are in equilibrium with the peridotite mineral assemblage, and that they formed in low redox conditions, comparable to those of the anhydrous mantle fragments of the same area (Fig. 8; Supplementary Table 2).

Handler Ridge. HR lherzolites record ΔFMQ values between −2.08 ± 0.18 and −1.07 ± 0.30, with many samples yielding ΔFMQ values between −1.4 and −1.3 (Fig. 8; Supplementary Table 2). As for the GP samples, these values are slightly lower compared with those previous obtained for this population (from −1.3 to +0.3 ΔFMQ: Pelorosso *et al.* 2017), for which we replicate the same explanation that these values reflect a pre-metasomatic condition.

Oxygen barometry of cumulate xenoliths

Mount Melbourne. Lherzolite BR56 from Baker Rocks records the highest f_{O_2} values (−0.87 ΔFMQ) for the BR population. Amphibole-bearing olivine-websterite 154G, on the other hand, records an oxygen fugacity value of about −2.07 ± 0.98 ΔFMQ, comparable to those of BR mantle rocks.

Handler Ridge. The oxygen fugacity values recorded by HR wehrlites are consistent with each other and are also in line with those of all northern Victoria Land cumulate rocks, lying well above the FMQ buffer. In detail, they vary from +1.31 ± 0.06 to +1.36 ± 0.05 ΔFMQ (Fig. 8), easily distinguishable from HR lherzolitic rocks.

Harrow Peaks. The f_{O_2} equilibration conditions of the Harrow Peaks cumulates as recorded in the coexisting mineral assemblage (from −2.24 ± 0.54 to +0.18 ± 0.72 ΔFMQ) (Fig. 8; Supplementary Table 2) confirm the previous results of Gentili *et al.* (2015) and Pelorosso *et al.* (2019). It is worth noting that these previous studies highlighted a strong discrepancy between the redox conditions recorded by the amphibole dehydration equilibrium (from +5 to +7 ΔFMQ) and those of the coexisting ultramafic mineral assemblage.

Mount Overlord. Few samples contain suitable parageneses for applying the conventional oxybarometers. The lack of crystallochemical data for amphiboles in the hornblendites prevents the use of the amphibole dehydrogenation model to estimate the ambient redox conditions (Bonadiman *et al.* 2014; Gentili *et al.* 2015). The OVP anhydrous wehrlite and clinopyroxenite assemblages yield f_{O_2} values ranging from −0.05 ± 0.30 to +0.04 ± 0.22 ΔFMQ, whereas the amphibole-bearing rocks record f_{O_2} values between −0.22 ± 0.55 and 0.76 ± 0.06 ΔFMQ (Fig. 8; Supplementary Table 2). These results suggest that OVP xenoliths equilibrated in a relatively oxidized environment.

Browning Pass. According to Perinelli *et al.* (2011), olivine-clinopyroxenites and wehrlites from this locality are fractionated products from a primitive magma similar to the OVP basanitic melt but with a lower water content. However, the BRP olivine-clinopyroxenites and wehrlites record markedly higher ambient redox conditions (ΔFMQ from +0.42 to +1.49 ± 0.08 ΔFMQ), suggesting that f_{O_2} conditions and water content are not strictly correlated (Fig. 8; Supplementary Table 2). As stated above, similar indications can also be obtained for some mantle sections where amphibole-free and amphibole-bearing xenoliths record perfectly comparable oxygen fugacity values (Bonadiman *et al.* 2014).

Discussion

Depletion event(s)

Most of the mantle xenoliths from GP, BR and HR follow the melting trends from lherzolites to harzburgites along anhydrous curves (Fig. 2), whereas a few samples from GP and, to a lesser extent, BR are displaced towards the orthopyroxene apex. This enrichment in modal orthopyroxene cannot be explained by any melting trend moving from the PM position (Johnson *et al.* 1990); thus indicating the operation of processes other than partial melting, such as post-melting peridotite–melt interactions (see the following subsection). In Supplementary Table 5, the degree of partial melting estimated on the basis of Al_2O_3 partitioning in orthopyroxene and clinopyroxene (Fig. 5) (Bonadiman and Coltorti 2011; Upton *et al.* 2011), and the Yb v. Y distribution in clinopyroxene (Fig. 9) (Zou 1998), are summarized, analogous to the treatment of the Patagonian SCLM by Melchiorre *et al.* (2020). In addition, the degree of partial melting recorded by the Fo content in olivine v. Cr# ($Cr_2O_3/(Cr_2O_3 + Al_2O_3)$ mol%) of spinel are shown in Figure 4 (Arai 1994). HREE

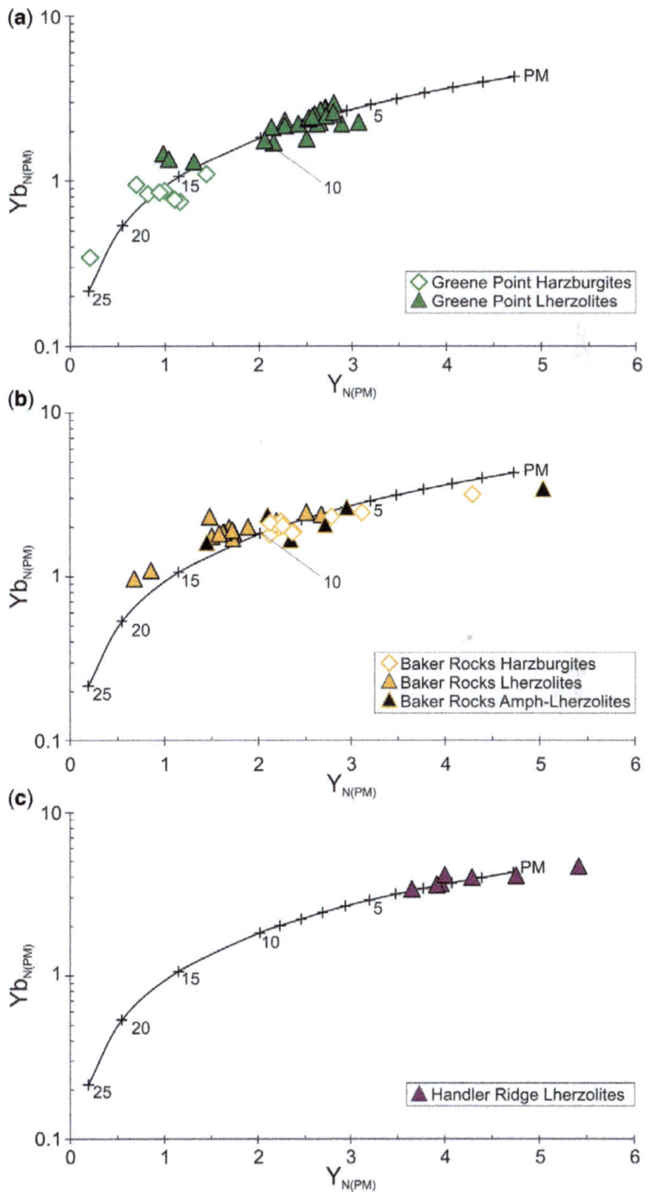

Fig. 9. Primitive-mantle (PM)-normalized Yb v. Y diagrams showing the composition of clinopyroxenes from the northern Victoria Land mantle xenoliths. (**a**) Greene Point harzburgites and lherzolites. (**b**) Baker Rocks harzburgites, lherzolites and amphibole-bearing lherzolites. (**c**) Handler Ridge lherzolites. Curves in (a), (b), and (c) refer to the melting model of Zou (1998), developed from a starting PM composition (Sun and McDonough 1989). Numbers along the curves correspond to the partial melting percentages. Amph, amphibole.

are more useful than LREE in estimating the degree of partial melting, since LREE are more prone to be affected by secondary textural features (i.e. metasomatism or refertilization) (Wood and Blundy 2001; Niu 2004; Bédard 2014). These different approaches give estimates consistent with each other, and record degrees of melting between 18 and 21% (on average) for GP rocks, and slightly lower (12–16% on average) values for BR xenoliths. HR has the most fertile lithologies, showing a degree of partial melting of less than 13% (Figs 5 & 9).

Further information on the depletion process can be derived from the REE contents of Cpx1. The $(La/Sm)_N$ v. $(Gd/Yb)_N$ diagram (Fig. 10) differentiates the same four groups of REE patterns of clinopyroxenes typical of northern Victoria Land xenoliths evidenced in Figure 7. The enriched and depleted clinopyroxenes are plotted in quadrants I and III, respectively, while the clinopyroxenes with $(La/Sm)_N < 1$ and $(Gd/Yb)_N > 1$ or with $(La/Sm)_N > 1$ and $(Gd/Yb)_N < 1$ fall in quadrants II and IV, respectively.

The great majority of GP xenoliths and a few BR samples fall in quadrant II, showing a hump-shaped pattern. This pattern is due to polybaric melting that starts in the garnet facies and terminates in the spinel facies (Perinelli *et al.* 2006). Alternatively, Pelorosso *et al.* (2016) proposed that the humped patterns were inherited from the interaction between a tholeiitic melt and depleted peridotite, envisaging a refertilization phenomenon with a high melt/rock ratio. This hypothesis is discussed in the following paragraph.

At both localities there are one or two samples with a spoon-shaped pattern (quadrant IV) and Yb_N as low as $2 \times Ch$ (Fig. 7d), comparable to those reported by Perinelli *et al.* (2006). They are likely to have originated by extreme depletion followed by a metasomatic event(s) that caused a strong enrichment in LREE. Whatever the original shape was, it is unlikely that a tholeiitic melt could be in equilibrium with clinopyroxene with such low HREEs. Therefore, it can be inferred that part of the BR and GP lithospheric mantle experienced partial melting starting in the garnet facies. According to Perinelli *et al.* (2006), a degree of partial melting of up to 15–20% for a lherzolite in the garnet stability field and subsequently 5–15% partial melting in the spinel facies can reproduce these HREE patterns, yielding Yb_N values down to 1 (e.g. sample GP6 of Perinelli *et al.* 2006, which has REE patterns comparable to our samples GP23 and GP81). The two samples GP23 and GP81 record a total degree of melting of up to 30–32% based on major elements in orthopyroxene (Supplementary Table 5). This extreme depletion was followed by a variable enrichment in LREE, which ultimately generated the spoon-shaped patterns. However, the number of samples that still record this feature is limited, probably because most were efficiently overprinted by subsequent refertilization/metasomatism events. Nevertheless, their presence indicates an extensive early partial melting event(s) that, according to Melchiorre *et al.* (2011), may have started in the Archean.

A rather large number of samples in BR, GP and HR is characterized by LREE-depleted patterns (Fig. 7b and quadrant III of Fig. 10), suggesting partial melting in the spinel facies, whose extent varies, on average, from 16 to 19% (Figs 4, 5 & 9; Supplementary Table 5). Most of them, however, reflect LREE enrichment by metasomatism, passing from the depleted quadrant III to the spoon-shaped quadrant IV (in Fig. 10). The flat patterns of MREE and HREE are a clear indication of melting in shallower (spinel facies) lithospheric mantle (e.g. Johnson *et al.* 1990; Bonadiman *et al.* 2005).

The overall framework for mantle depletion events suggests that the SCLM beneath northern Victoria Land was formed by the accretion of several domains, following a long evolution over a time interval between Archean (3.3 Ga) and Cretaceous times (120 Ma: Melchiorre *et al.* 2011).

Refertilization event

Several lines of evidence support the hypothesis that after various events of melt extraction, a refertilization event occurred in the northern Victoria Land area, which refertilized harzburgites back into more fertile lithotypes (Le Roux *et al.* 2009). In Figure 2, both BR and GP xenoliths form a trend towards higher modal abundance of orthopyroxene. The general lack of correlation between the modal content of silicate minerals and various melting-sensitive geochemical parameters suggests that the modal abundances are not the result of depletion event(s) only.

The refertilization hypothesis for the northern Victoria Land mantle segments was first introduced by Pelorosso *et al.* (2016) after a study of GP xenoliths. The occurrence of cumulate orthopyroxenites (i.e. sample HP163) among the HP xenoliths testifies to the occurrence of silica-saturated melts in the proximity of the mantle–crustal boundary (Pelorosso *et al.* 2017). However, the cumulate rocks found in the BRP and OVP areas (wehrlites or clinopyroxenites, with or without amphibole) are likely to have formed by crystallization of SiO_2-undersaturated melts, in which orthopyroxene cannot be an early liquidus phase. These melts are consistent with the Cenozoic alkaline magmas that carried the xenoliths to the surface in the Mount Melbourne district. Consequently, Pelorosso *et al.* (2019) proposed that most of the xenoliths from HP are cumulates from the Jurassic Ferrar tholeiites. The rare clinopyroxene of the HP orthopyroxenite (sample HP163), which coexists with $Fo_{80.5-84.5}$ olivine, has remarkably hump-shaped REE patterns. Similar REE profiles were observed in clinopyroxene of harzburgite HP166 (see fig. 6 of Pelorosso *et al.* 2019), which coexist with fertile (Fo_{88}) olivine, inconsistent with the residual nature of the harzburgite (Niu 2004; Herzberg and Asimow 2008). All these features support the petrogenesis of HP rocks by crystallization of a tholeiitic melt. The clinopyroxene REE patterns are comparable to those reported in figure 5e of Pelorosso *et al.* (2016) from GP lherzolites and represent the hump-shaped patterns seen in this study (Fig. 7a and quadrant II of Fig. 10).

A few analyses from BR also belong to this group, indicating that the infiltration of tholeiitic melts that led to the crystallization of HP cumulates also took place beneath both GP and BR. This is confirmed by Melchiorre *et al.* (2011), who documented the existence of Os-rich sulfides in BR rocks that yielded Jurassic minimum ages. These ages were determined in an amphibole-bearing lherzolite (sample BR56) containing Fo_{82} olivine, and are thus attributable to the cumulate group. This sample is most probably the product of an interaction between Ferrar tholeiitic melts and the mantle, similar to that reported for HP. This hypothesis is supported both by petrographical features and by oxybarometric estimates described in the previous paragraphs. Moreover, from the thermobarometric point of view, BR and GP mantle domains show identical f_{O_2} conditions, with GP having slightly higher temperatures and pressures.

The above-mentioned features suggest that the area of influence of the Ferrar refertilization event should also be extended to the SCLM domains beneath HP and BR in addition to GP. This appears reasonable when also taking into account the short distances (≤50 km) between these three localities and the Ferrar dolerite outcrops (Fig. 1).

Metasomatic event(s)

The last event that affected the SCLM beneath northern Victoria Land is the alkaline magmatism related to the opening of the WARS, which started during the Eocene and continued until the present. It first developed with the Meander

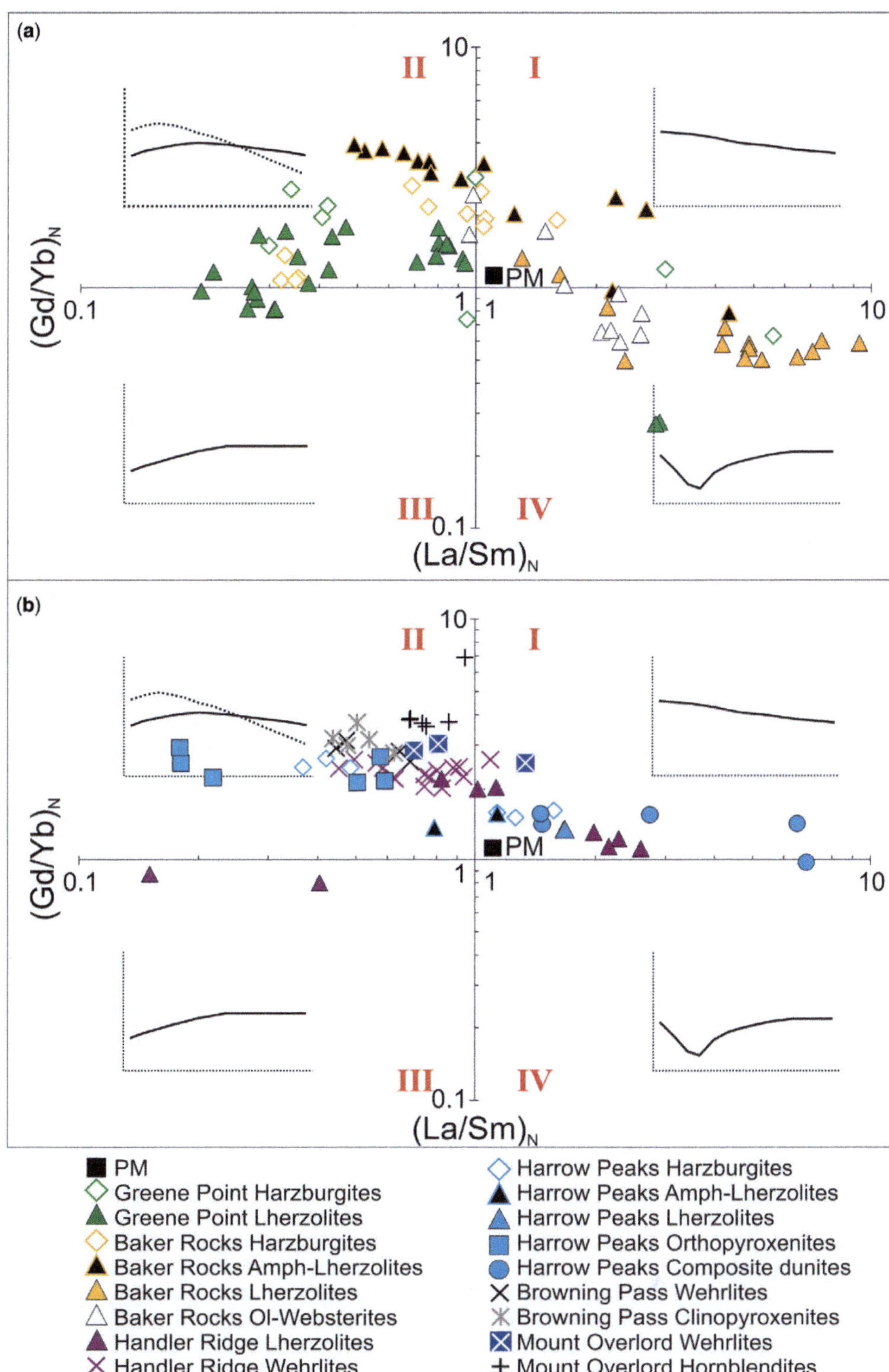

Fig. 10. Chondrite-normalized (N: Sun and McDonough 1989) Gd/Yb v. La/Sm diagram showing REE concentrations in clinopyroxene from northern Victoria Land mantle and cumulate xenoliths. (**a**) Clinopyroxene in Greene Point and Baker Rocks xenoliths. (**b**) Clinopyroxene in Handler Ridge, Harrow Peaks, Browning Pass and Mount Overlord samples. In each diagram, the intersection of the *x*- and *y*-axes at unit implies that each of the four quadrants (I–IV) corresponds to a distinct clinopyroxene REE pattern, as shown by the simplified shapes in each panel. Quadrant I includes clinopyroxenes with a LREE-enriched pattern (see also Fig. 7c). Quadrant II corresponds to clinopyroxenes with hump-shaped (solid line in the sketch) and convex-upward 'alkaline' patterns (dotted line in the sketch), easily distinguishable in terms of absolute abundances of HREE (see also Fig. 7a, c). Quadrant III includes clinopyroxene with a LREE-depleted pattern (see also Fig. 7b). Quadrant IV includes clinopyroxenes with a spoon-shaped pattern (see also Fig. 7d). The composition of clinopyroxene in primitive mantle (PM: Bonadiman *et al.* 2005) is also reported for comparison. Amph, amphibole; Ol, olivine.

intrusions, followed by the McMurdo volcanics (Hart *et al.* 1997; Rocchi *et al.* 2002). In BR and GP peridotite xenoliths, a sign of the metasomatic imprint related to the alkaline magmatism is provided by the oxygen isotope disequilibrium among minerals, accompanied by the core-to-rim trace element zoning of pyroxenes. These trace the metasomatic event around 40 Ma (Perinelli *et al.* 2006), in agreement with the youngest Re–Os ages determined on sulfides (i.e. 46 Ma) by Melchiorre *et al.* (2011). This fact is further supported by the presence within BR xenoliths of amphibole (and glass), which represent the last products of the metasomatic process triggered by the Cenozoic magmas. At HR, Pelorosso *et al.* (2017) proposed that over a period of 100–100 years complete overprinting of primary geochemical features could occur for crystals 1.5 mm in size. This seems an extremely short amount of time that, considering the average crystal size of the xenoliths, would have quickly eliminated both the primary depletion features and any indication of the Jurassic refertilization event. Around Mount Melbourne, the effects of the Cenozoic alkaline event are the most widespread, and are likely to have obliterated the effects of the Ferrar large magmatic event.

As reported by Coltorti *et al.* (2004) at BR, clinopyroxene underwent a remarkable chemical perturbation before amphibole growth. In terms of major elements, clinopyroxene is characterized by a progressive enrichment in Al_2O_3 (up to

12 wt% at constant Mg# of *c.* 91: see Fig. 6) (Coltorti *et al.* 2004) before transforming into amphibole. REE patterns change progressively from spoon-shaped to enriched patterns (Fig. 7): that is, from quadrant IV to quadrant I in Figure 10, irrespective of both the initial trend and the locality. Therefore, clinopyroxene from both GP and BR, starting from either quadrant II or IV, will move towards quadrant I of Figure 10. Clinopyroxene crystallizing from an alkaline melt tends to assume a hump-shaped pattern. In Figure 10, this corresponds to the spoon-shaped pattern of quadrant IV evolving into the humped-shaped patterns of quadrant II, crossing quadrant I. This is different to the origin of the strongly humped primary clinopyroxene, which is easily distinguished on chondrite-normalized values of La and Yb (Fig. 7a, c).

This REE trend, mostly represented at BR, is also recorded in Al_2O_3 v. Mg# trends in clinopyroxene of ultramafic rocks from both GP and HR (Fig. 6). This metasomatic episode caused a LREE enrichment in clinopyroxene before adding amphibole (and glass) to the pristine peridotite assemblage. This enrichment variably affected the previous clinopyroxene REE distributions, which were the results of polybaric and isobaric melting or refertilization:

1. clinopyroxene that suffered polybaric melting (i.e. from garnet to spinel facies) moved from quadrant II directly to quadrant I;
2. clinopyroxene refertilized by the Ferrar large magmatic event moved from quadrant II directly to quadrant I; and
3. variably depleted clinopyroxene in spinel peridotites from quadrant III shifted first towards quadrant IV and then to quadrant I; in some cases also reaching quadrant II (Fig. 10).

The third transition can be observed even within a single sample. It is evident that the last alkaline episode affected extremely heterogeneous mantle domains. These domains were characterized by the juxtaposition and intimate mingling of cumulate and mantle lithologies, as well as by the reaction products of these with various melts 'en route' to the surface. In this context, HP may represent a locality with an abundance of genetically different rocks. The amphibole at this locality is very similar to that found at Baker Rocks, both disseminated and in veins, although with slightly lower incompatible element contents (Coltorti *et al.* 2004; Pelorosso *et al.* 2019). It is highly likely that this amphibole bears witness to the final alkaline magmatic event, which gave rise to the HP neck. Similarly, the amphibole veins in sample 154G (BR) bear witness to the passage through the highly heterogeneous mantle of alkaline melts originating in the Mount Melbourne alkaline district.

At the moment, it is not possible to establish if amphibole growth was related to a distinct and younger alkaline metasomatism, distinct from the enrichment event that modified the pristine clinopyroxenes. These two processes could have been part of a unique continuous and progressive event, as modelled in Coltorti *et al.* (2004).

As pointed out by Pelorosso *et al.* (2016), BR and GP glass compositions are remarkably different for both major and trace element contents, and volatiles. Water contents in NAMs from the two localities are also quite different, with GP having the lowest values (Bonadiman *et al.* 2009). This fact suggests that substantially more depleted lithotypes were infiltrated by metasomatic melts, similar to those occurring at BR. The higher modal abundance of orthopyroxene could also explain the higher SiO_2 content of GP glasses (Coltorti *et al.* 2006). If this were the case, it would be expected that the clinopyroxene REE enrichment and the amphibole formation may be considered as a single event. The Na_2O/K_2O ratios of the two glass populations, however, are rather different. Together with the relatively high volatile content of BR and the older age of GP magmatic activity (Rocchi *et al.* 2002), this would suggest that the two events, although part of the same tectonomagmatic process, were separated in time. In this view, the ascent of hydrated alkaline melts in the SCLM beneath northern Victoria Land took place after the infiltration of highly alkaline and more anhydrous melts (Perinelli *et al.* 2011, 2017). This is testified by the occurrence of both BRP anhydrous and OVP hydrous cumulates (Perinelli *et al.* 2011, 2017). To our knowledge, no age determinations are available for these two outcrops but, from the morphological point of view, it is evident that Mount Overlord is the younger volcano, fostering the hypothesis that alkaline H_2O-rich magmatism followed a pristine less hydrous event.

Conclusions

The petrographical and geochemical features of mantle xenoliths from Baker Rocks, Greene Point and Handler Ridge, and of cumulate xenoliths at Harrow Peaks, Browning Pass and Mount Overlord, enabled us to reconstruct the main depletion and enrichment processes that took place in the subcontinental lithospheric mantle (SCLM) beneath northern Victoria Land. A similar evolution was identified for the two nearest and most studied mantle xenolith localities, namely Baker Rocks and Greene Point, although the former was more strongly affected by the final metasomatic event(s) that overprinted earlier chemical modifications.

The degree of partial melting determined by several independent methods gave comparable results. Lherzolitic and harzburgitic xenoliths record strong to mild depletion, with the highest degree of partial melting yielded at Greene Point (18–21% on average, with a maximum of 28%), the intermediate degree at Baker Rocks (12–16% on average, with a maximum of 19%) and the lowest at Handler Ridge (7–13% on average, with a maximum of 13%). The age of these depletion event(s), based on Re/Os ratios in sulfides, span from the Archean to the Cenozoic (Melchiorre *et al.* 2011). Relicts of polybaric melting that begin in the garnet facies and terminate in the spinel facies (Perinelli *et al.* 2006) are recorded in the Greene Point population, while most of the melting episodes recorded at Baker Rocks and Handler Ridge occurred in the spinel stability field.

A refertilization episode related to the Jurassic Ferrar large magmatic event was superimposed on these heterogeneous lithospheric mantle domains, especially at Greene Point and Baker Rocks. This event reintroduced orthopyroxene and clinopyroxene, with the latter characterized by a 'hump-shaped' pattern. Several lithologies, including an orthopyroxenite, olivine-websterites and a few lherzolites at both Harrow Peaks and Baker Rocks represent the cumulate counterparts of the Ferrar magmatic event.

Later, the Cenozoic alkaline magmatism responsible for the opening of the WARS introduced the last metasomatic enrichment(s) in the lithospheric domains. This was characterized by:

1. a progressive increase of the modal abundance of clinopyroxene;
2. an enrichment of its trace element contents;
3. the final appearance of amphibole, both disseminated and in veins; and
4. the formation of glass, just before the xenoliths were transported towards the surface.

Major and trace element compositions of clinopyroxene from cumulate rocks match the evolutionary enrichment path observed in clinopyroxene from mantle rocks. This supports the link between the last metasomatic episode observed within the mantle and the alkaline magmatism that gave rise to recent volcanoes such as Mount Melbourne or Mount Overlord.

Thermobarometric estimates suggest that Greene Point xenoliths record the highest temperatures and pressures, and the least oxidized conditions (T = 870–1059°C, P = 0.8–1.6 GPa and f_{O_2} values from −3.08 to −1.64 ΔFMQ). Within the mantle xenoliths, Baker Rocks and Handler Ridge tend to have more oxidized conditions at lower temperatures and pressures. Baker Rocks xenoliths yield temperatures of 940–1041°C, pressures of 0.7–1.1 GPa and f_{O_2} values from −2.85 to −0.87 ΔFMQ. Handler Ridge mantle rocks record temperatures of 790–1066°C, pressures of 0.7–1.0 GPa and f_{O_2} values from −2.08 to −1.07 ΔFMQ. Amphibole-bearing and amphibole-free parageneses at Baker Rocks, the only locality with amphibole-bearing mantle xenoliths, give similar oxygen fugacity (f_{O_2}) readings.

Baker Rocks and Handler Ridge cumulate lherzolites and wehrlites record temperatures between 968 and 1013°C, and oxygen fugacity ranging from −0.87 to +1.36 ΔFMQ. Harrow Peaks rocks record the shallowest equilibrium conditions (T = 893–959°C; P = 0.2–0.8 GPa), while Mount Overlord shows higher temperature (991–1106°C) and pressure values (0.8–1.5 GPa). Browning Pass has comparable temperatures (930–1170°C), although no pressure values could be determined. Oxygen fugacity is clearly distinct between Harrow Peaks (from −2.27 to +0.18 ΔFMQ) and Mount Overlord (from −0.22 to +0.76 ΔFMQ) or Browning Pass (from +0.42 to +1.49 ΔFMQ), with the last of these recording the highest oxidized conditions. The higher values are more representative of a crystallizing melt, while Baker Rocks and Harrow Peaks seem to represent a transition between mantle and cumulate values. Amongst the cumulate rocks, Mount Overlord is characterized by the widespread presence of amphibole, which, taking into account the high f_{O_2} value in Browning Pass, negates a strict correlation between amphibole stability, water activity and oxygen fugacity.

The H_2O content determined in nominally anhydrous minerals (NAMs) from Baker Rocks, Greene Point and Handler Ridge showed that the northern Victoria Land lithospheric mantle was rather anhydrous, with maximum whole-rock H_2O contents of 128 ppm. The alkaline metasomatic event, which added amphibole at BR, caused a remarkable increase in the whole-rock H_2O content, which ranges from 354 to 1120 ppm in lherzolites with disseminated amphibole and up to 1.42 wt% in the hornblenditic veins.

Acknowledgements Authors are grateful to Sergio Rocchi, Stephen Foley and Gerhard Wörner for their careful reviews, which significantly improved an earlier version of the manuscript. The Editor, Adam P. Martin, is acknowledged for his careful guidance.

Author contributions **MC**: conceptualization (lead), data curation (lead), funding acquisition (lead), methodology (lead), project administration (lead), software (supporting), supervision (lead), writing – original draft (lead), writing – review & editing (lead); **CB**: data curation (equal), formal analysis (equal), investigation (equal), methodology (lead), project administration (equal), writing – original draft (equal), writing – review & editing (supporting); **FC**: data curation (equal), formal analysis (supporting), resources (supporting), software (lead), writing – review & editing (equal); **BF**: data curation (lead), formal analysis (supporting), investigation (equal), methodology (equal), software (supporting), supervision (supporting), writing – original draft (equal), writing – review & editing (supporting); **PPG**: data curation (supporting), formal analysis (supporting), funding acquisition (equal), project administration (supporting), writing – review & editing (supporting); **BP**: data curation (equal), formal analysis (equal), investigation (equal), methodology (equal), software (equal), writing – review & editing (supporting); **CP**: data curation (equal), formal analysis (lead), investigation (lead), writing – original draft (supporting), writing – review & editing (supporting).

Funding This work was funded by the Italian National Research Programme with grant PRIN_2017 Project 20178LPCPW to M. Coltorti.

Data availability All data generated or analysed during this study are included in this published article (and its supplementary information files).

References

Arai, S. 1994. Characterization of spinel peridotites by olivine–spinel compositional relationships: Review and interpretation. *Chemical Geology*, **113**, 191–204, https://doi.org/10.1016/0009-2541(94)90066-3

Aviado, K.B., Rilling-Hall, S., Bryce, J.G. and Mukasa, S.B. 2015. Submarine and subaerial lavas in the West Antarctic Rift System: Temporal record of shifting magma source components from the lithosphere and asthenosphere. *Geochemistry, Geophysics, Geosystems*, **16**, 4344–4361, https://doi.org/10.1002/2015GC006076

Ballhaus, C., Berry, R.F. and Green, D.H. 1991. High pressure experimental calibration of the olivine–orthopyroxene–spinel oxygen geobarometer: implications for the oxidation state of the upper mantle. *Contributions to Mineralogy and Petrology*, **107**, 27–40, https://doi.org/10.1007/BF00311183

Beccaluva, L., Coltorti, M., Orsi, G., Saccani, E. and Siena, F. 1991. Nature and evolution of the sub-continental lithospheric mantle of Antarctica: evidence from ultramafic xenoliths of the Melbourne volcanic province (Northern Victoria Land, Antarctica). *Memorie della Società Geologica Italiana*, **46**, 353–370.

Beccaluva, L., Bonadiman, C., Coltorti, M., Salvini, L. and Siena, F. 2001. Depletion events, nature of metasomatizing agent and timing of enrichment processes in lithospheric mantle xenoliths from the Veneto Volcanic Province. *Journal of Petrology*, **42**, 173–188, https://doi.org/10.1093/petrology/42.1.173

Bédard, J.H. 2014. Parameterizations of calcic clinopyroxene–melt trace element partition coefficients. *Geochemistry, Geophysics, Geosystems*, **15**, 303–336, https://doi.org/10.1002/2013GC005112

Bénard, A., Woodland, A.B., Arculus, R.J., Nebel, O. and McAlpine, S.R. 2018. Variation in sub-arc mantle oxygen fugacity during partial melting recorded in refractory peridotite xenoliths from the West Bismarck Arc. *Chemical Geology*, **486**, 16–30, https://doi.org/10.1016/j.chemgeo.2018.03.004

Bonadiman, C. and Coltorti, M. 2011. Numerical modelling for peridotite phase melting trends in the SiO_2–Al_2O_3–FeO–MgO–CaO system at 2 GPa. *Mineralogical Magazine*, **75**, 548.

Bonadiman, C., Beccaluva, L., Coltorti, M. and Siena, F. 2005. Kimberlite-like metasomatism and 'garnet signature' in spinel–peridotite xenoliths from Sal, Cape Verde Archipelago: Relics of a subcontinental mantle domain within the Atlantic oceanic lithosphere? *Journal of Petrology*, **46**, 2465–2493, https://doi.org/10.1093/petrology/egi061

Bonadiman, C., Hao, Y., Coltorti, M., Dallai, L., Faccini, B., Huang, Y. and Xia, Q. 2009. Water contents of pyroxenes in intraplate lithospheric mantle. *European Journal of Mineralogy*, **21**, 637–647, https://doi.org/10.1127/0935-1221/2009/0021-1935

Bonadiman, C., Nazzareni, S., Coltorti, M., Comodi, P., Giuli, G. and Faccini, B. 2014. Crystal chemistry of amphiboles: implications for oxygen fugacity and water activity in lithospheric mantle beneath Victoria Land, Antarctica. *Contributions to Mineralogy and Petrology*, **167**, 984, https://doi.org/10.1007/s00410-014-0984-8

Brey, G.P. and Köhler, T. 1990. Geothermobarometry in four-phase lherzolites II. New thermobarometers, and practical assessment of existing thermobarometers. *Journal of Petrology*, **31**, 1353–1378, https://doi.org/10.1093/petrology/31.6.1353

Broadley, M.W., Ballentine, C.J., Chavrit, D., Dallai, L. and Burgess, R. 2016. Sedimentary halogens and noble gases within Western Antarctic xenoliths: Implications of extensive volatile recycling to the sub continental lithospheric mantle. *Geochimica et Cosmochimica Acta*, **176**, 139–156, https://doi.org/10.1016/j.gca.2015.12.013

Cabato, J.A., Stefano, C.J. and Mukasa, S.B. 2015. Volatile concentrations in olivine-hosted melt inclusions from the Columbia River flood basalts and associated lavas of the Oregon Plateau: implications for magma genesis. *Chemical Geology*, **392**, 59–73, https://doi.org/10.1016/j.chemgeo.2014.11.015

Chand, S., Radhakrishna, M. and Subrahmanyam, C. 2001. India–East Antarctica conjugate margins: rift-shear tectonic setting inferred from gravity and bathymetry data. *Earth and Planetary Science Letters*, **185**, 225–236, https://doi.org/10.1016/S0012-821X(00)00349-6

Coltorti, M., Bonadiman, C., Hinton, R.W., Siena, F. and Upton, B.G.J. 1999. Carbonatite metasomatism of the oceanic upper mantle: evidence from clinopyroxenes and glasses in ultramafic xenoliths of Grande Comore, Indian Ocean. *Journal of Petrology*, **40**, 133–165, https://doi.org/10.1093/petroj/40.1.133

Coltorti, M., Beccaluva, L., Bonadiman, C., Faccini, B., Ntaflos, T. and Siena, F. 2004. Amphibole genesis via metasomatic reaction with clinopyroxene in mantle xenoliths from Victoria Land, Antarctica. *Lithos*, **75**, 115–139, https://doi.org/10.1016/j.lithos.2003.12.021

Coltorti, M., Bonadiman, C., Faccini, B., Melchiorre, M., Ntaflos, T. and Siena, F. 2006. Mantle xenoliths from Northern Victoria land, Antarctica: evidence for heterogeneous lithospheric metasomatism. *Geochimica et Cosmochimica Acta*, **70**, A108–A108, https://doi.org/10.1016/j.gca.2006.06.130

Correale, A., Pelorosso, B., Rizzo, A.L., Coltorti, M., Italiano, F., Bonadiman, C. and Giacomoni, P.P. 2019. The nature of the West Antarctic Rift System as revealed by noble gases in mantle minerals. *Chemical Geology*, **524**, 104–118, https://doi.org/10.1016/j.chemgeo.2019.06.020

Cox, K.G. 1978. Flood basalts, subduction and the break-up of Gondwanaland. *Nature*, **274**, 47–49, https://doi.org/10.1038/274047a0

Davis, F.A., Cottrell, E., Birner, S.K., Warren, J.M. and Lopez, O.G. 2017. Revisiting the electron microprobe method of spinel–olivine–orthopyroxene oxybarometry applied to spinel peridotites. *American Mineralogist*, **102**, 421–435, https://doi.org/10.2138/am-2017-5823

Dick, H.J. and Bullen, T. 1984. Chromian spinel as a petrogenetic indicator in abyssal and alpine-type peridotites and spatially associated lavas. *Contributions to Mineralogy and Petrology*, **86**, 54–76, https://doi.org/10.1007/BF00373711

Elliot, D.H. and Fleming, T.H. 2018. The Ferrar Large Igneous Province: field and geochemical constraints on supra-crustal (high-level) emplacement of the magmatic system. *Geological Society, London, Special Publications*, **463**, 41–58, https://doi.org/10.1144/SP463.1

Estrada, S., Läufer, A., Eckelmann, K., Hofmann, M., Gärtner, A. and Linnemann, U. 2016. Continuous Neoproterozoic to Ordovician sedimentation at the East Gondwana margin – Implications from detrital zircons of the Ross Orogen in northern Victoria Land, Antarctica. *Gondwana Research*, **37**, 426–448, https://doi.org/10.1016/j.gr.2015.10.006

Fabriès, J. 1979. Spinel–olivine geothermometry in peridotites from ultramafic complexes. *Contributions to Mineralogy and Petrology*, **69**, 329–336, https://doi.org/10.1007/BF00372258

Fitzgerald, P.G. and Stump, E. 1997. Cretaceous and Cenozoic episodic denudation of the Transantarctic Mountains, Antarctica: New constraints from apatite fission track thermochronology in the Scott Glacier region. *Journal of Geophysical Research: Solid Earth*, **102**, 7747–7765, https://doi.org/10.1029/96JB03898

Foley, S.F. 2011. A reappraisal of redox melting in the Earth's mantle as a function of tectonic setting and time. *Journal of Petrology*, **52**, 1363–1391, https://doi.org/10.1093/petrology/egq061

Frost, B.R. 1991. Introduction to oxygen fugacity and its petrologic importance. *Reviews in Mineralogy and Geochemistry*, **25**, 1–9.

Gamble, J.A., McGibbon, F., Kyle, P.R., Menzies, M.A. and Kirsch, I. 1988. Metasomatized xenoliths from Foster Crater, Antarctica: implications for lithosphere structure and processes beneath the Transantarctic Mountains. *Journal of Petrology, Special Lithosphere Issue*, 109–138, https://doi.org/10.1093/petrology/Special_Volume.1.109

Gentili, S., Bonadiman, C., Biagioni, C., Comodi, P., Coltorti, M., Zucchini, A. and Ottolini, L. 2015. Oxo-amphiboles in mantle xenoliths: evidence for H_2O-rich melt interacting with the lithospheric mantle of Harrow Peaks (Northern Victoria Land, Antarctica). *Mineralogy and Petrology*, **109**, 741–759, https://doi.org/10.1007/s00710-015-0404-4

Giacomoni, P.P., Bonadiman, C., Casetta, F., Faccini, B., Ferlito, C., Ottolini, L., Zanetti, A. and Coltorti, M. 2020. Long-term storage of subduction-related volatiles in Northern Victoria Land lithospheric mantle: Insight from olivine-hosted melt inclusions from McMurdo basic lavas (Antarctica). *Lithos*, **378**, 105826, https://doi.org/10.1016/j.lithos.2020.105826

Grant, K., Ingrin, J., Lorand, J.P. and Dumas, P. 2007. Water partitioning between mantle minerals from peridotite xenoliths. *Contributions to Mineralogy and Petrology*, **154**, 15–34, https://doi.org/10.1007/s00410-006-0177-1

Greenfield, A.M.R., Ghent, E.D. and Russell, J.K. 2013. Geothermobarometry of spinel peridotites from southern British Columbia: implications for the thermal conditions in the upper mantle. *Canadian Journal of Earth Sciences*, **50**, 1019–1032, https://doi.org/10.1139/cjes-2013-0037

Harrington, H.J. 1958. Nomenclature of rock units in the Ross Sea region, Antarctica. *Nature*, **182**, 290–290, https://doi.org/10.1038/182290a0

Hart, S.R., Blusztajn, J., LeMasurier, W.E. and Rex, D.C. 1997. Hobbs Coast Cenozoic volcanism: implications for the West Antarctic rift system. *Chemical Geology*, **139**, 223–248, https://doi.org/10.1016/S0009-2541(97)00037-5

Hauri, E.H., Gaetani, G.A. and Green, T.H. 2004: Partitioning of H_2O between mantle minerals and silicate melts. *Geochimica et Cosmochimica Acta*, **68**, A33.

Herzberg, C. and Asimow, P.D. 2008. Petrology of some oceanic island basalts: PRIMELT2.XLS software for primary magma calculation. *Geochemistry, Geophysics, Geosystems*, **9**, 1–25, https://doi.org/10.1029/2008GC002057

Hornig, I., Wörner, G. and Zipfel, J. 1991. Lower crustal and mantle xenoliths from Mt. Melbourne Volcanic Field, Northern Victoria Land, Antarctica. *Memorie della Società Geologica Italiana*, **46**, 337–352.

Hudgins, T.R., Mukasa, S.B., Simon, A.C., Moore, G. and Barifaijo, E. 2015. Melt inclusion evidence for CO_2-rich melts beneath the western branch of the East African Rift: implications for long-term storage of volatiles in the deep lithospheric mantle. *Contributions to Mineralogy and Petrology*, **169**, 46, https://doi.org/10.1007/s00410-015-1140-9

Ingrin, J. and Blanchard, M. 2006. Diffusion of hydrogen in minerals. *Reviews in Mineralogy and Geochemistry*, **62**, 291–320, https://doi.org/10.2138/rmg.2006.62.13

Ionov, D.A., Doucet, L.S., Xu, Y., Golovin, A.V. and Oleinikov, O.B. 2018. Reworking of Archean mantle in the NE Siberian craton by carbonatite and silicate melt metasomatism: Evidence from a carbonate-bearing, dunite-to-websterite xenolith suite from the Obnazhennaya kimberlite. *Geochimica et Cosmochimica Acta*, **224**, 132–153, https://doi.org/10.1016/j.gca.2017.12.028

Ivanov, A.V., Mukasa, S.B. *et al.* 2018. Volatile concentrations in olivine-hosted melt inclusions from meimechite and melanephelinite lavas of the Siberian Traps Large Igneous Province:

Evidence for flux-related high-Ti, high-Mg magmatism. *Chemical Geology*, **483**, 442–462, https://doi.org/10.1016/j.chemgeo.2018.03.011

Jianping, L., Kornprobst, J. and Provost, A. 1995. Spinel as a chemical indicator during partial melting and subsolidus equilibration of mantle peridotite: experimental study and application in natural rocks. *Acta Geologica Sinica*, **2**, 6.

Johnson, K.T., Dick, H.J. and Shimizu, N. 1990. Melting in the oceanic upper mantle: An ion microprobe study of diopsides in abyssal peridotites. *Journal of Geophysical Research: Solid Earth*, **95**, 2661–2678, https://doi.org/10.1029/JB095iB03p02661

Kleinschmidt, G., Tessensohn, F. and Vetter, U. 1987. Paleozoic accretion at the Paleopacific margin of Antarctica. *Polarforschung*, **57**, 1–8.

Köhler, T.P. and Brey, G. 1990. Calcium exchange between olivine and clinopyroxene calibrated as a geothermobarometer for natural peridotites from 2 to 60 kb with applications. *Geochimica et Cosmochimica Acta*, **54**, 2375–2388, https://doi.org/10.1016/0016-7037(90)90226-B

Kyle, P.R. and Muncy, H.L. 1989. Geology and geochronology of McMurdo Volcanic Group rocks in the vicinity of Lake Morning, McMurdo Sound, Antarctica. *Antarctic Science*, **1**, 345–350, https://doi.org/10.1017/S0954102089000520

Lazarov, M., Woodland, A.B. and Brey, G.P. 2009. Thermal state and redox conditions of the Kaapvaal mantle: a study of xenoliths from the Finsch mine, South Africa. *Lithos*, **112**, 913–923, https://doi.org/10.1016/j.lithos.2009.03.035

LeMasurier, W.E. and Landis, C.A. 1996. Mantle-plume activity recorded by low-relief erosion surfaces in West Antarctica and New Zealand. *Geological Society of America Bulletin*, **108**, 1450–1466, https://doi.org/10.1130/0016-7606(1996)108<1450:MPARBL>2.3.CO;2

Le Roux, V., Bodinier, J.L., Alard, O., O'Reilly, S.Y. and Griffin, W.L. 2009. Isotopic decoupling during porous melt flow: A case-study in the Lherz peridotite. *Earth and Planetary Science Letters*, **279**, 76–85, https://doi.org/10.1016/j.epsl.2008.12.033

Litasov, K.D. and Ohtani, E. 2009. Solidus and phase relations of carbonated peridotite in the system $CaO–Al_2O_3–MgO–SiO_2–Na_2O–CO_2$ to the lower mantle depths. *Physics of the Earth and Planetary Interiors*, **177**, 46–58, https://doi.org/10.1016/j.pepi.2009.07.008

MacGregor, I.D. 2015. Empirical geothermometers and geothermobarometers for spinel peridotite phase assemblages. *International Geology Review*, **57**, 1940–1974, https://doi.org/10.1080/00206814.2015.1045307

Mallmann, G. and O'Neill, H.S.C. 2013. Calibration of an empirical thermometer and oxybarometer based on the partitioning of Sc, Y and V between olivine and silicate melt. *Journal of Petrology*, **54**, 933–949, https://doi.org/10.1093/petrology/egt001

Martin, A.P., Price, R.C., Cooper, A.F. and McCammon, C.A. 2015. Petrogenesis of the rifted southern Victoria Land lithospheric mantle, Antarctica, inferred from petrography, geochemistry, thermobarometry and oxybarometry of peridotite and pyroxenite xenoliths from the Mount Morning eruptive centre. *Journal of Petrology*, **56**, 193–226, https://doi.org/10.1093/petrology/egu075

Matjuschkin, V., Brey, G.P., Höfer, H.E. and Woodland, A.B. 2014. The influence of Fe^{3+} on garnet–orthopyroxene and garnet–olivine geothermometers. *Contributions to Mineralogy and Petrology*, **167**, 972, https://doi.org/10.1007/s00410-014-0972-z

Mattioli, G.S. and Wood, B.J. 1988. Magnetite activities across the $MgAl_2O_4–Fe_3O_4$ spinel join, with application to thermobarometric estimates of upper mantle oxygen fugacity. *Contributions to Mineralogy and Petrology*, **98**, 148–162, https://doi.org/10.1007/BF00402108

McIntosh, W.C. 2000. $^{40}Ar/^{39}Ar$ geochronology of tephra and volcanic clasts in CRP-2A, Victoria Land Basin, Antarctica. *Terra Antartica*, **7**, 621–630.

Melchiorre, M., Coltorti, M., Bonadiman, C., Faccini, B., O'Reilly, S.Y. and Pearson, N.J. 2011. The role of eclogite in the rift-related metasomatism and Cenozoic magmatism of Northern Victoria Land, Antarctica. *Lithos*, **124**, 319–330, https://doi.org/10.1016/j.lithos.2010.11.012

Melchiorre, M., Faccini, B., Grégoire, M., Benoit, M., Casetta, F. and Coltorti, M. 2020. Melting and metasomatism/refertilisation processes in the Patagonian sub-continental lithospheric mantle: a review. *Lithos*, **354**, 105324, https://doi.org/10.1016/j.lithos.2019.105324

Mercier, J.C.C. 1980. Single-pyroxene thermobarometry. *Tectonophysics*, **70**, 1–37, https://doi.org/10.1016/0040-1951(80)90019-0

Mercier, J.C.C. and Nicolas, A. 1975. Textures and fabrics of upper-mantle peridotites as illustrated by xenoliths from basalts. *Journal of Petrology*, **16**, 454–487, https://doi.org/10.1093/petrology/16.1.454

Mukasa, S.B. and Dalziel, I.W. 2000. Marie Byrd Land, West Antarctica: Evolution of Gondwana's Pacific margin constrained by zircon U–Pb geochronology and feldspar common-Pb isotopic compositions. *Geological Society of America Bulletin*, **112**, 611–627, https://doi.org/10.1130/0016-7606(2000)112<611:MBLWAE>2.0.CO;2

Müller, P., Schmidt-Thomé, M., Kreuzer, H., Tessensohn, F. and Vetter, U. 1991. Cenozoic peralkaline magmatism at the western margin of the Ross Sea, Antarctica. *Memorie della Società Geologica Italiana*, **46**, 315–336.

Nardini, I., Armienti, P., Rocchi, S., Dallai, L. and Harrison, D. 2009. Sr–Nd–Pb–He–O isotope and geochemical constraints on the genesis of Cenozoic magmas from the West Antarctic Rift. *Journal of Petrology*, **50**, 1359–1375, https://doi.org/10.1093/petrology/egn082

Nimis, P. and Taylor, W.R. 2000. Single clinopyroxene thermobarometry for garnet peridotites. Part I. Calibration and testing of a Cr-in-Cpx barometer and an enstatite-in-Cpx thermometer. *Contributions to Mineralogy and Petrology*, **139**, 541–554, https://doi.org/10.1007/s004100000156

Nimis, P. and Ulmer, P. 1998. Clinopyroxene geobarometry of magmatic rocks Part 1: An expanded structural geobarometer for anhydrous and hydrous, basic and ultrabasic systems. *Contributions to Mineralogy and Petrology*, **133**, 122–135, https://doi.org/10.1007/s004100050442

Niu, Y. 1997. Mantle melting and melt extraction processes beneath ocean ridges: evidence from abyssal peridotites. *Journal of Petrology*, **38**, 1047–1074, https://doi.org/10.1093/petroj/38.8.1047

Niu, Y. 2004. Bulk-rock major and trace element compositions of abyssal peridotites: implications for mantle melting, melt extraction and post-melting processes beneath mid-ocean ridges. *Journal of Petrology*, **45**, 2423–2458, https://doi.org/10.1093/petrology/egh068

O'Neill, H.S.C. and Wall, V.J. 1987. The olivine–orthopyroxene–spinel oxygen geobarometer, the nickel precipitation curve, and the oxygen fugacity of the Earth's upper mantle. *Journal of Petrology*, **28**, 1169–1191, https://doi.org/10.1093/petrology/28.6.1169

Panter, K.S., Castillo, P. *et al.* 2018. Melt origin across a rifted continental margin: a case for subduction-related metasomatic agents in the lithospheric source of alkaline basalt, NW Ross Sea, Antarctica. *Journal of Petrology*, **59**, 517–558, https://doi.org/10.1093/petrology/egy036

Pearce, J.A., Barker, P.F., Edwards, S.J., Parkinson, I.J. and Leat, P.T. 2000. Geochemistry and tectonic significance of peridotites from the South Sandwich arc–basin system, South Atlantic. *Contributions to Mineralogy and Petrology*, **139**, 36–53, https://doi.org/10.1007/s004100050572

Pelorosso, B., Bonadiman, C., Coltorti, M., Faccini, B., Melchiorre, M., Ntaflos, T. and Gregoire, M. 2016. Pervasive, tholeiitic refertilisation and heterogeneous metasomatism in Northern Victoria Land lithospheric mantle (Antarctica). *Lithos*, **248**, 493–505, https://doi.org/10.1016/j.lithos.2016.01.032

Pelorosso, B., Bonadiman, C. *et al.* 2017. Role of percolating melts in Antarctic subcontinental lithospheric mantle: New insights from Handler Ridge mantle xenoliths (northern Victoria Land,

Antarctica). *Geological Society of America Special Papers*, **526**, 133–150.

Pelorosso, B., Bonadiman, C., Ntaflos, T., Gregoire, M., Gentili, S., Zanetti, A. and Coltorti, M. 2019. An insight into the first stages of the Ferrar magmatism: ultramafic cumulates from Harrow Peaks, northern Victoria Land, Antarctica. *Contributions to Mineralogy and Petrology*, **174**, 44, https://doi.org/10.1007/s00410-019-1579-1

Perinelli, C., Armienti, P., Trigila, R. and Aurisicchio, C. 1998. Intergranular melt inclusions within ultramafic xenoliths from Baker Rocks and Greene Point volcanics (Northern Victoria Land, Antarctica). *Terra Antartica*, **5**, 217–233.

Perinelli, C., Armienti, P. and Dallai, L. 2006. Geochemical and O-isotope constraints on the evolution of lithospheric mantle in the Ross Sea rift area (Antarctica). *Contributions to Mineralogy and Petrology*, **151**, 245–266, https://doi.org/10.1007/s00410-006-0065-8

Perinelli, C., Armienti, P. and Dallai, L. 2011. Thermal evolution of the lithosphere in a rift environment as inferred from the geochemistry of mantle cumulates, Northern Victoria Land, Antarctica. *Journal of Petrology*, **52**, 665–690, https://doi.org/10.1093/petrology/egq099

Perinelli, C., Andreozzi, G.B., Conte, A.M., Oberti, R. and Armienti, P. 2012. Redox state of subcontinental lithospheric mantle and relationships with metasomatism: insights from spinel peridotites from northern Victoria Land (Antarctica). *Contributions to Mineralogy and Petrology*, **164**, 1053–1067, https://doi.org/10.1007/s00410-012-0788-7

Perinelli, C., Gaeta, M. and Armienti, P. 2017. Cumulate xenoliths from Mt. Overlord, northern Victoria Land, Antarctica: A window into high pressure storage and differentiation of mantle-derived basalts. *Lithos*, **268**, 225–239, https://doi.org/10.1016/j.lithos.2016.10.027

Pike, J.N. and Schwarzman, E.C. 1977. Classification of textures in ultramafic xenoliths. *The Journal of Geology*, **85**, 49–61, https://doi.org/10.1086/628268

Riley, T.R., Leat, P.T., Curtis, M.L., Millar, I.L., Duncan, R.A. and Fazel, A. 2005. Early–Middle Jurassic dolerite dykes from Western Dronning Maud Land (Antarctica): identifying mantle sources in the Karoo large igneous province. *Journal of Petrology*, **46**, 1489–1524, https://doi.org/10.1093/petrology/egi023

Rocchi, S., Armienti, P., D'Orazio, M., Tonarini, S., Wijbrans, J. and Di Vincenzo, G.D. 2002. Cenozoic magmatism in the western Ross Embayment: Role of mantle plume v. plate dynamics in the development of the West Antarctic Rift System. *Journal of Geophysical Research: Solid Earth*, **107**, 2195, https://doi.org/10.1029/2001JB000515

Rocchi, S., Armienti, P. and Di Vincenzo, G. 2005. No plume, no rift magmatism in the West Antarctic Rift. *Geological Society of America Special Papers*, **388**, 435–447.

Stefano, C.J., Mukasa, S.B., Andronikov, A. and Leeman, W.P. 2011. Water and other volatile systematics of olivine-hosted melt inclusions from the Yellowstone hotspot track. *Contributions to Mineralogy and Petrology*, **161**, 615–633, https://doi.org/10.1007/s00410-010-0553-8

Sun, S.-s. and McDonough, W.F. 1989. Chemical and isotopic systematics of oceanic basalts: implications for mantle composition and processes. *Geological Society, London, Special Publications*, **42**, 313–345, https://doi.org/10.1144/GSL.SP.1989.042.01.19

Taylor, W.R. 1998. An experimental test of some geothermometer and geobarometer formulations for upper mantle peridotites with application to the thermobarometry of fertile lherzolite and garnet websterite. *Neues Jahrbuch für Mineralogie – Abhandlungen*, **172**, 381–408, https://doi.org/10.1127/njma/172/1998/381

Tessensohn, F. and Wörner, G. 1991. The Ross Sea Rift System (Antarctica): structure, evolution and analogues. *In*: Thomson, M.R.A., Crame, J.A. and Thomson, J.W. (eds) *Geological Evolution of Antarctica*. Cambridge University Press, Cambridge, UK, 273–277.

Tonarini, S., Rocchi, S., Armienti, P. and Innocenti, F. 1997. Constraints on timing of Ross Sea rifting inferred from Cainozoic intrusions from northern Victoria Land, Antarctica. *In*: Ricci, C.A. (ed.) *The Antarctic Region: Geological Evolution and Processes*. Terra Antarctica, Siena, Italy, 511–521.

Uenver-Thiele, L., Woodland, A.B., Downes, H. and Altherr, R. 2014. Oxidation state of the lithospheric mantle below the Massif Central, France. *Journal of Petrology*, **55**, 2457–2480, https://doi.org/10.1093/petrology/egu063

Upton, B.G.J., Downes, H., Kirstein, L.A., Bonadiman, C., Hill, P.G. and Ntaflos, T. 2011. The lithospheric mantle and lower crust–mantle relationships under Scotland: a xenolithic perspective. *Journal of the Geological Society, London*, **168**, 873–886, https://doi.org/10.1144/0016-76492009-172

Wang, X.C., Wilde, S.A., Xu, B. and Pang, C.J. 2016. Origin of arc-like continental basalts: Implications for deep-Earth fluid cycling and tectonic discrimination. *Lithos*, **261**, 5–45, https://doi.org/10.1016/j.lithos.2015.12.014

Wells, P.R. 1977. Pyroxene thermometry in simple and complex systems. *Contributions to Mineralogy and Petrology*, **62**, 129–139, https://doi.org/10.1007/BF00372872

Wood, B.J. 1990. An experimental test of the spinel peridotite oxygen barometer. *Journal of Geophysical Research: Solid Earth*, **95**, 15 845–15 851, https://doi.org/10.1029/JB095iB10p15845

Wood, B.J. 1991. Oxygen barometry of spinel peridotites. *Reviews in Mineralogy*, **25**, 417–431.

Wood, B.J. and Blundy, J.D. 2001. The effect of cation charge on crystal–melt partitioning of trace elements. *Earth and Planetary Science Letters*, **188**, 59–71, https://doi.org/10.1016/S0012-821X(01)00294-1

Wood, B.J. and Virgo, D. 1989. Upper mantle oxidation state: Ferric iron contents of lherzolite spinels by ^{57}Fe Mössbauer spectroscopy and resultant oxygen fugacities. *Geochimica et Cosmochimica Acta*, **53**, 1277–1291, https://doi.org/10.1016/0016-7037(89)90062-8

Wood, B.J., Bryndzia, L.T. and Johnson, K.E. 1990. Mantle oxidation state and its relationship to tectonic environment and fluid speciation. *Science*, **248**, 337–345, https://doi.org/10.1126/science.248.4953.337

Woodland, A.B., Kornprobst, J. and Wood, B.J. 1992. Oxygen thermobarometry of orogenic lherzolite massifs. *Journal of Petrology*, **33**, 203–230, https://doi.org/10.1093/petrology/33.1.203

Woodland, A.B., Kornprobst, J., McPherson, E., Bodinier, J.-L. and Menzies, M.A. 1996. Metasomatic interactions in the lithospheric mantle: petrologic evidence from the Lherz massif, French Pyrenees. *Chemical Geology*, **134**, 83–112, https://doi.org/10.1016/S0009-2541(96)00082-4

Woodland, A.B., Kornprobst, J. and Tabit, A. 2006. Ferric iron in orogenic lherzolite massifs and controls of oxygen fugacity in the upper mantle. *Lithos*, **89**, 222–241, https://doi.org/10.1016/j.lithos.2005.12.014

Wörner, G. 1999. Lithospheric dynamics and mantle sources of alkaline magmatism of the Cenozoic West Antarctic Rift System. *Global and Planetary Change*, **23**, 61–77, https://doi.org/10.1016/S0921-8181(99)00051-X

Wörner, G., Fricke, A. and Burke, E.A.J. 1993. Fluid-inclusion studies on lower crustal gabbroic xenoliths from the Mt. Melbourne Volcanic field (Antarctica): Evidence for the post-crystallization uplift history during Cenozoic Ross Sea Rifting. *European Journal of Mineralogy*, **5**, 775–785, https://doi.org/10.1127/ejm/5/4/0775

Zipfel, J. and Wörner, G. 1992. Four- and five-phase peridotites from a continental rift system: evidence for upper mantle uplift and cooling at the Ross Sea margin (Antarctica). *Contributions to Mineralogy and Petrology*, **111**, 24–36, https://doi.org/10.1007/BF00296575

Zou, H. 1998. Trace element fractionation during modal and nonmodal dynamic melting and open-system melting: a mathematical treatment. *Geochimica et Cosmochimica Acta*, **62**, 1937–1945, https://doi.org/10.1016/S0016-7037(98)00115-X

Marie Byrd Land lithospheric mantle: a review of the xenolith record

Monica R. Handler[1]*, Richard J. Wysoczanski[2] and John A. Gamble[1]

[1]School of Geography, Environment and Earth Sciences, Te Herenga Waka–Victoria University of Wellington, New Zealand

[2]National Institute of Water and Atmospheric Research, Private Bag 14901, Wellington, New Zealand

MRH, 0000-0001-7095-0835; RJW, 0000-0002-7941-1608

*Correspondence: monica.handler@vuw.ac.nz

Abstract: The Marie Byrd Land (MBL) lithospheric mantle xenolith record comprises over 100 samples from a range of localities spanning both major crustal terranes that comprise MBL: Ross and Amundsen provinces. Coarse granular to porphyroclastic in texture, the xenoliths are predominantly Type I spinel-bearing lherzolites to harburgites, but include rare dunite and pyroxenite examples. Garnet is absent and no hydrous phases, such as amphibole or mica, have been reported to date, although traces of apatite may be present. Characterisation of the lithospheric mantle composition and its evolution however, is hampered by patchy and uneven geochemical analyses across the xenolith suite. Nonetheless, a picture emerges of a heterogeneous lithosphere beneath both Ross and Amundsen Provinces. Previously published and new data reported here are consistent with samples ranging from variably cryptically metasomatized residua from variable (10–25%) degrees of partial melt extraction to refertilized compositions. Limited isotopic data point to a complex history, providing evidence for both ancient Proterozoic lithospheric mantle and preservation of Ordovician events. The Sr–Nd–Pb composition of the sampled lithospheric mantle overlaps the common low-μ isotopic endmember identified in Cenozoic magmatism from MBL and the wider West Antarctic Rift System.

Supplementary material: Analytical methods and data tables are available at https://doi.org/10.6084/m9.figshare.c.5309814

Marie Byrd Land forms one of four major crustal blocks comprising the composite West Antarctica (Fig. 1a). For much of its evolution Marie Byrd Land was part of the paleo-Pacific Gondwana margin and in this setting experienced a protracted Paleozoic history defined by convergence and subduction (Fig. 1b). Currently however, it forms the northern flank of the West Antarctic Rift System (WARS), a continental rift complex that initiated in the late Cretaceous and now extends for over 2500 km along the eastern margin of the Antarctic continent from Cape Adare to the base of the Antarctic Peninsula (Fig. 1a; LeMasurier and Thomson 1990; Storey *et al.* 1999; Jordan *et al.* 2020). As part of the WARS, Marie Byrd Land has been a focus for intraplate volcanism dating back at least 37 Ma (Kyle and Muncy 1989; LeMasurier and Thomson 1990; Smellie *et al.* 2020; Panter *et al.* 2021).

Much of West Antarctic surface geology is obscured by extensive ice sheets, and over large tracts of country only the peaks of basanite-trachyte–phonolite stratovolcanoes (e.g. Mount Sidley, 4285 m) are exposed through the veneer of ice (e.g. Fig. 1). Surface exposure is particularly limited in Marie Byrd Land. However, basanitic scoria cones associated with and peripheral to the major Cenozoic stratovolcanoes are commonly host to diverse suites of xenoliths comprising upper-crustal (granitoids and metamorphic rocks), lower-crustal (mafic and felsic granulites) and mantle (spinel facies peridotites and pyroxenites) xenoliths, and megacrysts, thereby providing key insights into the composition and history of the deeper lithosphere.

Lithospheric mantle xenoliths have been described from several Marie Byrd Land Cenozoic volcanic centres in the Executive Committee Range, Usas Escarpment, and Fosdick Mountains, and from the coastal Mount Murphy (Fig. 1), although limited mineralogy and geochemical data have been reported to date (Wysoczanski and Gamble 1992; Wysoczanski 1993; Handler *et al.* 2003; Chatzaras *et al.* 2016; Cohen 2016). This contribution presents a review of the composition and petrogenesis of Marie Byrd Land lithospheric mantle as discerned from the record of mantle xenoliths entrained in Cenozoic magmas, summarizing the available literature data and reporting previously unpublished mineral and isotopic data on subsets of the xenoliths.

Geological setting

Prior to *c.* 85 Ma, Marie Byrd Land was contiguous with New Zealand (Zealandia) as part of the Gondwana margin, linking eastern Australia with the Antarctic Penninsula and South America (Jordan *et al.* 2020; Smellie *et al.* 2020). This margin was characterised by a prolonged period of convergence with magmatic arc(s) represented by Devonian to Late Cretaceous peraluminous, calc-alkaline series granites now exposed in coastal Marie Byrd Land (Larter *et al.* 2002). The granitoids intrude Neoproterozoic to lower Paleozoic metamorphic basement (Pankhurst *et al.* 1998; Mukasa and Dalziel 2000).

Marie Byrd Land basement has been divided into two terranes on the basis of differing granite Nd model ages: the Ross Province to the south (T_{DM} *c.* 1.5–1.3 Ga), and the Amundson Province to the north (T_{DM} *c.* 1.3–1.0 Ga) (Pankhurst *et al.* 1998) (Fig. 1). In central Marie Byrd Land, compositional differences in lower crustal xenoliths from the Executive Committee Range are consistent with the definition of these two crustal terranes and likely constrain their boundary to between mounts Sidley and Hampton (Wysoczanski *et al.* 1995; Fig. 1). Paleomagnetic data suggest that these two provinces have shared a common history since mid-Cretaceous times, prior to the separation of Marie Byrd Land and Zealandia (DiVenere *et al.* 1996; Luyendyk *et al.* 1996).

The age of the basement to Marie Byrd Land remains enigmatic however, with the only direct samples of lower crustal rocks proving resistant to dating techniques (Wysoczanski 1993; Wysoczanski *et al.* 1995). The oldest crustal unit exposed in the Ross Province is the Neoproterozoic to lower Paleozoic Swanson Formation, which has been correlated with metaturbidites in western New Zealand and Northern Victoria Land, (e.g. Fig. 1b) (Adams 1986; Pankhurst *et al.* 1998), and lower Paleozoic gneissic crystalline basement is exposed in the vicinity of Mount Murphy (Fig. 1a) (Pankhurst *et al.* 1998; Jordan *et al.* 2020). The Amundsen Province has been correlated with the Zealandia Median Batholith in New Zealand, with the oldest known crustal rocks Ordovician–Silurian calc-alkaline granitoids (e.g. Pankhurst *et al.* 1998; Smellie *et al.* 2020).

Proterozoic ages associated with the widespread granitoid suites – both whole-rock Nd model ages and U–Pb ages of

From: Martin, A. P. and van der Wal, W. (eds) 2023. *The Geochemistry and Geophysics of the Antarctic Mantle*. Geological Society, London, Memoirs, **56**, 83–100,
First published online 16 July 2021, https://doi.org/10.1144/M56-2020-17

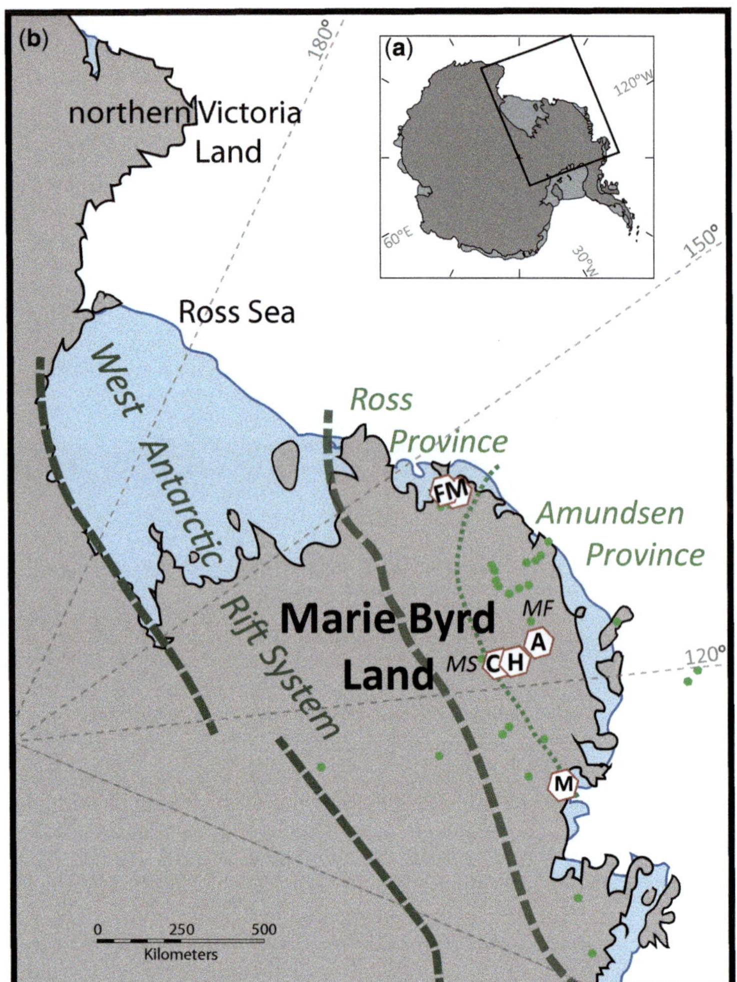

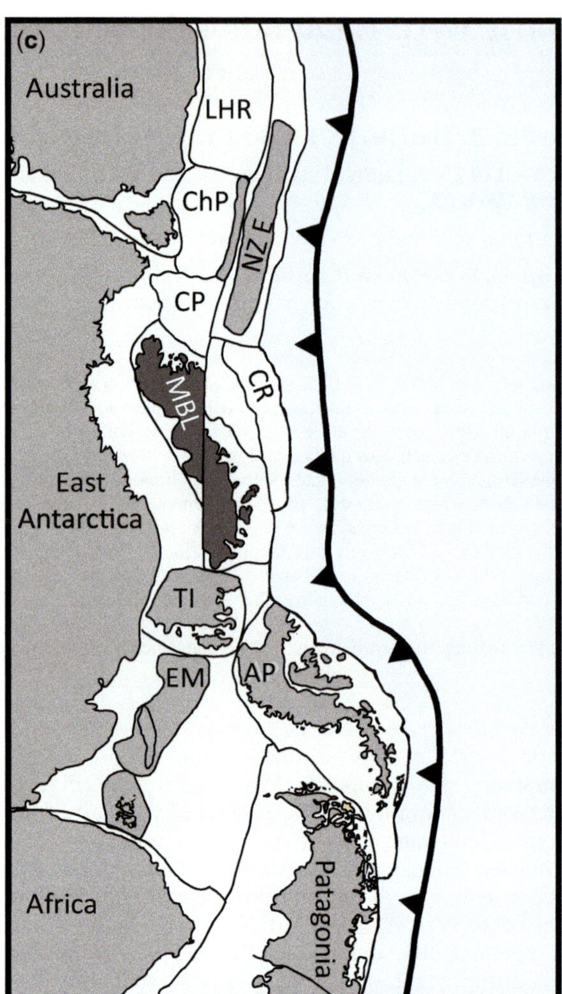

Fig. 1. (**a**) Outline map of Antarctica indicating area shown in b. (**b**) Map of West Antarctica showing Marie Byrd Land, with approximate boundaries of the West Antarctic Rift System (thick green dashed line) and Ross and Amundsen tectonostratigraphic provinces (green dotted line). The xenolith localities discussed in this paper are indicated by red-outlined polygons: FM, Fosdick Mountains; C, Mount Cumming and H, Mount Hampton, Executive Committee Range; A, Mount Aldaz, Usas Escarpment; M, Mount Murphy. Green dots indicate other Cenozoic volcanic centres within Marie Byrd Land; MF, Mountt Flint; MS, Mount Sidley. (**c**) Permian-Triassic reconstruction of Gondwana margin, after Elliot (2013): MBL, Marie Byrd Land; AP, Antarctic Peninsula; ChP, Challenger Plateau; CP, Campbell Plateau; CR, Chatham Rise; EM, Ellsworth Mountains; LHR, Lord Howe Rise; NZ E, eastern New Zealand; TI, Thurston Island. Grey regions indicate subaerial continental crust; pale green, subaqueous continental crust; pale blue, oceanic crust. Thick black line represents subduction plate boundary.

inherited zircon – have been used to suggest that Proterozoic basement underlies Marie Byrd Land's two provinces (e.g. Pankhurst *et al.* 1998; Mukasa and Dalziel 2000), apparently supported by Proterozoic Os model ages in mantle peridotite xenoliths from the Executive Committee Range (Handler *et al.* 2003). However, the granite model ages and zircon core ages could equally be explained by contributions from the thick lower Paleozoic turbidite deposits sourced from Proterozoic terranes (Pankhurst *et al.* 1998), and Proterozoic mantle has been reported from Phanerozoic crustal terranes, oceanic plateau and subduction zones elsewhere (e.g. Hassler and Shimizu 1998; Parkinson *et al.* 1998; McCoy-West *et al.* 2013). Lower crustal xenolith Pb isotope data are more consistent with Phanerozoic ages (Wysoczanski 1993).

The middle Cretaceous marked a period of extension (e.g. Weaver *et al.* 1992, 1994; Luyendyk *et al.* 1996) associated with the break-up of this portion of the Gondwana margin. Evidence for pre-break-up rifting offshore Marie Byrd Land has been documented from *c.* 96 Ma, with Marie Byrd Land-Zealandia break-up estimated *c.* 84–80 Ma (Wobbe *et al.* 2012; LeMasurier *et al.* 2016; Tulloch *et al.* 2019). Coastal Marie Byrd Land is characterised by basin and range style faulting, attenuated crust and alkaline volcanism (LeMasurier and Landis 1996; Heinemann *et al.* 1999; LeMasurier 2008; LeMasurier *et al.* 2016). Crustal thickness is estimated to be 22–33 km (Chaput *et al.* 2014) with the lithosphere extending to depths of 80–100 km (An *et al.* 2015).

The Cenozoic Marie Byrd Land intraplate volcanic province forms a 1000 × 550 km tectonomagmatic dome centered near Mount Flint (Fig. 1a). Active since *c.* 36 Ma, the volcanic province is dominantly mafic in nature with felsic volcanism apparent since *c.* 19 Ma (Hole and LeMasurier 1994; Panter 1997; Winberry and Anandakrishnan 2004; LeMasurier *et al.* 2016; Panter *et al.* 2021). All but two of 18 identified shield-like composite volcanoes are partially buried within the West Antarctic Ice Sheet with mainly the summit regions exposed (e.g. LeMasurier 2013), and additional volcanic centres are fully buried beneath the West Antarctic Ice Sheet (Van Wyk de Vries *et al.* 2017).

Xenolith locations and sample datasets

Lithospheric mantle xenoliths, typically 2–6 cm in diameter but ranging up to *c.* 20 cm, have been described from Mount Murphy (on the Walgreen Coast), Mount Aldaz (in the Usas

Table 1. *Xenolith localities and references*

Volcanic Centre	Latitude (S)	Longitude (W)	Reference/Source
Executive Committee Range			
Mount Cumming	76.667	125.820	Wysoczanski (1993); Handler *et al.* (2003); Chatzaras *et al.* (2016); Cohen (2016)
Mount Hampton	76.48	125.97	Wysoczanski (1993); Handler *et al.* (2003); Moreira and Madureira (2005)
Usas Escarpment			
Mount Aldaz	76.051	124.417	Wysoczanski (1993); Handler *et al.* (2003); Chatzaras *et al.* (2016); Cohen (2016)
Fosdick Mountains			
Mount Avers	76.481	145.396	Chatzaras *et al.* (2016); Cohen (2016)
Bird Bluff	76.504	144.598	Chatzaras *et al.* (2016); Cohen (2016)
Demas Bluff	76.568	144.853	Chatzaras *et al.* (2016); Cohen (2016)
Marujupu	76.508	145.670	Chatzaras *et al.* (2016); Cohen (2016)
Recess Nunatak	76.519	144.507	Chatzaras *et al.* (2016); Cohen (2016)
Walgreen Coast			
Mount Murphy	75.33	110.51	Wysoczanski (1993)

Escarpment), and several volcanic centres in the Executive Committee Range and Fosdick Mountains as part of two PhD and MSc theses (Wysoczanski 1993; Cohen 2016; Table 1). Moreover, additional geochemical, isotopic and/or textural and imaging data are reported in Handler *et al.* (2003), Chatzaras *et al.* (2016) and Chatzaras and Kruckenberg (in Press). Whilst limited to volcanic centres that protrude the ice sheet, the xenolith sites are sufficiently distributed to sample lithosphere beneath the Ross Province (Fosdick Mountains), the Amundsen Province (Mounts Cumming, Hampton and Aldaz) and in the vicinity of the boundary between the two (Mount Murphy; Mount Cumming) (Fig. 1).

Over 100 upper mantle xenoliths have been reported from across Marie Byrd Land, however the data available for the xenoliths is varied and uneven. The xenolith suites described by Cohen (2016) and Chatzaras *et al.* (2016) come dominantly, but not exclusively, from the Fosdick Mountains and provide a wealth of textural and mineral major element data. By contrast, data for the upper mantle xenolith suites reported by Wysoczanski (1993), whose main thesis focus was on lower crustal xenoliths, are largely limited to whole rock major element chemistry. Small subsets of the xenolith suites from the Executive Committee Range, Fosdick Mountains and Usas Escarpment have been analysed for mineral major and trace element data, or whole rock isotopes (Sr, Nd, Pb and Os), much of which has not been previously reported (the exception being Os isotopic data and limited major elements, in Handler *et al.* 2003). In addition, a spinel lherzolite from Mount Hampton (PK91005) was the focus of a study on cosmogenic helium and neon in older mantle samples (Moreira and Madureira 2005).

Petrography and mineral chemistry

The upper mantle xenolith suite is dominated by Type 1 peridotites (Frey and Prinz 1978), predominantly spinel lherzolite but also harzburgite and dunite compositions (Wysoczanski 1993; Cohen 2016). Rarer clinopyroxenite (Mount Hampton, Mount Aldaz, Demas Bluff), and websterite and wehrlite (Fosdick Mountains) xenoliths have also been reported (Wysoczanski 1993; Cohen 2016; Fig. 2). Clinopyroxenite xenoliths from Mount Hampton contain Cr-diopside chemically identical to those found as mm-scale veins cross-cutting lherzolite samples from the same locality and are interpreted to represent fragments of larger, cm- to dm-scale, veins (Wysoczanski 1993). Clinopyroxene-rich veins are also observed cross-cutting lherzolite xenoliths from the Usas Escarpment (e.g. MB69B; Fig. 3) and a clinopyroxenite xenolith has also been reported from this locality (AD6021-X01; Cohen 2016).

The sampled lithospheric mantle appears to be anhydrous with a mineral assemblage restricted to olivine + enstatite ± chrome-diopside + chrome-spinel; metasomatic hydrous phases, such as amphibole and mica, have not been reported in any xenoliths studied to date. Garnet is also absent.

The xenoliths have metamorphic textures and are dominantly coarse-grained granular and tabular, with porphyroclastic textures present but less common (e.g. Fig. 3; Wysoczanski 1993; Cohen 2016). Olivine and orthopyroxene are generally present as larger grains, commonly reaching up to 6 or 5 mm, respectively, and up to *c.* 9 mm in a few xenoliths from Demas Bluff, Fosdick Mountains (e.g. Cohen 2016). The average olivine and orthopyroxene grain sizes for individual samples vary from 100 to 2000 and 90–1200 μm, respectively (Wysoczanski 1993; Cohen 2016). Clinopyroxene is typically finer than olivine and orthopyroxene, except where veining is apparent, and is commonly associated with spinel. Average clinopyroxene grain size varies from *c.* 75–860 μm (Wysoczanski 1993; Cohen 2016).

Olivine porphyroclasts show strain features (dislocation boundaries and subgrains) and curvilinear boundaries,

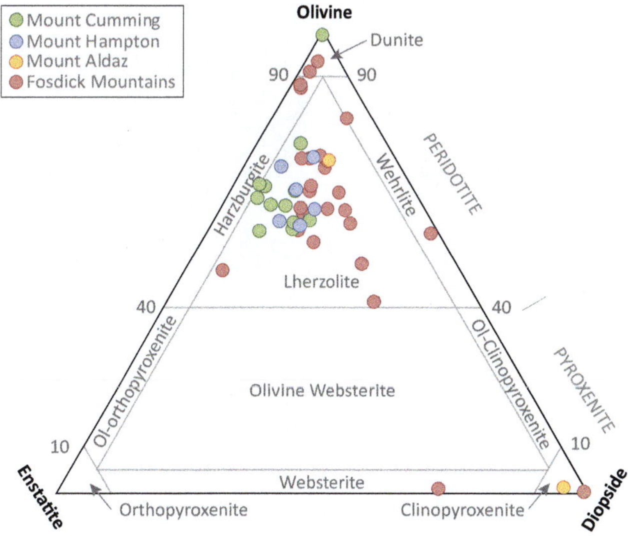

Fig. 2. Marie Byrd Land xenolith modal mineralogy on the ultramafic classification diagram (Streckeisen 1976), by locality. Data from Cohen (2016) and this study. The majority of xenoliths fall within the lherzolite peridotite field, but include harzburgite, dunite, wehrlite and clinopyroxenite samples.

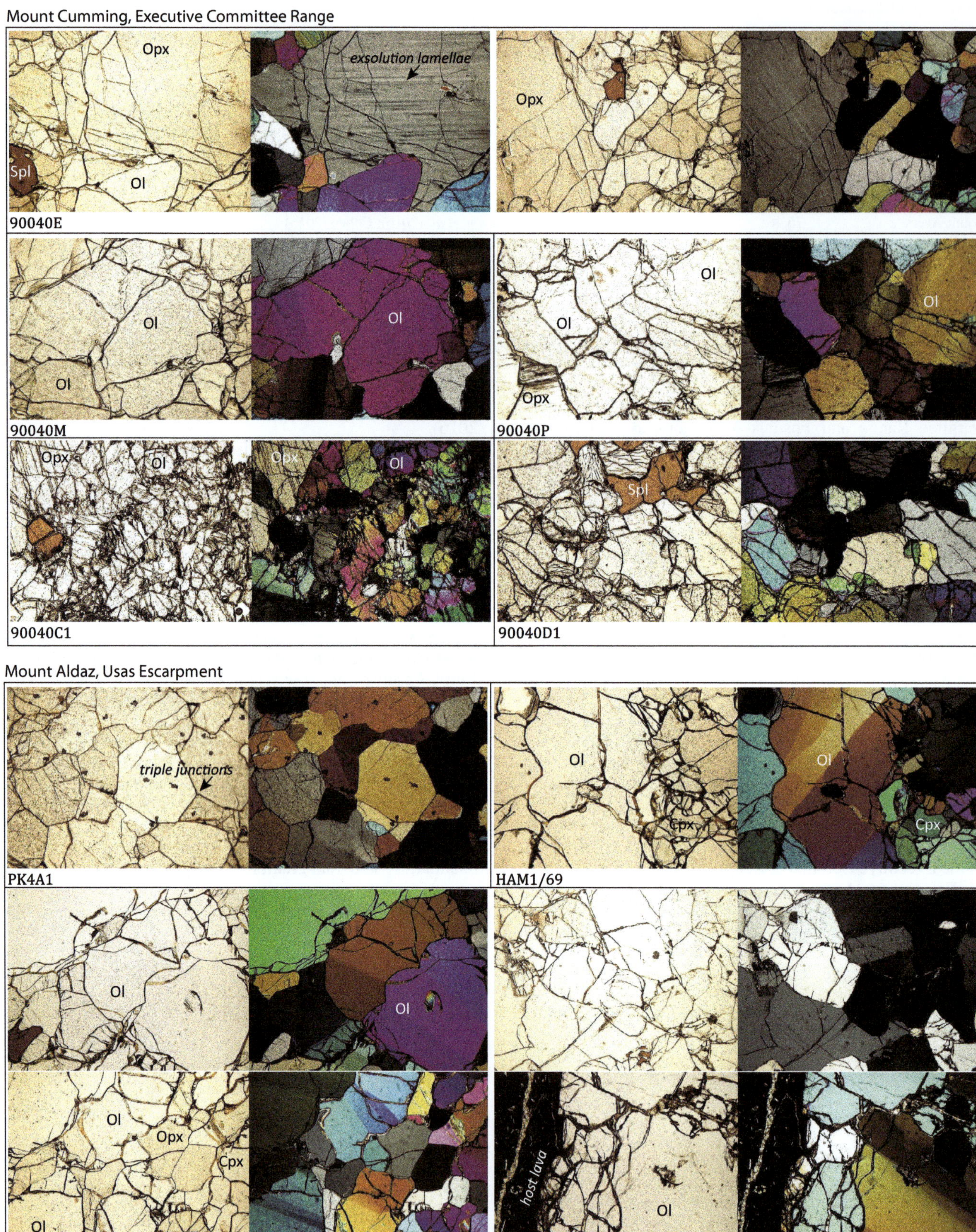

Fig. 3. Photomicrographs of representative xenoliths from Mount Cumming and Mount Hampton, Executive Committee Range, and Mount Aldaz, Usas Escarpment, showing textural and grain size variations. Textures are dominantly coarse granular (e.g. HAM1/69, Mt Hampton) but include well equilibrated, equigranular textures with 120° triple junctions (PK4A1, Mt Hampton) and samples exhibiting fine-grained deformation (90040C1, Mount Cumming). Exsolution lamellae are common in larger orthopyroxene grains (e.g. 90040E, Mount Cumming; MB69E, Mount Aldaz), and deformation bands in larger olivine grains (e.g. Ham1/69, Mount Hampton). MB69B, Mount Aldaz, contains a clinopyroxene-rich vein approximately 10 mm in diameter, with larger grains showing exsolution lamellae. All imaged areas have the same field of view, approximately 4 mm, and are shown as both plane and cross polar light images. Ol, Olivine; Opx, orthopyroxene; Cpx, clinopyroxene; Spl, spinel.

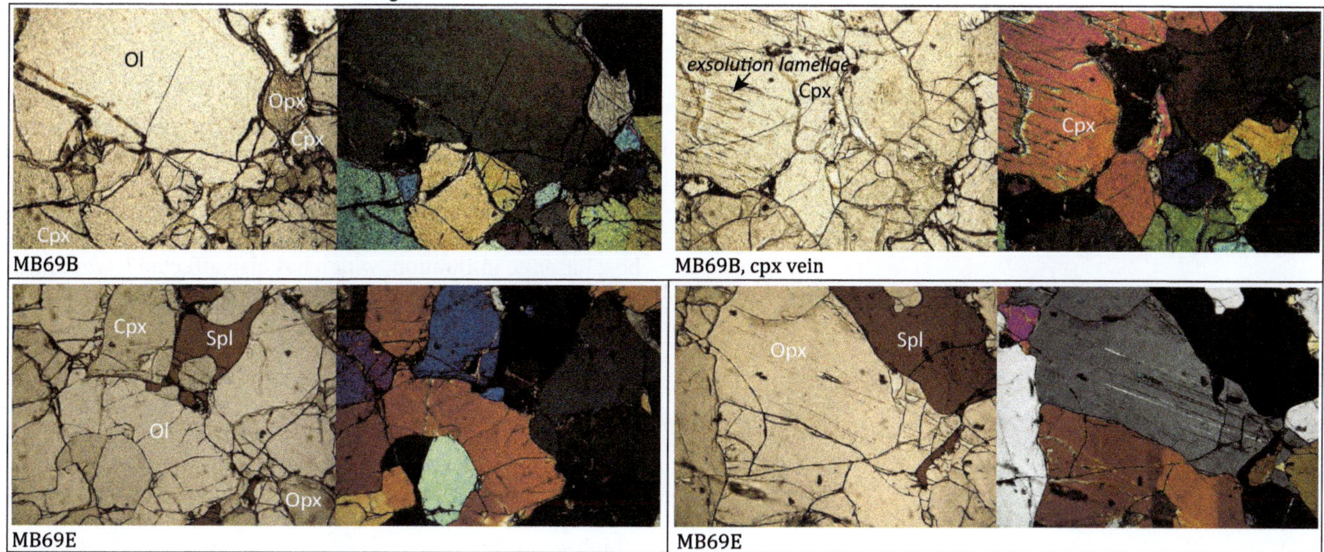

Fig. 3. *Continued.*

with kinkbands common in larger olivine grains across the suite (e.g. Fig. 3). Larger pyroxene grains can show deformation twinning or exsolution lamellae, including those within veins (e.g. MB69B vein, Fig. 3) and in pyroxenites. Smaller, recrystallised neoblasts are generally dislocation-free and commonly characterised by curvilinear boundaries (e.g. 90040E; Fig. 3), in places attaining 120° triple junctions (e.g. PK4A1; Fig. 3). Spinel grains are typically interstitial and randomly distributed, although spinel trails have been observed in some xenoliths from the Fosdick Mountains (Cohen 2016), as well as larger, 'holly-leaf' spinel grains. Detailed textural descriptions of the Fosdick Mountains suite can be found in Cohen (2016) including grain-size distributions, with deformation and crystallographic orientation analyses presented in Cohen (2016) and Chatzaras *et al.* (2016).

Mineral major element compositions (Fig. 4) are reported for 41 xenoliths in Chatzaras *et al.* (2016). Olivine and pyroxene compositions for an additional seven xenoliths are reported here (see Supplementary material), with representative analyses shown in Table 2.

Olivine compositions are Mg-rich with $100.Mg/(Mg + Fe^{2+})$ (Mg#) ranging from 88 to 92, within the range of mantle compositions, with the exception of one sample from Demas Bluff (FDM-DB02-X06, Mg# = 83.4; Chatzaras *et al.* 2016). All analysed olivine have low CaO [<0.2 weight percent (wt%)] and NiO ranges from 0.3–0.5 (wt%). Orthopyroxene compositions parallel those of coexisting olivine, with compositions showing a range in Mg# from 89 to 93. Clinopyroxene have Mg# ranging from 89 to 95 (excluding FDM-DB02-X06), and Cr_2O_3 = 0.1–1.9 wt%.

All peridotite samples for which both spinel and olivine data are available, aside from a dunite xenolith from Mount Cumming, plot on the olivine-spinel mantle array (Fig. 4) consistent with the suite representing residua from varying degrees of partial melting. Spinel Cr# [100.Cr/(Cr + Al)] ranges between *c.* 6–40. The Mount Cumming dunite contains two populations of spinel (high- and low-Cr) and has olivine–spinel compositions indicative of melt interaction rather than being residual from large degrees of partial melting (Fig. 4; Cohen 2016). Plots of Al_2O_3 and MgO are shown for both clinopyroxene and orthopyroxene for the xenolith suite (Fig. 4), with negative trends that are broadly compatible with up to 15–20% depletion of spinel-facies mantle for the bulk of the suite. In detail however, many of the clinopyroxene from the Fosdick Mountains scatter to higher Al_2O_3 contents (Fig. 4).

Trace element concentrations, as determined by laser ablation–inductively coupled plasma mass spectrometry or instrumental neutron activation analysis on mineral separates (see Supplementary material) for olivine, orthopyroxene and clinopyroxene for six peridotites are shown in chondrite-normalized (N) rare earth element (REE) diagrams (Fig. 5) and primitive-mantle normalized multielement diagrams (Fig. 6). Minerals have been analysed from xenoliths from two locations in the coastal ranges of Marie Byrd Land (West Recess Nunatak and Demas Bluff, Fosdick Mountains, Ross Province), and three from central Marie Byrd Land (Mount Hampton and Mount Cumming, Executive Committee Range, and Mount Aldaz, Usas Escarpment, Amundsen Province). Samples from the Fosdick Mountains and Mount Aldaz are light (L)REE depleted with convex upward REE patterns for the clinopyroxenes ($[Ce/Sm]_N$ *c.* 0.4) and generally flat heavy (H)REE ($[Sm/Yb]_N$ *c.* 1.0). Incompatible trace element plots of clinopyroxene from Demas Bluff and Recess Nunatak are very similar, showing depletions in the highly incompatible trace elements relative to the slightly less incompatible trace elements (Fig. 6). Both patterns show distinctive negative Ba- and Ti-anomalies. The LREE-depleted patterns are consistent with variable degrees of partial melt extraction, however despite preserving LREE-depleted patterns the samples show evidence of open system behaviour. The LREE (La + Ce and in some examples Nd), are out of equilibrium between coexisting clinopyroxene and orthopyroxene (Fig. 7), suggestive of cryptic metasomatism.

Clinopyroxene from Mounts Cumming and Hampton in the Executive Committee Range have variable REE patterns ranging from LREE depleted, similar to those of the Fosdick Mountains, to LREE enriched (Fig. 5). Clinopyroxene from Mount Cumming sample 1998 and Mount Hampton sample 2000/12 show convex upward REE patterns with $(Ce/Sm)_N = 0.7$ and 0.3 respectively (Supplementary file). Clinopyroxene from Mount Hampton sample 91005 and Mount Cumming sample 2000/2 have a concave upward REE pattern that is commonly interpreted as indicating metasomatic enrichment of LREE (e.g. McDonough and Frey 1989; Blusztajn and Shimizu 1994; Baker *et al.* 1998). Despite the very different REE profiles, the coexisting clinopyroxene and orthopyroxene in the two Mount Hampton xenoliths appear to be in equilibrium (Fig. 7). By contrast, La + Ce + Nd are strongly out of equilibrium in the LREE-enriched Mount Cumming xenolith (2000/2; Fig. 7). The multielement plots of clinopyroxene

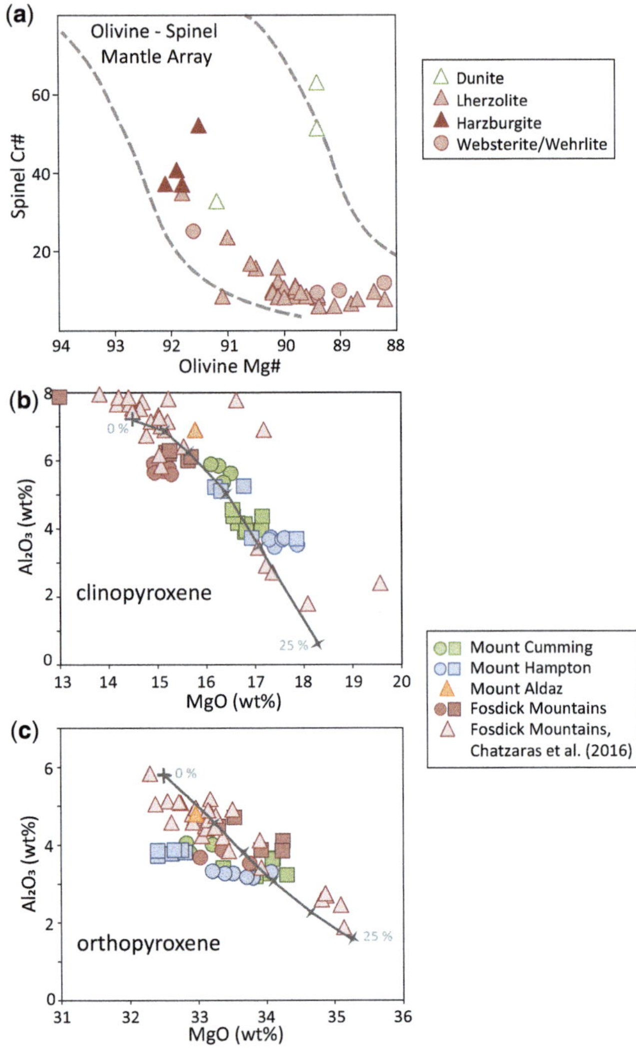

Fig. 4. Selected mineral major element chemistry. (**a**) spinel Cr# v. olivine Mg#, with the Olivine-Spinel Mantle Array of Arai (1994) indicated by dashed grey lines. Data are from Chatzaras *et al.* (2016), predominantly from Fosdick Mountains localities. (**b**) and (**c**) Al_2O_3 and MgO in clinopyroxene (**b**) and orthopyroxene (**c**) from lherzolite and harzburgite samples. Green symbols, Mount Cumming; blue, Mount Hampton; orange, Mount Aldaz; red, Fosdick Mountains. Triangles represent samples from Chatzaras *et al.* (2016); circles and squares represent analyses of multiple grains from two xenoliths from each of Mounts Hampton, Cumming, and the Fosdick Mountains (this study; see Supplementary material). Grey lines represent theoretical partial melting models from a primitive upper mantle source (McDonough and Sun 1995) following Upton *et al.* (2011). Ticks represent 5% partial melting increments through to 25%.

from the LREE enriched Mount Hampton and Mount Cumming samples (91005, 2000/2) also show complex patterns, with both showing marked Zr–Hf depletions, and enrichments in Th and U relative to the LREE-depleted samples from these sites (1998, 2000/12; Fig. 6). Both Mount Hampton samples show distinctive Ti anomalies, which are less evident in the samples from Mount Cumming.

Whole rock geochemical composition

Whole rock major element data are reported for 59 xenoliths from mounts Cumming (11) and Hampton (27) in the Executive Committee Range, Mount Aldaz (8), and Mount Murphy (13) with a limited range of trace element data reported for a subset of these (Wysoczanski 1993; see Supplementary material). Whole rock major element compositions are broadly consistent with the samples representing residual compositions following variable degrees of melt extraction (Fig. 8). Low P_2O_5 contents in most samples (<0.05 wt%) are consistent with an absence of secondary apatite (see Supplementary material), however one Mount Hampton peridotite (PKO) has 0.10 wt% P_2O_5, suggesting traces of apatite may be present. A number of the peridotites scatter to high FeO_{Total} contents, most notably the Mount Murphy suite, and to high Na_2O, suggestive of secondary processes. These may be related to host basanite infiltration, which has also affected lower crustal xenoliths (Wysoczanski 1993), or refertilization within the lithospheric mantle via interaction with Fe-rich melts.

Strontium, Nd and Pb isotopes have been measured for a suite of xenoliths from central Marie Byrd Land (mounts Hampton, Cumming and Aldaz). These are reported in Table 3 and plotted on standard covariation diagrams in Figure 9. On the Sr v. Nd isotope diagram, data from the Executive Committee Range and Usas Escarpment (uncorrected for age) plot in the depleted mantle field, clustering close to the HIMU ['high-μ', where $\mu = (^{238}U/^{204}Pb)_{t=0}$; Stracke *et al.* 2005] mantle end-member, as do data from Executive Committee Range lava flows. A single sample from the Usas Escarpment (MB69E) has highly depleted characteristics ($\varepsilon Nd = +22.9$; $^{87}Sr/^{86}Sr = 0.70191$). In $^{208}Pb/^{204}Pb$ v. $^{206}Pb/^{204}Pb$ space, the bulk of the data cluster around and below the Northern Hemisphere Reference Line (NHRL; Hart 1984), and overlap with the less radiogenic compositions of Marie Byrd Land mafic volcanic rocks (Fig. 9).

Osmium isotopic data for 17 xenoliths from mounts Aldaz, Hampton and Cumming were reported in Handler *et al.* (2003) with Re-Os data for an additional two samples from the Fosdick Mountains reported here (Table 3). The $^{187}Os/^{188}Os$ data for all but one xenolith range from depleted to values close to primitive upper mantle estimates (0.1162–0.1288), consistent with a time-integrated history of variable melt depletion (e.g. CaO ranging from 0.4–3.4 wt%). A single sample from Mount Aldaz however, has a superchondritic composition (MB69F; $^{187}Os/^{188}Os$ *c.* 0.132).

Petrogenesis and regional variability

Geothermobarometry

Chatzaras *et al.* (2016) reported two-pyroxene equilibrium temperatures for 41 xenoliths, predominantly from sites in the Fosdick Mountains (Fig. 10). Preferred temperature estimates were taken as the average of three two-pyroxene thermometers (Bertrand and Mercier 1985; Brey and Köhler 1990; Taylor 1998) calculated at 1.5 GPa, which were all in close agreement. In the absence of a reliable geobarometer for spinel peridotite compositions, pressures were estimated using a theoretical geotherm constructed such that all xenoliths were at pressures consistent with being below the crust and within the spinel stability field, and subparallel to the present day Ross Embayment geotherm of ten Brink *et al.* (1997) (Chatzaras *et al.* 2016).

Equilibration temperatures based on the coexisting clinopyroxene and orthopyroxene (Wells 1977; Brey and Köhler 1990; Putirka 2008; Liang *et al.* 2013) for an additional eight xenoliths are reported here in Table 4. Temperatures calculated using the Putirka (2008) thermometer are consistently higher than those calculated using the Wells (1977) or Brey and Köhler (1990) thermometers, although the Mount Hampton and Demas Bluff samples agree within uncertainties across the thermometers. The xenolith from West Recess Nunatak shows the highest discrepancy between temperature estimates,

Table 2. *Selected mineral compositions*

	Mount Cumming		Mount Hampton		Demas Bluff		W Recess Nunatak
	MC1999	2000/2	PK91005	2000/12	2000/8	1998/99	1998
Olivine							
SiO_2 (wt%)	41.67	41.34	41.40	43.07	41.21	42.09	41.94
TiO_2	0.06	0.02	0.06	0.05	0.05	0.06	0.17
Al_2O_3	0.05	0.02	0.06	0.05	0.04	0.02	0.04
FeO	9.46	8.44	8.38	8.93	8.86	9.01	8.68
MnO	0.18	0.10	0.21	0.12	0.13	0.18	0.19
MgO	49.16	49.85	49.97	49.55	48.94	50.30	49.95
CaO	0.11	0.07	0.11	0.12	0.05	0.05	0.12
Na_2O	0.02	0.01	0.09	0.04	0.05	0.02	0.05
K_2O	0.02	0.03	0.04	0.03	0.00	0.11	0.10
Cl	0.03		0.03		0.01	0.07	0.09
Cr_2O_3	0.11	0.04	0.04	0.09	0.04	0.12	0.06
NiO	0.37	0.10	0.40	0.12	0.08	0.41	0.44
Total	101.23	100.04	100.79	102.14	99.43	102.44	101.83
Mg#	90.3	91.3	91.2	90.8	90.8	90.9	91.2
Orthopyroxene							
SiO_2 (wt%)	55.13	56.31	55.97	56.40	55.12	56.48	56.81
TiO_2	0.33	0.07	0.09	0.13	0.12	0.18	0.27
Al_2O_3	3.93	3.38	3.23	3.80	3.77	3.99	3.62
FeO	6.11	5.37	5.30	5.68	5.33	5.62	5.62
MnO	0.17	0.13	0.17	0.15	0.13	0.20	0.14
MgO	32.97	33.94	33.61	32.60	32.68	33.99	33.70
CaO	0.99	0.68	0.99	0.98	1.22	0.63	0.61
Na_2O	0.07	0.06	0.11	0.12	0.15	0.11	0.12
K_2O	0.05	0.02	0.07	0.02	0.04	0.03	0.14
Cl	0.05		0.04			0.06	0.13
Cr_2O_3		0.69	0.77	0.67		0.56	0.60
NiO		0.09	0.15	0.06		0.15	0.17
Total	99.79	100.74	100.49	100.60	98.56	102.01	101.94
Mg#	90.6	91.8	91.9	91.1	91.6	91.5	91.4
Mg	88.85	90.65	90.12	89.33	89.39	90.42	90.37
Fe	9.23	8.05	7.97	8.73	8.18	8.39	8.46
Ca	1.92	1.30	1.91	1.94	2.43	1.20	1.17
Clinopyroxene							
SiO_2 (wt%)	50.44	53.00	52.61	52.82	52.50	52.67	53.04
TiO_2	1.01	0.21	0.05	0.26	0.43	0.53	0.48
Al_2O_3	5.68	4.17	3.63	4.70	5.71	6.31	5.55
FeO	3.29	2.04	2.44	2.61	2.29	2.70	2.30
MnO	0.18	0.10	0.14	0.13	0.11	0.15	0.12
MgO	16.32	16.82	17.52	16.72	15.08	15.01	15.17
CaO	20.65	21.87	21.08	20.56	20.24	21.14	21.14
Na_2O	0.83	1.00	0.72	1.10	1.97	1.80	2.00
K_2O	0.04	0.01	0.04	0.03	0.03	0.04	0.16
Cl	0.04		0.04		0.03	0.05	0.12
Cr_2O_3		1.20	1.15	1.40	1.64	1.10	1.55
NiO		0.06	0.13	0.06	0.05	0.08	0.12
Total	98.47	100.46	99.55	100.39	99.85	101.58	101.74
Mg#	89.9	93.6	92.8	91.9	92.1	90.8	92.2
Mg	49.44	49.93	51.46	50.71	48.77	47.32	47.93
Fe	5.58	3.40	4.01	4.45	4.16	4.77	4.08
Ca	44.98	46.68	44.52	44.83	47.07	47.90	48.00

suggesting that the analysed pyroxenes in this sample are not fully in equilibrium. Rare earth element data were obtained for coexisting clinopyroxene and orthopyroxene for five of the xenoliths, allowing the application of the Liang *et al.* (2013) REE geothermometer. For all five samples here, the middle (M) and HREE appear to be in equilibrium for the coexisting pyroxene pairs (Fig. 7) and calculated temperatures are in good agreement with the Brey and Köhler (1990) thermometer (Table 4). Notably, the West Recess Nunatak xenolith that produced the inconsistent temperature estimates amongst the major element thermometers also produced very large uncertainties on the REE in pyroxene thermometer estimate (965 ± 105°C).

The temperature estimates are also shown on Figure 10, using the Brey and Köhler (1990) temperatures and hypothetical geotherm of Chatzaras *et al.* (2016) for consistency between the two datasets. Excluding the West Recess Nunatak sample, calculated temperatures are broadly in the range 900–1100°C, within the range of temperatures determined for the larger Fosdick Mountain xenolith suite (Chatzaras *et al.* 2016).

There appear to be no obvious differences in temperature between porphyroclastic and protogranular textural varieties,

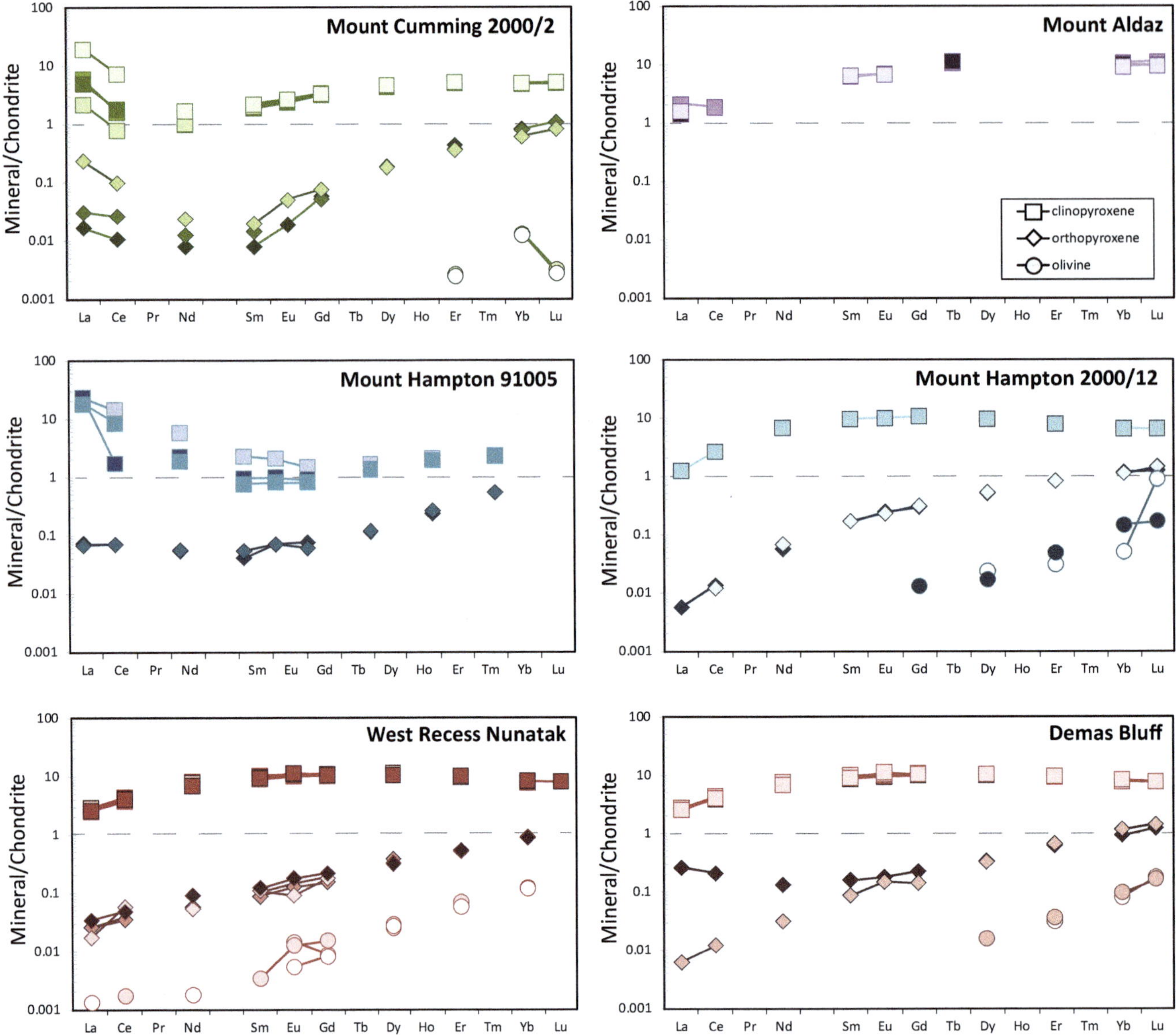

Fig. 5. Chondrite normalized (Anders and Grevasse 1989) REE diagrams for minerals from six xenoliths. Squares, clinopyroxene; diamonds, orthopyroxene; circles, olivine. Analyses of different grains from the same sample are indicated by different shading of symbols and lines.

nor by locality. The Demas Bluff xenolith suite, with the largest number of xenoliths analysed, shows the widest range in temperatures and is the only site to include samples with equilibration temperatures significantly higher than 1100°C. Overall however, the different locations have broadly similar temperature ranges, which are largely restricted to temperatures between 800 and 1050°C. This range is consistent with the exsolved nature of pyroxenes in many of the samples and with a generally 'cool' nature of stabilized subcontinental mantle lithosphere (e.g. McKenzie *et al.* 2007; Lee *et al.* 2010).

Melt depletion and refertilzsation

Whole rock major element compositions are broadly consistent with being residues from variable degrees of partial melting, however higher FeO_{Total} and Na_2O in some samples suggest secondary processes have also affected these samples (Fig. 8). This is particularly evident for the xenolith suite from Mount Murphy, which typically has FeO_{Total} >10 wt% as well as tending to lower Al_2O_3 relative to MgO than predicted for partial melt extraction from a primitive mantle-like source (Fig. 8). Major element trends very similar to those produced by melt extraction can also be produced by refertilization of strongly depleted peridotite (e.g. Van der Wal and Bodinier 1996; Bedini *et al.* 1997; Martin *et al.* 2014). Distinguishing the extent to which partial melt extraction or refertilization of depleted peridotite have affected the sampled Marie Byrd Land upper mantle is hampered by incomplete and patchy datasets for the xenoliths.

Whole rock FeO_{Total} and/or Na_2O data suggest refertilization may have affected at least some of the sampled lithospheric mantle, particularly that sampled at Mount Murphy. Trace element data are available for two Mount Murphy xenoliths, which have relatively low FeO_{Total} contents (8.9–9 wt% compared to others that are up to *c.* 14 wt%) for this locality (see Supplementary material). These have LREE enriched patterns ($[La/Yb]_N = 2.0–3.0$), consistent with interaction with moderately LREE-enriched melts.

Limited whole rock trace element data are available for a subset of the xenoliths (see Supplementary material). Of these, three lherzolites from Mount Aldaz (2) and Mount Cumming (1) have whole rock LREE-depleted patterns with $[La/Yb]_N = 0.15–0.69$ and Yb = 0.05–0.09 parts per million (ppm), consistent with residues from partial melting in spinel facies, however most analysed samples have elevated

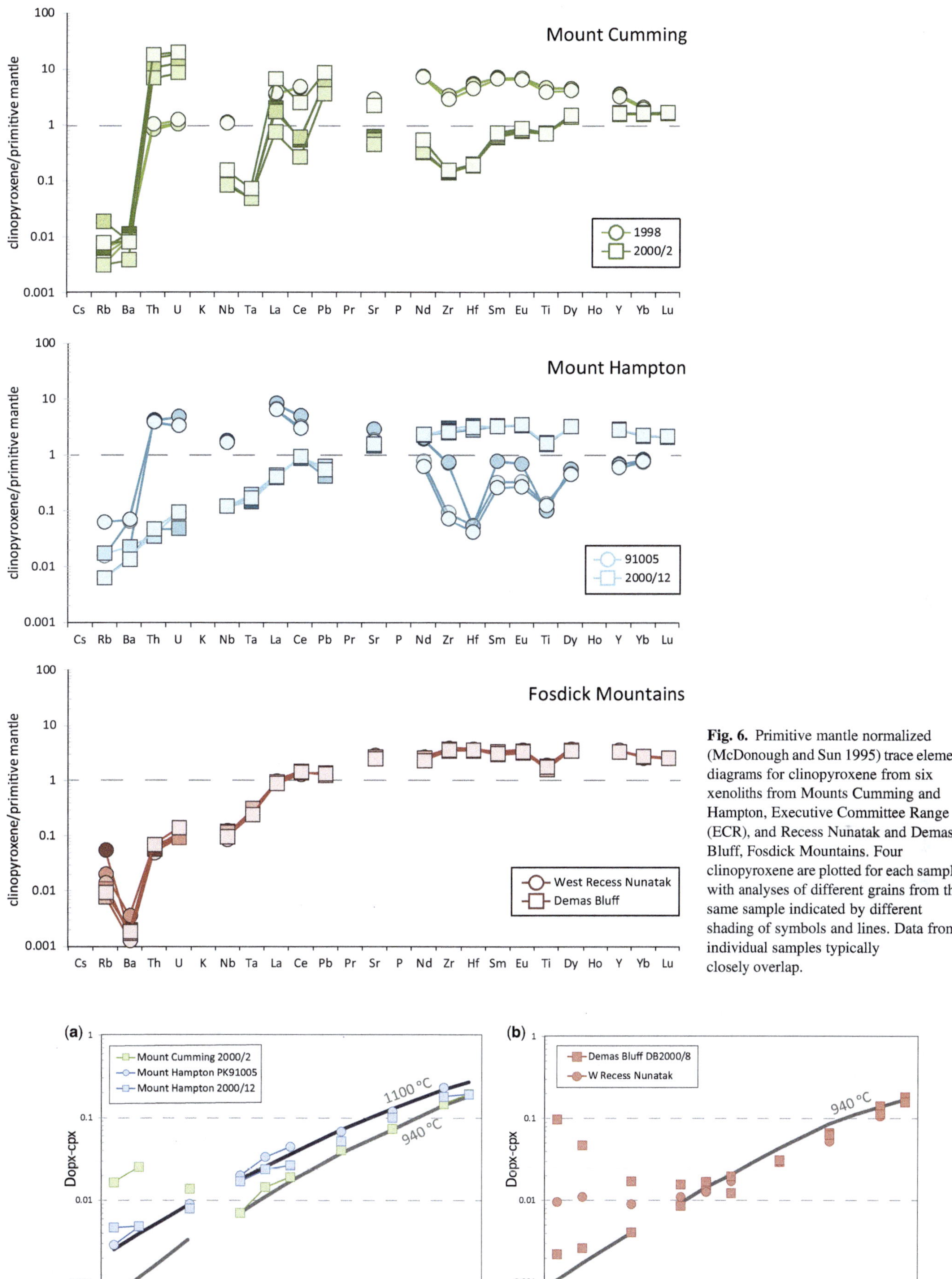

Fig. 6. Primitive mantle normalized (McDonough and Sun 1995) trace element diagrams for clinopyroxene from six xenoliths from Mounts Cumming and Hampton, Executive Committee Range (ECR), and Recess Nunatak and Demas Bluff, Fosdick Mountains. Four clinopyroxene are plotted for each sample, with analyses of different grains from the same sample indicated by different shading of symbols and lines. Data from individual samples typically closely overlap.

Fig. 7. Rare earth element partition coefficients (D) calculated for coexisting clinopyroxene and orthopyroxene. Thick lines represent calculated equilibrium partition coefficients at different temperatures for the xenolith compositions [(**a**) 940 and 1100°C (**b**) 940°C], following methods discussed in Liang *et al.* (2013). The H-MREE are in good agreement for all samples, with the LREE showing variable disequilibrium consistent with open system behaviour.

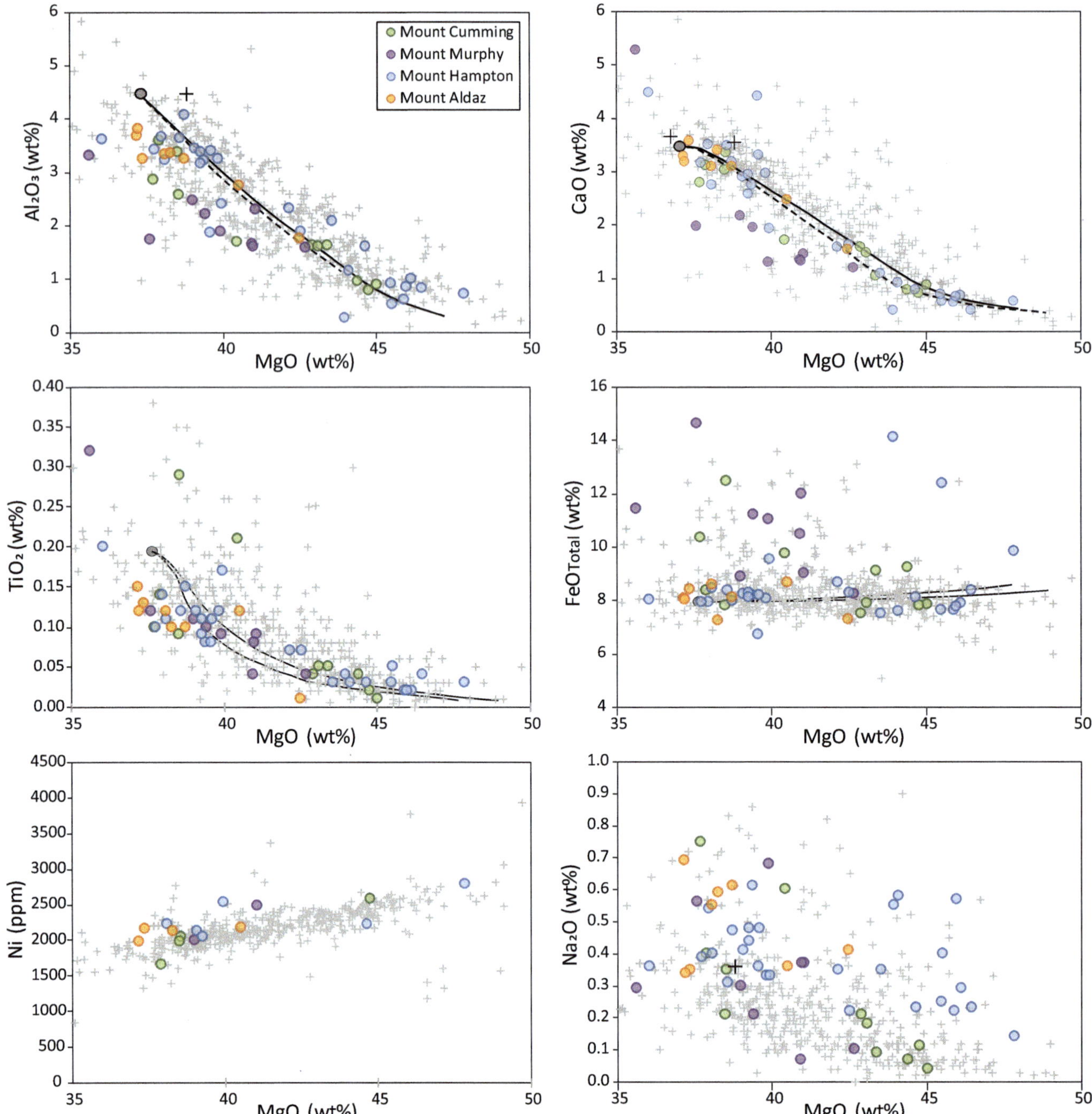

Fig. 8. Whole rock major element data (Wysoczanski 1993). Grey crosses represent continental spinel peridotite xenoliths from off-cratonic regions in Australia, China, Japan, Europe, Siberia, North America, Algeria and Tanzania (see Supplementary file for full reference list). (**a**)–(**d**), curves represent calculated residual compositions from incremental batch melting of a MORB source at 2.0 GPa (dashed) and 1.5 GPa (solid) using the MELTS thermodynamic approach, after Aldanmaz *et al.* (2009).

L-MREE consistent with enrichment processes (e.g. $[La/Yb]_N$ *c.* 1–5.9). The range of HREE are consistent with most of the lherzolite samples analysed having experienced <10% partial melt extraction in spinel facies (e.g. Yb = 0.26–0.42 ppm), with three samples from Mount Hampton and one from Mount Cumming preserving low Lu and Yb concentrations more consistent with up to *c.* 20% partial melt extraction (e.g. Yb = 0.02–0.09 ppm).

Pyroxene major element chemistry is also broadly consistent with the bulk of the suite being residual from < 20% partial melt extraction, with a few harzburgites and lherzolites from the Fosdick Mountains ranging to 25%, based on theoretical melting trends from a primitive upper mantle source (Upton *et al.* 2011; Fig. 4). In detail however, co-existing orthopyroxene and clinopyroxene suggest differing degrees of melt extraction for many of the xenolith suite. For example, clinopyroxenes from Fosdick Mountains lherzolite samples reported in Chatzaras *et al.* (2016) extend to higher Al_2O_3 and lower MgO contents than the primitive upper mantle starting composition, and the composition of Mount Hampton and Mount Cumming clinopyroxenes reported here suggest higher degrees of melting than their respective orthopyroxenes. Discrepancy between melt extraction estimates derived from coexisting pyroxenes in spinel lherzolite xenoliths elsewhere has been used to suggest a more complex history involving refertilization of strongly depleted peridotite (e.g. Upton *et al.* 2011). However, the compositions of pyroxenes from the two xenoliths from the Fosdick

Table 3. *Sr, Nd, Pb and Os isotopic data*

Sample	Al_2O_3 (wt%)	CaO (wt%)	Sr (ppm)	$^{87}Sr/^{86}Sr$	Nd (ppm)	Sm (ppm)	εNd	$^{143}Nd/^{144}Nd$	$^{147}Sm/^{144}Nd$	T_{DM} (Ma)	$^{206}Pb/^{204}Pb$	$^{207}Pb/^{204}Pb$	$^{208}Pb/^{204}Pb$	Re (ppb)	Os (ppb)	$^{187}Re/^{188}Os$	$^{187}Os/^{188}Os$	γOs	T_{RD}(Ma)	T_{MA}(Ma)
Executive Committee Range																				
Mount Hampton																				
PK1A	1.90	1.62	11.5	0.70284(3)	0.784	0.230	1.3	0.51275(1)	0.177	1861	19.5(6)	15.38(4)	38.50(12)	0.074	0.88	0.0783(8)	0.11907(9)	−7.5	1347	1648
PK4H	0.62	0.54	2	0.70307(4)	0.166	0.275	5.1	0.51294(4)	1.000	−48	17.85(5)	15.21(4)	35.9(3.5)	0.001	2.42	0.0011(1)	0.12064(9)	−6.3	1130	1132
PK4Z	1.15	0.91	2	0.70365(4)	0.031	0.010					19.12(1)	15.60(1)	38.69(2)	0.005	2.81	0.0122(1)	0.1214(2)	−5.7	1025	1055
PK4J1	2.33	1.58	3.2	0.70267(3)	0.291	0.096	8.2	0.51310(2)	0.200	1009	18.88(4)	15.33(4)	38.18(10)	0.061	1.41	0.206(2)	0.1214(1)	−5.7	1021	1966
														0.044	1.48	0.1483(15)	0.12151(9)	−5.6	1009	1544
PK4P1	4.07	3.17	4	0.70248(3)	0.454	0.251	9.9	0.51319(1)	0.334	−8	19.071(6)	15.566(5)	38.55(1)	0.174	2.49	0.337(3)	0.12768(7)	−0.8	144	694
PK5C	0.84	0.39	7	0.70309(3)							19.47(1)	15.52(1)	39.09(8)	0.256	2.13	0.584(6)	0.12809(7)	−0.5	86	–
					0.413	0.152	7.2	0.51305(1)	0.223	–	19.28(1)	15.55(1)	38.56(2)	0.350	2.24	0.529(5)	0.12823(8)	−0.4	66	–
PK5I	2.09	1.09									19.641(4)	15.636(4)	39.18(1)	0.000	0.88	0.0000	0.1219(1)	−5.3	953	953
PK5M	0.85	0.64	2	0.70349(3)										0.004	1.56	0.0119(1)	0.12064(9)	−6.3	1130	1162
PK4O	3.26	2.95												0.109	0.61	0.867(9)	0.1258(2)	−2.2	407	–
Mount Cumming																				
90040E	3.59	3.10	12	0.70279(4)		0.274	1.3	0.51270(2)	0.190	3142	19.104(6)	15.569(5)	38.2(4)	0.031	0.39	0.378(4)	0.1288(3)	0.1	–	–
90040M														0.044	0.82	0.261(3)	0.1272(2)	−1.2	215	555
90040K	1.62	1.57	8	0.70302(3)	0.308	0.066	4.5	0.51287(2)	0.130	592	19.561(3)	15.636(2)	38.69(1)	0.001	14.4	0.0004(1)	0.12263(5)	−4.7	853	853
											19.258(6)	15.648(5)	38.88(1)							
MC 1					0.577	0.179	5.9	0.51294(2)	0.187	1460	18.718(8)	15.552(6)	38.30(2)	0.045	2.95	0.0724(7)	0.11794(8)	−8.4	1503	1808
USAS Escarpment																				
Mount Aldaz																				
MB69A	3.68	3.29									19.201(8)	15.590(6)	38.73(2)	0.221	3.27	0.338(3)	0.12778(8)	−0.7	130	639
MB69E	3.24	3.08	3.0	0.70191(3)	0.209	0.141	22.9	0.51381(2)	0.407	488	18.34(1)	15.29(2)	37.34(5)	0.184	3.69	0.243(2)	0.12576(7)	−2.3	414	964
MB69F	3.36	3.38	6	0.70298(2)	0.382	0.146	6.1	0.51295(1)	0.318	–	19.172(4)	15.626(3)	38.564(8)	0.176	1.46	0.583(6)	0.1318(1)	2.4	–	1165
				0.70295(4)										0.077	1.51	0.246(3)	0.1333(2)	3.5	–	–
MB69H	2.76	2.47	10	0.70277(3)	0.350	0.165	8.9	0.51309(1)	0.285	–	18.611(4)	15.589	38.06(1)	0.101	3.19	0.152(2)	0.12466(9)	−3.1	569	885
											18.597(2)	15.571	37.987(4)							
Fosdick Mountains																				
Western Recesses															2.50		0.11623(8)	−9.7	1739	–
Demas Bluff														0.038	3.09	0.0538(5)	0.1222(1)	−5.0	911	1042

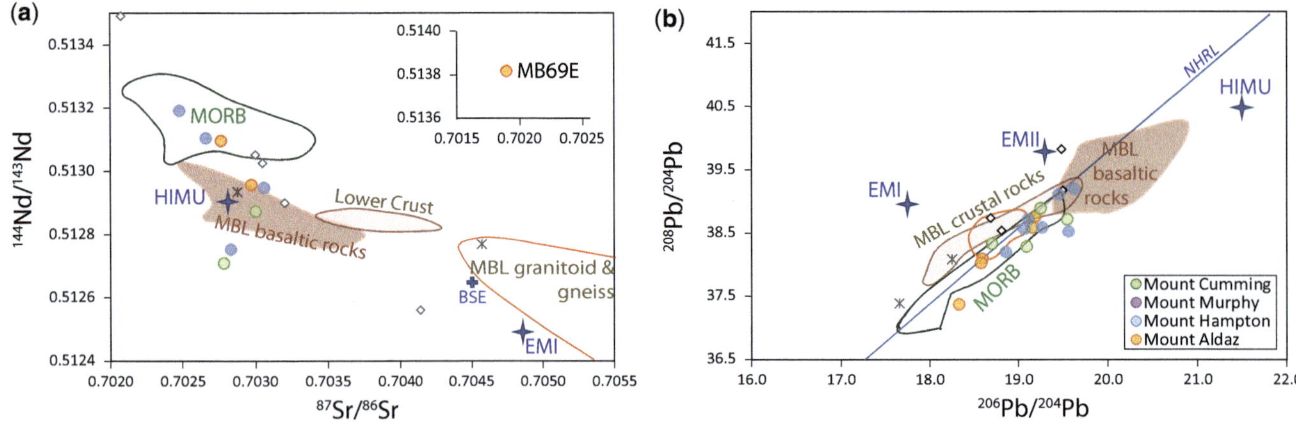

Fig. 9. Strontium, Nd and Pb isotopes for Marie Byrd Land (MBL) peridotites. Inset in (**a**) shows sample with extreme Sr and Nd composition that plots outside of the area shown in (**a**). Grey crosses indicate pyroxenites from Executive Committee Range, Marie Byrd Land (Wysoczanski *et al.* 1995), and open diamonds are peridotites from southern Victoria Land (Martin *et al.* 2013). Fields for Marie Byrd Land crustal and mafic volcanic rocks and MORB (mid-ocean ridge basalts) from Wysoczanski *et al.* (1995); Panter *et al.* (2000) and; LeMasurier *et al.* (2016). NRHL, northern hemisphere reference line. BSE, bulk silicate earth; EMI, enriched mantle I; EMII, enriched mantle II and; HIMU, 'high-μ' mantle endmember compositions (Stracke *et al.* 2005).

Mountains (this study) do provide consistent melt extraction estimates (Fig. 4).

Modelling of partial melting based on REE in clinopyroxene for six of the seven xenoliths that data are available for are broadly consistent with 2–15% fractional melting or 2–22% batch melting of a primitive upper mantle source in spinel facies (Fig. 11). One Mount Cumming xenolith (1998) has a distinctly HREE depleted, convex upward REE pattern suggestive of partial melting with garnet present (e.g. Bonadiman *et al.* 2005), with subsequent reequilibration within spinel facies in the lithosphere (e.g. $[Gd/Yb]_N$ *c.* 2.8 v. *c.* 0.6 at similar Yb contents; Fig. 11a and b). The Fosdick Mountains xenoliths and one Mount Hampton sample (2000/12) also have a less pronounced convex up REE pattern that suggests there may have been some influence of garnet during partial melting (e.g. Fosdick Mountains and Mount Hampton 2000/12 $[Gd/Yb]_N = 1.3–1.7$) perhaps in the first stages of polybaric melt extraction. Coupled Cr_2O_3 and CaO contents of clinopyroxene in these samples and across the wider xenolith suite, however, are generally consistent with spinel lherzolite rather than garnet peridotite precursor (e.g. Bonadiman *et al.* 2005).

Overall, the lithospheric mantle, as sampled, is heterogeneous both within and across localities and terranes, and likely preserves a mixture of variably cryptically metasomatized (e.g. LREE enriched) mantle residues, and more depleted samples that underwent subsequent refertilization to produce fertile lherzolite compositions.

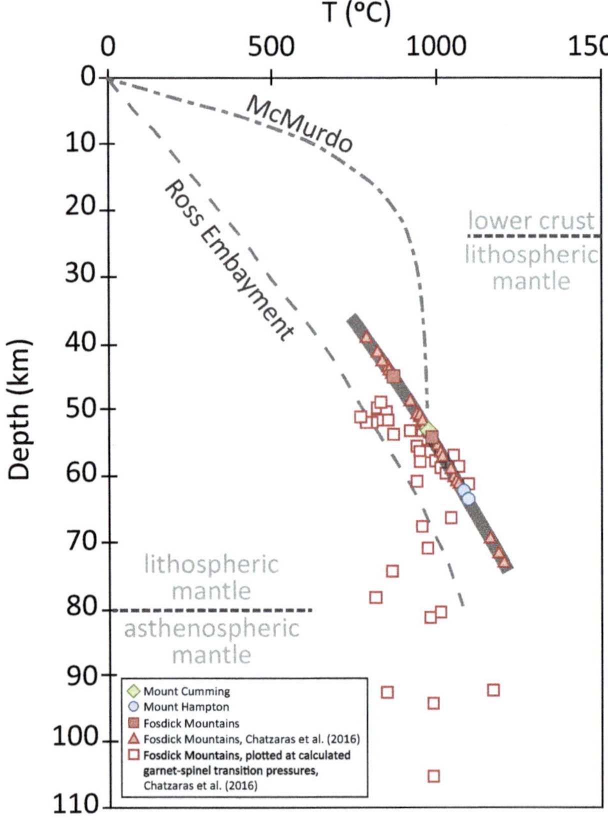

Fig. 10. Pressure-temperature diagram showing Marie Byrd Land peridotites, after Chatzaras *et al.* (2016). Temperatures are based on the Brey and Köhler (1990) thermometer and are plotted on a theoretical geotherm from Chatzaras *et al.* (2016; thick grey line). Also shown for comparison are the petrologic geotherm for McMurdo from Berg *et al.* (1989) and present-day Ross Embayment geotherm from ten Brink *et al.* (1997). Open red squares represent the calculated pressures of the spinel-garnet transition for each peridotite composition from Chatzaras *et al.* (2016), and provide a maximum pressure bound for those xenoliths used to constrain the theoretical geotherm.

Table 4. *Temperature estimates from two-pyroxene geothemometers*

Locality/ Sample	Major elements in pyroxene: Wells (1977) (°C)	BandK'90* (°C)	Putirka (2008) (°C)	REE in pyroxene Liang *et al.* (2013) (°C)
West Recess Nunatak	846	808	950	965 ± 105
Demas Bluff				
2000/8	908	876	959	926 ± 14
2000/3		930	1009	
Mount Hampton				
2000/12	1008	1048	1070	1088 ± 22
PK 91005	1023	1038	1065	1108 ± 16
Mount Cumming				
2000/2	965	929	992	957 ± 22
WA/4	973	1025	1006	

*BandK '90; Brey and Köhler (1990), calculated assuming a pressure of 1.5 GPa.

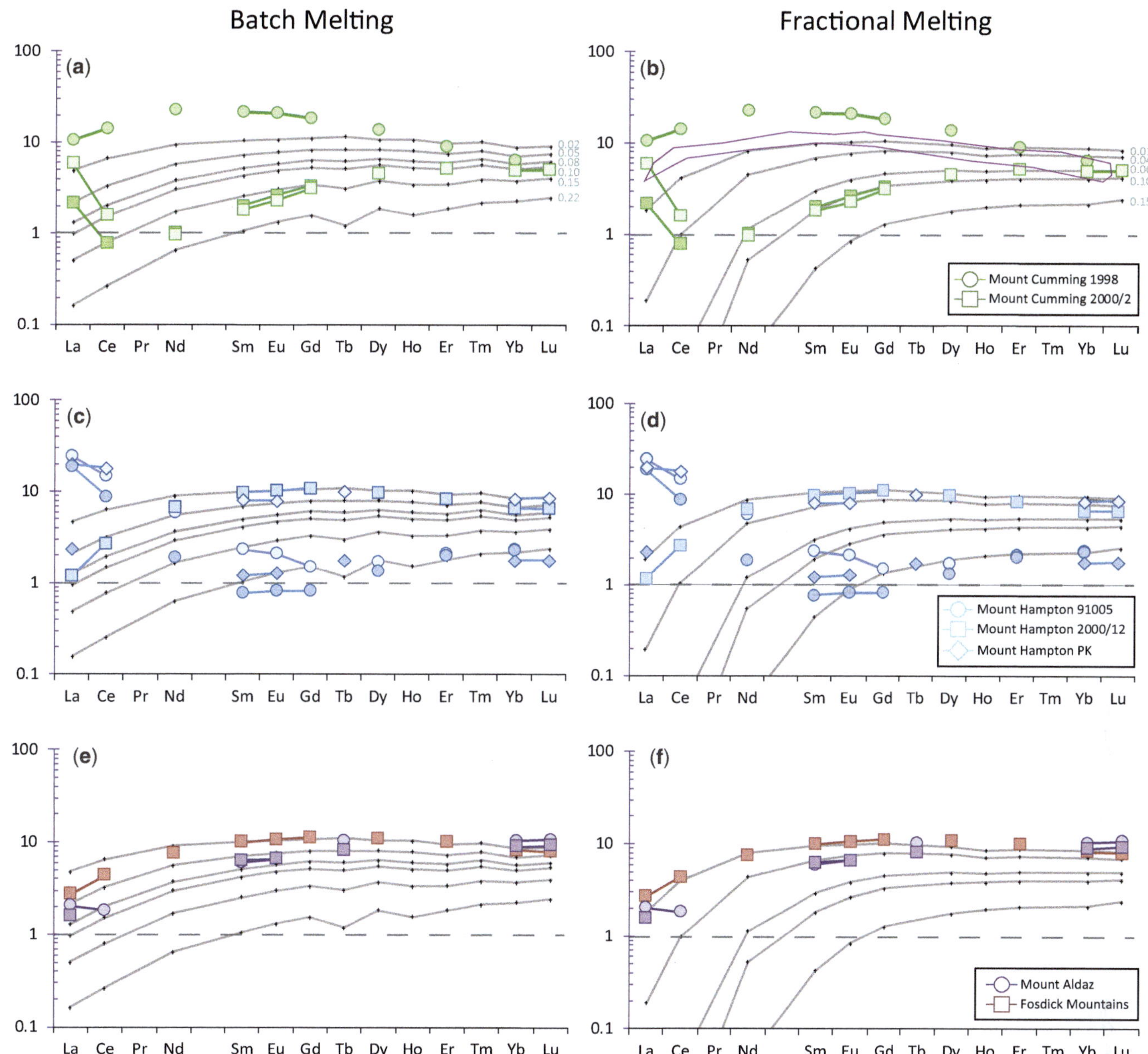

Fig. 11. Primitive mantle normalized REE patterns for selected clinopyroxenes overlain with melt extraction models for spinel facies partial melting of a primitive upper mantle source for both batch (**a, c, e**) and fractional (**b, d, f**) melting. Patterns for different grains from the same sample are indicated by different shading of symbols. The degrees of melt extraction are indicated in panels (**a**) and (**b**), and were calculated using melting modes and partition coefficients from Johnston and Schwab (2004), Kelemen *et al.* (1993) and references therein. Purple overlay in panel (**b**), shows field of clinopyroxene that underwent melting in garnet facies prior to re-equilibrating in spinel facies from Bonadiman *et al.* (2005). Primitive mantle composition from McDonough and Sun (1995).

A large number of the xenoliths from within the Ross Province (Fosdick Mountains) have pyroxene mineral compositions suggestive of refertilization of a harzburgitic source: high Al_2O_3 in many clinopyroxenes, and orthopyroxene and clinopyroxene major element compositions consistent with different estimates for degree of partial melt extraction. However, some samples preserve clinopyroxene and orthopyroxene chemistry consistent with larger (20–25%) degrees of partial melt extraction.

Within the Amundsen Province sites, samples from Executive Committee Range typically show smaller discrepancies between major element pyroxene partial melting estimates, and variable REE patterns broadly consistent with the major elements (e.g. 91005 higher degrees of melt extraction than 2000/12; Figs 4b, c, 11c and d). One sample from Mount Cumming retains a REE pattern in clinopyroxene suggestive of the influence of garnet in its melting history. By contrast, Mount Aldaz xenolith whole rock and clinopyroxene REE data are consistent with low degrees of partial melting in the spinel facies (e.g. Fig. 4e and f).

Only whole rock data are available for xenoliths from Mount Murphy, in the boundary region between the two provinces, and of all the xenoliths suites, the chemistry for these is the least consistent with a simple melt extraction history.

Metasomatism

Whilst there is an absence of secondary minerals reported for the xenolith suite, major and trace element data record interaction of the peridotites with metasomatic fluids and/or melts (Figs 4–8). Characterizing the nature and extent of these interactions is hampered by the limited trace element and isotopic data available. However, some constraints can be determined from the trace element profiles of clinopyroxene from the six samples from the Executive Committee Range and Fosdick Mountains (Fig. 6), which point to varied metasomatic agents influencing the lithospheric mantle.

Clinopyroxene from three spinel lherzolite samples show variable and elevated L-MREE patterns and enrichments in Th and U (1998 and 2000/2 from Mount Cumming, and 91005 from Mount Hampton; Fig. 6). Relative depletions in Zr, Hf and Ti, and Rb and Ba, will at least in part reflect that, unlike the REE, these elements partition more strongly into co-existing spinel than clinopyroxene (e.g. Grégoire *et al.* 2000; Bonadiman *et al.* 2005).

Relatively high Pb/Ce ratios for the incompatible trace element enriched Mount Cumming sample 2000/2 [$(Pb/Ce)_N$ *c.* 3–10] compared with the incompatible trace element depleted samples [Mount Hampton 2000/12 and Demas Bluff; $(Pb/Ce)_N$ *c.* 0.5–1], coupled with strong enrichments in Th, U and depleted Nb and Ta, are consistent with interaction with a melt or fluid bearing a subduction signature. By contrast, the incompatible trace element enrichment pattern in sample 91005 from Mount Hampton shows stronger enrichments in L-MREE and Nb, but lesser enrichments in Th and U (Ta and Pb were not measured), suggesting interaction with melts of a differing origin. This sample also has high Eu/Ti (Fig. 6), which is a signature attributed to interaction with carbonatitic melts (e.g. Yaxley *et al.* 1991; Scott *et al.* 2014; Martin *et al.* 2015). Notably, the Mount Cumming xenoliths with high Th, U and low Nb and Ta do not have elevated Eu/Ti.

Age and source characteristics

Re–Os data of xenoliths from the Amundsen Province sites (mounts Hampton, Cumming and Aldaz) provide mantle depletion age constraints consistent with Proterozoic stabilization of the lithosphere (1.3–1.1 Ga), and coincident with Nd model ages of the granitoids of the Amundsen Province (1.3–1.0 Ga, Pankhurst *et al.* 1998; Handler *et al.* 2003).

Two additional Re–Os data for samples from Fosdick Mountains, Ross Province, also return Proterozoic model ages, with one sample overlapping those from the Amundsen Province samples and the second significantly less radiogenic sample, yielding an older Re depletion age of *c.* 1.7 Ga (Table 3). This is even older than the Nd model ages of granitoids of the Ross Province (1.5–1.3 Ga), suggesting that the crust is younger than the mantle, which contains older Proterozoic foundations or perhaps remnants of older mantle caught up during continent formation. If the Nd model ages (and U–Pb ages of inherited zircon) in the granitoids reflect a recycled sediment source rather than crust formation ages, the discrepancy between mantle and crustal ages may be quite significant. Isolated, ancient mantle domains decoupled from younger overlying crust have been reported from other terranes along the eastern Gondwana margin, most notably from Zealandia (e.g. McCoy-West *et al.* 2013; Liu *et al.* 2015).

But not all of the xenolith isotopic data point to Proterozoic ages. In detail, the Mount Aldaz samples have isotopic systematics that are more consistent with an influence from the prolonged history of Paleozoic subduction known to have affected Marie Byrd Land. Three of the four xenoliths from Mount Aldaz have Os concentrations [3.2–3.7 parts per billion (ppb)] and isotopic compositions ($^{187}Os/^{188}Os$ *c.* 0.125–0.128; Handler *et al.* 2003) that span estimates of modern primitive and depleted upper mantle compositions (e.g. Meisel *et al.* 1996; Brandon *et al.* 2000; Day *et al.* 2017), producing relatively young Re-depletion model ages (570–130 Ma). One of these samples, MB69E, also preserves extremely depleted $^{87}Sr/^{86}Sr$ and $^{143}Nd/^{144}Nd$ compositions (Fig. 9a). This sample has a Nd depleted mantle model age of *c.* 488 Ma, broadly in agreement with its Re depletion age of *c.* 420 Ma. Both ages are consistent with material added to the lithospheric mantle during the Ordovician, a time of convergence and subduction magmatism along the Gondwana margin (e.g. Pankhurst *et al.* 1998). The fourth xenolith from Mount Aldaz (MB69F) has a radiogenic Os isotopic composition ($^{187}Os/^{188}Os$ *c.* 0.132), coupled with lower Os content (*c.* 1.5 ppb; Handler *et al.* 2003). These characteristics are consistent with radiogenic Os addition such as has been described for suprasubduction zone modified peridotites (e.g. Brandon *et al.* 1996) and chromites (Suzuki *et al.* 2011) elsewhere. In addition, these ages closely overlap the timing of suggested pyroxenite emplacement and metasomatism in the adjacent Victoria Land lithospheric mantle, as evidenced by ages for a clinopyroxenite xenolith from Foster Crater (McGibbon 1991) and eclogite and carbonatite emplacement in Victoria Land (Martin *et al.* 2015).

Aside from MB69E with its strongly depleted Sr and Nd isotopic composition, the sampled central Marie Byrd Land lithospheric mantle has a Sr-Nd-Pb isotopic composition that broadly overlaps an endmember identified in several studies of Cenozoic mafic volcanism across Marie Byrd Land and the Ross Sea region (Fig. 9). This isotopic endmember, variously described as low-μ or FOZO ('FOcal ZOne'; Stracke *et al.* 2005), appears to be a common component in much of the Cenozoic volcanism across the WARS (e.g. Hart *et al.* 1997; Panter *et al.* 2000; Aviado *et al.* 2015). Similar $^{87}Sr/^{86}Sr$ and $^{143}Nd/^{144}Nd$ compositions have also been reported for pyroxenite xenoliths from Marie Byrd Land (Wysoczanski *et al.* 1995) and upper mantle peridotite from Victoria Land (Martin *et al.* 2013), consistent with a widespread nature of this common signature in the lithospheric mantle across the wider region (e.g. Panter *et al.* 2000). The origin of this low-μ signature has been variously attributed to metasomatic enrichment of lithospheric mantle by both subduction and intraplate related magmatism (e.g. Day *et al.* 2019; Panter *et al.* 2000; LeMasurier *et al.* 2016). Although very limited, trace element data for clinopyroxene from Marie Byrd Land mantle xenoliths are consistent with either or both of these mechanisms influencing the sampled lithospheric mantle. Furthermore, as noted, the age and Os isotopic composition of some Mount Aldaz peridotites are consistent with lithospheric mantle influenced by subduction zone magmatism. However, there is no evidence for hydrous phases in the Marie Byrd Land lithospheric mantle sampled to date, as had been proposed from studies of volcanic rocks (e.g. Panter *et al.* 2000; LeMasurier *et al.* 2016) and the source of this hydrous component in the source of the Cenozoic magmatism in Marie Byrd Land remains elusive. This is in apparent contrast with lithospheric mantle xenolith suites from adjacent terranes, for example southern Victoria Land, where hydrous phases such as phlogopite and amphibole have been reported in lithospheric mantle xenoliths (e.g. Gamble *et al.* 1988; Martin *et al.* 2014; Day *et al.* 2019) although not all xenolith locations in southern Victoria Land contain hydrous mineral bearing xenoliths; the xenolith suite from Mount Morning is also anhydrous (Martin *et al.* 2015).

Cenozoic basaltic magmatism from Marie Byrd Land points to a mixing between this low-μ source and a HIMU endmember (e.g. Panter *et al.* 2000) that has been observed in magmatism more broadly in the southwest Pacific (e.g. McCoy-West *et al.* 2010; Gamble *et al.* 2018; Smellie *et al.* 2020). There is no evidence for a HIMU signature preserved in the sampled lithospheric mantle, nor in lower crustal or pyroxenite samples (Wysoczanski *et al.* 1995) analysed to date. This would be consistent with a compositionally layered mantle beneath Marie Byrd Land whereby a HIMU (fossil plume?) component underplates the lithospheric mantle, such as has been postulated in several studies (e.g. Weaver *et al.* 1994; Panter *et al.*

2000; Kipf *et al.* 2014). Direct samples of this mantle material have yet to be documented however.

Summary and conclusions

Cenozoic mafic magmatism has carried a large suite of lithospheric mantle xenoliths to the surface, from localities within both major terranes, the Ross and Amundsen provinces, that comprise Marie Byrd Land. The xenoliths are predominantly lherzolitic in composition, but include lesser amounts of harzburgite and rare dunite and pyroxenites. All are spinel-bearing; garnet is absent. No hydrous phases have been documented, although it is likely traces of apatite are present in at least a few samples. The sampled lithosphere is typically coarse granular to porphyroclastic, with pyroxenite veining present. Equilibration temperatures fall largely within 800–1050°C, with no apparent temperature variation with textural variety or by locality.

Characterisation of the lithospheric mantle and unravelling its age and evolution is hampered by patchy analytical datasets across the suite, and in particular, a lack of fully characterized samples from any single locality. Nonetheless, what emerges is a heterogeneous lithospheric mantle, likely preserving residues from various episodes and degrees of melt extraction together with refertilized material, and the influence of a variety of secondary cryptic metasomatic processes. Isotopically, the Marie Byrd Land lithospheric mantle is consistent with a low-μ (FOZO) component recognized in Cenozoic magmatism across the region. Limited Re–Os data suggest preservation of ancient, Proterozoic lithospheric mantle beneath both Ross and Amundsen provinces that may pre-date the overlying crust. However, isotopic data also indicate the influence of younger events, most notably Ordovician Os and Nd model ages for a lherzolite xenolith from the Usas Escarpment, Ross Province, that are consistent with a lithospheric mantle influenced by the protracted history of Paleozoic convergence and subduction.

Acknowledgements Xenoliths were collected during Antarctic expeditions supported by NSF-DPP, Antarctica New Zealand and VUWAE in 1982, 1983, 1984, 1988, 1989, 1990 and 1992, with the majority collected as an objective of the joint NZ – US – UK sponsored WAVE (West Antarctica Volcano Exploration) project that formed the basis of the Wysoczanski (1993) and Handler *et al.* (2003) studies. During the WAVE project field support from Bill McIntosh, Philip Kyle, Kurt Panter, Nelia Dunbar and John Smellie gave valuable assistance. Additional mineral data are reported here for samples donated by Chris Adams (Demas Bluff) and Dave Kimbrough (Recess Nunatak). Strontium, Nd and Pb isotopic data reported here were analysed at DTM, Carnegie Institution of Washington; MRH and RJW thank Rick Carlson, Mary Horan and Tim Mock for their support and assistance. This contribution has benefited from the very thoughtful and constructive comments of Alan Cooper, Adam Martin, and Kurt Panter, and we further thank Adam Martin for his editorial handling.

Author contributions **MRH**: formal analysis (lead), investigation (equal), writing – original draft (lead), writing – review and editing (equal); **RJW**: formal analysis (equal), investigation (equal), methodology (equal), writing – original draft (supporting), writing – review and editing (equal); **JAG**: formal analysis (equal), investigation (equal), methodology (equal), writing – original draft (supporting), writing – review and editing (equal).

Funding RW was funded by the Ministry of Business, Innovation and Employment (MBIE) Strategic Science Investment Fund (SSIF) programme 'Marine Geological Processes' (COPR2102).

Data availability All data generated or analysed during this study are included in this published article (and its supplementary files).

References

Adams, C.J. 1986. Geochronological studies of the Swanson Formation of Marie Byrd Land, West Antarctica, and correlation with northern Victoria Land, East Antarctica, and South Island, New Zealand. *New Zealand Journal of Geology and Geophysics*, **29**, 345–358, https://doi.org/10.1080/00288306.1986.10422157

Aldanmaz, E., Schmidt, M.W., Gourgaud, A. and Meisel, T. 2009. Mid-ocean ridge and suprasubduction geochemical signatures in spinel-peridotites from the Neothethyan ophiolites in SW Turkey: Implications for upper mantle melting processes. *Lithos*, **113**, 691–708, https://doi.org/10.1016/j.lithos.2009.03.010

An, M., Wiens, D. *et al.* 2015. S-velocity model and inferred Moho topography beneath the Antarctic Plate from rayleigh waves. *Journal of Geophysical Research: Solid Earth*, **120**, 359–383, https://doi.org/10.1002/2014JB011332

Arai, S. 1994. Characterization of spinel peridotites by olivine-spinel compositional relationships: review and interpretation. *Chemical Geology*, **113**, 191–204, https://doi.org/10.1016/0009-2541(94)90066-3

Anders, E. and Grevasse, N. 1989. Abundances of the elements: meteoric and solar. *Geochimica et Cosmochimica Acta*, **53**, 197–214, https://doi.org/10.1016/0016-7037(89)90286-X

Aviado, K.B., Rilling-Hall, S., Bryce, J.G. and Mukasa, S.B. 2015. Submarine and subaerial lavas in the West Antarctic Rift System: temporal record of shifting magma source components from the lithosphere and asthenosphere. *Geochemistry, Geophysics, Geosystems*, **16**, 4344–4361, https://doi.org/10.1002/2015GC006076

Baker, J.A., Chazot, G., Menzies, M. and Thirlwall, M. 1998. Metasomatism of the shallow mantle beneath Yemen by the Afar plume–implications for mantle plumes, flood volcanism and intraplate volcanism. *Geology*, **26**, 431–434, https://doi.org/10.1130/0091-7613(1998)026<0431:MOTSMB>2.3.CO;2

Bedini, R.M., Bodinier, J.-L., Dautria, J.-M. and Morten, L. 1997. Evolution of LILE-enriched small melt fractions in the lithospheric mantle: a case study from the East African Rift. *Earth and Planetary Science Letters*, **153**, 67–83, https://doi.org/10.1016/S0012-821X(97)00167-2

Berg, J.H., Moscati, R.J. and Herz, D.L. 1989. A petrologic geotherm from a continental rift in Antarctica. *Earth and Planetary Science Letters*, **93**, 98–108, https://doi.org/10.1016/0012-821X(89)90187-8

Bertrand, P. and Mercier, J.-C.C. 1985. The mutual solubility of coexisting ortho- and clinopyroxene: toward an absolute geothermometer for the natural system? *Earth and Planetary Science Letters*, **76**, 109–122, https://doi.org/10.1016/0012-821X(85)90152-9

Blusztajn, J. and Shimizu, N. 1994. The trace-element variations in clinopyroxenes from spinel peridotite xenoliths from southwest Poland. *Chemical Geology*, **111**, 227–243, https://doi.org/10.1016/0009-2541(94)90091-4

Bonadiman, C., Beccaluva, L., Coltorti, M. and Siena, F. 2005. Kimberlite-like metasomatism and 'garnet signature' in spinel-peridotite xenoliths from Sal, Cape Verde archipelago: relics of a subcontinental mantle domain with the Atlantic Ocean lithosphere? *Journal of Petrology*, **46**, 2465–2493, https://doi.org/10.1093/petrology/egi061

Brandon, A.D., Creaser, R.A., Shirey, S.B. and Carlson, R.W. 1996. Osmium recycling in subduction zones. *Science*, **272**, 861–863, https://doi.org/10.1126/science.272.5263.861

Brandon, A.D., Snow, J.E., Walker, R.J., Morgan, J.W. and Mock, T.D. 2000. ^{190}Pt–^{186}Os and ^{187}Re–^{187}Os systematics of abyssal peridotites. *Earth and Planetary Science Letters*, **177**, 319–335, https://doi.org/10.1016/S0012-821X(00)00044-3

Brey, G.P. and Köhler, T. 1990. Geothermometry in four-phase lherzolites: II New thermobarometers and practical assessment of existing thermobarometry. *Journal of Petrology*, **31**, 1352–1378.

Chaput, J., Aster, R.C. *et al.* 2014. The crustal thickness of West Antarctica. *Journal of Geophysical Research: Solid Earth*, **119**, 378–395, https://doi.org/10.1002/2013JB010642

Chatzaras, V., Krukenberg, S.C., Cohen, S.M., Medaris, L.G. Jr, Withers, A.C. and Bagley, B. 2016. Axial-type olivine crystallographic preffered orientations: the effect of strain geometry on mantle texture. *Journal of Geophysical Research: Solid Earth*, **121**, 4895–4922, https://doi.org/10.1002/2015JB012628

Chatzaras, V. and Kruckenberg, S.C. in Press. Effects of melt-percolation, refertilization, and deformation on upper mantle seismic anisotropy: constraints from peridotite xenoliths, Marie Byrd Land, West Antarctica. *Geological Society, London, Memoirs*, **56**, https://doi.org/10.1144/M56-2020-16

Cohen, S.M. 2016. *An assessment of heterogeneity within the lithospheric mantle, Marie Byrd Land, West Antarctica*. MSc thesis, Boston College.

Day, J.M.D., Walker, R.J. and Warren, J.M. 2017. $^{186}Os–^{187}Os$ and highly siderophile element abundance systematics of the mantle revealed by abyssal peridotites and Os-rich alloys. *Geochimica et Cosmochimica Acta*, **200**, 232–254, https://doi.org/10.1016/j.gca.2016.12.013

Day, J.M.D., Harvey, R.P. and Hilton, D.R. 2019. Melt-modified lithosphere beneath Ross Island and its role in the tectonomagmatic evolution of the West Antarctic Rift System. *Chemical Geology*, **518**, 45–54, https://doi.org/10.1016/j.chemgeo.2019.04.012

DiVenere, V., Kent, D.V. and Dalziel, I.W.D. 1996. Summary of palaeomagnetic results from West Antarctica: implications for the tectonic evolution of the Pacific margin of Gondwana during the Mesozoic. *Geological Society, London, Special Publications*, **108**, 31–43, https://doi.org/10.1144/GSL.SP.1996.108.01.03

Elliot, D.H. 2013. The geological and tectonic evolution of the Transantarctic Mountains: a review. *Geological Society, London, Special Publications*, **381**, 7–35, https://doi.org/10.1144/SP381.14

Frey, F.A. and Prinz, M. 1978. Ultramafic inclusions from San Carlos, Arizona: petrologic and geochemical data bearing on their petrogenesis. *Earth and Planetary Science Letters*, **38**, 129–176, https://doi.org/10.1016/0012-821X(78)90130-9

Gamble, J.A., McGibbon, F., Kyle, P.R., Menzies, M.A. and Kirsch, I. 1988. Metasomatised xenoliths from Foster Crater, Antarctica: implications for lithospheric structure and processes beneath the Transantarctic Mountain front. *Journal of Petrology*, Special Lithosphere Issue, 109–138, https://doi.org/10.1093/petrology/Special_Volume.1.109

Gamble, J.A., Adams, C.J., Morris, M.A., Wysoczanski, R.J., Handler, M.R. and Timm, C. 2018. The geochemistry and petrogenesis of Carnley Volcano, Auckland Islands, SW Pacific. *New Zealand Journal of Geology and Geophysics*, **61**, 480–497, https://doi.org/10.1080/00288306.2018.1505642

Grégoire, M., Moine, B.N., O'Reilly, S.Y., Cottin, J.Y. and Giret, A. 2000. Trace element residence and partitioning in mantle xenoliths metasomatized by highly alkaline, silicate- and carbonate-rich melts (Kerguelen Islands, Indian Ocean). *Journal of Petrology*, **41**, 477–509, https://doi.org/10.1093/petrology/41.4.477

Handler, M.R., Wysoczanski, R.J. and Gamble, J.A. 2003. Proterozoic lithosphere in Marie Byrd Land, West Antarctica: Re–Os systematics of spinel peridotite xenoliths. *Chemical Geology*, **196**, 131–145, https://doi.org/10.1016/S0009-2541(02)00410-2

Hart, S.R. 1984. A large-scale isotope anomaly in the Southern Hemisphere mantle. *Nature*, **309**, 753–757, https://doi.org/10.1038/309753a0

Hart, S.R., Blusztajn, J., LeMasurier, W.E. and Rex, D.C. 1997. Hobbs coast Cenozoic volcanism: implications for the West Antarctic rift system. *Chemical Geology*, **139**, 223–248, https://doi.org/10.1016/S0009-2541(97)00037-5

Hassler, D.R. and Shimizu, N. 1998. Osmium isotopic evidence for ancient subcontinental lithospheric mantle beneath the Kerguelen Islands, Southern Indian Ocean. *Science*, **280**, 418–421, https://doi.org/10.1126/science.280.5362.418

Heinemann, J., Stock, J., Clayton, R., Hafner, K., Cande, S. and Raymond, C. 1999. Constraints on the proposed Marie Byrd Land – Bellingshausen plate boundary from seismic reflection data. *Journal of Geophyscial Research*, **104**, 25,321–25,330, https://doi.org/10.1029/1998JB900079

Hole, M.J. and LaMasurier, W. 1994. Tectonic controls on the geochemical composition of Cenozoic, mafic alkaline volcanic rocks from West Antarctica. *Contributions to Mineralogy and Petrology*, **11**, 187–202, https://doi.org/10.1007/BF00286842

Johnston, A.D. and Schwab, B.E. 2004. Constraints on clinopyroxene/melt partitioning of REE, Rb, Sr, Ti, Cr, Zr and Nb during mantle melting: first insights from direct peridotite melting experiments at 1.0 Gpa. *Geochimica et Cosmochimica Acta*, **68**, 4949–4962, https://doi.org/10.1016/j.gca.2004.06.009

Jordan, T.A., Riley, T.R. and Siddoway, C.S. 2020. The geological history and evolution of West Antarctica. *Nature Reviews Earth and Environment*, **1**, 117–133, https://doi.org/10.1038/s43017-019-0013-6

Kelemen, P.B., Shimizu, N. and Dunn, T. 1993. Relative depletion of niobium in some arc magmas and the continental crust: partitioning of K, Nb, La and Ce during melt/rock reaction in the upper mantle. *Earth and Planetary Science Letters*, **120**, 111–134, https://doi.org/10.1016/0012-821X(93)90234-Z

Kipf, A., Hauff, F. *et al.* 2014. Seamounts off the West Antarctic margin: a case for non-hotspot driven intraplate volcanism. *Gondwana Research*, **25**, 1660–1679, https://doi.org/10.1016/j.gr.2013.06.013

Kyle, P.R. and Muncy, H.L. 1989. Geology and geochronology of McMurdo Volcanic Group rocks in the vicinity of Lake Morning, McMurdo Sound, Antarctica. *Antarctic Science*, **1**, 345–350, https://doi.org/10.1017/S0954102089000520

Larter, R.D., Cunningham, A.P., Barker, P.F., Gohl, K. and Nitsche, F.O. 2002. Tectonic evolution of the Paciic margin of Antarctica. I. Late Cretaceous tectonic reconstructions. *Journal of Geophysical Research*, **107**, 2345, https://doi.org/10.1029/2000JB000052

Lee, C.-T., Luffi, P. and Chin, E.J. 2010. Building and destroying continental mantle. *Annual Review of Earth and Planetary Sciences*, **39**, 59–90, https://doi.org/10.1146/annurev-earth-040610-133505

LeMasurier, W.E. 2008. Neogene extension and basin deepening in the West Antarctic rift inferred from comparisons with the East African rift and other analogs. *Geology*, **36**, 247–250, https://doi.org/10.1130/G24363A.1

LeMasurier, W. 2013. Shield volcanoes of Marie Byrd Land, West Antarctic rift: oceanic island similarities, continental signature, and tectonic controls. *Bulletin of Volcanology*, **75**, 726, https://doi.org/10.1007/s00445-013-0726-1

LeMasurier, W.E. and Landis, C.A. 1996. Mantle-plume activity recorded by low-relief erosion surfaces in West Antarctica and New Zealand. *Geological Society of America Bulletin*, **108**, 1450–1466, https://doi.org/10.1130/0016-7606(1996)108<1450:MPARBL>2.3.CO;2

LeMasurier, W.E. and Thomson, J.W. 1990. *In*: *Volcanoes of the Antarctic Plate and Southern Oceans*. American Geophysical Union.

LeMasurier, W.E., Choi, S.H., Hart, S.R., Mukasa, S. and Rogers, N. 2016. Reconciling the shadow of a subduction signature with rift geochemistry and tectonic environment in Eastern Marie Byrd Land, Antarctica. *Lithos*, **260**, 134–153, https://doi.org/10.1016/j.lithos.2016.05.018

Liang, Y., Sun, C. and Yao, L. 2013. A REE-in-two-pyroxene thermometer for mafic and ultramafic rocks. *Geochimica et Cosmochimica Acta*, **102**, 246–260, https://doi.org/10.1016/j.gca.2012.10.035

Liu, J., Scott, J.M., Martin, C.E. and Pearson, D.G. 2015. The longevity of Archean mantle residues in the convecting upper mantle and their role in young continent formation. *Earth and Planetary Science Letters*, **424**, 109–118, https://doi.org/10.1016/j.epsl.2015.05.027

Luyendyk, B., Cisowski, S., Smith, C., Richard, S. and Kimbrough, D. 1996. Paleomagnetic study of the northern Ford Ranges, western Marie Byrd Land, West Antarctica: motion between West and East Antarctica. *Tectonics*, **15**, 122–141, https://doi.org/10.1029/95TC02524

Martin, A.P., Cooper, A.F. and Price, R.C. 2013. Petrogenesis of Cenozoic, alkalic volcanic lineages at Mount Morning, West Antarctica and their entrained lithospheric mantle xenoliths: lithospheric v. asthenospheric mantle sources. *Geochimica et Cosmochimica Acta*, **122**, 127–152, https://doi.org/10.1016/j.gca.2013.08.025

Martin, A.P., Cooper, A.F. and Price, R.C. 2014. Increased mantle heat flow with on-going rifting of the West Antarctic rift system inferred from characterisation of plagioclase peridotite in the shallow Antarctic mantle. *Lithos*, **190–191**, 173–190, https://doi.org/10.1016/j.lithos.2013.12.012

Martin, A.P., Price, R.C., Cooper, A.F. and McCammon, C.A. 2015. Petrogenesis of the rifted southern Victoria Land lithospheric mantle, Antarctica, inferred from petrography, geochemistry, thermobarometry and oxybarometry of peridotite and pyroxenite xenoliths from the Mount Morning eruptive centre. *Journal of Petrology*, **56**, 193–226, https://doi.org/10.1093/petrology/egu075

McCoy-West, A.J., Baker, J.A., Faure, K. and Wysoczanski, R.J. 2010. Petrogenesis and origins of mid-Cretaceous continental intraplate volcanism in Marlborough, New Zealand: implications for the long-lived HIMU magmatic mega-province of SW Pacific. *Journal of Petrology*, **51**, 2003–2045, https://doi.org/10.1093/petrology/egq046

McCoy-West, A.J., Bennet, V.C., Puchtel, I.S. and Walker, R.J. 2013. Extreme persistence of cratonic lithosphere in the southwest Pacific: Paleoproterozoic Os isotopic signatures in Zealandia. *Geology*, **41**, 231–234, https://doi.org/10.1130/G33626.1

McDonough, W.F. and Frey, F.A. 1989. Rare Earth Elements in upper mantle rocks. *Reviews in Mineralogy*, **21**, 99–145.

McDonough, W.F. and Sun, S.-S. 1995. The composition of the Earth. *Chemical Geology*, **120**, 223–253, https://doi.org/10.1016/0009-2541(94)00140-4

McGibbon, F.M. 1991. Geochemistry and petrology of ultramafic xenoliths of the Erebus Volcanic Province. In: Thomson, M.R.A, Crame, J.A. and Thomson, J.W. (eds). *Geological Evolution of Antarctica-Proceedings of the 5th International Symposium on Antarctic Earth Sciences*. University of Cambridge Press, 317–321.

McKenzie, D., Jackson, J. and Priestley, K. 2007. Thermal structure of oceanic and continental lithosphere. *Earth and Planetary Science Letters*, **233**, 337–349, https://doi.org/10.1016/j.epsl.2005.02.005

Meisel, T., Walker, R.J. and Morgan, J.W. 1996. The osmium isotopic composition of the Earth's primitive upper mantle. *Nature*, **383**, 517–520, https://doi.org/10.1038/383517a0

Moreira, M. and Madureira, P. 2005. Cosmogenic helium and neon in 11 Myr old ultramafic xenoliths: consequences for mantle signatures in old samples. *Geochemistry, Geophysics, Geosystems*, https://doi.org/10.1029/2005GC000939

Mukasa, S.B. and Dalziel, I.W.D. 2000. Marie Byrd Land, West Antarctica: evolution of Gondwana's Pacific margin constrained by zircon U-Pb geochronology and feldspar common-Pb isotopic compositions. *Geological Society of America Bulletin*, **112**, 611–627, https://doi.org/10.1130/0016-7606(2000)112<611:MBLWAE>2.0.CO;2

Pankhurst, R.J., Weaver, S.D., Bradshaw, J.D., Storey, B.C. and Ireland, T.R. 1998. Geochronology and geochemistry of pre-Jurassic superterranes in Marie Byrd Land, Antarctica. *Journal of Geophysical Research*, **103**, 2529–2547, https://doi.org/10.1029/97JB02605

Panter, K.S. 1997. Petrogenesis of a phonolite-trachyte succession at Mount Sidley, Marie Byrd Land, Antarctica. *Journal of Petrology*, **38**, 1225–1253, https://doi.org/10.1093/petroj/38.9.1225

Panter, K.S., Hart, S.R., Kyle, P.R., Blusztajn, J. and Wilch, T. 2000. Geochemisty of late Cenozoic basalts from the Crary Mountains: characterization of mantle sources in Marie Byrd Land, Antarctica. *Chemical Geology*, **165**, 215–241, https://doi.org/10.1016/S0009-2541(99)00171-0

Panter, K.S., Wilch, T.I., Smellie, J.L., Kyle, P.R. and McIntosh, W.C. 2021. Marie Byrd Land and Ellsworth Land II Petrology. *In*: Smellie, J.L., Panter, K.S. and Geyer, A. (eds) *Volcanism in Antarctica: 200 Million Years of Subduction, Rifting and Continental Break-Up*. *Geological Society, London, Memoirs*, 55, https://doi.org/10.1144/M55-2019-50

Parkinson, I.J., Hawkesworth, C.J. and Cohen, A.S. 1998. Ancient mantle in a modern arc: osmium isotopes in Izu-Bonin-Mariana forearc peridotites. *Science*, **281**, 2011–2013, https://doi.org/10.1126/science.281.5385.2011

Putirka, K.D. 2008. Thermometers and barometers for volcanic systems. *In*: Putirka, K.D. and Tepley, F.J., III (eds) *Minerals, Inclusions and Volcanic Processes*. Mineralogical Society of America and Geochemical Society, Reviews in Mineralogy and Geochemistry, **69**, 61–120.

Scott, J.M., Hodgkinson, A., Palin, J.M., Waight, T.E., Van der Meer, Q.H.A. and Cooper, A.F. 2014. Ancient melt depletion overprinted by young carbonatitic metasomatism in the New Zealand lithospheric mantle. *Contributions to Mineralogy and Petrology*, **167**, 963, https://doi.org/10.1007/s00410-014-0963-0

Smellie, J.L., Martin, A.P., Panter, K.S., Kyle, P.R. and Geyer, A. 2020. Magmatism in Antarctica and its relation to Zealandia. *New Zealand Journal of Geology and Geophysics*, **63**, 578–588, https://doi.org/10.1080/00288306.2020.1781666

Storey, B.C., Leat, P.T., Weaver, S.D., Pankhurst, R.J., Bradshaw, J.D. and Kelley, S. 1999. Mantle plumes and Antarctica-New Zealand rifting: evidence from mid-Cretaceous mafic dykes. *Journal of the Geological Society*, **156**, 659–671, https://doi.org/10.1144/gsjgs.156.4.0659

Stracke, A., Hofmann, A.W. and Hart, S.R. 2005. FOZO, HIMU, and the rest of the mantle zoo. *Geochemistry, Geophysics, Geosystems*, https://doi.org/10.1029/2004GC000824

Streckeisen, A. 1976. To each plutonic rock its proper name. *Earth Science Reviews*, **12**, 1–33, https://doi.org/10.1016/0012-8252(76)90052-0

Suzuki, K., Senda, R. and Shimizu, K. 2011. Osmium behavior in a subduction system elucidated from chromian spinel in Bonin Island beach sands. *Geology*, **39**, 999–1002, https://doi.org/10.1130/G32044.1

Taylor, W.R. 1998. An experimental test of some geothermometer and geobarometer formulations for upper mantle peridotites with application to the thermobarometry of fertile lherzolite and garnet websterite. *Neues Jahrbuch fur Mineralogie*, **172**, 381–408, https://doi.org/10.1127/njma/172/1998/381

ten Brink, U.T., Hackney, R.I., Bannister, S., Stern, T.A. and Makovsky, Y. 1997. Uplift of the Transantarctic Mountains and the bedrock beneath the East Antarctic ice sheet. *Journal of Geophysical Research*, **102**, 27,603–27,621, https://doi.org/10.1029/97JB02483

Tulloch, A.J., Mortimer, N. *et al.* 2019. Reconnaissance basement geology and tectonics of South Zealandia. *Tectonics*, **38**, 516–551, https://doi.org/10.1029/2018TC005116

Upton, B.G.J., Downes, H., Kirstein, L.A., Bonadiman, C., Hill, P.G. and Ntaflos, T. 2011. The lithospheric mantle and lower crust-mantle relationships under Scotland: a xenolithic perspective. *Journal of the Geological Society, London*, **168**, 873–885, https://doi.org/10.1144/0016-76492009-172

Van der Wal, D. and Bodinier, J.-L. 1996. Origin of the recrystallisation front in the Ronda peridotite by km-scale pervasive porous melt flow. *Contributions to Mineralogy and Petrology*, **122**, 237–405.

van Wyk de Vries, M., Bingham, R.G. and Hein, A.S. 2017. A new volcanic province: an inventory of subglacial volcanoes in West Antarctica. *Geological Society, London, Special Publications*, **461**, 231–248, https://doi.org/10.1144/SP461.7

Weaver, S.D., Adams, C.J. and Pankhurst, R.J. 1992. Granites of Edward VII Peninsula, Marie Byrd Land: anorogenic magmatism related to Antarctic-New Zealand rifting. *Transactions of the Royal Society of Edinburgh*, **83**, 281–290, https://doi.org/10.1017/S0263593300007963

Weaver, S.D., Storey, B.C., Pankhurst, R.J., Mukasa, S.B., DiVenere, V.J. and Bradshaw, J.D. 1994. Antarctica-New Zealand rifting and Marie Byrd Land lithospheric magmatism linked to ridge subduction and mantle plume activity. *Geology*, **22**, 811–814, https://doi.org/10.1130/0091-7613(1994)022<0811:ANZRAM>2.3.CO;2

Wells, T.R.A. 1977. Pyroxene thermometry in simple and complex systems. *Contributions to Mineralogy and Petrology*, **62**, 129–139, https://doi.org/10.1007/BF00372872

Winberry, J.P. and Anandakrishnan, S. 2004. Crustal structure of the West Antarctic rift system and Marie Byrd Land hotspot. *Geology*, **32**, 977–980, https://doi.org/10.1130/G20768.1

Wobbe, F., Gohl, K., Chambord, A. and Sutherland, R. 2012. Structure and breakup history of the rifted margin of West Antarctica in relation to Cretaceous separation from Zealandia and Bellingshausen plate motion. *Geochemistry, Geophysics, Geosystems*, **13**, 1–19, https://doi.org/10.1029/2011GC003742

Wysoczanski, R.J. 1993. *Lithospheric xenoliths from Marie Byrd Land volcanic province, West Antarctica*. PhD thesis, Victoria University of Wellington, 476.

Wysoczanski, R.J. and Gamble, J.A. 1992. Xenoliths from the volcanic province of West Antarctica and implications for lithospheric structure and processes. *In*: Yoshida, Y., Kaminuma, K. and Shiraishi, K. (eds) *Recent progress in Antarctic Earth Science*. Terrapub, Tokyo, 273–277.

Wysoczanski, R.J., Gamble, J.A., Kyle, P.R. and Thirlwall, M.F. 1995. The petrology of lower crustal xenoliths from the Executive Committee Range, Marie Byrd Land Volcanic Provnice, West Antarctica. *Lithos*, **36**, 185–201, https://doi.org/10.1016/0024-4937(95)00017-8

Yaxley, G.M., Crawford, A.J. and Green, D.H. 1991. Evidence for carbonatite metasomatism in spinel peridotite xenoliths from western Victoria, Australia. *Earth and Planetary Science Letters*, **107**, 305–317, https://doi.org/10.1016/0012-821X(91)90078-V

Ultramafic mantle xenoliths in the Late Cenozoic volcanic rocks of the Antarctic Peninsula and Jones Mountains, West Antarctica

Philip T. Leat[1,2]*, Aidan J. Ross[3] and Sally A. Gibson[3]

[1]British Antarctic Survey, High Cross, Madingley Road, Cambridge CB3 0ET, UK

[2]School of Geography, Geology and the Environment, University of Leicester, University Road, Leicester LE1 7RH, UK

[3]Department of Earth Sciences, University of Cambridge, Downing Street, Cambridge CB2 3EQ, UK

*Correspondence: ptle@bas.ac.uk

Abstract: Abundant mantle-derived ultramafic xenoliths occur in Cenozoic (7.7–1.5 Ma) mafic alkaline volcanic rocks along the former active margin of West Antarctica, that extends from the northern Antarctic Peninsula to Jones Mountains. The xenoliths are restricted to post-subduction volcanic rocks that were emplaced in fore-arc or back-arc positions relative to the Mesozoic–Cenozoic Antarctic Peninsula volcanic arc. The xenoliths are spinel-bearing, include harzburgites, lherzolites, wehrlites and pyroxenites, and provide the only direct evidence of the composition of the lithospheric mantle underlying most of the margin. The harzburgites may be residues of melt extraction from the upper mantle (in a mid-ocean ridge type setting), that accreted to form oceanic lithosphere, which was then subsequently tectonically emplaced along the active Gondwana margin. An exposed highly depleted dunite–serpentinite upper mantle complex on Gibbs Island, South Shetland Islands, supports this interpretation. In contrast, pyroxenites, wehrlites and lherzolites reflect percolation of mafic alkaline melts through the lithospheric mantle. Volatile and incompatible trace element compositions imply that these interacting melts were related to the post-subduction magmatism which hosts the xenoliths. The scattered distribution of such magmatism and the history of accretion suggest that the dominant composition of sub-Antarctic Peninsula lithospheric mantle is likely to be harzburgitic.

Ultramafic mantle-derived xenoliths occur in Cenozoic–Quaternary mafic volcanic rocks in the Antarctic Peninsula region and also further west along the Pacific margin of West Antarctic in the Jones Mountains (Fig. 1). Apart from one outcrop of mantle rocks in Gibbs Island, South Shetland Islands, these xenoliths provide the only direct evidence of the composition of the lithospheric mantle underlying this part of West Antarctica. The host volcanic rocks (mainly basanites and alkali basalts) form scattered outcrops in the region, and were erupted after the cessation of a long period of subduction along the continental margin. Many of these volcanic rocks have been studied in detail and their geochemical and volcanological characteristics are generally well understood (Smellie *et al.* 1988, 2008; Hole *et al.* 1993; Košler *et al.* 2009; Hole 2021). Ultramafic xenoliths have been noted in these volcanic rocks since the 1960s (Craddock *et al.* 1964; Nelson 1966; Fleet 1968). Chemical data for some xenolith localities have been described in detail (Gibson *et al.* 2020), but for other localities, information is sparse. Petrological information on the xenoliths is mainly contained in abstract form (e.g. Ross *et al.* 2009; Gibson *et al.* 2010*a*, *b*; Calabozo *et al.* 2014; Rooks *et al.* 2015) and in unpublished theses (e.g. Ross 2008; Gibson 2012; Rooks 2016). The purpose of this chapter is to present the first overview of ultramafic xenolith localities in the Antarctic Peninsula and Jones Mountains, and examine the evidence they provide for the composition of the regional subcontinental lithospheric mantle.

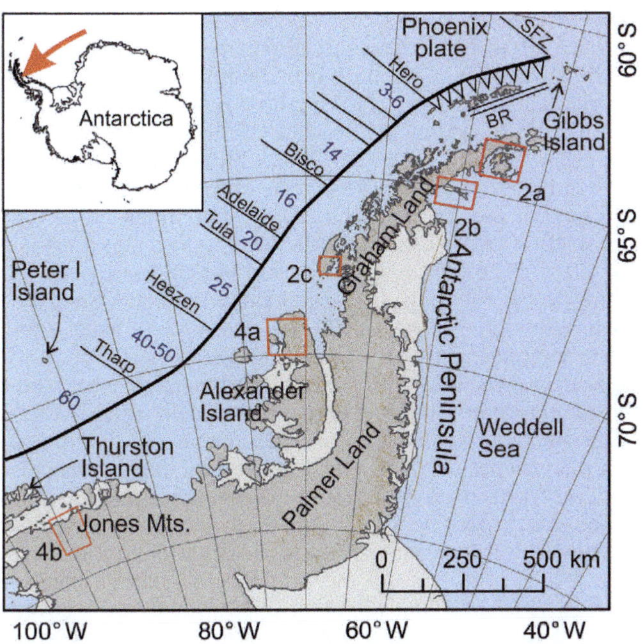

Fig. 1. Tectonic sketch map of the continental margin of West Antarctica. The solid heavy line is the site of subsequent ridge–trench collisions that now forms the continent–ocean boundary. Red boxes show areas of detailed maps in Figures 2 and 4. Fracture zones in the Antarctic ocean plate are indicated, with ages (in Ma) of ridge–trench collision for selected segments. SE-directed subduction of the Phoenix oceanic plate continues between the Hero and Shackleton (SFZ) fracture zones and is related to extension along the Bransfield rift (BR) (after Barker 1982; Hole *et al.* 1991; Larter and Barker 1991; Larter *et al.* 2002).

Tectonic setting and magmatic history

West Antarctica formed part of the margin between Gondwana and the Pacific Ocean, from the Early Paleozoic until the Mesozoic, and consists largely of magmatic and accreted rocks of this age (Storey and Garrett 1985; Pankhurst *et al.* 1993; Millar *et al.* 2002; Veevers and Saeed 2013). The oldest exposed rocks in West Antarctica are Mesoproterozoic gneisses at Haag Nunataks (zircon U–Pb age 1238 ± 4 Ma; Riley *et al.* 2020). These are of similar age to some detrital zircon populations in West Antarctica (Flowerdew *et al.* 2006; Veevers and Saeed 2013) and also Re–Os model ages of mantle xenoliths from Marie Byrd Land (Handler *et al.* 2003).

Zircon U–Pb data suggest that the oldest protolith ages of exposed crustal lithologies in the Antarctic Peninsula are Ordovician–Silurian (Millar *et al.* 2002; Riley *et al.* 2012*a*). Evidence from detrital zircons suggests that a volcanic arc existed in this part of West Antarctica from Late Carboniferous to Early Triassic times, with a peak in the Permian *c.* 276 Ma (Barbeau *et al.* 2009; Castillo *et al.* 2016). Triassic orthogneisses and granitoid plutonic rocks of this age range

From: Martin, A. P. and van der Wal, W. (eds) 2023. *The Geochemistry and Geophysics of the Antarctic Mantle*.
Geological Society, London, Memoirs, **56**, 101–114,
First published online 28 April 2021, https://doi.org/10.1144/M56-2019-44

have been identified in the Antarctic Peninsula (Millar *et al.* 2002; Riley *et al.* 2012*a*; Bastias *et al.* 2020). Widespread magmatism occurred along the southern Gondwana margin during Jurassic to Early Cenozoic times (Pankhurst 1982; Pankhurst *et al.* 1993; Leat *et al.* 1995; Tangeman *et al.* 1996; Burton-Johnson and Riley 2015; Riley *et al.* 2017; Zheng *et al.* 2018). Early–Middle Jurassic volcanism is concentrated along the eastern coast of the Antarctic Peninsula and formed part of the Chon Aike Province, a silicic large igneous province (LIP) related to the Karoo LIP (Pankhurst *et al.* 1998; Riley and Leat 2021). Early Cretaceous to Early Cenozoic volcanism was dominated by a volcanic arc with an axis trending along the current west coast from the South Shetland Islands to Thurston Island (Pankhurst *et al.* 1993; McCarron and Millar 1997; Haase *et al.* 2012; Riley *et al.* 2012*b*; Leat and Riley 2021).

Accreted units include the Trinity Peninsula Group. This forms much of the northern part of Graham Land and has structural and metamorphic characteristics of an accretionary complex (Storey and Garrett 1985). It comprises Carboniferous to Triassic metasediments that were deposited close to the active margin (Barbeau *et al.* 2009; Castillo *et al.* 2015). Outboard of the Trinity Peninsula Group, the Scotia Metamorphic Complex underlies parts of the South Shetland Islands and consists dominantly of deformed ocean floor sedimentary rocks. This is interpreted as an accretionary complex metamorphosed to amphibolite and blueschist facies during the mid-Cretaceous (Tanner *et al.* 1982; Grunow *et al.* 1992; Trouw *et al.* 2000). On Gibbs Island (Fig. 1), a dunite–serpentinite ultramafic complex is thought to represent oceanic mantle (De Wit *et al.* 1977; Lee *et al.* 2015). The ultramafic complex structurally overlies muscovite schists representing a metamorphosed volcano-sedimentary series, probably of Late Paleozoic depositional age (De Wit *et al.* 1977; Hervé *et al.* 1991). K–Ar and Ar–Ar ages for the schists range from 32 to 25 Ma, and are interpreted as uplift ages (Grunow *et al.* 1992; Kim *et al.* 1995). Olivines in dunites from the ultramafic complex have high Mg# (Mg# = Mg/Mg + Fe) values that range from 93.4 to 95.2 and the unit is interpreted as residual mantle formed in a suprasubduction zone setting (Lee *et al.* 2015). Another group of accreted rocks forms the LeMay Group which underlies central and western Alexander Island. This Group is dominated by sedimentary and volcaniclastic rocks, and includes Mesozoic accreted oceanic basalts and sediments, but there are no known exposed ultramafic oceanic mantle lithologies (Doubleday *et al.* 1994).

Cessation of the Early Cretaceous to Early Cenozoic subduction beneath the Antarctic Peninsula was caused by collision of the Antarctic–Phoenix oceanic spreading ridge with the subduction zone. The evolution of the cessation of subduction is determined from seafloor magnetic anomalies (Barker 1982; Larter and Barker 1991). The ridge–trench collision took place in a series of steps controlled by the position of transform offsets, younging from SW to NE along the margin (Fig. 1). As a result of the ridge–trench collision, a slab window progressively opened beneath the Antarctic Peninsula (Hole *et al.* 1991). Because the trailing oceanic lithosphere belonged to the same Antarctic Plate as the continental margin, the collision was followed by a reconfiguration of the margin as a non-active ocean–continent margin (Barker 1982; Larter and Barker 1991). Subduction is continuing beneath the South Shetland Islands, but at a much-reduced rate (Maldonado *et al.* 1994).

The Miocene to Recent volcanism along the Antarctic Peninsula reflects its complex tectonic setting. Mafic alkaline volcanic rocks were erupted in isolated locations between 7.7 and <1 Ma and their distribution mostly follows opening of the slab window (Hole 1988, 1990*a*; Hole *et al.* 1991). The mantle source of the mafic alkaline magmas may have been peridotite upwelling in the slab window (Hole *et al.* 1991) or subducted pyroxenite (Hole 2021). Nevertheless, in the north of the Antarctic Peninsula at James Ross Island, Miocene–Recent volcanism is thought to be controlled mainly by lithospheric extension (Hole *et al.* 1991; Košler *et al.* 2009) in a back-arc position relative to continuing subduction beneath the South Shetland Islands (Hole *et al.* 1991; Košler *et al.* 2009). At Jones Mountains, Miocene volcanism may be related to both extension and the presence of an underlying mantle plume (Hart *et al.* 1995).

The Miocene–Recent mafic alkaline volcanic rocks are, with the possible exception of an undated occurrence on Adelaide Island, the hosts of all ultramafic mantle xenoliths in the Antarctic Peninsula. Trace element and Rb–Sr–Pb isotopes indicate that these volcanic rocks have compositions similar to ocean island basalts (OIB), with dominantly asthenospheric mantle sources, and only minor contributions from subduction-modified lithospheric mantle (Hole 1988, 1990*a*, 2021; Hole *et al.* 1993; Hart *et al.* 1995; Košler *et al.* 2009).

Ultramafic xenolith localities

Ultramafic xenolith localities in the Antarctic Peninsula–Jones Mountains region of Antarctica are summarized in Table 1. The list is likely to be incomplete, however, owing to the limited investigation and reporting of many localities. Most of the xenoliths are hosted by Miocene–Pliocene alkali basaltic and basanitic lavas, dykes, and pyroclastic deposits. They are strikingly fresh, with largely unaltered green olivines. Absence of surface alteration probably reflects dominantly frozen subglacial conditions for all locations since eruption.

James Ross Island Volcanic Group

The James Ross Island Volcanic Group is situated to the west of Trinity Peninsula, at the northern tip of Graham Land (Fig. 1). The volcanic rocks overlie >7 km of Cretaceous–Early Cenozoic sedimentary rocks of the marine back-arc James Ross basin (Elliot 1988; Crame 2019). These sediments may overlie deformed metasediments of the mainly Carboniferous to Triassic Trinity Peninsula Group which crops out extensively west of James Ross Island (Fig. 2a; Castillo *et al.* 2015). The volcanism is situated in a back-arc position with respect to ongoing subduction along the South Shetland trench, although geochemically there is no significant subducted component in the magmas (Košler *et al.* 2009). The magmatism is thought to be predominantly related to back-arc lithospheric extension (Hole *et al.* 1991; Košler *et al.* 2009).

The James Ross Island Volcanic Group is distributed mainly in James Ross Island and several nearby islands (Fig. 2a). It covers an area of 4500 km^2 and has considerable thickness, estimated up to 1470 m based on field observations (Smellie 1999). Aeromagnetic data indicate a central magmatic feeder zone beneath the central and southern part of James Ross Island (Ghidella *et al.* 2013). James Ross Island itself is a composite shield volcano, with a highest point at Mount Haddington (*c.* 1630 m), dominated by lavas, tuffs, and palagonite breccias predominantly forming lava-fed deltas and tuff cones (Smellie *et al.* 1988, 2008). There are also exposed dykes, sills and plugs. Ar–Ar ages range from 6.16 to 0.13 Ma with most between 6.0 and 1.9 Ma. (Smellie *et al.* 2008). The James Ross Island Volcanic Group forms an alkali basalt–trachyte series, with rare basanites (Smellie 1987; Košler *et al.* 2009).

Ultramafic xenoliths have been identified in two areas of the James Ross Island Volcanic Group, in the Ulu Peninsula, and

Table 1. *Ultramafic xenolith localities in the Antarctic Peninsula and Jones Mountains*

Volcanic field	Location	Latitude	Longitude	Host rock	Host rock age (Ma)	Xenolith lithology	References
James Ross Island	Cape Lachman	63° 47′ S	57° 48′ W	Basalt lava	5.85	Dunite, wehrlite	Altunkaynak *et al.* (2018); Nývlt *et al.* (2011)
	Lachman Crags	63° 49′ S	57° 51′ W	Basalt lava	5.64–5.04		Altunkaynak *et al.* (2018); Smellie *et al.* (2008)
	Santa Martha Cove	63° 56′ S	57° 52′ W	Basalt lava or dyke	5.2–4.3	Ultramafic	Bastías *et al.* (2012); Calabozo *et al.* (2015)
	Croft Bay	64° 03′ S	57° 50′ W	Basaltic tuff	<6.16	Spinel harzburgite, spinel lherzolite	Sykes unpublished 1986; Smellie *et al.* (2008)
	Ekelöf Point	64° 13′ S	57° 12′ W	Alkali basalt dyke	<6.16	Spinel harzburgite, spinel lherzolite,	Nelson (1966); Keller and Strelin (1992); Calabozo *et al.* (2014)
	Rabot Point	64° 16′ S	57° 20′ W	Alkali basalt	<6.16	spinel pyroxenite	Calabozo *et al.* (2014)
	Lockyer Island	64° 27′ S	57° 36′ W	n.d.	<6.16	n.d.	Calabozo *et al.* (2014)
Seal Nunataks	Bruce Nunatak	65° 04′ S	60° 14′ W	Agglutinated alkali basalt scoria	1.5 ± 0.3	Spinel lherzolite	Fleet (1968); Gonzalez-Ferran (1983); Hole (1990*a*, *b*)
Adelaide Island	Wright Peninsula	67° 27′ S	68° 04′ W	Mafic dyke	<60	Spinel pyroxenite, spinel wehrlite, spinel lherzolite	Dewar (1970); Smellie *et al.* (1988); Ross (2008); Ross *et al.* (2009); Gibson (2012); Gibson *et al.* (2020)
Alexander Island	Mount Pinafore	69° 47′ S	70° 49′ W	Basanite lavas and tuffs	7.7 ± 0.6–7.3 ± 0.4	Spinel harzburgite	Smellie *et al.* (1988, 1993); Ross (2008); Gibson (2012)
	Hornpipe Heights	69° 51′ S	70° 43′ W	Basanite lavas and tuffs	2.7 ± 0.2–2.5 ± 0.8	Spinel lherzolite, spinel wehrlite, spinel harzburgite	Smellie *et al.* (1988); Hole (1990*b*); Ross (2008); Gibson (2012); Gibson *et al.* (2020)
Rothschild Island	Overton Peak	69° 40′ S	72° 01′ W	Basanite lavas and tuffs	5.4 ± 0.7	Spinel lherzolite, spinel wehrlite, spinel pyroxenites	Care (1980); Johnson *et al.* (2012); Ross (2008); Gibson (2012); Gibson *et al.* (2020)
	Southernmost Rothschild Island	69° 42′ S	72° 07′ W	Mafic alkaline dyke and tuffs	*c.* 5.4	n.d.	Care (1980)
Jones Mountains	Intrusion Point	73° 30.5′ S	94° 24′ W	Lava breccia	10–7	Spinel lherzolite	Craddock *et al.* (1964); Ross (2008); Ross *et al.* (2009)

n.d., no data.

along the southeastern coast of James Ross Island (Fig. 2a). In the northwestern part of the Ulu Peninsula, two subglacial lava-fed deltas crop out at Cape Lachman and Lachman Crags (Altunkaynak *et al.* 2018). Both are basaltic (Košler *et al.* 2009). Ar–Ar age determinations for the lava-fed deltas at Cape Lachman and the overlying Lachman Crags are 5.85 Ma (Nývlt *et al.* 2011) and 5.64–5.04 Ma (Smellie *et al.* 2008), respectively. Ultramafic xenoliths were described from the lavas of both of these lava-fed deltas by Altunkaynak *et al.* (2018), although they did not specify their relative abundances. They described the olivine rich (85–95%) xenoliths as mainly dunites and wehrlites with coarse-grained granular textures. West of Santa Martha Cove, a 600 m by 200 m zone of columnar-jointed basalt is described as lava or a dyke and being ‘rich in mantle xenoliths’ up to 2 cm in diameter (Bastías *et al.* 2012, page 365). Calabozo *et al.* (2015) also described a columnar-jointed lava bearing peridotite xenoliths in this location. No petrographic or chemical data for these xenoliths have been reported.

A xenolith locality within basaltic volcaniclastic deposits to the west of Croft Bay (Fig. 2a) was described by M.A. Sykes (unpublished information, British Antarctic Survey 1986). This is on the SE flank of Dobson Dome, and the volcaniclastic deposits may be related to this. The Dobson Dome is a satellite centre of the Mount Haddington stratocone and has been dated by Ar–Ar at <80 ka (Smellie *et al.* 2008). Of the two ultramafic xenoliths from material collected by Sykes, one is a granoblastic spinel harzburgite with grains up to 2 mm across. This sample (D8833.2C) contains olivine with Mg# of 89.9–91.4, orthopyroxene in which Mg# is 91.1–92.0 and has moderate Al_2O_3 contents (2.5–3.1 wt%), and clinopyroxene that is mostly Cr diopside ($Wo_{37\text{-}39}En_{57\text{-}59}Fs_{4\text{-}5}$). The spinel has a Cr# (Cr/Cr + Al) of 0.33–0.40 (P.T. Leat unpublished data 2019). The other sample (D8833.2A) is a spinel harzburgite infiltrated by pyroxenite veins, producing a bulk lherzolite composition (Fig. 3a). The harzburgite is granular with grains up to 2 mm across. The spinel-rich pyroxenite veins have a smaller grain size, up to 0.5 mm, suggesting that they may have formed via melt infiltration in the harzburgite.

Nelson (1966) first described ultramafic xenoliths along the southeastern coast of James Ross Island in a dyke 1.6 km to the NW of Ekelöf Point (Fig. 2a). This contains peridotite xenoliths that are 5–8 cm in diameter. The xenoliths consist of olivine and orthopyroxene, with minor amounts of clinopyroxene and spinel. Sample D4049.3, collected by Nelson (1966), is a coarse spinel harzburgite (Fig. 3b) with grains up to 3 mm across and strained olivine. Keller and Strelin (1992, page 22) also visited a site near Ekelöf Point, possibly the same location, and described xenoliths in a basaltic dyke as

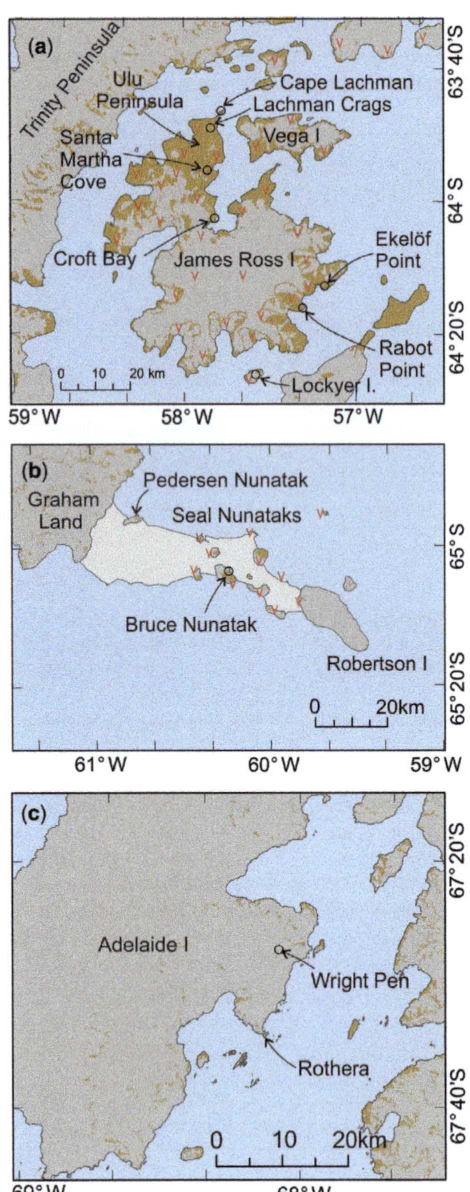

Fig. 2. Maps of ultramafic xenolith localities in the northern Antarctic Peninsula. Late Cenozoic to Quaternary alkaline volcanic rocks are indicated by red 'V' symbols (after Smellie *et al.* 1988), and ultramafic xenolith localities are indicated with a circle. (**a**) James Ross Island area. (**b**) Seal Nunataks. (**c**) Adelaide Island.

'predominantly spinel lherzolites' consisting of 50–95% olivine with clinopyroxene, orthopyroxene and spinel. Further details of the Ekelöf Point xenoliths were provided by Calabozo *et al.* (2014), who also described ultramafic xenoliths hosted in an alkali basalt at Rabot Point, and mentioned a third locality at Lockyer Island (for which no details were provided). Calabozo *et al.* (2014) describe the Ekelöf Point and Rabot Point xenoliths together as spinel-bearing and comprising, in order of abundance, harzburgites, lherzolites and pyroxenites. Textures are protogranular, with grains up to 10 mm, and with transitions to foliated porphyroclastic textures. Whole-rock major-element compositions of the xenoliths (MgO 44.5–47.6 wt%; Al_2O_3 0.6–1.3 wt%) indicate that they have depleted compositions (Calabozo *et al.* 2014). The depleted, harzburgitic character of the xenoliths was confirmed by Valladares Gaete (2017). Pyroxene-rich veins, which contain spinel, carbonate and sulfide, indicate melt or fluid infiltration and host replacement, consistent with observations of the Croft Bay samples (Calabozo *et al.* 2014). In the Ekelöf Point and Rabot Point xenolith suite, olivines have Mg# of 90.4–91.5 and clinopyroxenes are Cr-rich diopsides ($Wo_{36-50}En_{47-59}Fs_{2-4}$) similar to the Croft Bay harzburgite. Calabozo *et al.* (2014) obtained a pressure estimate of 2.3–2.7 GPa based on Cr content in spinel using the geobarometer of O'Neill (1981).

Seal Nunataks

The Seal Nunataks are situated to the east of Graham Land (Fig. 2b), within the Larsen Ice Shelf, *c.* 100 km south of James Ross Island. The volcanic rocks probably overlie back-arc Cretaceous marine sediments which crop out on Pedersen Nunatak and Robertson Island (Farquharson 1982). Basement rocks are not exposed within the nunataks. The basement in the adjacent part of western Graham Land consists of deformed dominantly Carboniferous to Triassic Trinity Peninsula Group metasediments (Fleet 1968; Castillo *et al.* 2015), and the broad negative magnetic anomaly around the Seal Nunataks suggests that they may be underlain by the metasediments (Johnson 1999). To the SW, in eastern Graham Land, Ordovician protolith ages have been determined using zircon U–Pb geochronology on dioritic gneisses (Riley *et al.* 2012*a*), indicating that Early Paleozoic crustal basement may also underlie the Nunataks.

The Seal Nunataks are some fourteen nunataks that occupy an area of 45 × 20 km and are formed by a group of hydrovolcanic and strombolian volcanic cones together with pillow and subaerial lavas (Gonzalez-Ferran 1983; Smellie *et al.* 1988; Hole 1990*a*; Smellie 1990, 1999). Compositionally, the volcanic rocks range from tholeiitic to alkaline basalt (Hole 1990*a*). Whole-rock K–Ar dates for the volcanism range from 4.0 to <0.1 Ma (Smellie *et al.* 1988; Hole 1990*a*, Smellie 1990) and suggest the volcanism may be related to rifting (Gonzalez-Ferran 1983) and opening of a slab window following cessations of subduction (Hole *et al.* 1991; Hole 2021).

Fleet (1968, page 40) described ultramafic xenoliths as 'frequently seen' in an agglomerate in the northern part of Bruce Nunatak (Fig. 2b). From the description of slumping within the deposit, it appears to be an agglutinate. The xenoliths were described as up to 5 mm across and enclosed in basaltic glass. The Bruce Nunatak host volcanic rock is alkali basalt in composition and has been dated at 1.5 ± 0.3 Ma (Smellie *et al.* 1988; Hole 1990*a*). According to the descriptions of Gonzalez-Ferran (1983), Smellie *et al.* (1988) and Smellie (1990), 'abundant' ultramafic xenoliths are contained in more dykes, lavas and tuffs in Seal Nunataks.

Fleet (1968) described one sample (D4692.2) from Bruce Nunatak as containing fresh, green olivine, clinopyroxene, orthopyroxene and picotite (spinel). Two thin sections from this xenolith are the only ultramafic samples identified by the authors from Seal Nunataks. They are spinel lherzolites with granular textures and grains up to 2 mm across (Fig. 3c).

Adelaide Island

Adelaide Island is situated within the Mesozoic–Early Cenozoic magmatic arc of the Antarctic Peninsula (Dewar 1970; Riley *et al.* 2012*b*). The oldest rock on and around Adelaide Island is a tonalite on Rigsby Island, zircon U–Pb dated at 156.4 ± 0.9 Ma (Jordan *et al.* 2014). On this Late Jurassic (or older) basement, a Late Jurassic to Early Cretaceous volcaniclastic and volcanic sequence was deposited which is thought to have formed in a fore-arc position (Riley *et al.* 2012*b*; Leat and Riley 2021). This succession was overlain during Late Cretaceous to Early Cenozoic times by volcanic arc rocks and intruded by mainly Early Cenozoic gabbroic

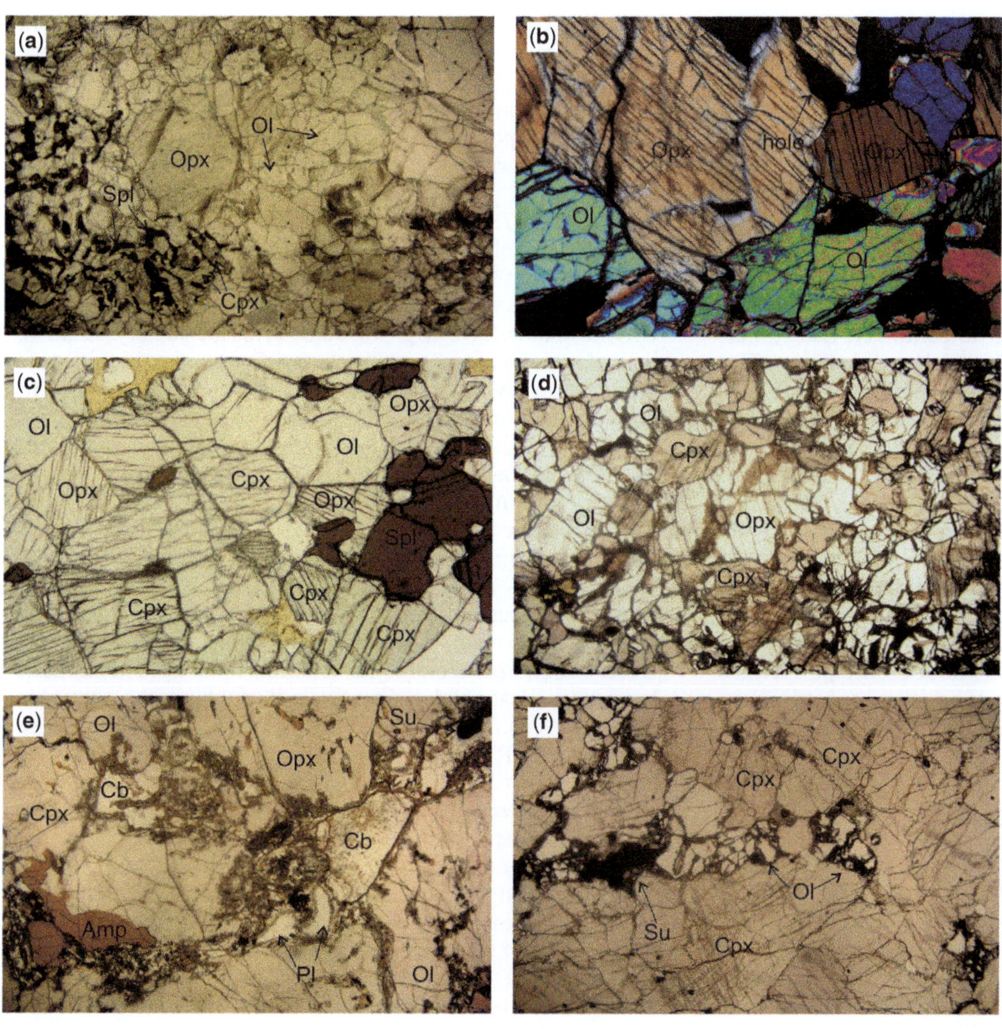

Fig. 3. Images of ultramafic xenoliths from the northern Antarctic Peninsula in thin section. (**a**) Sample D8833.2A from Croft Bay, James Ross Island. Image width 9 mm. Harzburgite host in the central and upper part of the image consists of colourless olivine and pale brown orthopyroxene invaded by a finer-grained vein rich in clinopyroxene, olivine and spinel. (**b**) Harzburgite D4049.3 from Ekelöf Point, James Ross Island. Image width 3 mm, crossed polars. (**c**) Lherzolite D4692.2 from Bruce Nunatak, Seal Nunataks. Image width 3 mm. (**d**) Lherzolite R3708.1A from Wright Peninsula, Adelaide Island. Image width 9 mm. (**e**) Lherzolite R5194.2q with abundant plagioclase and calcite from Wright Peninsula, Adelaide Island. Image width 9 mm. (**f**) Clinopyroxenite R5194.2F from Wright Peninsula, Adelaide Island. Image width 9 mm. Labels: Ol, olivine; Opx, orthopyroxene; Cpx, clinopyroxene; Spl, spinel; Cb, carbonate; Amp, amphibole; Pl, plagioclase; Su, sulfide.

to granodioritic plutons of the Adelaide Island Intrusive Suite (Pankhurst 1982; Griffiths and Oglethorpe 1998; Riley *et al.* 2012*b*; Jordan *et al.* 2014).

Ultramafic xenoliths occur in a mafic dyke in Wright Peninsula, in the southeastern part of the island, 13 km north of the British Antarctic Survey Rothera base (Fig. 2c). The dyke intrudes a granitic-granodioritic intrusion that has an Rb–Sr geochron age of 60 ± 3 Ma (Pankhurst 1982), which is within error of the 44–58 Ma age range for other intrusions of the Adelaide Island Intrusive suite (Riley *et al.* 2012*b*). Dewar (1970, p. 54–55) described the 0.9 m wide mafic dyke as forming part of a pipe intrusion. No analysis of the dyke itself is known to exist, although it is thought to be part of the Late Cenozoic post-subduction mafic magmatism of the Antarctic Peninsula (Smellie 1987; Smellie *et al.* 1988). Petrographically, this could be a calc-alkaline lamprophyre (Dewar 1970; Ross *et al.* 2009), but the fine-grained groundmass makes this identification uncertain. A post-subduction age would suggest that the dyke is younger than ridge–trench collision which was at 19.8 Ma alongside southern Adelaide Island, and 16.5 Ma for the central and northern parts of the island (Larter and Barker 1991). However, only a post-granite intrusion age of <60 Ma is certain.

The mafic dyke contains multiple ultramafic xenoliths up to 0.3 m in diameter (Dewar 1970). These xenoliths are highly varied. Nineteen examined xenoliths, including samples described by Ross (2008), Gibson (2012) and Gibson *et al.* (2020), are peridotites and pyroxenites, comprising eight lherzolites, three clinopyroxenites, three websterites, two olivine websterites, two wehrlites and a single olivine clinopyroxenite.

The Adelaide Island lherzolites have a coarse, granular texture with grains up to 12 mm across, and rarely have a weak foliation. They are spinel-bearing, and several also contain minor plagioclase, indicating that the xenoliths were derived from the spinel-plagioclase transition depth (Ross 2008). There is significant variation among the lherzolites. Sample R3708.1A contains olivine with Mg# 80.3–81.5, almost colourless orthopyroxene and pale brown, Ti-rich clinopyroxene (Fig. 3d). By contrast, sample R5194.2B has olivine with very low Mg# (74.6–76.2), with orthopyroxene, clinopyroxene, plagioclase and low-Cr spinel (Cr# 0.02) as well as Cu–Fe sulfides (Ross 2008; P.T. Leat, unpublished data 2019). A similar lherzolite sample (R5194.2q) contains amphibole, plagioclase and calcite (Fig. 3e), indicating significant metasomatism.

The Adelaide Island pyroxenites and wehrlites have coarse to granuloblastic textures. Wehrlite olivines have Mg# of 79.4–80.0, Ti-rich clinopyroxenes, rare orthopyroxene with moderate Al_2O_3 (4.9 wt%), low-Cr spinels (Cr# 0.04–0.09) and calcite (Ross 2008). Pyroxenite olivines have Mg# of 78–82 (i.e. slightly lower than those in lherzolite R3708.1A), and pale-brown augites (Fig. 3f) with high TiO_2 (1.0–2.0 wt%) and high Al_2O_3 (8.0 wt%) (Ross 2008; Gibson *et al.* 2020). Spinels have low Cr# (typically 0.02). Plagioclase, calcite and Fe–Ni and Cu–Fe sulfides are also present. The pyroxenites were interpreted as products of reaction of lithospheric peridotites with variably carbonate-rich, or silicate melts generated during subduction by Ross *et al.* (2009), and Gibson (2012). However, Gibson *et al.* (2020) concluded from the low Cl and Li, but high Ti, Nb, H and F of the pyroxenites that they crystallized from post-subduction alkaline melts.

Alexander Island

The basement of Alexander Island consists of an accretionary complex comprising Jurassic and Cretaceous trench-slope and trench-fill sediments, accreted Mesozoic ocean floor lavas and seamounts and pelagic sedimentary rocks. Together these form the LeMay Group (Burn 1984; Nell 1990; Doubleday *et al.* 1994). The accretionary complex formed on the western, fore-arc side of the Antarctic Peninsula volcanic arc. The LeMay Group is unconformably overlain by volcanic arc rocks and intruded by the calc-alkaline Rouen Mountains pluton of the same event, both related to continued Late Cretaceous to Early Tertiary east-directed subduction. The volcanic rocks and plutonic rocks have been dated by K–Ar and Ar–Ar techniques to be Late Cretaceous to Paleogene (80–46 Ma) (McCarron and Millar 1997). Ultramafic xenoliths are found in isolated outcrops of mafic alkaline volcanic rocks in the Elgar Uplands in northern Alexander Island that post-date the volcanic arc and cessation of subduction (Hole 1988; Smellie *et al.* 1988). Ultramafic xenoliths are described as being 'extremely common at all localities in the Elgar Uplands' (Smellie *et al.* 1988, page 29), and have been collected and described from two locations: at site KG3609 near Mount Pinafore and site KG3610 at Hornpipe Heights (Fig. 4a). Cessation of east-directed subduction beneath Alexander Island occurred at *c.* 25 Ma (Barker 1982), and the Late Cenozoic volcanism in northern Alexander Island is thought to be related to slab-window formation (Hole 1988; Hole *et al.* 1991).

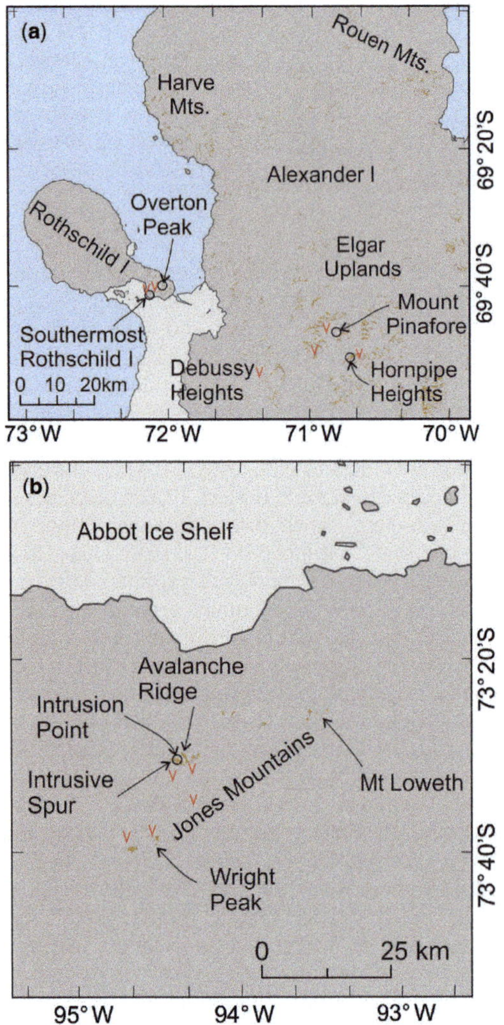

Fig. 4. Location maps of ultramafic xenolith localities in Alexander and Rothschild Islands and Jones Mountains. Symbols as in Figure 2. (**a**) Alexander and Rothschild islands. (**b**) Jones Mountains.

SE of the summit of Mount Pinafore (site KG3609), a 60 m thick, valley-ponded basanite lava flow hosting ultramafic xenoliths overlies volcaniclastic and diamictite deposits (Smellie *et al.* 1988, 1993). The lava was dated by K–Ar to be 7.7 ± 0.6–7.3 ± 0.4 Ma (Smellie *et al.* 1988). The only described xenolith from Mount Pinafore is spinel harzburgite sample KG3609.13 (Ross 2008; Gibson 2012), which has a coarse texture with a typical grain size of 0.6 mm (Fig. 5a). Olivine grains have Mg# of 90.3–91.6, are locally strained, with larger grains 7 mm across. Orthopyroxene is pale brown, with a Mg# of 91.7 and low Al_2O_3 (2.1–2.3 wt%). Clinopyroxene is a pale green diopside with moderate Cr_2O_3 and low TiO_2 (0.51 and 0.01 wt%, respectively) and a Mg# of 94.1. Spinel has a high Cr# (0.41). The geochemistry of this sample is summarized in Table 2.

At Hornpipe Heights (site KG3610), agglutinate, lava flows and tuffs form an isolated sequence up to *c.* 20 m thick (Smellie *et al.* 1988; Hole 1990*b*; Smellie 1999). Two lavas from the sequence have been dated by K–Ar at 2.5–2.7 Ma (Smellie *et al.* 1988). Compositionally, the lavas are basanites (Hole 1988). Spinel lherzolite xenoliths up to 6 cm in diameter have been described as common throughout the sequence (Hole 1990*b*). Six documented samples, including those described by Ross (2008), Gibson (2012) and Gibson *et al.* (2020), comprise three lherzolites, two harzburgites and a wehrlite. Harzburgite sample KG3610.7 has a coarse texture with grains up to 8 mm across, and has strongly serpentinized olivines. Harzburgite KG3610.10A has olivine with Mg# of 91.1–91.5, and orthopyroxene with Mg# of 91.8 and Al_2O_3 of 2.8 wt%. Clinopyroxene forms 4.7 mode% of this xenolith. It has a Mg# of 93.8 and is a Cr-diopside (1.1–1.4 wt% Cr_2O_3, 0.02 wt% TiO_2). The Cr# of the spinel (0.59) is the highest among the Antarctic Peninsula ultramafic xenoliths (Ross 2008; Gibson *et al.* 2020). Geochemical data for KG3610.10A are summarized in Table 2.

The three lherzolite samples from Hornpipe Heights have coarse textures with orthopyroxene grains up to 7 mm across. Olivines typically have Mg# of 86.4, brown orthopyroxenes with similar Mg# (typically of 87.3) have moderate-to-high Al_2O_3 of 3.2–4.5 wt%. Clinopyroxenes are pale green Cr-rich diopsides (0.9–1.0 wt% Cr_2O_3, 0.1–0.3 wt% TiO_2), with Mg# of 89.5. Spinels have Cr# of 0.14–0.18 (Ross 2008; Gibson 2012; Gibson *et al.* 2020). Two of the lherzolites contain metasomatic amphibole (Fig. 5b).

Spinel wehrlite sample KG3610.8 is dominated by large (2–7 mm) grains of strained olivine. Patches of fine-grained (*c.* 0.2 mm) olivine and clinopyroxene may represent metasomatic recrystallization along grain boundaries (Ross 2008; Gibson 2012). The olivines have Mg# of 86.2–87.6 and the clinopyroxenes have very high 1.2–2.7 wt% Cr_2O_3, i.e. they are Cr-diopsides.

Rothschild Island

The basement to Rothschild Island, which is situated west of northern Alexander Island (Fig. 4a), consists of sedimentary rocks of the LeMay Group accretionary complex intruded by dominantly intermediate calc-alkaline plutons (Care 1980). The plutons are similar to the Rouen Mountains pluton, which has been dated at 56 ± 3 Ma by zircon U–Pb geochronology (McCarron and Millar 1997). In the southeastern peninsula of the island, LeMay Group rocks are intruded and overlain by Late Cenozoic mafic dykes and volcanic rocks that have a similar relationship to cessation of subduction as in northern Alexander Island.

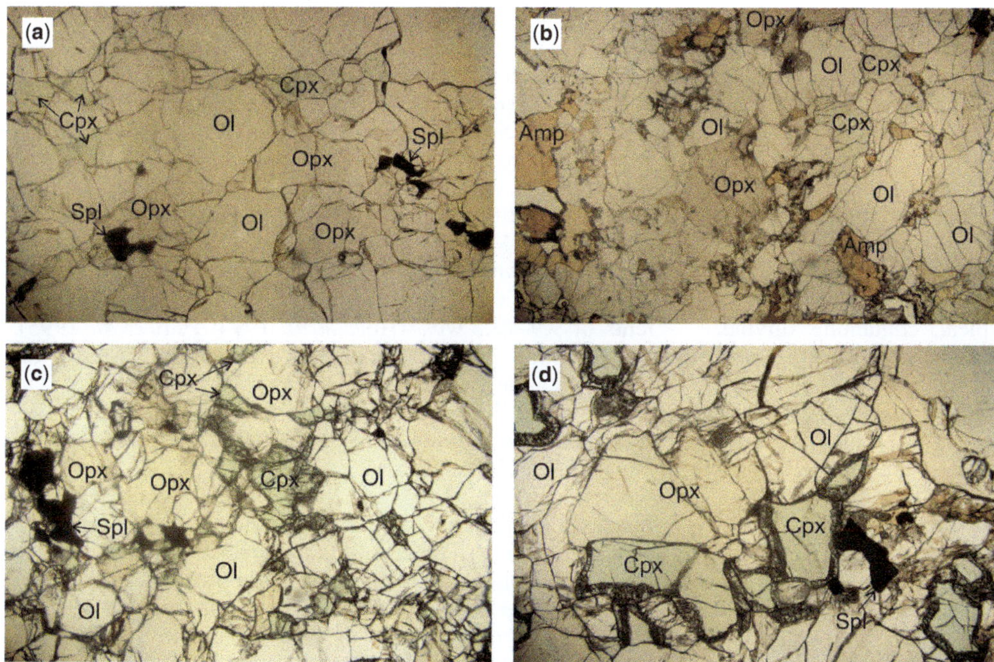

Fig. 5. Images of ultramafic xenoliths from the southern Antarctic Peninsula and Jones Mountains in thin section: (**a**) Harzburgite KG3609.13 from Mount Pinafore, Alexander Island. Image width 9 mm. (**b**) Lherzolite KG3610.1 from Hornpipe Heights, Alexander Island. Image width 9 mm. (**c**) Lherzolite L7201.1.12 from near Overton Peak, Rothschild Island. Image width 9 mm. (**d**) Lherzolite R3123.1 from Intrusion Point, Jones Mountains. Image width 9 mm. Labels as in Figure 3.

Cenozoic mafic alkaline rocks occur at two locations in Rothschild Island (Fig. 4a). The first, a 100 m high cliff near the southernmost point of the island, consists of Late Cenozoic bedded palagonite tuffs and volcaniclastic deposits cut by mafic dykes (Care 1980; Smellie *et al.* 1988; Smellie 1999). Both the tuffs and dykes are probably of similar age (5.4 Ma) to the Overton Peak location (Smellie 1999). They were described as containing 'nodules of ferromagnesian minerals' (Care 1980, page 106), although these nodules have not been further documented.

The second Rothschild Island location is approximately 2 km NW of Overton Peak and consists of two rounded hills covered by scree derived from palagonite tuffs (Fig. 6), presumably originally forming small cones (Care 1980; Smellie 1999). The surface is characterized by periglacial features and is scattered with glacially transported cobbles and boulders (Johnson *et al.* 2012). Minor lava flows and at least one mafic dyke have also been described at this locality (Care 1980; Thomson *et al.* 1990). The lavas are basanitic in composition (Smellie 1987; Hole 1988) and have been dated by K–Ar at 5.4 ± 0.7 Ma (Smellie *et al.* 1988). Ultramafic xenoliths are very abundant at this locality, most obviously scattered on

Table 2. *Representative compositions of Antarctic Peninsula upper mantle*

Harzburgite KG3610.10A Hornpipe Heights, Alexander Island					
	Ol	**Opx**	**Cpx**	**Spinel**	**Whole rock**
Mode	81.7	11.4	4.7	2.2	
SiO_2	41.53	56.86	54.59	0.06	42.98
Al_2O_3	0.01	2.83	3.50	22.68	1.03
MgO	49.89	34.28	15.83	16.08	45.78
FeO	8.52	5.46	1.86	10.32	7.93
CaO	0.03	0.43	20.31	0.18	1.03
H_2O ppm	20	181	307		51
Mg#	91.3	91.8	93.8	72.54	91.1
Cr#				0.59	
Harzburgite KG3609.13 Mt. Pinafore, Alexander Island					
	Ol	Opx	Cpx	Spinel	Whole rock
Mode	71.0	22.6	3.7	2.7	
SiO_2	41.24	57.23	54.34	0.11	44.23
Al_2O_3	0.01	2.29	2.07	34.0	1.52
MgO	49.02	33.87	17.21	16.8	43.55
FeO	8.88	5.61	1.94	12.2	7.97
CaO	0.06	0.62	22.9	0.02	1.03
Mg#	90.8	91.5	94.1	72.6	90.6
Cr#				0.41	
Lherzolite L7.201.1.67 Rothschild Island					
	Ol	Opx	Cpx	Spinel	Whole rock
Mode	56.0	27.8	14.6	1.6	
SiO_2	40.80	55.33	52.13	0.05	45.84
Al_2O_3	0.02	4.49	6.64	56.11	3.13
MgO	48.70	32.74	15.17	20.47	38.92
FeO	10.51	6.59	2.84	9.27	8.28
CaO	0.08	0.76	20.18	0.1	3.20
Mg#	89.2	89.9	90.5	79.7	89.3
Cr#				0.125	

Data are averaged analyses, with major oxides in wt%. Whole rock compositions are calculated from mineral compositions and modal proportions of phases. Data sources: Ross (2008), Gibson (2012), Gibson *et al.* (2020).

Fig. 6. Photograph of the mantle xenolith locality (site L7.201) near Overton Peak. The distant high, craggy outcrops of Overton Peak consist of metasediments of the LeMay Group. In the low outcrops of scree and patterned ground in the foreground, most of the black cobbles are mantle xenoliths coated by basalt. Granite and metasediments cobbles and boulders littering the surface are glacially deposited.

the slopes of the volcanic hills and also contained within dykes (Care 1980; Smellie *et al.* 1988; Thomson *et al.* 1990; Gibson 2012).

Although Care (1980) described the Rothschild Island xenoliths as spinel lherzolites, considerable compositional variation has since been identified. Of thirty-one xenoliths from the Overton Peak location, including those described by Ross (2008) and Gibson (2012), there are twenty lherzolites, four wehrlites, three olivine clinopyroxenites, two clinopyroxenites and two olivine websterites.

Lherzolites have a coarse texture with grains up to 7 mm across. Sample L7201.1.12 is typical, containing olivine, orthopyroxene, clinopyroxene and spinel, with a grain size of 0.5 mm (Fig. 5c). Olivine typically has moderate Mg# ranging from 87.7–90.6. Pale-brown orthopyroxenes have relatively high Al_2O_3 (typically 3.5–5.7 wt%). Clinopyroxenes (0.6–1.1 wt% Cr_2O_3, 0.3–1.0 wt% TiO_2) are mostly Cr-rich diopsides in the peridotites and augites in the pyroxenites. Spinels have Cr# of 0.11–0.22 (Ross 2008; Gibson 2012; Gibson *et al.* 2020). One lherzolite (sample L7.201.1.41) contains accessory metasomatic amphibole and phlogopite (Gibson 2012). Geochemical data for a representative lherzolite sample are given in Table 2.

Wehrlites contain olivine with Mg# of 79.3–84.0, and so are significantly more Fe-rich than the lherzolites. The clinopyroxenes are mostly Cr-rich diopsides and augites (0.6–0.8 wt % Cr_2O_3, 0.9–1.3 wt% TiO_2). Spinels in the wehrlites have low Cr# (0.05–0.11).

Pyroxenites typically have olivine with a similar Mg# (79.4–82.7) to those in the wehrlites but the clinopyroxenes are augites with low Cr_2O_3 (0.4 wt%) and high TiO_2 (1.3 wt %); some have 'spongy' rims caused by decompression melting. Spinels in the pyroxenites have very low Cr# (0.003–0.02) (Gibson 2012; Gibson *et al.* 2020).

The olivine websterite sample KG3919.9 is distinct in having olivine with a relatively high Mg# (89.6), high spinel Cr# (0.53), and Cr_2O_3-rich (2.2 wt%) clinopyroxenes (i.e. Cr-diopside; Ross 2008; Gibson 2012).

Jones Mountains

The Jones Mountains are situated on the Eights Coast and within the Thurston Island crustal block, one of several crustal blocks that comprise West Antarctica (Fig. 4b; Storey *et al.* 1988). This block forms part of the Pacific margin of Antarctica and is situated between the Antarctic Peninsula and Marie Byrd Land crustal blocks. Most outcrops in the crustal block are Carboniferous to Late Cretaceous intrusive and extrusive calc-alkaline rocks formed at the active continental margin (Leat *et al.* 1993; Pankhurst *et al.* 1993; Riley *et al.* 2017). An S-type granite in the Jones Mountains with a Rb–Sr geochron age of 198 ± 2 Ma has Nd T_{dm} model ages of *c.* 1330 Ma, suggesting the presence of non-exposed Mesoproterozoic crust (Pankhurst *et al.* 1993). Nelson and Cottle (2018) dated granite from a similar location using zircon U–Pb geochronology at between 208±1 and 215±1 Ma (Triassic). The presence of Mesoproterozoic basement is supported by Hf model ages measured on zircons on Paleozoic and Mesozoic igneous rocks from Thurston Island (Riley *et al.* 2017; Nelson and Cottle 2018). The age of cessation of subduction beneath the Jones Mountain is uncertain, but may have been at *c.* 61 Ma when the small Bellingshausen oceanic plate was incorporated into the Antarctic Plate (Larter *et al.* 2002).

Mafic alkaline volcanic rocks of Cenozoic age overlie the Paleozoic to Mesozoic basement with a pronounced unconformity in the Jones Mountains. The basement rocks are cut by a planar surface at about 750 m above sea-level overlain by glacial deposits (Rutford *et al.* 1968). The Cenozoic volcanic rocks are *c.* 500–700 m thick and consist of lavas, lava breccias and volcaniclastic rocks that are associated with the glacial deposits (Rutford *et al.* 1968, 1972; Rowley 1990; Hole *et al.* 1994). Several attempts to date the volcanic rocks by whole rock K–Ar methods have generated highly variable ages, perhaps due to extraneous radiogenic argon (Craddock *et al.* 1964; Rutford *et al.* 1968, 1972; Hole *et al.* 1994). An age of 10–7 Ma for the volcanism is, however, generally accepted. Compositionally, the volcanic rocks are mostly alkali basalts and basanites and have notably high MgO contents of 9.9–12.3 wt% (Hole *et al.* 1994; Hart *et al.* 1995). The eruption of the magmas was linked by Hole *et al.* (1994) to a mantle plume underlying Marie Byrd Land. Hart *et al.* (1995), however, suggested that the magmatism was rift-related and possibly linked to a different mantle plume beneath Peter I Island.

Craddock *et al.* (1964) and Rowley (1990) reported peridotite nodules in Cenozoic basaltic pyroclastic deposits in the Jones Mountains. The xenoliths were described as containing olivine, diopside, enstatite and spinel in proportions that suggest lherzolite compositions. One xenolith was collected in 1985 as part of a sample of lava breccia from a location known as Intrusion Point (unpublished information B.C. Storey, British Antarctic Survey 1985). The location is on the north-facing scarp between Avalanche Ridge and Intrusive Spur (Fig. 4b).

The Intrusion Point sample (R3123.1) is a spinel lherzolite (Ross 2008) and has a coarse texture, with some orthopyroxene grains up to 4 mm across (Fig. 5d). Olivines have average Mg# of 89.2. Orthopyroxenes have slightly higher Mg# of 89.6–90.5 and high Al_2O_3 contents (4.7 wt%). Clinopyroxenes are mostly bright-green Cr-diopsides (Cr_2O_3 0.8–0.9 wt %) and have well-formed 'spongy' rims probably caused by decompression melting. Spinels have Cr# of 0.10–0.11 (Ross 2008). Fe–Ni and Cu–Fe sulfides are present. This lherzolite sample is close in composition to fertile mantle (Ross 2008; Ross *et al.* 2009).

Discussion

Tectonic considerations and lithospheric mantle evolution

There have been many contributions to understanding the crustal development of the West Antarctic margin from northern Antarctic Peninsula to Thurston Island (e.g. Storey and Garrett 1985; Grunow *et al.* 1992; Tangeman *et al.* 1996; Vaughan and Storey 2000; Ferraccioli *et al.* 2006; Burton-Johnson and Riley 2015; Riley *et al.* 2017; Nelson and Cottle 2018; Zheng *et al.* 2018; Bastias *et al.* 2020). U–Pb zircon geochronology and isotopic model ages strongly suggest the presence of Mesoproterozoic, Neoproterozoic and Paleozoic crustal basement beneath much of the margin (Tangeman *et al.* 1996; Millar *et al.* 2002; Veevers and Saeed 2013; Riley *et al.* 2017; Zheng *et al.* 2018; Bastias *et al.* 2020). Accretion of oceanic igneous crust and associated volcaniclastic and oceanic sediments is also well documented (Hervé *et al.* 1991; Grunow *et al.* 1992; Doubleday *et al.* 1994; Lee *et al.* 2015), and much of northern Graham Land and all of western Alexander Island are underlain by accreted terranes (Fig. 7). In the 'terrane' model for the tectonic evolution of the margin (Vaughan and Storey 2000; Ferraccioli *et al.* 2006), large areas of western and central Antarctic Peninsula are interpreted as being underlain by terranes accreted during the Mesozoic. Alternative, 'autochthonous' models are not consistent with accretion of crustal terranes in Palmer Land during Mesozoic times (Storey and Garrett 1985;

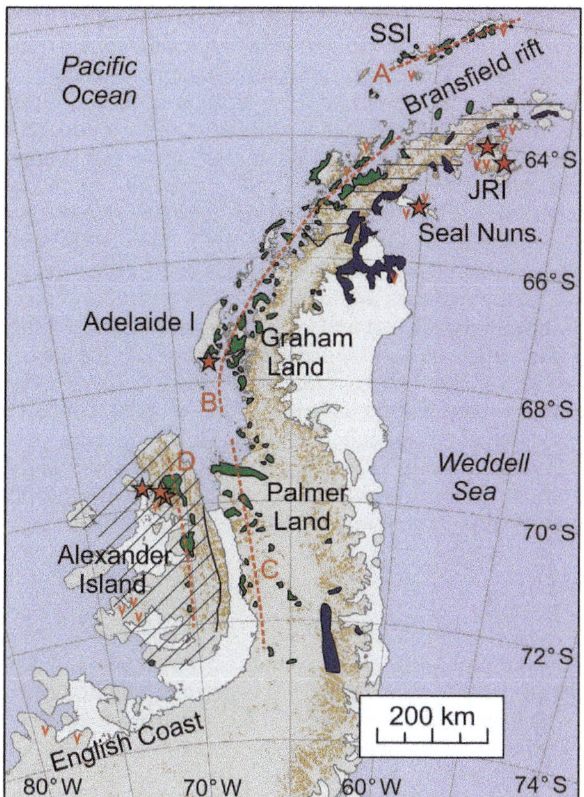

Fig. 7. Sketch geological map of the Antarctic Peninsula showing locations of ultramafic xenolith localities (stars) in relation to Late Cenozoic volcanic rocks (red 'V' ornament), volcanic rocks of the Early Cretaceous–Early Cenozoic volcanic arc (green ornament) and Early–Middle Jurassic Chon Aike volcanic rocks (blue ornament). The red dashed lines are the approximate axis of the volcanic arc in: **A**, South Shetland Islands; **B**, Graham Land; **C**, Early to mid-Cretaceous, Palmer Land; **D**, Late Cretaceous to Cenozoic, Alexander Island. Horizontal and diagonal rule are accreted units of, respectively, the Trinity Peninsula Group and the LeMay Group. SSI, South Shetland Islands; JRI, James Ross Island. Geological distributions after Smellie *et al.* (1988), Riley and Leat (2021), Leat and Riley (2021).

Burton-Johnson and Riley 2015), but are consistent with long-lived earlier Paleozoic or Proterozoic tectonic accretion along the palaeo-Pacific margin of Gondwana (Vaughan *et al.* 2005; Harley *et al.* 2013).

In contrast, the origin of the lithospheric mantle beneath the West Antarctic margin has not been investigated with the same intensity. Judging from the crustal evolution, it is likely that much of the lithospheric mantle beneath the Antarctic Peninsula formed in a long-lived active margin. The dunite–serpentinite ultramafic complex on Gibbs Island in the South Shetland Islands (Fig. 1; Lee *et al.* 2015) provides the only known evidence for such accretion of oceanic mantle exposed on the surface. A Rb–Sr isochron age of 287 ± 48 Ma for metamorphic rocks associated with the dunite suggest that this oceanic lithosphere formed during late Paleozoic times (Hervé *et al.* 1991; Grunow *et al.* 1992). Furthermore, Gibson *et al.* (2010*b*) and Gibson (2012) suggested that depleted mantle sampled by Antarctic Peninsula xenoliths may also have originated as accreted sub-oceanic lithosphere. Gibson (2012) interpreted Nd–Os isotopic data to suggest that the age of such mantle lithosphere could be as old as Archean or Paleoproterozoic, similar to ages obtained from ultramafic xenoliths from southern Patagonia (Mundl *et al.* 2015; Schilling *et al.* 2017).

Gibson (2012) estimated that the harzburgites formed by removal of a 20–30% melt fraction and that some lherzolites also have compositions which imply removal of melt fractions of up to 11%. These high degrees of melting are consistent with adiabatic decompression melting beneath a mid-ocean ridge type setting (Dick *et al.* 1984; Arai 1994). Calabozo *et al.* (2014) noted that harzburgites from the James Ross Island Volcanic Group had experienced a similar amount of melt removal. The higher melt fraction of 45% required to produce the Gibbs Island dunites is thought to have occurred in a subduction zone environment, where higher degrees of 'wet' melting are possible (Lee *et al.* 2015). Interestingly, dunites reported from the James Ross Island Volcanic Group by Altunkaynak *et al.* (2018) suggest that such highly depleted mantle may extend from beneath Gibbs Island to northern Graham Land. These relationships are illustrated in Figure 8, which shows that the Antarctic Peninsula lherzolites, harzburgites and dunites are distributed at progressively higher degrees of melting of fertile MORB mantle. The Jones Mountains lherzolite plots close to fertile MORB mantle in the figure and represents undepleted mantle, possibly related to mantle plumes beneath this part of West Antarctica (Hole *et al.* 1994; Hart *et al.* 1995).

Modal variations in mineralogy and pyroxene compositions of the Antarctic Peninsula and Jones Mountains xenoliths are summarized in Figure 9. Peridotites dominate in the James Ross Island Volcanic Group, and Alexander and Rothschild islands. In contrast, the Adelaide Island xenolith suite is dominated by pyroxenites and wehrlites that contain clinopyroxenes and orthopyroxenes with significantly lower Mg# than found in the peridotites. The pyroxenites and wehrlites from Adelaide, Alexander and Rothschild islands are interpreted as resulting from channelized flow of percolating mantle melts (Gibson 2012; Gibson *et al.* 2020), with the clinopyroxene-rich pyroxenites representing crystallized magmas. The magmatic origin is also reflected in the low olivine Mg# in the pyroxenites (Figs 8 and 10). Mineral trace element compositions and the presence of amphibole, carbonate and phlogopite in some samples suggest that many of the lherzolites have also been significantly modified from initially more depleted compositions, i.e. are metasomatized (Ross 2008; Gibson *et al.* 2010*a*, *b*; Gibson 2012), and similar enrichments are evident in James Ross Island peridotites (Calabozo *et al.* 2014). The origin of the melts that caused the

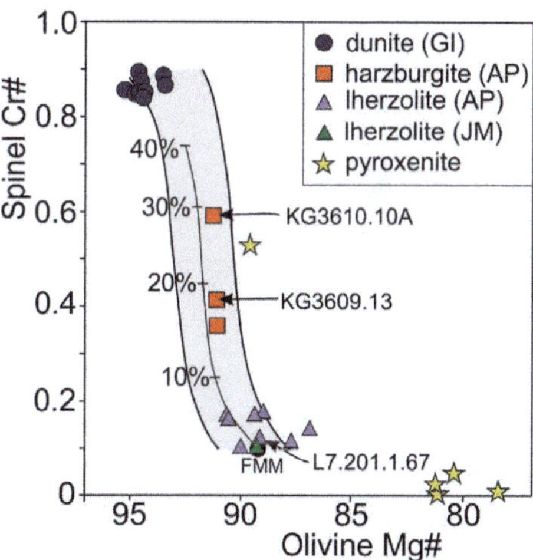

Fig. 8. Plot of spinel Cr# v. olivine Mg# for selected Antarctic Peninsula (AP) and Jones Mountains (JM) ultramafic xenoliths and dunites of Gibbs Island (GI) (after Gibson 2012 and Gibson *et al.* 2020). Field of the olivine–spinel mantle array (OSMA) and calculated residue compositions with degrees of melt removal (in percent) from fertile MORB mantle (FMM) indicated after Arai (1994). Antarctic data from Lee *et al.* (2015), P.T. Leat unpublished data (2019), Gibson *et al.* (2020).

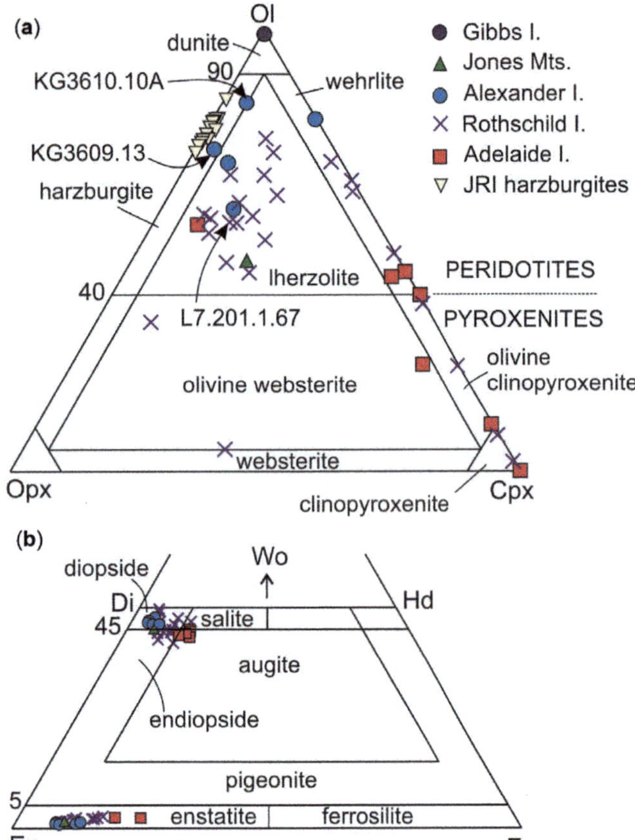

Fig. 9. (**a**) Modal mineralogy and (**b**) pyroxene compositions of Antarctic Peninsula and Jones Mountains ultramafic xenoliths and the Gibbs Island dunites. Data sources: Gibbs Island dunites, Lee *et al.* (2015); James Ross Island (JRI) harzburgites, Valladares Gaete (2017); Adelaide, Alexander and Rothschild islands and Jones Mountains, Ross (2008), Gibson (2012).

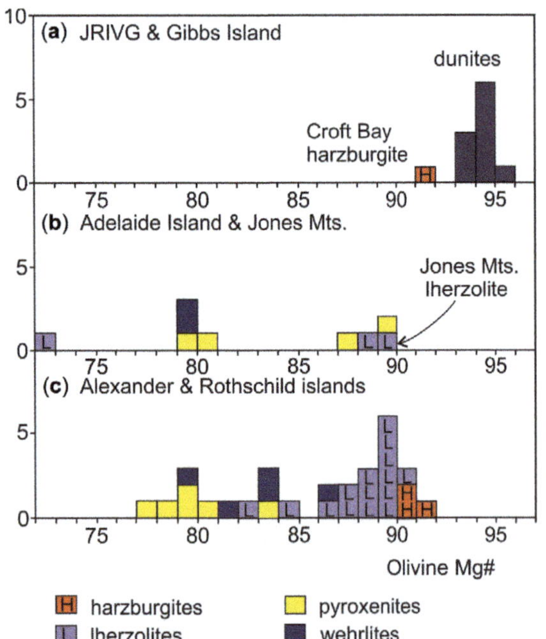

Fig. 10. Distribution of Mg# in olivines in Antarctic Peninsula and Jones Mountains ultramafic xenoliths and the Gibbs Island dunites. Each box represents one xenolith or dunite sample. (**a**) Croft Bay harzburgite, James Ross Island and Gibbs Island dunites. (**b**) Adelaide Island and Jones Mountains. (**c**) Alexander Island and Rothschild Island. Data sources: Lee *et al.* (2015), Ross (2008), Gibson (2012), PT Leat unpublished data (2019).

metasomatism have been variously identified using major and trace element data as associated with either subduction-related boninitic melts, carbonate-rich melts, adakite-like melts, or post-subduction alkaline magmas (Ross *et al.* 2009; Gibson *et al.* 2010*a*, *b*; Gibson 2012). However, new evidence for the composition of the metasomatizing agent has emerged from analyses of volatiles (H, F, Cl, Li, B) in olivines and pyroxenes in the Antarctic Peninsula xenoliths (Rooks *et al.* 2015; Rooks 2016; Gibson *et al.* 2020). This study shows that clinopyroxene is the main host of H and F. Fluorine contents of clinopyroxenes are high (up to 154 ppm; Gibson *et al.* 2020). By contrast, Li, B and Cl are present in low concentrations. This suggests that the magmas responsible for the crystallization of the pyroxenites and metasomatism of the peridotites were alkaline and not related to subduction. This supports the identification of a high-Nb component in the metasomatizing melts (Gibson 2012). The alkaline, high-F, high-Nb component is interpreted to be related to the post-subduction alkaline magmatism (Gibson *et al.* 2020).

Geotherms

Garnet is absent from all the Antarctic Peninsula and Jones Mountains mantle xenolith suites, therefore accurate pressure estimates cannot be determined. The two-pyroxene thermometer (Taylor 1998) indicates that spinel peridotites from Alexander Island and Rothschild Island have final equilibration temperatures between 850 and 1050°C, with the harzburgites yielding higher temperatures than lherzolites (Gibson *et al.* 2020). Gibson (2012) suggested that the mantle xenoliths from Adelaide, Alexander and Rothschild islands lay on similar pre-eruptive geotherms. Assuming normal asthenospheric mantle potential temperatures (1315°C), the thickness of the mechanical boundary layer is estimated to be approximately 70 km (Gibson *et al.* 2010*a*). Calabozo *et al.* (2014) estimated final equilibration pressures of 2.3–2.7 GPa (*c.* 75–85 km) for James Ross Island ultramafic xenoliths, based on Cr content of spinel. All these depth estimates are compatible with the *c.* 80 km thickness suggested for the Antarctic Peninsula lithosphere using S-wave velocities (An *et al.* 2015).

Typical mantle compositions for geophysical modelling

Selection of mantle compositions for use in geophysical models requires an understanding of the composition of the regional lithospheric mantle. The range of modal compositions sampled by Antarctic Peninsula mantle-derived xenoliths is large, ranging from highly depleted dunites and harzburgites to enriched pyroxenites (Fig. 9). This is reflected in the large range in Mg# of olivines, with values over 90 for depleted harzburgites, and typically 77–87 for wehrlites and pyroxenites from Adelaide, Alexander and Rothschild islands (Fig. 10). Several considerations indicate, however, that the dominant composition of the lithospheric mantle beneath the Antarctic Peninsula is depleted in clinopyroxene. Firstly, harzburgitic xenoliths are widely distributed, including in the James Ross Island Volcanic Group, Alexander Island and Rothschild Island. Secondly, these depleted mantle lithologies are interpreted to have originated by accretion of oceanic lithosphere during formation of the continental margin lithosphere (Gibson 2012). Such tectonic accretion is reflected in the known crustal evolution (Grunow *et al.* 1992; Doubleday *et al.* 1994; Vaughan *et al.* 2005). Harzburgitic compositions are thus interpreted to dominate the composition of the initially accreted continental lithosphere. Thirdly, more enriched lherzolite and pyroxenite compositions were a product of metasomatism by post-subduction alkaline melts. The distribution of post-subduction alkaline volcanism on the Antarctic Peninsula is shown in Figure 7. The occurrences are isolated and scattered mostly within both the fore-arc and back-arc relative to the Late Mesozoic–Early Cenozoic volcanic arc. Within the back-arc, the volcanism is interpreted to have been related to lithospheric extension at both the James Ross Island Volcanic Group and Seal Nunataks localities (Gonzalez-Ferran 1983; Hole *et al.* 1991; Košler *et al.* 2009). If the surface expression of the post-subduction alkaline volcanism reflects the generation of alkaline magmatism in the mantle, the generation of pyroxenites, wehrlites and enriched lherzolites would be similarly patchy and concentrated in extended areas on either flank of the spine of the Antarctic Peninsula. The mantle metasomatism associated with the alkaline magmatism may also be concentrated near the base of the lithosphere. Therefore, it is very unlikely that pyroxenite or related enriched compositions are typical of the whole Antarctic Peninsula lithospheric mantle.

Table 2 shows three peridotite compositions that together represent the likely range of typical bulk lithospheric mantle beneath the Antarctic Peninsula. Sample KG3610.10A, a harzburgite from Hornpipe Heights on Alexander Island represents a strongly depleted mantle end-member. It has a high modal amount of olivine (81.7%) with a high Mg# (91.3) compared to other Antarctic Peninsula mantle xenoliths (Figs 9a and 10). The co-existing spinels have a high Cr# which suggests this is a residue from 25 to 30% partial melting of fertile mantle (Fig. 8). The calculated whole rock composition of KG3610.10A has high MgO and low SiO_2, CaO and Al_2O_3. It has a whole rock H_2O content of 50 ppm, which is low relative to the values of 81–130 ppm determined for other Antarctic Peninsula peridotite xenoliths (Gibson *et al.* 2020). Sample KG3609.13 is a less depleted harzburgite that is modally more similar to other peridotite xenoliths from the Antarctic Peninsula and Jones Mountains (Fig. 9a). In Figure 8, spinels in KG3609.13 have a moderate Cr# which indicates that this is a residue from approximately 18% partial melting. The calculated whole rock composition for KG3609.13 has higher SiO_2 and Al_2O_3 and lower MgO than sample KG3610.10A and may be more typical of the regional lithospheric mantle. To represent more enriched typical mantle, data for sample L7.201.1.67, a lherzolite from Rothschild Island, are summarized in Table 2. This xenolith has 14.6 modal % clinopyroxene and plots within the cluster of lherzolites in Figure 9a. It consists of olivine, clinopyroxene, orthopyroxene and spinel, with no accessory phases. It has an olivine Mg# of 89.2, which is typical for Antarctic Peninsula lherzolites (Fig. 10). Lherzolite L7.201.1.67 has a spinel Cr# of 0.125, and plots close to fertile MORB mantle in Figure 8, suggesting that it has not been significantly depleted. It has not been analysed for volatiles, but likely has a whole rock water content with the range of 81–130 ppm determined for Antarctic Peninsula lherzolites (Gibson *et al.* 2020). We suggest that this sample represents the most enriched end-member that should realistically be used for modelling of regional Antarctic lithosphere, as more metasomatized lithologies are probably spatially restricted.

Conclusions

Volcanic rocks from the part of the Pacific margin of Antarctica that includes the Antarctic Peninsula and the Jones Mountains host a wide variety of fresh mantle xenoliths. These are spinel-bearing and typical of those from global off-craton continental mantle. Apart from one outcrop of a mantle-derived dunite–serpentinite complex in the South Shetland Islands, they provide the only direct evidence of the composition of the regional lithospheric mantle.

1. Mantle-derived xenoliths from the Antarctic Peninsula and Eights Coast (Jones Mountains) are compositionally varied and restricted to Cenozoic (7.7–1.5 Ma) mafic alkaline volcanic rocks that erupted in a post-subduction setting. The latter was linked to a Late Jurassic–Early Cenozoic continental margin volcanic arc.
2. Ultramafic xenolith localities related to the James Ross Island Volcanic Group have been described in two zones: in the Ulu Peninsula and along the southeastern coast of James Ross Island. The xenoliths are dominantly harzburgites (i.e. have depleted compositions) with localized pyroxene-rich regions formed by infiltration of metasomatic melts.
3. The composition of the mantle beneath Adelaide Island has been strongly modified by metasomatism, generating a pyroxenite–lherzolite–wehrlite suite with among the most enriched, Mg-poor compositions in the region.
4. Xenolith localities on Alexander Island and Rothschild Island formed in a fore-arc setting and have a wide range of compositions. These include: harzburgites that are interpreted to have originated as sub-oceanic lithosphere accreted to the margin; and lherzolites, pyroxenites, and wehrlites generated during metasomatism by alkaline magmas.
5. The wide distribution of harzburgite xenoliths, representing sub-oceanic lithospheric mantle accreted to the active margin, and the evidence for formation of the crust by tectonic accretion suggest that the dominant composition of the lithospheric mantle beneath the Antarctic Peninsula is harzburgitic. Such clinopyroxene-depleted compositions are recommended to be used in geophysical modelling. Generation of enriched mantle associated with metasomatic enrichment by mafic alkaline magmas was probably spatially restricted, and these compositions are unlikely to be widely represented in the lithospheric mantle beneath the continental margin.
6. A lherzolite sample from the Jones Mountains is compositionally similar to MORB-source mantle. The lithospheric mantle beneath this part of the Eights Coast appears to be less depleted than that below the Antarctic Peninsula, and may be related to mantle plume activity.

Acknowledgements We are grateful for constructive reviews by Adam Martin, Joaquin Bastias and an anonymous reviewer. We are grateful to Mari Whitelaw and Ieuan Hopkins for assistance with data on xenolith localities in British Antarctic Survey archives and databases, and to Laura Gerrish for generating the maps. We thank Hilary Blagbrough for help with locating thin sections and Teal Riley for assistance with SEM analysis. Electron microprobe analyses were undertaken in the Department of Earth Sciences at the University of Cambridge under the guidance of Chris Hayward. We gratefully acknowledge the significant contributions to understanding the evolution of the Antarctic Peninsula mantle made by Lydia Gibson and Eve Rooks.

Author contributions **PTL**: conceptualization (equal), formal analysis (supporting), investigation (equal), supervision (supporting), writing – original draft (lead), writing – review & editing (lead); **AJR**: data curation (equal), formal analysis (equal), investigation (equal), validation (supporting), writing – review & editing (supporting); **SAG**: conceptualization (lead), data curation (equal), formal analysis (equal), funding acquisition (lead), investigation (equal), project administration (lead), supervision (lead), writing – review & editing (supporting).

Funding Natural Environment Research Council (NE/G523404/1 and NE/K500884/1)to Sally A. Gibson.

Data availability Data used for this review were derived from published work and unpublished theses held at the University of Cambridge and the British Antarctic Survey supported by a small amount of unpublished data generated at the British Antarctic Survey. Some data is held on the British Antarctic Survey geochemical database held by the UK Polar Data Centre https://www.bas.ac.uk/data/uk-pdc/.

References

Altunkaynak, S., Aldanmaz, E., Güraslan, I.N., Çalişkanoğlu, A.Z., Ünal, A. and Nývlt, D. 2018. Lithostratigraphy and petrology of Lachman Crags and Cape Lachman lava-fed deltas, Ulu Peninsula, James Ross Island, north-eastern Antarctic Peninsula: preliminary results. *Czech Polar Reports*, **8**, 60–83, https://doi.org/10.5817/CPR2018-1-5

An, M., Wiens, D.A. *et al.* 2015. Temperature, lithosphere–asthenosphere boundary, and heat flux beneath the Antarctic plate inferred from seismic velocities. *Journal of Geophysical Research Solid Earth*, **120**, 8720–8742, https://doi.org/10.1002/2015JB011917

Arai, S. 1994. Characterization of spinel peridotites by olivine–spinel compositional relationships: review and interpretation. *Chemical Geology*, **113**, 191–204, https://doi.org/10.1016/0009-2541(94)90066-3

Barbeau, D.L., Davis, J.T., Murray, K.E., Valencia, V., Gehrels, G.E., Zahid, K.M. and Gombosi, D.J. 2009. Detrital-zircon geochronology of the metasedimentary rocks of north-western Graham Land. *Antarctic Science*, **22**, 65–78, https://doi.org/10.1017/S095410200999054X

Barker, P.F. 1982. The Cenozoic subduction history of the Pacific margin of the Antarctic Peninsula: ridge crest–trench interactions. *Journal of the Geological Society, London*, **139**, 787–801, https://doi.org/10.1144/gsjgs.139.6.0787

Bastías, J., Hervé, F., Fuentes, F., Schilling, M., Poiron, A., Gutiérrez, F. and Ramirez, C. 2012. Mantle xenoliths found at Santa Martha cove, James Ross Island, Antarctica. *13th Congreso Geológico Chileno, Actas*, Antofagasta, 365–367.

Bastias, J., Spikings, R. *et al.* 2020. The Gondwanan margin in West Antarctica: insights from Late Triassic magmatism of the Antarctic Peninsula. *Gondwana Research*, **81**, 1–20, https://doi.org/10.1016/j.gr.2019.10.018

Burn, R.W. 1984. The geology of the LeMay Group, Alexander Island. *British Antarctic Survey Scientific Reports*, **109**.

Burton-Johnson, A. and Riley, T.R. 2015. Autochthonous v. accreted terrane development of continental margins: a revised in situ tectonic history of the Antarctic Peninsula. *Journal of the Geological Society, London*, **172**, 822–835, https://doi.org/10.1144/jgs2014-110

Calabozo, F.M., Strelin, J.A., Keller, R., Brod, J.A. and Fuentes Iza, F. 2014. Mantle xenoliths from James Ross Island, Antarctic Peninsula: a preliminary whole rock and mineral chemistry study. *XIX Congreso Geológico Argentino*, Córdoba, T8-11.

Calabozo, F.M., Strelin, J.A., Orihashi, Y., Sumino, H. and Keller, R.A. 2015. Volcano-ice-sea interaction in the Cerro Santa Marta area, northwest James Ross Island, Antarctic Peninsula. *Journal of Volcanology and Geothermal Research*, **297**, 89–108, https://doi.org/10.1016/j.jvolgeores.2015.03.011

Care, B.W. 1980. The geology of Rothschild Island, northwest Alexander Island. *British Antarctic Survey Bulletin*, **50**, 87–112.

Castillo, P., Lacassie, J.P., Augustsson, C. and Hervé, F. 2015. Petrography and geochemistry of the Carboniferous-Triassic Trinity Peninsula Group, West Antarctica: implications for provenance and tectonic setting. *Geological Magazine*, **152**, 575–588, https://doi.org/10.1017/S0016756814000454

Castillo, P., Fanning, C.M., Hervé, F. and Lacassie, J.P. 2016. Characterisation and tracing of Permian magmatism in the south-western segment of the Gondwana margin: U–Pb age, Lu–Hf and O isotopic compositions of detrital zircons from metasedimentary complexes of northern Antarctic Peninsula and western Patagonia. *Gondwana Research*, **36**, 1–13, https://doi.org/10.1016/j.gr.2015.07.014

Craddock, C., Bastien, T.W. and Rutford, R.H. 1964. Geology of the Jones Mountains area. *In*: Adie, R.J. (ed.) *Antarctic Geology*. North Holland Publishing Company, Amsterdam, 171–187.

Crame, J.A. 2019. Paleobiological significance of the James Ross Basin. *Advances in Polar Science*, **30**, 165–177, https://doi.org/10.13679/j.advps.2018.0047

Dewar, G.J. 1970. The geology of Adelaide Island. *British Antarctic Survey Scientific Reports*, **57**.

De Wit, M.J., Dutch, S., Kligfield, R., Allen, R. and Stern, C. 1977. Deformation, serpentinization and emplacement of a dunite complex, Gibbs Island, South Shetland Islands: possible fracture zone tectonics. *Journal of Geology*, **85**, 745–762, https://doi.org/10.1086/628360

Dick, H.J.B., Fisher, R.L. and Bryan, W.B. 1984. Mineralogic variability of the uppermost mantle along mid-ocean ridges. *Earth and Planetary Science Letters*, **69**, 88–106, https://doi.org/10.1016/0012-821X(84)90076-1

Doubleday, P.A., Leat, P.T., Alabaster, T., Nell, P.A.R. and Tranter, T.H. 1994. Allochthonous oceanic basalts within the Mesozoic accretionary complex of Alexander Island, Antarctica: remnants of proto-Pacific oceanic crust. *Journal of the Geological Society, London*, **151**, 65–78, https://doi.org/10.1144/gsjgs.151.1.0065

Elliot, D.H. 1988. Tectonic setting and evolution of the James Ross Basin, northern Antarctic Peninsula. *Geological Society of America, Memoirs*, **169**, 541–555, https://doi.org/10.1130/MEM169-p541

Farquharson, G.W. 1982. Late Mesozoic sedimentation in the northern Antarctic Peninsula and its relationship to the southern Andes. *Journal of the Geological Society, London*, **139**, 721–727, https://doi.org/10.1144/gsjgs.139.6.0721

Ferraccioli, F., Jones, P.C., Vaughan, A.P.M. and Leat, P.T. 2006. New aerogeophysical view of the Antarctic Peninsula: More pieces, less puzzle. *Geophysical Research Letters*, **33**, L05310, https://doi.org/10.1029/2005GL024636

Flowerdew, M.J., Millar, I.L., Vaughan, A.P.M., Horstwood, M.S.A. and Fanning, C.M. 2006. The source of granitic gneisses and migmatites in the Antarctic Peninsula: a combined U–Pb SHRIMP and laser ablation Hf isotope study of complex zircons. *Contributions to Mineralogy and Petrology*, **151**, 751–768, https://doi.org/10.1007/s00410-006-0091-6

Fleet, M. 1968. The geology of the Oscar II Coast, Graham Land. *British Antarctic Survey Scientific Reports*, **59**.

Gibson, L.C. 2012. *Antarctic lithosphere architecture and evolution; direct constraints from mantle xenoliths*. PhD thesis, University of Cambridge.

Gibson, L., Gibson, S. and Leat, P. 2010*a*. Lithospheric mantle evolution above a subducting plate: direct constraints from Antarctic Peninsula spinel peridotite xenoliths. *EGU General Assembly*, Vienna, Austria, 2–7 May 2010, Abstracts p.10502.

Gibson, L.C., Gibson, S.A. and Leat, P.T. 2010*b*. Melt production and mantle refertilisation above a subduction zone: direct constraints from Antarctic Peninsula spinel-peridotite xenoliths. *American Geophysical Union Fall Meeting 2010*, San Francisco, USA, Abstract V11F-07.

Gibson, S.A., Rooks, E.E., Day, J.A., Petrone, C.M. and Leat, P.T. 2020. The role of sub-continental mantle as both 'sink' and 'source' in deep Earth volatile cycles. *Geochimica et Cosmochimica Acta*, **275**, 140–162, https://doi.org/10.1016/j.gca.2020.02.018

Ghidella, M.E., Zambrano, O.M., Ferraccioli, F., Lirio, J.M., Zakrajsek, A.F., Ferris, J. and Jordan, T.A. 2013. Analysis of James Ross Island volcanic complex and sedimentary basin based on high-resolution aeromagnetic data. *Tectonophysics*, **585**, 90–101, https://doi.org/10.1016/j.tecto.2012.06.039

Gonzalez-Ferran, O. 1983. The Seal Nunataks: an active volcanic group on the Larsen Ice Shelf, West Antarctica. *In*: Oliver, R.L., James, P.R. and Jago, J.B. (eds) *Antarctic Earth Science*. Australian Academy of Science, Canberra, 334–337.

Griffiths, C.J. and Oglethorpe, R.D.J. 1998. The stratigraphy and geochronology of Adelaide Island. *Antarctic Science*, **10**, 462–475, https://doi.org/10.1017/S095410209800056X

Grunow, A.M., Dalziel, I.W.D., Harrison, T.M. and Heizler, M.T. 1992. Structural geology and geochronology of subduction complexes along the margin of Gondwanaland: new data from the Antarctic Peninsula and southernmost Andes. *Geological Society of America Bulletin*, **104**, 1497–1514, https://doi.org/10.1130/0016-7606(1992)104<1497:SGAGOS>2.3.CO;2

Haase, K.M., Beier, C., Fretzdorff, S., Smellie, J.L. and Garbe-Schönberg, D. 2012. Magmatic evolution of the South Shetland Islands, Antarctica, and implications for continental crust formation. *Contributions to Mineralogy and Petrology*, **163**, 1103–1119, https://doi.org/10.1007/s00410-012-0719-7

Handler, M.R., Wysoczanski, R.J. and Gamble, J.A. 2003. Proterozoic lithosphere in Marie Byrd Land, West Antarctica: Re–Os systematics of spinel peridotite xenoliths. *Chemical Geology*, **196**, 131–145, https://doi.org/10.1016/S0009-2541(02)00410-2

Harley, S.L., Fitzsimons, I.C.W. and Zhao, Y. 2013. Antarctica and supercontinent evolution: historical perspectives, recent advances and unresolved issues. *Geological Society, London, Special Publications*, **383**, 1–34, https://doi.org/10.1144/SP383.9

Hart, S.R., Blusztajn, J. and Craddock, C. 1995. Cenozoic volcanism in Antarctica: Jones Mountains and Peter I Island. *Geochimica et Cosmochimica Acta*, **59**, 3379–3388, https://doi.org/10.1016/0016-7037(95)00212-I

Hervé, F., Loske, W., Miller, H. and Pankhurst, R.J. 1991. Chronology of provenance, deposition and metamorphism of deformed fore-arc sequences, southern Scotia arc. *In*: Thomson, M.R.A., Crame, J.A. and Thomson, J.W. (eds) *Geological Evolution of Antarctica*. Cambridge University Press, Cambridge, 429–435.

Hole, M.J. 1988. Post-subduction alkaline volcanism along the Antarctic Peninsula. *Journal of the Geological Society, London*, **145**, 985–998, https://doi.org/10.1144/gsjgs.145.6.0985

Hole, M.J. 1990*a*. Geochemical evolution of Pliocene–Recent post-subduction alkalic basalts from Seal Nunataks, Antarctic Peninsula. *Journal of Volcanology and Geothermal Research*, **40**, 149–167, https://doi.org/10.1016/0377-0273(90)90118-Y

Hole, M.J. 1990*b*. Hornpipe Heights. *In*: LeMasurier, W.E. and Thomson, J.W. (eds) *Volcanoes of the Antarctic Plate and Southern Oceans*. AGU, Antarctic Research Series, **48**, 271–272.

Hole, M.J. 2021. Antarctic Peninsula: petrology. *Geological Society, London, Memoirs*, **55**, https://doi.org/10.1144/M55-2018-40

Hole, M.J., Rogers, G., Saunders, A.D. and Storey, M. 1991. Relation between alkalic volcanism and slab-window formation. *Geology*, **19**, 657–660, https://doi.org/10.1130/0091-7613(1991)019<0657:RBAVAS>2.3.CO;2

Hole, M.J., Kempton, P.D. and Millar, I.L. 1993. Trace-element and isotopic characteristics of small-degree melts of the asthenosphere: evidence from alkalic basalts of the Antarctic Peninsula. *Chemical Geology*, **109**, 51–68, https://doi.org/10.1016/0009-2541(93)90061-M

Hole, M.J., Storey, B.C. and LeMasurier, W.E. 1994. Tectonic setting and geochemistry of Miocene alkalic basalts from the Jones Mountains, West Antarctic. *Antarctic Science*, **6**, 85–92, https://doi.org/10.1017/S0954102094000118

Johnson, A.C. 1999. Interpretation of new aeromagnetic anomaly data from the central Antarctic Peninsula. *Journal of Geophysical Research*, **104**, 5031–5046, https://doi.org/10.1029/1998JB900073

Johnson, J.S., Everest, J.D., Leat, P.T., Golledge, N.R., Rood, D.H. and Stuart, F.M. 2012. The deglacial history of NW Alexander Island, Antarctica, from surface exposure dating. *Quaternary Research*, **77**, 273–280, https://doi.org/10.1016/j.yqres.2011.11.012

Jordan, T.A., Neale, R.F. *et al.* 2014. Structure and evolution of Cenozoic arc magmatism on the Antarctic Peninsula; a high resolution aeromagnetic perspective. *Geophysical Journal International*, **198**, 1758–1774, https://doi.org/10.1093/gji/ggu233

Keller, R.A. and Strelin, J.A. 1992. Alkalic basalts and ultramafic xenoliths on James Ross Island, Antarctic Peninsula. *Antarctic Journal of the United States*, **27**, 22–23.

Kim, H., Cho, M., Lee, J.I. and Nagao, K. 1995. The K–Ar age of schists in Gibbs Island, Antarctica. *Korean Journal of Polar Research*, **6**, 91–96.

Košler, J., Magna, T., Mlčoch, B., Mixa, P. and Holub, F.V. 2009. Combined Sr, Nd, Pb and Li isotope geochemistry of alkaline lavas from northern James Ross Island (Antarctic Peninsula) and implications for back-arc magma formation. *Chemical Geology*, **258**, 207–218, https://doi.org/10.1016/j.chemgeo.2008.10.006

Larter, R.D. and Barker, P.F. 1991. Effects of ridge crest-trench interaction on Antarctic–Phoenix spreading: forces on a young subducting plate. *Journal of Geophysical Research*, **96**, 19583–19607, https://doi.org/10.1029/91JB02053

Larter, R.D., Cunningham, A.P., Barker, P.F., Gohl, K. and Nitsche, F.O. 2002. Tectonic evolution of the Pacific margin of Antarctica 1. Late Cretaceous tectonic reconstructions. *Journal of Geophysical Research*, **107**, 2345, https://doi.org/10.1029/2000JB000052

Leat, P.T. and Riley, T.R. 2021. Antarctic Peninsula and South Shetland Islands: volcanology. *Geological Society, London, Memoirs*, **55**, https://doi.org/10.1144/M55-2018-52

Leat, P.T., Storey, B.C. and Pankhurst, R.J. 1993. Geochemistry of Palaeozoic–Mesozoic Pacific rim orogenic magmatism, Thurston Island area, West Antarctica. *Antarctic Science*, **5**, 281–296, https://doi.org/10.1017/S0954102093000380

Leat, P.T., Scarrow, J.H. and Millar, I.L. 1995. On the Antarctic Peninsula batholith. *Geological Magazine*, **132**, 399–4127, https://doi.org/10.1017/S0016756800021464

Lee, J.I., Choi, S.H., Lee, J.-I. and Choe, M.Y. 2015. Petrogenesis of dunites from Gibbs Island, South Shetland Islands, Antarctica. *Geosciences Journal*, **19**, 33–44, https://doi.org/10.1007/s12303-014-0026-6

Maldonado, A., Larter, R.D. and Aldaya, F. 1994. Forearc tectonic evolution of the South Shetland margin, Antarctic Peninsula. *Tectonics*, **13**, 1345–1370, https://doi.org/10.1029/94TC01352

McCarron, J.J. and Millar, I.L. 1997. The age and stratigraphy of fore-arc magmatism on Alexander Island, Antarctica. *Geological Magazine*, **134**, 507–522, https://doi.org/10.1017/S0016756897007437

Millar, I.L., Pankhurst, R.J. and Fanning, C.M. 2002. Basement chronology of the Antarctic Peninsula: recurrent magmatism and anataxis in the Palaeozoic Gondwana margin. *Journal of the Geological Society, London*, **159**, 145–157, https://doi.org/10.1144/0016-764901-020

Mundl, A., Ntaflos, T., Ackerman, L., Bizimis, M., Bjerg, E.A. and Hauzenberger, C.A. 2015. Mesoproterozoic and Paleoproterozoic subcontinental lithospheric mantle domains beneath southern Patagonia: isotopic evidence for its connection to Africa and Antarctica. *Geology*, **43**, 39–42, https://doi.org/10.1130/G36344.1

Nell, P.A.R. 1990. Deformation in an accretionary melange Alexander Island, Antarctica. *Geological Society, London, Special Publications*, **54**, 405–416, https://doi.org/10.1144/GSL.SP.1990.054.01.37

Nelson, P.H.H. 1966. The James Ross Island Volcanic Group of north-east Graham Land. *British Antarctic Survey Scientific Reports*, **54**.

Nelson, D.A. and Cottle, J.M. 2018. The secular development of accretionary orogens: linking the Gondwana magmatic arc record of West Antarctica, Australia and South America. *Gondwana Research*, **63**, 15–33, https://doi.org/10.1016/j.gr.2018.06.002

Nývlt, D., Košler, J., Mlčoch, B., Mixa, P., Lisá, L., Bubík, M. and Hendriks, B.W.H. 2011. The Mendel Formation: evidence for Late Miocene climatic cyclicity at the northern tip of the Antarctic Peninsula. *Palaeogeography, Palaeoclimatology, Palaeoecology*, **299**, 363–384, https://doi.org/10.1016/j.palaeo.2010.11.017

O'Neill, H.St.C. 1981. The transition between spinel lherzolite and garnet lherzolite, and its use as a geobarometer. *Contributions to Mineralogy and Petrology*, **77**, 185–194, https://doi.org/10.1007/BF00636522

Pankhurst, R.J. 1982. Rb–Sr geochronology of Graham Land. *Journal of the Geological Society, London*, **139**, 701–711, https://doi.org/10.1144/gsjgs.139.6.0701

Pankhurst, R.J., Millar, I.L., Grunow, A.M. and Storey, B.C. 1993. The pre-Cenozoic magmatic history of the Thurston Island crustal block, West Antarctica. *Journal of Geophysical Research*, **98**, 11835–11849, https://doi.org/10.1029/93JB01157

Pankhurst, R.J., Leat, P.T., Sruoga, P., Rapela, C.W., Márquez, M., Storey, B.C. and Riley, T.R. 1998. The Chon Aike silicic province of Patagonia and related rocks in West Antarctica: a silicic large igneous province. *Journal of Volcanology and Geothermal Research*, **81**, 113–136, https://doi.org/10.1016/S0377-0273(97)00070-X

Riley, T.R. and Leat, P.T. 2021. Palmer Land and Graham Land volcanic groups (Antarctic Peninsula): volcanology. *Geological Society, London, Memoirs*, **55**, https://doi.org/10.1144/M55-2018-36

Riley, T.R., Flowerdew, M.J. and Whitehouse, M.J. 2012*a*. U–Pb ion-microprobe zircon geochronology from the basement inliers of eastern Graham Land, Antarctic Peninsula. *Journal of the Geological Society, London*, **169**, 381–393, https://doi.org/10.1144/0016-76492011-142

Riley, T.R., Flowerdew, M.J. and Whitehouse, M.J. 2012*b*. Chrono- and lithostratigraphy of a Mesozoic–Tertiary fore- to intra-arc basin: Adelaide Island, Antarctic Peninsula. *Geological Magazine*, **149**, 768–782, https://doi.org/10.1017/S0016756811001002

Riley, T.R., Flowerdew, M.J., Pankhurst, R.J., Leat, P.T., Millar, I.L., Fanning, C.M. and Whitehouse, M.J. 2017. A revised geochronology of Thurston Island, West Antarctica, and correlations along the proto-Pacific margin of Gondwana. *Antarctic Science*, **29**, 47–60, https://doi.org/10.1017/S0954102016000341

Riley, T.R., Flowerdew, M.J., Pankhurst, R.J., Millar, I.L. and Whitehouse, M.J. 2020. U–Pb zircon geochronology from Haag Nunataks, Coats Land and Shackleton Range (Antarctica): constraining the extent of juvenile Late Mesoproterozoic arc terranes. *Precambrian Research*, **340**, 105646, https://doi.org/10.1016/j.precamres.2020.105646

Rooks, E. 2016. *The subcontinental lithospheric mantle evolution of the southern Gondwana margin*. PhD thesis, University of Cambridge.

Rooks, E.E., Gibson, S.A., Leat, P.T. and Petrone, C.M. 2015. Compositionally controlled volatile content of nominally volatile-free minerals in the continental upper mantle of southern Gondwana (Patagonia & W. Antarctica). *American Geophysical Union Fall Meeting 2015*, San Francisco, USA, Abstract V51I-02.

Ross, A.J. 2008. *Compositional variability of the mantle beneath West Antarctica and its relationship to terrane tectonics: evidence from mantle xenoliths*. MSci dissertation, University of Cambridge.

Ross, A.J., Gibson, S.A., Leat, P.T. and Vaughan, A.P.M. 2009. Compositional variability of the mantle beneath West Antarctica and relationship to terrane tectonics: evidence from mantle xenoliths. *EGU General Assembly*, Vienna, Austria, 19–24 April 2009, Abstracts p. 10831.

Rowley, P.D. 1990. Jones Mountains. *In*: LeMasurier, W.E. and Thomson, J.W. (eds) *AGU, Antarctic Research Series*, **48**, 286–288.

Rutford, R.H., Craddock, C. and Bastien, T.W. 1968. Late Tertiary glaciation and sea-level changes in Antarctica. *Palaeogeography, Palaeoclimatology, Palaeoecology*, **5**, 15–39, https://doi.org/10.1016/0031-0182(68)90058-8

Rutford, R.H., Craddock, C., White, C.M. and Armstrong, R.L. 1972. Tertiary glaciation in the Jones Mountains. *In*: Adie, R.J. (ed.) *Antarctic Geology and Geophysics*. Universitetsforlaget, Oslo, 239–243.

Schilling, M.E., Carlson, R.W. *et al.* 2017. The origin of Patagonia revealed by Re–Os systematics of mantle xenoliths. *Precambrian Research*, **294**, 15–32, https://doi.org/10.1016/j.precamres.2017.03.008

Smellie, J.L. 1987. Geochemistry and tectonic setting of alkaline volcanic rocks in the Antarctic Peninsula: a review. *Journal of Volcanology and Geothermal Research*, **32**, 269–285, https://doi.org/10.1016/0377-0273(87)90048-5

Smellie, J.L. 1990. Seal Nunataks. *In*: LeMasurier, W.E. and Thomson, J.W. (eds) *AGU, Antarctic Research Series*, **48**, 349–351.

Smellie, J.L. 1999. Lithostratigraphy of Miocene–Recent, alkaline volcanic fields in the Antarctic Peninsula and eastern Ellsworth Land. *Antarctic Science*, **11**, 362–378, https://doi.org/10.1017/S0954102099000450

Smellie, J.L., Pankhurst, R.J., Hole, M.J. and Thomson, J.W. 1988. Age, distribution and eruptive conditions of Late Cenozoic alkaline volcanism in the Antarctic Peninsula and eastern Ellsworth Land: review. *British Antarctic Survey Bulletin*, **80**, 21–49.

Smellie, J.L., Hole, M.J. and Nell, P.A.R. 1993. Late Miocene valley-confined subglacial volcanism in northern Alexander Island, Antarctic Peninsula. *Bulletin of Volcanology*, **55**, 273–288, https://doi.org/10.1007/BF00624355

Smellie, J.L., Johnson, J.S., McIntosh, W.C., Esser, R., Gudmundsson, M.T., Hambrey, M.J. and van Wyk de Vries, B. 2008. Six million years of glacial history recorded in volcanic lithofacies of the James Ross Island Volcanic Group, Antarctica. *Palaeogeography, Palaeoclimatology, Palaeoecology*, **260**, 122–148, https://doi.org/10.1016/j.palaeo.2007.08.011

Storey, B.C. and Garrett, S.W. 1985. Crustal growth of the Antarctic Peninsula by accretion, magmatism and extension. *Geological Magazine*, **122**, 5–14, https://doi.org/10.1017/S0016756800034038

Storey, B.C., Dalziel, I.W.D., Garrett, S.W., Grunow, A.M., Pankhurst, R.J. and Vennum, W.R. 1988. West Antarctica in Gondwanaland: crustal blocks, reconstruction and breakup processes. *Tectonophysics*, **155**, 381–390.

Tangeman, J.A., Mukasa, S.B. and Grunow, A.M. 1996. Zircon U–Pb geochronology of plutonic rocks from the Antarctic Peninsula: confirmation of the presence of unexposed Paleozoic crust. *Tectonics*, **15**, 1309–1324, https://doi.org/10.1029/96TC00840

Tanner, P.W.G., Pankhurst, R.J. and Hyden, G. 1982. Radiometric evidence for the age of the subduction complex in the South Orkney and South Shetland islands, West Antarctica. *Journal of the Geological Society, London*, **139**, 683–690, https://doi.org/10.1144/gsjgs.139.6.0683

Taylor, W.R. 1998. An experimental test of some geothermometer and geobarometer formulations for upper mantle peridotites with application to the thermobarometers of fertile lherzolite and garnet websterite. *Neues Jahrbuch für Mineralogie Abhandlungen*, **172**, 381–408, https://doi.org/10.1127/njma/172/1998/381

Thomson, J.W., Rowley, P.D. and Hole, M.J. 1990. Rothschild Island. *In*: LeMasurier, W.E. and Thomson, J.W. (eds) *AGU, Antarctic Research Series*, **48**, 266–267.

Trouw, R.A.J., Passchier, C.W., Valeriano, C.M., Simões, L.S.A., Paciullo, F.V.P. and Riberio, A. 2000. Deformational evolution of a Cretaceous subduction complex: Elephant Island, South Shetland Islands, Antarctica. *Tectonophysics*, **319**, 93–110, https://doi.org/10.1016/S0040-1951(00)00021-4

Valladares Gaete, A.V. 2017. *Petrografía y geoquímica de xenolitos mantélicos de la Isla James Ross, Antártica*. Thesis, Universidad Andrés Bello, Santiago.

Vaughan, A.P.M. and Storey, B.C. 2000. The eastern Palmer Land shear zone: a new terrane accretion model for the Mesozoic development of the Antarctic Peninsula. *Journal of the Geological Society, London*, **157**, 1243–1256, https://doi.org/10.1144/jgs.157.6.1243

Vaughan, A.P.M., Leat, P.T. and Pankhurst, R.J. 2005. Terrane processes at the margins of Gondwana: introduction. *Geological Society, London, Special Publications*, **246**, 1–21, https://doi.org/10.1144/GSL.SP.2005.246.01.01

Veevers, J.J. and Saeed, A. 2013. Age and composition of Antarctic sub-glacial bedrock reflected by detrital zircons, erratics, and recycled microfossils in the Ellsworth Land–Antarctic Peninsula–Weddell Sea–Dronning Maud Land sector. *Gondwana Research*, **23**, 296–332, https://doi.org/10.1016/j.gr.2012.05.010

Zheng, G.G., Liu, X.C., Liu, S., Zhang, S.H. and Zhao, Y. 2018. Late Mesozoic–early Cretaceous intermediate-acid intrusive rocks from the Gerlache Strait area, Antarctic Peninsula: zircon U–Pb geochronology, petrogenesis and tectonic implications. *Lithos*, **312–313**, 204–222, https://doi.org/10.1016/j.lithos.2018.05.008

The subantarctic lithospheric mantle

Guillaume Delpech[1]*, James M. Scott[2], Michel Grégoire[3], Bertrand N. Moine[4], Dongxu Li[5], Jingao Liu[5], D. Graham Pearson[6], Quinten H. A. van der Meer[7], Tod E. Waight[8], Gilbert Michon[4], Damien Guillaume[4], Suzanne Y. O'Reilly[9], Jean-Yves Cottin[4] and André Giret[4†]

[1]Université Paris-Saclay, CNRS, GEOPS, 91405 Orsay Cedex, France

[2]Department of Geology, University of Otago, Dunedin 9054, New Zealand

[3]Géosciences Environnement Toulouse, OMP, CNRS, CNES, IRD, Toulouse III University 14 Avenue Édouard Belin, 31400 Toulouse, France

[4]Laboratoire Magmas et Volcans, CNRS, IRD, St Etienne and Clermont-Ferrand Universities, St Etienne, France

[5]State Key Laboratory of Geological Processes and Mineral Resources, China University of Geosciences, Beijing 100083, China

[6]Department of Earth and Atmospheric Sciences, University of Alberta, Edmonton, Alberta, Canada T6G 2E3

[7]Nordic Volcanological Centre, Institute of Earth Sciences, University of Iceland, Sturlugata 7, 101 Reykjavık, Iceland

[8]Department of Geosciences and Natural Resource Management (Geology Section), Copenhagen University Øster Voldgade 10, 1350 Copenhagen K, Denmark

[9]Australian Research Council Centre of Excellence for Core to Crust Fluid Systems (CCFS) and GEMOC Key Centre, Macquarie University, Sydney, NSW 2019, Australia

GD, 0000-0001-6496-0287; JMS, 0000-0001-5185-6261; MG, 0000-0001-9196-876X; BNM, 0000-0003-2780-2882; JL, 0000-0002-8800-5898; TEW, 0000-0003-2601-1202; SYO, 0000-0002-3883-5498

*Correspondence: guillaume.delpech@universite-paris-saclay.fr

[†]deceased

Abstract: We present a summary of peridotite in the Subantarctic (46–60° S) surrounding the Antarctic Plate. Peridotite xenoliths occur on the Kerguelen Islands and Auckland Islands. The Kerguelen Islands are underlain by a plume, whereas the Auckland Islands are part of continental Zealandia, which is a Gondwana-rifted fragment. Small amounts of serpentinized peridotite has been dredged from fracture zones on the Southeast Indian Ridge, Southwest Indian Ridge and Pacific Antarctic Ridge, and represent upwelled asthenosphere accreted to form lithosphere. Suprasubduction-zone peridotite was collected from two locations on the Sandwich Plate. Peridotites from most subantarctic occurrences are moderately to highly depleted, and many show signs of subsequent metasomatic enrichment. Os isotopes indicate that subantarctic continental and oceanic lithospheric mantle contains ancient fragments that underwent depletion long before formation of the overlying crust.

Supplementary material: Photomicrographs and mineralogical and geochemical data are available at https://doi.org/10.6084/m9.figshare.c.5424956

The Subantarctic is an area circumnavigating Earth between 46° S and 60° S (Fig. 1). It encompasses seven major tectonic plates (Antarctic, Australian, Pacific, Scotia, South Sandwich, South American and African) but, with the exception of the southern tip of South America and the southern portion of New Zealand, only small island groups breech the ocean surfaces. The islands, referred to as subantarctic islands, form 13 groups that range in area from 1.4 km^2 (Bounty) to 7215 km^2 (Kerguelen) (Fig. 1; Table 1); all are difficult to access and remote, and none have a permanent human population. The island groups are typically intraplate volcanoes (Weaver *et al.* 1987; Weis *et al.* 1993; Barling *et al.* 1994; Mahoney *et al.* 1996; Giret *et al.* 1997; Frey *et al.* 2000; Quilty and Wheller 2000; Leat *et al.* 2003; le Roex *et al.* 2012; Scott and Turnbull 2019), with the exception of the Zealandia granite-dominated Snares and Bounty island groups (e.g. Scott and Turnbull 2019), and Macquarie Island, which is a slice of exhumed oceanic lithosphere (Kamenetsky *et al.* 2000; Varne *et al.* 2000).

In this contribution, we summarize the characteristics of subantarctic lithospheric mantle in the Southern Ocean and the southern Atlantic and Indian oceans, as known from mantle peridotite. As will be discussed below, peridotite xenoliths have been found on three subantarctic island groups: the Kerguelen Islands and Heard Island in the southern Indian Ocean, and the Auckland Islands in the Southern Ocean. Exhumed abyssal peridotite is exposed on Macquarie Island in the Southern Ocean, and has been dredged from the Southwest Indian Ridge, Southeast Indian Ridge and America–Atlantic Ridge (Fig. 1), and suprasubduction-zone peridotite has been dredged and drilled from the South Sandwich arc in the Atlantic Ocean. Our synthesis of subantarctic lithospheric mantle peridotite is presented in three domains: the Indian Ocean, the Southern Ocean and the Atlantic Ocean. A very brief review of some locations was given in Nixon (1987).

Since the oldest oceanic crust surrounding Antarctica is Cretaceous, the fragmentation of Australia, India, Zealandia, South America and Africa from Antarctica has led to the dispersion of continental lithosphere and the formation of intervening areas of oceanic lithosphere by asthenosphere decompression. Therefore, the composition and evolution of rare fragments of peridotite from under the southern portions of the Indian, Southern and Atlantic oceans provide rare

From: Martin, A. P. and van der Wal, W. (eds) 2023. *The Geochemistry and Geophysics of the Antarctic Mantle.* Geological Society, London, Memoirs, **56**, 115–132,
First published online 1 September 2021, https://doi.org/10.1144/M56-2020-13

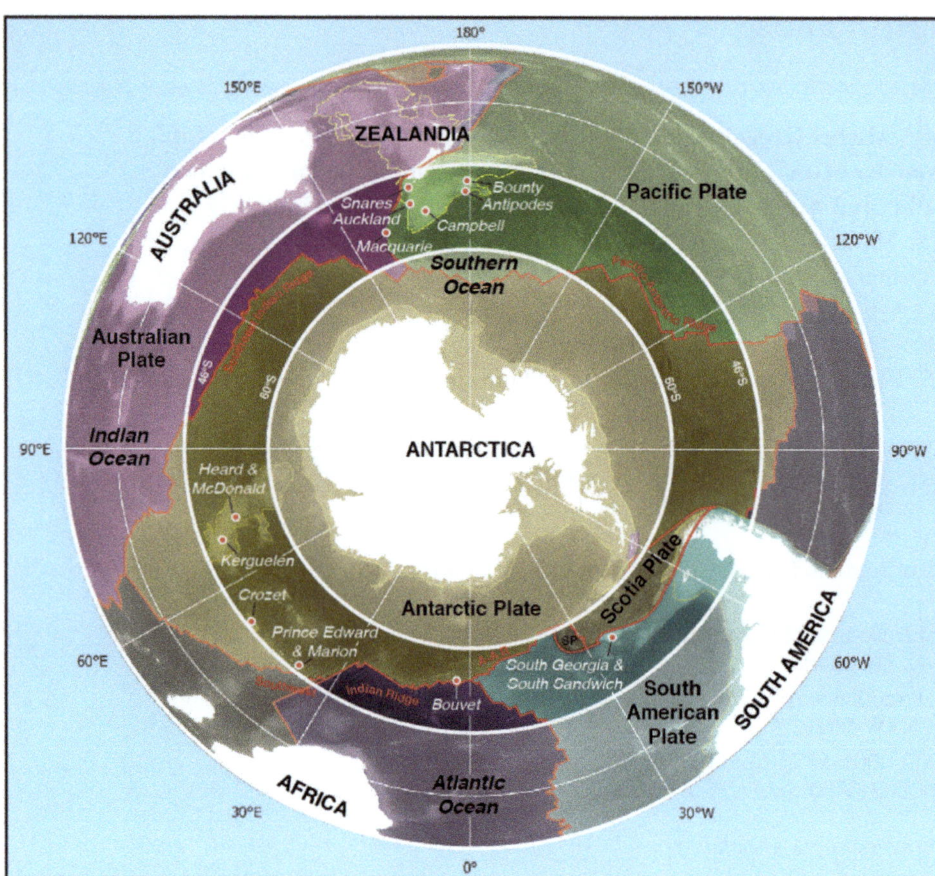

Fig. 1. Map illustrating the Subantarctic and the tectonic plates and oceans within this area. Mantle xenoliths occur on the Auckland Islands, Crozet Island, Heard Island and Kerguelen islands. Abyssal peridotite crops out on Macquarie Island, and has been dredged from the Southwest Indian Ridge and the America–Antarctica Ridge. Suprasubduction-zone peridotite occurs on the South Sandwich Arc. AAR, America–Antarctica Ridge; SP, Sandwich Plate.

insights into the evolution of fragments of dispersed Gondwana lithosphere, as well as the upwelled and accreted mantle that formerly resided deep beneath the continents but now represents the intervening oceanic lithosphere.

Subantarctic mantle under the southern Indian Ocean

The *c.* 7200 km^2 Kerguelen archipelago is the main island of the French subantarctic island group (Terres Australes et Antarctiques Françaises: TAAF) in the Indian Ocean and is located on the northern part of the submerged Kerguelen–Heard oceanic plateau (Fig. 2a). This plateau is the second largest oceanic large igneous province in the world (25 × 10^6 km^3) after the Ontong Java Plateau (Coffin and Eldholm 1993). The Kerguelen archipelago is the only place where mantle xenoliths have been intensively studied in the southern Indian Ocean. Rare peridotite xenoliths occur on Heard Island (O'Reilly pers. obs.) but no data are currently available, and the dunite xenoliths reported from Crozet by Nixon (1987) are actually igneous cumulates (Table 1).

The Kerguelen–Heard plateau formed in response to Kerguelen plume magmatism, which is closely linked to the Gondwana break-up in the southern hemisphere that started 120 myr ago (Coffin *et al.* 2002). According to geodynamic reconstructions and geochemical investigations (e.g. Royer and Coffin 1992; Weis and Frey 2002), the archipelago originated from interaction between the Kerguelen plume and the Southeast Indian Ridge at about 40 Ma. The ridge then progressively moved away from the hotspot, leaving the plume in its oceanic intraplate position since *c.* 25 Ma. The majority (80–85%) of the subaerial surface of the Kerguelen archipelago is covered by flood basalts that erupted from 30 to

Table 1. *Summary of subantarctic islands and mantle occurrences*

Subantarctic island group	Ocean	Latitude (°S)	Area (km^2)	Administrator	Mantle peridotite xenoliths?
South Georgia and South Sandwich	Atlantic	54	3903	UK	No but peridotite dredged and drilled
Bouvet	Atlantic	54	49	Norwegian	No
Heard and McDonald	Indian	53	368	Australia	Yes but only on Heard Island (S. O'Reilly unpublished data)
Crozet	Indian	46	352	French	No, dunites reported are cumulates
Kerguelen	Indian	49	7215	French	Yes
Prince Edward and Marion	Indian	46	335	South Africa	No
Macquarie	Southern	55	128	Australia	No but exhumed oceanic mantle
Antipodes	Southern	49	21	New Zealand	No
Auckland	Southern	51	626	New Zealand	Yes
Bounty	Southern	47	1.4	New Zealand	No
Campbell	Southern	52	113	New Zealand	No
Snares	Southern	48	3.5	New Zealand	No

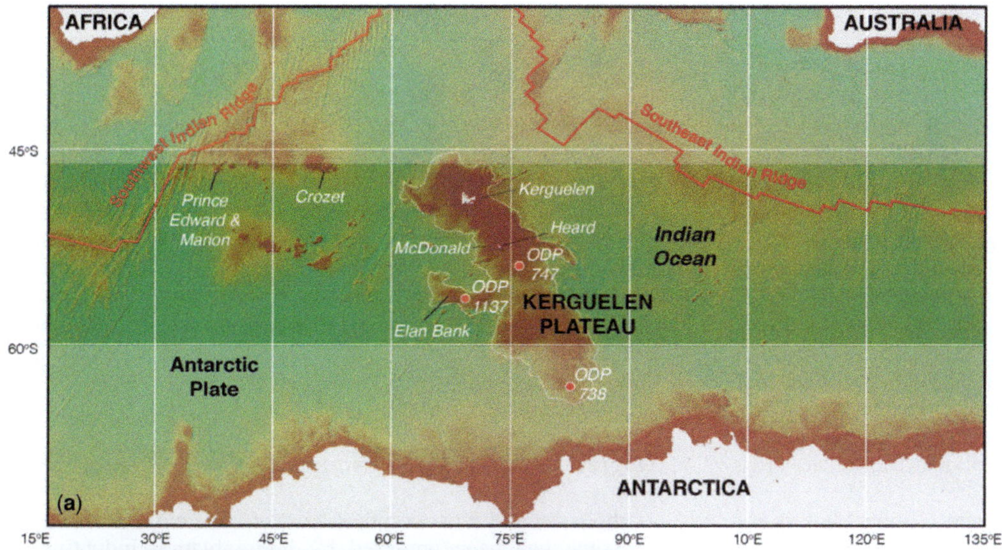

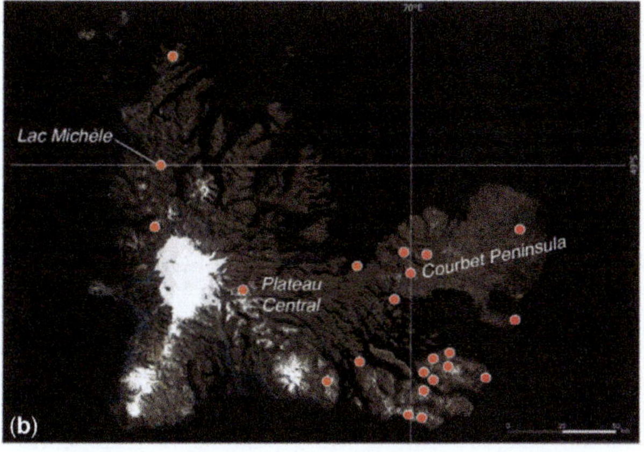

Fig. 2. (**a**) Bathymetric map showing the Indian Ocean subantarctic area and features discussed in the text. Red circles indicate the locations of ODP (Ocean Drilling Project) sites 1137, 747 and 738 where material with continental crust affinities have been found (see text for details). (**b**) The lower satellite image is of the Kerguelen archipelago, with known peridotite xenolith occurrences after Wasilewski *et al.* (2017).

25 Ma, and reached >1000 m in thickness in places (Giret 1993). However, small-volume disseminated Quaternary volcanic centres point to ongoing but limited magmatic activity. The volcanic rocks comprise an early tholeiitic-transitional series followed by alkaline and then highly alkaline magmas (Gautier *et al.* 1990; Weis *et al.* 1993; Yang *et al.* 1998; Frey *et al.* 2000; Doucet *et al.* 2002, 2005). The rest of the subaerial rocks are made up of plutonic rock types (*c.* 15%, comprising gabbros, syenites, quartz monzonites and rare granites) associated with volcanic systems (Lameyre *et al.* 1976; Dosso *et al.* 1979; Giret 1983; Scoates *et al.* 2008) and sedimentary rocks formed by fluvio-glacial erosion.

Kerguelen peridotite xenolith characterization

The earliest descriptions and collection of mantle xenoliths from Kerguelen were made by Aubert de la Rüe (1932) and Edwards (1938). Although xenoliths were subsequently mentioned regularly in publications (Talbot *et al.* 1963; Nougier 1970; McBirney and Aoki 1973; Giret 1983; Gautier 1987; Leyrit 1992), systematic study only really began in the early 1990s during the Kerguelen cartographical programmes initiated by André Giret and the TAAF. The results of these studies make Kerguelen one of the best-studied off-craton mantle xenolith occurrences on Earth (Grégoire *et al.* 1992, 1994, 1995, 1996, 1997, 1998, 2000*a*, *b*; Schiano *et al.* 1994; Mattielli 1996; Mattielli *et al.* 1996, 1999; Valbracht *et al.* 1996; Moine 2000; Moine *et al.* 2001, 2004; Delpech 2004; Delpech *et al.* 2004, 2012; Lorand *et al.* 2004; Bascou *et al.* 2008; Wasilewski *et al.* 2017). There are at least 22 mantle xenolith localities (Fig. 2b), all occurring within dykes, lava flows and breccia pipes of the youngest and most alkaline volcanic rocks. Most xenoliths have subrounded shapes, with sizes ranging up to 60 cm in diameter. Grégoire (1994) and Grégoire *et al.* (1995) subdivided the Kerguelen xenoliths into two main groups: Type I and Type II. The Type I xenoliths comprise spinel-bearing harzburgite (30%), dunite (10%) and associated composite rocks. These are commonly very fresh, except for a group of moderately serpentinized harzburgites from Lac Michèle (Wasilewski *et al.* 2017). The Type II xenoliths (50%) are metacumulates (peridotites, pyroxenites and metagabbros) recrystallized in the pressure–temperature conditions of the lower crust and upper mantle. The remaining 10% are hornblendites, biotitites and amphibole- or biotite-bearing clinopyroxenites with typical cumulative textures. In this contribution, we discuss published and new data for Type I xenoliths.

The Type I spinel-bearing harzburgite xenoliths mostly have coarse-grained textures (Harte 1977; see Supplementary Table 1 and Supplementary Fig. 1) that locally grade into porphyroclastic textures comprising fine-grained mosaics of olivine and orthopyroxene neoblasts surrounding elongate porphyroclasts. These common 'protogranular' harzburgites consist mostly of olivine and orthopyroxene with low modal contents of clinopyroxene (<5%) and spinel. Accessory phlogopite, amphibole, apatite, carbonate and feldspar occur in some samples. Although Cr-diopside is the main clinopyroxene (Grégoire 1994), many harzburgites also display 'poikilitic' Mg-augite crystals with magmatic twins and/or as spongy rims surrounding orthopyroxene, olivine and spinel, sometimes associated with phlogopite and more rarely with amphibole and apatite (Supplementary Fig. 1B, C). Poikilitic peridotites are often cut by veins of amphibole (pargasite) and/or phlogopite (Supplementary Fig. 1D). These two main types of peridotite, protogranular and poikilitic, are observed in almost all the xenolith localities in the archipelago but the protogranular samples always occur in greater abundance. Very rare clinopyroxene-poor lherzolites are present (Schiano *et al.* 1994; Mattielli *et al.* 1996; Valbracht *et al.* 1996; Grégoire *et al.* 2000*a*, *b*) and they are considered to belong to the poikilitic peridotite group. The third subgroup within the Type 1 Kerguelen peridotites – dunites – essentially consists of olivine with minor spinel and clinopyroxene. These rocks show equigranular or inequigranular textures (Supplementary Fig. 1E, F). The spinel and clinopyroxene modal contents are sometimes more abundant than in harzburgites (up to 3.5 and 7%, respectively) and the clinopyroxene is Mg-augite, as in poikilitic harzburgites. Accessory minerals are primary orthopyroxene, with secondary phlogopite, amphibole and carbonate disseminated in the olivine-rich matrix (Supplementary Fig. 1G, H) or occurring in veins of plagioclase, ilmenite, rutile, sulfides and carbonate cross-cutting xenoliths.

The protogranular harzburgites have olivine Mg# (where Mg# = 100 × Mg/(Mg + Fe)) >91, clinopyroxene Mg# >92 and spinel Cr# of *c.* 40–60 (where Cr# = 100 × Cr/(Cr + Al)), which is far more refractory than primitive mantle mineral compositions (Fig. 3f). Poikilitic peridotites commonly display olivine, clinopyroxene, orthopyroxene and spinel with lower Mg# and higher Fe, Ti, Al, Na in Mg-augite and orthopyroxene. Most clinopyroxene grains in poikilitic peridotites have major element abundances lower than the primitive mantle estimate, especially in Ca, Ti and Al, but they have higher Si, Na and Mg# associated with higher Mg# in olivine and Cr# in spinel (Fig. 3). Olivine and clinopyroxene in the dunites almost always have a lower Mg# (*c.* 90–85 and *c.* 92–87, respectively) than protogranular harzburgites, and are in the range of compositions of poikilitic harzburgites or even lower (Fig. 3). Clinopyroxene grains in these rocks have higher contents of Ca and Ti, but lower Al, compared to primitive mantle. Hydrous-mineral-bearing dunites contain olivine and clinopyroxene with a very similar Mg# to anhydrous dunites; however, their clinopyroxene has a higher Na content for similar ranges in Al, Ti and Ca. The spinel content in dunites displays a different range of compositions based on the presence or absence of hydrous minerals; in anhydrous dunites, spinel has lower Cr# (*c.* 17–49) than protogranular harzburgites but similar Mg# (*c.* 82–52), whereas hydrous dunites contain spinel with lower Mg# (*c.* 62–37) and higher Cr# (*c.* 31–69) than anhydrous dunites.

Two-pyroxene geothermometry (Brey *et al.* 1990) from co-existing pyroxenes indicates that the protogranular harzburgites equilibrated at between 845 and 1005°C (assuming a pressure of 1.5 GPa: Grégoire *et al.* 2000*b*). These temperatures are lower than those of the Lac Michèle harzburgites (1050–1200°C: Wasilewski *et al.* 2017) and the poikilitic peridotites (1015–1135°C). The range of calculated equilibrium temperatures in the dunites is difficult to determine due to the general absence of orthopyroxene but, where this mineral occurs with clinopyroxene, the temperature estimates are 940–1090°C.

Bascou *et al.* (2008) interpreted the lattice-preferred orientations (LPOs) of olivine and orthopyroxene in harzburgites as reflecting a deformation regime in axial compression or transpression. They also showed that the fabric strength of olivine progressively decreases from protogranular harzburgite to poikilitic peridotite to dunite, in response to prolonged melt–rock reactions. The petrophysical parameters (P-wave velocity (V_P) and density) of one Type I poikilitic harzburgite xenolith (harzburgite GM92-453), representative of the average major element composition of peridotite mantle xenoliths from Kerguelen, were determined by Grégoire *et al.* (2001). The measured and calculated V_P at a pressure of 0.9 GPa were 8.45 and 8.29 km s^{-1}, respectively. The seismic properties calculated from the LPOs of minerals indicate that metasomatism at a high melt/rock ratio lowers the V_P, and that the most significant difference between harzburgites and dunites corresponds to the distribution of S-wave anisotropy. Thus, metasomatism by the Kerguelen plume may have induced seismic heterogeneities in the lithospheric mantle (Bascou *et al.* 2008). The measured and calculated densities (3.30 and 3.34 g cm^{-3}, respectively) indicate that the lithospheric mantle beneath the Kerguelen archipelago, assuming that it is dominated by harzburgite as indicated by the xenoliths, should therefore be relatively buoyant and corroborates geophysical studies that interpret the occurrence of relatively low-density material down to a depth of about 80 km (e.g. Charvis *et al.* 1995).

Kerguelen xenolith major and trace element geochemistry

As the main and most common host of trace elements, clinopyroxene has been analysed by laser ablation inductively coupled plasma mass spectrometry (LA-ICP-MS) in most Kerguelen peridotites types. Clinopyroxene grains in protogranular harzburgites have heterogeneous rare earth element (REE) patterns that range from spoon-shaped patterns with heavy REE (HREE) and middle REE (MREE) contents of approximately 0.2 times chondrite to light REE-enriched (LREE) contents of more than 100 times CI-chondrite (Fig. 4a). Clinopyroxenes in some Lac Michèle protogranular harzburgites have even lower MREE and LREE contents down to 0.05 times CI-chondrite (Fig. 4b). Most protogranular clinopyroxene grains display spoon-shape REE patterns (Fig. 4a, b), and are accompanied by pronounced enrichments in Rb, Th, U and Pb ± Sr but depletions in Ba, Nb, Zr and Ti (not shown). In contrast, clinopyroxene grains in poikilitic peridotites display similar REE patterns that are enriched in LREEs (Fig. 4c), as well as in other very incompatible elements (Th, U and high field strength elements (HFSEs): see Grégoire *et al.* 2000*b*). These clinopyroxenes are also more enriched in the most incompatible elements (LREEs: Fig. 4c; HFSEs or large ion lithophile elements (LILEs)) (see Grégoire *et al.* 2000*b*) compared to the protogranular harzburgites. Clinopyroxene trace element contents in anhydrous dunites are similar to those in hydrous dunites (Fig. 4d), and, as a whole, the LREE- and MREE-enriched clinopyroxenes in dunites resemble those in poikilitic peridotites. Some clinopyroxenes in anhydrous dunites have flat REE patterns or show depletions in the most incompatible elements compared to MREEs and HREEs (LREEs: Fig. 5d; also LILEs and HFSEs) (see Grégoire *et al.* 2000*a*, *b* for details). Other samples show REE patterns with large enrichments in LREEs associated with high Th and U, and low Rb, Ba and HFSEs. Clinopyroxene in the phlogopite-amphibole-bearing dunite MG91-143 shows the highest enrichment in the most incompatible elements (Th and U) and large depletions in HFSEs (Nb, Ta, Zr, Hf and Ti). Clinopyroxene associated with carbonates in dunite GM92-140 displays lower enrichments of MREEs over LREEs and HREEs, and high contents of Th and U associated with a low HFSE content (Moine *et al.* 2004).

Due to their large size, bulk rock major and trace elements have been measured in most Kerguelen peridotite types (Supplementary Tables 2B and C). The protogranular, poikilitic and Lac Michèle peridotites have refractory bulk rock major element compositions (Fig. 5) with low CaO (<1.35 wt%), Al_2O_3 (<1.5 wt%) and Na_2O (<0.25 wt%). The Mg# of the protogranular (91.5–92.0) and Lac Michèle (92.0–93.0) harzburgites are commonly higher than that of poikilitic peridotites (88.0–91.5). Protogranular harzburgites have a low REE content and display spoon-shaped REE patterns

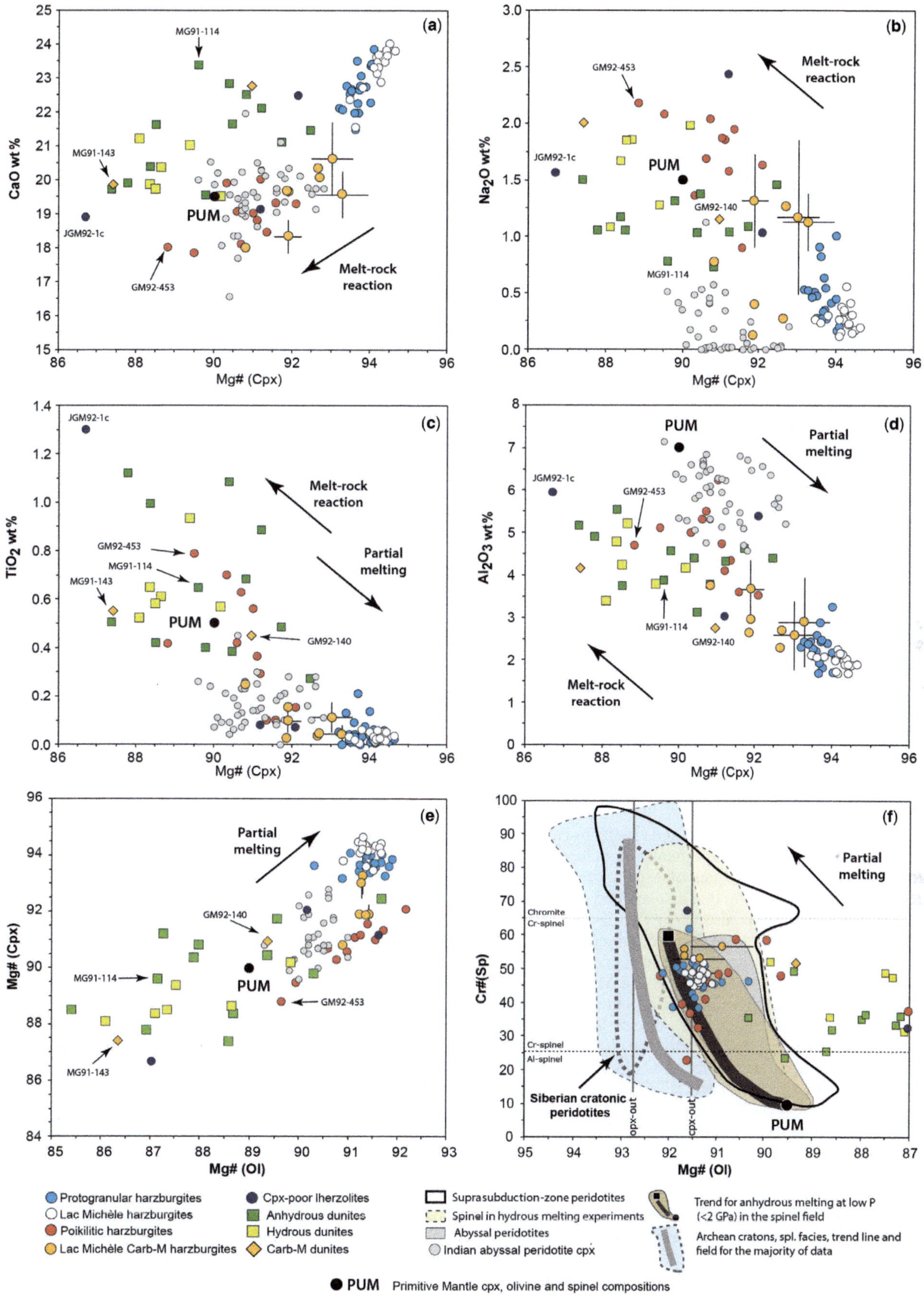

Fig. 3. Major element compositions of minerals in Kerguelen peridotite xenoliths (data after Grégoire 1994; Schiano *et al.* 1994; Grégoire *et al.* 1997, 2000*a*, *b*; Hassler 1999; Moine 2000; Moine *et al.* 2004; Delpech *et al.* 2004; Bascou *et al.* 2008; Wasilewski *et al.* 2017; this study). Ol, olivine; Cpx, clinopyroxene; Sp, spinel. Mg# (Ol, Cpx) and Cr# (Sp) are molar ratios of 100 × Mg/(Mg + Fe^{2+}) and 100 × Cr/(Cr + Al). Indian abyssal Cpx data are from Dick and Bullen (1984) and Dick (1989). PUM, primitive upper mantle estimates. Hydrous dunites contain hydrous minerals such as amphibole and/or phlogopite. Samples from this study are available in Supplementary Table 3. Carb-M samples refer to samples metasomatized by carbonatitic metasomatism (Delpech *et al.* 2004; Moine *et al.* 2004; Wasilewski *et al.* 2017). Error bars for some carbonate-rich samples show the extreme compositional major element variability in highly metasomatized samples. Spinel Cr# v. olivine Mg# and various updated fields plot from Scott *et al.* (2019).

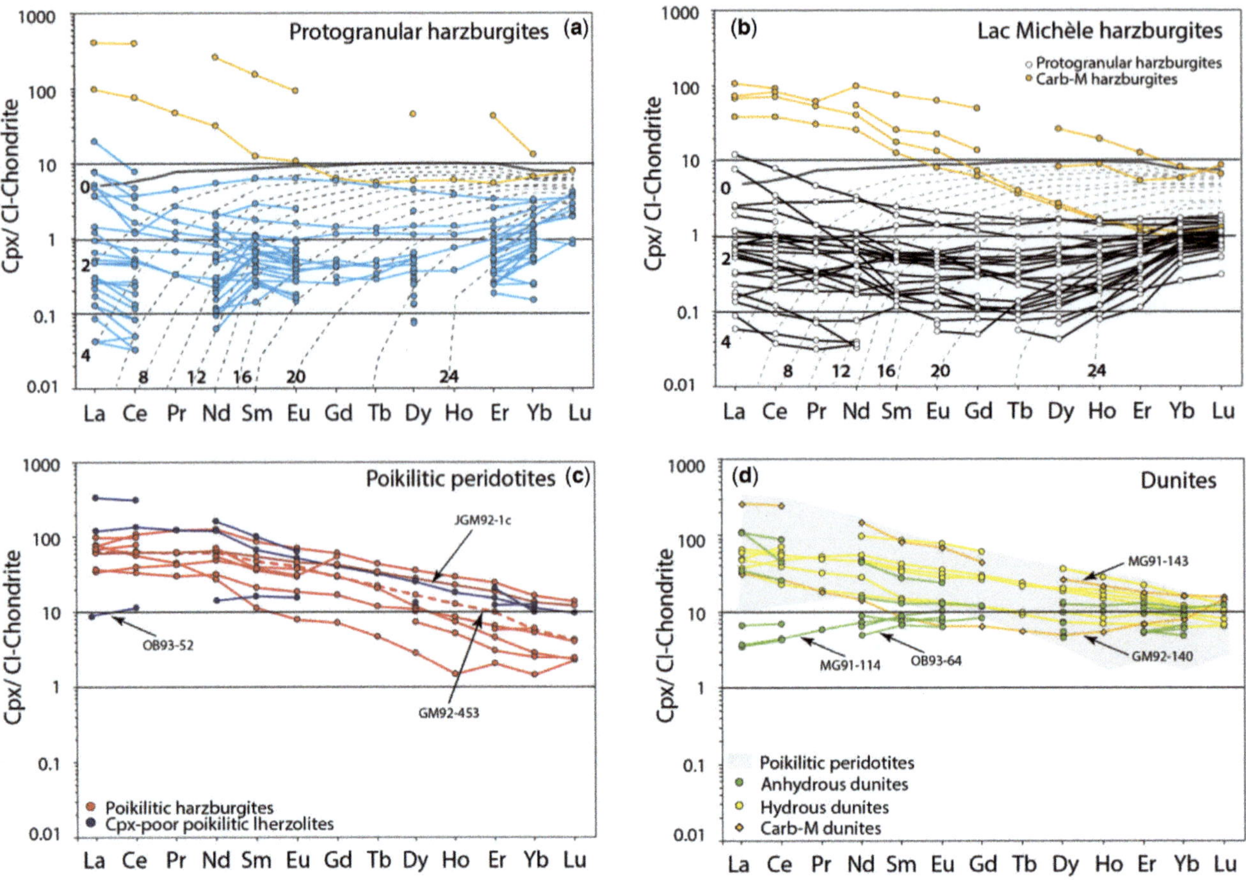

Fig. 4. Trace element and REE patterns of clinopyroxene in Kerguelen peridotite xenoliths. Data from Schiano *et al.* (1994), Mattielli *et al.* (1996, 1999), Hassler (1999), Grégoire *et al.* (2000*a*, *b*), Delpech (2004), Delpech *et al.* (2004), Moine *et al.* (2004), Bascou *et al.* (2008), Wasilewski *et al.* (2017) and this study. Samples from this study are reported in Supplementary Table 3A. Each REE spectrum is the average of several clinopyroxene (Cpx) core analyses, except for a few individual analyses from Hassler (1999). Carb-M samples refer to samples metasomatized by carbonatitic metasomatism (Delpech *et al.* 2004; Moine *et al.* 2004; Wasilewski *et al.* 2017). Melting curves were taken from Scott *et al.* (2016) and CI-chondrite values from McDonough and Sun (1995).

characterized by $(La/Sm)_N = 3.5–15$ and $(Sm/Yb)_N$ as low as 0.2 (subscript N indicated that the measured values are normalized to primitive-mantle values of McDonough and Sun 1995). Some of the harzburgites that display modal metasomatism by carbonate-rich melts also show LREE-enriched patterns $[La/Sm]_N$ varying between 2 and 6, which is similar to the poikilitic harzburgites (Fig. 4b) (Delpech *et al.* 2004). Poikilitic harzburgites have enriched LREE and MREE contents compared to HREEs, and their total REE contents are higher than protogranular harzburgites (Fig. 4c; see Grégoire *et al.* 2000*b*) with $(La/Sm)_N = 1.7–10.9$ and $(Sm/Yb)_N = 2.5–4.3$.

Dunite samples tend to exhibit low SiO_2 for a similar range in MgO compared to harzburgites (Fig. 5a), and low CaO (0.18–1.6 wt%), Na_2O (<0.27 wt%), Al_2O_3 (0.15–1.91 wt%: Fig. 5b), TiO_2 (0.03–0.09 wt%) contents but their Mg# (85–89, average of 88) is lower than most of the harzburgites (Fig. 5c). Dunite CaO and Al_2O_3 contents are sometimes higher than those in harzburgites (Fig. 5c), which reflects the relatively high modal content of clinopyroxene and spinel (Grégoire 1994). The dunites are characterized by variable REE shapes compared to harzburgites, with a combination of almost flat REE patterns, LREE-enriched patterns and upward-convex REE patterns (not shown: see Grégoire *et al.* 2000*b*).

Kerguelen peridotite isotopic compositions

Bulk-rock Os isotopic compositions from a restricted suite of harzburgite, lherzolite and dunite xenoliths ($^{187}Os/^{188}Os = 0.1189–0.1383$; $n = 19$: Hassler 1999; Hassler and Shimizu 1998) extend from unradiogenic values (0.1189) with rhenium-depletion model ages (T_{RD}) as old as 1.36 Ga to compositions more radiogenic than primitive upper mantle (PUM) (0.1296: Meisel *et al.* 2001). The radiogenic values are similar to Os isotope compositions of Kerguelen basalts, and may therefore represent the signature of the mantle accreted by the Kerguelen plume (Weis *et al.* 2000).

Sr–Nd ± Pb isotopes analysed for bulk-rock ($n = 13$) and clinopyroxene separates ($n = 10$) from harzburgites, dunites, two lherzolites and one clinopyroxenite show heterogeneous compositions (Fig. 6) (Mattielli *et al.* 1996, 1999; Hassler 1999). Most of the peridotite xenoliths have isotopic compositions in the range $^{87}Sr/^{86}Sr = 0.7050–0.7065$, $^{143}Nd/^{144}Nd = 0.5123–0.5127$ and $^{206}Pb/^{204}Pb = 18.0–18.5$; $^{207}Pb/^{204}Pb = 15.50–15.65$; $^{208}Pb/^{204}Pb = 38.5–39.2$. These values are distinct from basalts of the Southeast Indian Ridge, and rather similar to those of Kerguelen archipelago lavas or even more enriched compositions (Mattielli *et al.* 1996, 1999; Hassler 1999). Peridotites showing fingerprints of metasomatism by carbonatitic melts seem to extend to more enriched Sr–Nd isotopic compositions than those metasomatized by alkaline basaltic silicate melts. Their Sr–Nd isotope signature is similar to young (<10 Ma) Kerguelen volcanic rocks, and is comparable to the least evolved Heard Island lavas (Barling *et al.* 1994) (Fig. 6). A few samples show extreme Sr–Nd isotopic compositions; the most 'depleted' end of the spectrum is represented by one little-metasomatized harzburgite that has clinopyroxene with unradiogenic $^{87}Sr/^{86}Sr$ of 0.70329 (OB93-78: Hassler 1999), which falls in the field of Indian mid-ocean ridge basalt (MORB) (Fig. 6). Two samples (one anhydrous dunite and one

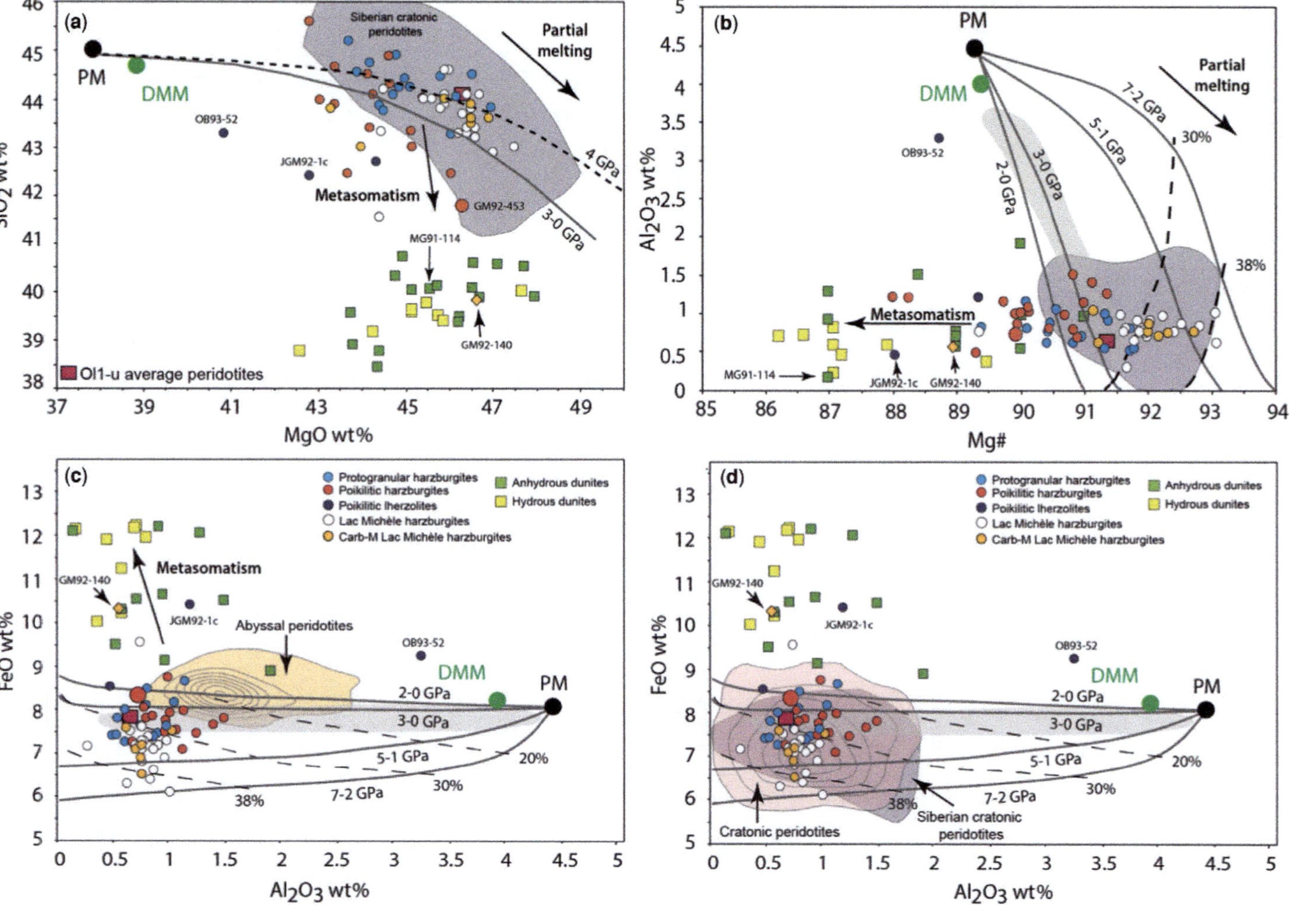

Fig. 5. Major element variations in bulk-rock peridotites xenoliths from the Kerguelen Islands (after Grégoire 1994; Grégoire *et al.* 1997, 2000*b*; Hassler 1999; Moine *et al.* 2000; this study). Samples from this study are available in Supplementary Table 3. Carb-M Lac Michèle harzburgites refer to samples metasomatized by carbonatitic metasomatism (Delpech *et al.* 2004; Wasilewski *et al.* 2017). The primitive mantle (PM) estimate is from McDonough and Sun (1995) and the depleted MORB mantle (DMM) estimate is from Workman and Hart (2005). Grey field: Horoman peridotites (Takazawa *et al.* 2000). Continuous black lines are residues of polybaric fractional melting at 2.0, 3. 0, 5.1 and 7.2 GPa (Herzberg 2004); thick dashed black lines correspond to 30 and 38% of polybaric fractional melting. Abyssal ($n = 446$) and cratonic ($n = 250$) peridotite density contours are from Wasilewski *et al.* (2017). Siberian cratonic peridotites are fertile off-craton garnet and spinel peridotite xenoliths from central Asia (see Wasilewski *et al.* 2017 for details). Average Ol1-u bulk peridotites (ultra-refractory oceanic island peridotites) are from Simon *et al.* (2008).

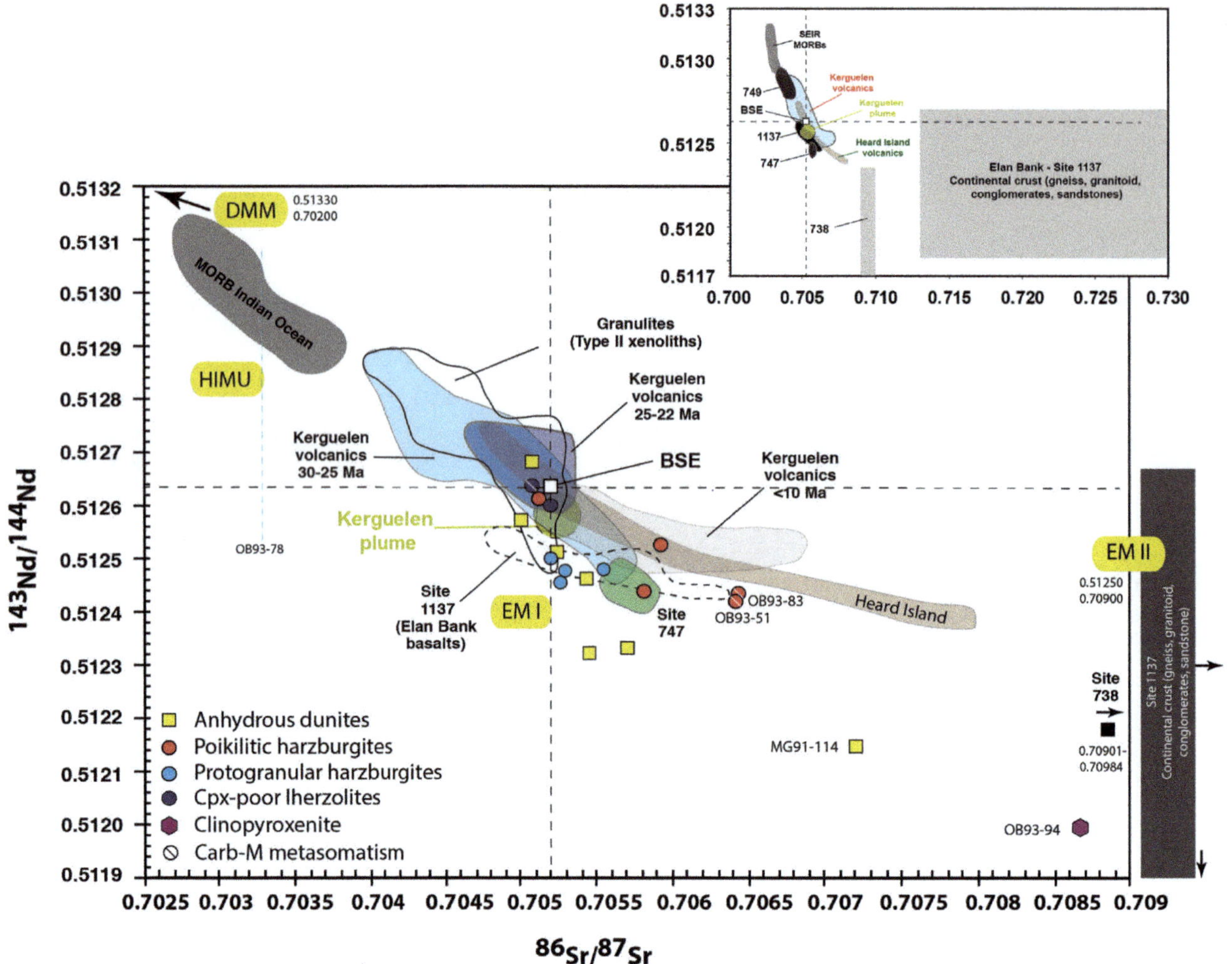

Fig. 6. Present-day Sr–Nd isotopic compositions for the Kerguelen xenolith suite. Analyses correspond either to bulk-rock or clinopyroxene separates (Mattielli *et al.* 1996, 1999; Hassler 1999). The fields for Kerguelen Type II xenoliths (granulites) are from Mattielli *et al.* (1996, 1999) and Hassler (1999). Fields for Kerguelen and Heard volcanic rocks and ODP Leg 183 (Site 1137 – Elan Bank), 120 (Site 747) and 119 (Site 738) are after http://georoc.mpch-mainz.gwdg.de/georoc/. Fields for sites 1137, 747 and 738 indicate plume-derived Cretaceous Kerguelen plateau volcanics with trace element and isotopic compositions indicative of contamination by continental-lithosphere-derived magmas, these geochemical fingerprints have not yet been found in Kerguelen Islands basalts. MORB, mid-ocean ridge basalt; SEIR MORB, South East Indian Ridge MORB; HIMU, high μ (high time-integrated U/Pb material); EM I and EM II are enriched mantle components I and II, BSE; Bulk Silicate Earth, see Zindler and Hart (1986) for details.

phlogopite (Phl)-bearing clinopyroxenite: Hassler 1999; Mattielli *et al.* 1999) show very radiogenic $^{87}Sr/^{86}Sr$ (0.70730 and 0.70869) and unradiogenic $^{143}Nd/^{144}Nd$ (0.51215 and 0.51199), and are isotopically very different from the Kerguelen basalts as they extend towards Sr–Nd isotopes found in continental crust material recovered at Elan Bank (Fig. 6; Frey *et al.* 2002) or volcanics highly contaminated by continental crust material (Site 738, Fig. 6). Such Sr–Nd isotope signatures indicate that some xenoliths interacted with, or were formed from, very evolved metasomatic melts or fluids percolating in the lithospheric mantle; either of plume origin or by interaction with recycled continental lithosphere (Mattielli *et al.* 1999). It has been argued in the literature that the typical enriched mantle 1 (EM1) isotope signature of the Kerguelen basalts reflect that of melts produced by the Kerguelen plume (Weis *et al.* 1993).

Synthesis of the Kerguelen peridotitic mantle lithosphere

The Type I Kerguelen xenoliths show the peridotitic lithospheric mantle beneath this archipelago to be largely a refractory domain. The very low clinopyroxene volume percentage, olivine Mg# up to 92 and Cr-rich spinel (Cr# >38), low HREE content and MREE in some clinopyroxenes, high bulk-rock MgO content, and low CaO, Al_2O_3, Na_2O, Fe_2O_3, S and platinum group element (PGE) content all indicate that the underlying mantle experienced partial melting in excess of 15–25% close to or beyond the exhaustion of clinopyroxene (Fig. 4) (Grégoire *et al.* 1997, 2000*a*, *b*; Lorand *et al.* 2004). Clinopyroxenes with moderate HREE contents in Figure 4a and b are consistent with melting in the spinel stability field at low pressure (<2 GPa); however, clinopyroxenes in Figure 4a and b with low (HREE/HREE) ratios (eg. Dy/Yb) and the lowest HREE (Er, Yb, Lu) contents, lower than those predicted by 24% melting at low pressure, may be explained by polybaric melting initiated at higher pressure (>2 GPa). Based on bulk-rock trace element modelling using melting equations of Walter (1998), Hassler (1999) showed that the REE abundances of a residual clinopyroxene after polybaric melting starting at 3 GPa and terminating at low pressure in the spinel facies is similar to those of clinopyroxenes in protogranular samples with the lowest HREE content (Fig. 4a). The most depleted clinopyroxenes in the Lac Michèle harzburgites (Fig. 4b), characterized by low HREE/HREE (e.g. Dy/Yb) ratios and low HREE contents, can also be explained in a similar way. This idea is supported by Wasilewski *et al.* (2017), who also interpreted the more refractory bulk compositions of the Lac Michèle harzburgites to record evidence of polybaric decompression fractional melting between 5 and 3 GPa, initiated at high pressure in the garnet stability field (Fig. 5b, d). The Kerguelen peridotites have petrographical, mineralogical and geochemical compositions that are different from abyssal peridotites (Fig. 5b) or the local Indian oceanic lithosphere (Figs 3 and 6) but are comparable to oceanic island peridotites worldwide (Fig. 5) (Simon *et al.* 2008). These authors suggested that depleted and, therefore, buoyant peridotite domains (Ol1-u peridotites of Simon *et al.* 2008), such as Kerguelen peridotites, were accreted to the oceanic lithosphere from the convecting mantle by the ascending plume. Simon *et al.* (2008) proposed that reworking of older oceanic or subduction-related mantle domains from the convecting mantle is also consistent with the occurrence of some rhenium-depletion model ages far older than the Kerguelen lithosphere. Conversely, Hassler and Shimizu (1998) interpreted these Proterozoic rhenium-depletion model ages as evidence for the occurrence of fragments of old subcontinental mantle domains that were incorporated into the Indian oceanic lithosphere during the Gondwana break-up. Based on their modal and bulk major element compositions, which resemble cratonic peridotites (Fig. 5a, b, d), the Lac Michèle harzburgites were considered by Wasilewski *et al.* (2017) in the same manner as in Hassler and Shimizu (1998). However, the low Mg# in olivine (91–92) for a given Cr# in spinel (Fig. 3f; or modal olivine contents in bulk rocks – not shown) compared to cratonic peridotites (Fig. 5) and the lack of Os isotope data for these peridotites do not yet allow a definite interpretation to made of their origin, especially considering that the existing Os isotope data on Kerguelen are within the overall range of data found in modern oceanic lithospheric mantle (Pearson *et al.* 2007; Chatterjee and Lassiter 2016; Day *et al.* 2017) or in oceanic plume settings (Simon *et al.* 2008).

The mineralogy and geochemistry of Kerguelen Type I peridotite xenoliths also require that the refractory mantle has experienced extensive post-depletion melt–rock reaction between ascending magmas (e.g. Schiano *et al.* 1994; Mattielli *et al.* 1996, 1999; Grégoire *et al.* 1997, 2000*a*, *b*; Moine *et al.* 2001, 2004; Delpech *et al.* 2004, 2012; Lorand *et al.* 2004). For instance, the occurrence of metasomatic minerals in peridotites (amphibole, phlogopite, carbonate and apatite) and the enrichment in LREEs (and very incompatible trace elements), Na, Fe, Ti and Al in some clinopyroxenes are indicative of melt–rock reactions (Figs 3 and 4). The metasomatic fingerprint of the Kerguelen protogranular harzburgites and poikilitic peridotites may result from the circulation of primary CO_2-bearing alkaline to high-alkaline silicate melts within the upper mantle closely linked to the magmatic activity of the Kerguelen plume (Grégoire *et al.* 2000*a*, *b*; Lorand *et al.* 2004; Delpech *et al.* 2012). The Sr–Nd–Pb isotopic characteristics of the mantle xenolith suite, similar to the Kerguelen basaltic lavas (Mattielli *et al.* 1996, 1999; Hassler 1999), indicate metasomatism by ascending plume-derived magmas on their way to the surface. Wall-rock interaction in the mantle with such ascending magmas at low to high melt/rock ratios resulted in different xenolith types (Grégoire *et al.* 1997, 2000*b*). Cryptic metasomatism at low melt/rock ratios affected the refractory protogranular harzburgites and caused a slight enrichment in the most incompatible trace elements (e.g. Fig. 4a, b). Poikilitic peridotites, however, experienced metasomatism at a higher melt/rock ratio, causing olivine and Mg-augite precipitation at the expense of orthopyroxene. As a result of the interaction with Fe-bearing alkaline silicate melts at high melt/rock ratios, minerals in the poikilitic rocks have lower Mg#, and their clinopyroxene has higher Al, Ti, Na and incompatible trace elements contents, but commonly less Ca than those of protogranular harzburgites (Fig. 3). Additionally, some of these Mg-augite grains also have trace element contents in near-equilibrium with alkaline basaltic silicate magmas erupted at the surface, supporting a genetic link with such magmas.

The dunites are considered to represent end products of the reaction of harzburgite with fluids at a high melt/rock ratio, causing total resorption of orthopyroxene, and crystallization of olivine and secondary clinopyroxene (Grégoire *et al.* 1997, 2000*b*). The common occurrence of composite xenoliths where dunite is the wall rock of small dykes of websterite, clinopyroxenite or hornblendite attest to a genetic relationship between dunite formation and magma percolation. Dunites with or without veins have a similar range of compositions, suggesting that all the Kerguelen dunites are end products of such reaction processes between a former harzburgitic protolith and a basaltic silicate melt (Grégoire *et al.* 1997, 2000*b*). This evolution is indicated in Figure 3 by the progressive decrease of Mg# of minerals in anhydrous or hydrous dunites, associated with increasing modal olivine contents, and an enrichment of Al, Ti and Na in clinopyroxene (Grégoire *et al.* 1997, 2000*b*). Some clinopyroxenes in dunites have trace element contents (Fig. 4) that are in near-equilibrium with various type of magmas erupted at the surface; ranging from early tholeiitic-transitional magmas to younger alkaline or highly alkaline magmas (see Grégoire *et al.* 1997, 2000*b*; Hassler 1999; Moine 2000). This is supported by their very variable Sr–Nd isotopes (Fig. 6) that cover a wide spectrum of Kerguelen volcanics but also extend well outside the field for Kerguelen volcanics (30–0 Ma) towards very evolved Sr–Nd isotopic compositions for a few samples. Mattielli *et al.* (1999) argued that the evolved Sr–Nd (and low Pb) signature of dunite MG91-114 (Fig. 6) reflects mixing of plume melts with melts derived from recycled continental crustal material. The radiogenic Os isotope compositions of wherlitic dunites (Hassler and Shimizu 1998) are, however, similar to the Os isotopes of the Kerguelen basalts. Hence, dunite bodies probably formed in the lithospheric mantle at different times in the history of the Kerguelen Islands as the geodynamic setting changed; their mineralogical and geochemical compositions show that they essentially equilibrated with metasomatic melts/fluids derived from the Kerguelen plume. The occurrence of hydrous minerals in dunites such as phlogopite and/or amphibole, as in some harzburgites, has been attributed to younger metasomatic events by small volumes of fluid-enriched alkaline or highly alkaline magmas, such as those forming the young lamprophyres at the surface (Hassler 1999; Grégoire *et al.* 2000*b*; Moine *et al.* 2001). For instance, Moine *et al.* (2001) showed that disseminated amphibole in dunitic wall rocks of hornblendite veins have geochemical compositions genetically related to those in the hornblendite veins; the latter having a trace element composition similar to young ultramafic silica-undersaturated highly alkaline lavas from Kerguelen (Moine *et al.* 2001). The occurrence of interstitial Mg-bearing calcites within pockets disseminated in some dunites (Moine *et al.* 2004; GM92-140: Fig. 4d; see also Supplementary Fig. 1G) that are very enriched in the most incompatible trace elements may reflect metasomatism by small melt fractions of carbonate-rich melts shortly before eruption of the xenoliths. In some metasomatized harzburgites, carbonate-rich, alkaline silicate-rich and CO_2 inclusions have been found physically connected forming trails along fracture planes, and this led Schiano *et al.* (1994) to suggest a genetic relationship between both types of metasomatic fluids. It may be postulated that the original fluid-bearing alkaline basaltic silicate melts formed by the Kerguelen plume that metasomatized the lithospheric mantle and formed poikilitic peridotites and dunites evolved into small volumes of volatile-rich melts following extensive percolation–reaction–crystallization processes. In this scenario, carbonate-rich melts/fluids may be formed by continuous reaction of an originally volatile-bearing alkaline silicate melt and do not require melting of a specific mantle source.

The Kerguelen lithospheric mantle records a complex and multi-stage evolution. The peridotite xenoliths indicate that they were formed by a high degree of melting, possibly in the garnet stability field. Their mineralogical and geochemical compositions were later sometimes strongly modified by the circulation of metasomatic fluids originating from the Kerguelen plume.

Subantarctic mantle under the Southern Ocean

There are three known occurrences of subantarctic mantle material exhumed in the Southern Ocean: the Auckland Islands, Macquarie Island and Campbell Island (Fig. 1; Table 1; see also Supplementary Table 3). The Auckland Islands comprise two intersecting glaciated Early–Middle Miocene intraplate shield volcanoes, the Ross and Carnley volcanoes, located just east of the steep descent to the adjacent oceanic-lithosphere-floored Eocene–Oligocene Emerald Basin (Fig. 7) (Scott and Turnbull 2019). The island group is dominated by Early Miocene basaltic–rhyolitic lava flows, tuffs, dykes, and basaltic and gabbroic plugs that erupted through the Zealandia continental crust (Wright 1967, 1968; Gamble *et al.* 2018; Scott and Turnbull 2019). Peridotite xenoliths have been collected and analysed from one location, Mount Eden, in the northern Ross Volcano (Scott *et al.* 2014*b*, 2019). Macquarie Island, on the other hand, is a slice of exhumed Miocene oceanic lithosphere located on the SW of the Emerald Basin on the transform boundary separating the Australian and Pacific plates (Fig. 7). It represents young oceanic lithosphere that formed after the 84 Ma Cretaceous separation of Zealandia and Australia. It is dominated by gabbro and mafic dykes, with serpentinized peridotite occurring in the northern portion (Goscombe and Everard 1998; Wertz 2003; Dijkstra *et al.* 2010). No peridotite has been found on the Campbell Island volcano, although coarse detrital xenocrystic zircon grains have been panned from clays. Trace elements, oxygen isotopes, and U–Pb and Hf isotopes indicate that these grains were derived from metasomatized mantle (van der Meer *et al.* 2019).

Auckland Islands peridotites

The inspected Auckland Islands mantle xenoliths are very fresh lherzolite and harzburgite, with the spinel textures, mineral chemistries and calculated equilibration indicating there to be two prominent types, herein referred to as Type 1 and Type 2. The Type 1 xenoliths have coarse textures (Harte 1977; Supplementary Fig. 2) except for the porphyroclastic AMED-3. Average olivine compositions for Type 1 peridotites range from Mg# = 89.8 to Mg# = 90.7. Their spinel grains form blebs or holly-leaf shapes (e.g. Supplementary Fig. 2A) and have Cr# = 10.4–16.1 (Supplementary Table 3A). In contrast to the Type 1 rocks, Type 2 xenoliths have a higher olivine Mg#, with average compositions ranging from 90.7–91.3. Spinel grains also have higher Cr# (22.9–31. 5) than the Type 1 peridotites, and form symplectitic textures with pyroxene (Supplementary Fig. 2B). There is one exception that does not fit either the Type 1 or Type 2 classification: AMED-11, which has an olivine Mg# of *c.* 87.5. Orthopyroxene in all Auckland Islands peridotites is zoned, with MgO decreasing, and Al_2O_3 and CaO increasing, towards the rims. Clinopyroxene CaO and Al_2O_3 decrease towards the rims, whereas MgO and FeO tend to increase. The rim-core trends in the Type 2 xenoliths are not quite as clear in clinopyroxene but, nonetheless, there is an increase in CaO (AMED-4 and AMED-5) and Al_2O_3 (AMED-5 and AMED-6). Water content measured by Fourier transform infrared spectrometry for olivine (calculated to be 6–26 ppm), clinopyroxene (60–264 ppm) and orthopyroxene (28–125 ppm) indicate the peridotites to be fairly dry (Li *et al.* 2018).

Geothermometric calculations reveal a complex cooling history and a stratification to the Auckland Islands mantle. Using the Taylor (1998) Fe–Mg exchange geothermometer (with temperatures calculated assuming pressure of 15 kbar, although varying the pressure by 5 kbar makes <30°C difference) on the cores of adjacent orthopyroxene and clinopyroxene grains shows the Type 1 peridotites to yield temperatures <950°C, whereas the Type 2 peridotites have temperatures >1050°C (Supplementary Table 3B). AMED-10, which has Type 1 chemistry, yields an intermediate temperature (1023°C), and the chemically anomalous AMED-11 has a high temperature (1134°C). If the Auckland Islands xenoliths were extracted along a single geotherm, then: (1) the Type 1 peridotites are from a shallower depth than the Type 2 peridotites; and (2) the large (*c.* 300°C) spread in temperatures for rocks that all belong to the spinel facies requires a heat flow of *c.* 70 mW m^{-2} (Fig. 8). This is similar to the Late Oligocene–Miocene heat flow calculated for elsewhere in Zealandia (Scott *et al.* 2014*a*, *b*). However, since clinopyroxene incorporates more CaO and less MgO at higher temperatures (Bertrand and Mercier 1985), the orthopyroxene Ca-rimward increase and clinopyroxene Ca-rimward decrease in adjacent grains in the Auckland Islands suite indicates that the mantle column also experienced a small temperature increase. Orthopyroxene analyses that fall within the compositional

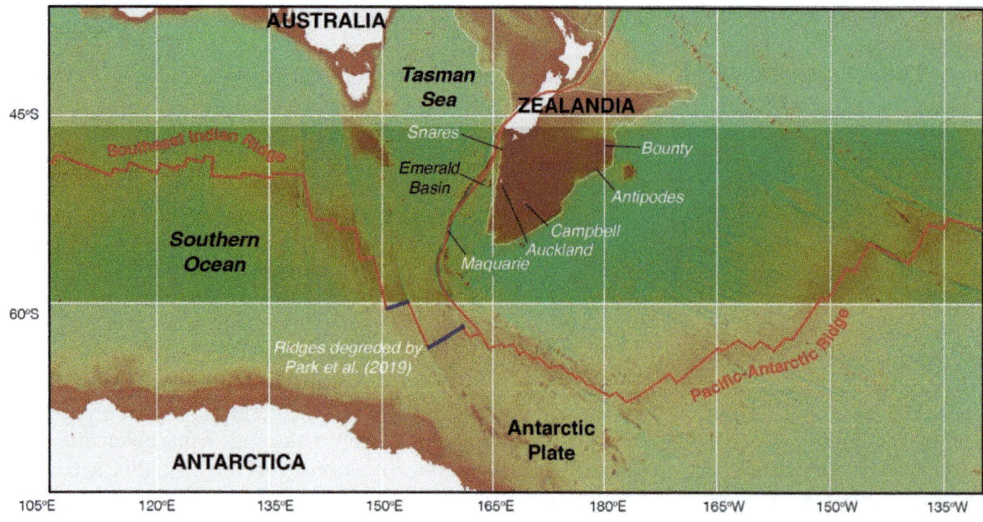

Fig. 7. Bathymetric map showing the Southern Ocean subantarctic and features discussed in the text.

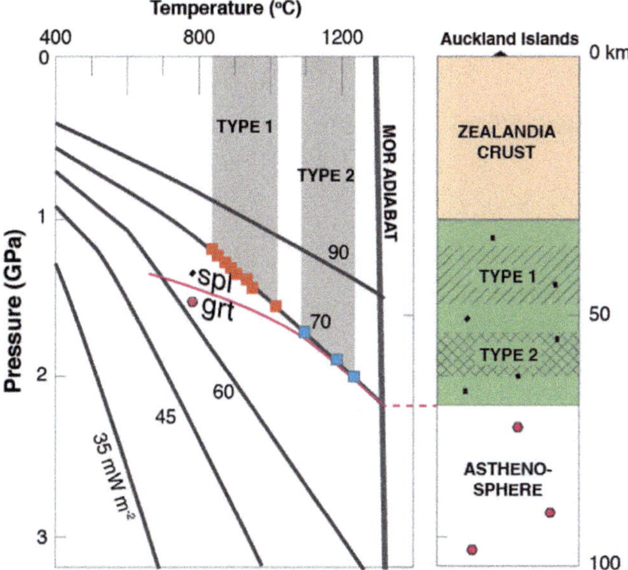

Fig. 8. Auckland Islands peridotite equilibration temperatures, projected onto the 70 mW m^{-2} geotherm of Hasterok and Chapman (2011). The spinel–garnet transition is for lherzolite (Klemme and O'Neill 2000) and a more Cr-rich bulk composition would increase the pressure of the transition. The column on the right indicates the inferred lithosphere thickness underneath the Auckland Islands at the Miocene time of xenolith entrainment.

bounds of the Witt-Eickschen and Seck (1991) Cr–Al–orthopyroxene geothermometer (AMED-6 and AMED-7; the other samples tend to have too much Al in the M1 site) indicate that the pyroxene rims record a temperature rise of up to 80°C. Although the changes could be due to heating of the xenoliths in the host magma, this would not have promoted such large diffusion profiles since the element exchange of Ca in clinopyroxene is slow (e.g. a 0.15 mm-wide profile at 1000°C would take more than 1 kyr: Zhang *et al.* 2010; Dalton *et al.* 2017).

Trace elements from the cores of clinopyroxene grains also support the interpretation of a stratified mantle beneath the Auckland Islands. The Type 1 clinopyroxenes form a tight compositional cluster with HREE concentrations around 10 times chondrite but low concentrations of LREE (Fig. 9a). When compared with theoretical compositions calculated for different degrees of melting (Scott *et al.* 2016), the Type 1 data deviate little from the area of low degree melting. In comparison, the clinopyroxenes from hotter Type 2 xenoliths have distinctly lower HREE contents, with the MREEs and LREEs showing a dramatic departure from the melting curves, which is consistent with enrichment of depleted residues by a LREE-bearing fluid (Fig. 9b). Elsewhere in Zealandia, this type of enrichment, coupled with Ti/Eu ratios, has been interpreted to result from Mesozoic carbonatitic or CO_2-rich silicate fluids (Scott *et al.* 2014*a*, *b*; McCoy-West *et al.* 2015, 2016; Scott *et al.* 2016; Dalton *et al.* 2017).

In an adiabatic melting column, the most depleted residues should be those that undergo the most decompression melting and are thus the shallowest in the end process of melting. This logic makes the configuration of a less-depleted mantle domain (Type 1) residing above a highly depleted mantle domain (Type 2) unusual, and would require some form of tectonic juxtaposition. An added complication is that the pyroxenes from both appear to record a slight temperature increase that is likely to predate the xenolith entrainment. This thermal increase may be due to the rifting and formation of oceanic lithosphere in the adjacent Emerald Basin (Fig. 7), which occurred in the Eocene–Oligocene before the intraplate Auckland Islands formed.

Macquarie Island peridotite

In contrast to the fresh Auckland Islands peridotites, the freshest Macquarie Island peridotites at best retain up to 60% primary minerals and most are thoroughly serpentinized (Wertz 2003; Dijkstra *et al.* 2010) (see Supplementary Fig. 2C). Olivine and orthopyroxene are commonly partially to totally altered to serpentine or bastite, respectively, with abundant magnetite and minor talc, amphibole and carbonate occurring as secondary phases. Although Dijkstra *et al.* (2010) showed images of relict olivine grains amongst serpentine, they provided no chemical analyses, and our own samples and those of Wertz (2003) have no olivine preserved. However, the Cr-rich nature of spinel (Cr# = 39–49: Wertz 2003; Dijkstra *et al.* 2010) indicates that these underwent moderate levels of melt extraction, which is consistent with the very low clinopyroxene modal percentage (<2%); the peridotites were therefore likely to have been harzburgitic prior to alteration. This interpretation is supported by the HREE concentrations of clinopyroxene grains (Dijkstra *et al.* 2010), which when compared to theoretical depleted clinopyroxene compositions indicate that the peridotite experienced from 20 to >25% partial melt depletion (Fig. 9c). However, the clinopyroxene analyses also reveal that the Macquarie Island mantle has been subsequently enriched in LREEs, with compositions varying widely within single samples (Wertz 2003; Dijkstra *et al.* 2010) (Fig. 9c). Equilibrium temperatures have not been calculated for the Macquarie Island peridotites, as it is not clear if the clinopyroxene and orthopyroxene grains are in equilibrium. Nonetheless, the rimward decrease in Al_2O_3 in orthopyroxene (Dijkstra *et al.* 2010) and diffusion rates of Al in orthopyroxene may mean that the mantle experienced slow cooling.

Auckland Islands and Macquarie Island isotopes

Os isotopes. Osmium isotopes are reported for 15 Auckland Islands peridotites, for which seven were reported by Scott *et al.* (2019) and eight are new (all are summarized in Supplementary Table 3B). The Auckland Islands Type 1 peridotites have $^{187}Os/^{188}Os$ values of 0.1222–0.1299 with Re-depletion modal ages (T_{RD}) ranging up to 0.87 Ga but clustering at *c.* 0.4 Ga (Fig. 10a). The Type 2 peridotites extend to slightly less radiogenic Os values than the Type 1 xenoliths, with $^{187}Os/^{188}Os = 0.1183–0.1248$ and Re-depletion Os model ages of 1.58–0.68 Ga. These results overlap with the Os isotope and model ages of other suites of peridotite erupted through the Zealandia continental lithosphere (Fig. 10a) (McCoy-West *et al.* 2013; Liu *et al.* 2015; Scott *et al.* 2019), although both Auckland Islands suites are at the more fertile Al_2O_3 end of the Zealandia spectrum (Fig. 10b). The Auckland Islands $^{187}Os/^{188}Os$ data are very similar to the oceanic lithospheric mantle associated with Kerguelen, the Dun Mountain Ophiolite Belt and modern abyssal peridotite (Fig. 10a). PGEs show that many of the Auckland Islands Type 1 xenoliths and one of the Type 2 xenoliths have higher Pd/Ir values than PUM (Fig. 10c; see also Supplementary Table 3B), which indicate that the PGEs in these samples have been disturbed via metasomatism. These Type 1 data are, however, similar to the ultramafic portion of the Dun Mountain Ophiolite Belt (Fig. 10a, b) (Scott *et al.* 2019), which is a Permian oceanic lithosphere accreted to Zealandia.

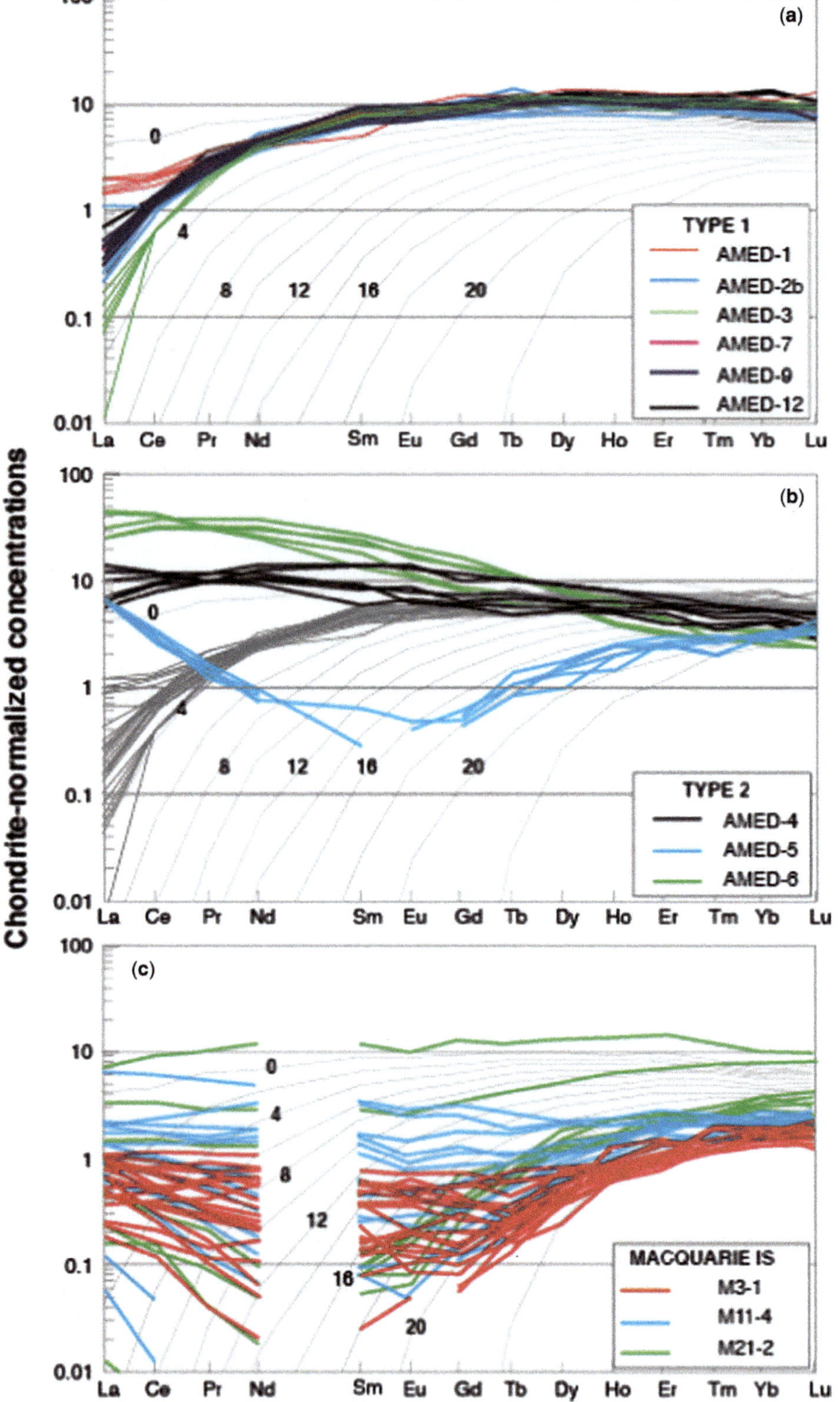

Fig. 9. Auckland Islands Type 1 (**a**) and Type 2 (**b**) and Macquarie Island (**c**) peridotite clinopyroxene REE data plotted against theoretical clinopyroxene compositions for different percentages of bulk peridotite melting. Auckland Islands data are from Scott *et al.* (2014*b*), and Macquarie Island data are from Dijkstra *et al.* (2010). Theoretical clinopyroxene compositions are from Scott *et al.* (2016). CI-chondrite is from Sun and McDonough (1989).

Despite Macquarie Island having a Miocene lithosphere stabilization age, the peridotites have $^{187}Os/^{188}Os = 0.1194–0.1227$ and ancient Re-depletion Os model ages of 1.23–0.73 Ga (Dijkstra *et al.* 2010; Fig. 10a; see also Supplementary Table 3B). These data are similar to Kerguelen, Auckland Islands Type 2 and abyssal peridotite data. An implication of the Auckland Islands, Kerguelen Islands and Macquarie Island Os isotope data is therefore that subantarctic mantle probably contains significant amounts of peridotite that is hundreds of millions to billions of years older than the overlying crust. Since peridotite on Macquarie Island represents decompressed asthenosphere accreted to form lithosphere, these anomalies must also be present in the convecting mantle in the Subantarctic.

Sr–Nd–Pb and Hf isotopes on the Auckland Islands or Macquarie Island are restricted to reconnaissance studies. A small dataset of four clinopyroxene $^{87}Sr/^{86}Sr$, $^{143}Nd/^{144}Nd$ and $^{206}Pb/^{204}Pb$, $^{207}Pb/^{204}Pb$ and $^{208}Pb/^{204}Pb$, and two $^{176}Hf/^{177}Hf$ analyses was described from the Auckland Islands by Scott *et al.* (2014*b*). The lowest $^{87}Sr/^{86}Sr$ (0.70231) and highest $^{143}Nd/^{144}Nd$ (0.51332) values occur in the Type 1 AMED-7 (see Supplementary Table 3B),

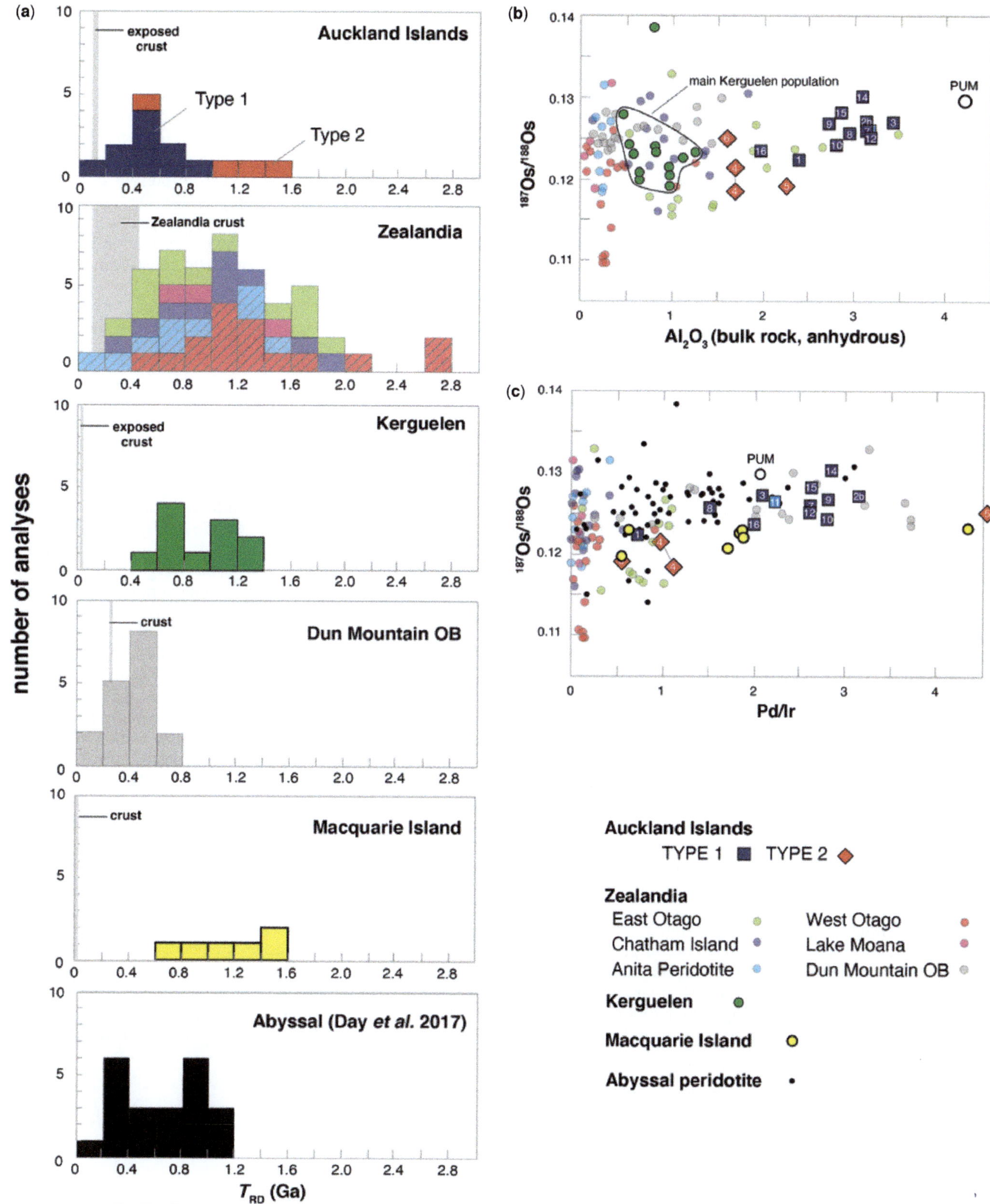

Fig. 10. Summaries of T_{RD} ages for subantarctic mantle peridotites. (**a**) Histograms for different locations. (**b**) Auckland Islands and Macquarie Island osmium v. bulk-rock Al_2O_3 and/or Pd/Ir. The Zealandia data are from McCoy-West *et al.* (2013), Liu *et al.* (2015) and Scott *et al.* (2019); the Kerguelen data are from Hassler (1999) and Hassler and Shimizu (1998); the Dun Mountain Ophiolite Belt (OB) data are from Scott *et al.* (2019); Macquarie Island data are from Dijkstra *et al.* (2010); and the abyssal peridotite data are from Day *et al.* (2017). The primitive upper mantle (PUM) value is from Meisel *et al.* (2001).

whereas the three analysed Type 2 xenoliths have slightly more radiogenic $^{87}Sr/^{86}Sr$ (0.70282–0.70293) and variable $^{143}Nd/^{144}Nd$ (0.51294–0.51325). Due to the low concentrations, Pb isotopes have only been obtained from clinopyroxene separates in the Type 2 rocks. These yielded $^{206}Pb/^{204}Pb = 19.5–20.2$, $^{207}Pb/^{204}Pb = 15.6–15.7$ and $^{208}Pb/^{204}Pb = 39.1–40.0$. ε_{Hf} data show Type 1 AMED-7 to have $\varepsilon_{Hf}(16\ Ma) = +22$ and Type 2 AMED-5 to have extremely radiogenic $\varepsilon_{Hf} = +85$, with this latter value being decoupled from the Nd and Sr isotopic record but within the range of abyssal peridotites from modern oceanic lithosphere (e.g. Stracke *et al.* 2011). The clinopyroxene Sr and Nd isotopes for clinopyroxene from six Macquarie Island peridotites fall between $^{87}Sr/^{86}Sr$ values of 0.70256–0.70322 and $^{143}Nd/^{144}Nd$ values of 0.51314–0.51305 (Dijkstra *et al.* 2010), and overlap with most of the Auckland Islands data (see Supplementary Table 3B). Furthermore, the data are also comparable to Sr–Nd–Pb–Hf data collected from peridotites erupted though continental lithosphere elsewhere in Zealandia (Scott *et al.* 2014*a*, *b*; McCoy-West *et al.* 2016; Dalton *et al.* 2017), which represent a combination of variable depletion ages overprinted by carbonatitic or related metasomatic melts.

Subantarctic lithospheric mantle under the Atlantic Ocean

Mantle peridotite xenoliths have not been found on Bouvet, South Georgia or the South Sandwich island groups in the Atlantic Ocean (Figs 1 and 11). However, samples of peridotite have been dredged from along the trench wall of the South Sandwich Arc, and at the intersection of the South Sandwich Trench with the South American and Antarctic plates (Pearce *et al.* 2000), as well as along fracture zones along subantarctic portions of the Southwest Indian Ridge and South America–Antarctic Ridge (e.g. Johnson *et al.* 1990; Snow and Dick 1995; Jaroslow *et al.* 1996; Hellebrand *et al.* 2001; Brunelli *et al.* 2003; Warren *et al.* 2009) (Fig. 11). Since these occurrences have already been described in detail by other workers and we have no new information to complement existing data, the reader is guided to the aforementioned publications and Warren (2016) for detailed information, and only a brief summary is given here. The abyssal peridotites in all cases are moderately to extensively serpentinized, with most estimated to have been harzburgite prior to hydration. Trace elements collected from abyssal peridotite clinopyroxene grains indicate the rocks to have mostly undergone moderate depletion consistent with decompression melting (Hellebrand *et al.* 2001; Brunelli *et al.* 2003; Warren 2016). Many of the clinopyroxenes also show some re-enrichment of LREE abundances. The South Sandwich peridotites represent suprasubduction-zone mantle, and therefore a distinct geodynamic setting to the peridotites found at mid-ocean ridge fracture zones. These peridotites are, in general, more melt-depleted than the abyssal peridotites, and are characterized as having spinels with very high Cr# at variable Ti contents, thought to indicate 'upgrading' of the Cr# in spinel by melt–rock reaction with a variety of melts including those of boninitic composition. Like the abyssal peridotites, the suprasubduction-zone peridotites appear to have been harzburgites prior to extensive serpentinization, and the clinopyroxene trace elements show moderate depletion variably overprinted by LREE enrichment (Pearce *et al.* 2000).

Osmium, Sr, Nd and Pb isotopes from abyssal peridotite sulfide and pyroxene minerals indicate that portions of the mantle beneath the subantarctic Southwest Indian Ridge were depleted billions of years before present day (Warren *et al.* 2009; Warren and Shirey 2012; Day *et al.* 2017). Thus, like the Southern Ocean and Kerguelen occurrences discussed above, the modern convecting mantle beneath the southern Atlantic in the Subantarctic must contain ancient domains embedded within younger mantle that stabilized as lithosphere during Mesozoic and more recent times.

Conclusions

Mantle peridotite occurs as xenoliths in intraplate basalts in the subantarctic (between 46 and 60° S) Kerguelen Islands and Heard Islands in the Indian Ocean, and on the Auckland Islands in the Southern Ocean, and represents samples of lithospheric mantle from beneath these locations. Little is known about the Heard Island mantle lithosphere but the Kerguelen Islands and Auckland Islands xenoliths were derived from lithosphere associated with the Kerguelen Plateau and Zealandia, respectively. The peridotite xenoliths are variably depleted, some to quite refractory levels, especially those from beneath the Kerguelen and South Sandwich islands, and they commonly contain complex metasomatic histories. The Auckland Islands mantle is distinctive as a result of it having been chemically stratified, with an apparently fertile cooler and shallower layer underlain by a more depleted but metasomatized hotter domain – although further work elsewhere may find this to be a common occurrence. Osmium ± Hf isotopes point to some of the Kerguelen Islands and Auckland Islands mantle fragments having undergone depletion hundreds of millions to billions of years before formation of the overlying and much younger crust. These ancient depletion events – as with those documented in abyssal peridotites from modern oceanic mantle – therefore cannot explain the timing of lithosphere formation as it is currently represented beneath these locations but are more likely to record evidence for more ancient events that produced residual mantle or minerals that were then incorporated into younger lithosphere.

Abyssal peridotite has been dredged from fracture zones along the subantarctic portions of the South America–Antarctica Ridge, the Southwest Indian Ridge and the

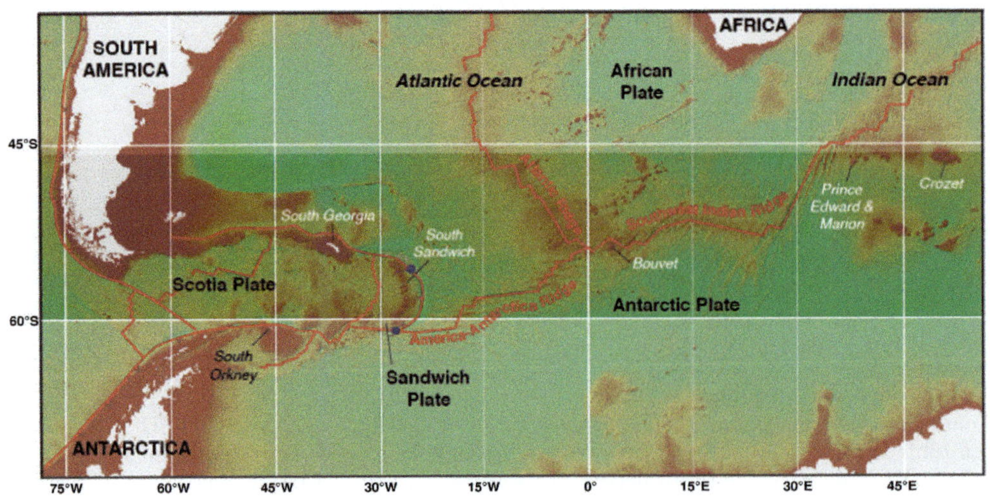

Fig. 11. South Atlantic Ocean subantarctic features with locations of mantle samples. Bathymetric depth decreases from red to green. Blue circles on the South Sandwich Arc indicate locations of dredged and drilled peridotite (Pearce *et al.* 2000).

Southeast Indian Ridge in the southern Atlantic Ocean and Indian Ocean. Although these abyssal peridotites are typically very serpentinized, most appear to have harzburgite protoliths. Serpentinized oceanic lithospheric peridotite has also been found exposed on Macquarie Island, which is an exhumed fracture zone that now forms part of the plate boundary separating the Pacific and Australian plates in the Southern Ocean. Like the subantarctic mantle lithosphere associated with Kerguelen and Zealandia, Os isotopic analyses show that the modern subantarctic oceanic upper mantle must contain embedded ancient peridotite fragments that have histories that long-predate the formation of the oceanic lithosphere.

Acknowledgements This is contribution 1528 from the ARC Centre of Excellence for Core to Crust Fluid Systems (http://www.ccfs.mq.edu.au) and 1402 in the GEMOC Key Centre (http://www.gemoc.mq.edu.au). S. Read (Otago) helped to generate the maps. Comments by A. Stracke and A. McCoy-West improved the paper.

Author contributions **DG**: conceptualization (lead), data curation (lead), supervision (lead), validation (lead), writing – original draft (lead), writing – review & editing (lead); **JMS**: conceptualization (lead), data curation (lead), supervision (lead), validation (lead), writing – original draft (lead), writing – review & editing (lead); **MG**: conceptualization (lead), supervision (lead), validation (lead), writing – original draft (lead), writing – review & editing (lead); **BNM**: writing – review & editing (supporting); **DL**: data curation (equal), formal analysis (equal), funding acquisition (equal), validation (lead), writing – review & editing (equal); **JL**: data curation (equal), formal analysis (equal), funding acquisition (equal), writing – original draft (equal), writing – review & editing (equal); **DGP**: data curation (equal), formal analysis (equal), funding acquisition (equal), writing – original draft (equal), writing – review & editing (equal); **QHAVDM**: data curation (equal), formal analysis (equal), writing – original draft (equal), writing – review & editing (equal); **TEW**: data curation (equal), formal analysis (equal), writing – original draft (equal), writing – review & editing (equal); **GM**: investigation (equal), supervision (equal); **DG**: conceptualization (supporting); **SYO**: conceptualization (equal), funding acquisition (equal), project administration (equal), supervision (equal); **J-YC**: conceptualization (equal), funding acquisition (equal), project administration (equal), supervision (equal), writing – review & editing (lead); **AG**: conceptualization (supporting).

Funding The Kerguelen archipelago research was funded by CNRS, St Etienne and Toulouse III Universities research programmes, and the Paul Emile Victor Institute field campaigns (IPEV, France). The analytical data were obtained using instrumentation funded by DEST Systemic Infrastructure Grants, ARC LIEF, NCRIS/AuScope, industry partners and Macquarie University. The Auckland Islands research was funded by a Foundation for Science Research and Technology Fellowship (contract UOOX1004) to J.M. Scott, and a Canada Excellence Research Chair Funding to DGP and a 1000 Youth Talents Programme to J. Liu.

Data availability All data generated or analysed during this study are included in this published article (and its supplementary information files).

References

Aubert de la Rüe, E. 1932. *Étude géologique et géographique de l'archipel de Kerguelen*. Revue de Géographie Physique et de la Géologie Dynamique.

Barling, J., Goldstein, S. and Nichols, I.A. 1994. Geochemistry of Heard Island (Southern Indian Ocean): characterization of an enriched mantle component and implications for enrichment of the Sub-Indian Ocean mantle. *Journal of Petrology*, **3**, 1017–1053.

Bascou, J., Delpech, G., Vauchez, A., Moine, B.N., Cottin, J.Y. and Barruol, G . 2008. An integrated study of microstructural, geochemical, and seismic properties of the lithospheric mantle above the Kerguelen plume (Indian Ocean). *Geochemistry, Geophysics, Geosystems*, **9**, Q04036, https://doi.org/10.1029/2007GC001879

Bertrand, P. and Mercier, J.C.C. 1985. The mutual solubility of coexisting ortho- and clinopyroxene: toward an absolute geothermometer for the natural system? *Earth and Planetary Science Letters*, **76**, 109–122, https://doi.org/10.1016/0012-821X(85)90152-9

Brey, G.P., Kohler, T. and Nickel, K.G. 1990. Geothermobarometry in Four-phase Lherzolites I. Experimental results from 10 to 60 kb. *Journal of Petrology*, **31**, 1313–1352.

Brunelli, D., Cipriani, A., Ottolini, L., Peyve, A. and Bonatti, E. 2003. Mantle peridotites from the Bouvet triple junction region, south Atlantic. *Terra Nova*, **15**, 194–203, https://doi.org/10.1046/j.1365-3121.2003.00482.x

Charvis, P., Recq, M., Operto, S., and Brefort, D. 1995. Deep structures of the northern Kerguelen plateau and hotspot related activity. *Geophysical Journal International*, **122**, 899–924, https://doi.org/10.1111/j.1365-246X.1995.tb06845.x

Chatterjee, R. and Lassiter, J.C. 2016. 186Os/188Os variations in upper mantle peridotites: constraints on the Pt/Os ratio of primitive upper mantle, and implications for late veneer accretion and mantle mixing timescales. *Chemical Geology*, **442**, 11–22, https://doi.org/10.1016/j.chemgeo.2016.08.033

Coffin, M.F. and Eldholm, O. 1993. Scratching the surface: Estimating dimensions of large igneous provinces. *Geology*, **21**, 515–518, https://doi.org/10.1130/0091-7613(1993)021<0515:STSEDO>2.3.CO;2

Coffin, M.F., Pringle, M.S., Duncan, R.A., Gladczenko, T.P., Storey, M., Müller, R.D. and Gahagan, L.A. 2002. Kerguelen hotspot magma output since 130 Ma. *Journal of Petrology*, **43**, 1121–1137, https://doi.org/10.1093/petrology/43.7.1121

Dalton, H.B., Scott, J.M. *et al.* 2017. Diffusion-zoned pyroxenes in an isotopically heterogeneous mantle lithosphere beneath the Dunedin Volcanic Group, New Zealand, and their implications for intraplate alkaline magma sources. *Lithosphere*, **9**, 463–475, https://doi.org/10.1130/L631.1

Day, J.M., Walker, R.J. and Warren, J.M. 2017. 186Os–187Os and highly siderophile element abundance systematics of the mantle revealed by abyssal peridotites and Os-rich alloys. *Geochimica et Cosmochimica Acta*, **200**, 232–254, https://doi.org/10.1016/j.gca.2016.12.013

Delpech, G. 2004. *Trace Element and Isotopic Fingerprints in Ultramafic Xenoliths from the Kerguelen Archipelago (South Indian Ocean)*. PhD thesis, Macquarie University, Sydney, Australia.

Delpech, G., Grégoire, M., O'Reilly, S.Y., Cottin, J.Y., Moine, B., Michon, G. and Giret, A. 2004. Feldspar from carbonate-rich silicate metasomatism in the shallow oceanic mantle under Kerguelen Islands (South Indian Ocean). *Lithos*, **75**, 209–237, https://doi.org/10.1016/j.lithos.2003.12.018

Delpech, G., Lorand, J.-P., Grégoire, M., Cottin, J.-Y. and O'Reilly, S.Y. 2012. In-situ geochemistry of sulfides in highly metasomatized mantle xenoliths from Kerguelen, southern Indian Ocean. *Lithos*, **154**, 296–314, https://doi.org/10.1016/j.lithos.2012.07.018

Dick, H.J.B. 1989. Abyssal peridotites, very slow spreading ridges and ocean magmatism. *Geological Society, London, Special Publications*, **42**, 71–105, https://doi.org/10.1144/GSL.SP.1989.042.01.06

Dick, H.J.B. and Bullen, T. 1984. Chromian spinel as a petrogenetic indicator in abyssal and alpine-type peridotites and spatially associated lavas. *Contribution to Mineral and Petrology*, **86**, 54–76.

Dijkstra, A.H., Sergeev, D.S., Spandler, C., Pettke, T., Meisel, T. and Cawood, P.A. 2010. Highly refractory peridotites on Macquarie Island and the case for anciently depleted domains in the Earth's mantle. *Journal of Petrology*, **51**, 469–493, https://doi.org/10.1093/petrology/egp084

Dosso, L., Vidal, P., Cantagrel, J.M., Lameyre, J., Marot, A. and Zimine, S. 1979. 'Kerguelen: Continental fragment or oceanic island?': Petrology and isotopic geochemistry evidence. *Earth and Planetary Science Letters*, **43**, 46–60, https://doi.org/10.1016/0012-821X(79)90154-7

Doucet, S., Weis, D., Scoates, J.S., Nicolaysen, K., Frey, F.A. and Giret, A. 2002. The depleted mantle component in Kerguelen archipelago basalts: petrogenesis of tholeiitic-transitional basalts from the Loranchet peninsula. *Journal of Petrology*, **43**, 1341–1366, https://doi.org/10.1093/petrology/43.7.1341

Doucet, S., Scoates, J.S., Weis, D. and Giret, A. 2005. Constraining the components of the Kerguelen mantle plume: a Hf–Pb–Sr–Nd isotopic study of picrites and high-MgO basalts from the Kerguelen Archipelago. *Geochemistry, Geophysics, Geosystems*, **6**, Q04007, https://doi.org/10.1029/2004GC000806

Edwards, A.B. 1938. *Tertiary lavas from the Kerguelen Archipelago*. B.A.N.Z.A.R.E. Report Series A, **5**(2), 72–100.

Frey, F.A., Weis, D., Yang, H.J., Nicolaysen, K., Leyrit, H. and Giret, A. 2000. Temporal geochemical trends in Kerguelen archipelago basalts: evidence for decreasing magma supply from the Kerguelen plume. *Chemical Geology*, **164**, 61– 80, https://doi.org/10.1016/S0009-2541(99)00144-8

Frey, F.A, Weis, D., Borissova, A.Y. and Xu, G. 2002. Involvement of continental crust in the formation of the Cretaceous Kerguelen Plateau: new perspective from ODP Leg 120 sites. *Journal of Petrology*, **43**, 1207–1239, https://doi.org/10.1093/petrology/43.7.1207

Gamble, J.A., Adams, C.J., Morris, P.A., Wysoczanski, R.J., Handler, M. and Timm, C. 2018. The geochemistry and petrogenesis of Carnley Volcano, Auckland Islands, SW Pacific. *New Zealand Journal of Geology and Geophysics*, **61**, 480–497, https://doi.org/10.1080/00288306.2018.1505642

Gautier, I. 1987. *Les Basaltes des îles Kerguelen (Terres Australes et Antarctiques Françaises)*. PhD thesis, Université de Paris VI, Paris, France.

Gautier, I., Weis, D., Mennessier, J.-P., Vidal, P., Giret, A. and Loubet, M. 1990. Petrology and geochemistry of the Kerguelen Archipelago basalts (South Indian Ocean): evolution of the mantle sources from ridge to intraplate position. *Earth and Planetary Science Letters*, **100**, 59–76, https://doi.org/10.1016/0012-821X(90)90176-X

Giret, A. 1983. *Le Plutonisme Océanique Intraplaque, Exemple des îles Kerguelen*. PhD thesis, Université de Paris VI, Paris, France; Bulletin CNFRA, Paris, **54**.

Giret, A. 1993. Les étapes magmatiques de l'édification des îles Kerguelen, Océan Indien. *Mémoire Société Géologique de France*, **163**, 273–282.

Giret, A., Grégoire, M., Cottin, J.Y. and Michon, G. 1997. Kerguelen, a third type of oceanic island? *In*: Ricci, C.A. (ed.) *The Antarctic Region: Geological Evolution and Processes*. Terra Antarctica, Siena, Italy, 735–741.

Goscombe, B.D. and Everard, J.L. 1998. *1:10 000 Geological Map of Macquarie Island*. Mineral Resources, Tasmania.

Grégoire, M. 1994. *Pétrologie des Enclaves Ultrabasiques et Basiques des Iles Kerguelen. Les Contraintes Minéralogiques et Thermobarométriques et Leurs Implications Géodynamiques*. PhD thesis, Université Jean Monnet, St Etienne, France.

Grégoire, M., Leyrit, H., Cottin, J.Y., Giret, A. and Mattielli, N. 1992. Les phases précoces et profondes du magmatisme des îles Kerguelen révélées par les enclaves basiques et ultrabasiques. *Comptes Rendus de l'Académie des Sciences, Paris*, **314**, 1203–1209.

Grégoire, M., Mattielli, N. *et al.* 1994. Oceanic mafic granulite xenoliths from the Kerguelen archipelago. *Nature*, **367**, 360–363, https://doi.org/10.1038/367360a0

Grégoire, M., Cottin, J.Y., Mattielli, N., Nicollet, C., Weis, D. and Giret, A. 1995. The Kerguelen archipelago: a hypothetic continental mafic protolith. *Terra Antarctica*, **2**, 1–6.

Grégoire, M., Cottin, J.Y., Giret, A., Mattielli, N. and Weis, D. 1996. Mantle–melt interactions and magmatic underplating beneath the Kerguelen oceanic islands revealed by ultrabasic and basic xenoliths. *In*: Demaiffe, D. (ed.) *Petrology and Geochemistry of Magmatic Suites of Rocks in the Continental and Oceanic Crust: A Volume Dedicated to Professor Jean Michot*. Université Libre de Bruxelles, 371–384.

Grégoire, M., Lorand, J.-P., Cottin, J.-Y., Giret, A., Mattielli, N. and Weis, D. 1997. Xenoliths evidence for a refractory oceanic mantle percolated by basaltic melts beneath the Kerguelen archipelago. *European Journal of Mineralogy*, **9**, 1085–1100, https://doi.org/10.1127/ejm/9/5/1085

Grégoire, M., Cottin, J.Y., Giret, A., Mattielli, N. and Weis, D. 1998. The meta-igneous granulite xenoliths from Kerguelen Archipelago: evidence of a continent nucleation in an oceanic setting. *Contributions to Mineralogy and Petrology*, **133**, 259–283, https://doi.org/10.1007/s004100050451

Grégoire, M., Lorand, J.P., O'Reilly, S.Y. and Cottin, J.-Y. 2000*a*. Armalcolite-bearing, Ti-rich metasomatic assemblages in harzburgitic xenoliths from the Kerguelen Islands: implications for the oceanic mantle budget of high-field strength elements. *Geochimica et Cosmochimica Acta*, **64**, 673–694, https://doi.org/10.1016/S0016-7037(99)00345-2

Grégoire, M., Moine, B.N., O'Reilly, S.Y., Cottin, J.Y. and Giret, A. 2000*b*. Trace element residence and partitioning in mantle xenoliths metasomatized by highly alkaline, silicate- and carbonate-rich melts (Kerguelen Islands, Indian Ocean). *Journal of Petrology*, **41**, 477–509, https://doi.org/10.1093/petrology/41.4.477

Grégoire, M., Jackson, I., O'Reilly, S.Y. and Cottin, J.-Y. 2001. The lithospheric mantle beneath the Kerguelen Islands (Indian Ocean): petrological and petrophysical characteristics of mantle mafic rock types and correlation with seismic profiles. *Contributions to Mineralogy and Petrology*, **142**, 244–259 https://doi.org/10.1007/s004100100289

Harte, B. 1977. Rock nomenclature with particular relation to deformation and recrystallization textures in olivine-bearing xenoliths. *The Journal of Geology*, **85**, 279–288, https://doi.org/10.1086/628299

Hassler, D.R. 1999. *Plume Lithosphere Interaction: Geochemical Evidence from Upper Mantle and Lower Crustal Xenoliths from the Kerguelen Islands*. PhD thesis, Massachusetts Institute of Technology (MIT) and the Woods Hole Oceanographic Institution (WHOI), Woods Hole, Massachusetts, USA.

Hassler, D.R., and Shimizu, N. 1998. Osmium isotopic evidence for ancient subcontinental lithospheric mantle beneath the Kerguelen Islands, Southern Indian Ocean. *Science*, **280**, 418–421, https://doi.org/10.1126/science.280.5362.418

Hasterok, D. and Chapman, D.S. 2011. Heat production and geotherms for the continental lithosphere. *Earth and Planetary Science Letters*, **307**, 59–70, https://doi.org/10.1016/j.epsl.2011.04.034

Hellebrand, E., Snow, J.E., Dick, H.J. and Hofmann, A.W. 2001. Coupled major and trace elements as indicators of the extent of melting in mid-ocean-ridge peridotites. *Nature*, **410**, 677–681, https://doi.org/10.1038/35070546

Herzberg, C. 2004. Geodynamic information in peridotite petrology. *Journal of Petrology*, **45**, 2507–2530, https://doi.org/10.1093/petrology/egh039

Jaroslow, G.E., Hirth, G. and Dick, H.J.B. 1996. Abyssal peridotite mylonites: implications for grain-size sensitive flow and strain localization in the oceanic lithosphere. *Tectonophysics*, **256**, 17–37, https://doi.org/10.1016/0040-1951(95)00163-8

Johnson, K.T., Dick, H.J. and Shimizu, N. 1990. Melting in the oceanic upper mantle: an ion microprobe study of diopsides in abyssal peridotites. *Journal of Geophysical Research: Solid Earth*, **95**, 2661–2678, https://doi.org/10.1029/JB095iB03p02661

Kamenetsky, V.S., Everard, J.L., Crawford, A.J., Varne, R., Eggins, S.M. and Lanyon, R. 2000. Enriched end-member of primitive MORB melts: petrology and geochemistry of glasses from Macquarie Island (SW Pacific). *Journal of Petrology*, **41**, 411–430, https://doi.org/10.1093/petrology/41.3.411

Klemme, S. and O'Neill, H.S. 2000. The near-solidus transition from garnet lherzolite to spinel lherzolite. *Contributions to Mineralogy and Petrology*, **138**, 237–248, https://doi.org/10.1007/s004100050560

Lameyre, J., Marot, A., Zimine, S., Cantagrel, J.M., Dosso, L. and Vidal, P. 1976. Chronological evolution of the Kerguelen islands syenite–granite ring complexes. *Nature*, **263**, 306–307, https://doi.org/10.1038/263306a0

Leat, P.T., Smellie, J.L., Millar, I.L. and Larter, R.D. 2003. Magmatism in the South Sandwich arc. Geological Society, London, Special Publications, **219**, 285–313, https://doi.org/10.1144/GSL.SP.2003.219.01.14

le Roex, A.P., Chevallier, L., Verwoerd, W.J. and Barends, R. 2012. Petrology and geochemistry of Marion and Prince Edward Islands, Southern Ocean: Magma chamber processes and source region characteristics. *Journal of Volcanology and Geothermal Research*, **223**, 11–28, https://doi.org/10.1016/j.jvolgeores.2012.01.009

Leyrit, H. 1992. *Kerguelen: Cartographie et Magmatologie des presqu'îles Jeanne d'Arc et Ronarc'h.* PhD thesis, Université Paris XI, Orsay, France.

Li, P., Scott, J.M., Liu, J. and Xia, Q.-K. 2018. Lateral H_2O variation in the Zealandia lithospheric mantle controls orogen width. *Earth and Planetary Science Letters*, **502**, 200–209, https://doi.org/10.1016/j.epsl.2018.09.004

Liu, J., Scott, J.M., Martin, C.E. and Pearson, D.G. 2015. The longevity of Archean mantle residues in the convecting upper mantle and their role in young continent formation. *Earth and Planetary Science Letters*, **424**, 109–118, https://doi.org/10.1016/j.epsl.2015.05.027

Lorand, J.-P., Delpech, G., Grégoire, M., Moine, B., O'Reilly, S.Y. and Cottin, J.Y. 2004. Platinum-group elements and the multistage metasomatic history of Kerguelen lithospheric mantle (South Indian Ocean). *Chemical Geology*, **208**, 195–215 https://doi.org/10.1016/j.chemgeo.2004.04.012

Mahoney, J.J., White, W.M., Upton, B.G.J., Neal, C.R. and Scrutton, R.A. 1996. Beyond EM-1: Lavas from Afanasy-Nikitin Rise and the Crozet Archipelago, Indian Ocean. *Geology*, **24**, 615–618.

Mattielli, N. 1996. *Magmatisme et métasomatisme associés au panache des Kerguelen : Contribution de la géochimie des enclaves basiques et ultrabasiques*. PhD Université Libre de Bruxelles.

Mattielli, N., Weis, D., Grégoire, M., Mennesier, J.P., Cottin, J.Y. and Giret, A. 1996. Kerguelen basic and ultrabasic xenoliths: evidence for long-lived Kerguelen hotspot activity. *Lithos*, **37**, 261–280, https://doi.org/10.1016/0024-4937(95)00040-2

Mattielli, N., Weis, D. *et al.* 1999. Evolution of heterogeneous lithospheric mantle in a plume environment beneath the Kerguelen Archipelago. *Journal of Petrology*, **40**, 1721–1744, https://doi.org/10.1093/petroj/40.11.1721

McBirney, A. and Aoki, K.I. 1973. Factors governing the stability of plagioclase at high-pressures as shown by spinel-gabbro xenoliths from the Kerguelen Archipelago. *American Mineralogist*, **58**, 271–276.

McCoy-West, A.J., Bennett, V.C., Puchtel, I.S. and Walker, R.J. 2013. Extreme persistence of cratonic lithosphere in the southwest Pacific: Paleoproterozoic Os isotopic signatures in Zealandia. *Geology*, **41**, 231–234, https://doi.org/10.1130/G33626.1

McCoy-West, A.J., Bennett, V.C., O'Neill, H.S.C., Hermann, J. and Puchtel, I.S. 2015. The interplay between melting, refertilization and carbonatite metasomatism in off-cratonic lithospheric mantle under Zealandia: an integrated major, trace and platinum group element study. *Journal of Petrology*, **56**, 563–604, https://doi.org/10.1093/petrology/egv011

McCoy-West, A.J., Bennett, V.C. and Amelin, Y. 2016. Rapid Cenozoic ingrowth of isotopic signatures simulating 'HIMU' in ancient lithospheric mantle: distinguishing source from process. *Geochimica et Cosmochimica Acta*, **187**, 79–101, https://doi.org/10.1016/j.gca.2016.05.013

McDonough, W.F., and Sun, S.-s. 1995. The composition of the Earth. *Chemical Geology*, **120**, 223–253, https://doi.org/10.1016/0009-2541(94)00140-4

Meisel, T., Walker, R.J., Irving, A.J., and Lorand, J.P. 2001. Osmium isotopic compositions of mantle xenoliths: a global perspective. *Geochimica et Cosmochimica Acta*, **65**, 1311–1323, https://doi.org/10.1016/S0016-7037(00)00566-4

Moine, B. 2000. *Volatile-bearing Ultramafic to Mafic Xenoliths from the Kerguelen Archipelago (Southern Indian Ocean), Fluids Migration and Mantle Metasomatism within Oceanic I-ntraplate Setting*. PhD thesis, Université Jean Monnet, St Etienne, France.

Moine, B.N., Cottin, J.Y., Sheppard, S.M.F., Grégoire, M., O'Reilly, S.Y. and Giret, A. 2000. Incompatible trace element and isotopic (D/H) characteristics of amphibole- and phlogopite-bearing ultramafic to mafic xenoliths from Kerguelen Islands (TAAF, South Indian Ocean). *European Journal of Mineralogy*, **12**, 761–777, https://doi.org/10.1127/ejm/12/4/0761

Moine, B., Grégoire, M., O'Reilly, S.Y., Sheppard, S.M.F. and Cottin, J.Y. 2001. High field strength element fractionation in the upper mantle: evidence from amphibole-rich composite mantle xenoliths from the Kerguelen Islands (Indian Ocean). *Journal of Petrology*, **42**, 2143–2167, https://doi.org/10.1093/petrology/42.11.2145

Moine, B., Grégoire, M. *et al.* 2004. Carbonatite melt in oceanic upper mantle beneath the Kerguelen Archipelago. *Lithos*, **75**, 239–252, https://doi.org/10.1016/j.lithos.2003.12.019

Nixon, P.H. 1987. *Mantle Xenoliths*. John Wiley & Sons, Chichester, UK.

Nougier, J. 1970. *Contribution à l'étude Géologique et Géomorphologique des îles Kerguelen*. PhD thesis, Université P. et M. Curie, Paris, France; Bulletin CNFRA, Paris, **27**.

Park, S-H., Langmuir, C. *et al.* 2019. An isotopically distinct Zealandia-Antarctic mantle domain in the Southern Ocean. *Nature Geoscience*, **12**, 206–214, https://doi.org/10.1038/s41561-018-0292-4

Pearce, J.A., Barker, P.F., Edwards, S.J., Parkinson, I.J. and Leat, P.T. 2000. Geochemistry and tectonic significance of peridotites from the South Sandwich arc–basin system, South Atlantic. *Contributions to Mineralogy and Petrology*, **139**, 36–53, https://doi.org/10.1007/s004100050572

Pearson, D.G., Parman, S.W. and Nowell, G.M. 2007. A link between large mantle melting events and continent growth seen in osmium isotopes. *Nature*, **449**, 202–205, https://doi.org/10.1038/nature06122

Quilty, P.G. and Wheller, G.E. 2000. Heard Island and the McDonald Islands: a window into the Kerguelen Plateau. *Papers and Proceedings of the Royal Society of Tasmania*, **133**(2) 1–12.

Royer, J.Y. and Coffin, M.F. 1992. Jurassic to Eocene plate tectonics reconstruction in the Kerguelen plateau region. *In*: Wise, S.W., Jr., Schlich, R. *et al.* (eds) *Proceedings of the Ocean Drilling Program, Scientific Results, Volume 120*. Ocean Drilling Program, College Station, TX, 917–928.

Schiano, P., Clocchiatti, R., Shimizu, N., Weis, D. and Mattielli, N. 1994. Cogenetic silica-rich and carbonate-rich melts trapped in mantle minerals in Kerguelen ultramafic xenoliths: Implications for metasomatized upper mantle. *Earth and Planetary Science Letters*, **123**, 167–178, https://doi.org/10.1016/0012-821X(94)90265-8

Scoates, J.S., Weis, D. *et al.* 2008. The Val Gabbro plutonic suite: a sub-volcanic intrusion emplaced at the end of flood basalt volcanism on the Kerguelen Archipelago. *Journal of Petrology*, **49**, 79–105, https://doi.org/10.1093/petrology/egm071

Scott, J.M. and Turnbull, I.M. 2019. Geology of New Zealand's Sub-Antarctic islands. *New Zealand Journal of Geology and Geophysics*, **623**, 291–317, https://doi.org/10.1080/00288306.2019.1600557

Scott, J.M., Hodgkinson, A., Palin, J.M., Waight, T.E., van der Meer, Q.H.A. and Cooper, A.F. 2014*a*. Ancient melt depletion overprinted by young carbonatitic metasomatism in the New Zealand lithospheric mantle. *Contributions to Mineralogy and Petrology*, **167**, 963, https://doi.org/10.1007/s00410-014-0963-0

Scott, J.M., Waight, T.E., van der Meer, Q.H.A., Palin, J.M., Cooper, A.F. and Münker, C. 2014*b*. Metasomatized ancient lithospheric mantle beneath the young Zealandia microcontinent and its role in HIMU-like intraplate magmatism. *Geochemistry, Geophysics, Geosystems*, **15**, 3477–3501, https://doi.org/10.1002/2014GC005300

Scott, J.M., Liu, J., Pearson, D.G. and Waight, T.E. 2016. Mantle depletion and metasomatism recorded in orthopyroxene in highly depleted peridotites. *Chemical Geology*, **441**, 280–291, https://doi.org/10.1016/j.chemgeo.2016.08.024

Scott, J.M., Liu, J. *et al.* 2019. Continent stabilisation by lateral accretion of subduction zone-processed depleted mantle residues; insights from Zealandia. *Earth and Planetary Science Letters*, **507**, 175–186, https://doi.org/10.1016/j.epsl.2018.11.039

Simon, N.S.C., Neumann, E-R., Bonadiman, C., Coltorti, M., Delpech, G., Grégoire, M. and Widom, E. 2008. Ultra-refractory domains in the oceanic mantle lithosphere sampled as mantle xenoliths at ocean islands. *Journal of Petrology*, **49**, 1223–1251, https://doi.org/10.1093/petrology/egn023

Snow, J.E. and Dick, H.J. 1995. Pervasive magnesium loss by marine weathering of peridotite. *Geochimica et Cosmochimica Acta*, **59**, 4219–4235, https://doi.org/10.1016/0016-7037(95)00239-V

Stracke, A., Snow, J.E., Hellebrand, E., Von Der Handt, A., Bourdon, B., Birbaum, K. and Günther, D. 2011. Abyssal peridotite Hf isotopes identify extreme mantle depletion. *Earth and Planetary Science Letters*, **308**, 359–368, https://doi.org/10.1016/j.epsl.2011.06.012

Sun, S.S. and McDonough, W.F. 1989. Chemical and isotopic systematics of oceanic basalts: implications for mantle composition and processes. *Geological Society, London, Special Publications*, **42**, 313–345, https://doi.org/10.1144/GSL.SP.1989.042.01.19

Takazawa, E., Frey, F.A., Shimizu, N. and Obata, M. 2000. Whole rock compositional variations in an upper mantle peridotite (Horoman, Hokkaido, Japan): are they consistent with a partial melting process. *Geochimica et Cosmochimica Acta*, **64**, 695–716, https://doi.org/10.1016/S0016-7037(99)00346-4

Talbot, J.L., Hobbs, B.E., Wilshire, H.G. and Sweatman, T.R. 1963. Xenoliths and xenocrysts from lavas of the Kerguelen archipelago. *American Mineralogist*, **48**, 159–179.

Taylor, W.R. 1998. An experimental test of some geothermometer and geobarometer formulations for upper mantle peridotites with application to the thermobarometry of fertile lherzolite and garnet websterite. *Neues Jahrbuch für Mineralogie – Abhandlungen*, **172**, 381–408, https://doi.org/10.1127/njma/172/1998/381

Valbracht, P.J., Honda, M., Matsumoto, T., Mattielli, N., McDougall, I., Ragettli, R. and Weis, D. 1996. Helium, neon, and argon systematics in Kerguelen ultramafic xenoliths: implications for mantle source signatures. *Earth and Planetary Science Letters*, **138**, 29–38, https://doi.org/10.1016/0012-821X(95)00226-3

van der Meer, Q.H., Scott, J.M., Serre, S.H., Whitehouse, M.J., Kristoffersen, M., Le Roux, P.J. and Pope, E.C. 2019. Low-$\delta^{18}O$ zircon xenocrysts in alkaline basalts; a window into the complex carbonatite–metasomatic history of the Zealandia lithospheric mantle. *Geochimica et Cosmochimica Acta*, **254**, 21–39, https://doi.org/10.1016/j.gca.2019.03.029

Varne, R., Brown, A.V. and Falloon, T. 2000. Macquarie Island: its geology, structural history, and the timing and tectonic setting of its N-MORB to E-MORB magmatism. *Geological Society of America Special Papers*, **349**, 301–320.

Walter, M.J., 1998. Melting of garnet peridotite and the origin of komatiite and depleted lithosphere. *Journal of Petrology*, **39**, 29–60, https://doi.org/10.1093/petroj/39.1.29

Warren, J.M. 2016. Global variations in abyssal peridotite compositions. *Lithos*, **248**, 193–219, https://doi.org/10.1016/j.lithos.2015.12.023

Warren, J.M. and Shirey, S.B. 2012. Lead and osmium isotopic constraints on the oceanic mantle from single abyssal peridotite sulfides. *Earth and Planetary Science Letters*, **359**, 279–293, https://doi.org/10.1016/j.epsl.2012.09.055

Warren, J.M., Shimizu, N., Sakaguchi, C., Dick, H.J. and Nakamura, E. 2009. An assessment of upper mantle heterogeneity based on abyssal peridotite isotopic compositions. *Journal of Geophysical Research: Solid Earth*, **114**, B12203, https://doi.org/10.1029/2008JB006186

Wasilewski, B., Doucet, L.S. *et al.* 2017. Ultra-refractory mantle within oceanic plateau: Petrology of the spinel harzburgites from Lac Michèle, Kerguelen Archipelago. *Lithos*, **272–273**, 336–349, https://doi.org/10.1016/j.lithos.2016.12.010

Weaver, B.L., Wood, D.A., Tarney, J. and Joron, J.L. 1987. Geochemistry of ocean island basalts from the South Atlantic: Ascension, Bouvet, St. Helena, Gough and Tristan da Cunha. *Geological Society, London, Special Publications*, **30**, 253–267, https://doi.org/10.1144/GSL.SP.1987.030.01.11

Weis, D. and Frey, F.A. 2002. Submarine basalts of the northern Plateau: interaction between the Kerguelen plume and the Southeast Indian Ridge revealed at ODP site 1140. *Journal of Petrology*, **45**, 1287–1309, https://doi.org/10.1093/petrology/43.7.1287

Weis, D., Frey, F.A., Leyrit, H. and Gautier, I. 1993. Kerguelen Archipelago revisited: geochemical and isotopic study of the SE provinces lavas. *Earth and Planetary Science Letters*, **118**, 101–119, https://doi.org/10.1016/0012-821X(93)90162-3

Weis, D., Shirey, S.B. and Frey, F.A. 2000. Re–Os systematics of Kerguelen plume basalts: enriched components and lower mantle source. *Eos, Transactions of the American Geophysical Union*, **81**, F1340.

Wertz, K.L. 2003. *From Seafloor Spreading to Uplift: The Structural and Geochemical Evolution of Macquarie Island on the Australian–Pacific Plate boundary*. PhD thesis, University of Texas at Austin, Austin, Texas, USA.

Witt-Eickschen, G. and Seck, H.A. 1991. Solubility of Ca and Al in orthopyroxene from spinel peridotite: an improved version of an empirical geothermometer. *Contributions to Mineralogy and Petrology*, **106**, 431–439, https://doi.org/10.1007/BF00321986

Workman, R.K. and Hart, S.R. 2005. Major and trace element composition of the depleted MORB mantle (DMM). *Earth and Planetary Science Letters*, **231**, 53–72, https://doi.org/10.1016/j.epsl.2004.12.005

Wright, J.B. 1967. Contributions to volcanic succession and petrology of Auckland Islands. 2. Upper parts of Ross Volcano. *Transactions of the Royal Society of New Zealand–Geology*, **5**, 71–86.

Wright, J.B. 1968. Contributions to volcanic succession and petrology of Aukland Islands, New Zealand. 3. Minor intrusives on Ross Volcano. *Transactions of the Royal Society of New Zealand–Geology*, **6**, 1–11.

Yang, H.J., Frey, F.A., Weis, D., Giret, A., Pyle, D. and Michon, G. 1998. Petrogenesis of the flood basalts forming the northern Kerguelen archipelago: implications for the Kerguelen plume. *Journal of Petrology*, **39**, 711–748, https://doi.org/10.1093/petroj/39.4.711

Zhang, X., Ganguly, J. and Ito, M. 2010. Ca–Mg diffusion in diopside: tracer and chemical inter-diffusion coefficients. *Contributions to Mineralogy and Petrology*, **159**, 175, https://doi.org/10.1007/s00410-009-0422-5

Zindler, A. and Hart, S. 1986. Chemical geodynamics. *Annual review of Earth and Planetary Sciences*, **14**, 493–571, https://doi.org/10.1146/annurev.ea.14.050186.002425

West Antarctic mantle deduced from mafic magmatism

Kurt S. Panter[1]* and Adam P. Martin[2]

[1]School of Earth, Environment and Society, Bowling Green State University, Bowling Green, OH 43403, USA

[2]GNS Science, Private Bag 1930, Dunedin, New Zealand

KSP, 0000-0002-0990-5880; APM, 0000-0002-4676-8344

*Correspondence: kpanter@bgsu.edu

Abstract: Distinct mantle compositions recorded in primitive West Antarctic magmatic rocks vary by tectonic setting and time. Deep asthenospheric mantle-plume sources or shallow metasomatized mantle sources may operate either coincidently or independently to supply melts for magmatism. For example, contemporaneous subduction and plume dynamics produced the Ferrar–Karoo Large Igneous Province; subduction-related melting followed by slab-rollback or melting of slab-hosted pyroxenite explains Antarctic Peninsula volcanism through time; Marie Byrd Land magmatism results from plume materials variably mixed with subduction-modified mantle; while magmatism in Victoria Land and western Ross Sea is best explained by plate dynamics and melting of asthenospheric and metasomatized lithospheric sources, and not by an upwelling plume. Element and isotopic ratios show a fundamental change between Marie Byrd Land and Victoria Land mantle domains. Specifically, Pb isotopes indicate that Victoria Land magmatism sources have a stronger focal zone (FOZO) mantle component, while Marie Byrd Land magmatism possesses more of the high μ = high $^{238}U/^{204}Pb$ (HIMU) mantle component that leads to high $^{206}Pb/^{204}Pb$ over time. The chemical and isotopic heterogeneity of relatively unfractionated igneous rocks in West Antarctica reflects fundamental differences in mantle domains and melting conditions. This mantle variability coincides with changes in crustal structure and composition, and has a geophysical signature that is manifest in seismic data and tomographic models.

Petrological studies of mafic igneous rocks from Antarctica yield crucial information about upper-mantle compositions and insights into the geodynamic evolution of the Antarctic Plate. Mantle-derived mafic magmas were intruded and erupted within several distinct tectonic settings over the past 200 million years (Smellie *et al.* 2020; Panter 2021), including continental break-up and the formation of the Ferrar and Karoo Large Igneous Provinces (LIPs), subduction to form the Antarctic Peninsula volcanic arc that has transitioned to post-subduction slab-rollback and slab-window magmatism, and broad extension resulting in widespread intraplate magmatism associated with the West Antarctic Rift System (WARS) – one of Earth's major continental rifts (Fig. 1). The diverse and dynamic tectonic history of Antarctica is revealed by the geochemical and isotopic diversity of igneous rock types, whose origins can ultimately be traced back to differences in mantle source (i.e. mineral mode, and geochemical and isotopic composition) and the conditions that promoted melting. The identification of different sources for magmatism (e.g. plume, upwelling asthenosphere and lithosphere) can provide important constraints on heat flux to the crust. Furthermore, information on the mineral mode for melting (e.g. anhydrous peridotite, pyroxenite and volatile-phase-bearing varieties) can contribute to our understanding of mantle rheology, which, in turn, can aid interpretations of seismic velocity and models for glacial isostatic adjustment.

The occurrence of igneous rocks representing primary melts in equilibrium with mantle peridotite are exceedingly rare and thus petrologists often rely on samples that are the least fractionated chemically (e.g. low SiO_2 and high MgO, Cr, Ni concentrations, and high olivine forsterite contents) to best tackle questions related to mantle source types and melting conditions. These mafic compositions that are used to reconcile magma origins in Antarctica include olivine tholeiite, basaltic andesite, alkaline and subalkaline basalts, basanite, hawaiite, trachybasalt, tephrite, and olivine nephelinite (Fig. 2a). The wide-ranging geochemical characteristics of Antarctic mafic magmatism permit discrimination by tectonic setting (Fig. 2b–d). Compositions plotted on tectonic discrimination diagrams fall consistently within fields that delimit two basic types: intraplate and convergent margin settings. Notably, compositions associated with the WARS and post-subduction compositions from the Antarctic Peninsula are tightly constrained to within-plate alkaline types. Conversely, compositions of the Ferrar–Karoo LIP and back-arc magmatism from the Antarctic Peninsula fall within both the arc (i.e. volcanic arc basalt (VAB) and island arc basalt (IAB)) and within-plate tholeiitic fields that encompass mid-ocean ridge basalt (MORB)-type compositions (Fig. 2b–d).

Major and trace element concentrations are also used to estimate mantle source enrichment or depletion (i.e. concentration of highly incompatible trace elements relative to undifferentiated or primitive mantle) and fertility (i.e. source mineralogy that dictates the degree of partial melting) to generate magma. Radiogenic isotopes, most commonly the Rb–Sr, Sm–Nd and U–Th–Pb systems, are used to characterize the long (hundreds of millions to billions of years) time-integrated history of different mantle end-member reservoirs (e.g. depleted mantle (DMM), enriched mantle (EM), high $\mu = {}^{238}U/^{204}Pb$ (HIMU), etc.) that have been defined by MORB and ocean island basalt (OIB) sample suites worldwide (Fig. 3). Results from radiogenic isotopes, coupled with oxygen isotopes and trace elements, are used to identify mantle source heterogeneities but are also important in identifying magmas that have been contaminated by crust during their ascent towards the surface.

Over the past 30 years, there has been significant progress in our understanding of the primary origins of magmas in Antarctica, facilitated by more comprehensive petrological data coverage along with better constraints on age (Smellie *et al.* 2021 and references therein). Other contributions include higher-resolution imaging of the architecture of the upper mantle and lithosphere, as well as finer-tuned plate reconstructions, all an outcome of extensive geological and geophysical campaigns (e.g. Pappa *et al.* 2019; Jordan *et al.* 2020; Lloyd *et al.* 2020; Martin and van der Wal 2022). Despite these advances, considerable debate remains. Petrological studies on intraplate alkaline magmatism associated with the WARS differ on whether magmas were produced within the asthenosphere or within metasomatized lithospheric mantle, and whether melting was facilitated by extension-enhanced edge-driven mantle flow or by mantle plumes (e.g. LeMasurier and Rex 1989; Kyle *et al.* 1992; Hole and LeMasurier 1994; Hart *et al.* 1995, 1997; Rocchi *et al.* 2002; Gaffney and Siddoway 2007; Nardini *et al.* 2009; Martin *et al.* 2013; LeMasurier *et al.* 2016; Phillips *et al.* 2018; Panter *et al.* 2018). Debate also

From: Martin, A. P. and van der Wal, W. (eds) 2023. *The Geochemistry and Geophysics of the Antarctic Mantle*.
Geological Society, London, Memoirs, **56**, 133–149,
First published online 28 September 2021, https://doi.org/10.1144/M56-2021-10

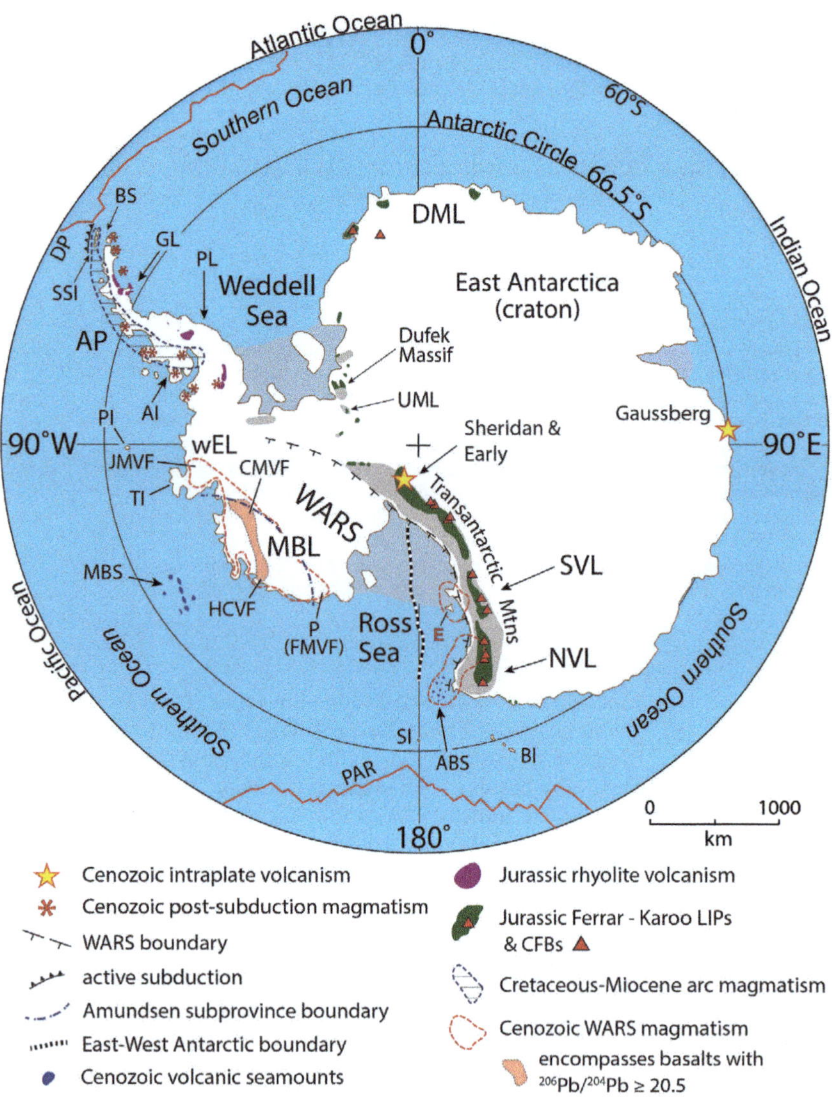

Fig. 1. Map of Antarctica showing the distribution of Mesozoic (Jurassic and Cretaceous) and Cenozoic tectonomagmatic provinces and volcanism (after Panter 2021). The distribution of Ferrar and Karoo LIPs and associated continental flood basalts (CFBs) are after Luttinen (2018) and Elliot and Fleming (2021). The locations of Jurassic rhyolite volcanism are from Riley and Leat (2021). The boundary of the Amundsen (seaward) and Ross (inland) geotectonic provinces of the West Antarctica Rift System (WARS) is after Jordan *et al.* (2020), and the geotectonic boundary between East and West Antarctica (cratonic margin province) is after Tinto *et al.* (2019) and Jordan *et al.* (2020). Cenozoic intraplate basalts with $^{206}Pb/^{204}Pb$ isotopic ratios ≥20.5 occur in four volcanic fields in Marie Byrd Land (MBL) and the region that encompasses them is outlined (from Panter *et al.* 2021*b* and references therein). Abbreviations: ABS, Adare Basin seamounts; AI, Alexander Island; AP, Antarctic Peninsula; BI, Balleny Islands; BS, Bransfield Strait; CMVF, Crary Mountains Volcanic Field; DML, Dronning Maud Land; DP, Drake Passage; E, Mount Erebus volcano on Ross Island; GL, Graham Land; HCVF, Hobbs Coast Volcanic Field; JMVF, Jones Mountains Volcanic Field; MBS, Marie Byrd seamounts; NVL, northern Victoria Land; P, Mount Perkins, which is part of the Fosdick Mountains Volcanic Field (FMVF); PAR, Pacific–Antarctic Ridge; PI, Peter I Island; PL, Palmer Land; SI, Scott Island; SSI, South Shetland Islands; SVL, southern Victoria Land; TI, Thurston Island; UML, ultramafic lamprophyres of the Ferrar LIP; wEL, western Ellsworth Land.

exists among petrologists who study intraplate magmatism closely associated with continental margin tectonics (i.e. Antarctic Peninsula, and the Jurassic Ferrar and Karoo LIPs). Magmatism in these areas has been explained by upwelling asthenosphere and plumes and/or upper mantle modified by subduction activity (e.g. Heinonen *et al.* 2014; Luttinen 2018; Choi *et al.* 2019; Elliot and Fleming 2021 and references therein).

In this chapter we provide a synthesis of mantle-source compositions that have been proposed for magmatism across West Antarctica, with reference to East Antarctica and sub-Antarctic island and seamount locations where relevant. It should be noted that this overview chapter is not a summary of the occurrence of mafic igneous rocks in Antarctica. Furthermore, all geochemical and isotopic data gathered from mafic igneous rocks in Antarctica that are discussed within this synthesis were procured from published studies, including compilations of Cenozoic samples from the WARS by Martin *et al.* (2021*b*), Panter *et al.* (2021*a*) and Rocchi and Smellie (2021). A detailed overview of mantle information recorded in igneous rocks from East Antarctica is provided by Foley *et al.* (2021) and is not repeated here. Because igneous activity has occurred intermittently from the Jurassic Period to the present and within a variety of tectonic settings, secular changes in mantle sources have been identified and offer a unique view of the evolution of the Antarctic Plate and the physical and chemical influences of plate dynamics on mantle domains that lie beneath.

Melt sources: deep origins

Sublithospheric mantle sources for magmatism include the melting of convecting upper mantle and emerging mantle plumes. Plumes are broadly defined as thermally buoyant mantle (i.e. having high Rayleigh numbers above that of ambient mantle) that rise as diapirs from deep thermochemical boundary layers (e.g. 410 km, 660 km or from ultralow-velocity zones in the lower mantle) and flatten at the base of the lithosphere where they decompressively melt. Geologically, mantle plumes can cause regional domal uplift, produce magmatism to form LIPs and long-lived linear volcanic ranges (e.g. Hawaiian–Emperor chain), and cause lithospheric thinning that can initiate rifting and continental break-ups. Geophysically, mantle plumes are defined by seismic tomography models that image low-velocity anomalies extending from the surface into the deep mantle (e.g. Montelli *et al.* 2004; Zhao 2007; French and Romanowicz 2015; Marignier *et al.* 2020). Geochemically, mantle plumes are characterized by OIBs with a range of compositions, erupted from oceanic island 'hotspot' volcanoes (e.g. Hofmann 1997; Stracke *et al.* 2005; Hawkesworth and Scherstén 2007; White 2010; Castillo 2015; Zhang *et al.* 2020) whose isotopic end members (HIMU, focus zone (FOZO), and enriched mantle types I (EMI) and II (EMII): Fig. 3) are considered to be the products of long-term recycling of subducted oceanic lithosphere and subduction-modified oceanic crust with or without a sedimentary cargo. However, in continental settings, especially

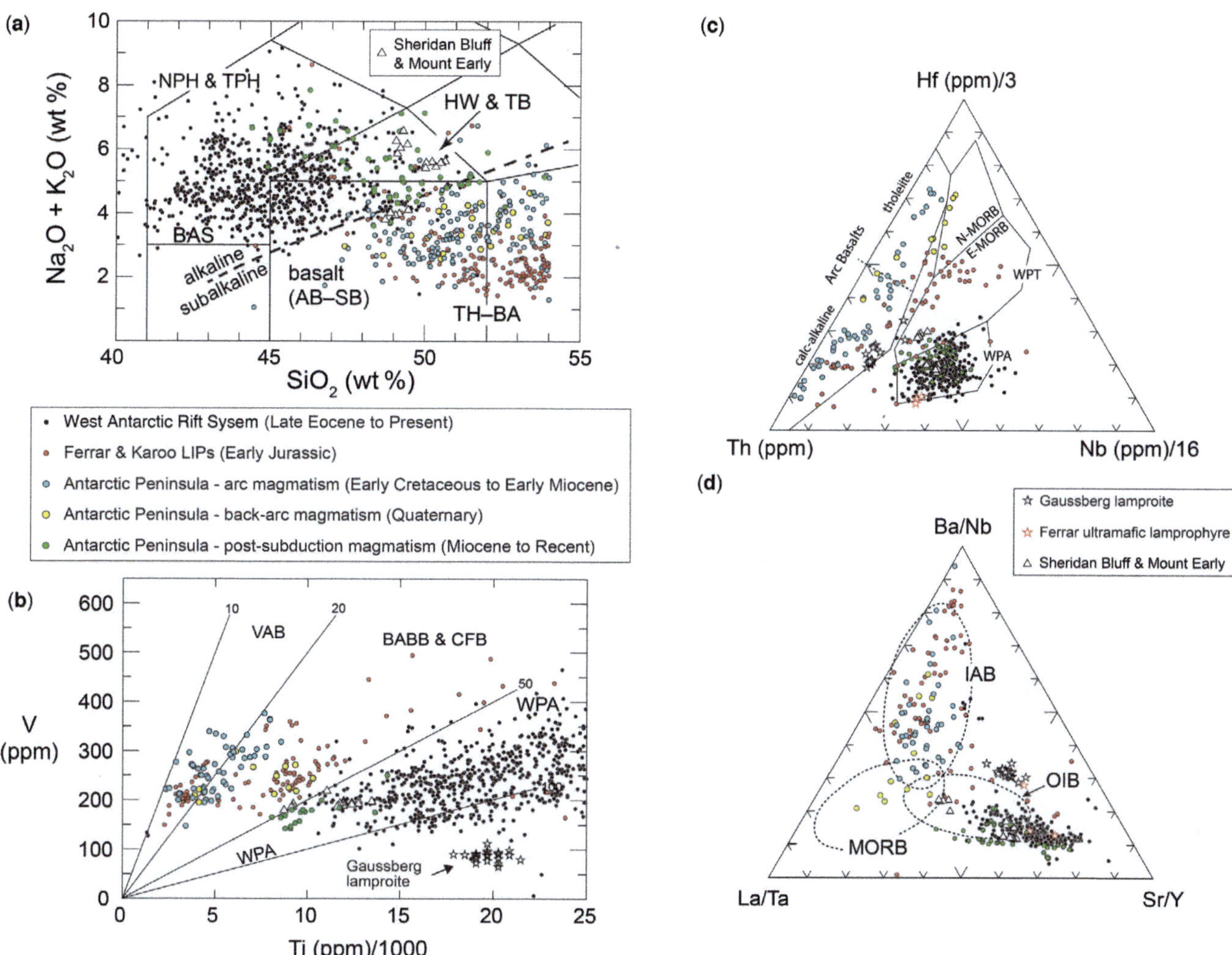

Fig. 2. Plots of mafic igneous rock compositions from Antarctica and their tectonomagmatic associations. Mafic samples are restricted in SiO_2 between 40 and 54 wt%, and MgO between 5 and 15 wt%. All analyses are normalized to a 100% volatile-free basis. (**a**) Total alkalis v. silica classification plot following the criteria of LeMaitre (2002). The thick dashed line delimits the subalkaline from the alkaline compositions (after Irvine and Baragar 1971). Field abbreviations are for nephelinite (NPH), tephrite (TPH), basanite (BAS), hawaiite (HW), trachybasalt (TB), tholeiite (TH), basaltic andesite (BA), alkaline basalt (AB) and subalkaline basalt (SB). Olivine-bearing dolerites of the Ferrar LIP plot with tholeiite and basaltic andesite compositions, and lamproite dykes from the Karoo LIP in Dronning Maud Land (Fig. 1) fall within the basanite classification field. Not plotted are the ultramafic lamprophyre dykes of the Ferrar LIP (Riley *et al.* 2003) with SiO_2 contents of *c.* 35 wt% and $Na_2O + K_2O$ <4 wt% (MgO >16 wt%) and lamproite pillow lavas from the Gaussberg volcano with SiO_2 contents of *c.* 50 wt% and $Na_2O + K_2O$ >12 wt% (MgO *c.* 8 wt%). (**b**) Mafic compositions plotted on a ppm Ti v. V diagram, with tectonomagmatic associations from Shervais (1982). Field labels between lines of equal proportions are volcanic arc basalt (VAB), back-arc basin basalt (BABB), continental flood basalt (CFB) and within-plate alkaline (WPA) basalt. (**c**) Mafic compositions plotted on the Th–Hf–Nb discrimination diagram of Wood (1980). Abbreviations: WPA, within-plate alkaline; WPT, within-plate tholeiite; N-MORB, normal mid-ocean ridge basalt; E-MORB, enriched mid-ocean ridge basalt. (**d**) A La/Ta–Ba/Nb–Sr/Y basalt discrimination diagram from Zhang *et al.* (2020). Zhang *et al.* (2020) employed GEOROC and PetDB global databases and big data statistical methods to define confidence ellipses (dashed) that discriminate island arc basalts (IAB) from mid-ocean ridge basalts (MORB) and ocean island basalts (OIB). Data sources for Antarctic mafic samples for the West Antarctic rift system are from Martin *et al.* (2021*b*), Panter *et al.* (2021*b*) and Rocchi and Smellie (2021), which are compilations of previously published and unpublished datasets for the Ross Sea and Marie Byrd Land regions of the rift. Please refer to these publications for the original data sources. Samples for the Ferrar and Karoo LIPs and arc magmatism from the Antarctic Peninsula are from the GEOROC database (http://georoc.mpch-mainz.gwdg.de/georoc/), samples for back-arc magmatism from the Bransfield Strait, Antarctic Peninsula are from Keller *et al.* (2002), and the compositions for post-subduction magmatism on the Antarctic Peninsula are from Hole (2021) and the GEOROC database. The compositions of ultramafic lamprophyre dykes from the Ferrar LIP are from Riley *et al.* (2003) and lamproite pillow lava compositions from Gaussberg in East Antarctica provided by Murphy *et al.* (2002). Basaltic compositions from Sheridan Bluff and Mount Early are from Panter *et al.* (2021*a*).

beneath thick cratons or inboard of active subduction zones, assigning a geochemical signature to a mantle plume is a complex undertaking that requires the unravelling of other potential lithospheric and sublithospheric influences on magma composition (e.g. Yellowstone hotspot: Hanan *et al.* 2008; Leeman *et al.* 2009; Stefano *et al.* 2019).

Mantle-plume sources have been proposed for Antarctic magmatism based on geological, geophysical and geochemical evidence. Magmatism that produced the Ferrar and Karoo LIPs and fostered Gondwana break-up, which separated Africa from Antarctica by the Late Jurassic, is one such case (Storey and Kyle 1997). Outcrops of the Middle Jurassic Ferrar and Karoo LIPs are intermittently exposed from northern Victoria Land along the Transantarctic Mountains to Dronning Maud Land (Fig. 1). The magmatism occurred over a very short time interval of less than 0.4 myr at *c.* 183 Ma (Svensen *et al.* 2012; Elliot and Fleming 2021) with an estimated volume of greater than 1×10^6 km^3 (Eldholm and Coffin 2000; Burgess *et al.* 2015). There has been extensive study and debate on the origin of Ferrar–Karoo magmatism stemming from its unique distribution and geochemistry relative to other LIPs (e.g. Marsh 2004; Elliot and Fleming

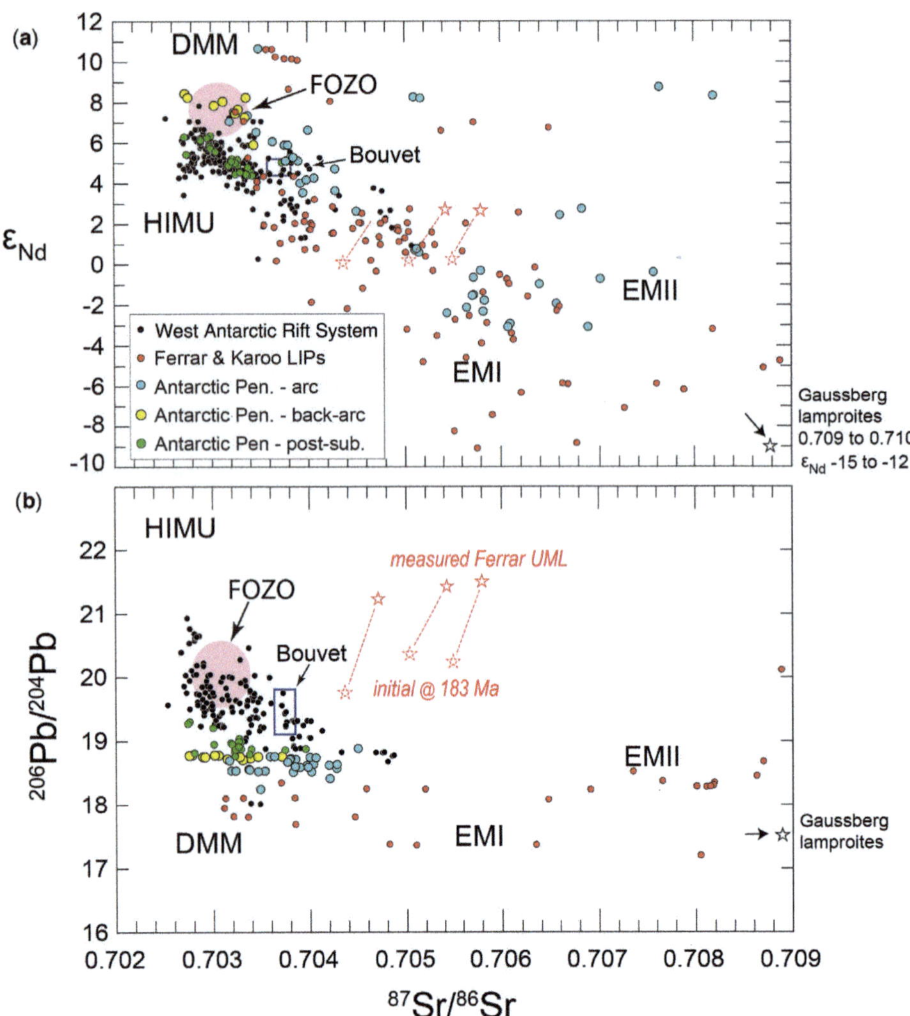

Fig. 3. Variations in $^{87}Sr/^{86}Sr$ v. (**a**) ε_{Nd} and (**b**) $^{206}Pb/^{204}Pb$ in Antarctic mafic igneous rocks. Data sources are the same as in Figure 2. Compositional restrictions in major element concentrations are also the same as in Figure 2 except for Ferrar–Karoo, which include some samples with higher SiO_2 contents (up to 62 wt%), and several samples with higher (up to 20 wt%) and lower (≥4 wt%) MgO contents. Mantle end-member compositions HIMU, DMM, EMI and EMII are plotted after Hofmann (2007), and FOZO (Hart *et al.* 1992) are after Stracke *et al.* (2005). The isotopic composition of the Bouvet OIB is from Riley *et al.* (2003 and references therein). The initial isotopic ratios for the Ferrar ultramafic lamprophyres (UML) are corrected to 183 Ma (Riley *et al.* 2003). All other isotopic data are plotted as measured values. Note that in (b) the Pb isotopic values are not available for samples of Antarctic Peninsula arc magmatism that have $^{87}Sr/^{86}Sr$ values >0.7046.

2021). In Antarctica, the long and narrow distribution of outcrops is explained by two different scenarios: (1) magmas are sourced from a single region of mantle melting followed by long-distance lateral migration within the crust; and (2) magmas were supplied from multiple, roughly aligned mantle-melt regions (i.e. 'linear source') and emplaced within the crust with restricted lateral migration (Elliot and Fleming 2021). Geochemically, mafic rocks of the Karoo and the Ferrar LIPs have mantle-like oxygen and osmium isotopic signatures; however, they also have enriched Sr, Nd and Pb isotopic values (Fig. 3), which, along with trace element patterns on normalized multi-element plots (not shown), suggest addition from continental crust (Luttinen 2018; Elliot and Fleming 2021). To explain both the geographical distribution (Fig. 1) and geochemistry, some researchers call upon a single mantle-plume source, now represented by compositions from the current Bouvet hotspot (Fig. 3) located in the South Atlantic Ocean, and long-distance transport of magma at various depths within the crust (e.g. Storey and Kyle 1997; Elliot *et al.* 1999; Riley *et al.* 2003; Vaughan and Storey 2007). Other researchers call for a linear zone of melting of subduction-modified mantle along the length of the palaeo-Pacific Gondwana margin (e.g. Hergt *et al.* 1991; Molzahn *et al.* 1996; Ivanov *et al.* 2017). In a recent study by Choi *et al.* (2019), an arc-like mantle source is favoured over a plume source based on platinum group elements abundances and Os isotope systematics. The authors suggest that rapid decompression of hydrated mantle materials parallel to the Gondwana margin subduction zone facilitated the large-volume and short-lived duration of Ferrar magmatism. Alternatively, a mixture of subduction-modified MORB-type mantle (pyroxenite source) and depleted ambient upper-mantle or plume(?) recycled MORB-type mantle (peridotite source) is used by Heinonen *et al.* (2014) to explain the Sr, Nd, Pb and Os isotopic and trace element compositions of mafic Karoo dykes in Dronning Maud Land. Luttinen (2018) proposed that melting occurred under the influence of both active subduction and an active mantle plume head. Here, subduction-modified mantle sources explain the Nb-depleted Karoo compositions found in Dronning Maud Land (Fig. 1) and, along with the distribution of similarly Nb-depleted compositions of the Ferrar LIP, the plume-influenced and subduction-influenced mantle regions of the palaeo-Pacific margin of Gondwana (Fig. 4). In summary, it is likely that both mantle-plume and subduction-modified mantle sources supplied Ferrar and Karoo LIP magmas based on the geological and geochemical evidence. This mixture of primary mantle-source types is reinforced by the distribution of Karoo–Ferrar mafic magma compositions across fields that define both arc basalts and within-plate tholeiitic flood basalts in Figure 2b–d.

Further disintegration of Gondwana in the Late Cretaceous separated the continental fragments that became Zealandia. The rapid transition (*c.* 110–80 Ma) from subduction to extension to seafloor spreading occurred without voluminous magmatism (i.e. no LIP formed) and did not produce any known volcanism in Antarctica (Smellie *et al.* 2020). Yet, there was relatively extensive intraplate volcanism in Zealandia occurring prior to and just after break-up between 99 and 79 Ma (Hoernle *et al.* 2020 and references therein). A large Late Cretaceous plume may have aided in this continental break-up event and may have provided a mantle reservoir – a 'fossil

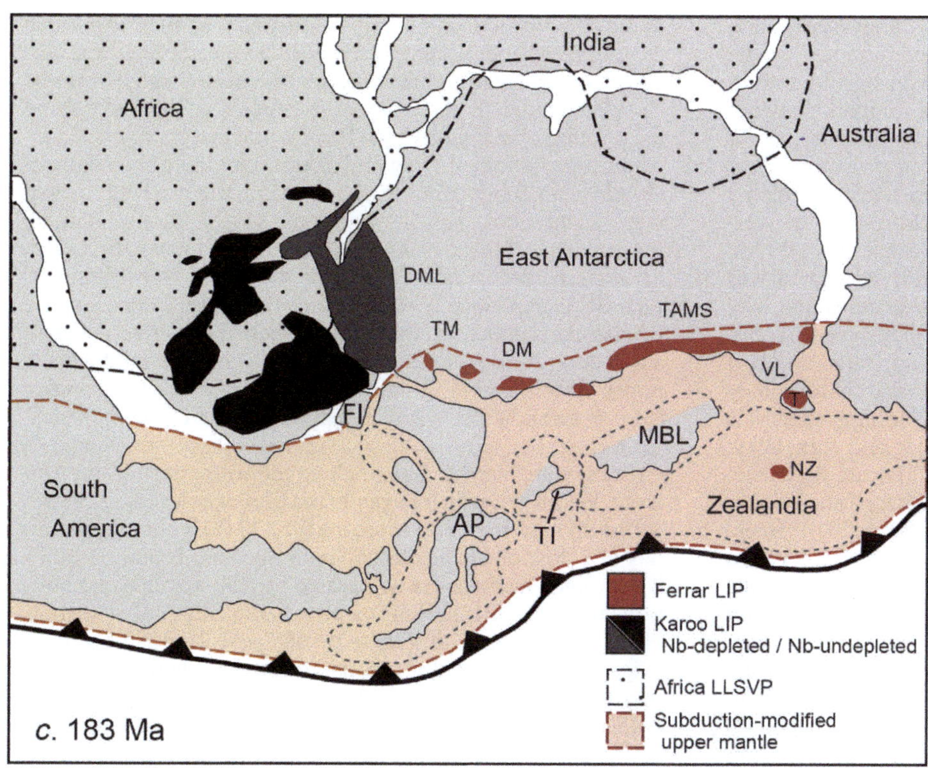

Fig. 4. Plate reconstruction at *c.* 183 Ma after Storey and Kyle (1997) showing for comparison the position of the Ferrar LIP after Smellie (2021) and the Karoo LIP after Luttinen (2018). The Karro LIP is divided into Nb-depleted (grey) and Nb-undepleted (black) regions, and together mark the location of a Jurassic plume head. Shown for comparison is the region were the shallow mantle is inferred to have been modified by subduction (Luttinen 2018) (in red) and the sub-African large low-shear velocity province (LLSVP) projected onto the surface from a depth of 2800 km (Luttinen 2018: stippled pattern). AP, Antarctic Peninsula; DM, Dufek Massif; DML, Dronning Maud Land; FI, Falkland Islands; MBL, Marie Byrd Land; NZ, New Zealand; T, Tasmania; TAMS, Transantarctic Mountains; TI, Thurston Island; TM, Theron Mountains; VL, Victoria Land.

plume source' for intraplate alkaline magmatism in West Antarctica throughout the Cenozoic (Lanyon *et al.* 1993; Weaver *et al.* 1994; Rocholl *et al.* 1995; Hart *et al.* 1997; Storey *et al.* 1999; Panter *et al.* 2000; Kipf *et al.* 2014; Park *et al.* 2019).

Apart from the magmatism along the Antarctic Peninsula and a few isolated volcanic centres in the southern Transantarctic Mountains (Mount Early and Sheridan Bluff) and in East Antarctica (Gaussberg), all other known Cenozoic igneous activity exposed on the continent is associated with the WARS (Fig. 1). The WARS initiated during the Late Cretaceous breakoff of Zealandia, and continued to develop with motion occurring between the East Antarctic craton and the geotectonic provinces of West Antarctica (Granot *et al.* 2013; Jordan *et al.* 2020; Storey and Granot 2021). The extension has produced a very broad region of thinned continental lithosphere (Behrendt *et al.* 1991; Behrendt 1999) that is 3000 km in length and 750–1000 km wide. Rift magmatism began approximately 50 myr ago (Tonarini *et al.* 1997; Rocchi *et al.* 2002) but only became abundant and widespread since the Middle Miocene (Martin *et al.* 2010). The WARS-related volcanism produced over 30 major polygenetic shield-like composite volcanoes (each $\geq$30–1800 km^3 of exposed volume above ice sheet or sea level) and numerous monogenetic volcanic fields, including seamounts in the southern Ross Sea (Aviado *et al.* 2015; Martin *et al.* 2021*b*) and in the oceanic Adare Basin (Fig. 1) (Panter *et al.* 2018). Mafic compositions from the WARS are almost exclusively alkaline (Fig. 2a) and consistently classify on tectonic discriminate diagrams as within-plate alkali basalt (Fig. 2b, c) that are OIB-like (Fig. 2d). Isotopically, WARS compositions indicate a mantle source with strong HIMU- and FOZO-like affinities (^{206}Pb/^{204}Pb $\cong$ 19–21, ^{87}Sr/^{86}Sr $\cong$ 0.7027–0.7040, $\varepsilon_{Nd} \cong$ 3–7: Fig. 3). The affinity of WARS magmatism with the geochemical and isotopic characteristics of ocean-island hotspot volcanoes has been the cornerstone of mantle-plume models. Plume activity beneath West Antarctica has also been based on the geographical distribution and age migration patterns of volcanism, as well as more recently by seismic tomography (e.g. Bredow and Steinberger 2021).

An active mantle plume beneath Marie Byrd Land (Fig. 1) has been used to explain the geochemistry of Cenozoic magmatism, along with regional doming and volcanic migration patterns (LeMasurier and Rex 1989; Hole and LeMasurier 1994; Hole *et al.* 1994; Hansen *et al.* 2014; Wiens *et al.* 2021). The volcanoes in Marie Byrd Land province are mostly located on a *c.* 1000 × 500 km structural dome (LeMasurier 2006) that lies on the north flank of the WARS and has been interpreted as a topographical expression of a mantle plume (LeMasurier and Landis 1996). The mean crustal thicknesses beneath the Marie Byrd Land dome are calculated from seismic data to be 28–33 km, which is 5–10 km thicker than in the rest of the WARS (Chaput *et al.* 2014; Ramirez *et al.* 2017; Shen *et al.* 2018; Wiens *et al.* 2021). However, the dome is not attributed to isostatic compensation of thicker crust but to a thermal anomaly beneath this volcanic province (Winberry and Anandakrishnan 2004). Marie Byrd Land primitive magma compositions have been explained by mixing between subduction-modified lithosphere (discussed below) and HIMU-like and/or FOZO-like mantle-plume materials (e.g. Hart *et al.* 1995, 1997; Panter *et al.* 2000, 2006; Finn *et al.* 2005; Gaffney and Siddoway 2007; LeMasurier *et al.* 2016). There has been a documented west–east gradient in isotopic and trace element composition in primitive volcanic rocks through Marie Byrd Land (LeMasurier *et al.* 2016) that continues into western Ellsworth Land with an interpreted distinct change in Pb isotopic and trace element composition (Hart *et al.* 1995; LeMasurier *et al.* 2016). In a recent compilation and review of basalt compositions from Marie Byrd Land and western Ellsworth Land (Fig. 1), Panter *et al.* (2021*b*), using an extended dataset of major and trace element concentrations and isotopic values (Sr, Nd and Pb), support this regional gradient and show a more extensive and gradual transition from central Marie Byrd Land (e.g. Hobbs Coast Volcanic Field (HCVF) and Crary Mountains Volcanic Field (CMVF): Fig. 1) east into Ellsworth Land (e.g. Jones Mountains Volcanic Field (JMVF)) and west to the Fosdick Mountains (e.g. Fosdick Mountains Volcanic Field (FMVF), Fig. 1). An additional observation is that the current known distribution of the most radiogenic Pb signatures ($^{206}Pb/^{204}Pb$ >20) found in Marie Byrd Land reveals a broadly linear feature orientated oblique to the main axis of the WARS (Fig. 1; also refer to figs 44 and 45 in Panter *et al.* 2021*b*) that overlaps with a broad eastward decrease in $^{143}Nd/^{144}Nd$ ratios, which LeMasurier *et al.* (2016) concluded, along with evidence from Sr isotopes and trace elements, to be the result of an increasing influence by subduction-modified mantle that was formed during convergence at the proto-Pacific margin of Gondwana. At *c.* 90 Ma the area encompassed by the radiogenic Pb signatures was roughly parallel with several other colinear features and may reflect continental strike-slip zones formed in response to oblique subduction prior to Gondwana break-up (refer to figs 4 and 5 in Eagles *et al.* 2004). Translithospheric faulting may have helped to localize upwelling HIMU plume materials for melting beneath the region (LeMasurier and Rex 1989; Panter *et al.* 1997). Geophysical studies indicate high heat flow (Seroussi *et al.* 2017; Shen *et al.* 2020; Pappa and Ebbing 2021) and slow seismic velocities (Hansen *et al.* 2014; Heeszel *et al.* 2016; Lloyd *et al.* 2020; Lucas *et al.* 2020; Wiens *et al.* 2021) that extend at least to the mantle transition zone (i.e. 400–600 km) beneath central Marie Byrd Land, thus supporting a mantle-plume influence on volcanism and tectonism.

The 'Erebus plume' was proposed by Kyle *et al.* (1992) and Esser *et al.* (2004) to explain the geochemistry, along with volcanic patterns and the generation of large volumes of phonolitic magmas from basanitic melts beneath the active Mount Erebus volcano (Fig. 1). The authors suggested that uplift and crustal extension was enhanced by plume buoyancy, and may explain the radial pattern of volcanism on Ross Island and a similar pattern centred on the Mount Discovery volcano *c.* 100 km to the south. More recent detailed geochemical and isotopic (Sr, Nd, Hf and Pb) studies of Mount Erebus and the rest of Ross Island by Sims *et al.* (2008) and Phillips *et al.* (2018) support a mantle-plume source. Phillips *et al.* (2018) modelled mixing between DMM and HIMU sources to explain Ross Island samples, and concluded that the HIMU isotopic signature originated from high time-integrated (Archean–Early Proterozoic) U/Pb and Th/Pb ratios from crustal materials that were recycled from the deep mantle. They and Emry *et al.* (2020) employed seismic tomography models, which suggested that a low-velocity zone exists to a depth of *c.* 1200 km and the mantle transition zone is thinner beneath Ross Island, supporting the deep plume hypothesis. Alternatively, Day *et al.* (2019) in analysing mantle xenoliths and lava flow rocks from Hut Point Peninsula on Ross Island defined $^3He/^4He$ ratios of 6.8 ± 0.3 R_A (2σ: where R_A is atmospheric $^3He/^4He$) beneath Ross Island that are distinct from high-$^3He/^4He$ plume mantle (e.g. ≥9 R_A: Class and Goldstein 2005). Nardini *et al.* (2009) came to a similar conclusion that northern Victoria Land volcanism with low mantle $^3He/^4He$ ratios (5.73–7.22 R_A) was a strong argument against involvement of a Cenozoic plume. This would be consistent with mantle images that do not show a deep (>150 km) root beneath Ross Island (e.g. Faccenna *et al.* 2008), and the observation that Ross Island and surrounding areas are a site of rifting rather than doming (Cooper *et al.* 2007; Martin *et al.* 2013).

Melt sources: shallow origins

Shallow melt sources have been proposed for magmatism in Antarctica, and include depleted upper-mantle sources (i.e. MORB-types), metasomatized lithospheric mantle, subducting slab and subduction-modified mantle wedge (i.e. asthenospheric mantle that lies above a subducting slab and below the overriding plate). The contributions from subduction-related processes to mantle sources are most clearly expressed in mafic compositions intimately associated with the progressive development of the Antarctic Peninsula volcanic arc. Yet, subduction dominated the tectonic regime episodically from the Neoproterozoic to the Late Cretaceous (*c.* 450 myr) along the pre-dispersal length of the Pan-Pacific margin of the Gondwana supercontinent (Cawood 2005), including the southern Gondwanan subduction zone and magmatic arc (Fig. 4) that was synchronous with the initial phase of Jurassic break-up magmatism (Rapela *et al.* 2005). As previously discussed, the regional geochemical characteristics of the Karoo and Ferrar LIPs is explained by Luttinen (2018) to be a consequence of magma production under the influence of active zones of subduction and plume upwelling. Subduction-influenced upper mantle may still reside beneath regions of Antarctica (Fig. 4) and could be a source tapped by Late Cenozoic volcanism (Fig. 1).

Continental arc volcanism along the west coast of the Antarctic Peninsula, which began in the Early Cretaceous and was active up until the Early Miocene (*c.* 23 Ma), continues to this day at a slow rate beneath the South Shetland Islands (Fig. 1). The erupted magmas are dominated by calc-alkaline compositions (e.g. basaltic andesite) and tholeiite (Fig. 2a), and show clear geochemical affinities to arc and back-arc settings (Fig. 2b–d). Isotopically, arc magmas were generated by melting of enriched mantle sources, as shown by mafic samples with high $^{87}Sr/^{86}Sr$ ratios and low ε_{Nd} values in Figure 3a. However, other arc magmas, along with back-arc compositions, reveal mantle sources that are much less enriched with much lower $^{87}Sr/^{86}Sr$ and $^{206}Pb/^{204}Pb$ ratios (Fig. 3b), and have higher measured ε_{Nd} values. Furthermore, samples with $^{87}Sr/^{86}Sr$ ratios of <0.7040 have Nd and Pb isotopic compositions generated from mantle that is more depleted than the post-subduction magmatism (discussed below). The variable origins of mafic compositions are explained by Leat and Riley (2021) as: (1) melting of variably depleted mantle-wedge material fluxed by hydrous fluids from the subducting slab (e.g. calc-alkaline series: South Shetland Islands); (2) partial melting of slab material and the equilibration of those melts under hydrous conditions within the mantle wedge prior to eruption (e.g. high-Mg andesite group: Alexander Island); (3) partial melting of garnet peridotite within the mantle wedge that incorporated melts of mafic slab material (e.g. adakitic group: South Shetland Islands); and (4) mantle partial melting that was triggered by arc extension or within a back-arc setting (e.g. high-Zr group: southern Graham Land–northern Palmer Land).

Since the Middle–Late Miocene, post-subduction volcanism produced extensive monogenetic volcanic fields, large (≥30 km^3) polygenetic shield volcanoes and small isolated centres scattered along the Antarctic Peninsula (Fig. 1). The geochemistry of mafic compositions is almost exclusively alkaline (Fig. 2a) and classify as being from a within-plate tectonic environment (Fig. 2b, c). Furthermore, they have a strong affinity to OIBs, as well as to intraplate mafic compositions from the WARS (Fig. 2c, d). The compositional resemblance to WARS basalts is also apparent with respect to Sr and Nd isotopes (Fig. 3a), although with Pb values that are much less radiogenic (Fig. 3b). The post-subduction intraplate volcanism in the Bransfield Strait is considered to be the result of extension related to the rollback of the Phoenix Plate, leading to partial melting of shallow mantle, to a high degree, with variable input of materials and fluids from the subducting slab (Haase and Beier 2021). The origin of the remaining post-subduction alkaline volcanism on the Antarctic Peninsula has been ascribed to back-arc extension or slab-window tectonics as a way to promote decompression melting of asthenospheric mantle. As an alternative to the slab-window hypothesis, Hole (2021) proposes that partial melting of slab-hosted pyroxenite can produce the geochemical characteristics of the post-subduction alkaline volcanism and may account for the lack of age progression of volcanism, as well as the relatively short time period of activity (i.e. predominantly ≤7 myr).

Submarine volcanism to the north of the South Shetland Islands in the Drake Passage (Fig. 1) records a transition in shallow mantle sources between the Late Miocene and the Pleistocene (Choe *et al.* 2007). Choe *et al.* (2007) found that prior to the shutdown of the Antarctic–Phoenix Ridge at 3.3 Ma (Livermore *et al.* 2000) partial melting of peridotite at shallow depths in the mantle produced tholeiitic basalts (6.4–3.5 Ma) with a normal (N)-MORB source affinity. After spreading ceased, mildly alkaline compositions were erupted (3.1–1.4 Ma) with an enriched (E)-MORB source affinity. Choe *et al.* (2007) suggested that the later phase of partial melting occurred to a smaller degree at greater depths and was facilitated by pyroxenite, possibly localized in veins (i.e. metasomatized mantle).

Metasomatized lithospheric mantle has been proposed as a shallow mantle reservoir for alkaline magmas erupted in continental settings (e.g. Stein *et al.* 1997; Jung *et al.* 2005; Panter *et al.* 2006; Ma *et al.* 2011; Mayer *et al.* 2014; Rooney *et al.* 2014, 2017; Scott *et al.* 2020), as well as oceanic settings (Pilet *et al.* 2008; Pilet 2015), and is prevalent in hypotheses on the origin of Cenozoic magmatism associated with the WARS (Hart *et al.* 1995; Rocchi *et al.* 2002; Panter *et al.* 2003, 2018; Gaffney and Siddoway 2007; Nardini *et al.* 2009; Rilling *et al.* 2009; Perinelli *et al.* 2011; Martin *et al.* 2013, 2015, 2021*b*; Aviado *et al.* 2015; LeMasurier *et al.* 2016; Correale *et al.* 2019; Day *et al.* 2019; Kim *et al.* 2019; Giacomoni *et al.* 2020). The foundations of this idea are to account for the relatively uniform geochemical and isotopic signatures (e.g. Figs 2c & 3) of rocks from three widely dispersed volcanic regions in West Antarctica (Fig. 1) and for the absence of voluminous tholeiitic magma series compositions that would be expected from the melting of a single plume head of the size required to encompass these areas. More specifically, the geochemical arguments for metasomatized sources supplying WARS magmatism are made based on their enriched concentrations of incompatible trace elements and relative depletion in potassium, as illustrated by mafic compositions plotted on mantle-normalized multi-element diagrams (Fig. 5). The negative K anomalies are considered a result of incomplete melting and retention of hydrous potassic minerals (amphibole ± phlogopite) in their sources. Mantle amphiboles are stable at temperatures of <1150°C (Mandler and Grove 2016) and therefore can reside only within the lithosphere. Lithospheric mantle xenoliths that contain amphibole (e.g. disseminated and vein amphibole, as well as their occurrence as reaction replacements of pyroxene) have been reported in both northern Victoria Land (Coltorti *et al.* 2004; Perinelli *et al.* 2006, 2011, 2017; Melchiorre *et al.* 2011; Broadley *et al.* 2016) and southern Victoria Land (Gamble and Kyle 1987; Gamble *et al.* 1988; Martin *et al.* 2014; Day *et al.* 2019). Except for rare occurrences of apatite, hydrous phases are absent in mantle xenoliths collected within the Marie Byrd Land Volcanic Province (Handler *et al.* 2021).

The cause of metasomatism has been attributed to several mechanisms including enrichments from subduction-derived melts and fluids, plume-derived melts and fluids, and extension-related auto-metasomatism. Evidence for metasomatism of West Antarctic lithosphere by subduction-zone processes is well documented in mantle xenoliths. Metasomatism of the lithosphere in northern Victoria Land by marine-derived volatiles is indicated by heavy halogen (Br and I) and noble gas (Ar, Kr and Xe) compositions liberated from fluid inclusions in olivine and pyroxene in peridotite xenoliths (Broadley *et al.* 2016). Broadley *et al.* (2016) suggested that the volatiles were released, perhaps at depths of up to 200 km (Sumino

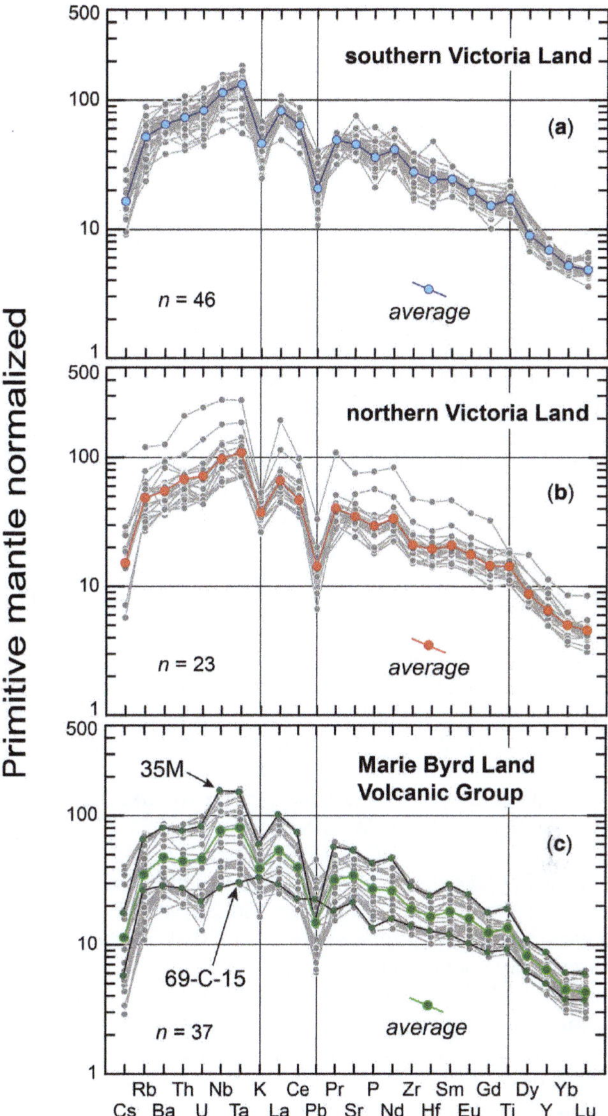

Fig. 5. A trace element comparison of mafic basaltic compositions within the West Antarctic Rift System modified from Panter (2021). Elemental concentrations are normalized to the primitive mantle values provided by McDonough and Sun (1995). Referenced datasets for (**a**) southern Victoria Land (Erebus Volcanic Province) are from Martin *et al.* (2021*b*), (**b**) northern Victoria Land (Hallett Volcanic Province and Melbourne Volcanic Province) from Rocchi and Smellie (2021) and (**c**) Marie Byrd Land and western Ellsworth Land (Marie Byrd Land Volcanic Group after Wilch *et al.* (2021)) are from Panter *et al.* (2021*b*). Labelled in (c) are basanite sample 35M (Hart *et al.* 1997) from the Hobbs Coast Volcanic Field and alkali basalt sample 69-C-15 (Hart *et al.* 1995) from the Jones Mountains Volcanic Field (Fig. 1).

et al. 2010; Kendrick *et al.* 2013), during subduction and were incorporated into the overlying mantle wedge and subcontinental lithospheric mantle beneath the Gondwana continental arc during the Paleozoic. In another noble gas study of mantle xenoliths from northern Victoria Land, Correale *et al.* (2019) found homogeneous $^3He/^4He$ (7.1 ± 0.4 R_A) and low $^4He/^{40}Ar^*$ ratios (<0.4) from fluid inclusions in olivine, pyroxene and amphibole that they interpret as a result of subcontinental lithospheric mantle being metasomatized by depleted asthenospheric melts (MORB-type). In southern Victoria Land, Day *et al.* (2019) concluded, based on a petrological, geochemical and isotopic (He and Os) study of glass- and amphibole-rich veins in peridotites, that the mantle lithosphere beneath Ross Island (approximating to Mount Erebus: Fig. 1) was metasomatized during the Cretaceous by subduction-related processes prior to Gondwana break-up. Also, within southern Victoria Land, a subduction-enriched component has been inferred from peridotite (EMI) and pyroxenite (EMII) xenoliths 30–100 km distant from Ross Island (Martin *et al.* 2014, 2015, 2021*a*), and in the primitive volcanic rocks of the Erebus Volcanic Province (Martin *et al.* 2013, 2021*b*). Whilst the age of metasomatism is not definitively constrained, existing whole-rock isochron dates on a phlogopite-bearing clinopyroxenite xenolith from the region (and the micas within it) give an age of 439 ± 5 Ma (McGibbon 1991), and an adjacent carbonatite dyke is dated at 531 ± 5 Ma (Hall *et al.* 1995) and may indicate that this was a time of general (carbonatite) metasomatism of the lithospheric mantle in this part of southern Victoria Land. Martin *et al.* (2014) inferred that alkaline metasomatism recorded in plagioclase-bearing spinel lherzolite xenoliths is coincident with alkaline magmatism in the region from *c.* 25 Ma. Handler *et al.* (2021) describe the effects of metasomatic processes on lithospheric mantle xenoliths from Marie Byrd Land, which, based on Os and Nd model ages, occurred between 570 and 130 Ma. These authors conclude that the geochemical and isotopic evidence, along with the ages, are consistent with a lithospheric mantle that was influenced by the long history of Paleozoic subduction that occurred beneath this region.

Geochemical and isotopic studies of mantle xenoliths and their basaltic hosts provide strong support for metasomatism caused by subduction-zone processes and that subduction-modified mantle is a common melt source for alkaline magmatism associated with the WARS (Hart *et al.* 1995; Martin *et al.* 2015; LeMasurier *et al.* 2016; Panter *et al.* 2021*b*; Coltorti *et al.* 2021; Handler *et al.* 2021). However, as has been proposed for the origin of the Karoo and Ferrar LIPs (discussed above), both deep plume sources and shallow metasomatic sources may coexist and supply melts for magmatism. Basalts from the CMVF in central Marie Byrd Land (Fig. 1) have geochemical and isotopic characteristics that led Panter *et al.* (2000) to conclude that mixing had occurred between a HIMU-like ($^{206}Pb/^{204}Pb$ >20.5) plume and a hydrous lower-μ mantle component. They propose a scenario in which a mantle plume was trapped and stored (‘fossilized’) beneath pre-existing metasomatized lithosphere prior to the Late Cretaceous break-up of Zealandia from West Antarctica. The geographical distribution of the most radiogenic Pb isotopic (HIMU-like) signatures are restricted to the middle portion of the region that comprises the Marie Byrd Land Volcanic Group (MBLVG: Wilch *et al.* 2021), which includes the HCVF (Hart *et al.* 1997) and the CMVF (Fig. 1), while the less radiogenic Pb (lower-μ source) signatures generally lie at the periphery of the MBLVG: that is, the JMVF and the FMVF (Fig. 1) (Hart *et al.* 1995; Gaffney and Siddoway 2007; LeMasurier *et al.* 2016; Panter *et al.* 2021*b*). In Figure 6, this relationship also correlates broadly with K/Ta ratios, and can be modelled by mixing between the HIMU and lower-μ end members within the MBLVG. The K/Ta ratios of mafic samples plotted in Figure 6 are proportional to the magnitude of the K anomalies (K/K*) shown on the mantle normalized multi-element diagram in Figure 5c. The relationship also exists for Ce/Pb ratios that are also proportional to Pb anomalies shown in Figure 5c, where samples with lower Ce/Pb ratios (*c.* <30) and higher Pb/Pb* values (*c.* >0.5: a value of 1 represents a linear alignment of Ce–Pb–Pr on multi-element diagrams) have less radiogenic Pb isotopic signatures. It is important to note in Figure 6 that the MBLVG samples with lower Pb isotopic ratios (lower-μ source) and lower K/Ta ratios (and lower Ce/Pb ratios, not shown) approach back-arc and post-subduction mafic compositions from the Antarctic Peninsula, which supports the existence of a subduction-modified mantle component beneath Marie Byrd Land and western Ellsworth Land to supply volcanism (e.g. Hart *et al.* 1995; Panter *et al.* 2006; LeMasurier *et al.* 2016).

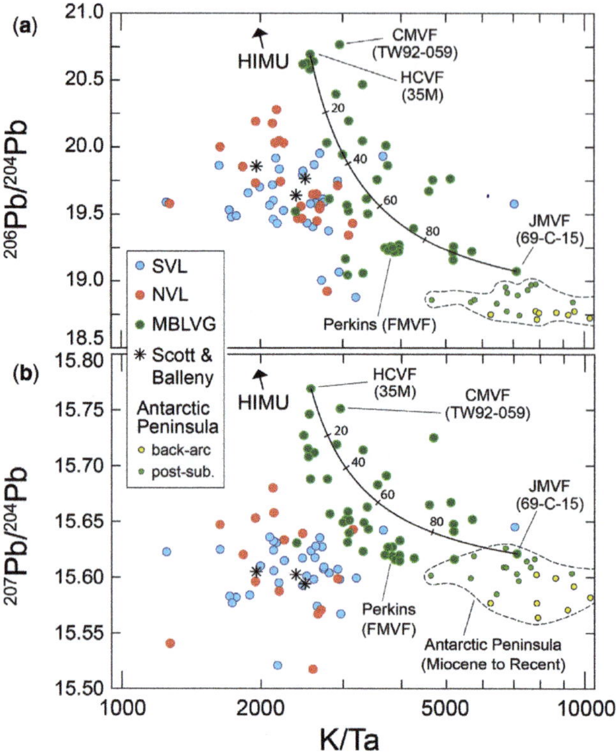

Fig. 6. Plots of K/Ta v. measured (**a**) $^{206}Pb/^{204}Pb$ and (**b**) $^{207}Pb/^{204}Pb$ ratios of mafic basaltic compositions from the West Antarctic Rift System (symbols are the same as those used in Fig. 5) and from the Antarctic Peninsula (back-arc and post-subduction magmatism; symbols and sources given in Fig. 2). Data from the Balleny and Scott islands, located north of the Ross Sea (Fig. 1), are from Hart *et al.* (1992) and Hart and Kyle (1994). MBLVG, Marie Byrd Land Volcanic Group; the other abbreviations are defined in Figures 1 and 3. Hypothetical curves of mixing between basanite (sample 35M) from the Hobbs Coast Volcanic Field (HCVF) and alkali basalt (sample 69-C-15) from the Jones Mountains Volcanic Field (JMVF) are marked at 20% increments. The curves illustrate that MBLVG compositions were likely to have been generated by the melting of a HIMU-like component (mantle plume?) and a lower-μ subduction-modified mantle component characterized by higher K/Ta (as well as higher K/Nb) and lower Ce/Pb ratios (*c.* <20). Alkali basalt sample TW92-059 (Panter *et al.* 2000) from the CMVF has the highest $^{206}Pb/^{204}Pb$ ratio (20.93) yet measured in West Antarctica. Also indicated are samples from Mount Perkins located within the Fosdick Mountains Volcanic Field (FMVF) (Gaffney and Siddoway 2007; Panter *et al.* 2021*b*). The FMVF and the JMVF lie approximately 1500 km apart on the periphery of the region that contains the MBLVG (Fig. 1).

In northern Victoria Land, Rocholl *et al.* (1995) proposed three mantle source components to explain the geochemistry and isotopic compositions of mafic continental volcanism in this region: DMM, fossilized plume head (HIMU) and enriched mantle (EM). The enriched mantle is considered to reside within the subcontinental lithosphere. The authors envisage a pre-rift mantle that was stratified in these components (i.e. with depth from EM to HIMU to DMM) and that during rift development the rising asthenosphere progressively replaced the overlying sources to supply DMM-type mantle in greater proportions within the rift relative to the rift shoulder. An origin from metasomatized lithosphere without any plume influence is proposed for continental and oceanic (Adare Basin seamounts: Fig. 1) volcanism from this region (Rocchi *et al.* 2002; Nardini *et al.* 2009; Panter *et al.* 2018; Rocchi and Smellie 2021). Models for metasomatic origins are explained by a multistage process that begins with

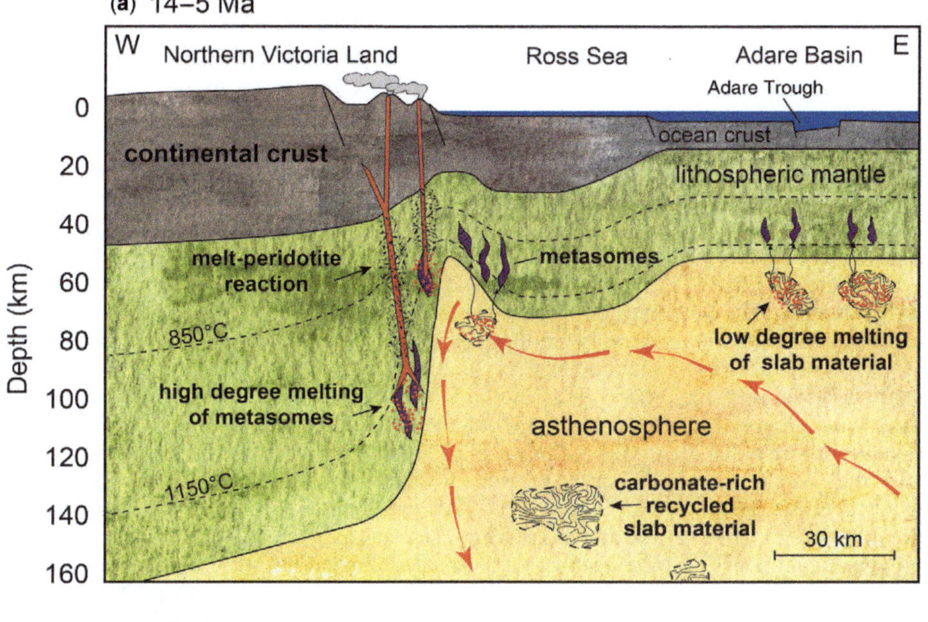

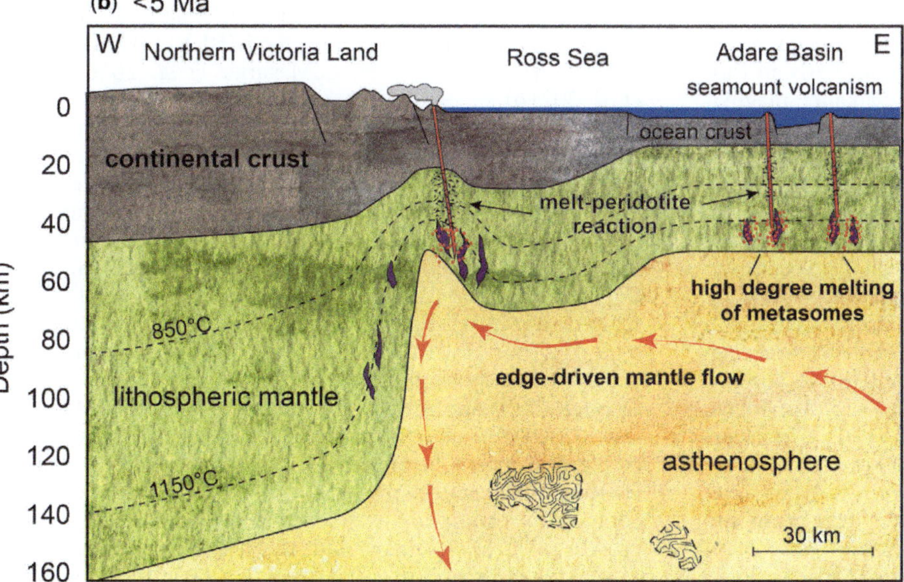

Fig. 7. Schematic model for the Late Oligocene–recent petrogenesis of alkaline volcanism in northern Victoria Land (Hallett Volcanic Province) and the oceanic Adare Basin seamounts (Fig. 1) modified from Panter *et al.* (2018). Plate architecture from craton to ocean in the northern Ross Sea region is shown for two time frames: 14–5 Ma and <5 Ma. Thinned lithosphere ('necked zone') beneath the rift boundary is considered to be a result of an earlier period (*c.* 80–40 Ma) of focused extension (Huerta and Harry 2007). The craton-directed edge-driven convective flow, depicted by red arrows, is considered to have been established in the Eocene (Faccenna *et al.* 2008). To explain the geochemistry and isotopic compositions of basalts and the time delay between rifting and volcanism, a multistage process is used: (**a**) decompression melting of upwelling subduction-derived materials produced carbonate-rich silicate liquids that rose and froze within the cooler lithosphere to form amphibole-rich veins ('metasomes'). Conductive heating at the base of the lithosphere by edge-driven flow eventually (*c.* 25 myr after focused extension) reached temperatures to melt metasomes (≥1150°C) and produce silica-undersaturated liquid. The reaction of this liquid with the surrounding peridotite modified the melt composition as it traversed the thicker continental plate and erupted to form large, elongated shield volcanoes along the continental coastline. (**b**) Thermal evolution of the lithosphere oceanwards reached the melting temperature of the earlier formed metasomes and produced the smaller Pliocene–Pleistocene volcanic islands (e.g. Possession Islands), and the seamounts located on the continental shelf and within the oceanic Adare Basin.

extension and/or transtension to initiate low degrees of melting of the asthenosphere (within either the garnet or the spinel stability fields). The melts rise and penetrate the base of the lithosphere to form a hydrous phase (pargasite-amphibole-and/or phlogopite-bearing cumulates) in veins (Fig. 7). Later, higher degrees of partial melting of these relatively small-volume, trace element and volatile-rich metasomatic veins, and the reaction of this silica-undersaturated liquid with peridotite (cf. Pilet *et al.* 2008; Pilet 2015) in the surrounding mantle during ascent (Fig. 7), is called upon to reproduce the major and trace elements, as well as the isotopic characteristics, of the mafic alkaline magmas from this region.

Overall, trace element concentrations of mafic compositions from the Ross Sea portion of the WARS (i.e. northern and southern Victoria Land (Fig. 1), which constitute the Western Ross Supergroup after Smellie and Martin 2021) show more restricted and uniform patterns relative to MBLVG samples on mantle-normalized plots, particularly with respect to elements Cs, K and Pb (Fig. 5a, b). The greater variability of the MBLVG mafic compositions may, in part, be controlled by partial melting processes (i.e. a greater range in the degree of melting) but it is also likely to be a consequence of differences in the proportions of plume and subduction-modified materials that are consumed during melting (e.g. Hart *et al.* 1995; LeMasurier *et al.* 2016). Another notable difference between MBLVG samples and those from the Ross Sea portion of the WARS in Figure 6 is that, on average, MBLVG compositions have higher K/Ta ratios and more radiogenic Pb isotope values. We propose that these compositional differences reflect a fundamental change in mantle-source domains somewhere between the Ross Sea portion of the WARS and the northern shoulder of the rift, which encompasses the MBLVG in Marie Byrd Land–western Ellsworth Land. It is possible that the change in mantle make-up may in some way be associated with the observed differences in crustal structure and composition across West Antarctica (Tinto *et al.* 2019; Jordan *et al.* 2020) (Fig. 1).

Secular changes in mantle domains across Antarctica

The past 200 million years of magmatism in Antarctica is intimately linked to its dynamic tectonic history. Over this period, igneous activity resulted from concurrent tectonic processes of continental fragmentation and subduction as the Gondwana supercontinent drifted southwards towards the Pole. Progressive disintegration of Gondwana changed the global continental configuration and eventually led to the creation of the Southern Ocean. By the Middle–Late Eocene, Australia had separated from Antarctica; and by the middle Oligocene, the opening of the Drake Passage established the Antarctic Circumpolar Current, isolating the continent of Antarctica from all other landmasses. Prior to continental rifting, the palaeo-Pacific margin of Gondwana, which consisted of conjugate continental blocks of Zealandia and what is now southeastern Australia and western Antarctica (Fig. 4), was impacted by subduction that occurred nearly continuously since the late Neoproterozoic. Contemporaneous subduction tectonics and an upwelling mantle plume are credited to have produced the Ferrar–Karoo LIP magmatism in the Jurassic (Heinonen *et al.* 2014; Luttinen 2018). Subduction shutdown progressively eastwards, beginning in the Early Cretaceous (Bradshaw 1989; Cawood 2005) and forming the Antarctic Peninsula volcanic arc by the Early Miocene (Leat and Riley 2021). Slow rates of subduction still occur at the Antarctic Peninsula's northernmost tip. Evidence from mafic igneous rocks and mantle xenoliths highlighted the widespread influence of long-lived subduction on upper-mantle sources for magmatism across West Antarctica.

Except for the Antarctic Peninsula and the isolated volcanoes of Gaussberg, Sheridan Bluff and Mount Early in East Antarctica (Fig. 1), all Cenozoic continental alkaline magmatism is associated with the post-subduction extensional tectonics of the WARS. The WARS was initiated during the break-up of Gondwana in the Late Cretaceous and developed in two main phases (e.g. Huerta and Harry 2007) ending in the Late Miocene (*c.* 11 Ma: Granot and Dyment 2018). However, most of the WARS magmatic activity is younger, occurring from the Middle Miocene (*c.* 14 Ma) through to the Holocene (Dunbar *et al.* 2021; Smellie and Martin 2021; Smellie and Rocchi 2021; Wilch *et al.* 2021). The significant time gap between tectonism and magmatism, as well as the absence of any Late Cretaceous–Cenozoic LIPs in West Antarctica, call into question a second plume-assisted Gondwana break-up event (e.g. Storey *et al.* 1999). However, Hoernle *et al.* (2020) made the case for a plume origin of Late Cretaceous (99–69 Ma) intraplate HIMU-like magmatism on and around the crustal blocks of Zealandia, which were landmasses outboard of Marie Byrd Land and Victoria Land (Fig. 4) before continental separation (>83 Ma). They propose that a Late Cretaceous plume rising beneath Zealandia impacted shortly after the collision between the oceanic Hikurangi Plateau and the active Gondwana margin (*c.* 110–100 Ma). Kipf *et al.* (2014) suggested that the plume beneath Zealandia underplated the adjacent oceanic lithosphere and, with seafloor spreading, became part of the Antarctica Plate. They contend that this source was melted by continental-insulation mantle flow in the Early Cenozoic (65–56 Ma) to produce the HIMU-like alkaline volcanism at the Marie Byrd seamounts (Fig. 1). Approximately 20–30 myr later in the Late Eocene–Early Oligocene, the earliest continental volcanism in Marie Byrd Land began (Wilch *et al.* 2021) and coincided with uplift of the Marie Byrd Land dome (LeMasurier and Landis 1996; LeMasurier 2006; Rocchi *et al.* 2006). The uplift may have occurred as much as 10–15 myr after the earliest volcanism, according to Spiegel *et al.* (2016), but this is still well before the onset of the phase of widespread and voluminous volcanic activity that occurred after 14 Ma (Wilch *et al.* 2021). The timing and spatial distribution of volcanism relative to domal uplift, the OIB HIMU-like compositions of erupted materials and the evidence for a low seismic velocity zone extending to at least the mantle transition zone (discussed previously) all support the influence of a mantle plume beneath this region. Whether there is an affiliation with the HIMU mantle imprinted upon the Late Cretaceous magmatism in Zealandia or the source reservoir for the Paleocene Marie Byrd seamount volcanism is uncertain. However, it is intriguing to note that tomographic models provided by Lloyd *et al.* (2020) show a broad region of slow shear-wave speeds beneath Marie Byrd Land that extend north of the coastline and underlie old oceanic lithosphere formed after the separation of Zealandia from Antarctica (see figs 6 and 11 in Wiens *et al.* 2021). This slow anomaly is relatively shallow (*c.* 75 km) beneath the Marie Byrd Land coastline but seawards connects at greater depths (*c.* 250 km) to a deeper (*c.* 400 km) low-velocity anomaly that is outboard of the Marie Byrd seamounts (Fig. 1). In addition, modelling of present-day bathymetic data (Sutherland *et al.* 2010) and residual basement topography (Wobbe *et al.* 2014) suggest that a lower-density upper mantle beneath this region may be long lived (*c.* 100 myr) in order to account for the subsidence history of the Campbell Plateau as Zealandia drifted away from West Antarctica. Setting aside the possible connections with Late Cretaceous magmatism in Zealandia, we conclude that melting of upwelling plume materials mixed with melts sourced from pre-existing subduction-modified mantle, in varying proportions, to account for the compositional array of Cenozoic magmatism that constitutes the MBLVG (Figs 5c & 6) (refer to Hart *et al.* 1995; LeMasurier *et al.* 2016; Panter *et al.* 2021*b*).

Applying a similar scenario to explain magmatism in the Ross Sea portion of the WARS is improbable. First, as discussed previously, mafic compositions in this region contrast with those of the MBLVG in that they have more uniform and restricted minor and trace element concentrations (e.g. Fig. 5) and have less of the HIMU-like Pb isotopic signature (Fig. 6). Tectonomagmatic relationships between the two regions also differ. For instance, magmatism in the Ross Sea region is not contemporaneous with Cenozoic uplift that occurred mostly in the Paleogene (Balestrieri *et al.* 2020; Goodge 2020 and references therein). In southern Victoria Land, localized domal uplift, which has been suggested to be plume initiated, is inferred to explain patterns of volcanism for Ross Island and nearby Mount Discovery. However, this is contrary to records for subsidence in this area caused by the growth of these volcanoes and the resultant loading of the crust over the past 5 myr (Naish *et al.* 2007; Johnston *et al.* 2008). Another complexity for plume models is how to account for the timing between magmatism and extension. Rilling *et al.* (2009) found that widespread Pliocene and Pleistocene volcanism along the Terror Rift in southern Victoria Land post-dates the major period of extension that occurred in the Miocene. Panter *et al.* (2018) also described significant stepwise time lags between major episodes of rifting and alkaline magmatism in the northwestern portion of the Ross Sea, which include plutons and volcanoes exposed in northern Victoria Land and the volcanic seamounts found within the oceanic Adare Basin (Fig. 1). In this region, the earliest alkaline magmatism, the Meander Intrusive Group (48–23 Ma), followed Gondwana break-up and a phase of broad WARS extension (105–80 Ma) by *c.* 30 myr. A focused phase of WARS extension (80–40 Ma) occurred *c.* 25 myr prior to the eruption of large alkaline shield volcanoes along the continental coastline (*c.* 14–5 Ma: Fig. 7a). In the final stage, monogenetic island and seamount volcanism located on the continental shelf and ocean floor of the Adare Basin (<5 Ma–<100 ka: Fig. 7b) followed extension and seafloor spreading (43–26 Ma) by *c.* 20 myr. The pattern of alkaline magmatism in northern Victoria Land has also shifted with time, indicating a change in tectonic conditions. Rocchi *et al.* (2002, 2005) and Rocchi and Smellie (2021) propose that the emplacement of Eocene–Oligocene plutons and dykes occurred along translithospheric dextral strike-slip faults, whereas younger volcanism (i.e. Miocene–recent) occurred along normal faults at the boundary of the WARS and above where the lithosphere has been locally thinned ('necked') by focused extension (Fig. 7).

Overall, the patterns and timing of alkaline magmatism in the Ross Sea portion of the WARS is best explained by plate dynamics and not by an upwelling plume or plumes. However, like Marie Byrd Land, a slow seismic wave anomaly also exists beneath the Ross Sea region (e.g. Heeszel *et al.* 2016; Shen *et al.* 2018; Lloyd *et al.* 2020; Wiens *et al.* 2021). The slow anomaly underlies areas of Cenozoic magmatism within the western Ross Sea and is almost continuous for nearly 3000 km from the southern Transantarctic Mountains, beneath the Sheridan Bluff and Mount Early volcanoes (Shen *et al.* 2017; Panter *et al.* 2021*a*), northwards to the Balleny Islands and further oceanwards where it is superimposed on transform fracture zones and the Pacific–Antarctic Ridge (Fig. 1; see also fig. 6 in Wiens *et al.* 2021). In contrast to the Marie Byrd Land anomaly, the mantle tomography of Lloyd *et al.* (2020) confined the slow anomaly to depths above 250 km (see fig. 6 in Wiens *et al.* 2021). This relatively shallow and linear seismic structure traces the continental margin defined by thinned WARS lithosphere against the thick cratonic lithosphere of East Antarctica. The juxtaposition of these features strongly suggests an origin controlled by Cenozoic extensional tectonics. It follows that the architecture of this lithospheric boundary promotes decompressive melting by passive asthenospheric upwelling and edge-driven mantle flow (Faccenna *et al.* 2008).

The uniformity of averaged trace element compositions (Fig. 5a, b), and the relatively narrow ranges in measured Sr (*c.* 0.7028–0.7038), Nd (*c.* 0.5128–0.5130) and Pb ($^{208}Pb/^{204}Pb$ *c.* 38.9–39.6: Fig. 6) isotopic ratios, strongly support a common origin for magmatism in the Ross Sea region of the WARS but one that is distinct from magmatism in the Marie Byrd Land region of the WARS. This distinction is further illustrated by data plotted on a Δ8/4Pb v. Δ7/4Pb diagram (Fig. 8). In Figure 8, the MBLVG samples define a linear array extending towards low Δ8/4Pb values of basalts

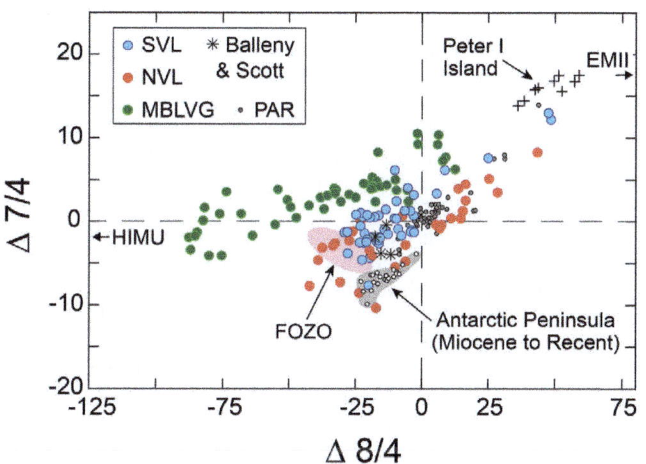

Fig. 8. Variation in Δ8/4Pb v. Δ7/4Pb for WARS mafic compositions in comparison with basalts from the Antarctic Peninsula (back-arc and post-subduction magmatism), nearby oceanic islands (Balleny, Scott and Peter I islands: Fig. 1) and the Pacific–Antarctic Ridge (PAR). Lead isotopic data used to calculate ΔPb values (i.e. variance of $^{207}Pb/^{204}Pb$ and $^{208}Pb/^{204}Pb$ ratios from the Northern Hemisphere Reference Line (NHRL), after Hart 1984) for PAR samples are from Ferguson and Klein (1993) and Vlastélic *et al.* (1999). Samples from Peter I Island are from Prestvik *et al.* (1990) and Kipf *et al.* (2014). Other sources of data are the same as in previous figures.

sourced from HIMU mantle, whereas the Ross Sea array cluster closer to the origin point, and with oceanic compositions from the Balleny and Scott islands and MORB from the Pacific–Antarctic Ridge. Basalts from the Balleny Islands (Fig. 1) are considered by some researchers to be plume sourced ('Balleny plume': e.g. Lanyon *et al.* 1993; Weaver *et al.* 1994; Storey 1995; Hart *et al.* 1997) and were used by Hart *et al.* (1992) to help to define the isotopic signature of the FOZO mantle end member, which has since been redefined by Stracke *et al.* (2005) and is shown in Figures 3 and 8. Stracke *et al.* (2005) concluded that FOZO is a ubiquitous small-scale component in MORB sources and is likely to be found throughout the entire mantle. Castillo (2015), on the other hand, proposed that FOZO is previously subducted oceanic lithospheric mantle and, hence, represents older, uppermost sections of MORB sources; such sources are inherently quite heterogeneous as they contain small-scale enriched components to begin with. Either way, the MORB–FOZO source mixture (cf. Stracke *et al.* 2005) or FOZO domains (cf. Castillo 2015) can explain most of the array for magmatism in the Ross Sea region, including alkaline volcanism on the Australian–Antarctic Ridge (Yi *et al.* 2020). In contrast, the HIMU component, which is most strongly represented in the Marie Byrd Land region, and EM components, which are present in sources for Cenozoic alkaline magmatism from both regions of the WARS, are most likely to be artefacts of subduction-zone processes including subducted altered oceanic crust or marine carbonates (for HIMU) and sediments (for EMII: Fig. 8), respectively.

Summary and conclusions

The petrological study of mafic igneous rocks is critical to our understanding of the upper mantle. In addition, key information about the tectonic environment in which mantle partial melts are generated is revealed by geochemical and isotopic data gathered from mafic compositions. Based on this perspective, we have highlighted the state of our current understanding of the tectonomagmatic origins of mafic igneous rocks in West Antarctica since the Triassic:

- Ferrar–Karoo LIP rock compositions generated in the Jurassic (*c.* 183 Ma) plot coincident with arc basalt and within-plate tholeiitic flood basalt fields (Fig. 2), and have enriched Sr (Fig. 3) and positive Δ8/4Pb and Δ7/4Pb values similar to Peter I Island basalt that trend towards EMII (Fig. 8). Melting is likely to have occurred as a result of active subduction and active mantle pluming at *c.* 183 Ma (Fig. 4), resulting in variable Nb depletions (Fig. 2c) that also reflect mantle-source heterogeneity at this time.
- Along the Antarctic Peninsula, the main period of subduction occurred from the Early Cretaceous to the Early Miocene and persists at a slow rate to the present day. The mantle source has high $^{87}Sr/^{86}Sr$ ratios and low ε_{Nd} values (Fig. 3a) that are consistent with melting of either an enriched mantle (EM) source in the subducted slab material or melting of subduction-modified mantle wedge materials, or both. Variation in the mantle source of the Antarctic Peninsula subduction-related magmatism – for example, as seen by relatively depleted $^{206}Pb/^{204}Pb$ and Sr isotope ratios in some samples (Fig. 3b) – is attributed to variations in the mantle source or episodic arc extension. Post-subduction magmatism along the Antarctic Peninsula between the Miocene and present day is related to slab-rollback, slab-window or melting of slab-hosted pyroxenite, as shown by the transition in composition from calc-alkaline to tholeiitic compositions from subduction-related magmatism to mostly alkaline magmatism post-subduction

(Fig. 2a). The chemical (Fig. 2) and isotopic (Sr–Nd: Fig. 3) composition of post-subduction magmatism is comparable to OIB or WARS primitive magmatism, although it is depleted with respect to isotopic Pb in WARS samples (Fig. 3b).

- In the Ross Sea and Victoria Land, primitive magmatic rocks have been emplaced since *c.* 50 Ma, with the majority of preserved volcanic activity occurring over the past 14 myr. Magmatism has occurred in a rift setting that is consistent with the alkaline whole-rock chemistry (Fig. 2a) and plotting in the within-plate field on trace element discrimination plots (Fig. 2b, c). The uniformity of averaged trace element compositions (Fig. 5) and the restricted range in isotopic composition (Figs 3 & 6) relative to other regions in this study supports a common mantle source that is characterized by a focal zone (FOZO) composition (Fig. 8). The trace element and isotopic patterns of primitive Victoria Land volcanic rocks and the timing of magmatism relative to major tectonic events is best explained by plate dynamics and not by an upwelling plume. In contrast, partial melting to produce Marie Byrd Land primitive volcanic rocks was induced by mantle-plume upwelling and mixed with melts from pre-existing subduction-modified mantle. This resulted in Pb isotopic ratios for Marie Byrd Land (Figs 6 & 8) that are distinct from either Victoria Land or the Antarctic Peninsula and represent a distinct mantle source.
- Through regional comparison in this study, mantle domains are defined by differences in chemical and isotopic signatures of mafic magmatism. These mantle-source differences reflect variability in bulk composition and mineral mode that will ultimately affect mantle rheology. The different mantle domains described here help to account for variations in geophysical studies of West Antarctica.

Acknowledgements This work includes support from many entities, most notably the US National Science Foundation, Antarctica New Zealand, GNS Science (New Zealand) and the British Antarctic Survey. We wish to thank handling editor Wouter van der Wal, and the constructive comments from John Gamble, Monica Handler, Pat Castillo and two anonymous reviewers that improved the paper.

Author contributions **KSP**: conceptualization (lead), writing – original draft (lead), writing – review & editing (equal); **APM**: conceptualization (supporting), writing – original draft (supporting), writing – review & editing (equal).

Funding This research received no specific grant from any funding agency in the public, commercial, or not-for-profit sectors.

Data availability All data generated or analysed during this study are included in this published article (and its supplementary information files).

References

Aviado, K.B., Rilling-Hall, S., Bryce, J.G. and Mukasa, S.B. 2015. Submarine and subaerial lavas in the West Antarctic Rift System: temporal record of shifting magma source components from the lithosphere and asthenosphere. *Geochemistry, Geophysics, Geosystems*, **16**, 4344–4361, https://doi.org/10.1002/2015GC006076

Balestrieri, M.L., Olivetti, V., Rossetti, F., Gautheron, C., Cattò, S. and Zattin, M. 2020. Topography, structural and exhumation history of the Admiralty Mountains region, northern Victoria Land, Antarctica. *Geoscience Frontiers*, **11**, 1841–1858, https://doi.org/10.1016/j.gsf.2020.01.018

Behrendt, J.C. 1999. Crustal and lithospheric structure of the West Antarctic Rift System from geophysical investigations – a review. *Global and Planetary Change*, **23**, 25–44, https://doi.org/10.1016/S0921-8181(99)00049-1

Behrendt, J.C., LeMasurier, W.E., Cooper, A.K., Tessensohn, F., Tréhu, A. and Damaske, D. 1991. Geophysical studies of the West Antarctic Rift System. *Tectonics*, **10**, 1257–1273, https://doi.org/10.1029/91TC00868

Bradshaw, J.D. 1989. Cretaceous geotectonic patterns in the New Zealand region. *Tectonics*, **8**, 803–820, https://doi.org/10.1029/TC008i004p00803

Bredow, E. and Steinberger, B. 2021. Mantle convection and possible mantle plumes beneath Antarctica – insights from geodynamic models and implications for topography. *Geological Society, London, Memoirs*, **56**, https://doi.org/10.1144/M56-2020-2

Broadley, M.W., Ballentine, C.J., Chavrit, D., Dallai, L. and Burgess, R. 2016. Sedimentary halogens and noble gases within Western Antarctic xenoliths: implications of extensive volatile recycling to the sub continental lithospheric mantle. *Geochimica et Cosmochimica Acta*, **176**, 139–156, https://doi.org/10.1016/j.gca.2015.12.013

Burgess, S.D., Bowring, S.A., Fleming, T.H. and Elliot, D.H. 2015. High-precision geochronology links the Ferrar large igneous province with early-Jurassic ocean anoxia and biotic crisis. *Earth and Planetary Science Letters*, **415**, 90–99, https://doi.org/10.1016/j.epsl.2015.01.037

Castillo, P.R. 2015. The recycling of marine carbonates and sources of HIMU and FOZO ocean island basalts. *Lithos*, **216–217**, 254–263, https://doi.org/10.1016/j.lithos.2014.12.005

Cawood, P.A. 2005. Terra Australis Orogen: Rodinia breakup and development of the Pacific and Iapetus margins of Gondwana during the Neoproterozoic and Paleozoic. *Earth-Science Reviews*, **69**, 249–279, https://doi.org/10.1016/j.earscirev.2004.09.001

Chaput, J., Aster, R.C. *et al.* 2014. The crustal thickness of West Antarctica. *Journal of Geophysical Research: Solid Earth*, **119**, 378–395, https://doi.org/10.1002/2013JB010642

Choe, W.H., Lee, J.I., Lee, M.J., Hur, S.D. and Jin, Y.K. 2007. Origin of E-MORB in a fossil spreading center: the Antarctic–Phoenix Ridge, Drake Passage, Antarctica. *Geochemical Journal*, **11**, 185–199.

Choi, S.H., Mukasa, S.B., Ravizza, G., Fleming, T.H., Marsh, B.D. and Bédard, J.H.J. 2019. Fossil subduction zone origin for magmas in the Ferrar Large Igneous Province, Antarctica: Evidence from PGE and Os isotope systematics in the Basement Sill of the McMurdo Dry Valleys. *Earth and Planetary Science Letters*, **506**, 507–519, https://doi.org/10.1016/j.epsl.2018.11.027

Class, C. and Goldstein, S.L. 2005. Evolution of helium isotopes in the Earth's mantle. *Nature*, **436**, 1107–1112, https://doi.org/10.1038/nature03930

Coltorti, M., Beccaluva, L., Bonadiman, C., Faccini, B., Ntaflos, T. and Siena, F. 2004. Amphibole genesis via metasomatic reaction with clinopyroxene in mantle xenoliths from Victoria Land, Antarctica. *Lithos*, **75**, 115–139, https://doi.org/10.1016/j.lithos.2003.12.021

Coltorti, M., Bonadiman, C., Casetta, F., Faccini, B., Giacomoni, P.P., Pelorosso, B. and Perinelli, C. 2021. Nature and evolution of the northern Victoria Land lithospheric mantle (Antarctica) as revealed by ultramafic xenoliths. *Geological Society, London, Memoirs*, **56**, https://doi.org/10.1144/M56-2020-11

Cooper, A.F., Adam, L.J., Coulter, R.F., Eby, G.N. and McIntosh, W.C. 2007. Geology, geochronology and geochemistry of a basanitic volcano, White Island, Ross Sea, Antarctica. *Journal of Volcanology and Geothermal Research*, **165**, 189–216, https://doi.org/10.1016/j.jvolgeores.2007.06.003

Correale, A., Pelorosso, B., Rizzo, A.L., Coltorti, M., Italiano, F., Bonadiman, C. and Giacomoni, P.P. 2019. The nature of the West Antarctic Rift Systemas revealed by noble gases in mantle minerals. *Chemical Geology*, **524**, 104–118, https://doi.org/10.1016/j.chemgeo.2019.06.020

Day, J.M.D., Harvey, R.P. and Hilton, D.R. 2019. Melt-modified lithosphere beneath Ross Island and its role in the tectonomagmatic evolution of the West Antarctic Rift System. *Chemical Geology*, **518**, 45–54, https://doi.org/10.1016/j.chemgeo.2019.04.012

Dunbar, N.W., Iverson, N.A., Smellie, J.L., McIntosh, W.C., Zimmerer, M.J. and Kyle, P.R. 2021. Active volcanoes in Marie Byrd Land. *Geological Society, London, Memoirs*, **55**, 759–783, https://doi.org/10.1144/M55-2019-29

Eagles, G., Gohl, K. and Larter, R.D. 2004. High-resolution animated tectonic reconstruction of the South Pacific and West Antarctic margin. *Geochemistry, Geophysics, Geosystems*, **5**, Q07002, https://doi.org/10.1029/2003GC000657

Eldholm, O. and Coffin, M.F. 2000. Large igneous provinces and plate tectonics. *American Geophysical Union Geophysical Monograph Series*, **121**, 309–326.

Elliot, D.H. and Fleming, T.H. 2021. Ferrar Large Igneous Province: petrology. *Geological Society, London, Memoirs*, **55**, 93–119, https://doi.org/10.1144/M55-2018-39

Elliot, D.H., Fleming, T.H., Kyle, P.R. and Foland, K.A. 1999. Long-distance transport of magmas in the Jurassic Ferrar Large Igneous Province, Antarctica. *Earth and Planetary Science Letters*, **167**, 87–104, https://doi.org/10.1016/S0012-821X(99)00023-0

Emry, E.L., Nyblade, A.A. *et al.* 2020. Prominent thermal anomalies in the mantle transition zone beneath the Transantarctic Mountains. *Geology*, **48**, 748–752, https://doi.org/10.1130/G47346.1

Esser, R.P., Kyle, P.R. and McIntosh, W.C. 2004. $^{40}Ar/^{39}Ar$ dating of the eruptive history of Mount Erebus, Antarctica: volcano evolution. *Bulletin of Volcanology*, **66**, 671–686, https://doi.org/10.1007/s00445-004-0354-x

Faccenna, C., Rossetti, F., Becker, T.W., Danesi, S. and Morelli, A. 2008. Recent extension driven by mantle upwelling beneath the Admiralty Mountains (East Antarctica). *Tectonics*, **27**, TC4015, https://doi.org/10.1029/2007TC002197

Ferguson, E.M. and Klein, E.M. 1993. Fresh basalts from the Pacific–Antarctic Ridge extend the Pacific geochemical province. *Nature*, **366**, 330–333, https://doi.org/10.1038/366330a0

Finn, C.A., Müller, R.D. and Panter, K.S. 2005. A Cenozoic diffuse alkaline magmatic province (DAMP) in the southwest Pacific without rift or plume origin. *Geochemistry, Geophysics, Geosystems*, **6**, Q02005, https://doi.org/10.1029/2004GC000723

Foley, S.F., Andronikov, A.V., Halpin, J.A., Daczko, N.R. and Jacob, D.E. 2021. Mantle rocks in East Antarctica. *Geological Society, London, Memoirs*, **56**, https://doi.org/10.1144/M56-2020-8

French, S.W. and Romanowicz, B.A. 2015. Broad plumes rooted at the base of the Earth's mantle beneath major hotspots. *Nature*, **525**, 95–99, https://doi.org/10.1038/nature14876

Gaffney, A.M. and Siddoway, C.S. 2007. Heterogeneous sources for Pleistocene lavas of Marie Byrd Land, Antarctica: new data from the SW Pacific diffuse alkaline magmatic province. *United States Geological Survey Open-File Report*, **2007-1047**, Extended Abstract 063.

Gamble, J.A. and Kyle, P.R. 1987. The origins of glass and amphibole in spinel wehrlite xenoliths from Foster Crater, McMurdo Volcanic Group, Antarctica. *Journal of Petrology*, **28**, 755–779, https://doi.org/10.1093/petrology/28.5.755

Gamble, J.A., McGibbon, F., Kyle, P.R., Menzies, M.A. and Kirsch, I. 1988. Metasomatised xenoliths from Foster Crater, Antarctica: implications for lithospheric structure and processes beneath the Transantarctic Mountain front. *Journal of Petrology*, **Special Volume**, Issue 1, 109–138.

Giacomoni, P.P., Bonadiman, C. *et al.* 2020. Long-term storage of subduction-related volatiels in Northern Victoria Land lithospheric mantle: Insight from olivine-hosted melt inclusions from McMurdo basic lavas (Antarctica). *Lithos*, **378–379**, 105826, https://doi.org/10.1016/j.lithos.2020.105826

Goodge, J.W. 2020. Geological and tectonic evolution of the Transantarctic Mountains, from ancient craton to recent enigma. *Gondwana Research*, **80**, 50–122, https://doi.org/10.1016/j.gr.2019.11.001

Granot, R. and Dyment, J. 2018. Late Cenozoic unification of East and West Antarctica. *Nature Communications*, **9**, 3189, https://doi.org/10.1038/s41467-018-05270-w

Granot, R., Cande, S.C., Stock, J.M. and Damaske, D. 2013. Revised Eocene–Oligocene kinematics for the West Antarctic rift system. *Geophysical Research Letters*, **40**, 279–284, https://doi.org/10.1029/2012GL054181

Hanan, B.B., Shervais, J.W. and Vetter, S.K. 2008. Yellowstone plume–continental lithosphere interaction beneath the Snake River Plain. *Geology*, **36**, 51–54, https://doi.org/10.1130/G23935A.1

Haase, K.M. and Beier, C. 2021. Bransfield Strait and James Ross Island: petrology. *Geological Society, London, Memoirs*, **55**, 285–301, https://doi.org/10.1144/M55-2018-37

Hall, C.E., Cooper, A.F. and Parkinson, D.L. 1995. Early Cambrian carbonatite in Antarctica. *Journal of the Geological Society, London*, **152**, 721–728, https://doi.org/10.1144/gsjgs.152.4.0721

Handler, M.R., Wysoczanski, R.J. and Gamble, J.A. 2021. Marie Byrd Land lithospheric mantle: A review of the xenolith record. *Geological Society, London, Memoirs*, **56**, https://doi.org/10.1144/M56-2020-17

Hansen, S.E., Graw, J.H. *et al.* 2014. Imaging the Antarctic mantle using adaptively parameterized P-wave tomography: evidence for heterogeneous structure beneath West Antarctica. *Earth and Planetary Science Letters*, **408**, 66–78, https://doi.org/10.1016/j.epsl.2014.09.043

Hart, S.R. 1984. A large-scale isotope anomaly in the Southern Hemisphere mantle. *Nature*, **309**, 753–757, https://doi.org/10.1038/309753a0

Hart, S.R. and Kyle, P.R. 1994. Geochemistry of McMurdo Group Volcanic Rocks. *Antarctic Journal of the United States*, **28**, 14–16.

Hart, S.R., Hauri, E.H., Oschmann, L.A. and Whitehead, J.A. 1992. Mantle plumes and entrainment: Isotopic evidence. *Science*, **256**, 517–520, https://doi.org/10.1126/science.256.5056.517

Hart, S.R., Blusztajn, J. and Craddock, C. 1995. Cenozoic volcanism in Antarctica; Jones Mountains and Peter I Island. *Geochimica et Cosmochimica Acta*, **59**, 3379–3388, https://doi.org/10.1016/0016-7037(95)00212-I

Hart, S.R., Blusztajn, J., LeMasurier, W.E. and Rex, D.C. 1997. Hobbs Coast Cenozoic volcanism: implications for the West Antarctic rift system. *Chemical Geology*, **139**, 223–248, https://doi.org/10.1016/S0009-2541(97)00037-5

Hawkesworth, C. and Scherstén, A. 2007. Mantle plumes and geochemisty. *Chemical Geology*, **241**, 319–331, https://doi.org/10.1016/j.chemgeo.2007.01.018

Heeszel, D.S., Wiens, D.A. *et al.* 2016. Upper mantle structure of central and West Antarctica from array analysis of Rayleigh wave phase velocities. *Journal of Geophysical Research: Solid Earth*, **121**, 1758–1775, https://doi.org/10.1002/2015jb012616

Heinonen, J.S., Carlson, R.W., Riley, T.R., Luttinen, A.V. and Horan, M.F. 2014. Subduction modified oceanic crust mixed with a depleted mantle reservoir in the sources of the Karoo continental flood basalt province. *Earth and Planetary Science Letters*, **394**, 229–241, https://doi.org/10.1016/j.epsl.2014.03.012

Hergt, J.M., Peate, D.W. and Hawkesworth, C.J. 1991. The petrogenesis of Mesozoic Gondwana low-Ti flood basalts. *Earth and Planetary Science Letters*, **105**, 134–148, https://doi.org/10.1016/0012-821X(91)90126-3

Hoernle, K., Timm, C. *et al.* 2020. Late Cretaceous (99–69 Ma) basaltic intraplate volcanism on and around Zealandia: tracing upper mantle geodynamics from Hikurangi Plateau collision to Gondwana breakup. *Earth and Planetary Science Letters*, **529**, https://doi.org/10.1016/j.epsl.2019.115864

Hofmann, A.W. 1997. Mantle geochemistry: the message from oceanic volcanism. *Nature*, **385**, 219–229.

Hofmann, A.W. 2007. Sampling mantle heterogeneity through oceanic basalts: Isotopes and trace elements. *In*: Holland, H.D. and Turekian, K.K. (eds) *Treatise on Geochemistry*. Pergamon, Oxford, UK, 1–44, https://doi.org/10.1016/B0-08-043751-6/02123-X

Hole, M.J. 2021. Antarctic Peninsula: petrology. *Geological Society, London, Memoirs*, **55**, 327–343, https://doi.org/10.1144/M55-2018-40

Hole, M.J. and LeMasurier, W.E. 1994. Tectonic controls on the geochemical composition of Cenozoic mafic alkaline volcanic rocks from West Antarctica. *Contributions to Mineralogy and Petrology*, **117**, 187–202, https://doi.org/10.1007/BF00286842

Hole, M.J., Storey, B.C. and LeMasurier, W.E. 1994. Tectonic setting and geochemistry of Miocene alkali basalts from the Jones Mountains, West Antarctica. *Antarctic Science*, **6**, 85–92, https://doi.org/10.1017/S0954102094000118

Huerta, A.D. and Harry, D.L. 2007. The transition from diffuse to focused extension: modelled evolution of the West Antarctic Rift system. *Earth and Planetary Science Letters*, **255**, 133–147, https://doi.org/10.1016/j.epsl.2006.12.011

Irvine, T.N. and Baragar, W.R.A. 1971. A guide to the chemical classification of the common volcanic rocks. *Canadian Journal of Earth Sciences*, **8**, 523–548, https://doi.org/10.1139/e71-055

Ivanov, A.V., Meffre, S., Thompson, J., Corfu, F., Kamenetskey, V.S., Kamenetsky, M.B. and Demonterova, E.I. 2017. Timing and genesis of the Karoo–Ferrar large igneous province: New high percision U–Pb data for Tasmania confirm short duration of the major magmatic pulse. *Chemical Geology*, **455**, 32–43, https://doi.org/10.1016/j.chemgeo.2016.10.008

Johnston, L., Wilson, G.S., Gorman, A.R., Henrys, S.A., Horgan, H., Clark, R. and Naish, T.R. 2008. Cenozoic basin evolution beneath the southern McMurdo Ice Shelf, Antarctica. *Global and Planetary Change*, **62**, 61–76, https://doi.org/10.1016/j.gloplacha.2007.11.004

Jordan, T.A., Riley, T.R. and Siddoway, C.S. 2020. The geological history and evolution of West Antarctica. *Nature Reviews Earth & Environment*, **1**, 117–133, https://doi.org/10.1038/s43017-019-0013-6

Jung, S., Pfänder, J.A., Brügmann, G. and Stracke, A. 2005. Sources of primitive alkaline volcanic rocks from the Central European Volcanic Province (Rhön, Germany) inferred from Hf, Os and Pb isotopes. *Contributions to Mineralogy and Petrology*, **150**, 546–559, https://doi.org/10.1007/s00410-005-0029-4

Keller, R.A., Fisk, M.R., Smellie, J.L., Strelin, J.A. and Lawver, L.A. 2002. Geochemistry of back arc basin volcanism in Bransfield Strait, Antarctica: subducted contributions and along axis variations. *Journal of Geophysical Research: Solid Earth*, **107**, 2171, https://doi.org/10.1029/2001JB000444

Kendrick, M.A., Honda, M., Pettke, T., Scambelluri, M., Phillips, D. and Giuliani, A. 2013. Subduction zone fluxes of halogens and noble gases in seafloor and forearc serpentinites. *Earth and Planetary Science Letters*, **365**, 86–96, https://doi.org/10.1016/j.epsl.2013.01.006

Kim, J., Park, J.-W., Lee, M.J., Lee, J.I. and Kyle, P.R. 2019. Evolution of alkalic magma systems: insight from coeval evolution of sodic and potassic fractionation lineages at The Pleiades volcanic complex, Antarctica. *Journal of Petrology*, **60**, 117–150, https://doi.org/10.1093/petrology/egy108

Kipf, A., Hauff, F. *et al.* 2014. Seamounts off the West Antarctic margin: a case for non-hotspot driven intraplate volcanism. *Gondwana Research*, **25**, 1660–1679, https://doi.org/10.1016/j.gr.2013.06.013

Kyle, P.R., Moore, J.A. and Thirlwall, M.F. 1992. Petrologic evolution of anorthoclase phonolite lavas at Mount Erebus, Ross Island, Antarctica. *Journal of Petrology*, **33**, 849–875, https://doi.org/10.1093/petrology/33.4.849

Lanyon, R., Varne, R. and Crawford, A.J. 1993. Tasmanian Tertiary basalts, the Balleny Plume, and opening of the Tasman Sea (southwest Pacific Ocean). *Geology* , **21**, 555–558, https://doi.org/10.1130/0091-7613(1993)021<0555:TTBTBP>2.3.CO;2

Leat, P.T. and Riley, T.R. 2021. Antarctic Peninsula and South Shetland Islands: petrology. *Geological Society, London, Memoirs*, **55**, 213–226, https://doi.org/10.1144/M55-2018-68

Leeman, W.P., Schutt, D.L. and Hughes, S.S. 2009. Thermal structure beneath the Snake River Plain: implications for the Yellowstone hot spot. *Journal of Volcanology and Geothermal Research*, **188**, 128–140.

LeMaitre, R.W. 2002. *Igneous Rocks: A Classification and Glossary of Terms: Recommendations of International Union of Geological Sciences Subcommission on the Systematics of Igneous Rocks*. Cambridge University Press, Cambridge, UK.

LeMasurier, W.E. 2006. What supports the Marie Byrd Land Dome? An evaluation of potential uplift mechanisms in a continental rift system. *In*: Fütterer, D.K., Damaske, D., Kleinschmidt, G., Miller, H. and Tessensohn, F. (eds) *Antarctica*. Springer, Berlin, 299–302.

LeMasurier, W.E. and Landis, C.A. 1996. Mantle-plume activity recorded by low-relief erosion surfaces in West Antarctica and New Zealand. *Geological Society of America Bulletin*, **108**, 1450–1466, https://doi.org/10.1130/0016-7606(1996)108<1450:MPARBL>2.3.CO;2

LeMasurier, W.E. and Rex, D.C. 1989. Evolution of linear volcanic ranges in Marie Byrd Land, West Antarctica. *Journal of Geophysical Research*, **94**, 7223–7236, https://doi.org/10.1029/JB094iB06p07223

LeMasurier, W.E., Choi, S.H., Hart, S.R., Mukasa, S.B. and Rogers, N.W. 2016. Reconciling the shadow of a subduction signature with rift geochemistry and tectonic environment in eastern Marie Byrd Land, Antarctica. *Lithos*, **260**, 134–153, https://doi.org/10.1016/j.lithos.2016.05.018

Livermore, R., Balanya, J.C. *et al.* 2000. Autopsy on a dead spreading center: The Phoenix Ridge, Drake Passage, Antarctica. *Geology*, **28**, 607–610, https://doi.org/10.1130/0091-7613(2000)28<607:AOADSC>2.0.CO;2

Lloyd, A.J., Wiens, D.A. *et al.* 2020. Seismic structure of the Antarctic upper mantle imaged with adjoint tomography. *Journal of Geophysical Research: Solid Earth*, **125**, https://doi.org/10.1029/2019JB017823

Lucas, E.M., Soto, D. *et al.* 2020. P- and S-wave velocity structure of central West Antarctica: implications for the tectonic evolution of the West Antarctic Rift System. *Earth and Planetary Science Letters*, **546**, 116437, https://doi.org/10.1016/j.epsl.2020.116437

Luttinen, A.V. 2018. Bilateral geochemical asymmetry in the Karoo large igneous province. *Scientific Reports*, **8**, 5223, https://doi.org/10.1038/s41598-018-23661-3

Ma, G.S.-K., Malpas, J., Xenophontos, C. and Chan, G.H.-N. 2011. Petrogenesis of latest Miocene–Quaternary continental intraplate volcanism along the northern Dead Sea Fault System (Al Ghab-Homs Volcanic Field), western Syria: evidence for lithosphere–asthenosphere interaction. *Journal of Petrology*, **52**, 401–430, https://doi.org/10.1093/petrology/egq085

Mandler, B.E. and Grove, T.L. 2016. Controls on the stability and composition of amphibole in the Earth's mantle. *Contributions to Mineralogy and Petrology*, **171**, 68, https://doi.org/10.1007/s00410-016-1281-5

Marignier, A., Ferreira, A.M.G. and Kitching, T. 2020. The probability of mantle plumes in global tomographic models. *Geochemistry, Geophysics, Geosystems*, **21**, e2020GC009276, https://doi.org/10.1029/2020GC009276

Marsh, B. 2004. A magmatic mush column rosetta stone: the McMurdo Dry Valleys of Antarctica. *Eos, Transactions of the American Geophysical Union*, **85**, 497–502, https://doi.org/10.1029/2004EO470001

Martin, A.P. and van der Wal, W. 2022. Introduction to the geochemistry and geophysics of the Antarctic mantle. *Geological Society, London, Memoirs*, **56**.

Martin, A.P., Cooper, A.F. and Dunlap, W.J. 2010. Geochronology of Mount Morning, Antarctica: two-phase evolution of a long-lived trachyte–basanite–phonolite eruptive center. *Bulletin of Volcanology*, **72**, 357–371, https://doi.org/10.1007/s00445-009-0319-1

Martin, A.P., Cooper, A.F. and Price, R.C. 2013. Petrogenesis of Cenozoic, alkalic volcanic lineages at Mount Morning, West Antarctica and their entrained lithospheric mantle xenoliths: lithospheric v. asthenospheric mantle sources. *Geochimica et Cosmochimica Acta*, **122**, 127–152, https://doi.org/10.1016/j.gca.2013.08.025

Martin, A.P., Cooper, A.F. and Price, R.C. 2014. Increased mantle heat flow with on-going rifting of the West Antarctic rift system inferred from characterisation of plagioclase peridotite in the shallow Antarctic mantle. *Lithos*, **190–191**, 173–190, https://doi.org/10.1016/j.lithos.2013.12.012

Martin, A.P., Price, R.C., Cooper, A.F. and McCammon, C.A. 2015. Petrogenesis of the rifted Southern Victoria Land lithospheric mantle, Antarctica, inferred from petrography, geochemistry, thermobarometry and oxybarometry of peridotite and pyroxenite xenoliths from the Mount Morning eruptive centre. *Journal of Petrology*, **56**, 193–226, https://doi.org/10.1093/petrology/egu075

Martin, A.P., Cooper, A.F., Price, R.C., Doherty, C.L. and Gamble, J.A. 2021*a*. A review of mantle xenoliths in volcanic rocks from southern Victoria Land, Antarctica. *Geological Society, London, Memoirs*, **56**, https://doi.org/10.1144/M56-2019-42

Martin, A.P., Cooper, A., Price, R., Kyle, P. and Gamble, J. 2021*b*. Erebus Volcanic Province: petrology. *Geological Society, London, Memoirs*, **55**, 447–489, https://doi.org/10.1144/M55-2018-80

Mayer, B., Jung, S., Romer, R.L., Pfänder, J.A., Klügel, A., Pack, A. and Gröner, E. 2014. Amphibole in alkaline basalts from intraplate settings: implications for the petrogenesis of alkaline lavas from the metasomatised lithospheric mantle. *Contributions to Mineralogy and Petrology*, **167**, 2095–2123, https://doi.org/10.1007/s00410-014-0989-3

McDonough, W.F. and Sun, S.-s. 1995. The composition of the earth. *Chemical Geology*, **120**, 223–253, https://doi.org/10.1016/0009-2541(94)00140-4

McGibbon, F.M. 1991. Geochemistry and petrology of ultramafic xenoliths of the Erebus Volcanic Province. *In*: Thomson, M.R.A., Crame, J.A. and Thomson, J.W. (eds) *Geological Evolution of Antarctica. Proceedings of the 5th International Symposium on Antarctic Earth Sciences*. Cambridge University Press, Cambridge, UK, 317–321.

Melchiorre, M., Coltorti, M., Bonadiman, C., Faccini, B., O'Reilly, S.Y. and Pearson, N.J. 2011. The role of eclogite in the rift-related metasomatism and Cenozoic magmatism of Northern Victoria Land, Antarctica. *Lithos*, **124**, 319–330, https://doi.org/10.1016/j.lithos.2010.11.012

Molzahn, M., Reisberg, L. and Wörner, G. 1996. Os, Sr, Nd, Pb, O isotope and trace element data from the Ferrar flood basalts, Antarctica: evidence for an enriched subcontinental lithospheric source. *Earth and Planetary Science Letters*, **144**, 529–546, https://doi.org/10.1016/S0012-821X(96)00178-1

Montelli, R., Nolet, G., Dahlen, F.A., Masters, G., Engdahl, E. and Hung, S.-H. 2004. Finite frequency tomography reveals a variety of plumes in the mantle. *Science*, **303**, 338–343, https://doi.org/10.1126/science.1092485

Murphy, D.T., Collerson, K.D. and Kamber, B.S. 2002. Lamproites from Gaussberg, Antarctica: possible transition zone melts of Archaean subducted sediments. *Journal of Petrology*, **43**, 981–1001, https://doi.org/10.1093/petrology/43.6.981

Naish, T., Powell, R., Levy, R. and the ANDRILL-SMS Science Team. 2007. Background to the ANDRILL McMurdo Ice Shelf Project, Antarctica. *Terra Antartica*, **14**, 121–130.

Nardini, I., Armienti, P., Rocchi, S., Dallai, L. and Harrison, D. 2009. Sr–Nd–Pb–He–O isotope and geochemical constraints on the genesis of Cenozoic magmas from the West Antarctic Rift. *Journal of Petrology*, **50**, 1359–1375, https://doi.org/10.1093/petrology/egn082

Panter, K.S. 2021. Antarctic volcanism: petrology and tectonomagmatic overview. *Geological Society, London, Memoirs*, **55**, 43–53, https://doi.org/10.1144/M55-2020-10

Panter, K.S., Kyle, P.R. and Smellie, J.L. 1997. Petrogenesis of a phonolite–trachyte succession at Mount Sidley, Marie Byrd Land, Antarctica. *Journal of Petrology*, **38**, 1225–1253, https://doi.org/10.1093/petroj/38.9.1225.

Panter, K.S., Hart, S.R., Kyle, P., Blusztajn, J. and Wilch, T. 2000. Geochemistry of Late Cenozoic basalts from the Crary Mountains: characterization of mantle sources in Marie Byrd Land, Antarctica. *Chemical Geology*, **165**, 215–241, https://doi.org/10.1016/S0009-2541(99)00171-0

Panter, K.S., Blusztajn, J., Wingrove, D., Hart, S. and Mattey, D. 2003. Sr, Nd, Pb, Os, O isotope, Major and trace element data from basalts, South Victoria Land, Antarctica: evidence for open-system processes in the evolution of mafic alkaline magmas. *General Assembly of the European Geosciences Union, Geophysical Research Abstracts*, **5**, 07583.

Panter, K.S., Blusztajn, J., Hart, S., Kyle, P., Esser, R. and McIntosh, W. 2006. The origin of HIMU in the SW Pacific: evidence from intraplate volcanism in Southern New Zealand and Subantarctic Islands. *Journal of Petrology*, **47**, 1673–1704, https://doi.org/10.1093/petrology/egl024

Panter, K.S., Castillo, P. *et al.* 2018. Melt origin across a rifted continental margin: a case for subduction-related metasomatic agents in the lithospheric source of alkaline basalt, northwest Ross Sea, Antarctica. *Journal of Petrology*, **59**, 517–558. https://doi.org/10.1093/petrology/egy036

Panter, K.S., Reindel, J. and Smellie, J.L. 2021*a*. Mount Early and Sheridan Bluff: petrology. *Geological Society, London, Memoirs*, **55**, 499–514, https://doi.org/10.1144/M55-2019-2

Panter, K.S., Wilch, T.I., Smellie, J.L., Kyle, P.R. and McIntosh, W.C. 2021*b*. Marie Byrd Land and Ellsworth Land: petrology. *Geological Society, London, Memoirs*, **55**, 577–614, https://doi.org/10.1144/M55-2019-50

Pappa, F. and Ebbing, J. 2021. Gravity, magnetics and geothermal heat flow of the Antarctic lithospheric crust and mantle. *Geological Society, London, Memoirs*, **56**, https://doi.org/10.1144/M56-2020-5

Pappa, F., Ebbing, J. and Ferraccioli, F. 2019. Moho depths of Antarctica: comparison of seismic, gravity, and isostatic results. *Geochemistry, Geophysics, Geosystems*, **20**, https://doi.org/10.1029/2018GC008111

Park, S.-H., Langmuir, C.H. *et al.* 2019. An isotopically distinct Zealandia–Antarctic mantle domain in the Southern Ocean. *Nature Geoscience*, **12**, 206–214, https://doi.org/10.1038/s41561-018-0292-4

Perinelli, C., Armienti, P. and Dallai, L. 2006. Geochemical and O-isotope constraints on the evolution of lithospheric mantle in the Ross Sea rift area (Antarctica). *Contributions to Mineralogy and Petrology*, **151**, 245–266, https://doi.org/10.1007/s00410-006-0065-8

Perinelli, C., Armienti, P. and Dallai, L. 2011. Thermal evolution of the lithosphere in a rift environment as inferred from the geochemistry of mantle cumulates, Northern Victoria Land, Antarctica. *Journal of Petrology*, **52**, 665–690, https://doi.org/10.1093/petrology/egq099

Perinelli, C., Gaeta, M. and Armienti, P. 2017. Cumulate xenoliths from Mt. Overlord, northern Victoria Land, Antarctica: a window into high pressure storage and differentiation of mantle-derived basalts. *Lithos*, **268–271**, 225–239, https://doi.org/10.1016/j.lithos.2016.10.027

Phillips, E.H., Sims, K.W.W. *et al.* 2018. The nature and evolution of mantle upwelling at Ross Island, Antarctica, with implications for the source of HIMU lavas. *Earth and Planetary Science Letters*, **498**, 38–53, https://doi.org/10.1016/j.epsl.2018.05.049

Pilet, S. 2015. Generation of low-silica alkaline lavas: Petrological constraints, models, and thermal implications. *Geological Society of America Special Papers*, **514**, 514–517.

Pilet, S., Baker, M.B. and Stolper, E.M. 2008. Metasomatized lithosphere and the origin of alkaline lavas. *Science*, **320**, 916–919, https://doi.org/10.1126/science.1156563

Prestvik, T., Barnes, C.G., Sundvoll, B. and Duncan, R.A. 1990. Petrology of Peter I Øy (Peter I Island), West Antarctica. *Journal of Volcanology and Geothermal Research*, **44**, 315–338, https://doi.org/10.1016/0377-0273(90)90025-B

Ramirez, C., Nyblade, A. *et al.* 2017. Crustal structure of the Transantarctic Mountains, Ellsworth Mountains and Marie Byrd Land, Antarctica: constraints on shear wave velocities, Poisson's ratios and Moho depths. *Geophysical Journal International*, **211**, 1328–1340, https://doi.org/10.1093/gji/ggx333

Rapela, C.W., Pankhurst, R.J., Fanning, C.M. and Hervé, F. 2005. Pacific subduction coeval with the Karoo mantle plume: the Early Jurassic Subcordilleran belt of northwestern Patagonia.

Geological Society, London, Special Publications, **246**, 217–239, https://doi.org/10.1144/GSL.SP.2005.246.01.07

Riley, P.T. and Leat, T.R. 2021. Antarctic Peninsula and South Shetland Islands: petrology. *Geological Society, London, Memoirs*, **55**, 213–226, https://doi.org/10.1144/M55-2018-68

Riley, T.R., Leat, P.T., Storey, B.C., Parkinson, I.J. and Miller, I.L. 2003. Ultramafic lamprophyres of the Ferrar large igneous province: evidence for a HIMU mantle component. *Lithos*, **66**, 63–76, https://doi.org/10.1016/S0024-4937(02)00213-X

Rilling, S., Mukasa, S., Wilson, T., Lawver, L. and Hall, C. 2009. New determinations of ^{40}Ar/^{39}Ar isotopic ages and flow volumes for Cenozoic volcanism in the Terror Rift, Ross Sea, Antarctica. *Journal of Geophysical Research: Solid Earth*, **114**, B12207, https://doi.org/10.1029/2009JB006303

Rocchi, S. and Smellie, J.L. 2021. Northern Victoria Land: petrology. *Geological Society, London, Memoirs*, **55**, 383–413, https://doi.org/10.1144/M55-2019-19

Rocchi, S., Armienti, P., D'Orazio, M., Tonarini, S., Wijbrans, J.R. and Di Vincenzo, G. 2002. Cenozoic magmatism in the western Ross Embayment: role of mantle plume v. plate dynamics in the development of the West Antarctic Rift System. *Journal of Geophysical Research: Solid Earth*, **107**, ECV 5-1–ECV 5-22, https://doi.org/10.1029/2001JB000515

Rocchi, S., Di Vincenzo, G. and Armienti, P. 2005. No plume, no rift magmatism in the West Antarctic rift. *Geological Society of America Special Papers*, **388**, 435–447.

Rocchi, S., LeMasurier, W.E. and Di Vincenzo, G. 2006. Oligocene to Holocene erosion and glacial history in Marie Byrd Land, West Antarctica, inferred from exhumation of the Dorrel Rock intrusive complex and from volcano morphologies. *Geological Society of America Bulletin*, **118**, 991–1005, https://doi.org/10.1130/B25675.1

Rocholl, A., Stein, M., Molzahn, M., Hart, S.R. and Wörner, G. 1995. Geochemical evolution of rift magmas by progressive tapping of a stratified mantle source beneath the Ross Sea Rift, Northern Victoria Land, Antarctica. *Earth and Planetary Science Letters*, **131**, 207–224, https://doi.org/10.1016/0012-821X(95)00024-7

Rooney, T.O., Nelson, W.R., Dosso, L., Furman, T. and Hanan, B. 2014. The role of continental lithosphere metasomes in the production of HIMU-like magmatism on the northeast African and Arabian plates. *Geology*, **42**, 419–422, https://doi.org/10.1130/G35216.1

Rooney, T.O., Nelson, W.R., Ayalew, D., Hanan, B., Yirgu, G. and Kappelman, J. 2017. Melting the lithosphere: Metasomes as a source for mantle-derived magmas. *Earth and Planetary Science Letters*, **461**, 105–118, https://doi.org/10.1016/j.epsl.2016.12.010

Scott, J.M., Pontesilli, A., Brenna, M., White, J.D., Giacalone, E., Palin, J.M. and Le Roux, P.J. 2020. The Dunedin Volcanic Group and a revised model for Zealandia's alkaline intraplate volcanism. *New Zealand Journal of Geology and Geophysics*, **63**, 510–529, https://doi.org/10.1080/00288306.2019.1707 695

Seroussi, H., Ivins, E.R., Wiens, D.A. and Bondzio, J. 2017. Influence of a West Antarctic mantle plume on ice sheet basal conditions. *Journal of Geophysical Research: Solid Earth*, **122**, 7127–7155, https://doi.org/10.1002/2017jb014423

Shen, W., Wiens, D.A. *et al.* 2017. Seismic evidence for lithospheric foundering beneath the southern Transantarctic Mountains, Antarctica. *Geology*, **46**, 71–74, https://doi.org/10.1130/G39555.1

Shen, W., Wiens, D.A. *et al.* 2018. The crust and upper mantle structure of central and West Antarctica from Bayesian inversion of Rayleigh wave and receiver functions. *Journal of Geophysical Research: Solid Earth*, **123**, 7824–7849, https://doi.org/10.1029/2017JB015346

Shen, W., Wiens, D.A., Lloyd, A.J. and Nyblade, A.A. 2020. A geothermal heat flux map of Antarctica empirically constrained by seismic structure. *Geophysical Research Letters*, **47**, e2020GL086955, https://doi.org/10.1029/2020GL086955

Shervais, J.W. 1982. Ti–V plots and the petrogenesis of modern and ophiolitic lavas. *Earth and Planetary Science Letters*, **59**, 101–118, https://doi.org/10.1016/0012-821X(82)90120-0

Sims, K.W.W., Blichert-Toft, J. *et al.* 2008. A Sr, Nd, Hf, and Pb isotope perspective on the genesis and long-term evolution of alkaline magmas from Erebus volcano, Antarctica. *Journal of Volcanology and Geothermal Research*, **177**, 606–618, https://doi.org/10.1016/j.jvolgeores.2007.08.006

Smellie, J.L. 2021. Antarctic volcanism: volcanology and palaeoenvironmental overview. *Geological Society, London, Memoirs*, **55**, 19–42, https://doi.org/10.1144/M55-2020-1

Smellie, J.L. and Martin, A.P. 2021. Erebus Volcanic Province: volcanology. *Geological Society, London, Memoirs*, **55**, 415–446, https://doi.org/10.1144/M55-2018-62

Smellie, J.L. and Rocchi, S. 2021. Northern Victoria Land: volcanology. *Geological Society, London, Memoirs*, **55**, 347–381, https://doi.org/10.1144/M55-2018-60

Smellie, J.L., Martin, A.P., Panter, K.S., Kyle, P.R. and Geyer, A. 2020. Magmatism in Antarctica and its relation to Zealandia. *New Zealand Journal of Geology and Geophysics*, **63**, 578–588, https://doi.org/10.1080/00288306.2020.1781666

Smellie, J.L., Panter, K.S. and Geyer, A. (eds) 2021. *Volcanism in Antarctica: 200 Million Years of Subduction, Rifting and Continental Break-Up. Geological Society, London, Memoirs*, **55**, https://doi.org/10.1144/M55

Spiegel, C., Lindow, J. *et al.* 2016. Tectonomagmatic evolution of Marie Byrd Land – Implications for Cenozoic rifting activity and onset of West Antarctic glaciation. *Global and Planetary Change*, **145**, 98–115, https://doi.org/10.1016/j.gloplacha.2016.08.013

Stefano, C.J., Mukasa, S.B. and Cabato, J.A. 2019. Elemental abundance patterns and Sr-, Nd- and Hf-isotope systematics for the Yellowstone hotspot and Columbia River basalts: bearing on petrogenesis. *Chemical Geology*, **513**, 44–53, https://doi.org/10.1016/j.chemgeo.2019.03.012

Stein, M., Navon, O. and Kessel, R. 1997. Chromatographic metasomatism of the Arabian–Nubian lithosphere. *Earth and Planetary Science Letters*, **152**, 75–91, https://doi.org/10.1016/S0012-821X(97)00156-8

Storey, B.C. 1995. The role of mantle plumes in continental breakup: case histories from Gondwanaland. *Nature*, **377**, 301–308, https://doi.org/10.1038/377301a0

Storey, B.C. and Granot, R. 2021. Tectonic history of Antarctica over the past 200 million years. *Geological Society, London, Memoirs*, **55**, 9–17, https://doi.org/10.1144/M55-2018-38

Storey, B.C. and Kyle, P. 1997. An active mantle mechanism for Gondwana breakup. *South African Journal of Geology*, **100**, 283–290.

Storey, B.C., Leat, P.T., Weaver, S.D., Pankhurst, R.J., Bradshaw, J.D. and Kelly, S. 1999. Mantle plumes and Antarctica–New Zealand rifting: evidence from mid-Cretaceous mafic dykes. *Journal of the Geological Society, London*, **156**, 659–671, https://doi.org/10.1144/gsjgs.156.4.0659

Stracke, A., Hofmann, A.W. and Hart, S.R. 2005. FOZO, HIMU, and the rest of the mantle zoo. *Geochemistry, Geophysics, Geosystems*, **6**, Q05007, https://doi.org/10.1029/2004GC000824

Sumino, H., Burgess, R., Mizukami, T., Wallis, S.R., Holland, G. and Ballentine, C.J. 2010. Seawater-derived noble gases and halogens preserved in exhumed mantle wedge peridotite. *Earth and Planetary Science Letters*, **294**, 163–172, https://doi.org/10.1016/j.epsl.2010.03.029

Sutherland, R., Spasojevic, S. and Gurnis, M. 2010. Mantle upwelling after Gondwana subduction death explains anomalous topography and subsidence histories of eastern New Zealand and West Antarctica. *Geology*, **38**, 155–158, https://doi.org/10.1130/G30613.1

Svensen, H., Corfu, F., Polteau, S., Hammer, Ø. and Planke, S. 2012. Rapid magma emplacement in the Karoo large igneous province. *Earth and Planetary Science Letters*, **325–326**, 1–9, https://doi.org/10.1016/j.epsl.2012.01.015

Tinto, K.J., Padman, L. *et al.* 2019. Ross Ice Shelf response to climate driven by the tectonic imprint on seafloor bathymetry. *Nature Geoscience*, **12**, 441–449, https://doi.org/10.1038/s41561-019-0370-2

Tonarini, S., Rocchi, S., Armienti, P. and Innocenti, F. 1997. Constraints on timing of Ross Sea rifting inferred from Cainozoic intrusions from northern Victoria Land, Antarctica. *In*: Ricci, C.A. (ed.) *The Antarctic Region: Geological Evolution and Processes*. Terra Antartica, Siena, Italy, 511–521.

Vaughan, A.P.M. and Storey, B.C. 2007. A new supercontinent self-destruct mechanism: evidence from the Late Triassic–Early Jurassic. *Journal of the Geological Society, London*, **164**, 383–392, https://doi.org/10.1144/0016-76492005-109

Vlastélic, I., Aslanian, D., Dosso, L., Bougault, H., Olivet, J.L. and Géli, L. 1999. Large-scale chemical and thermal division of the Pacific mantle. *Nature*, **399**, 345–350, https://doi.org/10.1038/20664

Weaver, S.D., Storey, B.C., Pankhurst, R.J., Mukasa, S.B., DiVenere, V.J. and Bradshaw, J.D. 1994. Antarctic–New Zealand rifting and Marie Byrd Land lithospheric magmatism linked to ridge subduction and mantle plume activity. *Geology*, **22**, 811–814, https://doi.org/10.1130/0091-7613(1994)022<0811:ANZRAM>2.3.CO;2

White, W.M. 2010. Oceanic island basalts and mantle plumes: the geochemical perspective. *Annual Review of Earth and Planetary Sciences*, **38**, 133–160, https://doi.org/10.1146/annurev-earth-040809-152450

Wiens, D.A., Shen, W. and Lloyd, A. 2021. The seismic structure of the Antarctic upper mantle. *Geological Society, London, Memoirs*, **56**, https://doi.org/10.1144/M56-2020-18

Wilch, T.I., McIntosh, W.C. and Panter, K.S. 2021. Marie Byrd Land and Ellsworth Land: volcanology. *Geological Society, London, Memoirs*, **55**, 515–576, https://doi.org/10.1144/M55-2019-39

Winberry, P.J. and Anandakrishnan, S. 2004. Crustal structure of the West Antarctic rift system and Marie Byrd Land hotspot. *Geology*, **32**, 977–980, https://doi.org/10.1130/G20768.1

Wobbe, F., Lindeque, A. and Gohl, K. 2014. Anomalous South Pacific lithosphere dynamics derived from new total sediment thickness estimates off the West Antarctic margin. *Global and Planetary Change*, **123**, 139–149, https://doi.org/10.1016/j.gloplacha.2014.09.006

Wood, D.A. 1980. The application of a Th–Hf–Ta diagram to problems of tectonomagmatic classification and to establishing the nature of crustal contamination of basaltic lavas of the British Tertiary volcanic province. *Earth and Planetary Science Letters*, **50**, 11–30, https://doi.org/10.1016/0012-821X(80)90116-8

Yi, S.-B., Lee, M.J. *et al.* 2020. Alkalic to tholeiitic magmatism near a mid-ocean ridge: petrogenesis of the KR1 Seamount Trail adjacent to the Australian–Antarctic Ridge. *International Geology Review*, https://doi.org/10.1080/00206814.2020.1756002

Zhang, Z., Li, S. *et al.* 2020. Plume interaction and mantle heterogeneity: a geochemical perspective. *Geoscience Frontiers*, **11**, 1571–1579, https://doi.org/10.1016/j.gsf.2020.02.009

Zhao, D. 2007. Seismic images under 60 hotspots: search for mantle plumes. *Gondwana Research*, **12**, 335–355, https://doi.org/10.1016/j.gr.2007.03.001

Effects of melt-percolation, refertilization and deformation on upper mantle seismic anisotropy: constraints from peridotite xenoliths, Marie Byrd Land, West Antarctica

Vasileios Chatzaras[1]* and Seth C. Kruckenberg[2]

[1]School of Geosciences, The University of Sydney, Sydney, NSW 2006, Australia

[2]Department of Earth and Environmental Sciences, Boston College, Chestnut Hill, MA 02467, USA

VC, 0000-0001-9759-4754; SCK, 0000-0001-7457-8778

*Correspondence: vasileios.chatzaras@sydney.edu.au

Abstract: We report on the petrology, microstructure and seismic properties of 44 peridotite xenoliths extracted from the upper mantle beneath Marie Byrd Land (MBL), West Antarctica. The aim of this work is to understand how melt-rock reaction, refertilization, and deformation affected the seismic properties (velocities, anisotropy) of the West Antarctic upper mantle, in the context of MBL tectonic evolution and West Antarctic Rift System formation. Modal compositions, mineral major element compositions, microstructures, and crystallographic preferred orientations (CPOs) provide evidence for diachronous reactive melt percolation and refertilization. Olivine shows three main CPO patterns, the A-type, axial-[010], and axial-[100] texture types. Average seismic properties of the MBL mantle lithosphere are mainly controlled by the strength of olivine crystallographic texture. Reactive melt percolation and refertilization likely modified seismic velocities and anisotropy, as is suggested by a systematic decrease in maximum P-wave and S-wave anisotropies with increasing modal abundance of pyroxene. At larger spatial scales, the seismic properties of the MBL mantle xenoliths are dominated by the anisotropy resulting from the A-type olivine CPO. Variations between individual volcanic centres, however, attest to spatial variations in the mantle structure, potentially related to 3D deformation and the prolonged tectonic history of MBL.

Supplementary material: Images of xenolith sections (Fig. S1), plots of crystallographic preferred orientations (Fig. S2) and seismic properties (Fig. S4) for all the xenoliths, plot of crystallographic preferred orientation strength with temperature (Fig. S3), geographic coordinates of volcanic centres (Table S1), microprobe mineral composition results (Table S2), and calculated synthetic seismic properties by modal composition (Table S3) are available at https://doi.org/10.6084/m9.figshare.c.5315261

West Antarctica has formed the active tectonic margin between East Antarctica and Palaeo-Pacific Ocean for almost half a billion years (Jordan *et al.* 2020). This region has had a significant impact on worldwide tectonism and climate for the last *c.* 120 m.y.. West Antarctica hosts one of the largest rift systems on Earth, the West Antarctic Rift System (WARS; e.g. Behrendt *et al.* 1996; LeMasurier 2008; Siddoway 2008) (Fig. 1), which may have affected Earth's palaeoclimate through CO_2 degassing (Brune *et al.* 2017). Constraining the seismic structure of the West Antarctic mantle (e.g. Ritzwoller *et al.* 2001; O'Donnell *et al.* 2017; Lloyd *et al.* 2019) is important for unravelling the dynamics of intracontinental extension within the WARS (e.g. Siddoway 2008). It is also key for mapping the three-dimensional viscosity structure of the upper mantle, required to accurately model the interaction between mantle dynamics and the evolution of the West Antarctic ice sheet, manifested by the glacial isostatic adjustment process (Van der Wal *et al.* 2015; Powell *et al.* 2020).

Seismic tomography and anisotropy provide important constraints on the thermal structure, viscosity, compositional heterogeneity, and flow pattern of the West Antarctic upper mantle. However, our understanding of how seismic properties of the West Antarctic upper mantle are affected by the deformation, petrological evolution and composition of mantle rocks, remains incomplete. Fast seismic velocities that extend down to 70–100 km depths beneath Marie Byrd Land (MBL) and the WARS (Heeszel *et al.* 2016; O'Donnell *et al.* 2019) are attributed to a melt-free and hydrogen deficient lithospheric mantle (O'Donnell *et al.* 2017) that was not reheated during WARS Cenozoic extension (White-Gaynor *et al.* 2019). Alternatively, the increased seismic wave speeds may correspond to Proterozoic continental fragments preserved in the MBL lithosphere (Handler *et al.* 2003; O'Donnell *et al.* 2019). Low shear velocities at 100–200 km depth beneath the broader WARS region are attributed to elevated temperatures in the upper mantle, potentially related to a mantle plume (Ritzwoller *et al.* 2001; Lloyd *et al.* 2015; Heeszel *et al.* 2016). A decrease in seismic wave speeds could also result from enrichment in volatiles and melt following slab detachment and sinking (Finn *et al.* 2005) or from compositional changes during melt-rock interaction, such as Fe enrichment in olivine (Tommasi *et al.* 2004), increase in orthopyroxene mode (Mainprice 1997), and increased oxygen fugacity (Cline *et al.* 2018).

Seismic anisotropy studies show that the polarization orientation of the fast shear waves in the WARS is perpendicular to the rift trend (Fig. 1) (Accardo *et al.* 2014), unlike other major continental rift systems with fast orientations parallel or oblique to the rift axis (Vauchez *et al.* 2000). Seismic anisotropy in the upper mantle is the result of olivine crystallographic preferred orientation (CPO) in response to mantle flow (Nicolas and Christensen 1987; Mainprice and Silver 1993). However, the interpretation of seismic anisotropy data in terms of mantle flow is not always straightforward. The geometry and magnitude of seismic anisotropy depend on the symmetry, orientation and strength of olivine CPO, which in turn may vary as a function of the deformation conditions (e.g. temperature, pressure, stress), strain history, deformation regime, modal and chemical composition (e.g. presence of pyroxene, water, melt), grain size, and deformation mechanism (Tommasi *et al.* 1999; Jung and Karato 2001; Holtzman *et al.* 2003; Karato *et al.* 2008; Boneh and Skemer 2014; Hansen *et al.* 2014*a*; Précigout and Hirth 2014; Chatzaras *et al.* 2016; Bernard *et al.* 2019; Kumamoto *et al.* 2019).

To illuminate the evolution of the subcontinental lithospheric mantle in the WARS, we use the rock record of mantle xenoliths extracted from the MBL upper mantle during Miocene and Pleistocene volcanic eruptions in the Fosdick Mountains, Executive Committee Range and the Usas Escarpment (Fig. 1). These xenoliths provide important information on the microstructural and petrological evolution of the upper mantle as a result of melt-rock interaction. We particularly focus on the effect the structural and

From: Martin, A. P. and van der Wal, W. (eds) 2023. *The Geochemistry and Geophysics of the Antarctic Mantle*. Geological Society, London, Memoirs, **56**, 151–180,
First published online 17 September 2021, https://doi.org/10.1144/M56-2020-16

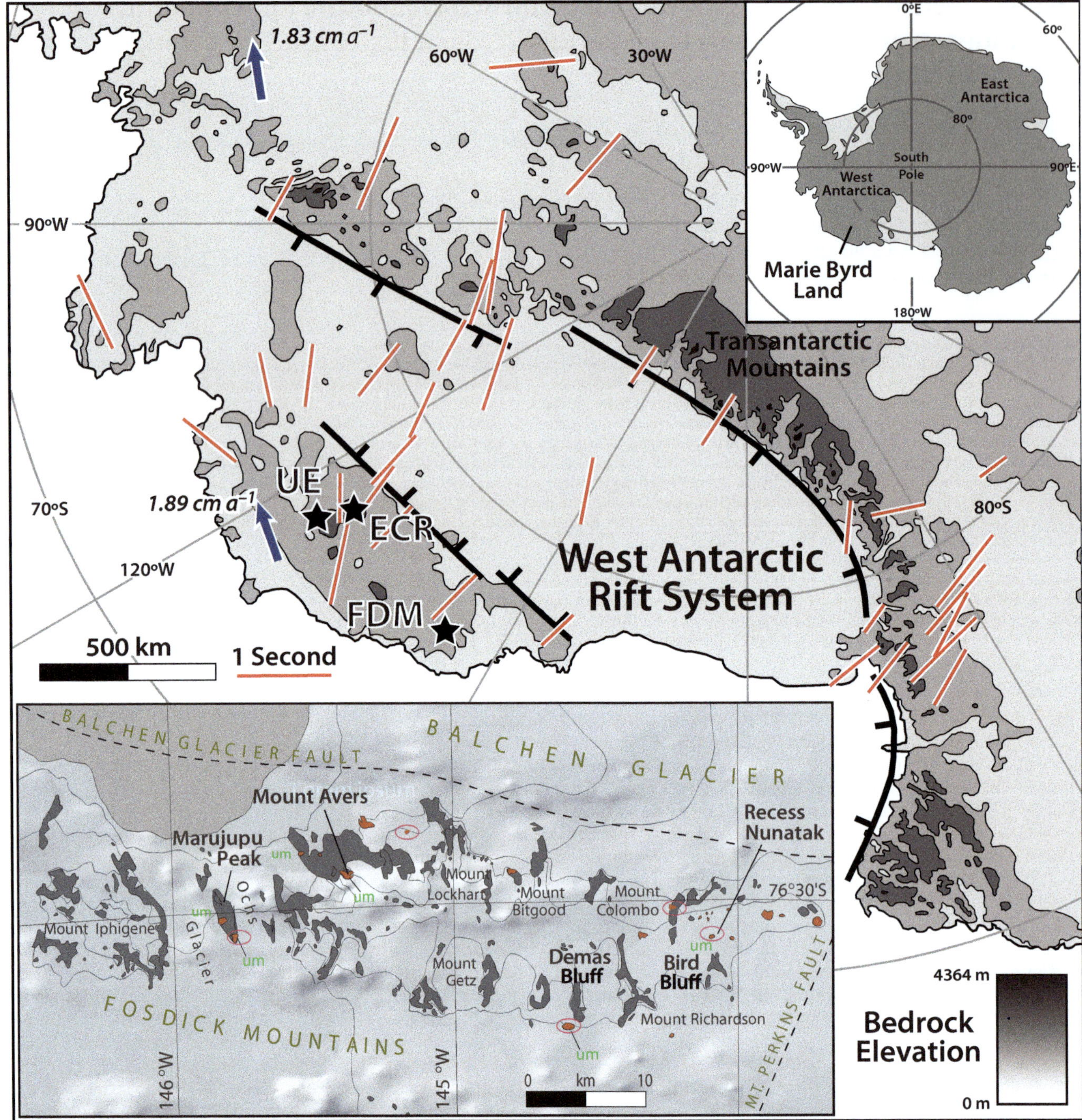

Fig. 1. Map of Antarctica showing major geographic features in relationship to the location of studied xenolith localities in the Fosdick Mountains (FDM; inset), the Executive Committee Range (ECR) and Usas Escarpment (UE). Grey contours correspond to surface elevations in West Antarctica (after Paulsen and Wilson 2010). Red vectors represent the polarization orientations of fast shear waves from Barklage *et al.* (2009) and Accardo *et al.* (2014) with the length of the vector being proportional to the splitting time. Solid blue arrows correspond to the direction of Antarctic absolute plate motion in the hotspot reference frame from Accardo *et al.* (2014). Inset shows the location of individual volcanic rock centres, denoted in orange, within the Fosdick Mountains (after Gaffney and Siddoway 2007). Sites with abundant ultramafic xenoliths are labelled 'um'. Volcanic vent bearing crustal xenoliths are indicated in pink.

compositional modification of the upper mantle has on its seismic properties. To the best of our knowledge, this is the first xenolith-based study of seismic anisotropy of the West Antarctic mantle.

Geological setting

West Antarctica developed along the tectonically active margin of Gondwana, between the East Antarctic craton and the subducting Palaeo-Pacific ocean (Jordan *et al.* 2020). The integration of results from geological and geophysical studies indicates that West Antarctica consists of several geological provinces (see Jordan *et al.* 2020 for a recent review). In this study, we focus on the WARS and MBL (Fig. 1).

The Marie Byrd Land dome and West Antarctic Rift System

This area experienced a complex geological history that includes: (1) Ordovician to Triassic ocean-continent convergence that resulted to thickening of the continental crust (Foden *et al.* 2006; Siddoway and Fanning 2009; Jordan *et al.* 2020); (2) Cretaceous transcurrent to oblique intracontinental extension, which lead to distributed crustal thinning, magmatism and the formation of the WARS (Rocchi *et al.* 2002; Siddoway *et al.* 2005; McFadden *et al.* 2010); and (3) widespread Cenozoic volcanism (LeMasurier 1990; Smellie *et al.* 2020*a*, *b*), associated with Oligocene to Miocene strike-slip motion (Müller *et al.* 2007) or oblique convergence (Granot *et al.* 2013), followed by mid-Miocene to present extension (LeMasurier and Rex 1989; Hart *et al.* 1997; Cande *et al.* 2000; Finn *et al.* 2005). GPS measurements indicate low strain rates between East and West Antarctica, which in addition to the low level of seismicity, suggest that the WARS is currently dormant or extends very slowly (Donnellan and Luyendyk 2004; Winberry and Anandakrishnan 2004).

The average lithospheric thickness is approximately 85 km across WARS and MBL (O'Donnell *et al.* 2019). The crust shows significant lateral thickness variations (Winberry and Anandakrishnan 2004; Chaput *et al.* 2014); crustal thickness decreases from 35–38 km at the MBL dome, to 28–30 km at the southern margin of the dome, and 22–30 km in the WARS. The thin crust at the southern flank of the MBL dome suggests that this region was subjected to significant thinning as part of the WARS, possibly during the diffuse Cretaceous rifting event (Winberry and Anandakrishnan 2004; LeMasurier 2008; Siddoway 2008).

Cenozoic volcanism and provenance of MBL and WARS subcontinental mantle

Cenozoic volcanism in West Antarctica is part of the diffuse alkaline magmatic province of the southwest Pacific, also encompassing Victoria Land, Australia, and New Zealand (Finn *et al.* 2005). Cenozoic igneous activity commenced at *c.* 37 Ma within MBL and is inferred to occur today beneath the West Antarctic ice sheet (e.g. LeMasurier 1990; Blankenship *et al.* 1993; Rocchi *et al.* 2002; Panter *et al.* 2021).

The geochemical composition of the erupted lavas can provide important constraints on mantle processes in the source area of the melt. The MBL Cenozoic alkaline basalts show incompatible trace element patterns similar to a high μ (HIMU; $\mu = {}^{206}Pb/{}^{204}Pb$), ocean island basalt-like mantle source (Panter and Martin 2021). They also have a wide range of isotopic compositions with both lower-μ and HIMU (μ >20.5) components, the latter being rather uncommon and geographically restricted (Hart *et al.* 1997; Panter *et al.* 1997, 2000). Basaltic lavas of *c.* 1.4 Ma in the Fosdick Mountains (western MBL) show variation in their isotopic composition between the volcanic centres, and bear no HIMU signature (Panter *et al.* 2000; Gaffney and Siddoway 2007). This observation implies heterogeneity in the mantle source component that feeds volcanism in MBL, with a HIMU-like signature of the mantle source being common but not exclusive (Gaffney and Siddoway 2007). Re–Os isotope measurements in peridotite xenoliths suggest the preservation of Proterozoic (*c.* 1.1 and >1.3 Ga) lithospheric mantle beneath central MBL (Handler *et al.* 2003, 2021).

The origin of the WARS and the Cenozoic igneous activity remain controversial. The mechanisms that have been proposed invoke: (1) presence of an active plume that underlies the WARS and MBL (LeMasurier and Rex 1989; Hole and LeMasurier 1994; Behrendt *et al.* 1996; Storey *et al.* 1999); (2) rift-related decompression melting within the WARS (Wörner 1999); (3) local decompression melting promoted by lithospheric-scale, strike-slip deformation (Salvini *et al.* 1997; Rocchi *et al.* 2005; Storti *et al.* 2007); (4) Jurassic plume-driven metasomatism of depleted upper mantle, followed by a Cretaceous HIMU-type mantle plume, trapped beneath the pre-existing metasomatized layer (Panter *et al.* 2000); and (5) Pre Late Cretaceous subduction-related metasomatism of the lithosphere, which led to the generation of geochemical reservoirs with HIMU-type signature, sampled from the alkali basalt melts that erupted during the Cenozoic to recent volcanism (Finn *et al.* 2005; Martin *et al.* 2013; Panter *et al.* 2018).

In support of the active mantle plume hypothesis beneath central West Antarctica is the Cenozoic volcanism in MBL, the isotopic signature of a HIMU mantle reservoir in the upper mantle, the topographic doming of MBL that is not isostatically supported by the crust (Chaput *et al.* 2014), and the radial distribution of the polarization direction of the fast shear wave (fast direction) in the Amundsen Sea region, which may indicate a mantle flow pattern away from the dome (Accardo *et al.* 2014).

A continuous zone of low seismic velocities at depths of 100–200 km extends along the coast of West Antarctica, from MBL to the Antarctic Peninsula (Danesi and Morelli 2001; Ritzwoller *et al.* 2001; Lloyd *et al.* 2019; Wiens *et al.* 2021). These slow shear wave speed anomalies are interpreted to indicate an upper mantle thermal anomaly (Lloyd *et al.* 2015; Shen *et al.* 2018). Alternatively, the anomalies may reflect the signature of a mantle metasomatized by subduction-related melts and fluids, as their strike is subparallel to the Palaeo-Pacific margin of Gondwana (O'Donnell *et al.* 2019). Beneath MBL, the slow shear wave speed anomalies are imaged to extend down to the mantle transition zone (600 km), suggesting that thermal anomalies in the upper mantle and the transition zone may be linked (Hansen *et al.* 2014*b*; Lloyd *et al.* 2019; Wiens *et al.* 2021). The distribution of slow shear wave speed anomalies with depth beneath MBL is consistent with the presence of a mantle plume (Hansen *et al.* 2014*b*; Lloyd *et al.* 2019; Wiens *et al.* 2021).

Shear wave splitting analysis provides important information regarding mantle processes associated with the origin of the WARS and the MBL dome (Accardo *et al.* 2014). Fast polarization directions in West Antarctica are oblique or at high angle to the direction of Antarctic absolute plate motion, suggesting that seismic anisotropy is not the result of shear due to plate motion over the mantle (Fig. 1). In the majority of West Antarctica, including the WARS, south MBL, and Transantarctic Mountains, fast polarization directions are at high angle to the strike of the WARS, i.e. parallel to the inferred extension orientation (Fig. 1). These fast polarization orientations are interpreted to result from lithospheric and asthenospheric mantle shearing frozen from the last major WARS rifting phase during the late Miocene or earlier (Accardo *et al.* 2014). Fast directions at the central part of the MBL dome are oblique to those under southern MBL and WARS (Accardo *et al.* 2014). The seismic anisotropy beneath the centre of the MBL dome could represent an older signature, inherited from the Proterozoic lithospheric mantle preserved beneath central MBL, consistent with past isotopic studies (e.g. Handler *et al.* 2003).

Marie Byrd Land xenoliths

The mantle xenoliths analysed in this study were sampled from seven volcanic centres in western and central MBL; five centres are located in the Fosdick Mountains (Marujupu Peak, Mount Avers, Demas Bluff, Bird Bluff, and Recess Nunatak), one in the Usas Escarpment (Mount Aldaz) and one in the Executive Committee Range (Mount Cumming) (Fig. 1 and Table S1). Mantle xenoliths were entrained in *c.* 1.4 Ma basaltic to basanitic cinder cones and flows in the Fosdick Mountains (Gaffney and Siddoway 2007), 10.4 Ma trachytes in Mount Cumming (LeMasurier and Rex 1989), and hawaiite lavas of 19.4 Ma in Mount Aldaz (LeMasurier and Rex 1982). The sampled xenoliths range between 3 and 15 cm in diameter, lack significant alteration, and have sharp contacts with the host basalts (Supplementary material, Fig. S1).

The average equilibration temperatures of the MBL xenoliths estimated from the three two-pyroxene geothermometers of Bertrand and Mercier (1985), Brey and Köhler (1990), and Taylor (1998) range from 780 to 1200°C, calculated at a pressure of 15 kbar (Chatzaras *et al.* 2016). This range of equilibration temperatures corresponds to extraction depths between 39 and 72 km. The extraction depths were estimated by projecting the pressure-dependent two-pyroxene temperatures on a geotherm constructed from the studied MBL xenoliths, in order to satisfy the stability of spinel in all the analysed samples (Chatzaras *et al.* 2016). Neither the McMurdo petrologic geotherm (Berg *et al.* 1989) nor the present-day Ross Embayment geotherm (ten Brink *et al.* 1997) seem to be appropriate for the MBL xenolith suite; the former would place half of the mantle xenoliths above the Moho, and the latter would require that the majority of the xenoliths contain garnet rather than spinel (Chatzaras *et al.* 2016, fig. 7b). The estimated depth of extraction corresponds to a region of intermediate shear wave velocities overlying a pronounced low velocity zone imaged beneath MBL (Ritzwoller *et al.* 2001; Shen *et al.* 2018; Lloyd *et al.* 2019; O'Donnell *et al.* 2019; Wiens *et al.* 2021).

Methods

We determined mineral major element compositions by means of electron probe microanalysis. We used electron backscatter diffraction to determine modal compositions, and the patterns of olivine and pyroxene crystallographic preferred orientations (CPOs). The CPO data are plotted relative to the spinel shape fabric (i.e. foliation and lineation), which was determined using X-ray computed tomography (Chatzaras *et al.* 2016). This workflow allows for an unbiased identification of the CPO symmetry and type, in cases where the xenolith shape fabric is not easy to identify. Seismic anisotropy properties of the peridotite xenoliths were calculated combining CPO and modal composition. Details for each of the methods applied are provided below. The spinel X-ray computed tomography, olivine CPO, and portion of the mineral major element composition results were previously presented in Chatzaras *et al.* (2016).

Electron probe microanalysis

The major element compositions of olivine, spinel, orthopyroxene, and clinopyroxene were analysed by wavelength-dispersion spectrometry with a Cameca SX50 instrument at the Department of Geoscience, University of Wisconsin-Madison. Operating conditions were 15 kV accelerating voltage, 20 nA beam current (Faraday cup), and beam diameter of 1 μm. Combinations of natural minerals and synthetic materials were used as standards for each mineral species, and data reduction was performed by Probe for Windows software, utilizing the $\varphi(\rho z)$ matrix correction of Armstrong (1988). Table S2 includes the major element compositions of the constituent minerals.

Electron backscatter diffraction

Crystallographic preferred orientations for all constituent mineral phases in the MBLxenoliths were collected by means of electron backscatter diffraction (EBSD) on polished thin sections. EBSD data were acquired on a Tescan Vega 3 LMU scanning electron microscope (SEM) equipped with a LaB_6 source and an Oxford Instruments Nordyls Max^2 EBSD detector housed within the Department of Earth and Environmental Sciences at Boston College. Typical SEM operating conditions were 20–100 nA for beam currents and an accelerating voltage of 30 kV. Large area maps of full thin sections (26 × 46 mm) were acquired using the Oxford Instruments AZtecHKL acquisition and analysis software. Mapping was performed at a step size of 7.5 μm to ensure a high density of crystallographic orientation data within individual grains. Indexing rates were typically >90%.

Post-acquisition data treatment involved processing of the EBSD raw maps using the HKL Channel5 software package. Processing included: (1) removal of isolated single pixels differing by more than 10° from their neighbours (i.e. wild spikes); and (2) assignment of the average orientation of neighbouring pixels to non-indexed pixels with eight nearest neighbours. The latter operation was iterated until no new pixels were filled and was repeated for non-indexed pixels with decreasing number of nearest neighbours as low as six. Microstructural maps were constructed from the processed EBSD datasets using version 5.5 of the MTEX Matlab toolbox for textural analysis. The abundance of the major constituent minerals (i.e. olivine, diopside and enstatite) was used to estimate the normalized modal compositions of the xenoliths, disregarding the non-indexed portions of the EBSD maps.

For grain reconstruction, a misorientation angle of 10° was used to define the lower limit of grain boundaries (Bachmann *et al.* 2011). One point per grain data (i.e. mean crystallographic orientation of each grain) were calculated for the reconstructed grains and the orientations of crystallographic axes were plotted on equal area, lower-hemisphere projections. The CPO data were plotted in the reference frame defined by the 3D spinel shape preferred orientation (i.e. relative to the spinel-defined foliation and lineation). We should note that thin sections in 30 samples were produced relative to the orientation of the spinel fabric ellipsoid determined using X-ray computed tomography, while thin sections from 14 samples were produced oblique to the spinel fabric due to restrictions imposed by small xenolith size (Chatzaras *et al.* 2016). In these 14 samples, the spinel SPO was determined by means of X-ray computed tomography from the rock billets used to make the thin sections, and the EBSD orientation data were rotated to the estimated spinel SPO reference framework.

To quantify the strength of CPO, we used the J-index (Bunge 1982) and M-index (Skemer *et al.* 2005). The J-index is calculated from the orientation distribution functions using MTEX (Bachmann *et al.* 2010) and has a value of 1 (random) to infinity (single crystal). The M-index is calculated from the distribution of uncorrelated misorientation axes and has a value of 0 (random) to 1 (single crystal). We used the BA-index (Mainprice *et al.* 2014) to quantify the tendency of olivine CPO toward end-member axial-[010] (BA = 0) or axial-[100] (BA = 1) symmetries. Calculated CPO strength indices are presented in Table 1.

Seismic anisotropy

Single-crystal elastic stiffness tensors defined for olivine (Abramson *et al.* 1997), orthopyroxene (Jackson *et al.* 2007) and clinopyroxene (Isaak *et al.* 2005) were used to determine the seismic anisotropy properties of the MBL xenoliths based on the crystallographic preferred orientation of all mineral phases and their modal abundance, as determined from full thin section EBSD maps (Mainprice 1990; Mainprice and

Table 1. *Modal composition, lithology, equilibration temperature, extraction depth, olivine CPO symmetry and CPO strength*

Sample	Phase Abundances (%)				Lithology	T (°C)	Depth (km)	Olivine				Orthopyroxene			Clinopyroxene		
	Ol	Opx	Cpx	Spl				n	BA	J	M	n	J	M	n	J	M
Mount Avers																	
FDM-AV01-X01	*59*	*22.9*	*15.7*	*2.4*	*Lherzolite*	*939*	*50*	*412*	*0.41*	*2.92*	*0.18*	*122*	*4.68*	*0.21*	*262*	*3.58*	*0.43*
Bird Bluff																	
FDM-BB01-X01	*55.5*	*1.3*	*42.1*	*1.1*	*Wehrlite*	*945*	*51*	*1698*	*0.42*	*1.39*	*0.07*	*93*	*5.84*	*0.31*	*1215*	*3.02*	*0.42*
FDM-BB02-X01	*53.2*	*23.9*	*20.9*	*2*	*Lherzolite*	*1053*	*60*	*903*	*0.33*	*1.43*	*0.02*	*453*	*1.99*	*0.05*	*712*	*2.54*	*0.43*
FDM-BB03-X01	*92.9*	*4.3*	*2.4*	*0.4*	*Dunite*	*856*	*45*	*263*	*0.1*	*4.07*	*0.21*	*129*	*3.35*	*0.24*	*83*	*6.72*	*0.44*
Demas Bluff																	
FDM-DB01-X01	*59.2*	*14.4*	*23.3*	*3.1*	*Lherzolite*	*1024*	*58*	*274*	*0.61*	*2.82*	*0.14*	*144*	*4.77*	*0.22*	*208*	*4.23*	*0.45*
FDM-DB02-X01	*69.6*	*14.4*	*15.1*	*0.9*	*Lherzolite*	*1183*	*71*	*620*	*0.77*	*3.05*	*0.2*	*308*	*2*	*0.05*	*357*	*3.99*	*0.43*
FDM-DB02-X02	*87.8*	*9.8*	*1.9*	*0.5*	*Harzburgite*	*978*	*54*	*111*	*0.5*	*8.95*	*0.36*	*53*	*9.28*	*0.44*	*60*	*14.06*	*0.69*
FDM-DB02-X03	*71.7*	*17.6*	*10.2*	*0.6*	*Lherzolite*	*999*	*55*	*91*	*0.33*	*4.54*	*0.3*	*26*	*16.75*	*0.56*	*63*	*10.58*	*0.57*
FDM-DB02-X04	*60.9*	*23*	*14.9*	*1.2*	*Lherzolite*	*933*	*50*	*103*	*0.33*	*3.93*	*0.2*	*39*	*8.09*	*0.46*	*88*	*7.83*	*0.43*
FDM-DB02-X05	*0.3*	*0.5*	*95*	*4.2*	*Clinopyroxenite*	*958*	*52*	–	–	–	–	*17*	*13.69*	*0.12*	*662*	*5.92*	*0.06*
FDM-DB02-X06	*48*	*44.6*	*7.3*	*0.1*	*Lherzolite*	*1198*	*72*	*117*	*0.43*	*10.43*	*0.29*	*127*	*4.27*	*0.22*	*629*	*10.14*	*0.59*
FDM-DB02-X08	*87.3*	*10.2*	*2.2*	*0.3*	*Harzburgite*	*803*	*41*	*107*	*0.49*	*7.07*	*0.37*	*48*	*10.51*	*0.57*	*9*	*61.01*	*0.9*
FDM-DB02-X10	*65.1*	*19.6*	*15.1*	*0.1*	*Lherzolite*	*1020*	*57*	*145*	*0.56*	*4.3*	*0.26*	*37*	*11.7*	*0.44*	*123*	*9.96*	*0.47*
FDM-DB02-X11	*71*	*12.9*	*15.2*	*0.8*	*Lherzolite*	*1039*	*59*	*224*	*0.36*	*3.54*	*0.19*	*89*	*5.14*	*0.29*	*206*	*5.4*	*0.48*
FDM-DB02-X12	*63.1*	*13.8*	*20.5*	*2.7*	*Lherzolite*	*1036*	*59*	*135*	*0.13*	*3.32*	*0.16*	*52*	*8.34*	*0.44*	*87*	*5.56*	*0.45*
FDM-DB02-X13	*71*	*13.7*	*12.8*	*2.5*	*Lherzolite*	*911*	*49*	*273*	*0.67*	*3.33*	*0.08*	*131*	*5.57*	*0.26*	*168*	*3.98*	*0.45*
FDM-DB03-X01	*89.8*	*6.9*	*1.9*	*1.4*	*Dunite*	*856*	*44*	*133*	*0.68*	*5.55*	*0.25*	*47*	*9.98*	*0.45*	*39*	*16.34*	*0.69*
FDM-DB03-X02	*54.2*	*22.1*	*19.1*	*4.5*	*Lherzolite*	*861*	*44*	*343*	*0.28*	*2.77*	*0.2*	*155*	*4.36*	*0.15*	*202*	*5.43*	*0.45*
FDM-DB03-X03	*58.7*	*21.3*	*17.6*	*2.4*	*Lherzolite*	*982*	*54*	*274*	*0.79*	*2.72*	*0.09*	*142*	*3.71*	*0.33*	*236*	*4.82*	*0.43*
FDM-DB03-X04	*76.8*	*12.5*	*9*	*2.7*	*Lherzolite*	*1002*	*56*	*149*	*0.38*	*9.88*	*0.13*	*59*	*11.66*	*0.61*	*29*	*18.91*	*0.73*
FDM-DB04-X01	*78.7*	*15*	*6.1*	*0.2*	*Lherzolite*	*991*	*55*	*117*	*0.25*	*4.03*	*0.2*	*58*	*5.96*	*0.28*	*68*	*10.29*	*0.5*
FDM-DB04-X02	*49.1*	*30.3*	*17.7*	*2.9*	*Lherzolite*	*984*	*54*	*208*	*0.31*	*3.37*	*0.16*	*106*	*4.78*	*0.18*	*168*	*4.15*	*0.45*
FDM-DB04-X03	*77.3*	*18.5*	*3.2*	*1*	*Harzburgite*	*1165*	*69*	*150*	*0.89*	*4.88*	*0.17*	*73*	*8.92*	*0.45*	*41*	*15.75*	*0.69*
FDM-DB04-X04	*78.7*	*19.5*	*1.6*	*0.2*	*Harzburgite*	*968*	*53*	*114*	*0.58*	*4.43*	*0.28*	*49*	*7.99*	*0.49*	*78*	*32.92*	*0.46*
Marujupu																	
FDM-MJ01-X01	*29.1*	*35.7*	*29.2*	*6*	*Ol. Websterite*	*898*	*48*	*408*	*0.68*	*2.5*	*0.05*	*197*	*7.39*	*0.09*	*247*	*4.82*	*0.05*
FDM-MJ01-X02	*64.9*	*15.2*	*17.5*	*2.4*	*Lherzolite*	*974*	*53*	*244*	*0.52*	*4*	*0.2*	*90*	*4.68*	*0.18*	*161*	*3.97*	*0.44*
FDM-MJ01-X03	*64.8*	*21.4*	*11.8*	*2*	*Lherzolite*	*929*	*50*	*220*	*0.49*	*2.7*	*0.08*	*86*	*4.88*	*0.26*	*85*	*7.09*	*0.48*
FDM-MJ01-X05	*69.4*	*8.4*	*18.1*	*4.1*	*Lherzolite*	*1014*	*57*	*163*	*0.35*	*3.76*	*0.16*	*75*	*10.27*	*0.34*	*163*	*6.11*	*0.48*
FDM-MJ01-X06	*45.7*	*29.4*	*23.9*	*1*	*Lherzolite*	*1070*	*61*	*1110*	*0.46*	*1.51*	*0.12*	*216*	*2.61*	*0.22*	*718*	*2.77*	*0.42*

(Continued)

Table 1. *Continued*

Sample	Phase Abundances (%)				Lithology	T (°C)	Depth (km)	Olivine				Orthopyroxene			Clinopyroxene		
	Ol	Opx	Cpx	Spl				n	BA	J	M	n	J	M	n	J	M
Recess Nunatak																	
FDM-RN01-X01	*87.8*	*1.2*	*9.6*	*1.4*	*Wehrlite*	*961*	*52*	*292*	*0.42*	*4.34*	*0.25*	*35*	*7.53*	*0.34*	*155*	*4.56*	*0.46*
FDM-RN02-X01	*61.2*	*14.4*	*22.4*	*2*	*Lherzolite*	*828*	*42*	*383*	*0.59*	*2.75*	*0.16*	*154*	*4.88*	*0.28*	*406*	*3.71*	*0.43*
FDM-RN03-X01	*84.1*	*9.3*	*6.2*	*0.5*	*Lherzolite*	*943*	*51*	*445*	*0.49*	*2.75*	*0.16*	*236*	*15.66*	*0.33*	*906*	*9.78*	*0.47*
FDM-RN04-X01	*53.2*	*27.1*	*18.4*	*1.4*	*Lherzolite*	*812*	*42*	*377*	*0.63*	*2.71*	*0.2*	*219*	*3.11*	*0.2*	*306*	*3.41*	*0.45*
Mount Avers – Bird Bluff																	
FDM-AVBB01	*59.8*	*17.8*	*19.9*	*2.5*	*Lherzolite*	*937*	*50*	*2681*	*0.15*	*2.19*	*0.12*	*896*	*1.56*	*0.06*	*1466*	*1.54*	*0.44*
FDM-AVBB02	*64.5*	*18.6*	*14.4*	*2.5*	*Lherzolite*	*779*	*39*	*630*	*0.19*	*1.97*	*0.11*	*195*	*2.97*	*0.16*	*331*	*2.63*	*0.42*
FDM-AVBB03	*0.7*	*24.2*	*62.9*	*12.2*	*Websterite*	*949*	*51*	–	–	–	–	*483*	*2.04*	*0.03*	*546*	*2.12*	*0.02*
FDM-AVBB04	*48.1*	*17.2*	*32*	*2.7*	*Lherzolite*	*822*	*42*	*579*	*0.24*	*1.88*	*0.09*	*216*	*3.32*	*0.15*	*425*	*4.2*	*0.44*
FDM-AVBB05	*71.1*	*16*	*11.3*	*1.6*	*Lherzolite*	*814*	*42*	*743*	*0.32*	*1.73*	*0.14*	*188*	*2.99*	*0.2*	*218*	*3.56*	*0.43*
FDM-AVBB06	*56.1*	*25.8*	*16.8*	*1.3*	*Lherzolite*	*940*	*50*	*911*	*0.26*	*1.9*	*0.09*	*298*	*2.68*	*0.1*	*475*	*2.52*	*0.42*
FDM-AVBB07	*79*	*4.6*	*14*	*2.4*	*Wehrlite*	*805*	*41*	*677*	*0.26*	*1.87*	*0.08*	*165*	*3.02*	*0.13*	*364*	*2.51*	*0.42*
FDM-AVBB08	*40.1*	*18.7*	*38.2*	*3*	*Lherzolite*	*832*	*43*	*831*	*0.24*	*3.27*	*0.12*	*296*	*2.95*	*0.16*	*559*	*5.88*	*0.39*
Mount Aldaz																	
AD6021-X01	*1.1*	*3.4*	*84.8*	*10.7*	*Clinopyroxenite*	*1017*	*56*	–	–	–	–	*14*	*29.38*	*0.34*	*134*	*26.23*	*0.21*
AD6021-X02	*71.4*	*12.5*	*15.4*	*0.7*	*Lherzolite*	*1084*	*63*	*312*	*0.27*	*3.54*	*0.20*	*132*	*4.24*	*0.20*	*283*	*3.50*	*0.44*
Mount Cumming																	
KSP89-181-X01	*99.3*	*0.00*	*0.2*	*0.6*	*Dunite*	*862 /995*	*42 /52*	*1354*	*0.80*	*4.34*	*0.34*	–	–	–	*46*	*12.82*	*0.50*

Phase abundances estimated from large area EBSD maps (Ol, olivine; Opx, orthopyroxene; Cpx, clinopyroxene; Spl, spinel); T, average temperature of three two-pyroxene geothermometers; n, number of grains analysed per phase; BA, BA-index; J, J-index; M, M-index.

Humbert 1994). Seismic properties were calculated using version 5.5 of the freely available MTEX toolbox for MATLAB using Voigt-Reuss-Hill averaging of single-crystal elastic constants at ambient conditions. Calculated seismic anisotropy parameters for individual samples are presented in Table 2 and Supplementary Figure S4. The seismic anisotropy data are plotted relative to the foliation and lineation defined by the shape-preferred orientation of spinel, determined by X-ray computed tomography (Chatzaras *et al.* 2016).

Average seismic properties were also calculated for the MBL xenoliths by summation of individual samples within a distinct region, volcanic centre, or based on common properties such as crystallographic texture symmetry. In order to construct an equally weighted average of seismic properties within a subset of samples that also reflected the variations in modal abundances of constituent minerals across the sample suite, crystallographic orientations were randomly selected from all phases within a sample such that each sample in the calculated subset contributed an equal number of crystallographic orientations to the averaged sample. The summation of individual samples was used to produce one million unique crystallographic orientations that were in turn used to calculate average seismic properties based on Voigt–Reuss–Hill averaging of single-crystal elastic constants, as described previously. Given that all samples within the xenolith suite share a common reference frame, defined by the shape-preferred orientation of spinel determined by X-ray computed tomography (Chatzaras *et al.* 2016), this calculation results in a maximum estimation of the seismic anisotropy for any given locality or texture symmetry, as it relies on the assumption of a common geographic orientation of foliation and lineation in all samples (Baptiste *et al.* 2015). The average seismic anisotropy properties of MBL xenolith localities are presented in Table 3. Seismic anisotropy properties based on the summation of distinct crystallographic texture symmetries are given in Table 4.

Results

Modal compositions

The MBL xenoliths are predominantly classified as lherzolite ($n = 30$), with lesser occurrences of harzburgite ($n = 4$), wehrlite ($n = 3$), dunite ($n = 3$), olivine websterite ($n = 1$), websterite ($n = 1$) and clinopyroxenite ($n = 2$) (Fig. 2a, Table 1). The upper mantle sampled beneath the volcanic centres is compositionally heterogeneous, comprised of lherzolite and one or more of the extreme compositions (i.e. wehrlite, websterite, clinopyroxenite, dunite, harzburgite). The xenoliths are fertile,

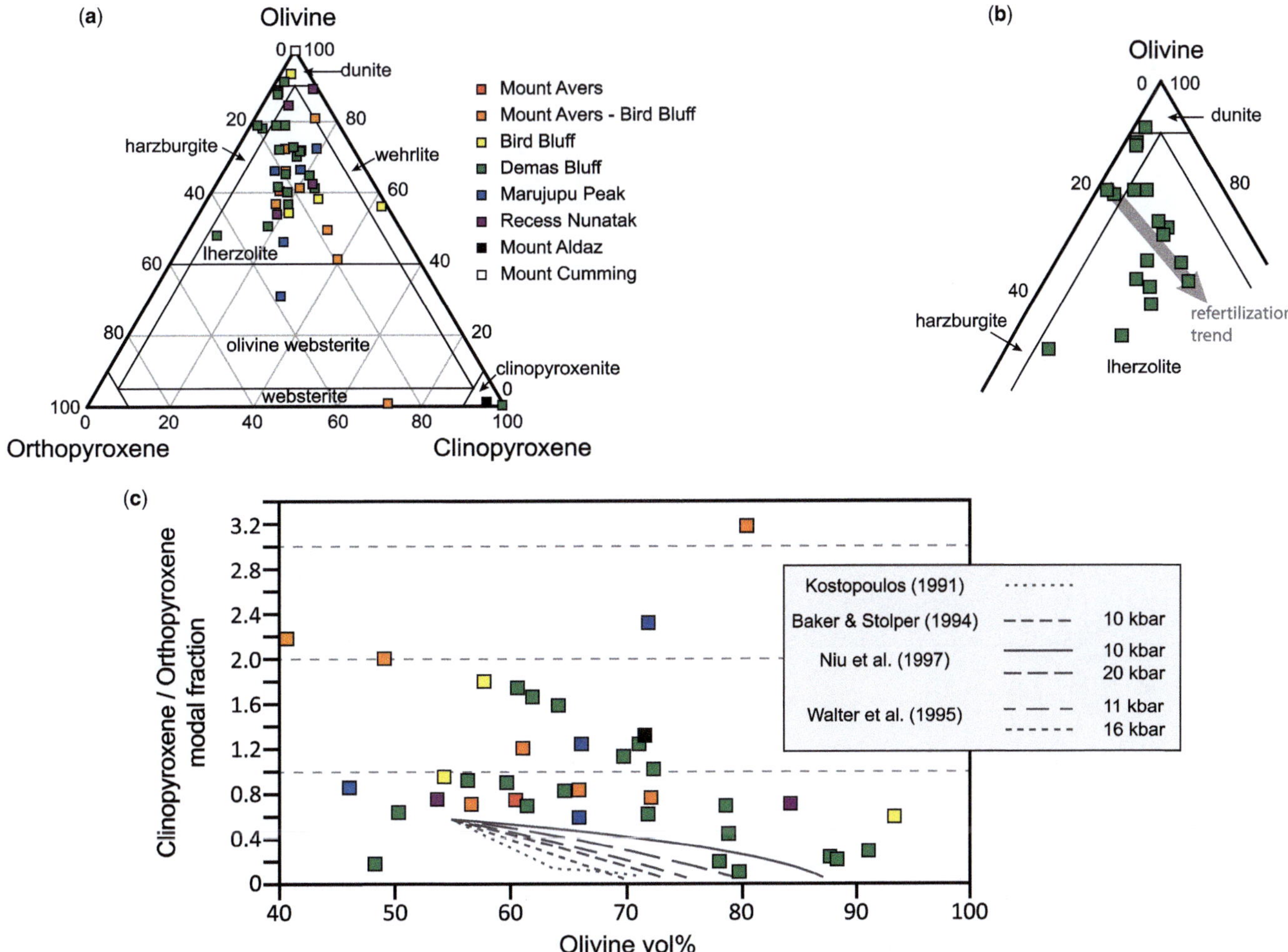

Fig. 2. Modal composition diagrams. (**a**) Ternary diagram showing the modal composition of all the analysed mantle xenoliths from Marie Byrd Land. (**b**) Ternary diagram showing the modal composition of the mantle xenoliths from the Demas Bluff volcanic centre. The grey arrow approximates the inferred refertilization trend, which may have produced the range in compositions seen in the Demas Bluff volcanic centre. (**c**) Ratio of clinopyroxene and orthopyroxene modal fractions as a function of olivine modal fraction. Dark grey curves correspond to the evolution of modal compositions predicted by different melting models for an initial modal composition of 55% olivine, 28% orthopyroxene, 15% clinopyroxene, and 2% spinel (from Zaffarana *et al.* 2014). Symbols similar to panel (a), vol% refers to volume percentage.

having average modal compositions of 55–71% olivine, 13–23% orthopyroxene, 15–24% clinopyroxene, and 1–2% spinel. High clinopyroxene mode could represent lower degrees of partial melting. However, the modal compositions of the MBL xenoliths depart from the modal compositions for partial melting trends predicted by melting models calculated for a starting ratio of clinopyroxene/orthopyroxene of 0.5 (Zaffarana *et al.* 2014), and indicate that the majority of the MBL xenoliths are enriched in clinopyroxene (Fig. 2c). Even for higher ratios of clinopyroxene/orthopyroxene (e.g. 0.8 or 1), a large number of xenoliths would still show clinopyroxene enrichment. Clinopyroxene enrichment could be the result of refertilization in response to reactions with percolating melts (Fig. 2b).

Mineral major element composition

The average olivine Mg number (Mg# = 100 × [Mg/(Mg + Fe)]) and spinel Cr number (Cr# = 100 × [Cr/(Cr + Al)]) in the MBL xenoliths range from 88 to 92, and from 6 to 63, respectively (Fig. 3a). The Mg# in olivine tends to increase with an increase in Cr# in coexisting spinel, a covariation that is normally observed in spinel peridotites (Arai 1994). In the MBL xenoliths, the covariation between olivine Mg# and spinel Cr# correlates with changes in modal composition, with the more clinopyroxene-rich xenoliths having low Mg# (<91) and Cr# (<17) values (Fig. 3a). With the exception of the dunite (KSP89-181-X01), all xenoliths plot within the olivine-spinel mantle array of Arai (1994), and are distributed

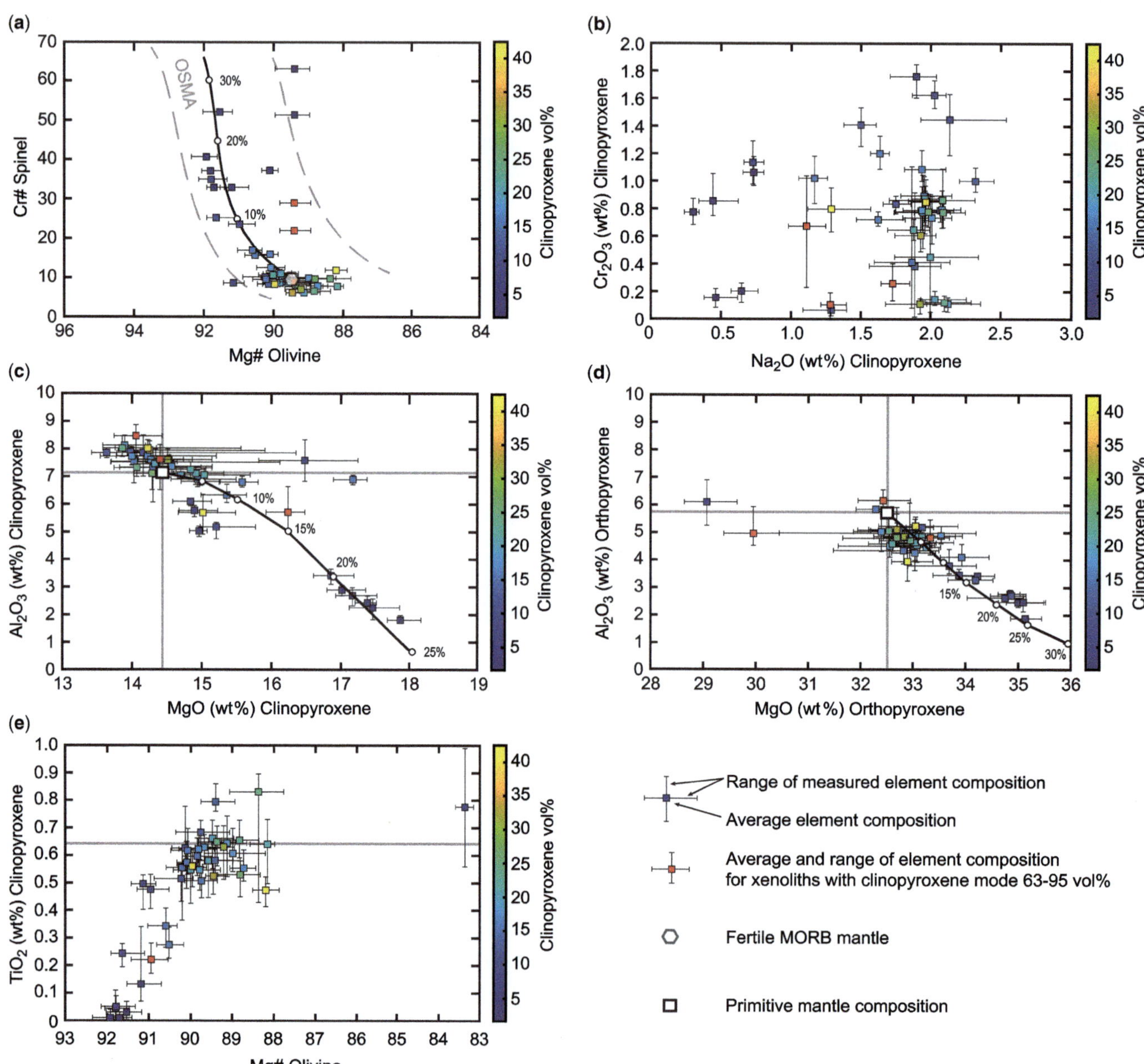

Fig. 3. Mineral major element chemistry of the Marie Byrd Land peridotite xenoliths. (**a**) Plot of Cr# in spinel v. Mg# in olivine. The olivine-spinel mantle array (OSMA) and the degrees of partial melting are from Arai (1994). The pentagon corresponds to the fertile MORB mantle (after Pearce *et al.* 2000). (**b**) Plot of Na_2O v. Cr_2O_3 in clinopyroxene. The majority of clinopyroxene grains are sodium rich (>1.5 wt%). (**c**), (**d**) Plot of Al_2O_3 v. MgO melting trends for clinopyroxene (c) and orthopyroxene (d). The Al_2O_3 and MgO contents of primitive mantle (white box and straight grey lines) and theoretical residual trend (curved black line) are from Upton *et al.* (2011) using the primitive mantle composition (spinel stability field) from McDonough and Sun (1995) and equations from Workman and Hart (2005). The white circles on curves indicate percentages of partial melting. (**e**) Plot of Mg# in olivine v. TiO_2 in the coexisting clinopyroxene. The straight grey line corresponds to the constant Ti content in clinopyroxene of refertilized peridotites of Le Roux *et al.* (2007). Abbreviation wt% refers to weight percent.

along the typical partial melting trend of a fertile mantle source. The more clinopyroxene-rich xenoliths plot near the composition of the fertile MORB mantle (Pearce *et al.* 2000) or are affected by very low degrees of partial melting (<5%). On the other hand, harzburgite, dunite, and clinopyroxene-poor lherzolite and wehrlite seem to be residues of relatively high degrees of partial melting (10–25%) (Fig. 3a).

Clinopyroxene shows a wide range in Al_2O_3 (1.8–8.5 wt %), MgO (13.6–17.9 wt%), Na_2O (0.30–2.32 wt%) and TiO_2 (0.01–0.83 wt%) content (Fig. 3b, c and e; Table S2). Orthopyroxene is also characterized by a large compositional variation in terms of Al_2O_3 (1.9–6.2 wt%) and MgO (32.3–35. 1 wt%) (Fig. 3d; Table S2). A clinopyroxenite and a clinopyroxene-poor lherzolite have orthopyroxene with anomalously low MgO (29.1 and 30.0 wt%) relative to the rest of the xenolith suite (Fig. 3d; Table S2). Pyroxene in clinopyroxene-poor lherzolite, harzburgite, and dunite show lower Al_2O_3 and higher MgO abundances than the more fertile peridotite (Fig. 3c and d). Two groups of clinopyroxene are identified on the basis of their major element composition, which shows correlation to lithology and clinopyroxene mode. A fertile group of lherzolite, wehrlite, websterite, and clinopyroxenite has low MgO (13.6–15.0 wt%) and high Al_2O_3 (6.8–8.5 wt%), Na_2O (1.50–2.32 wt%), and TiO_2 (0. 5–0.83 wt%) content, compared to a group of refractory peridotite, such as clinopyroxene-poor lherzolite, harzburgite, and dunite characterized by high MgO (16.9–17.9 wt%), and low Al_2O_3 (1.29–3.4 wt%), Na_2O (0.3–2.32 wt%), and TiO_2 (0. 01–0.13 wt%) content (Fig. 3b, d and e).

We used the theoretical melting trend of Upton *et al.* (2011) for Al_2O_3 and MgO in clinopyroxene and orthopyroxene to estimate the role of partial melting on the composition of pyroxene in the MBL xenoliths. Orthopyroxene in the MBL xenoliths follows the theoretical melting trend, indicating a degree of partial melting between 2% and 25% (Fig. 3d). Clinopyroxene only partially follows the melting trend and shows either a low degree of partial melting (less than 10% and predominantly less than 5%) or high degrees of melting (20–25%) with a gap in between (Fig. 3c). The range in the degree of partial melting estimated for the orthopyroxene (Fig. 3d) and spinel (Fig. 3a) is comparable but is not replicated by the estimations for the clinopyroxene. This discrepancy suggests that partial melting may not be the only process that influenced the Al–Mg distribution between the pyroxene in the MBL xenoliths.

In a plot of TiO_2 in clinopyroxene v. Mg# in the coexisting olivine, the clinopyroxene in the more fertile lherzolite and wehrlite shows a constant TiO_2 concentration at 0.65 and 0.8 wt% over a range of olivine Mg# values (Fig. 3d). The TiO_2 decreases rapidly with increasing Mg#, down to 0.01 wt% in the more refractory harzburgite and dunite (Fig. 3e). Based on Embey-Isztin (2016), the observed trend in the TiO_2 concentration in clinopyroxene is comparable to the trend in the TiO_2 concentration in clinopyroxene v. whole rocks in the Lherz peridotite, which is interpreted to indicate a refertilization process (Le Roux *et al.* 2007).

Microstructures

The MBL xenoliths have a coarse granular or tabular microstructure. Eight lherzolite xenoliths (FDM-AV01-X01, FDM-AVBB02, FDM-AVBB06, FDM-AVBB08, FDM-DB02-X03, FDM-DB02-X12, FDM-DB03-X04, FDM-MJ01-X03), however, show a coarse-grained porphyroclastic microstructure, characterized by bimodal olivine and orthopyroxene grain size distributions. In these xenoliths, olivine and orthopyroxene grains up to 10 mm coexist with a finer-grained (*c.* 100 µm) matrix composed of olivine, pyroxene, and spinel.

In the following, we summarize the main spinel and olivine microstructures previously described in Chatzaras *et al.* (2016), and then focus on the pyroxene microstructures and the microstructural relationships between the different phases. Spinel occurs as isolated mm-sized grains with euhedral to subhedral shape that mark the deformation fabric (Fig. 4c, d, i and j). The analysis of the 3D shape of spinel grains shows that the foliation plane containing the means of the long and intermediate axes of the spinel shape fabric ellipsoid, is oriented parallel to the 3D spatial distribution of spinel grains in layers. The trend of spinel grain long axes coincides with the trend of spinel trails (Fig. 4c). Interstitial, symplectitic, or complexly shaped irregular intergrowths of spinel with pyroxene are present but are less common (Fig. 4a).

Olivine grains tend to align their long axes parallel or at a small angle (<10°) to the spinel lineation, when observed in thin sections oriented normal to the spinel-defined foliation plane and parallel to the spinel lineation (Fig. 4b). Olivine shows intragrain deformation features including undulose extinction and well-developed subgrain boundaries that are oriented perpendicular, oblique (at high angle), or subparallel to the foliation (Fig. 4a–h, j and k). Interpenetrating olivine–olivine grain boundaries are present mainly in lherzolite (Fig. 4c and d) and suggest recrystallization, accommodated by grain boundary migration (Drury and Urai 1990). The occurrence of tabular grains of olivine with straight boundaries oriented subparallel to the lineation, four grain junctions, and diamond-shaped grains with boundaries at 20–50° to the foliation, suggest grain and phase boundary sliding (Ashby and Verrall 1973; Drury and Humphreys 1988; Ree 1994; Newman *et al.* 1999). Further, olivine grains may have gently curved to straight boundaries, which lead to polygonal grain shapes with approximately 120° triple junctions (Fig. 4j). Small rounded inclusions of pyroxene in olivine are present but are not common (Fig. 4c and g).

Orthopyroxene occurs as isolated grains that show two different morphologies. The first type includes large, millimetre-sized grains that may contain exsolution lamellae of clinopyroxene and spinel, as well as intragrain deformation features such as undulose extinction, subgrain boundaries, and deformation bands (Fig. 5a–c). In a small number of harzburgite and lherzolite xenoliths, the coarse orthopyroxene grains are elongated parallel to the spinel-defined lineation. In these harzburgite xenoliths, orthopyroxene–olivine grain boundaries are gently curved to straight, indicating a well-equilibrated microstructure (Fig. 4a). In the majority of the xenoliths, however, large orthopyroxene grains occur as irregularly shaped grains with either no obvious shaped preferred orientation, or elongation at high angle to the spinel and olivine lineation (Figs 4c, j and 5b). The irregularly shaped large orthopyroxene grains usually have concave boundaries with embayments filled by olivine (Figs 4c, e, j and 5c). Some of the large orthopyroxene grains seem to partially or fully enclose rounded olivine inclusions, which often show the same crystallographic orientation as neighbouring olivine grains (Figs 4b, f and 5c). These olivine inclusions could either represent true inclusions derived from a single coarse olivine grain that was replaced by orthopyroxene, or could result from sectioning of interpenetrating olivine and orthopyroxene grains (e.g. Soustelle *et al.* 2010; Morales and Tommasi 2011; Tommasi and Ishikawa 2014).

The second type of orthopyroxene grains are smaller in size and have subidiomorphic to xenomorphic shape. They show an interstitial habit, and form as cuspate-shaped grains in olivine triple junctions or as films along olivine–olivine grain boundaries (Fig. 4c and j). The interstitial orthopyroxene

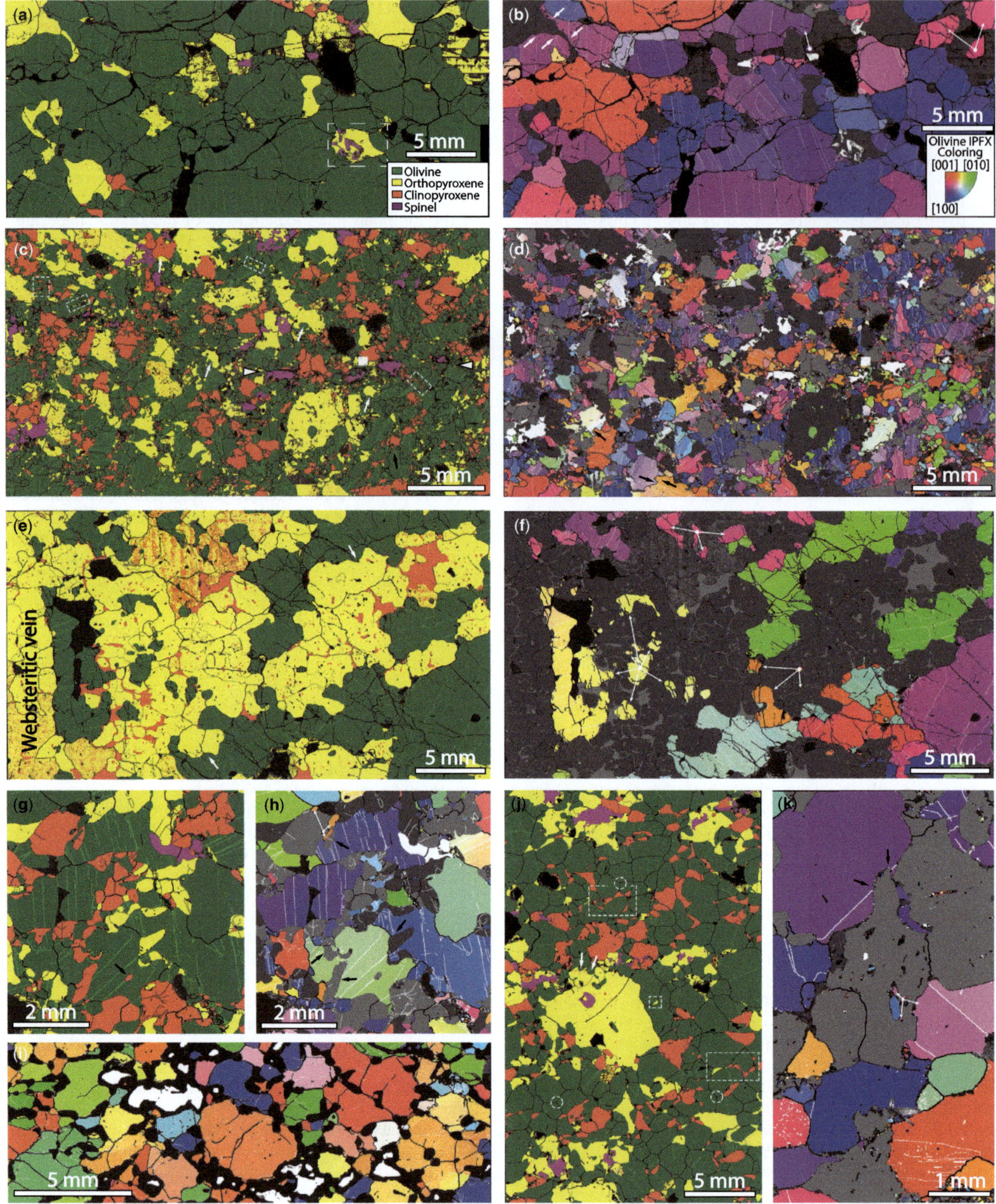

Fig. 4. EBSD maps illustrating the microstructure of the Marie Byrd Land xenoliths. All maps have upper edge parallel to the long axis (lineation) of the spinel shape fabric ellipsoid determined using X-ray computed tomography (see Chatzaras *et al.* 2016) with the foliation normal to the map. All the inverse pole figure maps of olivine and clinopyroxene crystallographic axis orientations are coloured relative to the spinel lineation (IPF-X colouring). In the phase maps, phase colouring is as follows: green, olivine; yellow, orthopyroxene; red, clinopyroxene; purple, spinel. In the combined EBSD phase and IPF-X maps, phase colouring is as follows: white, spinel; light grey, clinopyroxene; and dark grey, orthopyroxene. In all the maps, black lines correspond to grain boundaries and black areas to zero solutions (not-indexed domains). In the EBSD phase maps, subgrain boundaries are light green in olivine, olive green in orthopyroxene, and maroon in clinopyroxene. In the combined EBSD phase and IPF-X maps, subgrain boundaries are white for all phases. **(a)**, **(b)** EBSD phase map (a) and combined EBSD phase map and IPF-X map of olivine crystallographic axis orientations (b) for the harzburgite FDM-DB02-X02. The white box in (a) highlights an orthopyroxene–spinel symplectite. Note the gently curved nature of olivine–olivine grain boundaries and orthopyroxene–olivine phase boundaries. In (b), the white arrows (upper left corner) point to an orthopyroxene grain that cross-cuts olivine subgrain boundaries. The dots with arrows point to olivine inclusions (dots) within orthopyroxene grains with crystallographic orientation similar to neighbouring olivine grains (arrow heads). **(c)**, **(d)** EBSD

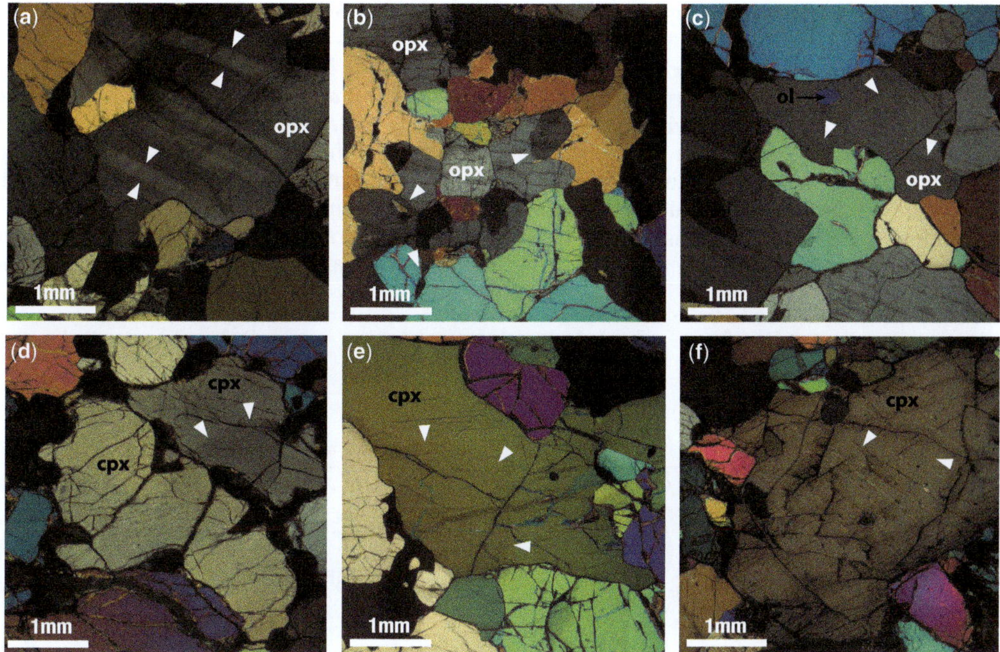

Fig. 5. Typical pyroxene microstructures in the Marie Byrd Land spinel peridotite xenoliths. In all photomicrographs lineation is parallel to the top edge and foliation is normal to the photomicrograph. All photomicrographs are under crossed-polarized light. (**a**) Coarse-grained websterite FDM-AVBB03 with granular microstructure. Orthopyroxene shows wide deformation bands (white triangles). (**b**) Orthopyroxene subgrain boundaries (white triangles) in the sparse orthopyroxene grains of wehrlite FDM-AVBB07. (**c**) Orthopyroxene subgrain boundaries (white triangles) and an olivine inclusion (black arrow) in lherzolite FDM-DB04-X01. (**d**) Clinopyroxene grains with subgrain boundaries (white triangles) in the granular clinopyroxenite FDM-DB02-X05. (**e**) Clinopyroxene with very subtle undulose extinction and weak subgrain boundaries (white arrows) in lherzolite FDM-DB03-X04. (**f**) Clinopyroxene with two sets of deformation bands in lherzolite FDM-MJ01-X05. Ol, olivine; Opx, orthopyroxene; Cpx, Clinopyroxene.

grains do not show exsolution lamellae nor evidence of intragrain deformation. Orthopyroxene grains of the second type cross-cut the subgrain boundaries of olivine grains (Fig. 4b and h), suggesting that the crystallization of at least some of orthopyroxene grains occurred after or during the late stages of olivine deformation.

Clinopyroxene occurs as isolated grains which, depending on the lithology, may show one or both of the two grain morphologies observed for orthopyroxene. Clinopyroxenite is composed of large clinopyroxene grains up to 4 mm in size that show evidence of intragrain deformation, such as undulose extinction, and subgrain boundaries (Figs 4i and 5d), but do not present any orthopyroxene exsolutions. In contrast to the clinopyroxenite, the lherzolite, wehrlite, websterite, and olivine websterite contain both large and interstitial clinopyroxene grains (Fig. 4c and j). The large clinopyroxene grains show similar intragrain deformation features with those in the clinopyroxenite but are characterized by more irregular shapes and by exsolution lamellae of orthopyroxene (Figs 4g, j and 5e, f). Some of the large clinopyroxene grains contain rounded olivine inclusions that may show similar crystallographic orientation as neighbouring olivine grains (Fig. 4h and k). The majority of the clinopyroxene, however, occur as subidiomorphic to xenomorphic shaped interstitial grains that may have concave grain boundaries with embayments filled by olivine or orthopyroxene, when in contact with each of the two phases. Clinopyroxene also occurs as films along olivine and orthopyroxene grain boundaries (Fig. 4c, e and j). Interstitial clinopyroxene grains show neither exsolutions nor indications of intragrain deformation. Both the large and the interstitial clinopyroxene may cross-cut the subgrain boundaries of olivine grains (Fig. 4h and k) suggesting that they crystallized during the late stages of olivine deformation and after olivine deformation, respectively. The harzburgite and dunite contain interstitial clinopyroxene with no evidence of intragrain deformation.

Fig. 4. (*Continued*). phase map (c) and combined EBSD phase map and IPF-X map of olivine crystallographic axis orientations (d) for the lherzolite FDM-AVBB01. The olivine and pyroxene show more irregular grain shapes compared to the harzburgite in (a). In (c), the white triangles mark a spinel trail with the elongation of the spinel grains parallel to the trace of the trail. White arrows highlight pyroxene with concave grain boundaries, and the white boxes include areas with interstitial pyroxene grains. The black arrows in (d) point to olivine grains with interpenetrating grain boundaries. (**e**), (**f**) EBSD phase map (e) and combined EBSD phase map and IPF-X map of olivine crystallographic axis orientations (f) for the xenolith FDM-DB02-X06. The xenolith is interpreted to be a coarse-grained dunite cross-cut by websteritic veins. Note the sharp boundary between the dunite and the websteritic vein on the left side of the maps, and the interstitial habit of clinopyroxene in the veins. White arrows in (e) point to orthopyroxene with concave grain boundaries. White dots with arrows in (f) as in (b). (**g**), (**h**) EBSD phase map (g) and combined EBSD phase map and IPF-X map of olivine crystallographic axis orientations (h) for the lherzolite FDM-BB02-X01. The black arrow in (g) shows a clinopyroxene inclusion within an olivine grain, while in (h) it points to pyroxene grains that cross-cut olivine subgrain boundaries. One of these clinopyroxene grains contains an olivine inclusion (dot) with similar crystallographic orientation as neighbouring olivine grains (arrowheads). (**i**) Combined EBSD phase map and IPF-X map of clinopyroxene crystallographic axis orientations for the clinopyroxenite FDM-DB02-X05. The formation of subgrain boundaries in clinopyroxene indicates intragrain deformation. Spinel is white and the black areas correspond to amorphous material. (**j**) EBSD phase map for the lherzolite FDM-AVBB02. White circles include olivine 120° triple junctions. White boxes include areas with interstitial pyroxene grains. (**k**) Combined EBSD phase map and IPF-X map of olivine crystallographic axis orientations for the wehrlite FDM-BB01-X01. The black arrows point to a clinopyroxene grain that cross-cuts an olivine subgrain boundary. The clinopyroxene grain also encloses olivine inclusions, which show the same crystallographic orientation as a neighbouring olivine grain (white dots with arrows).

Xenolith FDM-DB02-X06 consists of coarse-grained olivine and orthopyroxene rich domains with abrupt contacts (Fig. 4e and f). The orthopyroxene rich domains contain olivine inclusions with similar crystallographic orientation as neighbouring olivine grains, and cross-cut the olivine subgrain boundaries. We interpret the orthopyroxene rich domains as websteritic veins that cross-cut a coarse-grained dunite. Clinopyroxene in the websteritic veins has an interstitial habit being present both in grain and subgrain triple junctions, as well as forming grain and phase boundary films.

Crystallographic preferred orientations

Olivine in the MBL xenoliths preserves a range of CPO types (Chatzaras *et al.* 2016) (Tables 1 and 2, Fig. 6 and Fig. S2). We use the BA index to assign each olivine CPO to one of the three symmetries; the orthorhombic, axial-[100], or axial-[010] symmetry (Fig. 7a). Xenoliths with olivine orthorhombic symmetry show mainly the A-type CPO pattern, with point concentrations of [100] axes subparallel to the spinel lineation, [010] axes subparallel to the pole to the foliation, and

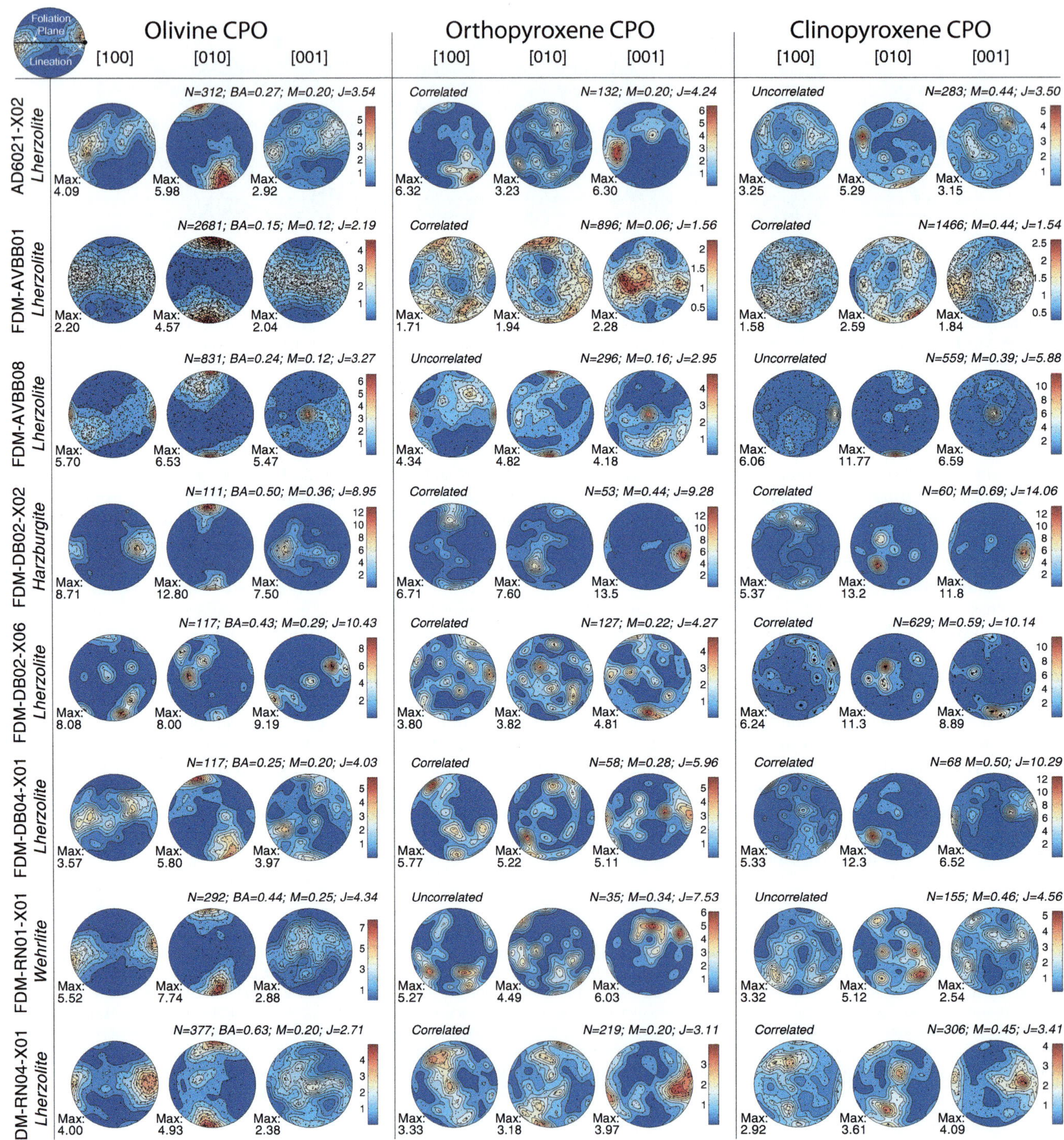

Fig. 6. Representative crystallographic preferred orientations of olivine, orthopyroxene and clinopyroxene grains. Crystallographic orientations are plotted as one point per grain data sets in lower hemisphere equal area projections, relative to the spinel fabric ellipsoid axes. Colour scales are for multiples of uniform distribution. For each analysed mineral in a sample, the number of grains analysed, as well the M, and J indices, are given, in addition to the BA index for olivine. For the full list of crystallographic orientations in all the Marie Byrd Land xenoliths see Figure S2.

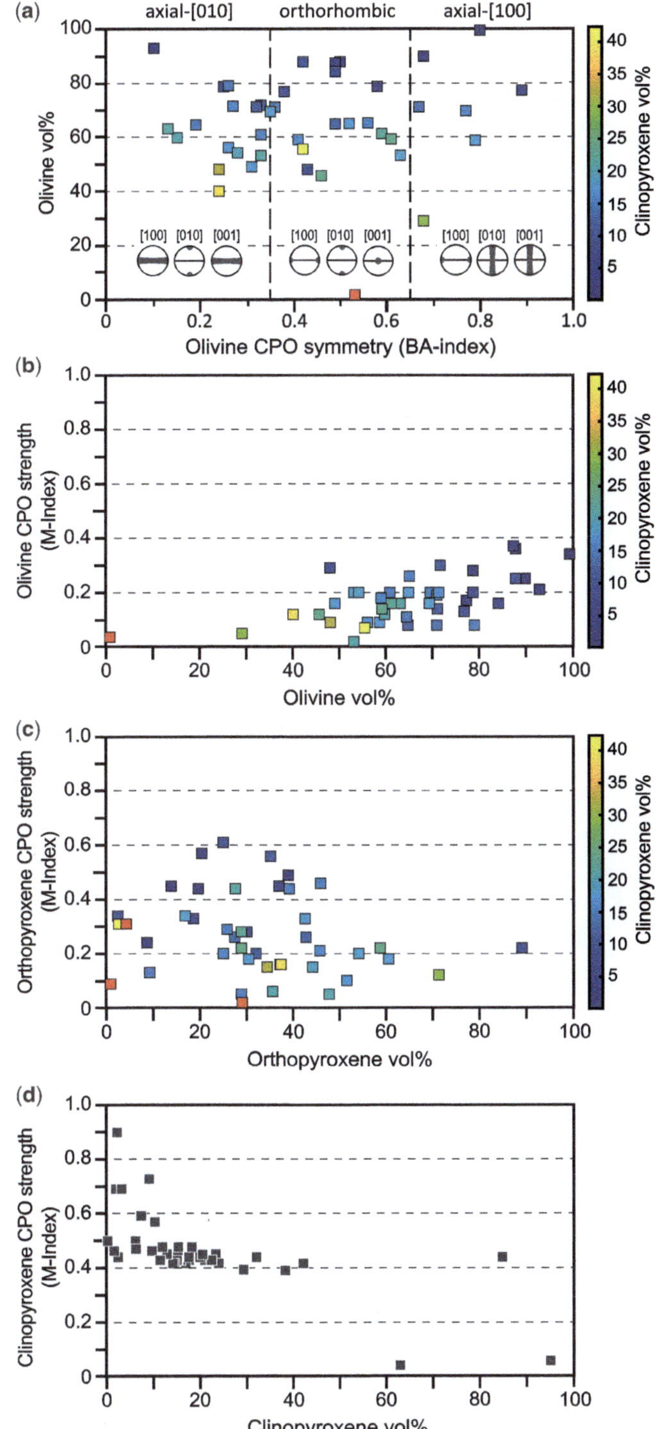

Fig. 7. (**a**) Plot of olivine CPO symmetry v. olivine modal composition [volume (vol) %]. (**b**)–(**d**) Relationship between CPO strength and modal composition (volume %) for olivine (b), orthopyroxene (c), and clinopyroxene (d). Data in (a–c) are colour coded for the clinopyroxene modal composition. Data in red are from xenoliths with clinopyroxene mode between 63 and 95 vol %.

[001] axes oriented at high angle to the lineation within the foliation plane (e.g. FDM-DB02-X02, FDM-RN01-X01 in Fig. 6). Xenoliths that show an axial-[100] symmetry have point concentrations of [100] axes near the lineation, [010] axes distributed along girdles at high angle to the lineation, and dispersed [001] axes leading to complex or random distributions (e.g. FDM-RN04-X01 in Fig. 6, and FDM-DB04-X03 in Fig. S2). Xenoliths with olivine axial-[010] symmetry are characterized by [010] axes with point concentrations at high angle to the foliation plane, while [100] and [001] axes form girdles within the foliation plane (e.g. FDM-AVBB01, FDM-AVBB08 in Fig. 6). The MBL xenoliths provide evidence that the symmetry of olivine CPO is primarily controlled by the geometry of the finite strain ellipsoid rather than the temperature, pressure, differential stress, and water content in the upper mantle (Chatzaras *et al.* 2016). Specifically, the orthorhombic (A-type), axial-[100] and axial-[010] CPOs form in relation to neutral, prolate and oblate fabric ellipsoids, respectively (Chatzaras *et al.* 2016).

Olivine CPO symmetry does not vary in a systematic way with olivine and clinopyroxene modal composition (Fig. 7a). There is, however, a tendency of axial-[010] patterns to develop at lower olivine modes (average of 63%), and higher clinopyroxene modes (average of 17%), compared to axial-[100] patterns (average olivine and clinopyroxene modes of 74% and 10%, respectively). The olivine CPO strength is moderate to low in most of the xenoliths. The M index ranges between 0.02 and 0.37, and the J index varies from 1.39 to 10.43 (Table 1). Olivine CPO strength increases with olivine mode (Fig. 7b) and does not vary systematically with equilibration temperature, and thus depth (Fig. S3).

Pyroxene CPOs show correlation to the olivine CPO in 22 xenoliths (Fig. 6 and Fig. S2). These xenoliths are primarily harzburgite and lherzolite with large pyroxene grains (Figs 4a and 5c; Fig. S2). Correlation between pyroxene and olivine CPO is expressed by concentration of pyroxene [001] axes subparallel to olivine [100] axes maximum. The weaker pyroxene [100] and/or [010] axes maxima are subparallel mainly to the olivine [010] axes and in some samples to olivine [001] axes maximum (Fig. 6: FDM-DB02-X02, FDM-DB04-X01, FDM-RN04-X01; Fig. S2). A weaker correlation between pyroxene and olivine CPO is observed in pyroxene-rich lherzolite with coexisting coarse-grained and fine-grained pyroxene grains (Figs 4c and 6: FDM-AVBB01). The pyroxene CPOs indicate dominant activation of the orthopyroxene (100)[001] slip system and the clinopyroxene (010)[001] and (100)[001] slip systems.

Xenolith FDM-DB02-X06, the dunite that is cross-cut by websteritic veins (composition averaged to lherzolite), exhibits a markedly different CPO from the rest of the xenoliths that have correlated olivine and pyroxene CPOs (i.e. subparallel olivine [100] and pyroxene [001] axes) (Fig. 6). Olivine [100] axes and pyroxene [001] axes are concentrated at high angle to the foliation, while olivine [001] and pyroxene [100] axes have maxima near the lineation. The olivine CPO suggests activation of the (100)[001] slip system; however, the pyroxene CPOs indicate [100] slip, which is not commonly observed in deformed peridotites. The olivine [100] and [001] axes, and pyroxene [001] axes are aligned with the traces of the two orientations of websteritic veins observed in the xenolith (Fig. 4e).

A combination of correlated orthopyroxene and olivine CPOs, and uncorrelated clinopyroxene and olivine CPOs (i.e. clinopyroxene [001] axes are not subparallel to olivine [100] axes) is observed in 9 xenoliths (Fig. 6 and Fig. S2). Orthopyroxene [001] axes show maximum concentration subparallel to olivine [100] axes near the lineation, while orthopyroxene [100] and/or [010] axes show alignment with olivine [010] axes, which are mainly oriented at high angle to the foliation (Fig. 6: FDM-AD6021-X02). Clinopyroxene, which primarily exhibits an interstitial habit, forms a non-random CPO that shows no correlation to the olivine CPO (Fig. 4j).

In 10 xenoliths, we observe no correlation between the olivine and pyroxene CPOs. The orthopyroxene and clinopyroxene CPOs are more dispersed from the corresponding olivine CPOs, and do not show well-developed patterns. In 3 lherzolite xenoliths, the orthopyroxene CPO is characterized by maximum concentrations of [001] axes at high angle to the

Table 2. *Seismic anisotropy parameters and olivine CPO type*

Volcanic centre	Sample	Vp Anisotropy (%)	Vs Max Anisotropy (%)	Vs1 Max Anisotropy (%)	Vs2 Max Anisotropy (%)	Vp/Vs1 Anisotropy (%)	Vp/Vs2 Anisotropy (%)	Olivine CPO Type
Mount Avers	*FDM-AV01-X01*	*7.58*	*5.94*	*2.80*	*3.73*	*5.38*	*4.79*	*A*
Bird Bluff	*FDM-BB01-X01*	*6.29*	*3.77*	*2.42*	*2.56*	*5.09*	*4.77*	*A*
Bird Bluff	*FDM-BB02-X01*	*2.08*	*1.60*	*0.84*	*1.33*	*1.84*	*1.88*	*random*
Bird Bluff	*FDM-BB03-X01*	*10.80*	*8.24*	*7.29*	*3.52*	*6.25*	*9.90*	*axial-[010]*
Demas Bluff	*FDM-DB01-X01*	*5.73*	*4.09*	*2.38*	*3.47*	*4.51*	*4.44*	*axial-[100]*
Demas Bluff	*FDM-DB02-X01*	*9.09*	*5.69*	*3.21*	*4.47*	*8.68*	*5.63*	*axial-[100]*
Demas Bluff	*FDM-DB02-X02*	*12.80*	*9.78*	*5.89*	*6.51*	*9.43*	*8.53*	*A*
Demas Bluff	*FDM-DB02-X03*	*9.28*	*6.35*	*2.53*	*5.42*	*8.76*	*4.93*	*axial-[010]*
Demas Bluff	*FDM-DB02-X04*	*6.32*	*5.18*	*4.28*	*2.53*	*4.56*	*6.03*	*n/a*
Demas Bluff	*FDM-DB02-X05**	*9.24*	*4.83*	*3.64*	*2.85*	*8.53*	*9.56*	*n/a**
Demas Bluff	*FDM-DB02-X06*	*3.96*	*4.26*	*2.07*	*2.87*	*3.48*	*4.98*	*C*
Demas Bluff	*FDM-DB02-X08*	*13.44*	*11.01*	*7.40*	*6.24*	*11.17*	*9.63*	*A*
Demas Bluff	*FDM-DB02-X10*	*8.61*	*6.44*	*2.92*	*5.15*	*7.75*	*4.79*	*A*
Demas Bluff	*FDM-DB02-X11*	*9.36*	*6.96*	*4.75*	*4.14*	*6.64*	*6.59*	*A*
Demas Bluff	*FDM-DB02-X12*	*6.87*	*6.15*	*4.97*	*3.57*	*4.49*	*6.85*	*axial-[010]*
Demas Bluff	*FDM-DB02-X13*	*5.53*	*3.47*	*1.66*	*2.76*	*5.15*	*3.86*	*axial-[100]*
Demas Bluff	*FDM-DB03-X01*	*12.94*	*10.39*	*6.88*	*5.69*	*12.45*	*9.13*	*axial-[100]*
Demas Bluff	*FDM-DB03-X02*	*6.64*	*5.22*	*4.38*	*2.65*	*3.92*	*5.69*	*axial-[010]*
Demas Bluff	*FDM-DB03-X03*	*5.21*	*3.92*	*1.66*	*3.14*	*4.51*	*2.96*	*axial-[100]*
Demas Bluff	*FDM-DB03-X04*	*6.24*	*5.11*	*2.89*	*4.53*	*7.54*	*3.59*	*n/a*
Demas Bluff	*FDM-DB04-X01*	*7.49*	*6.01*	*5.76*	*2.15*	*4.67*	*7.56*	*axial-[010]*
Demas Bluff	*FDM-DB04-X02*	*5.90*	*4.39*	*3.43*	*2.13*	*3.64*	*5.32*	*axial-[010]*
Demas Bluff	*FDM-DB04-X03*	*9.78*	*7.56*	*2.29*	*6.39*	*9.51*	*4.51*	*axial-[100]*
Demas Bluff	*FDM-DB04-X04*	*10.00*	*7.93*	*4.89*	*5.74*	*8.54*	*7.28*	*A*
Marujupu	*FDM-MJ01-X01**	*4.72*	*2.84*	*1.83*	*2.74*	*4.67*	*4.11*	*n/a**
Marujupu	*FDM-MJ01-X02*	*8.63*	*5.38*	*3.34*	*4.04*	*7.07*	*5.77*	*A*
Marujupu	*FDM-MJ01-X03*	*4.08*	*2.93*	*2.52*	*1.64*	*3.14*	*4.06*	*A*
Marujupu	*FDM-MJ01-X05*	*7.14*	*6.70*	*4.28*	*3.06*	*7.08*	*5.57*	*axial-[010]*
Marujupu	*FDM-MJ01-X06*	*5.18*	*3.58*	*1.69*	*3.11*	*4.96*	*3.20*	*A*
Recess Nunatak	*FDM-RN01-X01*	*12.43*	*7.89*	*4.29*	*5.47*	*8.83*	*8.35*	*A*
Recess Nunatak	*FDM-RN02-X01*	*8.57*	*5.73*	*2.33*	*4.54*	*7.84*	*4.70*	*A*
Recess Nunatak	*FDM-RN03-X01*	*9.65*	*6.31*	*2.47*	*5.08*	*7.84*	*5.34*	*B*
Recess Nunatak	*FDM-RN04-X01*	*7.66*	*5.44*	*2.67*	*4.60*	*6.67*	*4.51*	*axial-[100]*
Mount Avers – Bird Bluff	*FDM-AVBB01*	*5.26*	*3..15*	*2.62*	*1.61*	*3.20*	*4.82*	*axial-[010]*
Mount Avers – Bird Bluff	*FDM-AVBB02*	*6.18*	*5.11*	*3.18*	*2.49*	*3.99*	*4.53*	*axial-[010]*
Mount Avers – Bird Bluff	*FDM-AVBB03**	*2.23*	*2.28*	*1.89*	*1.54*	*3.50*	*3.29*	*n/a**
Mount Avers – Bird Bluff	*FDM-AVBB04*	*4.43*	*3.02*	*1.97*	*1.61*	*3.73*	*3.89*	*axial-[010]*
Mount Avers – Bird Bluff	*FDM-AVBB05*	*6.56*	*4.92*	*3.48*	*2.30*	*3.20*	*5.51*	*axial-[010]*
Mount Avers – Bird Bluff	*FDM-AVBB06*	*3.53*	*2.93*	*2.15*	*1.38*	*1.83*	*3.18*	*axial-[010]*
Mount Avers – Bird Bluff	*FDM-AVBB07*	*5.33*	*4.09*	*3.27*	*1.63*	*2.86*	*4.83*	*axial-[010]*
Mount Avers – Bird Bluff	*FDM-AVBB08*	*4.73*	*2.75*	*2.54*	*1.56*	*3.53*	*4.91*	*axial-[010]*
Mount Aldaz	*AD6021-X01**	*16.12*	*12.20*	*9.82*	*7.94*	*20.91*	*14.43*	*n/a**
Mount Aldaz	*AD6021-X02*	*8.48*	*6.09*	*4.33*	*2.93*	*5.29*	*6.49*	*axial-[010]*
Mount Cumming	*KSP89-181-X01*	*14.00*	*8.90*	*3.35*	*7.72*	*13.96*	*6.77*	*axial-[100]*

Samples and CPO type denoted with * correspond to samples with a low modal abundance of olivine (e.g. pyroxenite, websterite).

foliation and by concentrations of [100] and [010] axes near the foliation plane, either subparallel or at high angle to the lineation (Fig. S2: FDM-BB02-X01, FDM-DB02-X01, FDM-MJ01-X03). Lherzolite FDM-AVBB08 shows alignment of olivine, orthopyroxene, and clinopyroxene crystallographic axes, with [100] axes oriented parallel to the lineation, [010] axes perpendicular to the foliation, and [001] axes within the foliation and normal to the lineation orientation (Fig. 6). This CPO pattern suggests olivine slip on (010)[100], also supported by the concentration of olivine low angle

Table 3. *Weighted average seismic anisotropy by xenolith locality*

Location	Volcanic Centre	N samples averaged	Vp Anisotropy (%)	Vs Max Anisotropy (%)	Vs1 Max Anisotropy (%)	Vs2 Max Anisotropy (%)	Vp/Vs1 Anisotropy (%)	Vp/Vs2 Anisotropy (%)
Fosdick Mountains	*All localities*	*41*	*4.13*	*2.28*	*1.32*	*1.45*	*3.16*	*2.91*
	Mount Avers	*1*	*7.58*	*5.94*	*2.80*	*3.73*	*5.38*	*4.79*
	Bird Bluff	*3*	*3.46*	*2.01*	*1.16*	*1.37*	*3.42*	*3.30*
	Demas Bluff	*20*	*2.57*	*1.67*	*1.16*	*1.15*	*3.07*	*2.16*
	Marujupu	*5*	*1.17*	*1.22*	*0.93*	*0.70*	*1.37*	*1.14*
	Recess Nunatak	*4*	*6.80*	*4.39*	*3.31*	*1.45*	*3.78*	*5.93*
	Mount Avers – Bird Bluff	*8*	*3.38*	*2.11*	*1.96*	*0.33*	*1.74*	*3.30*
Usas Escarpment	*Mount Aldaz*	*2*	*10.01*	*6.46*	*4.36*	*3.92*	*8.31*	*7.22*
Executive Committee Range	*Mount Cumming*	*1*	*14.00*	*8.90*	*3.35*	*7.72*	*13.96*	*6.77*

misorientation axes near [001] (Fig. 9) (Chatzaras *et al.* 2016), but bears no relationship with the [001] slip in orthopyroxene that is commonly described in deformed mantle rocks (Jung *et al.* 2010; Soustelle *et al.* 2010).

The orthopyroxene CPO has a moderate to low strength (Fig. 7c). The M index ranges from 0.05 to 0.61, and the J index varies from 1.56 to 29.38 (Table 1; Fig. S2). High M index (>0.33) and J index (>15.66) values are an artifact of the low number (<100) of orthopyroxene grains in the studied xenoliths (Table 1). If we exclude these samples, the strength of the orthopyroxene CPO does not show a systematic variation with the orthopyroxene modal composition (Fig. 7c). The M and J index of clinopyroxene CPOs range from 0.02 to 0.9 and from 1.54 to 61.01, respectively (Table 1; Fig. S2). Similar to orthopyroxene, the strong clinopyroxene CPOs (M >0.59 and J >26.23) are artefacts associated with small numbers (<100) of clinopyroxene grains and are not considered in the following (Table 1). The M index decreases slightly from 0.5 to 0.4 with an increase in the clinopyroxene volume from 6% to 42%. The strength of both pyroxene CPOs does not show a systematic variation with the equilibration temperature (Fig. S3).

Seismic properties

Seismic properties of individual xenoliths. Anisotropic seismic properties for all individual MBL xenoliths (i.e. Fosdick Mountains, Usas Escarpment, and Executive Committee Range volcanic centres) are illustrated in Figures 8 and 9, Supplementary Figure S4, and are reported in Table 2. Seismic anisotropy values vary significantly among individual samples due to differences in CPO (distribution and intensity) and changes in the modal abundance of constituent mineral phases. Maximum P-wave propagation anisotropies (AVp_{Max}) for lherzolitic samples range between 2.1 and 10%, whereas corresponding maximum S-wave polarization anisotropies (AVs_{Max}) vary between 1.6 and 7%. For samples with low modal abundance of olivine (e.g. clinopyroxenite, olivine websterite), AVp_{Max} values range between 2.3–16.1%; AVs_{Max} in these samples vary between 2.3–12.2%. The maximum P-wave and S-wave anisotropies are positively correlated with increasing olivine modal abundance (Fig. 8d) but decrease with both increasing ortho- or clinopyroxene content (Fig. 8e and f). Exceptions are the two clinopyroxenites, characterized by high P-wave and S-wave anisotropy. AVp_{Max} and AVs_{Max} are best correlated with increasing intensity of the olivine crystallographic preferred orientation, as defined by the M-index of olivine (Fig. 8a). The Maximum P-wave anisotropy and the maximum S-wave polarization anisotropy also increase with increasing M-index in orthopyroxene, but is less correlated with clinopyroxene. The magnitude of the maximum P-wave propagation anisotropy and the polarization of S-waves are positively correlated (Fig. 8c), as has been observed in ot her xenolith studies (e.g. Tommasi and Mameri 2020). No significant correlation is observed between the P- or S-wave anisotropies and the olivine CPO symmetry, as characterized by the BA-index (Fig. 8b), in agreement with previous studies (Baptiste *et al.* 2015).

Despite the variability in CPO texture type and variations in the modal abundances of olivine, orthopyroxene and clinopyroxene, individual samples do share some common seismic properties. P-wave propagation is fastest parallel to the [100] axis of olivine. Given the presence of three dominant olivine CPO texture types in the xenolith suite, the fastest P-wave propagation direction is typically either parallel to lineation for orthorhombic A-type CPOs (e.g. Fig. 9: FDM-BB01-X01, FDM-DB02-X08) and axial-[100] (e.g. Fig. 9: FDM-DB02-X01, FDM-DB04-X03), or within the plane of the foliation at high-angle to lineation for orthorhombic B-type CPOs

Table 4. *Weighted average seismic anisotropy by olivine crystallographic texture type*

Olivine CPO texture type	N samples averaged	Vp Anisotropy (%)	Vs Max Anisotropy (%)	Vs1 Max Anisotropy (%)	Vs2 Max Anisotropy (%)	Vp/Vs1 Anisotropy (%)	Vp/Vs2 Anisotropy (%)
A-Type	*12*	*6.16*	*4.26*	*2.92*	*2.47*	*3.81*	*4.60*
Axial-[100]	*8*	*4.79*	*3.18*	*1.86*	*2.64*	*4.19*	*3.46*
Axial-[010]	*15*	*3.57*	*1.80*	*0.86*	*1.42*	*3.11*	*2.57*

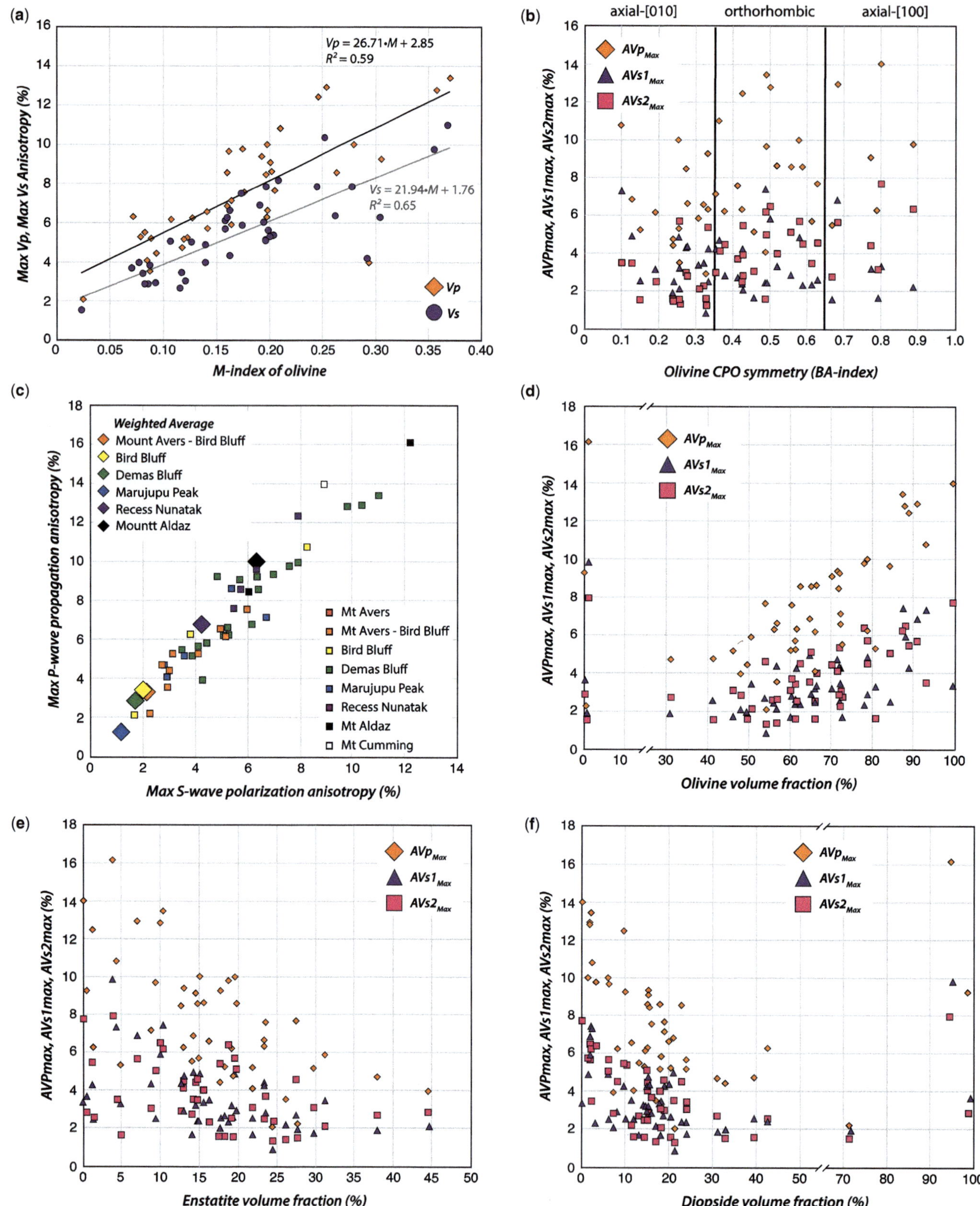

Fig. 8. Seismic anisotropy of individual samples. (**a**) Maximum P- and S-waves anisotropy v. olivine CPO strength. (**b**) Maximum P- and S-waves anisotropy v. olivine CPO symmetry. (**c**) Maximum P-wave anisotropy v. maximum S-wave anisotropy. (**d**) Maximum P-wave anisotropy, maximum S-wave anisotropy of fastest S-wave (S1) and slowest S-wave (S2) v. olivine volume fraction. (**e**) Maximum P-wave anisotropy, maximum S-wave anisotropy of fastest S-wave (S1) and slowest S-wave (S2) v. enstatite volume fraction. (**f**) Maximum P-wave anisotropy, maximum S-wave anisotropy of fastest S-wave (S1) and slowest S-wave (S2) v. diopside volume fraction.

(e.g. Fig. 9: FDM-RN03-X01) or subtending low-angles to lineation within the foliation plane (e.g. Fig. 9: FDM-AVBB05) for samples with axial-[010] CPO symmetry. In nearly all samples, P-wave propagation is slowest perpendicular to the foliation plane or in directions girdled about the lineation orientation (Fig. 9 and Fig. S4).

The distribution of the fastest split shear wave (S1) is typically coincident with the orientation of the maximum P-wave

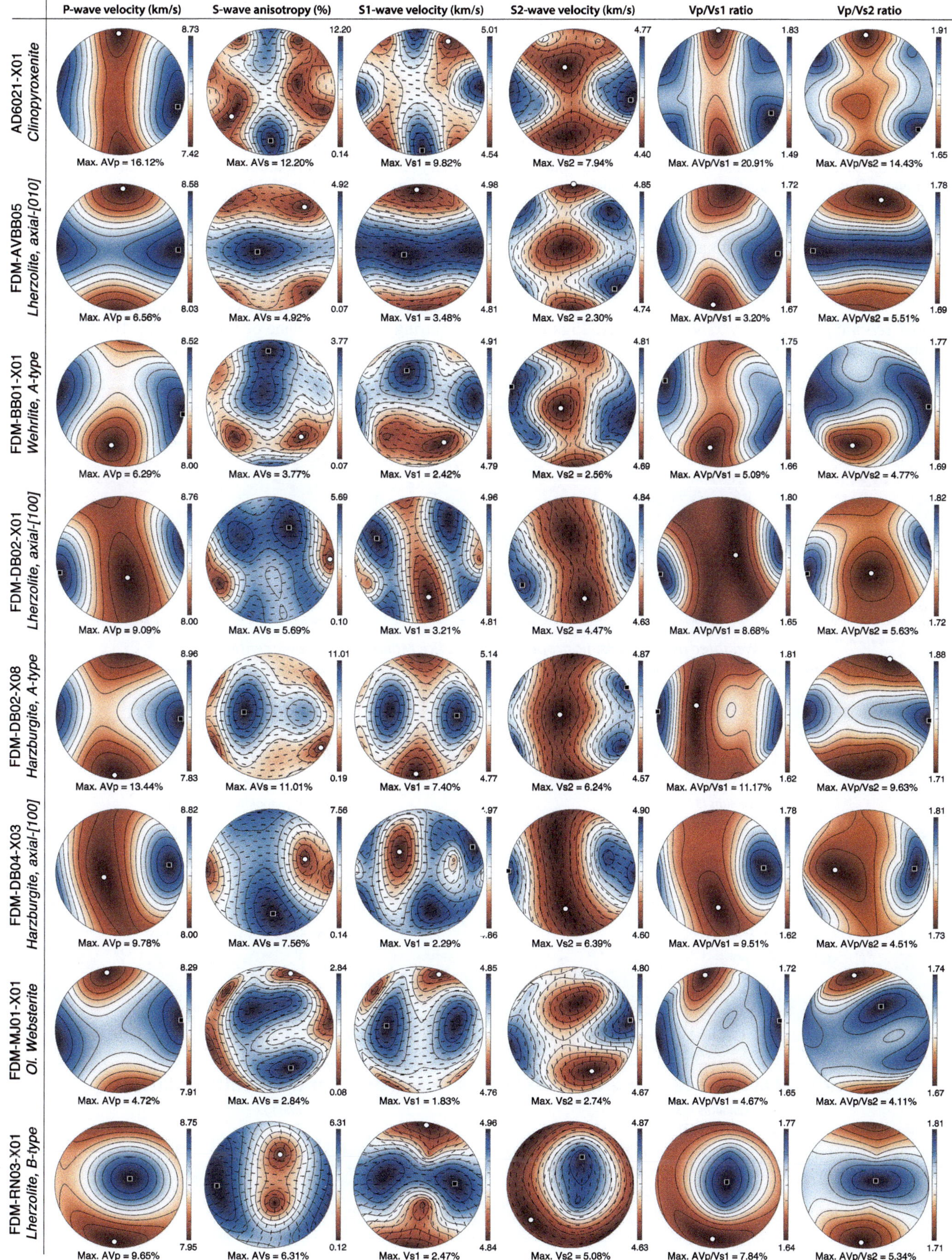

Fig. 9. Calculated seismic properties of individual xenoliths from Marie Byrd Land, West Antarctica. From left to right are displayed the 3D distributions of P-wave velocities (Vp), S-wave polarization anisotropy (AVs) and the orientation of fast shear wave polarization planes, S1 wave velocities (Vs1) and the orientation of fast shear wave polarization planes, S2 wave velocities (Vs2) and the orientation of slow shear wave polarization planes, Vp/Vs1 ratio, and Vp/Vs2 ratio. Black squares and white circles indicate maximum and minimum values, respectively. All plots are lower hemisphere equal area stereographic projections, displayed in the xenoliths shape fabric reference framework, as shown in Figure 6.

propagation direction, and therefore the orientation of the maximum concentration of olivine [100] axes (e.g. Fig. 9: FDM-AVBB05 and Fig. S4: FDM-BB03-X01). This correlation, however, begins to weaken with increasing pyroxene content (e.g. Fig. 9: clinopyroxenite sample AD6021-X01 and Fig. S4: FDM-DB02-X05). The propagation direction of the slowest S-wave (S2) commonly displays single maxima (e.g. Fig. 9: FDM-BB01-X01, FDM-DB02-X01) or multiple maxima (Fig. 9: FDM-AVBB05, FDM-DB02-X08) parallel to the fastest P-wave propagation direction, which typically lies parallel to lineation or within the foliation plane depending upon the olivine CPO symmetry observed. The three-dimensional pattern of the fastest and slowest Vp/Vs1 ratios mimics the distribution of fast and slow P-wave propagation directions in nearly all samples (Fig. 9 and Fig. S4).

Averaged seismic properties by xenolith locality. Figure 10 and Table 3 present the weighted average seismic properties for xenolith localities within the Fosdick Mountains, Usas Escarpment and Executive Committee Range of MBL. Weighted average seismic properties for individual volcanic centres, calculated by summation of individual samples (see methods), record significantly lower values for maximum P-wave propagation anisotropies (AVp_{Max}) and maximum S-wave polarization anisotropies (AVs_{Max}) than is observed in individual samples (Fig. 8c).

Crystallographic orientations, sampled at random from all phases within 41 individual xenoliths, were averaged in order to provide a composite description of the anisotropic seismic properties in the mantle beneath MBL. Maximum P-wave anisotropy is approximately 4.1%, corresponding to P-wave velocities ranging from 8 to 8.4 km s^{-1}. The fastest P-wave propagation direction is defined by well-developed maxima parallel to lineation with intermediate wave speeds occurring within the plane of foliation (Fig. 10). The slowest P-wave propagation direction occurs in the direction

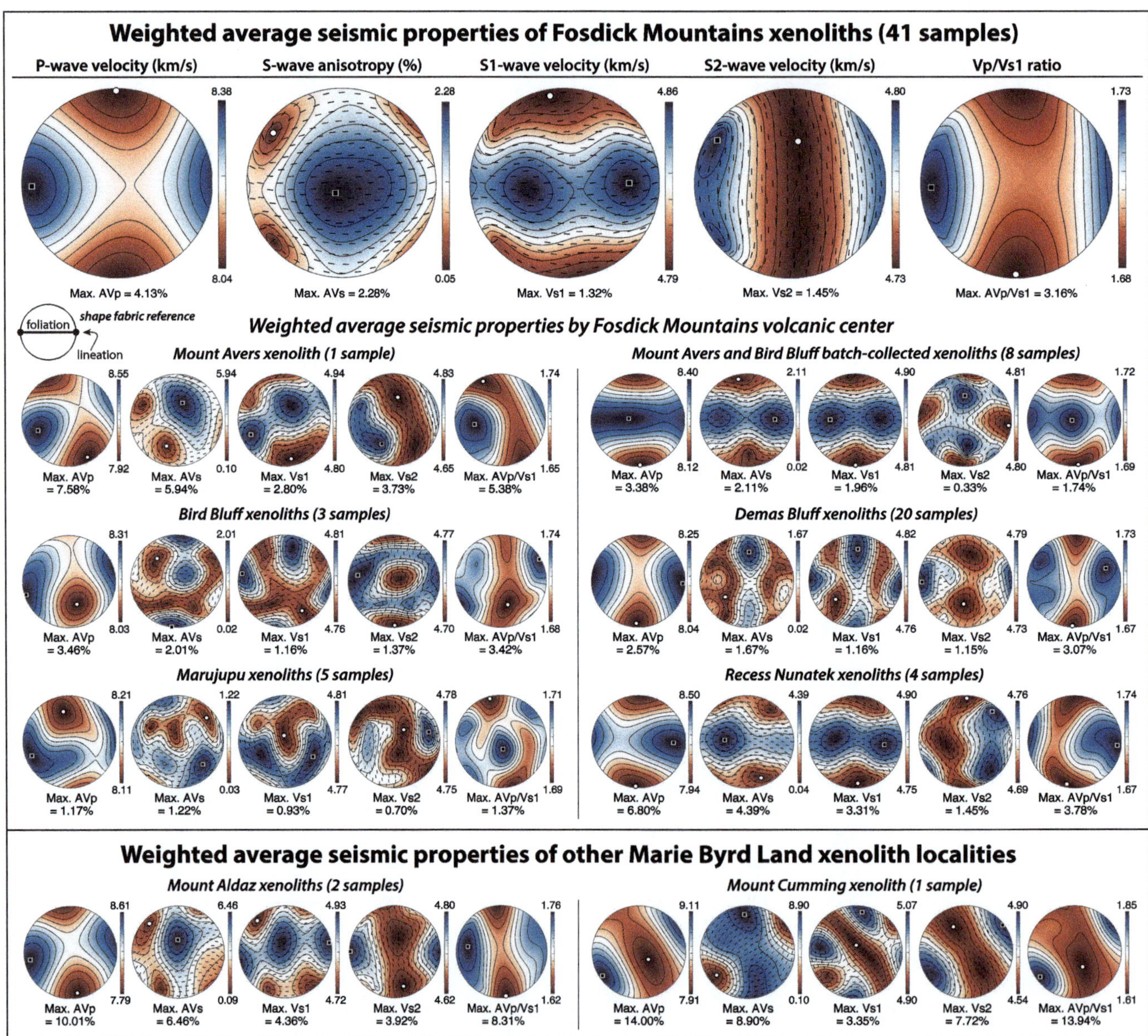

Fig. 10. Weighted average seismic properties for distinct regions within Marie Byrd Land and for individual volcanic centres within the Fosdick Mountains, calculated by summation of crystallographic orientations from constituent samples (e.g. 41 xenoliths for all Fosdick Mountains xenoliths; upper row) and subsequent Voigt-Reuss-Hill averaging of single-crystal elastic constants at ambient conditions (see Methods). From left to right are displayed the 3D distributions of P-wave velocities (Vp), S-wave polarization anisotropy (AVs) and the orientation of fast shear wave polarization planes, S1 wave velocities (Vs1) and the orientation of fast shear wave polarization planes, S2 wave velocities (Vs2) and the orientation of slow shear wave polarization planes, and Vp/Vs1 ratio. Black squares and white circles indicate maximum and minimum values, respectively. All plots are lower hemisphere equal area stereographic projections, displayed in the xenoliths shape fabric reference framework.

perpendicular to foliation. The maximum S-wave anisotropy is 2.3%, with the fastest S-wave speeds occurring within the plane of foliation perpendicular to lineation. The fastest shear wave (S1) is polarized parallel to the lineation orientation. S1-wave velocity is slowest in the plane perpendicular to foliation and fastest within the plane of foliation yielding dual maxima oriented approximately 45 degrees to the lineation. S2 waves are slowest in all propagation directions perpendicular to lineation. Vp/Vs1 ratios are fastest parallel to lineation, coincident with the distribution of the fastest P-wave velocities, and slowest in all directions perpendicular to lineation, similar to the slowest orientations of S2 waves. As discussed previously, weighted averages for individual volcanic centres record second-order variations of the seismic anisotropy pattern, owing to variations in the CPO symmetry, modal abundance of constituent mineral phases, and the total number of samples available for summation of average seismic properties (Fig. 10). For these reasons, we suggest that the weighted average properties of all 41 samples within the Fosdick Mountains represent the most robust estimate of the maximum average seismic anisotropy likely to be sampled by the propagation of seismic waves at larger (kilometre to hundreds of kilometres) spatial scales within the mantle underlying west MBL.

Averaged seismic properties by crystallographic texture type. All of the xenolith samples analysed are characterized by orthorhombic (A-type), axial-[100], or axial-[010] olivine CPO symmetries, with the exception of the four samples containing a low modal abundance of olivine (e.g. websterite, clinopyroxenite), two xenoliths with C-type and B-type CPO (FDM-DB02-X06, FDM-RN03-X01), and those with statistically random CPOs (Figs 6 and 7, Table 2). In order to determine the contribution of these crystallographic texture variations on anisotropic seismic properties, crystallographic orientations were sampled at random and averaged for all phases and samples that share a common olivine CPO symmetry.

The weighted average seismic properties for samples with orthorhombic (A-type) olivine CPO symmetry (Fig. 11) are nearly identical to those described previously for the weighted averaged properties of all Fosdick Mountains xenoliths (Fig. 10). The maximum P-wave propagation direction is aligned parallel to lineation and the slowest P-wave velocities occur perpendicular to foliation (Fig. 11). The maximum P-wave anisotropy is 6.2%. The maximum S-wave anisotropy is 4.3%. The fastest shear wave is polarized parallel to lineation with the fastest S1 wave speeds occurring as dual maxima within the plane of foliation 45 degrees from lineation. S2 waves define multiple maxima oriented 45 degrees from the lineation and the pole to foliation, with the slowest wave speeds occurring in girdles about the lineation orientation. Vp/Vs1 ratios are similarly slowest in all directions perpendicular to lineation and fastest parallel

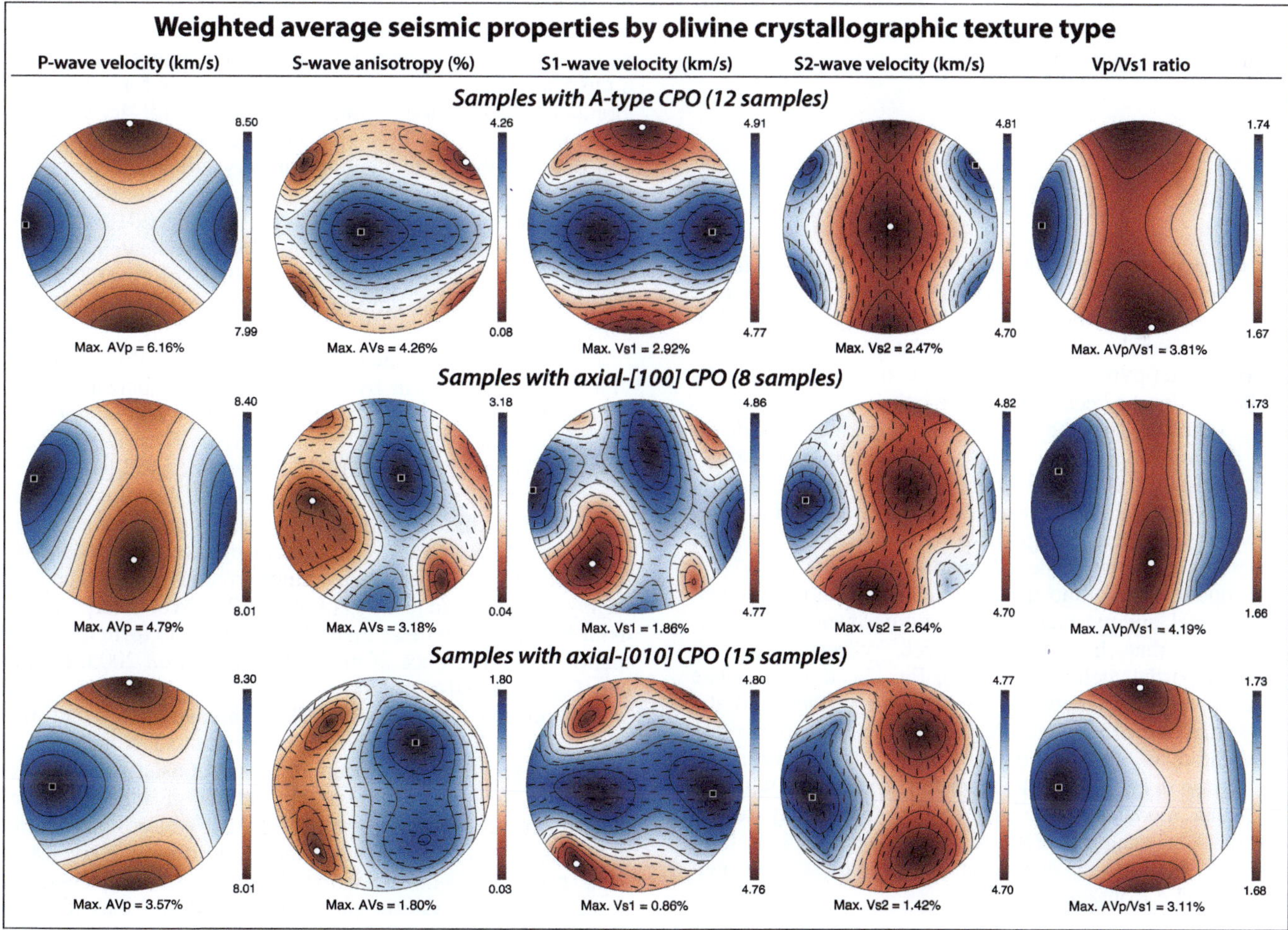

Fig. 11. Weighted average seismic properties calculated by summation of crystallographic orientations from xenoliths containing each of the three main olivine CPO texture types observed in the studied suite of the MBL xenoliths. From left to right are: the 3D distributions of P-wave velocities (Vp), S-wave polarization anisotropy (AVs) and the orientation of fast shear wave polarization planes, S1 wave velocities (Vs1) and the orientation of fast shear wave polarization planes, S2 wave velocities (Vs2) and the orientation of slow shear wave polarization planes, and Vp/Vs1 ratio. Black squares and white circles indicate maximum and minimum values, respectively. All plots are lower hemisphere equal area stereographic projections, displayed in the xenoliths shape fabric reference framework, as shown in Figures 6 and 10.

to lineation, coincident with the distribution of the fastest P-wave velocities.

Averaged seismic properties for samples with axial-[100] CPO symmetry share some similarities to those with orthorhombic symmetry in that the maximum propagation of P-wave, Vs2 waves, and the maximum Vp/Vs1 wave speeds are all fastest parallel to lineation (Fig. 11). In contrast, the slowest P-wave propagation directions occur in a girdled distribution about the lineation orientation, similar to the distribution observed for Vs2 and Vp/Vs1 wave speeds. The slowest P-waves are defined by maxima near the plane of the foliation and perpendicular to lineation (Fig. 11). The orientation of the slowest S1 waves is also oblique to both the foliation and lineation (Fig. 11). The maximum P-wave anisotropy for axial-[100] samples is 4.8%, whereas the maximum S-wave anisotropy is 3.2%.

The anisotropic seismic properties for samples that have an axial-[010] olivine CPO symmetry are the most distinctive in that the distributions of maxima are dramatically different than those observed for orthorhombic or axial-[100] weighted averages. P-wave velocities range from 8.0–8.3 km s^{-1}, corresponding to a maximum P-wave anisotropy of 3.6%. The fastest P-waves occur within the plane of the foliation, with the maximum P-wave velocities occurring in a well-defined maximum oriented approximately 30 degrees from the lineation orientation (Fig. 11). This orientation is also coincident with the direction of the maximum propagation for S2 waves and the maximum Vp/Vs1 ratio. P-wave speeds are slowest perpendicular to foliation. S1 waves are slowest oblique to the foliation pole and are fastest within the plane of foliation at approximately 30–45 degrees to lineation (Fig. 11).

Discussion

Origin of the Marie Byrd Land Mantle Xenoliths

The MBL peridotite xenoliths originate from the spinel-facies subcontinental lithospheric mantle beneath the southern part of West Antarctica. A deeper origin, from the garnet stability field, cannot be ruled out for some of the xenoliths; the presence of spinel-pyroxene symplectites in some xenoliths (Fig. 4a) implies the occurrence of pre-existing garnet. Garnet-bearing xenoliths may have experienced garnet breakdown during decompression prior to their equilibration in the spinel stability field (Morishita and Arai 2003), or cooling following a heating event that may have affected the MBL upper mantle, and which resulted in garnet-olivine reaction to form spinel-pyroxene symplectites (Medaris *et al.* 1997).

The modal content and the major element composition of olivine, spinel, and pyroxene in the clinopyroxene-poor xenoliths suggest that they were produced by relatively high degrees of partial melting (10–25%; Fig. 3a, c and d). The inverse relationships between the Mg# in olivine and the Cr# in spinel, and of MgO and Al_2O_3 contents in the pyroxene, are indicative of melt depletion (Johnson *et al.* 1990; Niu 2004; Martin *et al.* 2015). The more fertile xenoliths with clinopyroxene modal contents higher than 15%, may reflect either pristine mantle affected by low degrees of melting (<5%; Fig. 3a, c–d), or depleted mantle refertilized by melt percolation (Le Roux *et al.* 2007).

Melt percolation in the Marie Byrd Land lithospheric mantle

The variation in modal compositions, mineral major element compositions, and microstructural features observed in the MBL xenoliths suggest that the lithospheric mantle beneath the southern part of West Antarctica experienced some degree of reactive melt percolation and refertilization. Similar processes have been described in other upper mantle xenolith suites and exhumed peridotite massifs (Tommasi *et al.* 2004; Le Roux *et al.* 2007; Soustelle *et al.* 2010; Morales and Tommasi 2011; Tommasi and Ishikawa 2014; Zaffarana *et al.* 2014; Hidas *et al.* 2019). Our results are in agreement with xenolith studies suggesting melt-related metasomatism for the lithospheric mantle beneath West Antarctica (Martin *et al.* 2013; Doherty 2016; Day *et al.* 2019; Handler *et al.* 2021).

The following observations could be interpreted to indicate refertilization of the MBL upper mantle as a result of reaction with percolating melts: (1) enrichment in clinopyroxene leading to the formation of clinopyroxene-rich lherzolite, wehrlite, and clinopyroxenite as end-member products (Fig. 2b and c) implies crystallization of clinopyroxene ± olivine, at the expense of orthopyroxene, possibly due to a change in the silica-saturation of the melt (Lo Cascio *et al.* 2004; Morales and Tommasi 2011; Hidas *et al.* 2019); (2) discrepancy in the estimated degrees of partial melting for the two pyroxene, which implies that partial melting may not be the only process that influenced their Al_2O_3 and MgO content (Fig. 3c and d) (Martin *et al.* 2015; Pelorosso *et al.* 2017); (3) clinopyroxene with high Al_2O_3, Na_2O, and TiO_2 and low MgO content in a fertile group of lherzolite, wehrlite, websterite and clinopyroxenite (Fig. 3b, c and e) (Le Roux *et al.* 2007; Soustelle *et al.* 2010; Bodinier and Godard 2014); (4) 'secondary' origin of pyroxene, implied by the occurrence of subidiomorphic to xenomorphic grains with interstitial habit that either exhibit a cuspate shape in olivine triple junctions or form film-like elongated grains along grain boundaries (Fig. 4c, e and j) (Tommasi *et al.* 2008); (5) occurrence of pyroxene grains that contain olivine inclusions with crystallographic orientation similar to the neighbouring olivine grains (Fig. 4f, h and k) (Tommasi *et al.* 2004; Soustelle *et al.* 2010; Morales and Tommasi 2011); (6) formation of websteritic veins in dunite (Fig. 4e); (7) development of axial-[010] olivine CPO; and (8) development of an unusual orthopyroxene CPO with [001] axes oriented at high angle to the foliation.

The axial-[010] olivine CPO in the MBL xenoliths was previously shown to form by flattening strain (Chatzaras *et al.* 2016). Although overall valid, this relationship is not supported in a small number of xenoliths (e.g. FDM-DB03-X02, FDM-AVBB01, FDM-AVBB04), where the axial-[010] CPO forms in relation to prolate or neutral fabric ellipsoids (Chatzaras *et al.* 2016; fig. 14). These lherzolite xenoliths are enriched in clinopyroxene (Table 1), contain pyroxene of secondary origin (Fig. 4c), and their clinopyroxene is enriched in Al_2O_3, Na_2O and TiO_2 (Table S2). We interpret the axial-[010] CPO in these lherzolite xenoliths to form by deformation in the presence of melt (Holtzman *et al.* 2003; Higgie and Tommasi 2014).

The unusual orthopyroxene CPO, characterized by a [001] maximum at high angle to the foliation and concentration of [100] axes near the lineation, is not consistent with dislocation glide on any known slip systems in orthopyroxene, and may result from oriented crystallization of orthopyroxene in the presence of melt. Concentration of orthopyroxene [001] axes normal to the foliation is reported in high-strain peridotites in the Othris and Ronda peridotite massifs, and is attributed to a melt-facilitated reaction, leading to dissolution and precipitation of orthopyroxene (Dijkstra *et al.* 2002; Hidas *et al.* 2016). In the MBL xenoliths, this CPO is observed in lherzolite and in websteritic veins within dunite (Fig. S2: FDM-AVBB04, FDM-BB02-X01, FDM-DB02-X01, FDM-MJ01-X03; Table 1). Orthopyroxene grains in the lherzolite are elongated at high angle to the foliation, subparallel to the [001] axes. This relationship indicates correlation between the shape preferred orientation and CPO in orthopyroxene, and implies that the elongation of orthopyroxene grains could be controlled by oriented growth parallel to the [001] crystallographic axis.

Temporal relationships between melt percolation and deformation

The combined examination of the quenched microstructures and CPOs observed in the MBL xenoliths, allows identification of the dominant deformation processes and their temporal relationships to melt percolation in the MBL lithospheric mantle. Olivine microstructures indicate the simultaneous operation of dislocation creep, primarily exhibited by undulose extinction and subgrain boundaries in olivine grains, and microstructures indicative of grain and phase boundary sliding (this study; Chatzaras *et al.* 2016). The observed microstructures are consistent with the predictions of deformation mechanism maps constructed for the deformation conditions estimated for the MBL xenoliths, which indicate that deformation was primarily achieved by dislocation-accommodated grain boundary sliding (Chatzaras *et al.* 2016). The formation of 120° triple junctions between olivine grains that exhibit intragrain deformation features (Fig. 4j) suggests that dynamic recrystallization was followed by static recrystallization. The preservation of olivine intragrain deformation features and bimodal grain size distributions, indicates that microstructural modification due to static recrystallization was limited, even in samples deformed at low differential stresses; twenty two of the xenoliths were recrystallized at stresses $\leq$10 MPa (Chatzaras *et al.* 2016).

The relationship between olivine and pyroxene CPOs suggests diachronous melt percolation in the MBL upper mantle, relative to the deformation event that generated the olivine CPO. The correlation of olivine and pyroxene CPOs observed in 22 xenoliths (Fig. 6 and Fig. S2), indicates alignment of the dominant slip directions in the three minerals, suggesting coherent and synchronous deformation (Tommasi *et al.* 2004; Morales and Tommasi 2011; Tommasi and Ishikawa 2014; Zaffarana *et al.* 2014). This CPO correlation implies the pyroxene grains were present prior to or at the early stages of the deformation phase that produced the olivine CPO, and thus could have a primary or early-stage secondary origin, without the two interpretations being mutually exclusive. The clinopyroxene-rich lherzolite xenoliths (clinopyroxene mode >15%) seem to have experienced a complex history of melt-rock interaction and deformation, with: (1) primary, coarse pyroxene grains with undulose extinction, subgrain boundaries and shape alignment parallel to the spinel lineation, (2) finer-grained pyroxene showing intragrain deformation features that crystallized prior to or during mantle deformation and (3) secondary pyroxene that crystallized following the cessation of the deformation, as suggested by the interstitial habit, lack of intragrain deformation (e.g. FDM-AVBB08; Fig. 4c and d), and their cross-cutting relationships with olivine subgrain boundaries (e.g. FDM-BB01-X01; Fig. 4k). These relationships suggest that melt transport occurred prior, during, and after the CPO-forming deformation event in the xenoliths. Moreover, it implies that any post-deformation secondary crystallization of pyroxene may have resulted in a weakening of the pyroxene CPOs due to dispersion of the crystallographic axes distribution, but did not erase the pre-existing correlation with the olivine CPO (FDM-AVBB08; Fig. 6).

The correlation of olivine and orthopyroxene CPOs and lack of correlation between olivine and clinopyroxene CPOs in 9 xenoliths (Fig. 6 and Fig. S2) implies coherent deformation of olivine and orthopyroxene, and post-deformation crystallization of clinopyroxene (Tommasi *et al.* 2004; Soustelle *et al.* 2010; Morales and Tommasi 2011). Orthopyroxene could either have a primary origin or have crystallized from reactive melt percolation prior to the CPO-forming deformation event. Both types of orthopyroxene grains seem to be present in these 9 xenoliths. Large, millimetre-sized orthopyroxene grains with undulose extinction are interpreted as primary. Smaller, isolated grains that show elongation parallel to the lineation, have cuspate shapes, and show interstitial habits, are interpreted as secondary (Fig. 4j: FDM-AVBB02). The uncorrelated olivine and clinopyroxene CPOs (Fig. S1) suggest precipitation of the latter during post-deformation reactive melt percolation, consistent with the interstitial habit and the lack of evidence of intragrain deformation in the majority of the clinopyroxene grains (Fig. 4j: FDM-AVBB02). The occurrence of clinopyroxene film-like grains along straight olivine–olivine grain boundaries that meet at 120° triple junctions (FDM-AVBB02; Fig. 4j) implies that melt percolation and clinopyroxene precipitation possibly postdates static recrystallization of olivine. In support of this interpretation, clinopyroxene shows weak but non-random CPO (M index of 0.42–0.48), with distributions of crystallographic axes that do not indicate activation of any of the known slip systems in clinopyroxene (Van Roermund and Boland 1981; Ulrich and Mainprice 2005; Frets *et al.* 2012). We interpret the uncorrelated olivine and pyroxene (both orthopyroxene and clinopyroxene) CPOs in 10 MBL xenoliths to show post-deformation melt percolation and pyroxene precipitation (Soustelle *et al.* 2010; Zaffarana *et al.* 2014).

Effect of melt percolation and refertilization on seismic anisotropy

In order to evaluate the effect of modal compositional variations (e.g. increasing pyroxene content by melt percolation or refertilization) on the seismic properties of the MBL xenoliths, we calculated a series of synthetic anisotropy plots. Synthetic seismic properties were calculated using the weighted average crystallographic orientations for all Fosdick Mountains xenoliths and those comprising orthorhombic (A-type), axial-[100] or axial-[010] olivine CPO symmetries. From these datasets, which contained 1 million unique crystallographic orientations (see methods), crystallographic orientations were sampled at random from the natural data to produce orientations that reflected six synthetic modal compositions: 100% olivine; 80% olivine/20% pyroxene; 60% olivine/40% pyroxene; 40% olivine/60% pyroxene; 20% olivine/80% pyroxene; and 100% pyroxene. The ratio of orthopyroxene to clinopyroxene was assumed to be 1:1 for the calculation, which is justified based on the average modal proportions observed in the xenolith suite. The results of this analysis are shown in Figures 12 and 13.

For the summation of the Fosdick Mountains xenoliths, synthetic seismic properties vary as follows from olivine-rich ($Ol_{100}OPX_0CPX_0$) to olivine-free ($Ol_0OPX_{50}CPX_{50}$) compositions (Figs 12a and 13): (i) the maximum P-wave velocity decreases up to 5%, from 8.7 to 8.3 km s^{-1}, and the maximum S1-wave velocity decreases up to 3.7% from 5 to 4.8 km s^{-1}; (ii) the maximum P-wave and S-wave anisotropy decreases from 8.7% to 2.9% and from 6% to 1.7%, respectively; (iii) the P-wave propagation anisotropy transitions from fast lineation-parallel maxima with intermediate wave speeds occurring within the plane of foliation, to fast point maxima oblique to the lineation in the foliation plane, with girdled distributions of slow P-wave speeds oriented perpendicular to lineation; and (iv) the distribution of the fastest S-wave (S1)

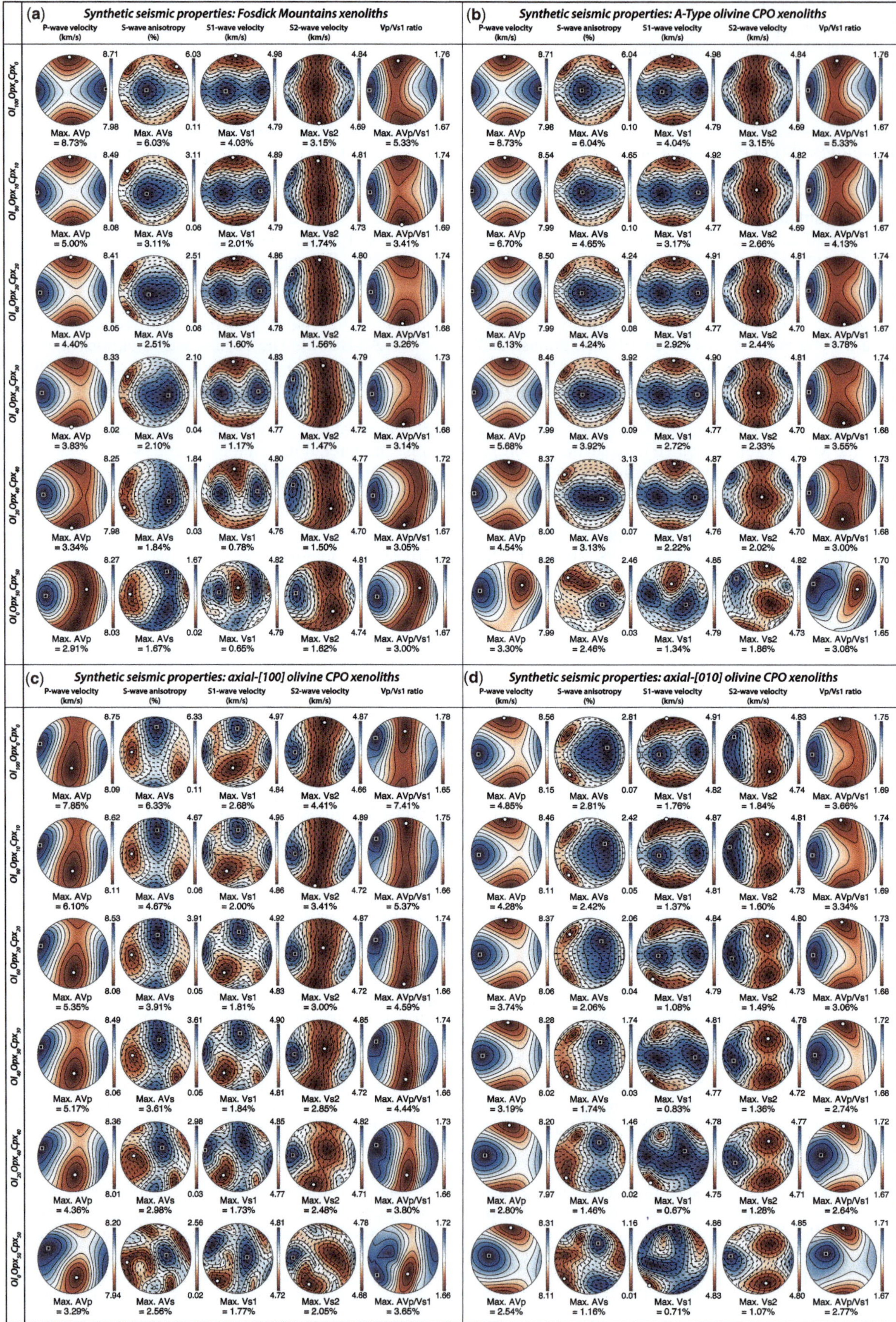

Fig. 12. Synthetic plots of seismic anisotropy properties calculated for a range of modal compositions using the weighted average crystallographic orientations for all Fosdick Mountains xenoliths (**a**) and those comprising A-type (**b**), axial-[100] (**c**) or axial-[010] (**d**) olivine CPO symmetries. Crystallographic orientations were sampled at random from composite natural datasets to produce orientations that reflected six synthetic modal compositions to evaluate the effect of changing composition on the seismic properties of the Marie Byrd Land xenoliths. All plots are lower hemisphere equal area stereographic projections, displayed in the xenoliths shape fabric reference framework, as shown in Figures 6 and 10.

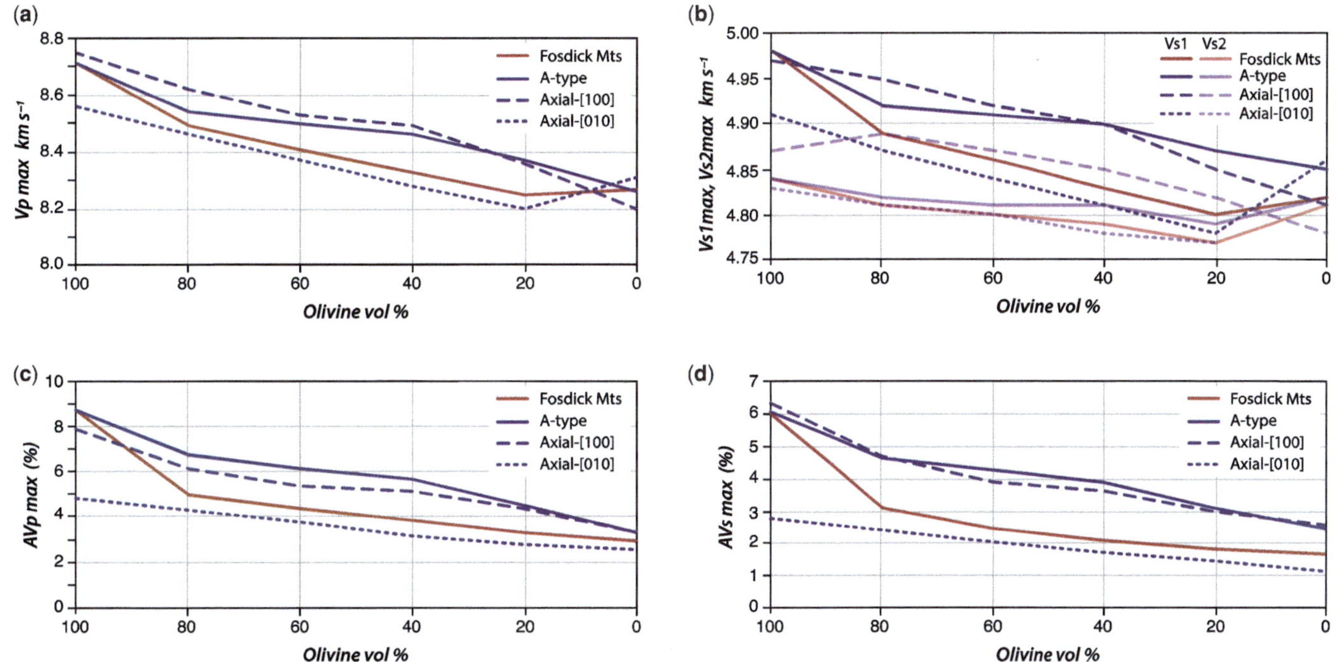

Fig. 13. Variation of synthetic seismic properties of all Fosdick Mountains xenoliths, and xenoliths with A-type, axial-[100], and axial-[010] olivine crystallographic texture, with modal composition. Relationship between olivine volume (vol) % and maximum P-wave velocity (**a**), maximum S1- and S2-wave velocity (**b**), maximum P-wave anisotropy (**c**) and maximum S-wave anisotropy (**d**).

defines multiple maxima within the plane of foliation up to modal compositions of approximately 20:80 olivine:pyroxene, while the fastest and slowest S1-wave maxima become distributed along a plane that is oblique to the foliation (Fig. 12a).

Changing modal composition in xenoliths with an olivine A-type CPO results in less drastic changes in the distribution of anisotropic seismic properties. A small decrease in the maximum P- and S1-wave propagation velocities is observed from 8.7 to 8.3 km s^{-1} (change of 5%), and from 5 to 4.9 km s^{-1} (change of 2.5%) from dunitic ($Ol_{100}Opx_0Cpx_0$) to pyroxenitic ($Ol_0Opx_{50}Cpx_{50}$) compositions (Figs 12b and 13a, b). A systematic change in both the maximum P-and S-wave propagation anisotropies is observed from 8.7 to 3.3% and 6 to 2.5%, respectively, for a similar change in the modal composition (Figs 12b and 13c, d). Synthetic seismic properties for axial-[100] samples show a decrease in the maximum P- and S-wave velocity of 6.5% (8.8 to 8.2 km s^{-1}) for P-waves, 3.3% (5 to 4.8 km s^{-1}) for S1-waves, and 2% (4.9 to 4.8 km s^{-1}) for S2-waves for a transition from a dunitic to a pyroxenitic composition (Figs 12c and 13a, b). The maximum P-wave anisotropy for the synthetic axial-[100] samples ranges from 7.9% to 3.3% with corresponding maximum S-wave propagation anisotropy ranging from 6.3% ($Ol_{100}Opx_0Cpx_0$) to 2.6% ($Ol_0Opx_{50}Cpx_{50}$) (Figs 12c and 13c, d). Finally, synthetic seismic properties for the summation of samples with axial-[010] samples yield a decrease in the P- and S1-wave velocities of 4.3% (8.6 to 8.2 km s^{-1}) and 2.7% (4.9 to 4.8 km s^{-1}), respectively, as well as a change in the maximum AVp from 4.9% to 2.5% and the maximum AVs from 2.8% to 1.2% with decreasing modal abundance of olivine (Figs 12d and 13).

The removal of olivine and the addition of orthopyroxene and clinopyroxene by melt infiltration, percolation, or refertilization may have thus modified the average anisotropic seismic properties of the MBL xenoliths. The P- and S1-wave velocities can be reduced up to 6.5% and 3.7%, respectively, while the respective wave anisotropies are considerably reduced (cf. Baptiste *et al.* 2015). Comparison of the weighted average seismic properties for the Fosdick Mountains (Fig. 10) with those calculated for the summation of various olivine CPO symmetries (Fig. 11), suggests that the seismic properties of the mantle underneath MBL are most dominated by the anisotropy resulting from the A-type olivine CPO. However, the abundance of axial-[010] olivine CPOs, which result in similar distributions of the fastest P- and S1-waves, likely enhances this anisotropy pattern. The strong correlation between the maximum P-wave and S-wave anisotropy with the M-index of olivine (Fig. 8a) implies that deformation intensity, as reflected in CPO development, is responsible for a significant portion of the observed seismic anisotropy. Reactive melt percolation and refertilization during deformation, which results in an increased modal abundance of pyroxene, may have also significantly contributed to the average seismic properties of the MBL xenoliths, as is suggested by geochemical trends and the systematic decrease in Vp, Vs1, AVp, AVs1 and AVs2 with increasing modal pyroxene (Figs 8d–f, 12 and 13).

Comparison with seismic wave velocities and anisotropies of tomography models

The mean P wave velocity of the average sample for the MBL xenoliths is 8.2 km s^{-1}, 2.5–3.8% higher than the Vp for the lithospheric mantle of West Antarctica estimated from Pnl travel time analysis (O'Donnell *et al.* 2017). The S wave tomography models point to vertical shear wave velocities (VSv) of 4.3 to 4.5 km s^{-1} between 40 and 70 km depth beneath MBL (Fig. 14) (Shen *et al.* 2018; O'Donnell *et al.* 2019). Assuming a sub-horizontal lineation and foliation in the uppermost mantle beneath MBL (see below), VSv ray paths would sample Vs2, which has a mean value of 4.8 km s^{-1} for the average sample of the MBL xenoliths. This value is significantly higher (5–10%) than VSv in tomography models (Shen *et al.* 2018; O'Donnell *et al.* 2019). Steeply propagating seismic waves would sample the low Vp/Vs ratios of 1.7 for the average MBL xenolith sample (Fig. 10).

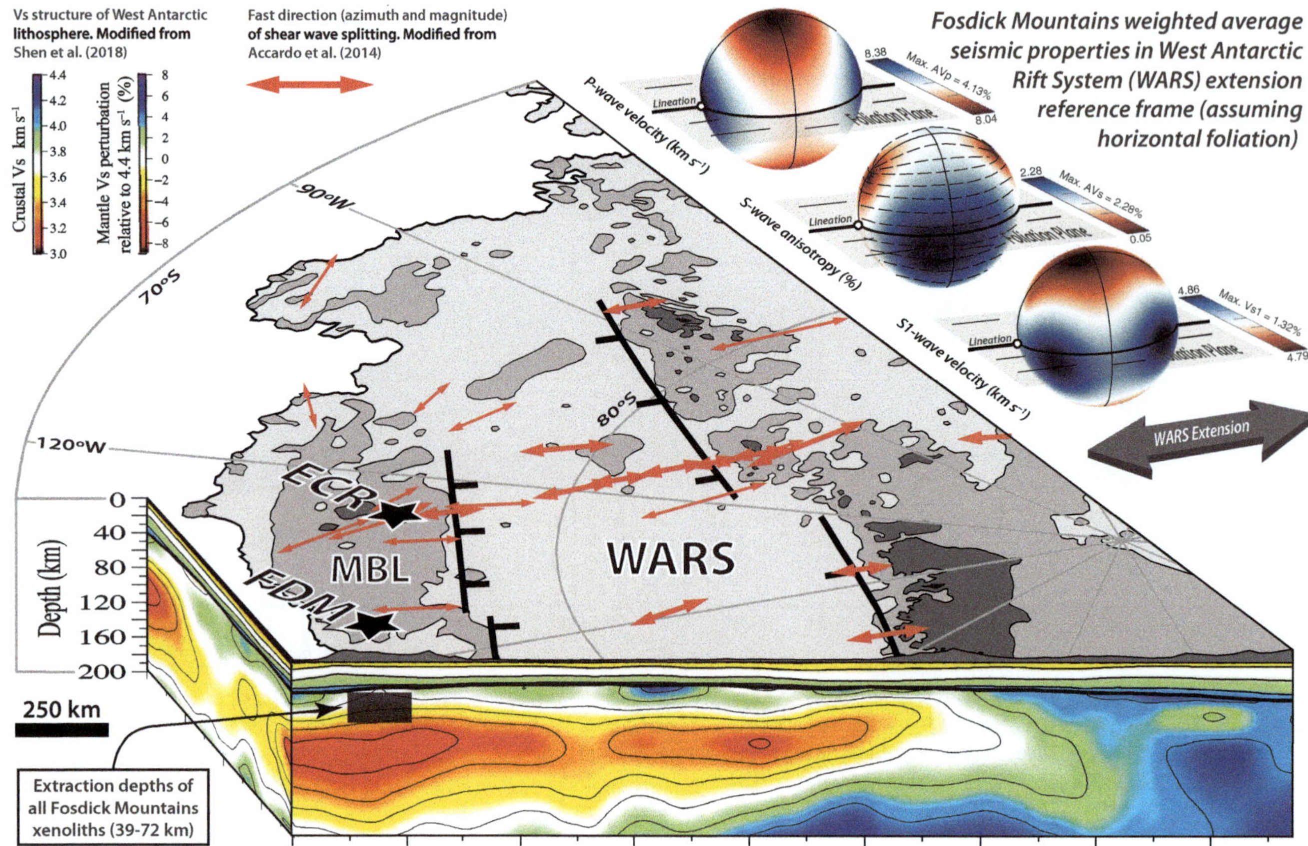

Fig. 14. Synoptic diagram showing the relationships between shear wave splitting (Accardo *et al.* 2014) and the crust and upper mantle velocity structure (after Shen *et al.* 2018), determined by geophysical studies, with respect to the approximate extraction depths of the Fosdick Mountains xenoliths and their weighted average seismic properties. Red arrows in map view show the fast direction (azimuth and magnitude) of shear wave splitting determined by Accardo *et al.* (2014), which show strong and consistent anisotropy throughout the West Antarctic Rift System (WARS) with fast axes subparallel to the inferred extension direction. Vertical cross-sections show the uppermost mantle Vs velocity structure beneath central and West Antarctica from Shen *et al.* (2018); crustal Vs is plotted with absolute values and the mantle is shown as a percentage perturbation relative to 4.4 km s^{-1}. The range of extraction depths for all Fosdick Mountains xenoliths (39–72 km) is shown by the grey box, which describes the relationship between the large-scale mantle velocity structure determined by geophysical methods and the region of upper mantle sampled by the MBL xenoliths. The three-dimensional pole figures at upper right show the interpreted relationship between the weighted average seismic properties of all Fosdick Mountains xenoliths and the fast axes of shear wave splitting, assuming a sub-horizontal tectonic foliation and a lineation developed sub-parallel to the direction of WARS extension. Note the presence of maximum P-wave propagation parallel to the inferred extension direction (and thus the <100> direction of olivine), slowest P-wave and S-wave in the vertical direction, and the maximum polarization of S1 waves within the foliation plane (i.e. horizontal plane). Grey contours in map view correspond to surface elevations in west Antarctica, with darker grey being higher elevation (after Paulsen and Wilson 2010). MBL, Marie Byrd Land; ECR, Executive Committee Range; FDM, Fosdick Mountains.

The upper mantle composition, microstructure, and thermal structure as described by the MBL xenoliths, is not consistent with the relatively low S wave velocities in the lithospheric mantle between 40 and 70 km depth, predicted by tomographic results (Shen *et al.* 2018; O'Donnell *et al.* 2019). Similar discrepancy, albeit smaller, between the S wave velocities derived by xenolith and tomographic studies is described by Demouchy *et al.* (2019) for the Tariat xeonliths, Mongolia. Seismic velocity varies with chemical and modal composition, hydrogen content, and presence of melt. There is a positive correlation between the shear wave velocity and Mg# in olivine (Lee 2003; Tommasi *et al.* 2004). The chemical composition of the MBL xenoliths is rather uniform with olivine Mg# in 90% of the xenoliths between 89 and 92 (Chatzaras *et al.* 2016). The pyroxene content has only a limited effect on shear wave velocity (Lee 2003; Soustelle *et al.* 2013). Thus, the chemical and modal composition of the xenoliths cannot account for the observed difference. High concentration of hydrogen embedded in point defects in the crystal structure of olivine was also suggested to decrease seismic wave velocities (Jacobsen 2006). Olivine in the MBL peridotites, however, is relatively dry (up to 31 ppm H_2O) (Chatzaras *et al.* 2016). The equilibrium temperatures in the majority of the MBL xenoliths are well below the wet solidus for peridotite, assumed due to a decrease in shear wave velocity due to the presence of melt.

Beneath West Antarctica, Ritzwoller *et al.* (2001) and O'Donnell *et al.* (2019) reported faster VSh than VSv, with average radial anisotropy (VSh – VSv)/VSv) in the uppermost mantle of 3–4%. For the average seismic properties of the MBL xenoliths, this observation suggests that mantle flow described by the peridotite lineation is sub-horizontal. In this configuration, the Vs1 and Vs2 calculated from the MBL xenoliths correspond to VSh and VSv, respectively. The radial anisotropy for the average seismic properties of the MBL xenoliths is up to 3%, and therefore is comparable with that predicted from seismological data.

Seismic structure and deformation of the lithospheric mantle beneath the West Antarctic Rift System

The summation of seismic properties of the Fosdick Mountains xenoliths shows polarization of the fastest shear wave (VS_1) parallel to the lineation or the flow direction (Fig. 10), which is parallel to the maximum concentration of olivine [100] axes (Fig. 6 and Fig. S2). This relationship suggests that shear wave splitting data can be used to map the horizontal component of the mantle flow, assuming the lithospheric mantle structures record mainly shearing along a sub-horizontal plane. The polarization orientation of the fast shear waves in the WARS and western MBL near the Fosdick Mountains, is perpendicular to the rift trend, as shown by shear wave splitting analysis (Fig. 14) (Accardo *et al.* 2014). This seismic anisotropy pattern implies that lithospheric thinning may be accommodated by plane strain deformation (pure shear with horizontal elongation, simple shear deformation, or a plane strain combination of the two), as asthenosphere rises beneath the rift axis and flows away, at high angle to the rift (McKenzie 1978; Wernicke 1981, 1985; Vauchez *et al.* 2000). Mantle flow is expected to produce an A-type olivine CPO with [100] axes oriented parallel to the elongation orientation at high angle to the rift axis, and [010] axes oriented near the pole to the sub-horizontal foliation plane (Vauchez *et al.* 2000). Comparison of Figures 10 and 11 suggests that the average seismic anisotropy calculated for all the Fosdick Mountains xenoliths is indeed dominated by the seismic anisotropy signature of the orthorhombic A-type olivine CPO (Fig. 14). Olivine A-type CPO in the upper mantle beneath MBL is shown to form in relation to plane strain deformation (Chatzaras *et al.* 2016), in agreement with rifting models for two-dimensional deformation (e.g. McKenzie 1978).

The average seismic properties calculated for the different volcanic centres in Fosdick Mountains further suggest that deformation in the lithospheric mantle is more complex and heterogeneous than the simple case of two-dimensional deformation described above. Upper mantle seismic anisotropy beneath Marujupu Peak seems to be dominated by the signature of the A-type olivine CPO, while the axial-[100] type dominates Demas Bluff and axial-[010] CPOs dominate xenoliths sourced from either Mount Avers or Bird Bluff (samples FDM-AVBB) (Figs 10 and 11). The A-type, axial-[100] and axial-[010] CPOs in the studied xenoliths, form in relation to plane strain, constriction and flattening, respectively (Chatzaras *et al.* 2016).

Given the observed heterogeneity at smaller spatial scales (i.e. between individual volcanic centres), it is possible that variations in the seismic anisotropy sourced from individual volcanic centres could be controlled by differences in local strain geometry, and thus the patterns of heterogeneous three-dimensional deformation in the upper mantle (e.g. Tommasi *et al.* 1999). This interpretation is supported by the protracted tectonic history of MBL, which involved active subduction and the development of a magmatic arc outboard from the East Antarctic Ross orogen, extreme crustal extension during the Cretaceous resulting in the formation of the Fosdick Mountains metamorphic core complex and associated strike-slip faults, and continued extension associated with the development of the West Antarctic Rift System and the rifting of the New Zealand continent (e.g. Jordan *et al.* 2020; Smellie *et al.* 2020*a*). However, despite the variations in the seismic properties of individual samples (Fig. 9 and Fig. S4) or those observed between individual volcanic centres or texture type (Figs 10 and 11, respectively), the weighted averaged properties of the MBL xenoliths yield seismic velocities and anisotropies that are in good agreement with those derived by geophysical studies. This result, combined with the observed patterns of shear wave splitting (Accardo *et al.* 2014) are interpreted to suggest that the seismic signature of the MBL xenoliths results primarily from the development of orthorhombic A-type olivine crystallographic preferred orientation contemporaneous with extension and active modification of the upper mantle beneath MBL associated with the development of the WARS.

Conclusions

Xenoliths sourced from volcanic centres throughout the Fosdick Mountains, Executive Committee Range, and Usas Escarpment of MBL preserve a unique record of the geochemical, microstructural, and textural changes within the lithospheric mantle of West Antarctica. Using an integrated study that combines analysis of major element mineral compositions, determination of CPO using electron backscatter diffraction and the calculation of seismic properties from CPO data, we demonstrate the effect of melt-rock reaction, refertilization and the corresponding changes in modal composition on the resultant rock microstructure and seismic properties of the MBL mantle lithosphere.

Variations in modal compositions, enrichment of clinopyroxene leading to the formation of clinopyroxene-rich lherzolite, discrepancies between the estimated degrees of partial melting inferred from geochemical trends in orthopyroxene and clinopyroxene, and microstructural relationships, indicate that the lithospheric mantle beneath MBL experienced a process of reactive melt percolation involving refertilization. Rock microstructures further elucidate the temporal relationships between melt percolation and deformation, suggesting a diachronous history of melt percolation in the MBL upper mantle relative to deformation that produced olivine CPOs.

Changes in modal composition resulting from melt percolation and refertilization in the MBL xenoliths, combined with the effects of heterogeneous deformation, resulted in compositionally and structurally heterogeneous xenoliths with CPO patterns dominated by one of three olivine texture types; either the orthorhombic A-type, axial-[100], or axial-[010] CPO. Variations in the modal abundance of constituent phases and the intensity of CPO development, as reflected by the J and M indices, resulted in corresponding variations in the seismic properties of the MBL xenoliths. A strong correlation between the maximum P-wave and S-wave anisotropy with olivine CPO strength implies that the latter is responsible for a significant portion of the observed seismic anisotropy. Reactive melt percolation and refertilization likely modified the average seismic properties of the MBL mantle lithosphere, as is suggested by a systematic decrease in maximum P-wave and S-wave anisotropies with increasing modal abundance of pyroxene. While the seismic properties of the MBL mantle xenoliths are heterogeneous with respect to individual samples, or between volcanic centres, at larger spatial scales (>100 km) they are most dominated by the anisotropy resulting from the A-type olivine CPO. The polarization of fast shear waves in the WARS oriented perpendicular to the rift trend suggest that the maximum P-wave anisotropy, and thus the alignment of olivine [100] axes and the lineation orientation, are parallel to the extension direction in the WARS. Assuming that WARS rifting and modification of the upper mantle lithosphere affected the majority of the MBL xenoliths, resulting in the production of a sub-horizontal tectonic foliation, the weighted average seismic properties of the MBL xenoliths that are dominated by A-type olivine textures suggest that lithospheric thinning was accommodated primarily by pure shear deformation. Variations within and amongst individual volcanic centres, however, attest to variations in local strain geometry and heterogeneous mantle deformation, likely resulting from the protracted history of deformation that has affected MBLand West Antarctica.

Acknowledgements We thank the United States Polar Rock Repository, Christine Siddoway, and Kurt Panter for providing us with xenolith samples. This manuscript greatly benefited from the constructive comments of Doug Wiens, an anonymous reviewer, and our editor, Adam Martin. We acknowledge the facilities and technical assistance of the Scanning Electron Microscope (SEM) Facility at Boston College, the electron microprobe laboratory at the University of Wisconsin–Madison, and Microscopy Australia at the Australian Centre for Microscopy and Microanalysis at the University of Sydney. We thank Shaina Cohen for assistance in the EBSD data collection, John Fournelle and Gordon Medaris (UW-Madison) for their advice and assistance with the microprobe analyses, and Brian Hess (UW-Madison) for the production of high-quality thin sections used in EBSD analysis. We also thank Gordon Medaris, Basil Tikoff, and Alexander Lusk for their feedback on an earlier version of the manuscript.

Author contributions **VC**: formal analysis (equal), visualization (equal), writing – original draft (lead); **SCK**: formal analysis (equal), funding acquisition (lead), visualization (equal), writing – original draft (equal).

Funding This work was funded by the National Science Foundation (NSF-ANT1246320, awarded to Seth C. Kruckenberg) and the University of Sydney (research support grant 199049, awarded to Vasileios Chatzaras).

Data availability The datasets generated or analysed during the current study are included in this published article and its supplementary files.

References

Abramson, E.H., Brown, J.M., Slutsky, L.J. and Zaug, J. 1997. The elastic constants of San Carlos olivine to 17 GPa. *Journal of Geophysical Research*, **102**, 12253–12263, https://doi.org/10.1029/97JB00682

Accardo, N.J., Wiens, D.A. *et al.* 2014. Upper mantle seismic anisotropy beneath the West Antarctic Rift System and surrounding region from shear wave splitting analysis. *Geophysical Journal International*, **198**, 414–429, https://doi.org/10.1093/gji/ggu117

Arai, S. 1994. Characterization of spinel peridotites by olivine-spinel compositional relationships: review and interpretation. *Chemical Geology*, **113**, 191–204, https://doi.org/10.1016/0009-2541(94)90066-3

Armstrong, J.T. 1988. Quantitative analysis of silicate and oxide materials: Comparison of Monte Carlo, ZaF, an φ(pz) procedures. *In*: Newbury, D.E. (ed.) *Proceedings of the 23rd Annual Conference of the Microbeam Analysis Society*, 239–246.

Ashby, M.F. and Verrall, R.A. 1973. Diffusion accommodated flow and superplasticity. *Acta Metallurgica*, **21**, 149–163, https://doi.org/10.1016/0001-6160(73)90057-6

Bachmann, F., Hielscher, R. and Schaeben, H. 2010. Texture analysis with MTEX-free and open source software toolbox. *Diffusion and Defect Data*, **160**, 63–68.

Bachmann, F., Hielscher, R. and Schaeben, H. 2011. Grain detection from 2d and 3d EBSD data—Specification of the MTEX algorithm. *Ultramicroscopy*, **111**, 1720–1733, https://doi.org/10.1016/j.ultramic.2011.08.002

Baker, B. and Stolper, E.M. 1994. Determining the composition of high-pressure mantle melts using diamond aggregates. *Geochimica et Cosmochimica Acta*, **58**, 2811–2827.

Baptiste, V., Tommasi, A., Vauchez, A., Demouchy, S. and Rudnick, R.L. 2015. Deformation, hydration, and anisotropy of the lithospheric mantle in an active rift: constraints from mantle xenoliths from the North Tanzanian Divergence of the East African Rift. *Tectonophysics*, **639**, 34–55, https://doi.org/10.1016/j.tecto.2014.11.011

Barklage, M., Wiens, D.A., Nyblade, A. and Anandakrishnan, S. 2009. Upper mantle seismic anisotropy of South Victoria Land and the Ross Sea coast, Antarctica from SKS and SKKS splitting analysis. *Geophysical Journal International*, **178**, 729–741, https://doi.org/10.1111/j.1365-246X.2009.04158.x

Behrendt, J.C., Saltus, R., Damaske, D., McCafferty, A., Finn, C.A., Blankenship, D. and Bell, R.E. 1996. Patterns of late Cenozoic volcanic and tectonic activity in the West Antarctic rift system revealed by aeromagnetic surveys. *Tectonics*, **15**, 660–676, https://doi.org/10.1029/95TC03500

Berg, J.H., Moscati, R.J. and Herz, D.L. 1989. A petrologic geotherm from a continental rift in Antarctica. *Earth and Planetary Science Letters*, **93**, 98–108, https://doi.org/10.1016/0012-821X(89)90187-8

Bernard, R.E., Behr, W.M., Becker, T.W. and Young, D.J. 2019. Relationships between olivine CPO and deformation parameters in naturally deformed rocks and implications for mantle seismic anisotropy. *Geochemistry, Geophysics, Geosystems*, **20**, 3469–3494, https://doi.org/10.1029/2019GC008289

Bertrand, P. and Mercier, J.-C.C. 1985. The mutual solubility of coexisting ortho- and clinopyroxene: toward an absolute geothermometer for the natural system? *Earth and Planetary Science Letters*, **76**, 109–122, https://doi.org/10.1016/0012-821X(85)90152-9

Blankenship, D.D., Bell, R.E., Hodge, S.M., Brozena, J.M., Behrendt, J.C. and Finn, C.A. 1993. Active volcanism beneath the West Antarctic Ice Sheet and implications for ice-sheet stability. *Nature*, **361**, 526–529, https://doi.org/10.1038/361526a0

Bodinier, J.-L. and Godard, M. 2014. Orogenic, ophiolitic, and abyssal peridotites. *In*: Carlson, R.W. (ed.) *Second Edition Treatise on Geochemistry*, **3**, 103–167.

Boneh, Y. and Skemer, P. 2014. The effect of deformation history on the evolution of olivine CPO. *Earth and Planetary Science Letters*, **406**, 213–222, https://doi.org/10.1016/j.epsl.2014.09.018

Brey, G.P. and Köhler, T. 1990. Geothermobarometry in four-phase lherzolites: II. New thermobarometers and practical assessment of existing thermobarometry. *Journal of Petrology*, **31**, 1352–1378.

Brune, S., Williams, S.E. and Müller, R.D. 2017. Potential links between continental rifting, CO_2 degassing and climate change through time. *Nature Geoscience*, **10**, 941–946, https://doi.org/10.1038/s41561-017-0003-6

Bunge, H.J. (ed.). 1982. *Texture Analysis in Materials Science*. Butterworths, Boston, Mass, 614, https://doi.org/10.1016/C2013-0-11769-2

Cande, S.C., Stock, J.M., Müller, R.D. and Ishihara, T. 2000. Cenozoic motion between East and West Antarctica. *Nature*, **404**, 145–150, https://doi.org/10.1038/35004501

Chaput, J., Aster, R.C. *et al.* 2014. The crustal thickness of West Antarctica. *Journal of Geophysical Research: Solid Earth*, **119**, 1–18.

Chatzaras, V., Kruckenberg, S.C., Cohen, S.M., Medaris, L.G. Jr, Withers, A.C. and Bagley, B. 2016. Axial-type olivine crystallographic preferred orientations: the effect of strain geometry on mantle texture. *Journal of Geophysical Research: Solid Earth*, **121**, https://doi.org/10.1002/2015JB012628

Cline, C.J., II, Faul, U.H., David, E.C., Berry, A.J. and Jackson, I. 2018. Redox-influenced seismic properties of upper-mantle olivine. *Nature*, **555**, 355–358, https://doi.org/10.1038/nature25764

Danesi, S. and Morelli, A. 2001. Structure of the upper mantle under the Antarctic plate from surface wave tomography. *Geophysical Research Letters*, **28**, 4395–4398, https://doi.org/10.1029/2001GL013431

Day, J.M.D., Harvey, R.P. and Hilton, D.R. 2019. Melt-modified lithosphere beneath Ross Island and its role in the tectonomagmatic evolution of the West Antarctic Rift System. *Chemical Geology*, **518**, 45–54, https://doi.org/10.1016/j.chemgeo.2019.04.012

Demouchy, S., Tommasi, A., Ionov, D., Higgie, K. and Carlson, R.W. 2019. Microstructures, water contents, and seismic properties of the mantle lithosphere beneath the northern limit of the

Hangay Dome, Mongolia. *Geochemistry, Geophysics, Geosystems*, **20**, 183–207, https://doi.org/10.1029/2018GC007931

Dijkstra, A.H., Drury, M.R., Vissers, R.L.M. and Newman, J. 2002. On the role of melt-rock reaction in mantle shear zone formation in the Othris Peridotite Massif (Greece). *Journal of Structural Geology*, **24**, 1431–1450, https://doi.org/10.1016/S0191-8141(01)00142-0

Doherty, C.L. 2016. *Multi stage evolution of the lithospheric mantle in the West Antarctic Rift System – a mantle xenolith study*. PhD, Columbia University, 228.

Donnellan, A. and Luyendyk, B.P. 2004. GPS evidence for a coherent Antarctic plate and for postglacial rebound in Marie Byrd Land. *Global and Planetary Change*, **42**, 305–311, https://doi.org/10.1016/j.gloplacha.2004.02.006

Drury, M.R. and Humphreys, F.J. 1988. Microstructural shear criteria associated with grain boundary sliding during ductile deformation. *Journal of Structural Geology*, **10**, 83–89, https://doi.org/10.1016/0191-8141(88)90130-7

Drury, M.R. and Urai, J.L. 1990. Deformation-related recrystallisation processes. *Tectonophysics*, **172**, 235–253, https://doi.org/10.1016/0040-1951(90)90033-5

Embey-Isztin, A. 2016. The role of melt depletion v. refertilization in the major element chemistry of four-phase spinel peridotite xenoliths. *Central European Geology*, **9**, 60–86, https://doi.org/10.1556/24.59.2016.002

Finn, C.A., Müller, R.D. and Panter, K.S. 2005. A Cenozoic diffuse alkaline magmatic province (DAMP) in the southwest Pacific without rift or plume origin. *Geochemistry, Geophysics, Geosystems*, **6**, https://doi.org/10.1029/2004GC000723

Foden, J., Elburg, M.A., Dougherty-Page, J. and Burtt, A. 2006. The timing and duration of the Delamerian Orogeny: correlation with the Ross Orogen and implications for Gondwana assembly. *Journal of Geology*, **114**, 189–210, https://doi.org/10.1086/499570

Frets, E., Tommasi, A., Garrido, C.J., Padrón-Navarta, J.A., Amri, I. and Targuisti, K. 2012. Deformation processes and rheology of pyroxenites under lithospheric mantle conditions. *Journal of Structural Geology*, **39**, 138–157, https://doi.org/10.1016/j.jsg.2012.02.019

Gaffney, A.M. and Siddoway, C.S. 2007. Heterogeneous sources for Pleistocene lavas of Marie Byrd Land, Antarctica: new data from the SW pacific diffuse alkaline magmatic province. *In*: Cooper, A.K., Raymond, C.R. *et al.* (eds) *Antarctica: A Keystone in a Changing World*. Online Proceedings for the Tenth International Symposium on Antarctic Earth Sciences, USGS Open-File Report 2007-1047, Extended abstract 063.

Granot, R., Cande, S.C., Stock, J.M. and Damaske, D. 2013. Revised Eocene-Oligocene kinematics for the West Antarctic rift system. *Geophysical Research Letters*, **40**, 279–284, https://doi.org/10.1029/2012GL054181

Handler, M.R., Wysoczanski, R.J. and Gamble, J.A. 2003. Proterozoic lithosphere in Marie Byrd Land, West Antarctica: Re–Os systematics of spinel peridotite xenoliths. *Chemical Geology*, **196**, 131–145, https://doi.org/10.1016/S0009-2541(02)00410-2

Handler, M.R., Wysoczanski, R.J. and Gamble, J.A. 2021. Marie Byrd Land lithospheric mantle: a review of the xenolith record. *Geological Society, London, Memoirs*, **56**, https://doi.org/10.1144/M56-2020-17

Hansen, L.N., Zimmerman, M.E. and Kohlstedt, D.L. 2014*a*. Protracted fabric evolution in olivine: implications for the relationship among strain, crystallographic fabric, and seismic anisotropy. *Earth and Planetary Science Letters*, **387**, 157–168, https://doi.org/10.1016/j.epsl.2013.11.009

Hansen, S.E., Graw, J.H. *et al.* 2014*b*. Imaging the Antarctic mantle using adaptively parameterized P-wave tomography: evidence for heterogeneous structure beneath West Antarctica. *Earth and Planetary Science Letters*, **408**, 66–78, https://doi.org/10.1016/j.epsl.2014.09.043

Hart, S.R., Blusztajn, J., LeMasurier, W.E., Rex, D.C., Hawkesworth, C.E. and Arndt, N.T.E. 1997. Hobbs Coast Cenozoic volcanism: implications for the West Antarctic rift system. *Chemical Geology*, **139**, 223–248, https://doi.org/10.1016/S0009-2541(97)00037-5

Heeszel, D.S., Wiens, D.A. *et al.* 2016. Upper mantle structure of central and West Antarctica from array analysis of Rayleigh wave phase velocities. *Journal of Geophysical Research: Solid Earth*, **121**, 1758–1775, https://doi.org/10.1002/2015JB012616

Hidas, K., Tommasi, A. *et al.* 2016. Fluid-assisted strain localization in the shallow subcontinental lithospheric mantle. *Lithos*, **262**, 636–650, https://doi.org/10.1016/j.lithos.2016.07.038

Hidas, K., Garrido, C.J. *et al.* 2019. Lithosphere tearing along STEP faults and synkinematic formation of lherzolite and wehrlite in the shallow subcontinental mantle. *Solid Earth*, **10**, 1099–1121, https://doi.org/10.5194/se-10-1099-2019

Higgie, K. and Tommasi, A. 2014. Deformation in a partially molten mantle: Constraints from plagioclase lherzolites from Lanzo, western Alps. *Tectonophysics*, **615–616**, 167–181, https://doi.org/10.1016/j.tecto.2014.01.007

Hole, M.J. and LeMasurier, W.E. 1994. Tectonic controls on the geochemical composition of Cenozoic, mafic alkaline volcanic rocks from West Antarctica. *Contributions to Mineralogy and Petrology*, **117**, 187–202, https://doi.org/10.1007/BF00286842

Holtzman, B.K., Kohlstedt, D.L., Zimmerman, M.E., Heidelbach, F., Hiraga, T. and Hustoft, J. 2003. Melt segregation and strain partitioning: implications for seismic anisotropy and mantle flow. *Science*, **301**, 1227–1230, https://doi.org/10.1126/science.1087132

Isaak, D.G., Ohno, I. and Lee, P.C. 2005. The elastic constants of monoclinic single-crystal chrome-diopside to 1300 K. *Physics and Chemistry of Minerals*, **32**, 691–699, https://doi.org/10.1007/s00269-005-0047-9

Jackson, J.M., Sinogeikin, S.V. and Bass, J.D. 2007. Sound velocities and single-crystal elasticity of orthoenstatite to 1073 K at ambient pressure. *Physics of the Earth and Planetary Interiors*, **161**, 1–12, https://doi.org/10.1016/j.pepi.2006.11.002

Jacobsen, S.D. 2006. Effect of water on the equation of state of nominally anhydrous minerals. *Reviews in Mineralogy and Geochemistry*, **62**, 321–342, https://doi.org/10.2138/rmg.2006.62.14

Johnson, K.T.M., Dick, H.J.B. and Shimizu, N. 1990. Melting in the oceanic upper mantle: an ion microprobe study of diopsides in abyssal peridotites. *Journal of Geophysical Research: Solid Earth*, **95**, 2661–2678, https://doi.org/10.1029/JB095iB03p02661

Jordan, T.A., Riley, T.R. and Siddoway, C.S. 2020. The geological history and evolution of West Antarctica. *Nature Reviews Earth and Environment*, **1**, 117–133, https://doi.org/10.1038/s43017-019-0013-6

Jung, H. and Karato, S. 2001. Water-induced fabric transitions in olivine. *Science*, **293**, 1460–1463, https://doi.org/10.1126/science.1062235

Jung, H., Park, M., Jung, S. and Lee, J. 2010. Lattice preferred orientation, water content, and seismic anisotropy of orthopyroxene. *Journal of Earth Science*, **21**, 555–568, https://doi.org/10.1007/s12583-010-0118-9

Karato, S., Jung, H., Katayama, I. and Skemer, P.A. 2008. Geodynamic significance of seismic anisotropy of the upper mantle: new insights from laboratory studies. *Annual Reviews of Earth and Planetary Science*, **36**, 59–95, https://doi.org/10.1146/annurev.earth.36.031207.124120

Kostopoulos, D.K. 1991. Melting of the shallow upper mantle: a new perspective. *Journal of Petrology*, **32**, 671–699, https://doi.org/10.1093/petrology/32.4.671

Kumamoto, K.M., Warren, J.M. and Hansen, L.N. 2019. Evolution of the Josephine Peridotite Shear Zones: 2. Influences on Olivine CPO Evolution. *Journal of Geophysical Research: Solid Earth*, **124**, 12763–12781, https://doi.org/10.1029/2019JB017968

Lee, C.-T.A. 2003. Compositional variation of density and seismic velocities in natural peridotites at STP conditions: Implications for seismic imaging of compositional heterogeneities in the upper mantle. *Journal of Geophysical Research*, **108**, 2441, https://doi.org/10.1029/2003JB002413

LeMasurier, W.E. 1990. Mount sidley. *In*: LeMasurier, W.E. and Thomson, J.W. (eds) *Volcanoes of the Antarctic Plate and Southern Oceans*. Antarctic Research Series, **48**, 203–206.

LeMasurier, W.E. 2008. Neogene extension and basin deepening in the West Antarctic rift inferred from comparisons with the East African rift and other analogs. *Geology*, **36**, 247–250, https://doi.org/10.1130/G24363A.1

LeMasurier, W.E. and Rex, D.C. 1982. Volcanic record of Cenozoic glacial history in Marie Byrd Land and Western Ellsworth Land: revised chronology and evaluation of tectonic factors. *In*: Craddock, C. (ed.) *Antarctic Geoscience*. University of Wisconsin Press, Madison, 725–734.

LeMasurier, W.E. and Rex, D.C. 1989. Evolution of linear volcanic ranges in Marie Byrd Land, West Antarctica. *Journal of Geophysical Research*, **94**, 7223–7236, https://doi.org/10.1029/JB094iB06p07223

Le Roux, V., Bodinier, J.L., Tommasi, A., Alard, O., Dautria, J.M., Vauchez, A. and Riches, A.J.V. 2007. The Lherz spinel lherzolite: refertilized rather than pristine mantle. *Earth and Planetary Science Letters*, **259**, 599–612, https://doi.org/10.1016/j.epsl.2007.05.026

Lloyd, A.J., Wiens, D.A. *et al.* 2015. A seismic transect across West Antarctica: evidence for mantle thermal anomalies beneath the Bentley Subglacial Trench and the Marie Byrd Land Dome. *Journal of Geophysical Research: Solid Earth*, **120**, 8439–8460, https://doi.org/10.1002/2015JB012455

Lloyd, A.J., Wiens, D.A. *et al.* 2019. Seismic structure of the Antarctic upper mantle imaged with adjoint tomography. *Journal of Geophysical Research: Solid Earth*, **124**, https://doi.org/10.1029/2019JB017823

Lo Cascio, M., Liang, Y. and Hess, P.C. 2004. Grain-scale processes during isothermal- isobaric melting of lherzolite. *Geophysical Research Letters*, **31**, L16605, https://doi.org/10.1029/2004GL020602

Mainprice, D. 1990. A FORTRAN program to calculate seismic anisotropy from the lattice preferred orientation of minerals. *Computers and Geoscience*, **16**, 385–393, https://doi.org/10.1016/0098-3004(90)90072-2

Mainprice, D. 1997. Modelling anisotropic seismic properties of partially molten rocks found at mid-ocean ridges. *Tectonophysics*, **279**, 161–179, https://doi.org/10.1016/S0040-1951(97)00122-4

Mainprice, D. and Humbert, M. 1994. Methods of calculating petrophysical properties from lattice preferred orientation data. *Surveys in Geophysics*, **15**, 575–592, https://doi.org/10.1007/BF00690175

Mainprice, D. and Silver, P.G. 1993. Interpretation of SKS waves using samples from the subcontinental lithosphere. *Physics of Earth and Planetary Science Interiors*, **78**, 257–280, https://doi.org/10.1016/0031-9201(93)90160-B

Mainprice, D., Bachmann, F., Hielscher, R. and Schaeben, H. 2014. Descriptive tools for the analysis of texture projects with large datasets using MTEX: strength, symmetry and components. *In*: Faulkner, D.R., Mariani, E. and Mecklenburgh, J. (eds) *Rock Deformation From Field, Experiments and Theory: A Volume in Honour of Ernie Rutter*. Geological Society, London, Special Publications, **409**, 223–249, https://doi.org/10.1144/SP409.8

Martin, A.P., Cooper, A.F. and Price, R.C. 2013, Petrogenesis of Cenozoic, alkalic volcanic lineages at Mount Morning, West Antarctica and their entrained lithospheric mantle xenoliths: lithospheric v. asthenospheric mantle sources. *Geochimica et Cosmochimica Acta*, **122**, 127–152, https://doi.org/10.1016/j.gca.2013.08.025

Martin, A.P., Price, R.C., Cooper, A.F. and McCammon, C.A. 2015. Petrogenesis of the rifted Southern Victoria Land lithospheric mantle, Antarctica, inferred from petrography, geochemistry, thermobarometry and oxybarometry of peridotite and pyroxenite xenoliths from the Mount Morning eruptive centre. *Journal of Petrology*, **56**, 193–226, https://doi.org/10.1093/petrology/egu075

McDonough, W.F. and Sun, S.-S. 1995. The composition of the Earth. *Chemical Geology*, **120**, 223–253, https://doi.org/10.1016/0009-2541(94)00140-4

McFadden, R.R., Siddoway, C.S., Teyssier, C. and Fanning, C.M. 2010. Cretaceous oblique extensional deformation and magma accumulation in the Fosdick Mountains migmatite-cored gneiss dome, West Antarctica. *Tectonics*, **29**, TC4022, https://doi.org/10.1029/2009TC002492

McKenzie, D. 1978. Some remarks on the development of sedimentary basins. *Earth and Planetary Science Letters*, **40**, 25–32, https://doi.org/10.1016/0012-821X(78)90071-7

Medaris, L.G., Fournelle, J.H., Wang, H.F. and Jelinek, E. 1997. Thermobarometry and reconstructed chemical compositions of spinel–pyroxene symplectites: evidence for pre-existing garnet in lherzolite xenolith from Czech Neogene lavas. *Russian Geology Geophysics*, **38**, 277–286.

Morales, L.F.G. and Tommasi, A. 2011. Composition, textures, seismic and thermal anisotropies of xenoliths from a thin and hot lithospheric mantle (Summit Lake, southern Canadian Cordillera). *Tectonophysics*, **507**, 1–15, https://doi.org/10.1016/j.tecto.2011.04.014

Morishita, T. and Arai, S. 2003. Evolution of spinel–pyroxene symplectite in spinel–lherzolites from the Horoman Complex, Japan. *Contributions to Mineralogy and Petrology*, **144**, 509–522, https://doi.org/10.1007/s00410-002-0417-y

Müller, R.D., Gohl, K., Cande, S.C., Goncharov, A. and Golynsky, A.V. 2007. Eocene to Miocene geometry of the West Antarctic rift system. *Australian Journal of Earth Sciences*, **54**, 1033–1045, https://doi.org/10.1080/08120090701615691

Newman, J., Lamb, W.M., Drury, M.R. and Vissers, R.L.M. 1999. Deformation processes in a peridotite shear zone: reaction-softening by an H_2O-deficient, continuous net transfer reaction. *Tectonophysics*, **303**, 193–222, https://doi.org/10.1016/S0040-1951(98)00259-5

Nicolas, A. and Christensen, N.I. 1987. Formation of anisotropy in upper mantle peridotites- a review. *Review of Geophysics*, **25**, 111–123.

Niu, Y. 2004. Bulk-rock major and trace element compositions of abyssal peridotites: implications for mantle melting, melt extraction and post-melting processes beneath mid-ocean ridges. *Journal of Petrology*, **45**, 2423–2458, https://doi.org/10.1093/petrology/egh068

Niu, Y., Langmuir, C.H. and Kinzler, R.J. 1997. The origin of abyssal peridotites: a new perspective. *Earth and Planetary Science Letters*, **152**, 251–265, https://doi.org/10.1016/S0012-821X(97)00119-2

O'Donnell, J.P., Selway, K. *et al.* 2017. The uppermost mantle seismic velocity and viscosity structure of central West Antarctica. *Earth and Planetary Science Letters*, **472**, 38–49, https://doi.org/10.1016/j.epsl.2017.05.016

O'Donnell, J.P., Stuart, G.W. *et al.* 2019. The uppermost mantle seismic velocity structure of West Antarctica from Rayleigh wave tomography: insights into tectonic structure and geothermal heat flow. *Earth and Planetary Science Letters*, **522**, 219–233, https://doi.org/10.1016/j.epsl.2019.06.024

Panter, K.S. and Martin, A. 2021. West Antarctic mantle deduced from mafic magmatism. *Geological Society, London, Memoirs*, **56**, https://doi.org/10.1144/M56-2021-10

Panter, K.S., Kyle, P.R. and Smellie, J.L. 1997. Petrogenesis of a phonolite–trachyte succession at Mount Sidley, Marie Byrd Land, Antarctica. *Journal of Petrology*, **38**, 1225–1253, https://doi.org/10.1093/petroj/38.9.1225

Panter, K.S., Hart, S.R., Kyle, P., Blusztanjn, J. and Wilch, T. 2000. Geochemistry of late Cenozoic basalts from the Crary Mountains: characterization of mantle sources in Marie Byrd Land, Antarctica. *Chemical Geology*, **165**, 215–241, https://doi.org/10.1016/S0009-2541(99)00171-0

Panter, K.S., Castillo, P. *et al.* 2018. Melt origin across a rifted continental margin: a case for subduction-related metasomatic agents in the lithospheric source of alkaline basalt, NW Ross Sea, Antarctica. *Journal of Petrology*, **59**, 517–558, https://doi.org/10.1093/petrology/egy036

Panter, K.S., Wilch, T.I., Smellie, J.L., Kyle, P.R. and McIntosh, W.C. 2021. Marie Byrd Land and Ellsworth Land: petrology. *Geological Society, London, Memoirs*, **55**, https://doi.org/10.1144/M55-2019-50

Paulsen, T.S. and Wilson, T.J. 2010. Evolution of Neogene volcanism and stress patterns in the glaciated West Antarctic Rift, Marie Byrd Land, Antarctica. *Journal of the Geological Society, London*, **167**, 401–416, https://doi.org/10.1144/0016-76492009-044

Pearce, J.A., Barker, P.F., Edwards, S.J., Parkinson, I.J. and Leat, P.T. 2000. Geochemistry and tectonic significance of peridotites from the South Sandwich arc-basin system, South Atlantic. *Contributions to Mineralogy and Petrology*, **139**, 36–53, https://doi.org/10.1007/s004100050572

Pelorosso, B., Bonadiman, C. *et al.* 2017. Role of percolating melts in Antarctic subcontinental lithospheric mantle: new insights from Handler Ridge mantle xenoliths (northern Victoria Land, Antarctica). *In*: Bianchini, G., Bodinier, J.-L., Braga, R. and Wilson, M. (eds) *The Crust-Mantle and Lithosphere-Asthenosphere Boundaries: Insights from Xenoliths, Orogenic Deep Sections, and Geophysical Studies*. Geological Society of America, Special Papers, **526**, 133–150.

Powell, E., Gomez, N., Hay, C., Latychev, K. and Mitrovica, J.X. 2020. Viscous effects in the solid earth response to modern Antarctic ice mass flux: implications for geodetic studies of WAIS stability in a warming world. *Journal of Climate*, **33**, 443–459, https://doi.org/10.1175/JCLI-D-19-0479.1

Précigout, J. and Hirth, G. 2014. B-type olivine fabric induced by grain boundary sliding. *Earth Planetary Science Letters*, **395**, 231–240, https://doi.org/10.1016/j.epsl.2014.03.052

Ree, J.H. 1994. Grain boundary sliding and development of grain boundary openings in experimentally deformed octachloropropane. *Journal of Structural Geology*, **16**, 403–418, https://doi.org/10.1016/0191-8141(94)90044-2

Ritzwoller, M.H., Shapiro, N.M., Levshin, A.L. and Leahy, G.M. 2001. Crustal and upper mantle structure beneath Antarctica and surrounding oceans. *Journal of Geophysical Research: Solid Earth*, **106**, B12, https://doi.org/10.1029/2001JB000179

Rocchi, S., Armienti, P., D'Orazio, M., Tonarini, S., Wijbrans, J. and Di Vincenzo, G. 2002. Cenozoic magmatism in the western Ross Embayment: role of mantle plume v. plate dynamics in the development of the West Antarctic rift system. *Journal of Geophysical Research*, **107**, B9, https://doi.org/10.129/2001JB000515

Rocchi, S., Armienti, P. and Di Vincenzo, G. 2005. No plume, no rift magmatism in the West Antarctic Rift. *In*: Foulger, G.R., Natland, J., Presnall, D. and Anderson, D.L. (eds) *Plates, Plumes, and Paradigms*. Geological Society of America, Special Paper, **388**, 435–447.

Salvini, F., Brancolini, G., Busetti, M., Storti, F., Mazzarini, F. and Coren, F. 1997. Cenozoic geodynamics of the Ross Sea region, Antarctica: crustal extension, intraplate strike-slip faulting, and tectonic inheritance. *Journal of Geophysical Research: Solid Earth*, **102**, 24669–24696, https://doi.org/10.1029/97JB01643

Shen, W., Wiens, D.A. *et al.* 2018. The crust and upper mantle structure of central and West Antarctica from Bayesian inversion of Rayleigh wave and receiver functions. *Journal of Geophysical Research: Solid Earth*, **123**, 7824–7849, https://doi.org/10.1029/2017JB015346

Siddoway, C.S. 2008. Tectonics of the West Antarctic rift system: new light on the history and dynamics of distributed intracontinental extension. *In*: Cooper, A.K., Barrett, P.J., Stagg, H., Storey, B., Stump, E., Wise, W. and the 10th ISAES editorial team (eds) *Antarctica: A Keystone in a Changing World*. National Academy of Sciences, The National Academies Press, 91–114.

Siddoway, C.S. and Fanning, C.M. 2009. Paleozoic tectonism on the East Gondwana margin: evidence from SHRIMP U–Pb zircon geochronology of a migmatite–granite complex in West Antarctica. *Tectonophysics*, **477**, 262–277, https://doi.org/10.1016/j.tecto.2009.04.021

Siddoway, C.S., Sass, III, L.C. and Esser, R. 2005. Kinematic history of Marie Byrd Land terrane, West Antarctica: direct evidence from cretaceous mafic dykes. *In*: Vaughan, A., Leat, P. and Pankhurst, R.J. (eds) *Terrane Processes at the Margin of Gondwana*. Geological Society, London, Special Publications, **246**, 417–438, https://doi.org/10.1144/GSL.SP.2005.246.01.17

Skemer, P., Katayama, I., Jiang, Z. and Karato, S. 2005. The misorientation index: development of a new method for calculating the strength of lattice-preferred orientation. *Tectonophysics*, **411**, 157–167, https://doi.org/10.1016/j.tecto.2005.08.023

Smellie, J.L., Martin, A.P., Panter, K.S., Kyle, P.R. and Geyer, A. 2020*a*. Magmatism in Antarctica and its relation to Zealandia. *New Zealand Journal of Geology and Geophysics*, **63**, 578–588, https://doi.org/10.1080/00288306.2020.1781666

Smellie, J.L., Panter, K.S. and Geyer, A. (eds). 2020*b*. *Volcanism in Antarctica: 200 Million Years of Subduction, Rifting and Continental Break-Up, 53*. Geological Society, London, Memoirs, **55**, https://doi.org/10.1144/M55

Soustelle, V., Tommasi, A., Demouchy, S. and Ionov, D.A. 2010. Deformation and fluid–rock interaction in the supra-subduction mantle: microstructures and water contents in peridotite xenoliths from the Avacha Volcano, Kamchatka. *Journal of Petrology*, **51**, 363–394, https://doi.org/10.1093/petrology/egp085

Soustelle, V., Tommasi, A., Demouchy, S. and Franz, L. 2013. Melt-rock interactions, deformation, hydration and seismic properties in the sub-arc lithospheric mantle inferred from xenoliths from seamounts near Lihir, Papua New Guinea. *Tectonophysics*, **608**, 330–345, https://doi.org/10.1016/j.tecto.2013.09.024

Storey, B., Leat, P.T., Weaver, S.D., Pankhurst, R.J., Bradshaw, J.D. and Kelley, S. 1999. Mantle plumes and Antarctica–New Zealand rifting: evidence from mid-Cretaceous mafic dykes. *Journal of the Geological Society, London*, **156**, 659–671, https://doi.org/10.1144/gsjgs.156.4.0659

Storti, F., Rossetti, F. and Salvini, F. 2007. Structural architecture and dis- placement accommodation mechanisms at the termination of the Priestley fault, northern Victoria Land, Antarctica. *Tectonophysics*, **341**, 141–161, https://doi.org/10.1016/S0040-1951(01)00198-6

Taylor, W.R. 1998. An experimental test of some geothermometer and geobarometer formulations for upper mantle peridotites with application to the thermobarometry of fertile lherzolite and garnet websterite. *Neues Jahrbuch für Geologie und Paläontologie - Abhandlungen*, **172**, 381–408, https://doi.org/10.1127/njma/172/1998/381

ten Brink, U.T., Hackney, R.I., Bannister, R.I.S., Stern, T.A. and Makovsky, Y. 1997. Uplift of the Transantarctic Mountains and the bedrock beneath the East Antarctic ice sheet. *Journal of Geophysical Research*, **102**, 603–627, https://doi.org/10.1029/97JB02483

Tommasi, A. and Ishikawa, A. 2014. Microstructures, composition, and seismic properties of the Ontong Java Plateau mantle root. *Geochemistry, Geophysics, Geosystems*, **15**, 4547–4569, https://doi.org/10.1002/2014GC005452

Tommasi, A. and Mameri, L. 2020. Textural and compositional changes in the lithospheric mantle atop the Hawaiian plume: consequences to seismic properties. *Earth and Space Science Open Archive*, https://doi.org/10.1002/essoar.10502936.1

Tommasi, A., Tikoff, B. and Vauchez, A. 1999. Upper mantle tectonics: three-dimensional deformation, olivine crystallographic fabrics and seismic properties. *Earth and Planetary Science Letters*, **168**, 173–186, https://doi.org/10.1016/S0012-821X(99)00046-1

Tommasi, A., Godard, M., Coromina, G., Dautria, J.-M. and Barsczus, H. 2004. Seismic anisotropy and compositionally induced velocity anomalies in the lithosphere above mantle plumes: a petrological and microstructural study of mantle xenoliths from French Polynesia. *Earth and Planetary Science Letters*, **227**, 539–556, https://doi.org/10.1016/j.epsl.2004.09.019

Tommasi, A., Vauchez, A. and Ionov, D.A. 2008. Deformation, static recrystallization, and reactive melt transport in shallow subcontinental mantle xenoliths (Tok Cenozoic volcanic field, SE Siberia). *Earth and Planetary Science Letters*, **272**, 65–77.

Ulrich, S. and Mainprice, D. 2005. Does cation ordering in omphacite influence development of lattice-preferred orientation? *Journal of Structural Geology*, **27**, 419–431, https://doi.org/10.1016/j.jsg.2004.11.003

Upton, B.G.J., Downes, H., Kirstein, L.A., Bonadiman, C., Hill, P.G. and Ntaflos, T. 2011. The lithospheric mantle and lower crust-mantle relationships under Scotland: A xenolithic perspective. *Journal of the Geological Society, London*, **168**, 873–886, https://doi.org/10.1144/0016-76492009-172

van der Wal, W., Whitehouse, P.L. and Schrama, E.J.O. 2015. Effect of GIA models with 3D composite mantle viscosity on GRACE mass balance estimates for Antarctica. *Earth and Planetary Science Letters*, **414**, 134–143, https://doi.org/10.1016/j.epsl.2015.01.001

Van Roermund, J. and Boland, H. 1981. The dislocation substructures of naturally deformed omphacites. *Tectonophysics*, **78**, 403–418, https://doi.org/10.1016/0040-1951(81)90022-6

Vauchez, A., Tommasi, A., Barruol, G. and Manumus, J. 2000. Uppermost mantle deforma- tion and seismic anisotropy in continental rifts. *Physics and Chemistry of the Earth*, **25**, 111–117, https://doi.org/10.1016/S1464-1895(00)00019-3

Walter, M.J., Sisson, T.W. and Presnall, D.C. 1995. A mass proportion method for calculating melting reactions and application to melting of model upper mantle lherzolite. *Earth and Planetary. Science. Letters*, **135**, 77–90, https://doi.org/10.1016/0012-821X(95)00148-6

Wernicke, B. 1981. Low angle normal faults in Basin and Range province. Nappe tectonics in an extending orogen. *Nature*, **291**, 645–648, https://doi.org/10.1038/291645a0

Wernicke, B. 1985. Uniform-sense normal simple shear of the continental lithosphere. *Canadian Journal of Earth Sciences*, **22**, 108–125, https://doi.org/10.1139/e85-009

White-Gaynor, A.L., Nyblade, A.A. *et al.* 2019. Heterogeneous upper mantle structure beneath the Ross Sea embayment and Marie Byrd Land, West Antarctica, revealed by P-wave tomography. *Earth and Planetary Science Letters*, **513**, 40–50, https://doi.org/10.1016/j.epsl.2019.02.013

Wiens, D.A., Shen, W. and Lloyd, A. 2021. The seismic structure of the Antarctic upper mantle. *Geological Society, London, Memoirs*, **56**, https://doi.org/10.1144/M56-2020-18

Winberry, J.P. and Anandakrishnan, R. 2004. Crustal structure of the West Antarctic rift system and Marie Byrd Land hotspot. *Geology*, **32**, 977–980, https://doi.org/10.1130/G20768.1

Workman, R.K. and Hart, S.R. 2005. Major and trace element composition of the depleted MORB mantle (DMM). *Earth and Planetary Science Letters*, **231**, 53–72, https://doi.org/10.1016/j.epsl.2004.12.005

Wörner, G. 1999. Lithospheric dynamics and mantle sources of alkaline magmatism of the Cenozoic West Antarctic Rift system. *Global and Planetary Change*, **23**, 61–77, https://doi.org/10.1016/S0921-8181(99)00051-X

Zaffarana, C., Tommasi, A., Vauchez, A. and Grégoire, M. 2014. Microstructures and seismic properties of south Patagonian mantle xenoliths (Gobernador Gregores and Pali Aike). *Tectonophysics*, **621**, 175–197, https://doi.org/10.1016/j.tecto.2014.02.017

Thermal regime and state of hydration of the Antarctic upper mantle from regional-scale electrical properties

Philip E. Wannamaker[1]*, John A. Stodt[2], Graham J. Hill[3,4], Virginie Maris[1] and Michal A. Kordy[1]
[1]University of Utah/Energy & Geoscience Institute, 423 Wakara Way, Suite 300, Salt Lake City, UT 84108, USA
[2]Numeric Resources LLC, PO Box 58195, Salt Lake City, UT 84158, USA
[3]Institute of Geophysics, Czech Academy of Sciences, Prague 4, Czech Republic
[4]Gateway Antarctica, University of Canterbury, Christchurch 8140, New Zealand
PEW, 0000-0001-5744-6474; GJH, 0000-0003-0058-3173
*Correspondence: pewanna@egi.utah.edu

Abstract: Large-scale electrical resistivity investigations of the Antarctic crust and upper mantle utilizing the magnetotelluric method (MT) are limited in number compared to temperate regions, but provide physical insights hard to achieve with other techniques. Key to the method's success are the instrumentation advances that allow microvolt (μV)-level measurements of the MT electric field in the face of mega-ohm (MΩ) contact resistances. Primarily in this chapter, we reanalyse existing data from three campaigns over the Antarctic interior using modern 3D non-linear inversion analysis, and offer additional geophysical conclusions and context beyond the original studies. A profile of MT soundings over the transitional Ellsworth–Whitmore block in central West Antarctica implies near-cratonic lithospheric geothermal conditions with interpreted graphite–sulfide horizons deformed along margins of high-grade silicate lithological blocks. Reanalysis of South Pole soundings confirms large-scale low resistivity spanning Moho depths that is consistent with limited seismic tomography and elevated crustal thermal regime inferences. Upper mantle under a presumed adiabatic thermal gradient below the Ross Ice Shelf near the central Transantarctic Mountains appears to be of a moderately hydrated state but not sufficient to induce melting. The degree of hydration there is comparable to that below the north-central Great Basin province of the western USA.

Supplementary material: pdf images of magnetotelluric sounding responses with inversion model fits for data sites from central West Antarctica (CWA), South Pole and central Transantarctic Mountains (CTAM) projects, plus a demonstration test of high-impedance pre-amplifier fidelity, are available at https://doi.org/10.6084/m9.figshare.c.5332327

Regional electrical resistivity investigations of the Antarctic crust and upper mantle using the magnetotelluric (MT) technique with resolution comparable to temperate region studies have advanced over the past two decades and yield physical insights difficult to obtain with other techniques (Wannamaker *et al.* 2017; Hill 2020). Resistivity (or its commonly considered inverse, conductivity) shows dependencies on temperature, hydration and fluid phases including melts in a manner that is complementary to other properties such as seismic velocity (Unsworth and Rondenay 2013; Tyburczy and Du Frane 2015). It also has demonstrated potential for terrane boundary and assembly analysis through deeply subsumed, formerly organic-bearing lithologies in Precambrian orogens with context provided by temperate terrain studies (e.g. Wannamaker *et al.* 2017; Wunderman *et al.* 2018).

Acquiring high-fidelity MT data, which derive from naturally occurring electromagnetic (EM) field variations caused by global solar processes, necessitates specialized instrumentation in the way of buffer amplifiers to counteract the high contact resistance of polar ice (firn) (Wannamaker *et al.* 2004, 2017; also see the Supplementary material). Extra attention also needs to be paid to possible violations of the MT method assumption of planar EM source fields due to the high magnetic latitudes where ionic disturbances may be channelled (Akasofu 1977, 2015), but this may be checked through comparisons of processed data at high and low activity times (Wannamaker *et al.* 2017; Hill 2020). Hence, the way forward to executing successful MT surveying and interpretation on the Antarctic continent is now largely established.

This paper focuses upon a more rigorous modelling of three existing MT datasets within the Antarctic interior using 3D non-linear inversion and an update of their implications in light of recent independent studies (Fig. 1). Two of these are a profile of 10 soundings in central West Antarctica (CWA) over lithosphere of enigmatic stability, and a similar-size profile near South Pole over lithosphere that has largely been considered stable. The third set is a slightly modified section view through the West to East Antarctic transition at the central Transantarctic Mountains (TAM) aimed at illuminating upper-mantle hydration especially below the Ross Ice Shelf (RIS) of West Antarctica. The large wave period bandwidth of the MT method allows a unified observation of tectonic processes and element transfer from upper-mantle source volumes through crustal residence and near-surface dispersal. As the observed MT responses here and in the Supplementary material show, significant 3D character exists in the on-diagonal impedance amplitudes and cross-strike vertical magnetic field variations (Chave and Jones 2012).

MT profiling in central West Antarctica: active rifting v. stable block

Multistage extension since the late Cretaceous has characterized West Antarctica generally, although current volcanic activity is concentrated in northern Victoria Land and Marie Byrd Peninsula (Hole and LeMasurier 1994; Behrendt 1999; LeMasurier 2008). The West Antarctic interior today is considered to be fairly cool and inactive, as signified by its low elevation, but local volcanic activity has been conjectured based upon aeromagnetic, altimetry and airborne radar data (Blankenship *et al.* 1993, 2001; Behrendt *et al.* 1994; LeMasurier 2008). A dedicated geophysical field camp, denoted Central West Antarctica (CWA), was established in the 1994–95 austral summer over the grid northern Byrd Subglacial Basin area as a component of the ANTALITH project (Blankenship *et al.* 2001; Dalziel and Lawver 2001). Located *c.* 100 km grid SE of a prominent geophysical anomaly suggestive of a volcanic edifice, activities included seismic reflection and refraction, gravity, and MT (Wannamaker *et al.* 1996; Clarke *et al.* 1997; Sen *et al.* 1998).

During this camp, 10 MT stations spaced 6 km apart in a profile orientated just west of grid north were acquired using

From: Martin, A. P. and van der Wal, W. (eds) 2023. *The Geochemistry and Geophysics of the Antarctic Mantle.*
Geological Society, London, Memoirs, **56**, 181–194,
First published online 28 April 2021, https://doi.org/10.1144/M56-2020-4

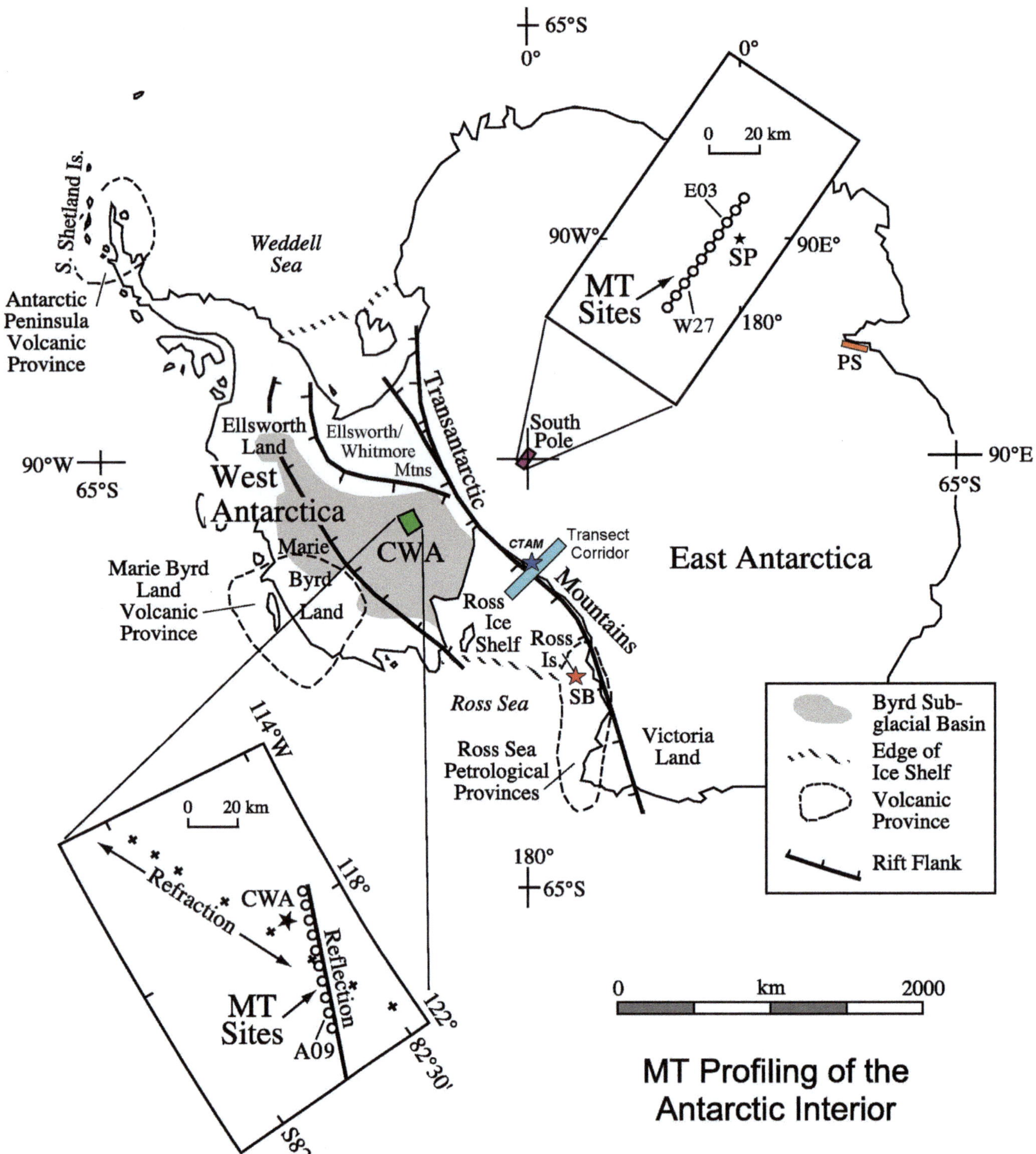

Fig. 1. Field site map for MT experiments of the Antarctic interior. The three surveys considered herein are at central West Antarctica (CWA), South Pole (SP) and the West to East Antarctic transition at the central Transantarctic Mountains (CTAM, cyan bar). Ross Island with Scott Base (SB) is denoted with a red star. The CWA and SP sites discussed in text are labelled in their insets. Grid north is up. The MT profiling of Peacock and Selway (2016) in northeastern East Antarctica is denoted with a light brown rectangle (PS). Map modified from LeMasurier and Rex (1991) and Wannamaker *et al.* (2004).

instrumentation constructed in-house at the University of Utah (Fig. 1). Profile sites were acquired in pairs, with synchronized MT time series from each transmitted via FM telemetry to a central recording hut. Perforated titanium metal sheets were used as electrodes at the ends of 50 m bipoles in a positive array, with the buffer amplifiers attached to the sheets via *c.* 30 cm-long ‘pigtail’ wires and banana plugs (for recent representative pictures, see Wannamaker *et al.* 2017; Hill 2020; see also the Supplementary material). The sites in each pair were cross-remote referenced, with spectral analysis via cascade decimation, and time series subsegments were sorted in quality according to spectral multiple coherence.

In the original study, the collected MT responses were forward-modelled using a finite-element algorithm developed in-house (Wannamaker *et al.* 1996, fig. 4). This provided a good fit to the data and resolved a lithosphere of generally high resistivity (2000–3000 Ω m) to *c.* 100 km depth, beyond which resistivity dropped modestly to a nominal value of *c.*

400 Ω m. These high values were taken to indicate only a subdued or fossil state of rifting in the study area. Embedded in the resistive host were two narrow concentrations of low resistivity in the middle crust located near the extremal ends of the profile, with a modest drop in resistivity south of centre. These were taken to have non-thermal causes such as local concentrations of graphite or sulfides caught up in subsequent orogeny. This interpretation of the MT results was the first geophysical argument that the thermal state below this area of West Antarctica is cold; the area is viewed now as lying on transitional Ellsworth–Whitmore Mountains lithosphere (Blankenship *et al.* 2001).

Although 2D forward modelling yielded insightful results, such an approach provides only a visually satisfying fit to a portion of the data and does not make use of the entire MT tensor. Hence, we have reanalysed these data using a fully 3D inversion involving all four elements of the MT impedance and both elements of the vertical magnetic field (tipper). The inversion is carried out using the deformable finite-element platform of Kordy *et al.* (2016*a*, *b*), which utilizes all direct solvers for the non-linear parameter step. The inverse solution seeks to fit the observations while simultaneously damping model roughness in a spatial slope sense. A wireframe view of the finite-element mesh appears in Figure 2. The starting model was a 600 Ω m half-space, which is close to the average of all observed apparent resistivities, below a fixed layer of ice 1500 m thick of 100 000 Ω m. Using error floors of $7.5\%|Z_{xy}-Z_{yx}|/2$ for all impedances and 0.1 for tipper elements, solution convergence is shown in Figure 3 achieving a normalized root-mean-square (nRMS) misfit of 1.38 where unity is ideal. An example sounding fit showing all MT data elements appears in Figure 4, with all soundings and fits provided in the Supplementary material. We note that the tipper elements were generally noisy in this set due mainly to wind and so did not have a large influence on the inversion.

The resultant model is depicted in Figures 5 and 6 in section and plan views. The sections show a reasonable resemblance to the original forward model of Wannamaker *et al.* (1996) within a smoothed framework. Low resistivity in the middle crust forms in a clumped texture with a general concentration near the distal ends of the profile. To some extent, the clumping may be a response to finer-scale anisotropy in the middle crust when the data are inverted assuming isotropic properties (e.g. Wannamaker 2005), so that the precise form and location of the clumps may depend on minor characteristics of the data. Our previous conclusion remains that these conductors represent graphitic or sulfide-bearing horizons metamorphosed and 'frozen' in earlier orogeny. The crustal conductors resolved near Prydz Bay, East Antarctica coastline, by Peacock and Selway (2016) (Fig. 1) would have similar causes, although the terrane assembly history differs. Since the study of Wannamaker *et al.* (1996), the importance of such elements worldwide in affecting resistivity structure and their potential for clarifying terrane assembly has developed (e.g. Boerner *et al.* 1996; Wunderman *et al.* 2018; and numerous others).

Multi-depth plan views in Figure 6 show the mid-crustal low resistivity approaching shallower depths in both directions offline. While some of this may be a consequence of inversion smoothing, the effect is also to boost the *xy*-mode of apparent resistivity as observed at upper-middle periods (0.1–1 s) (Fig. 4). At greater crustal depths, a rough alignment of low resistivity normal to the profile emerges. In general, because the MT data are restricted to a single profile, image portions off-profile become increasingly smooth and lack details. Even so, 3D inversion yields a useful sense of structural gradients or approximate locations, and the asymmetry of the models in plan views testifies to the presence of 3D effects.

The deeper-scale section view (Fig. 5) shows resistivities diminishing below *c.* 100 km in the upper mantle in accord with the original forward modelling. The resistivity decrease is slight, however, consistent with prior forward tests that a drop to 500 Ω m at 100 km suffices to simulate the subtle roll-over in apparent resistivity at the longest periods. Along an average mantle adiabat with potential temperature $T_p=$ 1600 K (Stixrude and Lithgow-Bertelloni 2011), temperature at 100 km depth should approach 1400°C such that even dry olivine-dominated mineralogy should have a bulk resistivity of *c.* 200 Ω m (Novella *et al.* 2017). Hence, we confirm with new analysis that the lithospheric thickness (depth to adiabat) in this area of CWA should be *c.* 100 km or more, consistent with seismic Rayleigh wave-velocity profile estimates for the Ellsworth–Whitmore block in O'Donnell *et al.* (2019), and represents near-stable tectonic conditions. We concur with

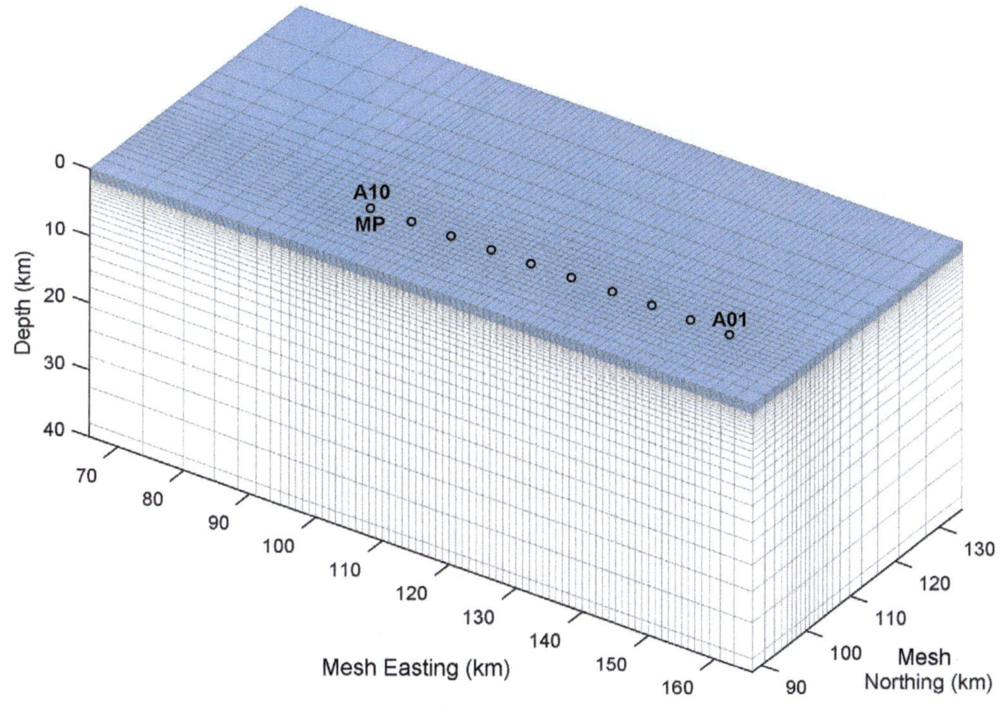

Fig. 2. Central portion of finite-element mesh used for inversion of the central West Antarctica MT profile. The ice layer is depicted in light blue. The location of site A10 is also that of the vertical upper-mantle profile (MP) of resistivity utilized in Figure 14.

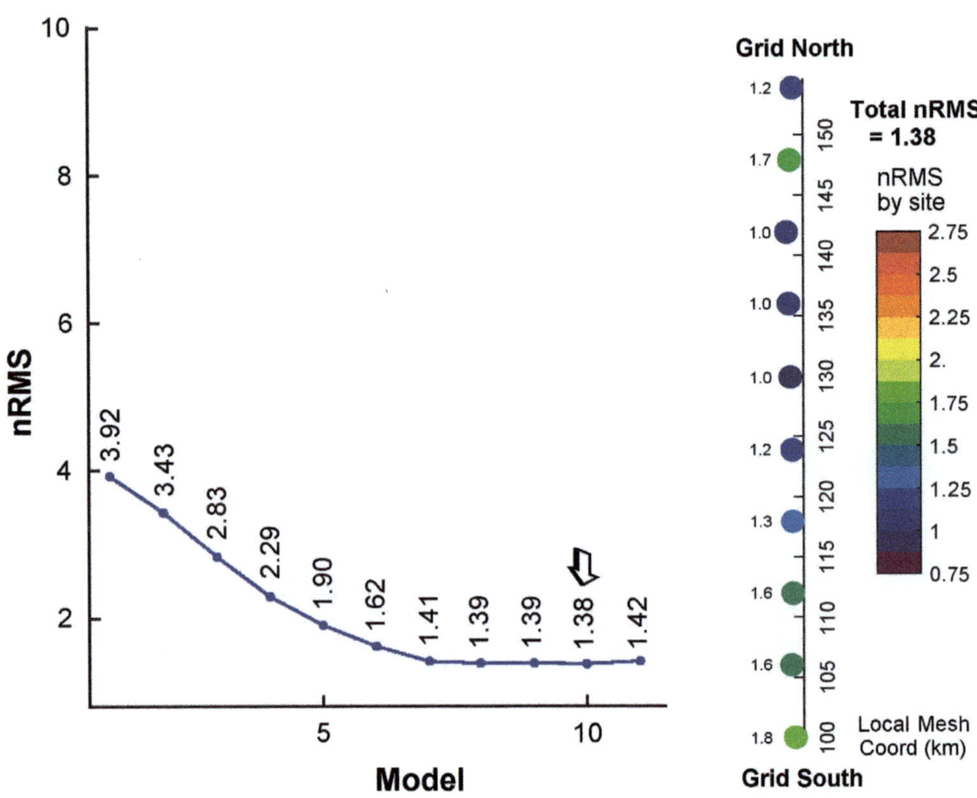

Fig. 3. Convergence plot for non-linear iterative inversion of the CWA MT profile. Model 10 is selected (arrow). Together with the nRMS for the entire dataset, nRMS values for each station are also plotted as coloured circles.

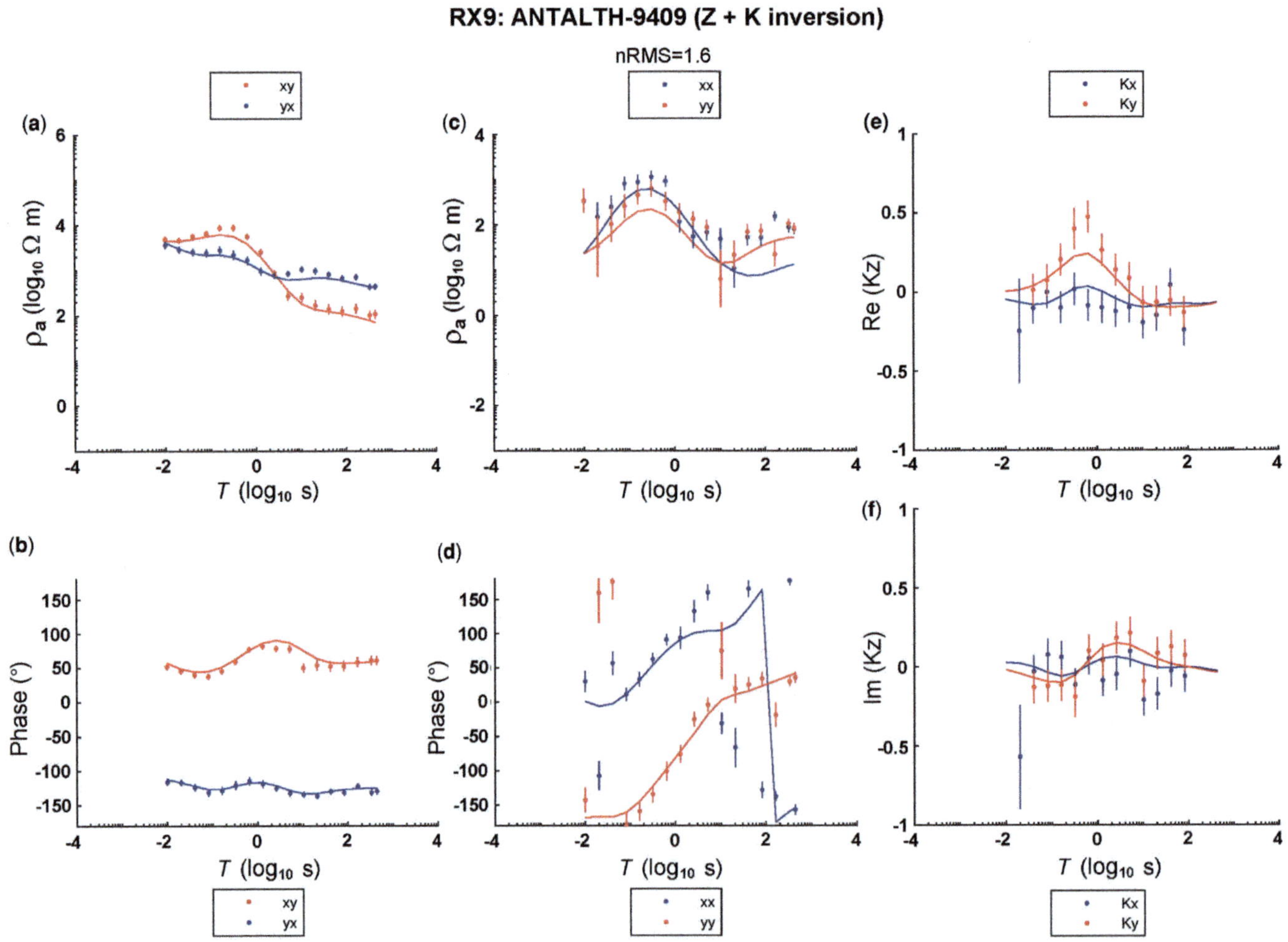

RX9: ANTALTH-9409 (Z + K inversion)

nRMS=1.6

Fig. 4. Example sounding data and computed response from the CWA inversion model of Figures 5 and 6. The *x*-direction is normal to the profile in the grid west direction, while the *y*-direction is along the profile towards grid north, approximately. This site is next to the grid south end of the profile (Wannamaker *et al.* 1996). The full suite of soundings and responses is presented in the Supplementary material.

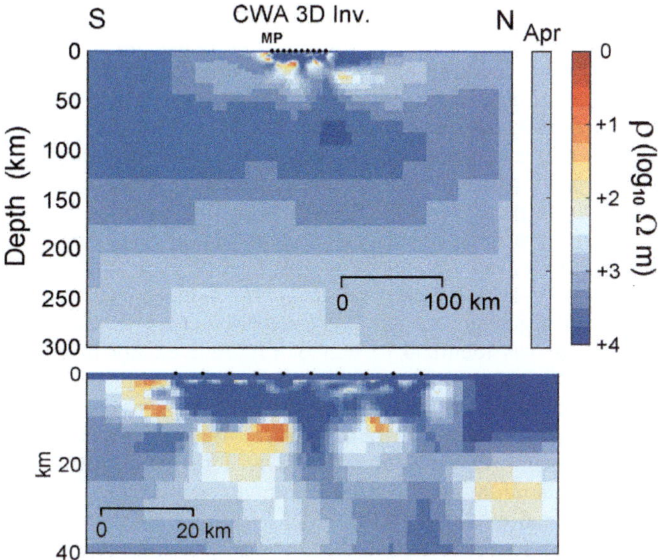

Fig. 5. Section view at two scales through the 3D inversion model of the central West Antarctica MT profile. Label MP denotes the location of the example vertical resistivity profile into the upper mantle for synthesis with other study areas shown in Figure 14. Column Apr denotes the starting and *a priori* model, which was a uniform 600 Ω m half-space below the ice layer.

other views that any late Cenozoic volcanism is restricted to occasional, narrow zones of weakness originating in Cretaceous rifting and reactivated during more recent activity (Hole and LeMasurier 1994; Blankenship *et al.* 2001). Additional MT data coverage to broaden the aperture and extend the period range is needed to probe the asthenosphere accurately here.

Active tectonic conditions beneath South Pole region suggested by MT

With a distance of over 400 km from the Transantarctic Mountains range front and no surface indication of Cenozoic activity, South Pole region in East Antarctica might be judged as a stable lithospheric regime of low heat flow. However, ice-stripped, rebounded basement elevations of much of East Antarctica are anomalously high (>1 km) suggesting some kind of active tectonic support (Bentley 1991). Until relatively recently, seismic tomography had difficulty resolving structure here due to sparse seismometer coverage and low earthquake activity. This caused tomographic models with a strong gradient in velocity beneath South Pole approximately parallel to the rangefront, with the lower velocities towards the West Antarctica side (Danesi and Morelli 2001; Ritzwoller *et al.* 2001).

To test whether a cratonic state characterizes South Pole region, MT profiling there was carried out in the 1997–98

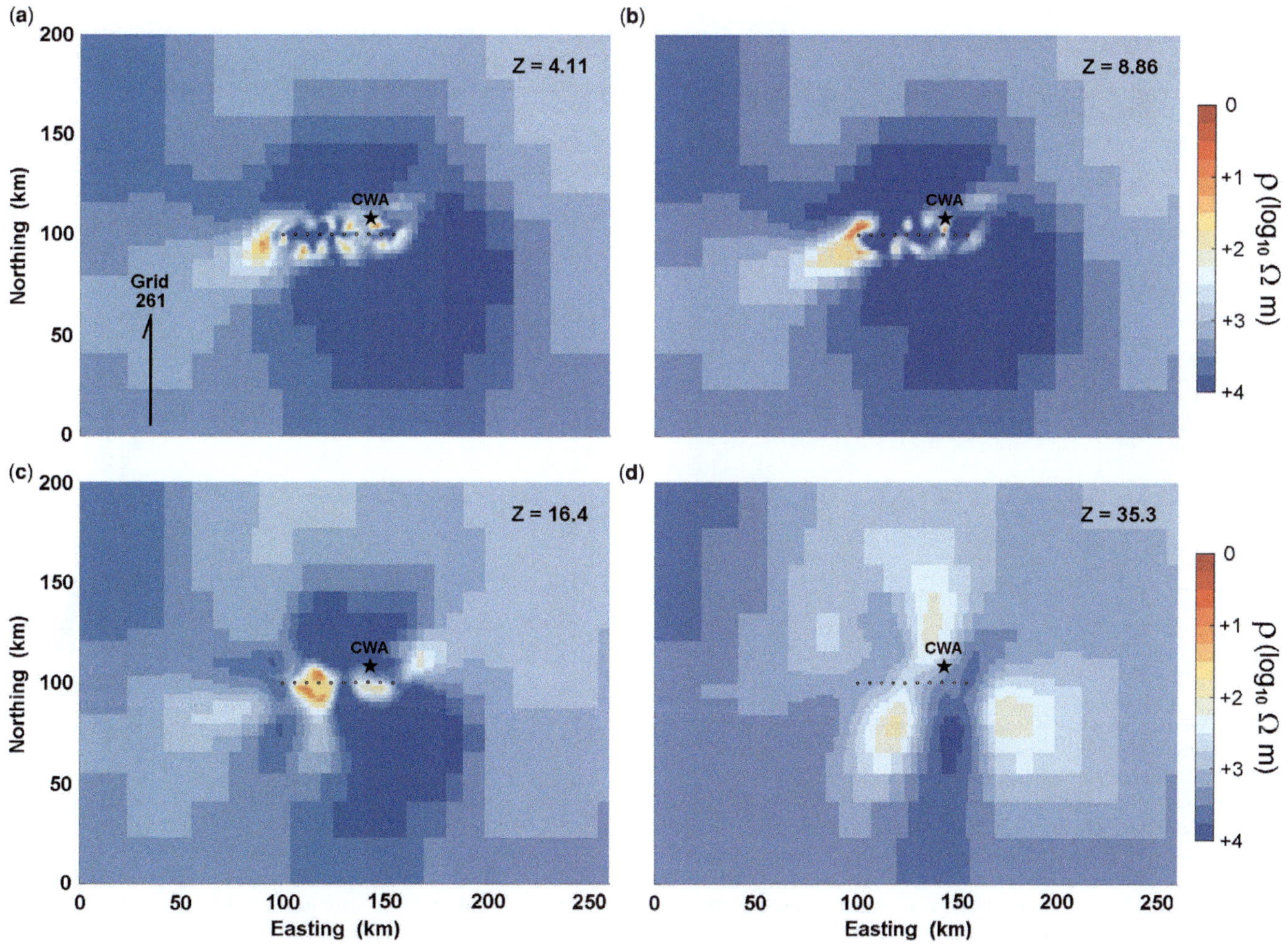

Fig. 6. Plan views at four depths through the 3D inversion model of the central West Antarctica MT profile. Depth labels are in km. Northing and easting values are local to the finite-element mesh. The grid north orientation is shown in (a).

austral summer (Wannamaker *et al.* 2004) (Fig. 1). The acquisition mode was essentially identical to that at CWA previously, having 10 sites spaced 6 km apart taken in pairs with time series telemetered to a central recording hut (Fig. 7). Recording was challenging in this campaign due to persistent wind-generated noise from statically charged firn particles blowing along the electric bipoles coupled with generally low resistivity at depth due to an apparent thick sedimentary layer under the ice, which reduces the strength of the electric field signal. Nevertheless, final data quality was good after long data stacking times.

In the original South Pole study, the collected MT responses were inverted using a 2D non-linear finite-element algorithm incorporating a direct Gauss–Newton parameter step (Wannamaker *et al.* 2004). The inversion emphasized the so-called transverse magnetic (TM) mode (electric current flowing along the profile) since it is generally accepted as less prone to 3D effects. One feature resolved was an undulatory, thin, low-resistivity layer at the base of the 2900 m-thick ice sheet taken to be Devonian Beacon Group clastic sediments that are widespread in East Antarctica. The layer explains the lack of seismic energy returns in early refraction experiments in South Pole area (Bentley 1991; Wannamaker *et al.* 2004). The deeper crust commonly showed higher resistivity as expected for mainly crystalline lithologies, but imaged below was a striking low-resistivity zone 20 km or more thick beginning in the lowermost crust and extending into the uppermost mantle. Wannamaker *et al.* (2004) concluded that cryptic enhanced thermal activity was taking place in the deepest crust below South Pole, which we believe to be the first interpretation of such a state.

We have reanalysed these data as well in a fully 3D inversion involving all four elements of the MT impedance and both elements of the vertical magnetic field (tipper) using the same algorithm as for CWA (Kordy *et al.* 2016*a*, *b*). The mesh was identical to that of Figure 2 except that the ice thickness was 2900 m, achieved by six additional layers of elements. Two starting models were considered: a 100 Ω m half-space, which is near the average of the data; and a smooth 1D variation from inverting a lateral average of the data along the profile (Wannamaker *et al.* 2004). Using error floors of $7.5\%|Z_{xy}–Z_{yx}|/2$ for all impedances and 0.05 for tipper elements, solution convergence for the half-space initial model is shown in Figure 8 achieving an nRMS misfit of 1.40 where unity is ideal. Example sounding fits showing all MT data elements appear in Figure 9. We note that the tipper elements were generally of high quality in this dataset and could contribute to the solution.

The resultant model cross-sections for the two starting models are depicted in Figure 10, and the plan views for the half-space starting model are shown in Figure 11. The section views show a reasonable resemblance to the original 2D inversion of Wannamaker *et al.* (2004) with some notable differences. There remains some conductive layering directly below the ice sheet that was previously correlated with Beacon Group sediments. The 3D inversion also shows low-resistivity 'blobs' residing in the middle crust, especially under the southwestern half of the profile, which also appeared in the original 2D inversion in a smoother form (Wannamaker *et al.* 2004). The precise location of fine details in this area has a modest dependence on the starting model (Fig. 10). They were suggested to be possible metamorphic graphitic material similar to that described for CWA, but this is not considered unique. Under the northeastern half of the section, higher resistivity

Fig. 7. (**a**) View of the central MT recording hut midway between the MT instrument sites at South Pole area, East Antarctica, during the 1997–98 field season. The hut would be pulled along the profile at the conclusion of paired recording using a Tucker Sno-Cat (pictured). The receiving radio mast is visible on the right. (**b**) View of a Nansen sled carrying MT instrument components at a nominal distance of 3 km from the central recording hut. The transmitting radio mast is visible on the right edge. Inset shows the high-impedance buffer preamplifier at that time used to treat the high contact resistance of the firn to electric field measurements. It is connected to the perforated titanium sheet electrode, the corner of which projects from the snow in this view. See Wannamaker *et al.* (2017) for other electrode photographs.

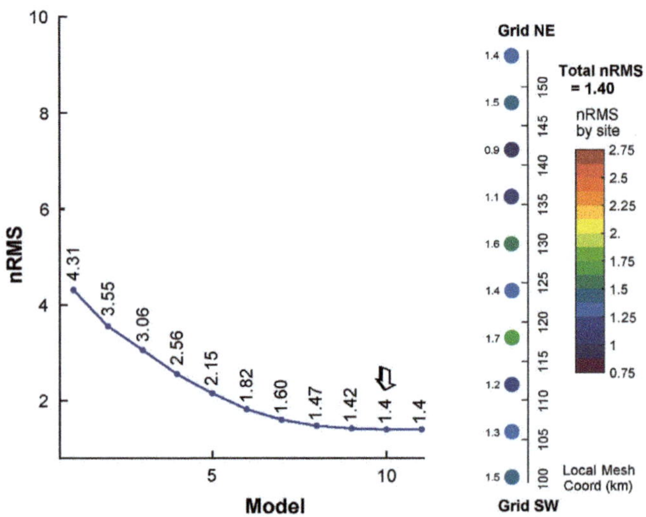

Fig. 8. Convergence plot for non-linear iterative inversion of the South Pole MT profile using the half-space initial model. Model 10 is selected (arrow). Together with the nRMS for the entire dataset, nRMS values for each station also also plotted as coloured circles. Convergence for the smooth 1D background starting/*a priori* model was similar but only achieved a final nRMS of 1.48.

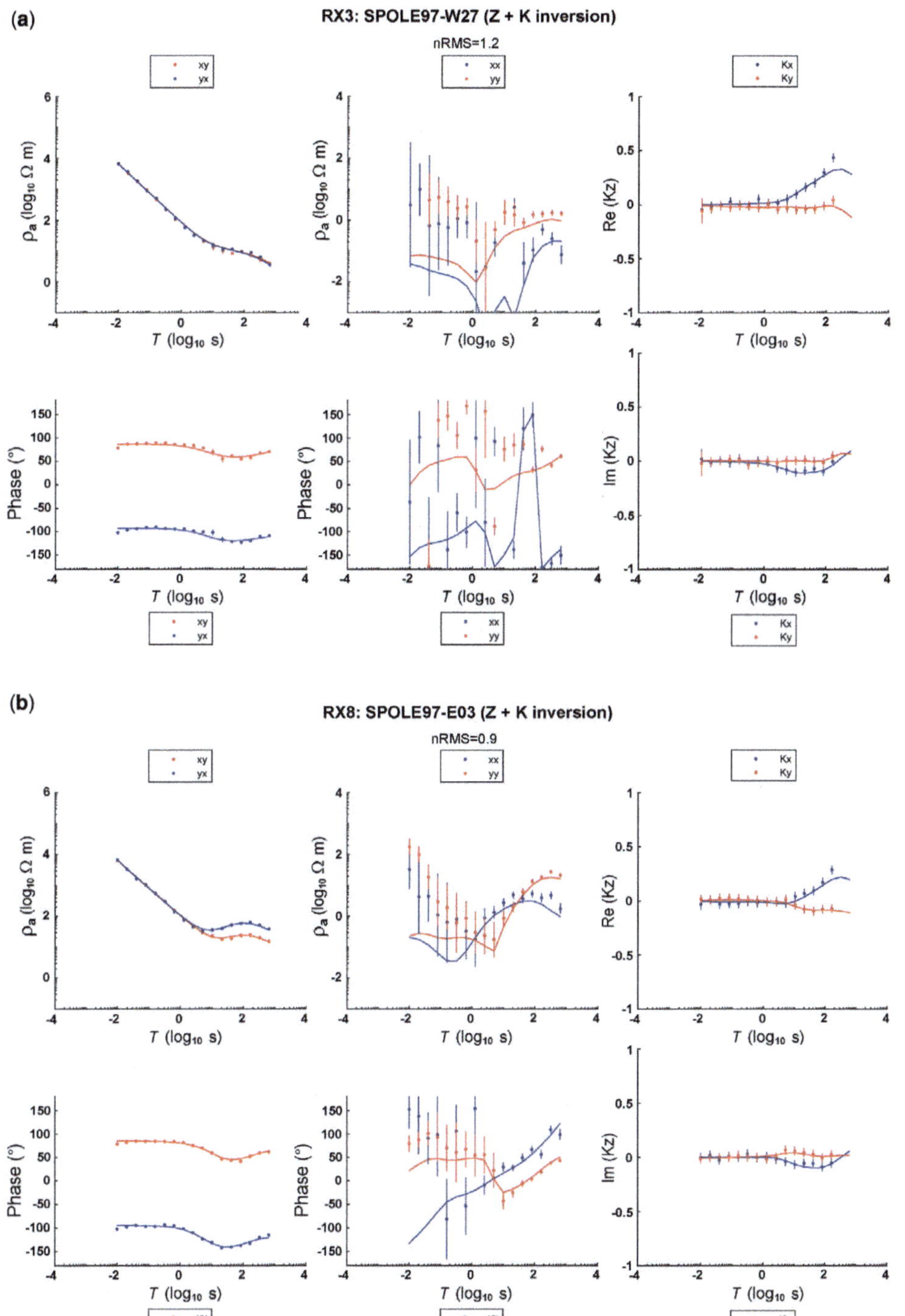

Fig. 9. Example sounding data and computed responses from South Pole inversion model of Figure 10 using the half-space starting model. The x-direction is normal to the profile in the grid NW direction, while the y-direction is along the profile towards grid NE, approximately. These sites are towards the grid SW and NE ends of the profile (Wannamaker *et al.* 2004). Note that the tipper response is almost entirely contained in the non-2D K_{zx}. The full suite of soundings and responses is presented in the the Supplementary material.

in the deeper levels of the crust appears as in the 2D model, taken to denote more purely silicate lithologies.

Most noteworthy we believe, the lowermost crust is of markedly low resistivity largely similar to the older 2D inversion model. There is some apparent upward dip to grid NE (away from the TAM) in the section view of Figure 10, although the requirement for this is not explicitly tested. Plan views with depth reveal that this low resistivity merges with a large conductor towards grid SE of the profile centre to effect a gross total dip towards the west (Fig. 11). High conductivity to the SE was inferred also by Wannamaker *et al.* (2004, fig. 10) through inversion of cross-strike tipper element K_{zx} constrained by an integrated impedance sounding. The 3D analysis suggests one large oblique low-resistivity feature. Limited aperture of our survey prevents further lateral extrapolation of the deep crustal structure.

The preferred explanation for the large lower-crustal conductor by Wannamaker *et al.* (2004) was cryptic thermal activity in the lowermost crust and uppermost mantle, specifically underplating of mafic melts and subsequent fluid release. This process was described in detail for the actively extensional Great Basin by Wannamaker *et al.* (2008). Given no evident crustal-scale extension below South Pole, a more obscure localized upper-mantle convection process would be acting. The identified presence of subglacial lakes below South Pole area suggests that heat flow may be at least 50 mW m^{-2} (Siegert 2017). Jordan *et al.* (2018) concluded from radar analysis of internal ice-sheet stratigraphy that a geothermal heat flux below South Pole of *c.* 120 mW m^{-2} was possible, at least locally, although they attributed it to heat-producing basement elements and hydrothermal focusing along fault zones.

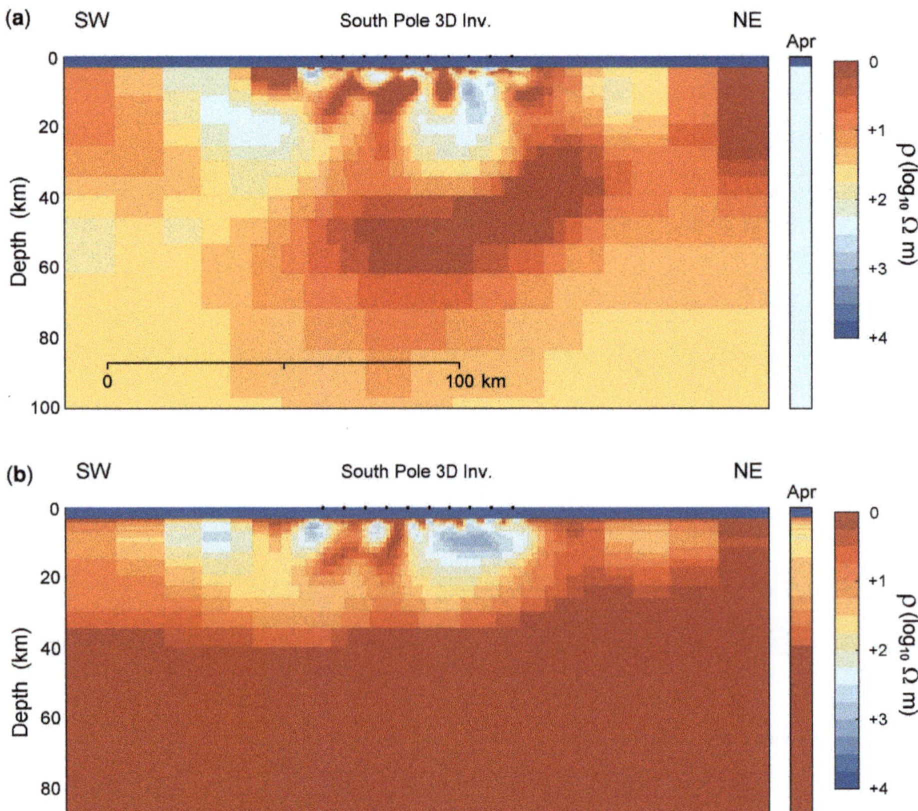

Fig. 10. Section views using two different starting/*a priori* models (Apr) through the 3D MT inversion model of South Pole region MT profile. Upper model (**a**) used a 100 Ω m half-space start, while the bottom model (**b**) used a smooth 1D start derived from 1D inversion of integrated TM mode impedance (Wannamaker *et al.* 2004).

The more recent establishment of several interior seismic arrays has improved velocity resolution substantially over the Antarctic continent. The P-wave tomography model of Hansen *et al.* (2014) shows a localized low-velocity zone in the upper mantle below South Pole. Combining S-wave teleseismic events with ambient noise recording and receiver functions, Shen *et al.* (2017) offer a model with low uppermost-mantle S-wave velocity under the entire southern TAM extending almost to South Pole. This was suggested to represent a modern lithospheric delamination event involving upwelling of asthenosphere to near Moho levels with possible melt production and underplating. If our resistivity interpretation is correct, the MT experiment was the first geophysical glimpse that a magmatic thermal event may be taking place

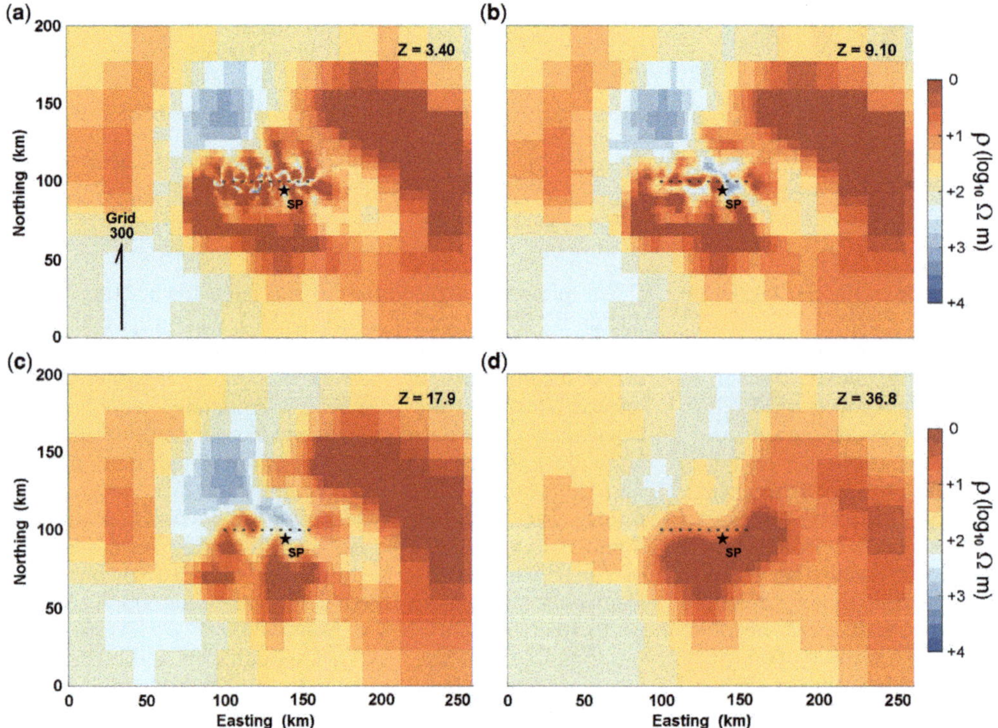

Fig. 11. Plan views at four depths through the 3D inversion model of South Pole region MT profile using the half-space starting model. The grid north orientation is denoted in (a).

under South Pole region. However, we would certainly like to have greater aperture of MT data coverage to better distinguish the contributions of thermally activated processes compared to fossil terrain suturing structure bearing graphite–sulfide metasediments (e.g. Boerner *et al.* 1996; Wunderman *et al.* 2018).

Evolution of upper-mantle resistivity structure across the central Transantarctic Mountains

The MT survey within the interior of the Antarctic continent having by far the greatest aperture while also obtaining good lateral sampling is the 550 km-long transect across the central Transantarctic Mountains (CTAM) acquired by Wannamaker *et al.* (2017) (Fig. 1). The survey sought to illuminate the tectonic transition from late Mesozoic–Cenozoic rifted West Antarctica beneath the RIS through the TAM rift shoulder and into cratonic East Antarctica. A particular goal was to compare the inferred physical state with that of currently extending margins of the US Great Basin, where thermal support below both the eastern and western rift shoulders has been inferred geophysically and volcanologically. The survey showed that the TAM lithosphere in this area (Law–Lennox-King glacier corridor adjacent to the Beardmore glacier) right up to the range front appears thick (*c.* 150 km) and that a rigid cantilever support mechanism is likelier than shallow thermal buoyancy as a support mechanism for the CTAM.

We have computed a re-inversion of the CTAM MT data here, with a slight alteration of the *a priori* and starting model. Data errors, mesh and inversion damping were otherwise identical to Wannamaker *et al.* (2017). With the longest wave periods being 4000–6000 s, penetration through most of the upper mantle is considered feasible. This time there has been no imposed drop in resistivity in the *a priori* model at the depth of 410 km. Such a drop as used by Wannamaker *et al.* (2017) considered earlier laboratory data on the olivine–wadsleyite transition, which have since been suggested to suffer contamination by iron loss to the platinum experimental specimen sleeve during measurement (Yoshino and Katsura 2013). The resultant model is almost identical except that the broadly lower resistivities of the asthenosphere below the RIS extend to depth somewhat more uniformly than in the model of Wannamaker *et al.* (2017). We use this newer model in our estimation of peridotite hydration in the next section of the paper. Otherwise, the model features in Figure 12 are essentially the same, with, for example, inference of focused rifting and magmatic underplating just grid SW of the CTAM front (RN) and probable Early Proterozoic graphitized metasediments (MS) in the deep crust of East Antarctica (Boerner *et al.* 1996; Wunderman *et al.* 2018). A primary characteristic is the apparent thick lithosphere under the TAM here, again, supporting the prior cantilever interpretation for uplift.

That conclusion by Wannamaker *et al.* (2017) seemed in contrast to the near-simultaneously published seismic shear-wave tomographic model of Shen *et al.* (2017), which showed low velocities in the shallowest mantle along a TAM corridor that projected towards South Pole to the grid NW of our MT transect. However, examination of the seismic model indicates that thicker, high-velocity TAM lithosphere could exist below the location of the MT study (see supplementary figure DR4b of Shen *et al.* 2017). Variability in the magnitude of mantle thermal uplift forces along the TAM can be expected (see also Brenn *et al.* 2017).

Both broader-scale seismic tomography (Heeszel *et al.* 2016) and the MT imaging (Fig. 12) suggest that lithospheric thickness continues to increase further into East Antarctica, although details of the MT structure cannot otherwise be resolved. The thickening appears consistent with continental-scale tectonic correlations based on reassembly models for Antarctica with Australia, potential field trends, and inferred East Antarctic basement rock ages. These correlations connect the Archean–Paleoproterozoic Gawler craton of Australia with the Mawson craton of Antarctica in deeper East Antarctica (Boger 2011; Goodge and Fanning 2016; Goodge *et al.* 2017).

Upper-mantle hydration below Western and Central Antarctica

One definition of the asthenosphere is a deeper zone of weak rheology and near-adiabatic temperature conditions in the upper mantle, separated mechanically from the overlying strong lithosphere allowing plate motions at varying scales (Turcotte and Schubert 2002; Meqbel *et al.* 2014). Increased temperature, presence of melt and degree of intracrystalline hydration all are factors that may promote weakness of peridotitic asthenosphere (e.g. Yoshino and Katsura 2013). Recent laboratory experiments have shown that seismic velocity

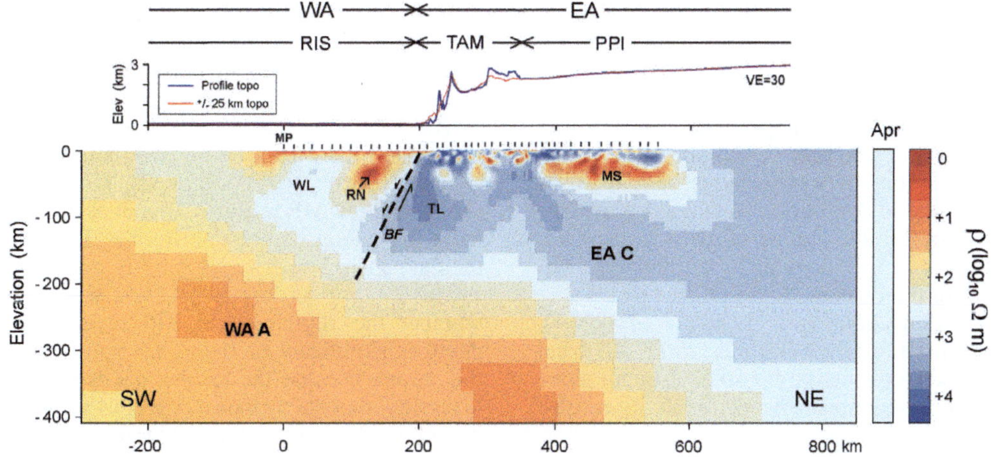

Fig. 12. CTAM 3D inversion section view along the main MT site profile. Label MP denotes the location of the example vertical resistivity profile through the upper mantle for synthesis with other study areas shown in Figure 14. Column Apr denotes the starting and *a priori* model, which was a uniform 300 Ω m half-space below the ice layer. Model features include West Antarctic asthenosphere (WA A), West Antarctic lithosphere (WL), rift necking (RN), Shackleton terrane boundary fault (BF), TAM lithosphere (TL), Early Proterozoic graphitized metasedimentary domain (MS) and East Antarctic craton (EA C). Physiographical divisions include the Ross Ice Shelf (RIS), Transantarctic Mountains (TAM) and Polar Plateau (PPl).

and attenuation are essentially negligibly affected by intracrystalline, solid-state hydration in otherwise nominally anhydrous minerals (NAMs) (Cline *et al.* 2018). However, electrical resistivity (conductivity) is strongly affected (Yoshino and Katsura 2013). This presents a unique opportunity for MT resistivity estimates to illuminate the degree of NAM hydration, especially in the asthenosphere that otherwise may not be possible without direct sampling.

Because electrical resistivity may be reduced strongly by the same three factors influencing rheology above, it is essential to apply likely or even plausible constraints to reduce non-uniqueness and address specific hypotheses. In particular, we ask whether the degree of peridotite NAM hydration under ambient temperatures is the maximum allowable before incipient water-undersaturated melting is triggered (Ardia *et al.* 2012). A similar question was asked by Naif (2018) for oceanic thermal regimes. For ambient temperature in the asthenosphere, we consider two adiabats with potential temperatures (T_p) of 1600 (1327°C) and 1800 K (1527°C) (Stixrude and Lithgow-Bertelloni 2011). The former is considered more globally representative than the latter, but a T_p of 1800 K has been invoked for deep plume-sourced upper mantle. For example, Plank and Forsyth (2016) suggested that this is the case for the eastern Great Basin, while plume contributions to upper-mantle melts have been advanced for Marie Byrd Land and central Victoria Land of West Antarctica (LeMasurier and Rex 1991; Phillips *et al.* 2018).

Water-undersaturated melting of peridotite v. pressure and temperature has been expressed according to the hydration degree of the bulk peridotite by Hirschmann *et al.* (2009), and results for a range of hydration are plotted in Figure 13 together with the two adiabats. The intersection of the adiabats with the 'damp' solidii defines the maximal amount of peridotite hydration possible just short of H_2O-undersaturated melting. These experimental data, together with laboratory results on the electrical conductivity of upper-mantle minerals as a function of temperature and hydration, allow us to predict upper-mantle conductivity (1/resistivity) v. depth along such maximally hydrated adiabats. We then may compare such model profiles with the field resistivities in the upper mantle from some of our Antarctica campaigns and elsewhere.

Of the upper-mantle minerals, the dominant olivine has undergone the most thorough measurement of conductivity under hydrated conditions. Using the peridotite mineral partitioning coefficients of Novella *et al.* (2014), nominally for a depleted MORB mantle (DMM), we estimate the weight ppm of H_2O in olivine along the adiabats (Fig. 13). Although different laboratories often have yielded diverse conductivity results v. P–T–H_2O, comprehension of olivine conductivity may be settling through application of the hydrogen self-diffusion technique (Novella *et al.* 2017). Given that partitioning diminishes towards the deeper upper mantle where the larger resistivity drops occur, and that there is not a strong difference in the conductivities between olivine and the two pyroxenes at asthenospheric conditions (Zhao and Yoshino 2016), we believe using the olivine standard is appropriate here. Hence, the precise mineralogical proportions of the model upper mantle are not crucial.

Hence, evaluation of equation (6) of Novella *et al.* (2017) for the hydrogen contents along the 1600 and 1800 K adiabats in Figure 13 yields model conductivity profiles v. depth plotted in Figure 14 that represent upper-mantle resistivity resulting from maximal hydration just short of melting. For reference, the predicted conductivity profiles for dry olivine from Du Frane *et al.* (2005) are also shown, as they were also used by Novella *et al.* (2017), and intersect the hydrated curves at the dry solidus (DSI) at shallower levels. We do not plot model curves for depths shallower than DSI because those will be in the quasi-dry melting regime as upwelling continues further, until heat losses are sufficient to establish the lithosphere. The dashed laboratory curves are for nominally isotropic olivine obtained through geometric mean averaging of conductivities along the three crystallographic axes (Novella *et al.* 2017). Both hydrated olivine adiabatic curves fall with depth to below 10 Ω m beyond *c.* 200 km, whereas dry solidus resistivities remain in the 40–200 Ω m range (Fig. 14). The deep hydrated values for the two adiabats differ modestly, only by a factor of about 2, because the higher T_p of the enhanced adiabat is countered by the lesser amount of permissible hydration before incipient melting (Fig. 13).

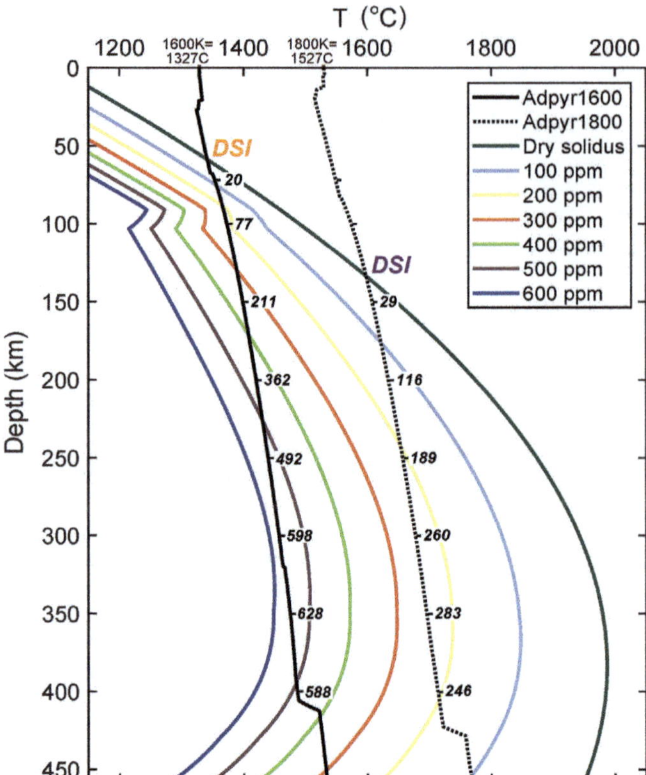

Fig. 13. Water-undersaturated peridotite solidii expressed as weight ppm in the bulk peridotite after Hirschmann *et al.* (2009) (coloured curves) with two model adiabats after Stixrude and Lithgow-Bertelloni (2011) (black solid and dotted curves). Numbers written at 50 km intervals along the adiabats are corresponding ppm contents of olivine according to the partition coefficients of Novella *et al.* (2014). DSI denotes the dry solidus intercept for each adiabat, used as pin points for the predicted adiabat resistivities in Figure 14. Tables of solidii and adiabatic temperatures were provided to P.E. Wannamaker by M. Hirschmann and by L. Stixrude.

The RIS upper-mantle profile from the CTAM project is plotted in Figure 14 for comparison and shows values falling to near 20 Ω m by 250 km depth. This location of the depth profile (below MP in Fig. 12) is chosen as it is relatively far from the strong contrast across the TAM but is still within the data span. Nowhere laterally in this depth range does the resistivity drop as low as 10 Ω m. It is difficult to evaluate whether the slight rise in RIS resistivity beyond 300 km depth is significant or represents some loss of resolution in the model, because the longest data period is 6000 s and the inversion imposes a modest degree of adherence to the starting model of 300 Ω m. Despite its large aperture, to firmly resolve the lowermost upper mantle would benefit from a profile about twice as long (*c.* 1000 km) and periods to 10 000 s. However, overall, the RIS depth profile is offset to greater resistivities relative to those of the hydrated standard adiabat by a factor of 3–4. Hence, solid-state NAM hydration does not appear to reach the maximum permitted short of H_2O-undersaturated

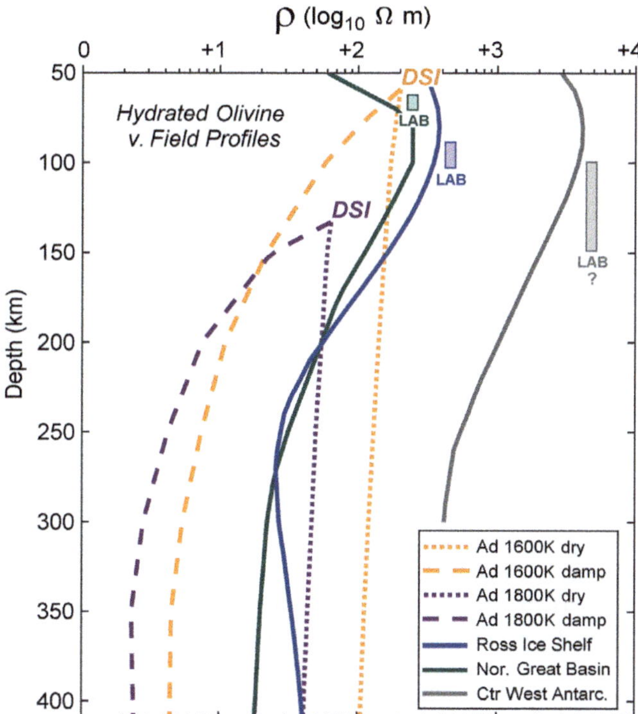

Fig. 14. Resistivity–depth curves of the adiabatic model for two potential temperatures (1600 and 1800 K) compared to the field resistivity profiles of CTAM and CWA. Predicted adiabat resistivities make use of the model olivine hydration values in Figure 13 and laboratory conductivity results of Novella *et al.* (2017). DSI denotes dry solidus intersection. Also plotted is the field profile of the central northern Great Basin of the western USA at latitude 41.5° N and longitude 117° W from Meqbel *et al.* (2014). Approximate lithosphere–asthenosphere boundary depth ranges, some with significant uncertainty, are included as lightly filled rectangular boxes labelled LAB.

melting for either adiabat, and appears sufficient to explain the RIS resistivity profile without invoking hydrous melt. Based on Figure 13, the H_2O content would be in the 75–200 ppm range below this portion of the RIS. This is in the range understood for oceanic asthenosphere (Hirschmann *et al.* 2009; Peslier *et al.* 2017) but petrology-based inferences for continental asthenosphere are lacking. Resistivities of 10–30 Ω m at depths >200 km in presumed asthenosphere beneath the Archean–Early Proterozoic shield of the north-central USA (Yang *et al.* 2015) are similar to our RIS and US Great Basin resistivities (Fig. 14), however, implying similar degrees of hydration (cf. Selway *et al.* 2019). Such a state of hydration may also occur at greater depths in the low resistivity beneath the thick lithosphere of the TAM in Figure 12

For further comparison, the field resistivity profile of the rifted, central northern Great Basin of the western USA at latitude 41.5° N and longitude 117° W from Meqbel *et al.* (2014) is plotted in Figure 14. It shows a shape comparable to that from the RIS, although the values decrease monotonically to 410 km depth perhaps in part due to increased resolution with their long-period (*c.* 10 000 s) Earthscope data. However, the northern Great Basin overall exhibits a similar apparent degree of hydration to the RIS. That the two upper-mantle profiles are so alike perhaps is surprising given the expected ongoing hydration of the Great Basin by protracted subduction of the Gorda-Juan de Fuca Plate (Busby 2013). This could imply that the state of hydration of the RIS upper mantle is long lived from events perhaps in Mesozoic or even Paleozoic time. Alternately, lateral flow of the deeper upper mantle may have occurred bringing hydrated asthenosphere from elsewhere. However, lithospheric xenoliths from subduction environments globally do not exhibit higher degrees of dissolved mineral hydration than those from other geological contexts (Demouchy and Bolfan-Casanova 2016).

For depths <100 km in Figure 14, the RIS resistivities reach or exceed *c.* 300 Ω m, and such values only increase northeastwards for the next 50–70 km. These suggest cooler lithospheric temperatures because the values are greater than those predicted from dry peridotite measurements at even the 1600 K adiabat (Fig. 13), although this is consistent with the regionally low elevations (LeMasurier 2008). No seismic stations are located close to this part of our profile and so an independent estimate of RIS depth to the lithosphere–asthenosphere boundary (LAB) is imprecise; we have shown the estimate from station SURP near the TAM range front (O'Donnell *et al.* 2019) (Fig. 14). Feature RN is interpreted to be local extensional magmatic underplating closer to the TAM range front (Wannamaker *et al.* 2017). This would be familial with the rollover in the Great Basin resistivity profile as one moves up to 50 km depth, expressing influence of its widespread, Moho-level magmatic underplating. LAB depth estimates for the Great Basin are shallow, in the 60–70 km depth range (Levander and Miller 2012) (Fig. 14). At depths shallower than the LAB, conductive geotherms passing to much lower temperatures than the adiabats would need to be modelled for Figure 13 to apply.

Uncertainties in physico-chemical state interpretation

At this point it seems appropriate to point out some analytical uncertainties. The water-undersaturated solidii in Figure 13 are based on experimental data points involving partitioning between crystal and melt water contents (Hirschmann *et al.* 2009), and updated analytical techniques are likely to refine the values (e.g. Novella *et al.* 2014; Naif 2018). Effects of other hydrated peridotite minerals, particularly pyroxenes (Zhao and Yoshino 2016), are not considered here; these appear to hold greater water contents than olivine, although this mainly is at depths <150 km (Novella *et al.* 2014). However, it is difficult in MT with resistive media to say more than the West Antarctic lithosphere (WL) simply is resistive, especially when lying below conductive layers such as the imaged seawater and possible marine sediments (Fig. 12) (e.g. Chave and Jones 2012). We have not used alternate laboratory data on hydrated olivine presented by Dai and Karato (2014) that would predict significantly higher conductivities in Figure 14 for a specific water content and temperature, although those would only strengthen the inference that field peridotite volumes are not hydrogen maximized. We also have not used the conductivity model of Gardes *et al.* (2014), which derives curves from a specific compendium of experiments and in itself is not based on a definite conduction species model. The hydrogen diffusion coefficient estimates of Novella *et al.* (2017) have accuracies within *c.* 20%, but the validity of a geometric mean to model hydrated crystal anisotropy that may exceed a factor of 10 should receive further study. We doubt, therefore, that differences between model mantle resistivity profiles and field estimates that are within a factor of 2 should be considered clearly significant at this time, but factors of 3–4 may be, so that the inference of broad-scale H_2O undersaturation is considered reasonable.

Because the predicted adiabatic resistivity profiles in Figure 14 are significantly lower than the field profiles, we do not see H_2O-undersaturated partial melts contributing to the lowered resistivity in these areas either of the RIS or the north-central Great Basin. If melts of any sort are contributing, then the inferred degree of NAM solid-state hydration would be less than our estimates above. Under dry conditions and a normal adiabat, melting would not occur until upwelling asthenosphere reached depths of *c.* 60 km (DSI in Fig. 13), whereas the field resistivities are much lower than unmelted dry peridotite essentially through the upper mantle (Fig. 14). Increasing the adiabat to a T_p of 1800 K only deepens the melt regime to *c.* 135 km (Fig. 13); raising T_p yet further is unrealistic. However, the presence of CO_2 can allow melts with much lower degrees of hydration, but such melts appear restricted to depths <*c.* 250 km (Dasgupta *et al.* 2013). Resistivity of these melts would be well under 1 Ω m (Sifre *et al.* 2015); nevertheless, for melt to reduce bulk peridotite resistivity enough to affect surface MT responses, it should be interconnected over regional distances. Disconnected melt may still have some effect on bulk seismic velocity, however, which could explain the weak low-velocity zones correlated with small amounts of melt modelled by O'Donnell *et al.* (2019) elsewhere in West Antarctica.

Also plotted in Figure 14 is a portion of the upper-mantle resistivity profile below the CWA MT transect over the grid northern Byrd Subglacial Basin from Figure 5. It is far more resistive than the RIS or Great Basin extensional regimes in keeping with its interpreted fossil state (Wannamaker *et al.* 1996; Blankenship *et al.* 2001). It is significantly more resistive than the model profile for dry peridotite and an average (1600 K) adiabat in Figure 13, and so certainly requires no hydration (see also O'Donnell *et al.* 2017). Clearly, the lithosphere in this project area is thick and relatively cold. However, the aperture of the profile being only 54 km could be prone to model distortion by heterogeneity of larger scale within which the dataset may sit. Specifically, if the profile lies within a block of dimensions of the order of 100 km that is more resistive than the regional average, then the resulting 3D inversion model may suffer some upward bias, especially towards depths of 100 km or more. For example, if the CWA upper-mantle model resistivities were inflated by a factor of *c.* 3, then lower model depths would be too great by sqrt(3) (*c.* 1.7). The MT profile straddles an aeromagnetic high anomaly (Wannamaker *et al.* 1996), although we do not know if this would represent lower or higher bulk resistivity, if at all. The conjectured LAB drawn in Figure 14 is deeper, similar to that below seismic station 1624 (O'Donnell *et al.* (2019), but essentially unresolvable from MT alone. A wider-ranging data campaign would be necessary to constrain this possibility. The much longer CTAM transect avoids such a difficulty through most of the upper mantle. Similarly thick and cold lithosphere could be conjectured for the Prydz Bay area survey of Peacock and Selway (2016) based directly on the age of terrane, but the period range of their data meant that they restricted their resistivity interpretation to the crust. Finally, due to both the aperture concerns and the MT signal penetration through the strong Moho conductor, we do not consider an upper-mantle resistivity profile at South Pole region. Additional data to significantly longer periods and over a wider area would be necessary. Moho-level low resistivity is more likely to be related to melt underplating and fluid release, although deeply underthrust fossil metasediments cannot be ruled out.

Conclusions

Electrical conductivity properties of the upper mantle can be revealing of the rheology and volatile transfer processes if interpreted using suitable constraints. With technology established to obtain high-fidelity MT data over polar ice sheets, inroads into upper-mantle electrical properties are developing. Apparent thick lithospheric domains of low electrical conductivity imply cold temperatures to depth and a resistance to deformation either from lateral (plate tectonic) or vertical (plume, rebound or flexural) forces (CWA, central TAM, East Antarctica). Sinuous conductors in the deep crust related to original continental margin graphite–sulfide stratigraphic deposition may assist in terrane reconstruction and correlation with domains on other continents as more MT data are acquired from Antarctica. Below the RIS region near the TAM in West Antarctica, bulk hydration of the upper-mantle NAMs seems significantly below the maximal amount allowed before peridotite undergoes incipient melting at standard adiabatic temperatures, although hydration could be locally channelized to greater concentrations. Small amounts of poorly interconnected melt, however, would not much affect the resistivity. The RIS resistivity depth profile is largely similar to that observed below the actively extensional northern Great Basin of the western USA. The active region profiles raise the question of how the asthenosphere is expressed electrically under continental extension; the upper depth limit of adiabatic conditions may lie in shallow upper mantle of lower hydration and so does not clearly express as a conductivity contrast. Further efforts to reduce uncertainty in the conductivity of mantle minerals and the peridotite solidus as functions of temperature and hydration through laboratory measurements and modelling is desirable.

Acknowledgements J. Stodt constructed the current version of the high-impedance buffer preamplifiers. We remember key early preamp design development by late engineer S. Olsen. V. Maris executed the 3D inversion and produced the model graphics. M. Kordy derived the finite-element equations and wrote the computer code for the 3D non-linear MT inversion. P.E. Wannamaker thanks M. Hirschmann for auxiliary calculations from his 2009 paper, and L. Stixrude and C. Lithgow-Bertelloni for a table of adiabatic temperature values from their 2011 paper that made Figure 13 possible. He also thanks J. Tyburczy for useful discussions on evaluating NAM conductivity. We gratefully acknowledge instrumentation support and in-field assistance over the years by Y. Ogawa of Tokyo Institute of Technology, by New Zealand GNS Science, and by the Polaris field instrument pool of Canada. Specific assistance in the CWA project by J. Johnston and W. Doerner, and in the CTAM project by Beth Murphy, is recognized. Generous and competent logistical support by the US Antarctic Program and by Antarctica New Zealand ensured success of the several field programmes. We acknowledge Dr Kate Selway and an anonymous reviewer for their remarks contributing to an improved paper. We also thank Editors Dr Adam Martin and Dr Wouter van der Wal for initiating this Memoir, inviting our submission and overseeing the review process.

Author contributions **PEW**: conceptualization (lead), formal analysis (lead), funding acquisition (lead), investigation (lead), methodology (lead), project administration (lead), supervision (lead), writing – original draft (lead), writing – review & editing (lead); **JAS**: methodology (equal), writing – review & editing (supporting); **GJH**: investigation (supporting), methodology (supporting), writing – review & editing (supporting); **VM**: data curation (supporting), formal analysis (equal), investigation (supporting), software (supporting), writing – review & editing (supporting); **MAK**: writing – review & editing (equal).

Funding The analysis and presentation of this paper were supported primarily by US National Science Foundation grant AES-1443522 to P.E. Wannamaker. Non-linear 3D MT inversion algorithm development was under US Department of Energy, Geothermal Technologies Office, contract DE-EE0002750 to P.E. Wannamaker. G.J. Hill acknowledges support from the Royal Society of New Zealand Marsden Fund award ASL-1301 for fieldwork participation, data analysis and manuscript preparation.

Data availability The CTAM dataset analysed during the current study is available in the NASA Global Change Master Directory, at http://gcmd.nasa.gov/getdif.htm?NSF-ANT08-38914. The CWA and South Pole datasets are available from P.E. Wannamaker upon request.

Correction notice An incorrect figure citation was included. This has been corrected.

References

Akasofu, S. (ed.) 1977. *Physics of Magnetospheric Substorms*. D. Reidel, Dordrecht, The Netherlands.

Akasofu, S. 2015. Auroral substorms as an electrical discharge phenomenon. *Progress in Earth and Planetary Science*, **2**, 20, https://doi.org/10.1186/s40645-015-0050-9

Ardia, P., Hirschmann, M.M., Withers, A.C. and Tenner, T.J. 2012. H_2O storage capacity of olivine at 5–8 GPa and consequences for dehydration partial melting of the upper mantle. *Earth and Planetary Science Letters*, **345**, 104–116, https://doi.org/10.1016/j.epsl.2012.05.038

Behrendt, J.D. 1999. Crustal and lithospheric structure of the West Antarctic Rift system from geophysical investigations – a review. *Global Planetary Change*, **23**, 25–44, https://doi.org/10.1016/S0921-8181(99)00049-1

Behrendt, J.D., Blankenship, D.D., Finn, C.A., Bell, R.E., Sweeney, R.E., Hodge, S.M. and Brozena, J.M. 1994. CASERTZ aeromagnetic data reveal late Cenozoic flood basalts (?) in the West Antarctic rift system. *Geology*, **22**, 527–530, https://doi.org/10.1130/0091-7613(1994)022<0527:CADRLC>2.3.CO;2

Bentley, C.R. 1991. Configuration and structure of the subglacial crust. *In*: Tingey, R.J. (ed.) *The Geology of Antarctica*. Clarendon Press, Oxford, UK, 335–364.

Blankenship, D.D., Bell, R.E., Hodge, S.M., Brozena, J.M., Behrendt, J.C. and Finn, C.A. 1993. Active volcanism beneath the West Antarctic ice sheet and implications for ice-sheet stability. *Nature*, **361**, 526–529, https://doi.org/10.1038/361526a0

Blankenship, D.D., Morse, D.L. *et al.* 2001. Geologic controls on the initiation of rapid basal motion for West Antarctic ice streams: A geophysical perspective including new airborne radar sounding and laser altimetry results. *American Geophysical Union Antarctic Research Series*, **77**, 105–121.

Boerner, D.E., Kurtz, R.D. and Craven, J.A. 1996. Electrical conductivity and Paleoproterozoic foredeeps. *Journal of Geophysical Research*, **101**, 13 775–13 791, https://doi.org/10.1029/96JB00171

Boger, S.D. 2011. Antarctica – before and after Gondwana. *Gondwana Research*, **19**, 335–371, https://doi.org/10.1016/j.gr.2010.09.003

Brenn, G.R., Hansen, S.E. and Park, Y. 2017. Variable thermal loading and flexural uplift along the Transantarctic Mountains, Antarctica. *Geology*, **45**, 463–466, https://doi.org/10.1130/G38784.1

Busby, C.J. 2013. Birth of a plate boundary at *c.* 12 Ma in the Ancestral Cascades arc, Walker Lane belt of California and Nevada. *Geosphere*, **9**, 1147–1160, https://doi.org/10.1130/GES00928.1

Chave, A.D. and Jones, A.G. (eds). 2012. *The Magnetotelluric Method: Theory and Practice*. Cambridge University Press, Cambridge, UK.

Clarke, T.S., Burkholder, P.D., Smithson, S.B. and Bentley, C.R. 1997. Optimum seismic shooting and recording parameters and a preliminary crustal model for the Byrd Subglacial Basin, Antarctica. *In*: Ricci, C.A. (ed.) *The Antarctic Region: Geological Evolution and Processes*. Terra Antarctica, Sienna, Italy, 485–493.

Cline, C.J.II, Faul, U.H., David, E.C., Berry, A.J. and Jackson, I. 2018. Redox-influenced seismic properties of upper-mantle olivine. *Nature*, **555**, 355–362, https://doi.org/10.1038/nature25764

Dai, L. and Karato, S.-I. 2014. High and highly anisotropic electrical conductivity of the asthenosphere due to hydrogen diffusion in olivine. *Earth and Planetary Science Letters*, **408**, 79–86, https://doi.org/10.1016/j.epsl.2014.10.003

Dalziel, I.W.D. and Lawver, L.A. 2001. The lithospheric setting of the West Antarctic ice sheet. *American Geophysical Union Antarctic Research Series*, **77**, 105–121.

Danesi, S. and Morelli, A. 2001. Structure of the upper mantle under the Antarctic Plate from surface wave tomography. *Geophysical Research Letters*, **28**, 4395–4398, https://doi.org/10.1029/2001GL013431

Dasgupta, R., Mallik, A., Tsuno, K., Withers, A.C., Hirth, G. and Hirschmann, M. 2013. Carbon-dioxide-rich silicate melt in the Earth's upper mantle. *Nature*, **493**, 211–216, https://doi.org/10.1038/nature11731

Demouchy, S. and Bolfan-Casanova, N. 2016. Distribution and transport of hydrogen in the lithospheric mantle: a review. *Lithos*, **240–243**, 402–425, https://doi.org/10.1016/j.lithos.2015.11.012

Du Frane, W.L., Roberts, J.J., Toffelmier, D.A. and Tyburczy, J.A. 2005. Anisotropy of electrical conductivity in dry olivine. *Geophysical Research Letters*, **32**, L24315, https://doi.org/10.1029/2005GL023879

Gardes, E., Gaillard, F. and Tarits, P. 2014. Toward a unied hydrous olivine electrical conductivity law. *Geochemistry, Geophysics, Geosystems*, **15**, 4984–5000, https://doi.org/10.1002/2014GC005496

Goodge, J.W. and Fanning, C.M. 2016. Mesoarchean and Paleoproterozoic history of the Nimrod Complex, central Transantarctic Mountains, Antarctica: Stratigraphic revisions and relation to the Mawson continent in East Gondwana. *Precambrian Research*, **285**, 242–271, https://doi.org/10.1016/j.precamres.2016.09.001

Goodge, J.W., Fanning, C.M., Fischer, C.M. and Vervoort, J.D. 2017. Proterozoic crustal evolution of central East Antarctica: Age and isotopic evidence from glacial igneous clasts, and links with Australia and Laurentia. *Precambrian Research*, **299**, 151–176, https://doi.org/10.1016/j.precamres.2017.07.026

Hansen, S.E., Graw, J.H. *et al.* 2014. Imaging the Antarctic mantle using adaptively parameterized P-wave tomography: Evidence for heterogeneous structure beneath West Antarctica. *Earth and Planetary Science Letters*, **408**, 66–78, https://doi.org/10.1016/j.epsl.2014.09.043

Heeszel, D.S., Wiens, D.A. *et al.* 2016. Upper mantle structure of central and West Antarctica from array analysis of Rayleigh wave phase velocities. *Journal of Geophysical Research: Solid Earth*, **121**, 1758–1775, https://doi.org/10.1002/2015JB01 2616

Hill, G.J. 2020. On the use of electromagnetics for Earth imaging of the polar regions. *Surveys in Geophysics*, **45**, 5–45, https://doi.org/10.1007/s10712-019-09570-8

Hirschmann, M.M., Tenner, T.J., Aubaud, C. and Withers, A.C. 2009. Dehydration melting of nominally anhydrous mantle: The primacy of partitioning. *Physics of the Earth and Planetary Interiors*, **176**, 54–68, https://doi.org/10.1016/j.pepi.2009.04.001

Hole, M.J. and LeMasurier, W.E. 1994. Tectonic controls on the geochemical composition of Cenozoic, mafic alkaline volcanic rocks from West Antarctica. *Contributions to Mineralogy and Petrology*, **117**, 187–202, https://doi.org/10.1007/BF002 86842

Jordan, T.A., Martin, C. *et al.* 2018. Anomalously high geothermal flux near the South Pole. *Scientific Reports*, **8**, 16785, https://doi.org/10.1038/s41598-018-35182-0

Kordy, M.A., Wannamaker, P.E., Maris, V., Cherkaev, E. and Hill, G.J. 2016*a*. Three-dimensional magnetotelluric inversion using deformed hexahedral edge finite elements and direct solvers parallelized on SMP computers, Part I: forward problem and parameter jacobians. *Geophysical Journal International*, **204**, 74–93, https://doi.org/10.1093/gji/ggv410

Kordy, M.A., Wannamaker, P.E., Maris, V., Cherkaev, E. and Hill, G.J. 2016*b*. Three-dimensional magnetotelluric inversion using deformed hexahedral edge finite elements and direct solvers parallelized on SMP computers, Part II: direct data-space inverse solution. *Geophysical Journal International*, **204**, 94–110, https://doi.org/10.1093/gji/ggv411

LeMasurier, W.E. 2008. Neogene extension and basin deepening in the West Antarctic rift inferred from comparisons with the East African rift and other analogs. *Geology*, **36**, 247–250, https://doi.org/10.1130/G24363A.1

LeMasurier, W.E. and Rex, D.C. 1991. The Marie Byrd Land volcanic province and its relation to the Cainozoic West Antarctic rift system. *In*: Tingey, R.J. (ed.) *The Geology of Antarctica*. Clarendon Press, Oxford, UK, 249–284.

Levander, A. and Miller, M.S. 2012. Evolutionary aspects of lithosphere discontinuity structure in the western U.S. *Geochemistry, Geophysics, Geosystems*, **13**, Q0AK07, https://doi.org/10.1029/2012GC004056

Meqbel, N.M., Egbert, G.D., Wannamaker, P.E., Kelbert, A. and Schultz, A. 2014. Deep electrical resistivity structure of the Pacific NW derived from 3-D inversion of Earthscope USArray magnetotelluric data. *Earth and Planetary Science Letters*, **402**, 290–304, https://doi.org/10.1016/j.epsl.2013.12.026

Naif, S. 2018. An upper bound on the electrical conductivity of hydrated oceanic mantle at the onset of dehydration melting. *Earth and Planetary Science Letters*, **482**, 357–366, https://doi.org/10.1016/j.epsl.2017.11.024

Novella, D., Frost, D.J., Hauri, E.H., Bureau, H., Raepsaet, C. and Mathilde, M. 2014. The distribution of H_2O between silicate melt and nominally anhydrous peridotite and the onset of hydrous melting in the deep upper mantle. *Earth and Planetary Science Letters*, **400**, 1–13, https://doi.org/10.1016/j.epsl.2014.05.006

Novella, D., Jacobsen, B., Weber, P.K., Tyburczy, J.A., Ryerson, F.J. and DuFrane, W.L. 2017. Hydrogen self-diffusion in single crystal olivine and electrical conductivity of the Earth's mantle. *Scientific Reports*, **7**, 5344, https://doi.org/10.1038/s41598-017-05113-6

O'Donnell, J.P., Selway, K. *et al.* 2017. The uppermost mantle seismic velocity structure and viscosity structure of central West Antarctica. *Earth and Planetary Science Letters*, **472**, 238–249, https://doi.org/10.1016/j.epsl.2017.05.016

O'Donnell, J.P., Stuart, G.W. *et al.* 2019. The uppermost mantle seismic velocity structure of West Antarctica from Rayleigh wave tomography: Insights into tectonic structure and geothermal heat flow. *Earth and Planetary Science Letters*, **522**, 219–233, https://doi.org/10.1016/j.epsl.2019.06.024

Peacock, J.R. and Selway, K. 2016. Magnetotelluric investigation of the Vestfold Hills and Rauer Group, East Antarctica. *Journal of Geophysical Research*, **121**, 2258–2273.

Peslier, A.H., Schönbächler, M., Busemann, H. and Karato, S.-I. 2017. Water in the Earth's interior: distribution and origin. *Space Science Review*, **212**, 743–810, https://doi.org/10.1007/s11214-017-0387-z

Phillips, E.H., Sims, K.W.W. *et al.* 2018. The nature and evolution of mantle upwelling at Ross Island, Antarctica, with implications for the source of HIMU lavas. *Earth and Planetary Science Letters*, **498**, 38–53, https://doi.org/10.1016/j.epsl.2018.05.049

Plank, T. and Forsyth, D.W. 2016. Thermal structure and melting conditions in the mantle beneath the Basin and Range province from seismology and petrology. *Geochemistry, Geophysics, Geosystems*, **17**, https://doi.org/10.1002/2015GC006205

Ritzwoller, M.H., Shapiro, N.M., Levshin, A.L. and Leahy, G.M. 2001. Crustal and upper mantle structure beneath Antarctica and surrounding oceans. *Journal of Geophysical Research*, **106**, 30 645–30 670, https://doi.org/10.1029/2001JB000179

Selway, K., O'Donnell, J.P. and Özaydin, S. 2019. Upper mantle melt distribution from petrologically constrained magnetotellurics. *Geochemistry, Geophysics, Geosystems*, **20**, 3328–3346, https://doi.org/10.1029/2019GC008227

Sen, V., Stoffa, P.L., Dalziel, I.W.D., Blankenship, D.D., Smith, A.M. and Anandakrishnan, S. 1998. Seismic surveys in central West Antarctica: data and processing examples from the ANT-ALITH field tests (1994–1995). *Terra Antarctica*, **5**, 761–772.

Shen, W., Wiens, D.A. *et al.* 2017. Seismic evidence for lithospheric foundering beneath the southern Transantarctic Mountains. *Antarctica. Geology*, **46**, 71–74, https://doi.org/10.1130/G39555.1

Siegert, M.J. 2017. A 60-year international history of Antarctic subglacial lake exploration. Geological Society, London, Special Publications, **461**, 7–21, https://doi.org/10.1144/SP461.5

Sifre, D., Hashim, L. and Gaillard, F. 2015. Effect of temperature, pressure and chemical compositions on the electrical conductivity of carbonated melts and its relationship with viscosity. *Chemical Geology*, **418**, 189–197, https://doi.org/10.1016/j.chemgeo.2014.09.022

Stixrude, L. and Lithgow-Bertelloni, C. 2011. Thermodynamics of mantle minerals – II. phase equilibria. *Geophysical Journal International*, **184**, 1180–1213, https://doi.org/10.1111/j.1365-246X.2010.04890.x

Turcotte, D.L. and Schubert, G. 2002. *Geodynamics*. Cambridge University Press, Cambridge, UK.

Tyburczy, J.A. and Du Frane, W.L. 2015. Properties of rocks and minerals – The electrical conductivity of rocks, minerals, and the Earth. *In*: Schubert, G. (ed.) *Treatise on Geophysics, Volume 2. Mineral Physics*. 2nd edn. Elsevier, Oxford, UK, 661–672.

Unsworth, M. and Rondenay, S. 2013. Mapping the distribution of fluids in the crust and lithospheric mantle utilizing geophysical methods. *In*: Harlov, D.E. and Austrheim, H. (eds) *Metasomatism and the Chemical Transformation of Rock*. Lecture Notes in Earth System Sciences. Springer, Berlin, 535–598, https://doi.org/10.1007/978-3-642-28394-9_13

Wannamaker, P.E. 2005. Anisotropy v. heterogeneity in continental solid earth electromagnetic studies: fundamental response characteristics and implications for physicochemical state: invited review paper. *Surveys in Geophysics*, **26**, 733–765, https://doi.org/10.1007/s10712-005-1832-1

Wannamaker, P.E., Stodt, J.A. and Olsen, S.L. 1996. Dormant state of rifting below the Byrd Subglacial Basin, West Antarctica, implied by magnetotelluric (MT) profiling. *Geophysical Research Letters*, **23**, 2983–2986, https://doi.org/10.1029/96GL02887

Wannamaker, P.E., Stodt, J.A., Pellerin, L., Olsen, S.L. and Hall, D.B. 2004. Structure and thermal regime beneath South Pole region, East Antarctica, from magnetotelluric measurements. *Geophysical Journal International*, **157**, 36–54, https://doi.org/10.1111/j.1365-246X.2004.02156.x

Wannamaker, P.E., Hasterok, D.P. *et al.* 2008. Lithospheric dismemberment and magmatic processes of the Great Basin–Colorado Plateau transition, Utah, implied from magnetotellurics. *Geochemistry, Geophysics, Geosystems*, **9**, Q05019, https://doi.org/10.1029/2007GC001886

Wannamaker, P.E., Hill, G.J. *et al.* 2017. Uplift of the central Transantarctic Mountains. *Nature Communications*, **8**, 1588, https://doi.org/10.1038/s41467-017-01577-2

Wunderman, R.L., Wannamaker, P.E. and Young, C.T. 2018. Architecture of the hidden Penokean terrane suture and Midcontinent Rift System overprint in eastern Minnesota and western Wisconsin from magnetotelluric profiling. *Lithosphere*, **10**, 291–300, https://doi.org/10.1130/L716.1

Yang, B., Egbert, G.D., Kelbert, A. and Meqbel, N.M. 2015. Three-dimensional electrical resistivity of the north-central USA from EarthScope long period magnetotelluric data. *Earth and Planetary Science Letters*, **422**, 87–93, https://doi.org/10.1016/j.epsl.2015.04.006

Yoshino, T. and Katsura, T. 2013. Electrical conductivity of mantle minerals: role of water conductivity anomalies. *Annual Reviews of Earth and Planetary Sciences*, **41**, 605–628, https://doi.org/10.1146/annurev-earth-050212-124022

Zhao, C. and Yoshino, T. 2016. Electrical conductivity of mantle clinopyroxene as a function of water content and its implication on electrical structure of uppermost mantle. *Earth and Planetary Science Letters*, **447**, 1–9, https://doi.org/10.1016/j.epsl.2016.04.028

The seismic structure of the Antarctic upper mantle

Douglas A. Wiens[1]*, Weisen Shen[2] and Andrew J. Lloyd[3]

[1]Department of Earth and Planetary Sciences, Washington University in St Louis, St Louis, MO 63130, USA

[2]Department of Geosciences, Stony Brook University, Stony Brook, NY 11794-2100, USA

[3]Lamont-Doherty Earth Observatory, Columbia University, Palisades, NY 10964, USA

DAW, 0000-0002-5169-4386; WS, 0000-0002-3766-632X; AJL, 0000-0002-2253-1342

*Correspondence: doug@wustl.edu

Abstract: The deployment of seismic stations and the development of ambient noise tomography as well as new analysis methods provide an opportunity for higher-resolution imaging of Antarctica. Here we review recent seismic structure models and describe their implications for the dynamics and history of the Antarctic upper mantle. Results show that most of East Antarctica is underlain by continental lithosphere to depths of approximately 200 km. The thickest lithosphere is found in a band 500–1000 km inboard from the Transantarctic Mountains, representing the continuation of cratonic lithosphere with Australian affinity beneath the ice. Dronning Maud Land and the Lambert Graben show much thinner lithosphere, consistent with Phanerozoic lithospheric disruption. The Transantarctic Mountains mark a sharp boundary between cratonic lithosphere and the warmer upper mantle of West Antarctica. In the southern Transantarctic Mountains, cratonic lithosphere has been replaced by warm asthenosphere, giving rise to Cenozoic volcanism and an elevated mountainous region. The Marie Byrd Land volcanic dome is underlain by slow seismic velocities extending through the transition zone, consistent with a mantle plume. Slow-velocity anomalies beneath the coast from the Amundsen Sea Embayment to the Antarctic Peninsula are likely to result from upwelling of warm asthenosphere during subduction of the Antarctic–Phoenix spreading centre.

Seismological imaging provides essential constraints on the physical processes and conditions shaping the upper mantle beneath Antarctica. Unlike other continents, the Antarctic mantle is overlain not only by the crust and sedimentary cover, but also by a thick ice sheet, which greatly limits geological mapping and sampling. Thus, seismological studies applied to Antarctica fulfill a unique role by providing information on poorly constrained geological processes that have formed the continent. The ice sheets also constitute large, temporally varying loads that cause deformation of the mantle and the land surface through glacial isostatic adjustment. Seismology is able to characterize lateral variations in the temperature profile of the upper mantle and crust, and thus constrain the geothermal heat flux (Shapiro and Ritzwoller 2004; Shen *et al.* 2020). These thermal properties control the response of the mantle and the land surface to glaciation (Ivins and Sammis 1995; Ivins and James 2005; van der Wal *et al.* 2015; Ivins *et al.* In press), and affect the future evolution of the ice sheet (Gomez *et al.* 2015; Whitehouse *et al.* 2019).

Seismographs were part of the scientific agenda for Antarctic exploration from the very first expeditions. A Milne seismograph was operated by Robert F. Scott's Discovery Expedition (1901–04) for more than a year at Hut Point, near modern-day McMurdo Station, and dozens of earthquakes were detected (Bernacchi and Milne 1908). A seismograph was installed at the South Pole base during the International Geophysical Year in 1957. However, despite these early efforts, it was not until the turn of the century that autonomous seismographs were installed in Antarctica, allowing the underlying structure of the continent to be revealed at greater resolution.

The last two decades have seen rapid progress both in seismological instrumentation in Antarctica and in seismic analysis methods to utilize these data. The results of these efforts are detailed seismological models describing the 3D structure of the Antarctic upper mantle, as well as crustal thickness. Here we review these models and briefly describe their implications for understanding the dynamics and history of the Antarctic upper mantle. This review concentrates on large-scale studies of the upper-mantle structure from analysis of passive seismic recording, while recognizing that active source seismic studies provide important details in some specific regions.

Seismic data and analysis methods

Antarctic seismic data

From the 1950s through to the 1990s, seismographs were largely restricted to a few permanently occupied bases. With the exception of South Pole, these were generally along the coasts, so the interior of Antarctica was relatively unexplored from a seismological standpoint. Thus, maps of Antarctic mantle structure from that time period showed very limited resolution compared to other parts of the world. Active source seismic refraction lines did provide estimates of crustal thickness and uppermost mantle P-wave velocity in several places (Bentley 1973; ten Brink *et al.* 1993; Leitchenkov and Kudryavtzev 1997; Trey *et al.* 1999).

The advent of technology for operating autonomous broadband seismographs in remote parts of the Antarctic interior has revolutionized seismic studies in Antarctica over the past two decades. Initial projects, including SEPA (Robertson Maurice *et al.* 2003), ANUBIS (Anandakrishnan *et al.* 2000), TAMSEIS (Lawrence *et al.* 2006*a*), SSCUA (Reading 2006) and a deployment in the Transantarctic Mountains (Bannister *et al.* 2003), generally operated stations only in the summer months. However, more recent large-scale deployments such as AGAP/GAMSEIS (Hansen *et al.* 2010), Ross Ice Shelf (Baker *et al.* 2019), TAMNNET (Hansen *et al.* 2015) and UKANET (O'Donnell *et al.* 2019*b*) have provided high-quality data throughout the year. In many cases the seismographs remained deployed for only about 2 years, although the POLENET/A-NET project (Accardo *et al.* 2014) has maintained some remote autonomous stations for over a decade. Figure 1 shows the distribution of Antarctic broadband seismic stations providing data for analyses discussed here. The combined distribution of seismic stations provides reasonably good coverage at the several hundred-kilometre level in West Antarctica; however, vast regions of East Antarctica remain without instrumentation.

From: Martin, A. P. and van der Wal, W. (eds) 2023. *The Geochemistry and Geophysics of the Antarctic Mantle*.
Geological Society, London, Memoirs, **56**, 195–212,
First published online 17 September 2021, https://doi.org/10.1144/M56-2020-18

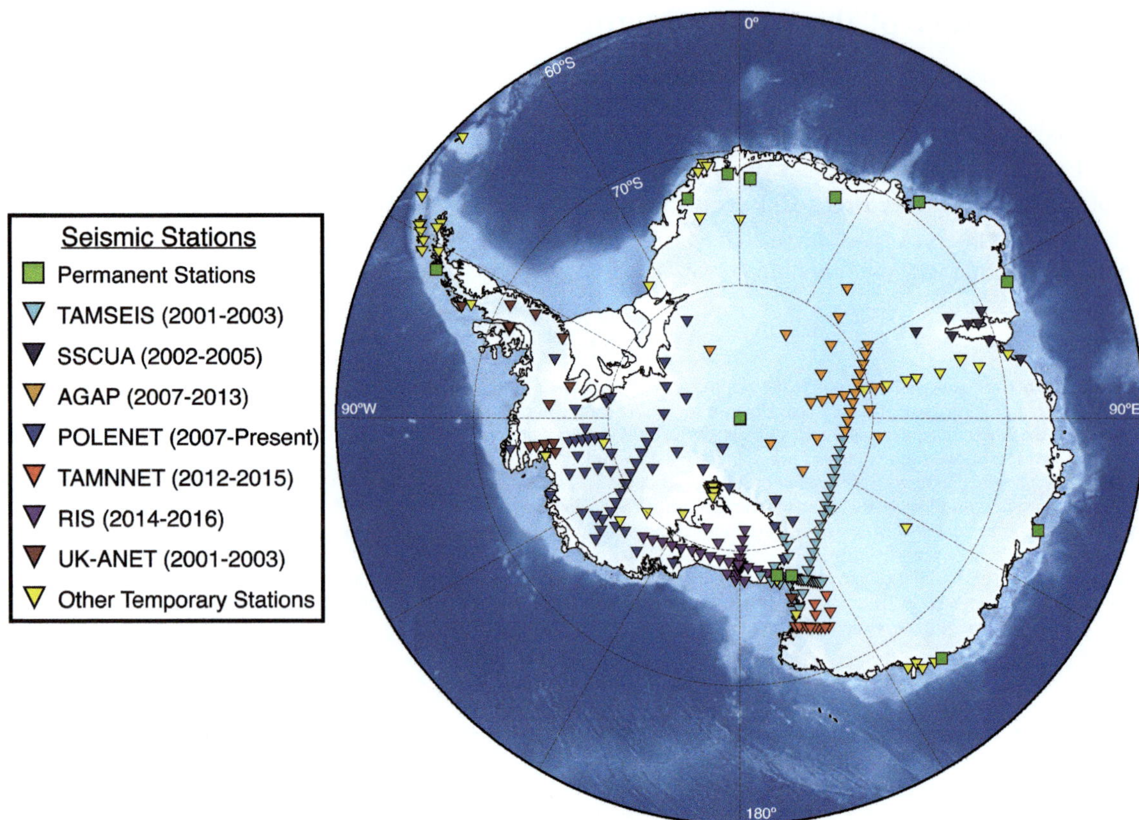

Fig. 1. Map of broadband seismic stations deployed in Antarctica and used in recent seismic studies. Permanent seismic stations, generally at Antarctic research stations, are shown as green boxes, and temporary seismic stations are shown as triangles. The key identifies some of the larger temporary seismic station deployments.

Body-wave tomography and receiver functions

P- and S-wave tomography using teleseismic arrivals can provide strong constraints on lateral velocity variations in the upper mantle of Antarctica (Watson *et al.* 2006; Lloyd *et al.* 2013, 2015; Hansen *et al.* 2014; Brenn *et al.* 2017; White-Gaynor *et al.* 2019; Lucas *et al.* 2020*b*). In these studies, arrival time anomalies across a seismic array are calculated from P- or S-waveforms using cross-correlation, and the travel time anomalies are then inverted for the velocity structure beneath the array. These studies typically yield detailed images of lateral velocity variations, but provide limited constraints on the depths of the velocity anomalies. Continent-wide body-wave tomography models suffer from highly variable resolution, with good resolution in areas with greater seismic station density, such as parts of West Antarctica, and poor resolution in regions of sparse station coverage.

Body-wave arrivals can also be analysed for structural interfaces or discontinuities below the seismic station using receiver function methodology. This method processes the horizontal and vertical components of body waves to enhance arrivals converted from S to P or P to S at structural interfaces (e.g. Ammon 1991). In general, P-wave receiver functions yield good resolution of Moho depth as well as shallow interfaces, and have provided estimates of crustal thickness for many locations around the continent (Bannister *et al.* 2003; Reading 2006; Chaput *et al.* 2014; Ramirez *et al.* 2017). For some areas with a thick ice sheet, S-wave receiver functions provide better results because the Moho conversion is not obscured by reverberations of seismic energy in the ice (Hansen *et al.* 2009, 2010; Ramirez *et al.* 2016). P-wave receiver functions are also used to study the depth of the 410 and 660 km discontinuities in the mantle, and thus infer the thickness of the transition zone (Reusch *et al.* 2008; Emry *et al.* 2015, 2020). The thickness of the transition zone allows identification of temperature anomalies at transition zone depths due to the different signs of the Clapyron slopes of the olivine to wadsleyite and ringwoodite to bridgemanite phase transitions denoted by these discontinuities (e.g. Bina and Helffrich 1994).

Surface waves from earthquakes and ambient noise

Surface-wave tomography can image the crust and upper mantle with superior depth resolution due to the relationship between surface-wave dispersion and velocity structure with depth. However, good resolution is generally limited to the upper 200–250 km of the Earth, and the resulting structures are usually quite smooth, without sharp interfaces. Traditional surface-wave methods involve analysing individual seismograms to determine the surface-wave group and phase velocity along the path from earthquake to receiver, and then performing a tomographic inversion at each period to generate phase and group velocity maps (Roult *et al.* 1994; Danesi and Morelli 2001; Ritzwoller *et al.* 2001). These maps can then be inverted to determine shear velocity at each location. The inclusion of higher-mode Rayleigh-wave measurements improves the resolution at depths of >200 km (Sieminski *et al.* 2003). In some cases, the surface-wave group velocity measurements may be inverted directly for shear velocity (An *et al.* 2015).

Denser arrays of seismographs on the Antarctic interior can be used to determine the phase velocity of Rayleigh waves from teleseismic earthquakes travelling across the continent (Lawrence *et al.* 2006*b*). In one implementation, the incoming Rayleigh waves are approximated as two interfering plane waves, with seismograms then analysed to determine phase velocity images as a function of period, which are then inverted for velocity structure (Forsyth and Li 2005; Heeszel *et al.* 2013, 2016). Surface-wave phase and group velocities can also be measured from seismic Green's functions derived by cross-correlating ambient noise at station pairs. This approach produces more accurate results than measurements from earthquake records at short periods (Shapiro *et al.* 2005; Pyle *et al.* 2010). Phase velocities determined by ambient noise at shorter periods (*c.* 8–45 s) are often combined with phase velocities from earthquakes at longer periods (*c.* 25–120 s) to provide phase velocities across a wide period band, and to constrain both shallow crustal and deeper mantle structure (Shen *et al.* 2018*a*; O'Donnell *et al.* 2019*b*). In areas with thinner crust, such as the Ross Embayment, the addition of ambient noise provides key constraints to the uppermost mantle structure.

Seismic anisotropy analysis

Seismic anisotropy, or the directional dependence of seismic velocities resulting from directional dependence of the elastic moduli, has the potential to reveal further information about mantle structure and processes. Seismic anisotropy is often a result of lattice-preferred orientation, which is the alignment of the crystallographic axes of anisotropic minerals by rock fabric or deformation. Alternatively, seismic anisotropy can result from shape-preferred orientation, or the alignment of geometrical objects of different seismic velocity, such as elongated minerals or larger objects such as dykes or sills. Studies of naturally deformed mantle xenoliths (Mainprice and Silver 1993; Chatzaras and Kruckenberg 2021), as well as laboratory studies of artificially deformed olivine aggregates (Zhang and Karato 1995), indicate that mantle seismic anisotropy generally results from lattice-preferred orientation. The fast axis of anisotropy is usually parallel to the extension direction or to the flow direction if there is a flow fabric, although other orientations are possible under conditions of high water or high stress (Karato *et al.* 2008). Thus, mantle anisotropy is commonly used to indicate the extension direction of mantle deformation or the direction of mantle flow.

Seismic observations cannot fully determine the general elastic tensor for anisotropic media, so several different techniques are used to constrain different aspects of anisotropy. Radial anisotropy (transverse isotropy) can be thought of as the difference between vertical (V_{SV}) and horizontally (V_{SH}) polarized shear-wave velocities, and is usually measured by jointly analysing Love and Rayleigh surface waves. The uppermost mantle shows positive radial anisotropy ($V_{SH}>V_{SV}$) in most places worldwide (Panning and Romanowicz 2006; Kustowski *et al.* 2008). Seismic models for Antarctica also show strong positive radial anisotropy (Ritzwoller *et al.* 2001; Lloyd 2018; Zhou *et al.* 2019; O'Donnell *et al.* 2019*a*), although the spatial pattern is not well resolved.

Azimuthal anisotropy can be evaluated by measuring the variation of P-, S- or surface-wave velocities with azimuth, or by shear-wave splitting measurements. Shear waves propagating through an anisotropic media are split along fast and slow vibration planes, allowing the fast direction and magnitude of anisotropy to be estimated. SKS and SKKS phases from distant earthquakes are particularly useful, since they eliminate the possibility that the anisotropy is near the source, and results are commonly interpreted as upper-mantle anisotropy beneath the receiver (Silver 1996; Savage 1999). Several SKS splitting studies have been carried out in Antarctica (Muller 2001; Bayer *et al.* 2007; Reading and Heintz 2008; Barklage *et al.* 2009; Accardo *et al.* 2014). Results show strong azimuthal anisotropy in parts of West Antarctica (Fig. 2), and indicate that the average upper-mantle azimuthal anisotropy of West Antarctica is greater than that of East Antarctica (Accardo *et al.* 2014; Lucas *et al.* 2020*a*). The anisotropy fast directions show a complex pattern related to the tectonic development of the region. The directions do not generally align with the velocity of the Antarctic Plate in an absolute reference frame, indicating that the anisotropy is not due to mantle shear from the continent moving relative to the deeper mantle (Accardo *et al.* 2014).

Bayesian joint inversion

The large-scale collection of broadband seismic data at regional or continental scales over the past two decades has enabled the joint analysis of multiple types of seismic observations. Local surface-wave dispersion properties and P-wave Moho conversion waveforms (i.e. receiver functions) are typical choices since they can be incorporated into a relatively simple local inversion for a 1D model (Julià *et al.* 2000; Chang *et al.* 2004; Lawrence and Wiens 2004). Among various realizations, the joint inversion under the Bayesian framework has been popular, since the associated uncertainties of the resulting 3D model can be quantified from Monte Carlo sampling.

For Antarctica, this approach has been applied to data collected by more than 200 seismic stations deployed between 1998 and 2017. Shen *et al.* (2018*b*) constructed a new seismic model for central and West Antarctica by jointly inverting Rayleigh-wave phase and group velocities along with P-wave receiver functions. In this work, ambient noise tomography is used to construct Rayleigh-wave phase and group velocity dispersion maps at relatively short periods (8–40 s), and teleseismic earthquake-derived phase velocity maps at longer periods (32–*c.* 140 s) are taken from Heeszel *et al.* (2016). Comparison between the two sets of phase velocity maps at 30 s presents a difference of -0.002 ± 0.027 km s^{-1} (<0.1% on average) for phase velocity in West and central Antarctica. This difference level is similar to the analogous comparison made in North America where seismic stations were denser, showing that high-quality ambient noise tomography can be applied to the remote continent.

These Rayleigh-wave phase and group velocity maps, together with P receiver function waveforms, were then used to construct a new 3D shear velocity model for the crust and uppermost mantle using a Bayesian Monte Carlo algorithm. Since the velocities are determined from Rayleigh waves, the maps show the velocity of vertically polarized shear waves (V_{SV}). Each Bayesian Monte Carlo inversion provides an ensemble of 1D models that fit both types of data. A final 3D model is then constructed by taking the average 1D models from the ensembles, and associated uncertainties are defined by the standard deviation of the ensembles. An example of the inversion procedure for a station in East Antarctica is shown in Figure 3, giving the fit of the preferred model to the surface-wave data (Fig. 3a) and P-wave receiver function (Fig. 3b). Incorporation of the P-wave receiver function into the inversion reduces the uncertainty in the Moho depth and crustal velocities, as shown by the comparison of the resulting structure and standard deviation (Fig. 3c, d) and prior and posterior distributions (Fig. 3f–h) for the joint inversion, as well as the inversion of surface-wave data alone. A similar Monte Carlo approach applied to Rayleigh- and Love-wave dispersions has helped to construct a 3D anisotropic model for the crust (Zhou *et al.* 2019).

The resulting 3D seismic model covers most of West Antarctica and parts of East Antarctica where there is adequate seismic station coverage for this method (Fig. 4), and extends down to the limit of good depth resolution with fundamental mode Rayleigh waves (about 200 km). The shear-wave

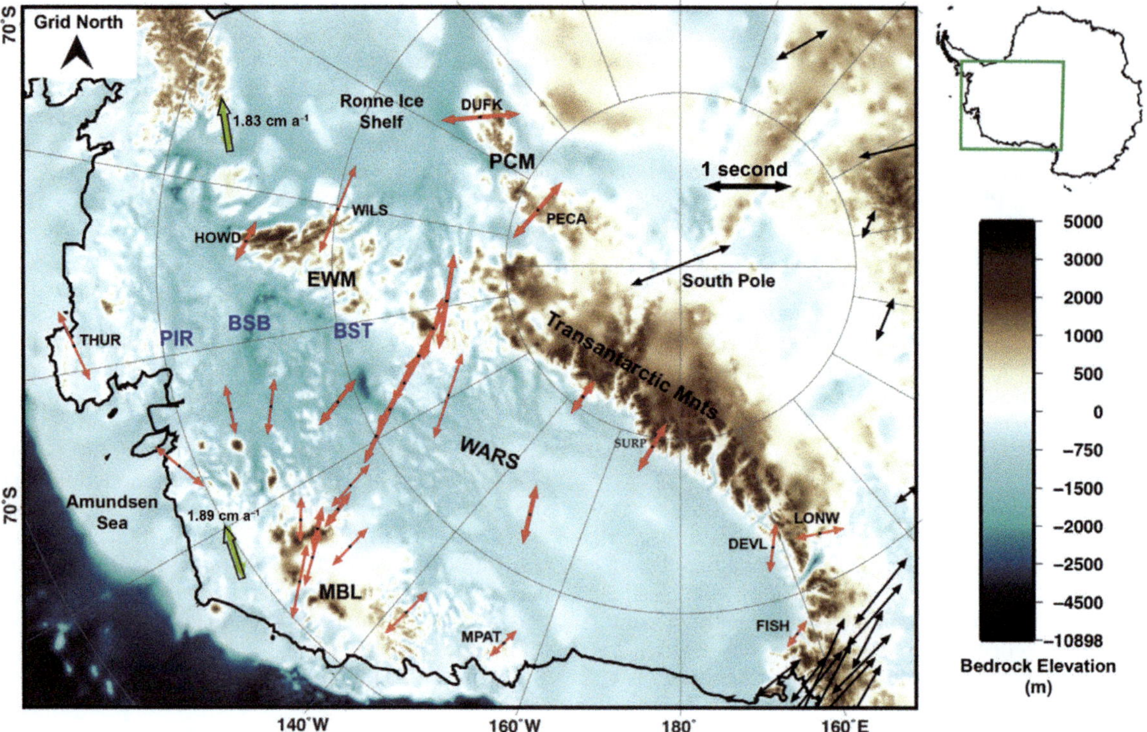

Fig. 2. Shear-wave splitting results for central West Antarctica from Accardo *et al.* (2014). The region shown is outlined in the green box on the inset map of Antarctica. Vector azimuths denote the average fast direction of splitting for arrivals at a given station. The length of the vector is proportional to the splitting time, with the splitting timescale indicated on the right. Thick red vectors represent the highest-quality measurements, thin red vectors represent the intermediate-quality measurements, and black vectors represent results assembled from other studies. Solid green arrows indicate the direction of Antarctic absolute plate motion in the hotspot reference frame. The background colour scale indicates bedrock elevation from Fretwell *et al.* (2013). Abbreviations: PIR, Pine Island Rift; BSB, Byrd Subglacial Basin; BST, Bentley Subglacial Trench; EWM, Ellsworth–Whitmore Mountains; WARS, West Antarctic Rift System.

velocity maps show a clear dichotomy of the tectonically active West Antarctica and the stable and ancient (Fig. 4). In addition, the model shows significant velocity anomalies within both West Antarctica and the central part of East Antarctica, which will be discussed later in this chapter.

Adjoint tomography

The use of 3D numerical wavefield simulations (Komatitsch and Vilotte 1998; Komatitsch and Tromp 1999) along with adjoint inversion methods (e.g. Tarantola 1984; Tromp *et al.* 2005) are proving to be a powerful tool for imaging Earth's seismological structure (e.g. Fichtner *et al.* 2009; Tape *et al.* 2010; Zhu *et al.* 2015; Bozdağ *et al.* 2016). Although computationally expensive, these iterative methods are advantageous because they account for many of the complexities associated with seismic-wave propagation in a complex 3D medium, and thus allow for the accurate and efficient determination of synthetic seismograms and sensitivity kernels for all body- and surface-wave arrivals (Komatitsch and Tromp 2002*a*, *b*). The ability to use a much larger portion of the seismic wavefield, in comparison to traditional seismic imaging, to accurately map out the observational sensitivities to Earth structure, and to avoid simplifying assumptions, such as the high frequency approximation of ray theory, permits higher-fidelity seismic images over large and poorly sampled regions.

Lloyd *et al.* (2020) used this iterative adjoint tomographic inversion to seismically image the entire Antarctic continent and the surrounding southern oceans to depths of 800 km at resolutions approaching that of regional studies. The resulting radial anisotropic model (ANT-20) was determined following 20 iterations, with a 3D starting model based on the global mantle model S362ANI (Kustowski *et al.* 2008) and a modified version of CRUST1.0 (see Lloyd *et al.* 2020). During each iteration, the model was updated to include increasingly shorter wavelength features based on travel time observations between observed and synthetic earthquake seismograms. These observations include P, S, Rayleigh and Love waves, including reflections and overtones, from 270 earthquakes recorded at 323 seismic stations, many of which are shown in Figure 1. Since ANT-20 uses both Love and Rayleigh waves to produce a radially anisotropic model, maps of velocity heterogeneity show the Voigt average shear velocity:

$$V_{S\_Voigt} = \sqrt{\frac{V_{SH}^2 + 2V_{SV}^2}{3}} \quad (1)$$

Because V_{SH} is generally faster than V_{SV} in the upper mantle, the Voigt average velocities are about 1–2% faster than V_{SV}.

The high computational cost of the adjoint inversion, *c.* 3 million CPU hours to produce ANT-20, limits the ability to assess the model's resolution or uncertainty. Thus, evaluation of resolution relies on proxies for data density and point-spread function tests. These tests show that structures appearing in ANT-20, south of 60° S, are reliably resolved within the upper mantle and transition zone (Lloyd *et al.* 2020). The length scales of the imaged features are limited by the smoothing employed in the inversion. The lateral smoothing length ranges from *c.* 140 km in the upper mantle to *c.* 340 km in the transition zone, while the vertical smoothing length is held fixed at *c.* 45 km at all depths throughout the inversion. These facts, combined with comparisons to regional seismic models (e.g. Heeszel *et al.* 2016; Shen *et al.* 2018*b*; O'Donnell *et al.* 2019*a*) and other geophysical observations, indicate that

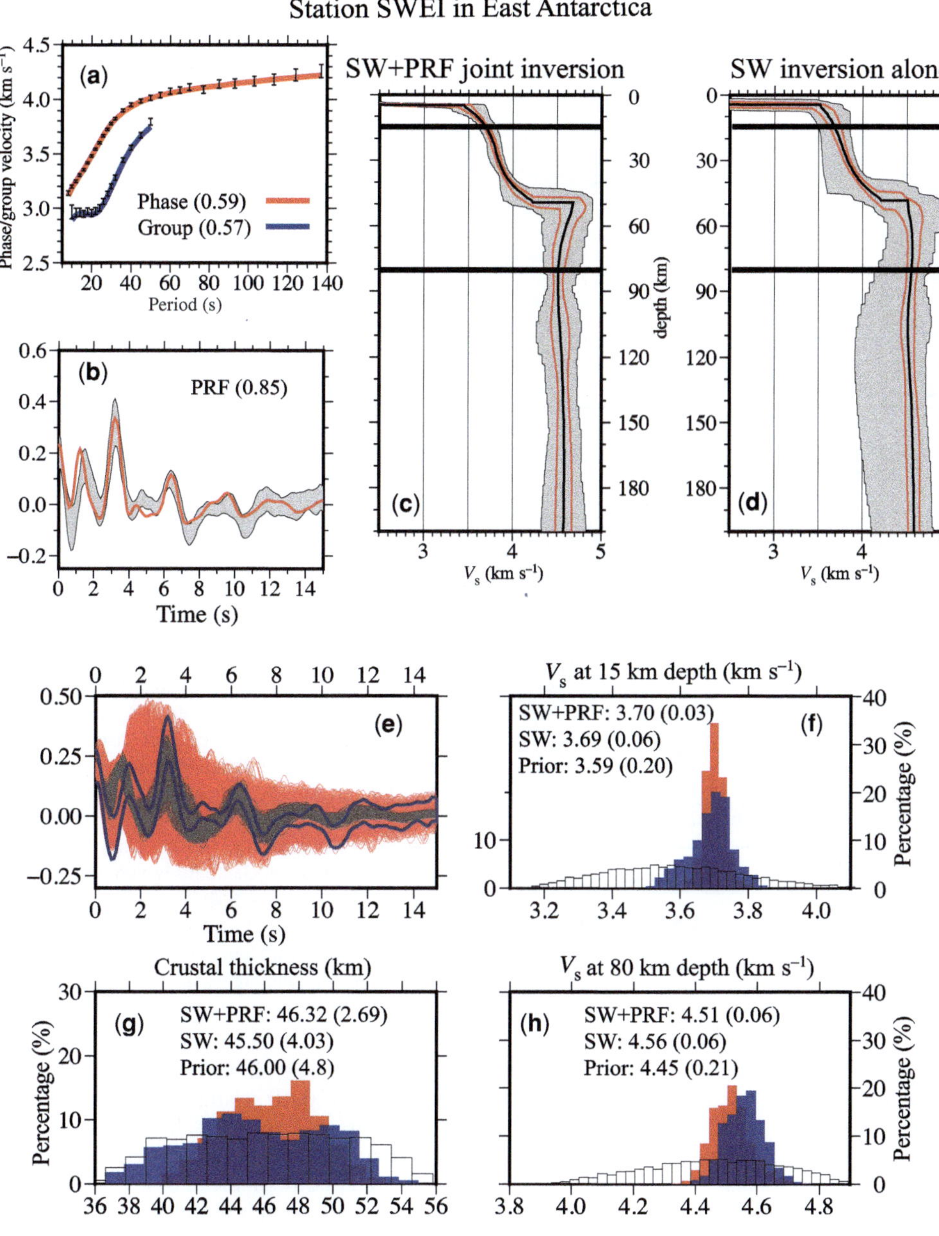

Fig. 3. Example of the joint inversion of surface-wave dispersion (SW) and P-wave receiver functions (PRF) for POLENET seismic station SWEI (Shen *et al.* 2018*b*). (**a**) and (**b**) Observed Rayleigh-wave phase and group velocity dispersion curves (a) and P-wave receiver function waveform (b) at station SWEI. The square root of the reduced χ^2 misfit for the preferred model is is given in parentheses. (**c**) and (**d**) Results from the joint inversion of both surface-wave and receiver function data (c) and surface-wave data alone (d). Grey corridors represent the 1D seismic model emsembles, with black and red lines showing the average shear-wave velocity (V_{SV}) and the standard deviation of the 1D model. (**e**) Predicted receiver function waveform from models in (c) and (d) are shown as red and grey waveforms. Observed receiver functions with uncertainties are shown as two thick blue lines. (**f**)–(**h**) Prior and posterior distributions of crustal V_{SV} (15 km: f), crustal thickness (g) and mantle V_{SV} (80 km: h). Prior distributions are shown by unfilled bars and joint inversion posterior distributions are shown in red, while the surface-wave inversion posterior distribution is in blue.

ANT-20 possesses similarly high resolution across the entire Antarctic region.

Upper-mantle structure of Antarctica's tectonic regions

East Antarctica

Antarctica is divided into two very different geological subcontinents, with East Antarctica (EA) representing largely Precambrian cratonic units and West Antarctica (WA) consisting of terranes formed or tectonically modified during the Mesozoic and Cenozoic. Most of EA shows crustal thicknesses that are typical for continental cratons (Pappa *et al.* 2019*b*; Szwillus *et al.* 2019), whereas WA shows thinner crust similar to crustal thicknesses found in regions of extended Phanerozoic continental crust (Fig. 5). Although Archean–early Paleozoic outcrops have been sampled and mapped near the coasts and in the Transantarctic Mountains (TAMS) (Tingey 1991; Fitzsimons 2000; Goodge *et al.* 2001), most of the EA interior is covered by thick ice sheets, greatly limiting our understanding of its geology. Thus, seismology and other geophysical techniques play an outsized role in constraining the formation and geological history of EA.

Most of EA is underlain by thick continental lithosphere, as defined by strong positive velocity anomalies relative to global averages, which extends to depths of >200 km in several places (Figs 6 & 7). Similar to mantle lithosphere beneath other cratons worldwide, this continental lithosphere represents ancient, cooled mantle that that may be isopycnic, and thus stable due to competing thermal and compositional buoyancy forces (Jordan 1981; Sleep 2005). The fast shear velocities arise predominantly from cold temperatures due to conductive cooling and to a lesser extent from a depleted mantle chemistry (e.g. Lee 2003; Schutt and Lesher 2006). For other continental cratons with extensive xenolith data, there is good agreement between mantle-temperature profiles inferred from mantle xenoliths and shear velocities, with both indicating thick, cold lithosphere extending to depths of >200 km (Priestley and McKenzie 2006).

Although most of EA shows seismic velocity anomalies of about 5–6% fast at depths of 75–150 km, greater variability in seismic velocity is found beneath the East Antarctic Highlands, stretching from western Dronning Maud Land to the Lambert Graben, containing several regions with mantle

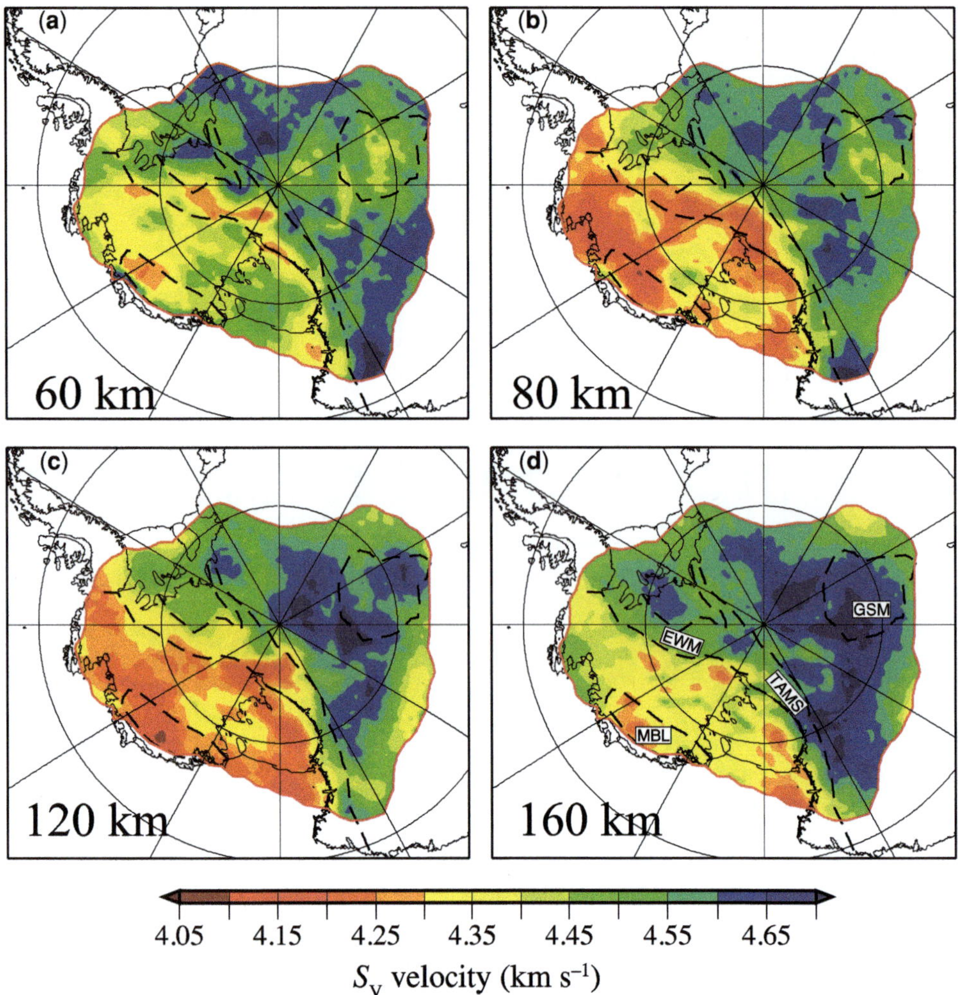

Fig. 4. Uppermost mantle V_{SV} structure from the Bayesian Monte Carlo joint inversion of surface-wave dispersion and receiver functions (Shen *et al.* 2018*b*). Maps show horizontal sections for depths of (**a**) 60 km, (**b**) 80 km, (**c**) 120 km and (**d**) 160 km. Dotted lines enclose the Gamburtsev Subglacial Mountains (GSM), Transantarctic Mountains (TAMS), Ellsworth–Whitmore Mountains (EWM) and Marie Byrd Land highlands (MBL).

velocities close to the global average (Fig. 6). This contrasts with several regions in central Antarctica that show fast-velocity anomalies as large as 7–8% and lithospheric thicknesses of >200 km. The variation in mantle lithospheric thickness and seismic velocity profiles allows us to understand better the distribution of lithospheric terrane ages within EA and to place constraints on the geological evolution of the continent.

The fastest seismic velocities and the thickest continental lithosphere, extending to depths of >200 km, occur along a band extending almost entirely across Antarctica 500–1000 km inboard from the TAMS (Fig. 7). Some of the thickest lithosphere is found inboard of the Miller Range in the TAMS where 3.15–3.05 Ga Nimrod Complex rocks confirm the existence of Archean-age terranes, although Mesoproterozoic units are also found in that region (Goodge and Fanning 1999, 2016), and glacial granitoid clasts show ages ranging from 2.01 to 1.06 Ga (Goodge *et al.* 2017). The thick lithosphere extends to the Shackleton Range, near the Weddell Embayment, where Paleoproterozoic rocks are preserved (Brommer *et al.* 1999; Will *et al.* 2009). Boger (2011) proposed that the 'Shackleton' and 'Nimrod' cratons are connected to late Archean 'Gawler Craton' rocks exposed in Australia (Swain *et al.* 2005) and along the corresponding Tierra Adélie coast of Antarctica (Oliver and Fanning 2002). The continuity of magnetic anomalies between the Antarctic and Australian coasts provides further evidence for the continuity of these cratonic regions prior to the break-up of Gondwana (Pappa and Ebbing 2021). The resulting 'Mawson Continent' extends entirely across EA from the coast on the Australian side to just SE of the Weddell Embayment. The distribution of the fastest seismic velocities and thickest lithosphere in the Lloyd *et al.* (2020) seismic velocity structure correlates well with the proposed extent of the Mawson Continent, supporting the idea that this region represents the late Archean–Paleoproterozoic cratonic nucleus of Gondwana.

Thick, cold lithosphere also extends beneath the Gamburtsev Subglacial Mountains, an enigmatic highland in central Antarctica with peaks reaching to 3000 m (Ferraccioli *et al.* 2011). Surface-wave (Heeszel *et al.* 2013) and body-wave (Lloyd *et al.* 2013) tomography show that the lateral variation in mantle velocity is modest, precluding significant Mesozoic or Cenozoic mantle tectonism or rejuvenation. A systematic comparison of Rayleigh-wave phase velocity curves worldwide shows that the Gamburtsev Mountains phase velocities are best matched by Archean and Paleoproterozoic regions, suggesting that the mountains are underlain by ancient continental lithosphere (Heeszel *et al.* 2013). Shen *et al.* (2018*a*) found that uppermost mantle velocities (Moho–100 km depth) below the Gamburtsev Mountains are anomalously low, by 2–4%, relative to expected lithospheric craton velocities. They are also lower than surrounding regions of EA and the deeper lithosphere. A detailed discussion in Shen *et al.* (2018*b*) demonstrated that this anomaly is unlikely to have a thermal origin but is most likely to be the signature of a compositionally anomalous body, perhaps remnant from a continental collision in Proterozoic or early Paleozoic time (Ferraccioli *et al.* 2011; An *et al.* 2015).

The mountain elevations are largely supported by crustal thicknesses of up to about 55 km (Hansen *et al.* 2010) (see Fig. 5). Initial crustal thickening is likely to have occurred through a compressional orogeny in the Neoproterozoic–

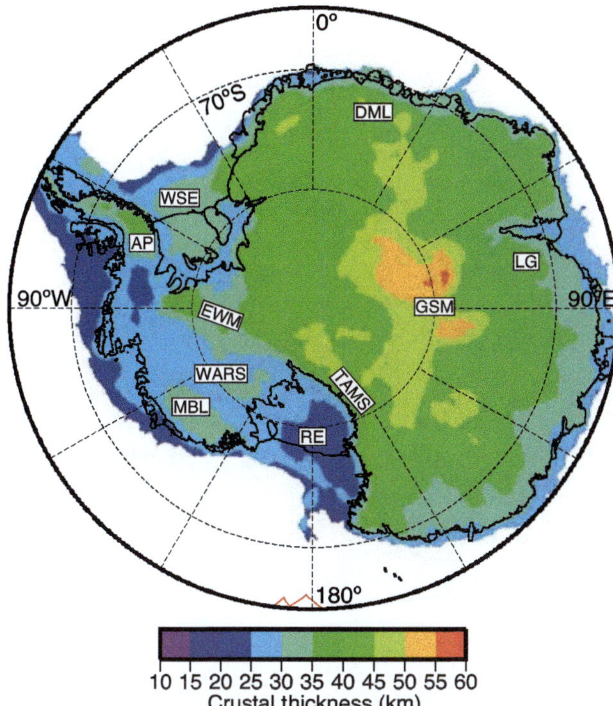

Fig. 5. Crustal thickness of Antarctica. The map is based on an inversion of seismic receiver functions and Rayleigh-wave velocities by Shen *et al.* (2018*b*) in West Antarctica and central Antarctica, and on GOCE satellite gravity constraints over the entire continent (Pappa *et al.* 2019*a*). The crustal thickness map based on gravity was corrected for a systematic crustal thickness offset with respect to the thicknesses derived from seismology. The resulting corrected gravity-based map was averaged with the seismic map in regions where the coverage overlapped. Abbreviations: AP, Antarctic Peninsula; EWM, Ellsworth–Whitmore Mountains; GSM, Gamburtsev Subglacial Mountains; LG, Lambert Graben; MBL, Marie Byrd Land; RE, Ross Embayment; TAMS, Transantarctic Mountains; WARS, West Antarctic Rift System; WSE, Weddell Sea Embayment.

earliest Paleozoic, which is consistent with detrital zircon studies (van de Flierdt *et al.* 2008), with additional uplift associated with the extension of the Lambert Graben in the late Paleozoic–early Mesozoic related to the early stages of Gondwana break-up (Phillips and Laufer 2009; Maritati *et al.* 2020). High elevations in this region are preserved through longer periods of geological time due to low erosion rates (Heeszel *et al.* 2013; Lloyd *et al.* 2013).

Much greater variability in lithospheric thickness is found beneath the highlands stretching from western Dronning Maud Land to the Lambert Graben (Figs 6 & 7). Thinner lithosphere, with thicknesses of the order of 100 km, is found throughout much of this region. The uppermost mantle beneath a portion of Dronning Maud Land is characterized by an absence of fast lithosphere, despite rock ages that extend from the Mesoproterozoic to the Early Paleozoic. The absence of lithosphere in a region of Paleozoic or earlier age suggests that the lithosphere has been tectonically destabilized or removed at a later time. Modest shear velocities at depths of 70–150 km suggest that the Precambrian-age lithosphere of this region has been removed, most likely by delamination (Lloyd *et al.* 2020). Unlike regions of Cenozoic delamination, which are characterized by low upper-mantle velocities as hot asthenosphere has replaced lithosphere (e.g. Levander *et al.* 2011; Shen *et al.* 2018*a*), uppermost mantle velocities in this region are near the global average. Thus, the asthenosphere that replaced the foundered lithosphere has already cooled, indicating a delamination event that occurred prior to the Cenozoic. Jacobs *et al.* (2008) suggested a delamination event occurring at *c.* 500 Ma to explain the late-tectonic granitoid intrusions found in Dronning Maud Land, consistent with the seismic structure.

The Lambert Graben and other neighbouring regions to the west represent an ancient terrane, with basement rocks ranging from Archean to earliest Paleozoic (Fitzsimons 2000), which also lacks thick continental lithosphere (Figs 6 & 7). Seismic structures in this region show that high lithospheric velocities are limited to depths of <75–100 km, suggesting the existence of thin lithosphere, in contrast to surrounding regions where high velocities extend to *c.* 200 km (Lloyd *et al.* 2020). The Lambert Graben and corresponding rift structures in India developed during Carboniferous–early Cretaceous extension associated with the break-up of Gondwana (Phillips and Laufer 2009). Xenolith suites from the Lambert Graben indicate a transition from a relatively cold to a warmer geotherm that is likely to have occurred during the late Paleozoic early rifting stages and Mesozoic formation of the graben (Foley *et al.* 2006, 2021). Thus, mantle geodynamic processes associated with this extension and rifting heated and destabilized the existing Precambrian lithosphere, thinning or removing it.

Transantarctic Mountains and adjacent rifts

The TAMS are traditionally viewed as the tectonic boundary between the thick Precambrian cratonic lithosphere of EA and the Phanerozoic lithosphere domains of WA, although there is some evidence that crust of EA affinity extends to the middle of the Ross Sea (Tinto *et al.* 2019). The 4000 km-long mountain range with peaks of up to *c.* 4000 m exhibits strong along-strike variations, varying from a narrow, rift-shoulder-like orogeny in the central TAMS to broad, 400 km-wide, plateau-like elevated areas in the southernmost and northernmost segments. These variations in topography reflect a complicated history of tectonism along the TAMS, and correlate with changes in the underlying upper-mantle structure, as imaged in the latest seismic models (Shen *et al.* 2018*a*; Lloyd *et al.* 2020).

Seismic tomography reveals a band of slow velocities extending from the Macquarie Triple Junction, through the volcanic Balleny Islands, the extinct Adare Trough spreading centre and along the front of the TAMS (Fig. 6). This pattern suggests that the TAMS and Terror Rift are part of a larger tectonic system linked to mantle geodynamic processes, stretching from the mid-ocean ridge to the southern TAMS. The Adare Trough, located just north of the Ross Embayment near the northern terminus of the TAMS, displays magnetic anomalies resulting from seafloor spreading from 46 to 26 Ma (Granot *et al.* 2013), with slow spreading continuing to 11 Ma and evidence of extensional faulting up to the present (Granot and Dyment 2018). The western side of the Ross Embayment accommodated almost 100 km of extension from 40 to 26 Ma associated with plate motion between EA and WA (Wilson and Luyendyk 2009; Granot *et al.* 2013; Davey *et al.* 2016). Very slow extension and rift sedimentation occurred since that time along the Terror Rift, which parallels the TAMS along the western Ross Embayment (Fielding *et al.* 2008; Martin and Cooper 2010; Granot and Dyment 2018). The slow wave speeds show that the upper mantle retains warm uppermost mantle temperatures caused by an extensional tectonic environment despite the fact that tectonic motion along this boundary has largely ceased, as shown by the absence of resolvable motion in GPS surveys (T. Wilson pers. comm.). Anomalously warm upper-mantle temperatures are also indicated for this region by mantle xenolith studies (e.g. Martin *et al.* 2021*a*).

The northern section of the TAMS mostly constitutes the highlands of northern Victoria Land (NVL). Surface- and

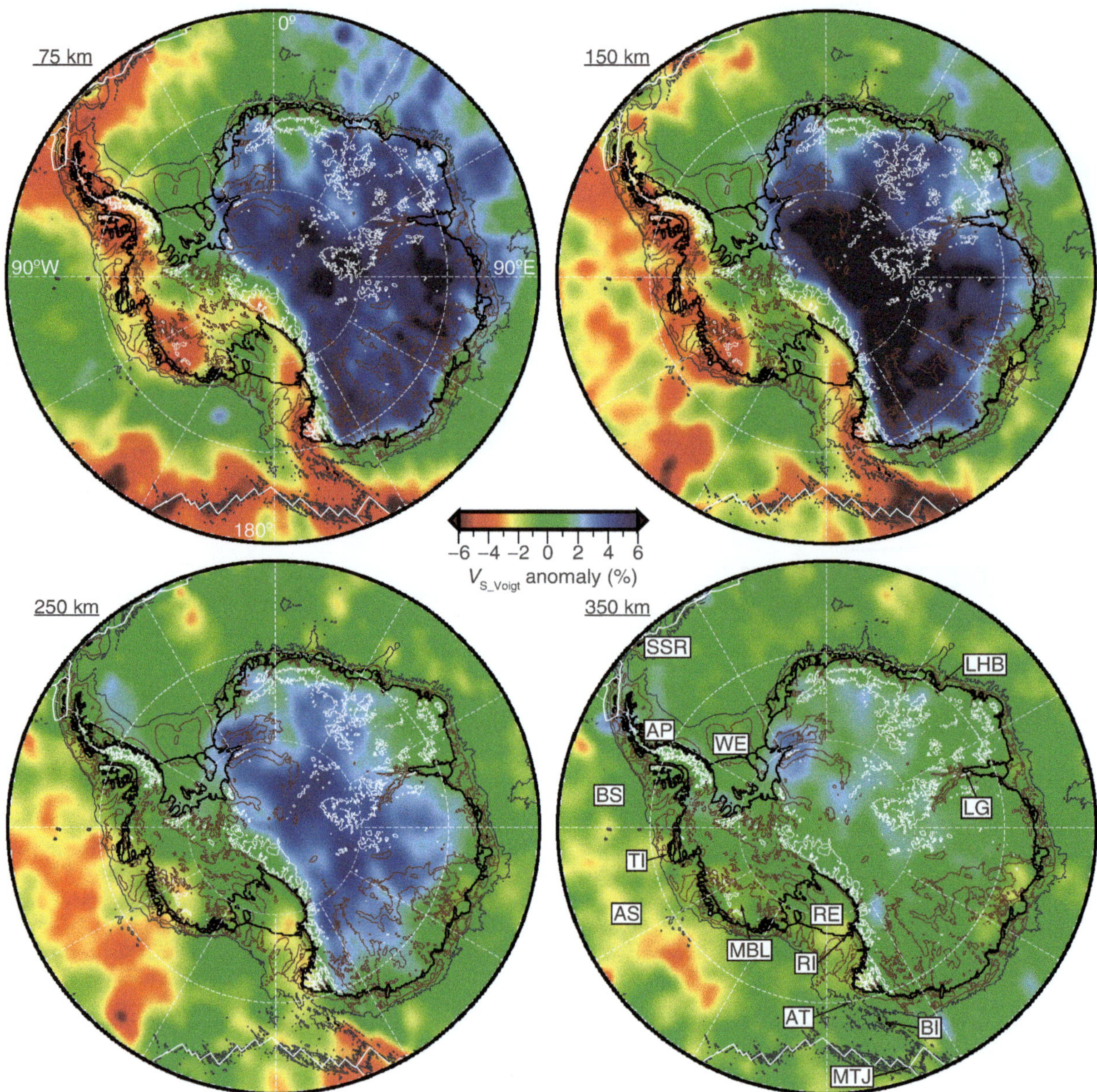

Fig. 6. Tomographic images of the shear-wave velocity structure of seismic model ANT-20 at depths of 75, 150, 250 and 350 km from Lloyd *et al.* (2020). Bathymetry and bedrock topography contours for 1000 m (thin white line) and −500 m (thin brown line), as well as −2500 m (thin dark grey line) in the oceans are also shown (Fretwell *et al.* 2013). Abbreviations: AT, Adare Trough; AS, Amundsen Sea; AP, Antarctic Peninsula; BI, Balleny Islands; BS, Bellingshausen Sea; LG, Lambert Graben; LHB, Lützow-Holm Bay; MTJ, Macquarie Triple Junction; MBL, Marie Byrd Land; RE, Ross Embayment; RI, Ross Island; SSR, South Scotia Ridge; TI, Thurston Island; WE, Weddell Embayment.

body-wave tomography using a regional temporary seismic array revealed that a slow-velocity anomaly in the upper mantle between Moho and 150 km depth is located near the coastline of NVL (Hansen *et al.* 2015; Graw *et al.* 2016; Brenn *et al.* 2017). The continental-scale model from Lloyd *et al.* (2020) shows a large slow-velocity region in the uppermost mantle beneath the high elevations of northernmost Victoria Land and the adjacent areas of the Ross Embayment that connect the Balleny Islands to the north and Ross Island to the south (Fig. 6). This extended region of low velocity in the upper mantle implies a strong thermal contribution to the uplift of NVL.

Detailed body- and surface-wave tomography using temporary seismic stations shows a sharp boundary between slow and fast upper mantle, near the crest of the TAMS near Ross Island (Lawrence *et al.* 2006*a*; Watson *et al.* 2006; Brenn *et al.* 2017; Shen *et al.* 2018*b*; White-Gaynor *et al.* 2019). There is also a sharp boundary along the TAMS between low mantle seismic attenuation in EA and high mantle attenuation beneath Ross Island and the Terror Rift (Lawrence *et al.* 2006*b*). The slow wave speeds in this region are consistent with the general absence of lithosphere and the high upper-mantle temperatures expected beneath a recently extending rift zone with a small amount of continuing decompression melting.

Active volcanism along the western coast of the Ross Embayment, mostly over the past 10 myr, extends from the northernmost tip of the coastline to just south of Ross Island (Kyle 1990). A mantle plume origin has been proposed for Mount Erebus, on Ross Island, based largely on petrological and geochemical data (Kyle *et al.* 1992; Phillips *et al.* 2018), although other studies associate the volcanism with mantle metasomatism from the long subduction history in this region (Day *et al.* 2019; Martin *et al.* 2021*b*). Adjoint tomography shows that Ross Island is underlain by a prominent slow-velocity anomaly extending down to the 410 km

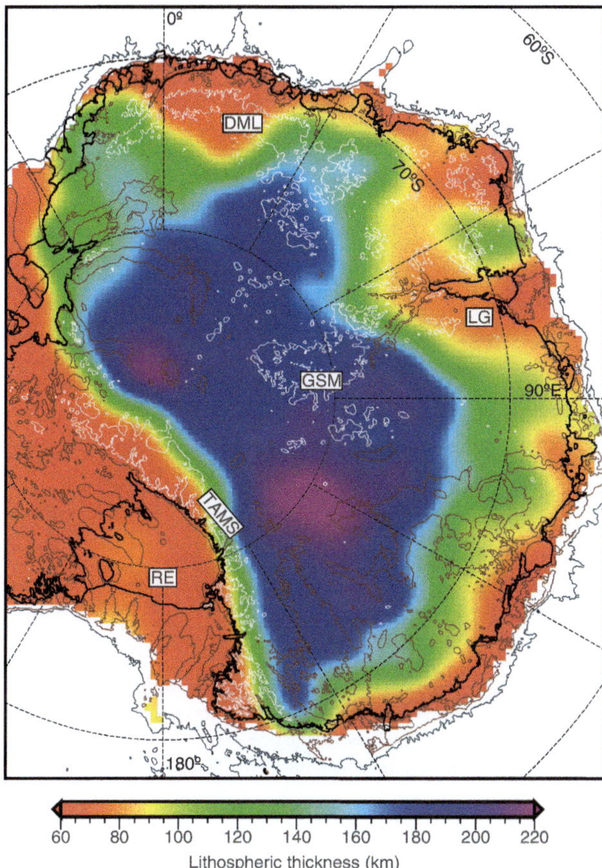

Fig. 7. Lithospheric thickness of East Antarctica based on the ANT-20 seismic model (Lloyd *et al.* 2020). Lithospheric thickness was calculated from velocity profiles assuming that the bottom of the lithosphere is located at the depth of the maximum negative velocity gradient above the low-velocity zone (van der Lee 2002). Geographical labels are the same as in Figure 5.

discontinuity but anomalies beneath this are low amplitude and indistinct (Lloyd *et al.* 2020). P-wave travel-time tomography also shows prominent upper-mantle slow-velocity anomalies but the anomalies diverge laterally in the transition zone (Hansen *et al.* 2014; White-Gaynor *et al.* 2019). This does not necessarily eliminate the plume model, since geodynamical modelling suggests that plumes may be highly tilted in the mid-mantle (Bredow and Steinberger 2021). Emry *et al.* (2020), using receiver functions, found that the mantle transition zone was anomalously thin near Mount Erebus and along the adjacent TAMS, suggesting a warm thermal anomaly in the transition zone. French and Romanowicz (2015), in a global tomographic study, showed little anomaly in the transition zone, a significant slow anomaly at depths of 800–1200 km and no anomalies in the lower mantle (Phillips *et al.* 2018). Thus, it is clear that the region is underlain by a significant thermal anomaly in the upper mantle but the existence of a classic mantle plume arising near the core–mantle boundary remains uncertain.

The southern section of the TAMS shows some unusual characteristics compared with other sections of the TAM. First, unlike the nearby central TAMS where the exposed high mountains are only *c.* 100 km wide, the southern TAMS has a plateau-like elevated region extending into EA for *c.* 400 km. Secondly, unlike the central and northern TAMS where volcanic rocks are only found on the West Antarctic Rift (WARS) side of the high mountains, Miocene-age (*c.* 20–15 Ma) volcanics at Sheridan Bluff and Mount Early (Fig. 8) are all located on the EA side of the peaks (Stump *et al.* 1980). Notably, recently found volcanic rocks in glacial deposits show ages of 25–17 Ma and can be traced to a magnetic anomaly 400 km into EA (Licht *et al.* 2018). Both the topographic and volcanic features pose difficulties to the flexural rift-shoulder mountain model as its support (see Paxman 2021 for a discussion of uplift mechanisms).

Using surface-wave tomography, Heeszel *et al.* (2016) found a slow uppermost mantle beneath the southern TAMS region. They hypothesized that the slow anomaly was either due to the reheating from the WARS activity related to the Terror Rift to the north, or evidence of a Cenozoic lithospheric delamination or destruction event. Shen *et al.* (2018*a*) further improved the images by combining data from Heeszel *et al.* (2016) together with short-period surface-wave velocity maps from ambient noise and receiver functions. The updated images show a low-velocity zone beneath the southern TAMS in the uppermost mantle (Fig. 8) on top of a dipping high-velocity zone (Fig. 9). They suggested that the deeper high-velocity zone represents a foundering lithosphere and the LVZ represents an upwelling asthenosphere, and the system is best interpreted by a lithosphere removal model (Shen *et al.* 2018*a*). Beneath the Thiel Mountains and Whitmore Mountains, the same study also identified an additional low seismic anomaly in the uppermost mantle, indicative of a thermal origin for these elevated areas in WA.

Although Shen *et al.* (2018*a*) showed that the lithosphere has been removed in the southern TAMS region and provided an estimate of the contribution to uplift from the mantle thermal effect due to the lithosphere removal, it is also clear that the TAMS is complicated and significant along-strike variation exists. Proposed uplift mechanisms for the TAMS include flexural uplift (ten Brink *et al.* 1997; Yamasaki *et al.* 2008; Wannamaker *et al.* 2017), thermal mantle support (Lawrence *et al.* 2006*c*; Brenn *et al.* 2017), and crustal thickness and density variations (Bialas *et al.* 2007; Huerta 2007). As an

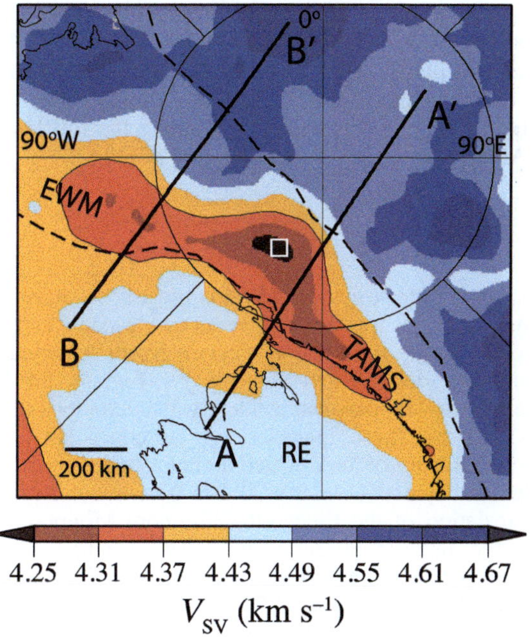

Fig. 8. Average V_S in the uppermost 50 km of the mantle beneath central Antarctica, showing the absence of cold, fast mantle lithosphere beneath the Transantarctic Mountains (Shen *et al.* 2018*a*). Black lines indicate the locations of the two vertical profiles shown in Figure 9. A small open box marks the approximate location of the Mount Early and Sheridan Bluff volcanism. The Ellsworth–Whitmore Mountains (EWM), the Transantarctic Mountains (TAMS) and the Ross Embayment (RE) are labelled. Lines A–A′ and B–B′ denote the location of the cross-sections shown in Figure 9.

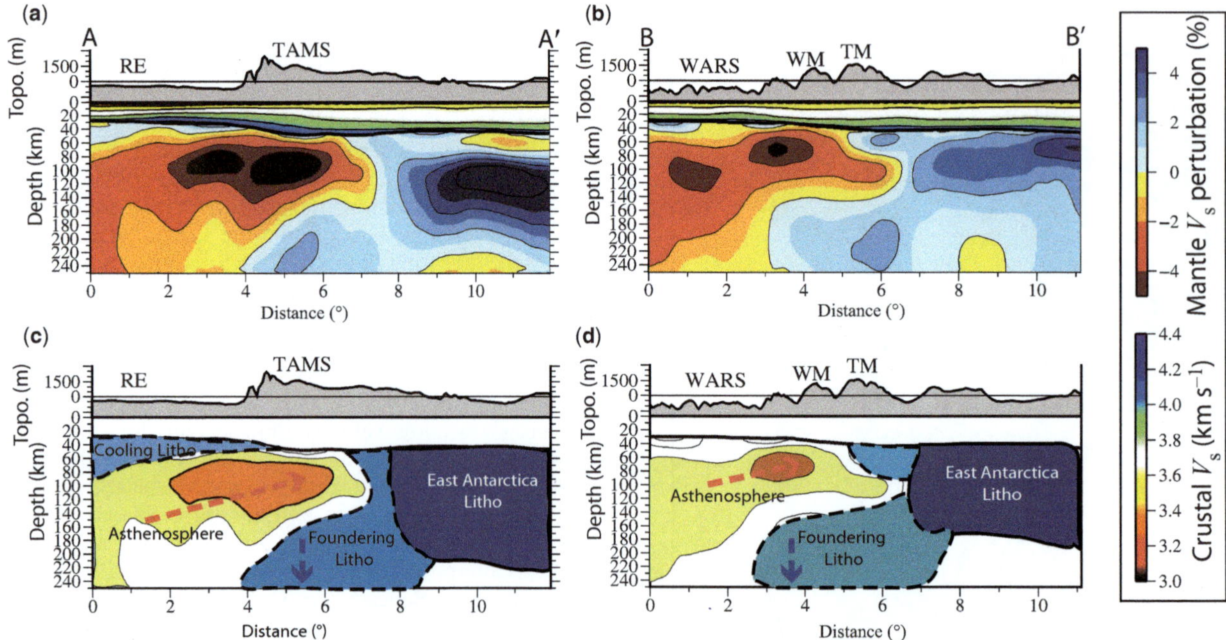

Fig. 9. Cross-sections beneath the southern Transantarctic Mountains and Whitmore Mountains along lines A–A′ and B–B′ in Figure 8, from Shen *et al.* (2018*a*). (**a**) and (**b**) show V_{SV}, and (**c**) and (**d**) show the corresponding interpretation. The elevation of the solid Earth surface in metres is shown above each plot. V_{SV} in the crust is plotted as an absolute value, and V_{SV} in the mantle is plotted as the per cent perturbation relative to the averaged 1D V_{SV} structure of the study region. Geographical features are marked by abbreviations: RE, Ross Embayment; TAMS, Transantarctic Mountains; TM, Thiel Mountains; WARS, West Antarctic Rift System; WM, Whitmore Mountains.

example of along-strike variation, a recent magnetotelluric (MT) study by Wannamaker *et al.* (2017) showed that in the central TAM, cratonic lithosphere extends nearly up to the TAMS range front. Cantilevered flexural uplift appears to be the preferred uplift mechanism and the thermal contribution from the mantle seems smaller in this region. These differences between the southern and central TAMS are consistent with the range morphology, in which the southern TAMS form a large plateau area, whereas the central TAMS form a much narrower mountain range. Indeed, such structural and morphological variability is evident along the full length of the TAMS (Shen *et al.* 2018*a*; Lloyd *et al.* 2020).

West Antarctica

The upper-mantle structure of WA is broadly similar to other regions worldwide that have experienced Mesozoic and Cenozoic tectonic activity, such as western North America. This structure includes shallow Moho depth (Fig. 5), thin lithosphere and slow upper-mantle shear wave velocities in many places (Fig. 6). These characteristics are consistent with the geological history of WA, which includes rifting in the Weddell Sea Embayment (Jordan *et al.* 2017) as well as the rotation and translation of the Ellsworth–Whitmore Mountain block during the Jurassic (Grunow *et al.* 1987), and Cretaceous extension between the WA crustal block and Zealandia (e.g. Siddoway 2008; Wobbe *et al.* 2012). The latter events culminated in the regional cessation of subduction along the margin of Marie Byrd Land. However, subduction continued from Thurston Island to the Antarctic Peninsula, where it gradually ceased from south to north during the Cenozoic (Eagles *et al.* 2004). The extended continental crust of the WARS underwent further episodes of focused extension during the Cenozoic (Granot *et al.* 2013; Davey *et al.* 2016).

The eastern Ross Embayment, central WA and the Weddell Sea show modest upper-mantle velocity anomalies, with velocities about 1–2% slower than the global average reference velocity (Figs 4, 6, 10 & 11). Mantle seismic velocities increase eastwards across the Ross Embayment, consistent with the increase in lithospheric age from the late Cenozoic rifting in the western Ross Embayment (Fielding *et al.* 2008) to early Cenozoic and Cretaceous lithosphere in the east (Wilson and Luyendyk 2009). Although aerogeophysical evidence suggests that the boundary between crust of EA and WA affinities occurs in the middle of the Ross Embayment (Tinto *et al.* 2019), from a seismic structure and tectonics standpoint the boundary occurs along the TAMS. Detailed surface-wave results show that the structure of the WARS consists of higher-velocity lithospheric mantle extending to depths of about 70–80 km, with lower velocities beneath (Heeszel *et al.* 2016; O'Donnell *et al.* 2017; Shen *et al.* 2018*b*), consistent with cooling since the major extensional episodes during the Mesozoic and early Cenozoic (Siddoway 2008; Granot *et al.* 2013). Some smaller regions show lower velocities, possibly delineating the locus of limited late Cenozoic extension (Lloyd *et al.* 2015). The Weddell Sea and areas beneath the Ronne Ice Shelf show upper-mantle velocities that are intermediate between the fast EA lithosphere and the lower velocities in other regions of WA (Figs 6 & 11). This is consistent with the age of the lithosphere in this region, dating to the Mesozoic opening of Gondwana (Jordan *et al.* 2013; 2017).

Shear-wave splitting measurements indicate large shear-wave splitting times (>1 s) in the southern WARS (Fig. 2). The splitting measurements show very consistent fast directions approximately perpendicular to the strike of the nearby Whitmore Mountains and other topographical features, and parallel to likely extension directions for the WARS opening. This strong upper-mantle azimuthal anisotropy is interpreted as resulting from lattice-preferred orientation induced by asthenospheric mantle strain associated with Cenozoic extension of the WARS (Accardo *et al.* 2014).

Marie Byrd Land (MBL), an elevated volcanic dome with 18 major subaerial shield and stratovolcanoes (LeMasurier and Rex 1989), is underlain by low mantle velocities (Figs 6

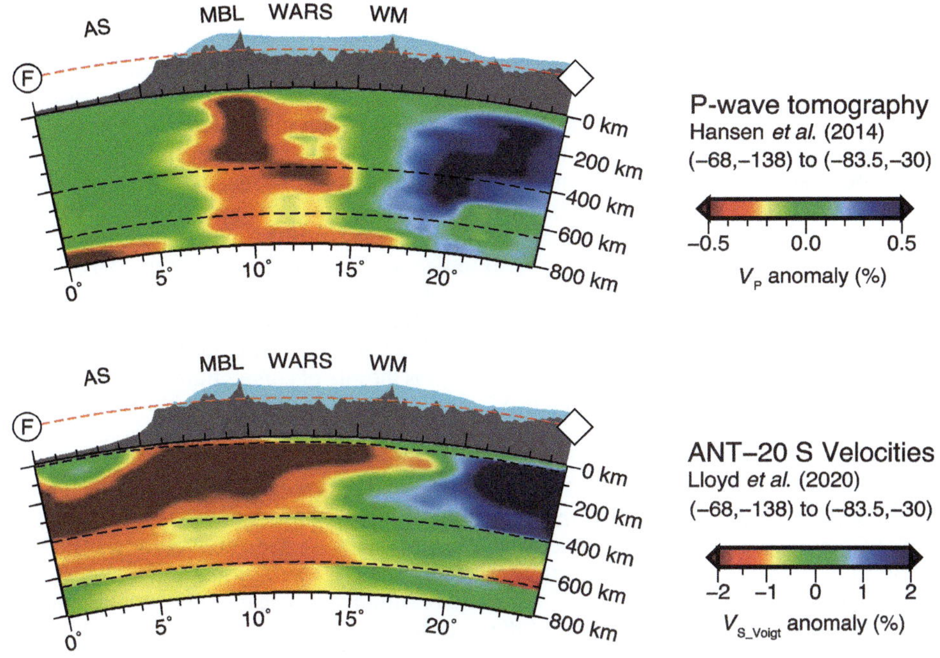

Fig. 10. Comparison of P-wave travel-time tomography (Hansen *et al.* 2014) with adjoint waveform tomography (Lloyd *et al.* 2020) results for a profile across the Amundsen Sea (AS), Marie Byrd Land (MBL), the West Antarctic Rift System (WARS) and the Whitmore Mountains (WM), along line F in Figure 11. The images show generally similar features in areas below the continent with good P-wave ray-path coverage (below MBL and East Antarctica) but anomalies are missing from the P-wave model in regions with little ray-path coverage, such as the Amundsen Sea coast. Both models show that slow-velocity anomalies beneath the Marie Byrd Land dome continue into the lower mantle, consistent with a mantle plume origin for the topography and volcanism.

& 10). Late Quaternary alkaline volcanism at Mount Berlin and Mount Takahae (Wilch *et al.* 1999), and an inferred subglacial magmatic system near Mount Waesche (Lough *et al.* 2013), demonstrate current volcanic activity. The crustal thickness of about 30–33 km (Fig. 5) is somewhat greater than the surrounding regions (*c.* 25–28 km) but this is not sufficient to explain the elevated topography, suggesting that the elevation is partially supported by a low-density thermal anomaly in the uppermost mantle (Chaput *et al.* 2014; Shen *et al.* 2018*a*). Several previous studies have proposed a mantle

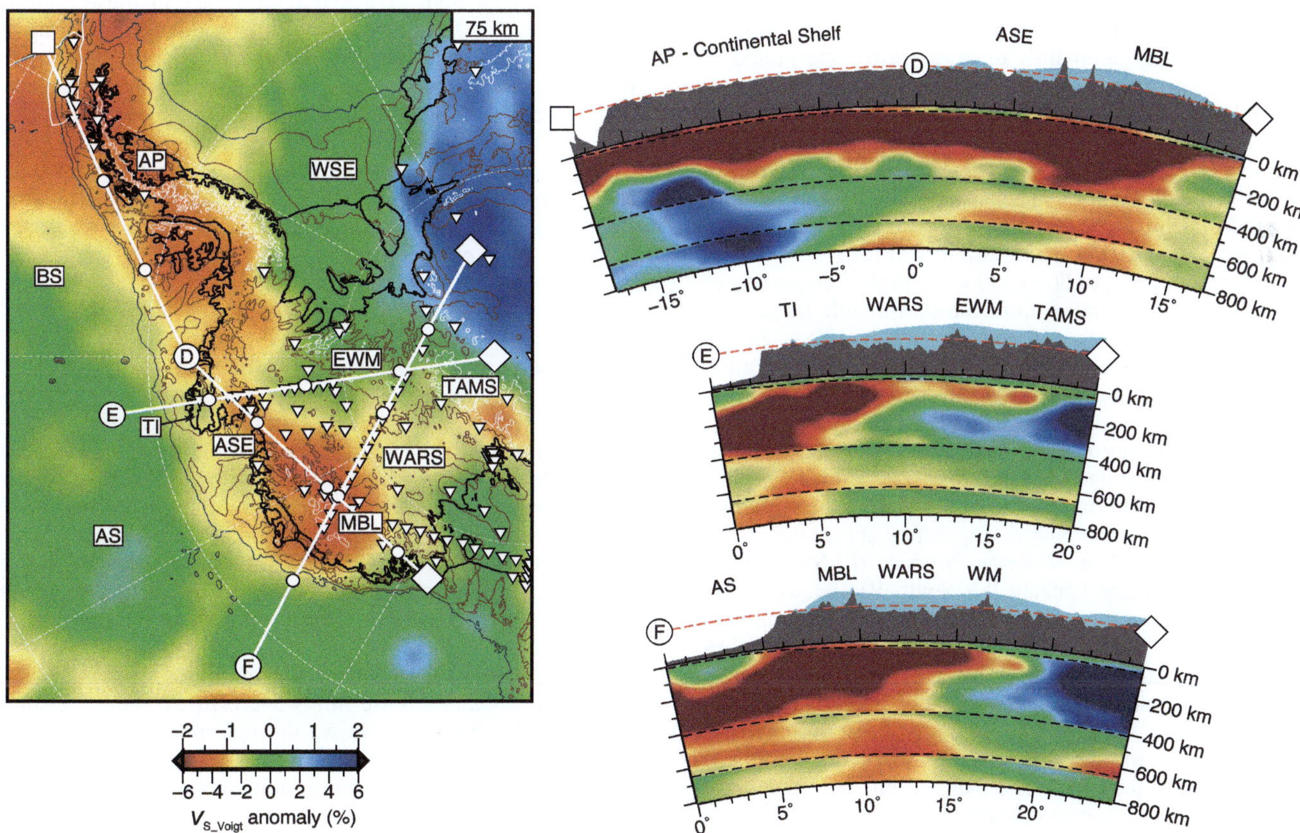

Fig. 11. Images of shear-wave speed structure beneath West Antarctica at 75 km depth and along profiles D, E and F from Lloyd *et al.* (2020). Horizontal and vertical slices have a shear-wave speed range of ±6 and ±2%, respectively. The horizontal slice depicts broadband seismic stations, wave-speed anomalies, bathymetry and topography as in Figure 6. Abbreviations: AS, Amundsen Sea; ASE, Amundsen Sea Embayment; AP, Antarctic Peninsula; BS, Bellingshausen Sea; EWM, Ellsworth–Whitmore Mountains; MBL, Marie Byrd Land; TAMS, Transantarctic Mountains; TI, Thurston Island; WARS, West Antarctic Rift System; WSE, Weddell Sea Embayment; WM, Whitmore Mountains.

plume origin for the MBL dome, based on the elevated topography and volcano chemistry and petrology (Hole and LeMasurier 1994; LeMasurier and Landis 1996; Behrendt 1999; Panter *et al.* 2000). Subaerial volcanic peaks are located along linear arrays, in many cases with age progressions, but directions of propagation are not consistent between difference lines, suggesting that the linear trends may be controlled by pre-existing fractures (LeMasurier and Rex 1989; Paulsen and Wilson 2010) rather than relative motion between the Antarctic Plate and the mantle plume. As an alternative to the mantle plume model, Finn *et al.* (2005) proposed that much of the volcanism of WA results from the volatile-enhanced melting of metasomatized lithosphere formed during Mesozoic subduction along this margin.

Both body-wave (Hansen *et al.* 2014; Lloyd *et al.* 2015) and surface-wave (Heeszel *et al.* 2016; Shen *et al.* 2018*b*) tomography studies show low velocities in the upper mantle beneath the MBL dome. The surface-wave results show the existence of a higher-velocity mantle lithosphere to depths of 60–80 km, underlain by lower seismic velocities. Methods with good sensitivity to mid-mantle depths generally show low-velocity anomalies beneath the MBL dome in the transition zone and upper part of the lower mantle. Figure 10 compares the results from P-wave tomography (Hansen *et al.* 2014) and adjoint tomography (Lloyd *et al.* 2020), with both results showing low velocities in the mid-mantle beneath the MBL dome. The P-wave tomography has almost no resolution of the coastal and oceanic regions due to a lack of seismic stations resulting in poor ray-path coverage but the adjoint tomography, which has good resolution, shows that the low upper-mantle velocities extend past the MBL coastline to a low-velocity anomaly at depths of 150–350 km beneath the Amundsen Sea.

The tomography results are consistent with a thermal plume extending at least from the mid-mantle to the surface beneath MBL. The velocity anomalies are most easily interpreted in terms of warmer mantle temperatures, as the effects of water on seismic velocities are uncertain and recent laboratory results suggest that water has an insignificant effect on upper-mantle seismic velocities (Cline *et al.* 2018). The magnitude of the shear velocity anomaly is consistent with a 150°C upper-mantle thermal anomaly, which can explain the excess topography (Lloyd *et al.* 2015). The continuity of the slow-velocity anomalies from depths of 80 to 800 km is highly suggestive of a mantle plume originating from the mid-mantle or deeper, which is consistent with geochemical evidence (e.g. Panter *et al.* 2000; Handler *et al.* 2021). However, the connection between the MBL anomaly and the offshore anomaly, as well as the diverse geographical trends of the volcanic lines, suggest the possibility of a more complicated geodynamical situation than indicated by the classical simple vertical plume model. Simulations of mantle plumes in global mantle-flow models suggest that plumes originating at the edge of the Pacific Large Low Shear Velocity Province (LLSVP) at the core–mantle boundary beneath the South Pacific would be tilted southwards by the large-scale mantle flow pattern, possibly explaining the MBL and offshore seismic anomalies (Bredow and Steinberger 2021).

Slow upper-mantle velocity anomalies extend along the coast from MBL through the Amundsen Sea Embayment and northwards to the tip of the Antarctic Peninsula (Fig. 11). Along the MBL and Amundsen Sea coastlines, slow velocities extend from the bottom of the thin lithosphere (70 km) to about 200 km depth, beneath late Pleistocene–recent volcanoes such as Mount Siple in MBL (Wilch *et al.* 1999) and Mount Hudson (Corr and Vaughan 2008) near the Amundsen Sea Embayment. The anomalies deepen to 150–300 km depth offshore beneath 90 Ma oceanic lithosphere formed during the separation of Zealandia from Antarctica (Eagles *et al.* 2004), where an extensive shallow bathymetric anomaly is found (Wobbe *et al.* 2014). The seismic and bathymetric anomalies may result from warm upper mantle near the region where spreading originated between Zealandia and Antarctica, in which the initiation of spreading has been attributed to a mantle plume (Weaver *et al.* 1994). The presence of low-viscosity hydrous mantle material resulting from the long-lived Mesozoic subduction zone that preceded the rifting along the northern Antarctic margin may also play a role (Finn *et al.* 2005; Sutherland *et al.* 2010).

Slow seismic anomalies beneath the Antarctic Peninsula are limited to depths of <200 km, and are underlain by fast velocities, in contrast to MBL and the Amundsen Sea regions (Fig. 11). This region was the site of subduction of the Phoenix Plate beneath Antarctica during the Mesozoic, with subduction ending from south to north during the Cenozoic as the Phoenix–Antarctic spreading centre was subducted (Eagles *et al.* 2004). The fast velocities found in the transition zone beneath the northern peninsula may represent the remnants of the subducted Phoenix slab. As the subducting ridge traversed northwards along the peninsula, it formed a slab window in its wake. Slab windows are generally accompanied by erosion of the overriding plate lithosphere and its replacement with hot upwelling asthenosphere (Groome and Thorkelson 2009). We interpret the slow seismic velocities extending along the entire length of the western margin of the peninsula as due to the effects of the slab window, consistent with the location and limited depth extent of the anomalies.

The slow mantle velocities beneath MBL, the Amundsen Sea Embayment and the Antarctic Peninsula imply warm upper-mantle temperatures, with important implications for ice sheet models and projections of the future of the Antarctic ice sheet. Heat-flow estimates, based on the seismic structure, show relatively high geothermal heat flux in these regions (Shen *et al.* 2020), influencing water production and drag at the base of the ice sheet (Pollard *et al.* 2005; Pattyn 2010). Warm mantle temperatures also imply low mantle viscosity, which predicts short glacial isostatic adjustment timescales that may have important influences on the evolution of the Antarctic ice sheet (Gomez *et al.* 2015; Whitehouse *et al.* 2019). The relationship between Antarctic seismic structure, mantle viscosity and glacial isostatic adjustment is discussed further in Ivins *et al.* (In press).

Conclusions and prospects for future work

The deployment of numerous seismic stations and the development of new seismic analysis tools over the last two decades have led to great improvements in our knowledge of the upper-mantle structure beneath Antarctica. Analysis of empirical Green's functions from ambient noise correlation allows much better resolution of surface-wave dispersion curves at short periods. Bayesian Monte Carlo inversion methods can be used to determine the structure that best fits constraints from multiple types of seismic data, and also provide uncertainty estimates. Adjoint inversion methods, combined with numerical calculations of full waveform synthetic seismograms in 3D structures, allows the determination of higher-resolution structure models of the entire Antarctic continent and surrounding regions throughout the upper mantle and transition zone.

Seismic structure models calculated with these methods now have sufficient resolution to provide important insights into the geological history of the continent. EA shows thick continental mantle lithosphere similar to Archean and Early Proterozoic terranes on other continents. The fast shear velocities of the lithosphere down to >200 km depth are interpreted as resulting from cold mantle temperatures, and to a lesser extent from chemical depletion, based on studies from other cratonic regions. The deepest and largest lithospheric velocity anomalies occur along a band extending almost entirely across Antarctica 500–1000 km inboard from the TAM, probably representing a 'Mawson Continent' that formed the core of ancient cratonic lithosphere around which other terranes accreted. Some other regions of EA, notably the Dronning Maud Land highlands and the Lambert Graben region, show much thinner lithosphere, indicating thermal modification of the lithosphere during the Phanerozoic.

The TAMS front represents a major boundary in the mantle between thick cratonic lithosphere in EA and much thinner lithosphere in WA. In the Ross Embayment region, it is flanked by the late Cenozoic rifts, which are underlain by an extensive low-velocity upper-mantle anomaly extending from the mid-ocean ridges north of Antarctica to the southern TAM. This velocity anomaly results from warm mantle along the trace of the rift, which persists even though extension has now apparently ceased. The TAMS show major structural variations along-strike, with the broad southern TAMS underlain at shallow mantle depths by low seismic velocities indicating an absence of lithospheric mantle, consistent with uplift as a result of lithospheric removal. In contrast, the much narrower central TAMS is underlain by a sharp boundary between slow WA and fast EA mantle, consistent with classical rift-shoulder or flexural uplift models.

The upper-mantle seismic structure of WA is dominated by large slow-velocity anomalies beneath central MBL, and along the Pacific coastline from MBL through the Antarctic Peninsula. The central MBL anomaly extends through the transition zone, and may represent the thermal anomaly resulting from a mantle plume. Slow velocities beneath the Amundsen Sea coast link to deeper anomalies offshore, suggesting a connection with deeper mantle processes to the north. Slow anomalies beneath the Antarctic Peninsula are limited to depths <200 km, and are likely to represent thermal anomalies resulting from the erosion of continental lithosphere and replacement by warm mantle during subduction of the Antarctic–Phoenix spreading centre.

Continuing work over the next decade should result in higher-resolution seismic images of the mantle structure due to improvements in both data collection and seismic analysis. Technological improvements should permit the deployment of denser arrays of seismic stations, as well as deployment of stations in regions without previous instrumentation. The development of much smaller and lighter autonomous seismic stations with more reliable power supplies should facilitate deployment in remote parts of the continent. Better and lower-cost satellite communications should allow real-time transmission of seismic data from remote sites, and should also reduce the cost of operation by reducing the need for maintenance visits. An international effort is needed to instrument the vast regions of EA that are so far without any seismograph deployments.

Improvements in data analysis will also allow advances to take place using existing data. Surface-wave data can be analysed to yield better constraints on anisotropy. Combined analysis of Love and Rayleigh waves will yield estimates on the variations in radial seismic anisotropy, the difference between vertically and horizontally polarized shear waves (O'Donnell *et al.* 2019*b*; Zhou *et al.* 2019). Analysis of Rayleigh-wave phase velocities for azimuthal anisotropy will allow resolution of the depth distribution of azimuthal anisotropy, which is poorly constrained by SKS analysis. The resulting maps of the lateral and depth distribution of azimuthal anisotropy will provide important constraints on the distribution and orientation of mantle fabric due to either past tectonic processes or to current mantle deformation and flow. Improved techniques will also increase the resolution of isotropic structural variations. Double difference tomography, which uses the differences in waveforms from the same earthquake recorded at different seismic stations, will improve resolution on the Antarctic continent by localizing the structure kernels in the vicinity of the stations (Yuan *et al.* 2016). The incorporation of Green's functions from ambient noise cross-correlation into adjoint tomography will also increase resolution (Chen *et al.* 2014; Liu *et al.* 2017). With the continued improvement of datasets and analysis methods, future seismic studies of the Antarctic mantle will provide important new insights into the geological history and current geodynamic processes of Antarctica.

Acknowledgements We thank the many individuals and organizations who have helped with the installation and maintenance of Antarctic seismic stations through the years. The seismic instruments were provided by the Incorporated Research Institutions for Seismology (IRIS) through the PASSCAL Polar Support Services. We thank Anya Reading, Fausto Ferraccioli, Adam Martin and Wouter van der Wal for helpful comments on an earlier draft.

The portable instrumentation and data management facilities of the IRIS Consortium are supported by the National Science Foundation's Seismological Facilities for the Advancement of Geoscience (SAGE) Award under Cooperative Support Agreement OPP-1851037.

Author contributions **DAW**: conceptualization (lead), funding acquisition (lead), investigation (equal), methodology (equal), project administration (lead), resources (lead), software (supporting), supervision (lead), validation (equal), visualization (equal), writing – original draft (lead), writing – review & editing (lead); **WS**: conceptualization (supporting), data curation (lead), formal analysis (lead), funding acquisition (supporting), investigation (equal), methodology (equal), project administration (supporting), resources (supporting), software (lead), validation (equal), visualization (equal), writing – original draft (supporting), writing – review & editing (supporting); **AJL**: conceptualization (supporting), data curation (lead), formal analysis (lead), funding acquisition (supporting), investigation (equal), methodology (equal), project administration (supporting), resources (supporting), software (lead), validation (equal), visualization (lead), writing – original draft (supporting), writing – review & editing (supporting).

Funding This work was funded by the United States National Science Foundation with grants PLR-1246712, OPP-1744883, OPP-1744889 and OPP-1945693 awarded to D.A. Wiens, and grant OPP-1945856 awarded to W. Shen.

Data availability Original data can be obtained from the IRIS Data Management Center (http://ds.iris.edu/ds/nodes/dmc/data/). The shear-wave velocity structures of the Bayesian Monte-Carlo and adjoint ANT-20 models can be obtained through the Incorporated Research Institutions for Seismology (IRIS) Earth Model Collaboration (http://ds.iris.edu/ds/products/emc/).

References

Accardo, N.J., Wiens, D.A. *et al.* 2014. Upper mantle seismic anisotropy beneath the West Antarctic Rift System and surrounding region from shear wave splitting analysis. *Geophysical Journal International*, **198**, 414–429, https://doi.org/10.1093/gji/ggu117

Ammon, C. 1991. The isolation of receiver effects from teleseismic P waveforms. *Bulletin of the Seismological Society of America*, **81**, 2504–2510.

An, M., Wiens, D.A. *et al.* 2015. S-velocity model and inferred Moho topography beneath the Antarctic Plate from Rayleigh waves. *Journal of Geophysical Research: Solid Earth*, **120**, 359–383, https://doi.org/10.1002/2014JB011332

Anandakrishnan, A., Voigt, D.E., Burkett, P.G. and Henry, R. 2000. Deployment of a broadband seismic network in west Antarctica. *Geophysical Research Letters*, **27**, 2053–2056, https://doi.org/10.1029/1999GL011189

Baker, M.G., Aster, R.C. *et al.* 2019. Seasonal and spatial variations in the ocean-coupled ambient wavefield of the Ross Ice Shelf. *Journal of Glaciology*, **65**, 912–925, https://doi.org/10.1017/jog.2019.64

Bannister, S., Yu, J., Leitner, B. and Kennett, B.L.N. 2003. Variations in crustal structure across the transition from West to East Antarctica, Southern Victoria Land. *Geophysical Journal International*, **155**, 870–884, https://doi.org/10.1111/j.1365-246X.2003.02094.x

Barklage, M., Wiens, D.A., Nyblade, A. and Anandakrishnan, S. 2009. Upper mantle seismic anisotropy of South Victoria Land and the Ross Sea coast, Antarctica from SKS and SKKS splitting analysis. *Geophysical Journal International*, **178**, 729–741, https://doi.org/10.1111/j.1365-246X.2009.04158.x

Bayer, B., Muller, C., Eaton, D.W. and Jokat, W. 2007. Seismic anisotropy beneath Dronning Maud Land, Antarctica, revealed by shear wave splitting. *Geophysical Journal International*, **171**, 339–351, https://doi.org/10.1111/j.1365-246X.2007.03519.x

Behrendt, J.C. 1999. Crustal and lithospheric structure of the West Antarctic Rift System from geophysical investigations – a review. *Global and Planetary Change*, **23**, 25–44.

Bentley, C. 1973. Crustal structure of Antarctica. *Tectonophysics*, **20**, 229–240, https://doi.org/10.1016/0040-1951(73)90112-1

Bernacchi, L.C. and Milne, J. 1908. Earthquakes and other earth movements recorded in the Antarctic regions, 1902–1903. *In*: *National Antarctic Expedition 1901–1904. Physical Observations*. The Royal Society, London, 37–96.

Bialas, R.W., Buck, W.R., Studinger, M. and Fitzgerald, P.G. 2007. Plateau collapse model for the Transantarctic Mountains–West Antarctic Rift System: insights from numerical experiments. *Geology*, **35**, 687–690, https://doi.org/10.1130/G23825A.1

Bina, C.R. and Helffrich, G. 1994. Phase transition Clapeyron slopes and transition zone seismic discontinuity topography. *Journal of Geophysical Research*, **99**, 15853–15860, https://doi.org/10.1029/94JB00462

Bredow, E. and Steinberger, B. 2021. Mantle convection and possible mantle plumes beneath Antarctica – insights from geodynamic models and implications for topography. *Geological Society, London, Memoirs*, **56**, https://doi.org/10.1144/M56-2020-2

Brenn, G.R., Hansen, S.E. and Park, Y. 2017. Variable thermal loading and flexural uplift along the Transantarctic Mountains, Antarctica. *Geology*, **45**, 463–466, https://doi.org/10.1130/G38784.1

Boger, S.D. 2011. Antarctica – before and after Gondwana. *Precambrian Research*, **19**, 335–371, https://doi.org/10.1016/j.gr.2010.09.003

Bozdağ, E., Peter, D. *et al.* 2016. Global adjoint tomography: first-generation model. *Geophysical Journal International*, **207**, 1739–1766, https://doi.org/10.1093/gji/ggw356

Brommer, A., Millar, I.L. and Zeh, A. 1999. Geochronology, structural geology and petrography of the northwestern La Grange Nunataks, Shackleton Range, Antarctica. *Terra Antarctica*, **6**, 269–278.

Chang, S.-J., Baag, C.-E. and Langston, C.A. 2004. Joint analysis of teleseismic receiver functions and surface wave dispersion using the genetic algorithm. *Bulletin of the Seismological Society of America*, **94**, 691–704, https://doi.org/10.1785/0120030110

Chaput, J., Aoster, R. *et al.* 2014. The crustal thickness of West Antarctica. *Journal of Geophysical Research: Solid Earth*, **119**, 378–395, https://doi.org/10.1002/2013JB010642

Chatzaras, V. and Kruckenberg, S.C. 2021. Effects of melt-percolation, refertilization, and deformation on upper mantle seismic anisotropy: constraints from peridotite xenoliths, Marie Byrd Land, West Antarctica. *Geological Society, London, Memoirs*, **56**, https://doi.org/10.1144/M56-2020-16

Chen, M., Huang Yao, H., van der Hilst, R. and Niu, F. 2014. Low wave speed zones in the crust beneath SE Tibet revealed by ambient noise adjoint tomography. *Geophysical Research Letters*, **41**, 334–340, https://doi.org/10.1002/2013GL058476

Cline, C.J., Faul, U.H., David, E.C., Berry, A.J. and Jackson, I. 2018. Redox-influenced seismic properties of upper-mantle olivine. *Nature*, **555**, 355–358, https://doi.org/10.1038/nature25764

Corr, H.F. and Vaughan, D.G. 2008. A recent volcanic eruption beneath the West Antarctic ice sheet. *Nature Geoscience*, **1**, 122–125, https://doi.org/10.1038/ngeo106

Danesi, S. and Morelli, A. 2001. Structure of the upper mantle under the Antarctic Plate from surface wave tomography. *Geophysical Research Letters*, **28**, 4395–4398, https://doi.org/10.1029/2001GL013431

Davey, F.J., Granot, R., Cande, S.C., Stock, J.M., Selvans, M. and Ferraccioli, F. 2016. Synchronous oceanic spreading and continental rifting in West Antarctica. *Geophysical Research Letters*, **43**, 6162–6169, https://doi.org/10.1002/2016GL069087

Day, J.M.D., Harvey, R.P. and Hilton, D.R. 2019. Melt-modified lithosphere beneath Ross Island and its role in the tectono-magmatic evolution of the West Antarctic Rift System. *Chemical Geology*, **518**, 45–54, https://doi.org/10.1016/j.chemgeo.2019.04.012

Eagles, G., Gohl, K. and Larter, R.D. 2004. High-resolution animated tectonic reconstruction of the South Pacific and West Antarctic margin. *Geochemistry, Geophysics, Geosystems*, **5**, Q07004, https://doi.org/10.1029/2003GC000657

Emry, E.L., Nyblade, A.A. *et al.* 2015. The mantle transition zone beneath West Antarctica: seismic evidence for hydration and thermal upwellings. *Geochemistry, Geophysics, Geosystems*, **15**, 40–58, https://doi.org/10.1002/2014GC005588

Emry, E.L., Nyblade, A.A. *et al.* 2020. Prominent thermal anomalies in the mantle transition zone beneath the Transantarctic Mountains. *Geology*, **48**, 748–752, https://doi.org/10.1130/G47346.1

Ferraccioli, F., Finn, C.A., Jordan, T.A., Bell, R.E., Anderson, L.M. and Damaske, D. 2011. East Antarctic rifting triggers uplift of the Gamburtsev Mountains. *Nature*, **479**, 388–392, https://doi.org/10.1038/nature10566

Fielding, C.R., Whittaker, J., Henrys, S.A., Wilson, T.J. and Naish, T.R. 2008. Seismic facies and stratigraphy of the Cenozoic succession in McMurdo Sound, Antarctica: implications for tectonic, climatic and glacial history. *Palaeogeography, Palaeoclimatology, Palaeoecology*, **260**, 8–29, https://doi.org/10.1016/j.palaeo.2007.08.016

Finn, C.A., Müller, R.D. and Panter, K.S. 2005. A Cenozoic diffuse alkaline magmatic province (DAMP) in the southwest Pacific without rift or plume origin. *Geochemistry, Geophysics, Geosystems*, **6**, Q02005, https://doi.org/10.1029/2004GC000723

Fichtner, A., Kennett, B.L., Igel, H. and Bunge, H.P. 2009. Full seismic waveform tomography for upper-mantle structure in the Australasian region using adjoint methods. *Geophysical Journal International*, **179**, 1703–1725, https://doi.org/10.1111/j.1365-246X.2009.04368.x

Fitzsimons, I.C.W. 2000. A review of tectonic events in the East Antarctic shield and their implications for Gondwana and earlier supercontinents. *Journal of African Earth Sciences*, **31**, 3–23, https://doi.org/10.1016/S0899-5362(00)00069-5

Foley, S.F., Andronikov, A.V., Jacob, D.E. and Melzer, S. 2006. Evidence from Antarctic mantle peridotite xenoliths for changes in mineralogy, geochemistry and geothermal gradients beneath a developing rift. *Geochimica et Cosmochimica Acta*, **70**, 3096–3120, https://doi.org/10.1016/j.gca.2006.03.010

Foley, S.F., Andronikov, A.V., Halpin, J.A., Daczko, N.R. and Jacob, D.E. 2021. Mantle rocks in East Antarctica. *Geological Society, London, Memoirs*, **56**, https://doi.org/10.1144/M56-2020-8

Forsyth, D.W. and Li, A. 2005. Array analysis of two-dimensional variations in surface wave phase velocity and azimuthal anisotropy in the presence of multipathing interference, in seismic earth: array analysis of broadband seismograms. *American Geophysical Union Geophysical Monograph Series*, **157**, 81–97.

French, S.W. and Romanowicz, B. 2015. Broad plumes rooted at the base of the Earth's mantle beneath major hotspots. *Nature*, **525**, 95–99, https://doi.org/10.1038/nature14876

Fretwell, P., Pritchard, H.D. *et al.* 2013. Bedmap2: improved ice bed, surface and thickness datasets for Antarctica. *The Cryosphere*, **7**, 375–393, https://doi.org/10.5194/tc-7-375-2013

Gomez, N., Pollard, D. and Holland, D. 2015. Sea-level feedback lowers projections of future Antarctic ice sheet mass loss. *Nature Communications*, **6**, 8798, https://doi.org/10.1038/ncomms9798

Goodge, J.W. and Fanning, C.M. 1999. 2.5 b.y. of punctuated Earth history as recorded in a single rock. *Geology*, **27**, 1007–1010, https://doi.org/10.1130/0091-7613(1999)027<1007:BYOPEH>2.3.CO;2

Goodge, J.W. and Fanning, C.M. 2016. Mesoarchean and Paleoproterozoic history of the Nimrod Complex, central Transantarctic Mountains, Antarctica: Stratigraphic revisions and relation to the Mawson Continent in East Gondwana. *Precambrian Research*, **285**, 242–271, https://doi.org/10.1016/j.precamres.2016.09.001

Goodge, J.W., Fanning, C.M. and Bennett, V.C. 2001. U–Pb evidence of c. 1.7 Ga crustal tectonism during the nimrod orogeny in the Transantarctic Mountains, Antarctica: implications for Proterozoic plate reconstructions. *Precambrian Research*, **112**, 261–288, https://doi.org/10.1016/S0301-9268(01)00193-0

Goodge, J.W., Fanning, C.M., Fisher, C.M. and Vervoort, J.D. 2017. Proterozoic crustal evolution of central East Antarctica: age and isotopic evidence from glacial igneous clasts, and links with Australia and Laurentia. *Precambrian Research*, **299**, 151–176, https://doi.org/10.1016/j.precamres.2017.07.026

Granot, R. and Dyment, J. 2018. Late Cenozoic unification of East and West Antarctica. *Nature Communications*, **9**, 3189, https://doi.org/10.1038/s41467-018-05270-w

Granot, R., Cande, S.C., Stock, J.M. and Damaske, D. 2013. Revised Eocene–Oligocene kinematics for the West Antarctic rift system. *Geophysical Research Letters*, **40**, 279–284, https://doi.org/10.1029/2012GL054181

Graw, J.H., Adams, A.N., Hansen, S.E., Wiens, D.A., Hackworth, L. and Park, Y. 2016. Upper mantle shear wave velocity structure beneath northern Victoria Land, Antarctica: volcanism and uplift in the northern Transantarctic Mountains. *Earth and Planetary Science Letters*, **449**, 48–60, https://doi.org/10.1016/j.epsl.2016.05.026

Groome, W.G. and Thorkelson, D.J. 2009. The three-dimensional thermo-mechanical signature of ridge subduction and slab window migration. *Tectonophysics*, **464**, 70–83, https://doi.org/10.1016/j.tecto.2008.07.003

Grunow, A.M., Kent, D.V. and Dalziel, I.W.D. 1987. Mesozoic evolution of West Antarctica and the Weddell Sea basin: new paleomagnetic constraints. *Earth and Planetary Science Letters*, **86**, 16–26, https://doi.org/10.1016/0012-821X(87)90184-1

Handler, M.R., Wysoczanski, R.J. and Gamble, J.A. 2021. Marie Byrd Land lithospheric mantle: A review of the xenolith record. *Geological Society, London, Memoirs*, **56**, https://doi.org/10.1144/M56-2020-17

Hansen, S.E., Julia, J., Nyblade, A.A., Pyle, M.L., Wiens, D.A. and Anandakrishnan, S. 2009. Using S wave receiver functions to estimate crustal structure beneath ice sheets: An application to the Transantarctic Mountains and East Antarctic craton. *Geochemistry, Geophysics, Geosystems*, **10**, Q08014, https://doi.org/10.1029/2009GC002576

Hansen, S.E., Nyblade, A.A., Heeszel, D.S., Wiens, D.A., Shore, P. and Kanao, M. 2010. Crustal structure of the Gamburtsev Mountains, East Antarctica, from S-wave receiver functions and Rayleigh wave phase velocities. *Earth and Planetary Science Letters*, **300**, 395–401, https://doi.org/10.1016/j.epsl.2010.10.022

Hansen, S.E., Graw, J.H. *et al.* 2014. Imaging the Antarctic mantle using adaptively parameterized P-wave tomography: evidence for heterogeneous structure beneath West Antarctica. *Earth and Planetary Science Letters*, **408**, 66–78, https://doi.org/10.1016/j.epsl.2014.09.043

Hansen, S.E., Reusch, A.M., Parker, T., Bloomquist, D.K., Carpenter, P., Graw, J.H. and Brenn, G.R. 2015. The Transantarctic Mountains Northern Network (TAMNNET): deployment and performance of a seismic array in Antarctica. *Seismological Research Letters*, **86**, 1636–1644, https://doi.org/10.1785/0220150117

Heeszel, D.S., Wiens, D.A., Nyblade, A.A., Hansen, S.E., Kanao, M., An, M. and Zhao, Y. 2013. Rayleigh wave constraints on the structure and tectonic history of the Gamburtsev Subglacial Mountains, East Antarctica. *Journal of Geophysical Research*, **118**, 2138–2153, https://doi.org/10.1002/jgrb.50171

Heeszel, D.S., Wiens, D.A. *et al.* 2016. Upper mantle structure of central and West Antarctica from array analysis of Rayleigh wave phase velocities. *Journal of Geophysical Research: Solid Earth*, **121**, 1758–1775, https://doi.org/10.1002/2015JB012616

Hole, M.J. and LeMasurier, W.E. 1994. Tectonic controls on the geochemical composition of Cenozoic mafic alkaline volcanic rocks from West Antarctica. *Contributions to Mineralogy and Petrology*, **117**, 187–202, https://doi.org/10.1007/BF00286842

Huerta, A.D. 2007. Lithospheric structure across the Transantarctic Mountains constrained by analysis of gravity and thermal structure. *United States Geological Survey Open-File Report*, **2007-1047**, https://doi.org/10.3133/of2007-1047.srp022

Ivins, E.R. and James, T.S. 2005. Antarctic glacial isostatic adjustment: a new assessment. *Antarctic Science*, **17**, 537–549, https://doi.org/10.1017/S0954102005002968

Ivins, E.R. and Sammis, C.G. 1995. On lateral viscosity contrast in the mantle and the rheology of low-frequency geodynamics. *Geophysical Journal International*, **123**, 305–322, https://doi.org/10.1111/j.1365-246X.1995.tb06856.x

Ivins, R.E., van der Wal, W., Wiens, D., Lloyd, A. and Caron, L. In press. Antarctic upper mantle rheology. Geological Society, London, Memoirs, **56**, https://doi.org/10.1144/M56-2020-19

Jacobs, J., Bingen, B., Thomas, R.J., Bauer, W., Wingate, M.T.D. and Feitio, P. 2008. Early Palaeozoic orogenic collapse and voluminous late-tectonic magmatism in Dronning Maud Land and Mozambique: insights into the partially delaminated orogenic root of the East African–Antarctic Orogen? *Geological Society, London, Special Publications*, **308**, 69–90, https://doi.org/10.1144/SP308.3

Jordan, T.A., Ferraccioli, F. *et al.* 2013. Inland extent of the Weddell Sea Rift imaged by new aerogeophysical data. *Tectonophysics*, **585**, 137–160, https://doi.org/10.1016/j.tecto.2012.09.010

Jordan, T.A., Ferraccioli, F. and Leat, P.T. 2017. New geophysical compilations link crustal block motion to Jurassic extension and strike-slip faulting in the Weddell Sea Rift System of West Antarctica. *Gondwana Research*, **42**, 29–48, https://doi.org/10.1016/j.gr.2016.09.009

Jordan, T.H. 1981. Continents as a chemical boundary layer. *Philosophical Transactions of the Royal Society A: Mathematical, Physical and Engineering Sciences*, **301**, 359–373, https://doi.org/10.1098/rsta.1981.0117

Julià, J., Ammon, C.J., Herrmann, R.B. and Correig, A.M. 2000. Joint inversion of receiver function and surface wave dispersion observations. *Geophysical Journal International*, **143**, 99–112, https://doi.org/10.1046/j.1365-246x.2000.00217.x

Karato, S., Jung, H., Katayama, I. and Skemer, P. 2008. Geodynamic significance of seismic anisotropy of the upper mantle: new insights from laboratory studies. *Annual Review of Earth and Planetary Sciences*, **36**, 59–95, https://doi.org/10.1146/annurev.earth.36.031207.124120

Komatitsch, D. and Tromp, J. 1999. Introduction to the spectral element method for three-dimensional seismic wave propagation. *Geophysical Journal International*, **139**, 806–822, https://doi.org/10.1046/j.1365-246x.1999.00967.x

Komatitsch, D. and Tromp, J. 2002*a*. Spectral-element simulations of global seismic wave propagation – I. Validation. *Geophysical Journal International*, **149**, 390–412, https://doi.org/10.1046/j.1365-246X.2002.01653.x

Komatitsch, D. and Tromp, J. 2002*b*. Spectral-element simulations of global seismic wave propagation – II. Three-dimensional models, oceans, rotation and self-gravitation. *Geophysical Journal International*, **150**, 303–318, https://doi.org/10.1046/j.1365-246X.2002.01716.x

Komatitsch, D. and Vilotte, J.P. 1998. The spectral element method: An efficient tool to simulate the seismic response of 2D and 3D geological structures. *Bulletin of the Seismological Society of America*, **88**, 368–392.

Kustowski, B., Ekström, G. and Dziewoński, A.M. 2008. Anisotropic shear-wave velocity structure of the Earth's mantle: a global model. *Journal of Geophysical Research: Solid Earth*, **113**, https://doi.org/10.1029/2007JB005169

Kyle, P.R. 1990. McMurdo Volcanic Group, western Ross Embayment: introduction. *American Geophysical Union Antarctic Research Series*, **48**, 19–25.

Kyle, P.R., Moore, J.A. and Thirlwall, M.F. 1992. Petrologic evolution of anorthoclase phonolite lavas at Mount Erebus, Ross Island, Antarctica. *Journal of Petrology*, **33**, 849–875, https://doi.org/10.1093/petrology/33.4.849

Lawrence, J.F. and Wiens, D.A. 2004. Combined receiver function and surface wave phase velocity inversion using a niching genetic algorithm: application to Patagonia. *Bulletin of the Seismological Society of America*, **94**, 977–987, https://doi.org/10.1785/0120030172

Lawrence, J.F., Wiens, D.A., Nyblade, A.A., Anandakrishnan, S., Shore, P.J. and Voigt, D. 2006*a*. Rayleigh wave phase velocity analysis of the Ross Sea. Transantarctic Mountains, and East Antarctica from a temporary seismograph array. *Journal of Geophysical Research: Solid Earth*, **111**, B06302, https://doi.org/10.1029/2005JB003812

Lawrence, J.F., Wiens, D.A., Nyblade, A.A., Anandakrishan, S., Shore, P.J. and Voigt, D. 2006*b*. Upper mantle thermal variations beneath the Transantarctic Mountains inferred from teleseismic S-wave attenuation. *Geophysical Research Letters*, **33**, L03303, https://doi.org/10.1029/2005GL024516

Lawrence, J.F., Wiens, D.A., Nyblade, A.A., Anandakrishnan, S., Shore, P.J. and Voigt, D. 2006*c*. Crust and upper mantle structure of the Transantarctic Mountains and surrounding regions from receiver functions, surface waves, and gravity: implications for uplift models. *Geochemistry, Geophysics, Geosystems*, **7**, Q10011, https://doi.org/10.1029/2006GC001282

Lee, C.-T.A. 2003. Compositional variation of density and seismic velocities in natural peridotites at STP conditions: implications for seismic imaging of compositional heterogeneities in the upper mantle. *Journal of Geophysical Research*, **108**, 2441, https://doi.org/10.1029/2003JB002413

Leitchenkov, G. and Kudryavtzev, G. 1997. Structure and origin of the Earth's Crust in the Weddell Sea Embayment (beneath the Front of the Filchner and Ronne Ice Shelves) from deep seismic sounding data. *Polarforschung*, **67**, 143–154.

LeMasurier, W.E. and Landis, C.A. 1996. Mantle-plume activity recorded by low-relief erosion surfaces in West Antarctica and New Zealand. *Geological Society of America Bulletin*, **108**, 1450–1466, https://doi.org/10.1130/0016-7606(1996)108<1450:MPARBL>2.3.CO;2

LeMasurier, W.E. and Rex, D.C. 1989. Evolution of linear volcanic ranges in Marie Byrd Land, West Antarctica. *Journal of Geophysical Research: Solid Earth*, **94**, 7223–7236, https://doi.org/10.1029/JB094iB06p07223

Levander, A., Schmandt, B. *et al.* 2011. Continuing Colorado Plateau uplift by delamination-style convective lithospheric downwelling. *Nature*, **472**, 461–465, https://doi.org/10.1038/nature10001

Licht, K.J., Groth, T., Townsend, J.P., Hennessy, A.J., Hemming, S.R., Flood, T.P. and Studinger, M. 2018. Evidence for extending anomalous Miocene volcanism at the edge of the East Antarctic craton. *Geophysical Research Letters*, **45**, 3009–3016, https://doi.org/10.1002/2018GL077237

Liu, Y., Niu, F., Chen, M. and Yang, W. 2017. 3-D crustal and uppermost mantle structure beneath NE China revealed by ambient noise adjoint tomography. *Earth and Planetary Science Letters*, **461**, 20–29, https://doi.org/10.1016/j.epsl.2016.12.029

Lloyd, A.J. 2018. *Seismic Tomography of Antarctica and the Southern Oceans: Regional and Continental Models from the Upper Mantle to the Transition Zone*. PhD thesis, Washington University in St Louis, St Louis, MO, USA.

Lloyd, A.J., Nyblade, A.A., Wiens, D.A., Hansen, S.E., Kanao, M., Shore, P.J. and Zhao, D. 2013. Upper mantle seismic structure beneath central East Antarctica from body wave tomography: implications for the origin of the Gamburtsev Subglacial Mountains. *Geochemistry, Geophysics, Geosystems*, **14**, 902–920, https://doi.org/10.1002/ggge.20098

Lloyd, A.J., Wiens, D.A. *et al.* 2015. A seismic transect across West Antarctica: Evidence for mantle thermal anomalies beneath the Bentley Subglacial Trench and the Marie Byrd Land Dome. *Journal of Geophysical Research: Solid Earth*, **120**, 8439–8460, https://doi.org/10.1002/2015JB012455

Lloyd, A.J., Wiens, D.A. *et al.* 2020. Seismic structure of the Antarctic upper mantle imaged with adjoint tomography. *Journal of Geophysical Research: Solid Earth*, **125**, https://doi.org/10.1029/2019JB017823

Lough, A.C., Wiens, D.A. *et al.* 2013. Seismic detection of an active subglacial magmatic complex in Marie Byrd Land, Antarctica. *Nature Geoscience*, **6**, 1031–1035, https://doi.org/10.1038/ngeo1992

Lucas, E.M., Nyblade, A. *et al.* 2020*a*. Upper mantle seismic anisotropy of Antarctica from shear wave splitting analysis. *AGU Fall Meeting Abstracts*, **2020**, T008-03, https://ui.adsabs.harvard.edu/#abs/2020AGUFMT008...03L/abstract

Lucas, E.M., Soto, D. *et al.* 2020*b*. P- and S-wave velocity structure of central West Antarctica: implications for the tectonic evolution of the West Antarctic Rift System. *Earth and Planetary Science Letters*, **546**, 116437, https://doi.org/10.1016/j.epsl.2020.116437

Mainprice, D. and Silver, P.G. 1993. Interpretation of SKS-waves using samples from the subcontinental lithosphere. *Physics of the Earth and Planetary Interiors*, **78**, 257–280, https://doi.org/10.1016/0031-9201(93)90160-B

Maritati, A., Danišík, M., Halpin, J.A., Whittaker, J.M. and Aitken, A.R.A. 2020. Pangea rifting shaped the East Antarctic landscape. *Tectonics*, **39**, e2020TC006180, https://doi.org/10.1029/2020TC006180

Martin, A.P. and Cooper, A.F. 2010. Post 3.9 Ma fault activity within the West Antarctic rift system: onshore evidence from Gandalf Ridge, Mount Morning eruptive centre, southern Victoria Land, Antarctica. *Antarctic Science*, **22**, 513–521, https://doi.org/10.1017/S095410201000026X

Martin, A.P., Cooper, A.F., Price, R.C., Doherty, C.L. and Gamble, J.A. 2021*a*. A review of mantle xenoliths in volcanic rocks from southern Victoria Land, Antarctica. *Geological Society, London, Memoirs*, **56**, https://doi.org/10.1144/M56-2019-42

Martin, A.P., Cooper, A.F., Price, R.C., Kyle, P.R. and Gamble, J.A. Smellie, J.L. 2021*b*. Erebus Volcanic Province: petrology. *Geological Society, London, Memoirs*, **55**, 447–489, https://doi.org/10.1144/M55-2018-80

Muller, C. 2001. Upper mantle seismic anisotropy beneath Antarctica and the Scotia Sea region. *Geophysical Journal International*, **147**, 105–122, https://doi.org/10.1046/j.1365-246X.2001.00517.x

O'Donnell, J.P., Selway, K. *et al.* 2017. The uppermost mantle seismic velocity and viscosity structure of central West Antarctica. *Earth and Planetary Science Letters*, **472**, 38–49, https://doi.org/10.1016/j.epsl.2017.05.016

O'Donnell, J.P., Brisbourne, A.M. *et al.* 2019*a*. Mapping crustal shear wave velocity structure and radial anisotropy beneath West Antarctica using seismic ambient noise. *Geochemistry, Geophysics, Geosystems*, **20**, 5014–5037, https://doi.org/10.1029/2019GC008459

O'Donnell, J.P., Stuart, G.W. *et al.* 2019*b*. The uppermost mantle seismic velocity structure of West Antarctica from Rayleigh wave tomography: insights into tectonic structure and geothermal heat flow. *Earth and Planetary Science Letters*, **522**, 219–233, https://doi.org/10.1016/j.epsl.2019.06.024

Oliver, R.L. and Fanning, C.M. 2002. Proterozoic geology east and southeast of Commonwealth Bay, George V Land, Antarctica, and its relationship to that of adjacent Gondwana terranes. *Royal Society of New Zealand Bulletin*, **35**, 51–58.

Panning, M. and Romanowicz, B. 2006. A three-dimensional radially anisotropic model of shear velocity in the mantle. *Geophysical Journal International*, **167**, 361–379, https://doi.org/10.1111/j.1365-246X.2006.03100.x

Panter, K.S., Hart, S.R., Kyle, P., Blusztanjn, J. and Wilch, T. 2000. Geochemistry of Late Cenozoic basalts from the Crary Mountains: characterization of mantle sources in Marie Byrd Land, Antarctica. *Chemical Geology*, **165**, 215–241, https://doi.org/10.1016/S0009-2541(99)00171-0

Pappa, F. and Ebbing, J. 2021. Gravity, magnetics and geothermal heat flow of the Antarctic lithospheric crust and mantle. *Geological Society, London, Memoirs*, **56**, https://doi.org/10.1144/M56-2020-5

Pappa, F., Ebbing, J. and Ferraccioli, F. 2019*a*. Moho depths of Antarctica: comparison of seismic, gravity, and isostatic results. *Geochemistry, Geophysics, Geosystems*, **20**, 1629–1645, https://doi.org/10.1029/2018GC008111

Pappa, F., Ebbing, J., Ferraccioli, F. and van der Wal, W. 2019*b*. Modeling satellite gravity gradient data to derive density, temperature, and viscosity structure of the antarctic lithosphere. *Journal of Geophysical Research: Solid Earth*, **124**, 12 053–12 076, https://doi.org/10.1029/2019JB017997

Pattyn, F. 2010. Antarctic subglacial conditions inferred from a hybrid ice sheet/ice stream model. *Earth and Planetary Science Letters*, **295**, 451–461, https://doi.org/10.1016/j.epsl.2010.04.025

Paulsen, T.S. and Wilson, T.J. 2010. Evolution of Neogene volcanism and stress patterns in the glaciated West Antarctic Rift, Marie Byrd Land, Antarctica. *Journal of the Geological Society, London*, **167**, 401–416, https://doi.org/10.1144/0016-76492009-044

Paxman, G.J.G. 2021. Antarctic palaeotopography. *Geological Society, London, Memoirs*, **56**, https://doi.org/10.1144/M56-2020-7

Phillips, E.H., Sims, K.W. *et al.* 2018. The nature and evolution of mantle upwelling at Ross Island, Antarctica, with implications for the source of HIMU lavas. *Earth and Planetary Science Letters*, **498**, 38–53, https://doi.org/10.1016/j.epsl.2018.05.049

Phillips, G. and Laufer, A.L. 2009. Brittle deformation relating to the Carboniferous–Cretaceous evolution of the Lambert Graben, East Antarctica: a precursor for Cenozoic relief development in an intraplate and glaciated region. *Tectonophysics*, **471**, 216–224, https://doi.org/10.1016/j.tecto.2009.02.012

Pollard, D., DeConto, R.M. and Nyblade, A.A. 2005. Sensitivity of Cenozoic Antarctic ice sheet variations to geothermal heat flux. *Global and Planetary Change*, **49**, 63–74, https://doi.org/10.1016/j.gloplacha.2005.05.003

Priestley, K. and McKenzie, D. 2006. The thermal structure of the lithosphere from shear wave velocities. *Earth and Planetary Science Letters*, **244**, 285–301, https://doi.org/10.1016/j.epsl.2006.01.008

Pyle, M.L., Wiens, D.A., Nyblade, A.A. and Anandakrishnan, S. 2010. Crustal structure of the Transantarctic Mountains near the Ross Sea from ambient seismic noise tomography. *Journal of Geophysical Research: Solid Earth*, **115**, B11310, https://doi.org/10.1029/2009JB007081

Ramirez, C., Nyblade, A. *et al.* 2016. Crustal and upper-mantle structure beneath ice-covered regions in Antarctica from S-wave receiver functions and implications for heat flow. *Geophysical Journal International*, **204**, 1636–1648, https://doi.org/10.1093/gji/ggv542

Ramirez, C., Nyblade, A. *et al.* 2017. Crustal structure of the Transantarctic Mountains, Ellsworth Mountains and Marie Byrd Land, Antarctica: constraints on shear wave velocities, Poisson's ratios and Moho depths. *Geophysical Journal International*, **211**, 1328–1340, https://doi.org/10.1093/gji/ggx333

Reading, A.M. 2006. The seismic structure of Precambrian and early Palaeozoic terranes in the Lambert Glacier region, East Antarctica. *Earth and Planetary Science Letters*, **244**, 44–57, https://doi.org/10.1016/j.epsl.2006.01.031

Reading, A.M. and Heintz, M. 2008. Seismic anisotropy of East Antarctica from shear-wave splitting: spatially varying contributions from lithospheric structural fabric and mantle flow? *Earth and Planetary Science Letters*, **268**, 433–443, https://doi.org/10.1016/j.epsl.2008.01.041

Reusch, A.M., Nyblade, A.A., Benoit, M.H., Wiens, D.A., Anandakrishnan, S., Voigt, D. and Shore, P.J. 2008. Mantle transition zone thickness beneath Ross Island, the Transantarctic Mountains, and East Antarctica. *Geophysical Research Letters*, **35**, https://doi.org/10.1029/2008GL033873

Ritzwoller, M.H., Shapiro, N.M., Levshin, A.L. and Leahy, G.M. 2001. Crustal and upper mantle structure beneath Antarctica and surrounding oceans. *Journal of Geophysical Research: Solid Earth*, **106**, 30 645–30 670, https://doi.org/10.1029/2001JB000179

Robertson Maurice, S.D., Wiens, D.A., Shore, P.J., Vera, E. and Dorman, L.M. 2003. Seismicity and tectonics of the South Shetland Islands and Bransfield Strait from a regional broadband seismograph deployment. *Journal of Geophysical Research: Solid Earth*, **108**, 2461, https://doi.org/10.1029/2003JB002416

Roult, G., Rouland, D. and Montagner, J.P. 1994. Antarctica II: Upper mantle structure from velocities and anisotropy. *Physics of the Earth and Planetary Interiors*, **84**, 33–57, https://doi.org/10.1016/0031-9201(94)90033-7

Savage, M.K. 1999. Seismic anisotropy and mantle deformation: what have we learned from shear wave splitting? *Reviews of Geophysics*, **37**, 65–106, https://doi.org/10.1029/98RG02075

Schutt, D.L. and Lesher, C.E. 2006. Effects of melt depletion on the density and seismic velocity of garnet and spinel lherzolite. *Journal of Geophysical Research: Solid Earth*, **111**, B05401, https://doi.org/10.1029/2003JB002950

Shapiro, N.M. and Ritzwoller, M.H. 2004. Inferring surface heat flux distributions guided by a global seismic model: Particular application to Antarctica. *Earth and Planetary Science Letters*, **223**, 213–224, https://doi.org/10.1016/j.epsl.2004.04.011

Shapiro, N.M., Campillo, M., Stehly, L. and Ritzwoller, M.H. 2005. High-resolution surface wave tomography from ambient seismic noise. *Science*, **307**, 1615–1618, https://doi.org/10.1126/science.1108339

Shen, W., Wiens, D.A. *et al.* 2018*a*. Seismic evidence for lithospheric foundering beneath the southern Transantarctic Mountains, Antarctica. *Geology*, **46**, 71–74, https://doi.org/10.1130/G39555.1

Shen, W., Wiens, D.A. *et al.* 2018*b*. The crust and upper mantle structure of central and West Antarctica from Bayesian inversion of Rayleigh wave and receiver functions. *Journal of Geophysical Research: Solid Earth*, **123**, 7824–7849, https://doi.org/10.1029/2017JB015346

Shen, W., Wiens, D.A., Lloyd, A. and Nyblade, A.A. 2020. A geothermal heat flux map of Antarctica empirically constrained by seismic structure. *Geophysical Research Letters*, **47**, e2020GL086955, https://doi.org/10.1029/2020GL086955

Siddoway, C.S. 2008. Tectonics of the West Antarctic Rift System: new light on the history and dynamics of distributed intracontinental extension. *In*: Cooper, A., Barrett, P.J. *et al.* (eds) *Antarctica: A Keystone in a Changing World.* National Academy of Sciences, Washington, DC, 91–114.

Sieminski, A., Debayle, E. and Lévêque, J.J. 2003. Seismic evidence for deep low-velocity anomalies in the transition zone beneath West Antarctica. *Earth and Planetary Science Letters*, **216**, 645–661, https://doi.org/10.1016/s0012-821x(03)00518-1

Silver, P.G. 1996. Seismic anisotropy beneath the continents: probing the depths of geology. *Annual Review of Earth and Planetary Sciences*, **24**, 385–432, https://doi.org/10.1146/annurev.earth.24.1.385

Sleep, N.H. 2005. Evolution of continental lithosphere. *Annual Review of Earth and Planetary Sciences*, **33**, 369–393, https://doi.org/10.1146/annurev.earth.33.092203.122643

Stump, E., Sheridan, M.F., Borg, S.G. and Sutter, J.F. 1980. Early Miocene subglacial basalts, the East Antarctic ice sheet, and uplift of the Transantarctic Mountains. *Science*, **207**, 757–759, https://doi.org/10.1126/science.207.4432.757

Sutherland, R., Spasojevic, S. and Gurnis, M. 2010. Mantle upwelling after Gondwana subduction death explains anomalous topography and subsidence histories of eastern New Zealand and West Antarctica. *Geology*, **38**, 155–158, https://doi.org/10.1130/G30613.1

Swain, G., Woodhouse, A., Hand, M., Barovich, K., Schwarz, M. and Fanning, C.M. 2005. Provenance and tectonic development of the late Archaean Gawler Craton, Australia; U–Pb zircon, geochemical and Sm–Nd isotopic implications. *Precambrian Research*, **141**, 106–136, https://doi.org/10.1016/j.precamres.2005.08.004

Szwillus, W., Afonso, J.C.C., Ebbing, J. and Mooney, W.D. 2019. Global crustal thickness and velocity structure from geostatistical analysis of seismic data. *Journal of Geophysical Research: Solid Earth*, **124**, 1626–1652, https://doi.org/10.1029/2018JB016593

Tape, C., Liu, Q., Maggi, A. and Tromp, J. 2010. Seismic tomography of the southern California crust based on spectral-element and adjoint methods. *Geophysical Journal International*, **180**, 433–462, https://doi.org/10.1111/j.1365-246X.2009.04429.x

Tarantola, A. 1984. Inversion of seismic reflection data in the acoustic approximation. *Geophysics*, **49**, 1259–1266, https://doi.org/10.1190/1.1441754

ten Brink, U.S., Bannister, S., Beaudoin, B.C. and Stern, T.A. 1993. Geophysical investigations of the tectonic boundary between East and West Antarctica. *Science*, **261**, 45–50, https://doi.org/10.1126/science.261.5117.45

ten Brink, U.S., Hackney, R.I., Bannister, S., Stern, T.A. and Makovsky, Y. 1997. Uplift of the Transantarctic Mountains and the bedrock beneath the East Antarctic ice sheet. *Journal of Geophysical Research: Solid Earth*, **102**, 27 603–27 621, https://doi.org/10.1029/97JB02483

Tingey, R.J. 1991. The regional geology of Archean and Proterozoic rocks in Antarctica. *In*: Tingey, R.J. (ed.) *The Geology of Antarctica*. Clarendon Press, Oxford, UK, 1–73.

Tinto, K.J., Padman, L. *et al.* 2019. Ross Ice Shelf response to climate driven by the tectonic imprint on seafloor bathymetry. *Nature Geoscience*, **12**, 441–449, https://doi.org/10.1038/s41561-019-0370-2

Trey, H., Cooper, A., Pellis, G., della Vedova, B., Cochrane, G., Brancolini, G. and Makris, J. 1999. Transect across the West Antarctic rift system in the Ross Sea, Antarctica. *Tectonophysics*, **301**, 61–74, https://doi.org/10.1016/S0040-1951(98)00155-3

Tromp, J., Tape, C. and Liu, Q. 2005. Seismic tomography, adjoint methods, time reversal and banana–doughnut kernels. *Geophysical Journal International*, **160**, 195–216, https://doi.org/10.1111/j.1365-246X.2004.02453.x

van de Flierdt, T., Hemming, S.R., Goldstein, S.L., Gehrels, G.E. and Cox, S.E. 2008. Evidence against a young volcanic origin of the Gamburtsev Subglacial Mountains, Antarctica. *Geophysical Research Letters*, **35**, L21303, https://doi.org/10.1029/2008GL035564

van der Lee, S. 2002. High-resolution estimates of lithospheric thickness from Missouri to Massachusetts, USA. *Earth and Planetary Science Letters*, **203**, 15–23, https://doi.org/10.1016/S0012-821X(02)00846-4

van der Wal, W., Whitehouse, P.L. and Schrama, E.J.O. 2015. Effect of GIA models with 3D composite mantle viscosity on GRACE mass balance estimates for Antarctica. *Earth and Planetary Science Letters*, **414**, 134–143, https://doi.org/10.1016/j.epsl.2015.01.001

Wannamaker, P., Hill, G. *et al.* 2017. Uplift of the central transantarctic mountains. *Nature Communications*, **8**, 1588, https://doi.org/10.1038/s41467-017-01577-2

Watson, T., Nyblade, A. *et al.* 2006. P and S velocity structure of the upper mantle beneath the Transantarctic Mountains, East Antarctic craton, and Ross Sea from travel time tomography. *Geochemistry, Geophysics, Geosystems*, **7**, Q07005, https://doi.org/10.1029/2005GC001238

Weaver, S.D., Storey, B.C., Pankhurst, R.J., Mukasa, S.B., DiVenere, V.J. and Bradshaw, J.D. 1994. Antarctica–New Zealand rifting and Marie Byrd Land lithospheric magmatism linked to ridge subduction and mantle plume activity. *Geology*, **22**, 811–814, https://doi.org/10.1130/0091-7613(1994)022<0811:ANZRAM>2.3.CO;2

White-Gaynor, A.L., Nyblade, A.A. *et al.* 2019. Heterogeneous upper mantle structure beneath the Ross Sea Embayment and Marie Byrd Land, West Antarctica, revealed by P-wave tomography. *Earth and Planetary Science Letters*, **513**, 40–50, https://doi.org/10.1016/j.epsl.2019.02.013

Whitehouse, P.L., Gomez, N., King, M.A. and Wiens, D.A. 2019. Solid Earth change and the evolution of the Antarctic Ice Sheet. *Nature Communications*, **10**, 503, https://doi.org/10.1038/s41467-018-08068-y

Wilch, T.I., McIntosh, W.C. and Dunbar, N.W. 1999. Late Quaternary volcanic activity in Marie Byrd Land: potential $^{40}Ar/^{39}Ar$-dated time horizons in West Antarctic ice and marine cores. *Geological Society of America Bulletin*, **111**, 1563–1580, https://doi.org/10.1130/0016-7606(1999)111<1563:LQVAIM>2.3.CO;2

Will, T.M., Zeh, A., Gerdes, A., Frimmel, H.E., Millar, I.L. and Schmädicke, E. 2009. Palaeoproterozoic to Palaeozoic magmatic and metamorphic events in theShackleton Range, East Antarctica: constraints from zircon andmonazite dating, and implications for the amalgamation of Gondwana. *Precambrian Research*, **182**, 25–45, https://doi.org/10.1016/j.precamres.2009.03.008

Wilson, D.S. and Luyendyk, B.P. 2009. West Antarctic paleotopography estimated at the Eocene-Oligocene climate transition. *Geophysical Research Letters*, **36**, L16302, https://doi.org/10.1029/2009GL039297

Wobbe, F., Gohl, K., Chambord, A. and Sutherland, R. 2012. Structure and breakup history of the rifted margin of West Antarctica in relation to Cretaceous separation from Zealandia and Bellingshausen plate motion. *Geochemistry, Geophysics, Geosystems*, **13**, Q04W12, https://doi.org/10.1029/2011GC003742

Wobbe, F., Lindeque, A. and Gohl, K. 2014. Anomalous South Pacific lithosphere dynamics derived from new total sediment thickness estimates off the West Antarctic margin. *Global and Planetary Change*, **123**, 139–149, https://doi.org/10.1016/j.gloplacha.2014.09.006

Yamasaki, T., Miura, H. and Nogi, Y. 2008. Numerical modelling study on the flexural uplift of the Transantarctic Mountains. *Geophysical Journal International*, **174**, 377–390, https://doi.org/10.1111/j.1365-246X.2008.03815.x

Yuan, Y.O., Simons, F.J. and Tromp, J. 2016. Double-difference adjoint seismic tomography. *Geophysical Journal International*, **206**, 1599–1618, https://doi.org/10.1093/gji/ggw233

Zhang, S. and Karato, S. 1995. Lattice preferred orientation of olivine aggregates deformed in simple shear. *Nature*, **375**, 774–777, https://doi.org/10.1038/375774a0

Zhou, Z., Wiens, D.A., Shen, W., Hansen, S., Aster, R. and Nyblade, A. 2019. Radial anisotropy of Antarctica from surface wave ambient noise tomography. *Seismological Research Letter*, **90**, 1047.

Zhu, H., Bozdağ, E. and Tromp, J. 2015. Seismic structure of the European upper mantle based on adjoint tomography. *Geophysical Journal International*, **201**, 18–52, https://doi.org/10.1093/gji/ggu492

Gravity, magnetics and geothermal heat flow of the Antarctic lithospheric crust and mantle

Folker Pappa* and Jörg Ebbing

Institute of Geosciences, Kiel University, Otto-Hahn-Platz 1, D-24118, Kiel, Germany

FP, 0000-0002-5170-4919; JE, 0000-0001-7492-5338

*Correspondence: folker.pappa@ifg.uni-kiel.de

Abstract: This chapter describes the application and coverage of gravity and magnetic data for Antarctica with emphasis on airborne and satellite models. Low resolution satellite data help to fill gaps between high-resolution airborne data. Satellite gravity data are best used to study broad-scale lithospheric architecture while airborne data, especially magnetic data, provide finer detail. We review examples of gravity and magnetic analysis and describe the possibilities and pitfalls for estimating the properties of the lithosphere as it relates to the mantle. This is followed by a discussion on geothermal heat flow and possible ways to combine different geophysical and petrological models for a better understanding of the Antarctic mantle.

Antarctica, despite its remoteness and size, has been the focus of systematic geophysical data acquisition from as early as the International Geophysical Year 1957–58. A wealth of modern airborne geophysical surveys, including airborne gravimetry and magnetics over largely unexplored Antarctic frontiers, such as the Gamburtsev Subglacial Mountains (Bell *et al.* 2011; Ferraccioli *et al.* 2011) and Wilkes Land in East Antarctica (Aitken *et al.* 2014) was stimulated by the International Polar Year 2007/2008 (Krupnik *et al.* 2011). Because of these efforts, more than two-thirds of the continent has been covered by airborne surveys, albeit with varying resolution and data quality, which is a challenge for data compilations.

Nevertheless, the first modern Antarctic-wide gravity data compilation was derived from 13 million data points and covered an area of 10 million km^2, which corresponds to 73% coverage of the continent (Scheinert *et al.* 2016). The first Antarctic magnetic anomaly compilation (ADMAP-1) was produced in 2001 from more than 1.5 million line-km of shipborne and airborne measurements (Golynsky *et al.* 2001), succeeded in 2018 by the second-generation Antarctic magnetic anomaly compilation (ADMAP-2) including now more than 3.5 million line-km of aeromagnetic and marine magnetic data that more than doubles the initial near-surface database (Golynsky *et al.* 2018).

Still, gaps in survey coverage exist, especially in the interior of the continent and satellite data can be used to fill these gaps. A number of missions (CHAMP, GRACE, GOCE, Swarm) have been launched by international agencies since the beginning of the century that begin to fill the gap between airborne data and earlier satellite data and which can be used to study the crust and upper mantle over the entire continent.

The applications of gravity and magnetic data are many and span from estimating variations in ice thickness or depth to the top of the crystalline basement, mapping geological units, the depth of the Moho boundary, to modelling the lithosphere thickness. Along with petrological studies, gravity and magnetic data can inform us about the thermal structure and composition of the lithosphere.

Geothermal heat flow at the base lithosphere is often interpreted from the 1300-degree isotherm (Eaton *et al.* 2009), with variable radiogenic heat-production in the crust related to rock type. To date, only a few direct estimates of geothermal heat flow are available over Antarctica, making geophysical models useful proxies that often rely on magnetic or gravity derived models, or a combination of those, with seismological results.

This chapter will provide a summary of gravity and magnetic data, discuss some applications of these data to understand the mantle and lithosphere, and provide a brief discussion of the state of the art with respect to geothermal heat flow estimates from geophysical models.

Gravity

Gravity observations in Antarctica: scales and strides

Gravity data allow detection and interpretation of subsurface structures related to density variations. Such features can, for instance, be subglacial or submarine troughs, sedimentary basins, sub-glacial volcanoes, intrusions, tectonic suture zones, the depth of the crust–mantle boundary (Moho discontinuity), or mantle density variations. The commonly used products for interpretation are the free-air gravity anomaly, the Bouguer gravity anomaly or the isostatic anomaly. Measured data are corrected for the station height and Earth's normal gravity, which also considers its ellipticity, which results in the free-air anomaly. If in addition the effects induced by topographic features with a constant standard rock density (2670 kg m^{-3}) are removed, one obtains the Bouguer gravity anomaly, which reveals density inhomogeneities in the subsurface (e.g. Nowell 1999). Typical example are the crustal roots under mountain chains that isostatically compensate topography. With additional information about subsurface density structures that are relevant for the isostatic behaviour of the solid Earth, one can compute their effect on the gravity field, which is referred to as the isostatic correction (e.g. Kaban *et al.* 2004). After applying the isostatic correction to the Bouguer gravity anomaly, the resulting, isostatic residual anomaly reveals deviations of the Earth from the state of isostatic equilibrium.

Gravimetric measurements can be taken on the ground with high precision, but require considerable logistical effort for remote regions to cover small areas relative to the size of the Antarctic continent. Therefore, mostly airborne and satellite surveys are acquired for Antarctica. In recent years, considerable progress has been made in both global Earth gravity field observations by satellite missions and regional airborne campaigns (e.g. Ferraccioli *et al.* 2011; Riedel *et al.* 2012; Aitken *et al.* 2014; Scheinert *et al.* 2016; Forsberg *et al.* 2018). The lateral resolution of these two techniques ranges from 130 to 80 km for static gravity models derived from satellite data (Mayer-Gürr *et al.* 2010, 2012) to less than 10 km for airborne data (e.g. Forsberg *et al.* 2011). Long-wavelength satellite data are particularly useful to study the deeper lithospheric and sub-lithospheric upper mantle structures. Aerogravimetry compilations often lack such information due to post-processing

From: Martin, A. P. and van der Wal, W. (eds) 2023. *The Geochemistry and Geophysics of the Antarctic Mantle*. Geological Society, London, Memoirs, **56**, 213–229,
First published online 28 April 2021, https://doi.org/10.1144/M56-2020-5

and combination procedures of individual flight lines and thus focus on small-scale and shallow features. Theoretically, the maximum wavelength detectable by airborne measurements is half the extent of the surveyed area, but it is often shorter due to processing steps such as levelling (e.g. Barzaghi *et al.* 2014). Aerogravimetric and satellite data are therefore combined in Earth gravitational models, in order to achieve a high degree of consistency together with high resolution where data are available (e.g. Forsberg 2015; Pail *et al.* 2018).

The twin-satellite mission GRACE (Gravity Recovery and Climate Experiment), orbiting Earth at *c.* 450 km height from 2002 to 2017, did not only measure the long wavelength signal of the gravitational field with high precision (Tapley *et al.* 2004) but also allowed us to track changes over time. In the years 2009–13 the complementary GOCE (Gravity field and steady-state Ocean Circulation Explorer) satellite mission mapped the gravity gradient field of the Earth at *c.* 255 km altitude (*c.* 225 km in the final stage) with its three-dimensional gradiometer (Rummel *et al.* 2011; Bouman *et al.* 2015). The gravity gradients are the second derivative of the gravitational potential, also known as the Marussi Tensor, consisting of six unique components. While the spatial resolution could be improved significantly to *c.* 80 km, GOCE's gravity gradient measurement has a poorer sensitivity at long wavelengths, which is why global Earth gravity models often combine GRACE and GOCE data to exploit the complementary advantages (e.g. Peidou and Pagiatakis 2019; Zingerle *et al.* 2019). Due to the orbit inclination, however, both satellite missions do not cover the entire Earth, which means that data gaps at the poles exist. In the case of GRACE, the observations extend to 89° (Reigber 2006), whereas for GOCE the polar gap covers regions beyond 83.5° S (Forsberg *et al.* 2011; Brockmann *et al.* 2014) and is as large as 1400 km in diameter (Scheinert *et al.* 2016). Recent efforts have been made in the course of the PolarGAP initiative to fill the gap with extensive airborne gravity measurements (Forsberg 2015).

Great advances have also been made in collecting and compiling airborne gravimetric data for the Antarctic continent and its adjacent oceanic areas. The AntGG compilation (Scheinert *et al.* 2016) covers 73% of Antarctica with a 10 km resolution grid and provides free-air and Bouguer gravity anomaly estimates (Fig. 1). Remaining gaps like the interior of Dronning Maud Land are being successively filled by additional surveys, which reveal unexplored geological features, help to

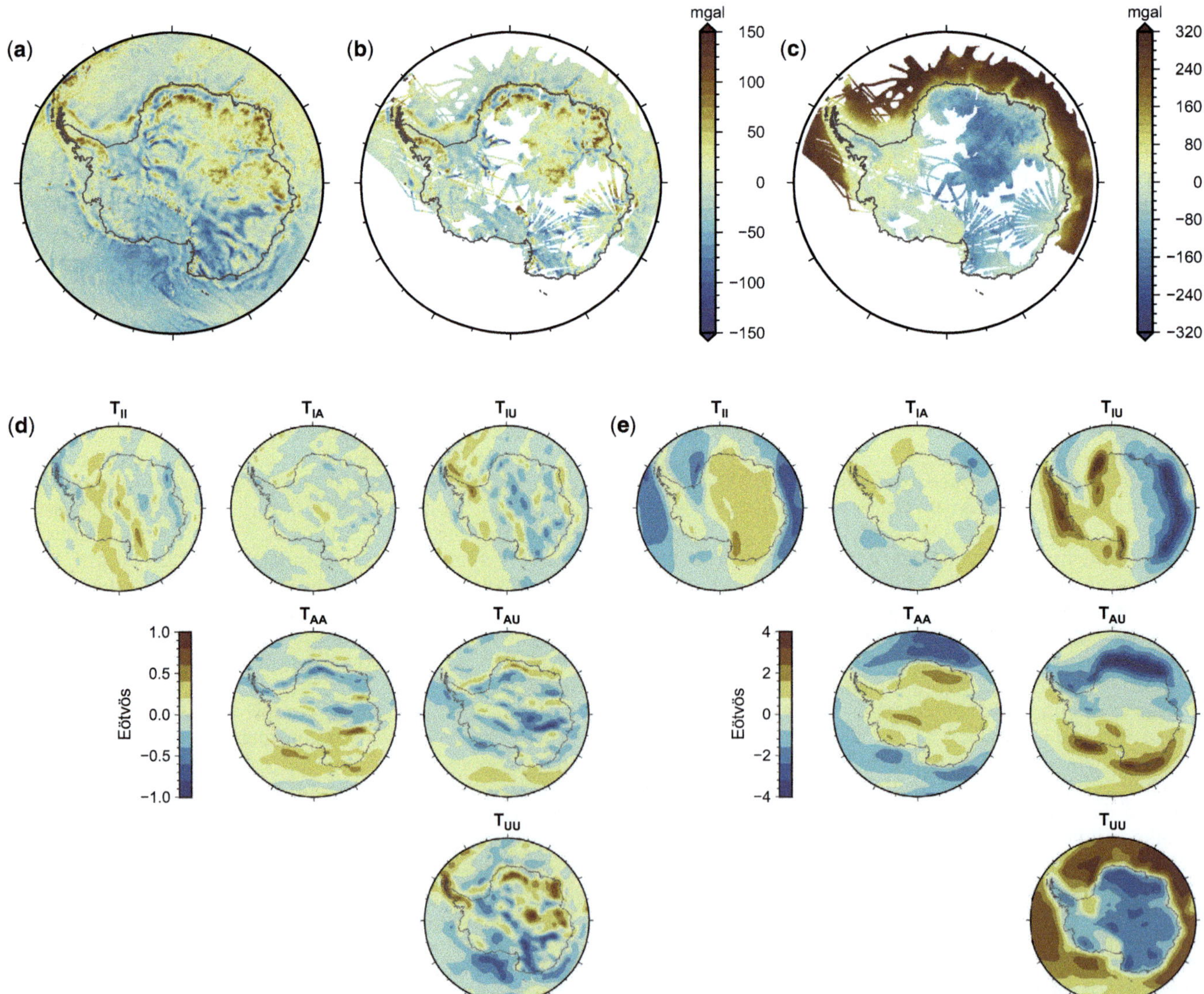

Fig. 1. Gravity data of Antarctica: (**a**) gravity anomaly of combined satellite model XGM2016 (Pail *et al.* 2018), (**b**) free-air and (**c**) Bouguer gravity anomalies from airborne survey data (Scheinert *et al.* 2016), (**d**) gravity gradient components from GOCE data (Bouman *et al.* 2016) at satellite altitude (225 km) and (**e**) topography-corrected gravity gradients. Gravity gradients are expressed in the IAU reference frame (Pappa *et al.* 2019*b*) with axes pointing to the Indian Ocean (I), Atlantic Ocean (A), and upward (U) direction.

evaluate existing gravity models, or solely confirm vintage data (Jordan *et al.* 2017; Yildiz *et al.* 2017; Forsberg *et al.* 2018). With the picture of Antarctica's gravitational signal getting successively clearer and more complete, large-scale tectonic features and crustal structures can be studied on a more robust basis. Furthermore, gravitational signatures can, together with other data like magnetic measurements, identify or assure the links between the Antarctic and its adjacent continents within the Gondwana, Rodinia and Columbia supercontinents (Aitken *et al.* 2016; Scheinert *et al.* 2016).

The Shape index: a gravitational field curvature product to study the lithospheric structure

Since satellite gravity measurements provide homogeneous data coverage over nearly the entire globe, one can search for similar characteristics in the gravitational field between regions in Antarctica and those in better-studied continents. Gravity gradients as measured by GOCE (Fig. 1) are particularly suited for this purpose due to their high sensitivity to density variations in lithospheric depth ranges (Bouman *et al.* 2016). In modelling of gravity gradients, the individual depth sensitivity of each gradient component can be used to support the modelling effort. For interpretation, it is, however, challenging to use multiple gravity gradient components simultaneously. Combining the gravity gradient components can simplify the interpretation and thus be of use for the broad scientific community.

An example of a gravity (gradient) data product that can be interpreted relatively easily is the shape index. Hereby, the gravity gradient tensor, is translated into an index value ranging from −1 for a bowl-like shape of the equipotential surface to +1 for a dome-like shape, which can be interpreted in terms of characteristics of the continental lithosphere. Ebbing *et al.* (2018) applied this technique on GOCE's global gravity gradient data after correcting them for topographic effects. The shape index indicates mass deficits in the lithosphere with negative values (bowls) and mass surplus with positive values (domes) and thus enhances lithospheric and intra-crustal density variations. In the global view (Fig. 2a), bowls can generally be correlated with cratonic areas or orogenic belts, e.g. the Kaapvaal Craton in South Africa and the Himalayas. In these

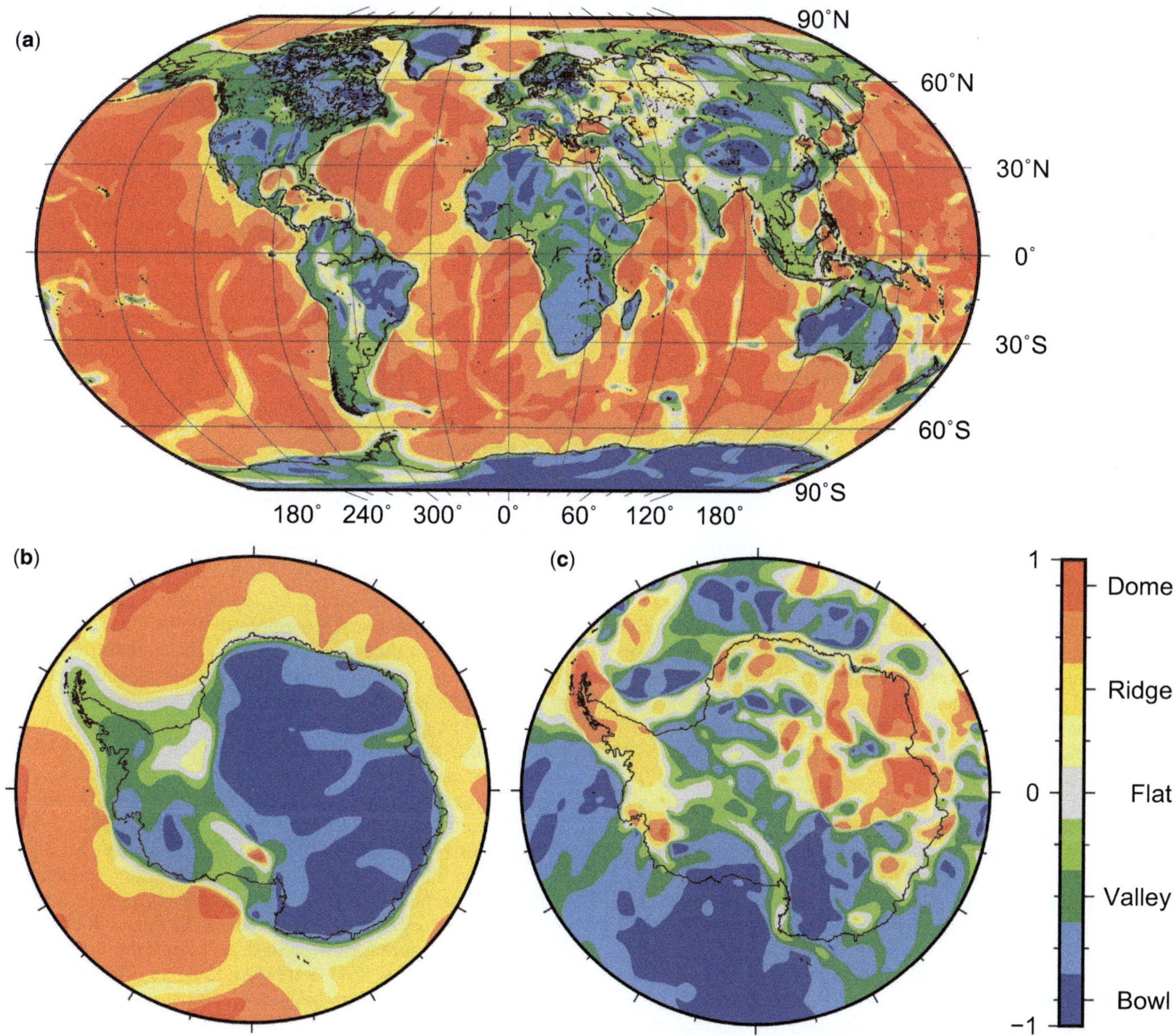

Fig. 2. Shape index based on GOCE gravity gradient data after topographic correction for (**a**) the Earth and (**b**) Antarctica (based on Ebbing *et al.* 2018). Panel (**c**) shows the shape index for Antarctica after additional correction for isostatic effects.

cases, mass deficits can be explained by depleted lithospheric mantle composition in cratonic settings and thick crustal roots beneath mountain ranges. On the other hand, the Andes exhibit a valley-like shape, most likely due to the additional masses of the cold subducted slab. Dome-like shapes indicate a positive shape index and generally correlate with oceanic lithosphere (e.g. the Pacific Ocean) and rifting (e.g. Red Sea).

In light of this global correlation, one can compare the shape index of Antarctica (Fig. 2b) with other regions of the world. A clear distinction between West Antarctica and East Antarctica is obvious, particularly pronounced through high shape index values in the West Antarctic Rift System and the Weddell Sea Rift, but also above the Antarctic Peninsula. Only Marie Byrd Land exhibits a bowl-like shape in West Antarctica. Underneath this region, a Cenozoic mantle plume has been proposed (LeMasurier and Rex 1989), which would likely imply relatively low densities of the upper mantle due to elevated temperatures. Independently observed low seismic velocities have been interpreted as a result of such a thermal anomaly (e.g. Lloyd *et al.* 2015; Wiens this volume, in review). The shape index can neither confirm nor disprove the existence of a mantle plume, but it is consistent with a thermal anomaly at depth (see Bredow and Steinberger (2021, this volume), for an in-depth discussion of the possible mantle plume).

East Antarctica, on the other hand, shows entirely negative shape index values, comparable to those of other cratonic regions on Earth. Here, the Lambert Graben stands out with an almost flat shape index, which likely reflects the tectonic setting of a failed rift system (e.g. Harrowfield *et al.* 2005). In addition to a topographic correction, the gravitational field can be reduced for isostatic effects in order to bring out the isostatic state. Figure 2c shows the shape index for Antarctica after applying a topographic reduction and subtracting the effect of an isostatically compensated Moho. The patterns are generally similar to the observed gravity anomaly (Fig. 1), though some differences can be identified. For instance, the Marie Byrd Land dome and the Antarctic Peninsula are more pronounced in the shape index. The Transantarctic Mountains, on the other hand, show a flat shape index, whereas their gravity anomaly is distinctly positive with more than 50 mGal. This can be interpreted as an expression of a missing crustal root beneath the Transantarctic Mountains (e.g. Hansen *et al.* 2016) and their still unresolved isostatic state and uplift history (e.g. Lisker *et al.* 2013).

Elastic lithosphere

Using a flexural uplift model, Stern and ten Brink (1989) early on could explain the apparent contradiction of a missing root under and the persistence of topography of the Transantarctic Mountains. Later studies used flexural uplift to study the architecture of the Ross Sea basin (Karner *et al.* 2005) or the interaction of the Wilkes basin and Transantarctic Mountains (Paxman *et al.* 2019*a*).

In a flexural uplift model, or regional isostasy, in comparison to conventional Airy or Pratt-isostasy, the elastic properties of the lithosphere are taken into account (e.g. Watts 2001). Hereby, in the simplest form, the load by the topography suppresses the elastic plate and the response is governed by the flexural rigidity of the plate. For simplicity, the flexural rigidity is often expressed by the effective elastic thickness that under the assumption of certain rheological parameters, describes the thickness of the idealized elastic plate and does not represent a physical boundary. Please note that the effective elastic thickness is not the same quantity as used in Glacial-Isostatic Adjustment modelling studies, where it describes the thickness of an idealized elastic layer.

Different techniques exist to calculate the flexural rigidity or effective elastic thickness, e.g. admittance or coherence (Watts 2001) or modifications of these (Paxman this volume). These methods all use sub-ice topography and the gravity field to derive a ratio of similarity between the two that reflects the elasticity of the lithosphere. McKenzie *et al.* (2015) or Chen *et al.* (2018) are two recent studies, that used satellite and a combined gravity model, respectively, to show that East and West Antarctica have significantly different properties. East Antarctica is generally expressed by a thick effective elastic thickness, while West Antarctica has thin effective elastic thickness. Paxman (2021, this volume) provides a comprehensive review of the elastic properties of the Antarctic continent.

Crustal thickness of Antarctica from gravity data

The Moho boundary separates crustal and mantle rocks and is, in general, associated with the most prominent density contrast in the lithosphere. Depending on the tectonic setting, the density contrast can vary between less than 200 kg m^{-3} in cratonic regions when eclogitization is involved, and more than 500 kg m^{-3} in extending basins or Phanerozoic orogens (Rabbel *et al.* 2013). Therefore, depth, geometry and density contrast at the Moho interface have a huge impact on the gravity field at a broad range of wavelengths (e.g. Sebera *et al.* 2018). Since geothermal heat flow, seismic velocities, and several other geophysical and geological parameters are strongly affected by the crustal thickness, knowledge about the Moho depth is of broad importance.

However, the observed gravity signal originates from several sources, and the free-air anomaly is largely affected by topography. In order to isolate the signal from the Moho boundary, effects from other sources, such as intra-crustal (sedimentary basins) and mantle density variations or dynamic forces as well as topography, ice and water need to be subtracted from the total signal. For Antarctica, the estimation of the sub-ice architecture is challenging. Nevertheless, huge progress has been made by establishing the Bedmap2 model (Fretwell *et al.* 2013), which describes the surface elevation, the ice thickness and the seafloor and sub-glacial bed elevation of the Antarctic and has been recently updated by the topographic model BedMachine Antarctica (Morlighem *et al.* 2020).

Since sedimentary rocks can have significantly lower densities than igneous or metamorphic rocks, it is crucial to incorporate them in a topographic reduction model. A compilation of offshore sedimentary basins in the Antarctic has recently been published (Straume *et al.* 2019), showing more than 12 km-thick sedimentary sequences in the Weddell Sea sector and up to 8 km in the Ross Sea region. The availability of data for sediment thickness is poorly constrained for onshore areas relative to offshore. Few studies exist that estimate the depth of East Antarctic inland sub-glacial basins from aerogeophysical surveys, but there are exceptions (e.g. Mishra *et al.* 1999; Bamber *et al.* 2006; Ferraccioli *et al.* 2011; Aitken *et al.* 2014; Frederick *et al.* 2016), predominantly in the Wilkes Land region, or seismic methods in Lake Vostok (Filina *et al.* 2008; Isanina *et al.* 2009) and its periphery (Studinger *et al.* 2003). It is, however, extremely challenging to establish a trustworthy model for thickness and density of East Antarctica's sedimentary basins from the available data, which would also have to account for different degrees of compaction, infill composition and age if it is supposed to be used in a gravimetric topographic reduction. In general, efforts to make sediment thickness maps of the Antarctic continent based on gravity and/or seismic estimates (e.g. Baranov *et al.* 2018; Haeger and Kaban 2019) still hold a high degree of uncertainty concerning the sediment's distribution, thickness and density.

Reasonably reliable data are currently only provided from active seismic surveys in offshore areas.

Multiple techniques exist to infer the Moho depth from gravimetric observations. The most widely used one is the Parker–Oldenburg scheme (Parker 1973; Oldenburg 1974). This approach mathematically relates the geometry of a two-dimensional interface with a vertical gravity anomaly and can thus be used to invert the inferred Bouguer anomaly of a particular region into a Moho depth model. It implies, however, a flat Earth approximation and requires presumptions of a certain (constant) density contrast and a reference depth for the Moho. Moreover, the data need to be given at Earth surface level, which means that low pass filtering and downward continuation must be performed on satellite gravity observations to prevent numerical instabilities and noise amplification. Other methods to estimate the Moho depth from the Bouguer gravity anomaly are, on the other hand, applicable in spherical coordinates and at satellite height (e.g. Moritz 1990).

Block *et al.* (2009) followed the Parker–Oldenburg approach and used GRACE satellite gravity data to estimate the crustal thickness of Antarctica, assuming a reference depth of 35 km and a density contrast of 500 kg m^{-3}. Their results showed a large contrast between the crustal thickness of West Antarctica with *c.* 30 km and East Antarctica with *c.* 40 km and a crustal root beneath the Gamburtsev Subglacial Mountains (GSM) of as much as 42 km. Seismological studies estimated the crustal thickness to be thinner (*c.* 20–30 km) in West Antarctica (e.g. Chaput *et al.* 2014; Ramirez *et al.* 2016; Shen *et al.* 2018) and much thicker (>50 km) beneath the GSM (e.g. Lawrence *et al.* 2006; Hansen *et al.* 2010; An *et al.* 2016). Even considering the rather high uncertainties of the first Bedmap model (Lythe and Vaughan 2001), the discrepancy suggests that a single density contrast and reference depth is not applicable for the whole Antarctic continent with its diverse tectonic provinces (Pappa *et al.* 2019*a*). Accounting for that, O'Donnell and Nyblade (2014) performed separate inversions of the Moho depth for West and East Antarctica and constrained the solutions with local seismological Moho depth estimates. Their results indicate a >50 km-thick crustal root beneath the GSM while a *c.* 23–27 km thick crust is suggested for most of West Antarctica. Integrating seismological findings in the inversion of gravity data enabled the authors to reduce the inherent ambiguity significantly. Despite these methodological improvements, however, it is still not viable to derive a Moho depth model for Antarctica from gravity with constraining seismic data (Pappa *et al.* 2019*a*), because the diverse tectonic regimes of the continent with different Moho density contrasts need to be accounted for. A thick and high-density crust can be similarly isostatically compensated as a thin and low-density crust. In this regard, gravity-inverted Moho depth models tend to underestimate the amplitude of crustal thickness variations.

Gravity-derived Moho depth models of Antarctica show similar morphologies because their methods are similar. In contrast to this, Moho models based solely on seismological data are notably diverse (e.g. Baranov and Morelli 2013; An *et al.* 2015*b*; Baranov *et al.* 2018; Haeger *et al.* 2019) and differ by more than 10 km in large areas of the continent. One reason is the variety of seismological methods applied (see also Wiens this volume, in review). Some studies suggest that S-wave receiver functions analysis is more reliable than P-wave receiver functions analysis due to the reverberation of seismic waves by the thick ice cover (e.g. Hansen *et al.* 2016). Another source for the disagreement of seismological models lies in the sparseness of seismic stations in Antarctica and the interpolation method for blank areas. Surface wave analyses can estimate the physical properties of the subsurface between seismic stations from dispersion curves, inherently leading to smooth results with averaged values. Numerical interpolation algorithms like the Kriging method, on the other hand, fill the blank areas based on local Moho depth information in the vicinity (Baranov and Morelli 2013; Haeger *et al.* 2019; Szwillus *et al.* 2019). Both methods can differ substantially in their results, but neither of them can be declared more correct than the other.

Figure 3a–f shows six Moho depth models derived with different methods. All models show the striking contrast in crustal thickness between West Antarctica and East Antarctica as well as some prominent features with thicker crust like the GSM, the southern Transantarctic Mountains, and (with the exception of the AN1-Moho model) Marie Byrd Land. The concept of vote maps (Shephard *et al.* 2017) can be used to illustrate the degree of consistency between models and to identify common features (Fig. 3g–i). This simple method counts the number of models that match a certain criterion. All presented Moho depth models agree on a >30 -km-thick crust in the whole of East Antarctica (see Fig. 3g). Likewise, large parts of West Antarctica consistently show a Moho depth shallower than 30 km, whereas the picture is diverse in the Weddell Sea region, the Antarctic Peninsula, and the vicinity of the Marie Byrd Land dome. In East Antarctica, on the other hand, only a few spots are present where at least a majority of the models exhibits a crust thicker than 40 km (Fig. 3i).

Using satellite gravity data to investigate the upper mantle structure of Antarctica

Joint interpretations of seismic findings and gravity data or isostasy (O'Donnell and Nyblade 2014; An *et al.* 2015*b*; Pappa *et al.* 2019*a*) suggest a laterally varying density contrast at the crust–mantle boundary, implying a heterogeneous upper mantle density structure beneath Antarctica. This is also indicated by seismic velocity models (e.g. Lloyd 2018). The dominating factors for both density and seismic properties of mantle rocks are temperature and composition. However, the relationships between these parameters are fundamentally different and partly non-linear or discontinuous due to mineral phase changes (e.g. Cammarano *et al.* 2003; Stixrude and Lithgow-Bertelloni 2005). For instance, the principle of isopycnicity (Jordan 1988) describes how density changes of mantle rocks due to temperatures can be compensated by opposite density changes due to compositional depletion (for discussion see also Kaban *et al.* 2003; Eaton and Perry 2013; Artemieva *et al.* 2019). Thermodynamic modelling of stable mineral assemblages (e.g. Connolly 2005) of mantle rocks under given pressure and temperature conditions can be used to establish lithospheric models that are self-consistent in terms of density, temperature and seismic velocities (e.g. Fullea *et al.* 2009, 2012). With this tool at hand, one can attempt to investigate the nature of the upper mantle of Antarctica by integrating seismological velocity models and satellite gravity data (Haeger *et al.* 2019). For this purpose, however, it is necessary to first separate the gravity signal originating from the lithospheric mantle as accurately as possible (e.g. Kaban *et al.* 2010). Hence, a crustal thickness and density model is needed, which is, as described above, subject to large uncertainties in the case of Antarctica. Haeger *et al.* (2019) used existing seismological information on Moho depth and seismic velocities in Antarctica to derive a crustal model through interpolation for a joint inversion of gravity and seismic tomography data with the aim of inferring the density, temperature, and composition of the Antarctic upper mantle. After applying the crustal correction, as described above, regions of low gravity residuals are associated with decreased

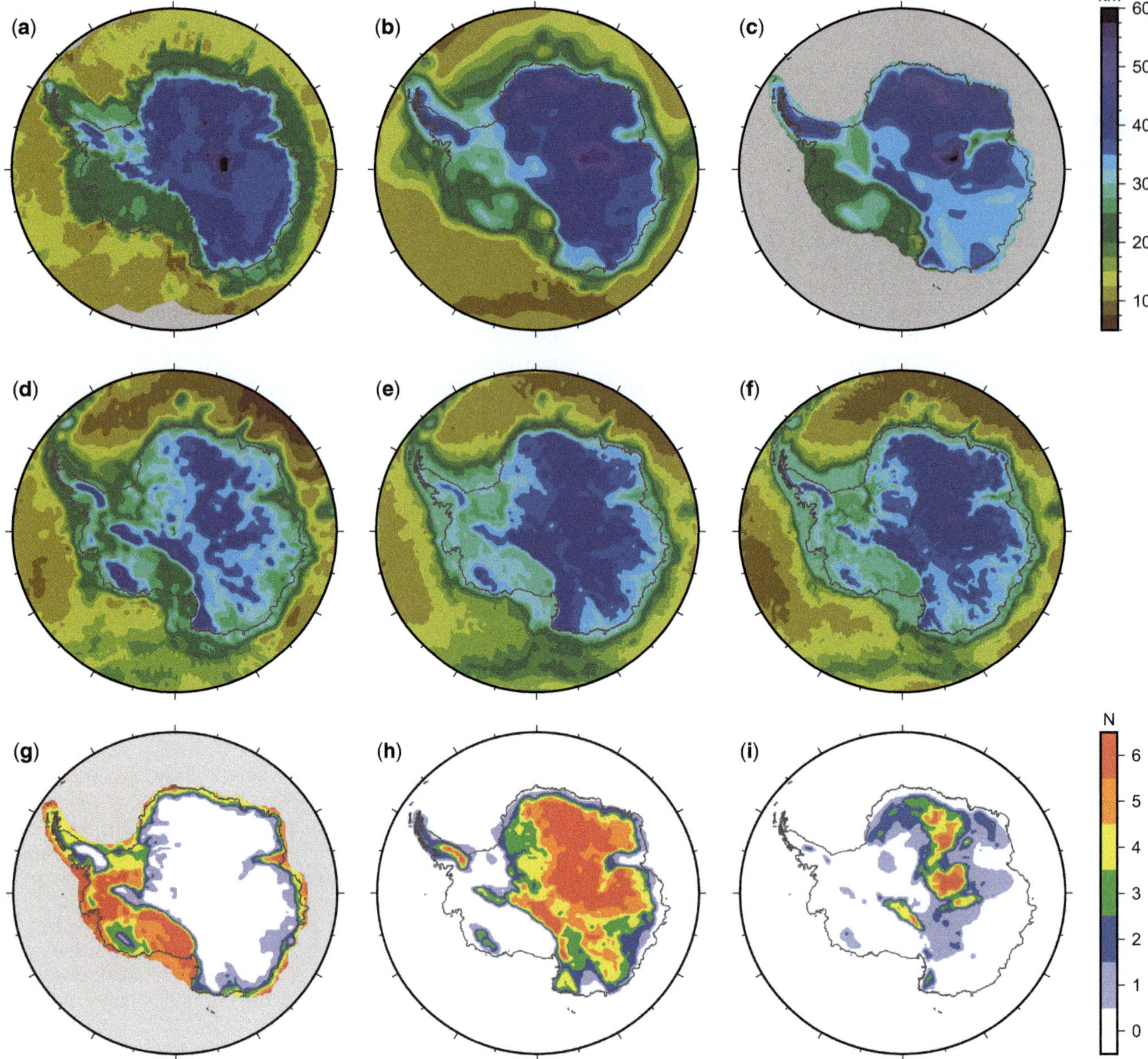

Fig. 3. **(a–f)** Moho depth models derived with different methods: (a) surface wave analysis, AN1-Moho (An *et al.* 2015*b*); (b) kriging interpolation of local seismologically derived Moho depth estimates filled with estimates from the AN1-Moho model (Haeger *et al.* 2019); (c) integrating seismic estimates, gravity data and isostasy (Baranov *et al.* 2018); (d) inversion of GOCE gravity gradient data, GEMMA (Reguzzoni and Sampietro 2015); (e) gravity inversion constrained with seismic estimates (Pappa *et al.* 2019*a*); (f) Airy-isostatic Moho depth model (Pappa *et al.* 2019*a*). **(g–i)** Vote-maps for different depth ranges illustrate the degree of agreement of the above models: (g) less than 30 km Moho depth; (h) more than 35 km; (i) more than 40 km. Colours indicate the numbers of models that agree for this depth range. For example, reddish areas show high agreement for a shallow Moho in the West Antarctic Rift System (g), whereas only a few spots exist in East Antarctica (i) where all models see a >40 km thick crust.

lithospheric mantle densities, potentially indicating a rock composition depleted in iron. Notwithstanding the poor data situation, the authors were able to show that large parts of East Antarctica's lithospheric mantle are likely strongly depleted in iron and thus enriched in magnesium (Fig. 4), indicated by a high magnesium number (Mg# = 100 Mg/(Mg + Fe)), which suggests cratonic settings of Proterozoic to Archean age. The fact, however, that different up-to-date seismological tomography models lead to inconsistent results in the inversion process demonstrates the need for further field campaigns to collect data, which will enable us to establish more robust models of Antarctica's lithospheric structure.

In order to investigate the characteristics of the upper mantle with gravity modelling, crustal corrections are commonly applied (e.g. Kaban *et al.* 2010; Haeger *et al.* 2019), which often rely on seismological models. If, however, a trade-off between crustal density, Moho geometry and mantle density is allowed in the modelling (e.g. Aitken *et al.* 2013), a reasonably consistent lithospheric model can be established with (partly) sparse seismic coverage. If, in contrast, a fixed crustal density and thickness model derived from seismic findings alone is used as a correction for gravity data, the potential errors in the gravity signal are likely larger than the gravity signal from mantle density variations. Such a seismologically derived crustal model should – if at all – be used extremely cautiously in gravity modelling. As long as the seismic findings considering the thickness and the density of the Antarctic crust remain ambiguous and, in consequence, seismological models cannot provide the degree of certainty required for gravity modelling, other strategies must be pursued in order

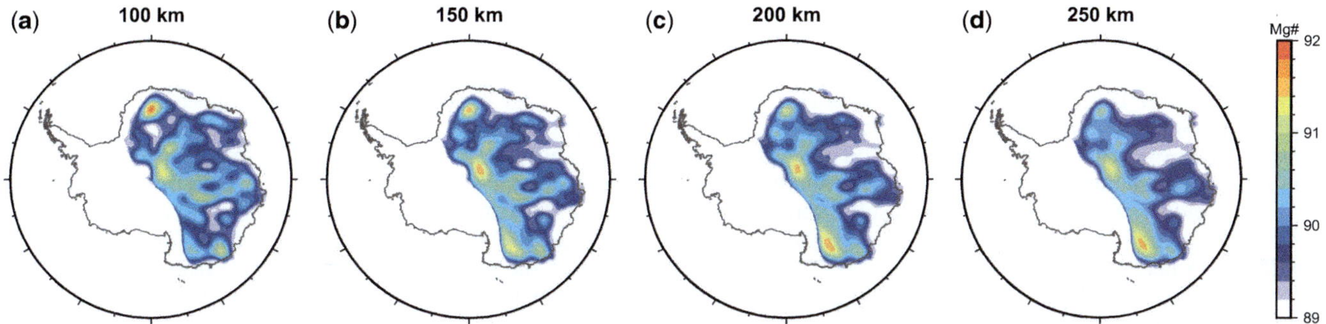

Fig. 4. Depletion variations in terms of Mg# estimated for (**a**) 100 km, (**b**) 150 km, (**c**) 200 km and (**d**) 250 km, based on the seismic S-wave velocity model SL2013sv (Schaeffer and Lebedev 2013) (after Haeger *et al.* 2019).

to investigate the architecture of the Antarctic lithosphere. One possibility is to model the lithosphere as a whole while considering all those parameters and processes that determine the observables for which data are available. For instance, the temperature strongly affects rock densities due to thermal expansion, the stable mineral assemblage, and the elastic behaviour and thus seismic velocities. Correspondingly, the thermal field of the lithosphere can reasonably be computed by solving the heat transfer equation, whereas the gravity field resulting from the modelled densities and the seismic velocities can be compared with measured data.

Despite the advantages, it is still necessary to make *a priori* assumptions about some parameters like crustal and mantle rock compositions, which are poorly explored or unknown for large parts of Antarctica. For establishing a lithospheric model of Antarctica, Pappa *et al.* (2019*b*) relied on petrological studies to define broad regions of different tectonothermal age (Fig. 5a), which were associated with varying degrees of depletion of lithospheric mantle composition. According to the pressure and temperature conditions inside the model, the stable mineral phases (and thus the density) of the respective peridotitic composition were inferred through the thermodynamic modelling software *Perple_X* (Connolly 2005). This method allows modelling of *in situ* densities for rocks under certain assumptions, such as the oxides considered in the system (in this case mantle rocks in the CaO–FeO–MgO–Al_2O_3–

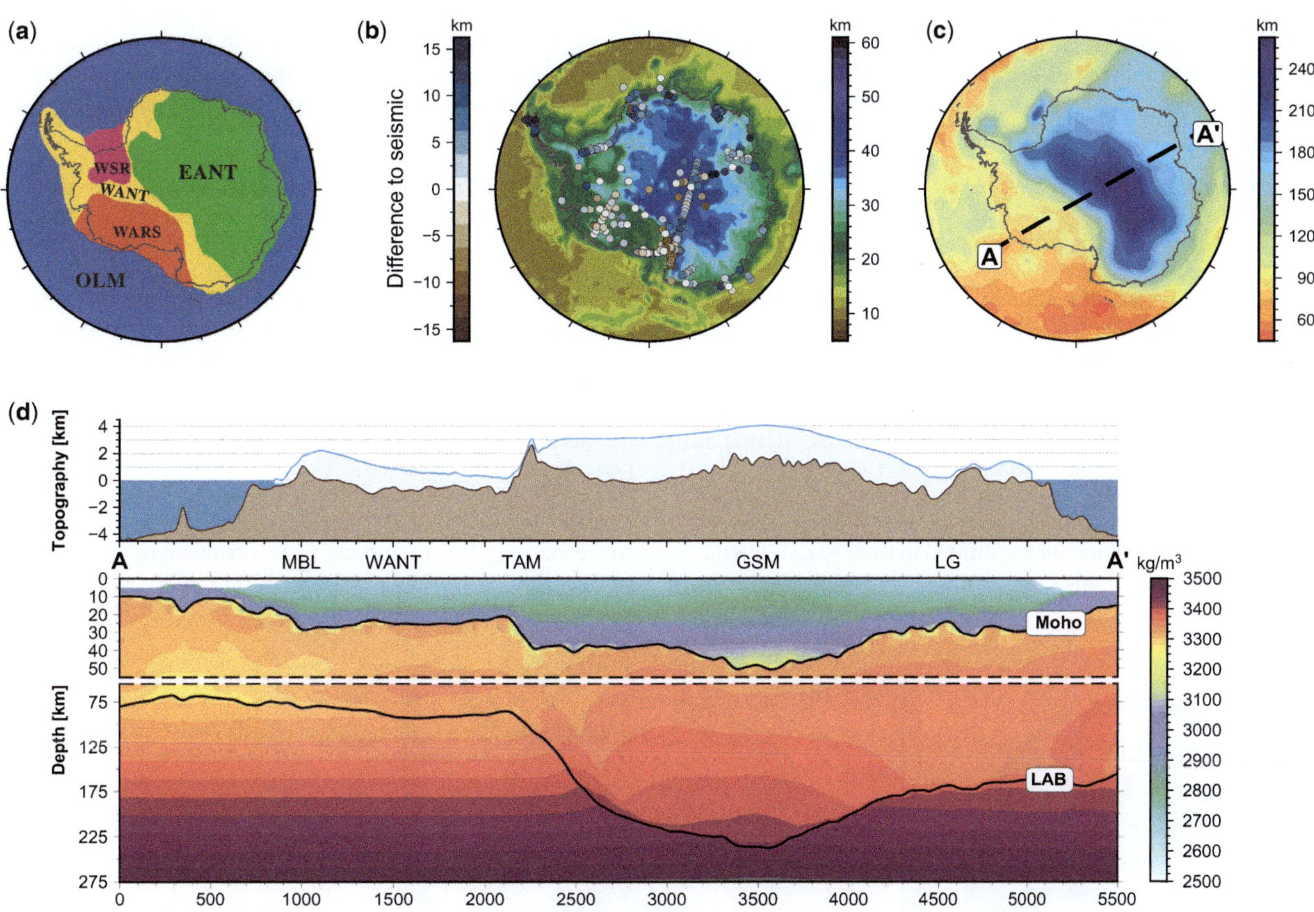

Fig. 5. (**a**) Domains of different lithospheric mantle composition in the model from Pappa *et al.* (2019*b*). (**b**) Resulting Moho depth with local seismic estimates as benchmark (coloured circles and left colour-bar). (**c**) Resulting total lithospheric thickness. (**d**) Density cross-section from A–A' (location on (c)). EANT, East Antarctica; GSM, Gamburtsev Subglacial Mountains; LAB, lithosphere–asthenosphere boundary; LG, Lambert Graben; MBL, Marie Byrd Land; OLM, oceanic lithospheric mantle; TAM, Transantarctic Mountains; WANT, West Antarctica; WARS, West Antarctic Rift System; WSR, Weddell Sea Rift.

SiO_2 system). The variety of modelled mineral phases is thus naturally limited. Additionally, representative global values were taken for crustal density and thermal parameters. With this initial set-up, the model was iteratively fitted to the observed topography by maintaining isostatic equilibrium and the satellite gravity gradient data measured by GOCE, always maintaining internal consistency in the relationships between all model properties. The main results of this study are new estimates of the Moho depth and the total lithospheric thickness of Antarctica (Fig. 5b and c).

Local seismic Moho depth estimates can serve as *a posteriori* benchmarks instead of using them as an input or constraint as described above. Since the density contrast at the Moho discontinuity in the lithospheric model is a function of temperature, pressure and rock properties, it can vary freely, which leads to a generally good fit of both the gravity gradient data and the seismic Moho depth estimates. In particular, the crustal root beneath the GSM is deeper (*c.* 54 km) in this model than in classical gravity inversions and has a high density of >3150 kg m^{-3}. A very small density contrast in this region was previously suggested by other studies (Ferraccioli *et al.* 2011; An *et al.* 2015*b*). The modelled total lithospheric thickness is largely similar to seismological estimates (An *et al.* 2015*a*) and likewise exhibits a stark contrast between thin lithosphere (*c.* 80 km) in West Antarctica and up to 260 km in East Antarctica. Having such a temperature model of the crust and upper mantle at hand, which agrees with topographic (isostatic), seismic and gravity gradient data, offers the opportunity to derive upper mantle viscosities (Pappa *et al.* 2019*b*). Especially in Antarctica, it is crucial to consider the three-dimensional variations of viscosity instead of using 1D models (Nield *et al.* 2018; Whitehouse *et al.* 2019) in estimating the glacial isostatic behaviour of the solid Earth, e.g. present-day uplift rates due to ice mass changes.

Magnetic data

Magnetic observations: scales and strides

Antarctica is covered by a number of aeromagnetic surveys as well as satellite magnetic models. Global models derived from single-satellite data have been obtained with accuracies better than a few nanotesla, but rather low resolution due to the dynamic behaviour of the external field, which strongly affects the measurements. The Earth's magnetic field has internal as well as external dynamic sources. Both vary over a wide range of time scales and separating them relies on their different temporal variations. Data from satellite missions have a uniform global coverage but are limited in their spatial resolution by the high altitudes. To distinguish variations in the lithospheric field from time-dependent contributions of the magnetosphere and ionosphere is especially challenging in polar regions (Langel and Hinze 1998).

A number of global magnetic satellite models have been released since the Ørsted satellite mission was launched in 1999 and joined by the CHAMP satellite mission in 2000. These satellites in low-Earth orbit provide the most effective means of mapping the long wavelengths of the magnetic field caused by the magnetization of the Earth's crust. For instance, the MF6 model was produced using the latest three years of measurements from the CHAMP flux-gate magnetometer (Maus *et al.* 2008). MF6 resolves the crustal magnetic field to spherical harmonic degree 120, corresponding to length scales down to 333 km. The follow-up model MF7 provided the crustal magnetic field to spherical harmonic degree 133, corresponding to length scales down to 300 km (Maus 2010).

In 2013, the Swarm satellite mission, a constellation of three satellites in three different polar orbits was launched. The satellites are located in two orbits, two as a pair in a lower orbit and one upper orbit, between 400 and 550 km altitude. One of the aims is to measure the lithospheric magnetic field to a resolution of 200 km. By combining the measurements of the strength and direction of the magnetic field, the vertical and horizontal derivatives can be calculated, which are less sensitive to time-dependent contributions of the magnetosphere and ionosphere (Kotsiaros and Olsen 2012) and therefore aid in the construction of lithospheric anomalies as in the model LCS-1 (Fig. 6a, Olsen *et al.* 2017). The model is improved in comparison to previous models (which were in good agreement up to around SH degree 85, i.e. down to 470 km wavelength); the lithospheric field contribution is now given from spherical harmonic degree 15 to 180 (*c.* 200 km length scales), but satellite data only constrain the part to spherical harmonic degree 135 (*c.* 300 km length scales). For the higher spectral range to degree 180, a certain correlation has been found with the Australian aeromagnetic compilation (Olsen *et al.* 2017). However, in polar regions, the noise to external field sources is significantly higher, making this part, before further detailed examination and validation, unreliable. As for all models, the spherical harmonic degrees up to degree 15 are omitted as this part is dominated by the normal, core field (ranging at the surface from 30 000 nT at the equator to 65 000 nT in polar regions), which masks the lithospheric field contribution.

In order to derive total magnetic field anomalies, called lithospheric magnetic anomalies, from aeromagnetic data, a reference field is subtracted from the measured field representing the normal core field. Routinely, the International Geomagnetic Reference Field (IGRF) for a certain epoch is used. The IGRF is a prediction of the behaviour of the magnetic normal field for an epoch of 5 years (e.g. Thébault *et al.* 2015) and has, especially for Antarctica, a low precision before the onset of satellite measurements. The Definitive Geomagnetic

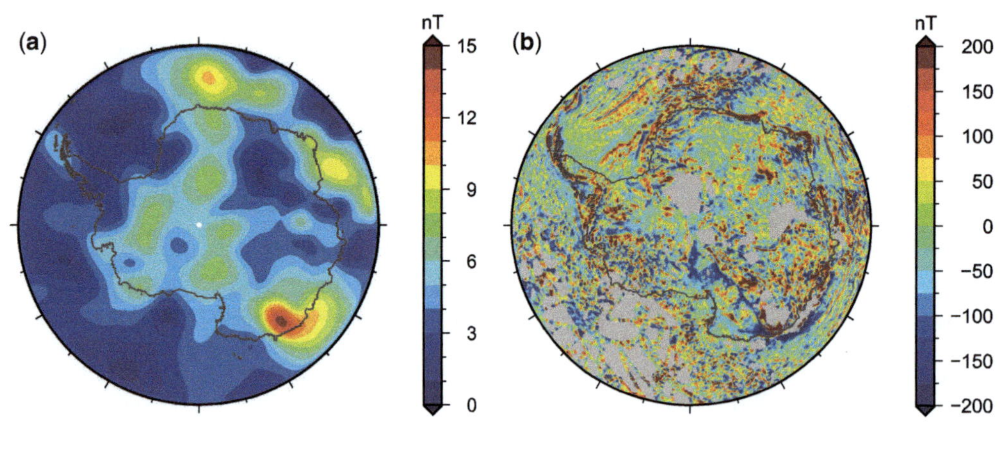

Fig. 6. Lithospheric magnetic anomaly of Antarctica: (**a**) satellite model LCS-1 (Olsen *et al.* 2017) defined at 400 km height; (**b**) ADMAP-2 compilation Golynsky *et al.* (2018) based on all available airborne surveys.

Reference Field (DGRF) is defined for these epochs based on the actual measurements, but data are seldom reprocessed to account for this, when this becomes available. Differences between IGRF and DGRF range in magnitude from 5 to 10 nT, which is low compared to the magnetic normal field, but can be high compared to lithospheric anomalies.

Between satellite and aeromagnetic data, a spectral gap still exists, which hopefully will be smaller with increasing lifetime of the Swarm mission, but also relates to the specifics of aeromagnetic surveys. The first controlling factor for aeromagnetic surveys is the extension, which is usually not larger than a few hundred kilometres. As seen in Figure 6b, Antarctica has an extensive coverage with aeromagnetic surveys and only a few gaps exist. However, the surveys have been acquired over a time-span of more than 50 years and can be divided in quality by data acquired before and after the availability of GPS for navigation (Golynsky *et al.* 2018). Flight lines before GPS have a relatively high uncertainty in navigation as other means have been used for referencing, for example topographic maps, while for later surveys the accuracy is relatively high. Inherent in aeromagnetic surveys is a low precision for the long-wavelength component. That relates to the size of the survey and routine processing techniques, e.g. correction for the reference field (IGRF/DGRF), levelling of datasets, diurnal correction and merging of individual surveys (e.g. Reeves 2005).

Levelling of the flight accounts for varying conditions during flight time such as diurnal variations (short-scale variations of the magnetic field) or varying altitude of the survey. Diurnal variations relate to changes in the magnetic field due to external sources, e.g. sometimes observable as polar lights. To correct for these time-varying effects, a survey has typically normal flight lines and orthogonal tie-lines of 2–5 times the spacing of the normal lines. Between lines and tie-lines, the difference is calculated at the cross-points and these differences are (after an outlier correction) distributed over the survey. After such statistical levelling, in compilations like ADMAP-2 (Golynsky *et al.* 2018) the individual surveys are merged, hereby correcting for the possible shifts in reference level and/or adjusting trends between the surveys. For Antarctic surveys, such a survey configuration has not always been possible as multiple data, which all have individual survey requirements, are acquired at the same time (e.g. Aitken *et al.* 2014; Paxman *et al.* 2019*b*). In such cases, the data are often filtered to a signal that corresponds to flight lines or can be levelled towards a regional or satellite grid, when available. However, here the spectral content of the data should overlap. All of these corrections do not exclude the overall usability for geological interpretation and affect mostly the long wavelengths (Golynsky *et al.* 2018). The ADMAP-2 compilation provides the most complete and coherent view of the magnetic properties of the Antarctic crust (Fig. 6).

Comparison between the satellite and aeromagnetic data (Fig. 6b) shows that the aeromagnetic surveys are able to image fine-scale crustal properties, while satellite data only provide an image of the broad-scale lithospheric properties, which is often omitted in geological interpretation. Aeromagnetic surveys often lack the long-wavelength component if the extension is not large enough. However, by processing steps affecting the trend, an artificial signal can easily be added unintendedly. Gaina *et al.* (2011) discuss the effect of replacing the long-wavelength component in aeromagnetic surveys by different bandwidth satellite data and show that the effect is significant and that, consequently, only satellite data can be analysed for the long-wavelength behaviour of the field.

Characterization of the magnetic lithosphere

With respect to the sources of the magnetic anomalies, the ADMAP-2 compilation reveals a wide variation of magnetic anomalies reflecting crustal terranes of different lithologies, ages, degrees of tectonic reworking and thermal attributes. The compilation also helps to map out Proterozoic–Archean cratons, Proterozoic–Paleozoic mobile belts, Paleozoic–Cenozoic magmatic arc systems, the boundary between East and West Antarctica, continent–ocean transitions and other regional Antarctic crustal features (Figs 6 and 7, Wilson and Ferraccioli this volume).

Figure 7 shows some examples of data products that can illuminate lithospheric properties. Interpretations of the sources of the magnetic anomalies are often carried out for smaller regions and on the survey-scale, often in combination with gravity analysis (e.g. Ferraccioli *et al.* 2009; Aitken *et al.* 2014). One method to help compare the sources of the gravity and magnetic field is to calculate the pseudo-gravity by applying the Poisson theorem (Blakely 1996). By converting the magnetic anomaly to a corresponding gravity anomaly, short-wavelength features are suppressed and the regional, large-scale features are enhanced. For example, the different characteristics of the West and East Antarctica lithosphere are now easily seen and can be used for further interpretation of the sub-glacial geology. Comparison to the gravity field allows discussion of whether the magnetic and gravity field have the same geological structure as source or, when anomalies do not correlate, whether they are only appearing in one of the datasets. For example, thick sedimentary sequences might not have a clear density contrast to the surrounding bedrock due to the sedimentary density increase with depth but will have a magnetic signature due to sedimentary rocks

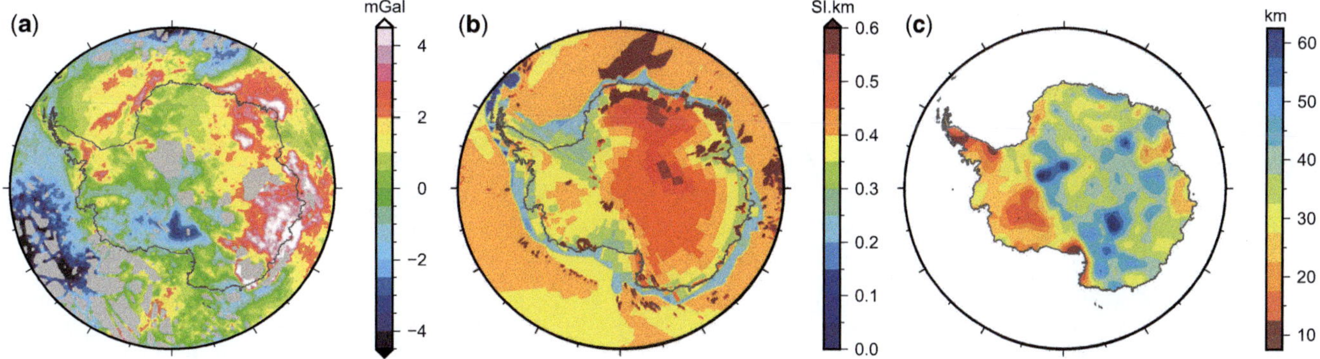

Fig. 7. Characteristics of the magnetic lithosphere: (**a**) pseudo-gravity calculated from ADMAP-2; (**b**) susceptibility of the lithosphere after the VIS model from Hemant and Maus (2005); (**c**) depth to deepest magnetic sources, after Martos *et al.* (2017).

typically having low magnetization relative to basement rocks, and are often regarded as magnetically transparent.

One of the few maps of lithospheric magnetization covering the Antarctic continent is presented by Hemant and Maus (2005). In their global study, the vertically integrated susceptibility (VIS) is calculated based on geological and tectonic maps of the world, laboratory susceptibility values of the occurring rock types and the seismic thickness of the crust. Unfortunately, the results for Antarctica reveal the poor quality and absence of such information and show the need to establish a database of rock properties to be able to update such products.

A different approach is not to invert for magnetization, but to estimate the deepest magnetic sources by assuming a piecewise constant susceptibility. This is of particular interest as the deepest sources can be associated with the depth to the Curie isotherm, hereby defining a thermal boundary. The Curie temperature of a ferromagnetic material is the temperature where the material loses its characteristic ferromagnetic ability: the ability to possess a net and spontaneous magnetization in the presence of an external inducing magnetic field. Rocks at higher temperatures will not generate a discernible magnetic signal. Magnetite, with a Curie temperature of 580°C (e.g. Hunt *et al.* 1995), is regarded as the dominant magnetic mineral in crustal rocks, while the Curie temperatures for felsic plutonic rocks can range from 400 to 550°C, related to variability in composition of titanomagnetites between substantially different rock types (e.g. basalt v. granite, Byerly and Stolt 1977). In most stable continental areas this temperature range includes the lower crust or upper mantle (e.g. McKenzie *et al.* 2005), although often the Moho depth is defined as a maximum depth for the Curie isotherm (Wasilewski and Mayhew 1992). However, at the Moho depth, temperatures might be well below the 580°C isotherm, especially in areas of tectonothermal cold cratons like East Antarctica. This implies that even the upper mantle can be magnetized, although the amplitudes of magnetization might be low.

Ferré *et al.* (2014) summarized a number of reasons why magnetization in the upper mantle might be reasonable, even though the classical view is that there is a lack of magnetic rocks in the mantle (e.g. Wasilewski and Mayhew 1992). Idoko *et al.* (2019) demonstrated by forward modelling based on assumed geotherms, the pressure dependence of the Curie temperature of magnetite, and the statistical distributions of rock magnetic data from mantle xenoliths, that the upper mantle magnetization potentially contributes up to 10% of the observed long-wavelength lithospheric magnetic anomalies.

In Antarctica, Fox Maule *et al.* (2005) and Martos *et al.* (2017) presented geothermal heat flow models based on analysis of magnetic data. Fox Maule *et al.* (2005) used one of the first magnetic satellite models (from the Ørsted mission) to calculate the deepest magnetic sources using some specific boundary conditions, for example a susceptibility of 0.035 (SI) and 0.040 (SI) for the continental and oceanic crust, respectively. The magnetic field used is filtered by spherical harmonic degree 14, as the lower spherical harmonic degrees are dominated by the core field. To circumvent this, the long-wavelength part of their model is taken from a crustal thickness model (Nataf and Ricard 1996), which is based on seismic and thermal data. This study was followed up by Martos *et al.* (2017), using ADMAP-1 (Golynksy *et al.* 2001) plus some additional new aeromagnetic surveys and by filling the gaps with satellite data. More specifically, the long-wavelength content (>150 km) of all aeromagnetic data was filtered and replaced with the satellite model MF7 (Maus 2010). In the characteristic power-spectrum, parts of the spectrum can be associated with a certain source depth, which can be estimated from the slope of the curve (e.g. Spector and Grant 1970; Tanaka *et al.* 1999). In their study, Martos *et al.* (2017) pre-defined a characteristic wavelength range both for East and West Antarctica separately, to calculate for a window size of 350 km the depth to the magnetic sources. Due to use of the window size, the results have to be treated carefully in areas where only satellite data are used as these are only reliable for a signal with wavelengths beyond 300 km.

Both the approach by Fox Maule *et al.* (2005) and Martos *et al.* (2017) assume a constant magnetization with depth, and Fox Maule *et al.* (2005) exclude depths larger than the Moho depth from a seismic model. Figure 7c shows the deepest magnetic sources from the analysis of Martos *et al.* (2017), which shows depths of more than 40 km in East Antarctica and areas of relatively shallow magnetic source depth in West Antarctica. Inversions of magnetic data to find the deepest magnetic source have been attempted previously in regions other than Antarctica in a number of studies (e.g. Bouligand *et al.* 2009; Ebbing *et al.* 2009; Tanaka 2017). Both Bouligand *et al.* (2009) and Ebbing *et al.* (2009) compared such depth estimates with thermal and structural information and discussed the uncertainties resulting, for example, from the fact that the deepest magnetic sources could be related to structural boundaries in the crust and not represent the depth of the Curie isotherm. Similar interpretations could be made for Antarctica using the pseudo-gravity map (Fig. 7a), which shows similar domains as the gravity field, since both the magnetization contrast and the geometry influence the magnetic field anomalies.

All this considered, at least in areas where the deepest magnetic source is located beneath the Moho depth, such depth estimates must be treated with care. In Figure 8b, we have therefore clipped the deepest magnetic sources for areas where they are located beneath the Moho depth. Somewhat surprisingly, the deepest magnetic sources are hereby clipped for half of the continent, indicating that a more complex magnetization model has to be developed in the future to address

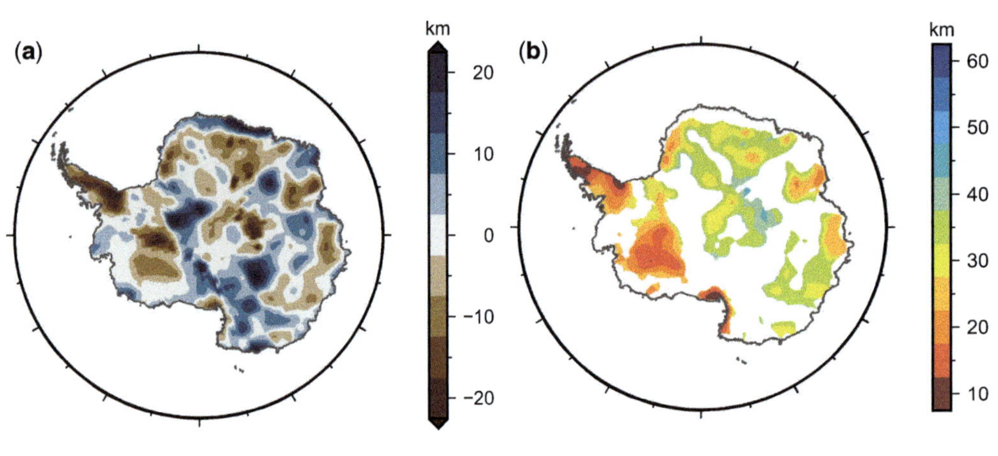

Fig. 8. (**a**) Difference between the depth to the deepest magnetic sources (Fig. 7c; Martos *et al.* 2017) and the crustal thickness (Fig. 3a; An *et al.* 2015*b*). (**b**) The deepest magnetic sources clipped for areas where the model is located beneath the Moho depth.

both the lateral and vertical variations in the magnetic parameters.

Antarctic magnetic data in a Gondwana framework

An application where magnetic data are often used is in linking Antarctica to its neighbours in a Gondwana framework. An example is shown in Figure 9 with the datasets of southern Africa, Antarctica and Australia. For example, Finn *et al.* (1999) and Aitken *et al.* (2014, 2016) use such reconstructions to discuss the structural setting of East Antarctica with respect to the structural setting of southern Australia and show that the magnetic data indicate the continuation of Australia under the ice of Antarctica. Similar studies have been carried out for other parts of Gondwana by Mieth and Jokat (2014), Ruppel *et al.* (2018) and Mueller and Jokat (2019).

Plate tectonic reconstructions have also been used by Pollett *et al.* (2019) to derive information on the thermal structure of Antarctica by interpolating thermal parameters from adjacent continents into Antarctica.

Geothermal heat flow

A short summary of models

The thermal structure of the continent is one of the most unknown parameters of Antarctica, but is critical for a variety of applications on different scales, e.g. ice-sheet modelling or glacial-isostatic adjustment (e.g. Burton-Johnson *et al.* 2020*a*, *b*). Despite some drilling initiatives, the remoteness and vastness of the continent limits the possibilities to achieve a good, homogeneous coverage of the entire continent for thermal information. Only very sparse heat flow measurements have been taken in boreholes into sub-glacial lake sediments (e.g. Fisher *et al.* 2015) or seafloor sedimentary strata (e.g. Schröder *et al.* 2011) or derived from basal temperature gradients in deep ice boreholes (e.g. Engelhardt 2004). Some of the recent drilling initiatives have measured values >120 mW m^{-2} in West Antarctica, far exceeding the heat flow normally measured over the continental crust. Such a high heat flow, if existing on a regional scale, is clearly a boundary condition that has to be considered in ice-sheet dynamics but might also indicate a thin and subsequently warm lithosphere, which might imply low viscosity influencing glacial isostatic adjustment.

A number of geothermal heat flow models are available for Antarctica as discussed in Van Liefferinge *et al.* (2018) and Burton-Johnson *et al.* (2020*a*). These models are derived from geophysical data that indirectly reflect the thermal structure and vary largely (Fig. 10). A full review of the data and models is provided in the two recent white papers from Burton-Johnson *et al.* (2020*a*, *b*), but in the absence of homogeneous coverage with heat flow measurements, different means have to be used. Pollett *et al.* (2019) assessed the heat flow over Antarctica by kriging interpolation of existing heat flow measurements, making use of the heat flow data of the continents adjacent to Antarctica. A motivation of the study was the disagreement of the existing models.

Magnetic analysis (e.g. Fox Maule *et al.* 2005; Martos *et al.* 2017) relies on estimates of the deepest magnetic source and has large uncertainty where the magnetic data quality is low and/or where the assumptions of a simple magnetic layer are not met. From the deepest sources the heat flow is then calculated by applying a one-dimensional thermal model assuming constant thermal parameters. Fox Maule *et al.* (2005) themselves estimate the error in their map of sub-glacial heat flow to be in the range of 21–27 mW m^{-2}, due to the limited resolution of a few hundred kilometres of the existing magnetic data (resulting from the satellite altitude), methodological uncertainties with associated errors and the use of a laterally constant crustal heat production. Martos *et al.* (2017) provide a considerably lower uncertainty based on their approach, which is, however, only valid where aeromagnetic survey data are used and the magnetic depth is not beneath

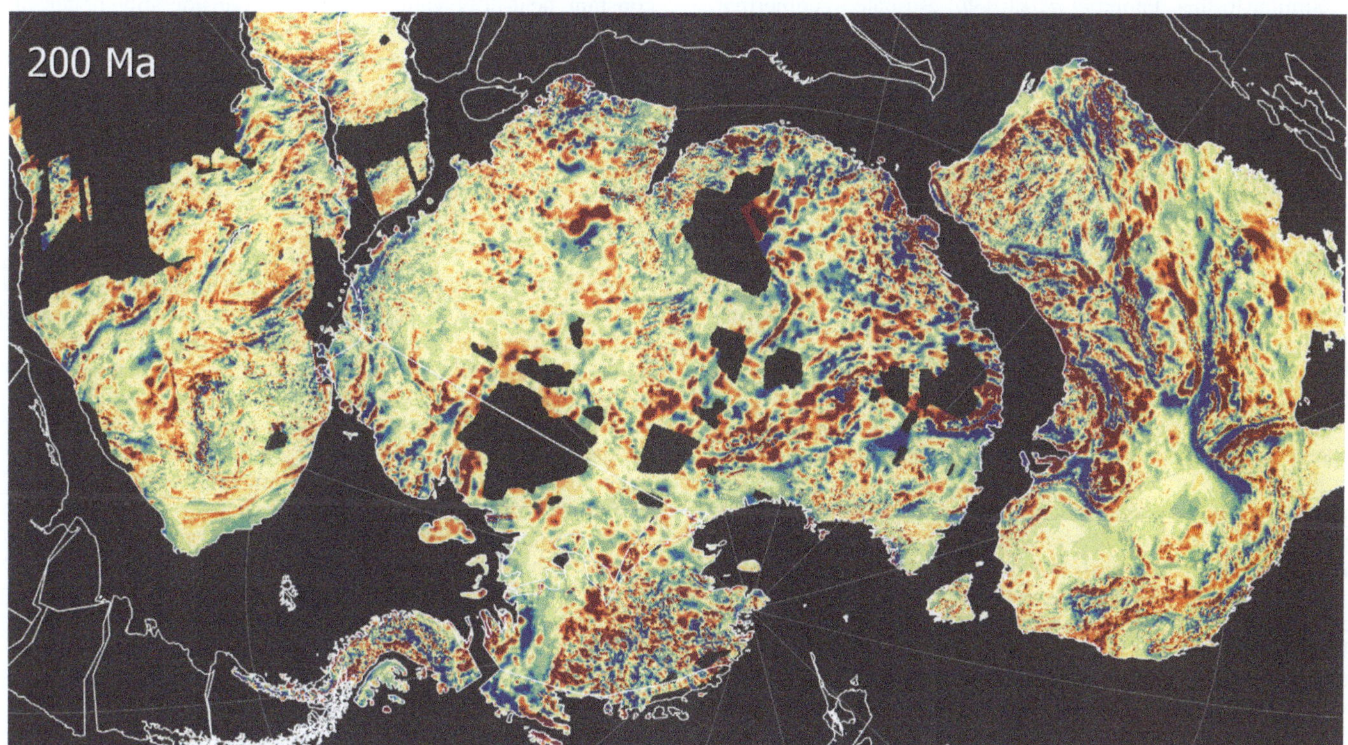

Fig. 9. Gondwana reconstruction of aeromagnetic anomaly maps. The datasets shown are the aeromagnetic compilations of South Africa (Stettler *et al.* 2000), Australia (Milligan *et al.* 2004) and ADMAP-2 (Golynsky *et al.* 2018) after conformation to the satellite model LCS-1.

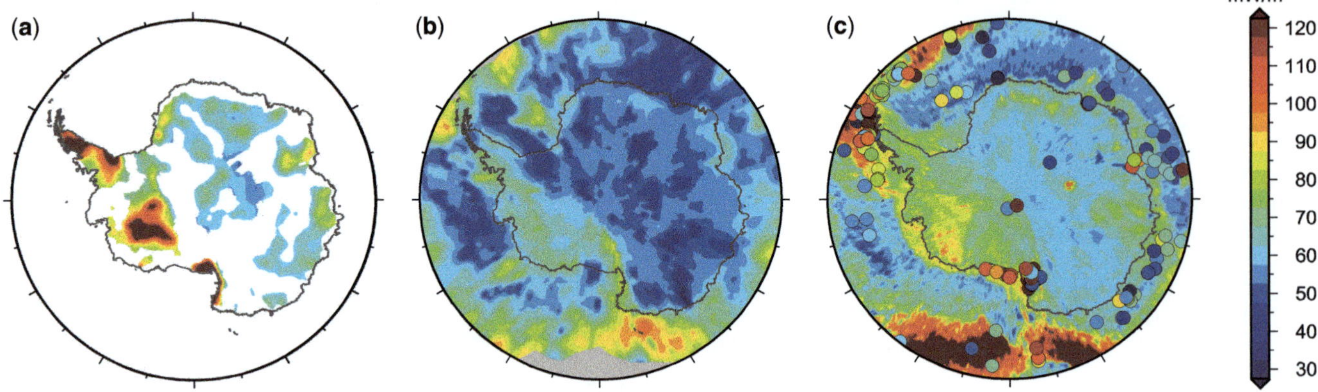

Fig. 10. Heat flow models for Antarctica: (**a**) Martos *et al.* (2017) clipped for areas, where the estimated deepest magnetic source is located beneath the Moho depth (compare Fig. 8b); (**b**) An *et al.* (2015*a*); (**c**) Lucazeau (2019) with measurements.

the Moho depth. In general, such estimates mainly reflect the changes in crustal properties.

Due to the absence of near-surface measurements and the uncertainties in magnetic data interpretation, heat flow has been alternatively determined from seismic data (e.g. Shapiro and Ritzwoller 2004; An *et al.* 2015*a*; Shen *et al.* 2020). Assuming that temperature is the dominant control on seismic velocity in the upper mantle, the temperature field in the lithosphere can be determined and with some assumptions the geothermal heat flow can be estimated. With an estimate of lithospheric heat production rates, a map of the sub-glacial heat flow can be produced. However, these models are also limited to a lateral resolution of >120 km and are insensitive to the lithospheric geotherm. Also, the lithosphere is treated as internally laterally uniform in its structure (only varying in thickness) and possessing low and homogeneous radiogenic heat production. Consequently, sub-glacial heat flow inversely correlates with lithospheric thickness with the highest values overlying the thinnest lithosphere.

Similar to this, Pappa *et al.* (2019*b*) estimated geothermal heat flow based on gravity modelling in combination with seismological constraints. These heat flow models, however, do not reflect small-scale patterns caused by variations in hydrothermal circulation, volcanic activity or crustal radioactive heat production, but mostly provide an assessment of the background heat flow from the mantle and an extrapolation of such values to the surface by assuming constant crustal parameters. A different approach is used by Macelloni *et al.* (2018) based on SMOS satellite data, who inverted geothermal heat flow against ice temperature profiles based on radar data. However, this approach requires specific boundary conditions such as thick ice cover and is therefore only valid in central East Antarctica. That explains some of the reasons for the differences between the heat flow models in Figure 10

To complicate the situation, Burton-Johnson *et al.* (2017) showed for the Antarctic Peninsula that the upper crust can contributes up to 70% to the geothermal heat flow. Therefore, models of geothermal heat flow should utilize a heterogeneous upper crust with variable radioactive heat production rates if they are to predict basal conditions of the ice-sheet accurately.

Alternative approaches have been proposed, for example machine learning (e.g. Rezvanbehbahani *et al.* 2017) or similarity methods (e.g. Lucazeau 2019). In both methods, a large number of geophysical models and geological data are used to statistically predict heat flow in areas of sparse measurement coverage. Both approaches are hereby first tested in areas where heat flow is known with relatively low uncertainty and later on applied to remote areas. In Figure 10c, the model from Lucazeau (2019) is shown derived as part of his new global heat flow model based on empirical correlations between geophysical and geological datasets. Such analysis can be done in a Gondwana framework to provide further input to derive the geothermal heat flow for East Antarctica. An example of this is Pollett *et al.* (2019), who assessed the heat flow over Antarctica by kriging interpolation of existing heat flow measurements of Antarctica and adjacent continents.

There is still significant disagreement between current geothermal heat flow models. Lösing *et al.* (2020) analysed with Bayesian inversion the trade-offs between the different geophysical parameters and showed that the current models cannot be reconciled within a plausible parameter range. However, that would be crucial in order to provide a useful assessment as input for modelling of Antarctic ice stream behaviour. For Greenland, Rogozhina *et al.* (2012) estimated the effect of such differences in geothermal heat flow models on the evolution of the Greenland ice-sheet. In summary, they concluded that the uncertainties of the geophysical models prohibit their use for ice-sheet modelling and that instead a constant heat-flow value provides a considerably better fit with the majority of the observational data. Yet, there is no geothermal heat flow model available for Antarctica to overcome this limitation.

Conclusions

A wealth of gravity and magnetic data from airborne surveys and satellite missions is available for Antarctica. While airborne data cover about two-thirds of the continent, their resolution and accuracy is variable. Satellite data offer homogeneous accuracy and limited resolution to help fill the gaps in airborne data, and aid interpretation of the lithospheric architecture. Satellite data, in combination with airborne data, or on their own, can be used to illuminate the structure, thermal state and composition of the lithosphere, especially when combined with seismological information. Studies based on these datasets agree with seismological data on the overall architecture of the Antarctic mantle and highlight the differences between East and West Antarctica in lithospheric thickness, mantle composition and geothermal heat flow. East Antarctica is cratonic, but the degree of complexity is modelled differently between recent studies (e.g. Ebbing *et al.* 2018; Haeger *et al.* 2019; Pappa *et al.* 2019*a*, *b*). Similar debates are ongoing for the makeup of West Antarctica. Gravity data can aid in testing competing hypotheses, while the magnetic data can provide details of the crustal configuration.

Geophysical models on scales relevant to understanding the dynamics of ice sheets can be improved by more detailed petrophysical information. This has worked well elsewhere, for example in Scandinavia Olesen *et al.* (2010) showed a susceptibility map based on rock measurements averaged over geological units aided interpretation and modelling. An important application of this in Antarctica is in understanding geothermal heat flow, which, on a local scale, is dominated by the thermal conductivity and heat production in the crust. Pollett *et al.* (2019) demonstrated how heat flow parameters can be used in a Gondwana framework, which for East Antarctica adds information from the adjacent continents of South Africa, India and Australia which could be better guided by incorporating geophysical data interpretation and petrophysical data. In this way, the different thermal history of the lithosphere after break-up could be considered, as could the interaction of the asthenosphere and lithosphere (see Bredow and Steinberger 2021, Wiens this volume in review, Hu *et al.* 2018).

Burton-Johnson *et al.* (2020*b*) summarized the challenges in providing accurate geothermal heat flow models with associated uncertainties and recommended future research directions. Coupled with modern approaches such as machine learning, one will be able to probe a wider range of geological and geophysical proxies to establish consistent integrated geophysical–petrological models of the solid Earth part of Antarctica. This is in line with the conclusions by Kennicutt *et al.* (2019) who summarized the need for improving and consolidating solid Earth studies in Antarctica. Both the need for multi-technique analyses combining seismological, petrological and potential field data has been expressed as well as open access to data and models to enhance our understanding of Antarctica.

Acknowledgements The authors thank Alan R.A. Aitken and one anonymous reviewer as well as the editors Wouter van der Wal and Adam P. Martin for their constructive contribution and criticism.

Author contributions **FP**: conceptualization (supporting), resources (equal), writing – original draft (equal), writing – review & editing (equal); **JE**: conceptualization (equal), methodology (lead), resources (equal), writing – original draft (equal), writing – review & editing (equal).

Funding This study was supported by the European Space Agency ESA by funding the projects GOCE + Antarctica and 3D Earth within the Support to Science Element (STSE) programme.

Data availability Data sharing is not applicable to this article as no datasets were generated or analysed during the current study.

References

Aitken, A.R.A., Salmon, M.L. and Kennett, B.L.N. 2013. Australia's Moho: a test of the usefulness of gravity modelling for the determination of Moho depth. *Tectonophysics*, **609**, 468–479, https://doi.org/10.1016/j.tecto.2012.06.049

Aitken, A.R.A., Young, D.A. *et al.* 2014. The subglacial geology of Wilkes Land, East Antarctica. *Geophysical Research Letters*, **41**, 2390–2400, https://doi.org/10.1002/2014GL059405

Aitken, A.R.A., Betts, P.G., Young, D.A., Blankenship, D.D., Roberts, J.L. and Siegert, M.J. 2016. The Australo-Antarctic Columbia to Gondwana transition. *Gondwana Research*, **29**, 136–152, https://doi.org/10.1016/j.gr.2014.10.019

An, M., Wiens, D.A. *et al.* 2015*a*. Temperature, lithosphere–asthenosphere boundary, and heat flux beneath the Antarctic Plate inferred from seismic velocities. *Journal of Geophysical Research: Solid Earth*, **120**, 8720–8742, https://doi.org/10.1002/2015JB011917

An, M., Wiens, D.A. *et al.* 2015*b*. S-velocity model and inferred Moho topography beneath the Antarctic Plate from Rayleigh waves. *Journal of Geophysical Research: Solid Earth*, **120**, 359–383, https://doi.org/10.1002/2014JB011332

An, M., Wiens, D.A. and Zhao, Y. 2016. A frozen collision belt beneath ice: an overview of seismic studies around the Gamburtsev Subglacial Mountains, East Antarctica. *Advances in Polar Science*, **27**, 78–89, https://doi.org/10.13679/j.advps.2016.2.00078

Artemieva, I.M., Thybo, H. and Cherepanova, Y. 2019. Isopycnicity of cratonic mantle restricted to kimberlite provinces. *Earth and Planetary Science Letters*, **505**, 13–19, https://doi.org/10.1016/j.epsl.2018.09.034

Bamber, J.L., Ferraccioli, F., Joughin, I., Shepherd, T., Rippin, D.M., Siegert, M.J. and Vaughan, D.G. 2006. East Antarctic ice stream tributary underlain by major sedimentary basin. *Geology*, **34**, 33–36, https://doi.org/10.1130/G22160.1

Baranov, A. and Morelli, A. 2013. The Moho depth map of the Antarctica region. *Tectonophysics*, **609**, 299–313, https://doi.org/10.1016/j.tecto.2012.12.023

Baranov, A., Tenzer, R. and Bagherbandi, M. 2018. Combined gravimetric–seismic crustal model for Antarctica. *Surveys in Geophysics*, **39**, 23–56, https://doi.org/10.1007/s10712-017-9423-5

Barzaghi, R., Betti, B., Carrion, D., Gentile, G., Maseroli, R. and Sacerdote, F. 2014. Orthometric correction and normal heights for Italian levelling network: a case study. *Applied Geomatics*, **6**, 17–25, https://doi.org/10.1007/s12518-013-0121-9

Bell, R.E., Ferraccioli, F. *et al.* 2011. Widespread persistent thickening of the East Antarctic Ice Sheet by freezing from the base. *Science*, **331**, 1592–1595, https://doi.org/10.1126/science.1200109

Blakely, R.J. 1996. *Potential Theory in Gravity and Magnetic Applications*. Cambridge University Press.

Block, A.E., Bell, R.E. and Studinger, M. 2009. Antarctic crustal thickness from satellite gravity: implications for the Transantarctic and Gamburtsev Subglacial Mountains. *Earth and Planetary Science Letters*, **288**, 194–203, https://doi.org/10.1016/j.epsl.2009.09.022

Bouligand, C., Glen, J.M.G. *et al.* 2009. Mapping Curie temperature depth in the western United States with a fractal model for crustal magnetization. *Journal of Geophysical Research*, **114**, B11104, https://doi.org/10.1029/2008JD011281

Bouman, J., Ebbing, J. *et al.* 2015. GOCE gravity gradient data for lithospheric modeling. *International Journal of Applied Earth Observation and Geoinformation*, **35**, 16–30, https://doi.org/10.1016/j.jag.2013.11.001

Bouman, J., Ebbing, J. *et al.* 2016. Satellite gravity gradient grids for geophysics. *Scientific Reports*, **6**, 21050, https://doi.org/10.1038/srep21050

Bredow, E. and Steinberger, B. 2021. Mantle convection and possible mantle plumes beneath Antarctica: insights from geodynamic models and implications for topography. *Geological Society, London, Memoirs*, **56**, https://doi.org/10.1144/M56-2020-2

Brockmann, J.M., Zehentner, N., Höck, E., Pail, R., Loth, I., Mayer-Gürr, T. and Schuh, W.-D. 2014. EGM_TIM_RL05: an independent geoid with centimeter accuracy purely based on the GOCE mission. *Geophysical Research Letters*, **41**, 8089–8099, https://doi.org/10.1002/2014GL061904

Burton-Johnson, A., Halpin, J.A., Whittaker, J.M., Graham, F.S. and Watson, S.J. 2017. A new heat flux model for the Antarctic Peninsula incorporating spatially variable upper crustal radiogenic heat production. *Geophysical Research Letters*, **44**, 5436–5446, https://doi.org/10.1002/2017GL073596

Burton-Johnson, A., Dziadek, R. and Martin, C. 2020*a*. Geothermal heat flow in Antarctica: current and future directions. *The Cryosphere Discussions*, https://doi.org/10.5194/tc-2020-59

Burton-Johnson, A., Dziadek, R. and The SERCE Geothermal Heat Flow Sub-Group 2020*b*. Antarctic Geothermal Heat Flow: Future research directions. White Paper, https://www.scar.org/scar-news/serce-news/scar-serce-ghf/

Byerly, P. and Stolt, R. 1977. An attempt to define the Curie point isotherm in northern and central Arizona. *Geophysics*, **42**, 1394–1400, https://doi.org/10.1190/1.1440800

Cammarano, F., Goes, S., Vacher, P. and Giardini, D. 2003. Inferring upper-mantle temperatures from seismic velocities. *Physics of the Earth and Planetary Interiors*, **138**, 197–222, https://doi.org/10.1016/S0031-9201(03)00156-0

Chaput, J., Aster, R.C. *et al.* 2014. The crustal thickness of West Antarctica. *Journal of Geophysical Research: Solid Earth*, **119**, 378–395, https://doi.org/10.1002/2013JB010642

Chen, B., Haeger, C., Kaban, M.K. and Petrunin, A.G. 2018. Variations of the effective elastic thickness reveal tectonic fragmentation of the Antarctic lithosphere. *Tectonophysics*, **746**, 412–424, https://doi.org/10.1016/j.tecto.2017.06.012

Connolly, J.A.D. 2005. Computation of phase equilibria by linear programming: a tool for geodynamic modeling and its application to subduction zone decarbonation. *Earth and Planetary Science Letters*, **236**, 524–541, https://doi.org/10.1016/j.epsl.2005.04.033

Eaton, D.W. and Perry, H.K.C. 2013. Ephemeral isopycnicity of cratonic mantle keels. *Nature Geoscience*, **6**, 967, https://doi.org/10.1038/ngeo1950

Eaton, D.W., Darbyshire, F., Evans, R.L., Grütter, H., Jones, A.G. and Yuan, X. 2009. The elusive lithosphere–asthenosphere boundary (LAB) beneath cratons. *Lithos*, **109**, 1–22, https://doi.org/10.1016/j.lithos.2008.05.009

Ebbing, J., Gernigon, L., Pascal, C., Olesen, O. and Osmundsen, P.T. 2009. A discussion of structural and thermal control of magnetic anomalies on the mid-Norwegian margin. *Geophysical Prospecting*, **57**, 665–681, https://doi.org/10.1111/j.1365-2478.2009.00800.x

Ebbing, J., Haas, P., Ferraccioli, F., Pappa, F., Szwillus, W. and Bouman, J. 2018. Earth tectonics as seen by GOCE - Enhanced satellite gravity gradient imaging. *Scientific Reports*, **8**, 16356, https://doi.org/10.1038/s41598-018-34733-9

Engelhardt, H. 2004, Thermal regime and dynamics of the West Antarctic ice sheet. *Annals of Glaciology*, **39**, 85–92, https://doi.org/10.3189/172756404781814203

Ferraccioli, F., Armadillo, E., Jordan, T., Bozzo, E. and Corr, H. 2009. Aeromagnetic exploration over the East Antarctic Ice Sheet: a new view of the Wilkes Subglacial Basin. *Tectonophysics*, **478**, 62–77, https://doi.org/10.1016/j.tecto.2009.03.013

Ferraccioli, F., Finn, C.A., Jordan, T.A., Bell, R.E., Anderson, L.M. and Damaske, D. 2011. East Antarctic rifting triggers uplift of the Gamburtsev Mountains. *Nature*, **479**, 388–392, https://doi.org/10.1038/nature10566

Ferré, E.C., Friedman, S.A. *et al.* 2014. Eight good reasons why the uppermost mantle could be magnetic. *Tectonophysics*, **624–625**, 3–14, https://doi.org/10.1016/j.tecto.2014.01.004

Filina, I.Y., Blankenship, D.D., Thoma, M., Lukin, V.V., Masolov, V.N. and Sen, M.K. 2008. New 3D bathymetry and sediment distribution in Lake Vostok: implication for pre-glacial origin and numerical modeling of the internal processes within the lake. *Earth and Planetary Science Letters*, **276**, 106–114, https://doi.org/10.1016/j.epsl.2008.09.012

Finn, C., Moore, D., Damaske, D. and Mackey, T. 1999. Aeromagnetic legacy of early Paleozoic subduction along the Pacific margin of Gondwana. *Geology*, **27**, 1087–1090, https://doi.org/10.1130/0091-7613(1999)027<1087:ALOEPS>2.3.CO;2

Fisher, A.T., Mankoff, K.D. *et al.* 2015. High geothermal heat flux measured below the West Antarctic Ice Sheet. *Science Advances*, **1**, e1500093, https://doi.org/10.1126/sciadv.1500093

Forsberg, R. 2015. GOCE and Antarctica. *In*: *5th International GOCE User Workshop*. ESA Special Publication, 3.

Forsberg, R., Olesen, A., Yildiz, H. and Tscherning, C.C. 2011. Polar Gravity Fields from GOCE and Airborne Gravity. *In*: *4th International GOCE User Workshop*. ESA Special Publication, 8.

Forsberg, R., Olesen, A.V. *et al.* 2018. Exploring the Recovery Lakes region and interior Dronning Maud Land, East Antarctica, with airborne gravity, magnetic and radar measurements. *Geological Society, London, Special Publications*, **461**, 23–34, https://doi.org/10.1144/SP461.17

Fox Maule, C., Purucker, M.E., Olsen, N. and Mosegaard, K. 2005. Heat flux anomalies in Antarctica revealed by satellite magnetic data. *Science*, **309**, 464–467, https://doi.org/10.1126/science.1106888

Frederick, B.C., Young, D.A., Blankenship, D.D., Richter, T.G., Kempf, S.D., Ferraccioli, F. and Siegert, M.J., 2016. Distribution of subglacial sediments across the Wilkes Subglacial Basin, East Antarctica. *Journal of Geophysical Research: Earth Surface*, **121**, 790–813, https://doi.org/10.1002/2015JF003760

Fretwell, P., Pritchard, H.D. *et al.* 2013. Bedmap2: improved ice bed, surface and thickness datasets for Antarctica. *The Cryosphere*, **7**, 375–393, https://doi.org/10.5194/tc-7-375-2013

Fullea, J., Afonso, J.C., Connolly, J.A.D., Fernàndez, M., Garcìa-Castellanos, D. and Zeyen, H. 2009. LitMod3D: an interactive 3-D software to model the thermal, compositional, density, rheological and seismological structure of the lithosphere and sublithospheric upper mantle. *Geochemistry, Geophysics, Geosystems*, **10**, https://doi.org/10.1029/2009GC002391

Fullea, J., Lebedev, S., Agius, M.R., Jones, A.G. and Afonso, J.C. 2012. Lithospheric structure in the Baikal–central Mongolia region from integrated geophysical-petrological inversion of surface-wave data and topographic elevation. *Geochemistry, Geophysics, Geosystems*, **13**, https://doi.org/10.1029/2012GC004138

Gaina, C., Werner, S.C., Saltus, R. and Maus, S. 2011. Circum-Arctic mapping project: new magnetic and gravity anomaly maps of the Arctic. *Geological Society, London, Memoirs*, **35**, 39–48, https://doi.org/10.1144/M35.3

Golynsky, A., Chiappini, M. *et al.* 2001. 'ADMAP – Magnetic Anomaly Map of the Antarctic,' 1:10 000 000 scale map. *In*: Morris, P. and von Frese, R. (eds) *BAS (Misc.)*. British Antarctic Survey, Cambridge, 10.

Golynsky, A.V., Ferraccioli, F. *et al.* 2018. New magnetic anomaly map of the Antarctic. *Geophysical Research Letters*, **45**, 6437–6449, https://doi.org/10.1029/2018GL078153

Haeger, C. and Kaban, M.K. 2019. Decompensative gravity anomalies reveal the structure of the upper crust of Antarctica. *Pure and Applied Geophysics*, **176**, 4401–4414, https://doi.org/10.1007/s00024-019-02212-5

Haeger, C., Kaban, M.K., Tesauro, M., Petrunin, A.G. and Mooney, W.D. 2019. 3-D density, thermal, and compositional model of the Antarctic lithosphere and implications for its evolution. *Geochemistry, Geophysics, Geosystems*, **20**, 688–707, https://doi.org/10.1029/2018GC008033

Hansen, S.E., Nyblade, A.A., Heeszel, D.S., Wiens, D.A., Shore, P. and Kanao, M. 2010. Crustal structure of the Gamburtsev Mountains, East Antarctica, from S-wave receiver functions and Rayleigh wave phase velocities. *Earth and Planetary Science Letters*, **300**, 395–401, https://doi.org/10.1016/j.epsl.2010.10.022

Hansen, S.E., Kenyon, L.M., Graw, J.H., Park, Y. and Nyblade, A.A. 2016. Crustal structure beneath the Northern Transantarctic Mountains and Wilkes Subglacial Basin: implications for tectonic origins. *Journal of Geophysical Research: Solid Earth*, **121**, 812–825, https://doi.org/10.1002/2015JB012325

Harrowfield, M., Holdgate, G.R., Wilson, C.J.L. and McLoughlin, S. 2005. Tectonic significance of the Lambert graben, East Antarctica: reconstructing the Gondwanan rift. *Geology*, **33**, 197–200, https://doi.org/10.1130/G21081.1

Hemant, K. and Maus, S. 2005. Geological modeling of the new CHAMP magnetic anomaly maps using a geographical information system technique. *Journal of Geophysical Research: Solid Earth*, **110**, https://doi.org/10.1029/2005JB003837

Hu, J., Liu, L., Faccenda, M., Zhou, Q., Fischer, K.M., Marshak, S. and Lundstrom, C. 2018. Modification of the Western Gondwana craton by plume–lithosphere interaction. *Nature Geoscience*, **11**, 203–210, https://doi.org/10.1038/s41561-018-0064-1

Hunt, C.P., Moskowitz, B.M., Banerjee, S.K. 1995. Magnetic properties of rocks and minerals. *In*: Ahrens, T.J. (ed.) *Rock Physics and Phase Relations. A Handbook of Physical Constants*. American Geophysical Union, 189–204.

Idoko, C.M., Conder, J.A., Ferré, E.C. and Filiberto, J. 2019. The potential contribution to long wavelength magnetic anomalies from the lithospheric mantle. *Physics of the Earth and Planetary Interiors*, **292**, 21–28, https://doi.org/10.1016/j.pepi.2019.05.002

Isanina, E.V., Krupnova, N.A., Popov, S.V., Masolov, V.N. and Lukin, V.V. 2009. Deep structure of the Vostok Basin, East Antarctica as deduced from seismological observations. *Geotectonics*, **43**, 221–225, https://doi.org/10.1134/S001685210903 0042

Jordan, T., Ferraccioli, F. *et al.* 2017. First complete regional view of the Pensacola-Pole Basin from PolarGAP radar data. *EGU General Assembly Conference Abstracts*, Vienna, Austria, 15902.

Jordan, T.H. 1988. Structure and formation of the continental tectosphere. *Journal of Petrology*, Special_Volume, 11–37, https://doi.org/10.1093/petrology/Special_Volume.1.11

Kaban, K.M., Schwintzer, P. and Reigber, C.H. 2004. A new isostatic model of the lithosphere and gravity field. *Journal of Geodesy*, **78**, 368–385, https://doi.org/10.1007/s00190-004-0401-6

Kaban, M.K., Schwintzer, P., Artemieva, I.M. and Mooney, W.D. 2003. Density of the continental roots: compositional and thermal contributions. *Earth and Planetary Science Letters*, **209**, 53–69, https://doi.org/10.1016/S0012-821X(03)00072-4

Kaban, M.K., Tesauro, M. and Cloetingh, S.A.P.L. 2010. An integrated gravity model for Europe's crust and upper mantle. *Earth and Planetary Science Letters*, **296**, 195–209, https://doi.org/10.1016/j.epsl.2010.04.041

Karner, G.D., Studinger, M. and Bell, R.E. 2005. Gravity anomalies of sedimentary basins and their mechanical implications: application to the Ross Sea basins, West Antarctica. *Earth and Planetary Science Letters*, **235**, 577–596, https://doi.org/10.1016/j.epsl.2005.04.016

Kennicutt, M.C. II, Bromwich, D. *et al.* 2019. Sustained Antarctic Research: a 21st century imperative. *One Earth*, **1**, 95–113, https://doi.org/10.1016/j.oneear.2019.08.014

Kotsiaros, S. and Olsen, N. 2012. The geomagnetic field gradient tensor. GEM. *International Journal on Geomathematics*, **3**, 297–314, https://doi.org/10.1007/s13137-012-0041-6

Krupnik, I., Allison, I. *et al.* (eds) 2011. *Understanding Earth's Polar Challenges: International Polar Year 2007–2008*, CCI Press, Edmonton, Alberta and University of the Arctic, Rovaniemi, Finland (printed version) and ICSU/WMO Joint Committee for International Polar Year 2007–2008.

Langel, R.A. and Hinze, W.J. 1998. *The Magnetic Field of the Earth's Lithosphere: The Satellite Perspective*. Cambridge University Press.

Lawrence, J.F., Wiens, D.A., Nyblade, A.A., Anandakrishnan, S., Shore, P.J. and Voigt, D. 2006. Rayleigh wave phase velocity analysis of the Ross Sea, Transantarctic Mountains, and East Antarctica from a temporary seismograph array. *Journal of Geophysical Research: Solid Earth*, **111**, https://doi.org/10.1029/2005JB003812

LeMasurier, W.E. and Rex, D.C. 1989. Evolution of linear volcanic ranges in Marie Byrd Land, West Antarctica. *Journal of Geophysical Research: Solid Earth*, **94**, 7223–7236, https://doi.org/10.1029/JB094iB06p07223

Lisker, F., Prenzel, J., Läufer, A. and Spiegel, C. 2013. The regional thermochronological record as evaluation criterion for uplift models of the Transantarctic Mountains. *EGU General Assembly Conference Abstracts, EGU2013-5985*, Vienna, Austria.

Lloyd, A.J. 2018. *Seismic Tomography of Antarctica and the Southern Oceans: Regional and Continental Models from the Upper Mantle to the Transition Zone*. Washington University, St. Louis, Missouri.

Lloyd, A.J., Wiens, D.A. *et al.* 2015. A seismic transect across West Antarctica: evidence for mantle thermal anomalies beneath the Bentley Subglacial Trench and the Marie Byrd Land Dome. *Journal of Geophysical Research: Solid Earth*, **120**, 8439–8460, https://doi.org/10.1002/2015JB012455

Lösing, M., Ebbing, J. and Szwillus, W. 2020. Geothermal heat flux in antarctica: assessing models and observations by Bayesian inversion. *Frontiers in Earth Science*, **8**, 105, https://doi.org/10.3389/feart.2020.00105

Lucazeau, F. 2019. Analysis and mapping of an updated terrestrial heat flow dataset. *Geochemistry, Geophysics, Geosystems*, **20**, 4001–4024, https://doi.org/10.1029/2019GC008389

Lythe, M.B. and Vaughan, D.G. 2001. BEDMAP: a new ice thickness and subglacial topographic model of Antarctica. *Journal of Geophysical Research: Solid Earth*, **106**, 11335–11351, https://doi.org/10.1029/2000JB900449

Macelloni, G. and Brogioni, M. *et al.* 2018. Cryorad: a low frequency wideband radiometer mission for the study of the cryosphere. IGARSS 2018–2018, *IEEE International Geoscience and Remote Sensing Symposium*, 1998–2000.

Martos, Y.M., Catalán, M., Jordan, T.A., Golynsky, A., Golynsky, D., Eagles, G. and Vaughan, D.G. 2017. Heat flux distribution of Antarctica unveiled. *Geophysical Research Letters*, **44**, 11,417–11,426, https://doi.org/10.1002/2017GL075609

Maus, S. 2010. Magnetic field model MF7, http://geomag.org/models/MF7.html

Maus, S., Yin, F. *et al.* 2008. Resolution of direction of oceanic magnetic lineations by the sixth-generation lithospheric magnetic field model from CHAMP satellite magnetic measurements. *Geochemistry Geophysics Geosystems*, **9**, Q07021, https://doi.org/10.1029/2008GC001949

Mayer-Gürr, T., Eicker, A., Kurtenbach, E. and Ilk, K.-H. 2010. ITG-GRACE: global static and temporal gravity field models from GRACE data. *In*: Flechtner, F.M., Gruber, T., Güntner, A., Mandea, M., Rothacher, M., Schöne, T. and Wickert, J. (eds) *System Earth via Geodetic-Geophysical Space Techniques*. Springer, Berlin, 159–168, https://doi.org/10.1007/978-3-642-10228-8_13

Mayer-Gürr, T., Rieser, D. *et al.* 2012. The new combined satellite only model GOCO03s. Venice.

McKenzie, D., Jackson, J. and Priestley, K. 2005. Thermal structure of oceanic and continental lithosphere. *Earth and Planetary Science Letters*, **233**, 337–349, https://doi.org/10.1016/j.epsl.2005.02.005

McKenzie, D., Yi, W. and Rummel, R. 2015. Estimates of Te for continental regions using GOCE gravity. *Earth and Planetary Science Letters*, **428**, 97–107, https://doi.org/10.1016/j.epsl.2015.07.036

Mieth, M. and Jokat, W. 2014. New aeromagnetic view of the geological fabric of southern Dronning Maud Land and Coats Land, East Antarctica. *Gondwana Research*, **25**, 358–367, https://doi.org/10.1016/j.gr.2013.04.003

Milligan, P.R., Franklin, R. and Ravat, D. 2004. A new generation magnetic anomaly grid database of Australia (MAGDA). *Preview*, **113**, 25–29.

Mishra, D.C., Sekhar, D.V.C., Raju, D.C.V. and Kumar, V.V. 1999. Crustal structure based on gravity–magnetic modelling constrained from seismic studies under Lambert Rift, Antarctica and Godavari and Mahanadi rifts, India and their interrelationship. *Earth and Planetary Science Letters*, **172**, 287–300, https://doi.org/10.1016/S0012-821X(99)00212-5

Moritz, H. 1990. The inverse Vening Meinesz problem in isostasy. *Geophysical Journal International*, **102**, 733–738, https://doi.org/10.1111/j.1365-246X.1990.tb04591.x

Morlighem, M., Rignot, E. *et al.* 2020. Deep glacial troughs and stabilizing ridges unveiled beneath the margins of the Antarctic ice sheet. *Nature Geoscience*, **13**, 132–137, https://doi.org/10.1038/s41561-019-0510-8

Mueller, C.O. and Jokat, W. 2019. The initial Gondwana break-up: a synthesis based on new potential field data of the Africa–Antarctica Corridor. *Tectonophysics*, **750**, 301–328, https://doi.org/10.1016/j.tecto.2018.11.008

Nataf, H.-C. and Ricard, Y. 1996. 3SMAC: an a priori tomo graphic model of the upper mantle based on geophysical modeling. *Physics of the Earth and Planetary Interiors*, **95**, 101–122, https://doi.org/10.1016/0031-9201(95)03105-7

Nield, G.A., Whitehouse, P.L., van der Wal, W., Blank, B., O'Donnell, J.P. and Stuart, G.W. 2018. The impact of lateral variations in lithospheric thickness on glacial isostatic adjustment in West Antarctica. *Geophysical Journal International*, **214**, 811–824, https://doi.org/10.1093/gji/ggy158

Nowell, D.A.G. 1999. Gravity terrain corrections – an overview. *Journal of Applied Geophysics*, **42**, 117–134, https://doi.org/10.1016/S0926-9851(99)00028-2

O'Donnell, J.P. and Nyblade, A.A. 2014. Antarctica's hypsometry and crustal thickness: implications for the origin of anomalous topography in East Antarctica. *Earth and Planetary Science Letters*, **388**, 143–155, https://doi.org/10.1016/j.epsl.2013.11.051

Oldenburg, D.W. 1974. The inversion and interpretation of gravity anomalies. *Geophysics*, **39**, 526–536, https://doi.org/10.1190/1.1440444

Olesen, O. and Brönner, M. *et al.* 2010. New aeromagnetic and gravity compilations from Norway and adjacent areas: methods and applications. *Geological Society, London, Petroleum Geology Conference Series*, **7**, 559, https://doi.org/10.1144/0070559

Olsen, N., Ravat, D., Finlay, C.C. and Kother, L.K. 2017. LCS-1: a high-resolution global model of the lithospheric magnetic field derived from CHAMP and Swarm satellite observations. *Geophysical Journal International*, **211**, 1461–1477, https://doi.org/10.1093/gji/ggx381

Pail, R., Fecher, T., Barnes, D., Factor, J.F., Holmes, S.A., Gruber, T. and Zingerle, P. 2018. Short note: the experimental geopotential model XGM2016. *Journal of Geodesy*, **92**, 443–451, https://doi.org/10.1007/s00190-017-1070-6

Pappa, F., Ebbing, J. and Ferraccioli, F. 2019*a*. Moho depths of antarctica: comparison of seismic, gravity, and isostatic results. *Geochemistry, Geophysics, Geosystems*, **20**, 1629–1645, https://doi.org/10.1029/2018GC008111

Pappa, F., Ebbing, J., Ferraccioli, F. and van der Wal, W. 2019*b*. Modeling satellite gravity gradient data to derive density, temperature, and viscosity structure of the antarctic lithosphere. *Journal of Geophysical Research: Solid Earth*, **124**, 12053–12076, https://doi.org/10.1029/2019JB017997

Parker, R.L. 1973. The rapid calculation of potential anomalies. *Geophysical Journal International*, **31**, 447–455, https://doi.org/10.1111/j.1365-246X.1973.tb06513.x

Paxman, G.J. 2021. Antarctic palaeotopography. *Geological Society, London, Memoirs*, **56**, https://doi.org/10.1144/M56-2020-7

Paxman, G.J., Jamieson, S.S., Ferraccioli, F., Bentley, M.J., Ross, N., Watts, A.B., … and Young, D.A. 2019*a*. The role of lithospheric flexure in the landscape evolution of the Wilkes Subglacial Basin and Transantarctic Mountains, East Antarctica. *Journal of Geophysical Research: Earth Surface*, **124**, 812–829, https://doi.org/10.1029/2018JF004705

Paxman, G.J., Jamieson, S.S. *et al.* 2019*b*. Subglacial geology and geomorphology of the Pensacola-Pole Basin, East Antarctica. *Geochemistry, Geophysics, Geosystems*, **20**, 2786–2807, https://doi.org/10.1029/2018GC008126

Peidou, A. and Pagiatakis, S. 2019. Gravity gradiometry with GRACE space missions: new opportunities for the geosciences. *Journal of Geophysical Research: Solid Earth*, **124**, 9130–9147, https://doi.org/10.1029/2018JB016382

Pollett, A., Hasterok, D., Raimondo, T., Halpin, J.A., Hand, M., Bendall, B. and McLaren, S. 2019. Heat flow in Southern Australia and connections with East Antarctica. *Geochemistry, Geophysics, Geosystems*, **20**, https://doi.org/10.1029/2019GC 008418

Rabbel, W., Kaban, M. and Tesauro, M. 2013. Contrasts of seismic velocity, density and strength across the Moho. *Tectonophysics*, **609**, 437–455, https://doi.org/10.1016/j.tecto.2013.06.020

Ramirez, C., Nyblade, A. *et al.* 2016. Crustal and upper-mantle structure beneath ice-covered regions in Antarctica from S-wave receiver functions and implications for heat flow. *Geophysical Journal International*, **204**, 1636–1648, https://doi.org/10.1093/gji/ggv542

Reeves, C. 2005. Aeromagnetic surveys: principles, practice and interpretation. *Geosoft.*, **155**.

Reguzzoni, M. and Sampietro, D. 2015. GEMMA: an Earth crustal model based on GOCE satellite data. *International Journal of Applied Earth Observation and Geoinformation*, **35**, 31–43, https://doi.org/10.1016/j.jag.2014.04.002

Reigber, C. 2006. A high resolution global gravity field model combining CHAMP and GRACE satellite mission and surface data: EIGEN-CG01C. *Scientific Technical Report*, Geoforschungszentrum Potsdam, https://doi.org/10.23689/fidgeo-564

Rezvanbehbahani, S., Stearns, L.A., Kadivar, A., Walker, J.D. and van der Veen, C.J. 2017. Predicting the geothermal heatflux in Greenland: a machine learning approach. *Geophysical Research Letters*, **44**, 12,271–12,279, https://doi.org/10.1002/2017GL075661

Riedel, S., Jokat, W. and Steinhage, D. 2012. Mapping tectonic provinces with airborne gravity and radar data in Dronning Maud Land, East Antarctica. *Geophysical Journal International*, **189**, 414–427, https://doi.org/10.1111/j.1365-246X.2012.05363.x

Rogozhina, I., Hagedoorn, J.M., Martinec, Z., Fleming, K., Soucek, O., Greve, R. and Thomas, M. 2012. Effects of uncertainties in the geothermal heat flux distribution on the Greenland Ice Sheet: an assessment of existing heat flow models. *Journal of Geophysical Research*, **117**, F02025, https://doi.org/10.1029/2011JF002098

Rummel, R., Yi, W. and Stummer, C. 2011. GOCE gravitational gradiometry. *Journal of Geodesy*, **85**, 777, https://doi.org/10.1007/s00190-011-0500-0

Ruppel, A., Jacobs, J., Eagles, G., Läufer, A. and Jokat, W. 2018. New geophysical data from a key region in East Antarctica: estimates for the spatial extent of the Tonian Oceanic Arc Super Terrane (TOAST). *Gondwana Research*, **59**, 97–107, https://doi.org/10.1016/j.gr.2018.02.019

Schaeffer, A.J. and Lebedev, S. 2013. Global shear speed structure of the upper mantle and transition zone. *Geophysical Journal International*, **194**, 417–449, https://doi.org/10.1093/gji/ggt095

Scheinert, M., Ferraccioli, F. *et al.* 2016. New Antarctic gravity anomaly grid for enhanced geodetic and geophysical studies in Antarctica. *Geophysical Research Letters*, **43**, 600–610, https://doi.org/10.1002/2015GL067439

Schröder, H., Paulsen, T. and Wonik, T. 2011. Thermal properties of the AND-2A borehole in the southern Victoria Land Basin, McMurdo Sound, Antarctica. *Geosphere*, **7**, 1324–1330, https://doi.org/10.1130/GES00690.1

Sebera, J., Haagmans, R., Floberghagen, R. and Ebbing, J. 2018. Gravity spectra from the density distribution of Earth's uppermost 435 km. *Surveys in Geophysics*, **39**, 227–244, https://doi.org/10.1007/s10712-017-9445-z

Shapiro, N.M. and Ritzwoller, M.H. 2004. Inferring surface heat flux distributions guided by a global seismic model: particular application to Antarctica. *Earth and Planetary Science Letters*, **223**, 213–224, https://doi.org/10.1016/j.epsl.2004.04.011

Shen, W., Wiens, D.A. *et al.* 2018. The crust and upper mantle structure of central and West Antarctica from Bayesian inversion of Rayleigh Wave and Receiver Functions. *Journal of Geophysical Research: Solid Earth*, **123**, 7824–7849, https://doi.org/10.1029/2017JB015346

Shen, W., Wiens, D.A., Lloyd, A.J. and Nyblade, A.A. 2020. A geothermal heat flux map of Antarctica empirically constrained by seismic structure. *Geophysical Research Letters*, **47**, e2020GL086955, https://doi.org/10.1029/2020GL086955

Shephard, G.E., Matthews, K.J., Hosseini, K. and Domeier, M. 2017. On the consistency of seismically imaged lower mantle slabs. *Scientific Reports*, **7**, 10976, https://doi.org/10.1038/s41598-017-11039-w

Spector, A. and Grant, F.S. 1970. Statistical models for interpreting aeromagnetic data. *Geophysics*, **35**, 293–302, https://doi.org/10.1190/1.1440092

Stern, T.A. and ten Brink, U.S. 1989. Flexural uplift of the Transantarctic Mountains. *Journal of Geophysical Research*, **94**, 10315–10330, https://doi.org/10.1029/JB094iB08p10315

Stettler, E.H., Fourie, C.J.S. and Cole, P. 2000. Total magnetic field intensity map of the Republic of South Africa (in 4 panels). *Council for Geoscience, Pretoria, 2000.*

Stixrude, L. and Lithgow-Bertelloni, C. 2005. Mineralogy and elasticity of the oceanic upper mantle: origin of the low-velocity zone. *Journal of Geophysical Research: Solid Earth*, **110**, 1978–2012, https://doi.org/10.1029/2004JB00 2965

Straume, E.O., Gaina, C. *et al.* 2019. GlobSed: updated total sediment thickness in the World's Oceans. *Geochemistry, Geophysics, Geosystems*, **20**, 1756–1772, https://doi.org/10.1029/2018G C008115

Studinger, M., Karner, G.D., Bell, R.E., Levin, V., Raymond, C.A. and Tikku, A.A. 2003. Geophysical models for the tectonic framework of the Lake Vostok region, East Antarctica. *Earth and Planetary Science Letters*, **216**, 663–677, https://doi.org/10.1016/S0012-821X(03)00548-X

Szwillus, W., Afonso, J.C., Ebbing, J. and Mooney, W.D. 2019. Global crustal thickness and velocity structure from geostatistical analysis of seismic data. *Journal of Geophysical Research: Solid Earth*, **124**, 1626–1652, https://doi.org/10.1029/2018JB016593

Tanaka, A. 2017. Global centroid distribution of magnetized layer from world digital magnetic anomaly map. *Tectonics*, **36**, 3248–3253, https://doi.org/10.1002/2017TC004770

Tanaka, A., Okubo, Y. and Matsubayashi, O. 1999. Curie point depth based on spectrum analysis of the magnetic anomaly data in East and Southeast Asia. *Tectonophysics*, **306**, 461–470, https://doi.org/10.1016/S0040-1951(99)00072-4

Tapley, B.D., Bettadpur, S., Ries, J.C., Thompson, P.F. and Watkins, M.M. 2004. GRACE measurements of mass variability in the earth system. *Science*, **305**, 503–505, https://doi.org/10.1126/science.1099192

Thébault, E., Finlay, C.C., Alken, P., Beggan, C.D., Canet, E., Chulliat, A. and Rother, M. 2015. Evaluation of candidate geomagnetic field models for IGRF-12. *Earth, Planets and Space*, **67**, 112, https://doi.org/10.1186/s40623-015-0273-4

Van Liefferinge, B., Pattyn, F., Cavitte, M.G.P., Karlsson, N.B., Young, D.A., Sutter, J. and Eisen, O. 2018. Promising Oldest Ice sites in East Antarctica based on thermodynamical modelling. *The Cryosphere*, **12**, 2773–2787, https://doi.org/10.5194/tc-12-2773-2018

Wasilewski, P.J. and Mayhew, M.A. 1992. The Moho as a magnetic boundary revisited. *Geophysical Research Letters*, **19**, 2259–2262, https://doi.org/10.1029/92GL01997

Watts, A.B. 2001. *Isostasy and Flexure of the Lithosphere*. Cambridge University Press.

Whitehouse, P.L., Gomez, N., King, M.A. and Wiens, D.A. 2019. Solid Earth change and the evolution of the Antarctic Ice Sheet. *Nature Communications*, **10**, 503, https://doi.org/10.1038/s41467-018-08068-y

Wiens, D. this volume, in review. The Seismic Structure of the Antarctic Upper Mantle. *Geological Society, Memoirs*, **56**.

Yildiz, H., Forsberg, R., Tscherning, C.C., Steinhage, D., Eagles, G. and Bouman, J. 2017. Upward continuation of Dome-C airborne gravity and comparison with GOCE gradients at orbit altitude in east Antarctica. *Studia Geophysica et Geodaetica*, **61**, 53–68, https://doi.org/10.1007/s11200-015-0634-2

Zingerle, P., Pail, R., Scheinert, M. and Schaller, T. 2019. Evaluation of terrestrial and airborne gravity data over Antarctica – a generic approach. *Journal of Geodetic Science*, **9**, 29, https://doi.org/10.1515/jogs-2019-0004

Antarctic palaeotopography

Guy J. G. Paxman[1,2]
[1]Department of Geography, Durham University, Durham, UK
[2]Lamont-Doherty Earth Observatory, Columbia University, New York, USA
0000-0003-1787-7442
gpaxman@ldeo.columbia.edu

Abstract: The development of a robust understanding of the response of the Antarctic Ice Sheet to present and projected future climatic change is a matter of key global societal importance. Numerical ice sheet models that simulate future ice sheet behaviour are typically evaluated with recourse to how well they reproduce past ice sheet behaviour, which is constrained by the geological record. However, subglacial topography, a key boundary condition in ice sheet models, has evolved significantly throughout Antarctica's glacial history. Since mantle processes play a fundamental role in the generation and modification of topography over geological timescales, an understanding of the interactions between the Antarctic mantle and palaeotopography is crucial for developing more accurate simulations of past ice sheet dynamics. This chapter provides a review of the influence of the Antarctic mantle on the long-term evolution of the subglacial landscape, through processes including structural inheritance, flexural isostatic adjustment, lithospheric cooling and thermal subsidence, volcanism and dynamic topography. The uncertainties associated with reconstructing these processes through time are discussed, as are important directions for future research and the implications of the evolving subglacial topography for the response of the Antarctic Ice Sheet to climatic and oceanographic change.

Overview

Unveiling the subglacial landscape of Antarctica

Antarctica is unique among Earth's continents in that the crust is almost entirely capped by a layer of ice up to 4 km thick. Today, *c.* 99.8% of the Antarctic landmass is concealed beneath the Antarctic Ice Sheet (AIS) (Burton-Johnson *et al.* 2016), and this figure was likely even larger during recent glacial maxima (Bentley *et al.* 2014). The near-total lack of exposure of the Antarctic 'land surface' presents major challenges for understanding the long-term history of the AIS.

The progressive mapping of Antarctica's subglacial topography has been driven by a substantial community effort since the 1970s (Drewry 1983; Lythe *et al.* 2001; Fretwell *et al.* 2013; Morlighem *et al.* 2020; Schroeder *et al.* 2019). Ground-based and airborne radio-echo sounding (RES) surveys are widely used to measure ice thickness in Antarctica. Although minor differences exist between the radar systems that have been employed in Antarctica, they share a common basic premise. An electronic signal is converted to radio waves, which are emitted from an antennae array mounted on the survey platform (e.g. aircraft). This radar pulse propagates through the air and ice sheet, and energy is reflected at interfaces with strong acoustic impedance contrasts (Fig. 1). The strongest reflections are from the air–ice and ice–bed interfaces; weaker reflections are also seen from internal layers within the ice. Upon their return to the carrier platform, the echoes are detected by a receiver and converted back to an electronic signal. The system then records the time delay between the pulse being emitted and received – the two-way travel time.

The ice surface and bed reflectors are extracted ('picked') from each radar trace (Fig. 1). The difference in the two-way travel times of these two reflectors gives the two-way travel time of the radar in the ice, which is multiplied by an assumed radar velocity in ice (*c.* 168 m μs^{-1}), with an additional small correction commonly made for the firn layer (the thin layer of snow and uncompacted ice at the top of the ice sheet, which does not generate strong radar reflections) to give the ice thickness. The difference between the ice surface elevation (which can be determined via laser altimetry) and the ice thickness gives the bed elevation. Location and altitude of the radar platform are determined using a global positioning system (GPS).

Up to 2013, *c.* 25 million ice thickness measurements (Fig. 1) had been acquired and incorporated into the Bedmap2 digital elevation model (DEM) of Antarctica's bedrock topography (Fig. 2) (Fretwell *et al.* 2013). However, approximately only one third of cells within Bedmap2 contain at least one real ice thickness measurement (the grid is at 5 km horizontal resolution), and some areas are >200 km away from the nearest real data point (Fretwell *et al.* 2013). Several multi-season field campaigns have commenced since the publication of Bedmap2, with the aim of filling some of these major data gaps (Fig. 1). This increased data coverage is reflected in the recently developed BedMachineAntarctica DEM (Morlighem *et al.* 2020).

Bed topography and ice sheet behaviour

Antarctica's subglacial topography is important to our understanding of AIS history. It has become increasingly recognized that subglacial topography exerts a fundamental influence on the dynamics of the overlying ice sheet (Wilson *et al.* 2013; Austermann *et al.* 2015; Gasson *et al.* 2015; Colleoni *et al.* 2018), and is therefore an important boundary condition in numerical ice sheet models. Ice sheet models aim to simulate the present dynamics of the AIS and predict the response of the ice sheet to future projected climatic change and, in turn, evaluate likely changes in global sea-levels (Mengel and Levermann 2014; Golledge *et al.* 2015; DeConto and Pollard 2016; Edwards *et al.* 2019). Since bed topography exerts a strong influence on ice sheet dynamics, it is important for any ice sheet model to incorporate an accurate DEM of present-day subglacial topography.

Of particular significance are 'marine ice sheets', which are grounded on bed situated below sea-level. Marine ice sheets that are grounded on an inland-dipping ('reverse sloping') bed are hypothesized to be particularly susceptible to perturbations in atmosphere and ocean temperatures (Mercer 1978; Thomas 1979). In this scenario, when the grounding line of the marine ice sheet retreats, it retreats into deeper water, leading to increased flotation, basal melting and ice discharge, resulting in further retreat and a self-sustained positive feedback (Schoof 2007). This process is known as marine ice sheet instability, and permits a highly non-linear response of the grounding line to initial climate and ocean forcing, creating the possibility of rapid retreat or even catastrophic collapse of marine-based ice sheets (Mercer 1978; Thomas 1979; Scherer *et al.* 1998; Schoof 2007; Joughin and Alley 2011). Ice sheet modelling studies incorporating marine ice sheet instability

From: Martin, A. P. and van der Wal, W. (eds) 2023. *The Geochemistry and Geophysics of the Antarctic Mantle.*
Geological Society, London, Memoirs, **56**, 231–251,
First published online 28 April 2021, https://doi.org/10.1144/M56-2020-7

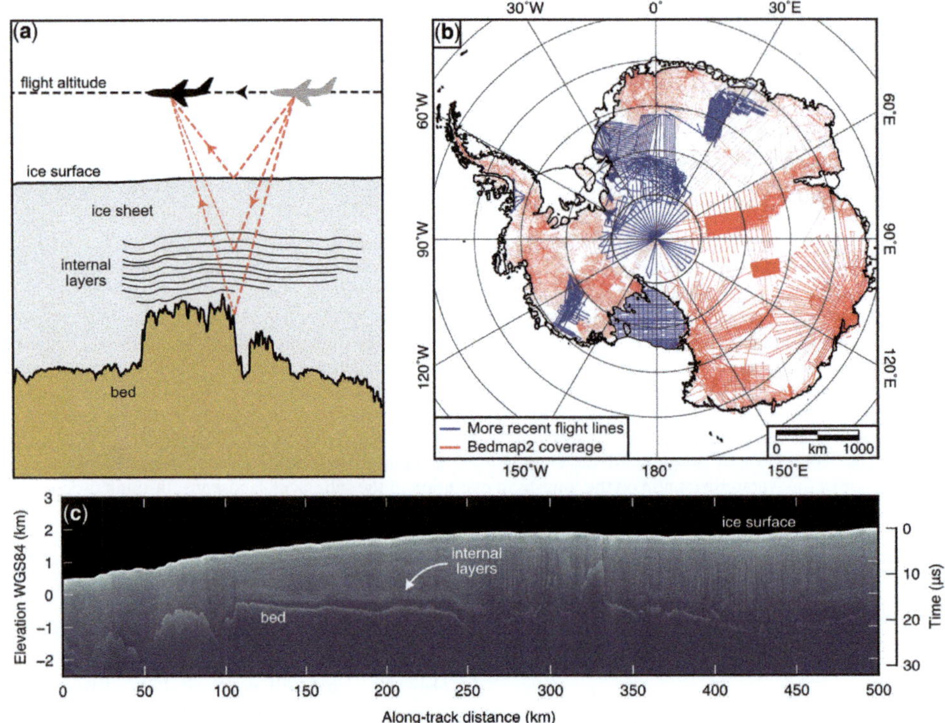

Fig. 1. Measurement of ice thickness and subglacial topography using radio-echo sounding (RES). (**a**) Schematic of an airborne RES survey. The aircraft emits a series of radar pulses (red dashed lines), which are reflected at interfaces including the ice surface, internal layers within the ice sheet, and the ice–bed interface. The amount of radar energy reflected depends on the acoustic impedance contrast at the interface. The aircraft then receives the return echoes, and their travel time is used to calculate the depth to the reflector. (**b**) Coverage of RES measurements in Antarctica. Red lines denote surveys that were included in the Bedmap2 compilation (Fretwell *et al.* 2013); blue lines denote more recently acquired survey data (Forsberg *et al.* 2018; Humbert *et al.* 2018; Jordan *et al.* 2018; Karlsson *et al.* 2018; Tinto *et al.* 2019). (**c**) Example of a radar trace (echogram) from Antarctica (Leuschen *et al.* 2016). Elevations are relative to the WGS84 ellipsoid. The y-axis is converted from time to depth assuming a radar velocity in ice of 168 m/μs. Reflectors from the ice surface and ice sheet bed are clearly visible, as are faint internal layer reflections within the ice sheet.

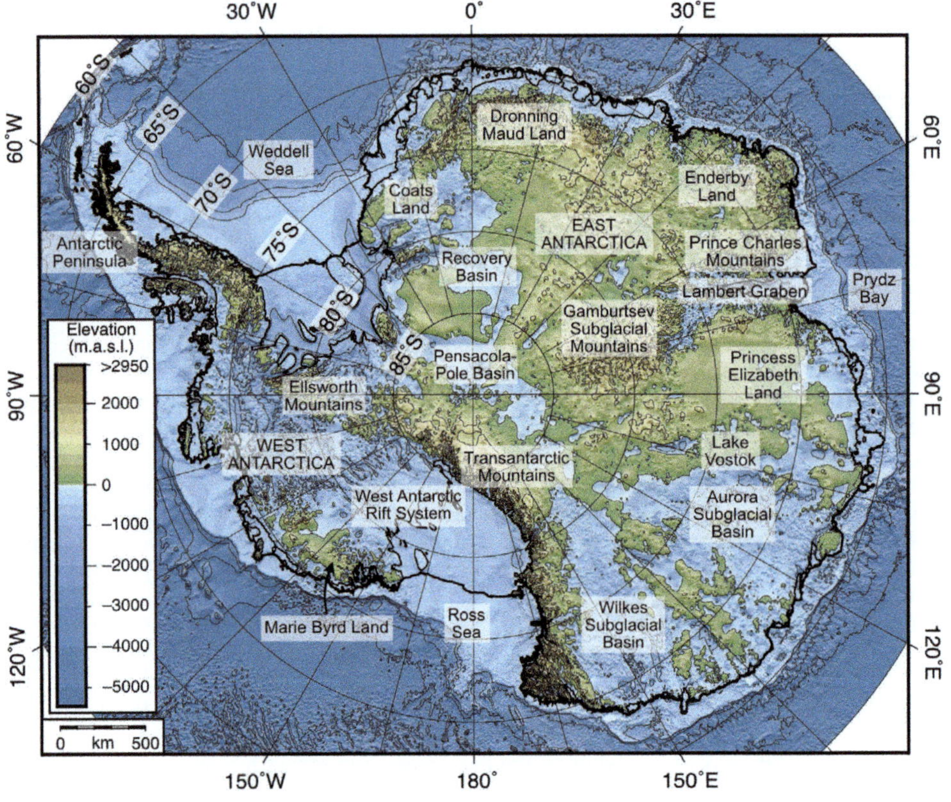

Fig. 2. Bed topography of Antarctica. Bed elevations and bathymetry are from the Bedmap2 dataset (Fretwell *et al.* 2013). Elevations are in metres above mean sea-level (m.a.s.l.) and contoured at 1000 m intervals. Major geographical and topographical features are labelled. The dark lines around the edge of the continent denote the modern grounding line and ice shelf calving front. Approximately 45% of the bed is presently situated below sea-level, including the extensive Recovery, Wilkes and Aurora subglacial basins in East Antarctica, wherein the deepest regions are *c.* 2 km below sea-level (Fretwell *et al.* 2013).

processes (Pollard *et al.* 2015) have predicted global sea-level rise of *c.* 1 m by the end of the current century (DeConto and Pollard 2016). However, probabilistic model projections indicate that these values may significantly overestimate the rate of ice mass loss and sea-level rise (Edwards *et al.* 2019). Moreover, a number of processes have been documented that may counteract marine ice sheet instability, such as slow ice shelf removal rates (Clerc *et al.* 2019), local isostatic uplift of the bed beneath the grounding line (Barletta *et al.* 2018; Kingslake *et al.* 2018), and deformation-induced relative sea-level change (Gomez *et al.* 2010).

Marine-based ice in Antarctica has the potential to contribute *c.* 23 m of global sea-level rise (Fretwell *et al.* 2013); the amount of warming required to induce significant retreat of marine-based parts of the ice sheet, and the likely rate and magnitude of the retreat, are therefore questions of crucial societal importance. In order to constrain these unknowns, attention has focused on records of AIS stability during past warm periods in Earth's history that may serve as analogues for future climate conditions. Ice sheet models are typically assessed by how well they can reproduce past ice sheet behaviour (as constrained from geological and sea-level records) before they can be confidently used to predict the response of the AIS to projected future climate scenarios. Since certain areas of the subglacial topography of Antarctica have been extensively modified since glacial inception *c.* 34 million years ago (Jamieson *et al.* 2010; Wilson *et al.* 2012; Paxman *et al.* 2019*b*; Pollard and DeConto 2019), it is unrealistic to use the modern bed topography when attempting to model early Antarctic ice sheets. Using a more realistic reconstruction of subglacial topography for past time intervals has the potential to increase the robustness of ice sheet model results and, in turn, (a) facilitate improved understanding of the response of past Antarctic ice sheets to climatic change (DeConto and Pollard 2003*a*; Wilson *et al.* 2013) and (b) engender greater confidence in projections of future ice sheet stability, mass loss and sea-level change.

Additionally, the morphology of subglacial topography is an important repository of information pertaining to past ice sheet dynamics. A landscape contains a record of the processes that have acted to shape it, and Antarctic geomorphology studies typically aim use age relationships and the morphology of landscape features to determine the history of the landscape and the processes that produced it. This analysis can reveal insights as to the spatial extent, dynamics and behaviour of past ice sheets, in particular during past warmer climate intervals potentially analogous to the future. For example, RES data have revealed subglacial landforms and areas of enhanced glacial erosion within the Aurora Subglacial Basin in East Antarctica, several hundred kilometres inland of the modern ice margin (Young *et al.* 2011; Aitken *et al.* 2016). Such landforms likely formed near the margin of a dynamic early ice sheet with a more restricted extent than at the present-day.

Antarctic glacial history

The timescales over which changes in Antarctic bed topography are most significant in this context are intrinsically linked to the intervals during which ice sheets have been present in the Southern Hemisphere. Marine oxygen isotope records indicate that the Earth has been experiencing a long-term cooling trend since the Early Eocene Climatic Optimum (*c.* 51 Ma; Fig. 3) (Zachos *et al.* 2001). The first appearance of continental-scale ice sheets in Antarctica is widely agreed to have occurred around the time of the Eocene–Oligocene boundary (*c.* 34 Ma) (Zachos *et al.* 2001; Coxall *et al.* 2005; Liu *et al.* 2009), which marks the transition between the 'greenhouse world' of the Cretaceous–Eocene and the 'icehouse world' that Earth has inhabited since the start of the Oligocene (Fig. 3). Stratigraphic records indicate that ice sheet growth occurred in a stepwise fashion across the Eocene–Oligocene transition, in concert with the marine oxygen isotope record (Coxall *et al.* 2005; Katz *et al.* 2008; Scher *et al.* 2011; Passchier *et al.* 2017). However, it is important to note that although this climate transition marks the first appearance of large-scale Southern Hemisphere ice sheets,

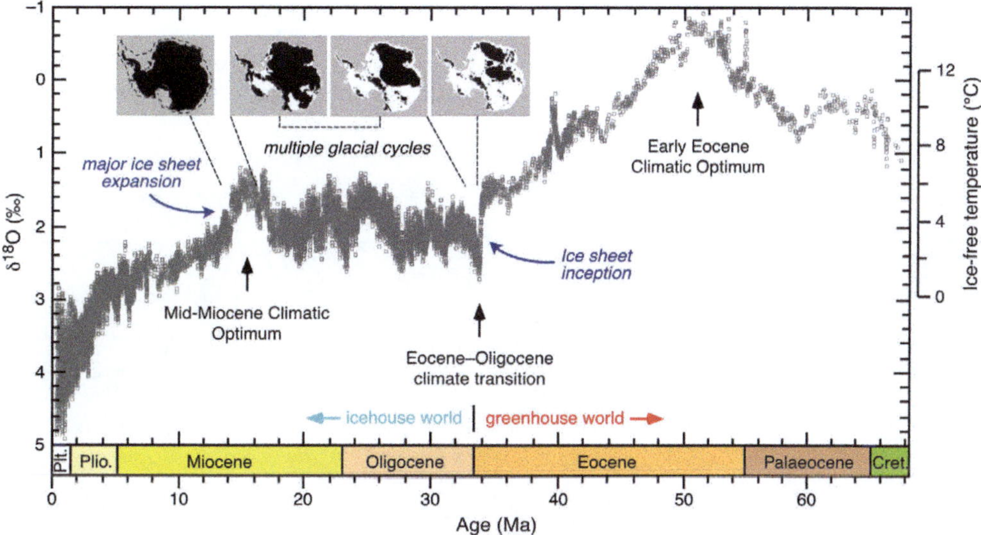

Fig. 3. Cenozoic ice sheet and climate history of Antarctica (modified from Zachos *et al.* 2008). Curve shows the Cenozoic stacked benthic foraminifera oxygen isotope record, based on deep-sea sediments obtained at Deep Sea Drilling Project and Ocean Drilling Program sites (Zachos *et al.* 2001, 2008). The raw data have been smoothed using a five-point running mean. $\delta^{18}O$ isotope values are given in parts per thousand (‰). The temperature scale is calculated for ice-free conditions, so only applies to the time interval preceding glaciation in Antarctica. The greenhouse-icehouse transition and the formation of large-scale ice sheets on Antarctica coincides with the step increase in $\delta^{18}O$ at the Eocene–Oligocene boundary (*c.* 34 Ma). The approximate ice extents for different stages (shaded black) are derived from ice sheet modelling (Jamieson *et al.* 2010). The Oligocene to middle Miocene witnessed numerous glacial cycles, with the ice sheet waxing and waning in concert with orbital cycles. Subsequent Miocene cooling allowed the ice sheet to advance to the continental shelf edge, perhaps on multiple occasions. Abbreviations: Cret. = Cretaceous; Plio. = Pliocene; Plt. = Pleistocene.

ephemeral ice sheets and/or mountain glaciers may have existed on Antarctica at an earlier stage (Carter *et al.* 2017).

The mechanisms behind the precise timing of glacial inception in Antarctica remain subject to continuing debate. However, it is widely believed that continental-scale glaciation on Antarctica was triggered by a drop in atmospheric pCO_2 levels below a threshold (DeConto and Pollard 2003*a*; DeConto *et al.* 2008; Pearson *et al.* 2009; Pagani *et al.* 2011) combined with associated internal feedbacks within the carbon cycle (Tripati *et al.* 2005; Zachos and Kump 2005; Reusch 2011; Scher *et al.* 2011). The opening/deepening of ocean gateways around Antarctica, in particular the Drake Passage and the Tasman Gateway, may have also played a role in the timing of glaciation (Kennett 1977; DeConto and Pollard 2003*b*).

Cyclicity in marine oxygen isotope and sedimentary records from drill cores around the Antarctic margin indicates that the early Antarctic ice sheets waxed and waned in concert with orbital cycles during the Oligocene and early Miocene (Hambrey *et al.* 1991; Zachos *et al.* 1997, 2001; Naish *et al.* 2001; Barrett 2008). Approximately 33 glacial cycles have been detected in this interval, and the sedimentary record suggests that meltwater was abundant, with conditions similar to present-day Greenland (Naish *et al.* 2001; Sugden and Jamieson 2018). Marine sediment cores from the Ross Sea revealed an erosional hiatus at the Oligocene–Miocene boundary (*c.* 23 Ma) indicative of a major episode of global cooling and ice sheet expansion (Naish *et al.* 2001), which has also been inferred from a contemporaneous benthic foraminiferal oxygen isotope shift that implies a significant increase in ice volume (Fig. 3) (Zachos *et al.* 2001). Following the mid-Miocene Climatic Optimum, a gradual cooling trend commenced at *c.* 14 Ma (Fig. 3). Decreasing air and ocean temperatures allowed the AIS to expand significantly under an arid polar climate (Zachos *et al.* 2001), and the oscillatory and more restricted ice masses evolved into a major and persistent continental ice sheet (Fig. 3) (DeConto and Pollard 2003*a*; Jamieson *et al.* 2010). The ice sheet extended to the continental shelf edge, perhaps on multiple occasions, before retreating and stabilizing with dimensions similar to that of the present-day (Sugden and Denton 2004).

This multi-faceted glacial history over the past 34 million years provides a large window of time for significant topographic and landscape evolution to occur in Antarctica. Efforts to understand palaeotopography in Antarctica have therefore primarily focused on the time period from glacial inception at *c.* 34 Ma to the present day (Jamieson and Sugden 2008; Wilson and Luyendyk 2009; Wilson *et al.* 2012; Paxman *et al.* 2019*b*). However, attempts are also ongoing to produce reconstructions of Antarctic palaeotopography during the early Cenozoic and Late Cretaceous, which will be important boundary conditions for the modelling of older ice masses, regional ocean and atmosphere circulation, and palaeoclimate.

Framework of landscape evolution in Antarctica

Landscape evolution reflects the competition between 'solid Earth' processes such as tectonics, which act to increase or decrease the elevation of the crust, and 'Earth surface' processes such as erosion and sedimentation, which modify it. In Antarctica, the mantle facilitates the generation of surface topography via:

1. The flexural isostatic response to load redistribution (via e.g. erosion, sedimentation, movement on faults), which is governed by the balance of forces in the crust and mantle lithosphere and causes uplift or subsidence of the Earth's surface.
2. Heating or cooling of the lithosphere, which alters the density of the crust/mantle lithosphere and drives a transient phase of surface uplift or subsidence. Heating can also induce volcanic activity, causing emplacement of magmatic material on the Earth's surface or intrusion of magmatic material within the crust.
3. Viscous normal tractions applied to the base of the lithosphere by density-driven convective flow within the mantle ('dynamic topography').

Each of these processes (Fig. 4) is to a greater or lesser extent influenced by the properties of the mantle; each is described in turn in the following sections of this chapter.

Isostasy

Elastic plate flexure

Isostatic adjustment occurs in response to the redistribution of loads on the surface of, or within, the lithosphere on

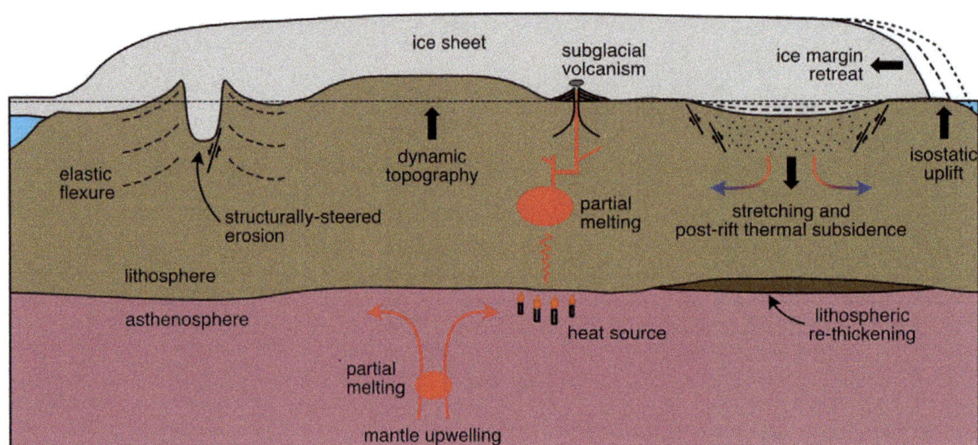

Fig. 4. Schematic summary of the interactions between mantle processes and topography beneath ice sheets. Changes in surface loading due to (e.g.) erosion or ice margin retreat induce a flexural isostatic response within the elastic lithosphere. Topographic subsidence occurs in response to lithospheric stretching and subsequent 'post-rift' cooling, re-thickening and densification, and commonly allows sediments to accumulate in the resulting basin. Deposition of sediments in turn triggers a flexural isostatic response, causing further subsidence. Heating of the lithosphere can induce partial melting, which leads to subglacial volcanism. Heating can also induce topographic uplift, as can normal tractions imparted on the base of the lithosphere by the convecting viscous mantle (dynamic topography).

wavelengths of order 100–1000 km. Landscape evolution processes involve the redistribution of surface and sub-surface material on these length scales, and therefore induce vertical surface displacement via the isostatic re-equilibration of the lithosphere. Moreover, isostasy is the process by which many solid Earth phenomena cause changes in surface topography, and is therefore the most ubiquitous link between mantle properties and landscape evolution processes (Fig. 4).

The flexural isostatic model states that surface and sub-surface loads are supported by a combination of (a) buoyancy forces imparted on the base of the crust by the mantle lithosphere, which arise due to the difference in density between crust and mantle (Airy isostasy), and (b) elastic bending forces within the lithosphere. The ratio of buoyancy forces and elastic bending forces is a function of the wavelength of the load and also the flexural rigidity of the lithosphere. The flexural response of the lithosphere to (un)loading over geological timescales (>10^6 years) may be computed by modelling the lithosphere as a thin elastic plate overlying an inviscid (non-viscous) substrate (the asthenospheric mantle) (Watts 2001; Watts *et al.* 2013). The general two-dimensional flexure equation giving the downward deflection (w) of the elastic plate in response to loading (h) is given by

$$\nabla^2[D(x, y)\nabla^2 w(x, y)] + (\rho_{\text{mantle}} - \rho_{\text{infill}})gw(x, y) = (\rho_{\text{load}} - \rho_{\text{displace}})gh(x, y) \quad (1)$$

where

$$D(x, y) = \frac{ET_e(x, y)^3}{12(1 - \nu^2)} \quad (2)$$

is the flexural rigidity of the lithosphere as a function of spatial dimensions x and y. E (Young's modulus; 100 GPa) and ν (Poisson's ratio; 0.25) are elastic constants and the density terms (ρ) depend on the particular loading or unloading scenario being considered. The densities of the load and the mantle are given by ρ_{load} and ρ_{mantle} respectively. The terms for ρ_{displace} and ρ_{infill} can be ignored for subaerial surface loads, but for subaqueous loads the density contrasts between (i) the load and the displaced water, and (ii) the mantle and the material filling in the flexural moat (e.g. water, sediment) must also be included.

The spatial pattern (amplitude and wavelength) of the flexural isostatic response to (un)loading is determined by the flexural rigidity (D) or its proxy, the effective elastic thickness (T_e) of the lithosphere (Watts 2001). T_e is a measure of the equivalent thickness the lithosphere would have if it behaved as a perfectly elastic plate overlying an inviscid fluid asthenospheric mantle, as is assumed by the thin elastic plate model (equation 2). In reality, the composition and rheology of the lithosphere is strongly heterogeneous, so T_e is more appropriately conceptualized as a proxy for the depth-integrated strength of the lithosphere (Watts and Burov 2003). An important implication of the flexure model is that provided the lithosphere has a finite rigidity/strength, vertical surface displacement occurs beyond the spatial limits of the (un)load (Fig. 4); flexural isostasy is therefore known as 'regional isostasy'.

Flexural rigidity of the Antarctic lithosphere

Flexurally rigid lithosphere is associated with long-wavelength, low-amplitude flexural displacement, whereas flexurally weak lithosphere is associated with short-wavelength, high-amplitude flexural displacement. Since the effective elastic thickness governs the pattern of flexural adjustment to load redistribution, the depth-integrated strength of the Antarctic lithospheric mantle is an important parameter in landscape evolution models. Attempts to estimate T_e in Antarctica can be divided into two broad categories.

1. Forward modelling techniques, whereby an observed manifestation of flexure such as a deflected/warped topographic surface and/or its associated gravity anomaly is compared to the predictions of an elastic plate model in which T_e can be adjusted. The value of T_e in the model that yields the best fit with the observed flexure is regarded as the estimate of T_e at the time of loading for the region under consideration. Such methods have recovered T_e values of 5 ± 5 km for parts of the West Antarctic Rift System (Jordan *et al.* 2010; Wenman *et al.* 2020), *c.* 5 km close to the front of the Transantarctic Mountains, increasing to *c.* 85–115 km beneath the Wilkes Subglacial Basin (Stern and ten Brink 1989; ten Brink *et al.* 1997; Stern *et al.* 2005), and 20–25 km in the Recovery Basin (Paxman *et al.* 2017) (Fig. 5a).
2. Inverse (spectral) modelling of observed gravity anomalies and subglacial topography. The two spectral parameters commonly used to estimate T_e from gravity and topography data are the free-air admittance and the Bouguer coherence. The free-air admittance is the linear transfer function between the free-air anomaly and the topography in the frequency domain, while the Bouguer coherence is essentially the square of the Pearson product-moment correlation coefficient between the Bouguer gravity anomaly and the topography, again computed in the frequency domain (Kirby 2014). The forms of these two spectral parameters as a function of wavelength can be used to estimate T_e, and have recovered values of *c.* 35 km in the Weddell Sea embayment (Studinger and Miller 1999), *c.* 10 km in the Gamburtsev Subglacial Mountains (Paxman *et al.* 2016) and *c.* 20 km in the Recovery Basin (Paxman *et al.* 2017) (Fig. 5b).

While the forward modelling methods are robust where a clear surface manifestation of flexure, such as tilted palaeoland surfaces, can be discerned (ten Brink *et al.* 1997; Paxman *et al.* 2017), reference surfaces with a known geometry prior to flexural warping are rarely identified in Antarctica given the lack of surface exposure. In addition, forward models recover the T_e pertaining to the time at which loading occurred, which is not necessarily equivalent to the contemporary T_e. The ability of T_e to evolve over geological timescales (e.g. Watts *et al.* 2013), has been documented for the rifted basins in the Ross Sea (Karner *et al.* 2005). The potential temporal variation of T_e adds an additional level of uncertainty in reconstructing flexure in Antarctica over long timescales.

By contrast, the inverse spectral methods can be applied over large areas, since most of Antarctica has fairly well-established topography (Fretwell *et al.* 2013) and gravity anomaly (Scheinert *et al.* 2016) grids. However, the validity of the results derived from the spectral methods, particularly the Bouguer coherence, has been questioned where there is a lack of clear coherence between the topography and the gravity anomaly, which may be caused by the presence of buried loads (i.e. loads with no surface topographic manifestation) or the modification of topography by erosion (McKenzie and Fairhead 1997; Watts 2001; Kirby 2014). It has been argued that the T_e estimates derived from the Bouguer coherence maybe consequently be biased towards high values and should be considered upper bounds rather than best estimates (McKenzie and Fairhead 1997). Moreover, in places of low RES coverage, Antarctic bed topography is often determined by inversion of gravity anomalies, which introduces circularity when T_e is estimated from the relationship *between* topography and gravity anomalies. These

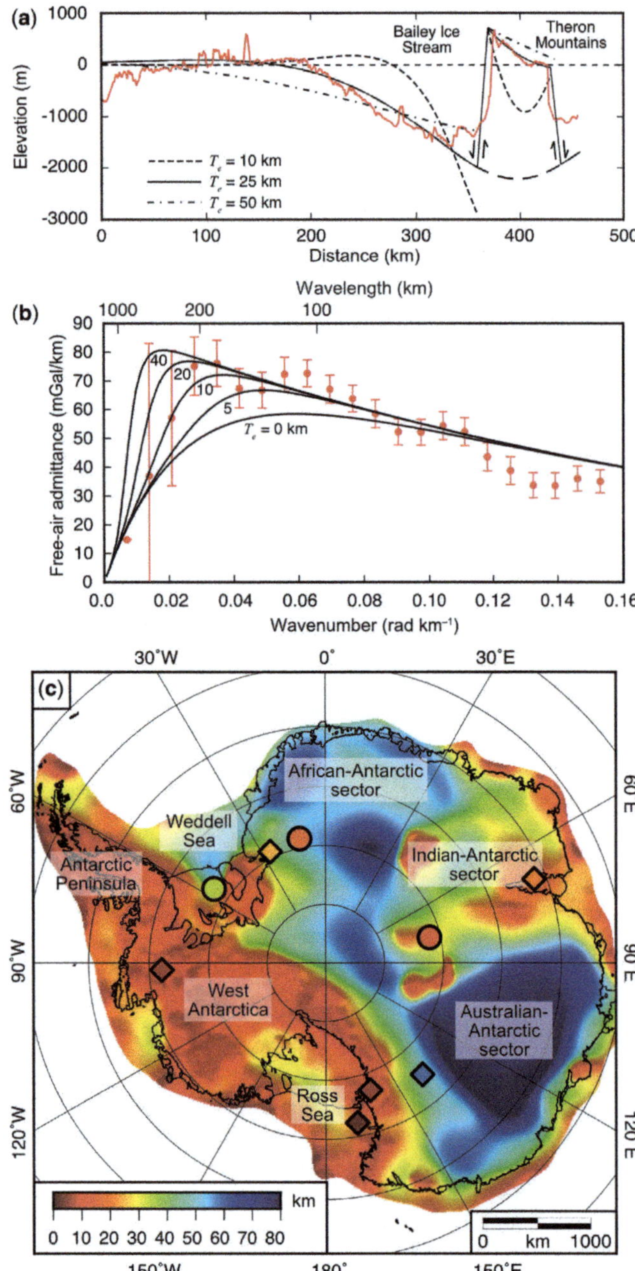

Fig. 5. Effective elastic thickness estimation in Antarctica. (**a**) Forward modelling of flexed topography along a profile crossing the Bailey Ice Stream and Theron Mountains, East Antarctica (modified from Paxman *et al.* 2017). Elastic plate models (black lines) indicate that the observed topography (red line) is best fit with a T_e value of 25 km. (**b**) Inverse (spectral) modelling of topography and gravity anomaly data for the region of East Antarctica including the profile in panel a. Comparison of the calculated free-air admittance (red circles with standard error bars) with model curves recovers a best fitting T_e of 18 km (modified from Paxman *et al.* 2017). (**c**) T_e map for Antarctica derived using a fan wavelet method to compute the Bouguer coherence (Chen *et al.* 2018). Coloured shapes denote published regional T_e estimates using forward (diamonds) and inverse (circles) modelling.

limitations mean that there is considerable uncertainty surrounding the use of inverse spectral methods to estimate T_e in Antarctica.

Fan wavelet techniques have recently been employed to determine variations in T_e across the entirety of Antarctica using the admittance and coherence methods (Ji *et al.* 2017; Chen *et al.* 2018). These methods show a strong contrast between predominantly high values in East Antarctica (T_e ~60–80 km) and low values in West Antarctica (T_e ~5–20 km) (Chen *et al.* 2018). However, the fan wavelet techniques also recover considerable variability within East Antarctica, with particularly high T_e values found in the Australian–Antarctic and Africa–Antarctic sectors, but relatively low values found in the Indian–Antarctic sector, including the Lambert Graben and Gamburtsev Subglacial Mountains (Fig. 5c) (Chen *et al.* 2018).

Spatial patterns in Antarctic T_e recovered by both forward and inverse models tend to show broad first-order agreement, albeit from a limited set of results (Fig. 5c). Significant uncertainty persists, however, as to the absolute values of T_e. The finding of a relatively weak West Antarctic lithosphere and a relatively rigid East Antarctic lithosphere may reflect the first-order differences between physical properties such as topography, crustal thickness, lithosphere thickness, lithosphere age, upper mantle viscosity, and heat flux in West and East Antarctica (An *et al.* 2015*a*, *b*). The differences in T_e *within* East Antarctica are indicative of spatial heterogeneities in the East Antarctic lithosphere, which may arise from the distinct thermotectonic histories of the different provinces (Chen *et al.* 2018). However, despite its importance in the context of long-term tectonic and topographic evolution, T_e in Antarctica remains poorly constrained.

Flexure and palaeotopography

Ice sheet loading. The most visible surface load in Antarctica today is the ice sheet itself (Fig. 6). Isostatic models indicate that complete removal of the modern ice sheet load would result in up to 1 km of bedrock uplift, and an average increase in bedrock elevation of *c.* 600 m (Paxman *et al.* 2019*b*). Any palaeotopography prior to Antarctic glaciation, or during a time at which the ice sheet was more restricted than today, was most likely characterized by higher average elevations. Under the modern ice sheet, *c.* 45% of the bed currently situated beneath grounded ice is situated below sea-level, whereas complete deglaciation would reduce this to *c.* 19% (Paxman *et al.* 2019*b*). Ice sheet loading therefore has a significant effect on the extent of marine-based ice, which is most vulnerable to potential rapid retreat (see section 1.2). For an in-depth review of glacial isostatic adjustment on decadal to multi-millennial timescales, see Nield and Barletta (this volume).

Erosional unloading. Parts of the subglacial landscape of Antarctica have also been significantly modified by glacial erosion (Taylor *et al.* 2004; Jamieson *et al.* 2005). Removal of crustal mass by erosion ('erosional unloading') induces an isostatic response from the lithosphere, causing uplift of the remaining topography (Fig. 6) (Molnar and England 1990). Selective linear erosion, whereby basal melting and erosion are concentrated beneath thick ice in glacial troughs, while neighbouring peaks remain preserved beneath non-erosive ice (Sugden and John 1976) and uplifted by the regional flexural adjustment to erosional unloading, is therefore optimal for creating substantial topographic relief.

Incision beneath major outlet glaciers dissecting mountain ranges around the Antarctic margin, such as the Beardmore and Byrd Glaciers in the Transantarctic Mountains (TAM) and the Recovery and Slessor Glaciers bounding the Shackleton Range, has produced subglacial troughs situated up to 3 km below sea-level and differences in elevation between mountain peaks and trough floors of up to *c.* 5 km over lateral distances of as little as 50 km (Fretwell *et al.* 2013; Morlighem *et al.* 2020). Elastic plate models show that the isostatic response to glacial incision can account for up to 2000 m of peak uplift in the central TAM and 1000 m in the Shackleton Range, which represents up to *c.* 50% of the total peak

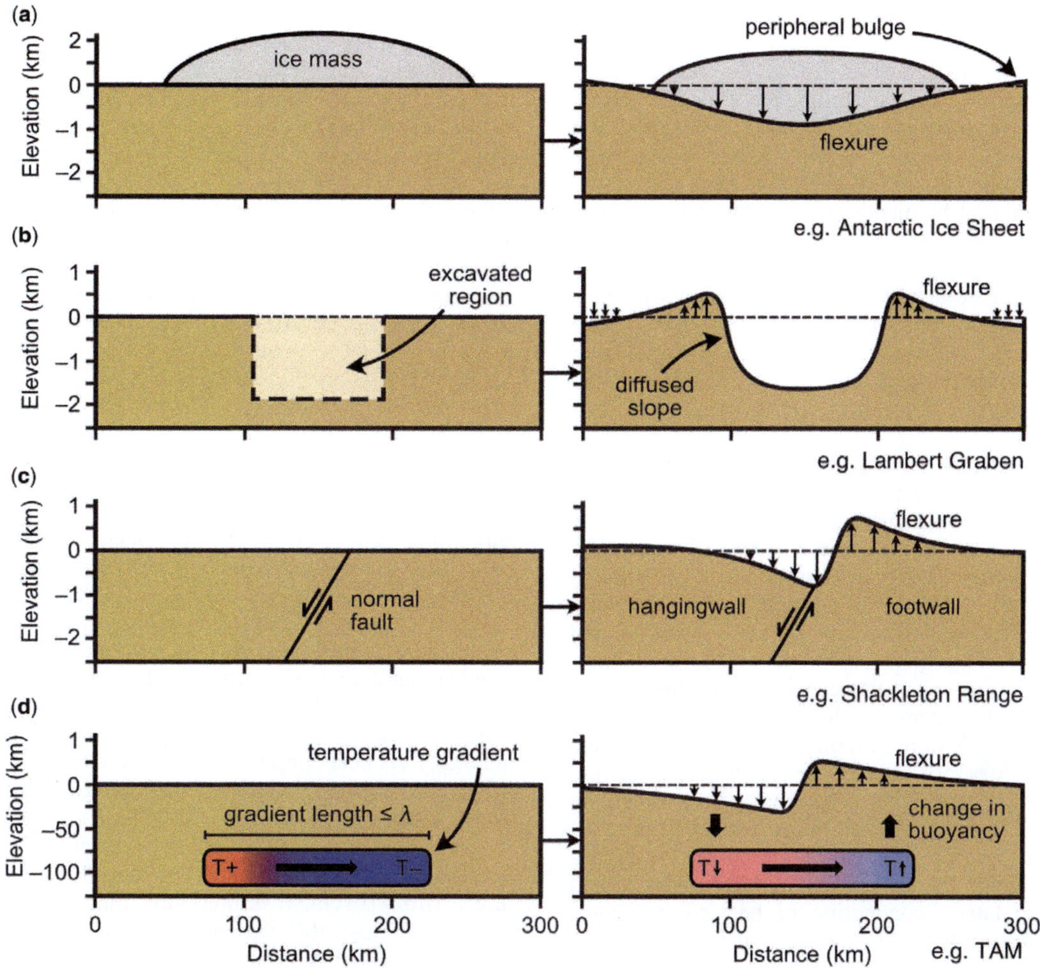

Fig. 6. Causes of lithospheric flexure in Antarctica. Diagrams are schematic, although representative horizontal and vertical scales are provided. (**a**) Ice loading, causing the land to be depressed beneath an ice mass, such as the AIS. Peripheral bulges are uplifted around the margin of the ice sheet. (**b**) Erosional unloading, caused by excavation of material by rivers or glaciers. The flexed topography exhibits uplifted flanks dipping gently away from the valley. Such topography is observed around numerous large outlet glaciers in Antarctica, such as the Lambert Glacier (White 2013). (**c**) Mechanical unloading, caused by shear along a normal fault plane. The footwall block is unloaded and flexurally uplifted, whereas in the hangingwall block surface material is replaced by mantle, causing loading and flexural subsidence (Weissel and Karner 1989). Such topography is observed along the flanks of major extensional fault systems, such as in the Shackleton Range (Paxman *et al.* 2017). (**d**) Thermal expansion (/ 'unloading'), caused by horizontal heat conduction within the lithosphere. Heat is conducted from areas of high temperature to areas of low temperature, and modifies the thermal buoyancy of the lithosphere, which induces a flexural response if the length of the thermal gradient ≤ the flexural wavelength of the lithosphere (λ). This process has been hypothesized to partially explain the uplift of the Transantarctic Mountains (TAM) (Stern and ten Brink 1989; ten Brink *et al.* 1997; Brenn *et al.* 2017).

elevations of the mountains (Stern *et al.* 2005; Paxman *et al.* 2017). In addition, flexural models, alongside geological observations such as uplifted fjordal sediments and detrital thermochronology, indicate that *c.* 3 km of incision within the Lambert Graben (Tochilin *et al.* 2012; Thomson *et al.* 2013; White 2013) has driven up to 1.4 km of post-Eocene flexural uplift along the graben flanks (Paxman *et al.* 2019*b*). In each of these examples, the high-amplitude flexural response is partly facilitated by relatively low regional T_e values (Fig. 5) and has helped to generate as much as 4 km of additional topographic relief.

Many outlet glaciers across Antarctica appear to follow the underlying tectonic structure, such as the Lambert Glacier above the Mesozoic Lambert Graben (Ferraccioli *et al.* 2011). Pre-existing fault systems and associated topographic relief were likely first exploited by pre-glacial river systems (Jamieson *et al.* 2014; Sugden and Jamieson 2018) and, in turn, by the early ice sheets, which progressively excavated the valleys to form deep subglacial troughs. Topographic steering became increasingly effective and ice flow and subglacial erosion became strongly focused within the troughs, whereas removal of mass by erosion caused the adjacent highlands to be flexurally uplifted to elevations where relatively thin, non-erosive ice predominates, preserving the mountain peaks (Kessler *et al.* 2008; Paxman *et al.* 2017). Significant topographic relief can therefore be generated in Antarctica through a combination of (a) the relative efficacy of glacial erosion (Koppes and Montgomery 2009), (b) peak preservation beneath non-erosive ice, and (c) flexural uplift induced by a relatively weak mantle lithosphere. Isostasy and lithospheric flexure can therefore drive and amplify feedbacks between tectonic structure, topography, climate and ice sheet behaviour.

Prior to glacial erosion and associated flexural uplift, the palaeotopography of Antarctica would have been characterized by considerably lower topographic relief, especially in the vicinity of the deep subglacial troughs near the coast, where fluvial valley networks could have only incised down to base level (Wilson *et al.* 2012; Thomson *et al.* 2013; Maritati *et al.* 2016; Paxman *et al.* 2017). The proportion of Antarctic topography situated below sea-level *c.* 34 Ma has been estimated at *c.* 6%, compared to *c.* 19% today (under ice-free conditions), with considerably less ice situated on deep, reverse-sloping beds (Paxman *et al.*

2019*b*). The overall effect of this change would be that earlier ice sheets were comparatively less susceptible to rapid, irreversible retreat associated with marine ice sheet instability processes. Prior to flexural uplift, the elevated terrain around the Antarctic margins would have been situated at lower average elevations at *c.* 34 Ma, which could have influenced the timing and scale of early ice sheet nucleation. By contrast, the central highlands of East Antarctica, where long-term erosion rates are low (Jamieson and Sugden 2008; Cox *et al.* 2010; Jamieson *et al.* 2010; Rose *et al.* 2013), and flexural responses to fluvial and alpine-style erosion are comparatively small (Paxman *et al.* 2016), have most likely remained relatively topographically stable over most of the past 34 myr.

Mechanical unloading and thermal expansion. Lithospheric flexure can also occur in response to mechanical unloading and thermal expansion. Mechanical unloading takes place in extensional settings (e.g. rift systems) when slip on a normal fault causes unloading of the footwall block by removal of the hanging wall; the result is flexural isostatic rebound and uplift of the footwall rift flank (Fig. 6) (Weissel and Karner 1989). Thermal expansion (sometimes termed 'thermal unloading') occurs when the density of the lithosphere is reduced owing to lateral/vertical conduction of heat, which can induce a flexural response if the length of the thermal gradient does not exceed the flexural wavelength of the lithosphere (Fig. 6). The TAM have been modelled as a flexurally-uplifted flank of the West Antarctic Rift System (WARS) arising from mechanical unloading on range-parallel normal faults bounding the WARS and thermal expansion due to lateral conduction of heat from higher-temperature mantle lithosphere beneath the WARS (see the 'West Antarctic Rift System' section) to lower-temperature mantle lithosphere beneath East Antarctica (Stern and ten Brink 1989; ten Brink *et al.* 1997; Lawrence *et al.* 2006).

However, the TAM chain exhibits considerable variation in elevation and width along its length, which cannot be easily explained by uniform flexure along the mountain front (Olivetti *et al.* 2018; Paxman *et al.* 2019*a*). This variation may in part be explained by the presence of heterogeneous thermal anomalies within the upper mantle beneath the TAM front, which may provide spatially variable buoyant thermal loads (Lawrence *et al.* 2006; Brenn *et al.* 2017). Particularly problematic however are the southern TAM, where a high plateau extends for more than 300 km into East Antarctica (Fretwell *et al.* 2013). It is difficult to envisage how edge-driven flexure at the front of the TAM alone can account for such a broadly distributed, long-wavelength pattern of uplift, even for a highly rigid lithosphere (Paxman *et al.* 2019*a*) (see also the 'Transantarctic Mountains' section).

The significance of the location of the TAM along the tectonic boundary between West and East Antarctica has been noted by a number of authors (ten Brink *et al.* 1997; van Wijk *et al.* 2008; Brenn *et al.* 2017; Ebbing *et al.* 2018). This boundary likely had a bearing, at least in part, on the uplift of the TAM. Numerical models of lithospheric extension indicate that the inherited tectonic boundary between West Antarctica and cratonic East Antarctica focused extensional deformation at the craton edge, resulting in a strongly asymmetrical pattern of TAM uplift (van Wijk *et al.* 2008). Although the mechanism for TAM uplift remains debated (see the 'Transantarctic Mountains' section), the presence of an inherited tectonic boundary and the contrasting lithospheric properties of East and West Antarctica have likely played a significant role in the tectonic and topographic evolution of the TAM.

Inherited fault systems may also have played an important role in the topographic uplift of highlands such as the Shackleton Range and Gamburtsev Subglacial Mountains, where modelling studies indicate that mechanical unloading along border faults may have contributed up to 1 km of flexural uplift (Ferraccioli *et al.* 2011; Paxman *et al.* 2017). In the case of the Gamburtsevs, fault activity may be related to rifting within the Mesozoic East Antarctic Rift System (Ferraccioli *et al.* 2011), while fault activity in the Shackleton Range region may have most recently occurred in association with Weddell Sea rifting during the Cretaceous (Krohne *et al.* 2016; Jordan *et al.* 2017).

Thermal subsidence

Continental extension and rifting are significant drivers of surface topographic subsidence. During the 'syn-rift' phase, stretching and faulting are associated with thinning of the crust and lithosphere, which in turn leads to surface subsidence in order to retain isostatic equilibrium. The extent of crustal thinning is quantified by the stretching factor (β), which is defined as the ratio of original crustal thickness to final crustal thickness. Rifting is also associated with elevated lithospheric heat flux; in the 'post-rift' phase, following the cessation of extensional faulting, the decay of the thermal anomaly via heat conduction causes cooling and densification of the lithosphere, which in turn drives subsidence of the Earth's surface (McKenzie 1978). Since lithospheric cooling is a transient process, thermal subsidence occurs over a protracted period of tens to hundreds of millions of years following a rifting event. It is therefore important to consider the effects of 'post-rift' thermal subsidence in the evolution of Antarctic palaeotopography since *c.* 34 Ma.

West Antarctic Rift System

The West Antarctic Rift System (WARS) is believed to extend from the Ross Sea embayment through West Antarctica to the Bellingshausen Sea (Fig. 7) (LeMasurier *et al.* 1990; LeMasurier and Landis 1996; Winberry and Anandakrishnan 2004; Jordan *et al.* 2010; Bingham *et al.* 2012). The presence of an extensive rift system beneath the West Antarctic Ice Sheet (WAIS) is highly significant, because the low-lying topography (in places as deep as 2.5 km below sea-level (Fig. 7)) is hypothesized to render the WAIS vulnerable to rapid retreat due to marine ice sheet instability processes (Mercer 1978; Thomas 1979; Schoof 2007; Joughin and Alley 2011). Moreover, the bedrock geology and crustal structure of the rift system may influence the on-going and future dynamics of the overlying glaciers (Studinger *et al.* 2001; Rignot *et al.* 2008; Bingham *et al.* 2012), which in turn affect the stability of the wider ice sheet.

Rifting in West Antarctica is thought to have commenced at *c.* 80–100 Ma (Siddoway *et al.* 2004), with the WARS subsequently experiencing 300–600 km of extension during the late Cretaceous and early Cenozoic (Behrendt *et al.* 1991; Cande *et al.* 2000; Wilson and Luyendyk 2009). Consequences of this extension were the separation of Marie Byrd Land from East Antarctica and the opening of the Ross Sea basin (Decesari *et al.* 2007). Relative motion between East and West Antarctica was historically believed to have ended abruptly in the late Oligocene on the basis of marine magnetic anomalies from the northern Ross Sea (Cande *et al.* 2000). However, more recently acquired magnetic anomaly data indicate that motion between East and West Antarctica may have persisted up to the mid-Miocene (Granot and Dyment 2018). Seismic and gravity

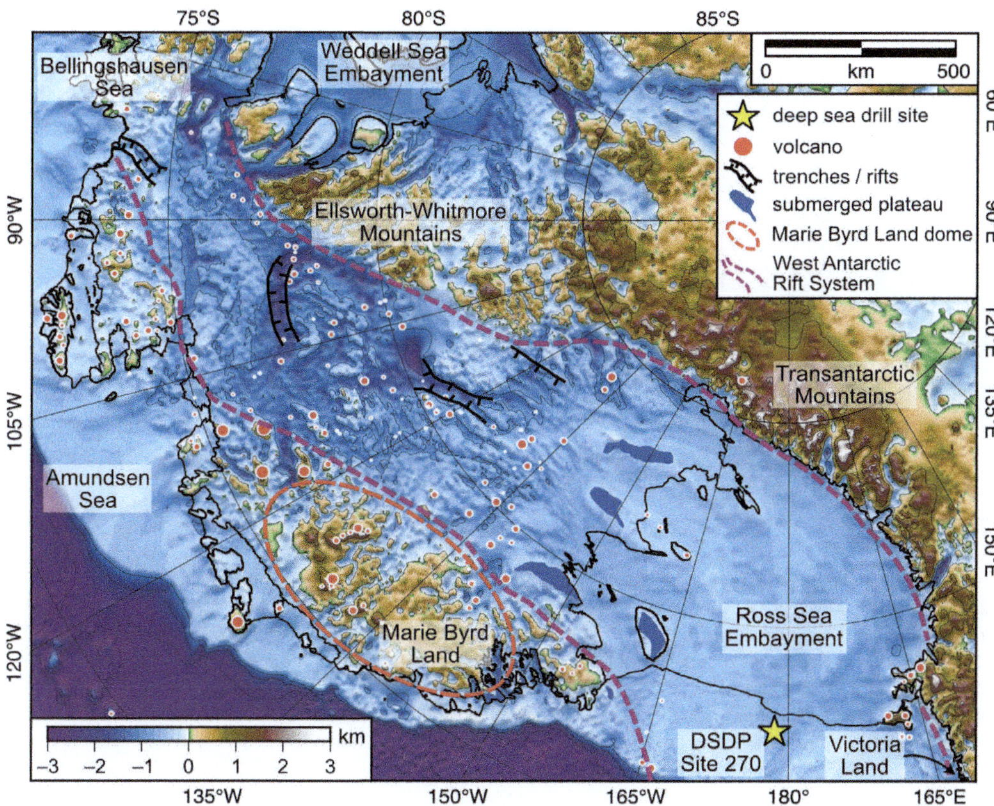

Fig. 7. West Antarctic Rift System. Present-day bed elevations are shown relative to mean sea-level (Paxman *et al.* 2019*b*). Contour interval is 1 km. Dashed magenta lines show the extent of the West Antarctic Rift System. An explanation of the symbology is provided in the figure legend. Red circles denoting volcanic edifices are proportional (4x true scale) to the basal diameter of the cone (van Wyk de Vries *et al.* 2018). Subglacial trenches and rift valleys (black outlines), as well as submerged flat-topped bedrock platforms (blue polygons) are imaged by radar and gravity/magnetic anomaly data (Bell *et al.* 1998; Studinger *et al.* 2001; Wilson and Luyendyk 2006; Bingham *et al.* 2012).

data have recovered crustal thicknesses as low as 21 km beneath the WARS (Winberry and Anandakrishnan 2004; Jordan *et al.* 2010; An *et al.* 2015*a*), indicating that the region has undergone significant crustal extension and thinning.

Rift-related magmatism appears to have commenced at *c.* 48 Ma in the form of large alkaline plutons exposed in northern Victoria Land (Rocchi *et al.* 2002), where airborne magnetic data suggest similar intrusions may be extensive (Ferraccioli *et al.* 2009). By contrast, Cenozoic intrusive magmatism in Marie Byrd Land began at *c.* 34 Ma (Rocchi *et al.* 2006). Effusive magmatism, in the form of Cenozoic volcanoes, has been documented in Victoria Land, Marie Byrd Land and across the WARS (LeMasurier *et al.* 1990). Dating of edifices and associated deposits indicates that the earliest volcanism commenced at *c.* 30 Ma, and a significant fraction of volcanic deposits date from after the mid-Miocene (*c.* 14 Ma) (Stump *et al.* 1980; LeMasurier *et al.* 1990; LeMasurier and Landis 1996; Rocchi *et al.* 2006; Shen *et al.* 2017). In addition to exposed magmatic rocks, aeromagnetic anomalies indicate that subglacial flood basalts may be preserved within the WARS (Behrendt *et al.* 1994). A recent inventory also indicates that there may be in excess of 138 subglacial volcanic cones across the WARS (van Wyk de Vries *et al.* 2018) (Fig. 7), supporting the presence of an extensive and diffuse alkaline magmatic province (Finn *et al.* 2005).

An implication of the thermotectonic history of the WARS is that the palaeotopography of West Antarctica was considerably higher prior to and during late Cretaceous and Cenozoic crustal extension and heating than at the present-day. This interpretation is supported by the presence of flat bedrock platforms of proposed Miocene age currently situated below sea-level in the Ross Sea Embayment (Fig. 7) (Wilson and Luyendyk 2006). These plateaus are interpreted as having formed by wave erosion at sea-level, yet are currently situated at water depths ranging from *c.* 150 to *c.* 350 m below sea-level, deepening southwards towards the TAM (Wilson and Luyendyk 2006). These observations imply regional subsidence of the Ross Sea Embayment since the Neogene, with a combination of younger and/or higher magnitude extension towards the TAM. In addition, coring at Deep Sea Drilling Project (DSDP) Site 270 in the Ross Sea (Fig. 7) recovered a terrestrial palaeosol several metres deep developed on a sedimentary breccia above basement gneiss at *c.* 1000 m below sea-level (Ford and Barrett 1975). The palaeosol is overlain by quartz sand and glauconitic greensand, which are dated at *c.* 26 ± 1.5 Ma and interpreted as having been deposited in a near-shore or shallow marine environment (Barrett 1975; De Santis *et al.* 1999; Kulhanek *et al.* 2019). This stratigraphy is consistent with a marine transgression and implies a significant amount of regional subsidence since the early Oligocene.

The first attempt to produce a model of post-Eocene thermal subsidence in the WARS assumed that the rift system comprises five non-overlapping regions, each representing a different phase of the regional extension history (Wilson and Luyendyk 2009). Each phase was assumed to have a uniform stretching factor and an instantaneous age of extension, permitting the use of a simple 1D cooling model (McKenzie 1978) to predict the amount of post-34 Ma thermal subsidence across the WARS. The amount of thermal subsidence (S) as a function of time (t) is given by (McKenzie 1978)

$$S(t) = \frac{4a\rho_m \alpha T_m}{\pi^2(\rho_m - \rho_i)} \frac{\beta}{\pi} \sin\left(\frac{\pi}{\beta}\right)(1 - e^{-t/\tau}) \tag{3}$$

with

$$\tau = \frac{a^2}{\pi^2 \kappa} \tag{4}$$

where a is the thickness of the lithosphere, α is the coefficient of thermal expansion, T_m is the temperature at the base of the lithosphere, κ is the thermal diffusivity of the lithosphere, ρ_m is the density of the lithospheric mantle, ρ_i is the density of the material infilling the surface subsidence, β is the stretching factor, and τ is thermal decay time constant.

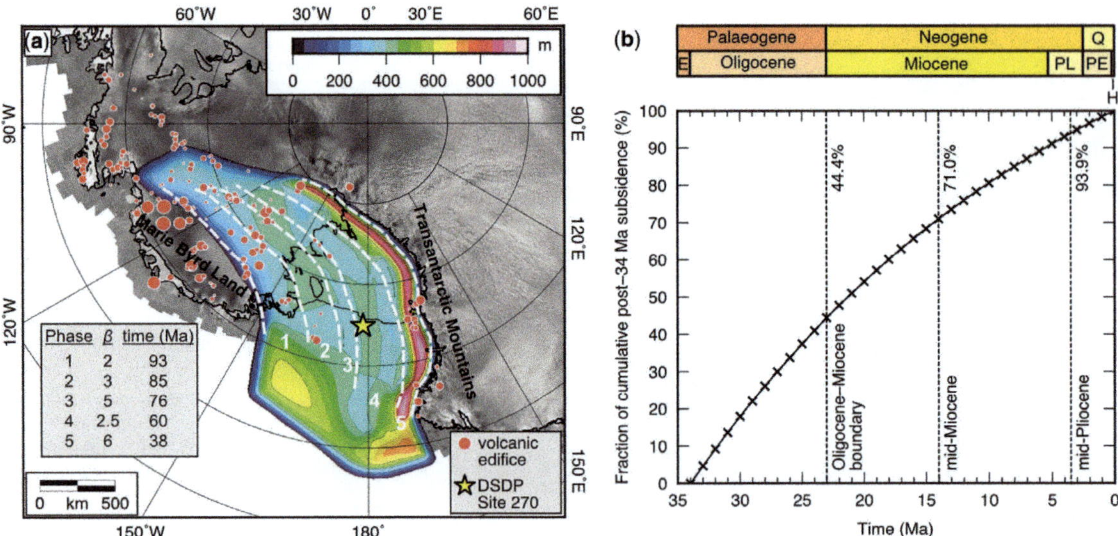

Fig. 8. Thermal subsidence in West Antarctica (modified from Paxman *et al.* 2019*b*). (**a**) Magnitude of post-34 Ma thermal subsidence associated with the West Antarctic Rift System (Wilson *et al.* 2012; Wilson and Luyendyk 2009). Dashed white lines mark modelled areas of uniform stretching factor (β) corresponding to five phases of extension in the WARS (Wilson and Luyendyk 2009). The inset table lists the β value and mean age of extension associated with each phase. Filled red circles denote volcanic edifices; circle size is proportional (4x true scale) to the basal diameter of the cone (van Wyk de Vries *et al.* 2018). Yellow star marks the location of Deep Sea Drilling Project (DSDP) Site 270 (Ford and Barrett 1975). (**b**) Thermal subsidence history since 34 Ma based on a simple 1D cooling model (McKenzie 1978). Cumulative subsidence is measured as a percentage of the total post-34 Ma subsidence; the magnitude of subsidence is spatially variable (see panel a). A simplified geological timescale is provided for reference. Abbreviations: E = Eocene; PL = Pliocene; PE = Pleistocene; H = Holocene; Q = Quaternary.

For their five phases of extension in the WARS, Wilson and Luyendyk (2009) used β factors from 2 to 6 and mean extension ages of 93 to 38 Ma (moving SW from Marie Byrd Land towards the TAM; Fig. 8). This model, with the locus of extension progressively shifting towards the TAM, predicts *c.* 500 m of post-34 Ma subsidence in the central WARS (Fig. 8) (Wilson and Luyendyk 2009). The above rifting parameters were constrained by the necessity to restore the terrestrial palaeosol at DSDP Site 270 to sea-level at *c.* 34 Ma, which required that extension in the central Ross Sea is younger than 70 Ma with a stretching factor of >2.0 (Wilson and Luyendyk 2009; Wilson *et al.* 2012). By correcting for thermal subsidence (following Wilson and Luyendyk 2009), erosion, flooding of the subsided area with water, and associated flexural isostatic responses, the reconstruction process of Wilson *et al.* (2012) produced a substantial upland feature in West Antarctica, with elevations of 500–1000 m above sea-level. The stretching factors also implied a pre-extension crustal thickness of *c.* 50 km in the WARS (Wilson and Luyendyk 2009). These results could be interpreted as lending to support to the suggestion that a significant highland plateau and thickened crust existed in West Antarctica prior to extension, of which the Transantarctic Mountains are a remnant (Bialas *et al.* 2007; Huerta 2007). Terrestrial and marine microfossil assemblages obtained via sub-ice drilling near the Siple Coast are indicative of low-elevation coastal plain and shallow marine palaeoenvironments during the Eocene (Coenen *et al.* 2019). This finding contradicts the presence of a uniformly high Siple Coast region during the Eocene (Wilson *et al.* 2012); an alternative scenario is that the palaeotopography was characterized by greater relief, with low-lying areas in the regions now occupied by ice streams and higher terrain in between.

A more recent reconstruction of Antarctic palaeotopography at the Eocene–Oligocene boundary topography (Paxman *et al.* 2019*b*) produced palaeo-elevations for the WARS up to 500 m lower than that of Wilson *et al.* (2012). This difference arises from a combination of updated observed offshore sediment volumes and modelled onshore erosion patterns, as well as updated constraints on the modern regional bedrock topography and T_e. Use of a simple 1D exponential decay model (McKenzie 1978) to compute the temporal history of thermal subsidence since *c.* 34 Ma (Equation 3; Fig. 8), predicted that DSDP Site 270 subsided below sea-level at *c.* 28 Ma (Paxman *et al.* 2019*b*), in good agreement with the recorded drill core stratigraphy (Leckie and Webb 1983; Kulhanek *et al.* 2019).

Since many subglacial volcanoes across the WARS may have been emplaced since *c.* 34 Ma, these edifices were removed from the recently updated *c.* 34 Ma palaeotopography (Paxman *et al.* 2019*b*). The approximate geometry of each volcanic edifice was determined using a recent inventory of the distribution, height, and diameter of subglacial volcanic cones across West Antarctica (Fig. 8) (van Wyk de Vries *et al.* 2018), and these cones were then subtracted from the palaeotopography (Paxman *et al.* 2019*b*). Since the average diameter of these volcanic cones is 21 km, which is smaller than the flexural length scale of the lithosphere, no flexural isostatic adjustment is required when they are removed from the topography. Moreover, any influence of the edifices (or lack thereof) on past ice sheet dynamics would be highly localized, and largely insignificant within continental-scale ice sheet models with spatial resolutions of >10 km.

Passive continental margins of East Antarctica

The transient effects of thermal subsidence are not solely pertinent to West Antarctica. During the early Mesozoic, East Antarctica was part of the supercontinent of Gondwana, and was situated in the centre of landmasses that constitute modern-day South America, Africa, Madagascar, India and Australia. Gondwana break-up began in the Jurassic *c.* 180–160 myr ago (Dalziel 1997) with the separation and movement northwards of South America and Africa away from East Antarctica. Break-up then proceeded clockwise around East Antarctica with the successive separation of India and then Australia (Boger 2011 and references therein). East Antarctica is now bounded by passive margins (Sugden and Jamieson 2018), which are separated from their conjugates in southern Africa, eastern India, and southern Australia by oceanic spreading ridges. These continents each contain Jurassic basalts and dolerite sills formed during volcanic activity associated with the break-up of Gondwana. In Antarctica, these rocks form the Ferrar Large Igneous Province (Elliot and Fleming 2004).

The East Antarctic passive margins would have experienced post-rift thermal subsidence, causing the continental shelves to subside (Hochmuth *et al.* 2020) and providing additional accommodation space for sediments being delivered to the continental margin via erosion and transport by ice or rivers (Jamieson *et al.* 2005, 2008). The break-up age of each sector of the Antarctic margin is well constrained from the age of the oldest sea floor (Müller *et al.* 2016), although the stretching factor and initial timing of continental rifting are more difficult to evaluate. A recent attempt to restore Antarctic margin palaeotopography (since *c.* 34 Ma) for the effects of residual post-rift cooling (Paxman *et al.* 2019*b*) adopted a 1D cooling model (McKenzie 1978) (equation 3) using the observed age of the sea floor (Müller *et al.* 2016) and a uniform β value of 4 across the widths of the continent–ocean transitions around Antarctica (Gohl 2008). The modelled magnitude of post-Eocene thermal subsidence around the margins of East Antarctica varies from *c.* 300 m for the earliest-formed continental shelves (e.g. offshore Dronning Maud Land), where separation occurred in the Jurassic, to up to *c.* 1200 m for the shelves that experienced separation as recently as the late Cretaceous (e.g. offshore George V Land) (Paxman *et al.* 2019*b*). However, no firm geological constraints that provide a record of post-Eocene thermal subsidence around the margin of East Antarctica have yet been documented.

Dynamic topography

Dynamic topography is a collective expression for changes in Earth's surface elevation caused by vertical tractions exerted on the base of the lithosphere by the convecting sub-lithospheric (viscous) mantle (Pekeris 1935; Hager *et al.* 1985; Flament *et al.* 2013; Molnar *et al.* 2015). Convective flow within the mantle is driven by the buoyancy effects of density anomalies, which are in turn caused by variations in temperature and/or composition. When this viscous flow interacts with the surface boundary, viscous normal stresses are generated, causing vertical deflection of the surface. Upward flow towards the surface leads to positive dynamic topography (surface uplift), whereas downward flow away from the surface leads to negative dynamic topography (surface subsidence) (Braun 2010).

Surface manifestations of dynamic topography can range from distributed long-wavelength continental tilting (Sandiford *et al.* 2009; Quigley *et al.* 2010; Richards *et al.* 2016), to more focused surface uplift above the impingement of a mantle plume on the base of the lithosphere (Behrendt *et al.* 1991) or subsidence above a subducting lithospheric slab (Whittaker *et al.* 2010). Ocean floor residual depth measurements from the Southern Ocean reveal dynamic topography amplitudes of up to 1 km and wavelengths of 10^3–10^4 km, with spatial variability in magnitude and direction (Hoggard *et al.* 2016, 2017). The temporal pattern of dynamic uplift/subsidence is typically derived from marine stratigraphy, with evidence for typical peak rates on the order of 100 m Ma^{-1} (Japsen *et al.* 2006; Al-Hajri *et al.* 2009; Czarnota *et al.* 2013; Richards *et al.* 2016). This implies that long-wavelength changes in bedrock elevation in Antarctica over multi-million year timescales may be attributed, at least in part, to changes in dynamic topography.

Marie Byrd Land

Marie Byrd Land is defined by a 1000 × 500 km dome-shaped volcanic province (Fig. 7) comprising a series of large shield volcanoes and an extensive sequence of alkali basalts up to *c.* 28–30 Ma in age (LeMasurier *et al.* 1990; Rocchi *et al.* 2006). The geochemistry of basalts in Marie Byrd Land and the WARS is comparable to those found on oceanic islands associated with well-established hot spot tracks (e.g. Hawaii) (LeMasurier and Rex 1989; Hole and LeMasurier 1994), supporting the suggestion of a mantle plume beneath the WARS (Behrendt *et al.* 1991) (see also Bredow and Steinberger 2021). In addition, several low-relief surfaces at scattered localities across Marie Byrd Land with elevations up to 2.7 km have been interpreted as remnants of a continuous, low-relief surface (LeMasurier and Landis 1996). It has been proposed that this erosion surface formed close to base level, and has been uplifted by up to *c.* 2.7 km since *c.* 28–30 Ma, accompanied by block faulting and alkali basalt volcanism, all of which may signal the impingement of a mantle plume on the base of the West Antarctic lithosphere (LeMasurier and Landis 1996).

However, the suggestion of a mantle plume beneath the WARS is controversial; the volume of magmatism imaged within the WARS by geophysical data implies time-averaged magma production rates much lower than those expected for a mantle plume-dominated setting, especially one with relatively thin lithosphere (Rocchi *et al.* 2002). Recent attempts to resolve this controversy have centred on the development of new seismic tomography models for the upper mantle beneath West Antarctica (An *et al.* 2015*b*; Lloyd *et al.* 2015; Shen *et al.* 2018; Bredow and Steinberger 2021; Wiens *et al.* this 2021). An approximately circular P- and S-wave low-velocity anomaly in the upper mantle would be indicative of a warm thermal anomaly, potentially related to a mantle plume. Such an anomaly is not clearly visible beneath central West Antarctica (An *et al.* 2015*a*), but may be resolved beneath Marie Byrd Land, although its depth extent is poorly constrained (Fig. 9) (Lloyd *et al.* 2015, 2020; Shen *et al.* 2018). Seismic wave velocities and receiver functions show that the crust beneath the high topography in Marie Byrd Land is up to 10 km thicker than beneath the WARS, indicating that this high topography is at least partially supported isostatically rather than entirely dynamically by a mantle plume (Shen *et al.* 2018). However, satellite gravity gradients show a bowl-shaped structure in Marie Byrd Land, possibly originating from a mantle plume source (Ebbing *et al.* 2018; Pappa and Ebbing 2021).

It is therefore possible, but not conclusively demonstrated, that the long-wavelength, high-elevation domal topography of Marie Byrd Land, anomalous when compared to the adjacent WARS, is in part supported by a hotspot-related thermal anomaly within the upper mantle. If this is the case, understanding the plume-induced topographic evolution of Marie Byrd Land is important for attempts to reconstruct regional palaeotopography. The beginning of Marie Byrd Land uplift was initially assumed to have been contemporaneous with early Oligocene magmatism (*c.* 28–30 Ma) (LeMasurier and Landis 1996), although older volcanic units may have been emplaced but subsequently eroded or not yet sampled. It is possible that dynamic uplift commenced earlier, in response to mantle upwelling during the Cretaceous and early Cenozoic after cessation of subduction following Gondwana break-up (Sutherland *et al.* 2010). Conversely, recent apatite fission track and (U–Th–Sm)/He thermochronology data have been used to tentatively suggest that denudation (and associated uplift) in Marie Byrd Land commenced in the early Miocene (*c.* 20 Ma) (Spiegel *et al.* 2016). However, most low-temperature cooling ages from Marie Byrd Land date to 100–60 Ma, and the 20 Ma age group is based on only two

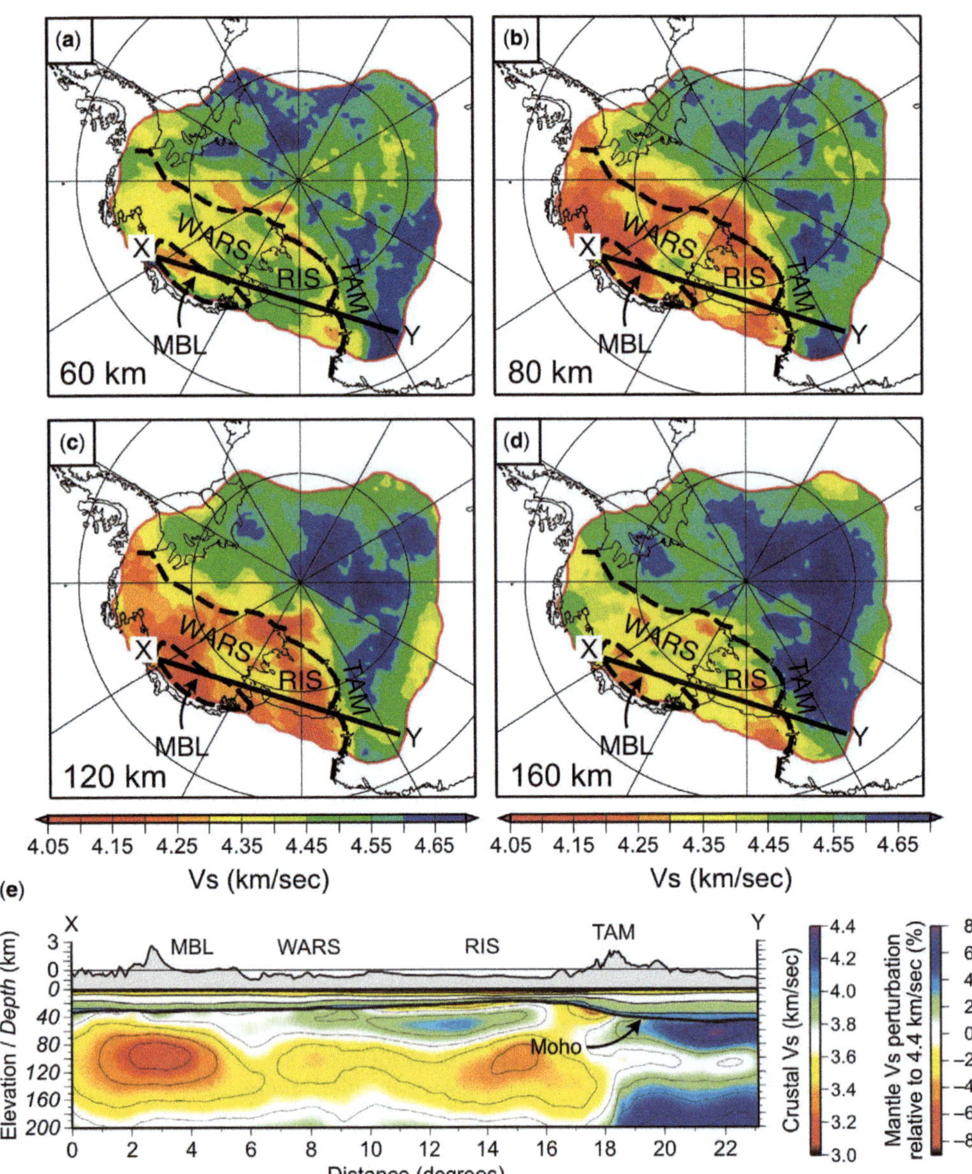

Fig. 9. S-wave velocity (Vs) model for central and West Antarctica (modified from Shen *et al.* 2018). (**a**)–(**d**) Map views of Vs at four depth slices in the tomography model (depths are relative to a reference datum in the upper crust). Dashed lines mark the approximate limits of Marie Byrd Land and the West Antarctic Rift System. (**e**) Vertical slice of the 3D tomography model of the crust and upper mantle along transect X–Y (location marked on above panels). The Vs of the crust is plotted as absolute values, whereas the Vs of the mantle is shown as percentage perturbation relative to 4.4 km/sec; the Moho is marked by the solid line in the pofile. MBL = Marie Byrd Land; RIS = Ross Ice Shelf; TAM = Transantarctic Mountains; WARS = West Antarctic Rift System.

rock samples from one location *c.* 250 km from the centre of the dome (Spiegel *et al.* 2016). Consequently, the questions of whether Marie Byrd Land was uplifted due to mantle plume activity, and the magnitude and timing of any dynamic uplift, remain matters of considerable uncertainty.

Large-scale continental glaciation in coastal West Antarctica would require areas of land situated at significant elevations above sea-level. Marie Byrd Land is therefore a potential candidate for ice sheet inception in West Antarctica. Ice sheet model simulations performed on the palaeotopographical reconstructions of Wilson *et al.* (2012) suggest that the high palaeo-elevations in Marie Byrd Land and central West Antarctica could have permitted the establishment of a significant WAIS at *c.* 34 Ma (Wilson *et al.* 2013). Significant ice still accumulates in the minimum *c.* 34 Ma palaeotopography of Wilson *et al.* (2012), which includes a correction for 1 km of post-34 Ma dynamic uplift in Marie Byrd Land. However, if uplift of up to 2.7 km (LeMasurier and Landis 1996) commenced at *c.* 20 Ma (Spiegel *et al.* 2016), Marie Byrd Land would have been low-lying at *c.* 34 Ma, and regional ice sheet inception may have occurred during the early Miocene at the earliest. This is turn presents a dilemma as to whether/where ice accumulated in West Antarctica during the early Oligocene, which has implications for global ice volumes, sea-level and ocean temperature at this time (Wilson *et al.* 2013). Since the magnitude of post-Eocene dynamic uplift across Marie Byrd Land may have been anywhere between 0 and 2.7 km, a high degree of uncertainty persists in our understanding of the topographic evolution and glacial history of Marie Byrd Land.

This case study highlights the potential importance of sub-plate processes such as the vertical tractions imposed on the base of the lithosphere by buoyant mantle plumes for the evolution of Antarctic palaeotopography, and in turn for Antarctic glacial history. As well as influencing past ice sheet dynamics via changes in topography, the presence of a mantle plume beneath Marie Byrd Land may also elevate the local geothermal heat flux to values of up to 150 mW m^{-2} (Seroussi *et al.* 2017) and, in turn, increase the rate of basal melting beneath the ice sheet, with consequences for ice sheet hydrology (Seroussi *et al.* 2017).

Transantarctic Mountains

As outlined in the 'Flexure and palaeotopography' section, the origin of the Transantarctic Mountains (TAM) remains a major open question in the history of Antarctica's subglacial topography. Some studies have proposed that the TAM are isostatically supported by a thickened crust, with gravity modelling implying crustal thicknesses *c.* 5–10 km greater than those beneath the Wilkes Subglacial Basin in the immediate hinterland (Studinger *et al.* 2004; Jordan *et al.* 2013). However, although recent seismic studies find a modest (*c.* 5 km) crustal root beneath the TAM, it does not appear to fully support the observed topography (Lawrence *et al.* 2006; Chaput *et al.* 2014; An *et al.* 2015*a*; Hansen *et al.* 2016; Ramirez *et al.* 2016). The absence of a fully compensating crustal root is more consistent with a flexural origin for the TAM. However, as discussed in the 'Mechanical unloading and thermal expansion' section, a problematic observation for the flexure model is that the southern TAM are particularly broad, with a high plateau extending for more than 300 km into East Antarctica (Fretwell *et al.* 2013), which is inconsistent with solely edge-driven flexure at the front of the TAM.

Recently constructed seismic tomographic images of the uppermost mantle beneath the southern TAM reveal a potential mantle source of the observed broad topographic plateau (Shen *et al.* 2017; Wiens *et al.* 2021). Seismically slow uppermost mantle beneath the southern TAM (Fig. 9) is indicative of a hot temperature anomaly capable of reducing the upper mantle density by *c.* 1–1.5%, which would provide sufficient buoyancy to support *c.* 1.5–2.0 km of surface elevation (Shen *et al.* 2017). Tomography images also reveal a fast seismic anomaly beneath the slow uppermost mantle (Fig. 9), which is interpreted as foundering lithosphere that is being replaced at shallow depths by a warmer, wedge-shaped body of asthenosphere (Shen *et al.* 2017, 2018). Cretaceous–Cenozoic rifting within the WARS may have caused thermal destabilization and ultimately delamination of pre-existing cold lithospheric mantle, thereby triggering lithospheric foundering (Shen *et al.* 2017). Better understanding of the timing of lithospheric foundering and buoyant uplift of the southern TAM plateau is required for more accurate reconstructions of the regional palaeotopography.

Wilkes Subglacial Basin

The Wilkes Subglacial Basin (WSB) is situated in the immediate hinterland of the TAM (Fig. 2). As a low-lying marine basin containing an ice volume equivalent to *c.* 5 m of global sea-level rise (Fretwell *et al.* 2013; Gasson *et al.* 2015), an understanding of ice sheet behaviour within the WSB during past warmer climates is of particular importance for predicting future ice sheet change in this sector of the East Antarctic Ice Sheet (EAIS). However, significant variation remains between numerical ice sheet model predictions of EAIS retreat within the WSB during past warm periods such as the mid-Pliocene (*c.* 3 Ma) (Pollard and DeConto 2009; Mengel and Levermann 2014; DeConto and Pollard 2016), at which time atmospheric CO_2 levels were similar to that of today (*c.* 400 ppm). Improved understanding of past ice sheet dynamics within the WSB relies in part on understanding the evolution of the subglacial topography of the basin (Paxman *et al.* 2018).

Austermann *et al.* (2015) used the 'ASPECT' mantle convection code (Bredow and Steinberger 2021) to model changes in dynamic topography in the WSB since the mid-Pliocene (Fig. 10). The present-day dynamic topography field was computed from the modern density and viscosity

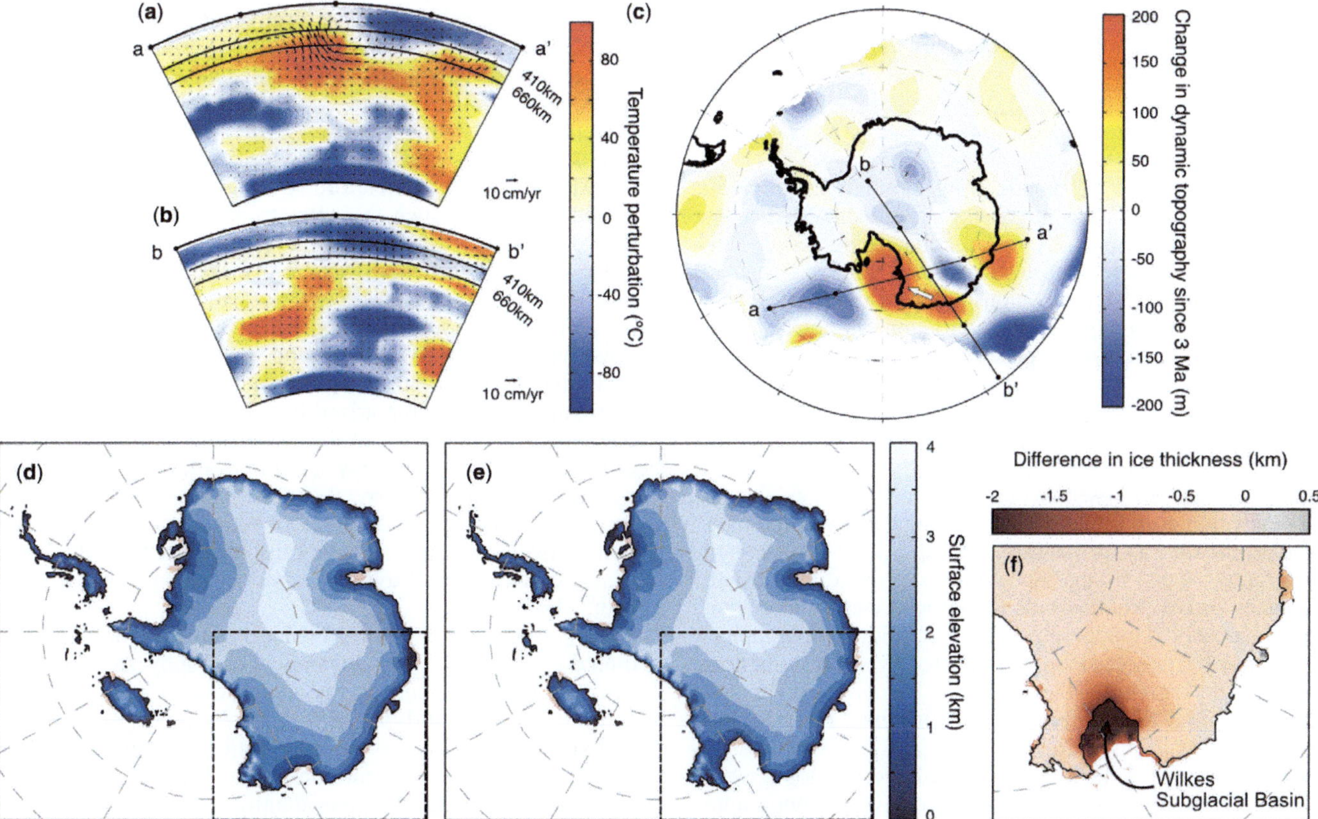

Fig. 10. Mantle flow, dynamic topography, and ice sheet modelling in the Wilkes Subglacial Basin (modified from Austermann *et al.* 2015). (**a**) and (**b**) Present-day mantle temperature fields (perturbations relative to the global average) and computed mantle flow vectors for transects a–a' and b–b' crossing the Wilkes Subglacial Basin (profile locations indicated in panel c). (**c**) Change in dynamic topography since 3 Ma caused by the motion of the Antarctic plate through the mantle flow field. The plate rotation vector in the Wilkes Subglacial Basin is marked by white arrow, and has a magnitude of 1.6 cm/yr. (**d**) and (**e**) AIS surface elevation computed using simulations for mid-Pliocene climates which (**d**) do not include, and (**e**) do include, changes in dynamic topography since 3 Ma (panel c). (**f**) Difference in ice sheet thickness in the Wilkes Subglacial Basin (panel extent marked by dashed boxes in panels d and e) between the simulations shown in panels d and e.

structure of the mantle (as constrained from seismic tomography and mineral physics). Changes in local dynamic topography were assumed to arise entirely from the motion of the Antarctic plate through this dynamic topography field (Austermann *et al.* 2015). This combined mantle flow and plate motion modelling approach predicted that the bed of the northern WSB (close to the present-day ice margin) was *c.* 100–200 m lower during the mid-Pliocene than today (Fig. 10) (Austermann *et al.* 2015). Incorporating this change in bed elevation into an ice sheet model (Pollard and DeConto 2009) resulted in an additional 200–560 km of grounding line retreat into the WSB under a mid-Pliocene climate (Fig. 10) (Austermann *et al.* 2015). The implication is that a relatively modest amount of dynamic uplift of the WSB since 3 Ma has acted to significantly increase the stability of this sector of the EAIS.

Austermann *et al.* (2015) also computed the change in dynamic topography across the entirety of Antarctica since 3 Ma (Fig. 10) and found that the amplitude of dynamic topography change in other marine-based sectors of the EAIS, such as the Aurora or Recovery subglacial basins, was relatively small by comparison to the WSB, with little impact on ice sheet behaviour. This study calculates the mantle flow field from the present-day density and viscosity structure of the mantle. Constraining these parameters – and the mantle flow field – in the geological past is particularly challenging, especially in Antarctica, and results in large uncertainties in predictions of changes in dynamic topography over multi-million-year timescales. Moreover, numerical mantle convection models tend to over-predict the amplitudes of dynamic topography measured from (e.g.) residual depth anomalies in the oceanic realm (Hoggard *et al.* 2016), indicating that mantle flow models alone cannot yet be relied upon to reliably correct for dynamic topography changes.

Antarctic-wide dynamic topography

Aside from the use of mantle flow simulations, separate attempts have been made to constrain continental-scale patterns of dynamic topography in Antarctica using geophysical datasets. For example, the seismic crustal thickness can be used to calculate surface elevations that would result if topography were entirely supported by variations in crustal thickness (Airy isostasy; 'Elastic plate flexure' section).

$$h_{\text{isostatic}} = \frac{(\rho_{\text{m}} - \rho_{\text{c}})}{\rho_{\text{m}}}(T_{\text{c}} - T_0) \tag{5}$$

where T_c is the observed crustal thickness, T_0 is the zero-elevation crustal thickness, ρ_m is the density of the mantle, ρ_c is the density of the crust, and $h_{\text{isostatic}}$ is the 'isostatic topography' – that is, the topography supported by isostatic buoyancy forces. The isostatic topography can be subtracted from the observed topography to give a 'residual topography', which may be, to some extent, supported by vertical tractions applied to the base of the lithosphere by the convecting mantle (Menard 1973; Cazenave *et al.* 1989; Forte *et al.* 1993; Molnar *et al.* 2015).

Comparison of Antarctica's observed bedrock topography (Fretwell *et al.* 2013) with the computed isostatic topography indicates that local crustal thickening beneath the highlands of the Gamburtsev Subglacial Mountains and Dronning Maud Land can support their differential elevation above the broader East Antarctic plateau (Fig. 11) (O'Donnell and Nyblade 2014). However, the modal elevations of the plateau itself are *c.* 700 m higher than can be accounted for based on its crustal thickness relative to that of the rest of East Antarctica (Fig. 11) (O'Donnell and Nyblade 2014). This Airy isostatic deficit beneath the East Antarctic plateau is also noted from Moho depths derived from gravity inversions (Pappa *et al.* 2019; Pappa and Ebbing this volume). Both studies speculate that this anomalous topography may be supported by a mid-to-lower mantle source, and is therefore 'dynamic' (O'Donnell and Nyblade 2014; Pappa *et al.* 2019).

However, calculations attempting to separate isostatic and dynamic topography are subject to a number of uncertainties:

1. Isostatic topography is calculated assuming an Airy model of isostasy, and so does not account for elastic plate flexure (see 'Elastic plate flexure' section).
2. Crustal thickness observations are often unreliable or associated with large uncertainties. In Antarctica, crustal thickness is either estimated from satellite gravity inversion models (Block *et al.* 2009; Pappa *et al.* 2019; Pappa and Ebbing 2021) or from seismic receiver functions (Hansen *et al.* 2010; An *et al.* 2015*a*; Wiens *et al.* 2021), which are typically associated with uncertainties of at least ±4 km, which translate into uncertainties of ±*c.* 700 m in the isostatic topography.
3. The density contrast between the lower crust and the lithospheric mantle is often also associated with large uncertainty.

Gravity anomalies, rather than crustal thickness measurements, are considered by several authors to offer the tightest constraints on dynamic topography in the continents (McKenzie 2010; Molnar *et al.* 2015). Numerical models of convective mantle flow indicate that the expected ratio between the 'dynamic' free-air anomaly and topography (the 'convective admittance') is *c.* 30 mGal km^{-1} for water/ice-covered terrain (McKenzie *et al.* 1974; McKenzie 1994). This 'rule' for converting free-air anomalies into dynamically supported topography is generally supported by limited observations in the oceanic realm (Winterbourne *et al.* 2014; Hoggard *et al.* 2017). The long-wavelength gravity field (>3300 km; Watts and Moore 2017) can therefore be scaled using the convective admittance to produce an estimate of dynamic topography (Fig. 11) (Hoggard *et al.* 2016, 2017).

However, this approach has not so far been used for Antarctica, and there are reasons to exercise caution with its application. Recent models of mantle convection show that the sensitivity kernels for gravity and topography are different, such that the relationship between 'dynamic' gravity and topography is not summarized by a constant admittance, but one which is a function of the wavelength and depth of the driving buoyancy anomaly (Colli *et al.* 2016). Sizeable dynamic topography may therefore exist without a corresponding gravity anomaly, and vice versa. Indeed, the correlation between residual topography (from crustal thickness measurements) and dynamic gravity is generally poor (Fig. 11).

The spatial and temporal evolution of the amplitude and direction of changes in dynamic topography in Antarctica since *c.* 34 Ma remain poorly understood. Development of an Antarctic palaeo-dynamic topography model will require an expanded coverage of seismic stations in Antarctica, and a reduction in the epistemic uncertainty within mantle flow models. Also required is the identification of readily dated geological markers that can constrain the magnitude and pattern of dynamic uplift or subsidence on length scales of hundreds to thousands of kilometres, although such exposure does not exist in Antarctica. These limitations and uncertainties become increasingly problematic when moving further back in time, which makes reconstructing changes in dynamic topography over the course of Antarctica's glacial history (since *c.* 34 Ma) particularly challenging.

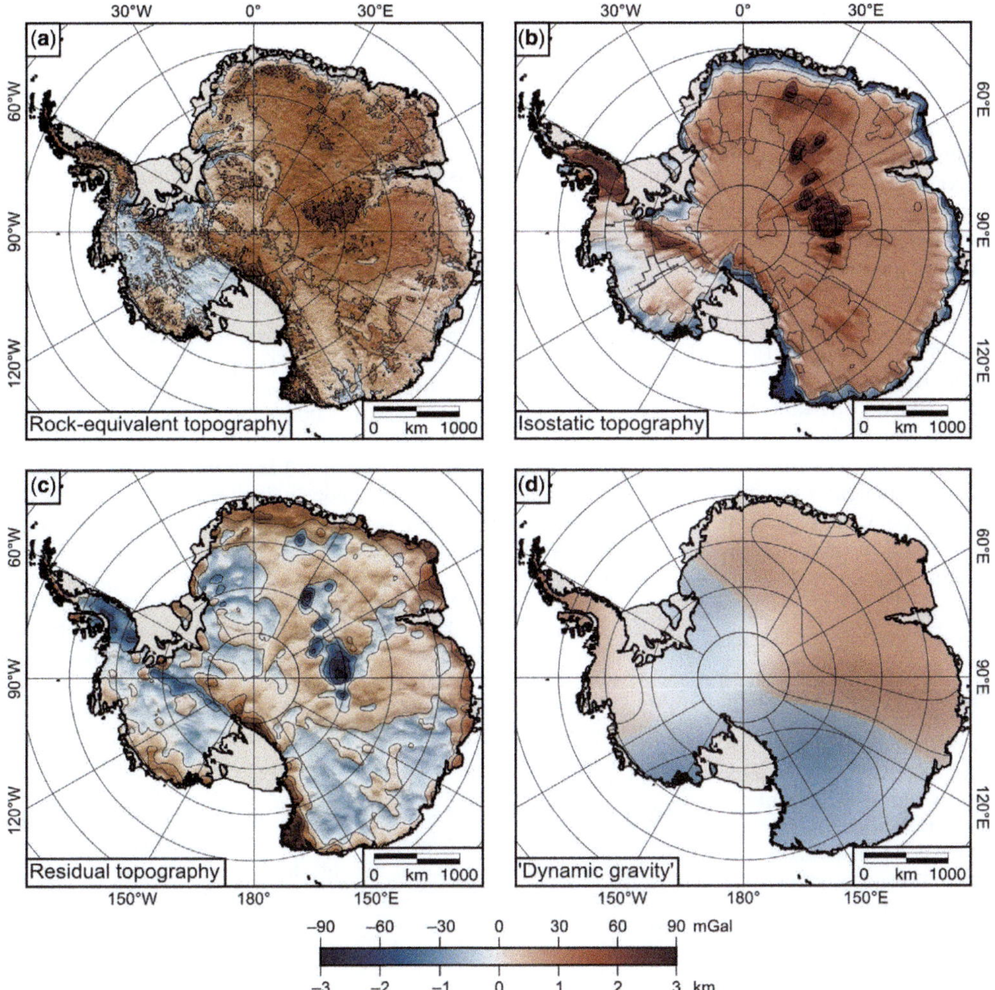

Fig. 11. Compensation of topography in Antarctica. (**a**) Present-day 'rock-equivalent' topography, whereby the ice sheet is converted into an equivalent mass of rock and added to the bed topography. Elevations are relative to mean sea-level. (**b**) Isostatic topography, calculated from the seismic crustal thickness assuming local compensation (Airy isostasy) and a zero-elevation crustal thickness of 35 km for East Antarctica and 25 km for West Antarctica (An *et al.* 2015*a*). (**c**) Residual topography, calculated by subtracting the isostatic topography from the rock-equivalent topography. This represents topography that is not in Airy isostatic balance. While this imbalance may be due to dynamic support of topography by the mantle, it may arise due to flexural isostasy or variations in crustal density. The topography has been median filtered at 100 km to remove short-wavelength features unlikely to arise from mantle convection. (**d**) Long-wavelength (>3300 km) GOCE satellite gravity anomaly (Bruinsma *et al.* 2014), which is scaled to dynamic topography using a uniform 'convective' admittance ratio of 30 mGal/km.

Conclusions

The preceding sections have illustrated the influences of the Antarctic mantle on palaeotopography. The 'Isostasy' section illustrated how lithospheric flexure has acted to increase (ice-free) bed elevations in regions subjected to unloading of the lithosphere via glacial erosion, extensional faulting, and conduction of heat within the lithosphere (Fig. 6). The third section, 'Thermal subsidence', highlighted the role of protracted lithospheric cooling in Antarctica, primarily in the West Antarctic Rift System. The result has been gradual topographic subsidence of up to 1 km since *c.* 34 Ma (Fig. 8) (Wilson and Luyendyk 2009). The fourth section, 'Dynamic topography', showed that it remains difficult to evaluate long-term changes in mantle convection and dynamic topography over geological timescales in Antarctica, despite recent modelling results from the Wilkes Subglacial Basin (Fig. 10) (Austermann *et al.* 2015).

The nature of the Antarctic mantle, and its role in the evolution of subglacial topography, has a number of implications for the past behaviour of the AIS.

1. Early Antarctic ice sheets likely exploited pre-existing tectonic structures such as fault systems and their associated topography. Subsequent excavation of deep subglacial troughs and associated flexural uplift of the trough flanks facilitated effective focussing of glacial flow and erosion within the troughs and contemporaneous preservation of the adjacent highlands beneath non-erosive ice. The patterns of glacial erosion and flexural uplift are therefore governed in part by pre-existing structural weaknesses within, and the flexural rigidity of, the Antarctic lithosphere.
2. Reconstructions of palaeotopography at *c.* 34 Ma have significantly less area below sea-level and much shallower marine basins than the modern topography (Wilson *et al.* 2012; Paxman *et al.* 2019*b*). This is largely due to bed overdeepening by glacial erosion (see above), and subsidence driven by post-rift lithospheric cooling. Widespread subsidence of up to 1 km in West Antarctica (Wilson and Luyendyk 2009; Wilson *et al.* 2012) has resulted in the development of an extensive area of low-lying, reverse-sloping topography beneath the WAIS. Landscape evolution models indicate that much of West Antarctica had subsided below sea-level by the mid-Miocene (*c.* 14 Ma) (Paxman *et al.* 2019*b*). The WAIS may therefore have become increasingly sensitive to marine ice sheet instability processes and associated rapid and irreversible losses of ice volume after this time. In addition, the progressive deepening of topography in East Antarctica by structurally steered glacial erosion during the past 34 Myr may have also facilitated an increase in the sensitivity of the EAIS to climate and ocean forcing (Paxman *et al.* 2020).
3. Changes in dynamic topography of as little as 100–200 m have potentially induced significant changes in ice sheet sensitivity in the Wilkes Subglacial Basin since as recently as the mid-Pliocene (*c.* 3 Ma) (Austermann *et al.* 2015). Considerable uncertainty remains, however, regarding the spatial and temporal evolution of Antarctic dynamic topography through the Cenozoic, with potentially major ramifications for ice sheet stability and sea-level change.

Overall, palaeotopographic reconstructions and ice sheet models suggest that the evolution of the subglacial topography of Antarctica has important implications for past Antarctic ice sheet behaviour (and in turn global sea-levels), in terms of the extent and rate of (i) ice growth during cold climate intervals, and (ii) ice loss during warm climate intervals (Wilson *et al.* 2013; Gasson *et al.* 2016; Colleoni *et al.* 2018). This modification of the subglacial landscape and associated impact on ice sheet behaviour can be attributed to interactions and feedbacks between surface, crustal, and mantle processes, such as glacial erosion, faulting, flexure, thermal subsidence, and dynamic topography. An improved understanding of mantle properties and processes, including in particular (a) a more robust model of the flexural rigidity of Antarctica's lithosphere, (b) further constraints on the pattern of rifting and thermal subsidence in West Antarctica and around the margins of East Antarctica, and (c) a better understanding of changes in mantle convection patterns and associated changes dynamic topography, on the evolution of Antarctic palaeotopography is therefore a key priority in the next generation of Antarctic landscape evolution and ice sheet models.

Acknowledgements Many of the topics described in this chapter are based on contributions from the ANTscape working group within the Past Antarctic Ice Sheet dynamics (PAIS) research programme of the Scientific Committee on Antarctic Research (SCAR). I would like to acknowledge the support of the major contributors to this programme, including Peter Barrett, Graeme Eagles, Karsten Gohl, Katharina Hochmuth, Stewart Jamieson, German Leitchenkov, Doug Wilson, and the numerous attendees of the various ANTscape workshops on Antarctic palaeotopography and palaeobathymetry that were instrumental in the development of several concepts reviewed in this chapter. I would also like to extend thanks to Mark Hoggard and Doug Wilson for their thorough and constructive reviews, which greatly improved the quality of this chapter.

Author contributions **GJGP**: writing – original draft (lead), writing – review & editing (lead).

Funding The author would like to acknowledge the support of a Natural Environment Research Council UK studentship (NE/L002590/1).

Data availability Data sharing is not applicable to this article as no datasets were generated or analysed during the current study.

References

Aitken, A.R.A., Roberts, J.L. *et al.* 2016. Repeated large-scale retreat and advance of Totten Glacier indicated by inland bed erosion. *Nature*, **533**, 385–389, https://doi.org/10.1038/nature17447

Al-Hajri, Y., White, N. and Fishwick, S. 2009. Scales of transient convective support beneath Africa. *Geology*, **37**, 883–886, https://doi.org/10.1130/G25703A.1

An, M., Wiens, D.A. *et al.* 2015*a*. S-velocity model and inferred Moho topography beneath the Antarctic Plate from Rayleigh waves. *Journal of Geophysical Research Solid Earth*, **120**, 359–383, https://doi.org/10.1002/2014JB011332

An, M., Wiens, D.A. *et al.* 2015*b*. Temperature, lithosphere–asthenosphere boundary, and heat flux beneath the Antarctic Plate inferred from seismic velocities. *Journal of Geophysical Research Solid Earth*, **120**, 8720–8742, https://doi.org/10.1002/2015JB011917

Austermann, J., Pollard, D. *et al.* 2015. The impact of dynamic topography change on Antarctic ice sheet stability during the mid-Pliocene warm period. *Geology*, **43**, 927–930, https://doi.org/10.1130/G36988.1

Barletta, V.R., Bevis, M. *et al.* 2018. Observed rapid bedrock uplift in Amundsen Sea Embayment promotes ice-sheet stability. *Science*, **360**, 1335–1339, https://doi.org/10.1126/science.aao1447

Barrett, P.J. 1975. Textural characteristics of Cenozoic preglacial and glacial sediments at Site 270, Ross Sea, Antarctica. *Initial Reports of the Deep Sea Drilling Project Leg 28*, **28**, 757–767.

Barrett, P.J. 2008. A history of antarctic cenozoic glaciation – view from the margin. *In*: Florindo, F. and Siegert, M.J. (eds) *Developments in Earth and Environmental Sciences*. Elsevier, 33–83, https://doi.org/10.1016/S1571-9197(08)00003-7

Behrendt, J.C., LeMasurier, W.E., Cooper, A.K., Tessensohn, F., Tréhu, A. and Damaske, D. 1991. Geophysical studies of the West Antarctic Rift System. *Tectonics*, **10**, 1257–1273, https://doi.org/10.1029/91TC00868

Behrendt, J.C., Blankenship, D.D., Finn, C.A., Bell, R.E., Sweeney, R.E., Hodge, S.M. and Brozena, J.M. 1994. CASERTZ aeromagnetic data reveal late Cenozoic flood basalts(?) in the West Antarctic rift system. *Geology*, **22**, 527, https://doi.org/10.1130/0091-7613(1994)022<0527:CADRLC>2.3.CO;2

Bell, R.E., Blankenship, D.D., Finn, C.A., Morse, D.L., Scambos, T.A., Brozena, J.M. and Hodge, S.M. 1998. Influence of subglacial geology on the onset of a West Antarctic ice stream from aerogeophysical observations. *Nature*, **394**, 58–62, https://doi.org/10.1038/27883

Bentley, M.J., O'Cofaigh, C. *et al.* 2014. A community-based geological reconstruction of Antarctic Ice Sheet deglaciation since the Last Glacial Maximum. *Quaternary Science Reviews*, **100**, 1–9, https://doi.org/10.1016/j.quascirev.2014.06.025

Bialas, R.W., Buck, W.R., Studinger, M. and Fitzgerald, P.G. 2007. Plateau collapse model for the Transantarctic Mountains-West Antarctic Rift System: insights from numerical experiments. *Geology*, **35**, 687–690, https://doi.org/10.1130/G23825A.1

Bingham, R.G., Ferraccioli, F., King, E.C., Larter, R.D., Pritchard, H.D., Smith, A.M. and Vaughan, D.G. 2012. Inland thinning of West Antarctic Ice Sheet steered along subglacial rifts. *Nature*, **487**, 468–471, https://doi.org/10.1038/nature11292

Block, A.E., Bell, R.E. and Studinger, M. 2009. Antarctic crustal thickness from satellite gravity: implications for the Transantarctic and Gamburtsev Subglacial Mountains. *Earth and Planetary Science Letters*, **288**, 194–203, https://doi.org/10.1016/j.epsl.2009.09.022

Boger, S.D. 2011. Antarctica – before and after Gondwana. *Gondwana Research*, **19**, 335–371, https://doi.org/10.1016/j.gr.2010.09.003

Braun, J. 2010. The many surface expressions of mantle dynamics. *Nature Geoscience*, **3**, 825–833, https://doi.org/10.1038/ngeo1020

Bredow, E. and Steinberger, B. 2021. Mantle convection and possible mantle plumes beneath Antarctica: insights from geodynamic models and implications for topography. *Geological Society, London, Memoirs*, **56**, https://doi.org/10.1144/M56-2020-2

Brenn, G.R., Hansen, S.E. and Park, Y. 2017. Variable thermal loading and flexural uplift along the Transantarctic Mountains, Antarctica. *Geology*, **45**, 463–466, https://doi.org/10.1130/G38784.1

Bruinsma, S.L., Förste, C. *et al.* 2014. ESA's satellite-only gravity field model via the direct approach based on all GOCE data. *Geophysical Research Letters*, **41**, 7508–7514, https://doi.org/10.1002/2014GL062045

Burton-Johnson, A., Black, M., Peter, T.F. and Kaluza-Gilbert, J. 2016. An automated methodology for differentiating rock from snow, clouds and sea in Antarctica from Landsat 8 imagery: a new rock outcrop map and area estimation for the entire Antarctic continent. *Cryosphere*, **10**, 1665–1677, https://doi.org/10.5194/tc-10-1665-2016

Cande, S.C., Stock, J.M., Müller, R.D. and Ishihara, T. 2000. Cenozoic motion between East and West Antarctica. *Nature*, **404**, 145–150, https://doi.org/10.1038/35004501

Carter, A., Riley, T.R., Hillenbrand, C.-D. and Rittner, M. 2017. Widespread Antarctic glaciation during the Late Eocene. *Earth and Planetary Science Letters*, **458**, 49–57, https://doi.org/10.1016/j.epsl.2016.10.045

Cazenave, A., Souriau, A. and Dominh, K. 1989. Global coupling of Earth surface topography with hotspots, geoid and mantle heterogeneities. *Nature*, **340**, 54–57, https://doi.org/10.1038/340054a0

Chaput, J., Aster, R.C. *et al.* 2014. The crustal thickness of West Antarctica. *Journal of Geophysical Research Solid Earth*, **119**, 378–395, https://doi.org/10.1002/2013JB010642

Chen, B., Haeger, C., Kaban, M.K. and Petrunin, A.G. 2018. Variations of the effective elastic thickness reveal tectonic fragmentation of the Antarctic lithosphere. *Tectonophysics*, **746**, 412–424, https://doi.org/10.1016/j.tecto.2017.06.012

Clerc, F., Minchew, B.M. and Behn, M.D. 2019. Marine ice cliff instability mitigated by slow removal of ice shelves. *Geophysical Research Letters*, **46**, 12108–12116, https://doi.org/10.1029/2019GL084183

Coenen, J.J., Scherer, R., Baudoin, P., Warny, S., Castañeda, I.S. and Askin, R. 2019. Paleogene marine and terrestrial development of the West Antarctic Rift System. *Geophysical Research Letters*, **47**, https://doi.org/10.1029/2019GL085281

Colleoni, F., De Santis, L. *et al.* 2018. Past continental shelf evolution increased Antarctic ice sheet sensitivity to climatic conditions. *Scientific Reports*, **8**, 11323, https://doi.org/10.1038/s41598-018-29718-7

Colli, L., Ghelichkhan, S. and Bunge, H.P. 2016. On the ratio of dynamic topography and gravity anomalies in a dynamic Earth. *Geophysical Research Letters*, **43**, 2510–2516, https://doi.org/10.1002/2016GL067929

Cox, S.E., Thomson, S.N., Reiners, P.W., Hemming, S.R. and van de Flierdt, T. 2010. Extremely low long-term erosion rates around the Gamburtsev Mountains in interior East Antarctica. *Geophysical Research Letters*, **37**, https://doi.org/10.1029/2010GL045106

Coxall, H.K., Wilson, P.A., Pälike, H., Lear, C.H. and Backman, J. 2005. Rapid stepwise onset of Antarctic glaciation and deeper calcite compensation in the Pacific Ocean. *Nature*, **433**, 53–57, https://doi.org/10.1038/nature03135

Czarnota, K., Hoggard, M.J., White, N. and Winterbourne, J., 2013. Spatial and temporal patterns of Cenozoic dynamic topography around Australia. *Geochemistry, Geophysics Geosystems*, **14**, 634–658, https://doi.org/10.1029/2012GC004392

Dalziel, I.W.D. 1997. Neoproterozoic-Paleozoic geography and tectonics: review, hypothesis, environmental speculation. *Geological Society of America Bulletin*, **109**, 16–42, https://doi.org/10.1130/0016-7606(1997)109<0016:ONPGAT>2.3.CO;2

De Santis, L., Prato, S., Brancolini, G., Lovo, M. and Torelli, L. 1999. The Eastern Ross Sea continental shelf during the Cenozoic: implications for the West Antarctic ice sheet development. *Global and Planetary Change*, **23**, 173–196, https://doi.org/10.1016/S0921-8181(99)00056-9

Decesari, R.C., Wilson, D.S., Luyendyk, B.P. and Faulkner, M. 2007. Cretaceous and Tertiary extension throughout the Ross Sea, Antarctica. *In*: Cooper, A.K. (ed.) *Antarctica: A Keystone in a Changing World – Online Proceedings of the 10th ISAES*. US Geological Survey. Open File Rep., 2007-1047, https://doi.org/10.3133/of2007-1047.srp098

DeConto, R.M. and Pollard, D. 2003*a*. Rapid Cenozoic glaciation of Antarctica induced by declining atmospheric CO_2. *Nature*, **421**, 245–249, https://doi.org/10.1038/nature01290

DeConto, R.M. and Pollard, D. 2003*b*. A coupled climate–ice sheet modeling approach to the Early Cenozoic history of the Antarctic ice sheet. *Palaeogeography Palaeoclimatology Palaeoecology*, **198**, 39–52, https://doi.org/10.1016/S0031-0182(03)00393-6

DeConto, R.M. and Pollard, D. 2016. Contribution of Antarctica to past and future sea-level rise. *Nature*, **531**, 591–597, https://doi.org/10.1038/nature17145

DeConto, R.M., Pollard, D., Wilson, P.A., Pälike, H., Lear, C.H. and Pagani, M. 2008. Thresholds for Cenozoic bipolar glaciation. *Nature*, **455**, 652–656, https://doi.org/10.1038/nature07337

Drewry, D.J. 1983. *Antarctica: Glaciological and Geophysical Folio*. Scott Polar Research Institute, University of Cambridge.

Ebbing, J., Haas, P., Ferraccioli, F., Pappa, F., Szwillus, W. and Bouman, J. 2018. Earth tectonics as seen by GOCE - Enhanced satellite gravity gradient imaging. *Scientific Reports*, **8**, 16356, https://doi.org/10.1038/s41598-018-34733-9

Edwards, T.L., Brandon, M.A. *et al.* 2019. Revisiting Antarctic ice loss due to marine ice-cliff instability. *Nature*, **566**, 58–64, https://doi.org/10.1038/s41586-019-0901-4

Elliot, D.H. and Fleming, T.H. 2004. Occurrence and dispersal of magmas in the Jurassic Ferrar Large Igneous Province, Antarctica. *Gondwana Research*, **7**, 223–237, https://doi.org/10.1016/S1342-937X(05)70322-1

Ferraccioli, F., Armadillo, E., Zunino, A., Bozzo, E., Rocchi, S. and Armienti, P. 2009. Magmatic and tectonic patterns over the Northern Victoria Land sector of the Transantarctic Mountains from new aeromagnetic imaging. *Tectonophysics*, **478**, 43–61, https://doi.org/10.1016/j.tecto.2008.11.028

Ferraccioli, F., Finn, C.A., Jordan, T.A., Bell, R.E., Anderson, L.M. and Damaske, D. 2011. East Antarctic rifting triggers uplift of the Gamburtsev Mountains. *Nature*, **479**, 388–392, https://doi.org/10.1038/nature10566

Finn, C.A., Müller, R.D. and Panter, K.S. 2005. A Cenozoic diffuse alkaline magmatic province (DAMP) in the southwest Pacific without rift or plume origin. *Geochemistry, Geophysics Geosystems*, **6**, https://doi.org/10.1029/2004GC000723

Flament, N., Gurnis, M. and Muller, R.D. 2013. A review of observations and models of dynamic topography. *Lithosphere*, **5**, 189–210, https://doi.org/10.1130/L245.1

Ford, A.B. and Barrett, P.J. 1975. Basement rocks of the south-central Ross Sea, Site 270, DSDP Leg 28. *Initial Reports of the Deep Sea Drilling Project Leg 28*, **28**, 861–868.

Forsberg, R., Olesen, A.V. *et al.* 2018. Exploring the Recovery Lakes region and interior Dronning Maud Land, East Antarctica, with airborne gravity, magnetic and radar measurements. *Geological Society, London, Special Publications*, **461**, https://doi.org/10.1144/SP461.17

Forte, A.M., Peltier, W.R., Dziewonski, A.M. and Woodward, R.L. 1993. Dynamic surface topography: a new interpretation based upon mantle flow models derived from seismic tomography. *Geophysical Research Letters*, **20**, 225–228, https://doi.org/10.1029/93GL00249

Fretwell, P., Pritchard, H.D. *et al.* 2013. Bedmap2: improved ice bed, surface and thickness datasets for Antarctica. *Cryosphere*, **7**, 375–393, https://doi.org/10.5194/tc-7-375-2013

Gasson, E., DeConto, R.M. and Pollard, D. 2015. Antarctic bedrock topography uncertainty and ice sheet stability. *Geophysical Research Letters*, **42**, 5372–5377, https://doi.org/10.1002/2015GL064322

Gasson, E., DeConto, R.M., Pollard, D. and Levy, R.H. 2016. Dynamic Antarctic ice sheet during the early to mid-Miocene. *Proceedings of the National Academy of Sciences*, **113**, 3459–3464, https://doi.org/10.1073/pnas.1516130113

Gohl, K. 2008. Antarctica's continent-ocean transitions: consequences for tectonic reconstructions. *In*: Cooper, A.K., Barrett, P.J., Stagg, H.M.J., Storey, B.C., Stump, E. and Wise, W. (eds) *Antarctica: A Keystone in a Changing World – Online Proceedings of the 10th ISAES*. The National Academies Press, Washington, DC, 29–38, https://doi.org/10.3133/of2007-1047.kp04

Golledge, N.R., Kowalewski, D.E., Naish, T.R., Levy, R.H., Fogwill, C.J. and Gasson, E.G.W. 2015. The multi-millennial Antarctic commitment to future sea-level rise. *Nature*, **526**, 421–425, https://doi.org/10.1038/nature15706

Gomez, N., Mitrovica, J.X., Huybers, P. and Clark, P.U. 2010. Sea level as a stabilizing factor for marine-ice-sheet grounding lines. *Nature Geoscience*, **3**, 850–853, https://doi.org/10.1038/ngeo1012

Granot, R. and Dyment, J. 2018. Late Cenozoic unification of East and West Antarctica. *Nature Communications*, **9**, 3189, https://doi.org/10.1038/s41467-018-05270-w

Hager, B.H., Clayton, R.W., Richards, M.A., Comer, R.P. and Dziewonski, A.M. 1985. Lower mantle heterogeneity, dynamic

topography and the geoid. *Nature*, **314**, 752–752, https://doi.org/10.1038/314752a0
Hambrey, M.J., Ehrmann, W.U. and Larsen, B. 1991. Cenozoic Glacial Record of the Prydz Bay Continental Shelf, East Antarctica. *Proceedings of the Ocean Drilling Program, Scientific Results*, **119**, 77–132, https://doi.org/10.2973/odp.proc.sr.119.200.1991
Hansen, S.E., Nyblade, A.A., Heeszel, D.S., Wiens, D.A., Shore, P. and Kanao, M. 2010. Crustal structure of the Gamburtsev Mountains, East Antarctica, from S-wave receiver functions and Rayleigh wave phase velocities. *Earth and Planetary Science Letters*, **300**, 395–401, https://doi.org/10.1016/j.epsl.2010.10.022
Hansen, S.E., Kenyon, L.M., Graw, J.H., Park, Y. and Nyblade, A.A. 2016. Crustal structure beneath the Northern Transantarctic Mountains and Wilkes Subglacial Basin: implications for tectonic origins. *Journal of Geophysical Research Solid Earth*, **121**, 812–825, https://doi.org/10.1002/2015JB012325
Hochmuth, K., Gohl, K. *et al.* 2020. The evolving paleobathymetry of the Circum-Antarctic Southern Ocean since 34 Ma: a key to understanding past cryosphere-ocean developments. *Geochemistry, Geophysics, Geosystems*, **21**.
Hoggard, M.J., White, N. and Al-Attar, D. 2016. Global dynamic topography observations reveal limited influence of large-scale mantle flow. *Nature Geoscience*, **9**, 456–463, https://doi.org/10.1038/ngeo2709
Hoggard, M.J., Winterbourne, J., Czarnota, K. and White, N. 2017. Oceanic residual depth measurements, the plate cooling model, and global dynamic topography. *Journal of Geophysical Research Solid Earth*, **122**, 2328–2372, https://doi.org/10.1002/2016JB013457
Hole, M.J. and LeMasurier, W.E. 1994. Tectonic controls on the geochemical composition of Cenozoic, mafic alkaline volcanic rocks from West Antarctica. *Contributions to Mineralogy and Petrology*, **117**, 187–202, https://doi.org/10.1007/BF00286842
Huerta, A.D. 2007. Byrd drainage system; evidence of a Mesozoic West Antarctic plateau. *In*: Cooper, A.K. and Raymond, C.R. (eds) *Antarctica: A Keystone in a Changing World – Online Proceedings of the 10th ISAES X*. USGS Open-File Report 2007-1047. Extended Abstract 091.
Humbert, A., Steinhage, D., Helm, V., Beyer, S. and Kleiner, T. 2018. Missing evidence of Widespread Subglacial Lakes at Recovery Glacier, Antarctica. *Journal of Geophysical Research Earth Surface*, **123**, 2802–2826, https://doi.org/10.1029/2017JF004591
Jamieson, S.S.R. and Sugden, D.E. 2008. Landscape evolution of Antarctica. *In*: Cooper, A.K., Barrett, P.J., Stagg, H., Storey, B., Stump, E. and Wise, W. (eds) *Antarctica: A Keystone in a Changing World*. The National Academies Press, Washington DC, 39–54, https://doi.org/10.1007/s13398-014-0173-7.2
Jamieson, S.S.R., Hulton, N.R.J., Sugden, D.E., Payne, A.J. and Taylor, J. 2005. Cenozoic landscape evolution of the Lambert basin, East Antarctica: the relative role of rivers and ice sheets. *Global and Planetary Change*, **45**, 35–49, https://doi.org/10.1016/j.gloplacha.2004.09.015
Jamieson, S.S.R., Hulton, N.R.J. and Hagdorn, M. 2008. Modelling landscape evolution under ice sheets. *Geomorphology*, **97**, 91–108, https://doi.org/10.1016/j.geomorph.2007.02.047
Jamieson, S.S.R., Sugden, D.E. and Hulton, N.R.J. 2010. The evolution of the subglacial landscape of Antarctica. *Earth and Planetary Science Letters*, **293**, 1–27, https://doi.org/10.1016/j.epsl.2010.02.012
Jamieson, S.S.R., Stokes, C.R. *et al.* 2014. The glacial geomorphology of the Antarctic ice sheet bed. *Antarctic Science*, **26**, 724–741, https://doi.org/10.1017/S0954102014000212
Japsen, P., Bonow, J.M., Green, P.F., Chalmers, J.A. and Lidmar-Bergström, K. 2006. Elevated, passive continental margins: long-term highs or Neogene uplifts? New evidence from West Greenland. *Earth and Planetary Science Letters*, **248**, 330–339, https://doi.org/10.1016/j.epsl.2006.05.036
Ji, F., Gao, J., Shen, Z., Zhang, Q. and Li, Y. 2017. Variations of the effective elastic thickness over the Ross Sea and Transantarctic Mountains and implications for their structure and tectonics. *Tectonophysics*, **717**, 127–138, https://doi.org/10.1016/j.tecto.2017.07.011
Jordan, T.A., Ferraccioli, F., Vaughan, D.G., Holt, J.W., Corr, H., Blankenship, D.D. and Diehl, T.M., 2010. Aerogravity evidence for major crustal thinning under the Pine Island Glacier region (West Antarctica). *Geological Society of America Bulletin*, **122**, 714–726, https://doi.org/10.1130/B26417.1
Jordan, T.A., Ferraccioli, F., Armadillo, E. and Bozzo, E. 2013. Crustal architecture of the Wilkes Subglacial Basin in East Antarctica, as revealed from airborne gravity data. *Tectonophysics*, **585**, 196–206, https://doi.org/10.1016/j.tecto.2012.06.041
Jordan, T.A., Ferraccioli, F. and Leat, P.T. 2017. New geophysical compilations link crustal block motion to Jurassic extension and strike-slip faulting in the Weddell Sea Rift System of West Antarctica. *Gondwana Research*, **42**, 29–48, https://doi.org/10.1016/j.gr.2016.09.009
Jordan, T.A., Martin, C. *et al.* 2018. Anomalously high geothermal flux near the South Pole. *Scientific Reports*, **8**, 16785, https://doi.org/10.1038/s41598-018-35182-0
Joughin, I. and Alley, R.B. 2011. Stability of the West Antarctic ice sheet in a warming world. *Nature Geoscience*, **4**, 506–513, https://doi.org/10.1038/ngeo1194
Karlsson, N.B., Binder, T., Eagles, G., Helm, V., Pattyn, F., Van Liefferinge, B. and Eisen, O. 2018. Glaciological characteristics in the Dome Fuji region and new assessment for 'Oldest Ice'. *Cryosphere*, **12**, 2413–2424, https://doi.org/10.5194/tc-12-2413-2018
Karner, G.D., Studinger, M. and Bell, R.E. 2005. Gravity anomalies of sedimentary basins and their mechanical implications: application to the Ross Sea basins, West Antarctica. *Earth and Planetary Science Letters*, **235**, 577–596, https://doi.org/10.1016/j.epsl.2005.04.016
Katz, M.E., Miller, K.G., Wright, J.D., Wade, B.S., Browning, J.V., Cramer, B.S. and Rosenthal, Y. 2008. Stepwise transition from the Eocene greenhouse to the Oligocene icehouse. *Nature Geoscience*, **1**, 329–334, https://doi.org/10.1038/ngeo179
Kennett, J.P. 1977. Cenozoic evolution of Antarctic glaciation, the circum-Antarctic Ocean, and their impact on global paleoceanography. *Journal of Geophysical Research*, **82**, 3843–3860, https://doi.org/10.1029/JC082i027p03843
Kessler, M.A., Anderson, R.S. and Briner, J.P. 2008. Fjord insertion into continental margins driven by topographic steering of ice. *Nature Geoscience*, **1**, 365–369, https://doi.org/10.1038/ngeo201
Kingslake, J., Scherer, R.P. *et al.* 2018. Extensive retreat and re-advance of the West Antarctic Ice Sheet during the Holocene. *Nature*, **558**, 430–434, https://doi.org/10.1038/s41586-018-0208-x
Kirby, J.F. 2014. Estimation of the effective elastic thickness of the lithosphere using inverse spectral methods: the state of the art. *Tectonophysics*, **631**, 87–116, https://doi.org/10.1016/j.tecto.2014.04.021
Koppes, M.N. and Montgomery, D.R. 2009. The relative efficacy of fluvial and glacial erosion over modern to orogenic timescales. *Nature Geoscience*, **2**, 644–647, https://doi.org/10.1038/ngeo616
Krohne, N., Lisker, F., Kleinschmidt, G., Klugel, A., Laufer, A., Estrada, S. and Spiegel, C. 2016. The Shackleton Range (East Antarctica): an alien block at the rim of Gondwana? *Geological Magazine*, **155**, 841–864, https://doi.org/10.1017/S0016756816001011
Kulhanek, D.K., Levy, R.H. *et al.* 2019. Revised chronostratigraphy of DSDP Site 270 and late Oligocene to early Miocene paleoecology of the Ross Sea sector of Antarctica. *Global and Planetary Change*, **178**, 46–64, https://doi.org/10.1016/j.gloplacha.2019.04.002
Lawrence, J.F., Wiens, D.A., Nyblade, A.A., Anandakrishnan, S., Shore, P.J. and Voigt, D. 2006. Crust and upper mantle structure of the Transantarctic Mountains and surrounding regions from

receiver functions, surface waves, and gravity: implications for uplift models. *Geochemistry, Geophysics Geosystems*, **7**, https://doi.org/10.1029/2006GC001282

Leckie, R.M. and Webb, P.-N. 1983. Late Oligocene-early Miocene glacial record of the Ross Sea, Antarctica: evidence from DSDP Site 270. *Geology*, **11**, 578–582, https://doi.org/10.1130/0091-7613(1983)11<578:LOMGRO>2.0.CO;2

LeMasurier, W.E. and Landis, C.A. 1996. Mantle-plume activity recorded by low-relief erosion surfaces in West Antarctica and New Zealand. *Geological Society of America Bulletin*, **108**, 1450–1466, https://doi.org/10.1130/0016-7606(1996)108<1450:MPARBL>2.3.CO;2

LeMasurier, W.E. and Rex, D.C. 1989. Evolution of linear volcanic ranges in Marie Byrd Land, West Antarctica. *Journal of Geophysical Research*, **94**, 7223, https://doi.org/10.1029/JB094iB06p07223

LeMasurier, W.E., Thomson, J.W., Baker, P.E., Kyle, P.R., Rowley, P.D., Smellie, J.L. and Verwoerd, W.J. (eds) 1990. *Volcanoes of the Antarctic Plate and Southern Oceans*. American Geophysical Union, *Antarctic Research Series*, Washington, DC, https://doi.org/10.1029/AR048

Leuschen, C., Gogineni, P., Rodriguez-Morales, F., Paden, J. and Allen, C. 2016. *IceBridge MCoRDS L2 Ice Thickness, Version 1*. NASA National Snow and Ice Data Center Distributed Active Archive Center, Boulder, Colorado, USA.

Liu, Z., Pagani, M. *et al.* 2009. Global cooling during the Eocene–Oligocene climate transition. *Science*, **323**, 1187–1190, https://doi.org/10.1126/science.1166368

Lloyd, A.J., Wiens, D.A. *et al.* 2015. A seismic transect across West Antarctica: evidence for mantle thermal anomalies beneath the Bentley Subglacial Trench and the Marie Byrd Land Dome. *Journal of Geophysical Research Solid Earth*, **120**, 8439–8460, https://doi.org/10.1002/2015JB012455

Lloyd, A.J., Wiens, D.A. *et al.* 2020. Seismic structure of the Antarctic upper mantle imaged with adjoint tomography. *Journal of Geophysical Research Solid Earth*, **125**, 2019JB017823, https://doi.org/10.1029/2019JB017823

Lythe, M., Vaughan, D.G. *et al.* 2001. BEDMAP: a new ice thickness and subglacial topographic model of Antarctica. *Journal of Geophysical Research*, **106**, 11335–11351, https://doi.org/10.1029/2000JB900449

Maritati, A., Aitken, A.R.A., Young, D.A., Roberts, J.L., Blankenship, D.D. and Siegert, M.J. 2016. The tectonic development and erosion of the Knox Subglacial Sedimentary Basin, East Antarctica. *Geophysical Research Letters*, **43**, 10728–10737, https://doi.org/10.1002/2016GL071063

McKenzie, D. 1978. Some remarks on development of sedimentary basins. *Earth and Planetary Science Letters*, **40**, 25–32, https://doi.org/10.1016/0012-821x(78)90071-7

McKenzie, D. 1994. *The Relationship between Topography and Gravity on Earth and Venus*. Icarus, https://doi.org/10.1006/icar.1994.1170

McKenzie, D. 2010. The influence of dynamically supported topography on estimates of Te. *Earth and Planetary Science Letters*, **295**, 127–138, https://doi.org/10.1016/j.epsl.2010.03.033

McKenzie, D. and Fairhead, D. 1997. Estimates of the effective elastic thickness of the continental lithosphere from Bouguer and free air gravity anomalies. *Journal of Geophysical Research*, **102**, 27523–27552, https://doi.org/10.1029/97JB02481

McKenzie, D.P., Roberts, J.M. and Weiss, N.O. 1974. Convection in the earth's mantle: towards a numerical simulation. *Journal of Fluid Mechanics*, **62**, 465–538, https://doi.org/10.1017/S0022112074000784

Menard, H.W. 1973. Depth anomalies and the bobbing motion of drifting islands. *Journal of Geophysical Research*, **78**, 5128–5137, https://doi.org/10.1029/jb078i023p05128

Mengel, M. and Levermann, A. 2014. Ice plug prevents irreversible discharge from East Antarctica. *Nature Climate Change*, **4**, 451–455, https://doi.org/10.1038/nclimate2226

Mercer, J.H. 1978. West Antarctic ice sheet and CO_2 greenhouse effect: a threat of disaster. *Nature*, **271**, 321–325, https://doi.org/10.1073/pnas.0703993104

Molnar, P. and England, P. 1990. Late Cenozoic uplift of mountain ranges and global climate change: chicken or egg? *Nature*, **346**, 29–34, https://doi.org/10.1038/346029a0

Molnar, P., England, P.C. and Jones, C.H. 2015. Mantle dynamics, isostasy, and the support of high terrain. *Journal of Geophysical Research Solid Earth*, **120**, 1–26, https://doi.org/10.1002/2014JB011724

Morlighem, M., Rignot, E. *et al.* 2020. Deep glacial troughs and stabilizing ridges unveiled beneath the margins of the Antarctic ice sheet. *Nature Geoscience*, **13**, 132–137, https://doi.org/10.1038/s41561-019-0510-8

Müller, R.D., Seton, M. *et al.* 2016. Ocean basin evolution and global-scale plate reorganization events since Pangea breakup. *Annual Review of Earth and Planetary Science*, **44**, 107–138, https://doi.org/10.1146/annurev-earth-060115-012211

Naish, T.R., Woolfe, K.J. *et al.* 2001. Orbitally induced oscillations in the East Antarctic ice sheet at the Oligocene/Miocene boundary. *Nature*, **413**, 719–723, https://doi.org/10.1038/35099534

O'Donnell, J.P. and Nyblade, A.A. 2014. Antarctica's hypsometry and crustal thickness: Implications for the origin of anomalous topography in East Antarctica. *Earth and Planetary Science Letters*, **388**, 143–155, https://doi.org/10.1016/j.epsl.2013.11.051

Olivetti, V., Rossetti, F., Balestrieri, M.L., Pace, D., Cornamusini, G. and Talarico, F. 2018. Variability in uplift, exhumation and crustal deformation along the Transantarctic Mountains front in southern Victoria Land, Antarctica. *Tectonophysics*, **745**, 229–244, https://doi.org/10.1016/j.tecto.2018.08.017

Pagani, M., Huber, M. *et al.* 2011. The role of carbon dioxide during the onset of Antarctic glaciation. *Science*, **334**, 1261–1264, https://doi.org/10.1126/science.1203909

Pappa, F. and Ebbing, J. 2021. Gravity, magnetics and geothermal heat flow of the Antarctic lithospheric crust and mantle. *Geological Society, London, Memoirs*, **56**, https://doi.org/10.1144/M56-2020-5

Pappa, F., Ebbing, J. and Ferraccioli, F. 2019. Moho depths of Antarctica: comparison of seismic, gravity, and isostatic results. *Geochemistry, Geophysics, Geosystems*, **20**, 1629–1645, https://doi.org/10.1029/2018GC008111

Passchier, S., Ciarletta, D.J., Miriagos, T.E., Bijl, P.K. and Bohaty, S.M. 2017. An Antarctic stratigraphic record of stepwise ice growth through the Eocene-Oligocene transition. *Geological Society of America Bulletin*, **129**, 318–330, https://doi.org/10.1130/B31482.1

Paxman, G.J.G., Watts, A.B., Ferraccioli, F., Jordan, T.A., Bell, R.E., Jamieson, S.S.R. and Finn, C.A. 2016. Erosion-driven uplift in the Gamburtsev Subglacial Mountains of East Antarctica. *Earth and Planetary Science Letters*, **452**, 1–14, https://doi.org/10.1016/j.epsl.2016.07.040

Paxman, G.J.G., Jamieson, S.S.R. *et al.* 2017. Uplift and tilting of the Shackleton Range in East Antarctica driven by glacial erosion and normal faulting. *Journal of Geophysical Research Solid Earth*, **122**, 2390–2408, https://doi.org/10.1002/2016JB01 3841

Paxman, G.J.G., Jamieson, S.S.R. *et al.* 2018. Bedrock erosion surfaces record former East Antarctic Ice Sheet extent. *Geophysical Research Letters*, **45**, 4114–4123, https://doi.org/10.1029/2018GL077268

Paxman, G.J.G., Jamieson, S.S.R. *et al.* 2019*a*. The role of lithospheric flexure in the landscape evolution of the Wilkes Subglacial Basin and Transantarctic Mountains, East Antarctica. *Journal of Geophysical Research Earth Surface*, **124**, 812–829, https://doi.org/10.1029/2018JF004705

Paxman, G.J.G., Jamieson, S.S.R., Hochmuth, K., Gohl, K., Bentley, M.J., Leitchenkov, G. and Ferraccioli, F. 2019*b*. Reconstructions of Antarctic topography since the Eocene–Oligocene boundary. *Palaeogeography Palaeoclimatology Palaeoecology*, **535**, 109346, https://doi.org/10.1016/j.palaeo.2019.109346

Paxman, G.J.G., Gasson, E.G.W., Jamieson, S.S.R., Bentley, M.J. and Ferraccioli, F. 2020. Long-term increase in Antarctic Ice Sheet vulnerability driven by bed topography evolution. *Geophysical Research Letters*, **47**.

Pearson, P.N., Foster, G.L. and Wade, B.S. 2009. Atmospheric carbon dioxide through the Eocene–Oligocene climate transition. *Nature*, **461**, 1110–1113, https://doi.org/10.1038/nature08447

Pekeris, C.L. 1935. Thermal convection in the interior of the earth. *Geophysics Journal International*, **3**, 343–367, https://doi.org/10.1111/j.1365-246X.1935.tb01742.x

Pollard, D. and DeConto, R.M. 2009. Modelling West Antarctic ice sheet growth and collapse through the past five million years. *Nature*, **458**, 329–332, https://doi.org/10.1038/nature07809

Pollard, D. and DeConto, R.M. 2019. Continuous simulations over the last 40 million years with a coupled Antarctic ice sheet-sediment model. *Palaeogeography Palaeoclimatology Palaeoecology*, **537**, 109374, https://doi.org/10.1016/J.PALAEO.2019.109374

Pollard, D., DeConto, R.M. and Alley, R.B. 2015. Potential Antarctic Ice Sheet retreat driven by hydrofracturing and ice cliff failure. *Earth and Planetary Science Letters*, **412**, 112–121, https://doi.org/10.1016/j.epsl.2014.12.035

Quigley, M.C., Clark, D. and Sandiford, M. 2010. Tectonic geomorphology of Australia. *Geological Society, London, Special Publications*, **346**, 243–265, https://doi.org/10.1144/SP346.13

Ramirez, C., Nyblade, A. *et al.* 2016. Crustal and upper-mantle structure beneath ice-covered regions in Antarctica from S-wave receiver functions and implications for heat flow. *Geophysics Journal International*, **204**, 1636–1648, https://doi.org/10.1093/gji/ggv542

Reusch, D.N. 2011. New Caledonian carbon sinks at the onset of Antarctic glaciation. *Geology*, **39**, 807–810, https://doi.org/10.1130/G31981.1

Richards, F.D., Hoggard, M.J. and White, N.J. 2016. Cenozoic epeirogeny of the Indian peninsula. *Geochemistry, Geophysics, Geosystems*, **17**, 4920–4954, https://doi.org/10.1002/2016GC006545

Rignot, E., Bamber, J.L., van den Broeke, M.R., Davis, C., Li, Y., van de Berg, W.J. and van Meijgaard, E. 2008. Recent Antarctic ice mass loss from radar interferometry and regional climate modelling. *Nature Geoscience*, **1**, 106–110, https://doi.org/10.1038/ngeo102

Rocchi, S., Armienti, P., D'Orazio, M., Tonarini, S., Wijbrans, J.R. and Di Vincenzo, G. 2002. Cenozoic magmatism in the western Ross Embayment: role of mantle plume v. plate dynamics in the development of the West Antarctic Rift System. *Journal of Geophysical Research Solid Earth*, **107**, https://doi.org/10.1029/2001JB000515

Rocchi, S., LeMasurier, W.E. and Di Vincenzo, G. 2006. Oligocene to Holocene erosion and glacial history in Marie Byrd Land, West Antarctica, inferred from exhumation of the Dorrel Rock intrusive complex and from volcano morphologies. *Geological Society of America Bulletin*, **118**, 991–1005, https://doi.org/10.1130/B25675.1

Rose, K.C., Ferraccioli, F. *et al.* 2013. Early East Antarctic Ice Sheet growth recorded in the landscape of the Gamburtsev Subglacial Mountains. *Earth and Planetary Science Letters*, **375**, 1–12, https://doi.org/10.1016/j.epsl.2013.03.053

Sandiford, M., Quigley, M., de Broekert, P. and Jakica, S. 2009. Tectonic framework for the Cenozoic cratonic basins of Australia. *Australian Journal of Earth Science*, **56**, S5–S18, https://doi.org/10.1080/08120090902870764

Scheinert, M., Ferraccioli, F. *et al.* 2016. New Antarctic gravity anomaly grid for enhanced geodetic and geophysical studies in Antarctica. *Geophysical Research Letters*, **43**, 600–610, https://doi.org/10.1002/2015GL067439

Scher, H.D., Bohaty, S.M., Zachos, J.C. and Delaney, M.L. 2011. Two-stepping into the icehouse: east Antarctic weathering during progressive ice-sheet expansion at the Eocene-Oligocene transition. *Geology*, **39**, 383–386, https://doi.org/10.1130/G31726.1

Scherer, R.P., Aldahan, A., Tulaczyk, S., Possnert, G., Engelhardt, H. and Kamb, B. 1998. Pleistocene collapse of the west Antarctic ice sheet. *Science*, **281**, 82–85, https://doi.org/10.1126/science.281.5373.82

Schoof, C. 2007. Ice sheet grounding line dynamics: steady states, stability, and hysteresis. *Journal of Geophysical Research*, **112**, F03S28, https://doi.org/10.1029/2006JF000664

Schroeder, D.M., Dowdeswell, J.A. *et al.* 2019. Multidecadal observations of the Antarctic ice sheet from restored analog radar records. *Proceedings of the National Academy of Sciences*, **116**, 18867–18873, https://doi.org/10.1073/pnas.1821646116

Seroussi, H., Ivins, E.R., Wiens, D.A. and Bondzio, J. 2017. Influence of a West Antarctic mantle plume on ice sheet basal conditions. *Journal of Geophysical Research Solid Earth*, **122**, 7127–7155, https://doi.org/10.1002/2017JB014423

Shen, W., Wiens, D.A. *et al.* 2017. Seismic evidence for lithospheric foundering beneath the southern Transantarctic Mountains, Antarctica. *Geology*, **46**, 1–4, https://doi.org/10.1130/G39555.1

Shen, W., Wiens, D.A. *et al.* 2018. The crust and upper mantle structure of central and west Antarctica from Bayesian inversion of Rayleigh wave and receiver functions. *Journal of Geophysical Research Solid Earth*, **123**, 7824–7849, https://doi.org/10.1029/2017JB015346

Siddoway, C.S., Richard, S.M., Fanning, C.M. and Luyendyk, B.P. 2004. Origin and emplacement of a middle Cretaceous gneiss dome, Fosdick Mountains, West Antarctica. *Geological Society of America Special Papers*, **380**, 267–294, https://doi.org/10.1130/0-8137-2380-9.267

Spiegel, C., Lindow, J. *et al.* 2016. Tectonomorphic evolution of Marie Byrd Land – implications for Cenozoic rifting activity and onset of West Antarctic glaciation. *Global and Planetary Change*, **145**, 98–115, https://doi.org/10.1016/j.gloplacha.2016.08.013

Stern, T.A. and ten Brink, U.S. 1989. Flexural uplift of the Transantarctic Mountains. *Journal of Geophysical Research*, **94**, 10315–10330, https://doi.org/10.1029/JB094iB08p10315

Stern, T.A., Baxter, A.K. and Barrett, P.J. 2005. Isostatic rebound due to glacial erosion within the Transantarctic Mountains. *Geology*, **33**, 221–224, https://doi.org/10.1130/G21068.1

Studinger, M. and Miller, H. 1999. Crustal structure of the Filchner-Ronne Shelf and Coats Land, Antarctica, from gravity and magnetic data: implications for the breakup of Gondwana. *Journal of Geophysical Research Solid Earth*, **104**, 20379–20394, https://doi.org/10.1029/1999JB900117

Studinger, M., Bell, R.E., Blankenship, D.D., Finn, C.A., Arko, R.A., Morse, D.L. and Joughin, I. 2001. Subglacial sediments: a regional geological template for ice flow in West Antarctica. *Geophysical Research Letters*, **28**, 3493–3496, https://doi.org/10.1029/2000GL011788

Studinger, M., Bell, R.E., Buck, W.R., Karner, G.D. and Blankenship, D.D. 2004. Sub-ice geology inland of the Transantarctic Mountains in light of new aerogeophysical data. *Earth and Planetary Science Letters*, **220**, 391–408, https://doi.org/10.1016/S0012-821X(04)00066-4

Stump, E., Sheridan, M.F., Borg, S.G. and Sutter, J.F. 1980. Early miocene subglacial basalts, the east Antarctic ice sheet, and uplift of the Transantarctic Mountains. *Science*, **207**, 757–759, https://doi.org/10.1126/science.207.4432.757

Sugden, D.E. and Denton, G.H. 2004. Cenozoic landscape evolution of the Convoy Range to Mackay Glacier area, Transantarctic Mountains: onshore to offshore synthesis. *Geological Society of America Bulletin*, **116**, 840–857, https://doi.org/10.1130/B25356.1

Sugden, D.E. and Jamieson, S.S.R. 2018. The pre-glacial landscape of Antarctica. *Scottish Geographical Journal*, **134**, 203–223, https://doi.org/10.1080/14702541.2018.1535090

Sugden, D.E. and John, B.S. 1976. *Glaciers and Landscape*. Edward Arnold, London.

Sutherland, R., Spasojevic, S. and Gurnis, M. 2010. Mantle upwelling after Gondwana subduction death explains anomalous topography and subsidence histories of eastern New Zealand and west Antarctica. *Geology*, **38**, 155–158, https://doi.org/10.1130/G30613.1

Taylor, J., Siegert, M.J., Payne, A.J. and Hubbard, B. 2004. Regional-scale bed roughness beneath ice masses: measurement and analysis. *Computers & Geosciences*, **30**, 899–908, https://doi.org/10.1016/j.cageo.2004.06.007

ten Brink, U.S., Hackney, R.I., Bannister, S., Stern, T.A. and Makovsky, Y. 1997. Uplift of the Transantarctic Mountains and the bedrock beneath the East Antarctic ice sheet. *Journal of Geophysical Research Solid Earth*, **102**, 27603–27621, https://doi.org/10.1029/97jb02483

Thomas, R.H. 1979. The dynamics of marine ice sheets. *Journal of Glaciology*, **24**, 167–177, https://doi.org/10.1017/S0022143000014726

Thomson, S.N., Reiners, P.W., Hemming, S.R. and Gehrels, G.E. 2013. The contribution of glacial erosion to shaping the hidden landscape of East Antarctica. *Nature Geoscience*, **6**, 203–207, https://doi.org/10.1038/ngeo1722

Tinto, K.J., Padman, L. *et al.* 2019. Ross Ice Shelf response to climate driven by the tectonic imprint on seafloor bathymetry. *Nature Geoscience*, **12**, 441–449, https://doi.org/10.1038/s41561-019-0370-2

Tochilin, C.J., Reiners, P.W., Thomson, S.N., Gehrels, G.E., Hemming, S.R. and Pierce, E.L. 2012. Erosional history of the Prydz Bay sector of East Antarctica from detrital apatite and zircon geo- and thermochronology multidating. *Geochemistry, Geophysics, Geosystems*, **13**, 1–21, https://doi.org/10.1029/2012GC004364

Tripati, A., Backman, J., Elderfield, H. and Ferretti, P. 2005. Eocene bipolar glaciation associated with global carbon cycle changes. *Nature*, **436**, 341–346, https://doi.org/10.1038/nature04289

van Wijk, J.W., Lawrence, J.F. and Driscoll, N.W. 2008. Formation of the Transantarctic Mountains related to extension of the West Antarctic Rift system. *Tectonophysics*, **458**, 117–126, https://doi.org/10.1016/j.tecto.2008.03.009

van Wyk de Vries, M., Bingham, R.G. and Hein, A.S. 2018. A new volcanic province: an inventory of subglacial volcanoes in West Antarctica. *Geological Society, London, Special Publications*, **461**, 231–248, https://doi.org/10.1144/SP461.7

Watts, A.B. 2001. *Isostasy and Flexure of the Lithosphere*. Cambridge University Press, Cambridge.

Watts, A.B. and Burov, E.B. 2003. Lithospheric strength and its relationship to the elastic and seismogenic layer thickness. *Earth and Planetary Science Letters*, **213**, 113–131, https://doi.org/10.1016/S0012-821X(03)00289-9

Watts, A.B. and Moore, J.D.P. 2017. Flexural isostasy: constraints from gravity and topography power spectra. *Journal of Geophysical Research Solid Earth*, **122**, 1–14, https://doi.org/10.1002/2017JB014571

Watts, A.B., Zhong, S.J. and Hunter, J. 2013. The behavior of the lithosphere on seismic to geologic timescales. *Annual Review of Earth and Planetary Science*, **41**, 443–468, https://doi.org/10.1146/annurev-earth-042711-105457

Weissel, J.K. and Karner, G.D. 1989. Flexural uplift of rift flanks due to mechanical unloading of the lithosphere during extension. *Journal of Geophysical Research Solid Earth*, **94**, 13919–13950, https://doi.org/10.1029/JB094iB10p13919

Wenman, C.P., Harry, D.L. and Jha, S. 2020. Post middle Miocene tectonomagmatic and stratigraphic evolution of the Victoria Land Basin, West Antarctica. *Geochemistry, Geophysics, Geosystems*, **21**, https://doi.org/10.1029/2019GC008568

White, D.A. 2013. Cenozoic landscape and ice drainage evolution in the Lambert Glacier-Amery Ice Shelf system. *Geological Society, London, Special Publications*, **381**, 151–165, https://doi.org/10.1144/SP381.15

Whittaker, J.M., Müller, R.D. and Gurnis, M. 2010. Development of the Australian–Antarctic depth anomaly. *Geochemistry, Geophysics, Geosystems*, **11**, https://doi.org/10.1029/2010G C003276

Wiens, D.A., Shen, W. and Lloyd, A. 2021. The seismic structure of the Antarctic upper mantle. *Geological Society, London, Memoirs*, **56**, https://doi.org/10.1144/M56-2020-18

Wilson, D.S. and Luyendyk, B.P. 2006. Bedrock platforms within the Ross Embayment, West Antarctica: hypotheses for ice sheet history, wave erosion, Cenozoic extension, and thermal subsidence. *Geochemistry, Geophysics Geosystems*, **7**, 1–23, https://doi.org/10.1029/2006GC001294

Wilson, D.S. and Luyendyk, B.P. 2009. West Antarctic paleotopography estimated at the Eocene-Oligocene climate transition. *Geophysical Research Letters*, **36**, L16302, https://doi.org/10.1029/2009GL039297

Wilson, D.S., Jamieson, S.S.R., Barrett, P.J., Leitchenkov, G., Gohl, K. and Larter, R.D. 2012. Antarctic topography at the Eocene-Oligocene boundary. *Palaeogeography Palaeoclimatology Palaeoecology*, **335–336**, 24–34, https://doi.org/10.1016/j.palaeo.2011.05.028

Wilson, D.S., Pollard, D., DeConto, R.M., Jamieson, S.S.R. and Luyendyk, B.P. 2013. Initiation of the West Antarctic Ice Sheet and estimates of total Antarctic ice volume in the earliest Oligocene. *Geophysical Research Letters*, **40**, 4305–4309, https://doi.org/10.1002/grl.50797

Winberry, J.P. and Anandakrishnan, S. 2004. Crustal structure of the West Antarctic rift system and Marie Byrd Land hotspot. *Geology*, **32**, 977–980, https://doi.org/10.1130/G20768.1

Winterbourne, J., White, N. and Crosby, A. 2014. Accurate measurements of residual topography from the oceanic realm. *Tectonics*, **33**, 982–1015, https://doi.org/10.1002/2013TC00 3372

Young, D.A., Wright, A.P. *et al.* 2011. A dynamic early East Antarctic Ice Sheet suggested by ice-covered fjord landscapes. *Nature*, **474**, 72–75, https://doi.org/10.1038/nature10114

Zachos, J. and Kump, L. 2005. Carbon cycle feedbacks and the initiation of Antarctic glaciation in the earliest Oligocene. *Global and Planetary Change*, **47**, 51–66, https://doi.org/10.1016/j.gloplacha.2005.01.001

Zachos, J.C., Flower, B.P. and Paul, H. 1997. Orbitally paced climate oscillations across the Oligocene/Miocene boundary. *Nature*, **388**, 567–570, https://doi.org/10.1038/41528

Zachos, J., Pagani, M., Sloan, L., Thomas, E. and Billups, K. 2001. Trends, rhythms, and aberrations in global climate 65 Ma to present. *Science*, **292**, 686–693, https://doi.org/10.1126/science.1059412

Zachos, J.C., Dickens, G.R. and Zeebe, R.E. 2008. An early Cenozoic perspective on greenhouse warming and carbon-cycle dynamics. *Nature*, **451**, 279–283, https://doi.org/10.1038/nature06588

Mantle convection and possible mantle plumes beneath Antarctica – insights from geodynamic models and implications for topography

Eva Bredow[1], Bernhard Steinberger[2,3]*, Rene Gassmöller[4] and Juliane Dannberg[4]

[1]Department of Geosciences, Kiel University, Otto-Hahn-Platz 1, 24118 Kiel, Germany

[2]GFZ German Research Centre for Geosciences, Telegrafenberg, 14473 Potsdam, Germany

[3]Centre for Earth Evolution and Dynamics, University of Oslo, PO Box 1028 Blindern, NO-0315 Oslo, Norway

[4]Department of Geological Sciences, University of Florida, Gainesville, Florida, USA

EB, 0000-0002-7111-1838; BS, 0000-0002-3643-3049; RG, 0000-0001-7098-8198; JD, 0000-0003-0357-7115

*Correspondence: bstein@gfz-potsdam.de

Abstract: This chapter describes large-scale mantle flow structures beneath Antarctica as derived from global seismic tomography models of the present-day state. In combination with plate reconstructions, the time-dependent pattern of palaeosubduction can be simulated and is shown from the rarely seen Antarctic perspective. Furthermore, a dynamic topography model demonstrates which kind and scales of surface manifestations can be expected as a direct and observable result of mantle convection. The last section of this chapter features an overview of the classical concept of deep-mantle plumes from a geodynamic point of view and how recent insights, mostly from seismic tomography, have changed the understanding of plume structures and dynamics over past decades. The long-standing and controversial hypothesis of a mantle plume beneath West Antarctica is summarized and addressed with geodynamic models, which estimate the excess heat flow of a potential plume at the bedrock surface. However, the predicted heat flow is small, while differences in surface heat-flux estimates are large; therefore, the results are not conclusive with regard to the existence of a West Antarctic mantle plume. Finally, it is shown that global mantle flow would cause the tilting of whole-mantle plume conduits beneath West Antarctica such that their base is predicted to be displaced about 20° northward relative to the surface position, closer to the southern margin of the Pacific Large Low-Shear Velocity Province.

Large-scale mantle flow beneath Antarctica

Mantle convection is the main mode of how heat, both from the Earth's initial formation and that continuously regenerated through radioactive decay, is transported from the deep Earth interior to near its surface (Schubert *et al.* 2001). An increase in temperature would lead to a reduction in viscosity, and it is standard theory that through a negative feedback the Earth will maintain a temperature such that it is sufficiently 'soft' to convect and loose its heat in this way (Tozer 1972). However, such self-regulation might be prevented due to the effect of mantle melting on viscosity (Korenaga 2016). Since the Earth's heat flux is not balanced by radiogenic heat production, the Earth is cooling with time (Korenaga 2008). Primary evidence is the greater prevalence of komatiites in the Archean, but radiogenic heating will be better constrained by future measurements of geoneutrino flux. The rheology of mantle materials is very poorly known; and even for the radial mantle viscosity structure, a wide variety of models has been proposed in recent years (e.g. Steinberger and Calderwood 2006; Čížková *et al.* 2012; Justo *et al.* 2015; Marquardt and Miyagi 2015; Roy and Peltier 2015; Rudolph *et al.* 2015; King 2016; Lau *et al.* 2016; Liu and Zhong 2016; Nakada *et al.* 2017). Hence, there are large uncertainties in the mantle flow structure. However, there are also certain observables that can be obtained as model output from mantle-flow computations and compared to observed values; in this way, flow structure can be better constrained. In particular, the large-scale geoid can be predicted quite successfully (Hager and Richards 1989), and therefore there is some confidence in at least models of the large-scale flow structure.

Tectonic plates are the surface expression of mantle convection. In particular, where plates converge, in most cases one plate will dive into the mantle, as it is cold and heavy, and the sinking plates are the essential drivers of plate tectonics (Forsyth and Uyeda 1975) and mantle convection (Davies 1977; Conrad and Lithgow-Bertelloni 2002). One way of reconstructing mantle structure and flow is hence based on plate reconstructions – where plates have been converging and sinking throughout geological history (Richards and Engebretson 1992; Ricard *et al.* 1993; Bunge *et al.* 2002; Shephard *et al.* 2012). Another main source of information on mantle structure is seismic tomography (e.g. Dziewonski and Woodhouse 1987; Becker and Boschi 2002; Grand 2002; Montelli *et al.* 2006; Schaeffer and Lebedev 2013; French and Romanowicz 2015; Hosseini *et al.* 2020): regions with fast seismic velocity anomalies are thought to be colder, denser and, hence, likely to be sinking, whereas regions with slow anomalies are hotter, less dense and buoyantly rising. If seismic anomalies are largely due to temperature variations, the latter can be computed based on mineral physics (Steinberger and Calderwood 2006; Fullea *et al.* 2009). Ideally, the mantle temperature and density structure inferred from tomography should closely match that inferred from subduction history. In practice, there is at least some similarity between them on the largest scales (Shephard *et al.* 2012; Steinberger *et al.* 2012).

On these largest scales, one feature that has probably been present for the last 300 million years or so is the 'Ring of Fire' – a ring of subduction zones surrounding the basin of the Pacific and its predecessor, the Panthalassic Ocean (Fig. 1) (Steinberger *et al.* 2012; Domeier and Torsvik 2019). As slabs in this circum-Pacific belt mostly cause sinking flow, there must be rising flow elsewhere. Hence, there are probably also two antipodal regions of rising flow, one beneath the Pacific and one beneath Africa, and mantle flow is dominated by a large-scale spherical harmonic of degree 2 (or quadrupolar) structure (Conrad *et al.* 2013). These regions of rising mantle flow roughly overlay the two large low-shear velocity provinces (LLSVPs) of the lowermost mantle. Figure 1 shows, from a south polar perspective, how this ring of subduction zones has been continuous across roughly the present-day location of Antarctica until around 80 Ma ago. Since then, southward subduction of the Phoenix Plate (which has now completely disappeared) has mostly stopped. Subduction continues on either side of it – north of New Zealand and beneath South America – but given that slabs probably take 200–300 million years to sink to the base of the mantle (van der Meer *et al.* 2010; Domeier *et al.* 2016), there is still a lot of sinking slab material beneath Antarctica. Hence, there is still an overall downward flow expected in the region.

From: Martin, A. P. and van der Wal, W. (eds) 2023. *The Geochemistry and Geophysics of the Antarctic Mantle.* Geological Society, London, Memoirs, **56**, 253–266,
First published online 9 September 2021, https://doi.org/10.1144/M56-2020-2

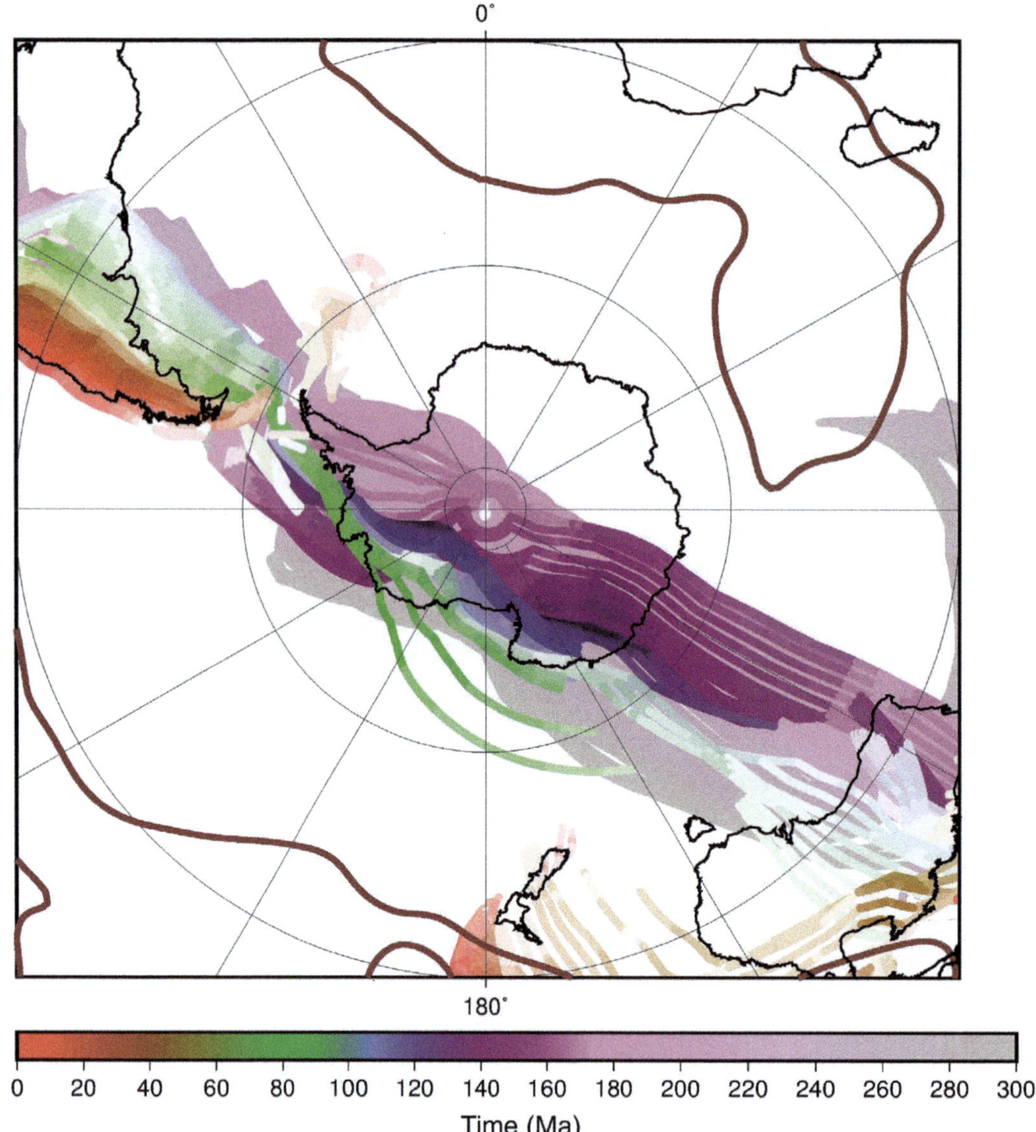

Fig. 1. Locations of palaeosubduction since 300 Ma from the Antarctic perspective. After 140 Ma, higher subduction rates correspond to darker colours. See Steinberger *et al.* (2012) for details, in particular their figure 2 for a complete colour scale that also specifies how darkness relates to subduction rates. The brown line shows the −1% contour in the lowermost layer of Smean (Becker and Boschi 2002), an average over three tomography models, as a proxy for LLSVP margins.

Figure 2 shows large-scale convective mantle flow (Hager and O'Connell 1979, 1981) based on a density model inferred from mantle tomography. Conversion from seismic velocity to density anomalies and radial viscosity structure is based on mineral physics; the latter additionally including constraints from geoid, heat flux and post-glacial rebound (Steinberger and Calderwood 2006). Overall, it conforms to the expectations outlined in the previous paragraph: there is a belt of downward flow crossing Antarctica. Flow is mostly towards it in the upper part of the mantle, and away from it towards the base of the mantle. The belt of downward flow tends to be narrower in the upper part of the mantle. Upward flow extends from the Pacific towards parts of West Antarctica, whereas most of East Antarctica is underlain by downward flow. Although this result is for one particular tomography model, the large-scale structure is has been quite consistently imaged throughout various recent tomography models. Hence, these features appear to be rather robust: even though there may still be slabs present in the upper *c.* 1000 km beneath Antarctica, especially beneath the Antarctic Peninsula (Lloyd 2018), upward flow may occur beneath the Ross Sea Embayment and Marie Byrd Land in the upper mantle, even above sinking slabs in the lower mantle. This could result if hot material has entered this region through horizontal flow in the upper mantle or tilted plume conduits (as discussed below), with its buoyancy counteracting the negative buoyancy of slabs beneath. A similar setting is also likely to be present in the western USA. Horizontal flow in the upper mantle, shown here at a depth of 650 km, also exhibits more small-scale structures – in particular, flow across East Antarctica, towards the West Antarctic upper-mantle upwelling. At a depth of 2650 km, viscosity is higher and small-scale flow structures are less evident.

Mantle convection and dynamic topography

Besides the geoid, dynamic topography – that is, how the lithosphere is pushed upwards above mantle upwellings and pulled down above downwellings (as sketched in Fig. 3) – is another important prediction from mantle flow models (Yang and Gurnis 2016; Steinberger *et al.* 2019*a*). Amplitudes are of the order of 1 km over regions extending for several

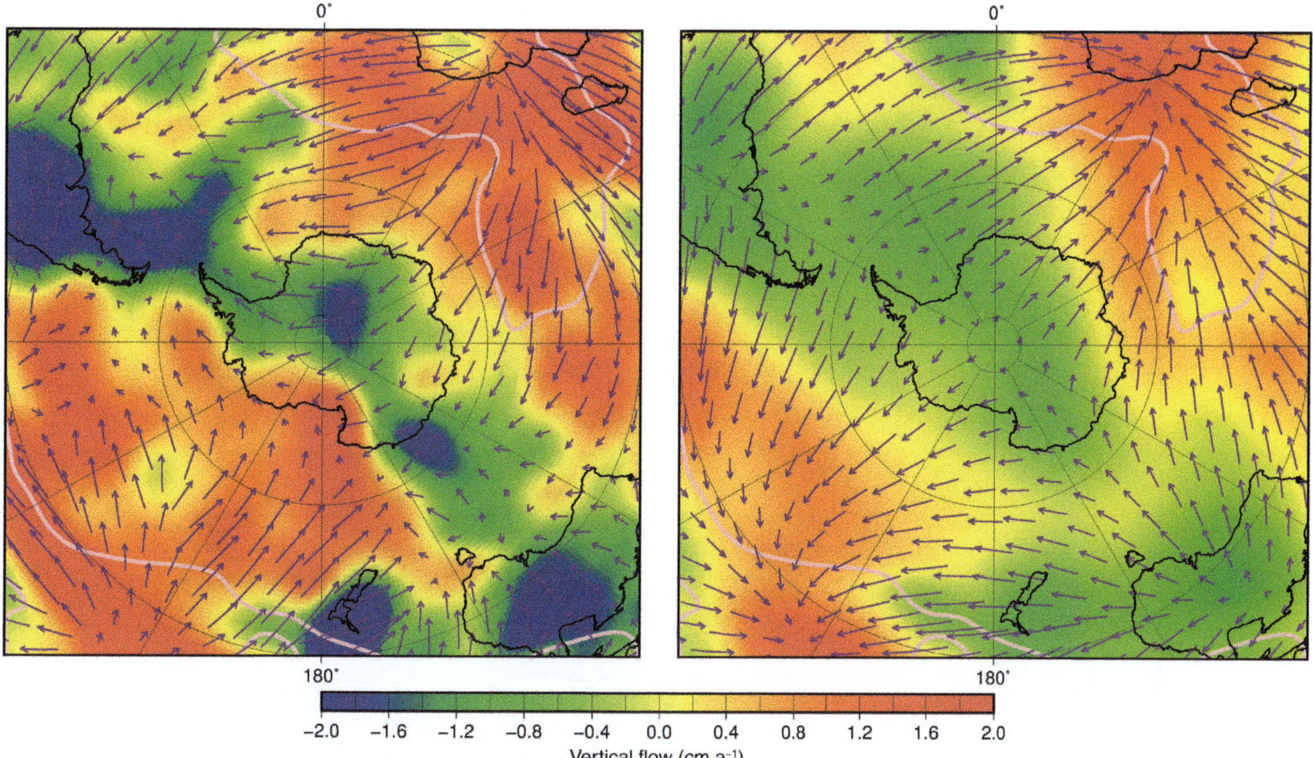

Fig. 2. Computed global mantle convection flow field from the Antarctic perspective, for S10mean (Doubrovine *et al.* 2016), an average over 10 tomography models, considering chemically distinct LLSVPs: see Steinberger *et al.* (2019*b*) for details. Depth slices are at 650 km (left) and 2650 km (right). Colours for vertical flow, arrows for horizontal flow. 1 cm a^{-1} corresponds to 4 degrees of arc arrow length. The pink line shows the −1% contour in the lowermost layer of Smean (Becker and Boschi 2002) as a proxy for LLSVP margins.

hundred to thousands of kilometres, reaching maximal amplitudes of 2–3 km in some regions. It can be compared to observed topography; however, the comparison is not straightforward, as most of the topography is sustained by crustal thickness variations that have to be corrected for and are uncertain. The corrected 'residual topography' (e.g. Hoggard *et al.* 2016), hence, has uncertainties of the order of 1 km. Figure 4 shows positive dynamic topography corresponding to upward flow in the upper mantle, beneath West Antarctica, whereas dynamic topography is mostly negative in East Antarctica, due to mostly downward flow beneath. This result

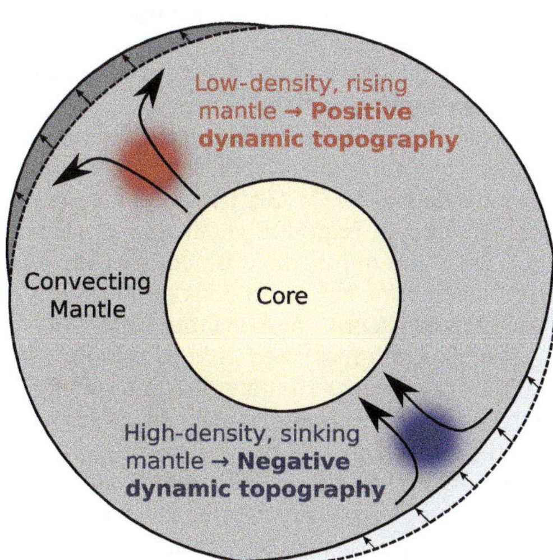

Fig. 3. Low- or high-density anomalies in the mantle induce upward or downward mantle flow, which affects the surface as positive or negative dynamic topography, respectively. Sketch after Braun (2010).

only considers variations of viscosity with depth; lateral viscosity variations (LVVs) are disregarded. Steinberger *et al.* (2019*a*) found that if LVVs due to temperature variations inferred from seismic tomography are considered, the dynamic topography pattern remains broadly similar, but the amplitude tends to be higher in continental regions because thicker continental lithosphere tends to couple more strongly to underlying mantle flow than thinner oceanic lithosphere. The presence of a plume, as discussed in the following 'Mantle plumes' section, could cause a pronounced plume-fed low-viscosity zone in the shallow asthenosphere, partly decoupling the lithosphere from the underlying mantle flow. Since such a decoupling layer is not present in our models, absolute amplitudes – in particular in plume-affected regions – could be somewhat too high in our models, while patterns are largely correct. In East Antarctica, modelled thick lithosphere leads to more pronounced negative dynamic topography, if LVVs are considered. This makes it even more at variance with observation-based estimates (Sleep 2006; Paxman 2021, this volume). The de-iced topography, which results from converting the ice sheet to an equivalent rock layer, is high in East Antarctica, and the inferred dynamic topography is positive. To explain this discrepancy, Sleep (2006) proposed a plume under East Antarctica that might even have contributed to triggering Oligocene glaciation, in addition to the effect of declining atmospheric CO_2 (DeConto and Pollard 2003). Ponded plume material below the lithosphere could cause dynamic uplift; while possibly neither the plume conduit nor the layer of ponded low-velocity material could be seismically imaged if they are rather thin. An alternative explanation for the high topography of East Antarctica could be that it has not eroded much since the last orogeny, and that its crust is thicker than in the models that are used to subtract isostatic topography. For more detail concerning specific Antarctic regions, see Paxman (2021, this volume).

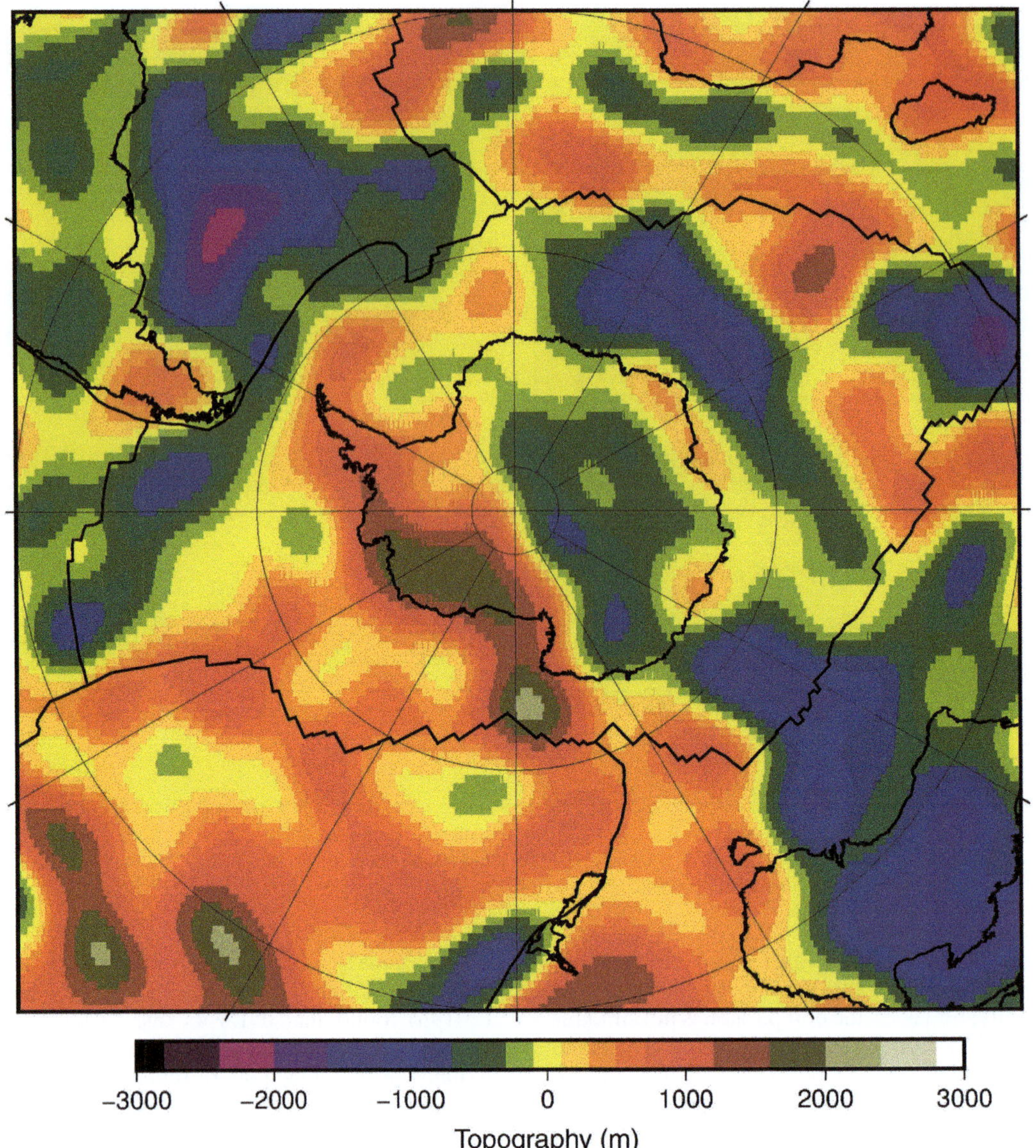

Fig. 4. A model of dynamic topography from Steinberger *et al.* (2019*a*), a case without lateral viscosity variations from the Antarctic perspective.

Mantle plumes

Classical and modern concepts of mantle plumes

Historically, the idea of steady plumes in the Earth's mantle started with surface observations in the Pacific Ocean, more precisely with the eye-catching Hawaiian–Emperor Seamount Chain. This strictly age-progressive line of seamounts, plateaus and islands is almost 6000 km long, linear (with the characteristic 60° bend) and ends close to Hawaii with its well-known volcanic activity. A stationary heat source within the mantle, above which the tectonic Pacific Plate moved slowly over time, provides a simple and elegant explanation for the origin of these impressive and long-lived surface features, and was first proposed by Wilson (1963).

This theory was later refined by Morgan (1971, 1972), who described mantle plumes as localized upwellings of hot, buoyant material rising from the core–mantle boundary (CMB) through the entire mantle up to the base of the lithosphere, where the material spreads laterally and pressure-release melting creates a volcanically active hotspot, such as Hawaii, at the surface. These central elements define what is nowadays referred to as the classical plume theory.

This theory has been revised and even entirely questioned (e.g. Anderson and Natland 2005; Foulger 2011) several times since its original formulation because mantle plumes are difficult to image. Therefore, the classical plume theory can neither be easily proved nor disproved.

However, evidence in favour of the existence of plumes has been provided by numerous laboratory experiments (e.g. Whitehead and Luther 1975; Griffiths and Campbell 1990) and numerical models (e.g. Farnetani and Richards 1995; van Keken 1997) that aim to investigate the basic principles of thermal convection and mantle dynamics (see Fig. 5). These studies consistently demonstrate that hot, buoyant upwellings (such as plumes) and cold, dense downwellings (such as subduction zones) are natural and dynamic counterparts of any convecting system, and can therefore also be expected in the Earth's mantle.

Concerning the shape of mantle plumes, both laboratory and numerical models indicate that thermal plumes initially consist of a large, spherical, plume head, subsequently supplied by a cylindrical, narrow and long plume tail (see Fig. 5); this head-and-tail structure results in two very different surface effects. Plume heads initiate voluminous eruptions that create gigantic flood basalt provinces within a relatively

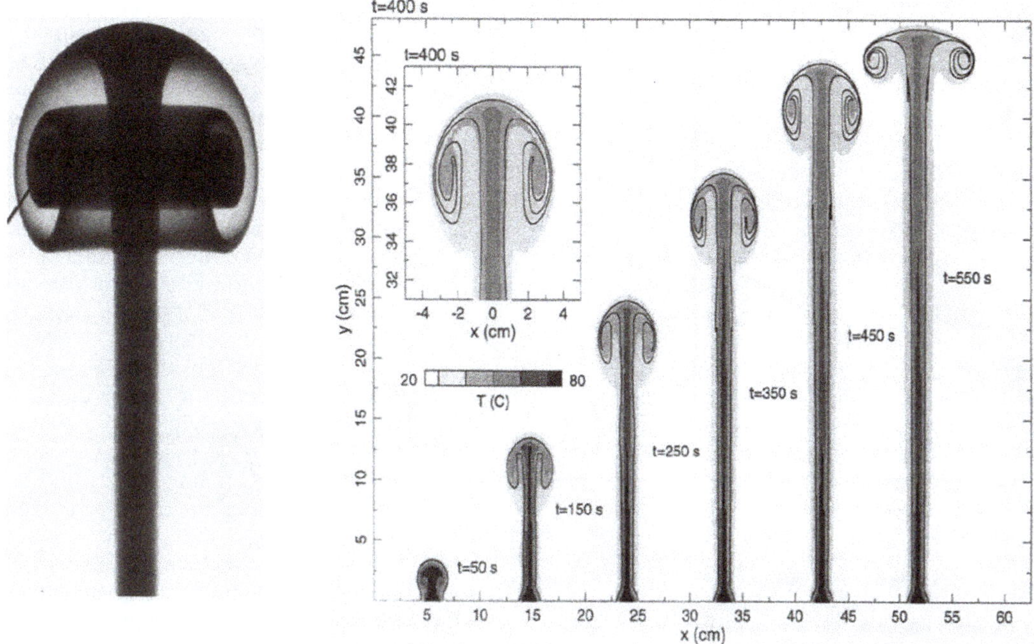

Fig. 5. Both laboratory experiments with highly viscous glucose syrup (left: from Griffiths and Campbell 1990) and 2D numerical models (right: from van Keken 1997, and with laboratory scaling) showed early on that upwelling plumes generally look like a mushroom: a large spherical head and a thin vertical tail. Our results in the 'Modelling plume conduits beneath Antarctica' subsection suggest that for a potential plume under West Antarctica, the head rise time translates to *c.* 30–60 million years, with large uncertainty, for mantle scales. The left image is reprinted from Griffiths and Campbell (1990, fig. 1c), with permission of Elsevier ©1990; the right image is reprinted from van Keken (1997, fig. 2), with permission of Elsevier ©1997.

short duration of a few million years (Large Igneous Provinces, abbreviated LIPs and defined by Bryan and Ernst (2008, p. 175) as 'magmatic provinces with areal extents >0.1 Mkm^2, igneous volumes >0.1 Mkm^3 and maximum lifespans of ~50 Myr that have intraplate tectonic settings or geochemical affinities, and are characterised by igneous pulse(s) of short duration (~1–5 Myr), during which a large proportion (>75%) of the total igneous volume has been emplaced'). Plume tails, conversely, can easily be active for more than a hundred million years, produce substantially less magma and create an age-progressive hotspot track when the lithosphere moves above the relatively stationary plume (Richards *et al.* 1989). Both plume heads and tails have reshaped substantial areas on the Earth's surface (see Fig. 6).

Note that apart from the interaction with the mobile tectonic plates, the amount of volcanic products produced also depends on the relief of the base of the lithosphere, since hot material can flow buoyantly upwards and pond beneath regions of thinner lithosphere. This process is known as upside-down drainage (Sleep 1997), and emphasizes the importance of considering local lithosphere thickness variations because melting and the associated hotspot do not necessarily occur vertically above the plume centre.

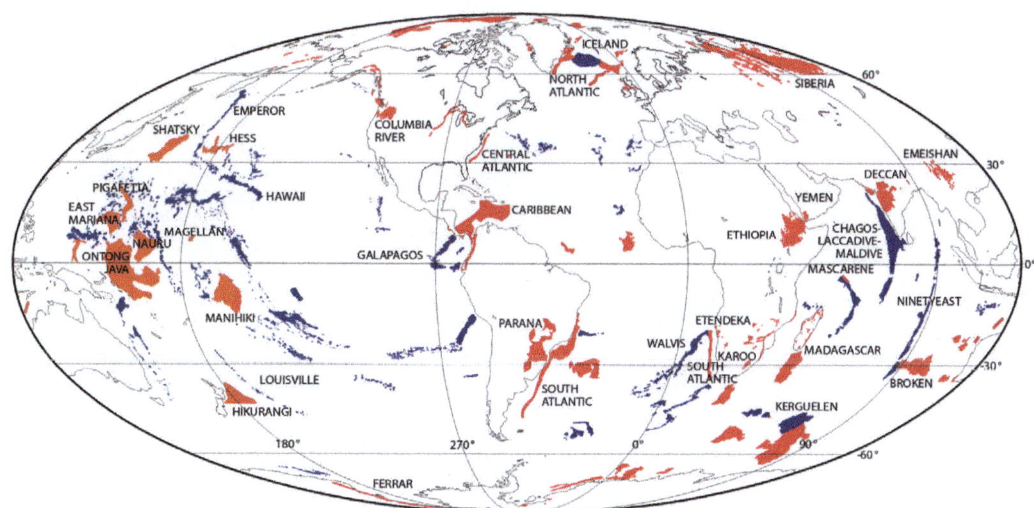

Fig. 6. Overview of LIPs (red) and hotspot tracks (blue), demonstrating the extent of the areas on Earth's surface that have been affected by plume heads and tails, respectively, over many millions of years, in some cases for more than 100 million years. The only onshore LIP marked in Antarctica is the *c.* 180 Ma Ferrar LIP, which follows the Transantarctic Mountains along almost 3500 km (Elliot and Fleming 2004). Offshore, there is also the Kerguelen LIP on the Antarctic Plate. The volcanic province in West Antarctica (see the 'West Antarctic mantle plume hypothesis' subsection) is not shown. Figure from Coffin *et al.* (2006) (licensed under CC BY 4.0; https://creativecommons.org/licenses/by/4.0/).

The term hotspot is rather vaguely defined as a localized surface region where volcanic activities take place over a long time and are independent of any plate tectonic processes (e.g. Schubert *et al.* 2001). Therefore, different records list different hotspots, usually between 40 and 50, depending on the applied criteria (e.g. Steinberger 2000; Courtillot *et al.* 2003; King and Adam 2014). Generally accepted factors for hotspots possibly fed by deep-mantle plumes are the occurrence of a LIP and a clearly age-progressive hotspot track in accordance with the reconstructed directions and velocities of the moving plates, as well as ongoing magmatic activities at the current hotspot location, surrounded by a broad topographical hotspot swell (e.g. Courtillot *et al.* 2003). Furthermore, the geochemical signature of hotspot-derived rock samples resembles that of ocean-island basalts while being distinctly different from other basalts produced at mid-ocean ridges (e.g. Moreira and Allègre 1998), and the plume requires a certain buoyancy to be able to ascend through the entire mantle (e.g. Steinberger and O'Connell 1998). Hotspots can be active for many tens of millions of years; for example, the Kerguelen hotspot has been persistently active for *c.* 130 million years (Coffin *et al.* 2002).

As mentioned previously, plumes are difficult to image. Seismic tomography is theoretically able to detect regions of reduced seismic velocities. The technique, however, is greatly limited by the available ray coverage and the rapidly decreasing resolution with depth, which make it extremely challenging to capture narrow plume conduits. A few years ago, French and Romanowicz (2015) provided the first convincing and long-awaited whole-mantle tomography images that do resolve continuous slow-velocity structures throughout the entire mantle (see Fig. 7 for their global distribution). The deep roots of the plumes at the CMB, however, appear to be much broader than expected and the highly deflected plume tails above approximately 1000 km depth deviate significantly from the classically predicted vertical conduits.

Additional geometrical discrepancies with respect to the classical plume concept comprise deformed, asymmetrical plume heads and highly tilted plume tails, which may result from an asymmetrical relief of the base of the lithosphere, interactions with the lithosphere moving above the plume, large-scale asthenospheric flow surrounding (and possibly deflecting) the plume or, in particular, interactions with nearby spreading ridges (e.g. as shown for the Réunion plume in the geodynamic models of Bredow *et al.* 2017 or the surface wave tomography model of Mazzullo *et al.* 2017).

Regarding the source location, plumes are assumed to start from instabilities at thermal boundary layers, such as the CMB. More precisely, reconstructed eruption sites of LIPs and present-day hotspot positions seem to indicate that deep plumes are generated at the margins of the African and Pacific LLSVPs (Torsvik *et al.* 2006; Burke *et al.* 2008), defined as the -1% shear wave velocity contour of the Smean composite tomography model (Becker and Boschi 2002). The choice of this specific contour is, however, rather arbitrary and the statistical significance of the spatial correlation can be questioned (Austermann *et al.* 2014). This means that the entire seismically slow zones in the lowermost mantle – and not just their margins – could be potential plume-generation zones, located approximately beneath the African continent and the central Pacific Ocean. Moreover, the global tomography model of Hosseini *et al.* (2020) also images the LLSVPs for the first time in a P-wave model, such that the term LLVP (without the 'shear') becomes more appropriate. Interestingly, the two large and continuous provinces that were consistently seen in previous shear-wave models appear in the P-wave model as numerous patches, which form an almost continuous global belt slightly south of the Equator. However, in any case, the regions in which deep plumes start their ascent through the mantle are located rather far away from Antarctica (see Fig. 7).

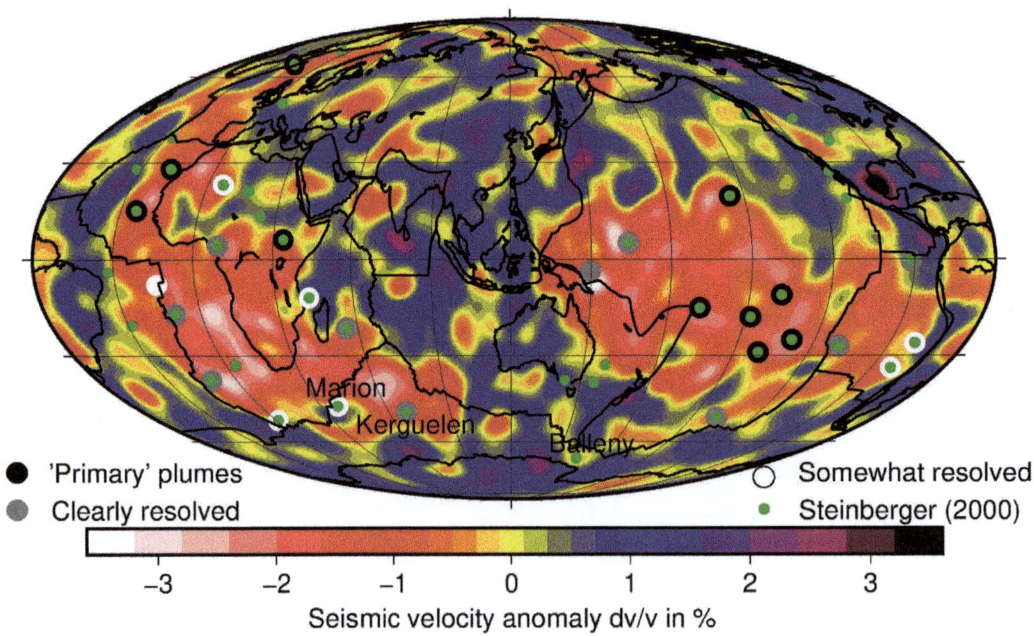

Fig. 7. Hotspot locations after Steinberger (2000) (all green dots) where the hotspots underlain by vertically continuous conduits in the lower mantle in the seismic tomography model of French and Romanowicz (2015) are marked as primary or clearly resolved plumes (black and grey circles, respectively). White circles are for somewhat resolved plumes. The background colours show the tomography model at a depth of 2800 km, which highlights the two zones of extremely slow anomalies, the African and Pacific LLSVPs (the two broad red areas). The only hotspot in the vicinity of Antarctica are the Balleny Islands (south of New Zealand), which lack even a somewhat resolved conduit and are far from the closest LLSVP. The Kerguelen and Marion plumes (clearly and somewhat resolved, respectively) are also beneath the Antarctic Plate, close to the southern margin of the African LLSVP but far from the continent. Figure redrawn after French and Romanowicz (2015).

West Antarctic mantle plume hypothesis

From a global geodynamic perspective, West Antarctica seems to be a rather unlikely location to observe the surface manifestations of a mantle plume (Sleep 2006) and, unsurprisingly, it has never been included in any global hotspot list so far (e.g. Steinberger 2000; Courtillot *et al.* 2003; King and Adam 2014; see also the previous subsection).

Nonetheless, a broad structural dome, resembling a hotspot swell, has been recognized in Marie Byrd Land, based on subglacial bedrock topography corrected for ice loading (Paxman *et al.* 2019). In addition, the geochemical characteristics of basaltic rocks throughout West Antarctica are similar to those of plume-related ocean island basalts and are most likely to originate from a depleted mantle source from depths of at least 100 km (see also Handler *et al.* 2021, this volume). Thus, the West Antarctic Ice Sheet might conceal an extensive LIP. Behrendt *et al.* (1992) were the first to link and explain these observations with a plume underneath West Antarctica, more precisely an ellipsoidal plume beneath the West Antarctic Rift System with a major axis about 3000 km in length. This suggested plume area comprises the entire Marie Byrd Land, the West Antarctic Rift System, the Ross Ice Shelf and even northern Victoria Land – exceeding by far the dimensions of the largest known plumes on Earth such as Hawaii or Iceland.

Another study, which was published in the same year, focused on the petrology of lavas from the still active Mount Erebus volcano on Ross Island (Kyle *et al.* 1992), which also seem to be derived from a depleted asthenospheric mantle source without much crustal contamination. The authors concluded that a relatively small plume centred beneath Mount Erebus with a diameter of 40 km and a rising rate of 6.5 cm a^{-1} would be sufficient to account for the estimated volume of volcanic material necessary to build Mount Erebus and the neighbouring volcanoes – values more within the range of classical plume parameters. However, there are also xenoliths representing young lithosphere (Day *et al.* 2019).

Ever since, the hypothesis of a mantle plume beneath West Antarctica (most often considered either beneath central Marie Byrd Land or Ross Island rather than beneath the entire West Antarctica) has been subject to detailed studies from various geoscientific disciplines. The most abundant indications at the surface (wherever rocks are exposed) are the widely spread basalts, which have been found throughout West Antarctica (LeMasurier and Rex 1989). Having been produced continuously over the past 30 million years, they do not follow a classical age-progressive hotspot track over hundreds of kilometres or even a single chain of volcanoes. This is not, however, a striking argument against a plume, considering that the Antarctic Plate has been virtually immobile over the past 85 million years (Larter *et al.* 2002).

Altogether, Marie Byrd Land hosts 18 large alkaline shield volcanoes, with volumes of up to 1800 km^3 and distributed over the approximately 500 × 800 km large tectonic dome (LeMasurier 2013). Additionally, a recent study found indications for up to 138 individual conical bedrock edifices beneath the thick ice sheets, based on combining aeromagnetic, aerogravity and satellite data (van Wyk de Vries *et al.* 2018). These potential volcanoes are distributed across the entire rift system, including the area of extended continental crust where no volcano had previously been reported. Whether a few or all of the conical bedrock topography edifices have a volcanic origin or not, the West Antarctic subglacial volcanic province is undoubtedly one of the largest volcanic provinces in the world. However, regarding the plume hypothesis, it remains uncertain if the volcanoes form a LIP, which means that conclusive proof for the surface manifestation of a plume head is still missing.

For the sake of completeness, it should be noted that there is also another, much older LIP in Antarctica: the Ferrar LIP (Elliot and Fleming 2004), emplaced at 183 Ma (Burgess *et al.* 2015) along 3500 km of the Transantarctic Mountain range (shown in Fig. 6). However, it is related neither to the volcanic province nor to the plume in West Antarctica.

As mentioned earlier, the geochemical signature of the volcanic rocks in West Antarctica can hardly be distinguished from ocean island basalts and seem to be derived from mantle depths (e.g. LeMasurier and Rex 1989; Martin *et al.* 2013; Panter *et al.* 2018), strongly suggesting the influence of a deep plume. Further geochemical evidence for mantle-plume components has been found in various studies of both Ross Island and Marie Byrd Land (Hole and LeMasurier 1994; Rocholl *et al.* 1995; Panter *et al.* 1997, 2000; Phillips *et al.* 2018). However, helium isotope data have been used to argue against a plume origin (Nardini *et al.* 2009; Day *et al.* 2019).

An increasing amount of seismic data provides evidence that large zones of slow seismic velocities exist both underneath Marie Byrd Land and Ross Island, and could be signs of possible plume structures (e.g. Accardo *et al.* 2014; Hansen *et al.* 2014; Lloyd *et al.* 2015; Shen *et al.* 2018; Wiens *et al.* 2021, this volume). Assuming that these seismic anomalies are temperature-driven, and not caused by the presence of fluids such as water or a different material composition, the potential plume structures can be followed at least down to the transition zone. At depths greater than 800 km, poor resolution impedes any clear findings. One of the most recent Antarctica-wide studies, by Lloyd (2018), concludes that the seismic anomaly beneath Marie Byrd Land (centred near Mount Sidley) may indeed indicate the existence of a mantle plume, whereas the presence of a potential plume beneath Mount Erebus remains elusive. Convincing proof for deep plume structures beneath West Antarctica is still lacking.

Another physical parameter indicating potential plume activities beneath West Antarctica is the elevated heat flux in West Antarctica, either measured at the bedrock surface or calculated in continent-wide heat-flux models (Pappa and Ebbing 2021, this volume) or inferred from mantle xenoliths (Coltorti *et al.* 2021, this volume; Handler *et al.* 2021, this volume; Martin *et al.* 2021, this volume). The surface heat flux has been estimated from global seismic models comprising the crust and upper mantle (Shapiro and Ritzwoller 2004), from satellite magnetic data (Fox Maule *et al.* 2005), from a continental shear velocity model (An *et al.* 2015) or from airborne magnetic data (Martos *et al.* 2017), as shown in Figure 8. Even though direct measurements are sparse and the models result in rather different value ranges and anomalies, the heat flux in West Antarctica is always distinctly elevated compared to the values above cratonic East Antarctica with its thick crust and lithosphere. More or less clearly pronounced anomalies appear in Marie Byrd Land and in the vicinity of Ross Island, following the Transantarctic Mountain chain, which could be interpreted as plume-related surface manifestations due to additional heat supply caused by plume material ponding beneath the lithosphere. This scenario has recently been tested in numerical models by Seroussi *et al.* (2017) in the context of the plume-induced heat flux at the base of the ice sheet, concluding that a plume with moderate parameters is certainly possible beneath Marie Byrd Land.

Additionally, the possible presence of a plume under East Antarctica (already discussed in the section 'Mantle convection and dynamic topography') needs to be considered here: plume material may spill across the Transantarctic Mountains from beneath the thicker East Antarctic lithosphere towards and beneath the thinner West Antarctic lithosphere (Sleep 2006) leading to pressure-release melting. Similar to beneath Africa (Ebinger and Sleep 1998), a single plume could

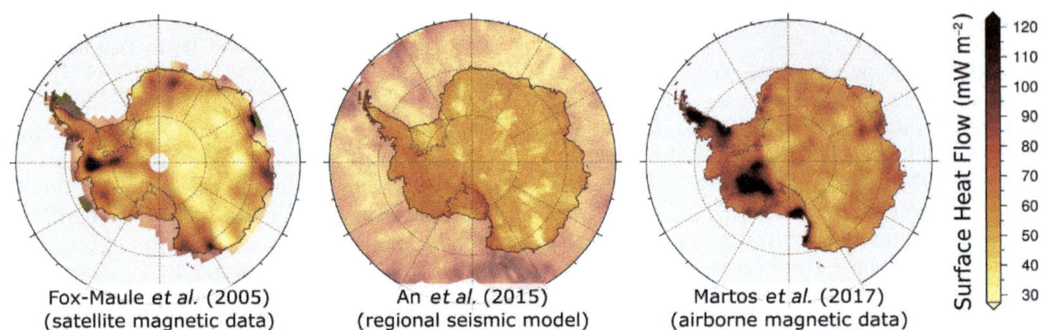

Fig. 8. Previous heat-flux estimates derived either from magnetic or seismic data consistently show an elevated heat flux beneath West Antarctica in contrast to East Antarctica (data from Fox Maule *et al.* 2005; An *et al.* 2015; Martos *et al.* 2017). Since the focus is on the continent, the colours in the oceanic areas are slightly dimmed.

possibly lead to distributed magmatism throughout a wide area, due to lateral flow and ponding of plume material in pre-existing zones of lithospheric thinning.

Putting together all indications for the presence of one or, even, two potential mantle plumes beneath West Antarctica, it becomes clear that without any direct evidence for a LIP, an age-progressive hotspot track or a possible plume origin at greater depths, it is practically impossible to constrain the dynamic history of the plume(s) without a considerable amount of speculation. To complicate the situation, there are a variety of alternative explanations for the intraplate volcanism, such as the presence of water or other fluids in the upper mantle instead of a thermal anomaly; and, in the case of Ross Island, there might also be edge-driven or edge-modulated convection effects (King 2007; Sleep 2007; Panter *et al.* 2018) at the pronounced step in the lithosphere thickness along the Transantarctic Mountain chain (van Wijk *et al.* 2008), or there might be rift- or transtension-related decompression melting (Rocchi *et al.* 2003, 2005; Cooper *et al.* 2007). Also, in contrast to the Marie Byrd Land dome, Ross island is located in a depression.

Geodynamic models of a West Antartic mantle plume

Although the existence of a mantle plume beneath the West Antarctic lithosphere remains uncertain (see the previous subsection for details), it is still useful to study geodynamic models of plumes beneath that region, and which predictions of such models could possibly be compared to observations.

Given the lack of time-dependent observations such as an age-progressive distribution of volcanoes, instantaneous models (which only consider the present-day situation and are de facto not truly ‘dynamic’) seem to be a reasonable choice. The comparatively well-confirmed anomalies of low seismic velocities in the crust and upper mantle can be used as constraints. Furthermore, the recently developed 3D model of the Antarctic lithosphere by Pappa *et al.* (2019) provides continuous regional information about the laterally varying crustal and lithospheric thickness. Model details can be found in Pappa and Ebbing (2021, this volume).

The mantle convection code ASPECT, short for Advanced Solver for Problems in Earth's ConvecTion (Kronbichler *et al.* 2012; Heister *et al.* 2017) and originally intended for simulations of time-dependent convection models, can be used to solve only the energy equation. Usually, the compressible Stokes equations also need to be solved for geodynamic models, but all time-dependent terms can be neglected if the material inside the model domain is not assumed to be moving. In this case, the energy equation is iteratively solved until the temperature field has reached a steady state.

As input parameters, the model requires the spatial distribution of temperatures and compositions. Each composition has a certain density, specific heat capacity, thermal conductivity and specific radiogenic heat production rate, all taken from the lithosphere model of Pappa *et al.* (2019). In order to test the plume hypothesis, a spherical hot anomaly (in the shape of a Gaussian spherical distribution) with a certain excess temperature and width is inserted into the model, either beneath Marie Byrd Land or beneath Ross Island (Fig. 9). The excess temperature is chosen in agreement with literature estimates for other plumes (i.e. between 100 and 250 K: e.g. Schilling 1991; Putirka 2008), whereas the width approximately fits the extent of the seismic anomalies (i.e. between 150 and 250 km: Lloyd 2018).

Besides a steady state for the temperatures, the model provides the heat flux at the bedrock surface beneath the ice sheets (an input parameter of major importance for Glacial Isostatic

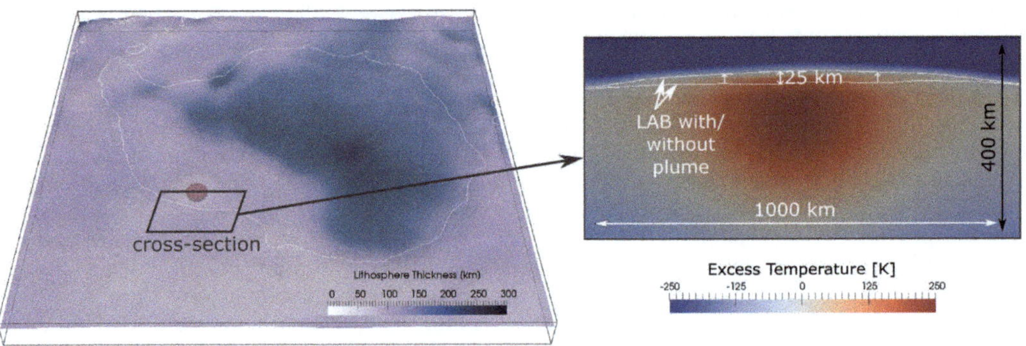

Fig. 9. Lithosphere thickness distribution in the ASPECT model with the inserted spherical thermal anomaly that simulates a plume beneath Marie Byrd Land with a maximum excess temperature of 250 K (left); the white line shows the outline of Antarctica. Cross-section of the thermal anomaly and the resulting uplift of the lithosphere–asthenosphere boundary (LAB) (right).

Adjustment models). The final model output represents an estimation of the additional heat flux that would be contributed by a plume-sized thermal anomaly either beneath Marie Byrd Land or beneath Ross Island.

The results show that the lithosphere–asthenosphere boundary (LAB) is affected by the plume anomaly over a maximum diameter of 1000 km and shifted upwards by up to 25 km (Fig. 9), in good agreement with previous models of plume–lithosphere interactions (Bredow *et al.* 2017). The additional heat-flux signature caused by the thermal (plume) anomaly reaches values of between 7.1 and 14.0 mW m^{-2}, with a diameter between 600 and 1290 km (Fig. 10). The shape of the plume heat-flux signature is not consistently circular, pointing out the importance of considering local lithosphere thickness variations. This is especially important in the case of the Ross Island plume, since it is very close to the sudden step in the lithosphere thickness along the Transantarctic Mountains. Altogether, the changes in the heat flux caused by the plume seem to be rather small. It should be noted, however, that these calculations only consider the conductive heat transfer and neglect any heat transport via volcanic activities that are definitively present in these areas.

Overall, the results do not disagree with previous studies of the surface heat flux, especially given the discrepancies between these studies (see Fig. 8). Evaluated solely from these models, neither position can be ruled out nor confirmed as a potential location for a plume.

It is noteworthy that the parameters used for the simulations have an impact on the results: for example, if the shape of the plume anomaly is pancake-like rather than spherical. The

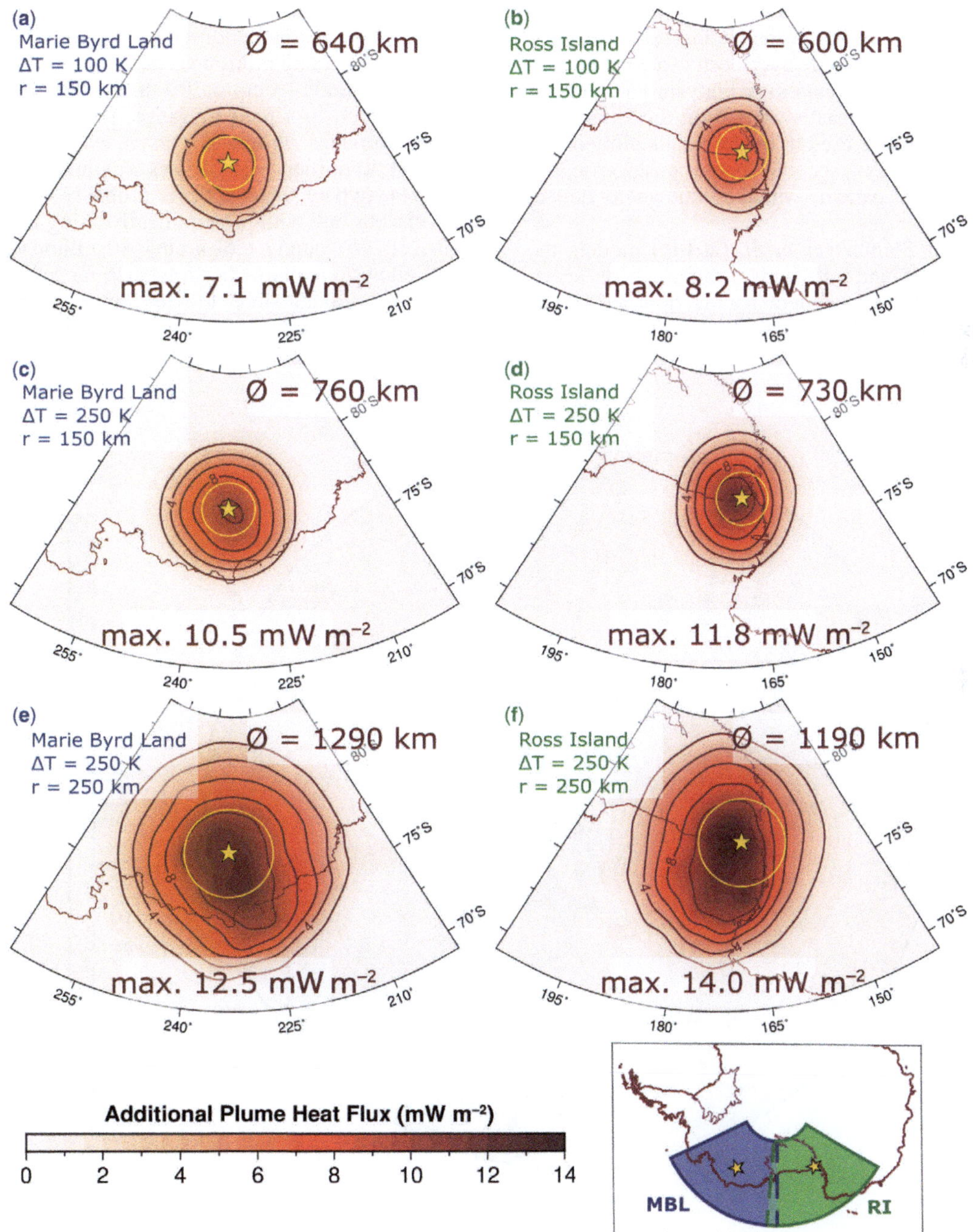

Fig. 10. Additional heat-flux signatures of a thermal plume with a radius of 150 km **(a)**–**(d)**, or 250 km **(e)** & **(f)**, variable plume excess temperatures of 100 K (a) & (b) or 250 K (c)–(f), and located beneath Marie Byrd Land (MBL) (a), (c) & (e) or Ross Island (RI) (b), (d) & (f). Yellow stars and circles denote the positions and outlines of the modelled thermal anomalies. The inset map shows the positions of the two model regions. The diameter is calculated as the average value of the 2 mW m^{-2} contour.

resulting difference, however, is most likely to be very small and since not even the presence of a plume can be confirmed, constraining its shape seems impossible at the moment and with this specific model set-up. Another critical parameter is the lithosphere model by Pappa *et al.* (2019), which is used as input configuration. The distribution of compositions and temperatures, in particular the depth of the LAB, certainly has an impact. Unfortunately, there is no similarly well-resolved model of Antarctica available currently and, hence, so far there is no alternative to the lithosphere model by Pappa *et al.* (2019).

Modelling plume conduits beneath Antarctica

Another approach to investigating a potential plume beneath West Antarctica with geodynamic models tackles the question of its origin. Assuming that the zones of low seismic velocities beneath Marie Byrd Land or Ross Island reflect thermal anomalies, this hot material must either have been heated at its current position via tectonic process(es) or buoyantly risen from greater depths (like a deep-mantle plume). In the latter case, geodynamic models can be used to assess the likelihood that a plume has risen towards a specific surface position within the global mantle-flow pattern, which is known to deflect plumes.

The procedure of Steinberger *et al.* (2019*b*) models the ascent of plume heads and tails with a certain rising speed and embedded in a time-dependent mantle flow field. The Smean tomography model (Becker and Boschi 2002) provides seismic velocities that can be converted into mantle densities (Steinberger and Calderwood 2006). Further input parameters are the radial viscosity structure (Steinberger and Calderwood 2006) and time-dependent plate motions (Torsvik *et al.* 2010), which enable the reconstruction of large-scale mantle flow. Time-dependence is also achieved by backward advecting the density heterogeneities. In the specific case of West Antarctica, the plume is assumed to have reached the surface at 30 Ma, approximately simultaneously with the onset of volcanic activities. Positions beneath both Marie Byrd Land and Ross Island have been tested.

The model result shows the present-day state of the plume conduit, with its deflection by mantle flow discussed earlier in the paper and the location in the D″ layer at the base of the mantle from which it needs to start rising in order to reach the chosen surface position. It also shows whether it is at all possible to generate a stable plume conduit beneath the surface position of interest. For Marie Byrd Land and Ross Island, stable plume conduits are possible if the plume head takes up to 60 or 30 million years, respectively, to rise through the entire mantle – but not if it takes any longer. Both values are, however, in a realistic range, positioned between other recent estimates. Whereas Torsvik *et al.* (2021) found rise times of 30 million years or less for plumes in the vicinity of LLSVPs (where the mantle is probably comparatively hot, less viscous and with predominantly rising flow), Steinberger *et al.* (2019*b*) found it took around 80 million years or more for the Yellowstone plume head, located far from LLSVPs in the vicinity of sinking slabs, to rise. Figure 11 shows the case in which each plume head needs 30 million years to ascend.

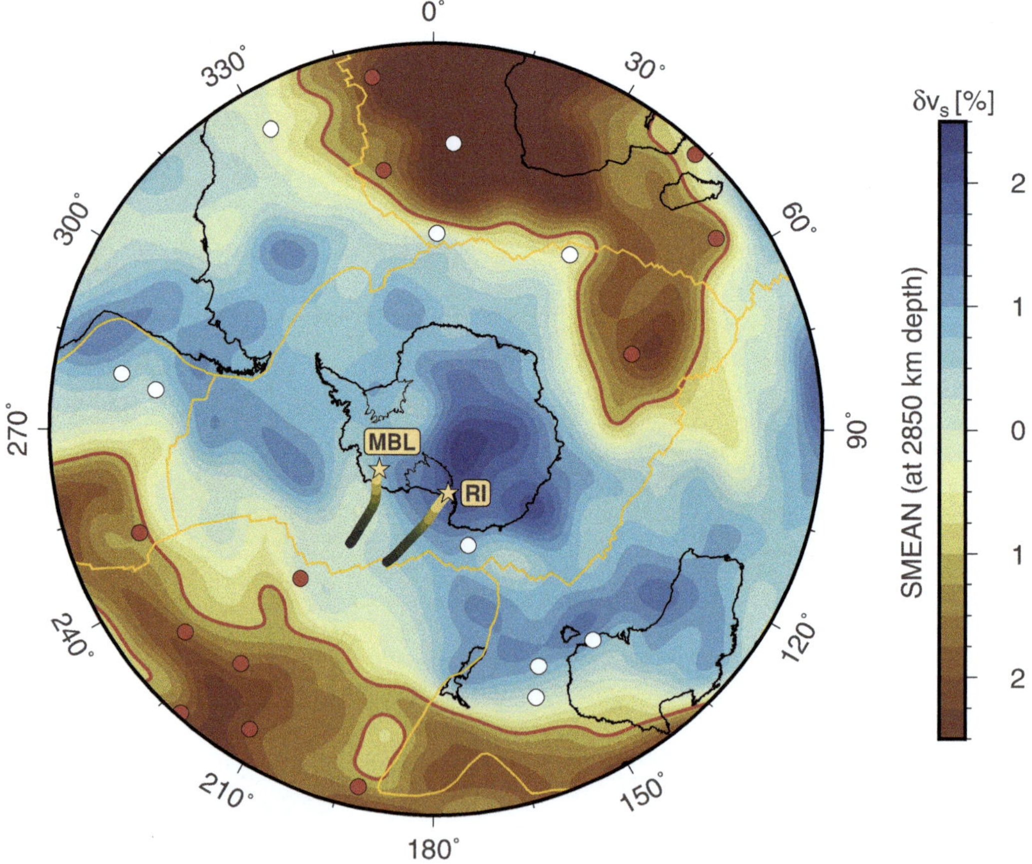

Fig. 11. Smean tomography model (Becker and Boschi 2002) at the CMB from the Antarctic perspective. The two LLSVPs are visible as brown areas, framed by the −1% shear-wave velocity contour (red line). All red and white circles denote the positions of the hotspots catalogued by Steinberger (2000), with red circles showing the primary and clearly resolved plumes detected by French and Romanowicz (2015). Yellow lines indicate plate boundaries. Stars show the locations of the potential plumes beneath Marie Byrd Land (MBL) or Ross Island (RI), underneath which the stability of whole-mantle plume conduits within the large-scale flow field was modelled. The pathways of the conduits are shown from the CMB up to the surface (colour-coded from dark to light green over depth).

Not surprisingly, and in accordance with seismic tomography images, the intrinsically vertical plume conduits are strongly tilted due to the complex flow pattern of the convecting mantle beneath Antarctica. The plumes start their ascent from the direction of the Pacific LLSVP; however, still far away from its margin or the known deep plumes (red circles).

Altogether, it can be summarized that it is not unlikely for hot material to flow towards and end up beneath Marie Byrd Land or Ross Island considering the directions and vigour of the mantle wind. A deep plume origin with a highly tilted plume conduit can therefore not be ruled out, in agreement with available seismic images.

The models are not suited, however, to finally confirming the existence of a whole-mantle plume. For that purpose, seismic tomography models with a higher resolution at greater depths are the most promising – if not the only – tool to provide a conclusive answer. Until then, the debate about a potential plume beneath West Antarctica will continue.

Acknowledgements The authors appreciate the constructive comments by Norm Sleep, an anonymous reviewer and the volume editor, Wouter van der Wal.

Author contributions **EB:** writing – original draft (equal); **BS:** writing – original draft (equal); **RG**: data curation (supporting), methodology (supporting), software (supporting); **JD**: data curation (supporting), methodology (supporting), software (supporting).

Funding This work was supported by the European Space Agency (ESA) as part of the Support to Science Element (STSE) '3D Earth – A Dynamic Living Planet'. The Computational Infrastructure for Geodynamics (geodynamics.org), which is funded by the National Science Foundation under awards EAR-0949446 and EAR-1550901, supported the development of ASPECT. BS acknowledges funding from the Centre of Excellence project 223272 through the Research Council of Norway and the innovation pool of the Helmholtz Association through the 'Advanced Earth System Modelling Capacity (ESM)' activity.

Data availability The datasets used to perform the modelling in the current study are available along with the original studies as referenced (Fox Maule *et al.* 2005; An *et al.* 2015; Martos *et al.* 2017; Pappa *et al.* 2019). The geodynamic models were computed with the open-source software ASPECT (https://aspect.geodynamics.org/). Large-scale mantle flow (Fig. 2), dynamic topography (Fig. 4) and plume conduits in the 'Modelling plume conduits beneath Antarctica' subsection were computed with code available from B. Steinberger upon request.

References

Accardo, N.J., Wiens, D.A. *et al.* 2014. Upper mantle seismic anisotropy beneath the West Antarctic Rift System and surrounding region from shear wave splitting analysis. *Geophysical Journal International*, **198**, 414–429, https://doi.org/10.1093/gji/ggu117

An, M., Wiens, D.A. *et al.* 2015. Temperature, lithosphere–asthenosphere boundary, and heat flux beneath the Antarctic Plate inferred from seismic velocities. *Journal of Geophysical Research: Solid Earth*, **120**, 8720–8742, https://doi.org/10.1002/2015JB011917

Anderson, D.L. and Natland, J.H. 2005. A brief history of the plume hypothesis and its competitors: concept and controversy. *Geological Society of America Special Papers*, **388**, 119–145.

Austermann, J., Kaye, B.T., Mitrovica, J.X. and Huybers, P. 2014. A statistical analysis of the correlation between large igneous provinces and lower mantle seismic structure. *Geophysical Journal International*, **197**, 1–9, https://doi.org/10.1093/gji/ggt500

Becker, T.W. and Boschi, L. 2002. A comparison of tomographic and geodynamic mantle models. *Geochemistry, Geophysics, Geosystems*, **3**, 1003, https://doi.org/10.1029/2001GC000168

Behrendt, J.C., LeMasurier, W. and Cooper, A.K. 1992. The West Antarctic Rift System – a propagating rift 'captured' by a mantle plume? *In*: Yoshida, K., Kaminuma, K. and Shiraishi, K. (eds) *Recent Progress in Antarctic Earth Science*. Terra Science, Tokyo, 315–322.

Braun, J. 2010. The many surface expressions of mantle dynamics. *Nature Geoscience*, **3**, 825–833, https://doi.org/10.1038/ngeo1020

Bredow, E., Steinberger, B., Gassmöller, R. and Dannberg, J. 2017. How plume–ridge interaction shapes the crustal thickness pattern of the Réunion hotspot track. *Geochemistry, Geophysics, Geosystems*, **18**, 2930–2948, https://doi.org/10.1002/2017GC006875

Bryan, S.E. and Ernst, R.E. 2008. Revised definition of Large Igneous Provinces (LIPs). *Earth-Science Reviews*, **86**, 175–202, https://doi.org/10.1016/j.earscirev.2007.08.008

Bunge, H.-P., Richards, M.A. and Baumgardner, J.R. 2002. Mantle-circulation models with sequential data assimilation: inferring present-day mantle structure from plate-motion histories. *Philosophical Transactions of the Royal Society A*, **360**, 2545–2567, https://doi.org/10.1098/rsta.2002.1080

Burgess, S.D., Bowring, S.A., Fleming, T.H. and Elliot, D.H. 2015. High-precision geochronology links the Ferrar large igneous province with early-Jurassic ocean anoxia and biotic crisis. *Earth and Planetary Science Letters*, **415**, 90–99, https://doi.org/10.1016/j.epsl.2015.01.037

Burke, K., Steinberger, B., Torsvik, T.H. and Smethurst, M.A. 2008. Plume Generation Zones at the margins of Large Low Shear Velocity Provinces on the core–mantle boundary. *Earth and Planetary Science Letters*, **265**, 49–60, https://doi.org/10.1016/j.epsl.2007.09.042

Čížková, H., van den Berg, A., Spakman, W. and Matyska, C. 2012. The viscosity of Earth's lower mantle inferred from sinking speed of subducted lithosphere. *Physics of the Earth and Planetary Interiors*, **200–201**, 56–62, https://doi.org/10.1016/j.pepi.2012.02.010

Coffin, M.F., Pringle, M., Duncan, R., Gladczenko, T., Storey, M., Müller, R. and Gahagan, L. 2002. Kerguelen hotspot magma output since 130 Ma. *Journal of Petrology*, **43**, 1121–1137, https://doi.org/10.1093/petrology/43.7.1121

Coffin, M.F., Duncan, R.A. *et al.* 2006. Large igneous provinces and scientific ocean drilling: Status quo and a look ahead. *Oceanography*, **19**, 150–160, https://doi.org/10.5670/oceanog.2006.13

Coltorti, M., Bonadiman, C., Casetta, F., Faccini, B., Giacomoni, P.P., Pelorosso, B. and Perinelli, C. 2021. Nature and evolution of the northern Victoria Land lithospheric mantle (Antarctica) as revealed by ultramafic xenoliths. *Geological Society, London, Memoirs*, **56**, https://doi.org/10.1144/M56-2020-11

Conrad, C.P. and Lithgow-Bertelloni, C. 2002. How mantle slabs drive plate tectonics. *Science*, **298**, 207–209, https://doi.org/10.1126/science.1074161

Conrad, C.P., Steinberger, B. and Torsvik, T.H. 2013. Stability of active mantle upwelling revealed by net characteristics of plate tectonics. *Nature*, **498**, 479–482, https://doi.org/10.1038/nature12203

Cooper, A.F., Adam, L.J., Coulter, R.F., Eby, G.N. and McIntosh, W.C. 2007. Geology, geochronology and geochemistry of a basanitic volcano, White Island, Ross Sea, Antarctica. *Journal of Volcanology and Geothermal Research*, **165**, 189–216, https://doi.org/10.1016/j.jvolgeores.2007.06.003

Courtillot, V., Davaille, A., Besse, J. and Stock, J. 2003. Three distinct types of hotspots in the Earth's mantle. *Earth and Planetary Science Letters*, **205**, 295–308, https://doi.org/10.1016/S0012-821X(02)01048-8

Davies, G.F. 1977. Whole-mantle convection and plate tectonics. *Geophysical Journal of the Royal Astronomical Society*, **49**, 459–486, https://doi.org/10.1111/j.1365-246X.1977.tb03717.x

Day, J.M.D., Harvey, R.P. and Hilton, D.R. 2019. Melt-modified lithosphere beneath Ross Island and its role in the tectono-magmatic evolution of the West Antarctic Rift System, *Chemical Geology*, **518**, 45–54, https://doi.org/10.1016/j.chemgeo.2019.04.012

DeConto, R.M. and Pollard, D. 2003. Rapid Cenozoic glaciation of Antarctica induced by declining atmospheric CO_2. *Nature*, **421**, 245–249, https://doi.org/10.1038/nature01290

Domeier, M. and Torsvik, T.H. 2019. Full-plate modelling in pre-Jurassic time. *Geological Magazine*, **156**, 261–280, https://doi.org/10.1017/S0016756817001005

Domeier, M., Doubrovine, P.V., Torsvik, T.H., Spakman, W. and Bull, A.L. 2016. Global correlation of lower mantle structure and past subduction. *Geophysical Research Letters*, **43**, 4945–4953, https://doi.org/10.1002/2016GL068827

Doubrovine, P.V., Steinberger, B. and Torsvik, T.H. 2016. A failure to reject: Testing the correlation between large igneous provinces and deep mantle structures with EDF statistics. *Geochemistry, Geophysics, Geosystems*, **17**, 1130–1163, https://doi.org/10.1002/2015GC006044

Dziewonski, A.M. and Woodhouse, J.H. 1987. Global images of the Earth's interior. *Science*, **236**, 37–48, https://doi.org/10.1126/science.236.4797.37

Ebinger, C.J. and Sleep, N. 1998. Cenozoic magmatism throughout East Africa resulting from impact of a single plume. *Nature*, **395**, 788–791, https://doi.org/10.1038/27417

Elliot, D.H. and Fleming, T.H. 2004. Occurrence and dispersal of magmas in the Jurassic Ferrar Large Igneous Province, Antarctica. *Gondwana Research*, **7**, 223–237, https://doi.org/10.1016/S1342-937X(05)70322-1

Farnetani, D.G. and Richards, M.A. 1995. Thermal entrainment and melting in mantle plumes. *Earth and Planetary Science Letters*, **136**, 251–267, https://doi.org/10.1016/0012-821X(95)00158-9

Forsyth, D. and Uyeda, S. 1975. On the relative importance of the driving forces of plate motion. *Geophysical Journal of the Royal Astronomical Society*, **43**, 163–200, https://doi.org/10.1111/j.1365-246X.1975.tb00631.x

Foulger, G.R. 2011. *Plates vs Plumes: A Geological Controversy*. John Wiley & Sons, Chichester, UK.

Fox Maule, C., Purucker, M.E., Olsen, N. and Mosegaard, K. 2005. Heat flux anomalies in Antarctica revealed by satellite magnetic data. *Science*, **309**, 464–467, https://doi.org/10.1126/science.1106888

French, S.W. and Romanowicz, B. 2015. Broad plumes rooted at the base of the Earth's mantle beneath major hotspots. *Nature*, **525**, 95–99, https://doi.org/10.1038/nature14876

Fullea, J., Afonso, J.C., Connolly, J., Fernandez, M., García-Castellanos, D. and Zeyen, H. 2009. LitMod3D: An interactive 3-D software to model the thermal, compositional, density, seismological, and rheological structure of the lithosphere and sublithospheric upper mantle. *Geochemistry, Geophysics, Geosystems*, **10**, Q08019, https://doi.org/10.1029/2009GC002391

Grand, S.P. 2002. Mantle shear-wave tomography and the fate of subducted slabs. *Philosophical Transactions of the Royal Society of London A: Mathematical, Physical and Engineering Sciences*, **360**, 2475–2491, https://doi.org/10.1098/rsta.2002.1077

Griffiths, R.W. and Campbell, I.H. 1990. Stirring and structure in mantle starting plumes. *Earth and Planetary Science Letters*, **99**, 66–78, https://doi.org/10.1016/0012-821X(90)90071-5

Hager, B.H. and O'Connell, R.J. 1979. Kinematic models of large-scale flow in the Earth's mantle. *Journal of Geophysical Research: Solid Earth*, **84**, 1031–1048, https://doi.org/10.1029/JB084iB03p01031

Hager, B.H. and O'Connell, R.J. 1981. A simple global model of plate dynamics and mantle convection. *Journal of Geophysical Research: Solid Earth*, **86**, 4843–4867, https://doi.org/10.1029/JB086iB06p04843

Hager, B.H. and Richards, M.A. 1989. Long-wavelength variations in Earth's geoid: physical models and dynamical implications. *Philosophical Transactions of the Royal Society of London A: Mathematical, Physical and Engineering Sciences*, **328**, 309–327, https://doi.org/10.1098/rsta.1989.0038

Handler, M.R., Wysoczanski, R.J. and Gamble, J.A. 2021. Marie Byrd Land lithospheric mantle: a review of the xenolith record. *Geological Society, London, Memoirs*, **56**, https://doi.org/10.1144/M56-2020-17

Hansen, S.E., Graw, J.H. *et al.* 2014. Imaging the Antarctic mantle using adaptively parameterized p-wave tomography: Evidence for heterogeneous structure beneath West Antarctica. *Earth and Planetary Science Letters*, **408**, 66–78, https://doi.org/10.1016/j.epsl.2014.09.043

Heister, T., Dannberg, J., Gassmöller, R. and Bangerth, W. 2017. High accuracy mantle convection simulation through modern numerical methods. II: realistic models and problems. *Geophysical Journal International*, **210**, 833–851, https://doi.org/10.1093/gji/ggx195

Hoggard, M.J., White, N. and Al-Attar, D. 2016. Global dynamic topography observations reveal limited influence of large-scale mantle flow. *Nature Geoscience*, **9**, 456–463, https://doi.org/10.1038/ngeo2709

Hole, M.J. and LeMasurier, W.E. 1994. Tectonic controls on the geochemical composition of Cenozoic, mafic alkaline volcanic rocks from West Antarctica. *Contributions to Mineralogy and Petrology*, **117**, 187–202, https://doi.org/10.1007/BF00286842

Hosseini, K., Sigloch, K., Tsekhmistrenko, M., Zaheri, A., Nissen-Meyer, T. and Igel, H. 2020. Global mantle structure from multifrequency tomography using *P*, *PP* and *P*-diffracted waves. *Geophysical Journal International*, **220**, 96–141, https://doi.org/10.1093/gji/ggz394

Justo, J., Morra, G. and Yuen, D. 2015. Viscosity undulations in the lower mantle: The dynamical role of iron spin transition. *Earth and Planetary Science Letters*, **421**, 20–26, https://doi.org/10.1016/j.epsl.2015.03.013

King, S.D. 2007. Hotspots and edge-driven convection. *Geology*, **35**, 223, https://doi.org/10.1130/G23291A.1

King, S.D. 2016. An evolving view of transition zone and midmantle viscosity. *Geochemistry, Geophysics, Geosystems*, **17**, 1234–1237, https://doi.org/10.1002/2016GC006279

King, S.D. and Adam, C. 2014. Hotspot swells revisited. *Physics of the Earth and Planetary Interiors*, **235**, 66–83, https://doi.org/10.1016/j.pepi.2014.07.006

Korenaga, J. 2008. Urey ratio and the structure and evolution of Earth's mantle. *Reviews of Geophysics*, **46**, RG2007, https://doi.org/10.1029/2007RG000241

Korenaga, J. 2016. Can mantle convection be self-regulated? *Science Advances*, **235**, 66–83, https://doi.org/10.1126/sciadv.1601168

Kronbichler, M., Heister, T. and Bangerth, W. 2012. High accuracy mantle convection simulation through modern numerical methods. *Geophysical Journal International*, **191**, 12–29, https://doi.org/10.1111/j.1365-246X.2012.05609.x

Kyle, P., Moore, J. and Thirlwall, M. 1992. Petrologic evolution of anorthoclase phonolite lavas at Mount Erebus, Ross Island, Antarctica. *Journal of Petrology*, **33**, 849–875, https://doi.org/10.1093/petrology/33.4.849

Larter, R.D., Cunningham, A.P., Barker, P.F., Gohl, K. and Nitsche, F.O. 2002. Tectonic evolution of the Pacific margin of Antarctica 1. Late Cretaceous tectonic reconstructions. *Journal of Geophysical Research: Solid Earth*, **107**, 2345, https://doi.org/10.1029/2000JB000052

Lau, H., Mitrovica, J., Austermann, J., Crawford, O., Al-Attar, D. and Latychev, K. 2016. Inferences of mantle viscosity based on ice age data sets: radial structure. *Journal of Geophysical Research: Solid Earth*, **121**, 6991–7012, https://doi.org/10.1002/2016JB013043

LeMasurier, W. 2013. Shield volcanoes of Marie Byrd Land, West Antarctic Rift: oceanic island similarities, continental signature, and tectonic controls. *Bulletin of Volcanology*, **75**, 726, https://doi.org/10.1007/s00445-013-0726-1

LeMasurier, W. and Rex, D. 1989. Evolution of linear volcanic ranges in Marie Byrd Land, West Antarctica. *Journal of Geophysical Research: Solid Earth*, **94**, 7223–7236, https://doi.org/10.1029/JB094iB06p07223

Liu, X. and Zhong, S. 2016. Constraining mantle viscosity structure for a thermochemical mantle using the geoid observation. *Geochemistry, Geophysics, Geosystems*, **17**, 895–913, https://doi.org/10.1002/2015GC006161

Lloyd, A.J. 2018. *Seismic Tomography of Antarctica and the Southern Oceans: Regional and Continental Models from the Upper Mantle to the Transition Zone*. PhD thesis, Washington University in St Louis, St Louis, Missouri, USA.

Lloyd, A.J., Wiens, D.A. *et al.* 2015. A seismic transect across West Antarctica: Evidence for mantle thermal anomalies beneath the Bentley Subglacial Trench and the Marie Byrd Land Dome. *Journal of Geophysical Research: Solid Earth*, **120**, 8439–8460, https://doi.org/10.1002/2015JB012455

Marquardt, H. and Miyagi, L. 2015. Slab stagnation in the shallow lower mantle linked to an increase in mantle viscosity. *Nature Geoscience*, **8**, 311–314, https://doi.org/10.1038/ngeo2393

Martin, A.P., Cooper, A.F. and Price, R.C. 2013. Petrogenesis of Cenozoic, alkalic volcanic lineages at Mount Morning, West Antarctica and their entrained lithospheric mantle xenoliths: Lithospheric v. asthenospheric mantle sources. *Geochimica et Cosmochimica Acta*, **122**, 127–152, https://doi.org/10.1016/j.gca.2013.08.025

Martin, A.P., Cooper, A.F., Price, R.C., Doherty, C.L. and Gamble, J.A. 2021. A review of mantle xenoliths in volcanic rocks from southern Victoria Land, Antarctica. *Geological Society, London, Memoirs*, **56**, https://doi.org/10.1144/M56-2019-42

Martos, Y.M., Catalán, M., Jordan, T.A., Golynsky, A., Golynsky, D., Eagles, G. and Vaughan, D.G. 2017. Heat flux distribution of Antarctica unveiled. *Geophysical Research Letters*, **44**, 11 417–11 426, https://doi.org/10.1002/2017GL075609

Mazzullo, A., Stutzmann, E., Montagner, J.-P., Kiselev, S., Maurya, S., Barruol, G. and Sigloch, K. 2017. Anisotropic tomography around Réunion Island from Rayleigh waves. *Journal of Geophysical Research: Solid Earth*, **122**, 9132–9148, https://doi.org/10.1002/2017JB014354

Montelli, R., Nolet, G., Dahlen, F. and Masters, G. 2006. A catalogue of deep mantle plumes: New results from finite-frequency tomography. *Geochemistry, Geophysics, Geosystems*, **7**, Q11007, https://doi.org/10.1029/2006GC001248

Moreira, M. and Allègre, C.J. 1998. Helium–neon systematics and the structure of the mantle. *Chemical Geology*, **147**, 53–59, https://doi.org/10.1016/S0009-2541(97)00171-X

Morgan, W.J. 1971. Convection plumes in the lower mantle. *Nature*, **230**, 42–43, https://doi.org/10.1038/230042a0

Morgan, W.J. 1972. Deep mantle convection plumes and plate motions. *AAPG Bulletin*, **56**, 203–213.

Nakada, M., Okuno, J. and Irie, Y. 2017. Inference of viscosity jump at 670 km depth and lower mantle viscosity structure from GIA observations. *Geophysical Journal International*, **212**, 2206–2225, https://doi.org/10.1093/gji/ggx519

Nardini, I., Armienti, P., Rocchi, S., Dallai, L. and Harrison, D. 2009. Sr–Nd–Pb–He–O isotope and geochemical constraints on the genesis of Cenozoic magmas from the West Antarctic Rift. *Journal of Petrology*, **50**, 1359–1375, https://doi.org/10.1093/petrology/egn082

Panter, K.S., Kyle, P.R. and Smellie, J.L. 1997. Petrogenesis of a phonolite–trachyte succession at Mount Sidley, Marie Byrd Land, Antarctica. *Journal of Petrology*, **38**, 1225–1253, https://doi.org/10.1093/petroj/38.9.1225

Panter, K.S., Hart, S.R., Kyle, P., Blusztanjn, J. and Wilch, T. 2000. Geochemistry of late Cenozoic basalts from the Crary Mountains: characterization of mantle sources in Marie Byrd Land, Antarctica. *Chemical Geology*, **165**, 215–241, https://doi.org/10.1016/S0009-2541(99)00171-0

Panter, K.S., Castillo, P. *et al.* 2018. Melt origin across a rifted continental margin: a case for subduction-related metasomatic agents in the lithospheric source of alkaline basalt, northwest Ross Sea, Antarctica. *Journal of Petrology*, **59**, 517–558, https://doi.org/10.1093/petrology/egy036

Pappa, F. and Ebbing, J. 2021. Gravity, magnetics and geothermal heat flow of the Antarctic lithospheric crust and mantle. *Geological Society, London, Memoirs*, **56**, https://doi.org/10.1144/M56-2020-5

Pappa, F., Ebbing, J., Ferraccioli, F. and van der Wal, W. 2019. Modeling satellite gravity gradient data to derive density, temperature, and viscosity structure of the Antarctic lithosphere. *Journal of Geophysical Research: Solid Earth*, **124**, 12 053–12 076, https://doi.org/10.1029/2019JB017997

Paxman, G.J.G. 2021. Antarctic palaeotopography. *Geological Society, London, Memoirs*, **56**, https://doi.org/10.1144/M56-2020-7

Paxman, G., Jamieson, S., Hochmuth, K., Gohl, K., Bentley, M., Leitchenkov, G. and Ferraccioli, F. 2019. Reconstructions of Antarctic topography since the Eocene–Oligocene boundary. *Palaeogeography, Palaeoclimatology, Palaeoecology*, **535**, 109346, https://doi.org/10.1016/j.palaeo.2019.109346

Phillips, E.H., Sims, K.W. *et al.* 2018. The nature and evolution of mantle upwelling at Ross Island, Antarctica, with implications for the source of HIMU lavas. *Earth and Planetary Science Letters*, **498**, 38–53, https://doi.org/10.1016/j.epsl.2018.05.049

Putirka, K. 2008. Excess temperatures at ocean islands: Implications for mantle layering and convection. *Geology*, **36**, 283–286, https://doi.org/10.1130/G24615A.1

Ricard, Y., Richards, M., Lithgow-Bertelloni, C. and Le Stunff, Y. 1993. A geodynamic model of mantle density heterogeneity. *Journal of Geophysical Research: Solid Earth*, **98**, 21 895–21 909, https://doi.org/10.1029/93JB02216

Richards, M.A. and Engebretson, D.C. 1992. Large-scale mantle convection and the history of subduction. *Nature*, **355**, 437–330, https://doi.org/10.1038/355437a0

Richards, M.A., Duncan, R.A. and Courtillot, V.E. 1989. Flood basalts and hot-spot tracks: Plume heads and tails. *Science*, **246**, 103–107, https://doi.org/10.1126/science.246.4926.103

Rocchi, S., Storti, F., Di Vincenzo, G. and Rosetti, F. 2003. Intraplate strike-slip tectonics as an alternative to mantle plume activity for the Cenozoic rift magmatism in the Ross Sea region, Antarctica. *Geological Society, London, Special Publications*, **210**, 145–158, https://doi.org/10.1144/GSL.SP.2003.210.01.09

Rocchi, S., Armienti, P. and Di Vincenzo, G. 2005. No plume, no rift magmatism in the West Antarctic Rift. *Geological Society of America Special Papers*, **388**, 435–447.

Rocholl, A., Stein, M., Molzahn, M., Hart, S. and Wörner, G. 1995. Geochemical evolution of rift magmas by progressive tapping of a stratified mantle source beneath the Ross Sea Rift, Northern Victoria Land, Antarctica. *Earth and Planetary Science Letters*, **131**, 207–224, https://doi.org/10.1016/0012-821X(95)00024-7

Roy, K. and Peltier, W. 2015. Glacial isostatic adjustment, relative sea level history and mantle viscosity: reconciling relative sea level model predictions for the us east coast with geological constraints. *Geophysical Journal International*, **201**, 1156–1181, https://doi.org/10.1093/gji/ggv066

Rudolph, M.L., Lekić, V. and Lithgow-Bertelloni, C. 2015. Viscosity jump in Earth's mid-mantle. *Science*, **350**, 1349–1352, https://doi.org/10.1126/science.aad1929

Schaeffer, A. and Lebedev, S. 2013. Global shear speed structure of the upper mantle and transition zone. *Geophysical Journal International*, **194**, 417–449, https://doi.org/10.1093/gji/ggt095

Schilling, J.-G. 1991. Fluxes and excess temperatures of mantle plumes inferred from their interaction with migrating mid-ocean ridges. *Nature*, **352**, 397–403, https://doi.org/10.1038/352397a0

Schubert, G., Turcotte, D.L. and Olson, P. 2001. *Mantle Convection in the Earth and Planets*. Cambridge University Press, Cambridge, UK.

Seroussi, H., Ivins, E.R., Wiens, D.A. and Bondzio, J. 2017. Influence of a West Antarctic mantle plume on ice sheet basal conditions. *Journal of Geophysical Research: Solid Earth*, **122**, 7127–7155, https://doi.org/10.1002/2017JB014423

Shapiro, N.M. and Ritzwoller, M.H. 2004. Inferring surface heat flux distributions guided by a global seismic model: particular application to Antarctica. *Earth and Planetary Science Letters*, **223**, 213–224, https://doi.org/10.1016/j.epsl.2004.04.011

Shen, W., Wiens, D.A. *et al.* 2018. The crust and upper mantle structure of Central and West Antarctica from Bayesian inversion of Rayleigh wave and receiver functions. *Journal of Geophysical Research: Solid Earth*, **123**, 7824–7849, https://doi.org/10.1029/2017JB015346

Shephard, G., Bunge, H.-P., Schuberth, B., Müller, R., Talsma, A., Moder, C. and Landgrebe, T. 2012. Testing absolute plate reference frames and the implications for the generation of geodynamic mantle heterogeneity structure. *Earth and Planetary Science Letters*, **317–318**, 204–217, https://doi.org/10.1016/j.epsl.2011.11.027

Sleep, N.H. 1997. Lateral flow and ponding of starting plume material. *Journal of Geophysical Research: Solid Earth*, **102**, 10 001–10 012, https://doi.org/10.1029/97JB00551

Sleep, N.H. 2006. Mantle plumes from top to bottom. *Earth-Science Reviews*, **77**, 231–271, https://doi.org/10.1016/j.earscirev.2006.03.007

Sleep, N.H. 2007. Edge-modulated stagnant-lid convection and volcanic passive margins. *Geochemistry, Geophysics, Geosystems*, **8**, Q12004, https://doi.org/10.1029/2007GC001672

Steinberger, B. 2000. Plumes in a convecting mantle: Models and observations for individual hotspots. *Journal of Geophysical Research: Solid Earth*, **105**, 11 127–11 152, https://doi.org/10.1029/1999JB900398

Steinberger, B. and Calderwood, A. 2006. Models of large-scale viscous flow in the Earth's mantle with constraints from mineral physics and surface observations. *Geophysical Journal International*, **167**, 1461–1481, https://doi.org/10.1111/j.1365-246X.2006.03131.x

Steinberger, B. and O'Connell, R.J. 1998. Advection of plumes in mantle flow: implications for hotspot motion, mantle viscosity and plume distribution. *Geophysical Journal International*, **132**, 412–434, https://doi.org/10.1046/j.1365-246x.1998.00447.x

Steinberger, B., Torsvik, T.H. and Becker, T.W. 2012. Subduction to the lower mantle – a comparison between geodynamic and tomographic models. *Solid Earth*, **3**, 415–432, https://doi.org/10.5194/se-3-415-2012

Steinberger, B., Conrad, C., Osei Tutu, A. and Hoggard, M. 2019*a*. On the amplitude of dynamic topography at spherical harmonic degree two. *Tectonophysics*, **760**, 221–228, https://doi.org/10.1016/j.tecto.2017.11.032

Steinberger, B., Nelson, P., Grand, S. and Wang, W. 2019*b*. Yellowstone plume conduit tilt caused by large-scale mantle flow. *Geochemistry, Geophysics, Geosystems*, **20**, 5896–5912, https://doi.org/10.1029/2019GC008490

Torsvik, T.H., Smethurst, M.A., Burke, K. and Steinberger, B. 2006. Large igneous provinces generated from the margins of the large low-velocity provinces in the deep mantle. *Geophysical Journal International*, **167**, 1447–1460, https://doi.org/10.1111/j.1365-246X.2006.03158.x

Torsvik, T.H., Steinberger, B., Gurnis, M. and Gaina, C. 2010. Plate tectonics and net lithosphere rotation over the past 150 My. *Earth and Planetary Science Letters*, **291**, 106–112, https://doi.org/10.1016/j.epsl.2009.12.055

Torsvik,T.H., Svensen, H.H., Steinberger, B., Royer, D.L., Jerram, D.A., Jones, M.T. and Domeier, M. 2021. Connecting the deep Earth and the atmosphere. *American Geophysical Union Geophysical Monograph*, **263**, 413–453, https://doi.org/10.1002/9781119528609.ch16

Tozer, D.C. 1972. The present thermal state of the terrestrial planets. *Physics of the Earth and Planetary Interiors*, **6**, 182–197, https://doi.org/10.1016/0031-9201(72)90052-0

van der Meer, D.G., Spakman, W., van Hinsbergen, D.J.J., Amaru, M.L. and Torsvik, T.H. 2010. Towards absolute plate motions constrained by lower-mantle slab remnants. *Nature Geoscience*, **3**, 36–40, https://doi.org/10.1038/ngeo708

van Keken, P. 1997. Evolution of starting mantle plumes: a comparison between numerical and laboratory models. *Earth and Planetary Science Letters*, **148**, 1–11, https://doi.org/10.1016/S0012-821X(97)00042-3

van Wijk, J., Lawrence, J. and Driscoll, N. 2008. Formation of the Transantarctic Mountains related to extension of the West Antarctic Rift System. *Tectonophysics*, **458**, 117–126, https://doi.org/10.1016/j.tecto.2008.03.009

van Wyk de Vries, M.V.W., Bingham, R.G. and Hein, A.S. 2018. A new volcanic province: an inventory of subglacial volcanoes in West Antarctica. *Geological Society, London, Special Publications*, **461**, 231–248, https://doi.org/10.1144/SP461.7

Whitehead, J.A. and Luther, D.S. 1975. Dynamics of laboratory diapir and plume models. *Journal of Geophysical Research*, **80**, 705–717, https://doi.org/10.1029/JB080i005p00705

Wiens, A.D., Shen, W. and Lloyd, A. 2021. The seismic structure of the Antarctic upper mantle. *Geological Society, London, Memoirs*, **56**, https://doi.org/10.1144/M56-2020-18

Wilson, J.T. 1963. A possible origin of the Hawaiian Islands. *Canadian Journal of Physics*, **41**, 863–870, https://doi.org/10.1139/p63-094

Yang, T. and Gurnis, M. 2016. Dynamic topography, gravity and the role of lateral viscosity variations from inversion of global mantle flow. *Geophysical Journal International*, **207**, 1186–1202, https://doi.org/10.1093/gji/ggw335

Antarctic upper mantle rheology

E. R. Ivins[1]*, W. van der Wal[2], D. A. Wiens[3], A. J. Lloyd[4] and L. Caron[1,5]

[1]Jet Propulsion Laboratory, California Institute of Technology, Pasadena, CA 91109, USA

[2]Faculty of Aerospace Engineering, Delft University of Technology, Delft, The Netherlands

[3]Department of Earth and Planetary Sciences, Washington University, St Louis, MO 63130, USA

[4]Lamont Doherty Earth Observatory, Columbia University, Palisades, NY 10964, USA

[5]JIFRESSE, University of California at Los Angeles, Los Angeles, CA 90095, USA

ERI, 0000-0003-0148-357X; WW, 0000-0001-8030-9080; DAW, 0000-0002-5169-4386; AJL, 0000-0002-2253-1342; LC, 0000-0001-8946-1222

*Correspondence: erik.r.ivins@jpl.nasa.gov

Abstract: The Antarctic mantle and lithosphere are known to have large lateral contrasts in seismic velocity and tectonic history. These contrasts suggest differences in the response timescale of mantle flow across the continent, similar to those documented between the northeastern and southwestern upper mantle of North America. Glacial isostatic adjustment and geodynamical modelling rely on independent estimates of lateral variability in effective viscosity. Recent improvements in imaging techniques and the distribution of seismic stations now allow resolution of both lateral and vertical variability of seismic velocity, making detailed inferences about lateral viscosity variations possible. Geodetic and palaeosea-level investigations of Antarctica provide quantitative ways of independently assessing the 3D mantle viscosity structure. While observational and causal connections between inferred lateral viscosity variability and seismic velocity changes are qualitatively reconciled, significant improvements in the quantitative relations between effective viscosity anomalies and those imaged by P- and S-wave tomography have remained elusive. Here we describe several methods for estimating effective viscosity from S-wave velocity. We then present and compare maps of the viscosity variability beneath Antarctica based on the recent S-wave velocity model ANT-20 using three different approaches.

Solid Earth rheology is that discipline which deals with the laws that govern the deformation of rocks under the high-temperature and -pressure conditions of the mantle. Observations linked to rheology require input from many disciplines in the geological and geophysical sciences, and examples include composition, temperature and pressure conditions from xenoliths, elastic parameters from seismic wave propagation, and density variations derived from gravity, topography and magnetic anomaly data (e.g. Scheinert *et al.* 2016; Martos *et al.* 2017; Pappa *et al.* 2019). Simulations of mantle circulation (Bredow *et al.* 2021, this volume), glacial isostatic adjustment (GIA), post-seismic deformation (Barletta *et al.* 2022, this volume) and their surface expressions require quantifying large variations in mantle rheology. For several areas of the world we have some quantitative bounds on lateral variability in upper mantle viscosity, such as the contrast between Fennoscandia and Iceland (Wang *et al.* 2008). Antarctica has long been known to have contrasting mantle and lithosphere properties between east and west, somewhat akin to the contrasts between the Precambrian craton of eastern Canada and the geologically younger crust and mantle of western North America (Crittenden 1963; Austermann *et al.* 2020).

A broader goal of this chapter is to make quantitative connections between flow laws derived in the rock deformation laboratory and the integrative power of the datasets for constraining crustal motion and Earth structure using a network of global positioning system (GPS) and broadband seismic instruments. A period of intense data collection began after the 4th International Polar Year of 2007 and was organized as the POLENET project. Much of this integrated dataset may be brought to bear on understanding the nature of the Antarctic lithosphere (e.g. Nield *et al.* 2018; Shen *et al.* 2018; Scheinert *et al.* 2021, this volume). An overview of seismic imaging of the Antarctic crust, lithosphere and upper mantle is given in the chapter by Wiens *et al.* (2021, this volume) that includes tectonophysical interpretations, thus complementing the study of Earth rheology. The reader may also find the recent review by Jordan *et al.* (2020) to be useful. However, in this chapter we focus on the mantle response to changes in surface ice loading over centennial to millennial timescales with steady-state high-temperature creep. The rate of strain associated with this creep may be causally connected to both vertical uplift and horizontal crustal motion rates that have been measured using bedrock GPS (e.g. James and Ivins 1995; Groh *et al.* 2012; Barletta *et al.* 2018). This motion also occurs where it cannot be measured, deep below the surface of the ice sheet, over extensive regions of the vast ice-covered interior. Such extensive continental-scale mantle and crustal motion is important to model since the vertical component creates a time-varying gravity field. The modelled GIA signal needs to be removed from gravimetry measurements, such that accurate ice mass-balance trends for the Antarctic ice sheet and its contribution to ongoing sea-level rise may be retrieved using space gravity mission data (e.g. Caron and Ivins 2020).

Early surface-wave studies in Antarctica (e.g. Knopoff and Vane 1978; Danesi and Morelli 2001; Ritzwoller *et al.* 2001) revealed the existence of a major contrast in the structure of the uppermost 300 km of the mantle between East and West Antarctica. The boundary between these deep structural elements is roughly coincident with the surface expression of the Transantarctic Mountains (TAM: see Fig. 1). The seismic data retrieved since 2007 are of sufficiently high quality that we can begin to quantify the lateral heterogeneity in the viscous response of the mantle to ice-mass surface-load changes. Lateral heterogeneity in the mantle is important for reconciling observed and predicted crustal motion in GIA models (Powell *et al.* 2020), and is an important aspect for determining the stability of the ice sheet over a changing bed topography, both in the past and in the future (Gomez *et al.* 2018). Lateral variability in elastic-wave velocity (see Fig. 1) is related to lateral temperature variations. Using the temperature-dependence of elastic constants and anelasticity determined by theory and experiment, we can try to relate seismic velocity change to changes in effective mantle viscosity, since the latter are known to be strongly dependent on temperature.

The focus of this chapter is to examine the feasibility of developing causal interconnections between seismic tomographic imaging, mantle viscosity retrieved from Maxwellian 1D GIA models and flow laws for steady-state creep quantified

From: Martin, A. P. and van der Wal, W. (eds) 2023. *The Geochemistry and Geophysics of the Antarctic Mantle*. Geological Society, London, Memoirs, **56**, 267–294,
First published online 17 November 2021, https://doi.org/10.1144/M56-2020-19

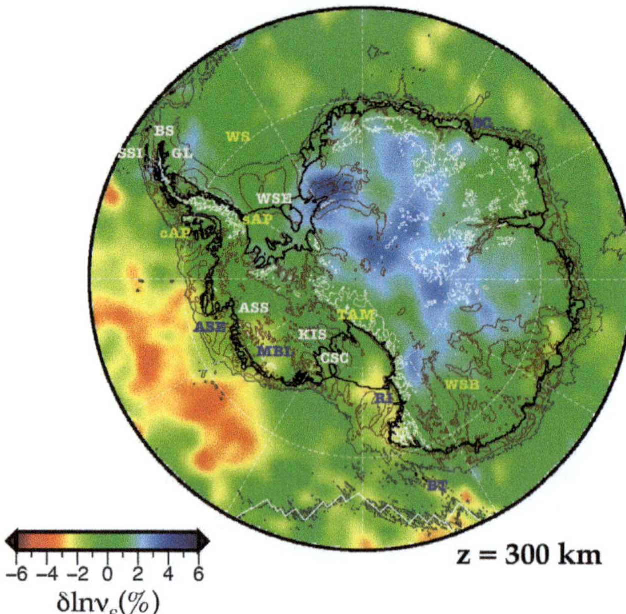

Fig. 1. Deviation in shear-wave velocity from the global average at 300 km depth. This is reported as the Voigt average of a transversely anisotropic shear-wave solution described by Lloyd *et al.* (2020) and by Wiens *et al.* (2021, this volume). Also shown are site names referred to throughout the chapter. The locations are: ASE, Amundsen Sea Embayment; ASS, Amundsen Sea sector; BS, Bransfield Strait; BT, Balleny Trough; cAP, central Antarctic Peninsula; CSC, Central Siple Coast; GL, Graham Land; KIS, Kamb Ice Stream; MBL, Marie Byrd Land; RI, Ross Island; sAP, southern Antarctic Peninsula; SC, Syowa Coast; SSI, South Shetland Islands; TAM, Transantarctic Mountains; WS, Weddell Sea; WSE, Weddell Sea Embayment; WSB, Wilkes Subglacial Basin. Bathymetry is contoured at −2500, −1000 and −500 m. White contours show ice topography. Thick white lines show plate boundaries.

by high-pressure and -temperature laboratory experiments. This is a rather grand challenge, and may have consequences for regions other than Antarctica. However, here we restrict our study only to a single tomographic S-wave imaging study (Lloyd *et al.* 2020) for Antarctica and restrict ourselves to the upper mantle. The feasibility of establishing causal interconnection is tested by conducting an intercomparison of three methods at three different depths. The methods for each are documented in the remaining pages of this chapter.

Table 1. *Intercomparison of techniques for scaling Antarctic upper mantle* $\log_{10}\eta(r, \theta, \phi)$*

Approach No.	Rock microphysics	Laboratory experiments
1†	Diffusion controlled	Dry olivine
2†	Grain-boundary dislocation sliding	Dry olivine
3†	Multiple microphysics	Olivine sliding + dislocation + diffusion
1‡	Dislocation controlled	1.5% H_2O by wt ringwoodite
2‡	Dislocation controlled	1.5% H_2O by wt ringwoodite
3‡	Multiple microphysics	Climb and glide disloclocation, *c.* 2.0% H_2O ringwoodite

*$T_0(r)$ is common to all three methods.
† z = 150 and 300 km depth of maps.
‡ z = 550 km depth of maps.

In Table 1 we give a short synopsis of some of the differences in the methods. Later we refer to these as Approaches 1, 2 and 3 (see the first column in Table 1). Not only is the selection of laboratory flow laws generally different, so are the implementation steps yielding anelastic slowing of the shear-wave velocity. The latter are necessary to derive the appropriate thermal deviation $\delta T(r, \theta, \phi)$ from tomography. We anticipate from the outset that there will be large differences between the predictions. These should manifest themselves as differences in mapped $\log_{10}\Delta\eta(r, \theta, \phi)$ for the Antarctic mantle and in comparisons to the inferences from 1D GIA models compatible with geodetic data.

The scope of study in this chapter is limited by the fact that Antarctica is an ice-sheet-covered continent with a logistically difficult environment in which to obtain bedrock GPS, relative sea-level (RSL) data and maintain seismic station operation. We are perhaps blessed by the fact that geochronological and ice-core constraints on the ice-sheet history during the late-Quaternary place tight constraints on the millennial time-scale ice-load history (e.g. Bentley *et al.* 2014; Anderson *et al.* 2014; Whitehouse *et al.* 2017), a critical ingredient for establishing viable GIA predictions. Neither a comprehensive 3D compressional (P-wave) map can be combined with S-wave mapping (e.g. Goes *et al.* 2000), nor does an attenuation map of the upper mantle for Antarctica exist (e.g. Lawrence and Prieto 2011). This fact diminishes the capability of retrieving higher-fidelity mapping of $\delta T(r, \theta, \phi)$ that can be produced in upper mantle environments where seismic station coverage is denser in both space and time (e.g. Goes *et al.* 2000). Nonetheless, in this chapter we demonstrate that a coherent rationale exists for quantitatively connecting Antarctic seismic tomography to GIA-determined mantle viscosity and laboratory flow laws. This inference may have an impact on the construction of future models of glacial isostatic adjustment with lateral variation in Earth structure (GIA-LV) and for formulating paradigms for their uncertainty.

In this chapter, we will first review the factors controlling mantle rheology as well as seismic-wave velocities in the mantle. We then discuss a general computational strategy for estimating 3D mantle viscosity from seismic observations and present new maps of viscosity for the Antarctic upper mantle derived in three different ways. In our conclusions, we attempt to draw out robust predictions that are common to the three methods. A common starting point for our estimates is the ANT-20 S-wave velocity model (Lloyd *et al.* 2020). This model takes advantage of recent increases in seismic station density and computational techniques to derive higher-resolution images of mantle structure using full waveform adjoint tomography (Wiens *et al.* 2021, this volume). The focus of this chapter is on the most basic elements of the large uncertainties encountered in deriving viscosity from seismic models. We identify and detail a number of the relevant quantitative elastic and plastic deformation parameters, but assess their relative uncertainty in a more qualitative way. Our main conclusions will, hopefully, offer clarity and guidance for further efforts to reduce this uncertainty, and thereby improve our estimation of mantle rheology.

Background

Deformation in the mantle is controlled by the mobility of defects in the crystal structure, either within or at the boundaries of rock grains (Kohlstedt and Hansen 2015; Masuti *et al.* 2019). Therefore, this chapter begins with a brief synopsis of some of the microphysical models derived from laboratory deformation experiments. Such experiments reveal the same defect structures that are found in field samples of mantle rocks (e.g. Matsyuk *et al.* 1998; Johanesen and Platt 2015).

While the essence of rock flow at mantle temperature and pressure is to be found microscopically, all geodynamic simulations rely on the continuum hypothesis. This hypothesis asserts that the constitutive law governing the relations between stress and strain, and their time derivatives, are independent of scale. While a computational convenience is to describe mantle flow controlled by a constitutive equation that has an effective Newtonian viscosity, a large body of experimental results suggests a non-Newtonian behaviour. Hence, any discussion of Antarctic mantle rheology must probe this issue. Detailed descriptions of the experimental basis of the laws derived from studies of the mechanical strength of mantle rocks at high temperature and pressure is well beyond the scope of this chapter.

Overall, there exists a gap between laboratory-based flow laws and the rheology obtained by models of geodetically measured solid Earth deformation. This gap may only be reconciled by placing both methodologies into a rigorous statistical framework that will define a subset of the flow laws and model predictions that are mutually compatible (e.g. Jain *et al.* 2019). It is hoped that by writing this chapter we can lay some groundwork that might further this aim for the Antarctic mantle.

The role of experiments

Figure 2 shows a classic representation of the deformation diagram used in study of mantle rheology to define regimes of control by microscopic mechanisms as they vary with stress, strain rate, temperature and/or grain size. Here we show one example of a deformation map produced by Karato (2010) for olivine, a rock type that dominates the mantle above 410 km depth. The portion of the deformation diagram in Figure 2 that shows stress values below 100 MPa corresponds to the regime in which GIA-generated stresses occur. GIA stresses may reach 10 MPa in Antarctica near the lithosphere–asthenosphere boundary for realistic ice loading and

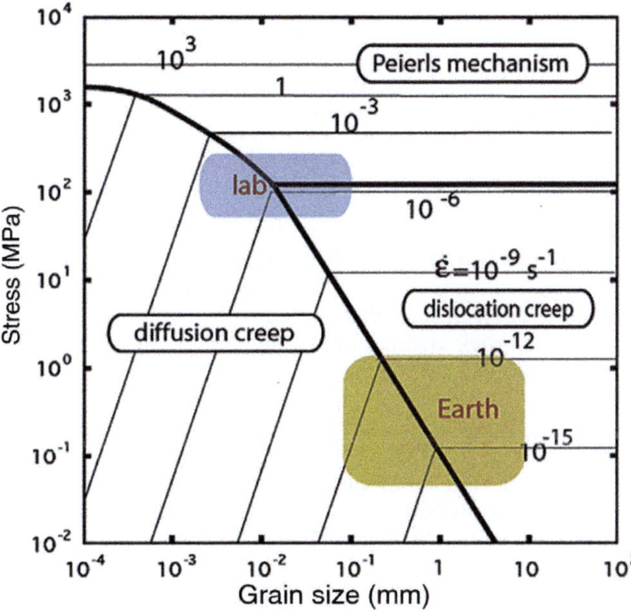

Fig. 2. Deformation maps of differential stress as a function of grain size for olivine at a pressure of 7 GPa and temperature of 1700 K. In the Earth's convecting mantle, grain size may be highly variable (Dannberg *et al.* 2017). Reprinted from Karato (2010), with permission from Elsevier.

unloading (e.g. Ivins *et al.* 2003). In this regime, two microscopic mechanisms may control effective rheology: diffusion or dislocation creep. Later in this chapter, both of these mechanisms are to play a role in the scaling to S-wave seismic tomography.

There are caveats that are well known when applying experimentally derived flow laws to geodynamic and GIA models. An obvious caveat is the difference between mantle strain rates ($\mathcal{O}\ 10^{-15} - 10^{-12}\mathrm{s}^{-1}$, where $\mathcal{O}$ indicates the 'order of magnitude') and those of the laboratory ($\mathcal{O}\ 10^{-6} - 10^{-3}\mathrm{s}^{-1}$) (see the light contour lines in Fig. 2). There is the additional caveat that the principal shear stresses are higher in the laboratory (see Fig. 2). Some experiments at higher strain rate ($\mathcal{O}\ 10^{-3}\mathrm{s}^{-1}$) and higher stress ($\mathcal{O}$ 80 – 100MPa) are relevant to shear localization in the lithosphere (e.g. Hansen *et al.* 2012). In Figure 2, the upper portion shows a region of the deformation map (deviatoric stress v. rock grain size) that corresponds to the material response that involves a plastic yield stress. This type of deformation may occur at relatively low temperature and high stress. In this regime, on a microscopic level, dislocation glide may dominate, a mechanism generally known as Peierls mechanism. This mechanism may be important to deep lithospheric shear zones (Kameyama *et al.* 1999). Such rheology has not been accounted for in GIA models and we do not treat this rheology in this chapter.

Constraints provided by seismology can be linked to temperature- and pressure-dependent creep laws for mantle rock aggregates composed of typical mantle minerals (olivine, pyroxene and Al-silicate). A modicum of understanding of the physics involved in both elastic wave propagation *and* dislocation and impurity diffusion within the rock grain interior and at grain boundaries is required. These will be briefly reviewed in this paper, along with some of the caveats that accompany these linkages. Furthermore, we assemble the current state-of-the-art interconnections between the high-pressure and -temperature thermodynamics and solid-state relations that allow us to quantify the effective mantle viscosity.

Steady-state v. transient laboratory experiments

The experimental set-ups for steady-state and transient creep are described by Paterson and Olgaard (2000), Karato (2010) and Kohlstedt and Hansen (2015). Two fundamental quantities must be extracted from such experiments for our application. These are the activation enthalpy for steady-state and transient, or anelastic, creep, H^* and H^*_{D}, respectively. The latter activation parameter was introduced into the interpretation of seismic tomography by Karato and Spetzler (1990) and Karato (1993). There are two types of high-pressure and -temperature mechanical strength laboratory experiments. Mantle flow models typically use experimental results that report 'steady-state' creep laws from static microstrain devices (e.g. Karato 2010; Kohlstedt and Hansen 2015), while models of seismic attenuation employ forced-oscillation devices over a range of oscillation frequencies (e.g. Berckhemer *et al.* 1982; Jackson 2019). Attaining a true steady state in the static experiments, wherein the rate of creation and destruction of microscopic elements at grain boundaries and within grains reaches equilibrium, is one of the challenges for developing creep laws that are acceptable in modelling mantle flow. As a rock deformation experiment begins, those microscopic elements that are capable of being mobilized under an imposed shear stress move relatively quickly. After a time, these elements (e.g. gliding or climbing dislocations and grain-boundary slip) may become blocked by obstacles, including their own bending or tangling, and, as a consequence, the creep rate of the rock aggregate slows. Technically, the macroscopic phenomenon is referred to as strain hardening (e.g. Peltier *et al.* 1980) and formal connections may be made between the damping of seismic waves to the microscopic viscoelastic strain (e.g. Minster and Anderson 1980; Jackson 2019). The mechanisms controlling the transition from transient to steady-state creep are numerous (e.g. Hanson and Spetzler 1994). While complex, these are linked to the anelastic properties of mantle rocks (e.g. Faul and Jackson 2015; Jackson 2019; Masuti *et al.* 2019).

Some general geophysical observations

Although our understanding of the physics of creep at upper and lower mantle conditions is limited, there is a clear observational connection between regions of low seismic-wave velocity and rapid isostatic response measured by geodetic uplift and/or rapid post-glacial land emergence (e.g. Sigmundsson 1991). While accounting for lateral heterogeneity in mantle deformation has been addressed over many decades (Wu 1999), today both higher-resolution seismic tomography and more dense and accurate geodetic networks are reporting data, meaning that the study of lateral heterogeneity in mantle rheology is evolving into a new frontier in solid Earth sciences (e.g. Austermann *et al.* 2013; Lau *et al.* 2018; Huang *et al.* 2019). Estimates of viscosity based on seismic-wave speeds and high-temperature creep experiments can be tested using geodetic data that are capable of constraining GIA or post-seismic modelled crustal motions. This enhances our knowledge of regional mantle rheological structure, and produces more confident viscosity estimates than in continental areas that lack such geodetic data constraints.

Scaling S-wave velocity to viscosity: retrospective for GIA

The general study of GIA has a substantial impact on how we interpret the rheology of the Earth's mantle and lithosphere over millennial to million year timescales. The preponderance of the GIA models that are used to simulate gravitational viscoelastic flow over the ice ages assume a radially stratified mantle that is approximated as a 3D linear Maxwell viscoelastic material in tensor form (e.g. Peltier *et al.* 1981). In this model rheological framework, the Earth responds viscously and elastically over long and short timescales, respectively. Its success in giving rational explanation to both geodetic (e.g. Milne *et al.* 2004) and RSL data (e.g. Lambeck *et al.* 2014) has meant that extensions to Maxwellian set-ups that contain non-linear power-law viscosity (e.g. Gasperini *et al.* 2004) generally fall outside of the GIA model paradigm most commonly used in cryospheric sciences, by geodesists and in the palaeo-oceanography community (e.g. Wahr *et al.* 2000; Whitehouse 2018; Dobslaw *et al.* 2020).

The structural roots of modelling GIA with laterally varying Earth structure (GIA-LV) have a number of commonalities. Seismic tomography provides elastic-wave velocity in 3D space (r, θ, ϕ), where the three spatial variables are radial position, co-latitude and longitude, respectively. A systematic way to describe these data is in the form $\delta v_S(r, \theta, \phi)/v_{S_0}(r)$ where the numerator of this expression, δv_S, represents the deviation of velocity from a mean value, v_{S_0}, that depends only the radial position. Both v_{S_0} and δv_S are related to absolute temperature. The temperature-dependence of the elastic constants governing wave propagation are known through theoretical and semi-empirical methods (e.g. Kumazawa and Anderson 1969). All of the GIA-LV models developed to date, including those that incorporate non-linear power-law viscosity, have assumed that the deviations from the mean are caused by a temperature deviation $\delta T(r, \theta, \phi)$ away from a radially dependent mean $T_0(r)$. Paulson *et al.* (2005), for example, assumed that the deviations in δv_S were linearly related to lateral density anomalies $\delta\rho$, such that:

$$d\ln v_S(r, \theta, \phi) = 3.33 d\ln\rho(r, \theta, \phi) \quad (1)$$

where here we use the logarithmic notation for the ratio of lateral deviation to the mean (see Karato 2008, p. 372). The factor 3.33 is one that may be deduced from high-temperature and -pressure mineral physics (e.g. Anderson 1995; Stixrude and Lithgow-Bertelloni 2010; Stixrude and Jeanloz 2015). (We elaborate on the physical relations for this factor later in this chapter.) Paulson *et al.* (2005) also assumed that the relationship of seismic velocity to temperature is completed by additionally assuming that equation (1) is divided by the thermal expansivity, α_{th}; thus, deriving the spatial dependence of the thermal deviation $\delta T(r, \theta, \phi)$ from the mean $T_0(r)$. The reader may note that this is the method adopted by A *et al.* (2013) for GIA-LV predictions for Antarctica. Both a nominal radial average temperature and its aspherical deviation were assumed to be known, and are written in the form:

$$T(r, \theta, \phi) = T_0(r) + \delta T(r, \theta, \phi) \quad (2)$$

and entered the viscosity relationship as:

$$\eta(r, \theta, \phi) = A_0 \exp[g T_m(r)/(T_0(r) + \delta T(r, \theta, \phi))] \quad (3)$$

where there is a prefactor, A_0, a dimensionless activation energy, g, and $T_m(r)$ is the melting temperature (e.g. Poirier 2000). Thus, all of the ingredients for performing finite-element model (FEM) solutions of the time-dependent sixth-order partial differential equation system for GIA-LV are defined.

Latychev *et al.* (2005) also developed a numerical finite-volume method for GIA-LV that also used equations (1) and (2), but made no assumption about the details of the governing rheology other than the Arrhenius-dependency given by the temperature in the exponential term. Therein, a viscosity ratio is defined as:

$$\Delta\eta \equiv \eta(r, \theta, \phi)/\eta_0(r) = \exp[-\varepsilon \delta T(r, \theta, \phi)] \quad (4)$$

where the parameter ε is used to adjust the total lateral variability in viscosity assumed in any GIA-LV computer simulation. A typical value for this adjustment parameter is $\varepsilon \approx 0.4\ K^{-1}$ (e.g. Austermann *et al.* 2013). Like Paulson *et al.* (2005), Latychev *et al.* (2005) assumed:

$$\delta T(r, \theta, \phi) = -\frac{1}{\alpha_{th}} d\ln\rho(r, \theta, \phi). \quad (5)$$

Latychev *et al.* (2005) and Paulson *et al.* (2005) developed simple and flexible scaling relations, but which are not closely tied to rheological laws that are currently in debate in the experimental rock deformation community (e.g. Kohlstedt and Hansen 2015). In contrast, the FEM formulation for GIA-LV by Wu (2005) sought to expand the scaling method so that it explicitly accounted for activation enthalpy of high-temperature creep, its decomposition into activation energy and activation volume, along with the pressure dependence. This was accomplished by adopting the formulation of Ivins and Sammis (1995). The latter method also forced the globally averaged radially-dependent viscosity, $\eta_0(r)$, to be defined by the same rheological dependencies. This guarantees an internal self-consistency, such that the rheological parameters could be tested against the most robust among competing 1-D GIA inversions for mantle viscosity. The formulation proposed by Ivins and Sammis (1995) used the same assumptions for the relationship of elastic shear-wave velocity to density and temperature as in equations (1), (2) and (5), although it more rigorously accounted for the Grüneisen parameter, a fundamental property of theoretical thermal elasticity (e.g. Isaak *et al.* 1992; Stacey and Hodgkinson 2019). These relationships will be explicitly given in the 'Methods for Approach 1' and 'Methods for Approach 2' sections later in this chapter.

Wu (2005) adopted the $\log_{10}$ representation of the viscosity field and employed the relationship to the activation enthalpy parameter, H^*, as proposed by Ivins and Sammis (1995). This representation can be written as the ratio:

$$\Delta\eta = \frac{\exp[a/(T_0 + \delta T)]}{\exp[a/T_0]} \tag{6}$$

with $a \equiv H^*/R_G$ and R_G being the universal gas constant. Upon taking $\log_{10}$ of both sides of equation (6), this becomes:

$$\frac{\log_{10}\Delta\eta}{\log_{10}e} = \frac{a\delta T}{(T_0 + \delta T)} \tag{7}$$

clearly a non-linear relationship between viscosity and lateral thermal anomaly. Ivins and Sammis (1995) further demonstrated with several global shear-wave seismic tomographic models that it was feasible to linearize this relationship by using a Taylor expansion about $\delta T = 0$:

$$\frac{\log_{10}\Delta\eta}{a\log_{10}e} \approx -\frac{\delta T}{T_0^2} + \text{higher-order terms.} \tag{8}$$

By truncating the expansion at a single term, we arrive at the form employed in Wu (2005) and to be used in our intercomparison of methods. The approximation is also inherent to the strategy of Latychev *et al.* (2005). In Appendix A of this chapter, we use variability up to the maximum temperature differences found in the Antarctic thermal model of An *et al.* (2015) at a depth of 240 km to give a quantitative evaluation of this approximation. Although Wu (2005) used this form, a later GIA-LV model (Wang *et al.* 2008) extended the relation for connecting δv_S and δT to include anelastic corrections that were earlier advocated by Karato (1993). It is this anelastically corrected form that is the starting point of our intercomparative study entertained later in this chapter.

Microphysics

It has long been understood that deformation mechanisms in ceramic materials at high temperature are controlled by the diffusion of ionic impurities (point defects) or by slip due to dislocations (linear or planar defects) in the crystal lattice. Ceramic materials are distinguished by their covalent and ionic atomic bonding, usually involving one or more oxygen atoms. This intracrystalline bonding allows experiments to be performed on certain analogue materials with crystal structure such as rock materials at mantle temperatures and pressures, thus relevant to both thermal elasticity and creep. This section discusses the main mechanisms, as well as a generalized flow law and the main rationale used to approximate an 'effective' viscosity.

Steady-state flow law and temperature-dependent anelasticity

The diffusion of point defects and the slip of dislocations are the microscopic manifestations of diffusion and dislocation creep, respectively. Both creep mechanisms are controlled at some stage by diffusion and, hence, by temperature. The strain rate, $\dot{\epsilon}$, in the presence of a background stress, σ, is then exponentially dependent on the local mantle temperature, T, typical of all thermal activation processes (e.g. Kittel 2004; Cressler and Moen 2012). The form is:

$$\dot{\epsilon} = \tilde{f}(\dot{\epsilon}_0)\sigma^n \exp(-H^*/R_G T) \tag{9}$$

where H^* is the activation enthalpy associated with diffusivity in solids (Poirier 2000; Karato 2008), R_G the universal gas constant (8.314×10^{-3} kJ mol^{-1} K^{-1}), n is a semi-empirical exponent ($1 \leq n \leq 5$) and $\tilde{f}(\dot{\epsilon}_0)$ a constant coefficient with units of s^{-1}Pa^{-n}, which is often tagged to a reference strain rate $\dot{\epsilon}_0$, or by other specific features of the microphysics operating. The temperature-dependence derives from the Maxwell–Boltzmann statistics governing the ionic–molecular interactions at the crystalline level; hence, equation (9) must have the temperature expressed in Kelvin (e.g, Kittel 2004). In laboratory settings, high confining pressures are required to stabilize the sample while reaching measurable strain rates. A tractable parameter to solve for is H^* at a series of temperature values. In this chapter we shall make use of two activation enthalpies: one for steady-state creep and the other for anelastic and transient (viscoelastic) strain processes in mantle rock. These activation enthalpies are decomposed into two terms:

$$H^* = E^* + PV^* \tag{10}$$

and

$$H_D^* = E_D^* + PV_D^* \tag{11}$$

where P is pressure, and pairs E^*, E_D^* and V^*, V_D^* are the activation energies and volume for the steady-state and transient creep states, respectively. Each activation parameter will be unique to the different creep mechanisms that operate within a crystalline rock structure, either within or at the boundaries of the rock grains. Generally, E^* is more straightforward to recover from laboratory experiments than values of V^*, and similarly for their transient counterparts, E_D^* and V_D^*. While the former may be recovered at relatively lower confining pressures, experiments determining V^* must accommodate stepwise increases in pressure while carefully accounting for all other factors influencing creep mobility (Dixon and Durham 2018).

It is to be understood that these activation parameters are arguments in exponential functions multiplying T^{-1}. Hence, they represent the fact that there is a strong temperature-dependence both to steady-state creep and to seismic dissipation and dispersion (e.g. Wiens *et al.* 2008). The latter viscoelastic properties are essential for properly developing transfer functions between S-wave velocity and mantle viscosity.

The normal procedure for determining E_D^* from experimental results is to fix V_D^* to roughly the values determined by steady-state creep experiments ($V_D^* \sim V^* \approx 15 \pm 5 \times 10^{-6}$m^{3}mol^{-1}) (e.g. Jackson and Faul 2010; Dixon and Durham 2018). For estimating the temperature-dependent anelastic influence on shear-wave velocity, we will assume pairs E_D^*, V_D^*, corresponding to the experimental results of Jackson and Faul (2010).

If the constitutive laws only involved these two thermodynamic variables, while probably non-unique, the problem of mapping seismic velocity to viscosity using the methods of either Ivins and Sammis (1995) or Wang *et al.* (2008) is quite tractable, provided that both the temperature and the rate-controlling creep mechanism are decipherable. However, not only is it difficult to decipher the rate-controlling mechanism, but the pre-exponential constant $\tilde{f}(\dot{\epsilon}_0)$ is dependent on both grain size and volatile impurities, such as water or free ions of hydrogen and oxygen. While current controversy exists over the realistic magnitude of the effect of volatile impurities, it may be treated in a formal way by including an additional pre-exponential term: $f_{H_2O}^{\bar{r}}(P, T)$, where f is the fugacity of the respective impurity in the rock material (Karato 2008). It also represents the partial pressure of the invasive atomic species in the crystal structure (here $\bar{r}$ is an experimentally determined exponent). The larger these impurity effects are, the more intractable becomes building a relation between seismic imaging in the mantle and lateral variations in effective viscosity, since impurities in the mantle may have spatial dependencies that do not follow, or even mimic, those of the spatial dependencies of $T(r, \theta, \phi)$. Therefore, they are briefly discussed in the following subsection.

Role of water and grain size

Small grain sizes tend to promote material transport by grain-boundary diffusion (Karato and Wu 1993), whereas large grain size promotes deformation by dislocation creep. Also, experiments with mantle olivine and synthetic anorthosite indicate that trace amounts of structurally bound water have a pronounced weakening effect (Mackwell *et al.* 1985; Hirth and Kohlstedt 1996; Dimanov *et al.* 1999; Mei and Kohlstedt 2000), thus indicating the importance of water-weakening in increasing the mobility of rock in the lithosphere and upper mantle environment. A wet sample of rock is defined when there are roughly 500–3000 H per 10^6 Si and a dry sample at one order of magnitude smaller fraction H content (1 ppm $H/Si \approx 10^{-5}$ wt% of water in olivine, or 5×10^{-3}–3×10^{-2} wt% H_2O). Proper quantification of the effect of water-weakening at mantle temperatures and pressures remains controversial (Fei *et al.* 2013). Generally, volatile impurities, such as water, are a subclass of point defects. Other atomic or molecular species may also be important in controlling mobility.

In order to account for grain size in the constitutive equations, $\tilde{f}(\dot{\epsilon}_0)$ is set equal to Ad^{-p} (Karato *et al.* 1986; Dimanov *et al.* 1999), and for both grain size and wetness content to $Ad^{-p}f_{H_2O}^{q}$, where the units and interpretation of the constant A are dependent on the inclusion, or exclusion, of the effects. Here p and q are experimentally determined exponents. Possibly the most complete prefactor would be to set:

$$\tilde{f}(\dot{\epsilon}_0) = Ad^{-p}f_{H_2O}^{q}\exp(\bar{\alpha}\bar{\phi}) \tag{12}$$

which would additionally account for the influence of the fraction of partial melt content, $\bar{\phi}$, ($\bar{\alpha}$ being a constant) on solid deformation (Hirth and Kohlstedt 2004). One of our approaches uses the prefactor in the form of equation (12). Lateral heterogeneity in grain size (Dannberg *et al.* 2017) and water content (O'Donnell *et al.* 2017) might be important for Antarctica. Examining these latter causes is warranted should we discover especially poor resemblance of effective viscosity scaled to shear-wave velocity maps, $v_S(r, \theta, \phi)$.

It is worth noting that both structurally bound water and grain size are likely to affect seismic velocities, although quantitative relationships are uncertain and, thus, difficult to implement formally in viscosity estimates. Water is proposed to reduce seismic velocities by increasing anelasticity, through a similar mechanism to its effect on viscosity (Karato 2003). Recent laboratory experiments indicate a small direct seismic velocity perturbation due to the presence of water (Cline *et al.* 2018), yet a substantial sensitivity to oxygen fugacity, f_{O_2}, a property that often positively correlates with water content. Rock mechanics experiments do show a significant dependence of anelasticity on grain size at upper mantle temperatures, suggesting that small grain size can lower seismic velocity (Jackson and Faul 2010). The effect of water and small grain size both reduce seismic velocity and mantle viscosity. Thus, they both qualitatively simulate the effects of higher temperatures and both act to reduce mechanical strength. Ignoring the effects of water and grain size in converting seismic velocities to viscosity will at least lead to viscosity anomalies of the correct sign, if not the correct magnitude.

Deformation mechanisms in mantle conditions

Simulation of upper mantle conditions in the laboratory are limited by confining pressure, limited temperature ranges and relatively high strain rates that must be studied in the laboratory. Nonetheless, there are advantages in that experiments have been performed on natural or synthetic rocks of either the major constituents of the mantle or their analogues. Sampling of upper mantle rocks under a microprobe reveal those subgrain-scale dislocation sites that are associated with ceramic high-temperature creep mobility. If diffusion creep dominates deformation in the upper mantle, then $n \approx 1$ in equation (9), an assumption often employed in GIA models. However, this requires the upper mantle to have small grain sizes throughout, contrary to observations (Karato and Wu 1993; Dannberg *et al.* 2017; Coltorti *et al.* 2021, this volume). For large grain size, or large stress, dislocation creep dominates, as in Figures 2 and 3. An alternative mechanism, which has been studied extensively over the past several years is grain-boundary sliding (GBS) (Hirth and Kohlstedt 2004; Hansen *et al.* 2011). While this mechanism may be controlled by diffusion for small grain size, experiments with naturally occurring olivine reveal that dislocation-accommodated GBS can dominate creep with $n \approx 2.9$ at upper mantle conditions and realistic grain size ($d \sim$ 1–10 mm). Figure 3 shows the dominant deformation mechanisms in a stress–temperature diagram for cases of small (0.01 mm) and large (1 mm) grain sizes. As an example, consider the contour line corresponding to a strain rate of 10^{-15} s^{-1} for an olivine with a 1 mm grain size. At high temperature and low stress, diffusion creep will dominate. With increased stress, dislocation creep will become more important. The stress at which the transition occurs will be higher when grain size is decreased. For very small grain sizes (10 μm), diffusion creep will give way to GBS when stress is increased.

Continuum approach and effective viscosity

In the continuum point of view taken in geodynamic flow simulations, material behaviour can be well described by the Maxwell model (e.g. Ivins *et al.* 1982; Karato 2008). In the Maxwell model, elastic deformation occurs instantaneously, while at relatively long timescales movement (≥1–10 kyr) is described by slow flow in a viscous fluid. GIA codes bookkeep the elastic deformation, as this is important for RSL and Little Ice Age geodetic computation (Ivins *et al.* 2000; Sabadini *et al.* 2016), but they are usually ignored in geodynamic models wherein the total viscous strain vastly exceeds that of the elastic components over tens of millions of years, or greater. The Maxwell model is generally faithful to the microphysical processes; it corresponds to a situation in which defects can move indefinitely after an initial elastic strain (Karato and Spetzler 1990; Karato 2010). To use the Maxwell model, a viscosity parameter is needed. The viscosity in such models can be linked to both the laboratory-based laws and to seismically based mantle models by defining a so-called effective viscosity.

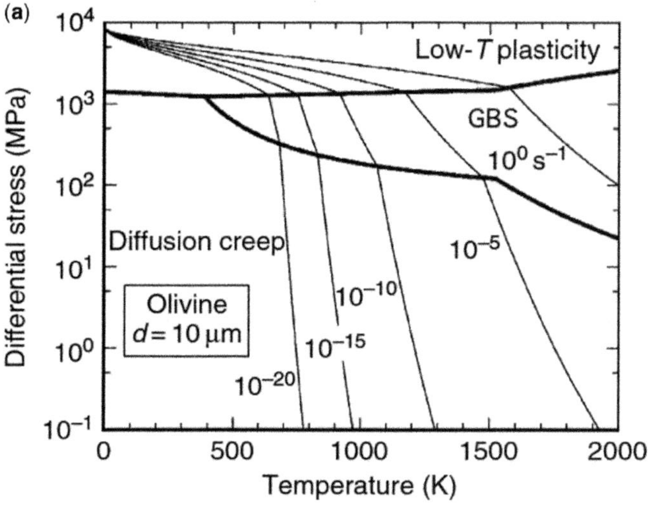

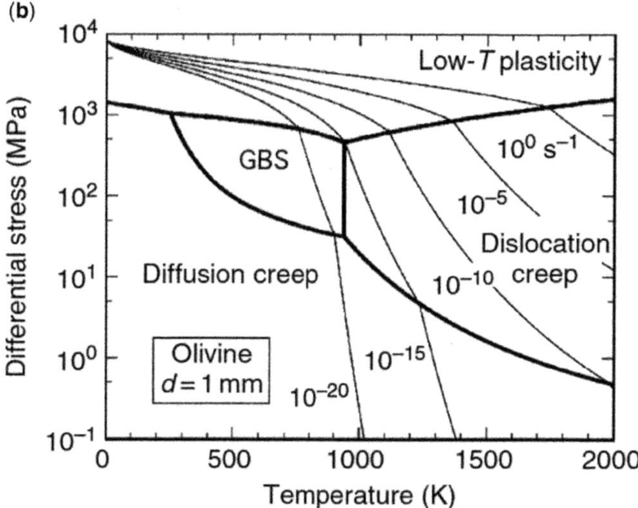

Fig. 3. Deformation maps of differential stress as a function of temperature for dunite, with grain sizes of (**a**) 10 µm and (**b**) 1 mm. Light contour lines show constant strain rate. GBS, grain-boundary sliding. Most observed grain sizes corresponding to the upper mantle are reflected better by (b). Modified from Kohlstedt (2007), with permission from Elsevier.

Choice of effective viscosity relation

The advantage of defining an effective viscosity, $\bar{\eta}$, is that it can be conveniently linked to the laboratory-based laws, to seismically based mantle models (Doubrovine *et al.* 2012) and to the boundary value problems typically employed in the GIA problem (Spada *et al.* 2011); three different forms that have been proposed by Poirier (2000) and Ranalli (2001). Each, however, require some flow variable to be fixed. The three relationships yielding an effective viscosity are given below. At constant local background strain rate:

$$\bar{\eta} = \eta_{\dot{\epsilon}} = \frac{1}{2}\tilde{f}_{\dot{\epsilon}}\dot{\epsilon}^{(1-n)/n}\exp\left(H^*/nR_{\mathrm{G}}T\right) \qquad (13)$$

while at constant background stress:

$$\bar{\eta} = \eta_{\sigma} = \frac{1}{2}\tilde{f}_{\sigma}\sigma^{(1-n)}\exp\left(H^*/R_{\mathrm{G}}T\right). \qquad (14)$$

Finally, at constant rate of energy dissipated in the flow:

$$\bar{\eta} = \eta_{\Phi} = \frac{1}{2}\tilde{f}_{\Phi}\Phi^{(1-n)/(1+n)}\exp\left[2H^*/(n+1)R_{\mathrm{G}}T\right] \qquad (15)$$

where $\Phi \equiv \dot{\epsilon}\sigma$, and the constant factors $\tilde{f}_{\dot{\epsilon}}$, $\tilde{f}_{\sigma}$ and $\tilde{f}_{\Phi}$ are functions of the experimental fits involving water or oxygen fugacity, grain size, partial melt and the power-law exponent n (e.g. see equation 12). The reader may consult Ranalli (2001) or Karato (2008, equations 19.3a–c) for the corresponding details. The Maxwell assumption so often used in GIA modelling is that either $n = 1$ or that one of the above relations offers a link to laboratory data. We have already mentioned that is possible to formulate Maxwellian models with $n > 1$. Should the laboratory experiments be well fit with $n = 1$ (diffusion creep), then the classic Newtonian viscosity, η, defined with the equation

$$\eta \equiv \frac{\sigma}{2\dot{\epsilon}_0} \qquad (16)$$

applies, and there are no ambiguities. Models using any approximations represented by equations (13–15) (e.g. Ivins 1989; Zhu 2014) are not as tightly linked to the laboratory experiments as are mantle-flow modelling that directly incorporate equation (9), and perhaps equation (12), into the computations (e.g. Wu 2002; Dal Forno *et al.* 2005).

Since the time of Norman Haskell's work in the 1930s, the simple Newtonian viscosity, η, is the sought after parameter for the mantle interior in various forward-inverse GIA and mantle-flow models. In what follows in this chapter, we will be using the effective viscosity as defined by equation (14). This will avoid employing the power-law exponent n for dislocation creep into the argument of the exponential in the flow law. We remark that in doing so, we choose the approximation that will tend to maximize the lateral variability predicted from the ANT-20 seismic tomography model.

Maxwell models: why are they relatively successful?

Considering that most laboratory experiments yield $n \sim 3$, it might come as a surprise that the Newtonian assumption, when inserted into the Maxwell constitutive relation, does fairly well at numerically simulating mantle geoid and GIA. There can be many reasons for this, which are beyond the scope of our discussion. Limitations of the datasets for constraining geodynamic models are just one obvious source. Two popular ideas are proposed as logical explanations. First, is that the flow law actually is Newtonian, a prediction shown in the two deformation diagrams of Figure 3 labelled as diffusion creep. Second, is a suggestion offered by Karato and Wu (1993) that mantle flow is self-regulating, via modifications of grain size. Here, the idea is that stress increases during the rise of ice loading/unloading phases with the consequence that grain size is reduced. Such reduction moves the local mantle creep state to the left in the deformation diagram of Figure 3, closer to the boundary between diffusional ($n = 1$) and dislocation ($n \sim 3$) dominant mechanisms. This might allow the operative microscopic plasticity to hover quite close to the boundaries between competing solid-state creep mechanisms (see the solid lines in Fig. 3 that separate the diffusion, dislocation and grain-boundary sliding regimes).

Viscosity estimates for Antarctica

As is clear from the previous section, viscosity involves many parameters that are themselves uncertain. Past studies of laterally varying viscosity beneath Antarctica (e.g. Kaufmann *et al.* 2005; A *et al.* 2013; van der Wal *et al.* 2015; Gomez *et al.* 2018) have all agreed on a larger East Antarctic viscosity, owing to the faster seismic velocities imaged there. The results of these studies, however, substantially differ in the predicted magnitudes. Although, the resolution of the mantle seismic structure has improved through the years, many of the differences may be caused by methods employed in transforming seismic velocity to viscosity.

In the several subsections that follow, we estimate Antarctic viscosity in three different ways. In the first approach, viscosity perturbations are calculated directly from seismic models according to the approach of Wu *et al.* (2013). These use the temperature derivatives of shear velocity from Karato (2008) to estimate a temperature anomaly and to estimate a viscosity anomaly relative to a reference viscosity model. The second method uses the structural form derived by Ivins and Sammis (1995), but is extended to capture the anelastic correction using the theoretical development of Karato and Karki (2001) and Karato (2008). The third method has similarities to the first two approaches, but assumes a complete representation of the rock experimental results, including the non-thermodynamic prefactors. Both the first and third approaches will rely on temperature derivatives as explicitly computed by Karato (2008). In all approaches, we assume that global average 1D seismic velocity and mantle temperature are models that are relatively well known. In addition, Approaches 1 and 2 assume that radial 1D viscosity models, $\eta_0(r)$, for the upper mantle are well determined. They then compute a 3D viscosity model based on perturbations to these global averages based on a straightforward conversion from seismic velocity to effective viscosity. Naturally, in each case of Approach 1 and Approach 2, $\eta_0(r)$ must also be logically linked to the flow law assumed. To give clarity to the formulations of the three methods, we give some of the requisite fundamentals of the classical thermodynamics and elasticity theory. To better elucidate the comparison and the uncertainties, the three methods all start from the same seismic structure model of Lloyd *et al.* (2020) (Wiens *et al.* 2021, this volume) termed ANT-20, which is the highest-resolution continental-scale model available. Figure 4 shows the ANT-20 $\delta \ln v_S(r, \theta, \phi)$ field at depths of 150 and 550 km. Together with the map in Figure 1, these show the lateral variations in shear-wave velocity common to all three approaches for estimating effective viscosity that will be discussed in the next four subsections of this chapter.

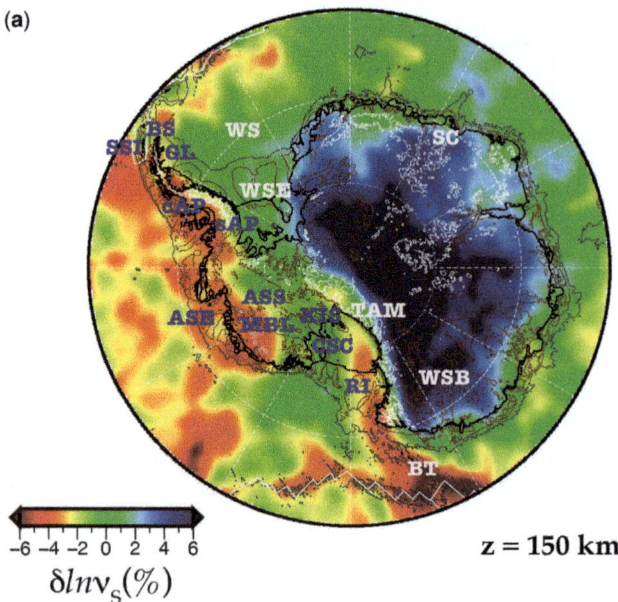

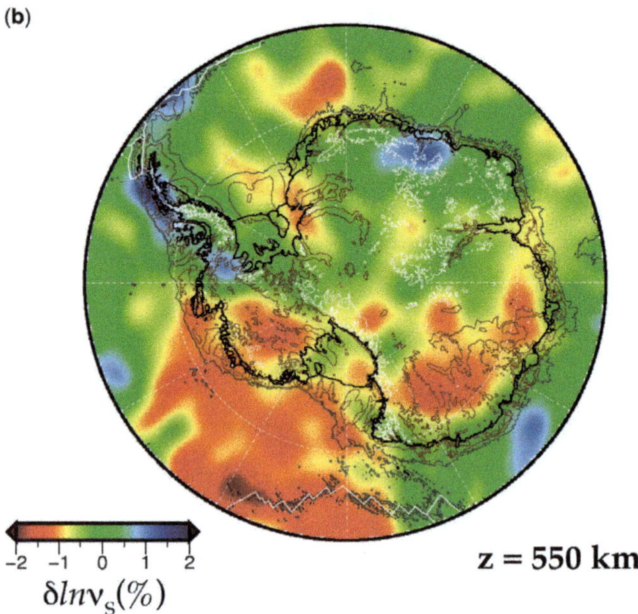

Fig. 4. Shear-wave velocity in ANT-20 as a deviation from global average (STW105) at (**a**) 150 and (**b**) 550 km depth. Note the differences in the colour scale bar. Refer to Figure 1 for the place name abbreviations in map (a) and additional information.

Relations between elastic and seismic parameters

Seismic velocity in isotropic rock media is related to density, ρ, elastic bulk modulus, κ, and shear modulus, μ, by:

$$v_S = \left[\frac{\mu}{\rho}\right]^{1/2} \quad (17)$$

$$v_P = \left[\frac{\kappa + \frac{4}{3}\mu}{\rho}\right]^{1/2}. \quad (18)$$

Thus, both shear (v_S) and compressional (v_P) seismic velocities are related to density. It is often useful to consider seismic velocities as anomalies relative to a 1D radial background (reference) model. For example, the change in shear velocity, δv_S ($v_S - v_{S_0}$), relative to the background model, v_{S_0}, is expressed as the ratio $\frac{\delta v_S}{v_{S_0}}$ for such anomalies. This quantity is accurately approximated using the natural logarithm, and doing so is advantageous for our purposes:

$$\delta \ln v_S \approx \frac{\delta v_S(r, \phi, \theta)}{v_{S_0}(r)} \quad (19)$$

and

$$\delta \ln v_P \approx \frac{\delta v_P(r, \phi, \theta)}{v_{P_0}(r)}. \quad (20)$$

In the following we focus on the shear-wave velocities, as shear-velocity models generally show the highest continent-wide resolution in the Antarctic upper mantle. Anderson (1987) derived relations from thermodynamic and solid-state physics principles that interrelate seismic velocity anomalies and density anomalies. One goal of linking global seismic mapping and geodynamical modelling is to constrain the ratio of variable seismic velocity to variable density, $\delta \ln v_S / \delta \ln \rho$ (e.g. Dziewonski *et al.* 1993). In theory, inversion of such ratios can be linked to thermodynamic quantities derived from pressure and temperature derivatives of elastic parameters of mantle mineralogy. For interpreting seismic velocity anomalies and converting them to viscosity, a fundamental interrelation used by Ivins and Sammis (1995) was:

$$\frac{d \ln v_S}{d \ln \rho} = \frac{1}{2}\left[\frac{\partial \ln \mu}{\partial \ln \rho} - 1\right]. \qquad (21)$$

The ratio appearing on the right-hand side of equation (21) is the Grüneisen parameter for the shear modulus, c_1:

$$c_1 \equiv \frac{\partial \ln \mu}{\partial \ln \rho}. \qquad (22)$$

c_1 is identical to δ_μ as defined in the monograph by Karato (2008, equation 4.35b). The thermal Grüneisen parameter is defined as a ratio of vibrational frequencies to molecular volume, γ_μ:

$$c_1 = 2\gamma_\mu + \frac{1}{3}. \qquad (23)$$

The relations we have presented in this subsection establish the basic theory for relating seismic-wave velocity to the Grüneisen parameter, a fundamental property of the thermodynamic state of crystalline rock at high temperatures and pressures. In the next subsection we show how this is a key link to temperature derivatives of the shear-wave velocity. Here we strictly employ the thermodynamic and elasticity approach that is advocated by D. L. Anderson (1988), O. L. Anderson and Isaak (1995), Karato (2008), Stixrude and Jeanloz (2015), Stacey and Hodgkinson (2019) and others. We must also be cognizant of the uncertainties associated with the thermal elastic constants of mineral physics at the high temperatures and pressures of the mantle. For the reader unfamiliar with thermal and seismic equations of state, but who wants to delve further into this subject for its practical ramifications (e.g. Núñez-Valdez *et al.* 2013), understanding the interrelationship between elastic constants and thermodynamic variables is quite important.

Thermal parameters

A key element of any geodynamical model is the thermally induced buoyancy. For our purposes, one element of this buoyancy, the thermal expansivity $\alpha_{th} \equiv -\partial \ln \rho / \partial T$ is a simple elastic parameter that can be derived by mineral physics (e.g. Anderson *et al.* 1992; Katsura *et al.* 2010) and it expresses the volume changes that occur due to heat content. The radial dependency of α_{th} can be adopted, for example, from Anderson *et al.* (1992) for the upper mantle and from Chopelas and Boehler (1992) for the lower mantle. Here we select the depth-dependence derived by Katsura *et al.* (2010). Our goal is to isolate anomalous heat content through its influence on shear-wave speed. Thus, the essential parameters are the thermal expansivity, α_{th}, and c_1 or γ_μ. These parameters were the essence of the Ivins and Sammis (1995) method and implicitly enter in all of the approaches in this chapter.

To describe anomalous heat content it is important to have a 3D description of the temperature distribution in the mantle. We define anomalous temperature as $\delta T(r, \theta, \phi)$, that part of the temperature that deviates from the globally based spherically symmetrical part. Stated formally: $\delta T(r, \theta, \phi) \equiv T - T_0(r)$, wherein $T_0(r)$ is the global spherically averaged temperature at radius r. This decomposition is standard procedure in computing free convection in planetary interiors (e.g. Durney 1968). To infer effective viscosity, all approaches require, implicitly or explicitly, a reference temperature profile in order to specify a radial variation of viscosity and a distinct laterally varying viscosity. Under the approximations developed in Ivins and Sammis (1995) these viscosity and temperature fields may be expanded in orthogonal functions. Hence, the relations for the viscosity and seismic velocity fields can both to be expanded in spherical harmonics and interrelated in the spectral domain. We do not, however, give the details of this procedure in this chapter. Table 2 gives the $T_0(r)$ values used for all three approaches where we also show non-global averages at 150 km depth.

The elastic parameters can be combined to form the desired derivative that relates seismic velocity anomalies to temperature anomalies:

$$\frac{\partial \ln v_s^{(ah)}}{\partial T} = -\frac{1}{2}\alpha_{th}(c_1 - 1) \qquad (24)$$

where superscript ah denotes anharmonic, meaning that the elastic lateral variation in velocity results from thermal expansion and associated changes in elastic structure (see equation 4.40a of Karato 2008).

What follows is a discussion of the three approaches to deriving lateral viscosity variations from a seismic model. This requires an estimate of the global average temperature, T_0, at each depth. Three depths are chosen for the intercomparison: z = 150, 300 and 550 km. Because the reference seismic and viscosity models are global, it is essential that the reference temperatures represent global averages. The 150 km depth lies within the conducting lithosphere for some locations. Thus, for T_0 at 150 km, we use the global average estimate of 1473 ± 40 K from Stacey and Davis (2008). Should non-global T_0 values for relatively hot or cold subcontinental mantle be assumed (see Table 2), they will seriously bias the predictions of deviations from a purely radially dependent viscosity. We employ estimates of the isentropic, convecting, variation with depth for our value of T_0 for the deeper depths, arriving at 1722 ± 42 K for 300 km depth and 1930 ± 50 K for 550 km (see figs 3 and 4 in Katsura *et al.* 2010). Karato and Karki (2001) pointed out the potential pitfalls of deriving seismic constraints on either mantle temperature and/or driving buoyancy in geodynamic models without *a priori* correction for anelasticity. We return to this issue in the section discussing Approach 2.

Table 2. *Temperature T_0 assumed at 150, 300 and 550 km depths*

Depth (km)	T_0 (K)	Method
150 (hot continental)*	1680 ± 20	Jaupart *et al.* (2015) ($q_{GHF} = 90\,\mathrm{mW\,m^{-2}}$)†
150 (cold continental)*	1283 ± 20	McKenzie *et al.* (2005) ($q_{GHF} = 58.6\,\mathrm{mW\,m^{-2}}$)
150 (global average)	1473 ± 40	Stacey and Davis (2008)
300	1722 ± 42	Katsura *et al.* (2010) (isentrope varying with depth)
550	1930 ± 50	Katsura *et al.* (2010) (computed temperature)

*Hot and cold values are for illustration and are not used in the computations of viscosity.

†q_{GHF} is the corresponding regional surface geothermal heat flux.

Road map for the three approaches

Karato and Spetzler (1990) and Karato (1993) demonstrated that all realistic mantle rock mineral assemblages will have elastic-wave velocities diminished by thermally activated defect processes, so-called anelastic dispersion (e.g. Jackson *et al.* 2002). This must be accounted for in seismic studies, and the three approaches generally do so in different ways. Some elements of the approaches are summarized in Tables 1 and 3. Approach 1 uses a method that was partly described in the maps of the lateral variation in mantle viscosity derived using a combination of the scaling described by Ivins and Sammis (1995) with the modifications described by Wu *et al.* (2013). As a consequence, a number of features of Approach 1 and Approach 2 are similar, and we shall attempt not to be redundant in describing these overlapping features. Each method does apply an anelastic correction, as first suggested by Karato (1993). The latter is a necessary feature of deriving variable temperatures from seismic tomography. However, the detailed application of this correction differs between approaches. For Approach 1 and Approach 3, the anelastic correction simply follows from the table recommended by Karato (2008, table 20.2, p. 376). For Approach 2 we assemble the components of the anelastic correction from the general relationships for diffusion-controlled energy loss as described by Karato and Karki (2001) and in the monograph of Karato (2008, equations 18.5, 18.6, 20.12 and 20.13). We also rederived the asymptotic relationships of Karato (2008, equation 20.15b and discussion). This revealed that the asymptotics come with the added caveat that the frequency-dependent power-law exponent (α) must be restricted to $\alpha \leq 0.3$. Without the asymptotics, the root of the anelastic theory and its temperature dependence is essentially identical to that employed by Guerri *et al.* (2016).

For our intercomparison exercise we generally assume rather different flow laws to be operative. Approach 1 assumes diffusion creep as described and quantified by Hirth and Kohlstedt (2004), and Approach 2 assumes GBS dislocation creep as described and quantified by Hansen *et al.* (2011). Readers that may be interested in the derivation that leads to the linear relations between $\mathrm{dln}v_S$ and δT, and the viscosity variability that we describe below, $\log_{10}\Delta\eta(r, \phi, \theta)$, may wish to consult the paper by Ivins and Sammis (1995) and Appendix A of this chapter.

Approach 3 is unique in that it uses flow laws in a composite form, combining olivine dislocation, diffusion and GBS micromechanics for calculations at 150 and 300 km depths, and similarly for ringwoodite at 550 km for dislocation and diffusion. Other differences that distinguish Approach 3 are summarized in Tables 1 and 3. Approach 3 uses thermoelastic plus anelastic scaling to $\delta\mathrm{ln}v_S$ that is identical to Approach 1.

Table 3. *Guide for the three intercomparative approaches**

Parameter	Rule	Subcommonality†
$T_0(r)$	Common	1, 2 and 3
$\frac{\mathrm{d}v_S(r)}{\mathrm{d}T}$	Differ	1 and 3
$H^*(r)$	Differ	1 and 2 at $z = 550$ km
$\bar{\eta}$	Common	Equations (4) and (14)
$\bar{\eta}_0(r)^‡$	Differ	None

*Differences in anelastic correction are discussed in the text.
†Numbers refer to different approaches.
‡See equation (4).

Inference from modelling geodetic data

The construction of trends from the time series of geodetic observations of the movement of the crust for the past three decades (e.g. Dietrich *et al.* 2004; Bevis *et al.* 2009) provides essential information for the Antarctic GIA modelling community. Inferences of mantle viscosity beneath Antarctica from geodetic observations are useful for fine-tuning, and even calibrating, the parameters used in mapping seismic velocity to viscosity. Unfortunately, these estimates generally require some assumption about ice-load history, which is not well understood, and thus different estimates for the same region can vary significantly. Nield *et al.* (2014) used GPS-measured uplift following the 2002 break-up of the Larsen B Ice Shelf to constrain the upper mantle viscosity beneath the northern Antarctic Peninsula to about 1×10^{18} Pa s or less. A more recent GIA analysis by Samrat *et al.* (2020) also favours a quite low shallow mantle viscosity in this region (0.3×10^{18}–3×10^{18} Pa s) and a modest upper mantle viscosity of 4×10^{20} Pa s deeper than 400 km depth. For the northern Antarctic Peninsula, Ivins *et al.* (2011) used a different GPS dataset, and Gravity Recovery and Climate Experiment (GRACE) data. The modelled GIA uplift also included an ice model with historical retreat (*c.* 100–150 years) and determined a best-fit mantle viscosity of 4×10^{19}–7×10^{19} Pa s. Bradley *et al.* (2015) and Wolstencroft *et al.* (2015) estimated upper mantle viscosity at 1×10^{20}–3×10^{20} Pa s across the southernmost Antarctic Peninsula and western Weddell Sea based on GPS measurements and improved ice-load history. Zhao *et al.* (2017) also modelled regional mantle viscosity in the central Peninsula and found values in excess of 2×10^{19} Pa s with an ice-loss history documented back to 1966 using remote-sensing data. Late Holocene beach uplift data was modelled by Simms *et al.* (2012) for the South Shetland Islands glacial advance and retreat, and associated viscoelastic GIA response. They determined a best-fit upper mantle viscosity value of 1.0×10^{18} Pa s, applicable to the asthenospheric mantle adjacent to the actively spreading Bransfield Strait Basin (e.g. Lawver *et al.* 1995). In summary, estimates of upper mantle viscosity beneath the Antarctic Peninsula range from less than 10^{18} to about 3×10^{20} Pa s, with lower values generally predominating in the northern and western areas.

A few viscosity estimates are also available for other areas of West Antarctica. Nield *et al.* (2016) suggested, based on observed responses to the stagnation of the Kamb Ice Stream, that upper mantle viscosity beneath the Central Siple Coast cannot be less than 1×10^{20} Pa s. Barletta *et al.* (2018) modelled extremely rapid GPS uplift rates in the Amundsen Sea Embayment region. After corrections for the elastic response from current ice-mass loss are applied, an upper mantle viscosity that is lower than the canonical upper mantle values (2–6×10^{20} Pa s) can be inferred. Their best-fitting model revealed a mantle viscosity of 4×10^{18} Pa s from 65 to 200 km depth, underlain by 1.6×10^{19} Pa s from 200to 400 km depth and 2.5×10^{19} Pa s from 400 to 660 km. Using a shorter ice-unloading history than Barletta *et al.* (2018), a GIA model proposed by Powell *et al.* (2020) that employed seismic scaling, suggests that viscosity as low as several times 10^{18} Pa s in the shallow upper mantle and several times 10^{19} Pa s down to a 900 km depth might be responsible for the rapid GPS uplift. Inevitably, these estimates of Antarctic mantle viscosity will improve as our constraints on past ice change become more robust (e.g. Nichols *et al.* 2019; Johnson *et al.* 2020).

We caution the reader that here we have discussed the ability of Maxwell Earth models, with a single parameter, $\bar{\eta}$, for steady-state viscosity per radial layer, to simulate Antarctic GPS crustal-motion trend data. It is possible that more sophisticated models involving higher-order viscoelasticity, of the type used to model torsion experiments, may also be appropriate to satisfying the geodetic datasets (e.g. Faul and Jackson 2015; Lau and Holtzman 2019; Ivins *et al.* 2020; Adhikari *et al.* 2021).

In the following three sections we describe the methods and calibration for each of the three intercomparative approaches. Approach 2 also presents a disambiguation of the anelastic correction parameters. Here, analytical representation facilitates studying the sensitivity to all parameters. It is important to provide the reader with a concrete set of principles that the intercomparisons obey. Table 3 provides some additional rules that we follow. These supplement those presented in Table 1. The next three sections further detail those rules in the intercomparison that are defined in Table 3. Before we proceed, the background, or globally averaged effective viscosity at depth $\bar{\eta}_0(r)$ should be explained, as these are different for each approach. For Approach 1, $\bar{\eta}_0(r)$ takes on a constant value of 2×10^{20} Pa s, a value found for the upper mantle using the Antarctic GIA model of Ivins *et al.* (2013). Approach 2 tests the rheological law assumed at a depth of $z = 225$ km and compares it to the most reliable estimates that are available from the northern hemisphere. In Approach 2, $\bar{\eta}_0(r)$ varies with the complete rheological law. Finally, for Approach 3 the value of $\bar{\eta}_0(r)$ is neither explicitly computed or assumed, and its effective value follows from the complete composite rheology that will later be described.

Approach 1

Methods for Approach 1

We now pursue computation of the ratio $\Delta\eta(r, \theta, \phi) \equiv \bar{\eta}(r, \phi, \theta)/\overline{\eta_0}(r)$ as a fundamental measure of global viscosity variation. Introducing the parameters on the right-hand side of equation (25), Ivins and Sammis (1995) proposed the simple relationship:

$$\log_{10}\Delta\eta(r, \theta, \phi) = M\frac{2a}{\alpha_{\text{th}}T_0^2}\frac{1}{(c_1 - 1)}\frac{\delta v_{\text{S}}(r, \theta, \phi)}{v_{\text{S}_0}} \tag{25}$$

and assumed that the observed S-wave anomaly $\delta v_{\text{S}}/v_{\text{S}_0} \equiv \delta\ln v_{\text{S}}$ could be equated to the anharmonic (ah) velocity change derived from mineral physics and elasticity. The relation uses $M \equiv \log_{10}e$ and $a \equiv (E^* + PV^*)/R_{\text{G}}$. Here we recognize the exponent from the strain-rate equation (9) or the effective viscosity, η_σ, as in equation (14). Provided that one recognizes the qualifications for writing equation (14), the approximation (equation 25) is not particularly tied to diffusional creep laws, as it is equally valid for any dislocational creep law. It is assumed that there is an r, θ, ϕ dependence associated with $\log_{10}\Delta\eta$ and δv_{S}, but only radial dependency with the other parameters. Clearly, the assumption of material homogeneity is implied in equations (9), (14) and (12), and is also implied in the practical application of the partial derivative relations derived from the elasticity relations for a realistic Earth model.

Wu *et al.* (2013) introduced a formulation that adds the anharmonic contribution of equation (24) to the contribution from anelastic dispersion:

$$[\partial\ln v_{\text{S}}/\partial T]_{\text{total}} \equiv \frac{\partial\ln v_{\text{S}}^{(\text{ah})}}{\partial T} + \frac{\partial\ln v_{\text{S}}^{(\text{an})}}{\partial T}. \tag{26}$$

This results in:

$$\log_{10}\Delta\eta(r, \phi, \theta) = -\frac{M\beta}{[\partial\ln v_{\text{S}}/\partial T]_{\text{total}}}\frac{a}{T_0^2}\frac{\delta v_{\text{S}}}{v_{\text{S}_0}} \tag{27}$$

with β discussed below. Karato (2008) provided approximations for the two derivatives on the right-hand side of equation (26) with values given in his table 20.2.

Computing viscosity using this method requires choices of global reference models for shear velocity, temperature and viscosity. Since the ANT-20 Antarctic seismic model was developed by Lloyd *et al.* (2020) from the global 3D model 362ANI, which was based on the 1D model STW105 (Kustowski *et al.* 2008), we use STW105 as the shear-velocity reference model. The temperatures are common to all three methods and are discussed in the previous section. Conductive temperature in the lithosphere is from a thermal boundary layer geotherm of Stacey and Davis (2008). The intersection of this profile and that of an adiabatic mantle occurs at about 200 km where we interpolate the geotherm onto the adiabatic temperature gradient of Katsura *et al.* (2010).

The computation requires rheological parameters for the construction of a in equation (25). For the activation energy, E^*, Hirth and Kohlstedt (2004) suggested a value of 375 kJ mol^{-1} for either wet or dry diffusion creep. While the activation volume, V^*, is poorly determined from experimental work, Hirth and Kohlstedt (2004) suggested that it could range from 2×10^{-6} to 10×10^{-6} m^3 mol^{-1} for dry diffusion creep and $V^* \leq 20 \times 10^{-6}$ m^3 mol^{-1} for wet diffusion creep, so we use 5×10^{-6} m^3 mol^{-1} in this calculation. These values are used for olivine mineralogy as it forms the dominant mineral phase above 410 km depth. At a depth of 550 km, we assume ringwoodite mineral assemblage, and hence laboratory experiments with this silicate phase are needed. Experiments by Fei *et al.* (2017) determined dislocation annealing-based mobility and its Arrhenius-dependence for both wet ringwoodite and bridgmanite at the appropriate mantle temperature and pressure. Here we use their estimates of activation enthalpy and the same value for activation volume as in Fei *et al.* (2017) to arrive at an estimate of the effective viscosity at 550 km depth.

The conversion of mantle shear velocity to mantle viscosity relies on the assumption that temperature variations are responsible for variations in both shear velocity and viscosity. However, other factors, particularly variations in mineralogy and volatile content, will also influence shear velocity. Here, the factor β ranges from 0 to 1, and is meant to account for effects on shear-wave velocity other than temperature (Wang *et al.* 2008). In the GIA analysis of Wu *et al.* (2013), β is used to relax the amplitude of the scale factor in fits to GIA-related datasets. In GIA analysis it forms a way of assessing the relative role of temperature in comparison to other factors affecting both the seismic velocity and $\log_{10}\Delta\eta$.

Calibration of Approach 1

The recent Maxwell viscoelastic models for GIA developed during the past decade in Antarctica are used for calibration (see the 'Inference from modelling geodetic data' subsection earlier in this chapter). These model-based estimates are a potential source for placing realistic constraints on the parameter choices for the viscosity conversion. The extremely low upper mantle viscosity estimates of 1×10^{18}–4×10^{18} Pa s in the northern Antarctic Peninsula and the Amundsen Sea Embayment can only be approximately fitted using the IJ05-R2 upper mantle reference viscosity model of Ivins *et al.* (2013), which has the lowest continent-wide 1D upper mantle viscosity (2×10^{20} Pa s). Providing an acceptable match to these GIA-modelled observations also requires that shear-velocity variations result entirely from thermal variations (i.e. $\beta = 1$), at least in West Antarctica. A values of β close to 1 is compatible with evidence from compositional modelling which indicates that temperature variations have the greatest effect on mantle shear velocities, as realistic compositional variability produces only modest changes in seismic velocity (e.g. Goes *et al.* 2000; Lee 2003) and melt extraction produces only minor velocity increases in the upper mantle (Afonso and Schutt 2012).

A complete absence of compositional influence on shear velocities is unreasonable in East Antarctica, a craton that is underlain by thick continental lithosphere. On a global comparative basis, such lithosphere has been shown to be depleted, with high Mg/Fe ratios producing fast seismic velocity anomalies (e.g. Jordan 1981; van der Lee 2001; Lee 2003). Thus, we use $\beta = 1$, but the high mantle shear velocities in the upper reaches of the East Antarctic mantle are reduced by an appropriate factor prior to viscosity conversion. Assuming that depletion affects mantle shear velocity at the 2% level, Approach 1 makes a depletion correction by subtracting up to 1.7% from East Antarctic velocity anomalies greater than 3%, with the correction increasing with increasing velocity anomaly. The magnitude of this correction is based on both xenolith data from outside of Antarctica (Lee 2003) and petrological modelling (Afonso and Schutt 2012).

Clarification of the role of globally averaged T_0 is important. Thermal convection in the upper mantle causes the globally averaged temperature gradient to deviate from an isentropic geotherm. However, using a common temperature for $T_0(r)$ allows internally consistent results to be employed that come from high-pressure experimental and theoretical thermal elasticity. In Table 2 we give values for T_0 that span relatively hot and cold subcontinental mantle environments and a global average estimate. This is important because in the shallow upper mantle, where the anelastic influence on $\delta \ln v_S$ is largest, the total $\log_{10}\Delta\eta$ predicted is sensitive to the choice of geotherms, as shown in Figure 5 where we employ the 'hot' and 'cold' examples alongside the global average (see Table 2) as a function of the activation enthalpy for anelastic mobility.

Approach 2

Methods for Approach 2

A second method is now introduced that builds on the formulation of Ivins and Sammis (1995) as written in equation (25),

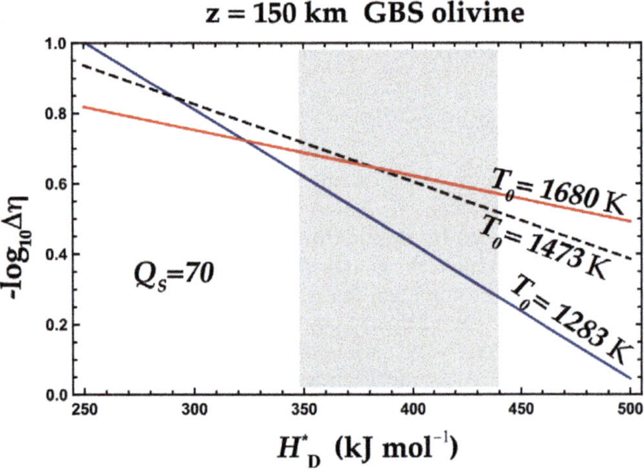

Fig. 5. – $\log_{10}\Delta\eta$ v. H_D^* at 150 km depth and rheology used in Approach 2. In this example $\delta \ln v_S = -0.02$. The red and blue solid lines refer to the geotherms referenced as 'hot' and 'cold' in Table 2. The grey zone (most probable range) assumes $H_D^* = \xi H^*$ with $0.8 \leq \xi \leq 1.0$, the rule of thumb suggested by Karato (2008). The black dashed line shows the result for a value of T_0 that is assumed in this chapter for all three intercomparative approaches for $z = 150$ km.

but accounts for anelastic effects, as represented in the second term on the right-hand side of equation (26). Here we will disambiguate the components of the anharmonic ('ah') and anelastic ('an') contributions to the velocity anomaly, to aid in deciphering sources of error and bias.

The observed shear-wave velocity anomaly can be partitioned as:

$$\delta \ln v_S \equiv \delta \ln v_S^{(ah)} + \delta \ln v_S^{(an)}. \tag{28}$$

All of the approaches essentially assume that $\delta \ln v_S^{(ah)}$ has a thermal origin.

One might ask logically: to zeroth order, what is the temperature deviation from spherically averaged value that is caused by anharmonicity? A zeroth-order estimate of the thermal anomaly is:

$$\delta T = -\frac{1}{\alpha_{th}} \frac{2}{(c_1 - 1)} \delta \ln v_S. \tag{29}$$

We should note that derivations of the theoretical anharmonic estimates of δT, as in equation (29), depend on the exact flavour of the experimental and theoretical representations used, but at a basic level each must deal with the Grüneisen parameter and the temperature derivatives of the elastic moduli (e.g. Isaak *et al.* 1992; Núñez-Valdez *et al.* 2013).

The anelastic effect depends on the seismic shear-wave attenuation, Q_S. The physics describing this attenuation is more nuanced than is the description of seismic wave velocity using the theory of thermal elasticity. At the grain-boundary level and interior to rock grains, some strain during seismic wave propagation involves motions that cannot be described with elasticity because they involve microstructural imperfections. Although the causal mechanical interactions at grain boundaries and of interior dislocation distortions and point defect motions may be numerous, one feature is clear: they are highly sensitive to temperature through thermally activated processes (e.g. Schwarz and Granato 1975). Hence, these processes are also described by an Arrhenius temperature-dependence with an activation enthalpy, H_D^* (e.g. Karato and Spetzler 1990; Gribb and Cooper 1998).

Karato (1993) found an approximate solution for the amplitude of the incremental diminution in velocity that is a function of this temperature anomaly and which is related to the observed shear-wave seismic Q_S and the activation enthalpy:

$$\delta \ln v_S^{(an)}|_{\delta Q} = -\frac{H_D^*}{\pi R_G T_0 Q_S} \frac{\delta T}{T_0}. \tag{30}$$

Equation (30) shows the strongly non-linear relationship between seismic velocity and the thermal anomaly ratio $\delta T/T_0$. This particular form is strictly valid for the case of frequency-independent Q_S. Such a lack of dependency is consistent with the construction of ANT-20 using the Durek and Ekström (1996) mantle Q_S model. A rederivation of the frequency-dependent equivalent, however, shows that the approximation by equation (30) is nonetheless accurate to 10% provided that the power exponent of the frequency-dependence is less than 1/2, consistent with experimental estimates (Jackson and Faul 2010; Faul and Jackson 2015). We term this H_D^* in equation (30) the activation enthalpy of attenuation. We note that equation (30) is consistent with the anelasticity effect on S-wave velocity as given by Takei *et al.* (2014, equation 26, therein) for high-temperature background dissipation. This allows a first-order anelastic correction to be applied to the observed S-wave anomaly map. The anharmonic anomalies can be calculated from the observed S-wave anomalies ($\delta \ln \nu_S$) as:

$$\delta \ln \nu_S^{(ah)} = \delta \ln \nu_S - \delta \ln \nu_S^{(an)}$$
$$= \delta \ln \nu_S \left[1 - \frac{2H_D^*}{\pi R_G T_0^2 Q_S \alpha_{th}(c_1 - 1)} \right]. \quad (31)$$

This result is essentially all that is needed to employ the method originally proposed by Ivins and Sammis (1995), which lacked corrections for temperature-dependent modulus dispersion (anelasticity). With this correction, we simply note that:

$$\frac{\delta \nu_S}{\nu_{S_0}} \rightarrow \delta \ln \nu_S^{(ah)} \quad (32)$$

in equation (25), and no other change needs to be applied in generating maps of $\log_{10}\Delta\eta(r, \phi, \theta)$.

The advantage of computing the velocity anomaly explicitly with equation (31) is that the dependence on the seismic quality factor, Q_S, can be appropriately tailored to any attenuation model, whereas this cannot occur when employing the derivatives $[\partial \ln \nu_S/\partial T]_{total}$ tabulated in Karato (2008). There is little fundamental difference in the way that Approach 1 and Approach 2 treat anelasticity. However, for Approach 1, all of the anelastic corrections are captured in the single term $[\partial \ln \nu_S/\partial T]_{total}$ and these may be recovered from a table of values supplied by Karato (2008). For Approach 2, this term is disambiguated into its constituent parts, thus aiding in developing formal uncertainties as they relate to the formal uncertainties of those constituent parameters.

Parameter sensitivity

A large body of experimental results quantify the spectrum of transient relaxation mechanics associated with thermally activated anelastic dispersion for olivine aggregates at high temperature ($1173 \leq T \leq 1473$ K) and pressure (0.2 GPa) (e.g. Jackson *et al.* 2002, 2004, 2014; Jackson and Faul 2010; Faul and Jackson 2015). Included are least-square fits to the experimental parameters on ensembles of samples strained under oscillatory conditions. These yield values for the activation energy for anelastic dispersion, E_D^* and, together with an estimate of the activation volume, V_D^*, we can assemble a large number of estimates of H_D^*.

There is considerable sensitivity of the computed values of $\log_{10}\Delta\eta$ at our depths of interest to c_1 and H_D^*, and we therefore sample a range of probable values. In Figures 6 and 7, we demonstrate the important role of H_D^* in determining the total lateral variation in viscosity for a 2% $\delta \nu_S$ seismic anomaly. The grey zone represents the breadth of values recommended by Karato (2008) and which satisfy laboratory values reported by Jackson and Faul (2010). The vertical dash-dot lines in Figures 6 show a value of H_D^* also determined by Jackson *et al.* (2014) using an alternative parameterization of torsional experimental data and modelling of deformation in dry olivine. The latter quantified $E_D^* = 259 \pm 25\,\text{kJ mol}^{-1}$ with $V_D^* = 10 \times 10^{-6}\text{m}^3\,\text{mol}^{-1}$. Clearly, the latter parameterization has a large impact on the interpretation of $\log_{10}\Delta\eta$ in terms of the observed lateral variability in S-wave velocity. A logical deduction from Figures 6 and 7 is that progress in theoretical and experimental interpretations of seismic anelasticity does impact our ability to make predictions of $\log_{10}\Delta\eta$ in Antarctica.

Calibration of Approach 2

The calibration of Approach 2 uses only past inferences of effective upper mantle viscosity in the northern hemisphere. The quality and abundance of the RSL and geodetic data in Fennoscandia makes it an ideal location for calibration, considering that our reference parameterization is global in nature. Studies that rely on the exponential decay curves of Holocene near-field uplift are especially important in Fennoscandia and Hudson Bay (e.g. Mitrovica 1996; Kuchar *et al.* 2019). The strongest sensitivity of these studies to the effective mantle viscosity is in the mid–upper mantle depth range. An ensemble of Fennoscandian ice-sheet reconstructions by Nordman *et al.* (2015) demonstrated that the GIA model solutions for upper mantle viscosity using the sea-level emergence record along the river valley of Ångermanälven on the western coast of the Gulf of Bothnia are decoupled from uncertainty in the ice-loading model. Such decoupling is critical to recovering tight bounds on the upper mantle viscosity, since the exponential decay of the rate of land emergence can be tightly coupled to the viscosity (e.g. Klemann and Wolf 2005). This fact

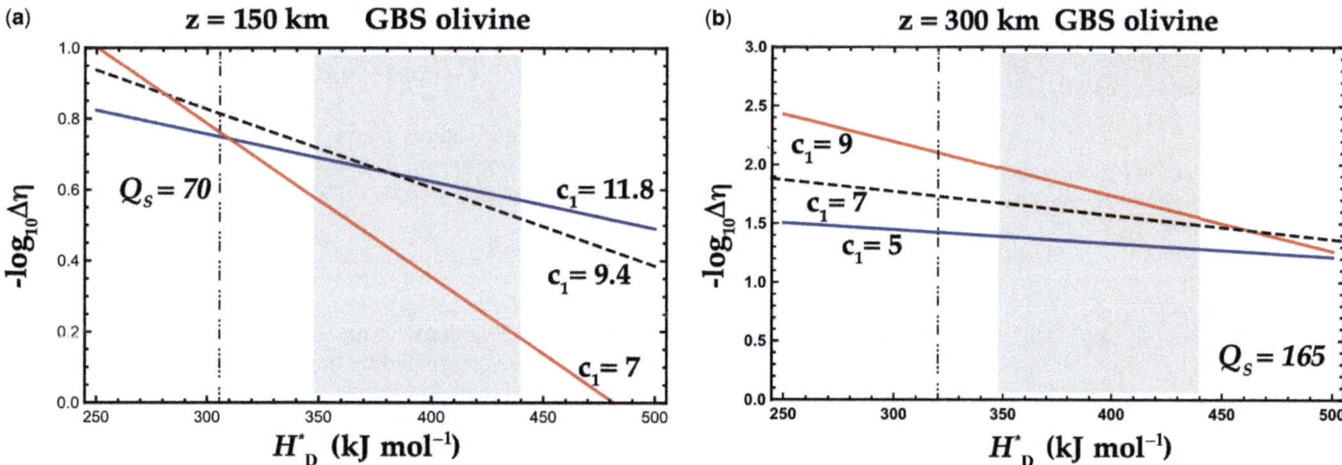

Fig. 6. $-\log_{10}\Delta\eta$ v. H_D^* at (**a**) $z = 150$ km and (**b**) $z = 300$ km depths for Approach 2. The S-wave velocity anomaly is −2%. A full range of possible H_D^* are shown, but the grey area shows the range of values inferred from Jackson and Faul (2010). The black dashed line in (a) and (b) are for our preferred c_1 at 150 and 300 km, respectively. The vertical dashed-double dot line shows the activation enthalpy calculated in experiments of Jackson *et al.* (2014) with an absorption peak centred at a period of 14.1 s and a more nuanced grain-size sensitivity. Parameters are given in Table 4. Approach 2 predictive maps use H_D^* within the grey shaded region.

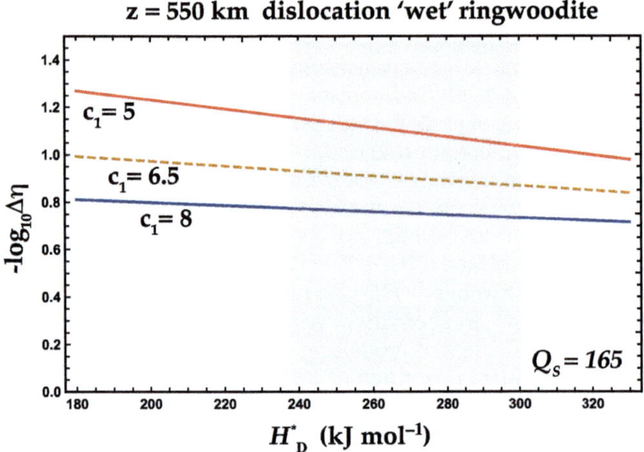

Fig. 7. – $\log_{10}\Delta\eta$ v. H^*_D at z = 550 km with differing c_1 at that depth. The S-wave velocity anomaly is −2%. The orange dashed line is our preferred estimate for mapping at 550 km depth. Other parameters are given in Table 4. Note that there are no experimental results on H^*_D for ringwoodite, so the grey zone (most probable range) assumes $H^*_D = \xi H^*$ with $0.8 \le \xi \le 1.0$, as is assumed in Figures 5 and 6. A steady-state creep value is used for H^* for ringwoodite as determined by Fei *et al.* (2017).

motivates calibrating the $\eta_0(r)$ to the experimentally developed flow law in Approach 2.

We place the activation enthalpies from experiments by Hansen *et al.* (2011) that determined the flow law for GBS dislocation mechanism into the approximation for effective viscosity using equation (14), resulting in:

$$\bar{\eta} = \frac{1}{2A_{\rm GBS}}\sigma^{1-n_{\rm GBS}} d^{p_{\rm GBS}} \exp[H^*_{\rm GBS}/R_{\rm G}T] \quad (33)$$

In Hansen *et al.* (2011), experimental data were fitted with $A_{\rm GBS} = 10^{4.8} \pm 0.8\ \mu{\rm m}^{p_{\rm GBS}}{\rm MPa}^{-n_{\rm GBS}}{\rm s}$, $n_{\rm GBS} = 2.9 \pm 0.3$, $p_{\rm GBS} = 0.7 \pm 0.1$ and $H^*_{\rm GBS} = 445 \pm 20\,{\rm kJ\,mol^{-1}}$.

At an average background deviatoric flow stress σ = 0.5 MPa and a mantle depth of 220 km, where $T_0 \simeq 1500$ K (Davies *et al.* 2012), we can obtain quite plausible values for $\bar{\eta}$. Towards this aim, the value of activation volume, V^* = 18 × 10^{-6}m^3 mol^{-1} , suggested by Hansen *et al.* (2011) is also employed. At a depth of 220 km, hydrostatic pressure is roughly 7.1 GPa and accounting for the confining pressures of the experiments (300 MPa), the effective activation energy can be determined to be $E^* \simeq 439.6$ kJ mol^{-1}. We find with a grain size, d, of 4.5 mm, as suggested for application to the mantle by Hansen *et al.* (2011), and the appropriate hydrostatic pressure at this depth, then $\bar{\eta}_{220\rm km} = 5.56 \times 10^{20}$ Pa s. This is a value quite consistent with what we know from GIA data-based inferences in central Canada and the Gulf of Bothnia (e.g. Wolf *et al.* 2006; Nordman *et al.* 2015), sites of relatively cold lithosphere. We note that GBS rheology is not assumed in Approach 1, so comparisons between the predictions of Approach 1 and Approach 2 may reveal biases that are tied specifically to the assumption of flow mechanism.

To derive an estimate of the effective viscosity at 550 km depth, we use the parameters derived by Fei *et al.* (2017) for hydrous ringwoodite. Using the activation enthalpy determined for ringwoodite (and activation volume, temperature and pressure estimate, see Table 4), we can estimate viscosity only by somewhat arbitrarily selecting additional parameters that have been derived in experiments for olivine, specifically using equation (12) with A = 1600, d = 25.5 mm, $\bar{r} = 1.2$, p = 3, $\alpha = \phi = 0$ and n = 3.5, as σ = 0.2 MPa (cf. Hirth and Kohlstedt 2004). With these, we calculate $\bar{\eta}_{550\rm km} = 6.0 \times 10^{20}$ Pa s, quite close to the most probable global value based on a Bayesian statistical GIA analysis using global RSL and GPS vertical trend datasets by Caron *et al.* (2018). The estimates that we have made calibrate H^* and temperature at the depth estimations we employ for $\log_{10}\Delta\eta$. Ultimately, the predictions should also be tested against Antarctic GIA inferences, as in Approach 1.

Table 4. *Parameters for $\log_{10}\Delta\eta$ in Approach 2*

z (km)	$\alpha_{\rm th}$[1] (10^{-5} K)	P (GPa)	c_1[2]	E^* (kJ mol^{-1})	V^*[3] (10^{-6} m^3 mol^{-1})	$H^*_{\rm D}$[4] (kJ mol^{-1})
150	4.05 ± 0.36	4.78	9.4 ± 2.4	444.2 ± 20[5]	18.0[5]	394 ± 45
300	3.02 ± 0.30	9.8	7.0 ± 2.0	444.2 ± 20[5]	18.0[5]	394 ± 45
550	2.60 ± 0.26	19.1	6.5 ± 2.0	250.4 ± 30[6]	2.6[6]	269 ± 31

[1]Katsura *et al.* (2010).
[2]Isaak *et al.* (1992) and Anderson (1995).
[3]No error estimates cited in the reports.
[4]Jackson and Faul (2010).
[5]Hansen *et al.* (2011).
[6]Wang *et al.* (2017).

Uncertainty quantification

We give predictions of the formal uncertainties in lateral variations of viscosity in Figure 8. The uncertainties are constructed from a combination of equations (25) and (29–32), yielding:

$$\log_{10}\Delta\eta = \frac{2MH^*}{R_{\rm G}\alpha_{\rm th}T_0^2(c_1-1)}\delta\ln\nu_{\rm S}\left[1 - \frac{2H^*_{\rm D}}{\pi R_{\rm G}\alpha_{\rm th}T_0^2 Q_{\rm S}(c_1-1)}\right] \quad (34)$$

and then taking partials with respect to each of the parameters. Note that in writing equation (34) all the parameters that have model uncertainties are revealed. Assuming that H^*, $H^*_{\rm D}$, $\alpha_{\rm th}$, T_0 and c_1 are the important contributors to the uncertainty of the viscosity variations, and that they follow a Gaussian distribution, we can calculate the uncertainty of $\log_{10}\Delta\eta$ as:

$$\sigma_{\log_{10}\Delta\eta} = \sqrt{\sum_{\sigma_x=\{H^*,H^*_{\rm D},\alpha_{\rm th},T_0,c_1\}}\left(\frac{\partial\log_{10}\Delta\eta}{\partial x}\sigma_x\right)^2} \quad (35)$$

with partial derivatives formed from equation (34) and the symbol σ_x representing any one of the error estimates of the corresponding parameters in the sum. For example:

$$\frac{\partial\log_{10}\Delta\eta}{\partial H^*_{\rm D}} = -\frac{4MH^*}{\pi Q_{\rm S}[R_{\rm G}(c_1-1)\alpha_{\rm th}]^2T_0^4}\delta\ln\nu_{\rm S}. \quad (36)$$

We give parameters assumed for the maps in Figure 8 in Table 4. These are guided by the parameter studies conducted as in Figures 6 and 7 (see also Fig. 9). Note that the estimates provide uncertainties for the viscosity conversion, but do not take into account any uncertainties in the shear velocities in the seismic model ANT-20. Furthermore, larger uncertainties than we can quantify here may be associated with unknown variability in composition, grain size or water. Deep uncertainty is also associated with the arbitrary selection of the effective Newtonian viscosity $\bar{\eta}$ from among the various constant background forms of equations (13), (14) and (15).

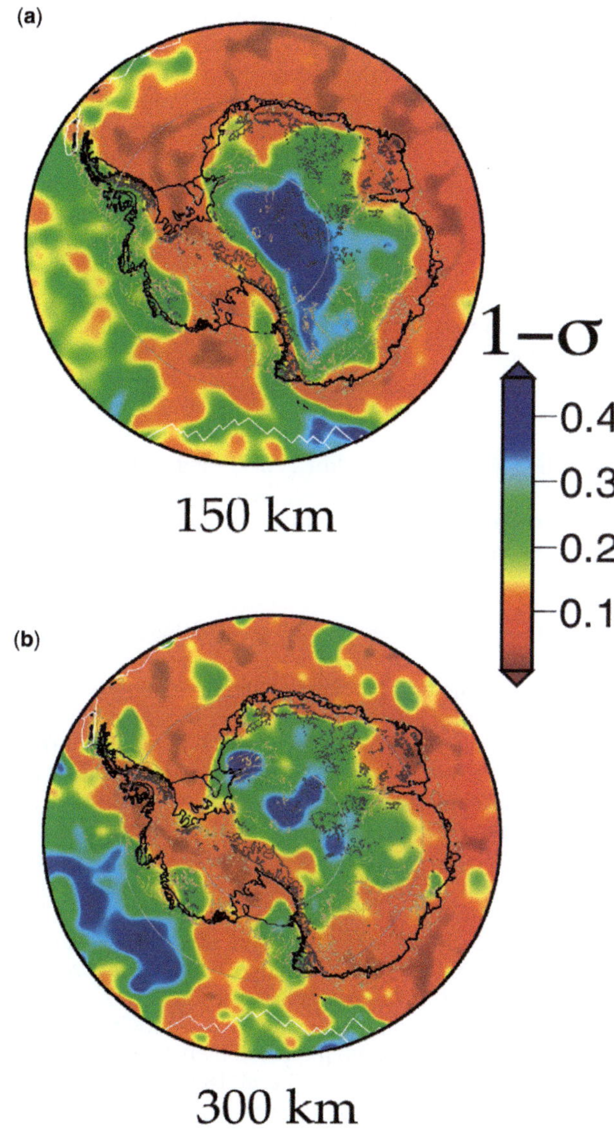

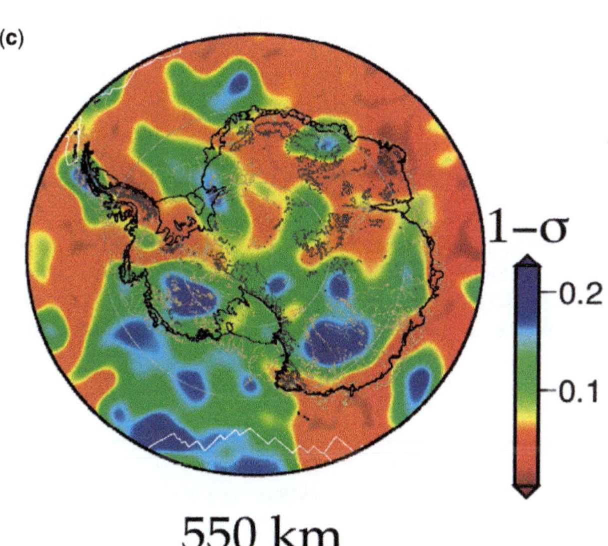

Fig. 8. Maps of the $1\!-\!\sigma$ uncertainty propagated in estimation of $\log_{10}\eta$ at the depths of (**a**) 150, (**b**) 300 and (**c**) 550 km for Approach 2. Table 4 gives parameter values along with uncertainty estimates. Units are in viscosity deviation from dimensionless $\Delta\log_{10}\eta$. A reduced colour threshold scale for (c) aids in revealing a different pattern of uncertainty at 550 km.

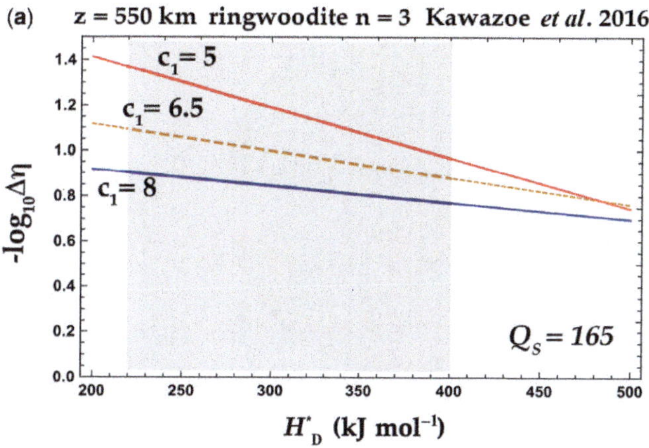

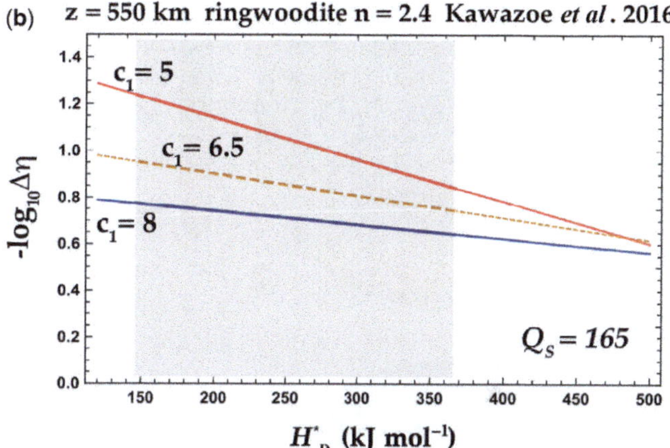

Fig. 9. $\log_{10}\Delta\eta$ v. H_D^* at the depths of 550 km with fit (**a**) $n = 3$ (fixed) and (**b**) $n = 2.4$. $H^* = 345$ kJ mol^{-1} and our preferred value for anelastically related activation enthalpy is with $\xi = 0.9$, or $H_\text{D}^* = 310.5$ kJ mol − 1 in (a). For (b) $H^* = 279$ kJ mol^{-1} and our preferred value at the same $\xi = 0.9$ is $H_\text{D}^* = 251.1$ kJ mol^{-1}. Other aspects are the same as in Figures 6 and 7.

Figure 8 allows a view of the formal uncertainties derived from the sum of an enumerable set of parameter uncertainties. Note that the larger uncertainties are obtained for regions with larger-amplitude velocity anomalies. The continuing advances in the model interpretations of creep and anelasticity from static creep and torsion experiments, respectively, must temper any overinterpretation of these uncertainty maps.

Approach 3

Methods for Approach 3

The roots of Approach 3 have been applied previously to Antarctic rebound by van der Wal *et al.* (2015). The temperature, $T_0 + \delta T$, is obtained following the description in the subsection 'Thermal parameters' and then inserted into a fully robust olivine flow law. The flow law allows diffusion, intragranular dislocation creep and GBS creep mechanics to operate simultaneously, each contributing to the macroscopic mantle strain (e.g. Ranalli 2001; Barnhoorn *et al.* 2011; Kohlstedt and Hansen 2015).

Dislocation creep is non-linearly related to stress, as has already been discussed. Viscosity, therefore, also carries this non-linear dependence, as is readily deduced from equations (9) and (16). van der Wal *et al.* (2013) formulated the composite rheology with the intention of solving time-dependent GIA loading–unloading problems in which effective viscosity

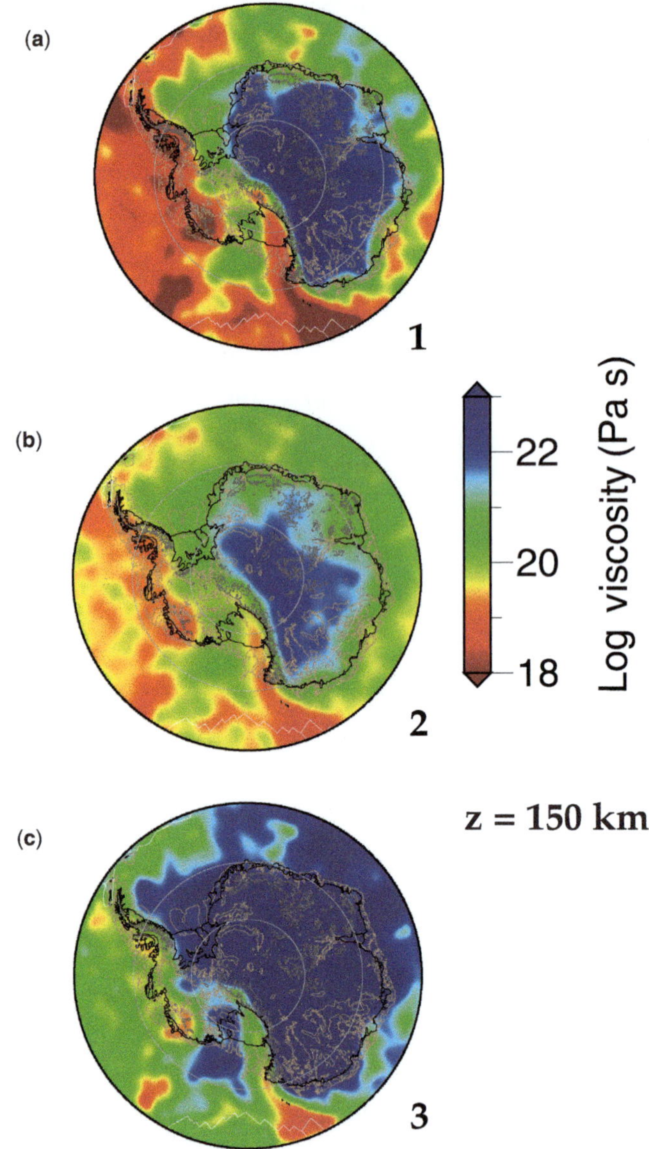

Fig. 10. Maps of $\log_{10}\eta$ at the depth of 150 km for three approaches. Approach 1 assumes $\beta = 1$ and the flow law for dry olivine from Hirth and Kohlstedt (2004) using $E^* = 375$ kJ mol^{-1} and $V^* = 5 \times 10^{-6}$m^3 mol^{-1}. Approach 2 uses parameters as shown in Table 4. Approach 3 assumes diffusion and the dislocation activation parameters taken from Hirth and Kohlstedt (2004) for dry olivine with a grain size of 4 mm and stress of 0.1 MPa. Composite component GBS parameters for Approach 3 are discussed in the text.

evolves in time, as does the deviatoric stress and strain rate. However, here we desire a simpler approach, one that would facilitate intercomparison with the predictions of Approaches 1 and 2. We provide estimates for a constant stress state. Using the seismic velocity anomaly model ANT-20, the relative change in viscosity is computed as in equations (26) and (27). The spherically averaged global background temperatures used are identical to Approach 1 and Approach 2, as described in Table 2. The derivatives of the seismic velocity anomaly to temperature anomaly are provided as a function of depth in Karato (2008), thus accounting for both the anelastic and anharmonic contributions to S-wave velocity perturbations.

Unique to Approach 3 is that we add this microscopic contribution to the strain rate resulting from diffusion, dislocation and GBS mobility in the following way:

$$\dot{\epsilon}_{\mathrm{tot}} = \dot{\epsilon}_{\mathrm{dif}} + \dot{\epsilon}_{\mathrm{disc}} + \dot{\epsilon}_{\mathrm{GBS}} \tag{37}$$

wherein the subscripts ‘tot’, ‘dif’, ‘disc’ and ‘GBS’ refer to total strain rate and the diffusional, dislocation and GBS contributions, respectively. We note that we cannot employ a simple form like in equation (33), but rather we need forms such as:

$$\dot{\epsilon}_{\mathrm{GBS}} = A_{\mathrm{GBS}}\sigma^{n_{\mathrm{GBS}}} d^{-p_{\mathrm{GBS}}} \exp[-(E^*_{\mathrm{GBS}} + PV^*_{\mathrm{GBS}})/R_{\mathrm{G}}T] \tag{38}$$

with the appropriate constants found in Hansen *et al.* (2011). Similar relations apply to the dislocation and diffusional component of the strain rate. To complete the evaluation of effective viscosity, the resultant $\dot{\epsilon}_{\mathrm{tot}}$ is employed in equation (16) with a constant stress selected. The grain size and stress are taken as constants and set to 4 mm and 0.1 MPa, respectively, for the purposes of our map computation in Figures 10c and 11c. The experiment-based flow parameters of Approach 3 are summarized in Table 5.

As in Approach 1 and Approach 2, we assume that the ringwoodite phase dominates at 550 km depth. Here, however, we select experimental results published by Kawazoe *et al.* (2016). In Figure 9 we show the lateral variability in viscosity that is predicted if we assume the parameters that are derived from experiments with ringwoodite at 18 GPa by Kawazoe *et al.* (2016). The experiments of Fei *et al.* (2017) used an annealing technique to obtain activation enthalpy values for dislocation mobility at temperatures and pressures relevant to the 550 km depth. However, annealing experiments do not apply a controlled shear stress and, hence, are not capable of determining a flow law. By contrast, the deformation experiments by Kawazoe *et al.* (2016) allowed a dislocation creep law to be derived, but with rather large uncertainties on the activation enthalpy. For a fit to the data in which the power exponent n was fixed to 3, $H^* = 345 \pm 90$ kJ mol^{-1}; and when n was simultaneously solved for using the data, $H^* = 279 \pm 105$ kJ mol^{-1}, with $n = 2.4 \pm 0.7$. In Figure 9 we see that for a 2% S-wave anomaly, a discrepancy of no more than about 20% arises from the two differing fits to the data of Kawazoe *et al.* (2016). At 550 km depth, it is also probable

Table 5. *Flow parameters used in Approach 3*

Depth (km)	E^*_{diff} † (kJ mol^{-1})	E^*_{disc} † (kJ mol^{-1})	E^*_{GBS} ‡ (kJ mol^{-1})	V^*_{diff} † (10^{-6} m^3 mol^{-1})	V^*_{disc} § (10^{-6} m^3 mol^{-1})	V^*_{GBS} § (10^{-6} m^3 mol^{-1})	σ (MPa)	H^*_{diff} ‖ (kJ mol^{-1})	H^*_{disc} ‖ (kJ mol^{-1})
150	375	530	445	5.0	3.0	12.0	0.1		
300	375	530	445	5.0	3.0	3.0	0.1		
550							0.002	402	345

†Hirth and Kohlstedt (2004).
‡Hansen *et al.* (2011).
§Kohlstedt and Hansen (2015),
‖Kawazoe *et al.* (2016).
Stress exponents (n) are: olivine dislocation, 3.5 (Hirth and Kohlstedt 2004), ringwoodite, 3 (Kawazoe *et al.* 2016); GBS, 2.9 (Hansen *et al.* 2011).

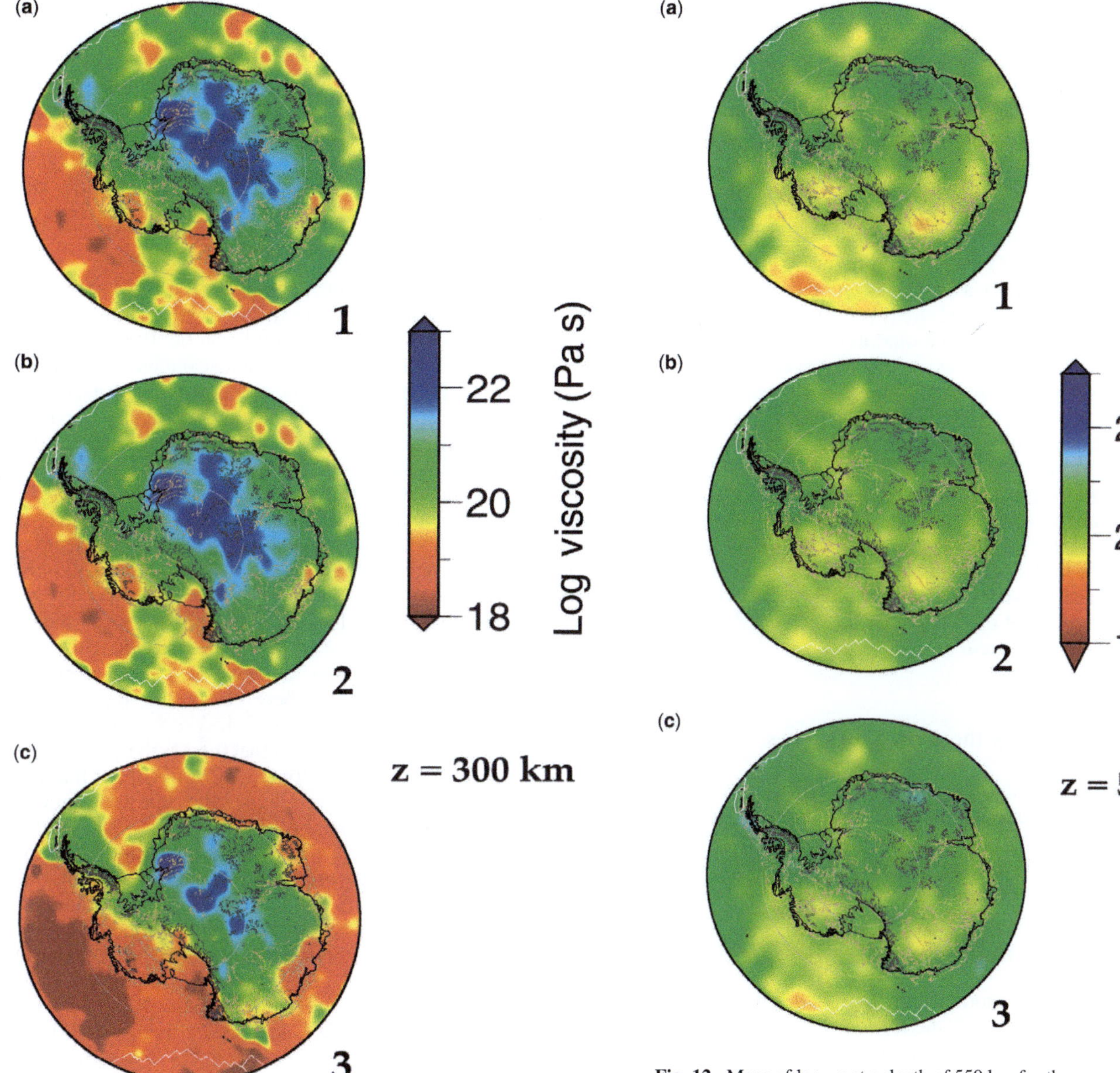

Fig. 11. Maps of $\log_{10}\eta$ at a depth of 300 km for the three approaches. Parameters are as in Figure 10 with adjustments for temperature and pressure as in Tables 1 and 2, respectively. Note the close, but serendipidous, similarity of the predicted viscosity in (**a**) and (**b**).

Fig. 12. Maps of $\log_{10}\eta$ at a depth of 550 km for three approaches. Viscosity for Approach 1 in (**a**) assumes $\beta = 1$. While assuming differing anelastic corrections, both Approach 1 and Approach 2 use the flow law for hydrous ringwoodite from Fei *et al.* (2017), with $E^* = 250.4$ kJ mol^{-1} and $V^* = 2.6 \times 10^{-6}$m^3 mol^{-1}. Approach 2 in (**b**) uses parameters as shown in Table 4. Approach 3 in (**c**) assumes $A = 10^{-0.2}$ s^{-1} MPa^n for dislocation creep (Kawazoe *et al.* 2016) and $A = 10^{-13.2}$ s^{-1} MPa^n for diffusion in ringwoodite (Shimojuku *et al.* 2009). The activation enthalpy assumed for dislocation and diffusion creep at 550 km in (c) are $H^* =$ 345 kJ mol^{-1} and $H^* = 402$ kJ mol^{-1}, respectively. No GBS is assumed at this depth.

that diffusion at grain boundaries contributes, and Shimojuku *et al.* (2009) provided diffusion constants for Si and O atomic species with activation enthalpies at 402 and 246 kJ mol^{-1}, respectively. We can use the former slowest diffusing species as rate controlling for diffusion creep in ringwoodite and form an estimate for the composite strain rate limited to $\dot{\epsilon}_{\text{dif}}$ and $\dot{\epsilon}_{\text{disc}}$ in equation (37). We note that the dislocation component dominates any reasonable estimate of the total strain rate and, therefore, primarily reflects the estimate of the viscosity that is mapped in Figure 12c

Unlike Approaches 1 and 2, wherein the full forms of the experimental flow laws are used for calibration purposes, the map views generated for Approach 3 for depths of 150 and 300 km attempt to use explicitly the experimentally determined prefactor and grain size (Figs 10c and 11c). For the ringwoodite mineralogy, it is more difficult to generate prefactors directly from the experiments and there are no experiments that collectively reveal the grain-size dependence. In Figure 12c, stress was set to 0.002 MPa and a self-consistent prefactor was derived.

Calibration of Approach 3

The largest uncertainties in obtaining effective viscosity from Approach 3 comes from grain size, water content and stress (Hirth and Kohlstedt 2004; O'Donnell *et al.* 2017). Effective viscosity is very sensitive to these values because no average viscosity is assumed in the calculation of effective viscosity. To establish that the approach is realistically retrieving effective viscosity, model predictions can be compared to GIA data. The general composite approach with both linear (diffusion control) and non-linear (dislocation control) constitutive relations is employed in computing predictions of vertical land motion and compared to GPS observations in Antarctica (van der Wal *et al.* 2015), Fennoscandian RSL and GPS crustal motion data (e.g. Tushingham and Peltier 1992;

Lidberg *et al.* 2010), and GRACE gravity trends (e.g. van der Wal *et al.* 2013). The computations are also used to estimate the GIA correction for the present-day mass balance of the Antarctic and Greenland ice sheets (e.g. Bouman *et al.* 2014; Xu *et al.* 2016). In general, a rheology with grain sizes of 4 mm or larger resulted in better fits to the uplift and gravity trend datasets. In contrast, a lower viscosity associated with smaller grain size and wet rheology predicts GIA signatures that underestimate the observed uplift rates. The dry rheology with 4 mm grain size is the preferred model in van der Wal *et al.* (2013) and is used here as the reference model. A stress of 0.5 MPa is assumed, which is a representative stress level for GIA-induced stress beneath the lithosphere locations underneath and near the margin of an ice sheet (van der Wal *et al.* 2010). Mantle stress levels, however, depend not only on the stress induced by GIA, but also on mantle convection (Bredow *et al.* 2021, this volume). There are some important caveats regarding our preferred values of grain size and water content. First, it is cautioned that these results are obtained with ice-loading histories that are implicitly based on 1D Earth profiles of linear viscosity. Progress with ice histories based on composite rheology is under investigation (Huang *et al.* 2019). Second, on the basis of xenolithic rocks recovered in Antarctica (see Chatzaras and Kruckenberg 2021, this volume), small grain sizes and wet olivine cannot be excluded (e.g. van der Wal *et al.* 2015; Chatzaras *et al.* 2016).

Map views predicted by the three approaches

Viscosity variation in $\log_{10}$ units in plan view is predicted using the three approaches at the three intercomparative upper mantle depths. The predictions are given in Figures 10, 11 and 12.

Synopsis of results for Approach 1

The resulting predictive model from Approach 1 is generally in approximate agreement with the ice-load geodetic viscosity estimates mentioned earlier (see Figures 10a, 11a and 12a, which show the three depth predictions). For example, in the Amundsen Sea Embayment, the modelled effective viscosity for the asthenosphere shallower than 200 km is 5×10^{18} Pa s, compared to 4×10^{18} Pa s estimated by Barletta *et al.* (2018). The modelled values for the rest of the upper mantle are somewhat higher than in Barletta *et al.* (2018), but still within a half of an order of magnitude. We note that in a GIA-LV model, Powell *et al.* (2020) inferred a viscosity in this same region of 2×10^{18} Pa s. The prediction of Approach 1 is unable to match the extremely low viscosity (*c.* 1×10^{18} Pa s) estimated for the northern Antarctic Peninsula and near Graham Land, by Simms *et al.* (2012) and Nield *et al.* (2014), yielding 7×10^{18} Pa s for 65–200 km depth,and higher viscosity values at greater upper mantle depths. At a depth of 550 km, lateral variability is further reduced, and predicted effective viscosity is close to 10^{20} Pa s. At this depth two regions, one in West Antarctica and the other in East Antarctica beneath the Wilkes Subglacial Basin, have a viscosity predicted to be close to 4×10^{19}–8×10^{19} Pa s.

Synopsis of results for Approach 2

Figure 10b shows the predictions for Approach 2 at 150 km depth. Here much of East Antarctica has an effective viscosity, η_σ, of slightly above 10^{21} Pa s, rising to just above 10^{22} Pa s in a cratonic core region that extends 1100 km from the South Pole into the interior of East Antarctica and along the Wilkes Subglacial Basin. At 150 km beneath West Antarctica, η_σ is predicted to be between about 2×10^{19} and 10^{20} Pa s. A viscosity as low as 10^{18} Pa s is predicted SW of the Balleny Trough in the Indian Ocean and 20° latitude north of Marie Byrd Land in the Pacific Ocean sector, each being far from the Antarctic continent. The $1-\sigma$ uncertainties propagated from the experimentally based parameters might allow for the West Antarctic mantle effective viscosity to be lowered to just below 10^{19} Pa s.

At 300 km depth (Fig. 11b) the East Antarctic pattern is similar to 150 km. However, at this depth a large swath of the subcontinent extending from the coast adjacent to Wilkes Subglacial Basin to the southern Antarctic Peninsula and Weddell Sea region is essentially Fennoscandian in character ($\eta_0 \sim 4 \times 10^{20}$ Pa s). Along this swath our $\bar{\eta}$ (η_σ) takes on values near 3×10^{20} -8×10^{20} Pa s. A portion of coastal West Antarctica and Marie Byrd Land may be in the range 0.9–10×10^{19} Pa s, not too different in value from 150 km depth prediction, but now of smaller overall regional extent. Again, the $1-\sigma$ uncertainties could allow predictions to descend to values slightly below 10^{19} Pa s in a small section of the mantle beneath Marie Byrd Land. (At this 300 km depth beneath Marie Byrd Land, P-wave tomography by Lucas *et al.* (2020) shows this region has a robust slow anomaly relative to surrounding West Antarctic mantle). The variability predicted at 550 km depth (Fig. 12b) is much reduced, with viscosity everywhere below 10^{21} Pa s, and is similar to that predicted by Approach 1.

We anticipate for all approaches that there will be large viscosity contrasts located at the upper half of the upper mantle, for it is here where convective temperature and partial melts may express themselves strongly (e.g. Yuen *et al.* 1993; Vacher *et al.* 1996). Indeed, we see that about four orders of magnitude characterize the prediction shown for each part of Figure 10 at 150 km depth when considering both continent and oceanic regions in map view. Excluding variability beneath oceanic crust in Figure 10, the variability reaches approximately three orders of magnitude. The largest variation occurs between coastal Marie Byrd Land, a region known for active volcanism, and the East Antarctic craton. At 300 km depth a magnitude variability of four orders is also reached if we include oceanic mantle. Hot oceanic upper mantle may be the site of highly channelized flow, at least in some interpretations of $\delta \ln v_S$ at this depth (e.g. French *et al.* 2013). Our focus, however, is on mantle beneath the continent of Antarctica where it is possible to retrieve corroborative viscosity information using bedrock geodesy.

At a depth of 550 km we predict that lateral variability is much reduced and, in fact, seems characterized by 1–1.5 orders of magnitude. This is caused by the smaller activation enthalpy for ringwoodite (see Table 4). We cautiously note that while Fei *et al.* (2017) discovered substantial variability, possibly from water content, we have not modelled this possible origin for lateral variability in rheology, for seismic models like ANT-20 are unlikely to deliver the required quantitative information.

Confining the assessment to West Antarctic mantle, the contrast in effective viscosity reduces to a little less than three orders at 150 km and 300 km depths. We carefully note that examination of Figure 6a reveals that a reduction in the anelastic activation enthalpy, H_D^*, from 394 kJ mol^{-1}, assumed for the predictive maps, to 307 kJ mol^{-1}, a value preferred in the summary of Jackson *et al.* (2014), increases the $\log_{10}$ variability by a factor of 1.56 at 150 km depth. Regardless of which value is assumed, Approach 2 predicts strong contrasts that will influence various ice history–geodetic solutions for mantle viscosity estimated through comparison to GPS crustal motions. This is because the regional viscosity has an exponential control on the isostatic decay time. Note that the preliminary $1-\sigma$ errors at 150, 300 and 550 km do not alter this conclusion.

Synopsis of results for Approach 3

The predictions using Approach 3 are distinctive in that they are constructed with the assumption of composite micromechanics and explicit adoption of the corresponding flow laws, at least at depths of 150 and 300 km. At a depth of 150 km the cratonic mantle beneath East Antarctica is predicted to be 10^{22}–10^{23} Pa s, and possibly higher in some locations (Fig. 10c). This high viscosity extends well into the surrounding oceanic environment. A small portion of the Antarctic Peninsula and Marie Byrd Land is predicted to be near 10^{19} Pa s, and for a small region well offshore beneath the Balleny Trough the effective viscosity is predicted to be a few times 10^{18} Pa s. At this same depth there is a three orders of magnitude gradient in viscosity predicted on-continent between the southern Transantarctic Mountains and Marie Byrd Land.

Viscosity at 300 km depth is predicted to be a few times 10^{18} Pa s in oceanic regions (Fig. 11c). This prediction of sub-oceanic mantle viscosity is a ubiquitous feature, except in the Weddell Sea region. The Marie Byrd Land viscosity prediction is close to 5×10^{18} Pa s. A narrow corridor of viscosity beneath West Antarctica extends from northern Victoria Land to Thurston Island and west of the Bellingshausen Sea that is below about 6×10^{19} Pa s. A somewhat parallel band of Fennoscandian-like viscosity extends from just south of the Balleny Trough to the western Weddell Sea and northern Antarctic Peninsula. Also, East Antarctica is predicted to have a distinctively Fennoscandian character from the coast to 500–1000 km into the interior from 0° E to 90° E. At 550 km depth, Approach 3 predicts two pockets of relatively low viscosity ($\geq 3 \times 10^{19}$ Pa s) beneath Marie Byrd Land and Wilkes Subglacial Basin (Fig. 12c). Nearly all other regions are near Fennoscandian values, except for small pockets just west of the Syowa Coast and in the Antarctic Peninsula region where viscosity values are predicted to be in the range 10^{21}–10^{22} Pa s range.

ANT-20-based viscosity v. model inferences

It is useful to display the GIA model results and the predictions of scaled effective viscosity at 150 and 550 km depths alongside one another, as is done in Table 6. Here we have not listed GIA models and our predictions at 300 km depth because the cases of 1D model viscosity at that depth are identical to the value assumed at 150 km. In Table 6, only the study by Powell *et al.* (2020) assumed a GIA-LV (3D) model. Differences between GIA model predictions can arise from differences in the details of the GPS time series employed, the ice-load history or depth parameterization. Given the underlying deep uncertainty in parameterizing the scaling factor *and* the fact that the values of the viscosity that best fit geodetic data depends on the selection of the model thickness of the uppermost mantle layer (e.g. Hu and Freymueller 2019), we caution against overinterpreting Table 6.

Assessment using ice history–geodetic estimates of viscosity

Given the simplicity of our assumption that seismic tomography can be used to infer temperature, and therefore an effective viscosity, it should not be too surprising to find deficiencies in correlating mapped 1D viscosity to mapped shear-wave speed. Lateral heterogeneity in water or melt content, chemistry or anisotropy may also influence the viscosity and seismic velocity fields, and we have not attempted to account for these factors. However, when comparing our results from the three approaches we find general agreement with geodetic and palaeosea-level inferences of effective viscosity from GIA modelling. Basically, all three methods predict high viscosity beneath East Antarctic cratonic lithosphere and low viscosity beneath West Antarctica, with the centres of predicted low viscosity beneath Ross Island and Marie Byrd Land, sites of extensive Quaternary volcanism (e.g. Kyle 1990; Wilch *et al.* 1999; Smellie *et al.* 2021).

Low-viscosity regions are also imaged along the Amundsen Sea Coast and the Antarctic Peninsula. Furthermore, there are vast and continuous swaths of mid-level upper mantle where Fennoscandian-like viscosity (e.g. Mitrovica 1996; Lidberg *et al.* 2010; Nordman *et al.* 2015) is predicted, consistent

Table 6. *Viscosity (10^{19} Pa s): scaling to $\delta \ln v_S$ v. modelled GIA with geodetic data support*

Region*	Approach 1	Approach 2	Approach 3	GIA viscosity	Study
			z = 150 km		
GL	0.6–3	2–6	9–20	7–9, 0.06–0.2, 0.03–0.3	Ivins *et al.* (2011)†, Nield *et al.* (2014)‡, Samrat *et al.* (2020)
SSI	2–6	3–7	10–30	0.1–0.2	Simms *et al.* (2012)
cAP	0.6–3	2–6	9–20	2	Zhao *et al.* (2017)
sAP	6–30	9–40	9–40	10–30	Wolstencroft *et al.* (2015)
KIS	6–20	10–30	40–80	≥10	Nield *et al.* (2016)
ASE	0.6–0.9	1–4	6–20	0.2, 0.4	Barletta *et al.* (2018), Powell *et al.* (2020)
WSE	7–40	10–50	100–1000	10, 10–30	Bradley *et al.* (2015), Wolstencroft *et al.* (2015)
WSB	20–60	40–80	200–2000	40	Amalvict *et al.* (2009)
			z = 550 km		
GL	20–60	20–60	70–100	7–9, 40, 40	Ivins *et al.* (2011), Nield *et al.* (2014), Samrat *et al.* (2020)
SSI	30–70	20–60	40–90	0.1–0.2	Simms *et al.* (2012)
cAP	30–70	30–70	70–100	4	Zhao *et al.* (2017)
sAP	40–80	40–80	50–90	10–30	Wolstencroft *et al.* (2015)
KIS	20–60	20–60	20–60	≥10	Nield *et al.* (2016)
ASE	4–9	6–20	4–9	1.5	Barletta *et al.* (2018)
WSE	9–70	8–50	40–80	10, 10–30	Bradley *et al.* (2015), Wolstencroft *et al.* (2015)
WSB	5–20	10–50	10–50	40	Amalvict *et al.* (2009)

*Refer to the labels on the maps of Figures 1 and 4a.
†Half-space mantle with ice loss since 1930.
‡Mantle constant to 400 km depth and twentieth-century mass gain.

with inferences of a number of continent-wide 1D GIA models (e.g. Whitehouse *et al.* 2012; Ivins *et al.* 2013; Argus *et al.* 2014). Among the most notable failures of the continent-wide 1D GIA models is their inability to predict large present-day vertical uplift on bedrock sites in the Amundsen Sea sector (e.g. Groh *et al.* 2012), several hundred kilometres from the Voigt-averaged S-wave anomaly minimum of ANT-20 beneath Marie Byrd Land. A low-viscosity region is predicted in this general region by each of the three approaches at 150 and 300 km. This is quite generally in accord with inferences of low viscosity that come from regional GIA modelling (e.g. Barletta *et al.* 2018; Kachuck *et al.* 2020; Powell *et al.* 2020). As we have discussed in reference to each of the predictions from the three approaches, there is discord on the magnitude of the inferred viscosity minimum. We do not speculate further on the causes of such a discrepancy. The list might include seismology, poor ice reconstruction, insufficient layering, low-temperature plasticity, transient creep or subtle effects of the heterogeneity itself. However, a positive result of our study is that the viscosity contrasts inferred from scaling seismic tomography to temperature, and then to viscosity do correlate well with the patterns of upper mantle viscosity inferred from GIA models that are well rooted in geodetic observation.

Scrutiny of the Antarctic Peninsula

The northern Antarctic Peninsula (Graham Land) is one region that rather poorly correlates with the ANT-20 scaling to viscosity in at least two of the three approaches, as each are limited to treating lateral variability as dominated only by thermal control. The most optimistic correlation is predicted by Approach 1 at $z = 150$ km (see Fig. 10a). Models by Simms *et al.* (2012) and Nield *et al.* (2014) infer shallow upper mantle viscosity in the range of 0.7×10^{18}–2×10^{18} Pa s. The viscosity we present in this chapter based upon scaling to ANT-20 is nearly an order of magnitude higher than this at all depths we compute and for all approaches, presenting a challenge for interpretation.

In dealing with this enigma, three candidate causes are of primary consideration. First, there are potential changes in mantle grain size and volatile content caused by a youthful ridge subduction tectonics that characterize the northernmost Peninsula upper mantle environment (e.g. Bercovici *et al.* 2015). Secondly, is a resolution issue: a sharp variation in viscosity near the Bransfield Strait spreading centre might be captured in the geodetic data, but smeared out in the Voigt-average ANT-20 model. Generally, the model has approximately 150–200 km lateral resolution and about 50 km depth resolution at 150 km depth, but this resolution diminishes with depth. Thirdly is the possibility that both the GIA forward model set-ups and the experimental flow laws assume 'steady-state' properties, the former from the Maxwell viscoelastic assumption and the latter from creep experiments that have reached steady state, and therefore lack any transient behaviour. We are not yet in a position to distinguish between these three possibilities, and any further elaboration should be tempered by the fact that a higher-viscosity solution was derived from geodetic data by Ivins *et al.* (2011). The latter are seemingly compatible with two of the three predictions at 150 and 300 km depth, as these are between 1×10^{19} and 9×10^{19} Pa s (cf. Figs 10a, b & 11a–c). Additional seismic imaging of the upper mantle in this complex region might shed further light on these issues (e.g. Park *et al.* 2012).

Discussion and conclusions

The three intercomparative approaches that we have used to convert the ANT-20 S-wave tomography of the Antarctic upper mantle to viscosity has been an exercise in testing for coherency in the resulting maps of $\log_{10}\Delta\eta(r, \theta, \phi)$, noting the considerable differences in the underlying assumptions and procedures. The contrasts in the approaches are substantial, as can be deduced from the summary Tables 1 and 3. In fact, what we have sampled are methods that are in current use, and therefore give the reader a sense of the spread in results that underlie any transfer function for S-wave images to $\log_{10}\Delta\eta(r, \theta, \phi)$ using laboratory-based steady-state flow laws. A more systematic study might be designed in the future that would consider a somewhat more broadened scope than that which we have developed here.

A key part of this chapter on rheology has been to revisit the issue of anelasticity in deriving quantitative relationships between lateral variation in effective mantle viscosity and S-wave velocity anomalies. Central to the physics of anelasticity are the grain-scale deformation processes associated with seismic wave motion. During the past two decades, torsional oscillation studies of GBS and static creep deformation in olivine have revealed that diffusionally assisted interionic mobility is a key feature of seismic attenuation (e.g. Jackson 2000; Marquardt and Faul 2018). Here we have taken the basic approach outlined in the monograph of Karato (2008). As the process of seismic wave attenuation involves grain-boundary assisted diffusion, an activation enthalpy of a transient creep, H_D^*, is a key controlling parameter. Our assessment employs experiments on olivine samples that are derived from torsional oscillation experimental work (e.g. Jackson *et al.* 2002, 2014; Jackson and Faul 2010; Faul and Jackson 2015). However, these experiments are performed under an array of conditions (e.g. grain sizes, pressures, temperatures and synthetic preparation procedures) and, hence, there is a relatively broad range of H_D^* that may apply to the upper mantle beneath Antarctica.

Our results reveal that among the competing parameters for calculations at $z = 150$ and 300 km for a 2% S-wave anomaly, H_D^* can have from a half to a full order of magnitude influence on the $\log_{10}\Delta\eta$ that is predicted. Approaches 1 and 3 have selected values for the anelastic correction that are taken directly from the numerical values as a function of depth as supplied by Karato (2008). Karato (2008) cautioned about the sensitivity to the background Q_S model. However, we find that these cause negligible differences between the predictions from the three different approaches for Antarctic upper mantle. The background Q_S model used in construction of ANT-20 ($Q_S = 70$) differs only modestly from the model presented by Karato (2008) with $Q_S = 80$ at $z = 150$ km. In Figure 13 we explore the prediction for $\delta \ln v_S = -6\%$, a value characterizing the minimum S-wave velocity anomaly in ANT-20. Similar comparisons at depths of 300 and 550 km led to the same conclusions: that the differences in these particular Q_S models ($\mathcal{O}10\%$) play a negligible role. However, if the background Q_S model is larger by a factor of 2, then the inverted S-wave anomaly values increase (e.g. Lloyd *et al.* 2020, fig. S7). In addition, the influence on the anelastic correction would then be smaller, resulting in a larger prediction of $\log_{10}\Delta\eta(r, \theta, \phi)$ associated with a thermal perturbation $\delta T(r, \theta, \phi)$.

In Figure 13 we relax the rule of thumb suggested by Karato (2008) ($0.8 \leq \xi \leq 1$) in the relation $H_D^* \approx \xi H^*$. Jackson *et al.* (2014) suggested a relaxation in the lower bound on ξ to allow $\xi \approx 0.5$. Such lowering of ξ is also consistent with the analysis of McCarthy *et al.* (2011). A reduction to this value does have some consequence. For example, the map of Figure 10b shows viscosity variability weaker than that allowed in Figure 13. At $H_D^* = 200$ kJ mol^{-1} (the lowest value of the shaded region in Fig. 13), $\log_{10}\Delta\eta$ would be raised by about a half an order of magnitude with respect to values in Figure 10b if using this lower activation enthalpy for thermally activated anelasticity.

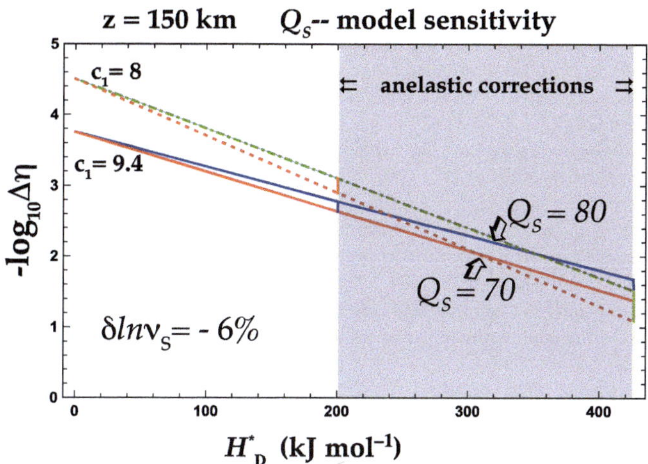

Fig. 13. $\log_{10}\Delta\eta$ v. H^*_D at the depth of 150 km. All parameters are as in Figure 6a with values referred to Table 4. Q_S models: the one used in ANT-20 is QL6 of Durek and Ekström (1996) and the other QLM9 by Lawrence and Wysession (2006) was used in table 20.2 of Karato (2008). The differences in prediction are small. The dash and dash-dot lines are for $c_1 = 8$, and the solid lines are for $c_1 = 9.4$. The full range of H^*_D explored are those reported in Jackson and Faul (2010). Expansion of the lower range to account for lower H^*_D is inferred from McCarthy *et al.* (2011) and Jackson *et al.* (2014).

Potentially, ambiguities in the range of acceptable H^*_D values are a possible source of bias in the results delivered by any one of the approaches examined in this chapter. In Figure 13 we see that the maximum diminution of viscosity at 6% S-wave anomaly is three orders of magnitude, but that four orders is never achieved, provided the anelastic correction is applied. Our chapter concludes that future experimental and theoretical progress for improved constraints on anelastic H^*_D will aid in better constraining *a priori* lateral viscosity models derived from seismology.

One method that can account for the different creep mechanisms that correspond to laboratory-derived flow laws is to abandon the use of viscosity as a single controlling parameter, and employ the creep law of equation (9) directly into the initial-value boundary-value problem (IVBVP) code structure that solves geodynamic and GIA model set-ups. This, in fact, has been done in GIA models (e.g. Wu 1999; Zhong *et al.* 2003). Diffusion, dislocation and GBS creep can operate simultaneously in the mantle (Ranalli 2001; Kohlstedt and Hansen 2015), and a composite rheology has been used in mantle convection (e.g. van den Berg *et al.* 1995; Becker 2006; Dannberg *et al.* 2017) and in GIA models (e.g. van der Wal *et al.* 2013; Huang *et al.* 2019). In this case, a mean effective viscosity could only be defined as a retrospective summary of the class of full numerical simulations, should the fully non-linear constitutive equations be employed. As our goal has been to discuss the rheology of the mantle beneath the Antarctic ice sheet that can be derived from a comprehensive S-wave tomographic image, we must acknowledge this limitation. Tractable relations between S-wave anomaly and the single Newtonian-like effective viscosity make bold approximations in order to be easily compared to viscosity derived from ice history–geodetic reconstructions of GIA (e.g. Ivins *et al.* 2011; Nield *et al.* 2014; Barletta *et al.* 2018; Powell *et al.* 2020). The latter models are nearly all based on a Maxwell constitutive assumption and report a Newtonian viscosity value. Hence, the scaling methods explored in this paper have a practical utilitarian value since the preponderance of GIA models are 1D and assume a Newtonian viscous element in a Maxwell rheology. An inherent goal of developing coherent transfer functions between tomography and viscosity is to develop more advanced 3D structures for use in GIA models. For example, in a study of the Amundsen Sea Embayment, Powell *et al.* (2020) showed that the sensitivity of horizontal GPS station motions to 3D structure is substantial. For regions peripheral to the locus of ice-mass change, 3D structure can also influence vertical motions since bulge migration involves lateral transport of mantle material (Klemann *et al.* 2007).

We conclude that the basic philosophy of deriving scaling relationships is a coherent and rationale one. With quantification and propagation of laboratory-derived and seismically derived uncertainty, we are also able to describe some of the uncertainties in the scaling to effective regional viscosity. A future challenge may be to explore more fully various poorly understood biases, and to understand better the grain-size dependency and impurity effects in the deeper upper mantle for wadsleyite and ringwoodite mineralogy. The issue of lower mantle viscosity just below the 660 km discontinuity has important implications for the space gravimetric interpretation of present-day ice-mass changes in East Antarctica, as shown recently by Caron and Ivins (2020). The coming decade will see the development of increasingly sophisticated numerical models for GIA-LV, possibly even those that will employ adjoint methods (e.g. Crawford *et al.* 2018). Critical to these advanced 3D approaches is having high confidence in a starting model that is well rooted spatially. For this purpose, seismic mapping in the mantle will be a valuable tool. Among a number of caveats we have not touched on, the most important is the *other* implication of the mobility processes associated with the activation parameter H^*_D: transient rheology. In studies of post-seismic relaxation it is now widely acknowledged that this is a critical component of deformation models (e.g. Lau and Holtzman 2019; Ivins *et al.* 2020). Transient rheology could help to explain the high amplitudes of geodetically determined responses to decade timescale ice-unloading events in Antarctica. How to deal with this short timescale rheology will be a major future challenge.

Acknowledgements We dedicate this chapter to the memory of Orson L. Anderson and Donald L. Anderson. We thank the editor, Adam Martin, Pippa Whitehouse and two anonymous reviewers for quite thorough and helpful suggestions.

Author contributions **ERI**: conceptualization (equal), data curation (supporting), formal analysis (lead), funding acquisition (lead), investigation (lead), methodology (equal), project administration (lead), resources (equal), software (equal), supervision (lead), validation (lead), visualization (equal), writing – original draft (lead), writing – review & editing (equal); **WVDW**: conceptualization (equal), data curation (equal), formal analysis (equal), funding acquisition (equal), investigation (equal), methodology (equal), project administration (equal), resources (equal), software (equal), supervision (equal), validation (equal), visualization (supporting), writing – original draft (supporting), writing – review & editing (supporting); **DAW**: conceptualization (equal), data curation (equal), formal analysis (equal), funding acquisition (equal), investigation (equal), methodology (equal), project administration (equal), resources (equal), software (equal), supervision (equal), validation (equal), visualization (equal), writing – original draft (equal), writing – review & editing (equal); **AJL**: conceptualization (supporting), data curation (equal), formal analysis (supporting), funding acquisition (supporting), investigation (supporting), methodology (supporting), project administration (supporting), resources (equal), software (supporting), supervision (supporting), validation (supporting), visualization (supporting), writing – original draft (supporting), writing – review & editing (supporting); **LC**: conceptualization (supporting), data curation (supporting), formal analysis (supporting), funding acquisition (supporting), investigation (supporting), methodology (supporting), project administration (supporting), resources (supporting), software (supporting), supervision (supporting), validation (supporting), visualization (supporting), writing – original draft (supporting), writing – review & editing (supporting)

Funding This research was carried out at the Jet Propulsion Laboratory (JPL), California Institute of Technology, under a contract with the National Aeronautics and Space Administration (NASA) and funded through NASA Post-doctoral Program, and through NASA's Earth Surface and Interior Focus Area Program (grant NNH15ZDA001N-ESI to E.R. Ivins), the Sea-Level Change Science Team (grant NNH16ZDA001N-SLCT to E.R. Ivins), the GRACE Follow-On Science Team (grant NNH19ZDA001N-GRACEFO to E.R. Ivins) and the NASA Cryosphere Program (grant NNH13ZDA001N-SLR to E.R. Ivins).

Data availability The data that support the findings of this study are available from Jet Propulsion Laboratory, California Institute of Technology, but restrictions apply to the availability of these data, which were used under licence for the current study, and so are not publicly available. Data are, however, available from the authors upon reasonable request and with permission of Jet Propulsion Laboratory, California Institute of Technology.

Appendix A: A linearization of the thermal perturbation

In this Appendix we show the level of bias that is inherent to a linearization of δT in equation (7). The approximation is technically unnecessary, but it eases a direct use of spherical harmonics for simultaneously treating global tomography models and semi-analytic formulations of the GIA-LV problem (e.g. Martinec 2000; Tromp and Mitrovica 2000; Tanaka *et al.* 2011). Spherical harmonic representation of the field $\log_{10}\Delta\eta(r, \theta, \phi)$ is then smoothed with the same resolution and fidelity as the seismic tomographic map of $\delta v_S(r, \theta, \phi)/v_{S_0}$ (e.g. Kaufmann *et al.* 2005). Additional terms in equation (8) include

$$\frac{\log_{10}\Delta\eta}{a\log_{10}e} \approx -\frac{\delta T}{T_0^2} - \frac{\delta T^2}{T_0^3} - \frac{\delta T^3}{T_0^4} + \text{higher-order terms.} \quad \text{(A1)}$$

Comparison of the exact relationship (equation 7) and two levels of approximation are shown in Figure A1. Also shown here (dotted curve overlying the yellow dash-dots) is the method of Latychev *et al.* (2005), which is also a first-order approximation. The study by An *et al.* (2015) used a Rayleigh-wave dispersion analysis to determine shear-wave structure. It is of note that if we adjust the parameter ε of Latychev *et al.* (2005) in equation (4) to a value of 0.067 K^{-1} the predictions become identical to the first-order approximation that employs physics into the parameterization. This contrasts to the ε value assumed in the global modelling by Austermann *et al.* (2013), $\varepsilon = 0.04\ K^{-1}$, and to one of the values used for Antarctica by Hay *et al.* (2017) ($\varepsilon = 0.037\ K^{-1}$).

In Figure A2 we show the positions from the steady-state thermal model computed by An *et al.* (2015) that were used in constructing Figure A1. Our method employed in this chapter uses a different set of relationships to compute temperature as a vehicle to estimate lateral viscosity contrasts. Nonetheless, we use Meijian An's model as an independent method for estimating δT and T_0 for the Antarctic mantle. It therefore provides an independent way of testing the linear approximation for $\log_{10}\Delta\eta$ widely assumed in the GIA-LV modelling community (e.g. Hay *et al.* 2017; Powell *et al.* 2020).

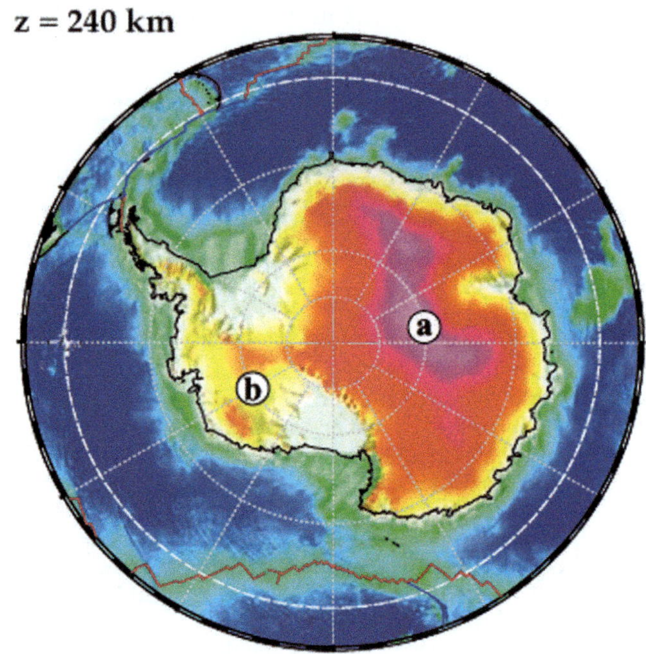

Fig. A2. Positions (a and b) of the thermal model computed in An *et al.* (2015) superimposed on a colour topography map of Antarctica. Figure A1 assumes values of $T_0 + \delta T$ from the model at $z = 240$ km. The temperature contrast retrieved at the positions in the model ($2\delta T$) is 285 K. Modified from An *et al.* (2015), with permission from John Wiley and Sons.

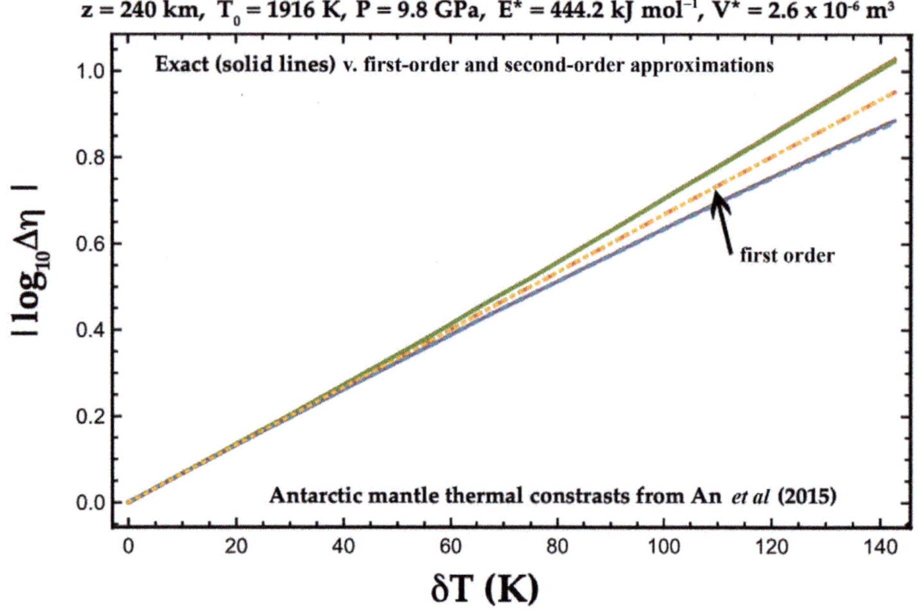

Fig. A1. Test of the linearization of δT as in equation (8) v. the exact form from equation (7). For the exact and second-order approximation (see equation A1), the influence of δT is asymmetrical with respect to the sign of δT. The green and blue colours indicate negative and positive temperature deviations, respectively. Solid lines indicate no approximation, while light dashed and lines are for second-order approximation (only visible for the positive δT case and overlap in the case of negative δT). For the orange dash, the linear approximation (first order) does not exhibit this asymmetry so that $\pm\,\delta T$ are coincident. We also plot the Latychev *et al.* (2005) method with the red dotted curve which here overlies the first-order approximation.

References

A, G., Wahr, J. and Zhong, S. 2013. Computations of the viscoelastic response of a 3-D compressible earth to surface loading: an application to glacial isostatic adjustment in Antarctica and Canada. *Geophysical Journal International*, **192**, 557–572, https://doi.org/10.1093/gji/ggs030

Adhikari, S., Milne, G.A. *et al.* 2021. Decadal to centennial timescale mantle viscosity inferred from modern crustal uplift rates in Greenland. *Geophysical Research Letters,* **48**, e2021GL094040, https://doi.org/10.1029/2021GL094040

Afonso, J.C. and Schutt, D.L. 2012. The effects of polybaric partial melting on density and seismic velocities of mantle restites. *Lithos*, **134–135**, 289–303, https://doi.org/10.1016/j.lithos.2012.01.009

Amalvict, M., Willis, P., Wöppelmann, G., Ivins, E.R., Bouin, M., Testut, L. and Hinderer, J. 2009. Isostatic stability of the East Antarctic station Dumont d'Urville from long-term geodetic observations and geophysical models. *Polar Research*, **28**, 193–202, https://doi.org/10.3402/polar.v28i2.6112

An, M., Wiens, D.A. *et al.* 2015. Temperature, lithosphere–asthenosphere boundary, and heat flux beneath the Antarctic Plate inferred from seismic velocities. *Journal of Geophysical Research: Solid Earth*, **120**, 8720–8742, https://doi.org/10.1002/2015JB011917

Anderson, D.L. 1987. A seismic equation of state II. shear properties and thermodynamics of the lower mantle. *Physics of the Earth and Planetary Interiors*, **45**, 307–323, https://doi.org/10.1016/0031-9201(87)90039-2

Anderson, D.L. 1988. Temperature and pressure derivatives of elastic constants with application to the mantle. *Journal of Geophysical Research: Solid Earth*, **93**, 4688–4700, https://doi.org/10.1029/JB093iB05p04688

Anderson, J.B., Conway, H. *et al.* 2014. Ross Sea paleo-ice sheet drainage and deglacial history during and since the LGM. *Quaternary Science Reviews*, **100**, 31–54, https://doi.org/10.1016/j.quascirev.2013.08.020

Anderson, O.L. 1995. *Equations of State of Solids for Geophysics and Ceramic Sciences*. Oxford University Press, Oxford, UK.

Anderson, O.L. and Isaak, D.J. 1995. Seismic tomography and geodynamics. *In*: Ahrens, T.J. (ed.) *Mineral Physics and Crystallography: A Handbook of Physical Constants*. American Geophysical Union Reference Shelf, **2**, 64–96.

Anderson, O.L., Oda, H. and Isaak, D. 1992. A model for the computation of thermal expansivity at high compression and high temperatures: MgO as an example. *Geophysical Research Letters*, **19**, 1987–1990, https://doi.org/10.1029/92GL02145

Argus, D.F., Peltier, W.R., Drummond, R. and Moore, A.W. 2014. The Antarctica component of postglacial rebound model ICE-6G_C (VM5a) based on GPS positioning, exposure age dating of ice thicknesses, and relative sea level histories. *Geophysical Journal International*, **198**, 537–563, https://doi.org/10.1093/gji/ggu140

Austermann, J., Mitrovica, J.X., Latychev, K. and Milne, G.A. 2013. Barbados-based estimate of ice volume at last glacial maximum affected by subducted plate. *Nature Geoscience*, **6**, 553–557, https://doi.org/10.1038/ngeo1859

Austermann, J., Chen, C., Lau, H., Maloof, A. and Latychev, K. 2020. Constraints on mantle viscosity and Laurentide ice sheet evolution from pluvial paleolake shorelines in the western United States. *Earth and Planetary Science Letters*, **532**, 116006, https://doi.org/10.1016/j.epsl.2019.116006

Barletta, V.R., Bevis, M. *et al.* 2018. Observed rapid bedrock uplift in Amundsen Sea Embayment promotes ice-sheet stability. *Science*, **360**, 1335–1339, https://doi.org/10.1126/science.aao1447

Barletta, V.R., Nield, G.A., van der Wal, W. and van Calcar, C.J. 2022. Glacial isostatic adjustment and postseismic deformation in Antarctica. *Geological Society, London, Memoirs,* **56**, in press.

Barnhoorn, A., van der Wal, W., Vermeersen, B.L. and Drury, M.R. 2011. Lateral, radial, and temporal variations in upper mantle viscosity and rheology under Scandinavia. *Geochemistry, Geophysics, Geosystems*, **12**, Q01007, https://doi.org/10.1029/2010GC003290

Becker, T.W. 2006. On the effect of temperature and strain-rate dependent viscosity on global mantle flow, net rotation, and plate-driving forces. *Geophysical Journal International*, **167**, 943–957, https://doi.org/10.1111/j.1365-246X.2006.03172.x

Bentley, M.J., Cofaigh, C. *et al.* 2014. A community-based geological reconstruction of antarctic Ice sheet deglaciation since the last glacial maximum. *Quaternary Science Reviews*, **100**, 1–9, https://doi.org/10.1016/j.quascirev.2014.06.025

Berckhemer, H., Kampfmann, W., Aulbach, E. and Schmeling, H. 1982. Shear modulus and Q of forsterite and dunite near partial melting from forced-oscillation experiments. *Physics of the Earth and Planetary Interiors*, **29**, 30–41, https://doi.org/10.1016/0031-9201(82)90135-2

Bercovici, D., Schubert, G. and Ricard, Y. 2015. Abrupt tectonics and rapid slab detachment with grain damage. *Proceedings of the National Academy of Sciences of the United States of America*, **112**, 1287–1291, https://doi.org/10.1073/pnas.1415473112

Bevis, M., Kendrick, E. *et al.* 2009. Geodetic measurements of vertical crustal velocity in West Antarctica and the implications for ice mass balance. *Geochemistry, Geophysics, Geosystems*, **10**, Q10005, https://doi.org/10.1029/2009GC002642

Bouman, J., Fuchs, M., Ivins, E.R., van der Wal, W., Schrama, E., Visser, P. and Horwath, M. 2014. Antarctic outlet glacier mass change resolved at basin scale from satellite gravity gradiometry. *Geophysical Research Letters*, **41**, 5919–5926, https://doi.org/10.1002/2014GL060637

Bradley, S.L., Hindmarsh, R.C., Whitehouse, P.L., Bentley, M.J. and King, M.A. 2015. Low post-glacial rebound rates in the Weddell Sea due to late Holocene ice-sheet readvance. *Earth and Planetary Science Letters*, **413**, 79–89, https://doi.org/10.1016/j.epsl.2014.12.039

Bredow, E., Steinberger, B., Gassmöller, R. and Dannberg, J. 2021. Mantle convection and possible mantle plumes beneath Antarctica – insights from geodynamic models and implications for topography. *Geological Society, London, Memoirs*, **56**, https://doi.org/10.1144/M56-2020-2

Caron, L. and Ivins, E.R. 2020. A baseline Antarctic GIA correction for space gravimetry. *Earth and Planetary Science Letters*, **531**, 115957, https://doi.org/10.1016/j.epsl.2019.115957

Caron, L., Ivins, E.R., Larour, E., Adhikari, S., Nilsson, J. and Blewitt, G. 2018. GIA model statistics for GRACE hydrology, cryosphere, and ocean science. *Geophysical Research Letters*, **45**, 2203–2212, https://doi.org/10.1002/2017GL076644

Chatzaras, V. and Kruckenberg, S.C. 2021. Effects of melt-percolation, refertilization and deformation on upper mantle seismic anisotropy: constraints from peridotite xenoliths, Marie Byrd Land, West Antarctica. *Geological Society, London, Memoirs*, **56**, https://doi.org/10.1144/M56-2020-16

Chatzaras, V., Kruckenberg, S.C., Cohen, S.M., Medaris, L.G., Jr, Withers, A.C. and Bagley, B. 2016. Axial-type olivine crystallographic preferred orientations: The effect of strain geometry on mantle texture. *Journal of Geophysical Research: Solid Earth*, **121**, 4895–4922, https://doi.org/10.1002/2015JB012628

Chopelas, A. and Boehler, R. 1992. Thermal expansivity in the lower mantle. *Geophysical Research Letters*, **19**, 1983–1986, https://doi.org/10.1029/92GL02144

Cline, C.J., II, Faul, U.H., David, E.C., Berry, A.J. and Jackson, I. 2018. Redox-influenced seismic properties of upper-mantle olivine. *Nature*, **555**, 355–358, https://doi.org/10.1038/nature25764

Coltorti, M., Bonadiman, C., Casetta, F., Faccini, B., Giacomoni, P.P., Pelorosso, B. and Perinelli, C. 2021. Nature and evolution of the northern Victoria Land lithospheric mantle (Antarctica) as revealed by ultramafic xenoliths. *Geological Society, London, Memoirs*, **56**, https://doi.org/10.1144/M56-2020-11

Crawford, O., Al-Attar, D., Tromp, J., Mitrovica, J.X., Austermann, J. and Lau, H.C.P. 2018. Quantifying the sensitivity of post-glacial sea level change to laterally varying viscosity.

Geophysical Journal International, **214**, 1324–1363, https://doi.org/10.1093/gji/ggy184

Cressler, J.M. and Moen, K.A. 2012. Physics of temperature and temperature's role in carrier transport. *In*: Cressler, J.M. and Mantooth, H.A. (eds) *Extreme Environments Electronics*. CRC Press, Boca Raton, FL, 61–70.

Crittenden, M.J. 1963. *New Data on the Isostatic Deformation of Lake Bonneville*. United States Geological Survey Professional Papers, **454-E**.

Dal Forno, G., Gasperini, P. and Boschi, E. 2005. Linear or nonlinear rheology in the mantle: a 3D finite-element approach to postglacial rebound modeling. *Journal of Geodynamics*, **39**, 183–195, https://doi.org/10.1016/j.jog.2004.08.008

Danesi, S. and Morelli, A. 2001. Structure of the upper mantle under the antarctic plate from surface wave tomography. *Geophysical Research Letters*, **28**, 4395–4398, https://doi.org/10.1029/2001GL013431

Dannberg, J., Eilon, Z., Faul, U., Gassmöller, R., Moulik, P. and Myhill, R. 2017. The importance of grain size to mantle dynamics and seismological observations. *Geochemistry, Geophysics, Geosystems*, **18**, 3034–3061, https://doi.org/10.1002/2017GC006944

Davies, D.R., Goes, S., Davies, J.H., Schuberth, B., Bunge, H.-P. and Ritsema, J. 2012. Reconciling dynamic and seismic models of Earth's lower mantle: The dominant role of thermal heterogeneity. *Earth and Planetary Science Letters*, **353**, 253–269, https://doi.org/10.1016/j.epsl.2012.08.016

Dietrich, R., Rülke, A. *et al.* 2004. Plate kinematics and deformation status of the Antarctic Peninsula based on GPS. *Global and Planetary Change*, **42**, 313–321, https://doi.org/10.1016/j.gloplacha.2003.12.003

Dimanov, A., Dresen, G., Xiao, X. and Wirth, R. 1999. Grain boundary diffusion creep of synthetic anorthite aggregates: The effect of water. *Journal of Geophysical Research: Solid Earth*, **104**, 10 483–10 497, https://doi.org/10.1029/1998JB900113

Dixon, N.A. and Durham, W.B. 2018. Measurement of activation volume for creep of dry olivine at upper-mantle conditions. *Journal of Geophysical Research: Solid Earth*, **123**, 8459–8473, https://doi.org/10.1029/2018JB015853

Dobslaw, H., Dill, R. *et al.* 2020. Gravitationally consistent mean barystatic sea level rise from leakage-corrected monthly grace data. *Journal of Geophysical Research: Solid Earth*, **125**, e2020JB020923, https://doi.org/10.1029/2020JB020923

Doubrovine, P.V., Steinberger, B. and Torsvik, T.H. 2012. Absolute plate motions in a reference frame defined by moving hot spots in the Pacific, Atlantic, and Indian oceans. *Journal of Geophysical Research: Solid Earth*, **117**, B09101, https://doi.org/10.1029/2011JB009072

Durek, J.J. and Ekström, G. 1996. A radial model of anelasticity consistent with long-period surface-wave attenuation. *Bulletin of the Seismological Society of America*, **86**, 144–158.

Durney, B. 1968. Convective spherical shell: I. No rotation. *Journal of the Atmospheric Sciences*, **25**, 372–380, https://doi.org/10.1175/1520-0469(1968)025<0372:CSSINR>2.0.CO;2

Dziewonski, A.M., Forte, A.M., Su, W. and Woodward, R.L. 1993. Seismic tomography and geodynamics. *American Geophysical Union, Geophysical Monograph Series*, **76**, 67–105, https://doi.org/10.1029/GM076p0067

Faul, U. and Jackson, I. 2015. Transient creep and strain energy dissipation: An experimental perspective. *Annual Review of Earth and Planetary Sciences*, **43**, 541–569, https://doi.org/10.1146/annurev-earth-060313-054732

Fei, H., Wiedenbeck, M., Yamazaki, D. and Katsura, T. 2013. Small effect of water on upper-mantle rheology based on silicon self-diffusion coefficients. *Nature*, **498**, 213–215, https://doi.org/10.1038/nature12193

Fei, H., Yamazaki, D., Sakurai, M., Miyajima, N., Ohfuji, H., Katsura, T. and Yamamoto, T. 2017. A nearly water-saturated mantle transition zone inferred from mineral viscosity. *Science Advances*, **3**, https://doi.org/10.1126/sciadv.1603024

French, S., Lekić, V. and Romanowicz, B. 2013. Waveform tomography reveals channeled flow at the base of the oceanic asthenosphere. *Science*, **342**, 227–230, https://doi.org/10.1126/science.1241514

Gasperini, P., Dal Forno, G. and Boschi, E. 2004. Linear or non-linear rheology in the Earth's mantle: the prevalence of power-law creep in the postglacial isostatic readjustment of Laurentia. *Geophysical Journal International*, **157**, 1297–1302, https://doi.org/10.1111/j.1365-246X.2004.02319.x

Goes, S., Govers, R. and Vacher, P. 2000. Shallow mantle temperatures under Europe from *P* and *S* wave tomography. *Journal of Geophysical Research: Solid Earth*, **105**, 11 153–11 169, https://doi.org/10.1029/1999JB900300

Gomez, N., Latychev, K. and Pollard, D. 2018. A coupled ice sheet-sea level model incorporating 3D earth structure: Variations in Antarctica during the last deglacial retreat. *Journal of Climate*, **31**, 4041–4054, https://doi.org/10.1175/JCLI-D-17-0352.1

Gribb, T.T. and Cooper, R.F. 1998. Low-frequency shear attenuation in polycrystalline olivine: Grain boundary diffusion and the physical significance of the Andrade model for viscoelastic rheology. *Journal of Geophysical Research: Solid Earth*, **103**, 27 267–27 279, https://doi.org/10.1029/98JB02786

Groh, A., Ewert, H. *et al.* 2012. An investigation of Glacial Isostatic Adjustment over the Amundsen Sea sector, West Antarctica. *Global and Planetary Change*, **98–99**, 45–53, https://doi.org/10.1016/j.gloplacha.2012.08.001

Guerri, M., Cammarano, F. and Tackley, P. 2016. Modelling Earth's surface topography: Decomposition of the static and dynamic components. *Physics of the Earth and Planetary Interiors*, **261**, 172–186, https://doi.org/10.1016/j.pepi.2016.10.009

Hansen, L.N., Zimmerman, M.E. and Kohlstedt, D.L. 2011. Grain boundary sliding in San Carlos olivine: Flow law parameters and crystallographic-preferred orientation. *Journal of Geophysical Research: Solid Earth*, **116**, B08201, https://doi.org/10.1029/2011JB008220

Hansen, L.N., Zimmerman, M.E. and Kohlstedt, D.L. 2012. The influence of microstructure on deformation of olivine in the grain-boundary sliding regime. *Journal of Geophysical Research: Solid Earth*, **117**, B09201, https://doi.org/10.1029/2012JB009305

Hanson, D.R. and Spetzler, H.A. 1994. Transient creep in natural and synthetic, iron-bearing olivine single crystals: Mechanical results and dislocation microstructures. *Tectonophysics*, **235**, 293–315, https://doi.org/10.1016/0040-1951(94)90191-0

Hay, C.C., Lau, H.C.P. *et al.* 2017. Sea level fingerprints in a region of complex Earth structure: The case of WAIS. *Journal of Climate*, **30**, 1881–1892, https://doi.org/10.1175/JCLI-D-16-0388.1

Hirth, G. and Kohlstedt, D.L. 1996. Water in the oceanic upper mantle: implications for rheology, melt extraction and the evolution of the lithosphere. *Earth and Planetary Science Letters*, **144**, 93–108, https://doi.org/10.1016/0012-821X(96)00154-9

Hirth, G. and Kohlstedt, D. 2004. Rheology of the upper mantle. *American Geophysical Union, Geophysical Monograph Series*, **138**, 83–106.

Hu, Y. and Freymueller, J.T. 2019. Geodetic observations of time-variable glacial isostatic adjustment in southeast Alaska and its implications for earth rheology. *Journal of Geophysical Research: Solid Earth*, **124**, 9870–9889, https://doi.org/10.1029/2018JB017028

Huang, P., Wu, P. and Steffen, H. 2019. In search of an ice history that is consistent with composite rheology in glacial isostatic adjustment modelling. *Earth and Planetary Science Letters*, **517**, 26–37, https://doi.org/10.1016/j.epsl.2019.04.011

Isaak, D.G., Anderson, O.L. and Cohen, R.E. 1992. The relationship between shear and compressional velocities at high pressures: reconciliation of seismic tomography and mineral physics. *Geophysical Research Letters*, **19**, 741–744, https://doi.org/10.1029/92GL00774

Ivins, E.R. 1989. New aspects of rotational dynamics within the North American-pacific ductile shear zone. *American Geophysical Union, Geophysical Monograph Series*, **50**, 179–201, https://doi.org/10.1029/GM050p0179

Ivins, E.R. and Sammis, C.G. 1995. On lateral viscosity contrast in the mantle and the rheology of low-frequency geodynamics.

Geophysical Journal International, **123**, 305–322, https://doi.org/10.1111/j.1365-246X.1995.tb06856.x

Ivins, E.R., Unti, T.W.J. and Phillips, R.J. 1982. Large Prandtl number finite-amplitude thermal convection with Maxwell viscoelasticity. *Geophysical & Astrophysical Fluid Dynamics*, **22**, 103–132, https://doi.org/10.1080/03091928208221739

Ivins, E.R., Raymond, C.A. and James, T.S. 2000. The influence of 5000 year-old and younger glacial mass variability on present-day crustal rebound in the Antarctic Peninsula. *Earth, Planets & Space*, **152**, 1023–1029, https://doi.org/10.1186/BF03352325

Ivins, E.R., James, T.S. and Klemann, V. 2003. Glacial isostatic stress shadowing by the Antarctic ice sheet. *Journal of Geophysical Research: Solid Earth*, **108**, 2560, https://doi.org/10.1029/2002JB002182

Ivins, E.R., Watkins, M.M., Yuan, D.-N., Dietrich, R., Casassa, G. and Rülke, A. 2011. On-land ice loss and glacial isostatic adjustment at the Drake Passage: 2003–2009. *Journal of Geophysical Research: Solid Earth*, **116**, B02403, https://doi.org/10.1029/2010JB007607

Ivins, E.R., James, T.S., Wahr, J., O. Schrama, E.J., Landerer, F.W. and Simon, K.M. 2013. Antarctic contribution to sea level rise observed by GRACE with improved GIA correction. *Journal of Geophysical Research: Solid Earth*, **118**, 3126–3141, https://doi.org/10.1002/jgrb.50208

Ivins, E.R., Caron, L., Adhikari, S., Larour, E. and Scheinert, M. 2020. A linear viscoelasticity for decadal to centennial time scale mantle deformation. *Reports on Progress in Physics*, **83**, 106801, https://doi.org/10.1088/1361-6633/aba346

Jackson, I. 2000. Laboratory measurement of seismic wave dispersion and attenuation: recent progress. *American Geophysical Union, Geophysical Monograph Series*, **117**, 265–289.

Jackson, I. 2019. Viscoelastic behaviour from complementary forced-oscillation and microcreep tests. *Minerals*, **9**, 721, https://doi.org/10.3390/min9120721

Jackson, I. and Faul, U.H. 2010. Grainsize-sensitive viscoelastic relaxation in olivine: Towards a robust laboratory-based model for seismological application. *Physics of the Earth and Planetary Interiors*, **183**, 151–163, https://doi.org/10.1016/j.pepi.2010.09.005

Jackson, I., Fitz Gerald, J.D., Faul, U.H. and Tan, B.H. 2002. Grain-size-sensitive seismic wave attenuation in polycrystalline olivine. *Journal of Geophysical Research: Solid Earth*, **107**, ECV 5-1–ECV 5-16, https://doi.org/10.1029/2001JB001225

Jackson, I., Faul, U.H., Fitz Gerald, J.D. and Tan, B.H. 2004. Shear wave attenuation and dispersion in melt-bearing olivine polycrystals: 1. Specimen fabrication and mechanical testing. *Journal of Geophysical Research: Solid Earth*, **109**, B06201, https://doi.org/10.1029/2003JB002406

Jackson, I., Faul, U.H. and Skelton, R. 2014. Elastically accommodated grain-boundary sliding: New insights from experiment and modeling. *Physics of the Earth and Planetary Interiors*, **228**, 203–210, https://doi.org/10.1016/j.pepi.2013.11.014

Jain, C., Korenaga, J. and Karato, S. 2019. Global analysis of experimental data on the rheology of olivine aggregates. *Journal of Geophysical Research: Solid Earth*, **124**, 310–334, https://doi.org/10.1029/2018JB016558

James, T.S. and Ivins, E.R. 1995. Present-day Antarctic ice mass changes and crustal motion. *Geophysical Research Letters*, **22**, 973–976, https://doi.org/10.1029/94GL02800

Jaupart, C., Labrosse, S., Lucazeau, F. and Mareschal, J.-C. 2015. Temperatures, heat, and energy in the mantle of the Earth. *In*: Schubert, G. (ed.) *Treatise on Geophysics, Volume 2*. 2nd edn. Elsevier, Oxford, UK, 223–270.

Johanesen, K.E. and Platt, J.P. 2015. Rheology, microstructure, and fabric in a large scale mantle shear zone, Ronda Peridotite, southern Spain. *Journal of Structural Geology*, **73**, 1–17, https://doi.org/10.1016/j.jsg.2015.01.007

Johnson, J.S., Roberts, S.J. *et al.* 2020. Deglaciation of Pope Glacier implies widespread early Holocene ice sheet thinning in the Amundsen Sea sector of Antarctica. *Earth and Planetary Science Letters*, **548**, 116501, https://doi.org/10.1016/j.epsl.2020.116501

Jordan, T. 1981. Continents as a chemical boundary layer. *Philosophical Transactions of the Royal Society A: Mathematical, Physical and Engineering Sciences*, **301**, 359–373, https://doi.org/10.1098/rsta.1981.0117

Jordan, T., Riley, T. and Siddoway, C. 2020. The geological history and evolution of West Antarctica. *Nature Reviews Earth & Environment*, **1**, 117–133, https://doi.org/10.1038/s43017-019-0013-6

Kachuck, S.B., Martin, D.F., Bassis, J.N. and Price, S.F. 2020. Rapid viscoelastic deformation slows marine ice sheet instability at Pine Island Glacier. *Geophysical Research Letters*, **47**, e2019GL086446, https://doi.org/10.1029/2019GL086446

Kameyama, M., Yuen, D.A. and Karato, S. 1999. Thermal–mechanical effects of low-temperature plasticity (the Peierls mechanism) on the deformation of a viscoelastic shear zone. *Earth and Planetary Science Letters*, **168**, 159–172, https://doi.org/10.1016/S0012-821X(99)00040-0

Karato, S.-I. 1993. Importance of anelasticity in the interpretation of seismic tomography. *Geophysical Research Letters*, **20**, 1623–1626, https://doi.org/10.1029/93GL01767

Karato, S.-I. 2003. Mapping water content in the mantle. *American Geophysical Union, Geophysical Monograph Series*, **138**, 135–152, https://doi.org/10.1029/138GM08

Karato, S.-I. 2008. *Deformation of Earth Materials: An Introduction to the Rheology of Solid Earth*. Cambridge University Press, Cambridge, UK.

Karato, S.-I. 2010. Rheology of the Earth's mantle: A historical review. *Gondwana Research*, **18**, 17–45, https://doi.org/10.1016/j.gr.2010.03.004

Karato, S.-I. and Karki, B. 2001. Origin of lateral variation of seismic wave velocities and density in the deep mantle. *Journal of Geophysical Research: Solid Earth*, **106**, 21 771–21 783, https://doi.org/10.1029/2001JB000214

Karato, S.-I. and Spetzler, H. 1990. Defect microdynamics in minerals and solid-state mechanisms of seismic wave attenuation and velocity dispersion in the mantle. *Reviews of Geophysics*, **28**, 399–421, https://doi.org/10.1029/RG028i004p00399

Karato, S.-I. and Wu, P. 1993. Rheology of the upper mantle: A synthesis. *Science*, **260**, 771–778, https://doi.org/10.1126/science.260.5109.771

Karato, S.-I., Paterson, M.S. and FitzGerald, J.D. 1986. Rheology of synthetic olivine aggregates: Influence of grain size and water. *Journal of Geophysical Research: Solid Earth*, **91**, 8151–8176, https://doi.org/10.1029/JB091iB08p08151

Katsura, T., Yoneda, A., Yamazaki, D., Yoshino, T. and Ito, E. 2010. Adiabatic temperature profile in the mantle. *Physics of the Earth and Planetary Interiors*, **183**, 212–218, https://doi.org/10.1016/j.pepi.2010.07.001

Kaufmann, G., Wu, P. and Ivins, E.R. 2005. Lateral viscosity variations beneath Antarctica and their implications on regional rebound motions and seismotectonics. *Journal of Geodynamics*, **39**, 165–181, https://doi.org/10.1016/j.jog.2004.08.009

Kawazoe, T., Nishihara, Y. *et al.* 2016. Creep strength of ringwoodite measured at pressure–temperature conditions of the lower part of the mantle transition zone using a deformation-DIA apparatus. *Earth and Planetary Science Letters*, **454**, 10–19, https://doi.org/10.1016/j.epsl.2016.08.011

Kittel, C. 2004. *Introduction to Solid State Physics*. 8th edn. Wiley, Hoboken, NJ.

Klemann, V. and Wolf, D. 2005. The eustatic reduction of shoreline diagrams: implications for the inference of relaxation-rate spectra and the viscosity stratification below Fennoscandia. *Geophysical Journal International*, **162**, 249–256, https://doi.org/10.1111/j.1365-246X.2005.02637.x

Klemann, V., Ivins, E.R., Martinec, Z. and Wolf, D. 2007. Models of active glacial isostasy roofing warm subduction: Case of the South Patagonian Ice Field. *Journal of Geophysical Research: Solid Earth*, **112**, B09405, https://doi.org/10.1029/2006JB004818

Knopoff, L. and Vane, G. 1978. Age of East Antarctica from surface wave dispersion. *Pure and Applied Geophysics*, **117**, 806–815, https://doi.org/10.1007/BF00879981

Kohlstedt, D.L. 2007. Properties of rocks and minerals – constitutive equations, rheological behavior, and viscosity of rocks. *In*: Schubert, G. (ed.) *Treatise on Geophysics, Volume 2*. Elsevier, Oxford, UK, 389–417, https://doi.org/10.1016/B978-0444 52748-6.00043-2

Kohlstedt, D.L. and Hansen, L.N. 2015. Constitutive equations, rheological behavior, and viscosity of rocks. *In*: Schubert, G. (ed.) *Treatise on Geophysics, Volume 2*. 2nd edn. Elsevier, Oxford, UK, 441–472, https://doi.org/10.1016/B978-044452748-6.00043-2

Kuchar, J., Milne, G., Hill, A., Tarasov, L. and Nordman, M. 2019. An investigation into the sensitivity of postglacial decay times to uncertainty in the adopted ice history. *Geophysical Journal International*, **220**, 1172–1186, https://doi.org/10.1093/gji/ggz512

Kumazawa, M. and Anderson, O.L. 1969. Elastic moduli, pressure derivatives, and temperature derivatives of single-crystal olivine and single-crystal forsterite. *Journal of Geophysical Research*, **74**, 5961–5972, https://doi.org/10.1029/JB074i025p05961

Kustowski, B., Ekstrom, G. and Dziewonski, A.M. 2008. Anisotropic shear-wave velocity structure of the earth's mantle: A global model. *Journal of Geophysical Research: Solid Earth*, **113**, B06306, https://doi.org/10.1029/2007JB005169

Kyle, P. 1990. McMurdo Volcanic Group, western Ross Embayment: Introduction. *American Geophysical Union Antarctic Research Series*, **48**, 19–25.

Lambeck, K., Rouby, H., Purcell, A., Sun, Y. and Sambridge, M. 2014. Sea level and global ice volumes from the last glacial maximum to the holocene. *Proceedings of the National Academy of Sciences of the United States of America*, **111**, 15 296–15 303, https://doi.org/10.1073/pnas.1411762111

Latychev, K., Mitrovica, J.X., Tromp, J., Tamisiea, M.E., Komatitsch, D. and Christara, C.C. 2005. Glacial isostatic adjustment on 3-D Earth models: a finite-volume formulation. *Geophysical Journal International*, **161**, 421–444, https://doi.org/10.1111/j.1365-246X.2005.02536.x

Lau, H.C.P. and Holtzman, B.K. 2019. 'Measures of dissipation in viscoelastic media' extended: Toward continuous characterization across very broad geophysical time scales. *Geophysical Research Letters*, **46**, 9544–9553, https://doi.org/10.1029/2019GL083529

Lau, H.C.P., Austermann, J., Mitrovica, J.X., Crawford, O., Al-Attar, D. and Latychev, K. 2018. Inferences of mantle viscosity based on ice age data sets: The bias in radial viscosity profiles due to the neglect of laterally heterogeneous viscosity structure. *Journal of Geophysical Research: Solid Earth*, **123**, 7237–7252, https://doi.org/10.1029/2018JB015740

Lawrence, J.F. and Prieto, G.A. 2011. Attenuation tomography of the western United States from ambient seismic noise. *Journal of Geophysical Research: Solid Earth*, **116**, B06302, https://doi.org/10.1029/2010JB007836

Lawrence, J.F. and Wysession, M.E. 2006. QLM9: A new radial quality factor (Q_μ) model for the lower mantle. *Earth and Planetary Science Letters*, **241**, 962–971, https://doi.org/10.1016/j.epsl.2005.10.030

Lawver, L., Keller, R., Fisk, M. and Strelin, J. 1995. Bransfield strait, antarctic peninsula active extension behind a dead arc. *In*: Taylor, B. (ed.) *Backarc Basins*. Springer, Boston, MA, 315–342, https://doi.org/10.1007/978-1-4615-1843-3_8

Lee, C.-T. 2003. Compositional variation of density and seismic velocities in natural peridotites at STP conditions: Implications for seismic imaging of compositional heterogeneities in the upper mantle. *Journal of Geophysical Research: Solid Earth*, **108**, 2441, https://doi.org/10.1029/2003JB002413

Lidberg, M., Johansson, J.M., Scherneck, H.-G. and Milne, G.A. 2010. Recent results based on continuous GPS observations of the GIA process in Fennoscandia from BIFROST. *Journal of Geodynamics*, **50**, 8–18, https://doi.org/10.1016/j.jog.2009.11.010

Lloyd, A.J., Wiens, D.A. *et al.* 2020. Seismic structure of the Antarctic upper mantle imaged with adjoint tomography. *Journal of Geophysical Research: Solid Earth*, **125**, https://doi.org/10.1029/2019JB017823

Lucas, E.M., Soto, D. *et al.* 2020. P- and S-wave velocity structure of central West Antarctica: Implications for the tectonic evolution of the West Antarctic Rift System. *Earth and Planetary Science Letters*, **546**, 116437, https://doi.org/10.1016/j.epsl.2020.116437

Mackwell, S.J., Kohlstedt, D.L. and Paterson, M.S. 1985. The role of water in the deformation of olivine single crystals. *Journal of Geophysical Research: Solid Earth*, **90**, 11 319–11 333, https://doi.org/10.1029/JB090iB13p11319

Marquardt, K. and Faul, U. 2018. The structure and composition of olivine grain boundaries: 40 years of studies, status and current developments. *Physics and Chemistry of Minerals*, **45**, 139–172, https://doi.org/10.1007/s00269-017-0935-9

Martinec, Z. 2000. Spectral-finite element approach to three-dimensional viscoelastic relaxation in a spherical Earth. *Geophysical Journal International*, **142**, 117–141, https://doi.org/10.1046/j.1365-246x.2000.00138.x

Martos, Y.M., Catalán, M., Jordan, T.A., Golynsky, A., Golynsky, D., Eagles, G. and Vaughan, D.G. 2017. Heat flux distribution of Antarctica unveiled. *Geophysical Research Letters*, **44**, 11 417–11 426, https://doi.org/10.1002/2017GL075609

Masuti, S., ichiro Karato, S., Girard, J. and Barbot, S.D. 2019. Anisotropic high-temperature creep in hydrous olivine single crystals and its geodynamic implications. *Physics of the Earth and Planetary Interiors*, **290**, 1–9, https://doi.org/10.1016/j.pepi.2019.03.002

Matsyuk, S., Langer, K. and Hösch, A. 1998. Hydroxyl defects in garnets from mantle xenoliths in kimberlites of the Siberian platform. *Contributions to Mineralogy and Petrology*, **132**, 163–179, https://doi.org/10.1007/s004100050414

McCarthy, C., Takei, Y. and Hiraga, T. 2011. Experimental study of attenuation and dispersion over a broad frequency range: 2. The universal scaling of polycrystalline materials. *Journal of Geophysical Research: Solid Earth*, **116**, B09207, https://doi.org/10.1029/2011JB008384

McKenzie, D., Jackson, J. and Priestley, K. 2005. Thermal structure of oceanic and continental lithosphere. *Earth and Planetary Science Letters*, **233**, 337–349, https://doi.org/10.1016/j.epsl.2005.02.005

Mei, S. and Kohlstedt, D.L. 2000. Influence of water on plastic deformation of olivine aggregates: 2. Dislocation creep regime. *Journal of Geophysical Research: Solid Earth*, **105**, 21 471–21 481, https://doi.org/10.1029/2000JB900180

Milne, G.A., Mitrovica, J.X., Scherneck, H.-G., Davis, J.L., Johansson, J.M., Koivula, H. and Vermeer, M. 2004. Continuous GPS measurements of postglacial adjustment in Fennoscandia: 2. Modeling results. *Journal of Geophysical Research: Solid Earth*, **109**, B02412, https://doi.org/10.1029/2003JB0 02619

Minster, J.B. and Anderson, D.L. 1980. Dislocations and nonelastic processes in the mantle. *Journal of Geophysical Research: Solid Earth*, **85**, 6347–6352, https://doi.org/10.1029/JB085iB11p06347

Mitrovica, J.X. 1996. Haskell [1935] revisited. *Journal of Geophysical Research: Solid Earth*, **101**, 555–569, https://doi.org/10.1029/95JB03208

Nichols, K.A., Goehring, B.M., Balco, G., Johnson, J.S., Hein, A.S. and Todd, C. 2019. New last glacial maximum ice thickness constraints for the Weddell Sea Embayment, Antarctica. *The Cryosphere*, **13**, 2935–2951, https://doi.org/10.5194/tc-13-2935-2019

Nield, G.A., Barletta, V.R. *et al.* 2014. Rapid bedrock uplift in the Antarctic Peninsula explained by viscoelastic response to recent ice unloading. *Earth and Planetary Science Letters*, **397**, 32–41, https://doi.org/10.1016/j.epsl.2014.04.019

Nield, G.A., Whitehouse, P.L., King, M.A. and Clarke, P.J. 2016. Glacial isostatic adjustment in response to changing late Holocene behaviour of ice streams on the Siple Coast, West Antarctica. *Geophysical Journal International*, **205**, 1–21, https://doi.org/10.1093/gji/ggv532

Nield, G.A., Whitehouse, P.L., van der Wal, W., Blank, B., O'Donnell, J.P. and Stuart, G.W. 2018. The impact of lateral variations in lithospheric thickness on glacial isostatic adjustment in West Antarctica. *Geophysical Journal International*, **214**, 811–824, https://doi.org/10.1093/gji/ggy158

Nordman, M., Milne, G. and Tarasov, L. 2015. Reappraisal of the Ångerman river decay time estimate and its application to determine uncertainty in Earth viscosity structure. *Geophysical Journal International*, **201**, 811–822, https://doi.org/10.1093/gji/ggv051

Núñez-Valdez, M., Wu, Z., Yu, Y.G. and Wentzcovitch, R.M. 2013. Thermal elasticity of $(Fe_X,Mg_{1-X})_2SiO_4$ olivine and wadsleyite. *Geophysical Research Letters*, **40**, 290–294, https://doi.org/10.1002/grl.50131

O'Donnell, J., Selway, K. *et al.* 2017. The uppermost mantle seismic velocity and viscosity structure of central West Antarctica. *Earth and Planetary Science Letters*, **472**, 38–49, https://doi.org/10.1016/j.epsl.2017.05.016

Pappa, F., Ebbing, J., Ferraccioli, F. and van der Wal, W. 2019. Modeling satellite gravity gradient data to derive density, temperature, and viscosity structure of the Antarctic lithosphere. *Journal of Geophysical Research: Solid Earth*, **124**, 12 053–12 076, https://doi.org/10.1029/2019JB017997

Park, Y., Kim, K.-H., Lee, J., Yoo, H.J. and Plasencia, L., M.P. 2012. *P*-wave velocity structure beneath the northern Antarctic Peninsula: evidence of a steeply subducting slab and a deep-rooted low-velocity anomaly beneath the central Bransfield Basin. *Geophysical Journal International*, **191**, 932–938, https://doi.org/10.1111/j.1365-246X.2012.05684.x

Paterson, M. and Olgaard, D. 2000. Rock deformation tests to large shear strains in torsion. *Journal of Structural Geology*, **22**, 1341–1358, https://doi.org/10.1016/S0191-8141(00)00042-0

Paulson, A., Zhong, S. and Wahr, J. 2005. Modelling post-glacial rebound with lateral viscosity variations. *Geophysical Journal International*, **163**, 357–371, https://doi.org/10.1111/j.1365-246X.2005.02645.x

Peltier, W.R., Yuen, D.A. and Wu, P. 1980. Postglacial rebound and transient rheology. *Geophysical Research Letters*, **7**, 733–736, https://doi.org/10.1029/GL007i010p00733

Peltier, W.R., Wu, P. and Yuen, D.A. 1981. The viscosities of the Earth's mantle. *American Geophysical Union Geodynamics Series*, **4**, 59–77, https://agupubs.onlinelibrary.wiley.com/doi/abs/10.1029/GD004p0059

Poirier, J.-P. 2000. *Introduction to the Physics of the Earth's Interior*. Cambridge University Press, Cambridge, UK.

Powell, E., Gomez, N., Hay, C., Latychev, K. and Mitrovica, J.X. 2020. Viscous effects in the solid earth response to modern Antarctic ice mass flux: Implications for geodetic studies of WAIS stability in a warming world. *Journal of Climate*, **33**, 443–459, https://doi.org/10.1175/JCLI-D-19-0479.1

Ranalli, G. 2001. Mantle rheology: radial and lateral viscosity variations inferred from microphysical creep laws. *Journal of Geodynamics*, **32**, 425–444, https://doi.org/10.1016/S0264-3707(01)00042-4

Ritzwoller, M.H., Shapiro, N.M., Levshin, A.L. and Leahy, G.M. 2001. Crustal and upper mantle structure beneath Antarctica and surrounding oceans. *Journal of Geophysical Research: Solid Earth*, **106**, 30 645–30 670, https://doi.org/10.1029/2001JB000179

Sabadini, R., Vermeersen, B. and Cambiotti, G. 2016. *Global Dynamics of the Earth: Applications of Viscoelastic Relaxation Theory to Solid-Earth and Planetary Geophysics*. 2nd edn. Springer, Dordrecht, The Netherlands.

Samrat, N.H., King, M.A., Watson, C., Hooper, A., Chen, X., Barletta, V.R. and Bordoni, A. 2020. Reduced ice mass loss and three-dimensional viscoelastic deformation in northern Antarctic peninsula inferred from GPS. *Geophysical Journal International*, **222**, 1013–1022, https://doi.org/10.1093/gji/ggaa229

Scheinert, M., Ferraccioli, F. *et al.* 2016. New Antarctic gravity anomaly grid for enhanced geodetic and geophysical studies in Antarctica. *Geophysical Research Letters*, **43**, 600–610, https://doi.org/10.1002/2015GL067439

Scheinert, M., Engels, O., Schrama, E.J.O., van der Wal, W. and Horwarth, M. 2021. Geodetic observations for constraining mantle processes in Antarctica. *Geological Society, London, Memoirs*, **56**, https://doi.org/10.1144/M56-2021-22

Schwarz, R. and Granato, A. 1975. Theory of damping due to thermally assisted unpinning of dislocations. *Physical Review Letters*, **34**, 1174–1177, https://doi.org/10.1103/PhysRevLett.34.1174

Shen, W., Wiens, D.A. *et al.* 2018. The crust and upper mantle structure of central and West Antarctica from Bayesian inversion of Rayleigh wave and receiver functions. *Journal of Geophysical Research: Solid Earth*, **123**, 7824–7849, https://doi.org/10.1029/2017JB015346

Shimojuku, A., Kubo, T., Ohtani, E., Nakamura, T., Okazaki, R., Dohmen, R. and Chakraborty, S. 2009. Si and O diffusion in $(Mg, Fe)_2SiO_4$ wadsleyite and ringwoodite and its implications for the rheology of the mantle transition zone. *Earth and Planetary Science Letters*, **284**, 103–112, https://doi.org/10.1016/j.epsl.2009.04.014

Sigmundsson, F. 1991. Post-glacial rebound and asthenosphere viscosity in Iceland. *Geophysical Research Letters*, **18**, 1131–1134, https://doi.org/10.1029/91GL01342

Simms, A.R., Ivins, E.R., DeWitt, R., Kouremenos, P. and Simkins, L.M. 2012. Timing of the most recent Neoglacial advance and retreat in the South Shetland Islands, Antarctic peninsula: insights from raised beaches and Holocene uplift rates. *Quaternary Science Review*, **47**, 41–55, https://doi.org/10.1016/j.quascirev.2012.05.013

Smellie, J.L., Panter, K.S. and Geyer, A. 2021. Introduction to volcanism in Antarctica: 200 million years of subduction, rifting and continental break-up. *Geological Society, London, Memoirs*, **55**, 1–6, https://doi.org/10.1144/M55-2020-14

Spada, G., Colleoni, F. and Ruggieri, G. 2011. Shallow upper mantle rheology and secular ice sheet fluctuations. *Tectonophysics*, **511**, 89–98, https://doi.org/10.1016/j.tecto.2009.12.020

Stacey, F.D. and Davis, P. 2008. *Physics of the Earth*. Cambridge University Press, Cambridge, UK.

Stacey, F.D. and Hodgkinson, J.H. 2019. Thermodynamics with the Grüneisen parameter: Fundamentals and applications to high pressure physics and geophysics. *Physics of the Earth and Planetary Interiors*, **286**, 42–68, https://doi.org/10.1016/j.pepi.2018.10.006

Stixrude, L. and Jeanloz, R. 2015. Constraints on seismic models from other disciplines – constraints from mineral physics on seismological models. *In*: Schubert, G. (ed.) *Treatise on Geophysics, Volume 2*. 2nd edn. Elsevier, Oxford, UK, 829–852, https://doi.org/10.1016/B978-0-444-53802-4.00022-1

Stixrude, L. and Lithgow-Bertelloni, C. 2010. Thermodynamics of the Earth's mantle. *Reviews in Mineralogy and Geochemistry*, **71**, 465–484, https://doi.org/10.2138/rmg.2010.71.21

Takei, Y., Karasawa, F. and Yamauchi, H. 2014. Temperature, grain size, and chemical controls on polycrystal anelasticity over a broad frequency range extending into the seismic range. *Journal of Geophysical Research: Solid Earth*, **119**, 5414–5443, https://doi.org/10.1002/2014JB011146

Tanaka, Y., Klemann, V., Martinec, Z. and Riva, R.E.M. 2011. Spectral-finite element approach to viscoelastic relaxation in a spherical compressible Earth: application to GIA modelling. *Geophysical Journal International*, **184**, 220–234, https://doi.org/10.1111/j.1365-246X.2010.04854.x

Tromp, J. and Mitrovica, J.X. 2000. Surface loading of a viscoelastic planet, III. Aspherical models. *Geophysical Journal International*, **140**, 425–441, https://doi.org/10.1046/j.1365-246x.2000.00027.x

Tushingham, A.M. and Peltier, W.R. 1992. Validation of the ICE-3G model of Würm-Wisconsin Deglaciation using a global data base of relative sea level histories. *Journal of Geophysical Research: Solid Earth*, **97**, 3285–3304, https://doi.org/10.1029/91JB02176

Vacher, P., Mocquet, A. and Sotin, C. 1996. Comparison between tomographic structures and models of convection in the upper mantle. *Geophysical Journal International*, **124**, 45–56, https://doi.org/10.1111/j.1365-246X.1996.tb06351.x

van den Berg, A.P., Yuen, D.A. and van Keken, P.E. 1995. Rheological transition in mantle convection with a composite temperature-dependent, non-Newtonian and Newtonian rheology. *Earth and Planetary Science Letters*, **129**, 249–260, https://doi.org/10.1016/0012-821X(94)00246-U

van der Lee, S. 2001. Deep below North America. *Science*, **294**, 1297–1298, https://doi.org/10.1126/science.1066319

van der Wal, W., Wu, P., Wang, H. and Sideris, M.G. 2010. Sea levels and uplift rate from composite rheology in glacial isostatic adjustment modeling. *Journal of Geodynamics*, **50**, 38–48, https://doi.org/10.1016/j.jog.2010.01.006

van der Wal, W., Barnhoorn, A., Stocchi, P., Gradmann, S., Wu, P., Drury, M. and Vermeersen, B. 2013. Glacial isostatic adjustment model with composite 3-D Earth rheology for Fennoscandia. *Geophysical Journal International*, **194**, 61–77, https://doi.org/10.1093/gji/ggt099

van der Wal, W., Whitehouse, P.L. and Schrama, E.J. 2015. Effect of GIA models with 3D composite mantle viscosity on GRACE mass balance estimates for Antarctica. *Earth and Planetary Science Letters*, **414**, 134–143, https://doi.org/10.1016/j.epsl.2015.01.001

Wahr, J., Wingham, D. and Bentley, C. 2000. A method of combining ICESat and GRACE satellite data to constrain Antarctic mass balance. *Journal of Geophysical Research: Solid Earth*, **105**, 16 279–16 294, https://doi.org/10.1029/2000JB900113

Wang, H., Wu, P. and van der Wal, W. 2008. Using postglacial sea level, crustal velocities and gravity-rate-of-change to constrain the influence of thermal effects on mantle lateral heterogeneities. *Journal of Geodynamics*, **46**, 104–117, https://doi.org/10.1016/j.jog.2008.03.003

Wang, L., Blaha, S., Kawazoe, T., Miyajima, N. and Katsura, T. 2017. Identical activation volumes of dislocation mobility in the [100](010) and [001](010) slip systems in natural olivine. *Geophysical Research Letters*, **44**, 2687–2692, https://doi.org/10.1002/2017GL073070

Whitehouse, P.L. 2018. Glacial isostatic adjustment modelling: historical perspectives, recent advances, and future directions. *Earth Surface Dynamics*, **6**, 401–429, https://doi.org/10.5194/esurf-6-401-2018

Whitehouse, P.L., Bentley, M.J., Milne, G.A., King, M.A. and Thomas, I.D. 2012. A new glacial isostatic adjustment model for Antarctica: calibrated and tested using observations of relative sea-level change and present-day uplift rates. *Geophysical Journal International*, **190**, 1464–1482, https://doi.org/10.1111/j.1365-246X.2012.05557.x

Whitehouse, P.L., Bentley, M.J., Vieli, A., Jamieson, S.S.R., Hein, A.S. and Sugden, D.E. 2017. Controls on Last Glacial Maximum ice extent in the Weddell Sea embayment, Antarctica. *Journal of Geophysical Research: Earth Surface*, **122**, 371–397, https://doi.org/10.1002/2016JF004121

Wiens, D.A., Conder, J.A. and Faul, U.H. 2008. The seismic structure and dynamics of the mantle wedge. *Annual Review of Earth and Planetary Sciences*, **36**, 421–455, https://doi.org/10.1146/annurev.earth.33.092203.122633

Wiens, D.A., Shen, W. and Lloyd, A.J. 2021. The seismic structure of the Antarctic upper mantle. *Geological Society, London, Memoirs*, **56**, https://doi.org/10.1144/M56-2020-18

Wilch, T., McIntosh, W. and Dunbar, N. 1999. Late Quaternary volcanic activity in Marie Byrd Land: Potential ^{40}Ar/^{39}Ar-dated time horizons in West Antarctic ice and marine cores. *Geological Society of America Bulletin*, **111**, 1563–1580, https://doi.org/10.1130/0016-7606(1999)111<1563:LQVAIM> 2.3.CO;2

Wolf, D., Klemann, V., Wünsch, J. and Zhang, F.-P. 2006. A reanalysis and reinterpretation of geodetic and geological evidence of glacial-isostatic adjustment in the Churchill region, Hudson Bay. *Surveys in Geophysics*, **27**, 19–61, https://doi.org/10.1007/s10712-005-0641-x

Wolstencroft, M., King, M.A. *et al.* 2015. Uplift rates from a new high-density GPS network in Palmer Land indicate significant late Holocene ice loss in the southwestern Weddell Sea. *Geophysical Journal International*, **203**, 737–754, https://doi.org/10.1093/gji/ggv327

Wu, P. 1999. Modelling postglacial sea levels with power-law rheology and a realistic ice model in the absence of ambient tectonic stress. *Geophysical Journal International*, **139**, 691–702, https://doi.org/10.1046/j.1365-246x.1999.00965.x

Wu, P. 2002. Effects of nonlinear rheology on degree 2 harmonic deformation in a spherical self-gravitating Earth. *Geophysical Research Letters*, **29**, 40-1–40-4, https://doi.org/10.1029/2001GL014109

Wu, P. 2005. Effects of lateral variations in lithospheric thickness and mantle viscosity on glacially induced surface motion in Laurentia. *Earth and Planetary Science Letters*, **235**, 549–563, https://doi.org/10.1016/j.epsl.2005.04.038

Wu, P., Wang, H. and Steffen, H. 2013. The role of thermal effect on mantle seismic anomalies under Laurentia and Fennoscandia from observations of glacial isostatic adjustment. *Geophysical Journal International*, **192**, 7–17, https://doi.org/10.1093/gji/ggs009

Xu, Z., Schrama, E.J., van der Wal, W., van den Broeke, M. and Enderlin, E.M. 2016. Improved GRACE regional mass balance estimates of the Greenland Ice sheet cross-validated with the input–output method. *The Cryosphere*, **10**, 895–912, https://doi.org/10.5194/tc-10-895-2016

Yuen, D.A., Hansen, U., Zhao, W., Vincent, A.P. and Malevsky, A.V. 1993. Hard turbulent thermal convection and thermal evolution of the mantle. *Journal of Geophysical Research: Planets*, **98**, 5355–5373, https://doi.org/10.1029/92JE02725

Zhao, C., King, M.A., Watson, C.S., Barletta, V.R., Bordoni, A., Dell, M. and Whitehouse, P.L. 2017. Rapid ice unloading in the Fleming Glacier region, southern Antarctic Peninsula, and its effect on bedrock uplift rates. *Earth and Planetary Science Letters*, **473**, 164–176, https://doi.org/10.1016/j.epsl.2017.06.002

Zhong, S., Paulson, A. and Wahr, J. 2003. Three-dimensional finite-element modelling of Earth's viscoelastic deformation: effects of lateral variations in lithospheric thickness. *Geophysical Journal International*, **155**, 679–695, https://doi.org/10.1046/j.1365-246X.2003.02084.x

Zhu, T. 2014. Tomography-based mantle flow beneath Mongolia-baikal area. *Physics of the Earth and Planetary Interiors*, **237**, 40–50, https://doi.org/10.1016/j.pepi.2014.10.003

Geodetic observations for constraining mantle processes in Antarctica

Mirko Scheinert[1]*, Olga Engels[2], Ernst J. O. Schrama[3], Wouter van der Wal[3,4] and Martin Horwath[1]

[1]Institut für Planetare Geodäsie, Technische Universität Dresden, Helmholtzstraße 10, 01069 Dresden, Germany

[2]Institute of Geodesy and Geoinformation, University of Bonn, Nussallee 17, 53115 Bonn, Germany

[3]Faculty of Aerospace Engineering, Delft University of Technology, Kluyverweg 1, 2629 HS Delft, The Netherlands

[4]Faculty of Civil Engineering, Delft University of Technology, Stevinweg 1, 2628 CN Delft, The Netherlands

MS, 0000-0002-0892-8941; OE, 0000-0001-8521-3475; EJOS, 0000-0002-9654-4082; WW, 0000-0001-8030-9080; MH, 0000-0001-5797-244X

*Correspondence: Mirko.Scheinert@tu-dresden.de

Abstract: Geodynamic processes in Antarctica such as glacial isostatic adjustment (GIA) and post-seismic deformation are measured by geodetic observations such as global navigation satellite systems (GNSS) and satellite gravimetry. GNSS measurements have comprised both continuous measurements and episodic measurements since the mid-1990s. The estimated velocities typically reach an accuracy of 1 mm a^{-1} for horizontal velocities and 2 mm a^{-1} for vertical velocities. However, the elastic deformation due to present-day ice-load change needs to be considered accordingly.

Space gravimetry derives mass changes from small variations in the inter-satellite distance of a pair of satellites, starting with the GRACE (Gravity Recovery and Climate Experiment) satellite mission in 2002 and continuing with the GRACE-FO (GRACE Follow-On) mission launched in 2018. The spatial resolution of the measurements is low (about 300 km) but the measurement error is homogeneous across Antarctica. The estimated trends contain signals from ice-mass change, and local and global GIA signals. To combine the strengths of the individual datasets, statistical combinations of GNSS, GRACE and satellite altimetry data have been developed. These combinations rely on realistic error estimates and assumptions of snow density. Nevertheless, they capture signals that are missing from geodynamic forward models such as the large uplift in the Amundsen Sea sector caused by a low-viscous response to century-scale ice-mass changes.

Several geodynamic processes occurring in the Earth's mantle manifest in geodetic observations in Antarctica such as global navigation satellite systems (GNSS: comprising GPS, GLONASS, Galileo, BeiDou and further systems), time-variable gravity and satellite altimetry. These include surface-loading processes from past ice-thickness changes, post-seismic deformation and possible dynamic topography. The geodetic observations are becoming more numerous because of continuous coverage from satellite measurements and increasing efforts to place GNSS antennas on bedrock in Antarctica. At the same time, measurements are becoming more accurate with increasing instrument precision and better corrections (King *et al.* 2010). Moreover, geodynamic processes are visible as secular trends, which means that with longer time series from GNSS or satellite measurements the signal/noise ratio of geodynamic signals increases. The dominant problem in observing geodynamic processes with geodetic observations is that several processes are mixed; uplift or gravity rates are influenced by current ice unloading, past ice-thickness changes and post-seismic changes. Therefore, several studies have been dedicated to disentangling the processes by using models as well as further non-geodetic data (e.g. firn compaction: Ligtenberg *et al.* 2011; rheological parameters based on seismic tomography: van der Wal *et al.* 2015). Recent studies have had success in constraining the glacial isostatic adjustment (GIA) process and post-seismic deformation (Barletta *et al.* 2022, this volume). This chapter provides an overview of the main geodetic data types, their strengths and weaknesses, and main results.

Measurements collected by permanently or temporarily installed GNSS receivers are most valuable since they allow coordinates and coordinate changes to be accurately inferred, and, thus, the time-variable geometry of the Earth's surface to be determined (King *et al.* 2010). For our purpose, we are interested in geodetic GNSS equipment with the antenna stably placed on bedrock and the receiver permanently recording data (permanent station) or in certain observation periods (campaign station). An overview of measurements and insights is presented in the following section. Time-variable gravity data from the GRACE (Gravity Recovery and Climate Experiment) and GRACE-FO (GRACE Follow-On) missions are discussed in the 'Temporal gravity field variations from satellite gravimetry' section, while the static gravity measurements are discussed by Pappa and Ebbing (2021, this volume). Further constraints for modelling GIA may be inferred from relative sea-level curves where they can be unequivocally identified and dated (e.g. Peltier 1998). However, Antarctica near-field relative sea-level data are scarce and are not able to provide a similar level of information as GNSS (e.g. Ivins and James 2005).

To address the problem of superposition of processes, several studies have produced combinations of measurements, each of which has a different sensitivity to surface and solid Earth processes (e.g. Riva *et al.* 2009; Groh *et al.* 2014; Gunter *et al.* 2014; Martín-Español *et al.* 2016*b*). These measurements can include GRACE-derived temporal gravity changes, surface changes from radar and/or laser altimetry, and position changes from GNSS. The combinations increased the constraints on the GIA process and have become known as 'empirical' or 'inverse' GIA models. These will be reviewed in the 'Results on mantle-related processes from geodetic data combination' section. The chapter concludes with a summary of the state of the art and an outlook for possible improvements in the field of geodetic measurements of mantle processes in Antarctica.

Geodetic GNSS measurements to constrain the modelling of ice-induced Earth deformation

Past and recent Antarctic ice-mass changes form the loading and unloading of the Earth's crust. These load changes cause crustal deformations that can be observed by geodetic measurement techniques, especially by GNSS. The response of the solid Earth depends on the timescales of the load and the related rheological properties of the Earth's interior (van der Wal *et al.* 2015; Ivins *et al.* 2020; Barletta *et al.* 2022, this volume). On short timescales (seismic waves), the response is driven primarily by elastic properties; and on

From: Martin, A. P. and van der Wal, W. (eds) 2023. *The Geochemistry and Geophysics of the Antarctic Mantle.* Geological Society, London, Memoirs, **56**, 295–313,

First published online 3 December 2021, https://doi.org/10.1144/M56-2021-22

longer timescales, from decades to thousands of years, anelastic and viscoelastic properties become more important. The ductile behaviour of the lithospheric and upper mantle that is a function of the pressure–temperature distribution, especially in the asthenosphere, govern the response on these longer timescales (Ivins *et al.* 2021, this volume). The slow, time-dependent deformation is known as creep, and dominates the strength of the material below the brittle crust where material strength is determined by the resistance to fracturing. To understand measured crustal deformation and to come up with informed predictions, suitable numerical models have to be set up and their parameters have to be determined. These rheological parameters include the effective elastic thickness of the lithosphere (Burov 2011) and the viscosities of the upper mantle (especially asthenosphere) and lower mantle, respectively, as well as their lateral variations. They can be determined based on seismic-wave tomography and laboratory experiments on material behaviour, but such inferences suffer from a large uncertainty (Ivins *et al.* 2021, this volume). Therefore, constraints on the models are needed.

GNSS observations are used to accurately and consistently determine coordinates and coordinate changes of specially marked points in the bedrock. However, in Antarctica these observations are limited to a few places where bedrock is accessible. It has to be noted that for a long time GPS was the only available system while further satellite navigation systems reached full operation mode and/or civil access (much) later. Today, more systems are available (e.g. GLONASS (Russia), Galileo (Europe) and BeiDou (China)), and modern receivers featuring several hundred receiving channels are capable of recording a multitude of GNSS signals. The resulting coordinate changes, especially vertical velocities or deformation rates, provide independent data at the Earth's surface and, thus, can serve as direct constraints in the modelling.

Short history and types of geodetic GNSS observations in Antarctica

The first, German-led, geodetic GNSS projects started in 1995 in the region of the Antarctic Peninsula and in Dronning Maud Land, East Antarctica (Dietrich 1996, 2000; Dietrich *et al.* 2001, 2004). These measurements formed the nucleus for the Scientific Committee on Antarctic Research (SCAR: https://scar.org/) Epoch Crustal Movement Campaigns, in which many international institutions have since contributed data. The SCAR GNSS database (https://data1.geo.tu-dresden.de/scar) was set-up and is maintained at TU Dresden. In these 'early' GPS years the observations were carried out only as campaign-style or epoch measurements (Figs 1 & 2). One observation epoch could comprise several days of quasi-continuous measurements from which one coordinate solution was computed. Thus, by combining several epochs (observed in different years), coordinate change rates can be estimated. However, it is only possible to deduce linear rates with reasonable accuracy; any non-linear (e.g. seasonal) signals cannot be derived. Similar activities were performed, for example, in the Transantarctic Mountains (1996–2001: Raymond *et al.* 2004; project TAMDEF, first field season 1996–97: Vazquez Becerra 2009), Marie Byrd Land, West Antarctica (1999–2002: Donnellan and Luyendyk 2004) and North Victoria Land (project VLNDEF: Zanutta *et al.* 2017, 2018). Tregoning *et al.* (1999, 2000) started to install continuously operating GPS receivers at Beaver Lake in 1998 and extended the installation in subsequent years to study vertical crustal movement in the region of Lambert Glacier, East Antarctica.

Around the turn of the century, the setting-up of a significant number of permanently recording geodetic GNSS sites was intensified. For this to succeed, the especially challenging task of independent power supply had to be solved. Each permanent site has to be equipped with a large battery pack to provide sufficient electrical charge, together with solar panels and wind generators, for recharging the batteries (Figs 3 & 4). Attempts to use fuel-cell technology were also made in these early years (Tregoning *et al.* 2000). Only in the case of a GNSS site situated near a year-round Antarctic station can a direct power link be utilized, as is the case for the International GNSS Service (IGS: https://www.igs.org) sites SANAE IV, Syowa, McMurdo, Casey, Davis, Mawson, Dumont d'Urville, Palmer, Rothera and O'Higgins (cf. Fig. 5). A significant boost to deploy and run epoch and permanent geodetic GNSS sites in Antarctica was initiated by the International

Fig. 1. Example of a GNSS campaign site: Clark Island, Pine Island Bay, West Antarctica. The GNSS antenna is fixed to the bedrock. Receiver, battery and electronics are stowed in an aluminium box, while the solar panel is used to recharge the battery. Image credits: M. Scheinert.

Fig. 2. Example of a GNSS campaign site: Backer Islands, Pine Island Bay, West Antarctica. Image credits: M. Scheinert.

Polar Year (IPY) 2007–09. The IPY project POLENET was the nucleus for a number of new projects, among which is the USA-led ANET project (Bevis *et al.* 2009; Barletta *et al.* 2018). Here, UNAVCO (University NAVSTAR Consortium: https://www.unavco.org) plays a decisive role in not only providing hardware and engineering expertise for the set-up and maintenance of GNSS sites, but also in operating one of the largest data archives including Antarctic GNSS data. Many national Antarctic programmes have also provided logistic support to carry out these operations in Antarctica.

Reference frame, technique-related issues and accuracy issues

In the data analysis, the linkage of GNSS position measurements to a proper reference system has to be realized (realization of the so-called 'geodetic datum') in order to maintain a practical and accessible solution of the origin, axes and scale of the terrestrial reference frame. In the differential GNSS (DGNSS) method, this is principally fulfilled by the inclusion of reference sites that are IGS sites within Antarctica and at the adjacent continents, and the application of a suitable six- or seven-parameter Helmert transformation. In the precise point positioning (PPP) technique, the role of these fiducial points is taken over by precise ephemeris of the GNSS satellite orbits. For both techniques, and especially in the processing of GNSS data usually spanning time periods of more than 1 year, further data and models, and precise and consistent orbit coordinates as well as clock parameters, have to be introduced: for example, from consistent reprocessing (Rülke *et al.* 2008; Rebischung *et al.* 2016; Villiger and Dach 2020, p. 11). Further parameters introduced into the analysis are phase centre offsets of satellite and receiver

Fig. 3. Example of a GNSS permanent site: Kottas Mountains, western Dronning Maud Land, East Antarctica. The GNSS antenna is installed on the left outcrop (see also Fig. 4). This GNSS site is co-located with a seismometer station of the Alfred Wegener Institute, Germany. On the right outcrop, solar panels and wind generators are mounted. Image credits: E. Buchta.

Fig. 4. Example of a GNSS permanent site: Kottas Mountains, western Dronning Maud Land, East Antarctica (see also Fig. 3). Image credits: E. Buchta.

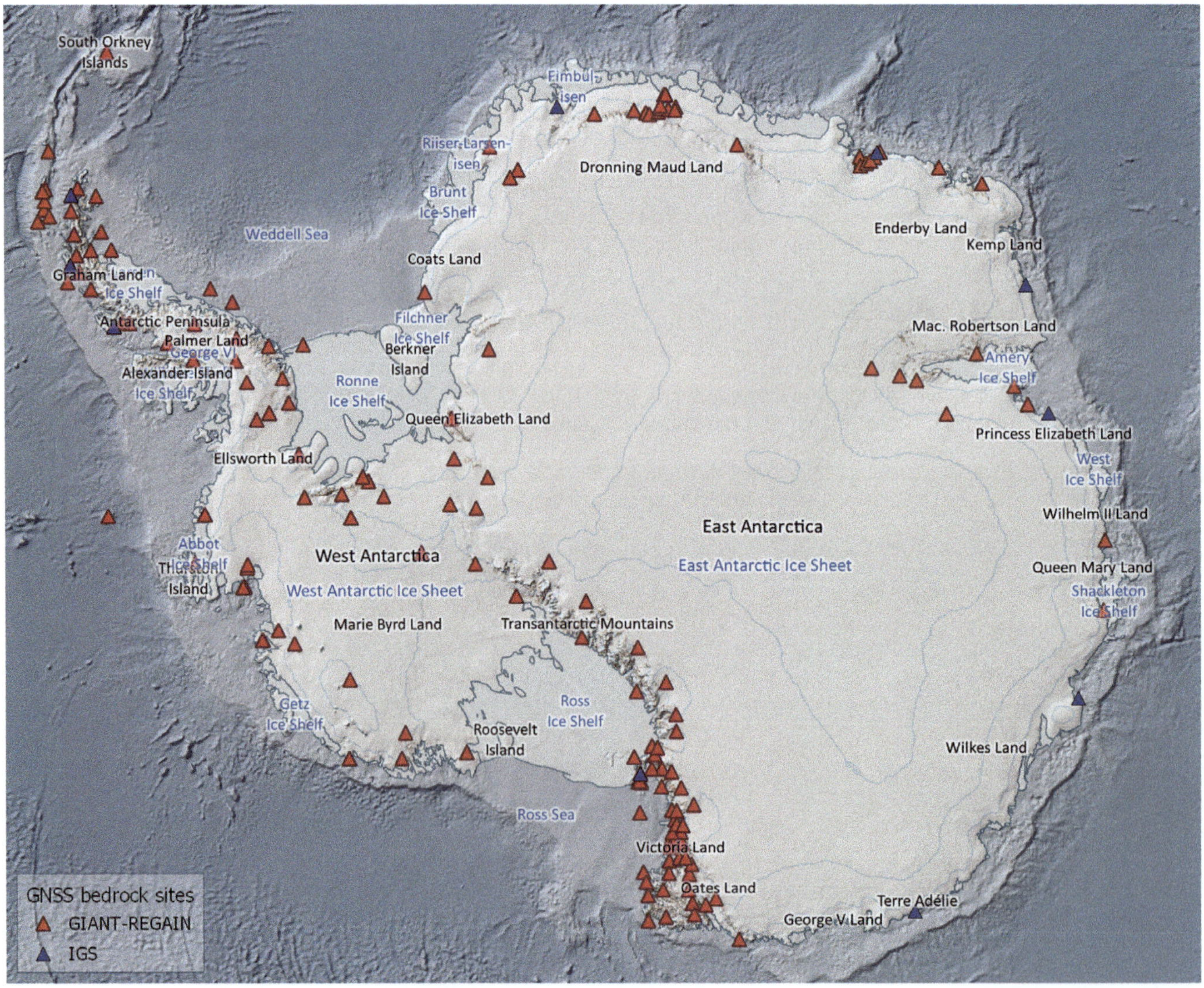

Fig. 5. Distribution of permanent and episodic GNSS sites in Antarctica used in the GIANT-REGAIN reprocessing (red triangles). Blue triangles denote IGS sites.

antennas, and tropospheric and ionospheric signal propagation delay. Geodynamic phenomena are likewise corrected for using models for solid Earth tides, ocean tidal loading and atmospheric loading; see King *et al.* (2010) for a comprehensive overview. In addition, King and Santamaría-Gómez (2016) and Turner *et al.* (2020) emphasize the problem of the separability of the (horizontal) GIA and plate motion from GNSS data. Therefore, GNSS-inferred horizontal velocities have to be used with caution when constraining GIA models in Antarctica.

A particular problem for GNSS in regions of high latitude can be the accumulation of snow on the antenna, which leads to perceived subsidence. In Antarctica, snow was found inside GNSS antennas equipped with radoms (Konfal *et al.* 2016), and the action of blocking the entry of snow led to substantial changes in position time series (Wilson *et al.* 2019). The effect was investigated by Koulali and Clarke (2020), who found that the accumulation of snow depends on local weather conditions and was greatest for cold regions with high wind speed. They showed a substantial effect of snow accumulation on annual signal, but the effect on the estimated trend is not yet known. However, this effect together with the effects of GNSS equipment change or further effects (such as earthquakes) that may prevent an undisturbed time series being obtained have to be taken into account in the estimation of the coordinate velocities. One proven strategy is to mark such events and to estimate the respective velocity only for distinct time periods (where the assumption of linearity is valid). In that respect it has to be emphasized that metadata (such as information on antenna height) are essential. However, often metadata are not, or only incompletely, accessible, which is a further challenge for reliable GNSS analysis.

The recent realization of the (International) Terrestrial Reference Frame (ITRF) is given by ITRF2014 (Altamimi *et al.* 2016), which is based on four basic geodetic space techniques (VLBI (Very Long Baseline Interferometry: see https://vlbi.org), SLR (satellite laser ranging; see https://ilrs.gsfc.nasa.gov), DORIS (Doppler Orbitography and Radiopositioning Integrated by Satellite, see https://ids-doris.org/) and GNSS). If GNSS coordinates and coordinate changes are linked to ITRF2014, they refer to its origin, which is defined as the centre of mass of the entire Earth system (CM). This is true in a long-time sense (or at a secular timescale) since the ITRF2014 origin is tied to the average CM as realized by SLR data (Altamimi *et al.* 2016). Some GIA models refer to CM as well (Spada 2017; Caron *et al.* 2018). However, more often GIA models refer to the centre of mass of the solid Earth (CE) (King *et al.* 2010; Whitehouse 2018). CE experiences a change of its trajectory in the inertial space, while CM is stationary with respect to the inertial system and is, therefore, accessible by orbiting satellites (Blewitt 2003). The effect of the relative velocity of CM with respect to CE on uplift rates has been assessed to be in the range of 0.1 (Thomas *et al.* 2011) to 0.5 mm a^{-1} (Argus *et al.* 2014). Apart from differences in origin definitions used by models and geodetic solutions, different geodetic solutions may refer to different realizations of the CM origin (see the ‘Results on mantle-related processes from geodetic data combination’ section).

The IGS provides a GNSS-only solution for the ITRF based on the reprocessing of global GNSS reference sites back to 1994, together with orbit ephemerides and further data (Rebischung *et al.* 2016), named the IGS14 reference frame, and which has recently been updated to IGb14. The position accuracy of these GNSS reference sites is on average ±1.5 mm for the horizontal components and ±4 mm for the vertical component (Rebischung *et al.* 2016). The rate of reference frame origin could be determined with an uncertainty of ±0.3 mm a^{-1} from the scatter of the contributing analysis centres (Rebischung *et al.* 2016). This level of uncertainty was confirmed by Riddell *et al.* (2017) when investigating weekly translations of the SLR solutions with respect to the ITRF2014. Adopting a combined power law and white noise model, Riddell *et al.* (2017) concluded that the uncertainty in the rates of SLR X, Y and Z translations is ±0.13, ±0.17 and ±0.33 mm a^{-1}, respectively. This represents an improvement of nearly 30% for the Z component in comparison to the results that Argus (2012) obtained for the previous ITRF2008 solution. Hence, the (long-term) stability of the origin realization was considerably improved in the ITRF2014 solution (Altamimi *et al.* 2016), which provides a decisive precondition for the analysis of GNSS data that may now cover more than 25 years.

To infer coordinate change rates (velocities) for permanent observations methods of time-series analysis can be applied to the daily GNSS position solutions. This allows researchers to adjust simultaneously for different signal parts and noise models. The flexibility of the stochastic modelling is much lower in the case of epoch solutions. The time-series analysis often consists of fitting the coordinate time series with a trajectory model composed of a linear trend, annual and semi-annual signal parts. Thereby, a suitable noise model has to be adopted that usually includes white and coloured (flicker) noise (Williams 2003). Respective software solutions incorporate these noise models (CATS: Williams 2008; HECTOR: Bos *et al.* 2013) and allow a robust trend determination to be performed, even with gaps and jumps in the time series (MIDAS: Blewitt *et al.* 2016). For most sites, coordinate changes can be determined with a precision of ±0.5 mm a^{-1} and better. In terms of accuracy, horizontal velocities can be determined at ±1 mm a^{-1}, while the accuracy of vertical velocities is worse by a factor of 2: that is, at the level of ±2 mm a^{-1} (King *et al.* 2010; Thomas *et al.* 2011; Sasgen *et al.* 2018). The precision may improve with the length of the coordinate time series (King *et al.* 2010). A common mode error (CME) analysis may further improve the precision of the coordinate velocity time series, especially when inferred by PPP, as was shown for the vertical component by Liu *et al.* (2018). CME considers the spatial correlation caused by unmodelled geophysical effects such as atmospheric or non-tidal loading and remaining systematic errors. To investigate the non-Gaussian features of the data, Liu *et al.* (2018) adopted the independent component analysis (ICA), an advancement of the principal component analysis (PCA), taking higher-order non-Gaussian statistics into account. Estimates based on ICA are found to show a stronger correlation with geophysical effects than PCA. After applying spatiotemporal filtering, the mean RMS of the GNSS-observed vertical velocities could be reduced by about 40%, indicating the gain of precision. Likewise, the agreement with predictions of four GIA models could be improved by a mean reduction of 0.9 mm a^{-1} in terms of weighted RMS (Liu *et al.* 2018).

For epoch observations (and, to a certain extent, also for permanent solutions) usually a further uncertainty of 5–10 mm is considered in addition to the precision of daily position solutions to come up with suitable accuracy measures. In this way, residual effects are accounted for: for example, those caused by differences in the site set-up and by uncertainties in the realization of the geodetic datum, especially in the determination of the origin (e.g. Argus *et al.* 2014). The accuracy measures for the resulting trends are inferred by variance propagation (depending on the number of observation epochs and the length of the time span), leading to an accuracy typically of about 4–8 mm a^{-1} for vertical velocities (Rülke *et al.* 2015). However, it could be shown that after a time span of at least 7 years the inferred (linear) rates have similar accuracies, as in the case of continuous observations (Rülke 2009).

Correction for the elastic effect

The GNSS-inferred coordinate change rates reflect the deformation due to instantaneous loading effects and processes governed by the mantle, such as GIA and post-seismic deformation. In order to separate out the mantle contribution, the GNSS-inferred coordinate change rates are commonly corrected for the elastic effect. This starts with the assumption that present-day ice-mass change exerts a load that is immediate or, at least, of comparably short time. Furthermore, it is assumed that at these timescales the response is governed by the elastic properties of the solid Earth. The present-day ice-mass change, for example, is estimated from ice-height observations of satellite altimetry (e.g. ERS-1/2, Envisat ICESat and CryoSat-2: cf. Schröder *et al.* 2019*a*, depending on the time span of the GNSS observations) together with a reasonable assumption on the pattern of firn and ice density (e.g. Groh *et al.* 2012) or applying the mass-budget method based on climate modelling and radar measurements of ice flux across grounding lines (e.g. King *et al.* 2010). In general, the calculation of this elastic effect follows Farrell (1972), where load Love numbers and, subsequently, Green's functions are calculated for a standard 1D Earth model like PREM (preliminary reference Earth model) (Dziewonski and Anderson 1981). Then, the computation can be done in the spectral (i.e. spherical harmonic) domain, while also taking the regional sea-level response into account (e.g. Mitrovica *et al.* 2001). High-resolution computations are necessary, as the measured uplift rates can be sensitive to loading changes from individual glaciers (Barletta *et al.* 2018).

The elastic response might also be sensitive to inhomogeneities in the structure of the lithospheric crust, which are not taken into account in the standard Love number formulation in the 1D Earth model. This sensitivity is pronounced when more localized load signals occur (Dill *et al.* 2015), which is the case for regions of considerable ice-mass loss such as the large outlet glaciers in Amundsen Sea Embayment, West Antarctica, but also for glacial streams in East Antarctica like Jutulstraumen (Dronning Maud Land) or Lambert Glacier, Totten Glacier or Moscow University Glacier (Wilkes Land). Changes in the lithospheric structure might account for variations of up to 25% of the vertical elastic deformation (Dill *et al.* 2015). An open problem is how the elastic and rheological properties can be derived for Antarctica and how they should be taken into account in a consistent way both in the modelling of the instantaneous elastic effect and in the GIA effect. This problem is being approached by further contributions in this Memoir (Ivins *et al.* 2021, this volume; Wiens *et al.* 2021, this volume).

In principle, elastic parameters derived from seismic wave speeds (e.g. Young's modulus and density) are sufficient to calculate elastic effects in surface-loading methods and can be checked by microphysical measurements of lithospheric rocks. Information on elastic properties is usually derived from seismic velocity anomalies. An *et al.* (2015*a*) used about 120 broadband seismographs for calculating a 3D S-velocity model of the Antarctic lithosphere and depth estimates of the Mohorovičić (Moho) discontinuity. From the average ratio between lithospheric mantle and crustal densities, they concluded that the lithospheric mantle is of Archean age. This has also been inferred by isotopic measurements of lithospheric mantle xenoliths (Handler *et al.* 2021, this volume; Martin *et al.* 2021, this volume). The thickest crust is located along the East Antarctic subglacial mountain range (*c.* 60 km in thickness close to Dome A: An *et al.* 2015*a*, figs 1 & 6). Furthermore, in a second paper, An *et al.* (2015*b*) determined the depth of the lithosphere–asthenosphere boundary (LAB). They found East Antarctica to have a 'thick lithosphere similar to other stable cratons' (up to 250 km between Dome A and C) (An *et al.* 2015*b*, p. 8720), while West Antarctic comprises thin crust and lithosphere 'similar to … modern subduction-related rift systems' (An *et al.* 2015*b*, p. 8720). These findings were partially confirmed by Lloyd *et al.* (2020), but were also readjusted to come up with a more detailed picture. In a comprehensive study, Lloyd *et al.* (2020) used adjoint tomography along with data of more than 300 broadband seismometers in Antarctica to image the structure beneath Antarctica, especially the upper mantle and transition zone. In East Antarctica, they found fast shear-wave speed anomalies that mirror 'cold thick lithospheric roots that are characteristic of cratons and Proterozoic fold belts' (Lloyd *et al.* 2020, p. 20). However, thicker lithosphere is suggested only for the interior of East Antarctica, while thinner lithosphere is inferred along the coast, and

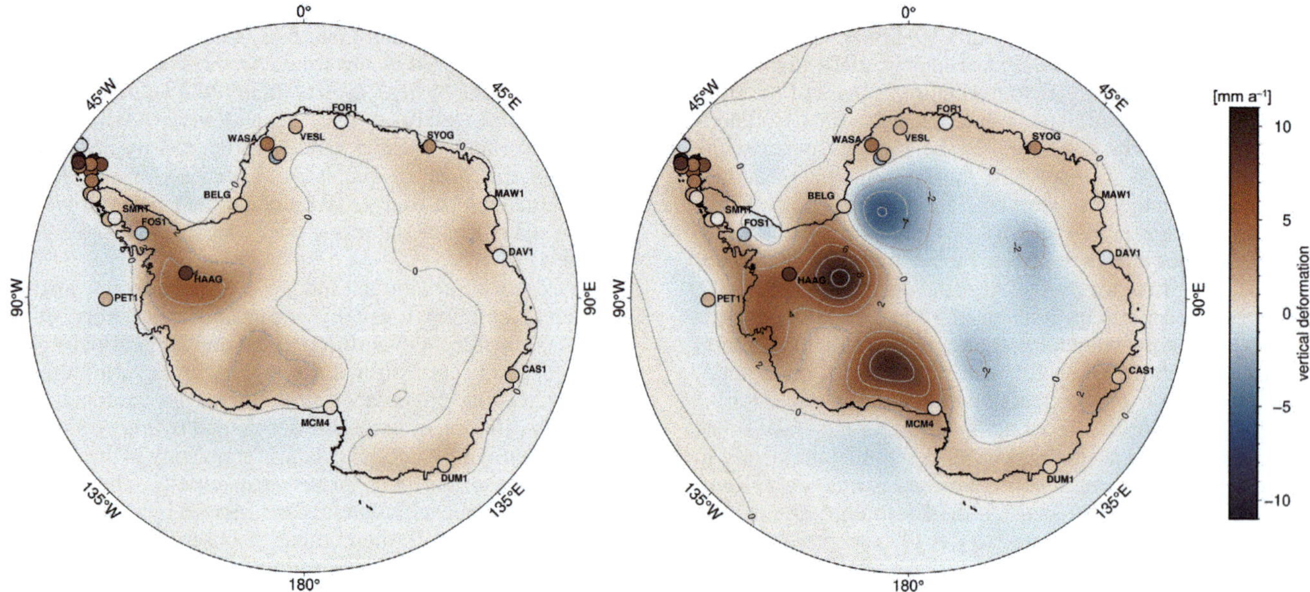

Fig. 6. Vertical deformation rates inferred for selected permanent and campaign GNSS sites in Antarctica (according to Rülke *et al.* 2015, table 1). The GNSS results were corrected for the present-day elastic uplift. In the background, vertical deformation rates are plotted as predicted by respective GIA models: (left) IJ05_R2 (Ivins *et al.* 2013); (right) W12a (Whitehouse *et al.* 2012).

along the sector from western Dronning Maud Land to Enderby and Kemp Land and the Lambert Graben in particular there is a 'much greater variability in lithospheric thickness' (Lloyd *et al.* 2020, p. 20) that may extend for several hundred kilometres inland.

However, many results and hypotheses are still inconsistent or partly contradictory (see Pappa and Ebbing 2021, this volume; Wiens *et al.* 2021, this volume). Gravity data take a more prominent role to estimate crustal thickness and LAB because of the sparse seismic data coverage (Pappa and Ebbing 2021, this volume). Further details on these issues can be found in this Memoir (e.g. Martin *et al.* 2021; Ivins *et al.* 2021; Wiens *et al.* 2021, all this volume). It should be noted that the different methods do not yield the same lithosphere or effective elastic thickness. In seismic studies the LAB is often based on a temperature boundary, whereas for GIA it is based on a transition from elastic to viscous behaviour (e.g. see Nield *et al.* 2018). Steinberger and Becker (2018) concluded that compositional variations in the mantle occur mainly in the upper 150 km. Their model yields lithospheric thickness of up to 250 km for cratons. Effective elastic lithosphere thickness is 'typically about a factor two lower' (Steinberger and Becker 2018, p. 337).

Although it has been argued for a long time that the Antarctic continent is composed of two distinct and contrasting geological provinces (e.g. Adie 1962), it is timely to step back from a uniform picture of one 'East Antarctic unit' (Mieth and Jokat 2014; Jacobs *et al.* 2015; Foley *et al.* 2021, this volume), which is also reflected by the findings of Lloyd *et al.* (2020) (see above).

A further aspect not fully explored yet is the interlinkage between the timescale of the load application and the time-dependent rheological properties of the Earth that govern the deformation response. This is reflected in the following discussion: on the one hand, even at such short timescales as those of semi-diurnal and diurnal Earth tides, effects such as anelasticity are significant and, thus, lead to a deviation from a pure elastic response. Bos *et al.* (2015) explained the discrepancy between GNSS-observed M_2 tidal loading signal and its prediction using ocean tide loading models based on the isotropic PREM (Dziewonski and Anderson 1981) by accounting for the elastic properties of the upper mantle and an anelastic dispersion in the asthenosphere. Bos *et al.* (2015) assumed that dissipation, which is responsible for a reduction in shear modulus, was taking place in an absorption band from seismic up to tidal wavelengths. Ivins *et al.* (2020) emphasized that in the transition from seismic to tidal and even longer wavelengths (up to a typical Maxwell time of *c.* 100 years) deformation has to be explained by taking anelasticity into account and the transition to transient and finally steady-state creep. The latter statement applies, on the other hand, to the problem of timescales for which 'present-day' ice-mass change can be regarded in the elastic framework only. Geodetic GNSS observations in Antarctica cover time periods of more than 25 years (see above), long enough to see the viscous effects due to loading (Nield *et al.* 2014). Ivins *et al.* (2020) discussed how crustal motion induced by glacial loading and unloading on decadal timescales is governed by mantle properties and has to be explained by viscoelastic response.

Thus, it becomes clear that geodetic GNSS observations provide a powerful tool to test rheological properties of the mantle and timescale dependency of the deformation. However, the challenge is to identify those locations on Earth that can serve as a suitable environment for natural experiments. All available processes should be explored such as ocean tidal loading (Bos *et al.* 2015) and body tide deformation (Kang *et al.* 2015) or immediate or catastrophic load changes (e.g. Marderwald *et al.* 2020). One may also focus on regions with fast-retreating glaciers or accelerated mass loss where the elastic response (in the 'classical sense', as described above) makes up more than half of the total observed deformation, as in the case of southern Patagonia (Lange *et al.* 2014) or (parts of) the Greenland ice sheet (Bevis *et al.* 2012, 2019; Nielsen *et al.* 2013). The insights from such experiments have to be transferred to the case of Antarctica in order to investigate their capability to separate and better explain the different effects and timescales of ice-load-driven solid Earth deformation.

Results of geodetic GNSS observations: deformation of the Earth's crust due to GIA

In the framework of the SCAR Epoch Crustal Movement Campaigns Antarctic-wide consistent reanalyses of all data together with IGS data were performed in 2008 and 2015 (Dietrich and Rülke 2008; Rülke *et al.* 2008, 2015). The GNSS-inferred coordinate change rates were corrected for the immediate elastic effect (see the discussion above), and the residual GNSS rates were taken as a ground-truth observation of GIA. However, the residual vertical uplift rates in particular exhibit large differences to the modelled GIA rates according to recent GIA models (Whitehouse *et al.* 2012; Ivins *et al.* 2013) (see Fig. 6). These findings are generally confirmed by investigations by King *et al.* (2010), Argus *et al.* (2011) and Thomas *et al.* (2011). Even if GNSS-inferred deformation rates (especially vertical rates) were used to constrain the modelling of GIA in Antarctica, as in the cases of IJ05_R2 (Ivins *et al.* 2013), ICE6G_C (VM5a) (Argus *et al.* 2014), W12a (Whitehouse *et al.* 2012) and Caron *et al.* (2018), the scatter between the models and the misfit of modelled rates to GNSS-inferred rates are still substantial. Data–model misfit reaches more than 10 mm a^{-1} in the Amundsen Sea Embayment and several millimetres per year in East Antarctica (see also Martín-Español *et al.* 2016*a*), which are comparable to or larger than errors in GNSS-derived trends of 3–4 mm a^{-1} in the Amundsen Sea sector (Martín-Español *et al.* 2016*b*).

Several regional investigations should be highlighted. Bevis *et al.* (2009) presented the first results of the West Antarctic GPS Network (WAGN), which consisted of campaign-style and partly continuous observations. Later, all WAGN sites were upgraded to continuously recording GNSS to form part of the POLENET/ANET project. Bevis *et al.* (2009) found that the GIA models used for comparison over-predicted vertical uplift rates. Zanutta *et al.* (2017, 2018) presented results of a long-term study in North Victoria Land where relatively low uplift rates confirm GIA model predictions but are also subject to tectonic effects. However, the effect of two main tectonic lineaments can only be seen in residual horizontal velocities.

Relatively large uplift rates of more than 10 mm a^{-1} were measured in the Antarctic Peninsula region. Nield *et al.* (2012, 2014) used GNSS data from stations that had mostly been installed since 2009. They carried out viscoelastic modelling to explain the non-linearity in the GNSS time series in the Antarctic Peninsula, and found a combination of elastic and viscoelastic responses fitted the observations. For this, the upper-mantle viscosity had to be constrained to relatively low values (of the order of 10^{18} Pa s). Low viscosity is one of the main findings of the GNSS measurements. However, several questions remain open relating, for example, to the effective elastic thickness of the lithosphere and the range of upper-mantle viscosities (see above). Nield *et al.* (2014) even questioned the validity of (linear) Maxwell rheology (which has been widely used as a standard rheology in GIA modelling). Thus, Nield *et al.* (2018) investigated the impact of a laterally varying elastic lithospheric thickness in contrast to power-law rheology allowing a viscous lithosphere. Here, the upper-mantle viscosity was not as low as applied by Nield *et al.* (2014). The latter model leads to larger uplift amplitudes and shorter wavelengths, which may be important in better capturing GNSS-inferred deformation that exhibits steeper gradients such as in West Antarctica. Wolstencroft *et al.* (2015) investigated GNSS observations to constrain GIA modelling in the southern Antarctic Peninsula. They localized the largest misfit in the SW Weddell Sea.

The largest uplift rates in Antarctica were measured in the Amundsen Sea sector at the continuously recording ANET sites (Barletta *et al.* 2018), as well as at GNSS epoch sites of TU Dresden (Groh *et al.* 2012, 2014). Including observations of the latest field campaign in early 2017, the TU Dresden group found uplift rates, corrected for the instantaneous elastic effect, of several centimetres per year up to about 4.5 cm a^{-1} (Busch *et al.* 2017). These results are largely underpredicted by the present GIA models. In their regional GIA model refinement, Barletta *et al.* (2018) concluded on a weak shallow upper mantle (from lithospheric base down to 200 km depth) with a viscosity of 4×10^{18} Pa s, a stiffer and deeper upper mantle (200–400 km) of 1.6×10^{19} Pa s, and a transition zone (400–670 km) of 2.5×10^{19} Pa s. To overcome limitations in the Maxwell model, the GNSS-inferred velocities in the Amundsen Sea sector could potentially form useful constraints on viscosity estimates based on seismic anomalies and creep experiments (Ivins *et al.* this volume; Ivins *et al.* 2020, 2021, this volume). Ivins *et al.* (2020) proposed to utilize the extended Burgers model (EBM) in GIA modelling. The EBM might explain rapid uplift even on short timescales such as decades, and enhances GIA uplift predicted by the (classical) Maxwell model by a factor of up to 2.5 (Ivins *et al.* 2020). This attention to shorter timescales has brought to the forefront the largest uncertainty in the modelling, namely the unknown ice history over the last centuries. This problem plays a smaller role in East Antarctica, where viscosity is higher and recent loading does not affect present-day uplift rates as many rates observed in East Antarctica are generally smaller (a few millimetres per year) showing small uplift or even subsidence near the coast. The misfit to GIA model predictions, however, can still be of the same order (Rülke *et al.* 2015) (Fig. 6) at some sites near the coast.

As already stated above (Nield *et al.* 2014), one may investigate non-linear relations between stress and strain rates in contrast to the treatment of the mantle as a Newtonian fluid. This leads to power-law creep equations, where the strain rate varies with time even under constant stress. More important is the fact that mantle viscosity varies with stress (or strain rate). Non-Newtonian flow behaviour has been found to be common in silicate polycrystals even at low stresses but high temperature (Ranalli 1995). However, it is difficult to use GNSS observations to constrain a more realistic (i.e. 3D) mantle rheology together with an ice-history model not starting from an *a priori* (1D) mantle viscosity profile (van der Wal *et al.* 2013). Taking into account the results of laboratory experiments on olivine, the most prominent constituent of the mantle, van der Wal *et al.* (2013) inferred diffusion and dislocation creep flow laws where the latter leads to non-Newtonian flow. However, they stated a difficulty in fitting GNSS uplift rates and sea-level data simultaneously, which may be due to uncertainties in the inferred flow law or not considering mantle minerals other than olivine (van der Wal *et al.* 2013). van der Wal *et al.* (2015) extended this study to the case of Antarctica, acknowledging that seismic studies inferred large viscosity variations in the Antarctic mantle. As applied elsewhere (e.g. Fennoscandia), a finite-element model was extended to also include power-law creep. Global GIA models (ICE-5G: Peltier 2004; W12a: Whitehouse *et al.* 2012) were 'tuned to fit constraints in the northern hemisphere' (van der Wal *et al.* 2015, p. 134) and, thereafter, compared to GNSS uplift rates taken from Argus *et al.* (2014). It could be shown that the models incorporating a 3D Earth structure and composite rheology give smaller misfits between observed and predicted uplift rates, both for W12a and for ICE-5G. Even so, introducing GNSS uplift rates as the only constraints cannot explain to what extent larger variations in the Earth structure beneath Antarctica exist that would certainly increase the influence of 3D rheology (van der Wal *et al.* 2015). However, the impact of modelling 3D rheology leads to variations in mass-balance estimates of the Antarctic ice sheet that are larger than the measurement error.

Thus, improved results can be expected when longer GNSS observation sets become available. The interior of East Antarctica lacks bedrock for realizing GNSS observations; in this case, a combination of measurement techniques may help to constrain GIA models, as will be detailed in the 'Results on mantle-related processes from geodetic data combination' section later in this chapter.

Temporal gravity field variations from satellite gravimetry

Temporal gravity can be measured with terrestrial instruments or with satellites. Although temporal variations have been measured from orbit variations for decades (e.g. Cheng *et al.* 1989), the accurate inter-satellite ranging measurements of the GRACE mission led to a breakthrough in observing mass changes. The GRACE mission is based on inter-satellite microwave ranging, whereby the distance between two satellites that follow each other is measured precisely. The mission operated between 2002 and 2017, and provides a monthly snapshot of the Earth's gravity field (e.g. Tapley *et al.* 2004; Wouters *et al.* 2014). A year after the end of this mission the GRACE-FO mission was launched. The mission aimed for a continuation of the time series, but also carries a novel laser instrument for inter-satellite ranging that has a higher accuracy (Koch *et al.* 2018). The first results showed that the accuracy of the GRACE-FO data was comparable to that of the GRACE data (Landerer *et al.* 2020). However, an accelerometer on one of the satellites had to be switched off, which means that some of the non-gravitational forces cannot be measured accurately. By transferring the accelerations from the leading satellite to the follower, the non-gravitational forces were mitigated. Currently, there is no indication of a bias between the time series of the two missions (Landerer *et al.* 2020), and a time series of mass-balance estimates in Antarctica between both missions appears to be consistent (Velicogna *et al.* 2020).

Three official GRACE science centres currently provide monthly spherical harmonic solutions, so-called level-2 data, with two versions being offered up to maximum spherical harmonic degree and order 60 and degree 96. The spatial resolutions corresponding to a certain maximum spherical harmonic degree L can be approximated by $\pi R_E/L$ where R_E is the mean radius of the Earth. The resolution is limited by the fact that the GRACE and GRACE-FO satellites fly at an altitude of around 500 km, which dampens the gravity signal compared to airborne or terrestrial sensors. The resolution also depends on measurement errors and errors of the applied correction models, which are briefly addressed below. As a result, with the GRACE/GRACE-FO time series one can identify mass-change processes that vary on a timescale of 30 days or more with a spatial resolution of approximately 200–300 km.

Spherical harmonic coefficients of the gravity field form the standard gravity-field products (level-2 data). The satellite gravimetry science product line is set-up in such a way that a user can always undo corrections to go back to an earlier state in the product chain. Important corrections include so-called de-aliasing models for removing temporal gravity signals that originate from non-tidal atmospheric and oceanic mass variations with timescales shorter than the 30 days production scheme. The de-aliasing products are based on a numerical weather prediction model and ocean bottom pressure output from a global ocean circulation model (Dobslaw *et al.* 2017). Uncertainties in the de-aliasing models still contribute a significant part of the GRACE measurement error (Kvas and Mayer-Gürr 2019). GRACE/GRACE FO measurements of very-long-wavelength signals are less accurate. Therefore, the C_{20} coefficient is usually replaced by a solution from satellite laser ranging (Dahle 2020). The same holds for the C_{30} coefficient, which also influences the Antarctic mass balance (Loomis *et al.* 2019; Velicogna *et al.* 2020).

The error in the GRACE data comprises sensor errors and errors of the correction models. Because of the sampling of the GRACE mission along a polar ground track, these form a characteristic pattern of north–south-orientated bands or stripes (Swenson and Wahr 2006). Because of undersampling of the high-frequency geophysical signal, errors in the correction models end up in the Earth's flattening term (C_{20}) and other low degree and order terms, but also spherical harmonic order around 15 and multiples thereof. This finally results in the equivalent water-height product being affected by aliasing with periods of 100 days up to several years (Seo *et al.* 2008). Part of the north–south-orientated errors can also be understood as the interaction of the shifting ground-track pattern with the low-frequency quasi-periodic gravitational signal (Peidou and Pagiatakis 2020).

As a result of this complex interplay of measurement errors, correction models and sampling, it has proven difficult to provide an accurate statistical description of the errors. The north–south error signal is a result of correlation in the data, but off-diagonal terms in the full covariance matrices did not provide corresponding information (Wahr *et al.* 2006). Therefore, most studies used the standard deviation of least-squares residuals as a proxy for the error (e.g. Chen *et al.* 2006; Velicogna and Wahr 2006). However, autocorrelation can result in an underestimation of the error by a factor of 2 or more (Horwath and Dietrich 2009; Williams *et al.* 2014). Because of the errors mentioned above, filtering is generally applied. Filters follow different approaches: (i) smoothing the short-wavelength with weights derived from a Gaussian function (e.g. Velicogna and Wahr 2006); (ii) applying an anisotropic filter such as by Kusche *et al.* (2009); (iii) applying statistical tests to spherical harmonic coefficients (Sasgen *et al.* 2007); or (iv) separating signals by empirical orthogonal function analysis (King *et al.* 2012). Because the exact noise properties are unknown, it is difficult to optimize the filter settings. However, the GIA signal is mostly linear across the measurement period, which means that the strength of the signal increases in an estimation of the trend. Less additional filtering is necessary as a result, with recent studies using a 250 km half-width of the filter (Velicogna *et al.* 2020), which results in higher-resolution mass-change estimates (Velicogna *et al.* 2020), whereas earlier results used a half-width of 800 km (Chen *et al.* 2006). This also reflects improvements in the background model corrections that reduce the noise at smaller spatial wavelengths.

An improvement in resolution was also demonstrated by using data from the GOCE (Gravity field and steady-state Ocean Circulation Explorer) satellite mission. This satellite carried a gradiometer that accurately measures the accelerations between proof masses levitating inside the satellite. The mission was designed for accurate measurement of the static gravity field. Its measurements were not very sensitive to gravity changes in the spatial scale of the ice-mass change. Combined with a relatively short lifetime of 3 years, GOCE was not expected to contribute to time-variable gravity estimates. However, a combination of GOCE and GRACE improved the spatial resolution of the mass changes in the Amundsen Sea sector (Bouman *et al.* 2014). Further improvement in resolution is possible by terrestrial gravity measurements. Terrestrial measurements of time-variable gravity can observe solid-Earth processes such as volcano inflation, solid Earth tides, slow slip events, GIA or co- and post-seismic signals (Van Camp *et al.* 2017). Instruments can be divided into absolute gravimeters that measure the absolute value of gravity and relative gravimeters that measure gravity differences with respect to a reference point or acquire time series. In Antarctica, absolute measurements were realized at several sites (Mäkinen *et al.* 2007). The measurements are sensitive to the uplift resulting from GIA, but the analysis of 5 years of data acquired at Syowa showed that the GIA signal is still obscured by changes in atmospheric mass and snow changes (Aoyama *et al.* 2016). Therefore, observations of solid Earth mass changes in Antarctica mostly rely on data from the GRACE and GRACE-FO mission.

Spherical harmonic coefficient sets provided by the GRACE data-processing centres can be synthesized to produce gravity changes on a map, or can be transformed to changes in equivalent water-layer thickness, which assumes that all gravity change originates from a layer of water at the Earth's surface. Water-layer-thickness changes are also provided by GRACE processing centres and other institutes as so-called level-3 products. Mascons provide an alternative to spherical harmonic solutions, which are based on regional mass concentrations obtained from a direct inversion of the inter-satellite distances for predefined regions on the Earth (Loomis *et al.* 2019). The mascon technique can also be applied to the spherical harmonic coefficients. This is documented, for instance, by Schrama *et al.* (2014), where it is used to obtain ice-mass-change estimates for basins in Antarctica. Leakage forms one of the potential error sources in surface-mass-change estimates, based either on synthesized equivalent water-layer grids or on mascons. Leakage refers to errors in the spatial attribution of mass changes associated with observed gravity field changes (Swenson and Wahr 2002; Horwath and Dietrich 2009). Different strategies for mitigating or correcting leakage effects have been developed based on *a priori* information on the signal location, patterns or covariance (e.g. King *et al.* 2012; Landerer and Swenson 2012; Harig and Simons 2015). Mascon approaches particularly lend themselves to incorporating such information (Wiese *et al.* 2016; Loomis *et al.* 2019).

Most of the global signals in GRACE level-2 data originate from changes in water storage. However, solid Earth signals are also visible, the main ones being GIA (e.g. Tamisiea *et al.* 2007), co-seismic signals and post-seismic signals (e.g. Chen *et al.* 2007; Han *et al.* 2016). Time-variable gravity from other mantle-flow processes is currently at the edge of detection of the GRACE mission (Ghelichkhan *et al.* 2018). In large parts of Antarctica, the increasing gravity signal is dominantly due to GIA. Post-seismic relaxation could potentially form a significant signal if it is also detected in GNSS data (King and Santamaría-Gómez 2016; Barletta *et al.* 2022, this volume). Ice-mass changes form a loading effect to which the solid Earth responds, but this can be taken into account by elastic effect only as described in the 'Geodetic GNSS measurements to constrain the modelling of ice-induced Earth deformation' section. Because the gravitational effect due to deformation is much smaller than the mass change itself, the elastic effect is also smaller than in GNSS uplift rates. It can be satisfactorily modelled by using Green functions based on a reference Earth model (see the section on 'Geodetic GNSS measurements to constrain the modelling of ice-induced Earth deformation').

The difficulty in interpreting gravity data is that it also contains other signals, so that corrections are required to be able to discern the signals of interest, even after the de-aliasing models. In Antarctica, the main signal of interest is the ice-mass loss in West Antarctica and parts of East Antarctica. Therefore, the solid Earth signal due to GIA is corrected for by subtracting gravity changes predicted from numerical models from the GRACE data (e.g. Velicogna and Wahr 2006; Ivins *et al.* 2013; Sasgen *et al.* 2013). An ensemble of forward GIA models was used to quantify uncertainty yielding error bars larger than the signal (Barletta *et al.* 2008). More realistic estimates can be obtained by using constraints on the GIA process in Antarctica such as ice margins and sea-level data in a Bayesian approach (Caron *et al.* 2018). This yields standard deviations of 10 mm a^{-1} water equivalent in large parts of West Antarctica and coastal areas in East Antarctica, compared to a maximum signal in West Antarctica of 160 mm a^{-1} (Ivins *et al.* 2013). It is important to realize that both GIA and satellite gravimetry are global in nature, and GIA outside Antarctica contributes to the mass changes signal within Antarctica. This effect is estimated to be between 17 and 42 Gt a^{-1} (Caron and Ivins 2020), which is significant compared to 160 Gt a^{-1} attributed to ice-mass loss over the period 1992–2017 (Shepherd *et al.* 2018).

Another approach to address signal separation is to combine GRACE data with complementary data, as will be discussed in the next section. Furthermore, a separation in the time domain is possible. The GIA signal is linear over the timescale of the GRACE mission; therefore, the acceleration signal will not be affected by the GIA (Velicogna 2009). An exception may be seen in areas where the viscosity is so low that the GIA process can contribute to an acceleration over a short time period (Barletta *et al.* 2022, this volume).

The Ice Sheet Mass Balance Inter-comparison Exercise (IMBIE) is aimed at summarizing the ice-sheet mass balance for Greenland and Antarctica (Shepherd *et al.* 2018, 2020). In both papers the GIA signal is reduced from observations using GIA models. As noted before, the uncertainty in the GIA correction is significant, and depends on the knowledge of past ice thickness and rheology. The largest uncertainty is due to the unknown ice thickness. This is reflected in the results presented by, for instance, Schrama *et al.* (2014), who discussed mass changes for the Antarctic ice sheet between 2003 and 2013 based on GIA models adopting different ice histories. They found the Antarctic ice-mass balance using GIA models (at this time new) IJ05_R2 (Ivins *et al.* 2013) and W12a (Whitehouse *et al.* 2012) to be half that inferred using the ICE-5G-based models (Peltier 2004).

Figure 7 (see also Shepherd *et al.* 2018) shows the mass change of the Antarctic ice sheet including subdomains based on a number of techniques including spaceborne gravimetry. The gravimetric estimates are in turn based on a variety of contributions to IMBIE where different processing approaches were applied based on the options mentioned above. It was found that the total melt of land ice between 1992 and 2017 contributed 7.6 ± 3.9 mm to sea-level rise. The GIA correction in East Antarctica suffers from a poor availability of constraints. In other regions, the high uplift rates observed by GNSS are not captured in the GIA models used in the IMBIE study, as these did not include ice-thickness changes more recent than the last few hundred years. An improvement on the latter aspect is the use of GIA models, which represent the regional properties of the solid Earth (e.g. Barletta *et al.* 2018). However, information on last millennial ice-sheet thickness necessary for such modelling is not available continent-wide. Models that consider variations in viscosity already exist (Geruo A. *et al.* 2013; van der Wal *et al.* 2015; Gomez *et al.* 2018), but the uncertainty in viscosity derived from seismic information is still large (Ivins *et al.* 2021, this volume).

Results on mantle-related processes from a combination of geodetic data

Traditionally, GIA is modelled using two key inputs: ice-loading history; and a model of solid Earth properties and structure (Barletta *et al.* 2022, this volume). GIA rates modelled in this way are called forward-modelled GIA. An alternative way to derive GIA is through combining different geodetic observational techniques, such as gravimetry ('Temporal gravity field variations from satellite gravimetry' section), altimetry and/or GNSS displacements (sse the introduction). In Antarctica, the satellite gravity missions GRACE and GRACE-FO observe a superposition of signals mostly originating from two different sources: mass variations within the surface layer; and GIA-induced mass variations ('Temporal gravity field variations from satellite gravimetry' section). The surface layer contains variations of ice and

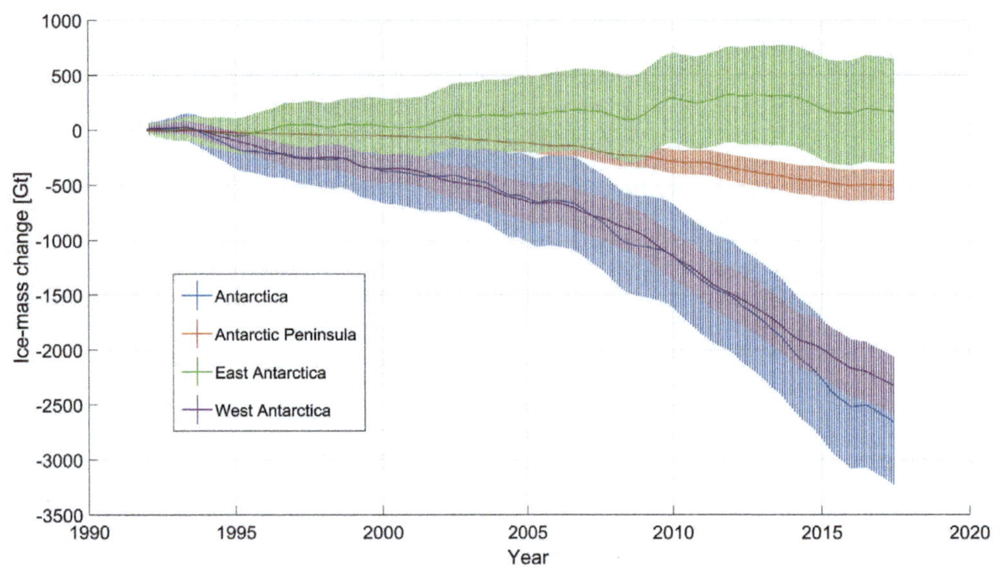

Fig. 7. Cumulative mass change of the Antarctic ice sheet, separated in West and East Antarctica, Antarctic Peninsula and the entire ice sheet, including 1σ error bars (unit: Gigaton). The figure is based on the data published by Shepherd *et al.* (2018) and https://www.imbie.org (last accessed: 8 September 2021), and corresponds to Shepherd *et al.* (2018, fig. 2).

firn, whereby the term 'firn' is used to describe all associated processes such as snowfall, melt and firn compaction (Ligtenberg *et al.* 2011). The variations within the surface layer will usually be and hereafter termed 'ice-mass changes'. GRACE and GRACE-FO have a different sensitivity to ice-mass changes than to GIA-induced mass changes due to the large differences of the corresponding densities ('Geodetic GNSS measurements to constrain the modelling of ice-induced Earth deformation' section).

In contrast to *mass* changes measured by gravimetry missions, altimeters track *height* changes at high spatial resolution. Height changes associated with surface processes are typically much larger than those caused by GIA. Thus, they are more easily detectable by altimeters due to a better signal/noise ratio. However, the altimetry signal contains the height changes related to both GIA and surface processes.

Dedicated ice radar satellite altimetry missions include Envisat (2002–12) or CryoSat-2 (2010–), and the laser altimetry missions ICESat (2003–09) and ICESat-2 (2018–). Radar altimeters provide the longest observational record of these two types. They are not sensitive to clouds (as is the case for laser altimeters) and, therefore, can measure under all atmospheric conditions. However, they have difficulties in regions of large surface slopes (Brenner *et al.* 2007). Furthermore, since the radar signal penetrates snow, an accurate model for surface penetration is required. When converting altimetry height changes into mass changes, both radar and laser altimetry data should be first corrected for firn compaction because the corresponding height changes are not associated with a mass change. In addition, this conversion requires assumptions about the density of snow and ice, which represents the largest source of uncertainty when using altimetry data for deriving mass changes. The processing of input data for a regional inversion of geodetic observations in Antarctica is discussed in detail in Sasgen *et al.* (2018). In the summer of 2020, a so-called CryoSat-2/ICESat-2 (CRYO2ICE) resonance campaign took place to improve the accuracy of ice-sheet elevation time series, in which both satellites are almost simultaneously passing the same areas providing, for the first time, radar and laser measurements of the same geophysical condition (https://earth.esa.int/eogateway/missions/cryosat/cryo2ice).

As discussed in the 'Geodetic GNSS measurements to constrain the modelling of ice-induced Earth deformation' section, GNSS observations represent a valuable constraint on GIA. However, they need to be corrected for elastic deformation (see 'Geodetic GNSS measurements to constrain the modelling of ice-induced Earth deformation'). Thus, for the computation of the elastic signal, an accurate knowledge of ice-mass changes with high spatial and temporal resolution is required. In addition, when using GNSS to constrain GIA, one has to keep in mind the fact that GNSS-inferred site velocities usually refer to CM (see the subsection 'Reference frame, technique-related issues and accuracy issues') and, therefore, contain global GIA. This means that the GNSS signal over the Antarctic ice sheet contains: (i) Antarctic GIA; (ii) Antarctic elastic deformation; and (iii) long-wavelength signal from non-Antarctic sources, which are mostly driven by GIA in the northern hemisphere (Caron and Ivins 2020).

The GIA-derived rates based on observations are labelled 'empirical' to guarantee a clear separation from the forward-modelled GIA rates. Sometimes, empirically derived GIA rates are also called 'inverse' (Martín-Español *et al.* 2016*a*). The methods developed to assess Antarctic GIA can be arranged in four categories:

1. Forward modelling of GIA based on ice-load history and assumed properties and structure of the solid Earth (e.g. Peltier 2004; Ivins and James 2005).
2. Forward modelling of GIA while constraining it using geodetic observations such as GNSS (e.g. Nield *et al.* 2014, 2016; Barletta *et al.* 2018) or GRACE and GPS (Sasgen *et al.* 2013).
3. Estimating GIA empirically using complementary geodetic observations while constraining it using forward models. For this, Schoen *et al.* (2015), Zammit-Mangion *et al.* (2015) and Martín-Español *et al.* (2016*b*) used a hierarchical Bayesian framework. Sasgen *et al.* (2017) extended their approach by including local viscoelastic response functions (Barletta *et al.* 2022, this volume) to account for lateral variations in the Earth structure.
4. Estimating empirical GIA from complementary geodetic observations independent of *a priori* assumptions about the Earth's viscosity structure and ice-loading history (e.g. Riva *et al.* 2009; Gunter *et al.* 2014; Engels *et al.* 2018; Willen *et al.* 2020).

Independence is a very important aspect as it allows forward-modelled and empirically derived GIA to be validated, providing insights into the underlying geophysics. Therefore, in this section, the main emphasis will be put on the fourth category. The first and second categories are discussed in the chapter by (Barletta *et al.* 2022, this volume). It is important to mention here that GIA encompasses a range of observables including solid Earth deformation and deformation of the shape of the geoid. In the following, unless stated otherwise, the rates of surface deformation associated with GIA will be discussed.

The concept of deriving the *empirical* GIA was first introduced by Wahr *et al.* (2000). Prior to the launch of the GRACE and ICESat satellite missions, Wahr *et al.* (2000) suggested a combined approach based on simulated altimetry and gravimetry data to simultaneously solve for GIA and ice-mass changes. Velicogna and Wahr (2002) extended this approach by additionally considering simulated continuous GNSS observations to account for time-varying density within the surface layer. However, the first real data implementation was performed by Riva *et al.* (2009) for the whole Antarctic ice sheet based on 5 years of ICESat and GRACE data. Their results demonstrated that GIA (Fig. 8a) and surface processes could indeed be separated by combining the two complementary geodetic observations.

Following Riva *et al.* (2009), Groh *et al.* (2012) focused their research on the Amundsen Sea sector in West Antarctica for which they derived GIA and ice-mass changes for the 6 year observation period utilizing GRACE and ICESat observations. Their results suggested a strong viscoelastic uplift confirmed (at that time) by only two seasonal GNSS campaigns at three sites located in this region.

In the time since the work by Riva *et al.* (2009), additional data from a regional atmospheric climate model (RACMO) were made available to the scientific community. Two products of the RACMO are usually utilized: the firn densification model (FDM: Ligtenberg *et al.* 2011) and the surface mass balance (SMB: Lenaerts *et al.* 2012), which describe temporal processes within the firn layer in terms of height and mass changes, respectively. Both the SMB and FDM temporal variations arc inherently linked. To improve the methodology of Riva *et al.* (2009), Gunter *et al.* (2014) incorporated the information provided by RACMO into the gravimetry/altimetry combination approach. After establishing a calibration approach of the empirically estimated results to a low-precipitation zone in East Antarctica, their derived GIA consistently outperformed forward-modelled results when compared to GNSS displacements, as was shown by Gunter *et al.* (2014) and Wolstencroft *et al.* (2015). Moreover, the high GIA uplift rates in the Amundsen Sea sector that were first suggested by Groh *et al.* (2012) could be confirmed and its spatial pattern could be demonstrated (Fig. 8b). Gunter

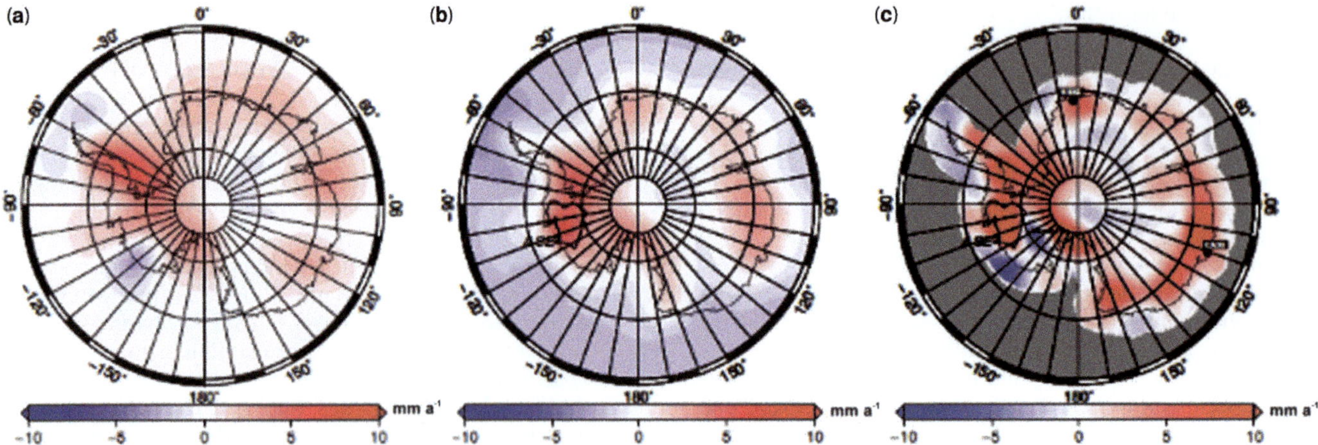

Fig. 8. GIA derived from: (**a**) Riva *et al.* (2009) using the gravimetry–altimetry combination approach; (**b**) Gunter *et al.* (2014), who additionally incorporated variations within the firn layer using climate data from RACMO and found strong uplift in the Amundsen Sea sector; and (**c**) Engels *et al.* (2018), who improved the combination approach to achieve high spatial resolution of the estimated signals.

et al. (2014) suggested that the disagreement between forward-modelled and empirically derived GIA rates in this region could be caused by the fact that the forward GIA models do not typically consider ice-load changes over the last 1000 years, which may, however, be a significant signal.

After allowing the spatial GIA length scale to vary from 500 km over West Antarctica and linearly increasing up to 1700 km over East Antarctica, as opposed to using a fixed GIA length scale by Schoen *et al.* (2015), Martín-Español *et al.* (2016*b*) also derived a strong GIA signal in the Amundsen Sea sector. Considering the weak Earth structure in the determination of utilized viscoelastic response functions, Sasgen *et al.* (2017) concluded that the large uplift in the Amundsen Sea sector is likely to have been caused by a rapid viscoelastic response to more recent ice retreat and thinning. A statistically significant purely data-driven strong GIA uplift was also retrieved by Engels *et al.* (2018). They suggested an approach that simultaneously (i) removes the correlated noise in gravity data (see the 'Geodetic GNSS measurements to constrain the modelling of ice-induced Earth deformation' section) and (ii) consistently combines them with high-resolution data from the altimetry–RACMO combination, yielding high spatial resolution of the empirically estimated GIA (Fig. 8c) and ice-mass change signals.

Empirical GIA models for Antarctica do not contain all long-wavelength signals from GIA or loading outside Antarctica. Global inversions can provide useful constraints providing a robust long-wavelength GIA signal as well as a regional GIA signal, albeit with lower accuracy. Such global inversion requires additional constraints. Over the ocean, researchers use ocean bottom pressure from ocean models (e.g. Wu *et al.* 2010) or sea-level observations from radar altimetry (Rietbroek *et al.* 2016) as additional input. Recently Jiang *et al.* (2021) obtained a global inverse GIA model with a rate of change of C_{20} that matched that of the GIA models and estimates from satellite laser ranging, while at the same time showing the enhanced GIA signal in areas of low viscosity such as the Amundsen Sea sector.

Recent GIA models that consider late Holocene ice-mass changes support the large GIA-induced uplift in the Amundsen Sea sector, which is known for its relatively thin lithosphere and low mantle viscosity (Nield *et al.* 2016; Barletta *et al.* 2018; Gomez *et al.* 2018). According to Barletta *et al.* (2018), who used GNSS observations to constrain mantle viscosity, the forward-modelled GIA-induced mass changes over that region for the GRACE time period range between 13.5 and 19.4 Gt a^{-1}. The empirical estimate by Engels *et al.* (2018) of 17.8 ± 0.7 Gt a^{-1}, which is completely independent of GNSS uplift rates used by Barletta *et al.* (2018), compares very well with the forward-modelled GIA. It is remarkable that empirically estimated and forward-modelled GIA solutions start to converge in this region, which is very important in light of current discussions regarding the stability of the Antarctic ice sheet and its contribution to sea-level rise (Engels *et al.* 2018; Larour *et al.* 2019).

The low mantle viscosity over the Amundsen Sea sector not only yields a localized GIA signal but also shortens the GIA response timescale from a century to decades. The GIA response is even predicted to accelerate with time, potentially preventing the complete collapse of the West Antarctic ice sheet (Barletta *et al.* 2018). The high empirical and forward-modelled GIA uplift rates in the Amundsen Sea sector provide arguments for a much more dynamic Antarctic ice sheet and, thus, questioning the assumption of a linear GIA signal that has been made in all earlier studies. Evidence of a time-variable GIA trend is also found in regions outside Antarctica, for instance when analysing geodetic observations in SE Alaska from 1992 to 2012 (Hu and Freymueller 2019). However, the open question is whether the non-linear GIA signal modelled by Barletta *et al.* (2018) in West Antarctica can be verified over a time period longer than a decade using a purely data-driven approach. For this, Willen *et al.* (2020) have undertaken a first attempt towards deriving monthly GIA-induced mass changes.

Summary and future research directions

Accurate geodetic data can constrain the uplift rate and the mass-change rate from mantle processes in Antarctica. Despite the logistical difficulty, GNSS observations have been collected since the 1990s. Satellite techniques provide the largest share of observations because they cover the entire continent apart from a small cap around the South Pole due to the inclination of the respective missions. Radar altimetry comprises the longest time period of the geodetic missions. However, altimetric height changes mainly reflect the change in ice thickness, while the GIA component in the height change is at least one order of magnitude smaller. Therefore, this chapter focused on GNSS and satellite gravimetry. However, altimetry plays an important role in untangling the ice-mass change signal in satellite gravimetry, and empirical GIA models can be seen as an independent product that can constrain the modelling of GIA.

GNSS measurements were started by campaigns, but more and more commonly continuous GNSS stations have been put into place since the turn of the century, especially since the IPY (2007–09). The remote environment necessitated installing independent power supplies from solar and wind generators, and protection from weather conditions. Standard GNSS processing methodology is followed, such as aligning the solutions to the terrestrial references by including IGS reference stations from adjacent continents. The accuracy achieved for horizontal velocities is typically 1 mm a^{-1}, while the accuracy of the vertical velocity is at the level of 2 mm a^{-1} (King *et al.* 2010; Thomas *et al.* 2011; Sasgen *et al.* 2018). This accuracy is sufficient to observe the solid Earth signal in many regions of Antarctica, although non-linear motion can often be better characterized than linear motion. However, identifying the viscous signals requires knowledge of the elastic loading, which becomes the largest error source when using GNSS to constrain GIA models. Still, GNSS provides the only constraint in large parts of Antarctica. The large uplift rates observed in the Antarctic Peninsula and the Amundsen Sea Embayment have motivated an increased effort to model GIA in low-viscosity regions. Similar to studies in Alaska, Iceland and Patagonia, it is now clear that parts of West Antarctica and the Antarctic Peninsula must be underlain by a low-viscous mantle. This is an important constraint on 3D viscosity maps that are created from seismic models (Ivins *et al.* 2021, this volume).

In order to come up with a consistent and homogeneous solution for coordinates and coordinate changes of bedrock sites in Antarctica, the GIANT-REGAIN (Geodynamics In ANTarctica based on REprocessing GNSS DAta INitiative) project compiles the most complete dataset of geodetic GNSS measurements recorded at bedrock sites in Antarctica. First results of this processing can be expected in 2022.

The success of improving constraints on the solid Earth response depends on the *continuation* of geodetic GNSS observations in Antarctica in order to extend the time span, as well as on the realization of *continuous* measurements in order to detect annual and non-linear signals. Observing the uplift over a longer time period will provide vital constraints to improve understanding of the evolution of the Antarctic ice sheet and its interaction with the solid Earth. Other research topics that can be addressed with longer time series are the importance of anelastic behaviour for timescales beyond instantaneous loading. Longer time series also help to reduce the error in horizontal velocities and to better separate the plate motion signal. The GIA signal in horizontal velocities is much smaller than in vertical uplift rates, but they are more sensitive to viscosity and possibly a viscosity contrast (Hermans *et al.* 2018). The analysis of horizontal velocities near the Ross Sea demonstrated the potential for constraining the viscosity contrast across that region (Konfal *et al.* 2018).

Satellite gravimetry using GRACE and GRACE-FO provides mass-change measurements with a temporal resolution of 1 month and a spatial resolution of around 300 km. With ground tracks passing near the poles, the coverage of

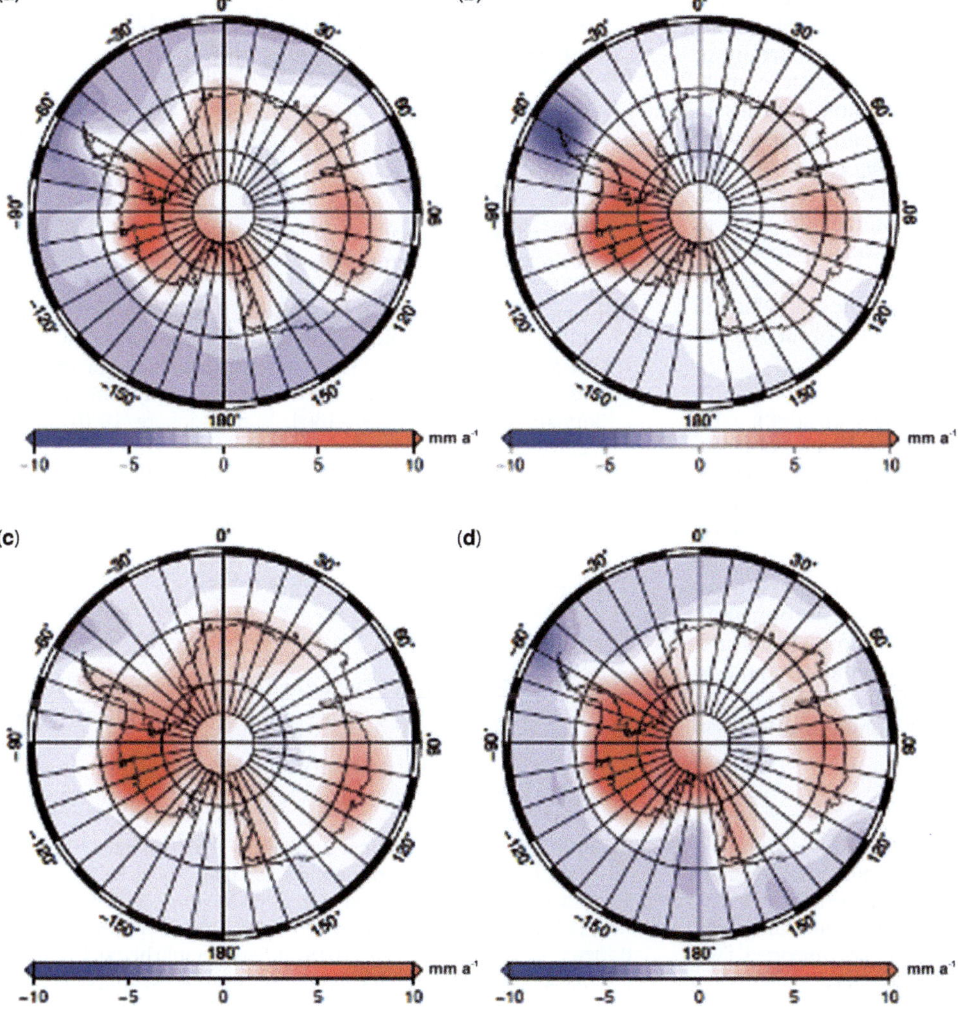

Fig. 9. (**a**) GIA estimated by Gunter *et al.* (2014). GIA estimated over the same time span and applying exactly the same methodology as Gunter *et al.* (2014), but using a new dataset for (**b**) satellite altimetry (publicly available: cf. Schröder *et al.* 2019*b*); (**c**) RACMO2.3p2 (provided by M. van den Broeke and P. Kuipers Munneke, Utrecht University, The Netherlands); and (**d**) satellite gravimetry (publicly available: cf. Mayer-Gürr *et al.* 2018).

Antarctica is much better than that of low-latitude regions. However, corrections models (e.g. for atmospheric pressure) can be less accurate in Antarctica and correlated errors manifesting as north–south stripes are still present. As for GNSS, the signal separation between current ice-mass changes and the readjustment to past mass changes remains the greatest problem. It is usually the ice-mass changes that are the focus of interest. However, satellite gravimetry needs to be corrected for GIA in order to obtain the real ice-mass changes, with this effect being the greatest source of noise in trend estimates. This demonstrates the importance of improved realism and accuracy in numerical GIA models taking into account lateral and temporal changes in viscosity, for example, and extending ice melt further back in time. The consistency between mass changes derived from satellite gravimetry as well as from radar altimetry or input–output mass changes provides a validation of GIA models. However, the problem of uncertainty in GIA predictions can only be solved by including further datasets. Apart from the empirical GIA modelling based on geodetic data, constraints on the GIA process (e.g. in the form of ice-margin data, measurements on bedrock and relative sea-level curves) are crucial.

For deriving empirical GIA estimates, the quality and properties of the input datasets are pivotal to the combination methodology. Following the approach introduced by Gunter *et al.* (2014) in their sensitivity study of Antarctic GIA, Willen *et al.* (2020) concluded that the estimated GIA is strongly sensitive to input datasets (as can be expected for a combination of real data). The sensitivity of the estimated GIA to different datasets is clearly visible in Figure 9 (reproduced from Willen *et al.* 2020). Apart from the data themselves, the input includes assumptions on the density of snow, ice and the upper mantle. Although several attempts have been made to improve the combination approach, a thorough validation of all datasets involved is still missing. Especially in light of large interannual variations in the Antarctic climate (Ligtenberg *et al.* 2012) and the possible non-linearity in the GIA signal (Barletta *et al.* 2018), a sophisticated approach for time-series analysis, such as suggested by Engels (2020), could be used to analyse and validate temporal variations of all input data prior to combining them: (i) to improve our understanding of the evolution of Antarctic geophysical processes; (ii) to identify significant disagreements; and (iii) to adjust the uncertainties correspondingly, which is crucial when combining data from different observational techniques.

Extended time series yielded by satellite gravimetry (GRACE-FO), altimetry (ICESat-2) and GNSS, along with improved methodologies, will increase the spatiotemporal resolution and the reliability of empirically derived GIA. Moreover, data-driven GIA should be extensively compared against independent GIA models that consider ice-mass changes in the late Holocene in addition to the last glacial maximum, as well as a wide range of plausible Earth parameters based on seismic, gravity and petrological information. Only a close collaboration of modellers, geoscientists and remote-sensing experts can help to increase the accuracy of the derived GIA models, which are so crucial in quantifying Antarctic ice-mass loss and in predicting feedback effects on future changes to ice dynamics.

Acknowledgements The successful realization of *in-situ* measurements in Antarctica is always a result of a collaborative effort of many people both in science and logistics as well as of many nations. Therefore, we would like to acknowledge the fruitful and constructive co-operation with all colleagues who supported field work and were involved in the research in Antarctic geosciences. Especially we would like to emphasize the collaboration within the Scientific Committee on Antarctic Research Expert Group on 'Geodetic Infrastructure in Antarctica' and the International Association of Geodesy Subcommission 1.3f 'Regional Reference Frame in Antarctica'. We would like to thank the editor, Adam Martin, as well as Matt King and an anonymous reviewer for their valuable comments which helped to improve the manuscript.

Author contributions **MS**: writing – original draft (lead), writing – review & editing (lead); **OE**: writing – original draft (equal), writing – review & editing (supporting); **EJOS**: writing – review & editing (supporting); **WVDW**: writing – review & editing (supporting); **MH**: writing – review & editing (supporting).

Funding We gratefully acknowledge the funding granted over many years by the different national funding agencies and national Antarctic programmes..

Data availability Data sharing is not applicable to this article as no datasets were generated or analysed during the current study.

References

Adie, R.J. 1962. The geology of Antarctica. *American Geophysical Union Geophysical Monograph Series*, **7**, 26–39, https://doi.org/10.1029/GM007p0026

Altamimi, Z., Rebischung, P., Métivier, L. and Collilieux, X. 2016. ITRF2014: a new release of the International Terrestrial Reference Frame modeling nonlinear station motions. *Journal of Geophysical Research: Solid Earth*, **121**, 6109–6131, https://doi.org/10.1002/2016JB013098

An, M., Wiens, D.A. *et al.* 2015*a*. S-velocity model and inferred Moho topography beneath the Antarctic plate from Rayleigh waves. *Journal of Geophysical Research: Solid Earth*, **120**, 359–383, https://doi.org/10.1002/2014JB011332

An, M., Wiens, D.A. *et al.* 2015*b*.Temperature, lithosphere–asthenosphere boundary, and heat flux beneath the Antarctic plate inferred from seismic velocities. *Journal of Geophysical Research: Solid Earth*, **120**, 8720–8742, https://doi.org/10.1002/2015JB011917

Aoyama, Y., Doi, K., Ikeda, H., Hayakawa, H. and Shibuya, K. 2016. Five years' gravity observation with the superconducting gravimeter OSG# 058 at Syowa Station, East Antarctica: gravitational effects of accumulated snow mass. *Geophysical Journal International*, **205**, 1290–1304, https://doi.org/10.1093/gji/ggw078

Argus, D.F. 2012. Uncertainty in the velocity between the mass center and surface of earth. *Journal of Geophysical Research: Solid Earth*, **117**, B10405, https://doi.org/10.1029/2012JB009196

Argus, D.F., Blewitt, G., Peltier, W.R. and Kreemer, C. 2011. Rise of the Ellsworth mountains and parts of the East Antarctic coast observed with GPS. *Geophysical Research Letters*, **38**, L16303, https://doi.org/10.1029/2011GL048025

Argus, D.F., Peltier, W.R., Drummond, R. and Moore, A.W. 2014. The Antarctica component of postglacial rebound model ICE-6G_C (VM5a) based on GPS positioning, exposure age dating of ice thicknesses, and relative sea level histories. *Geophysical Journal International*, **198**, 537–563, https://doi.org/10.1093/gji/ggu140

Barletta, V.R., Sabadini, R. and Bordoni, A. 2008. Isolating the PGR signal in the GRACE data: impact on mass balance estimates in Antarctica and Greenland. *Geophysical Journal International*, **172**, 18–30, https://doi.org/10.1111/j.1365-246X.2007.03630.x

Barletta, V.R., Bevis, M. *et al.* 2018. Observed rapid bedrock uplift in Amandsen Sea embayment promotes ice-sheet stability. *Science*, **360**, 1335–1339, https://doi.org/10.1126/science.aao1447

Barletta, V.R., Nield, G.A., van der Wal, W. and van Calcar, C.J. 2022. Glacial isostatic adjustment and postseismic deformation in Antarctica. *Geological Society, London, Memoirs*, **56**, in press.

Bevis, M., Kendrick, E. *et al.* 2009. Geodetic measurements of vertical crustal velocity in West Antarctica and the implications for ice mass balance. *Geochemistry, Geophysics, Geosystems*, **10**, Q10005, https://doi.org/10.1029/2009GC002642

Bevis, M., Wahr, J. *et al.* 2012. Bedrock displacements in Greenland manifest ice mass variations, climate cycles and climate change. *Proceedings of the National Academy of Sciences of the United States of America*, **109**, 11944–11948, https://doi.org/10.1073/pnas.1204664109

Bevis, M.C., Harig, S.A. *et al.* 2019. Accelerating changes in ice mass within Greenland, and the ice sheet's sensitivity to atmospheric forcing. *Proceedings of the National Academy of Sciences of the United States of America*, **116**, 1934–1939, https://doi.org/10.1073/pnas.1806562116

Blewitt, G. 2003. Self-consistency in reference frames, geocenter definition, and surface loading of the solid earth. *Journal of Geophysical Research: Solid Earth*, **108**, 2103, https://doi.org/10.1029/2002JB002082

Blewitt, G., Kreemer, C., Hammond, W.C. and Gazeaux, J. 2016. MIDAS robust trend estimator for accurate GPS station velocities without step detection. *Journal of Geophysical Research: Solid Earth*, **121**, 2054–2068, https://doi.org/10.1002/2015JB012552

Brenner, A.C., DiMarzio, J.P. and Zwally, H.J. 2007. Precision and accuracy of satellite radar and laser altimeter data over the continental ice sheets. *IEEE Transactions on Geoscience and Remote Sensing*, **45**, 321–331, https://doi.org/10.1109/TGRS.2006.887172

Bos, M.S., Fernandes, R.M.S., Williams, S.D.P. and Bastos, L. 2013. Fast error analysis of continuous GNSS observations with missing data. *Journal of Geodesy*, **87**, 351–360, https://doi.org/10.1007/s00190-012-0605-0

Bos, M.S., Penna, N.T., Baker, T.F. and Clarke, P.J. 2015. Ocean tide loading displacements in Western Europe: 2. GPS-observed anelastic dispersion in the asthenosphere. *Journal of Geophysical Research: Solid Earth*, **120**, 6540–6557, https://doi.org/10.1002/2015JB011884

Bouman, J., Fuchs, M., Ivins, E.V., Van Der Wal, W., Schrama, E., Visser, P.N.A.M. and Horwath, M. 2014. Antarctic outlet glacier mass change resolved at basin scale from satellite gravity gradiometry. *Geophysical Research Letters*, **41**, 5919–5926, https://doi.org/10.1002/2014GL060637

Burov, E.B. 2011. Rheology and strength of the lithosphere. *Marine and Petroleum Geology*, **28**, 1402–1443, https://doi.org/10.1016/j.marpetgeo.2011.05.008

Busch, P., Scheinert, M. *et al.* 2017. GNSS measurements in West Antarctica to reconcile glacial-isostatic adjustment. Presented at the IAG/SCAR-SERCE Workshop on Glacial Isostatic Adjustment and Elastic Deformation, 5–7 September 2017, Reykjavik, Iceland.

Caron, L. and Ivins, E.R. 2020. A baseline Antarctic GIA correction for space gravimetry. *Earth and Planetary Science Letters*, **531**, 115957, https://doi.org/10.1016/j.epsl.2019.115957

Caron, L., Ivins, E.R., Larour, E., Adhikari, S., Nilsson, J. and Blewitt, G. 2018. GIA model statistics for GRACE hydrology, cryosphere, and ocean science. *Geophysical Research Letters*, **45**, 2203–2212, https://doi.org/10.1002/2017GL076644

Chen, J.L., Wilson, C.R., Blankenship, D.D. and Tapley, B.D. 2006. Antarctic mass rates from GRACE. *Geophysical Research Letters*, **33**, L11502, https://doi.org/10.1029/2006GL026369

Chen, J.L., Wilson, C.R., Tapley, B.D. and Grand, S. 2007. GRACE detects coseismic and postseismic deformation from the Sumatra-Andaman earthquake. *Geophysical Research Letters*, **34**, L13302, https://doi.org/10.1029/2007GL030356

Cheng, M.K., Eanes, R.J., Shum, C.K., Schutz, B.E. and Tapley, B.D. 1989. Temporal variations in low degree zonal harmonics from Starlette orbit analysis. *Geophysical Research Letters*, **16**, 393–396, https://doi.org/10.1029/GL016i005p00393

Dahle, C. 2020. *Release Notes for GFZ GRACE-FO Level-2 Products – version RL06*. GFZ German Research Centre for Geosciences, Potsdam Germany.

Dietrich, R. (ed.) 1996. *The Geodetic Antarctic Project GAP95. German Contributions to the SCAR 95 Epoch Campaign. Deutsche Geodätische Kommission, B 304*. Bayerischen Akademie der Wissenschaften, München, Germany.

Dietrich, R. (ed.) 2000. *Deutsche Beiträge zu GPS-Kampagnen des Scientific Committee on Antarctic Research (SCAR) 1995–1998, Deutsche Geodätische Kommission, B310*. Bayerischen Akademie der Wissenschaften, München, Germany.

Dietrich, R. and Rülke, A. 2008. A precise reference frame for Antarctica from SCAR GPS campaign data and some geophysical implications. *In*: Capra, A. and Dietrich, R. (eds) *Geodetic and Geophysical Observations in Antarctica*. Springer, Berlin, 1–10.

Dietrich, R., Dach, R. *et al.* 2001. ITRF coordinates and plate velocities from repeated GPS campaigns in Antarctica – an analysis based on different individual solutions. *Journal of Geodesy*, **74**, 756–766, https://doi.org/10.1007/s001900000147

Dietrich, R., Rülke, A. *et al.* 2004. Plate kinematics and deformation status of the Antarctic Peninsula based on GPS. *Global and Planetary Change*, **42**, 313–321, https://doi.org/10.1016/j.gloplacha.2003.12.003

Dill, R., Klemann, V., Martinec, Z. and Tesauro, M. 2015. Applying local Green's functions to study the influence of the crustal structure on hydrological loading displacements. *Journal of Geodynamics*, **88**, 14–22, https://doi.org/10.1016/j.jog.2015.04.005

Dobslaw, H., Bergmann-Wolf, I. *et al.* 2017. A new high-resolution model of non-tidal atmosphere and ocean mass variability for de-aliasing of satellite gravity observations: AOD1B RL06. *Geophysical Journal International*, **211**, 263–269, https://doi.org/10.1093/gji/ggx302

Donnellan, A. and Luyendyk, B.P. 2004. GPS evidence for a coherent Antarctic plate and for postglacial rebound in Marie Byrd Land. *Global and Planetary Change*, **42**, 305–311, https://doi.org/10.1016/j.gloplacha.2004.02.006

Dziewonski, A.M. and Anderson, D.L. 1981. Preliminary reference earth model. *Physics of the Earth and Planetary Interiors*, **25**, 297–356, https://doi.org/10.1016/0031-9201(81)90046-7

Engels, O. 2020. Stochastic modelling of geophysical signal constituents within a kalman filter framework. *In*: Montillet, J.P. and Bos, M. (eds) *Geodetic Time Series Analysis in Earth Sciences*. Springer Geophysics. Springer, Cham, Switzerland, 239–260, https://doi.org/10.1007/978-3-030-21718-1_8

Engels, O., Gunter, B.C., Riva, R.E.M. and Klees, R. 2018. Separating geophysical signals using GRACE and high-resolution data: A case study in Antarctica. *Geophysical Research Letters*, **45**, 12 340–12 349, https://doi.org/10.1029/2018GL079670

Farrell, W.E. 1972. Deformation of the earth by surface loads. *Review of Geophysics*, **10**, 761–797, https://doi.org/10.1029/RG010i003p00761

Foley, S.F., Andronikov, A.V., Halpin, J.A., Daczko, N.R. and Jacob, D.E. 2021. Mantle rocks in East Antarctica. *Geological Society, London, Memoirs*, **56**, https://doi.org/10.1144/M56-2020-8

Geruo, A., Wahr, J. and Zhong, S. 2013. Computations of the viscoelastic response of a 3-D compressible Earth to surface loading: an application to Glacial Isostatic Adjustment in Antarctica and Canada. *Geophysical Journal International*, **192**, 557–572, https://doi.org/10.1093/gji/ggs030

Ghelichkhan, S., Murböck, M., Colli, L., Pail, R. and Bunge, H.P. 2018. On the observability of epeirogenic movement in current and future gravity missions. *Gondwana Research*, **53**, 273–284, https://doi.org/10.1016/j.gr.2017.04.016

Gomez, N., Latychev, K. and Pollard, D. 2018. A coupled ice sheet–sea level model incorporating 3D Earth structure: Variations in Antarctica during the last deglacial retreat. *Journal of Climate*, **31**, 4041–4054, https://doi.org/10.1175/JCLI-D-17-0352.1

Groh, A., Ewert, H. *et al.* 2012. An investigation of glacial isostatic adjustment over the Amundsen Sea sector, West Antarctica. *Global and Planetary Change*, **98–99**, 45–53, https://doi.org/10.1016/j.gloplacha.2012.08.001

Groh, A., Ewert, H. *et al.* 2014. Mass, volume and velocity of the Antarctic ice sheet: present-day changes and error effects. *Surveys in Geophysics*, **35**, 1481–1505, https://doi.org/10.1007/s10712-014-9286-y

Gunter, B., Didova, O. *et al.* 2014. Empirical estimation of present-day antarctic glacial isostatic adjustment and ice mass change. *The Cryosphere*, **8**, 743–760, https://doi.org/10.5194/tc-8-743-2014

Han, S.C., Sauber, J. and Pollitz, F. 2016. Postseismic gravity change after the 2006–2007 great earthquake doublet and constraints on the asthenosphere structure in the central Kuril Islands. *Geophysical Research Letters*, **43**, 3169–3177, https://doi.org/10.1002/2016GL068167

Handler, M.R., Wysoczanski, R.J. and Gamble, J.A. 2021. Marie Byrd Land lithospheric mantle: a review of the xenolith record. *Geological Society, London, Memoirs*, **56**, https://doi.org/10.1144/M56-2020-17

Harig, C. and Simons, F.J. 2015. Accelerated West Antarctic ice mass loss continues to outpace East Antarctic gains. *Earth and Planetary Science Letters*, **415**, 134–141, https://doi.org/10.1016/j.epsl.2015.01.029

Hermans, T.H.J., van der Wal, W. and Broerse, T. 2018. Reversal of the direction of horizontal velocities induced by GIA as a function of mantle viscosity. *Geophysical Research Letters*, **45**, 9597–9604, https://doi.org/10.1029/2018GL078533

Horwath, M. and Dietrich, R. 2009. Signal and error in mass change inferences from GRACE: the case of Antarctica. *Geophysical Journal International*, **177**, 849–864, https://doi.org/10.1111/j.1365-246X.2009.04139.x

Hu, Y. and Freymueller, J.T. 2019. Geodetic observations of time variable glacial isostatic adjustment in Southeast Alaska and its implications for Earth rheology. *Journal of Geophysical Research: Solid Earth*, **124**, 9870–9889, https://doi.org/10.1029/2018JB017028

Ivins, E.R. and James, T.S. 2005. Antarctic glacial isostatic adjustment: a new assessment. *Antarctic Science*, **17**, 541–553, https://doi.org/10.1017/S0954102005002968

Ivins, E.R., James, T.S., Wahr, J., Schrama, E.J.O., Landerer, F.W. and Simon, K.M. 2013. Antarctic contribution to sea level rise observed by GRACE with improved GIA correction. *Journal of Geophysical Research: Solid Earth*, **118**, 3126–3141, https://doi.org/10.1002/jgrb.50208

Ivins, E.R., Caron, L., Adhikari, S., Larour, E. and Scheinert, M. 2020. A linear viscoelasticity for decadal to centennial time scale mantle deformation. *Reports on Progress in Physics*, **83**, 106801, https://doi.org/10.1088/1361-6633/aba346

Ivins, E.R., van der Wal, W., Wiens, D.A., Lloyd, A.J. and Caron, L. 2021. Antarctic upper-mantle rheology. *Geological Society, London, Memoirs*, **56**, https://doi.org/10.1144/M56-2020-19

Jacobs, J., Elburg, M. *et al.* 2015. Two distinct Late Mesoproterozoic/Early Neoproterozoic basement provinces in central/eastern Dronning Maud Land, East Antarctica: The missing link, 15–21° E. *Precambrian Research*, **265**, 249–272, https://doi.org/10.1016/j.precamres.2015.05.003

Jiang, Y., Wu, X., van den Broeke, M.R., Munneke, P.K., Simonsen, S.B., van der Wal, W. and Vermeersen, B.L. 2021. Assessing global present-day surface mass transport and glacial isostatic adjustment from inversion of geodetic observations. *Journal of Geophysical Research: Solid Earth*, **126**, e2020JB020713, https://doi.org/10.1029/2020JB020713

Kang, K., Wahr, J., Heflin, M. and Desai, S. 2015. Stacking global GPS verticals and horizontals to solve for the fortnightly and monthly body tides: implications for mantle anelasticity. *Journal of Geophysical Research: Solid Earth*, **120**, 1787–1803, https://doi.org/10.1002/2014JB011572

King, M.A. and Santamaría-Gómez, A. 2016. Ongoing deformation of Antarctica following recent Great Earthquakes. *Geophysical Research Letters*, **43**, 1918–1927, https://doi.org/10.1002/2016GL067773

King, M.A., Altamimi, Z. *et al.* 2010. Improved constraints on models of glacial isostatic adjustment: a review of the contribution of ground-based geodetic observations. *Surveys in Geophysics*, **31**, 465–507, https://doi.org/10.1007/s10712-010-9100-4

King, M.A., Bingham, R.J., Moore, P., Whitehouse, P.L., Bentley, M.J. and Milne, G.A. 2012. Lower satellite-gravimetry estimates of Antarctic sea-level contribution. *Nature*, **491**, 586–589, https://doi.org/10.1038/nature11621

Koch, A., Sanjuan, J., Gohlke, M., Mahrdt, C., Brause, N., Braxmaier, C. and Heinzel, G. 2018. Line of sight calibration for the laser ranging interferometer on-board the GRACE follow-on mission: on-ground experimental validation. *Optics Express*, **26**, 25892–25908, https://doi.org/10.1364/OE.26.025892

Konfal, S., Kendrick, E., Saddler, D., Wilson, T. and Bevis, M. 2016. Crustal velocity solutions in polar environments: GPS position errors caused by ice in antennas. Presented at the XXXIV SCAR Meeting and Open Science Conference, 19–31 August 2016, Kuala Lumpur, Malaysia.

Konfal, S.A., Wilson, T.J.W. *et al.* 2018. Utilizing GPS to investigate past ice mass change in the Ross Sea region, Antarctica. *AGU Fall Meeting Abstracts*, **2018**, G43B-0716, https://ui.adsabs.harvard.edu/abs/2018AGUFM.G43B0716K

Koulali, A. and Clarke, P.J. 2020. Effect of antenna snow intrusion on vertical GPS position time series in Antarctica. *Journal of Geodesy*, **94**, 101, https://doi.org/10.1007/s00190-020-01403-6

Kusche, J., Schmidt, R., Petrovic, S. and Rietbroek, R. 2009. Decorrelated GRACE time-variable gravity solutions by GFZ, and their validation using a hydrological model. *Journal of Geodesy*, **83**, 903–913, https://doi.org/10.1007/s00190-009-0308-3

Kvas, A. and Mayer-Gürr, T. 2019. GRACE gravity field recovery with background model uncertainties. *Journal of Geodesy*, **93**, 2543–2552, https://doi.org/10.1007/s00190-019-01314-1

Landerer, F.W. and Swenson, S.C. 2012. Accuracy of scaled GRACE terrestrial water storage estimates. *Water Resources Research*, **48**, W04531, https://doi.org/10.1029/2011WR011453, w04531

Landerer, F.W., Flechtner, F.M. *et al.* 2020. Extending the global mass change data record: GRACE follow-on instrument and science data performance. *Geophysical Research Letters*, **47**, e2020GL088306, https://doi.org/10.1029/2020GL088306

Lange, H., Casassa, G. *et al.* 2014. Observed crustal uplift near the Southern Patagonian Icefield constrains improved viscoelastic Earth models. *Geophysical Research Letters*, **41**, 805–812, https://doi.org/10.1002/2013GL058419

Larour, E., Seroussi, H., Adhikari, S., Ivins, E., Caron, L., Morlighem, M. and Schlegel, N. 2019. Slowdown in Antarctic mass loss from solid Earth and sea-level feedbacks. *Science*, **364**, 6444, https://doi.org/10.1126/science.aav7908

Lenaerts, J.T.M., van den Broeke, M.R., van de Berg, W.J., van Meijgaard, E. and Kuipers Munneke, P. 2012. A new, high-resolution surface mass balance map of Antarctica 1979–2010) based on regional atmospheric climate modeling. *Geophysical Research Letters*, **39**, L04501, https://doi.org/10.1029/2011GL050713

Ligtenberg, S.R.M., Helsen, M.M. and van den Broeke, M.R. 2011. An improved semi-empirical model for the densification of Antarctic firn. *The Cryosphere*, **5**, 809–819, https://doi.org/10.5194/tc-5-809-2011

Ligtenberg, S.R.M., Horwath, M., van den Broeke, M.R. and Legrésy, B. 2012. Quantifying the seasonal 'breathing' of the Antarctic ice sheet. *Geophysica. Research Letters*, **39**, L23501, https://doi.org/10.1029/2012GL053628

Liu, B., King, M. and Dai, W. 2018. Common mode error in Antarctic GPS coordinate time-series on its effect on bedrock-uplift estimates. *Geophysical Journal International*, **214**, 1652–1664, https://doi.org/10.1093/gji/ggy217

Lloyd, A.J., Wiens, D.A. *et al.* 2020. Seismic structure of the Antarctic upper mantle imaged with adjoint tomography. *Journal of Geophysical Research: Solid Earth*, **125**, https://doi.org/10.1029/2019JB017823

Loomis, B.D., Luthcke, S.B. and Sabaka, T.J. 2019. Regularization and error characterization of GRACE mascons. *Journal of Geodesy*, **93**, 1381–1398, https://doi.org/10.1007/s00190-019-01252-y

Mäkinen, J., Amalvict, M., Shibuya, K. and Fukuda, Y. 2007. Absolute gravimetry in Antarctica: status and prospects. *Journal of Geodynamics*, **43**, 339–357, https://doi.org/10.1016/j.jog.2006.08.002

Marderwald, E.R., Aragon Paz, J.M.A. *et al.* 2020. Perito Moreno Glacier dam rupture – a recurrent natural experiment to probe solid-earth elasticity. *Journal of South American Earth Sciences*, **104**, 102904, https://doi.org/10.1016/j.jsames.2020.102904

Martin, A.P., Cooper, A.F., Price, R.C., Doherty, C.L. and Gamble, J.A. 2021. A review of mantle xenoliths in volcanic rocks from southern Victoria Land, Antarctica. *Geological Society, London, Memoirs*, **56**, https://doi.org/10.1144/M56-2019-42

Martín-Español, A., King, M.A., Zammit-Mangion, A., Andrews, B., Moore, P. and Bamber, J.L. 2016*a*. An assessment of forward and inverse GIA solutions for Antarctica. *Journal of Geophysical Research: Solid Earth*, **121**, 6947–6965, https://doi.org/10.1002/2016JB013154

Martín-Español, A., Zammit-Mangion, A. *et al.* 2016*b*. Spatial and temporal Antarctic Ice Sheet mass trends, glacio-isostatic adjustment, and surface processes from a joint inversion of satellite altimeter, gravity, and GPS data. *Journal of Geophysical Research: Earth Surface*, **121**, 182–200, https://doi.org/10.1002/2015JF003550

Mayer-Gürr, T., Behzadpur, S., Ellmer, M., Kvas, A., Klinger, B., Strasser, S. and Zehentner, N. 2018. *ITSG-Grace2018 – Monthly, Daily and Static Gravity Field Solutions from GRACE*. GFZ Data Services, https://doi.org/10.5880/ICGEM.2018.003; model data also available at https://www.tugraz.at/institute/ifg/downloads/gravity-field-models/

Mieth, M. and Jokat, W. 2014. New aeromagnetic view of the geological fabric of southern Dronning Maud Land and Coats Land, East Antarctica. *Gondwana Research*, **25**, 358–367, https://doi.org/10.1016/j.gr.2013.04.003

Mitrovica, J.X., Milne, G.A. and Davis, J.L. 2001. Glacial isostatic adjustment on a rotating earth. *Geophysical Journal International*, **147**, 562–578, https://doi.org/10.1046/j.1365-246x.2001.01550.x

Nield, G.A., Whitehouse, P.L., King, M.A., Clarke, P.J. and Bentley, M.J. 2012. Increased ice loading in the Antarctic Peninsula since the 1850s and its effect on glacial isostatic adjustment. *Geophysical Research Letters*, **39**, L17504, https://doi.org/10.1029/2012gl052559

Nield, G.A., Barletta, V.R. *et al.* 2014. Rapid bedrock uplift in the Antarctic Peninsula explained by viscoelastic response to recent ice unloading. *Earth and Planetary Science Letters*, **397**, 32–41, https://doi.org/10.1016/j.epsl.2014.04.019

Nield, G.A., Whitehouse, P.L., King, M.A. and Clarke, P.J. 2016. Glacial isostatic adjustment in response to changing Late Holocene behaviour of ice streams on the Siple Coast, West Antarctica. *Geophysical Journal International*, **205**, 1–21, https://doi.org/10.1093/gji/ggv532

Nield, G.A., Whitehouse, P.L., van der Wal, W., Blank, B., O'Donnell, J.P. and Stuart, G.W. 2018. The impact of lateral variations in lithospheric thickness on glacial isostatic adjustment in West Antarctica. *Geophysical Journal International*, **214**, 811–824, https://doi.org/10.1093/gji/ggy158

Nielsen, K., Khan, S.A., Spada, G., Wahr, J., Bevis, M., Liu, L. and van Dam, T. 2013. Vertical and horizontal surface displacements near Jakobshavn Isbræ driven by melt-induced and dynamic ice loss. *Journal of Geophysical Research: Solid Earth*, **118**, 1837–1844, https://doi.org/10.1002/jgrb.50145

Pappa, F. and Ebbing, J. 2021. Gravity, magnetics and geothermal heat flow of the Antarctic lithospheric crust and mantle. *Geological Society, London, Memoirs*, **56**, https://doi.org/10.1144/M56-2020-5

Peidou, A. and Pagiatakis, S. 2020. Stripe mystery in GRACE geopotential models revealed. *Geophysical Research Letters*, **47**, e2019GL085497, https://doi.org/10.1029/2019GL085497

Peltier, W.R. 1998. Postglacial variations in the level of the sea: Implications for climate dynamics and solid-Earth geophysics. *Reviews of Geophysics*, **36**, 603–689, https://doi.org/10.1029/98RG02638

Peltier, W.R. 2004. Global glacial isostasy and the surface of the ice-age earth: the ICE-5G (VM2) model and GRACE. *Annual Review of Earth and Planetary Science*, **32**, 111–149, https://doi.org/10.1146/annurev.earth.32.082503.144359

Ranalli, G. 1995. *Rheology of the Earth*. 2nd edn. Chapman & Hall, London.

Raymond, C.A., Ivins, E.R., Heflin, M.B. and James, T.S. 2004. Quasi-continuous global positioning system measurements of glacial isostatic deformation in the Northern Transantarctic Mountains. *Global and Planetary Change*, **42**, 295–303, https://doi.org/10.1016/j.gloplacha.2003.11.013

Rebischung, P., Altamimi, Z., Ray, J. and Garayt, B. 2016. The IGS contribution to ITRF2014. *Journal of Geodesy*, **90**, 611–630, https://doi.org/10.1007/s00190-016-0897-6

Riddell, A.R., King, M.A., Watson, C.S., Sun, Y., Riva, R.E.M. and Rietbroek, R. 2017. Uncertainty in geocenter estimates in the context of ITRF2014. *Journal of Geophysical Research: Solid Earth*, **122**, 4020–4032, https://doi.org/10.1002/2016JB013698

Rietbroek, R., Brunnabend, S.E., Kusche, J., Schröter, J. and Dahle, C. 2016. Revisiting the contemporary sea-level budget on global and regional scales. *Proceedings of the National Academy of Sciences of the United States of America*, **113**, 1504–1509, https://doi.org/10.1073/pnas.1519132113

Riva, R.E.M., Gunter, B.C. *et al.* 2009. Glacial isostatic adjustment over Antarctica from combined ICESat and GRACE satellite data. *Earth and Planetary Science Letters*, **288**, 516–523, https://doi.org/10.1016/j.epsl.2009.10.013

Rülke, A. 2009. *Zur Realisierung eines terrestrischen Referenzsystems in globalen und regionalen GPS-Netzen*. PhD thesis, TU Dresden, http://nbn-resolving.de/urn:nbn:de:bsz:14-qucosa-24543

Rülke, A., Dietrich, R., Fritsche, M., Rothacher, M. and Steigenberger, P. 2008. Realization of the Terrestrial Reference System by a reprocessed global GPS network. *Journal of Geophysical Research: Solid Earth*, **113**, B08403, https://doi.org/10.1029/2007JB005231

Rülke, A., Dietrich, R. *et al.* 2015. The Antarctic regional GPS network densification: status and results. *In*: Rizos, C. and Willis, P. (eds) *IAG 150 Years*. International Association of Geodesy Symposia, **143**. Springer, Cham, Switzerland, 133–139, https://doi.org/10.1007/1345_2015_79

Sasgen, I., Martinec, Z. and Fleming, K. 2007. Regional ice-mass changes and glacial-isostatic adjustment in Antarctica from GRACE. *Earth and Planetary Science Letters*, **264**, 391–401, https://doi.org/10.1016/j.epsl.2007.09.029

Sasgen, I., Konrad, H., Ivins, E.R., van den Broeke, M.R., Bamber, J.L., Martinec, Z. and Klemann, V. 2013. Antarctic ice-mass balance 2003 to 2012: regional reanalysis of GRACE satellite gravimetry measurements with improved estimate of glacial-isostatic adjustment based on GPS uplift rates. *The Cryosphere*, **7**, 1499–1512, https://doi.org/10.5194/tc-7-1499-2013

Sasgen, I., Martín-Español, A. *et al.* 2017. Joint inversion estimate of regional glacial isostatic adjustment in Antarctica considering a lateral varying earth structure (ESA STSE project REGINA). *Geophysical Journal International*, **211**, 1534–1553, https://doi.org/10.1093/gji/ggx368

Sasgen, I., Martín-Español, A. *et al.* 2018. Altimetry, gravimetry, GPS and viscoelastic modeling data for the joint inversion for glacial isostatic adjustment in Antarctica (ESA STSE project REGINA). *Earth System Science Data*, **10**, 493–523, https://doi.org/10.5194/essd-10-493-2018

Schoen, N., Zammit-Mangion, A., Rougier, J., Flament, T., Rémy, F., Luthcke, S. and Bamber, J. 2015. Simultaneous solution for mass trends on the West Antarctic Ice Sheet. *The Cryosphere*, **9**, 805–819, https://doi.org/10.5194/tc-9-805-2015

Schrama, E.J.O., Wouters, B. and Rietbroek, R. 2014. A mascon approach to assess ice sheet and glacier mass balances and their uncertainties from GRACE data. *Journal of Geophysical Research: Solid Earth*, **119**, 6048–6066, https://doi.org/10.1002/2013JB010923

Schröder, L., Horwath, M., Dietrich, R., Helm, V., van den Broeke, M.R. and Ligtenberg, S.R.M. 2019*a*. Four decades of Antarctic surface elevation changes from multi-mission satellite altimetry. *The Cryosphere*, **13**, 427–449, https://doi.org/10.5194/tc-13-427-2019

Schröder, L., Horwath, M., Dietrich, R., Helm, V., van den Broeke, M.R., and Ligtenberg, S.R.M. 2019*b*. Gridded surface elevation changes from multi-mission satellite altimetry 1978–2017. *PANGAEA*, https://doi.org/10.1594/PANGAEA.897390

Seo, K.W., Wilson, C.R., Chen, J. and Waliser, D.E. 2008. GRACE's spatial aliasing error. *Geophysical Journal International*, **172**, 41–48, https://doi.org/10.1111/j.1365-246X.2007.03611.x

Shepherd, A., Ivins, E. *et al.* 2018. Mass balance of the Antarctic Ice Sheet from 1992 to 2017. *Nature*, **558**, 219–222, https://doi.org/10.1038/s41586-018-0179-y

Shepherd, A., Ivins, E. *et al.* 2020. Mass balance of the Greenland Ice Sheet from 1992 to 2018. *Nature*, **579**, 233–239, https://doi.org/10.1038/s41586-019-1855-2

Spada, G. 2017. Glacial isostatic adjustment and contemporary sea level rise: an overview. *Surveys in Geophysics*, **38**, 153–185, https://doi.org/10.1007/s10712-016-9379-x

Steinberger, B. and Becker, T.W. 2018. A comparison of lithospheric thickness models. *Tectonophysics*, **746**, 325–338, https://doi.org/10.1016/j.tecto.2016.08.001

Swenson, S. and Wahr, J. 2002. Methods for inferring regional surface-mass anomalies from Gravity Recovery and Climate Experiment (GRACE) measurements of time-variable gravity. *Journal of Geophysical Research: Solid Earth*, **107**, 2193, https://doi.org/10.1029/2001JB000576

Swenson, S. and Wahr, J. 2006. Post-processing removal of correlated errors in GRACE data. *Geophysical Research Letters*, **33**, L08402, https://doi.org/10.1029/2005GL025285

Tamisiea, M.E., Mitrovica, J.X. and Davis, J.L. 2007. GRACE gravity data constrain ancient ice geometries and continental dynamics over Laurentia. *Science*, **316**, 881–883, https://doi.org/10.1126/science.1137157

Tapley, B.D., Bettadpur, S., Ries, J.C., Thompson, P.F. and Watkins, M.M. 2004. GRACE measurements of mass variability in the Earth system. *Science*, **305**, 503–505, https://doi.org/10.1126/science.1099192

Thomas, I.D., King, M.A. *et al.* 2011. Widespread low rates of Antarctic glacial isostatic adjustment revealed by GPS observations. *Geophysical Research Letters*, **38**, L22302, https://doi.org/10.1029/2011GL049277

Tregoning, P., Twilley, B., Hendy, M. and Zwartz, D. 1999. Monitoring isostatic rebound in Antarctica using continuous remote GPS observations. *GPS Solutions*, **2**, 70–75, https://doi.org/10.1007/PL00012759

Tregoning, P., Welsh, A., McQueen, H. and Lambeck, K. 2000. The search for postglacial rebound near the Lambert Glacier, Antarctica. *Earth, Planets and Space*, **52**, 1037–1041, https://doi.org/10.1186/BF03352327

Turner, R.J., Reading, A.M. and King, M.A. 2020. Separation of tectonic and local components of horizontal GPS station velocities: a case study for glacial isostatic adjustment in East Antarctica. *Geophysical Journal International*, **222**, 1555–1569, https://doi.org/10.1093/gji/ggaa265

Van Camp, M., de Viron, O., Watlet, A., Meurers, B., Francis, O. and Caudron, C. 2017. Geophysics from terrestrial time-variable gravity measurements. *Reviews of Geophysics*, **55**, 938–992. https://doi.org/10.1002/2017RG000566

van der Wal, W., Barnhoorn, A., Stocchi, P., Gradmann, S., Wu, P., Drury, M. and Vermeersen, B. 2013. Glacial isostatic adjustment model with composite 3-D Earth rheology for Fennoscandia. *Geophysical Journal International*, **194**, 61–77, https://doi.org/10.1093/gji/ggt099

van der Wal, W., Whitehouse, P.L. and Schrama, E.J.O. 2015. Effect of GIA models with 3D composite mantle viscosity on GRACE mass balance estimates for Antarctica. *Earth and Planetary Science Letters*, **414**, 134–143, https://doi.org/10.1016/j.epsl.2015.01.001

Vazquez Becerra, G.E. 2009. *Geodesy in Antarctica: A Pilot Study Based on the TAMDEF GPS Network, Victoria Land, Antarctica*. Ohio State University, Geodetic Science and Surveying Report 492, https://kb.osu.edu/handle/1811/78631

Velicogna, I. 2009. Increasing rates of ice mass loss from the Greenland and Antarctic ice sheets revealed by GRACE. *Geophysical Research Letters*, **36**, L19503, https://doi.org/10.1029/2009GL040222

Velicogna, I. and Wahr, J. 2002. A method for separating Antarctic postglacial rebound and ice mass balance using future ICESat geoscience laser altimeter system, Gravity Recovery and Climate Experiment, and GPS satellite data. *Journal of Geophysical Research: Solid Earth*, **107**, 2263, https://doi.org/10.1029/2001JB000708

Velicogna, I. and Wahr, J. 2006. Measurements of time-variable gravity show mass loss in Antarctica. *Science*, **311**, 1754–1756, https://doi.org/10.1126/science.1123785

Velicogna, I., Mohajerani, Y.E. *et al.* 2020. Continuity of ice sheet mass loss in Greenland and Antarctica from the GRACE and GRACE follow-on missions. *Geophysical Research Letters*, **47**, e2020GL087291, https://doi.org/10.1029/2020GL087291

Villiger, A. and Dach, R. (eds) 2020. *International GNSS Service: Technical Report 2019*. IGS Central Bureau and University of Bern Open Publishing, https://doi.org/10.7892/boris.144003

Wahr, J., Wingham, D. and Bentley, C. 2000. A method of combining ICESat and GRACE satellite data to constrain Antarctic mass balance. *Journal of Geophysical Research: Solid Earth*, **105**, 16 279–16 294, https://doi.org/10.1029/2000JB900113

Wahr, J., Swenson, S. and Velicogna, I. 2006. Accuracy of GRACE mass estimates. *Geophysical Research Letters*, **33**, L06401, https://doi.org/10.1029/2005GL025305

Whitehouse, P.L., Bentley, M.J., Milne, G.A., King, M.A. and Thomas, I.D. 2012. A new glacial isostatic adjustment model for Antarctica: calibrated and tested using observations of relative sea-level change and present-day uplift rates. *Geophysical Journal International*, **190**, 1464–1482, https://doi.org/10.1111/j.1365-246X.2012.05557.x

Whitehouse, P.L. 2018. Glacial isostatic adjustment modelling: historical perspectives, recent advances, and future directions. *Earth Surface Dynamics*, **6**, 401–429, https://doi.org/10.5194/esurf-6-401-2018

Wiens, D.A., Shen, W. and Lloyd, A.J. 2021. The seismic structure of the Antarctic upper mantle. *Geological Society, London, Memoirs*, **56**, https://doi.org/10.1144/M56-2020-18

Wiese, D.N., Landerer, F.W. and Watkins, M.M. 2016. Quantifying and reducing leakage errors in the JPL RL05M GRACE mascon solution. *Water Resources Research*, **52**, 7490–7502, https://doi.org/10.1002/2016WR019344

Willen, M.O., Horwath, M., Schröder, L., Groh, A., Ligtenberg, S.R.M., Kuipers Munneke, P. and van den Broeke, M.R. 2020. Sensitivity of inverse glacial isostatic adjustment estimates over Antarctica. *The Cryosphere*, **14**, 349–366, https://doi.org/10.5194/tc-14-349-2020, https://tc.copernicus.org/articles/14/349/2020/

Williams, D.P. 2003. The effect of coloured noise on the uncertainties of rates estimated from geodetic time series. *Journal of Geodesy*, **76**, 483–494, https://doi.org/10.1007/s00190-002-0283-4

Williams, D.P. 2008. CATS: GPS coordinate time series analysis software. *GPS Solutions*, **12**, 147–153, https://doi.org/10.1007/s10291-007-0086-4

Williams, S.D.P., Moore, P., King, M.A. and Whitehouse, P.L. 2014. Revisiting GRACE Antarctic ice mass trends and accelerations considering autocorrelation. *Earth and Planetary Science Letters*, **385**, 12–21, https://doi.org/10.1016/j.epsl.2013.10.016

Wilson, T., Bevis, M. *et al.* 2019. Understanding themismatch between measured and model-predicted crustal motions across West Antarctica: Insights from POLENET-ANET GPS results. Presented at the Workshop on Glacial Isostatic Adjustment, Ice Sheets, and Sea-level Change – Observations, Analysis, and Modelling, 24–26 September 2019, Ottawa, Canada.

Wolstencroft, M., King, M.A. *et al.* 2015. Uplift rates from a new high-density GPS network in Palmer Land indicate significant late Holocene ice loss in the southwestern Weddell Sea. *Geophysical Journal International*, **203**, 737–754, https://doi.org/10.1093/gji/ggv327

Wouters, B., Bonin, J.A., Chambers, D.P., Riva, R.E., Sasgen, I. and Wahr, J. 2014. GRACE, time-varying gravity, Earth system dynamics and climate change. *Reports on Progress in Physics*, **77**, 116801, https://doi.org/10.1088/0034-4885/77/11/116801

Wu, X., Heflin, M.B. *et al.* 2010. Simultaneous estimation of global present-day water transport and glacial isostatic adjustment. *Nature Geoscience*, **3**, 642–646, https://doi.org/10.1038/ngeo938

Zammit-Mangion, A., Bamber, J.L., Schoen, N.W. and Rougier, J.C. 2015. A data-driven approach for assessing ice-sheet mass balance in space and time. *Annals of Glaciology*, **56**, 175–183, https://doi.org/10.3189/2015AoG70A021

Zanutta, A., Negusini, M. *et al.* 2017. Monitoring geodynamic activity in the Victoria Land, East Antarctica: evidence from GNSS measurements. *Journal of Geodynamics*, **110**, 31–42, https://doi.org/10.1016/j.jog.2017.07.008

Zanutta, A., Negusini, M. *et al.* 2018. New geodetic and gravimetric maps to infer geodynamics of Antarctica with insights on Victoria Land. *Remote Sensing*, **10**, 1608, https://doi.org/10.3390/rs10101608

Glacial isostatic adjustment and post-seismic deformation in Antarctica

Wouter van der Wal[1,2*], Valentina Barletta[3], Grace Nield[4,5] and Caroline van Calcar[1,6]

[1]Faculty of Aerospace Engineering, Delft University of Technology, Kluyverweg 1, 2629 HS Delft, The Netherlands

[2]Department of Geosciences and Remote Sensing, Delft University of Technology, Stevinweg 1, 2628 CN Delft, The Netherlands

[3]DTU Space, National Space Institute, Geodynamics Department, Technical University of Denmark, Elektrovej, 328, 016, 2800 Kongens Lyngby, Denmark

[4]Department of Geography, Durham University, Lower Mountjoy, South Road, Durham DH1 3LE, UK

[5]School of Geography, Planning, and Spatial Sciences, University of Tasmania, Hobart, Tasmania 7001, Australia

[6]Institute for Marine and Atmospheric Research Utrecht, Utrecht University, Utrecht, 3508 TA, The Netherlands

WvdW, 0000-0001-8030-9080

*Correspondence: w.vanderwal@tudelft.nl

Abstract: This chapter reviews glacial isostatic adjustment (GIA) and post-seismic deformation in Antarctica. It discusses numerical models and their inputs, and observations and inferences that have been made from them. Both processes are controlled by mantle viscosity but their forcings are different. Ongoing GIA induced by the loss of ice since the last glacial maximum (LGM) could have amounted to 5–15 m of global sea-level rise. However, mantle viscosity is so low in parts of West Antarctica (*c.* 10^{18} Pa s) that changes in ice thickness over the last centuries and decades have controlled the current uplift rates there. The uplift due to GIA has promoted ice-sheet stability since the LGM, and in West Antarctica GIA is a significant negative feedback on the current decline of the ice sheet. Post-seismic deformation following the 1998 earthquake near the Balleny Islands south of New Zealand has been detected in global navigation satellite system (GNSS) data and compared to model outputs. The best-fitting viscosity for this area is *c.* 10^{19} Pa s, similar to GIA-based estimates for the Antarctic Peninsula. Future work should focus on unifying descriptions of viscosity across geodynamic models, and integrating information from seismic, gravity, experimental and geological data.

The mantle in Antarctica plays an important role in changing the elevation of the bedrock through forces applied to the base of the lithosphere, including convective flow of the mantle (dynamic topography), tectonics, heating and cooling of the lithosphere, and load redistribution at the surface (Paxman *et al.* 2019; Paxman 2021, this volume). Loading of the surface can be caused by changes in ice-sheet and snow thickness, sedimentation and erosion, sea-level changes, and changes in pressure due to oceanic and atmospheric currents and tides. Here we focus on the secular changes induced by the changing surface loading of the ice sheet and the stress redistribution caused by earthquakes. The adjustment of the Earth following growing or melting ice sheets is termed glacial isostatic adjustment (GIA) and is mostly measurable on a timescale from decades to millennia.

The adjustment of the Earth's crust and mantle after slip on the fault plane of an earthquake is called post-seismic relaxation; it is mostly measurable on a time frame from years to decades. Both are governed by the resistance to flow of the mantle, and numerical models rely on a similar description of the rheology of the mantle. Both processes can lead to vertical and horizontal motion that is linear or quasi-linear in global navigation satellite system (GNSS) time series (Scheinert *et al.* 2021, this volume). Post-seismic deformation is only detectable following large earthquakes, and only a few of these have occurred recently enough for them to have been measured in GNSS time series. In agreement with Scheinert *et al.* (2021, this volume), we will use the general term GNSS, which encompasses all global systems, even though most GNSS data on Antarctica are obtained from GPS receivers.

The response of the solid Earth to loading change can be instantaneous with respect to the forcing, in which case it is governed by the elastic properties of the crust and mantle that can be characterized, for example, by Young's modulus. The elastic deformation due to present-day surface-load changes is commonly applied as a correction to GNSS measurements (Scheinert *et al.* 2021, this volume). The solid Earth also shows a delayed response to the forcing, driven by the viscoelastic properties of the mantle. In this chapter, we focus on this viscoelastic behaviour. Modelling of the response of the solid Earth generally includes the elastic effect that occurs instantaneously at the moment of loading, as well as the delayed viscous effect (e.g. Peltier 1974).

Antarctica is particularly interesting for GIA studies as there are substantial variations in the Earth's structure beneath East and West Antarctica, as shown by the lower than average seismic velocities beneath West Antarctica. To first order, low seismic velocities correspond to high temperatures (e.g. Goes and van der Lee 2002). Therefore, West Antarctica is inferred to have a warmer and, hence, weaker mantle, while the East Antarctic Craton is underlain by a colder and stiffer mantle (Berg *et al.* 1989; Morelli and Danesi 2004; Hansen et al. 2014; Martin *et al.* 2014; An *et al.* 2015; Burton-Johnson *et al.* 2020; Wiens *et al.* 2021, this volume). Thus, it is expected that GIA and post-seismic deformation would proceed at a different rate in West Antarctica than in East Antarctica, and simulations should take this into account.

The following section of this chapter is dedicated to GIA. A brief introduction to the topic is given and the physics associated with the processes are discussed, together with the main results derived from existing GIA models. Readers familiar with GIA processes and models may wish to skip this part. For more details, the review paper by Whitehouse (2018) may be consulted. The section continues with a review of results of GIA models in Antarctica, focusing on deglaciation since the last glacial maximum (LGM) and inferences of mantle viscosity. Particular attention is paid to the interaction between GIA and ice-sheet dynamics. Post-seismic deformation is discussed in the third section of this chapter: specifically, the occurrence of earthquakes, a description of mantle flow and the results of post-seismic deformation following the 1998 earthquake. Finally, conclusions are presented and potential future work for GIA and post-seismic deformation are outlined.

From: Martin, A. P. and van der Wal, W. (eds) 2023. *The Geochemistry and Geophysics of the Antarctic Mantle*.
Geological Society, London, Memoirs, **56**, 315–341,
First published online 10 November 2022, https://doi.org/10.1144/M56-2022-13

GIA

The growing concern about climate change is generating great interest in the rate at which ice-covered regions are losing mass and transferring water to the oceans. Estimates of current ice-mass changes are produced by analyses that incorporate geodetic and glaciological observations (Shepherd *et al.* 2018). However, to predict future ice-sheet melt we also need physical models of ice dynamics. In this context, the behaviour of the solid Earth is particularly important. Ice change triggers GIA, which in turns affects the ice dynamics. For example, a rising coastal region can lead to a displacement of the grounding line, affecting the way in which the oceans interact with land-based ice (Thomas 1979; Oerlemans 1980).

The GIA process is not only relevant for ice sheet–solid Earth feedback, as the GIA signal is present in geodetic observations of present-day uplift and gravity change rate. For example, in GRACE (Gravity Recovery and Climate Experiment) satellite measurements of mass change in Antarctica, GIA makes up about one-third of the total signal (Shepherd *et al.* 2018). Therefore, in order to recover the signal due to the present-day ice melt, GIA must be removed from the observations. The deformation of the solid Earth can be detected on the surface by GNSS data on rock outcrops adjacent to the ice sheets (Scheinert *et al.* 2021, this volume). Many of such observations are available in Greenland, for example (Bevis *et al.* 2012), but observations in Antarctica are much more spread out due to the harsh terrain and because most of the bedrock is covered in ice. Therefore, to obtain spatial patterns of surface deformation, as well as to correct other relevant climate indicators for GIA, numerical simulations are used together with available observations (e.g. Whitehouse *et al.* 2019). With the increasing coverage of geodetic measurements, inversion techniques can exploit different sensitivities to changes in ice, snow and the solid Earth in order to separate out the measured signals (see Martín-Español *et al.* 2016*a*; Scheinert *et al.* 2021, this volume).

The GIA process is also an interesting topic of study in itself. The growth and decay of an ice sheet can be seen as a large deformation experiment. The response of the Earth's surface is governed by the viscosity of the underlying mantle, so measuring the GIA can be seen as a way to measure viscosity at depth. As viscosity is the parameter that controls the speed of motion in the Earth's mantle, which drives plate tectonics, constraining viscosity is an important contribution of GIA research. However, since viscosity affects several processes in the Earth, such as mantle convection (e.g. Bredow *et al.* 2022, this volume). GIA studies are not the only way to infer the viscosity of the Earth's interior. GIA research complements other methods for estimating viscosity: for example, through seismic models (e.g. Lloyd *et al.* 2020; Wiens *et al.* 2021, this volume) and laboratory experiments on mantle rocks (Hirth and Kohlstedt 2004), combined with information on rock composition and local mantle conditions derived from xenoliths (e.g. Martin 2021, this volume). How viscosity is derived for use in GIA models is described in Ivins *et al.* (2021, this volume).

This section presents methods for modelling GIA, discusses the main inputs and results of this modelling, and the observational constraints. It starts with a brief general history of GIA in general, after which the state of the art on GIA modelling in Antarctica is presented. In terms of data, the focus is on relative sea level (RSL: the difference between the surface of the sea and the ocean bottom), as the geodetic data are already described in Scheinert *et al.* (2021, this volume). Concerning numerical models, attention is paid to the description of the models themselves and their inputs, with a focus on ice thickness, as rheology is covered in Ivins *et al.* (2021, this volume). The focus is on the last glacial cycle because this influences present-day geodetic observations and more observations of the GIA process are available for this period. GIA projections for the Antarctic Ice Sheet are briefly addressed and the section concludes with pointers for future research.

Brief history of GIA research

What follows is a brief summary of the history of the concept of GIA up to the point that numerical models were used extensively from the 1980s. A more extensive history can be found in Ekman (1991), Steffen and Wu (2011) and Whitehouse (2018). Historically, the concept of GIA is associated with Fennoscandia, specifically a small Swedish village on an island in the Baltic Sea, where in 1731 Anders Celsius was invited to explain why a group of rocks in the coastal area was apparently slowly rising above the level of the sea (Ekman 2013). Even though Celsius and his contemporaries did not realize it at the time, they were observing land uplift and a sea-level readjustment caused by deglaciation since the LGM (*c.* 20 kyr ago). Given the strategic role of these rocks, which had been used for hunting seals for centuries, empirical records of their height with respect to sea level had been kept for generations, allowing the phenomenon to be tracked that would otherwise have been difficult to notice in one life time. All the important elements of GIA were already present: the uplift of the Earth's surface, the role of RSL as a reference for the Earth's deformation and the speed (or, rather, the slowness) of the phenomenon.

Celsius was able to compute the uplift rate of the rock; however, he was not able to provide the correct explanation, which was suggested first in 1865 by Thomas Jamieson who argued that 'the enormous weight of ice thrown upon the land may have something to do with this (land level) depression' (Jamieson 1865, p. 178). Until the pioneering work of Haskell (1935), there had been no attempt at quantitative modelling. Haskell used a simplified model that neglected the crust but considered a viscous layer underlain by a rigid layer on which the load was applied. To constrain his model, he used records of past changes in RSL. Assuming a certain thickness for the upper layer, he estimated the viscosity of this viscous layer to be around 10^{21} Pa s. The number, although based on a model that is incomplete, is remarkably close to the modern estimates of the average viscosity for the upper mantle (Mitrovica 1996). A viscosity of this magnitude translates into a characteristic relaxation timescale of thousands of years.

The success of Jamieson's idea of a vertical viscous response of the Earth to deglaciation was one of the main arguments used by Wegener in his book *The Origin of Continents and Oceans* (first published in 1915 in German; Wegener 1915) in favour of his new hypothesis of continental drift. He argued that if vertical viscosity-regulated motions could occur, it could also be possible that horizontal motions could be sustained. Thus, the idea of GIA predates, and even contributed to, the idea of continental drift.

New and more reliable methods to date sedimentary changes (palynology and, later, carbon dating) substantially improved and significantly expanded the sea-level records, stimulating the development of new GIA models. In the 1970s, the first spherically symmetrical, self-gravitating, models were developed (Peltier 1974; Cathles 1975). Spherically symmetrical means that the Earth model consists of concentric layers with constant properties, while self-gravity refers to the inclusion in the model of changes in gravity due to deformation. This can be considered the beginning of modern GIA modelling. A lot of effort went into reproducing the increasingly precise sea-level records with ice-history reconstructions and inferring the Earth's viscoelastic properties for distinct layers as opposed to a single average value. This required a detailed understanding of the relation between the spatial characteristics of the load and the sensitivity of the response of the different components of the model such as the lithosphere thickness and the viscosity profile.

Parallel to the studies on the solid Earth, ideas developed on the adjustment of sea level, the concept of 'eustasy'. Edward Suess (1888) described the idea that sea level is at a minimum during a glaciation when water from the oceans gets stored on land in the form of ice, and that it then reaches a maximum in the interglacial periods when the ice sheets retreat. However, this concept results in a uniform sea-level variation during the glacial oscillations, treating the ocean basin as a giant bathtub with extremely steep boundaries between water and land that is filled or depleted as needed. This is in contradiction with what records of sea level show. Palaeoreconstructions of sea level based on geological observations and debris dating that cover several millennia show a much more varied behaviour: sea-level records at the coasts close to the regions covered in ice at the LGM consistently show a sea-level fall in Fennoscandia and North America (Walcott 1972). Clearly, there is another process acting. The analogy of the bathtub fails because it neglects gravity, which dominates on a global scale. Moving large masses around changes the gravity and equipotential surfaces that determine sea level at rest. The inclusion of gravitational forces explains how the melting of a large ice sheet reduces gravity in a region around the shrinking ice sheet, so that the equipotential surface is lowered and a local sea-level decrease is obtained (Farrell and Clark 1976; Clark *et al.* 1978). Changes in the gravity field and changes in sea level cause deformation of the Earth, and are important to include in GIA models.

To get an idea of the magnitude of deformation, we quantify the GIA process in some key areas. Huge continental ice sheets of more than 3 km in thickness covered Canada, north-western Europe (Fennoscandia) and West Antarctica at the LGM when they started melting *c.* 20 kyr ago. Melting was essentially completed 7 kyr ago and resulted in a net rise of the mean sea level over the global ocean of about 130 m between LGM and present day (Clark and Tarasov 2014). This net increase in water is called a eustatic sea-level rise and it reached a rate of 40 mm a^{-1} during the deglaciation period (Lambeck *et al.* 2014). Due to the lag in Earth deformation having a characteristic timescale of thousands of years, the solid Earth surface in former ice-covered regions is currently uplifting by up to 10 mm a^{-1}. Just beyond the ice sheets, forebulges emerged during the growth of former ice sheets (see a schematic representation in Fig. 1), which currently subside at a rate of several millimetres per year, as detected by GNSS instruments (e.g. Schumacher *et al.* 2018). Thus, the bedrock deformation under the present-day ice sheets differs per region, depending on the historical and current loading and unloading of ice.

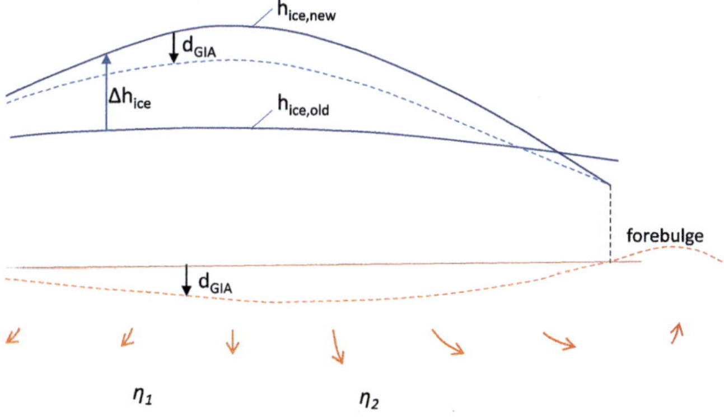

Fig. 1. Schematic representation of GIA following an increase in ice thickness of an ice sheet resting on a bedrock that has two different viscosities. The ice thickness increases by Δh_{ice}. As a result of this increase and past increases in ice thickness, the bedrock deforms by d_{GIA} with a forebulge forming outside the area of ice increase. Because viscosity η_2 is assumed to be smaller than η_1, deformation is larger under the right half of the ice sheet in a given time.

GIA physics simplified

When a load is placed on the Earth, after initial elastic displacement, pressure is distributed into the mantle, which deforms in a way that can be compared to deformation of memory foam or a sponge after pressing it: a slow readjustment to equilibrium controlled by the dynamic viscosity parameter, η, that is usually used for viscous flow (Fig. 1). Measurable deformation some time after the change in load occurs if the viscosity is low enough for rocks to creep on timescales of a glacial cycle or shorter. The numerical models used to compute GIA and post-seismic relaxation are generally viscoelastic, meaning that they include a viscous component as well as instantaneous elastic changes in deformation and in the gravitational field (and, therefore, in sea level). The Earth's interior, down to the fluid outer core, can be considered viscoelastic but the top layer is essentially elastic and is termed the lithosphere (not shown separately in Fig. 1). The flexural rigidity of the lithosphere acts as a dampening filter for both the forcing and the consequent deformation. The thinner the lithosphere, the more spatially concentrated is the deformation induced on the mantle; the thicker the lithosphere, the more distributed is the deformation. This can be used to infer the thickness of the lithosphere when the load's spatial and temporal histories are known. Finally, brittle deformation also occurs within the lithosphere, causing fractures under stress associated with surface loading, but this plays a smaller role in relieving overall stresses. This can be understood from the observation that GIA-induced earthquakes have a small effect on nearby faults (Brandes *et al.* 2015).

The subsidence below a thickened ice sheet must be compensated by uplift elsewhere because although the Earth is compressible, its volume does not change significantly. The basic shape of a GIA pattern is a concentric pattern of deformation, the so-called footprint of the ice sheet, with uplift in a band around the ice load – the forebulge (Fig. 1). Upon melting of the ice sheets, meltwater flows into the oceans and the overall mass distribution on the Earth's surface changes. This results in changes in the gravitational pull and a sea-level fall near the ice sheet that were mentioned in the previous subsection. The spatial pattern of sea-level change that results from ice melt has been labelled 'sea-level fingerprint' (Mitrovica *et al.* 2009), and includes gravitational effects and elastic deformation due to ice melt. In addition, the meltwater acts as a load that is redistributed across the globe, and which deforms the ocean basins. This leads to continental levering, which is the subsidence of the ocean floor and a corresponding uplift of the neighbouring continent as a result of the addition of meltwater (Walcott 1972; Mitrovica and Milne 2002). Also relevant to the sea-level change is the concept of ocean syphoning, where water flows from equatorial regions towards forebulges that are subsiding below the ocean, especially around Canada and Antarctica (Mitrovica and Peltier 1991).

Further improvements in sea-level modelling account for shifting land–ocean boundaries (Johnston 1993; Peltier 1994) and grounded ice that can become floating as sea level rises (e.g. Milne *et al.* 1999). Finally, the change in ice load and deformation of the Earth affect the Earth's moment

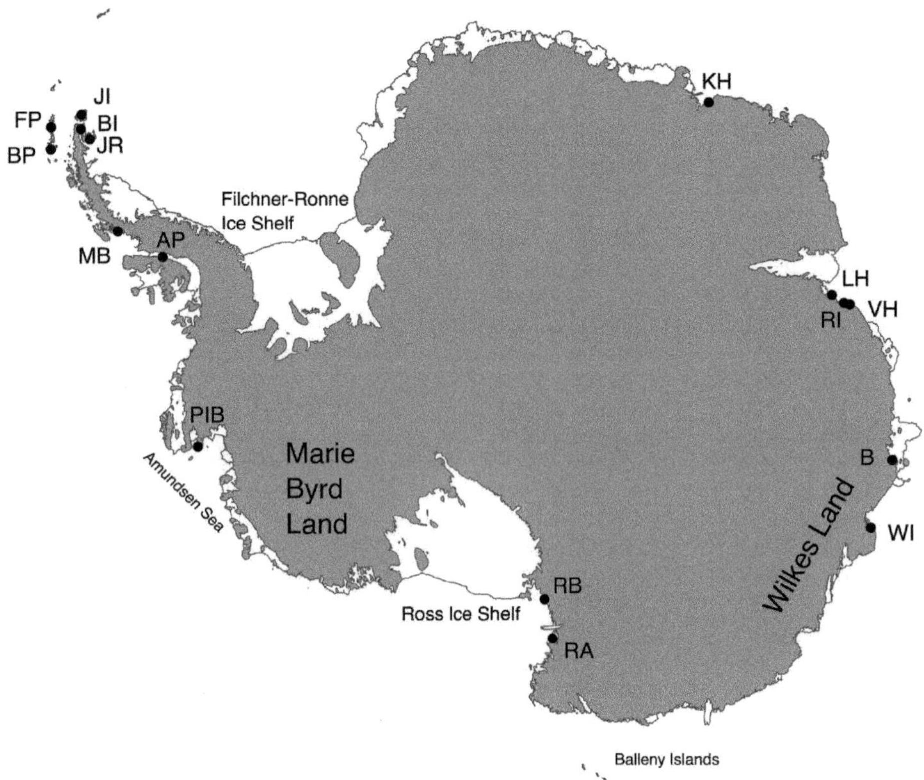

Fig. 2. Relative sea-level data in Antarctica compiled in Whitehouse *et al.* (2012*a*) (largely based on Bassett *et al.* 2007) and sites from Okuno and Miura (2013). Locations going along the coastline clockwise starting at KH: KH, Kizahasi Hama; LH, Larsemann Hills; RI, Rauer Islands; VH, Vestfold Hills; B, Bunger; WI, Windmill Island; RA, Ross A (Terra Nova Bay); RB, Ross B (Cape Roberts); PIB, Pine Island Bay; AP, Ablation Point; MB, Marguerite Bay; BP, Byers Peninsula; FP, Fildes Peninsula; JI, Joinville Island; BI, Beak Island; JR, James Ross. The 200 largest grounded regions are patched grey, and Antarctica's coast (including ice shelves) is shown in blue. These are plotted with MATLAB routines from Antmap (Greene *et al.* 2017) based on Bedmap2 (Fretwell *et al.* 2013).

of inertia, which changes the direction and magnitude of the rotational vector of the Earth. This in turn changes the rotational potential, which induces deformation of the solid Earth as well as changes in sea level (Bills and James 1996; Milne and Mitrovica 1998).

Considering the discussion of GIA above, it is clear why records of sea level at different locations on the globe can be used to gain knowledge on the history of deglaciation and on the Earth's physical properties. In the following subsection, we discuss observations available in Antarctica, including sea-level records.

GIA observations in Antarctica

Constraints on GIA are much more sparse in Antarctica than in other continents but the quantity and the quality has been steadily improving, so that a robust picture appears of some aspects of GIA in Antarctica. In this subsection, we discuss palaeosea-level data; geodetic data are reviewed in Scheinert *et al.* (2021, this volume). In addition to these two types of data, there are a few other manifestations of the GIA process that could be considered as observable. Post-glacial rebound is hypothesized to have contributed to the 1988 Antarctic earthquake south of the Balleny Islands (see their location in Fig. 2) (Tsuboi *et al.* 2000). Seismic events in the Ellsworth Mountains are thought to be related to ongoing GIA-induced stress, acting on pre-existing structural weaknesses (Lucas *et al.* 2021). Stress orientation obtained from boreholes indicates that significant GIA stress is present that could have triggered other earthquakes (Ivins *et al.* 2003). Glacial-induced earthquakes have also been suggested as contributing to sediment slides in the Miocene in the Western Wilkes Land margin (Donda *et al.* 2008). These hypotheses are interesting to explore in future modelling but as of yet do not yield useful constraints on GIA models.

Sea-level indicators. Information on sea level can come from the dating of organic material such as shells and seaweed in raised beaches. A clear raised beach is not always visible as the land rise and corresponding sea-level change happens gradually. Some findings indicate a minimum height for sea level, such as shells: sea level must have been higher than the location at which they are deposited (e.g. Bentley *et al.* 2005). Some findings represent a maximum sea level as they are deposited subaerially, such as seal hairs and penguin droppings (Bassett *et al.* 2007). Sometimes sea-level tilt can be derived, which provides a stronger constraint on spatial sea-level changes (Konfal *et al.* 2013).

The most accurate data are taken from so-called isolation basins. When land is rising, bodies of water that lose their connection to the sea transition from marine to lacustrine. This transition is recorded within the sediment deposited in these water bodies, and can be detected in sediment core studies and dated. The reverse can also occur: a transition from lacustrine to marine sediments indicates a sea-level rise.

A number of sea-level indicator data in Antarctica has been reported in recent literature; their location is shown in Figure 2. Bassett *et al.* (2007) compiled data from eight locations; their location is shown in Figure 2, which was extended in Whitehouse *et al.* (2012*a*) to 14 sites that were also used by Argus *et al.* (2014). Okuno and Miura (2013) used a subset of these data and included an extra site in East Antarctica (Bunger). Furthermore, raised beaches in the northern part of the Antarctic Peninsula have been dated to the late Holocene, indicating increasing sea-level fall during that time (Simms *et al.* 2018). A reconstruction in Joinville Island at the tip of the Antarctic Peninsula shows a sea-level fall in the last 3 kyr, with significant changes in rates (Zurbuchen and Simms 2019). Records in East Antarctica have been extended by Hodgson *et al.* (2016) and Verleyen *et al.* (2017), with the latter pointing out the possibility of neotectonics contributing to vertical deformation.

The amount of sea-level data is sparse, and there are no data available between the LGM and 12 ka before present (BP). However, in several locations beach deposits and lake and sea sediments can be used to infer sea level before the LGM (e.g. Nakada *et al.* 2000). They indicate a sea level close to present, which could only have been achieved with excessive ice (due to the self-gravity of the ice sheet), that would place the glacial maximum well before 20 ka (Ishiwa *et al.* 2021).

Interpretation of the GIA observations require numerical models to compute uplift rates and gravity changes for a given Earth model and ice-sheet thickness history, which can then be compared to the observations. Results of numerical GIA models are discussed in the following subsection.

Numerical models and results

The earliest GIA models solved the differential equations describing the physical phenomena involved, starting with the conservation of momentum and mass (Cathles 1975). The traditional way of solving them is by means of the normal mode method (Peltier 1974; Wu and Peltier 1982). This method is suitable for a layered Earth with radially varying viscoelastic parameters. The normal mode method succeeds in computing the response for a spherical body with concentric layers with constant parameters (we will refer to this as a 1D model), where the response is the change in deformation and the change in gravity. For each layer, the elastic parameters (density and Young's modulus) and viscosity need to be specified. For ice-dynamics models, the GIA component is often simulated by a model that is conceptually simpler than the normal mode method, the so-called elastic lithosphere and relaxing asthenosphere (ELRA) model, which will be addressed in the 'Coupling of ice to solid Earth: feedbacks and processes' subsection later in this section. The asthenosphere is the upper part of the upper mantle below the lithosphere. For the ELRA model, parameters are also constant for the two layers involved.

The normal mode theory assumes linear viscoelasticity, which means that stress is proportional to strain at a given time and the response to different stresses can be summed (Findley *et al.* 1976). Peltier (1974) and Wu and Peltier (1982), along with many others following their methodology, assume a Maxwell rheology (e.g. see Sabadini *et al.* 2016), which is a particular form of linear viscoelastic rheology where the elastic deformation occurs first upon the application of a force, which can be the increase in weight of the ice sheet, after which viscous deformation takes over. Other linear viscoelastic rheologies have been employed in GIA models: for example, the Burger's rheology, which uses a separate viscosity for the short-term initial and the long-term steady-state responses (Yuen and Peltier 1982; Sabadini *et al.* 1985). This rheology is commonly used in post-seismic deformation (see 'Conclusions and future work' later in this chapter). It has not been used widely to explain GIA observations since the first studies in the 1980s but it has received more attention recently (Caron *et al.* 2017; Ivins *et al.* 2020), not least because the behaviour in deformation experiments points towards Burger's rheology (Faul and Jackson 2015). Non-linear rheology, in which the strain rate depends non-linearly on stress, has also been included early on in GIA modelling (e.g. Nakada 1983; Wu 1992), with more recent use in GIA models informed by laboratory experiments on mantle rocks (e.g. van der Wal *et al.* 2013).

Model results produced by the normal mode method can explain GIA observations to a large extent for a global average viscosity profile (e.g. Peltier 2004; Lambeck *et al.* 2014; Lau *et al.* 2016), although studies do not all agree on the same radial viscosity profile, partly because of non-uniqueness in the inversion: different viscosity profiles can explain the same subset of observations to within the noise level (Paulson *et al.* 2007). Viscosity in the real Earth can vary by orders of magnitude but the normal mode method can still be used for applications in regions where viscosity variations underneath and near the ice sheet are small enough that viscosity can be approximated by an average value that varies with depth only.

For several regions, including Antarctica, viscosity cannot be assumed to be constant beneath the ice-covered region, and the viscosity, along with other Earth material properties in the model, needs to change with latitude and longitude. To model the effect of such lateral variations in viscosity, numerical methods are generally applied in which the Earth is discretized into small elements, finite elements or finite volumes. Global spherical Earth models that can incorporate lateral variations in viscoelastic parameters have been developed since the 2000s (Martinec 2000; Zhong *et al.* 2003; Wu 2004; Latychev *et al.* 2005). These models are sometimes referred to as 3D GIA models, where 3D refers to the variations in the Earth model input rather than the domain of the model itself. Most knowledge on GIA still comes from global 1D models as these have been around longer, and because 3D models have a long computation time and a potentially large number of degrees of freedom requiring additional constraints.

The largest uncertainty in GIA models stems from the unknown ice thickness through time. For several regions, ice sheets have left marks in the landscape from which the ice extent can be contoured and dated. However, in Antarctica, the ice sheet still covers most of the land and changes in the thickness of the ice sheet have left little indication on the landscape, unless particular glaciers were confined to a mountain valley that show the erosional limits (trimlines: Denton *et al.* 1992; Bentley and Anderson 1998) or there is other geological evidence: for example, from glacial and post-glacial deposits (e.g. Brook *et al.* 1993; Mackintosh *et al.* 2014) or glaciovolcanology (Smellie *et al.* 2009, 2011). An overview of studies on the large-scale changes in the Antarctic Ice Sheet since the LGM is given in the following subsection.

For the GIA-based method of inferring ice-sheet history, a forward GIA model is used that assumes a certain viscosity profile. However, the mantle viscosity is often derived from GIA studies that assume a certain ice history. This circularity is reduced by the different data available and the different sensitivities they have to the input Earth model parameters. For example, depending on the spatial scale of the load, the depth at which the mantle responds the strongest changes: smaller ice sheets, such as in Scotland, affected a shallower part of the mantle, while the largest ice sheets, such as the Laurentide Ice Sheet, excited a response in the lower mantle. This characteristic can therefore be used to employ observations around different ice sheets to constrain viscosity at different depths of the mantle. This qualitative idea has been confirmed by computing 'sensitivity kernels' from numerical models (Peltier 1976; Wu 2006; Lau *et al.* 2016) that show which depth of the mantle has the most impact on a certain observable. A sensitivity kernel can be computed with a forward GIA model. A small change in the viscosity in a small sublayer is applied and the change in the observable is computed, all with respect to a certain reference viscosity profile. By doing this for many sublayers, a plot can be created of the effect in the observable v. the depth at which the viscosity change is applied. An example for the Scandinavian Ice Sheet is shown in Figure 3. In this case, the observable is the present-day change in gravity but the curves would be similar for present-day uplift rate. Note that the magnitude itself is not relevant as it scales with the magnitude in the viscosity change but the relative magnitude is interesting. The sign is positive in the figure, meaning that the change in viscosity increases the gravity rate. However, a negative sign could also be achieved because both a viscosity that becomes very low and a viscosity that becomes very high will reduce gravity and uplift rates. The sensitivity with depth is directly dependent on the viscosity: for a profile with higher viscosity in

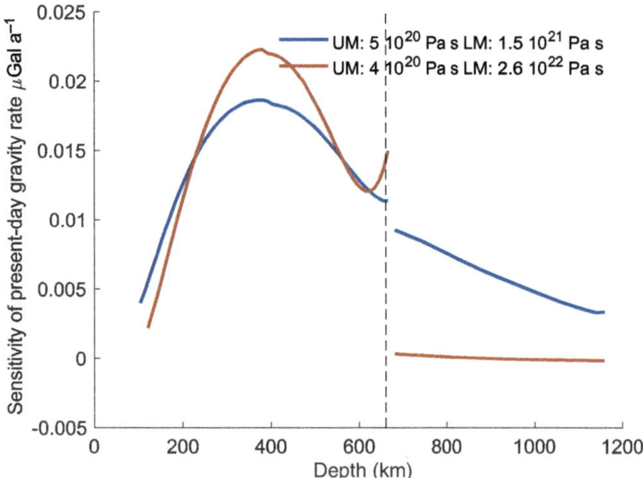

Fig. 3. Sensitivity of the present-day gravity rate in the centre of the Scandinavian ice sheet to viscosity at depth for two viscosity profiles. UM, upper mantle, above 670 km; LM, lower mantle, between 670 km and the core. The ice-sheet model used is ICE-5Gv1.2 (Peltier 2004). Source: Reproduced from van der Wal *et al.* (2011).

the lower mantle (orange line), more of the sensitivity is concentrated in the upper mantle. For Antarctica, the change in ice-sheet thickness since LGM in West Antarctica is similar to the thickness of the Scandinavian Ice Sheet but with a smaller spatial extent. Therefore, the sensitivity is expected to be concentrated at a shallower layer compared to what is shown in the figure.

A further reduction in the circularity of the problem is possible by recognizing that relaxation after the ice has disappeared mostly depends on the viscosity and not on the ice history (Wieczerkowski *et al.* 1999). Furthermore, combining and integrating sea-level records with GNSS measurements of uplift rate and gravity variation measurements from GRACE and GRACE-FO (GRACE Follow-On) satellite measurements reduces the uncertainty, although unique inversion for viscosity in three layers or more is challenging, even with a combination of these data (Paulson *et al.* 2007).

Despite the limitations, the GIA approach allows an estimation of Earth properties of the lithosphere and mantle that complements information coming from other geological, geophysical and geodynamic studies. Global GIA models predict RSL and other observations from all over the world, taking into account the global effect of the water redistribution by computing the sea-level change that is consistent with the loading and solid Earth deformation. Modern analyses indicate that the global average value for upper-mantle viscosity lies close to the value suggested by Haskell in 1935 (Mitrovica 1996) as stated in the earlier subsection on 'Brief history of GIA research'. A refinement of this result indicates that viscosity increases from the upper to the lower mantle by a factor of 2–3 (e.g. Peltier 2004) or more (e.g. Kaufmann and Lambeck 2002). The characteristics of the GIA process are such that both the low-viscosity and the high-viscosity solutions are allowed by the sea-level data (Lambeck *et al.* 2014) but the ambiguity in the lower-mantle viscosity can be reduced by including constraints on the Antarctic ice volume (Caron *et al.* 2017).

The estimates discussed above represent an average value for viscosity. They will provide incorrect predictions of RSL change and modern uplift rates in areas where the actual viscosity structure greatly deviates from the global average, such as in West Antarctica. The solid Earth's internal properties and Earth's loading history are interdependent in GIA inversions and both are unknown at a fine scale even if a global picture emerges. All of these problems exist in Antarctic GIA studies but steady progress with increased measurements and model refinements has led to GIA results for Antarctica that are gaining acceptance.

In 3D GIA models, the larger number of degrees of freedom combined with a longer computation time prohibit simulating a large parameter space. For this reason, uncertainty in 3D model predictions is not well quantified. However, 3D GIA models provide higher accuracy over 1D models for several applications. Some key sensitivities of GIA models can be summarized to facilitate interpretation of model results. First, the bedrock response at a certain location depends on the ice-sheet forcing in the surrounding area. The stress is transported from the area of loading downwards into the mantle but also outwards where it also deforms the surface. The observed response is a function of viscosity along this stress path (Paulson *et al.* 2005). As stress is largest beneath the load, viscosity beneath the load has a large control on the bedrock response but viscosity outside the loading area is also relevant. For example, near the edge of the ice sheet, the uplift rate is most sensitive to the local viscosity as opposed to the viscosity beneath the load (Wu 2006). In fact, this sensitivity is as strong as the sensitivity of the uplift rate to the viscosity in the centre of the ice sheet. RSL data outside the ice cap have an even stronger sensitivity to the viscosity outside the ice sheet; sensitivity to viscosity at a certain depth can have an opposite sign between two locations near the observation point (Crawford *et al.* 2018). This shows that RSL data have more resolving power inside the Earth than the uplift rate data, which demonstrates the importance of RSL data for GIA inversions. The RSL data was discussed in an earlier subsection.

Ice-sheet history in Antarctica

Applying a forward GIA model requires the ice thickness through time to be known. To reconstruct such a history, glacial geological constraints, ice-dynamics modelling and GIA observations are used, or a combination of these (e.g. Bentley 1999; Briggs and Tarasov 2013; Siegert *et al.* 2022). In the case of ice dynamics, ice-sheet behaviour is simulated using climatic conditions as input (e.g. Huybrechts 2002). In a GIA-based scheme, ice thickness is adjusted until a good fit is obtained with palaeo sea-level records, possibly including geodetic data (e.g. Peltier 2004). Several approaches lie in between: a simplified ice model with a regional tuning parameter (Lambeck *et al.* 2017; Gowan *et al.* 2021) or averages of an ensemble of ice-physics-based realizations can be employed (Tarasov *et al.* 2012).

Geological inferences of ice thickness. Comprehensive geological and ice-core data have become available for the deglaciation since the LGM. The extent of the ice sheet can be derived from ice-marginal features such as moraines, subglacial features and glacial deposits, which can be dated using different techniques (e.g. Whitehouse *et al.* 2012*a*; Siegert *et al.* 2022). An important dating technique is cosmogenic nuclide surface exposure dating. Nuclides form in rocks as a result of bombardment by cosmic rays. The longer a rock is exposed, the more nuclides it will contain. Therefore, counting the nuclides gives an indication of the time that the rock has been exposed, which can pinpoint the formation of a moraine or other features (Davis 2020). The technique, however, is affected by GIA itself, as the elevation of the sample affects nuclide production (e.g. Jones *et al.* 2019).

A second dating technique is optically stimulated luminescence dating, which measures the remaining electrons created from naturally present radioactive material after the electrons are 'reset' by exposure to sunlight before ice crystals are deposited. Where organic material is present, carbon dating can be used: for example, organic material in sediments or of animal remains at raised beaches. The latter usually provides a minimum age: whenever animals were present there was no ice. However, carbon dating is affected by natural variations in the background rate and, therefore, the ages have to be calibrated to calendar ages before they can be compared with model predictions.

Other data include mapping of grounding lines by seismic reflection profiles and sonar (Whitehouse *et al.* 2012*a*), and detection of grounding-line retreat in the transition of sediments (Briggs and Tarasov 2013). Such data are sparse in East Antarctica (Arndt *et al.* 2013) but exposure dating of glacial features has been performed for a number of locations on the coast of East Antarctica (Mackintosh *et al.* 2014). In the interior, constraints are more difficult to obtain. The isotope distribution in ice cores can be a proxy for the ice-accumulation rate, from which changes in ice thickness can be obtained (e.g. Parrenin *et al.* 2007). This requires an ice-flow model but also an accurate model for bedrock elevation and its changes (e.g. Siddall *et al.* 2012).

Finally, glaciovolcanism offers insights into ice-sheet parameters, such as age, thickness and surface elevation (Smellie 2018). The contact of lava with ice produces meltwater when the volcanic cone reaches the bottom of the ice sheet. The interaction of meltwater and lava produces a recognizable stratigraphic unit. Therefore, the thickness of the layer thus produced approximates to the thickness of the prevailing ice sheet. The elevation of the layer can also indicate the thermal regime, whether ice is cold or wet based (e.g. Smellie *et al.* 2011). The advantage of glaciovolcanic evidence is that volcanic deposits are thick and are resistant to erosion but the time resolution can be low as it depends on eruption frequency. Applications in Antarctic include the Antarctic Peninsula Ice Sheet in the Pliocene (Smellie *et al.* 2009; Davies *et al.* 2012), and the thermal regime of the East Antarctic Ice Sheet between 12 Ma and present (Smellie *et al.* 2011). The compilation by the RAISED Consortium in 2014 provided time-slice maps of grounding-line position and ice-sheet thickness at 5 kyr intervals (Bentley *et al.* 2014; see also Siegert *et al.* 2022). The results agreed with earlier findings of a total volume of Antarctic ice loss since the LGM of below 10 m sea-level equivalent (SLE). The amount of retreat varies across Antarctica, with the largest retreat found in the marine-based sector of the Ross and Filchner–Ronne ice shelves (for their location see Fig. 2). The timing of retreat also varied, from early retreat in the Antarctic Peninsula and the Amundsen Sea sector to late retreat in Marie Byrd Land (for the locations see Fig. 2). However, ice also advanced in the late Holocene (e.g. in the Antarctic Peninsula: Simms *et al.* 2021), as possibility already hinted at in Hollin (1962). In East Antarctica, ice extended to the shelf at the LGM but was probably thinner in the interior (Mackintosh *et al.* 2014). Recently, the role of ocean–ice interaction has received more attention. This interaction could provide a means for parts of East Antarctica to be susceptible to instability caused by the ocean ‘eating’ into an ice sheet on deepening topography (Schoof 2007; Kawamata *et al.* 2020).

Ice-sheet dynamics. To recreate an ice sheet, knowledge of the dynamics of an ice sheet is helpful but an ice sheet is subject to many forces and interfaces that are difficult to take into account. In particular, the interactions of an ice sheet and shelves with bedrock in subglacial water transport and with the ocean are crucial but complex, and are an active area of research (Siegert and Golledge 2022). Therefore, a unique representation of the ice sheet through time cannot come from ice dynamics alone. Proxies for total ice volume and evidence of the maximum ice extent can help to limit the range of ice-sheet realizations. However, at the same time it is difficult to create a model of ice dynamics that matches available data (Briggs and Tarasov 2013). Moreover, high spatial resolution in the models and in the input data such as topography is required to represent ice flow accurately, which is not yet feasible (Colleoni *et al.* 2018). What ice-sheet-dynamics models can provide is a shape for the ice sheet that is physically more realistic than ice sheets that are tuned to sea-level data or ice-extent data only. They can also provide uncertainty estimates based on physical parameters (e.g. Albrecht *et al.* 2020*b*).

A widely used Antarctic Ice Sheet reconstruction is CLIMAP (Hughes 1981; Stuiver *et al.* 1981), on which several GIA models were based (Nakiboglu *et al.* 1983; Wu and Peltier 1983). It consists of a 2D ice model with a basal shear parameter – the main degree of freedom. The total volume of ice melted since the LGM amounted to 24 m SLE. The Antarctic-wide reconstruction by Huybrechts (1990) showed in more detail that the Antarctic Ice Sheet behaved similar in time to the northern hemisphere ice sheets, and that most of the ice loss occurred on West Antarctica. Denton and Hughes (2002) obtained a value of 14 m SLE for Antarctic Ice Sheet loss since the LGM. Early models were not yet able to reproduce the processes that act on the grounding line such as damage, and calving and basal melting (Pattyn *et al.* 2017). These processes are very relevant for the post-LGM ice history in Antarctica as fast-flowing glaciers in response to ocean forcing can drain large glacier catchments (Golledge *et al.* 2012). More recent models do not result in an ice sheet extending to the continental shelf; they indicate thinner ice sheets with a smaller contribution to the period of fast sea-level rise known as Meltwater Pulse 1A, which occurred *c.* 14 kyr ago (Albrecht *et al.* 2020*b*).

Figure 4a shows the time history of the Antarctic Ice Sheet volume since the LGM for several recent ice models. Figure 4b indicates the total contribution of the Antarctic Ice Sheet to global sea level since the LGM (based on fig. 11 of Albrecht *et al.* 2020*b*). Note that some studies used volume of grounded ice above flotation (VAF), rather than total volume of grounded ice, which gives a smaller SLE. Ice thickness above flotation is what contributes to the loading of the Earth’s surface (see Goelzer *et al.* 2020). It can be seen that studies which simulate the dynamic behaviour of the Antarctic Ice Sheet yield the largest variations. Variations also stem from the timing of the LGM being unclear (Bentley *et al.* 2014), varying, for example, from 25 ka BP (Briggs *et al.* 2014) to around 16 ka BP (Maris *et al.* 2014), although most studies assume little decline in that period. In general, ice-sheet melting and retreat is not synchronous across Antarctica (Bentley *et al.* 2014). Some sections, such as the Antarctic Peninsula and Amundsen Sea sector, responded faster to temperature increase than others such as the Ross and Filchner Ronne ice shelves and Marie Byrd Land. There is also evidence for a minimum ice extent several thousands of years BP, and growth thereafter (e.g. Siegert *et al.* 2022). Such behaviour will by itself lead to subsidence, which complicates the interpretation of GIA observations.

GIA-based ice-sheet reconstruction. Ice-sheet thickness determines land uplift and the shape of the sea surface (see the earlier subsection on ‘GIA physics simplified’), and, therefore, sea-level estimates can be used to infer ice thickness. The first constraint is that the total volume of loss of land-based ice should match the global mean sea-level rise. A further refinement in locating past land ice is possible by combining sea-level indicators located near the centre and edges of the former ice sheet and in the far field. This requires a GIA solver that can provide predictions of sea-level change. When such a solver is used, the ice inferences are no longer independent of the

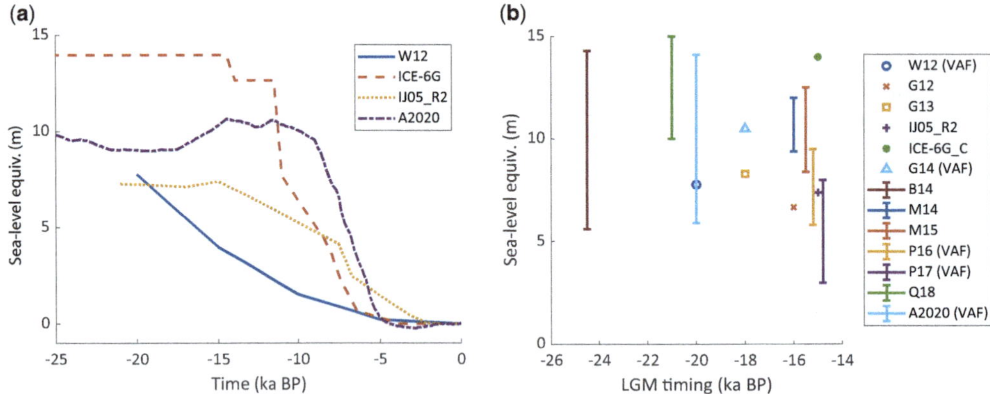

Fig. 4. (**a**) Contribution of the melting of the Antarctic Ice Sheet to global sea-level rise since the LGM for different ice models or studies of LGM ice cover. W12, Whitehouse *et al.* (2012*a*); ICE-6G, Argus *et al.* 2014; IJ05_R2, Ivins *et al.* 2013; A2020, Albrecht 2019; Albrecht *et al.* 2020*b*). The change in ice thickness is converted to volume of water using ice and ocean densities of 917 and 1000 kg m^{-3} and then dividing by the present-day ocean area of 361 × 10^6 km^2 to obtain SLE. For Whitehouse *et al.* (2012*a*), the output from the Glimmer ice-sheet model at 5 kyr intervals is used, from which the area above flotation is calculated using topography consistent with the ice-sheet model. (**b**) Total contribution of the melting of the Antarctic Ice Sheet to global sea-level rise since the LGM, as well as the LGM timing for selected studies. For some studies a range is provided, for others only a single number is available. Numbers are taken from Albrecht *et al.* (2020*b*), except where authors reported a different value or range in the original paper. W12, Whitehouse *et al.* (2012*a*); G12, Golledge *et al.* (2012); G13, Golledge *et al.* (2013); IJ05_R2, Ivins *et al.* 2013; ICE6G_C, Argus *et al.* (2014); G14, Golledge *et al.* (2014); B14, Briggs *et al.* (2014); M14, Maris *et al.* (2014); M15, Maris *et al.* (2015); P16, Pollard *et al.* (2016); P17, Pollard *et al.* (2017); Q18, Quiquet *et al.* (2018); A20, Albrecht *et al.* (2020*b*). VAF indicates that grounded ice above flotation is used to compute the sea-level contribution.

assumed mantle viscosity or other rheological parameters. A further limiting factor is that sea-level indicators are only found in Antarctica in select places near the edge of some of the ice sheets or shelves (see the earlier subsection on 'GIA observations in Antarctica'). Therefore, due to the lack of local constraints, in global GIA models until 2010 Antarctica has mostly been used to match the total sea-level change during the deglaciation (e.g. Lambeck *et al.* 2000). Wu and Peltier (1983) used an ice sheet by Clark and Lingle (1979) in their ICE-2 global reconstruction model; the timing was later adjusted by Peltier (1988). Nakiboglu *et al.* (1983) took estimates of the total melt volume and created a more detailed distribution of ice over the Weddell and Ross seas. The CLIMAP model was revised downwards in terms of volume by Nakada *et al.* (2000), whose range of 6–17 m SLE is closer to recent estimates. James and Ivins (1998) provided simulations of the uplift rate that could be tested by GPS data, which became available in the years thereafter (see Scheinert *et al.* 2021, this volume). Since then, reconstructions have relied more upon ice extent, thickness change, ice-dynamics estimates (Whitehouse *et al.* 2012*a*; Ivins *et al.* 2013; Argus *et al.* 2014) and GPS uplift rates (Argus *et al.* 2014).

Ice sheet histories used in GIA models. In the following we discuss ice-sheet histories that are widely used in GIA models, especially those that are used to correct GRACE measurements of mass change: W12 (Whitehouse *et al.* 2012*a*), IJ05-R2 (Ivins *et al.* 2013) and ICE-6G_C (Argus *et al.* 2014). We discuss the model of Albrecht *et al.* (2020*b*) as a recent example of a ice-dynamics model, specifically the best scoring simulation as shown in figure 15 of that paper.

Whitehouse *et al.* (2012*a*) assembled all the pertinent glacial-geological data and used a shallow ice numerical code to examine a spectrum of possible ice-change scenarios. Ice-thickness changes between 20 ka BP and the present as provided by this model are plotted in Figure 5a. The greatest changes can be seen in the marine-based sectors in the Ross and Weddell seas, followed by the coastal regions of West Antarctica. In East Antarctica, ice thickness increased during this period but actual constraints in East Antarctica are scarce. Compared to an ice-dynamics model such as that of Golledge *et al.* (2012), it has a greater increase in ice thickness in East Antarctica and a larger reduction in ice thickness in the Amundsen Sea sector.

IJ05_R2 (Ivins *et al.* 2013) is a revision of the IJ05 ice model (Ivins and James 2005) that is widely used to correct GRACE data. It does not include flow dynamics; instead, ice heights are adjusted to fit glacial-geological data, mostly the same as that used for W12. However, a choice was made to create the largest possible ice-thickness changes and the thickest and youngest LGM ice sheet to obtain an upper range for the GIA correction for Antarctica. In creating the ice-

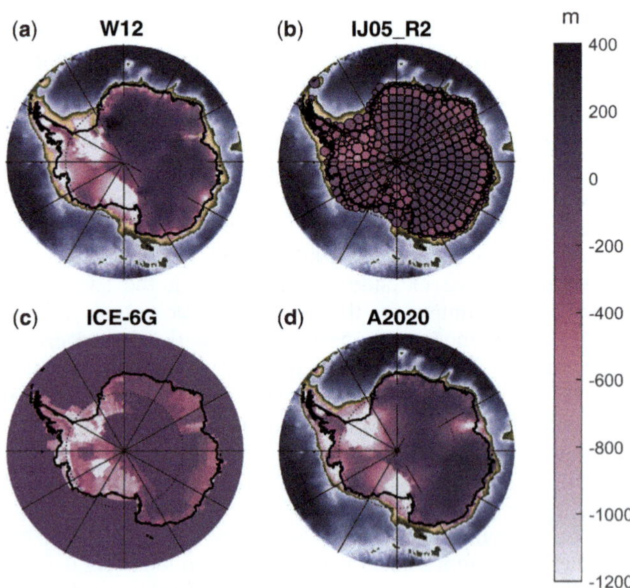

Fig. 5. Ice-thickness change between the LGM and the present for four different ice histories. LGM is taken to be 20 ka BP unless otherwise noted. Outside Antarctica, where no ice thickness existed at the LGM, the Bedmachine topography (Morlighem *et al.* 2020) is plotted. (**a**) W12 model (Whitehouse *et al.* 2012*a*). (**b**) IJ05_R2 (Ivins *et al.* 2013) with the LGM taken to be 21 ka BP. (**c**) ICE-6G_C (Argus *et al.* 2014). (**d**) A2020 (Albrecht *et al.* 2020*b*).

thickness values, the present-day ice and ocean is taken as reference: that is, only ice that is higher than the local water level corrected for the density difference, so-called ice above flotation, is outputted, as this ice thickness is the effective load in a GIA model. IJ05_R2 arrives at 7.5 m SLE that melted since 15 ka BP.

ICE-5G (version 1.2) (Peltier 2004) used limited regional constraints and ice-dome histories inferred from far-field sea-level models and records, arriving at a figure of 17.5 m SLE. This was reduced to 13.6 m SLE in ICE-6G_C (Argus *et al.* 2014). The ICE-6G_C model has been tested for consistency with an ice-dynamics simulation, by nudging the ice dynamics to the ICE-6G_C model (Stuhne and Peltier 2015). The result is a model that has a smoother shape and, for the most part, has lower ice-surface heights than ICE-6G_C. Overall, the large volume of ICE6G_C is at the high end of the range of ice-sheet reconstructions shown in Figure 3. It is also large compared to the compilation of studies in Simms *et al.* (2019), which gives an average contribution of Antarctica of 9.9 ± 1.7 m SLE, and compared to ice-sheet reconstructions that are used in GIA simulations such as W12 (Whitehouse *et al.* 2012*a*), IJ05_R2 (Ivins *et al.* 2013), and Briggs *et al.* (2014). A possible explanation is that ICE-6G_C attributes a large Antarctic contribution to the period of rapid sea-level rise known as Meltwater Pulse 1A (14.7–13.5 ka), while ice-dynamics studies Antarctic-wide generally find a smaller contribution (Mackintosh *et al.* 2011; Briggs *et al.* 2014), also supported by ice-sheet limiting data (Bentley *et al.* 2014).

A2020 (Albrecht *et al.* 2020*b*) use the Parallel Ice Sheet Model (Winkelmann *et al.* 2011) to simulate two glacial cycles. Uncertainty is quantified by varying parameters for internal ice dynamics, precipitation and mantle viscosity. The simulations are scored against present-day observations of grounded and ice-shelf areas, ice thickness, grounding-line location, and uplift rates. The model that achieved the best aggregated score shows ice thickness since the LGM to be reduced by more than 2000 m in the major shelves in West Antarctica, while ice grew by hundreds of metres in East Antarctica, similar to W12. The total volume of ice reduction is 9.7 m SLE, which is in between W12/IJ05_R2 and ICE-6G_C (Fig. 4a), and similar to other ice-dynamics-based models (Fig. 4b). A difference with the earlier discussed models is that the minimum ice volume is reached before present, with readvance and thickening occurring after 3 ka BP (e.g. at Siple Coast). Contrary to Argus *et al.* (2014), it finds that rapid Antarctica ice-sheet loss seems to be a consequence rather than a source of the meltwater pulses.

GIA model results and inferences of mantle structure

The trade-off between mantle properties and ice reconstruction is more pertinent in Antarctica, where the ice history is much less known than for other large Late Quaternary ice sheets. For a uniform viscosity and a given ice retreat, which does not vary too much spatially, GIA model simulations will show the largest uplift where the magnitude of ice change was largest: that is, at the Ross and Filchner–Ronne ice shelves, which have seen a large degree of ice retreat since 15 ka BP (Martín-Español *et al.* 2016*b*). In accordance with this relation between uplift rate and maximum ice-thickness change, the largest uplift rates are found for the ICE-6G_C (Argus *et al.* 2014) reconstruction, followed by the W12 (Whitehouse *et al.* 2012*a*) and IJ05_R2 (Ivins *et al.* 2013) ice histories (see uplift rate compilations in King 2013; Martín-Español *et al.* 2016*b*; Whitehouse *et al.* 2019). Comparison with GNSS uplift rates is not straightforward and requires correction of the GNSS uplift rates for elastic loading effects (Thomas *et al.* 2011; Scheinert *et al.* 2021, this volume), and also separation of the plate velocity and GIA models (King *et al.* 2016), as well as correcting for possible remaining tectonic effects.

GIA models based on ice models that are tuned to match GNSS or sea-level data still find an upper-mantle viscosity very close to the average global viscosity such as W12 (Whitehouse *et al.* 2012*a*), which finds a best-fit upper-mantle viscosity of *c.* 10^{21} Pa s, or IJ05_R2 (Ivins *et al.* 2013), which led to a one order of magnitude smaller best-fit viscosity (*c.* 2×10^{20} Pa s). These are the models that are applied most widely in correcting GRACE gravity measurements for the GIA signal in Antarctica (Shepherd *et al.* 2018; Scheinert *et al.* 2021, this volume).

While the GIA forward models can offer physical insight, their accuracy becomes a function of the many input parameters. Therefore, for some purposes, data-based approaches are used, which are termed ‘empirical’ or ‘inverse’ GIA models. For Antarctica, they are based on geodetic data such as GRACE time-variable gravity, satellite altimetry and GNSS (see Scheinert *et al.* 2021, this volume). Note that these are not the same as inversions of GIA models for Earth model parameters. Although the approaches are based largely on data and their error estimates, they are not entirely free from assumptions about the solid Earth such as the density of the solid Earth mass change that causes a GIA gravity change (Wahr *et al.* 2000; Riva *et al.* 2009) and an *a priori* set length-scale of GIA (Martín-Español *et al.* 2016*a*).

The forward GIA models cannot predict all of the signals seen in geodetic observations and in the empirical GIA models in West Antarctica. For instance, in a joint analysis of time-variable gravity (GRACE) data and altimetry data (Ice, Cloud, and Land Elevation Satellite (ICESat)), Groh *et al.* (2012) found a significant uplift signal of more than 20 mm a^{-1} in the Amundsen Sector in West Antarctica that is not predicted by any GIA model. The W12 (Whitehouse *et al.* 2012*a*) ice model, which accounts for ice changes in that sector during the deglaciation occurring since 20 ka BP, resulted in an uplift of at most 5 mm a^{-1} in that region. The discrepancy can be explained by the shortcomings in models that assume almost no ice-loading changes after 5 ka BP, in agreement with small change in global mean sea level during that time. However, it was later realized that the ice sheet in Antarctica responded to local climatic changes throughout the late Holocene.

A second reason for the discrepancy between data and models is that the ice in Antarctica covers a whole continent, from the East Antarctic thick crust to the West Antarctic thinner one, which cannot be truly represented by a single viscosity profile. As stated earlier, substantial differences exist in the Earth's structure between East and West Antarctica: warmer and weak mantle beneath West Antarctica and colder and stiffer mantle beneath the East Antarctic Craton (Berg *et al.* 1989; Morelli and Danesi 2004; Hansen *et al.* 2014; Martin *et al.* 2014; An *et al.* 2015; Burton-Johnson *et al.* 2020; Wiens *et al.* 2021, this volume). East Antarctica shows high seismic velocities down to a depth of 200 km, correlating with old structures, although velocities at the coast are close to the global average of around 150 km and deeper. The Ross and Weddell seas have modest positive anomalies, and the slowest velocities are found in the part of West Antarctica where Cenozoic rifting and extension took place, particularly in Marie Byrd Land, the Amundsen Sea sector and the Antarctic Peninsula. The slow velocities do not extend very deep beneath the Antarctic Peninsula, where fast anomalies are found in the transition zone, but for Marie Byrd Land they do extend deeper. Seismic velocity can be used to calculate viscosity anomalies but with several assumptions and with uncertainty resulting from unknown conversion factors and the varying quality of the seismic model. Therefore, other sources of information are necessary to infer mantle structure, as reviewed in this volume. For Antarctica, these can include gravity and magnetic data (Pappa and Ebbing 2021, this volume), geology (Panter and Martin 2021, this volume) or mantle xenoliths (Handler *et al.* 2021, this volume; Martin 2021, this volume).

The high mantle viscosity used in traditional GIA models makes the predicted effect of ice changes within the last few hundred years negligible. In several regions on Earth, adjustment is much faster; the response to the LGM load has all but relaxed and the current uplift is caused by deglaciation events in the last few centuries. This is the case for glaciated regions near ridges such as Iceland (e.g. Sigmundsson 1991), and near subduction zones such as Alaska (e.g. Larsen *et al.* 2005) and Patagonia (e.g. Dietrich *et al.* 2010), where mantle conditions clearly deviate from the global average. For these areas, local modelling is required combined with a local deglaciation history for the last few centuries. The possibility of a large signal due to local, recent ice unloading in Antarctica was hinted at in Ivins *et al.* (2000). In the last decade, the availability of GNSS data has enabled the development of regional GIA models that include changes in ice thickness over the last few centuries. Regional studies can provide a constraint on regional upper-mantle viscosities that together can lay out a picture of a varying mantle structure, which can be compared to other observations of the Earth's structure,

The best-fitting upper-mantle viscosity from regional GIA studies that can be compared to the geophysical image of Antarctica's mantle are discussed in the following. To match the observed uplift rate data, the regional studies need to include elastic deformation due to current ice-thickness changes and the response to recent ice-loading changes in a period that is sometimes called the Little Ice Age, although confusingly this period is not the same for different regions on Earth. Nield *et al.* (2014) analysed GPS uplift histories in the northern Antarctic Peninsula and found that geodetic uplift rates greatly exceeded their estimates for the Earth's purely elastic response to contemporary ice-mass changes, implying a significant viscous component of uplift. They concluded that the observed uplift rates after subtracting elastic deformation could be explained by invoking upper-mantle viscosities in the range 0.6×10^{18}–2×10^{18} Pa s. A similar range of 0.3×10^{18}–3×10^{18} Pa s was found in a more recent analysis by Samrat *et al.* (2020). Best-fit upper-mantle viscosity ranges are presented as coloured ellipses encompassing the respective study regions in Figure 6 on top of 3D viscosity maps from Ivins *et al.* (2021, this volume) at two depths in the upper mantle. Note that the tip of the Antarctic Peninsula does not have the lowest viscosity based on the 3D viscosity maps, especially at a depth of 300 km, which indicates the uncertainty in translating seismic velocities to viscosity.

The best-fit viscosities obtained can also be compared to the tectonic setting of the northern Antarctic Peninsula. Along the western Peninsula, a subduction zone, which shut down progressively from south to north between 50 and 4 Ma (Barker 1982; Larter and Barker 1991; Eagles *et al.* 2004), remains active along the South Shetland Trench and has an active back-arc basin (Lawver *et al.* 1996). The Bransfield Straight, separating the South Shetland Islands from the northern tip of the peninsula, is undergoing extension and there is evidence of active volcanism (Barker and Austin 1998; Smellie 2021). This means that a viscosity below the global average is not unexpected there. The low viscosities were confirmed by Samrat *et al.* (2021) in the northern Marguerite Bay, with viscosities found to range from 0.1×10^{18} to 9×10^{18} Pa s. This matches the slow seismic wave speeds along the Western Antarctic Peninsula (see the background viscosity map and Wiens *et al.* 2021, this volume). In the southern part of the Antarctic Peninsula, viscosity appears to be higher (Zhao *et al.* 2017), in agreement with past subduction that ended from south to north. Uplift rates due to the thinning of the Wordie Ice Shelf match the small GPS uplift rates observed there only for a lower bound of viscosity of 2×10^{19} Pa s. This larger viscosity implies that viscosity increases by an order of magnitude over 500 km.

Other examples include studies in the Weddell Sea, where the uplift rate allows tentative conclusions on the timing of retreat. Wolstencroft *et al.* (2015) argued for a later retreat than assumed in the Antarctic-wide ice reconstructions (Whitehouse *et al.* 2012*a*; Ivins *et al.* 2013; Argus *et al.* 2014), and Bradley *et al.* (2015) argued for retreat from the 6 ka grounding line and subsequent readvance. Wolstencroft *et al.* (2015) found best agreement with an upper-mantle viscosity somewhat larger than in the rest of the Antarctic Peninsula, although the evidence is not very conclusive. In East Antarctica, few data are available; a recent study by Hattori *et al.* (2021) showed uplift rates in Lützow-Holm Bay, East Antarctica, that agree with some of the global GIA models.

Barletta *et al.* (2018) analysed geodetically observed uplift rates in the Amundsen Sea Embayment, West Antarctica, by

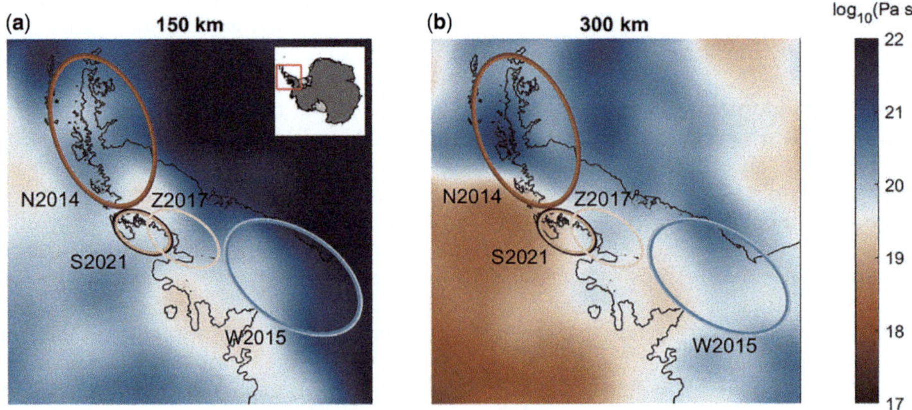

Fig. 6. 1D upper-mantle viscosity constraints $\log_{10}$(Pa s) from GIA studies in the Antarctic Peninsula. The location of the ellipse outlines roughly the region for which the viscosity constraint is obtained; the colour denotes the viscosity; the outer ellipse is the lower bound of the best-fit viscosity range and the inner ellipse is the upper bound. N2014, Nield *et al.* (2014); Z2017, Zhao *et al.* (2017), who only provide a minimum viscosity; W2015, Wolstencroft *et al.* (2015); S2021 (Samrat *et al.* 2021). (**a**) The background is the 3D viscosity map from Approach 3 in Ivins *et al.* (2021, this volume): i.e. derived from the Antarctic seismic model of Lloyd *et al.* (2020) for dry olivine with a 4 mm grain size and stress of 0.1 MPa at 150 km depth (**a**) and at 300 km depth (**b**).

modelling the horizontal and vertical displacement of all Antarctic GPS Network (ANET) stations in the Amundsen sector. The rate of ice-mass loss there is the major part of the ice-mass loss of the entire Antarctic Ice Sheet. Ice loss since 1890 in combination with satellite-observed ice loss and a shallow upper-mantle viscosity of 4×10^{18} Pa s gave best fit to the GPS uplift rates. Despite the fact that this region is far removed from any plate boundary, it is underlain by slow seismic velocities. There are also indications of geologically recent (and, perhaps, locally active) rifting in much of West Antarctica, including the Amundsen sector (Wobbe *et al.* 2012). An example are the indications of active volcanism (Smellie *et al.* 2021), and Cenozoic rifting in much of West Antarctica, and limited evidence that points to rift-related displacements of Oligocene age between the Thurston Island–Eights Coast and Marie Byrd Land crustal blocks flanking the Amundsen sector (Kalberg *et al.* 2015; Spiegel *et al.* 2016). This supports the finding of Barletta *et al.* (2018) that mantle viscosity in this region deviates significantly from the global average.

The regional studies are well suited to fit uplift rates but they fall short in representing the continental-scale signal seen by the GRACE mission as there is a contribution from the far-away ice sheets through the global sea-level variations and associated continental levering. In terms of mass change, this contribution is almost 40% of the post-LGM signal (Caron and Ivins 2020). Moreover, regional GIA models in Antarctica do not simulate the effect of neighbouring regions with a different viscosity, which is represented in Figure 1 as deformation in the mantle that is not symmetrical with respect to a vertical cross through the centre of the ice sheet. Thus, 3D GIA models should probably be used for some parts of Antarctica despite their longer computation time. A possible faster solution is the combination of multiple regional 1D solutions, as in Hartmann *et al.* (2020), or an ELRA model with regional parameters (Coulon *et al.* 2021) but their accuracy should be further investigated.

3D GIA studies require viscosity maps as input, which can be derived from seismic velocity anomalies, as reviewed in Ivins *et al.* (2021, this volume). Therefore, an anomaly in seismic velocities can be converted to an anomaly in temperature, which can be related to anomalies in logarithmic viscosity using scaling factors based on laboratory measurements. GIA studies using 3D models have mostly followed this approach, with more recent models taking advantage of more seismic data and full-waveform inversion that yield more accurate velocity estimates. Ivins *et al.* (2021, this volume) discuss and quantify some of the uncertainties: the scaling factors, the background temperature profile and other parameters in the creep laws of mantle rocks, resulting in a standard deviation of 1 in logarithmic viscosity, mainly in areas with high viscosity. This is not yet the full uncertainty range: seismic velocity anomalies have uncertainties that are hard to quantify. Furthermore, water content can affect seismic velocities directly and can also affect deformation for a given temperature. Water content varies within the Antarctic mantle (Martin 2021, this volume).

Thus, results of 3D models should be assessed with this uncertainty in mind. However, it should also be realized that because of the way that 3D viscosity maps for GIA studies are created, the 3D GIA models do not have many more degrees of freedom in fitting GIA observations than 1D GIA models. Mostly, it is assumed that viscosity can be obtained by scaling seismic velocity perturbations (Kaufmann *et al.* 2005; Geruo *et al.* 2013; Hay *et al.* 2017; Powell *et al.* 2020), as discussed by Ivins *et al.* (2021, this volume). There are a number of choices for parameters that change the amplitude of this spatial pattern, in addition to assumed background temperature or background viscosity, but the spatial pattern is determined by the seismic model used. A somewhat different approach translates the velocity anomalies into temperature anomalies, which are entered into a mantle flow law (van der Wal *et al.* 2015; King *et al.* 2016; O'Donnell *et al.* 2017). In this case, viscosity can also depend on the stress in the mantle, as will be discussed later. Even so, the spatial pattern in effective viscosity will be determined mostly by the pattern of the seismic velocity anomalies. This means that a better fit of 3D models compared to 1D models should not be expected automatically when using spatially homogeneous scaling parameters. 3D GIA models by themselves allow variations in viscosity for each element in the model, and some applications in the future might require fitting spatial variations in parameters directly, in which case the number of degrees of freedom will be much greater and inversions become more problematic. In this case, other information such as gravity anomalies (Pappa and Ebbing 2021, this volume) and geologically derived constraints can also be helpful (Burton-Johnson *et al.* 2020; Martin 2021, this volume; Martin *et al.* 2022, this volume).

An approach that can be followed to reduce uncertainty in 3D viscosity maps is to calibrate the 3D viscosity estimates with regional 1D studies. Viscosity values that can serve as such anchor points are compiled in table 6 of Ivins *et al.* (2021, this volume) and in figure 1 of Lau *et al.* (2021). Ivins *et al.* (2021, this volume) conclude that rheology derived from micromechanics and seismology compares well with viscosities derived from GIA models for several regions. In particular, the low viscosity in the Antarctic Peninsula can be replicated. A selection of studies for the Antarctic Peninsula is shown in Figure 6 together with a 3D viscosity estimate from Ivins *et al.* (2021, this volume). An important caveat is that it is not immediately clear if the best-fit 1D viscosity should be seen as an average over a certain part of the mantle. The question becomes: To what region and depth is a certain observable sensitive? We know from the sensitivity kernels discussed earlier that larger ice sheets are sensitive to deeper structure. However, the sensitivity itself depends on the viscosity profile. Sensitivity is concentrated in areas of low viscosity in the case of a viscosity contrast. In that case, the low viscosities have a larger weight in determining what 1D viscosity corresponds best to the 3D viscosity pattern. It could be advocated that a viscosity representative of the process should be weighted by strain rate or by internal work (Christensen 1984). Blank *et al.* (2021) used the approach of averaging only over those elements where the stress reaches a threshold value, relative to the maximum stress.

In 1D models, the lithospheric thickness is an independent input parameter. It is difficult to quantify lithosphere thickness based on other studies because the lithosphere is defined differently in different fields. For GIA, a suitable definition is the top part of the Earth that deforms elastically on the timescale of the glacial loading considered (e.g. Nield *et al.* 2018). This is different to a definition of lithosphere based on seismic wave speed (e.g. Wiens *et al.* 2021, this volume), from the elastic layer that is considered in gravity data (e.g. Pappa and Ebbing 2021, this volume) or the chemical-compositional-melt boundary considered from mantle xenolith data (Coltorti *et al.* 2021, this volume), see also Rychert *et al.* (2020). A further consequence of the GIA definition of effective lithosphere is that its thickness changes in time; for the same 3D Earth structure, the effective lithosphere for will be thicker when studying a response on the timescale of centuries than for a timescale of thousands of years. An advantage of using 3D viscosity is that it is no longer necessary to prescribe the lithosphere thickness *a priori*.

The influence of lateral viscosity on observables in Antarctica can be determined by comparing the outcome of 3D and 1D GIA models. The effect on uplift rate was found to be around 1–2 mm a^{-1} (Kaufmann *et al.* 2005; Geruo *et al.* 2013), which is relatively small, especially considering the larger ice sheets used in these studies compared to latest Antarctic Ice Sheet reconstructions. A change in the location of the maximum uplift rates can also be observed (e.g. comparing fig. 5b and c of Kaufmann *et al.* 2005 to their fig. 4b and c; van der Wal *et al.* 2015). This can be explained by the different time sensitivities: a higher viscosity will be sensitive to earlier ice-load changes that could have occurred at a different location. For the mass balance of the Antarctic Ice Sheet observed by the GRACE mission, van der Wal *et al.* (2015) found an effect of 3D viscosity of up to 20 Gt a^{-1} compared to current ice-mass loss of 160 Gt a^{-1}. Blank *et al.* (2021) showed that lateral changes in viscosity cannot as yet be detected in GPS measurements in the Amundsen Sea sector. Focusing on the spatial pattern of uplift rates, Nield *et al.* (2018) observed that the 3D models result in a higher amplitude and shorter wavelength, which is due to the thinner effective lithosphere that can warp easier.

For horizontal velocities, the effect of 3D viscosity variations is larger than for vertical uplift rates, as found by sensitivity studies (Wu 2006). The effect of viscosity variations is such that the horizontal displacement rate can be reversed depending on viscosity, as the horizontal rates are a resultant of two opposite motions: the lithosphere that flexes such that points move away from the former ice sheet, while the underlying mantle flows towards the ice-covered region (Hermans *et al.* 2018). For Antarctica, the stiff cratonic root of East Antarctica is thought to promote flow in the weaker West Antarctic mantle (Kaufmann *et al.* 2005), which might be visible in horizontal GNSS observations there (Konfal *et al.* 2018). An enhanced effect of 3D viscosity on horizontal velocities is also found for uplift caused by ongoing ice melt in West Antarctica (Powell *et al.* 2020).

Finally, in terms of RSL data, Gomez *et al.* (2018) found an effect of 3D rheology of tens of metres across Antarctica, well above the measurement accuracy. The sensitivity of RSL data depends greatly on the ice history. Bagge *et al.* (2021) found a large sensitivity to lateral variations in the Ross Sea, and they explain this by the relatively thin ice sheet compared to the ice sheets in classical GIA areas, which makes the response more sensitive to shallower layers where larger lateral variations exist.

Apart from lateral variations, the predictions in Antarctica are also sensitive to the presence of a power-law rheology, which results in stress-dependent rheology (Nield *et al.* 2018; Blank *et al.* 2021). Such a power-law rheology represents behaviour that is observed in deformation experiments: steady-state creep proceeds faster when stress is higher. This implies that effective viscosity depends on stress and changes in time: effective viscosity is lower where and when stress is high. This local weakening leads to larger peaks in uplift rate (Nield *et al.* 2018; Blank *et al.* 2021). During the glacial cycle, the change in viscosity can be up to two orders of magnitude (Barnhoorn *et al.* 2011). That has an important implication, namely that the viscosity that is valid for the glacial cycle is not necessarily the same viscosity that holds for a more recent loading process, or a longer term process such as mantle convection (Bredow *et al.* 2022, this volume). Figure 7 shows the change in viscosity in the Amundsen Sea sector as a result of stress from post-LGM ice loss for two different 3D viscosity models. The model in the top row has lower temperatures but, because the grain size is smaller, viscosity is on average lower than the bottom model (further details are given in Blank *et al.* 2021). The change in viscosity as a result of the post-LGM decline of the West Antarctic Ice Sheet ranges by up to two orders of magnitude for the model with a stiff rheology (top row) and a long 'memory' (Fig. 7), although the change is greatest for shallow regions where viscosity is so high that it can be considered to be part of the lithosphere.

Note that the stress-dependent rheology is not the same as transient rheology. Experiments show anelastic or transient rheology before steady-state creep is reached (Faul and Jackson 2015). In its simplest form, transient rheology can be parametrized by a short-term and long-term viscosity (e.g. Caron *et al.* 2017). Such a rheology is used more often in post-seismic deformation studies and will be discussed in the following 'Post-seismic deformation' section. The description of transient rheology can be more general than short-term v. long-term viscosity; experiments show that the rheology, in general, depends on the frequency of loading (Ivins *et al.* 2020). Accounting for this frequency dependence, Lau *et al.* (2021) calculated a complex viscosity for Antarctic regions for a mantle composition based on olivine and with a dependence on grain size. At a frequency between long-term and short-term GIA, the transient process is shown to make a large contribution (Lau *et al.* 2021). This also leads to an effective viscosity that changes in time, which is conceptually different from a stress-dependent rheology.

Coupling of ice to solid Earth: feedbacks and processes

As noted in the introduction to this chapter, the deformation of the solid Earth affects the dynamics and growth of the ice sheet on top of it. Thus, solid Earth deformation is important for the

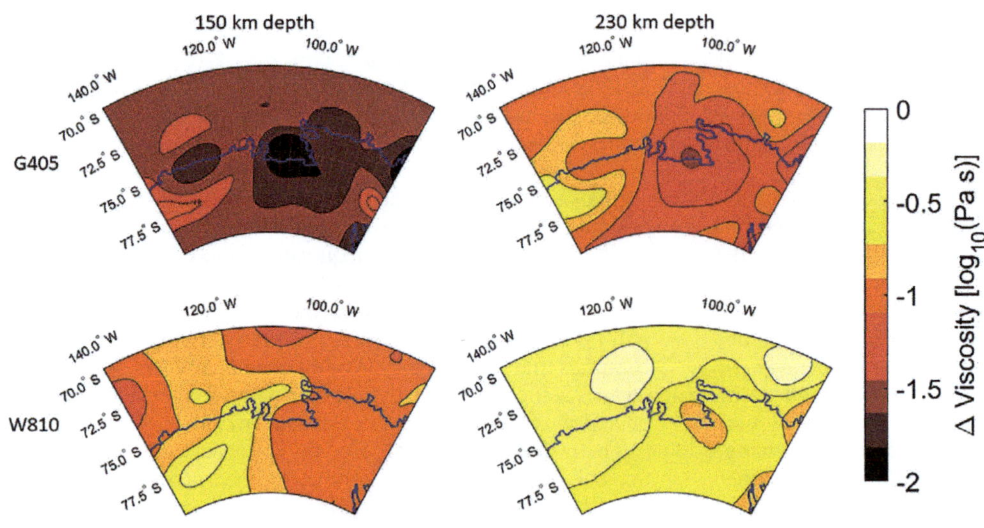

Fig. 7. The change in viscosity as a result of stress from the last glacial cycle at the start of the simulation for the Amundsen Sea Embayment in AD 1900. G405 is a 3D viscosity model based on a gravity-derived lithosphere model for Antarctica (Pappa *et al.* 2019) using an olivine flow law with a grain size of 4 mm and water content of 500 ppm. W810 is based on a global temperature model (an early realization of WINTERC-G of Fullea *et al.* 2021) using an olivine flow law with a grain size of 8 mm and 1000 ppm water content. Source: adapted from Blank *et al.* (2021).

ice sheet in the last glacial cycle but also for scenarios of future Antarctic Ice Sheet decline. In this subsection, the main effects are discussed, how they are modelled and what the implications are of variations in mantle viscosity.

GIA influences ice dynamics in different ways. A straightforward effect is that it brings the ice surface to a different altitude (schematized in Fig. 1) and hence to a different atmospheric temperature that might promote or inhibit melting. However, in Antarctica, temperatures are low everywhere and the greatest influence of the evolving topography induced by GIA is on the grounding-line position of the ice sheet. The grounding line lies at the transition between grounded ice and floating ice. Its position is influenced by the local ice thickness and the local sea level. A rise in sea level is associated with a grounding line that moves inwards (Hollin 1962). As the ice sheet gets thinner, the grounding line retreats inland and the amount of floating ice increases. Ice discharge scales with ice thickness at the grounding line. Therefore, if the depth of the bedrock increases inland, the ice-thinning rate increases more, a phenomenon termed ‘marine ice-sheet instability’ (Weertman 1974; Schoof 2007). As a result of a decreased ice load, the bedrock beneath the ice sheet uplifts, which causes the grounding line to shift back towards the ocean, leading to a decrease in melt and a reduction in discharge if topography reduces outwards. The main effect of GIA is therefore a negative feedback that stabilizes the Antarctic Ice Sheet, specifically at regions where the bedrock uplift is fast (e.g. Parizek and Alley 2004; Kachuck *et al.* 2020). When this coupling is considered, the palaeo-ice volume is 1–2 m SLE smaller (Whitehouse 2018). The uplift is also called upon to explain a readvancing grounding line in the late Holocene (e.g. Kingslake *et al.* 2018) and as a slowing effect on the Amundsen Sea sector ice-sheet retreat (Johnson *et al.* 2020).

A second effect related to GIA that modulates the ice dynamics is the sea-level fall due to a decrease in the gravitational attraction of a melting ice sheet. The sea-level fall will lead to an outward-moving grounding line and a decreased flux across the grounding line, once again promoting a negative feedback (Gomez *et al.* 2010). However, sea level can also change in response to other ice sheets or processes. The sea-level forcing due to northern hemisphere ice sheets was shown to increase the amount of ice loss since the LGM, and hence the size of the LGM ice sheet (Gomez *et al.* 2020). Furthermore, smaller feedback effects of GIA in Antarctica are an altered bed slope that could stabilize the ice sheet even more and the rise of pinning points at ice shelves as a result of bedrock uplift (Adhikari *et al.* 2014; Whitehouse *et al.* 2019). Pinning points, which can arise due to natural topography roughness (e.g. tectonic or volcanic features) or can be built by the ice sheet itself (Colleoni *et al.* 2022), can act as ‘brakes’ on the flow of the glacier. In reality, all feedback effects work together. For example, in the Weddell Sea, ice regrounded in the late Holocene is likely to have occurred due to the rebound of a topographical feature that formed an ice rise (Siegert *et al.* 2019). This was helped by the lowering of the water depth and smaller self-attraction.

To compute the solid Earth deformation, ice-dynamics models generally employ a fast approximation of GIA using an ELRA model to simulate GIA (e.g. Golledge *et al.* 2014; Maris *et al.* 2014; Pollard *et al.* 2016). In this model, it is assumed that the Earth, consisting of two flat layers, responds to a certain change in ice load by deforming exponentially towards the equilibrium displacement for that particular ice height (Le Meur and Huybrechts 1996). The time needed to reach equilibrium, the relaxation time, is typically taken to be 3 kyr for the Antarctic continent, which resulted in a present-day topography that was in agreement with that computed with a more realistic GIA model (Le Meur and Huybrechts 1996). More realistic GIA models include the physics of a layered Earth that differs from the ELRA model in several aspects. Even for a two-layered Earth, the relaxation-time response depends on wavelength, while for the ELRA model it does not. In addition, the forebulge is more pronounced in a GIA model and GIA models allow the inclusion of more layers. The difference in the LGM ice volume between coupling with an ELRA or GIA model can be more than 2 m SLE (Whitehouse *et al.* 2019). An intermediate case in between the ELRA and GIA models is the viscoelastic half-space, where relaxation time depends on the size of the ice sheet (e.g. Bueler *et al.* 2007; Garbe *et al.* 2020). This has been used by Albrecht *et al.* (2020*a*) to show that low mantle viscosity fits better to the observed ice flow, while higher viscosity fits better to GPS uplift-rate data.

Several ice-sheet models have been developed that are coupled to 1D GIA models: for example, GIA models based on the normal-mode method (e.g. Gomez *et al.* 2012; de Boer *et al.* 2014) or the spectral finite-element method (Konrad *et al.* 2014, 2015). Important parameters in the coupling are the spatial resolution of the ice-sheet model with which the model represents the bedrock and grounding line, and the time resolution, especially in areas with low viscosity and fast response. The ice-sheet model requires an initial topography, for which present-day topography is used, corrected with the deformation caused by GIA over the glacial cycle (Gomez *et al.* 2013). An example of the topographical deformation for a 3D GIA model based on the W12 ice-loading history (Whitehouse *et al.* 2012*a*, b) and olivine rheology (4 mm grain size as in Ivins *et al.* 2021, this volume, Approach 3) resulted in maximum displacements of 130 m for the Weddell Sea and 117 m for the Ross Sea.

To include the effects of GIA on sea level through the change in gravity, GIA feedback has also been modelled using coupled ice dynamics–sea level models (Gomez *et al.* 2013; de Boer *et al.* 2014; Konrad *et al.* 2015; Pollard *et al.* 2017; Larour *et al.* 2019). Here the sea-level model computes both the bedrock deformation and the RSL change following a change in the equipotential surface, and forwards this to the ice-sheet model. Although coupled ice sheet–sea level models are computationally more expensive than the ELRA model, a varying temporal resolution can be used to decrease computation time by 50% (Han *et al.* 2022). A disadvantage of such an algorithm is that, with the usual sea-level algorithm for every time step, the entire ice history needs to be convolved with the Earth’s response.

Ice-sheet models have also been coupled to 1D GIA models that are more consistent with seismological and geological evidence of the structure under the Antarctic Ice Sheet (Gomez *et al.* 2015). Less ice loss is predicted when using a structure consistent with the rheology under the West Antarctic Ice Sheet compared to using a structure derived from globally distributed ice-age datasets. DeConto *et al.* (2021) showed the limited effects of a relatively low mantle viscosity on Antarctic ice retreat in the coming two centuries using a coupled ice sheet–1D GIA model. However, tailoring the model for a certain region cannot always replace a model with true lateral variations, given the large region to which the ice sheet is sensitive and the large variations in viscosity across the Transantarctic Mountains and within the Antarctic Peninsula. Predictions of sea-level fall as a result of a West Antarctic melt event increase by 20 and 50% in the coming decades compared to purely elastic deformation when using a 1D and 3D rheology, respectively (Hay *et al.* 2017). It should be noted that the 1D viscosity in this study was relatively high (10^{21} Pa s), which subdues the response of the 1D model. On the other hand, due to the uncertainties in the conversion from seismic velocity to viscosity anomalies, the prediction of sea-level fall can be even larger for different 3D models. This shows the importance of the tuning of the model to the local Earth properties, both for 1D and 3D models. To know whether a 3D model is really necessary or if 1D models are sufficient, locally tuned 1D models should be used, as for example, in Blank *et al.* (2021).

Full 3D GIA models have large computation times of the order of days to weeks. Therefore, shortcuts have been proposed: the ELRA model with laterally varying relaxation time (Oude Egbrink 2017) or laterally varying flexural rigidity, or both (Coulon *et al.* 2021). While these can simulate the differences in relaxation time that exists, for example, between West and East Antarctica, they cannot capture the full physics of stress transfer from a region of higher to lower viscosity or vice versa. To study the effect of lateral variations on the evolution of the Antarctic Ice Sheet, Gomez *et al.* (2018) coupled an ice sheet model to a 3D GIA model for the past 40 kyr by iterating the initial topography until the present-day topography was achieved. Significant differences in ice loading were found; the grounding line retreated less during the deglaciation phase when using a 3D rheology tuned for Antarctica compared to using a 1D rheology not tuned for (part of) Antarctica. During the deglaciation phase, ice thickness was up to 1 km thicker in the Ross Sea when using 3D GIA, although still far from the present-day ice sheet. The enhanced flexibility of 3D model makes it suitable for those complex regions but its uncertainty is large without validation against data such as RSL or present-day ice cover.

The coupling is also relevant at other timescales, such as the Eocene–Oligocene transition (34–33.5 Ma BP), the Last Interglacial (130–116 ka) and future Antarctic ice loss. These will be discussed briefly in the following. In the Eocene–Oligocene transition, the Antarctic Ice Sheet emerged or became more prominent (Miller *et al.* 1991; Lear *et al.* 2000). At that time, the same physics applies as in the Pleistocene ice sheet; a growing ice sheet attracts and raises sea level. The resulting GIA effects modulate the ice sheets as the forebulge progresses through the edges of the Antarctic continental shelf (Stocchi *et al.* 2013). Sea level at this epoch, recorded in drill cores, reflects the competing effects of attracting sea level from the growing ice sheet, and the forebulge from the growing East Antarctic Ice Sheet (Galeotti *et al.* 2016). For Pliocene studies, the role of GIA is that of correcting sea-level measurements, as well as providing corrections on the initial topography on which the ice sheet formed (e.g. Rovere *et al.* 2014). As these studies can be probably considered preliminary from a GIA modelling point of view, there is room for improvement by including recent insights into the mantle structure of Antarctica.

The last interglacial saw smaller ice sheets on Antarctica, which provides a pessimistic window on the stability of the Antarctic Ice Sheet in a warming climate. Estimates for sea level at the last interglacial show a much higher sea level, of which 3–4 m SLE might have come from extra melting of the West Antarctic Ice Sheet beyond its present configuration (e.g. Turney *et al.* 2020). The sea-level change is partly controlled by GIA, with several decimetres found to have been determined by lateral variations in viscosity as a result of meltwater expulsion from rapid bedrock uplift due to low viscosity in West Antarctica (Pan *et al.* 2021; Powell *et al.* 2021).

The low viscosity will also act to slow the future decline of the West Antarctic Ice Sheet. Although earlier simulations concluded that the effect of bedrock uplift was minor (Bamber *et al.* 2009; Mitrovica *et al.* 2009), it is enhanced for low viscosity (Konrad *et al.* 2015; Barletta *et al.* 2018). For West Antarctic Ice Sheet collapse, the deformation of a 3D model reaches more than 50% of the elastic deformation after 100 years (Hay *et al.* 2017). Larour *et al.* (2019) found that uplift has a small effect (*c.* 1% in 2100), increasing to 27% after 500 years, which makes it a relevant process to take into account in future scenarios for the Antarctic Ice Sheet. To do so, spatial resolution is important, in particular near the grounding line where the coupling manifests. A resolution of 1 km is suggested for elastic effects (Larour *et al.* 2019) and a value of about 4 km is suggested for modelling the viscous effects (Wan *et al.* 2022). These resolutions can be reached by 3D GIA models but with increased computation time compared to standard GIA simulations. When many scenarios should be simulated, regional (flat) GIA models or ELRA models with spatially varying parameters can be a pragmatic choice but it is not yet clear how much they deviate from full 3D spherical GIA models.

Post-seismic deformation

Earthquakes are the result of the sudden release of built-up strain on a fault plane within the Earth's lithosphere. Most earthquakes occur at tectonic plate boundaries and are associated with the relative motion of the plates as they slowly move past each other (transform plate boundary), collide (convergent plate boundary) or separate (divergent plate boundary). Part of the interface at the plate boundary may become locked due to frictional resistance, preventing the two plates from moving past each other and resulting in an accumulation of strain. At a critical point in time, the accumulated strain becomes larger than the frictional resistance, the fault unlocks and the plate interface suddenly moves, causing an earthquake and a change in stress in the surrounding crust and mantle.

A small number of earthquakes also occur in the interior of a tectonic plate, termed 'intraplate earthquakes', although these earthquakes are not as common as those occurring at plate boundaries. The cause of intraplate earthquakes is not well understood but it has been suggested that they may occur due to pre-existing structures or weaknesses within the tectonic plate (Stein and Mazzotti 2007).

Earthquakes cause deformation at the Earth's surface from several processes (see Fig. 8) that can be categorized as coseismic or post-seismic, depending on whether they occur at the time of, or following, the earthquake. Coseismic displacement is the instantaneous slip that occurs on the fault plane at the time of the earthquake and the associated elastic response of the Earth in the surrounding region. Following this initial coseismic displacement, the Earth undergoes subsequent post-

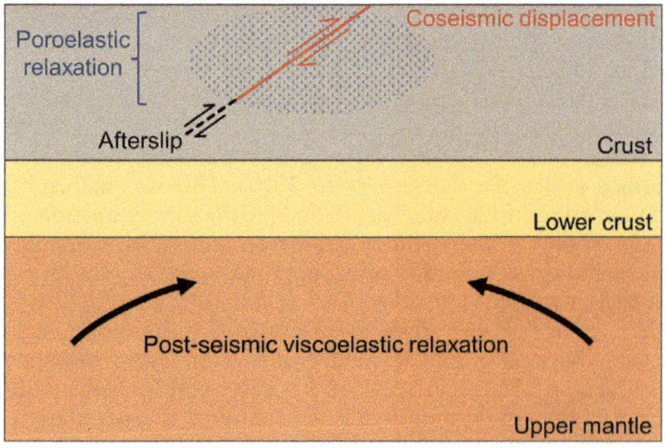

Fig. 8. Schematic representation of earthquake deformation processes. Fault-plane and coseismic displacement is shown in red, with deep afterslip shown as a black dashed line. The blue shaded region indicates poroelastic relaxation and black arrows in the upper-mantle show where post-seismic relaxation may occur.

seismic deformation caused by several processes (Wright 2016). Afterslip, which is the subsequent slow slip on the fault plane controlled by friction, usually occurs above or below the location of seismic rupture (e.g. Fig. 8). Changes in fluid pressure in the shallow upper crust close to the fault (e.g. Fig. 8) can cause poroelastic relaxation to occur in the months following an earthquake. Poroelastic relaxation takes into account the stress change due to movement of a fluid through a porous solid matrix. Finally, viscoelastic relaxation within the lower crust and upper mantle occurs as the viscous material responds to stress changes. Because this second process occurs at depth, it produces a longer wavelength signal and, hence, is the dominant cause of surface deformation in the far field (typically more than 300 km from the earthquake). Post-seismic viscoelastic deformation from large earthquakes (greater than magnitude 9) can be observed over 1000 km away from the epicentre (Shao *et al.* 2016) and can be sustained over several decades. For example, deformation from the 1960 Chile earthquake, with a magnitude of 9.5, was still being observed *c.* 40 years after the event (Khazaradze and Klotz 2003).

In this section, we will focus on post-seismic viscoelastic deformation, explore what it might reveal about the underlying mantle and outline some results for Antarctica where the study of post-seismic deformation is still in its infancy.

Post-seismic deformation, observations and models

The sudden large stress change associated with an earthquake causes viscoelastic relaxation of the lower crust and upper mantle over the years and decades following an earthquake. The magnitude, spatial pattern and time evolution of the deformation is dependent on the size of the earthquake, the geometry of the fault plane and the rheological properties of the Earth.

Post-seismic viscoelastic deformation can be observed by increasingly dense networks of geodetic measurements such as GPS (e.g. Freed *et al.* 2012) and InSAR (Interferometric Synthetic Aperture Radar: e.g. Wang and Fialko 2018), motivating the development of models to help to interpret geodetically observed deformation (see also Scheinert *et al.* 2021, this volume) and to fill in the gaps between observations in space and time. Through a process of varying model inputs and comparing model predictions to observations, constraints can be placed on the likely Earth structure (Pollitz 2005) or rheological model (Freed *et al.* 2012). In particular, the viscosity of the mantle has a dominant control over the response of the Earth to an earthquake.

Modelling techniques used for post-seismic deformation include the normal-mode method that is also widely applied in GIA modelling. The widely used VISCO1D program of Pollitz (1997) has been used in Antarctica (King and Santamaría-Gómez 2016) but it can only deal with radially varying parameters that might not be appropriate for a large part of Antarctica. Finite-element methods exist with half-space geometry (e.g. Freed *et al.* 2012) or with spherical geometry (Hu and Wang 2012; Agata *et al.* 2019). The latter are not global and do not include self-gravity, as the normal-mode method does, which can be relevant for large earthquakes. Recently, a global spherical finite-element method was presented that does include self-gravity (Nield *et al.* 2022), and has capabilities to include different rheological models and the sea-level equation, which is relevant for earthquakes under the ocean (e.g. Broerse *et al.* 2011).

Estimating post-seismic viscoelastic deformation via modelling techniques requires detailed knowledge of the earthquake such as location, depth, fault geometry and slip parameters. This information can often be found from studies of fault inversions, whereby slip properties are deduced through the inversion of the seismic observations (e.g. Ji *et al.* 2002; Hayes 2017) or from geodetic inversions (e.g. Lohman and Simons 2005; Feng *et al.* 2010). The properties of the Earth (e.g. mantle viscosity and elastic properties) and rheological model (e.g. Maxwell or Burger rheology: see Ivins *et al.* 2021, this volume), which are used as input to a forward model of post-seismic viscoelastic deformation, are then varied and the predicted displacements tested against geodetic observations, similar to GIA studies (see the 'GIA model results and inferences of mantle structure' subsection in the previous section). Alternatively, the Earth properties can be inverted using gradient-based optimization with the required gradients computed efficiently using the adjoint method (Crawford *et al.* 2017).

The rheological model can be of particular importance in the study of post-seismic deformation. The short timescale of stress change involved means that transient effects, where the short-term mantle viscosity is lower than the long-term steady-state viscosity, are important (Ranalli 1995). Several studies have shown that a transient component of deformation in addition to a steady-state viscosity is required to explain observations, specifically Burger's rheology (Pollitz and Thatcher 2010) or transient power-law rheology (Freed *et al.* 2010, 2012). This is in contrast to GIA, where non-linear or transient rheologies have been used but do not always improve the fit (van der Wal *et al.* 2015; Caron *et al.* 2017).

Insights from other regions

The study of post-seismic deformation in Antarctica is still in its infancy; however, there are many studies that have successfully constrained mantle properties or rheology in other regions of the world. The 1960 magnitude 9.5 earthquake in Chile was the largest earthquake to have ever been recorded. It occurred at the convergent plate boundary where the Nazca Plate is subducting below the South American Plate. The earthquake occurred long before the geodetic measurements of the Chile subduction zone commenced in 1993; however, the signature of the post-seismic deformation is still visible in observations 35 years after the earthquake (Hu *et al.* 2004). Several studies have constrained the continental upper-mantle viscosity in this region to 2×10^{19}–3×10^{19} Pa s using a 3D linear Maxwell viscoelastic model (Khazaradze *et al.* 2002; Hu *et al.* 2004). Sun *et al.* (2018) also explored the possibility of a Burger's rheology for the 1960 event; however, due to the length of time between the earthquake and the beginning of the GPS measurements, it was not possible for them to constrain a value for the transient viscosity. Their results for the steady-state viscosity of the mantle wedge above the subducting slab (2×10^{19} Pa s), constrained with GPS observations, agree with studies mentioned above.

In Alaska, a magnitude 7.9 strike-slip earthquake occurred on 3 November 2002. GPS stations recorded deformation rates of up to 300 mm a^{-1} during the month following the earthquake and up to 100 mm a^{-1} during the subsequent 18 months (Pollitz 2005). Pollitz (2005) interpreted the two distinct deformation rates as being indicative of a Burger's rheology with an initial transient viscosity and a long-term steady-state viscosity. Using a forward model to calculate displacement and then comparing the results to GPS observations, Pollitz (2005) was able to determine the range of possible Earth structures that best fit the data, finding upper-mantle viscosities of 1×10^{17} (transient) and 2.5×10^{18} Pa s (steady state).

The approach used by the aforementioned studies to solve for a value of mantle viscosity does not consider the environmental conditions such as temperature, pressure, grain size or water content. Freed *et al.* (2010) took a different approach by using a laboratory-derived steady-state power-law rheology, which is a function of the different environmental parameters, and investigated whether it could explain the GPS-observed post-seismic response to the 1999, magnitude 7.1, Hector Mine earthquake in California. They found that they could not replicate the observed surface displacement from 7 years of GPS data at 55 locations in the far field (>50 km from the earthquake rupture) with the steady-state flow law and suggested that a transient phase is also required: that is, where deformation is enhanced immediately after the earthquake compared to the steady-state response.

Freed *et al.* (2012) developed this idea by formulating a constitutive relationship combining transient and steady-state flow, which is a function of the environmental parameters, to predict the spatial and temporal post-seismic viscoelastic deformation. They found that they could fit the GPS data with a transient phase of approximately 1 year and with a viscosity around 10 times lower than the steady-state viscosity. Because this flow law is a function of ambient conditions, in theory it can be applied to any earthquake in any region so long as the environmental parameters are known.

Results for Antarctica

Antarctica lies within the Antarctic Plate and comprises two geologically distinct regions: the old cratons of East Antarctica and the active rift system of West Antarctica. The majority of the Antarctic Plate is connected to the surrounding tectonic plates by spreading rifts (Reading 2007). The northern Antarctic Peninsula, lying close to the plate boundary, is tectonically complex as discussed in the earlier subsection on 'GIA model results and inferences of mantle structure'. The active tectonic setting of the Antarctic Peninsula means that it is likely to have a lower mantle viscosity than other regions of Antarctica, as discussed in the 'GIA model results and inferences of mantle structure' subsection (Ivins *et al.* 2011; Nield *et al.* 2014), which may result in a more pronounced post-seismic response.

Seismicity generally occurs at a low level around the plate boundaries of the Antarctic Plate, except for two regions. Earthquakes with a magnitude greater than 7 have been recorded at the plate boundary between the Antarctic Plate and the Scotia Plate (Fig. 9); two strike-slip earthquakes have occurred at this location since 2003 (magnitude 7.6 in 2003 and magnitude 7.7 in 2013: Ye *et al.* 2014). The Macquarie Ridge Complex, where the Antarctic Plate meets the Australian and Pacific plates (Fig. 9), has also experienced three earthquakes of magnitude 7.7 or greater in the past *c.* 100 years (Watson *et al.* 2010).

Continental Antarctica was once thought to be aseismic (i.e. very few earthquakes occurred in this region) and it was not until the 1980s that an earthquake was definitively shown to have occurred here (Adams *et al.* 1985; Adams and Akoto 1986). The magnitude 4.5 earthquake occurred on 4 November 1982, *c.* 1200 km from the coast of Dronning Maud Land, and was detected by five seismic stations. Despite this confirmation of the existence of intraplate earthquakes, the continent of Antarctica was still thought to experience fewer intraplate earthquakes than other equivalent regions possibly due to the weight of the overlying ice sheet (Johnston 1987). However, the lack of recorded seismic activity was, in fact, due to the small number of seismic stations operating in this region rather than the absence of earthquakes. Following the installation of seismographs in East Antarctica during the International Polar Year, Lough *et al.* (2018) showed that during 2009 intraplate earthquakes occurred at the same frequency and magnitude as in other cratons. Given the lack of recorded earthquakes, it is no surprise that the study of post-seismic deformation in Antarctica is not well established compared with other regions of the world.

Earthquakes in continental Antarctica occur in several regions such as the Transantarctic Mountains and around the coastline (Reading 2007). The most notable intraplate earthquake occurred near Balleny Islands, an area of high seismicity close to the plate boundary south of New Zealand. This magnitude 8.1 earthquake occurred on 25 March 1998, 500 km from the Antarctic coastline (Nettles *et al.* 1999; Henry *et al.* 2000) and is the subject of the only published study on post-seismic deformation of Antarctica (King and Santamaría-Gómez 2016).

King and Santamaría-Gómez (2016) sought to place constraints on the properties of the Antarctic mantle by constraining a model of post-seismic viscoelastic deformation using geodetic measurements. They estimated initial coseismic displacement from continuous GPS observations made at the Dumont d'Urville Station located 600 km from the 1998

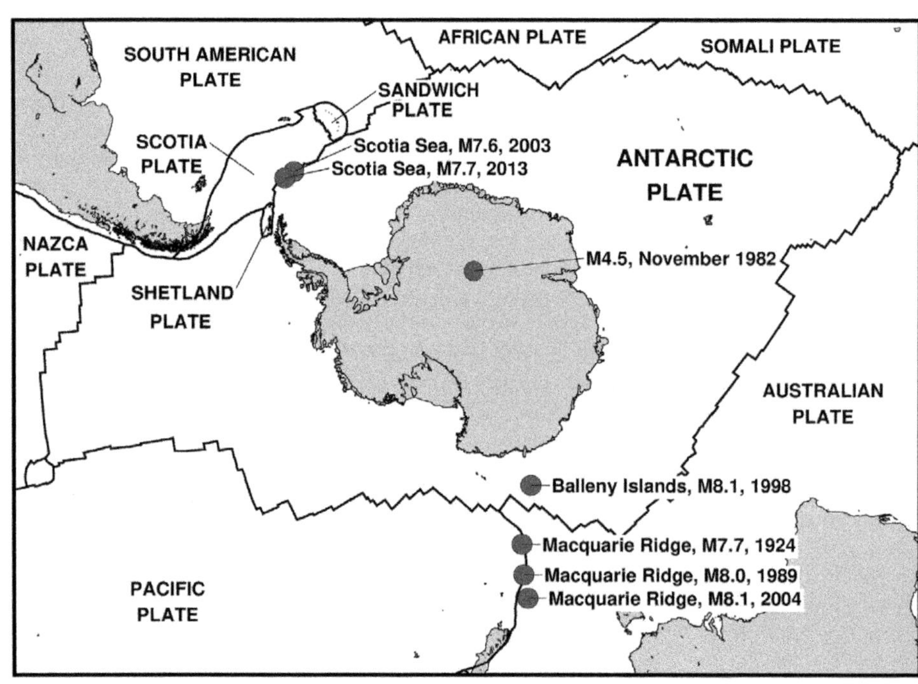

Fig. 9. Map of Antarctica with plate boundaries and the location of earthquakes mentioned in the text.

earthquake, and used this displacement to tune fault parameters such as geometry and slip magnitude. The resulting fault parameters provided input to a model of post-seismic viscoelastic deformation. The authors also considered the magnitude 8.1 2004 earthquake that occurred at the Macquarie Ridge (Fig. 9). Whilst the coseismic offset from this earthquake is small (submillimetre) at all Antarctic sites, there may be a post-seismic deformation signal present at Dumont d'Urville, as indicated in the GPS time series.

They used the VISCO1D program with a suite of layered, spherically symmetrical Earth models to predict post-seismic deformation from both the 1998 and 2004 earthquakes, and compared model outputs to bedrock coordinate time series from continuous GPS, campaign GPS, and Doppler Orbitography and Radiopositioning Integrated by Satellite (DORIS) data from the Dumont d'Urville Station. The Earth models consisted of a purely elastic lithosphere, and asthenosphere and upper mantle layers with biviscous Burger's rheology that has a transient viscosity one order of magnitude lower than the steady-state viscosity. The thickness of each layer was varied but the base of the upper mantle was fixed at a depth of 670 km.

When comparing modelled with observed horizontal motions, the authors found a best-fitting Earth model with a 90 km-thick lithosphere, a 130 km-thick asthenosphere with steady-state viscosity of 1.2×10^{19} Pa s and an upper-mantle steady-state viscosity of 1.4×10^{19} Pa s (at a depth of between 220 and 670 km). However, King and Santamaría-Gómez (2016) could not find any Earth structure that satisfactorily reproduced both the observed vertical deformation and the horizontal deformation simultaneously, and suggested that using a 3D Earth structure may solve at least part of this issue. This result is important because it may be indicative of the presence of lateral variations in the Earth structure, particularly a contrast between oceanic and continental mantle viscosity, as has also been hypothesized from isotopic studies of volcanic rock in this region (Panter *et al.* 2018). The study of post-seismic deformation clearly has the potential to constrain mantle structure in Antarctica.

Conclusions and future work

In this section, the main conclusions of glacial isostatic adjustment (GIA) and post-seismic studies in Antarctica are summarized and possible future research is outlined in three sections: the contribution of GIA and post-seismic deformation to knowledge of the mantle, future research in the field of GIA, and future research in the field of post-seismic deformation.

GIA is mainly due to the shrinking of the ice sheet since the last glacial maximum (LGM: *c.* 20 kyr ago). The total volume of ice that disappeared was for a long time mainly constrained by assigning to the Antarctic Ice Sheet the amount of ice needed to reconcile the contribution to the sea level from the known land ice masses on the other continents with the total global sea-level change. This resulted in a large ice loss from Antarctica of 25 m sea-level equivalent (SLE). Relying on an indirect constraint from the global sea level was necessary due to the lack of direct and indirect evidence from Antarctica itself. This situation has now improved, and, more recently, geodetic and seismic data and improved ice-dynamics models have led to lower estimates of 5–15 m of global SLE. Melting occurred asynchronously, mostly starting at *c.* 15 ka BP. Most of the ice disappeared from the Ross and Filchner-Ronne ice shelf areas. For East Antarctica, some models show an increase in ice thickness since the LGM. During the late Holocene, readvance occurred in some places, possibly aided by rebound due to GIA following a retreating ice sheet. Closer to present, melting after the Little Ice Age occurred in parts of West Antarctica such as the Amundsen Sea Embayment.

Post-seismic deformation occurs due to plate motion that generates large (magnitude >7) earthquakes. Earthquakes of sufficient magnitude occurred at plate margins north of the Antarctic Peninsula and south of New Zealand. Data about this phenomenon have been lacking in the region as insufficient GPS receivers were in place to detect most of the large earthquakes. A 1998 earthquake was both large enough and sufficiently covered by GPS measurements for post-seismic motion to be detected at the level of 2 mm a^{-1}.

GNSS receivers measure 3D deformation and show uplift of centimetres per year in the Amundsen Sea sector. Smaller velocities in the range of millimetres per year in areas such as the Filchner-Ronne and Ross ice shelves can be related to GIA. Horizontal velocities are harder to relate to GIA models, except in the Amundsen Sea sector. GIA is recorded in palaeosea-level observations at a dozen locations near the Antarctic coastline, showing tens of metres of uplift. Data in East Antarctica are quite sparse relative to West Antarctica, and data from both are infrequent relative to other continents. In general, continental-scale GIA models for Antarctica are relatively unconstrained but GNSS data help to make regional GIA studies more constrained.

GIA and post-seismic deformation use similar geophysical models. These require elastic and viscous parameters as input. In most GIA models, a Maxwell rheology is used where elastic deformation is followed by steady-state viscous deformation. In post-seismic models, a fast, short-term viscosity is found to be necessary, which leads to a Burger's rheology with a short-term and a long-term viscous element. GIA models estimate an average viscosity of 10^{20}–10^{21} Pa s for the Antarctic upper mantle. Regional studies show low viscosity ($<10^{18}$ Pa s) in the northern Antarctic Peninsula, increasing to 10^{20} Pa s in the southern Antarctic Peninsula. This increasing viscosity agrees with subduction that terminates from south to north. The post-seismic deformation following the 1998 earthquake near the Balleny Islands that has been detected and compared to model outputs leads to a best-fitting viscosity of 1.2×10^{19} Pa s in the shallow, and 1.4×10^{19} Pa s in the deeper upper mantle, similar to GIA-based estimates for the Antarctic Peninsula. Low viscosity (4×10^{18} Pa s) is also found in the Amundsen Sea sector, an area with geologically recent rifting and volcanism. These viscosity inferences agree with maps of 3D viscosity created from seismic velocity anomalies that are converted to viscosity anomalies. In fact, the argument should be reversed: the uncertainty in 3D viscosity maps can be considered so high that regional studies (both geophysical and geological) are needed to provide a constraint on (parts of) such viscosity maps.

The timing of the Earth's response to changes in loading is driven by the local mantle viscosity. Low viscosities (10^{17}–10^{18} Pa s) in the Antarctic upper mantle can be translated into a relaxation time in the years to decades range, much faster than in classical GIA areas, but similar to GIA in areas near rifts or subduction such as Iceland, Alaska and Patagonia. The low viscosity also guarantees a strong influence on the ice sheets in Antarctica by changing the local sea level: post-glacial uplift raises land ice from the sea and reduces the ice–ocean contact area and, hence, melt. The instability of an inward-moving grounding line in combination with topography that deepens inland can be somewhat reduced by GIA. This process can be accounted for with schematic models that require a single relaxation time as a parameter, or with a viscoelastic half-space model, or full-fledged 3D GIA simulations. Accounting for this feedback with the most accurate model is shown to affect the total volume of the LGM Antarctic Ice Sheet by 1–3 m SLE and to affect the grounding-line position by hundreds of kilometres in the marine-grounded ice sheets in the Ross and Weddell seas.

Outlook on contribution of GIA and post-seismic deformation to knowledge of the mantle

Simulations of GIA and post-seismic deformation depend on mantle viscosity. However, their different sensitivities and timescales mean that results from one study are not directly comparable to mantle viscosity structure in another study because different regions and timescales are sampled. For example, even if the Earth structure is identical across different simulations, the effective lithosphere for shorter loading phenomena is thicker because the mantle below has had less time to adjust viscously (Nield *et al.* 2018; Lau *et al.* 2021). A step forward is the use of seismic models, flow laws, geological observations and material parameters to create mantle profiles that can be used and tested in different geodynamic studies, and where lithosphere thickness is implicit in the 3D viscosity distribution.

The use of seismic velocity anomalies as a proxy for viscosity variations raises the issues of accuracy of conversion of seismic velocity anomalies to temperature and viscosity. The question whether temperature or composition explains a seismic velocity anomaly cannot be answered *a priori* and requires other information that could come from xenoliths, which record grain size, water content and chemistry from which pressure and temperature conditions can be calculated that are relevant in the flow of the main mantle material olivine (Martin 2021, this volume). These show heterogeneity on all scales. There is evidence in Antarctica for the occurrence of large grain sizes that would increase viscosity but also for the presence of water in mantle rocks, which has a weakening effect. Besides isolated findings of mantle rocks, the mantle structure can also be elucidated by the geological history and the tectonic link to surrounding continents. The observed high heat flow in West Antarctica could provide a constraint on thermal modelling of the mantle, although heat-flow measurements are sparse (Burton-Johnson *et al.* 2020; Pappa and Ebbing 2021, this volume). Gravity-field variations provide a strong test of mantle structure. Thermal anomalies as used in geodynamic models can be converted to density and gravity anomalies with less uncertainty than in the conversion to viscosity. The resulting gravity-field variations must match the observed gravity, which, for the long wavelengths that contain the mantle signal, is known with high accuracy (Pappa and Ebbing 2021, this volume).

The translation from material parameters and conditions of the mantle to viscosity requires flow laws (i.e. relations between stress and strain rate). Experiments show that even for the same material, the effective viscosity is different for different processes because the bedrock relaxation is not in steady state (i.e. transient rheology) or because one of the mantle-flow mechanisms is strongly dependent on stress (i.e. dislocation creep). For the above reasons, Lau *et al.* (2021) argued that the standard description of viscosity should be replaced by frequency dependence of viscosity, which would lead to a different effective viscosity for LGM ice-sheet unloading than for the recent changes in the Antarctic Peninsula. It will be a large challenge to create a framework that can unify geological observations, the seismic measurements with the obtained viscosity in West Antarctica and the average viscosities found in 1D GIA studies.

The emerging area of post-seismic deformation research has the potential to reveal information about the mantle in Antarctica, providing complementary results to GIA as the processes span different timescales and affect different regions of the mantle. Post-seismic deformation occurs over years to decades, predominantly in the shallow upper mantle, whereas GIA may occur over hundreds to thousands of years invoking a response at much greater depths. There are a few select regions (such as the northern Antarctic Peninsula) where the timescale of these processes overlaps, presenting an opportunity to use post-seismic deformation and GIA to constrain the lithosphere and mantle.

Generalizing this idea further, the same Earth model should be applied for regional recent loading and continent-wide LGM ice sheets, post-seismic deformation and mantle convection, and be compared to the geophysical, geological and geodetic observations of these processes. In particular, the horizontal velocities hold promise because of their different sensitivities (Hermans *et al.* 2018) and availabilities near areas of viscosity contrast (Konfal *et al.* 2018). Integrating results from regional studies and other processes in 3D viscosity maps is not straightforward; it is not known *a priori* how the GIA process samples the 3D Earth (Powell *et al.* 2020). Sensitivity studies can enlighten the depth sensitivity, as well as the relevant horizontal extent, so that the regional GIA or post-seismic studies can provide unbiased anchor points for 3D effective viscosity models (see table 6 in Ivins *et al.* 2021, this volume). Integrating more geodynamic processes and their observations provide more constraints that can be assimilated. This complicates and lengthens simulations and interpretations but in the end is the best way to unveil the true Earth structure.

Future research on GIA in Antarctica

The construction of the ice history is the main uncertainty in GIA model predictions. Ice dynamics cannot yet provide an ice-sheet history with large confidence: for example, knowledge of subglacial processes and interaction between ice and ocean requires improvement (Siegert and Golledge 2022). In addition, more efforts are necessary to estimate ice-thickness variations in the last millennium. It has been demonstrated that several regions in West Antarctica, and also at the coast of East Antarctica, have lower upper-mantle viscosity, responding on a timescale of years to hundreds of years, which means their uplift is sensitive to the late Holocene or even the last decades of ice-sheet evolution. Furthermore, there is evidence for grounding-line readvance, which complicates the picture in most GIA models of monotonous retreat (Kingslake *et al.* 2018; Simms *et al.* 2021). Deepening and integrating this knowledge with the last LGM maximum ice history will be a major step forward for GIA modelling. At the same time, improvements in the physical description of ice-sheet behaviour will go hand in hand with improved predictions for future scenarios of decline of the Antarctic Ice Sheet.

The picture of the Antarctic mantle that emerged from seismic data and mantle xenolith and volcanic rock studies shows variations in mantle structure from West Antarctica to East Antarctica and within subregions of Antarctica such as the Antarctic Peninsula. However, there are uncertainties in the seismic models themselves (Wiens *et al.* 2021, this volume), as well as in the conversion to viscosity (Ivins *et al.* 2021, this volume). Moreover, seismic data have limited resolution in the lower mantle, where a significant part of the GIA sensitivity could reside. Global models show small anomalies in the lower mantle but Lloyd *et al.* (2020) showed an anomaly below Marie Bird Land that continues into the lower mantle (Wiens *et al.* 2021, this volume). Improved lower-mantle seismic velocity anomalies in other regions will lead to improved GIA predictions. Further constraints on viscosity are necessary. If flow laws are used to derive effective viscosity, Antarctica water content and grain size are usually the largest uncertainties (van der Wal *et al.* 2015; King *et al.* 2016; Blank *et al.* 2021). Water content, however, can be measured in xenoliths or igneous olivine from which trends on spatial variations in water content within Antarctica could be deduced that could be applied in GIA models (Martin 2021, this volume).

With the current knowledge of the distinct domains in East and West Antarctica, the few studies that interpreted sea-level records before the LGM could be revisited. On such time scales, infill of sediments can form a considerable load. Sediment deposition mapped offshore also offers constraints on where erosion occurred and, hence, where ice-sheet flow originated (Naish *et al.* 2022). The latter could require coupling between ice-sheet erosion models such as done by Jamieson *et al.* (2010) to model the evolution of subglacial topography. On longer timescales, the evolution of the topography due to sedimentation and vertical motion due to dynamic topography and tectonics should also be taken into account, as the topography controls ice-sheet flow (Whitehouse *et al.* 2019; Colleoni *et al.* 2022).

All of the above play a large role in West Antarctica where most of the rapid ice melt occurs, and which sees low viscosity and large variations in tectonic regimes. However, East Antarctica holds most of the freshwater and there are several regions in East Antarctica where glaciers are thinning fast (Rignot 2006). East Antarctica is much less covered by seismic or geodetic data, for obvious logistic reasons. This lack of knowledge presents considerable uncertainty in understanding the evolution and future of the Antarctica Ice Sheet and associated sea-level changes. Therefore, improved data coverage in East Antarctica is desirable, albeit challenging and costly.

Future research on post-seismic deformation in Antarctica

The study of post-seismic viscoelastic deformation in Antarctica is still in its early stages but King and Santamaría-Gómez (2016) have shown that the continent is, or has been, affected by widespread deformation following the 1998 earthquake near the Balleny Islands. This demonstrates the potential for future post-seismic studies to reveal information about Antarctica's mantle, particularly in terms of 3D Earth properties that may be needed to explain observations. Future studies will be likely to focus on other large earthquakes, particularly those that have occurred during the era of geodetic observation such as the 2003 and 2013 Scotia Sea earthquakes that occurred close to the northern Antarctic Peninsula. Furthermore, the possibility of a transient phase should be investigated where the mantle responds more quickly over a shorter timescale immediately following the sudden change in stress, as has proved necessary to explain the response to earthquakes in other regions (Pollitz 2005; Freed *et al.* 2012).

High-quality and plentiful geodetic data are key to being able to constrain Earth properties from post-seismic deformation, as studies have done for other regions of the world. With around 75 continuous GPS stations operating across Antarctica, the situation has vastly improved from the sparse arrays in operation at the time of the 1998 Balleny Islands and 2003 Scotia Sea earthquakes. This shows the importance of maintaining and building on existing networks, such as the Antarctic-Network (A-NET) project (Wilson *et al.* 2020), which will provide more observations to constrain models. Furthermore, improved models of post-seismic viscoelastic deformation, which include lateral variations in Earth properties, will be needed in regions where traditional 1D Earth structures (10^{20}–10^{21} Pa s upper-mantle viscosity: see the subsection on 'GIA model results and inferences of mantle structure') cannot match observations in all components of deformation.

Acknowledgements The authors thank the two reviewers for their comments, which helped to improve the manuscript, and Pippa Whitehouse for providing data and information on the W12 ice model. Several of the figures used the mapping tools of Greene *et al.* (2017), the colour maps of Crameri (2018) and the Generic Mapping Tools (Wessel *et al.* 2019).

Competing interest The authors declare that they have no known competing financial interests or personal relationships that could have appeared to influence the work reported in this chapter.

Author contributions **WVDW**: writing – original draft (lead), writing – review & editing (equal); **VB**: writing – original draft (supporting), writing – review & editing (equal); **GN**: writing – original draft (supporting), writing – review & editing (equal); **CVC**: writing – original draft (supporting), writing – review & editing (equal).

Funding The work is partially funded by projects GOCE + Antarctica and 3-D Earth funded by the European Space Agency (ESA) as a Support to Science Element.

Data availability Data sharing is not applicable to this article as no datasets were generated or analysed during the current study.

References

Adams, R.D. and Akoto, A.M. 1986. Earthquakes in continental Antarctica. *Journal of Geodynamics*, **6**, 263–270, https://doi.org/10.1016/0264-3707(86)90043-8

Adams, R.D., Hughes, A.A. and Zhang, B.M. 1985. A confirmed earthquake in continental Antarctica. *Geophysical Journal International*, **81**, 489–492, https://doi.org/10.1111/j.1365-246X.1985.tb06416.x

Adhikari, S., Ivins, E.R., Larour, E., Seroussi, H., Morlighem, M. and Nowicki, S. 2014. Future Antarctic bed topography and its implications for ice sheet dynamics. *Solid Earth*, **5**, 569–584, https://doi.org/10.5194/se-5-569-2014

Agata, R., Barbot, S.D. *et al.* 2019. Rapid mantle flow with power-law creep explains deformation after the 2011 Tohoku mega-quake. *Nature Communications*, **10**, 1385, https://doi.org/10.1038/s41467-019-08984-7

Albrecht, T. 2019. PISM parameter ensemble analysis of Antarctic Ice Sheet glacial cycle simulations, PANGAEA, https://doi.org/10.1594/PANGAEA.909728

Albrecht, T., Winkelmann, R. and Levermann, A. 2020*a*. Glacial-cycle simulations of the Antarctic Ice Sheet with the Parallel Ice Sheet Model (PISM) – Part 1: Boundary conditions and climatic forcing. *The Cryosphere*, **14**, 599–632, https://doi.org/10.5194/tc-14-599-2020

Albrecht, T., Winkelmann, R. and Levermann, A. 2020*b*. Glacial-cycle simulations of the Antarctic Ice Sheet with the Parallel Ice Sheet Model (PISM) – Part 2: Parameter ensemble analysis. *The Cryosphere*, **14**, 633–656, https://doi.org/10.5194/tc-14-633-2020

An, M., Wiens, D.A. *et al.* 2015. Temperature, lithosphere-asthenosphere boundary, and heat flux beneath the Antarctic Plate inferred from seismic velocities. *Journal of Geophysical Research: Solid Earth*, **120**, 8720–8742, https://doi.org/10.1002/2015JB011917

Argus, D.F., Peltier, W.R., Drummond, R. and Moore, A.W. 2014. The Antarctica component of postglacial rebound model ICE-6G_C (VM5a) based on GPS positioning, exposure age dating of ice thicknesses, and relative sea level histories. *Geophysical Journal International*, **198**, 537–563, https://doi.org/10.1093/gji/ggu140

Arndt, J.E., Schenke, H.W. *et al.* 2013. The International Bathymetric Chart of the Southern Ocean (IBCSO) Version 1.0 – A new bathymetric compilation covering circum-Antarctic waters. *Geophysical Research Letters*, **40**, 3111–3117, https://doi.org/10.1002/grl.50413

Bagge, M., Klemann, V., Steinberger, B., Latinović, M. and Thomas, M. 2021. Glacial-isostatic adjustment models using geodynamically constrained 3D earth structures. *Geochemistry, Geophysics, Geosystems*, **22**, e2021GC009853, https://doi.org/10.1029/2021GC009853

Bamber, J.L., Riva, R.E.M., Vermeersen, B.L.A. and LeBrocq, A.M. 2009. Reassessment of the potential sea-level rise from a collapse of the West Antarctic Ice Sheet. *Science*, **324**, 901–903, https://doi.org/10.1126/science.1169335

Barker, D.H.N. and Austin, J.A. Jr 1998. Rift propagation, detachment faulting, and associated magmatism in Bransfield Strait, Antarctic Peninsula. *Journal of Geophysical Research: Solid Earth*, **103**, 24017–24043, https://doi.org/10.1029/98JB01117

Barker, P. 1982. The Cenozoic subduction history of the Pacific margin of the Antarctic Peninsula: ridge crest–trench interactions. *Journal of the Geological Society, London*, **139**, 787–801, https://doi.org/10.1144/gsjgs.139.6.0787

Barletta, V.R., Bevis, M. *et al.* 2018. Observed rapid bedrock uplift in Amundsen Sea Embayment promotes ice-sheet stability. *Science*, **360**, 1335–1339, https://doi.org/10.1126/science.aao1447

Barnhoorn, A., van der Wal, W., Vermeersen, B.L.A. and Drury, M.R. 2011. Lateral, radial, and temporal variations in upper mantle viscosity and rheology under Scandinavia. *Geochemistry, Geophysics, Geosystems*, **12**, Q01007, https://doi.org/10.1029/2010GC003290

Bassett, S.E., Milne, G.A., Bentley, M.J. and Huybrechts, P. 2007. Modelling Antarctic sea-level data to explore the possibility of a dominant Antarctic contribution to meltwater pulse IA. *Quaternary Science Reviews*, **26**, 2113–2127, https://doi.org/10.1016/j.quascirev.2007.06.011

Bentley, M.J. 1999. Volume of Antarctic Ice at the Last Glacial Maximum, and its impact on global sea level change. *Quaternary Science Reviews*, **18**, 1569–1595, https://doi.org/10.1016/S0277-3791(98)00118-8

Bentley, M.J. and Anderson, J.B. 1998. Glacial and marine geological evidence for the ice sheet configuration in the Weddell Sea–Antarctic Peninsula region during the Last Glacial Maximum. *Antarctic Science*, **10**, 309–325, https://doi.org/10.1017/S0954102098000388

Bentley, M.J., Hodgson, D.A., Smith, J.A. and Cox, N.J. 2005. Relative sea level curves for the South Shetland Islands and Marguerite Bay, Antarctic Peninsula. *Quaternary Science Reviews*, **24**, 1203–1216, https://doi.org/10.1016/j.quascirev.2004.10.004

Bentley, M.J., Ó Cofaigh, C. *et al.* 2014. A community-based geological reconstruction of Antarctic Ice Sheet deglaciation since the Last Glacial Maximum. *Quaternary Science Reviews*, **100**, 1–9, https://doi.org/10.1016/j.quascirev.2014.06.025

Berg, J.H., Moscati, R.J. and Herz, D.L. 1989. A petrologic geotherm from a continental rift in Antarctica. *Earth and Planetary Science Letters*, **93**, 98–108, https://doi.org/10.1016/0012-821X(89)90187-8

Bevis, M., Wahr, J. *et al.* 2012. Bedrock displacements in Greenland manifest ice mass variations, climate cycles and climate change. *Proceedings of the National Academy of Sciences of the United States of America*, **109**, 11 944–11 948, https://doi.org/10.1073/pnas.1204664109

Bills, B.G. and James, T.S. 1996. Late Quaternary variations in relative sea level due to glacial cycle polar wander. *Geophysical Research Letters*, **23**, 3023–3026, https://doi.org/10.1029/96GL02886

Blank, B., Barletta, V., Hu, H., Pappa, F. and van der Wal, W. 2021. Effect of lateral and stress-dependent viscosity variations on GIA induced uplift rates in the Amundsen Sea embayment. *Geochemistry, Geophysics, Geosystems*, **22**, e2021GC009807, https://doi.org/10.1029/2021GC009807

Bradley, S.L., Hindmarsh, R.C.A., Whitehouse, P.L., Bentley, M.J. and King, M.A. 2015. Low post-glacial rebound rates in the Weddell Sea due to Late Holocene ice-sheet readvance. *Earth and Planetary Science Letters*, **413**, 79–89, https://doi.org/10.1016/j.epsl.2014.12.039

Brandes, C., Steffen, H., Steffen, R. and Wu, P. 2015. Intraplate seismicity in northern Central Europe is induced by the last glaciation. *Geology*, **43**, 611–614, https://doi.org/10.1130/G36710.1

Bredow, E., Steinberger, B., Gassmöller, R. and Dannberg, J. 2022. Mantle convection and possible mantle plumes beneath Antarctica – insights from geodynamic models and implications for topography. *Geological Society, London, Memoirs*, **56**, https://doi.org/10.1144/M56-2020-2

Briggs, R.D. and Tarasov, L. 2013. How to evaluate model-derived deglaciation chronologies: a case study using Antarctica. *Quaternary Science Reviews*, **63**, 109–127, https://doi.org/10.1016/j.quascirev.2012.11.021

Briggs, R.D., Pollard, D. and Tarasov, L. 2014. A data-constrained large ensemble analysis of Antarctic evolution since the Eemian. *Quaternary Science Reviews*, **103**, 91–115, https://doi.org/10.1016/j.quascirev.2014.09.003

Broerse, D.B.T., Vermeersen, L.L.A., Riva, R.E.M. and van der Wal, W. 2011. Ocean contribution to co-seismic crustal deformation and geoid anomalies: Application to the 2004 December 26 Sumatra–Andaman earthquake. *Earth and Planetary Science Letters*, **305**, 341–349, https://doi.org/10.1016/j.epsl.2011.03.011

Brook, E.J., Kurz, M.D., Ackert, R.P., Denton, G.H., Brown, E.T., Raisbeck, G.M. and Yiou, F. 1993. Chronology of Taylor Glacier advances in Arena Valley, Antarctica, using *in situ* cosmogenic ^{3}He and ^{10}Be. *Quaternary Research*, **39**, 11–23, https://doi.org/10.1006/qres.1993.1002

Bueler, E., Lingle, C.S. and Brown, J. 2007. Fast computation of a viscoelastic deformable Earth model for ice-sheet simulations. *Annals of Glaciology*, **46**, 97–105, https://doi.org/10.3189/172756407782871567

Burton-Johnson, A., Dziadek, R. *et al.* 2020. *Antarctic Geothermal Heat Flow: Future Research Directions*. SCAR-SERCE White Paper. Scientific Committee on Antarctic Research (SCAR), Cambridge, UK, https://scar.org/scar-library/search/science-4/research-programmes/serce/5454-scar-serce-white-paper-on-antarctic-geothermal-heat-flow/

Caron, L. and Ivins, E.R. 2020. A baseline Antarctic GIA correction for space gravimetry. *Earth and Planetary Science Letters*, **531**, 115957, https://doi.org/10.1016/j.epsl.2019.115957

Caron, L., Métivier, L., Greff-Lefftz, M., Fleitout, L. and Rouby, H. 2017. Inverting Glacial Isostatic Adjustment signal using Bayesian framework and two linearly relaxing rheologies. *Geophysical Journal International*, **209**, 1126–1147, https://doi.org/10.1093/gji/ggx083

Cathles, L.M. 1975. *Viscosity of the Earth's Mantle*. Princeton Legacy Library, Princeton, NJ.

Christensen, U. 1984. Convection with pressure- and temperature-dependent non-Newtonian rheology. *Geophysical Journal International*, **77**, 343–384, https://doi.org/10.1111/j.1365-246X.1984.tb01939.x

Clark, J.A. and Lingle, C.S. 1979. Predicted relative sea-level changes (18,000 years B.P. to present) caused by late-glacial retreat of the Antarctic Ice Sheet. *Quaternary Research*, **11**, 279–298, https://doi.org/10.1016/0033-5894(79)90076-0

Clark, J.A., Farrell, W.E. and Peltier, W.R. 1978. Global changes in postglacial sea level: A numerical calculation. *Quaternary Research*, **9**, 265–287, https://doi.org/10.1016/0033-5894(78)90033-9

Clark, P.U. and Tarasov, L. 2014. Closing the sea level budget at the Last Glacial Maximum. *Proceedings of the National Academy of Sciences of the United States of America*, **111**, 15 861–15 862, https://doi.org/10.1073/pnas.1418970111

Colleoni, F., De Santis, L. *et al.* 2018. Spatio-temporal variability of processes across Antarctic ice-bed–ocean interfaces. *Nature Communications*, **9**, 2289, https://doi.org/10.1038/s41467-018-04583-0

Colleoni, F., De Santis, L. *et al.* 2022. Past Antarctic ice sheet dynamics (PAIS) and implications for future sea-level change. *In*: Florindo, F., Siegart, M., De Santis, L. and Naish, T. (eds) *Antarctic Climate Evolution*. Elsevier, 689–768, https://doi.org/10.1016/B978-0-12-819109-5.00010-4

Coltorti, M., Bonadiman, C., Casetta, F., Faccini, B., Giacomoni, P.P., Pelorosso, B. and Perinelli, C. 2021. Nature and evolution of the northern Victoria Land lithospheric mantle (Antarctica) as revealed by ultramafic xenoliths. *Geological Society, London, Memoirs*, **56**, https://doi.org/10.1144/M56-2020-11

Coulon, V., Bulthuis, K., Whitehouse, P.L., Sun, S., Haubner, K., Zipf, L. and Pattyn, F. 2021. Contrasting response of West and East Antarctic ice sheets to glacial isostatic adjustment. *Journal of Geophysical Research: Earth Surface*, **126**, e2020JF006003, https://doi.org/10.1029/2020JF006003

Crameri, F. 2018. Geodynamic diagnostics, scientific visualisation and StagLab 3.0. *Geoscientific Model Development*, **11**, 2541–2562, https://doi.org/10.5194/gmd-11-2541-2018

Crawford, O., Al-Attar, D., Tromp, J. and Mitrovica, J.X. 2017. Forward and inverse modelling of post-seismic deformation. *Geophysical Journal International*, **208**, 845–876, https://doi.org/10.1093/gji/ggw414

Crawford, O., Al-Attar, D., Tromp, J., Mitrovica, J.X., Austermann, J. and Lau, H.C.P. 2018. Quantifying the sensitivity of post-glacial sea level change to laterally varying viscosity. *Geophysical Journal International*, **214**, 1324–1363, https://doi.org/10.1093/gji/ggy184

Davies, B.J., Hambrey, M.J., Smellie, J.L., Carrivick, J.L. and Glasser, N.F. 2012. Antarctic Peninsula Ice Sheet evolution during the Cenozoic Era. *Quaternary Science Reviews*, **31**, 30–66, https://doi.org/10.1016/j.quascirev.2011.10.012

Davis, B. 2020. Cosmogenic nuclide dating. AntarcticGlaciers.org, http://www.antarcticglaciers.org/glacial-geology/dating-glacial-sediments-2/cosmogenic_nuclide_datin/ [last accessed April 2022].

de Boer, B., Stocchi, P. and van de Wal, R.S.W. 2014. A fully coupled 3-D ice-sheet–sea-level model: algorithm and applications. *Geoscientific Model Development*, **7**, 2141–2156, https://doi.org/10.5194/gmd-7-2141-2014

DeConto, R.M., Pollard, D. *et al.* 2021. The Paris Climate Agreement and future sea-level rise from Antarctica. *Nature*, **593**, 83–89, https://doi.org/10.1038/s41586-021-03427-0

Denton, G.H. and Hughes, T.J. 2002. Reconstructing the Antarctic Ice Sheet at the Last Glacial Maximum. *Quaternary Science Reviews*, **21**, 193–202, https://doi.org/10.1016/S0277-3791(01)00090-7

Denton, G.H., Bockheim, J.G., Rutford, R.H., Andersen, B.G., Webers, G.F., Craddock, C. and Splettstoesser, J.F. 1992. Glacial history of the Ellsworth Mountains, West Antarctica. *Geological Society of America Memoirs*, **170**, 403–432, https://doi.org/10.1130/MEM170-p403

Dietrich, R., Ivins, E.R., Casassa, G., Lange, H., Wendt, J. and Fritsche, M. 2010. Rapid crustal uplift in Patagonia due to enhanced ice loss. *Earth and Planetary Science Letters*, **289**, 22–29, https://doi.org/10.1016/j.epsl.2009.10.021

Donda, F., O'Brien, P.E., De Santis, L., Rebesco, M. and Brancolini, G. 2008. Mass wasting processes in the Western Wilkes Land margin: Possible implications for East Antarctic glacial history. *Palaeogeography, Palaeoclimatology, Palaeoecology*, **260**, 77–91, https://doi.org/10.1016/j.palaeo.2007.08.008

Eagles, G., Gohl, K. and Larter, R.D. 2004. High-resolution animated tectonic reconstruction of the South Pacific and West Antarctic Margin. *Geochemistry, Geophysics, Geosystems*, **5**, Q07002, https://doi.org/10.1029/2003GC000657

Ekman, M. 1991. A concise history of postglacial land uplift research (from its beginning to 1950). *Terra Nova*, **3**, 358–365, https://doi.org/10.1111/j.1365-3121.1991.tb00163.x

Ekman, M. 2013. *An Investigation of Celsius' Pioneering Determination of the Fennoscandian Land Uplift Rate, and of His Mean Sea Level Mark.* Small Publications in Historical Geophysics, **25**.

Farrell, W.E. and Clark, J.A. 1976. On postglacial sea level. *Geophysical Journal International*, **46**, 647–667, https://doi.org/10.1111/j.1365-246X.1976.tb01252.x

Faul, U. and Jackson, I. 2015. Transient creep and strain energy dissipation: An experimental perspective. *Annual Review of Earth and Planetary Sciences*, **43**, 541–569, https://doi.org/10.1146/annurev-earth-060313-054732

Feng, G., Hetland, E.A., Ding, X., Li, Z. and Zhang, L. 2010. Coseismic fault slip of the 2008 M_w 7.9 Wenchuan earthquake estimated from InSAR and GPS measurements. *Geophysical Research Letters*, **37**, L01302, https://doi.org/10.1029/2009GL041213

Findley, W.N., Lai, J.S. and Onaran, K. 1976. *Creep and Relaxation of Nonlinear Viscoelastic Materials: With an Introduction to Linear Viscoelasticity*. North Holland, Amsterdam.

Freed, A.M., Herring, T. and Bürgmann, R. 2010. Steady-state laboratory flow laws alone fail to explain postseismic observations. *Earth and Planetary Science Letters*, **300**, 1–10, https://doi.org/10.1016/j.epsl.2010.10.005

Freed, A.M., Hirth, G. and Behn, M.D. 2012. Using short-term post-seismic displacements to infer the ambient deformation conditions of the upper mantle. *Journal of Geophysical Research: Solid Earth*, **117**, B01409, https://doi.org/10.1029/2011JB008562

Fretwell, P., Pritchard, H.D. *et al.* 2013. Bedmap2: improved ice bed, surface and thickness datasets for Antarctica. *The Cryosphere*, **7**, 375–393, https://doi.org/10.5194/tc-7-375-2013

Fullea, J., Lebedev, S., Martinec, Z. and Celli, N.L. 2021. WINTERC-G: mapping the upper mantle thermochemical heterogeneity from coupled geophysical–petrological inversion of seismic waveforms, heat flow, surface elevation and gravity satellite data. *Geophysical Journal International*, **226**, 146–191, https://doi.org/10.1093/gji/ggab094

Galeotti, S., DeConto, R. *et al.* 2016. Antarctic Ice Sheet variability across the Eocene–Oligocene boundary climate transition. *Science*, **352**, 76–80, https://doi.org/10.1126/science.aab0669

Garbe, J., Albrecht, T., Levermann, A., Donges, J.F. and Winkelmann, R. 2020. The hysteresis of the Antarctic Ice Sheet. *Nature*, **585**, 538–544, https://doi.org/10.1038/s41586-020-2727-5

Geruo, A., Wahr, J. and Zhong, S. 2013. Computations of the viscoelastic response of a 3-D compressible Earth to surface loading: an application to Glacial Isostatic Adjustment in Antarctica and Canada. *Geophysical Journal International*, **192**, 557–572, https://doi.org/10.1093/gji/ggs030

Goelzer, H., Coulon, V., Pattyn, F., de Boer, B. and van de Wal, R. 2020. Brief communication: On calculating the sea-level contribution in marine ice-sheet models. *The Cryosphere*, **14**, 833–840, https://doi.org/10.5194/tc-14-833-2020

Goes, S. and van der Lee, S. 2002. Thermal structure of the North American uppermost mantle inferred from seismic tomography. *Journal of Geophysical Research: Solid Earth*, **107**, ETG 2-1–ETG 2-13, https://doi.org/10.1029/2000JB000049

Golledge, N.R., Fogwill, C.J., Mackintosh, A.N. and Buckley, K.M. 2012. Dynamics of the last glacial maximum Antarctic ice-sheet and its response to ocean forcing. *Proceedings of the National Academy of Sciences of the United States of America*, **109**, 16 052–16 056, https://doi.org/10.1073/pnas.1205385109

Golledge, N.R., Levy, R.H. *et al.* 2013. Glaciology and geological signature of the Last Glacial Maximum Antarctic ice sheet. *Quaternary Science Reviews*, **78**, 225–247, https://doi.org/10.1016/j.quascirev.2013.08.011

Golledge, N.R., Menviel, L., Carter, L., Fogwill, C.J., England, M.H., Cortese, G. and Levy, R.H. 2014. Antarctic contribution to meltwater pulse 1A from reduced Southern Ocean overturning. *Nature Communications*, **5**, 5107, https://doi.org/10.1038/ncomms6107

Gomez, N., Mitrovica, J.X., Huybers, P. and Clark, P.U. 2010. Sea level as a stabilizing factor for marine-ice-sheet grounding lines. *Nature Geoscience*, **3**, 850–853, https://doi.org/10.1038/ngeo1012

Gomez, N., Pollard, D., Mitrovica, J.X., Huybers, P. and Clark, P.U. 2012. Evolution of a coupled marine ice sheet–sea level model. *Journal of Geophysical Research: Earth Surface*, **117**, F01013, https://doi.org/10.1029/2011JF002128

Gomez, N., Pollard, D. and Mitrovica, J.X. 2013. A 3-D coupled ice sheet – sea level model applied to Antarctica through the last

40 ky. *Earth and Planetary Science Letters*, **384**, 88–99, https://doi.org/10.1016/j.epsl.2013.09.042

Gomez, N., Pollard, D. and Holland, D. 2015. Sea-level feedback lowers projections of future Antarctic Ice-Sheet mass loss. *Nature Communications*, **6**, 8798, https://doi.org/10.1038/ncomms9798

Gomez, N., Latychev, K. and Pollard, D. 2018. A coupled ice sheet–sea level model incorporating 3D Earth structure: Variations in Antarctica during the last deglacial retreat. *Journal of Climate*, **31**, 4041–4054, https://doi.org/10.1175/jcli-d-17-0352.1

Gomez, N., Weber, M.E., Clark, P.U., Mitrovica, J.X. and Han, H.K. 2020. Antarctic ice dynamics amplified by Northern Hemisphere sea-level forcing. *Nature*, **587**, 600–604, https://doi.org/10.1038/s41586-020-2916-2

Gowan, E.J., Zhang, X. *et al.* 2021. A new global ice sheet reconstruction for the past 80 000 years. *Nature Communications*, **12**, 1199, https://doi.org/10.1038/s41467-021-21469-w

Greene, C.A., Gwyther, D.E. and Blankenship, D.D. 2017. Antarctic Mapping Tools for Matlab. *Computers & Geosciences*, **104**, 151–157, https://doi.org/10.1016/j.cageo.2016.08.003

Groh, A., Ewert, H. *et al.* 2012. An investigation of Glacial Isostatic Adjustment over the Amundsen Sea sector, West Antarctica. *Global and Planetary Change*, **98-99**, 45–53, https://doi.org/10.1016/j.gloplacha.2012.08.001

Han, H.K., Gomez, N. and Wan, J.X.W. 2022. Capturing the interactions between ice sheets, sea level and the solid Earth on a range of timescales: a new 'time window' algorithm. *Geoscientific Model Development*, **15**, 1355–1373, https://doi.org/10.5194/gmd-15-1355-2022

Handler, M.R., Wysoczanski, R.J. and Gamble, J.A. 2021. Marie Byrd Land lithospheric mantle: a review of the xenolith record. *Geological Society, London, Memoirs*, **56**, https://doi.org/10.1144/M56-2020-17

Hansen, S.E., Graw, J.H. *et al.* 2014. Imaging the Antarctic mantle using adaptively parameterized P-wave tomography: Evidence for heterogeneous structure beneath West Antarctica. *Earth and Planetary Science Letters*, **408**, 66–78, https://doi.org/10.1016/j.epsl.2014.09.043

Hartmann, R., Ebbing, J. and Conrad, C.P. 2020. A Multiple 1D Earth Approach (M1DEA) to account for lateral viscosity variations in solutions of the sea level equation: An application for glacial isostatic adjustment by Antarctic deglaciation. *Journal of Geodynamics*, **135**, 101695, https://doi.org/10.1016/j.jog.2020.101695

Haskell, N. 1935. The motion of a viscous fluid under a surface load. *Physics*, **6**, 265–269, https://doi.org/10.1063/1.1745329

Hattori, A., Aoyama, Y., Okuno, J. and Doi, K. 2021. GNSS observations of GIA-induced crustal deformation in Lützow-Holm Bay, East Antarctica. *Geophysical Research Letters*, **48**, e2021GL093479, https://doi.org/10.1029/2021GL093479

Hay, C.C., Lau, H.C.P. *et al.* 2017. Sea level fingerprints in a region of complex Earth structure: The case of WAIS. *Journal of Climate*, **30**, 1881–1892, https://doi.org/10.1175/jcli-d-16-0388.1

Hayes, G.P. 2017. The finite, kinematic rupture properties of great-sized earthquakes since 1990. *Earth and Planetary Science Letters*, **468**, 94–100, https://doi.org/10.1016/j.epsl.2017.04.003

Henry, C., Das, S. and Woodhouse, J.H. 2000. The great March 25, 1998, Antarctic Plate earthquake: Moment tensor and rupture history. *Journal of Geophysical Research: Solid Earth*, **105**, 16 097–16 118, https://doi.org/10.1029/2000JB900077

Hermans, T.H.J., van der Wal, W. and Broerse, T. 2018. Reversal of the direction of horizontal velocities induced by GIA as a function of mantle viscosity. *Geophysical Research Letters*, **45**, 9597–9604, https://doi.org/10.1029/2018GL078533

Hirth, G. and Kohlstedt, D. 2004. Rheology of the upper mantle and the mantle wedge: A view from the experimentalists. *American Geophysical Union Geophysical Monograph Series*, **138**, 83–105, https://doi.org/10.1029/138GM06

Hodgson, D.A., Whitehouse, P.L. *et al.* 2016. Rapid early Holocene sea-level rise in Prydz Bay, East Antarctica. *Global and Planetary Change*, **139**, 128–140, https://doi.org/10.1016/j.gloplacha.2015.12.020

Hollin, J.T. 1962. On the glacial history of Antarctica. *Journal of Glaciology*, **4**, 173–195, https://doi.org/10.3189/S0022143000027386

Hu, Y. and Wang, K. 2012. Spherical-Earth finite element model of short-term postseismic deformation following the 2004 Sumatra earthquake. *Journal of Geophysical Research: Solid Earth*, **117**, B05404, https://doi.org/10.1029/2012JB009153

Hu, Y., Wang, K., He, J., Klotz, J. and Khazaradze, G. 2004. Three-dimensional viscoelastic finite element model for postseismic deformation of the great 1960 Chile earthquake. *Journal of Geophysical Research: Solid Earth*, **109**, B12403, https://doi.org/10.1029/2004JB003163

Hughes, T.J. 1981. The last great ice sheets: A global view. *In*: Denton, G.H. and Hughes, T.J. (eds) *The Last Great Ice Sheets*. Wiley, New York, 263–317.

Huybrechts, P. 1990. The Antarctic ice sheet during the last glacial–interglacial cycle: a three-dimensional experiment. *Annals of Glaciology*, **14**, 115–119, https://doi.org/10.3189/S0260305500008387

Huybrechts, P. 2002. Sea-level changes at the LGM from ice-dynamic reconstructions of the Greenland and Antarctic ice sheets during the glacial cycles. *Quaternary Science Reviews*, **21**, 203–231, https://doi.org/10.1016/S0277-3791(01)00082-8

Ishiwa, T., Okuno, J.I. and Suganuma, Y. 2021. Excess ice loads in the Indian Ocean sector of East Antarctica during the last glacial period. *Geology*, **49**, 1182–1186, https://doi.org/10.1130/g48830.1

Ivins, E.R. and James, T.S. 2005. Antarctic glacial isostatic adjustment: a new assessment. *Antarctic Science*, **17**, 541–553, https://doi.org/10.1017/S0954102005002968

Ivins, E.R., Raymond, C.A. and James, T.S. 2000. The influence of 5000 year-old and younger glacial mass variability on present-day crustal rebound in the Antarctic Peninsula. *Earth, Planets and Space*, **52**, 1023–1029, https://doi.org/10.1186/BF03352325

Ivins, E.R., James, T.S. and Klemann, V. 2003. Glacial isostatic stress shadowing by the Antarctic ice sheet. *Journal of Geophysical Research: Solid Earth*, **108**, 2560, https://doi.org/10.1029/2002JB002182

Ivins, E.R., Watkins, M.M., Yuan, D.-N., Dietrich, R., Casassa, G. and Rülke, A. 2011. On-land ice loss and glacial isostatic adjustment at the Drake Passage: 2003–2009. *Journal of Geophysical Research: Solid Earth*, **116**, B02403, https://doi.org/10.1029/2010JB007607

Ivins, E.R., James, T.S., Wahr, J.O., Schrama, E.J., Landerer, F.W. and Simon, K.M. 2013. Antarctic contribution to sea level rise observed by GRACE with improved GIA correction. *Journal of Geophysical Research: Solid Earth*, **118**, 3126–3141, https://doi.org/10.1002/jgrb.50208

Ivins, E.R., Caron, L., Adhikari, S., Larour, E. and Scheinert, M. 2020. A linear viscoelasticity for decadal to centennial time scale mantle deformation. *Reports on Progress in Physics*, **83**, 106801, https://doi.org/10.1088/1361-6633/aba346

Ivins, E.R., van der Wal, W., Wiens, D.A., Lloyd, A.J. and Caron, L. 2021. Antarctic upper mantle rheology. *Geological Society, London, Memoirs*, **56**, https://doi.org/10.1144/M56-2020-19

James, T.S. and Ivins, E.R. 1998. Predictions of Antarctic crustal motions driven by present-day ice sheet evolution and by isostatic memory of the Last Glacial Maximum. *Journal of Geophysical Research: Solid Earth*, **103**, 4993–5017, https://doi.org/10.1029/97JB03539

Jamieson, S.S.R., Sugden, D.E. and Hulton, N.R.J. 2010. The evolution of the subglacial landscape of Antarctica. *Earth and Planetary Science Letters*, **293**, 1–27, https://doi.org/10.1016/j.epsl.2010.02.012

Jamieson, T.F. 1865. On the history of the last geological changes in Scotland. *Quarterly Journal of the Geological Society, London*, **21**, 161–204, https://doi.org/10.1144/gsl.Jgs.1865.021.01-02.24

Ji, C., Wald, D.J. and Helmberger, D.V. 2002. Source description of the 1999 Hector Mine, California, earthquake, part I: Wavelet domain inversion theory and resolution analysis. *Bulletin of the Seismological Society of America*, **92**, 1192–1207, https://doi.org/10.1785/0120000916

Johnson, J.S., Roberts, S.J. *et al.* 2020. Deglaciation of Pope Glacier implies widespread early Holocene ice sheet thinning in the Amundsen Sea sector of Antarctica. *Earth and Planetary Science Letters*, **548**, 116501, https://doi.org/10.1016/j.epsl.2020.116501

Johnston, A.C. 1987. Suppression of earthquakes by large continental ice sheets. *Nature*, **330**, 467–469, https://doi.org/10.1038/330467a0

Johnston, P. 1993. The effect of spatially non-uniform water loads on prediction of sea-level change. *Geophysical Journal International*, **114**, 615–634, https://doi.org/10.1111/j.1365-246X.1993.tb06992.x

Jones, R.S., Whitehouse, P.L., Bentley, M.J., Small, D. and Dalton, A.S. 2019. Impact of glacial isostatic adjustment on cosmogenic surface-exposure dating. *Quaternary Science Reviews*, **212**, 206–212, https://doi.org/10.1016/j.quascirev.2019.03.012

Kachuck, S.B., Martin, D.F., Bassis, J.N. and Price, S.F. 2020. Rapid viscoelastic deformation slows marine ice sheet instability at Pine Island Glacier. *Geophysical Research Letters*, **47**, e2019GL086446, https://doi.org/10.1029/2019GL086446

Kalberg, T., Gohl, K., Eagles, G. and Spiegel, C. 2015. Rift processes and crustal structure of the Amundsen Sea Embayment, West Antarctica, from 3D potential field modelling. *Marine Geophysical Research*, **36**, 263–279, https://doi.org/10.1007/s11001-015-9261-0

Kaufmann, G. and Lambeck, K. 2002. Glacial isostatic adjustment and the radial viscosity profile from inverse modeling. *Journal of Geophysical Research: Solid Earth*, **107**, ETG 5-1–ETG 5-15, https://doi.org/10.1029/2001JB000941

Kaufmann, G., Wu, P. and Ivins, E.R. 2005. Lateral viscosity variations beneath Antarctica and their implications on regional rebound motions and seismotectonics. *Journal of Geodynamics*, **39**, 165–181, https://doi.org/10.1016/j.jog.2004.08.009

Kawamata, M., Suganuma, Y., Doi, K., Misawa, K., Hirabayashi, M., Hattori, A. and Sawagaki, T. 2020. Abrupt Holocene ice-sheet thinning along the southern Soya Coast, Lützow-Holm Bay, East Antarctica, revealed by glacial geomorphology and surface exposure dating. *Quaternary Science Reviews*, **247**, 106540, https://doi.org/10.1016/j.quascirev.2020.106540

Khazaradze, G. and Klotz, J. 2003. Short- and long-term effects of GPS measured crustal deformation rates along the south central Andes. *Journal of Geophysical Research: Solid Earth*, **108**, 2289, https://doi.org/10.1029/2002JB001879

Khazaradze, G., Wang, K., Klotz, J., Hu, Y. and He, J. 2002. Prolonged post-seismic deformation of the 1960 great Chile earthquake and implications for mantle rheology. *Geophysical Research Letters*, **29**, 7-1–7-4, https://doi.org/10.1029/2002GL015986

King, M.A. 2013. Progress in modelling and observing Antarctic glacial isostatic adjustment. *Astronomy & Geophysics*, **54**, 4.33–34.38, https://doi.org/10.1093/astrogeo/att122

King, M.A. and Santamaría-Gómez, A. 2016. Ongoing deformation of Antarctica following recent Great Earthquakes. *Geophysical Research Letters*, **43**, 1918–1927, https://doi.org/10.1002/2016GL067773

King, M.A., Whitehouse, P.L. and van der Wal, W. 2016. Incomplete separability of Antarctic plate rotation from glacial isostatic adjustment deformation within geodetic observations. *Geophysical Journal International*, **204**, 324–330, https://doi.org/10.1093/gji/ggv461

Kingslake, J., Scherer, R.P. *et al.* 2018. Extensive retreat and re-advance of the West Antarctic Ice Sheet during the Holocene. *Nature*, **558**, 430–434, https://doi.org/10.1038/s41586-018-0208-x

Konfal, S.A., Wilson, T.J. *et al.* 2013. Palaeoshoreline records of glacial isostatic adjustment in the Dry Valleys region, Antarctica. *Geological Society, London, Special Publications*, **381**, 455–467, https://doi.org/10.1144/SP381.26

Konfal, S.A., Wilson, T.J. *et al.* 2018. Utilizing GPS to investigate past ice mass change in the Ross Sea region, Antarctica. Abstract #G43B-0716 presented at the 2018 AGU Fall Meeting, 10–14 December 2018, Washington, DC.

Konrad, H., Thoma, M., Sasgen, I., Klemann, V., Grosfeld, K., Barbi, D. and Martinec, Z. 2014. The deformational response of a viscoelastic solid Earth model coupled to a thermomechanical ice sheet model. *Surveys in Geophysics*, **35**, 1441–1458, https://doi.org/10.1007/s10712-013-9257-8

Konrad, H., Sasgen, I., Pollard, D. and Klemann, V. 2015. Potential of the solid-Earth response for limiting long-term West Antarctic Ice Sheet retreat in a warming climate. *Earth and Planetary Science Letters*, **432**, 254–264, https://doi.org/10.1016/j.epsl.2015.10.008

Lambeck, K., Yokoyama, Y., Johnston, P. and Purcell, A. 2000. Global ice volumes at the Last Glacial Maximum and early Lateglacial. *Earth and Planetary Science Letters*, **181**, 513–527, https://doi.org/10.1016/S0012-821X(00)00223-5

Lambeck, K., Rouby, H., Purcell, A., Sun, Y. and Sambridge, M. 2014. Sea level and global ice volumes from the Last Glacial Maximum to the Holocene. *Proceedings of the National Academy of Sciences of the United States of America*, **111**, 15 296–15 303, https://doi.org/10.1073/pnas.1411762111

Lambeck, K., Purcell, A. and Zhao, S. 2017. The North American Late Wisconsin ice sheet and mantle viscosity from glacial rebound analyses. *Quaternary Science Reviews*, **158**, 172–210, https://doi.org/10.1016/j.quascirev.2016.11.033

Larour, E., Seroussi, H., Adhikari, S., Ivins, E., Caron, L., Morlighem, M. and Schlegel, N. 2019. Slowdown in Antarctic mass loss from solid Earth and sea-level feedbacks. *Science*, **364**, eaav7908, https://doi.org/10.1126/science.aav7908

Larsen, C.F., Motyka, R.J., Freymueller, J.T., Echelmeyer, K.A. and Ivins, E.R. 2005. Rapid viscoelastic uplift in southeast Alaska caused by post-Little Ice Age glacial retreat. *Earth and Planetary Science Letters*, **237**, 548–560, https://doi.org/10.1016/j.epsl.2005.06.032

Larter, R.D. and Barker, P.F. 1991. Effects of ridge crest–trench interaction on Antarctic–Phoenix Spreading: Forces on a young subducting plate. *Journal of Geophysical Research: Solid Earth*, **96**, 19 583–19 607, https://doi.org/10.1029/91JB02053

Latychev, K., Mitrovica, J.X., Tromp, J., Tamisiea, M.E., Komatitsch, D. and Christara, C.C. 2005. Glacial isostatic adjustment on 3-D Earth models: a finite-volume formulation. *Geophysical Journal International*, **161**, 421–444, https://doi.org/10.1111/j.1365-246X.2005.02536.x

Lau, H.C.P., Mitrovica, J.X., Austermann, J., Crawford, O., Al-Attar, D. and Latychev, K. 2016. Inferences of mantle viscosity based on ice age data sets: Radial structure. *Journal of Geophysical Research: Solid Earth*, **121**, 6991–7012, https://doi.org/10.1002/2016JB013043

Lau, H.C.P., Austermann, J., Holtzman, B.K., Havlin, C., Lloyd, A.J., Book, C. and Hopper, E. 2021. Frequency dependent mantle viscoelasticity via the complex viscosity: Cases from Antarctica. *Journal of Geophysical Research: Solid Earth*, **126**, e2021JB022622, https://doi.org/10.1029/2021JB022622

Lawver, L.A., Sloan, B.J. *et al.* 1996. Distributed, active extension in Bransfield basin Antarctic Peninsula: Evidence from multibeam bathymetry. *GSA Today*, **6**, 1–6.

Lear, C.H., Elderfield, H. and Wilson, P.A. 2000. Cenozoic deep-sea temperatures and global ice volumes from Mg/Ca in benthic foraminiferal calcite. *Science*, **287**, 269–272, https://doi.org/10.1126/science.287.5451.269

Le Meur, E. and Huybrechts, P. 1996. A comparison of different ways of dealing with isostasy: examples from modelling the Antarctic ice sheet during the last glacial cycle. *Annals of Glaciology*, **23**, 309–317, https://doi.org/10.3189/S0260305500013586

Lloyd, A.J., Wiens, D.A. *et al.* 2020. Seismic structure of the Antarctic upper mantle imaged with adjoint tomography. *Journal of Geophysical Research: Solid Earth*, **125**, https://doi.org/10.1029/2019JB017823

Lohman, R.B. and Simons, M. 2005. Some thoughts on the use of InSAR data to constrain models of surface deformation: Noise structure and data downsampling. *Geochemistry,*

Geophysics, Geosystems, **6**, Q01007, https://doi.org/10.1029/2004GC000841

Lough, A.C., Wiens, D.A. and Nyblade, A. 2018. Reactivation of ancient Antarctic rift zones by intraplate seismicity. *Nature Geoscience*, **11**, 515–519, https://doi.org/10.1038/s41561-018-0140-6

Lucas, E.M., Nyblade, A.A. *et al.* 2021. Seismicity and Pn velocity structure of central West Antarctica. *Geochemistry, Geophysics, Geosystems*, **22**, e2020GC009471, https://doi.org/10.1029/2020GC009471

Mackintosh, A., Golledge, N. *et al.* 2011. Retreat of the East Antarctic ice sheet during the last glacial termination. *Nature Geoscience*, **4**, 195–202, https://doi.org/10.1038/ngeo1061

Mackintosh, A.N., Verleyen, E. *et al.* 2014. Retreat history of the East Antarctic Ice Sheet since the Last Glacial Maximum. *Quaternary Science Reviews*, **100**, 10–30, https://doi.org/10.1016/j.quascirev.2013.07.024

Maris, M.N.A., de Boer, B., Ligtenberg, S.R.M., Crucifix, M., van de Berg, W.J. and Oerlemans, J. 2014. Modelling the evolution of the Antarctic ice sheet since the last interglacial. *The Cryosphere*, **8**, 1347–1360, https://doi.org/10.5194/tc-8-1347-2014

Maris, M.N.A., van Wessem, J.M., van de Berg, W.J., de Boer, B. and Oerlemans, J. 2015. A model study of the effect of climate and sea-level change on the evolution of the Antarctic Ice Sheet from the Last Glacial Maximum to 2100. *Climate Dynamics*, **45**, 837–851, https://doi.org/10.1007/s00382-014-2317-z

Martin, A.P. 2021. A review of the composition and chemistry of peridotite mantle xenoliths in volcanic rocks from Antarctica and their relevance to petrological and geophysical models for the lithospheric mantle. *Geological Society, London, Memoirs*, **56**, https://doi.org/10.1144/M56-2021-26

Martin, A.P., Cooper, A.F. and Price, R.C. 2014. Increased mantle heat flow with on-going rifting of the West Antarctic rift system inferred from characterisation of plagioclase peridotite in the shallow Antarctic mantle. *Lithos*, **190–191**, 173–190, https://doi.org/10.1016/j.lithos.2013.12.012

Martin, A.P., van der Wal, W. and de Boer, B. 2022. An introduction to the geochemistry and geophysics of the Antarctic mantle. *Geological Society, London, Memoirs*, **56**, https://doi.org/10.1144/M56-2022-21

Martinec, Z. 2000. Spectral–finite element approach to three-dimensional viscoelastic relaxation in a spherical earth. *Geophysical Journal International*, **142**, 117–141, https://doi.org/10.1046/j.1365-246x.2000.00138.x

Martín-Español, A., King, M.A., Zammit-Mangion, A., Andrews, S.B., Moore, P. and Bamber, J.L. 2016*a*. An assessment of forward and inverse GIA solutions for Antarctica. *Journal of Geophysical Research: Solid Earth*, **121**, 6947–6965, https://doi.org/10.1002/2016JB013154

Martín-Español, A., Zammit-Mangion, A. *et al.* 2016*b*. Spatial and temporal Antarctic Ice Sheet mass trends, glacio-isostatic adjustment, and surface processes from a joint inversion of satellite altimeter, gravity, and GPS data. *Journal of Geophysical Research: Earth Surface*, **121**, 182–200, https://doi.org/10.1002/2015JF003550

Miller, K.G., Wright, J.D. and Fairbanks, R.G. 1991. Unlocking the Ice House: Oligocene–Miocene oxygen isotopes, eustasy, and margin erosion. *Journal of Geophysical Research: Solid Earth*, **96**, 6829–6848, https://doi.org/10.1029/90JB02015

Milne, G.A. and Mitrovica, J.X. 1998. Postglacial sea-level change on a rotating Earth. *Geophysical Journal International*, **133**, 1–19, https://doi.org/10.1046/j.1365-246X.1998.1331455.x

Milne, G.A., Mitrovica, J.X. and Davis, J.L. 1999. Near-field hydro-isostasy: the implementation of a revised sea-level equation. *Geophysical Journal International*, **139**, 464–482, https://doi.org/10.1046/j.1365-246x.1999.00971.x

Mitrovica, J.X. 1996. Haskell [1935] revisited. *Journal of Geophysical Research: Solid Earth*, **101**, 555–569, https://doi.org/10.1029/95JB03208

Mitrovica, J.X. and Peltier, W.R. 1991. On postglacial geoid subsidence over the equatorial oceans. *Journal of Geophysical Research: Solid Earth*, **96**, 20053–20071, https://doi.org/10.1029/91JB01284

Mitrovica, J.X. and Milne, G.A. 2002. On the origin of late Holocene sea-level highstands within equatorial ocean basins. *Quaternary Science Reviews*, **21**, 2179–2190, https://doi.org/10.1016/S0277-3791(02)00080-X

Mitrovica, J.X., Gomez, N. and Clark, P.U. 2009. The sea-level fingerprint of West Antarctic collapse. *Science*, **323**, 753–753, https://doi.org/10.1126/science.1166510

Morelli, A. and Danesi, S. 2004. Seismological imaging of the Antarctic continental lithosphere: a review. *Global and Planetary Change*, **42**, 155–165, https://doi.org/10.1016/j.gloplacha.2003.12.005

Morlighem, M., Rignot, E. *et al.* 2020. Deep glacial troughs and stabilizing ridges unveiled beneath the margins of the Antarctic ice sheet. *Nature Geoscience*, **13**, 132–137, https://doi.org/10.1038/s41561-019-0510-8

Naish, T.R., Duncan, B. *et al.* 2022. Antarctic Ice Sheet dynamics during the Late Oligocene and Early Miocene: climatic conundrums revisited. *In*: Florindo, F., Siegert, M., Santis, L.D. and Naish, T. (eds) *Antarctic Climate Evolution*. 2nd edn. Elsevier, Amsterdam, 363–387, https://doi.org/10.1016/B978-0-12-819109-5.00013-X

Nakada, M. 1983. Rheological structure of the earth's mantle derived from glacial rebound in Laurentide. *Journal of Physics of the Earth*, **31**, 349–386, https://doi.org/10.4294/jpe1952.31.349

Nakada, M., Kimura, R., Okuno, J., Moriwaki, K., Miura, H. and Maemoku, H. 2000. Late Pleistocene and Holocene melting history of the Antarctic ice sheet derived from sea-level variations. *Marine Geology*, **167**, 85–103, https://doi.org/10.1016/S0025-3227(00)00018-9

Nakiboglu, S.M., Lambeck, K. and Aharon, P. 1983. Postglacial sea-levels in the Pacific: Implications with respect to deglaciation regime and local tectonics. *Tectonophysics*, **91**, 335–358, https://doi.org/10.1016/0040-1951(83)90049-5

Nettles, M., Wallace, T.C. and Beck, S.L. 1999. The March 25, 1998 Antarctic Plate Earthquake. *Geophysical Research Letters*, **26**, 2097–2100, https://doi.org/10.1029/1999GL900387

Nield, G.A., Barletta, V.R. *et al.* 2014. Rapid bedrock uplift in the Antarctic Peninsula explained by viscoelastic response to recent ice unloading. *Earth and Planetary Science Letters*, **397**, 32–41, https://doi.org/10.1016/j.epsl.2014.04.019

Nield, G.A., Whitehouse, P.L., van der Wal, W., Blank, B., O'Donnell, J.P. and Stuart, G.W. 2018. The impact of lateral variations in lithospheric thickness on glacial isostatic adjustment in West Antarctica. *Geophysical Journal International*, **214**, 811–824, https://doi.org/10.1093/gji/ggy158

Nield, G.A., King, M.A., Steffen, R. and Blank, B. 2022. A global, spherical finite-element model for post-seismic deformation using Abaqus. *Geoscientific Model Development*, **15**, 2489–2503, https://doi.org/10.5194/gmd-15-2489-2022

O'Donnell, J.P., Selway, K. *et al.* 2017. The uppermost mantle seismic velocity and viscosity structure of central West Antarctica. *Earth and Planetary Science Letters*, **472**, 38–49, https://doi.org/10.1016/j.epsl.2017.05.016

Oerlemans, J. 1980. Model experiments on the 100,000-yr glacial cycle. *Nature*, **287**, 430–432, https://doi.org/10.1038/287430a0

Okuno, J.I. and Miura, H. 2013. Last deglacial relative sea level variations in Antarctica derived from glacial isostatic adjustment modelling. *Geoscience Frontiers*, **4**, 623–632, https://doi.org/10.1016/j.gsf.2012.11.004

Oude Egbrink, D. 2017. *Modelling the Last Glacial Ice Sheet on Antarctica with Laterally Varying Relaxation Time*. MSc dissertation, Delft University of Technology, Delft, The Netherlands.

Pan, L., Powell, E.M. *et al.* 2021. Rapid postglacial rebound amplifies global sea level rise following West Antarctic Ice Sheet collapse. *Science Advances*, **7**, eabf7787, https://doi.org/10.1126/sciadv.abf7787

Panter, K.S. and Martin, A.P. 2021. West Antarctic mantle deduced from mafic magmatism. *Geological Society, London, Memoirs*, **56**, https://doi.org/10.1144/M56-2021-10

Panter, K.S., Castillo, P. *et al.* 2018. Melt origin across a rifted continental margin: A case for subduction-related metasomatic agents in the lithospheric source of alkaline basalt, NW Ross

Sea, Antarctica. *Journal of Petrology*, **59**, 517–558, https://doi.org/10.1093/petrology/egy036

Pappa, F. and Ebbing, J. 2021. Gravity, magnetics and geothermal heat flow of the Antarctic lithospheric crust and mantle. *Geological Society, London, Memoirs*, **56**, https://doi.org/10.1144/M56-2020-5

Pappa, F., Ebbing, J., Ferraccioli, F. and van der Wal, W. 2019. Modeling satellite gravity gradient data to derive density, temperature, and viscosity structure of the Antarctic lithosphere. *Journal of Geophysical Research: Solid Earth*, **124**, 12 053–12 076, https://doi.org/10.1029/2019JB017997

Parizek, B.R. and Alley, R.B. 2004. Ice thickness and isostatic imbalances in the Ross Embayment, West Antarctica: model results. *Global and Planetary Change*, **42**, 265–278, https://doi.org/10.1016/j.gloplacha.2003.09.005

Parrenin, F., Dreyfus, G. *et al.* 2007. 1-D-ice flow modelling at EPICA Dome C and Dome Fuji, East Antarctica. *Climate of the Past*, **3**, 243–259, https://doi.org/10.5194/cp-3-243-2007

Pattyn, F., Favier, L., Sun, S. and Durand, G. 2017. Progress in numerical modeling of Antarctic ice-sheet dynamics. *Current Climate Change Reports*, **3**, 174–184, https://doi.org/10.1007/s40641-017-0069-7

Paulson, A., Zhong, S. and Wahr, J. 2005. Modelling post-glacial rebound with lateral viscosity variations. *Geophysical Journal International*, **163**, 357–371, https://doi.org/10.1111/j.1365-246X.2005.02645.x

Paulson, A., Zhong, S. and Wahr, J. 2007. Limitations on the inversion for mantle viscosity from postglacial rebound. *Geophysical Journal International*, **168**, 1195–1209, https://doi.org/10.1111/j.1365-246X.2006.03222.x

Paxman, G.J.G. 2021. Antarctic palaeotopography. *Geological Society, London, Memoirs*, **56**, https://doi.org/10.1144/M56-2020-7

Paxman, G.J.G., Jamieson, S.S.R., Hochmuth, K., Gohl, K., Bentley, M.J., Leitchenkov, G. and Ferraccioli, F. 2019. Reconstructions of Antarctic topography since the Eocene–Oligocene boundary. *Palaeogeography, Palaeoclimatology, Palaeoecology*, **535**, 109346, https://doi.org/10.1016/j.palaeo.2019.109346

Peltier, W.R. 1974. The impulse response of a Maxwell Earth. *Reviews of Geophysics*, **12**, 649–669, https://doi.org/10.1029/RG012i004p00649

Peltier, W.R. 1976. Glacial-isostatic adjustment – II. The inverse problem. *Geophysical Journal International*, **46**, 669–705, https://doi.org/10.1111/j.1365-246X.1976.tb01253.x

Peltier, W.R. 1988. Lithospheric thickness, Antarctic deglaciation history, and ocean basin discretization effects in a global model of postglacial sea level change: A summary of some sources of nonuniqueness. *Quaternary Research*, **29**, 93–112, https://doi.org/10.1016/0033-5894(88)90054-3

Peltier, W.R. 1994. Ice Age paleotopography. *Science*, **265**, 195–201, https://doi.org/10.1126/science.265.5169.195

Peltier, W.R. 2004. Global glacial isostasy and the surface of the ice-age earth: the ICE-5G (VM2) model and GRACE. *Annual Review of Earth and Planetary Sciences*, **32**, 111–149, https://doi.org/10.1146/annurev.earth.32.082503.144359

Pollard, D., Chang, W., Haran, M., Applegate, P. and DeConto, R. 2016. Large ensemble modeling of the last deglacial retreat of the West Antarctic Ice Sheet: comparison of simple and advanced statistical techniques. *Geoscientific Model Development*, **9**, 1697–1723, https://doi.org/10.5194/gmd-9-1697-2016

Pollard, D., Gomez, N. and Deconto, R.M. 2017. Variations of the Antarctic Ice Sheet in a coupled ice sheet–Earth–sea level model: Sensitivity to viscoelastic Earth properties. *Journal of Geophysical Research: Earth Surface*, **122**, 2124–2138, https://doi.org/10.1002/2017JF004371

Pollitz, F.F. 1997. Gravitational viscoelastic postseismic relaxation on a layered spherical Earth. *Journal of Geophysical Research: Solid Earth*, **102**, 17 921–17 941, https://doi.org/10.1029/97JB01277

Pollitz, F.F. 2005. Transient rheology of the upper mantle beneath central Alaska inferred from the crustal velocity field following the 2002 Denali earthquake. *Journal of Geophysical Research B: Solid Earth*, **110**, 1–16, https://doi.org/10.1029/2005JB003672

Pollitz, F.F. and Thatcher, W. 2010. On the resolution of shallow mantle viscosity structure using postearthquake relaxation data: Application to the 1999 Hector Mine, California, earthquake. *Journal of Geophysical Research: Solid Earth*, **115**, B10412, https://doi.org/10.1029/2010JB007405

Powell, E., Gomez, N., Hay, C., Latychev, K. and Mitrovica, J.X. 2020. Viscous effects in the solid Earth response to modern Antarctic ice mass flux: Implications for geodetic studies of WAIS stability in a warming world. *Journal of Climate*, **33**, 443–459, https://doi.org/10.1175/jcli-d-19-0479.1

Powell, E.M., Pan, L., Hoggard, M.J., Latychev, K., Gomez, N., Austermann, J. and Mitrovica, J.X. 2021. The impact of 3-D Earth structure on far-field sea level following interglacial West Antarctic Ice Sheet collapse. *Quaternary Science Reviews*, **273**, 107256, https://doi.org/10.1016/j.quascirev.2021.107256

Quiquet, A., Dumas, C., Ritz, C., Peyaud, V. and Roche, D.M. 2018. The GRISLI ice sheet model (version 2.0): calibration and validation for multi-millennial changes of the Antarctic ice sheet. *Geoscientific Model Development*, **11**, 5003–5025, https://doi.org/10.5194/gmd-11-5003-2018

Ranalli, G. 1995. *Rheology of the Earth*. Springer, Dordrecht, The Netherlands.

Reading, A.M. 2007. The seismicity of the Antarctic plate. *Geological Society of America Special Papers*, **425**, 285–298, https://doi.org/10.1130/2007.2425(18)

Rignot, E. 2006. Changes in ice dynamics and mass balance of the Antarctic ice sheet. *Philosophical Transactions of the Royal Society A: Mathematical, Physical and Engineering Sciences*, **364**, 1637–1655, https://doi.org/10.1098/rsta.2006.1793

Riva, R.E.M., Gunter, B.C. *et al.* 2009. Glacial Isostatic Adjustment over Antarctica from combined ICESat and GRACE satellite data. *Earth and Planetary Science Letters*, **288**, 516–523, https://doi.org/10.1016/j.epsl.2009.10.013

Rovere, A., Raymo, M.E., Mitrovica, J.X., Hearty, P.J., O'Leary, M.J. and Inglis, J.D. 2014. The Mid-Pliocene sea-level conundrum: Glacial isostasy, eustasy and dynamic topography. *Earth and Planetary Science Letters*, **387**, 27–33, https://doi.org/10.1016/j.epsl.2013.10.030

Rychert, C.A., Harmon, N., Constable, S. and Wang, S. 2020. The Nature of the Lithosphere-Asthenosphere Boundary. *Journal of Geophysical Research: Solid Earth*, **125**, e2018JB016463, https://doi.org/10.1029/2018JB016463

Sabadini, R., Yuen, D.A. and Gasperini, P. 1985. The effects of transient rheology on the interpretation of lower mantle viscosity. *Geophysical Research Letters*, **12**, 361–364, https://doi.org/10.1029/GL012i006p00361

Sabadini, R., Vermeersen, B. and Cambiotti, G. 2016. *Global Dynamics of the Earth: Applications of Viscoelastic Relaxation Theory to Solid-Earth and Planetary Geophysics*. Springer, Dordrecht, The Netherlands.

Samrat, N.H., King, M.A., Watson, C., Hooper, A., Chen, X., Barletta, V.R. and Bordoni, A. 2020. Reduced ice mass loss and three-dimensional viscoelastic deformation in northern Antarctic Peninsula inferred from GPS. *Geophysical Journal International*, **222**, 1013–1022, https://doi.org/10.1093/gji/ggaa229

Samrat, N.H., King, M.A., Watson, C., Hay, A., Barletta, V.R. and Bordoni, A. 2021. Upper mantle viscosity underneath northern Marguerite Bay, Antarctic Peninsula constrained by bedrock uplift and ice mass variability. *Geophysical Research Letters*, **48**, e2021GL097065, https://doi.org/10.1029/2021GL097065

Scheinert, M., Engels, O., Schrama, E.J.O., van der Wal, W. and Horwath, M. 2021. Geodetic observations for constraining mantle processes in Antarctica. *Geological Society, London, Memoirs*, **56**, https://doi.org/10.1144/M56-2021-22

Schoof, C. 2007. Ice sheet grounding line dynamics: Steady states, stability, and hysteresis. *Journal of Geophysical Research: Earth Surface*, **112**, F03S28, https://doi.org/10.1029/2006JF000664

Schumacher, M., King, M.A., Rougier, J., Sha, Z., Khan, S.A. and Bamber, J.L. 2018. A new global GPS data set for testing and improving modelled GIA uplift rates. *Geophysical Journal International*, **214**, 2164–2176, https://doi.org/10.1093/gji/ggy235

Shao, Z., Zhan, W., Zhang, L. and Xu, J. 2016. Analysis of the far-field co-seismic and post-seismic responses caused by the 2011 M_W 9.0 Tohoku-Oki earthquake. *Pure and Applied Geophysics*, **173**, 411–424, https://doi.org/10.1007/s00024-015-1131-9

Shepherd, A., Ivins, E. *et al.* 2018. Mass balance of the Antarctic Ice Sheet from 1992 to 2017. *Nature*, **558**, 219–222, https://doi.org/10.1038/s41586-018-0179-y

Siddall, M., Milne, G.A. and Masson-Delmotte, V. 2012. Uncertainties in elevation changes and their impact on Antarctic temperature records since the end of the last glacial period. *Earth and Planetary Science Letters*, **315–316**, 12–23, https://doi.org/10.1016/j.epsl.2011.04.032

Siegert, M. and Golledge, N.R. 2022. Advances in numerical modelling of the Antarctic ice sheet. *In*: Florindo, F., Siegert, M., Santis, L.D. and Naish, T. (eds) *Antarctic Climate Evolution*. 2nd edn. Elsevier, Amsterdam, 199–218, https://doi.org/10.1016/B978-0-12-819109-5.00006-2

Siegert, M., Hein, A.S., White, D.A., Gore, D.B., De Santis, L. and Hillenbrand, C.-D. 2022. Antarctic Ice Sheet changes since the Last Glacial Maximum. *In*: Florindo, F., Siegert, M., Santis, L.D. and Naish, T. (eds) *Antarctic Climate Evolution*. 2nd edn. Elsevier, Amsterdam, 623–687, https://doi.org/10.1016/B978-0-12-819109-5.00002-5

Siegert, M.J., Kingslake, J. *et al.* 2019. Major ice sheet change in the Weddell Sea sector of West Antarctica over the last 5,000 years. *Reviews of Geophysics*, **57**, 1197–1223, https://doi.org/10.1029/2019RG000651

Sigmundsson, F. 1991. Post-glacial rebound and asthenosphere viscosity in Iceland. *Geophysical Research Letters*, **18**, 1131–1134, https://doi.org/10.1029/91GL01342

Simms, A.R., Whitehouse, P.L., Simkins, L.M., Nield, G., DeWitt, R. and Bentley, M.J. 2018. Late Holocene relative sea levels near Palmer Station, northern Antarctic Peninsula, strongly controlled by late Holocene ice-mass changes. *Quaternary Science Reviews*, **199**, 49–59, https://doi.org/10.1016/j.quascirev.2018.09.017

Simms, A.R., Lisiecki, L., Gebbie, G., Whitehouse, P.L. and Clark, J.F. 2019. Balancing the last glacial maximum (LGM) sea-level budget. *Quaternary Science Reviews*, **205**, 143–153, https://doi.org/10.1016/j.quascirev.2018.12.018

Simms, A.R., Bentley, M.J., Simkins, L.M., Zurbuchen, J., Reynolds, L.C., DeWitt, R. and Thomas, E.R. 2021. Evidence for a 'Little Ice Age' glacial advance within the Antarctic Peninsula – Examples from glacially-overrun raised beaches. *Quaternary Science Reviews*, **271**, 107195, https://doi.org/10.1016/j.quascirev.2021.107195

Smellie, J.L. 2018. Glaciovolcanism: A 21st century proxy for palaeo-ice. *In*: Menzies, J. and van der Meer, J.J.M. (eds) *Past Glacial Environments*. 2nd edn. Elsevier, Amsterdam, 335–375, https://doi.org/10.1016/B978-0-08-100524-8.00010-5

Smellie, J.L. 2021. Bransfield Strait and James Ross Island: volcanology. *Geological Society, London, Memoirs*, **55**, 227–284, https://doi.org/10.1144/m55-2018-58

Smellie, J.L., Haywood, A.M., Hillenbrand, C.-D., Lunt, D.J. and Valdes, P.J. 2009. Nature of the Antarctic Peninsula Ice Sheet during the Pliocene: Geological evidence and modelling results compared. *Earth-Science Reviews*, **94**, 79–94, https://doi.org/10.1016/j.earscirev.2009.03.005

Smellie, J.L., Rocchi, S., Gemelli, M., Di Vincenzo, G. and Armienti, P. 2011. A thin predominantly cold-based Late Miocene East Antarctic ice sheet inferred from glaciovolcanic sequences in northern Victoria Land, Antarctica. *Palaeogeography, Palaeoclimatology, Palaeoecology*, **307**, 129–149, https://doi.org/10.1016/j.palaeo.2011.05.008

Smellie, J.L., Panter, K.S. and Geyer, A. 2021. Introduction to volcanism in Antarctica: 200 million years of subduction, rifting and continental break-up. *Geological Society, London, Memoirs*, **55**, 1–6, https://doi.org/10.1144/m55-2020-14

Spiegel, C., Lindow, J. *et al.* 2016. Tectonomorphic evolution of Marie Byrd Land – Implications for Cenozoic rifting activity and onset of West Antarctic glaciation. *Global and Planetary Change*, **145**, 98–115, https://doi.org/10.1016/j.gloplacha.2016.08.013

Steffen, H. and Wu, P. 2011. Glacial isostatic adjustment in Fennoscandia – A review of data and modeling. *Journal of Geodynamics*, **52**, 169–204, https://doi.org/10.1016/j.jog.2011.03.002

Stein, S. and Mazzotti, S. 2007. *Continental Intraplate Earthquakes: Science, Hazard, and Policy Issues*. Geological Society of America Special Papers, **425**, https://doi.org/10.1130/SPE425

Stocchi, P., Escutia, C. *et al.* 2013. Relative sea-level rise around East Antarctica during Oligocene glaciation. *Nature Geoscience*, **6**, 380–384, https://doi.org/10.1038/ngeo1783

Stuhne, G.R. and Peltier, W.R. 2015. Reconciling the ICE-6G_C reconstruction of glacial chronology with ice sheet dynamics: The cases of Greenland and Antarctica. *Journal of Geophysical Research: Earth Surface*, **120**, 1841–1865, https://doi.org/10.1002/2015JF003580

Stuiver, M., Denton, G.H., Hughes, T.J. and Fastook, J.L. 1981. History of the marine ice sheet in West Antarctica during the last glaciation: a working hypothesis. *In*: Denton, G.H. and Hughes, T.J. (eds) *The Last Great Ice Sheets*. Wiley-Interscience, New York, 319–436.

Sun, T., Wang, K. and He, J. 2018. Crustal deformation following great subduction earthquakes controlled by earthquake size and mantle rheology. *Journal of Geophysical Research: Solid Earth*, **123**, 5323–5345, https://doi.org/10.1029/2017JB01 5242

Suess, E. 1888. *Das Antlitz der Erde, Vol. 2, Part 3: Die Meere der Erde*. F. Tempsky, Vienna.

Tarasov, L., Dyke, A.S., Neal, R.M. and Peltier, W.R. 2012. A data-calibrated distribution of deglacial chronologies for the North American ice complex from glaciological modeling. *Earth and Planetary Science Letters*, **315-316**, 30–40, https://doi.org/10.1016/j.epsl.2011.09.010

Thomas, I.D., King, M.A. *et al.* 2011. Widespread low rates of Antarctic glacial isostatic adjustment revealed by GPS observations. *Geophysical Research Letters*, **38**, https://doi.org/10.1029/2011GL049277

Thomas, R.H. 1979. The dynamics of marine ice sheets. *Journal of Glaciology*, **24**, 167–177, https://doi.org/10.3189/S0022143000014726

Tsuboi, S., Kikuchi, M., Yamanaka, Y. and Kanao, M. 2000. The March 25, 1998 Antarctic Earthquake: Great earthquake caused by postglacial rebound. *Earth, Planets and Space*, **52**, 133–136, https://doi.org/10.1186/BF03351621

Turney, C.S.M., Fogwill, C.J. *et al.* 2020. Early Last Interglacial ocean warming drove substantial ice mass loss from Antarctica. *Proceedings of the National Academy of Sciences of the United States of America*, **117**, 3996–4006, https://doi.org/10.1073/pnas.1902469117

van der Wal, W., Kurtenbach, E., Kusche, J. and Vermeersen, B. 2011. Radial and tangential gravity rates from GRACE in areas of glacial isostatic adjustment. *Geophysical Journal International*, **187**, 797–812, https://doi.org/10.1111/j.1365-246X.2011.05206.x

van der Wal, W., Barnhoorn, A., Stocchi, P., Gradmann, S., Wu, P., Drury, M. and Vermeersen, B. 2013. Glacial isostatic adjustment model with composite 3-D Earth rheology for Fennoscandia. *Geophysical Journal International*, **194**, 61–77, https://doi.org/10.1093/gji/ggt099

van der Wal, W., Whitehouse, P.L. and Schrama, E.J.O. 2015. Effect of GIA models with 3D composite mantle viscosity on GRACE mass balance estimates for Antarctica. *Earth and Planetary Science Letters*, **414**, 134–143, https://doi.org/10.1016/j.epsl.2015.01.001

Verleyen, E., Tavernier, I. *et al.* 2017. Ice sheet retreat and glacio-isostatic adjustment in Lützow-Holm Bay, East Antarctica. *Quaternary Science Reviews*, **169**, 85–98, https://doi.org/10.1016/j.quascirev.2017.06.003

Wahr, J., Wingham, D. and Bentley, C. 2000. A method of combining ICESat and GRACE satellite data to constrain Antarctic mass balance. *Journal of Geophysical Research: Solid Earth*, **105**, 16 279–16 294, https://doi.org/10.1029/2000JB900113

Walcott, R.I. 1972. Late Quaternary vertical movements in eastern North America: Quantitative evidence of glacio-isostatic rebound. *Reviews of Geophysics*, **10**, 849–884, https://doi.org/10.1029/RG010i004p00849

Wan, J.X.W., Gomez, N., Latychev, K. and Han, H.K. 2022. Resolving glacial isostatic adjustment (GIA) in response to modern and future ice loss at marine grounding lines in West Antarctica. *The Cryosphere*, **16**, 2203–2223, https://doi.org/10.5194/tc-16-2203-2022

Wang, K. and Fialko, Y. 2018. Observations and modeling of coseismic and postseismic deformation due to the 2015 M_w 7.8 Gorkha (Nepal) earthquake. *Journal of Geophysical Research: Solid Earth*, **123**, 761–779, https://doi.org/10.1002/2017JB014620

Watson, C., Burgette, R. *et al.* 2010. Twentieth century constraints on sea level change and earthquake deformation at Macquarie Island. *Geophysical Journal International*, **182**, 781–796, https://doi.org/10.1111/j.1365-246X.2010.04640.x

Weertman, J. 1974. Stability of the junction of an ice sheet and an ice shelf. *Journal of Glaciology*, **13**, 3–11, https://doi.org/10.3189/S0022143000023327

Wegener, A. 1915. *The Origin of Continents and Oceans*. Vieweg, Braunschweig, Germany [in German].

Wessel, P., Luis, J.F., Uieda, L., Scharroo, R., Wobbe, F., Smith, W.H.F. and Tian, D. 2019. The Generic Mapping Tools version 6. *Geochemistry, Geophysics, Geosystems*, **20**, 5556–5564, https://doi.org/10.1029/2019GC008515

Whitehouse, P.L. 2018. Glacial isostatic adjustment modelling: historical perspectives, recent advances, and future directions. *Earth Surface Dynamics*, **6**, 401–429, https://doi.org/10.5194/esurf-6-401-2018

Whitehouse, P.L., Bentley, M.J. and Le Brocq, A.M. 2012*a*. A deglacial model for Antarctica: geological constraints and glaciological modelling as a basis for a new model of Antarctic glacial isostatic adjustment. *Quaternary Science Reviews*, **32**, 1–24, https://doi.org/10.1016/j.quascirev.2011.11.016

Whitehouse, P.L., Bentley, M.J., Milne, G.A., King, M.A. and Thomas, I.D. 2012*b*. A new glacial isostatic adjustment model for Antarctica: calibrated and tested using observations of relative sea-level change and present-day uplift rates. *Geophysical Journal International*, **190**, 1464–1482, https://doi.org/10.1111/j.1365-246X.2012.05557.x

Whitehouse, P.L., Gomez, N., King, M.A. and Wiens, D.A. 2019. Solid Earth change and the evolution of the Antarctic Ice Sheet. *Nature Communications*, **10**, 503, https://doi.org/10.1038/s41467-018-08068-y

Wieczerkowski, K., Mitrovica, J.X. and Wolf, D. 1999. A revised relaxation-time spectrum for Fennoscandia. *Geophysical Journal International*, **139**, 69–86, https://doi.org/10.1046/j.1365-246X.1999.00924.x

Wiens, D.A., Shen, W. and Lloyd, A.J. 2021. The seismic structure of the Antarctic upper mantle. *Geological Society, London, Memoirs*, **56**, https://doi.org/10.1144/M56-2020-18

Wilson, T.J., Anandakrishnan, S. *et al.* 2020. Antarctic Network–Polar Earth Observing Network: Achievements from ten years of autonomous measurements. Abstract #C032-002 presented at the 2020 AGU Fall Meeting, 1–17 December 2020, virtual meeting.

Winkelmann, R., Martin, M.A., Haseloff, M., Albrecht, T., Bueler, E., Khroulev, C. and Levermann, A. 2011. The Potsdam Parallel Ice Sheet Model (PISM-PIK) – Part 1: Model description. *The Cryosphere*, **5**, 715–726, https://doi.org/10.5194/tc-5-715-2011

Wobbe, F., Gohl, K., Chambord, A. and Sutherland, R. 2012. Structure and breakup history of the rifted margin of West Antarctica in relation to Cretaceous separation from Zealandia and Bellingshausen plate motion. *Geochemistry, Geophysics, Geosystems*, **13**, Q04W12, https://doi.org/10.1029/2011GC0 03742

Wolstencroft, M., King, M.A. *et al.* 2015. Uplift rates from a new high-density GPS network in Palmer Land indicate significant late Holocene ice loss in the southwestern Weddell Sea. *Geophysical Journal International*, **203**, 737–754, https://doi.org/10.1093/gji/ggv327

Wright, T.J. 2016. The earthquake deformation cycle. *Astronomy & Geophysics*, **57**, 4.20–24.26, https://doi.org/10.1093/astrogeo/atw148

Wu, P. 1992. Deformation of an incompressible viscoelastic flat earth with powerlaw creep: a finite element approach. *Geophysical Journal International*, **108**, 35–51, https://doi.org/10.1111/j.1365-246X.1992.tb00837.x

Wu, P. 2004. Using commercial finite element packages for the study of earth deformations, sea levels and the state of stress. *Geophysical Journal International*, **158**, 401–408, https://doi.org/10.1111/j.1365-246X.2004.02338.x

Wu, P. 2006. Sensitivity of relative sea levels and crustal velocities in Laurentide to radial and lateral viscosity variations in the mantle. *Geophysical Journal International*, **165**, 401–413, https://doi.org/10.1111/j.1365-246X.2006.02960.x

Wu, P. and Peltier, W.R. 1982. Viscous gravitational relaxation. *Geophysical Journal International*, **70**, 435–485, https://doi.org/10.1111/j.1365-246X.1982.tb04976.x

Wu, P. and Peltier, W.R. 1983. Glacial isostatic adjustment and the free air gravity anomaly as a constraint on deep mantle viscosity. *Geophysical Journal International*, **74**, 377–449, https://doi.org/10.1111/j.1365-246X.1983.tb01884.x

Ye, L., Lay, T. *et al.* 2014. Complementary slip distributions of the August 4, 2003 M_w 7.6 and November 17, 2013 M_w 7.8 South Scotia Ridge earthquakes. *Earth and Planetary Science Letters*, **401**, 215–226, https://doi.org/10.1016/j.epsl.2014.06.007

Yuen, D.A. and Peltier, W.R. 1982. Normal modes of the viscoelastic earth. *Geophysical Journal International*, **69**, 495–526, https://doi.org/10.1111/j.1365-246X.1982.tb04962.x

Zhao, C., King, M.A., Watson, C.S., Barletta, V.R., Bordoni, A., Dell, M. and Whitehouse, P.L. 2017. Rapid ice unloading in the Fleming Glacier region, southern Antarctic Peninsula, and its effect on bedrock uplift rates. *Earth and Planetary Science Letters*, **473**, 164–176, https://doi.org/10.1016/j.epsl.2017.06.002

Zhong, S., Paulson, A. and Wahr, J. 2003. Three-dimensional finite-element modelling of Earth's viscoelastic deformation: effects of lateral variations in lithospheric thickness. *Geophysical Journal International*, **155**, 679–695, https://doi.org/10.1046/j.1365-246X.2003.02084.x

Zurbuchen, J. and Simms, A.R. 2019. Late Holocene ice-mass changes recorded in a relative sea-level record from Joinville Island, Antarctica. *Geology*, **47**, 1064–1068, https://doi.org/10.1130/g46649.1

A review of the composition and chemistry of peridotite mantle xenoliths in volcanic rocks from Antarctica and their relevance to petrological and geophysical models for the lithospheric mantle

Adam P. Martin

GNS Science, Private Bag 1930, Dunedin, New Zealand

0000-0002-4676-8344

a.martin@gns.cri.nz

Abstract: This chapter reviews the geochemistry and petrology of mantle peridotite xenoliths from across Antarctica, including parameters that are of most relevance to geophysical studies. This Memoir is the first time such a complete overview of the chemistry of Antarctic mantle xenoliths has been available and Antarctica should no longer be the ignored continent in studies of mantle xenoliths in volcanic rocks. Xenoliths indicate that the chemistry, heat flow and water content of the Antarctic lithospheric mantle varies regionally at scales of one to thousands of kilometres. The prevalence of variability in xenoliths suggests that the Antarctic mantle is ubiquitously heterogeneous. This has important, yet unquantified, implications for interpreting geophysical data and for reference Earth models used in Antarctic geophysical studies. Information about and interpretations of Antarctic mantle xenoliths can be linked to studies from once adjacent continental blocks in Africa, India, Australia, New Zealand and South America. Together, this can improve understanding of the mantle contribution to glacial isostatic adjustment and geodynamic models to show how the Antarctic mantle fits with adjacent continents in the puzzle of lithospheric blocks. Numerous, fundamental and important research questions remain unanswered making further study of the Antarctic mantle an exciting prospect for future research.

The contributions in this Memoir constitute the most complete review of rocks with mantle affinity ever undertaken for Antarctica and the sub-Antarctic islands. This fills an important gap in the global literature on mantle geochemistry, and allows, perhaps for the first time, the sum of mantle rock occurrences from Antarctica to be considered. It builds upon important work from regional studies across the continent (e.g. Kyle *et al.* 1987; Nixon 1987*a*; Gamble *et al.* 1988; Zipfel and Wörner 1992; Grégoire *et al.* 2000; Handler *et al.* 2003; Coltorti *et al.* 2004; Foley *et al.* 2006; Martin *et al.* 2015) and the relevant chapters in this memoir are accompanied by exhaustive references.

This chapter focuses on peridotite mantle xenoliths, which are fragments of the lithospheric mantle brought to the Antarctic surface in alkaline or ultramafic igneous rocks. Mantle xenoliths reflect a snapshot of the mantle at the time they were entrained in their magmatic hosts. In the case of Antarctica, studies of tectonically emplaced mantle rocks or abyssal peridotites have attracted less attention than mantle xenoliths and although a review of these occurrences is needed, it is beyond the scope of this chapter. For information about the inferences drawn about the Antarctic mantle from the study of volcanic rocks, the reader is referred to the recent, authoritative volume on volcanism in Antarctica (Smellie *et al.* 2021). Elsewhere in the current Memoir, relevant chapters include those on relatively undifferentiated rocks of East Antarctica (Foley *et al.* 2021, this volume), Cenozoic volcanic rocks of West Antarctica (Panter and Martin 2021, this volume) and the sub-Antarctic Islands (Delpech *et al.* 2021, this volume).

Antarctica is composed of a mosaic of crustal blocks in West Antarctica and cratons in East Antarctica (Fig. 1; Martin and van der Wal, 2022, this volume). In West Antarctica, mantle xenoliths are known to occur in primitive, Cenozoic alkaline rocks of Victoria Land and Marie Byrd Land, and relatively undifferentiated rocks of alkaline and possibly calc-alkaline lamprophyre affinity, in the Antarctic Peninsula (Fig. 2). In the sub-Antarctic islands, xenoliths are reported from Kerguelen Island and Heard Island in the Indian Ocean, and Auckland Island in the Southern Ocean. The sub-Antarctic island occurrences are in relatively undifferentiated, Cenozoic alkaline igneous rocks. In East Antarctica, mantle rocks are known from only three localities, all in the Indo-Antarctic region (Fig. 1). These occurrences include the Mesozoic alkaline-ultramafic igneous rocks at Jetty Peninsula, Proterozoic ultramafic igneous rocks at Vestfold Hills and a single peridotite xenolith in leucite lamproite rocks of the Quaternary Gaussberg Volcano (Fig. 2).

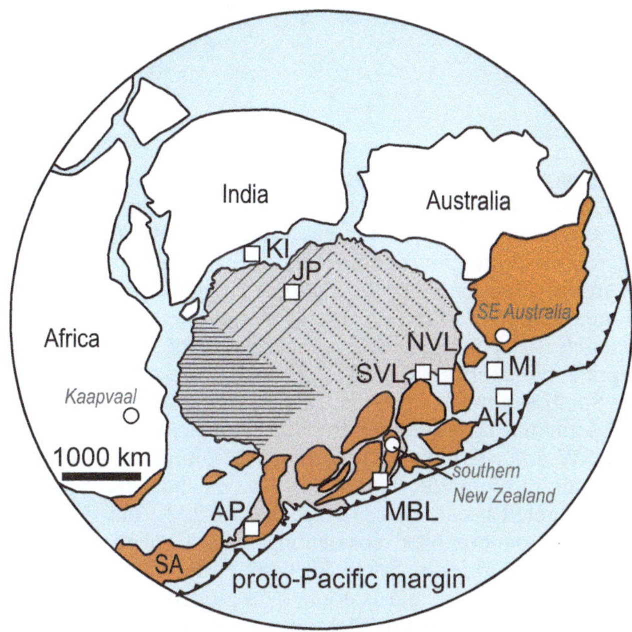

Fig. 1. Diagram showing locations discussed relative to their approximate pre-breakup positions in Gondwana, including Phanerozoic rocks in orange and older rocks in white (Storey and Kyle 1997). Grey shading shows Antarctica. The different hatching patterns in East Antarctica are from Foley *et al.* (2021, this volume) and they show correlations of first-order tectonic domains between Antarctica and Africa (horizontal lines), India (inclined lines) and Australia (inclined dotted lines). The thick black line with triangles indicates the inferred subduction zone prior to continental breakup. Localities discussed in the text with Antarctic mantle xenoliths (square symbols) are: AkI, Auckland Island; AP, Antarctic Peninsula; KI, Kerguelen Island; JP, Jetty Peninsula; MBL, Marie Byrd Land; MI, Macquarie Island; NVL, north Victoria Land; SA, South America; SVL, south Victoria Land. Also shown (as circles) are Kaapvaal, southern New Zealand, and southeastern (SE) Australia localities discussed in the text.

From: Martin, A. P. and van der Wal, W. (eds) 2023. *The Geochemistry and Geophysics of the Antarctic Mantle*. Geological Society, London, Memoirs, **56**, 343–354,
First published online 22 November 2021, https://doi.org/10.1144/M56-2021-26

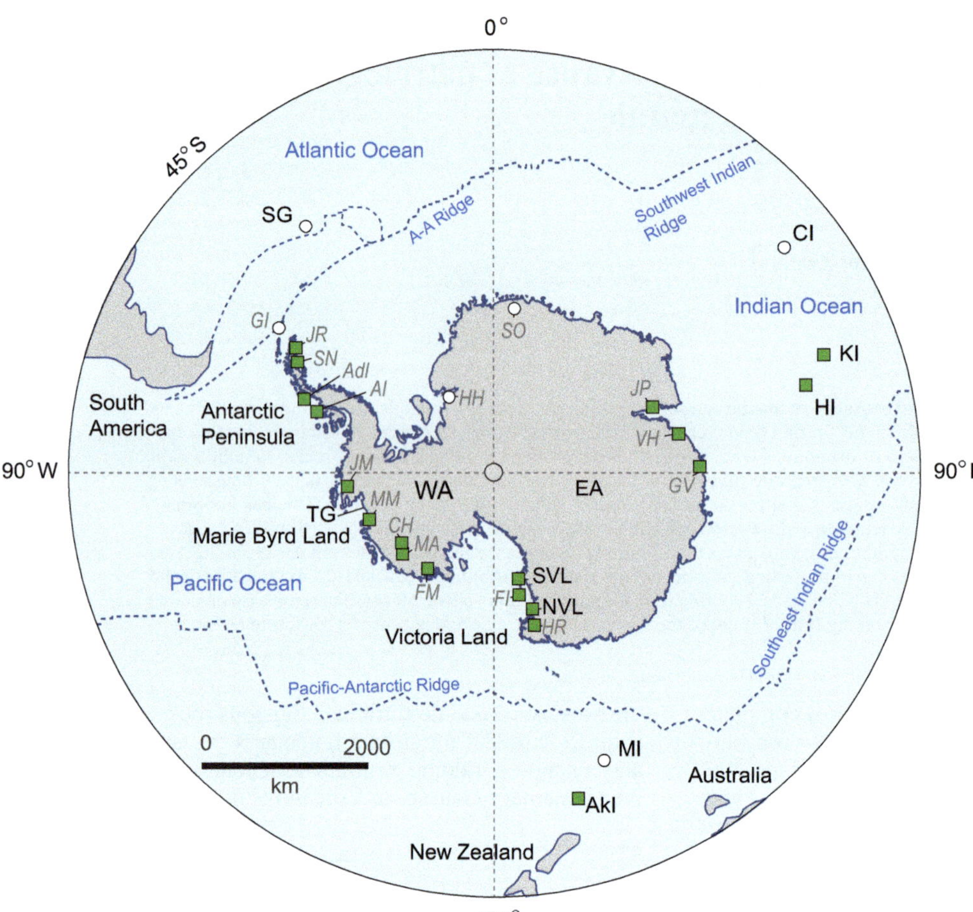

Fig. 2. Location of key areas and tectonic and regional context for mantle xenolith localities around Antarctica. Refer to the original references (Table 1) for regional- to local-scale information. Mantle xenolith locations are shown as green squares. Other mantle or cumulate rock locations are shown as white circles. A-A Ridge, America–Atlantic Ridge; AdI, Adelaide Island; AI, Alexander Island; AkI, Auckland Island; CH, Mount Cumming and Mount Hampton; CI, Crozet Island; EA, East Antarctica; FI, Franklin Island; FM, Fosdick Mountains; GI, Gibbs Island; GV, Gaussberg Volcano; HH, Haskard Highlands; HI, Heard Island; HR, Handler Ridge; JM, Jones Mountains; JP, Jetty Peninsula; JR, James Ross Island; KI, Kerguelen Island; MA, Mount Aldaz; MI, Macquarie Island; MM, Mount Murphy; NVL, north Victoria Land; SG, South Georgia and South Sandwich islands; SN, Seal Nunataks; SO, Schirmacher Oasis; SVL, south Victoria Land; TG, Thwaites Glacier; VH, Vestfold Hills; WA, West Antarctica.

For completeness, the locality of mantle rocks other than xenoliths are summarized here, but as this chapter focuses on xenoliths they will not be further discussed. Variably altered peridotite blocks are found at Schirmacher Oasis and as supracrustal lenses in the Haskard Highlands of East Antarctica (Fig. 2). Exhumed abyssal peridotite occurs on the sub-Antarctic Macquarie Island in the Southern Ocean and as supra-subduction zone peridotite drilled and dredged from the South Sandwich arc in the Atlantic Ocean (Fig. 2). Abyssal peridotite has been dredged from the America-Atlantic Ridge and SW and SE Indian Ridge (Fig. 2). Ultramafic xenoliths originally reported from Crozet Island have been re-interpreted as cumulates (Delpech *et al.* 2021, this volume). An ultramafic complex representing oceanic mantle is found on Gibbs Island in the South Shetland Island chain (Fig. 2).

The geochemical and isotopic data plotted in this chapter are from references in this Memoir and selected supporting literature, as summarized in Table 1. The reader is referred to references in Table 1 for detailed regional overviews. The choice of what to plot in this chapter was severely limited to the handful of elements and isotopes commonly reported from all regions of the continent. Abundance data for even very common and useful elements, like Pb, are not always available and it is for this reason that a wider range of plots was not used. Nonetheless, using what is available some important, continent-wide inferences can be made.

Viscosity is used to model glacial isostatic adjustment, post-seismic deformation (Barletta *et al.* 2022, this volume) and other geodynamic models such as mantle convection (Bredow *et al.* 2021, this volume). Viscosity can be derived from strain rate through the following relationship (e.g. Ranalli 1995):

$$\text{viscosity} = \text{stress}/(2 \times \text{strain rate}) \quad (1)$$

and strain rate depends on several parameters as observed in laboratory experiments:

$$\text{strain rate } (\dot{\varepsilon}) = A\sigma^n\, d^{-p}\, fH_2O^r \exp(\alpha\phi)\exp^{[-(E*+PV*)/RT]} \quad (2)$$

where A is a constant, n is the stress exponent, d is the grain size, p is the grain size exponent, fH_2O is water fugacity, r is the water fugacity exponent, ϕ is melt fraction, α is a constant, $E*$ is the activation energy, $V*$ is the activation volume, R is the gas constant and T is absolute temperature for olivine aggregates (Hirth and Kohlstedt 2003). Parameters important to viscosity (equation 1), via strain rate (equation 2), will be discussed, namely mantle grain size, heat flow, water content and chemistry as recorded in mantle xenoliths. Each parameter in equation (2) will affect viscosity. In particular, natural variation in parameters, such as grain size or water content, as recorded in mantle xenoliths will affect the rate of glacial isostatic adjustment and can be applied at the continental scale and sub-continental scale. This chapter discusses the composition and formation of Antarctic mantle xenoliths in the context of the Antarctic and global mantle. The implications of these data and interpretations are explored and recommendations are made for future work.

Petrology

Mantle xenoliths in Antarctica range in maximum dimension from sub-millimetre observed in thin section, up to >1 m. The majority of mantle xenoliths in Antarctica are spinel-bearing peridotites, with the abundance of harzburgite approximately

Table 1. *Sources for mantle xenolith data used in this chapter*

Location	Authors
East Antarctica	Foley *et al.* (2006, 2021, this volume)
South Victoria Land	Martin *et al.* (2021*a*, this volume)
North Victoria Land	Coltorti *et al.* (2021, this volume)
Marie Byrd Land	Handler *et al.* (2021, this volume)
	Chatzaras and Kruckenberg (2021, this volume)
	Chatzaras *et al.* (2016)
Antarctic Peninsula	Leat *et al.* (2021, this volume)
	Gaete (2017)
	Gibson *et al.* (2020)
Subantarctic Islands	Delpech *et al.* (2021, this volume)
	Grégoire *et al.* (2000)
	Scott *et al.* (2014)

equal to lherzolite, and both more abundant than dunite (this Memoir; pers. obs.). Garnet is completely absent from mantle xenoliths in Antarctica, except in the Jetty Peninsula occurrence (Figs 1 & 2). By comparison to garnet, plagioclase is much more common which is atypical, globally (Martin *et al.* 2014*b*). Plagioclase-bearing spinel peridotite is reported from Victoria Land (Zipfel and Wörner 1992; Martin *et al.* 2014*a*), the Antarctic Peninsula (Leat *et al.* 2021, this volume), Kerguelen Island (Delpech *et al.* 2004, 2021, this volume) and Vestfold Hills (Andronikov *et al.* 1994; Foley *et al.* 2021, this volume) in East Antarctica. Pyroxenite xenoliths and mantle-derived megacrysts are near-ubiquitous where mantle peridotite xenoliths are reported, albeit in significantly lower abundances, and they are much more rarely described.

A full range of mantle textures (Harte 1977) is observed from coarse (rare) to porphyroclastic, tabular granuloblastic or equant granuloblastic. A macroscopic fabric is evident in mantle xenoliths at several localities, and mantle anisotropy is dealt with specifically by Chatzaras and Kruckenberg (2021, this volume; also see Chatzaras *et al.* 2016). This is important as the crystallographic preferred orientation of olivine controls much of the seismic anisotropy and velocity (Chatzaras and Kruckenberg 2021, this volume). Where inequigranular textures occur, porphyroclasts of olivine and orthopyroxene are common, with clinopyroxene rarely being present as porphyroclasts. Frequently, multiple generations of clinopyroxene are present, as are exsolution lamellae in pyroxene. Neoblasts are typically olivine, orthopyroxene, clinopyroxene (multiple generations) plus an alumina-phase (typically spinel) showing triple-point junctions between grains. Accessory minerals vary within and between regions. For example, neither amphibole nor mica is reported from Marie Byrd Land (Handler *et al.* 2021, this volume), but these phases are common, though not ubiquitous, in southern Victoria Land (Gamble and Kyle 1987; Cooper *et al.* 2007) (Martin *et al.* 2021*a*, this volume – south Victoria Land). However, accessory minerals can include amphibole, phlogopite, apatite, carbonate and sulfide.

Grain size

In the development of geophysical and geological models for the mantle, incorrect grain size can affect the magnitude of viscosity values calculated. Grain size in Antarctic mantle xenoliths is reasonably consistent across the continent. Inequigranular textures are common, with porphyroclasts being between 2 and 10 mm, and neoblasts between 0.1 and 1 mm. A statistical treatment of grain size by Chatzaras *et al.* (2016) gave a representative grain size in spinel peridotite from Marie Byrd Land as 0.6 mm, which is consistent with petrographic observations made by authors in this Memoir. The Marie Byryd Land samples come from the Fosdick Mountains, Mount Aldaz and Mount Cumming (Fig. 2). When plotted onto a reference model for average grain size (Fig. 3) used by Dannberg *et al.* (2017), the average Antarctic grain size (0.6 mm) overlaps with the reference model at depths between 20 and 100 km. It is perhaps interesting that the maximum porphyroblast diameters in Antarctic xenoliths (≤10 mm), which represent relict grain sizes from greater depths, are frequently greater than the 4 mm maximum grain size inferred in the Dannberg *et al.* (2017) reference model (Fig. 3).

Heat flow

There is now a substantial body of work on petrologically-determined lithospheric geotherms in Antarctica (Fig. 4). The information from mantle xenoliths records pressures and temperatures at the time and depth of entrainment, and frequently there is evidence of multiple, older events also recorded in the rocks (e.g. Wörner and Zipfel 1996; Foley *et al.* 2006; Martin *et al.* 2014*a*). In Antarctic rift settings there is evidence of a cooler, pre- or early-rift geotherm (Fig. 4; East Antarctica stage 1; south Victoria Land) that becomes hotter with rift development (Fig. 4; East Antarctica stage 2, south Victoria Land late-stage overprint).

The modern-day Ross Embayment geotherm (Fig. 4) is calculated using measured heat flow in the Ross Sea of 71 mW m^{-2} (Blackman *et al.* 1987) and sediment conductivity of 1.6 W m^{-1} K^{-1}, and an assumption that the geotherm has been in a steady-state since 70 Ma (see ten Brink *et al.* 1997 for details of calculations). The mantle xenolith data used in constructing the south Victoria Land geotherm were entrained in igneous rocks erupted and intruded <1 Ma (Martin *et al.* 2015). Petrographic evidence from these samples is not consistent with a steady-state thermal history. This indicates that heat flow close to the Transantarctic Mountain front may be hotter by up to 30 mW m^{-2} than modelled by ten Brink *et al.* (1997), but within error of heat flow empirically constrained by

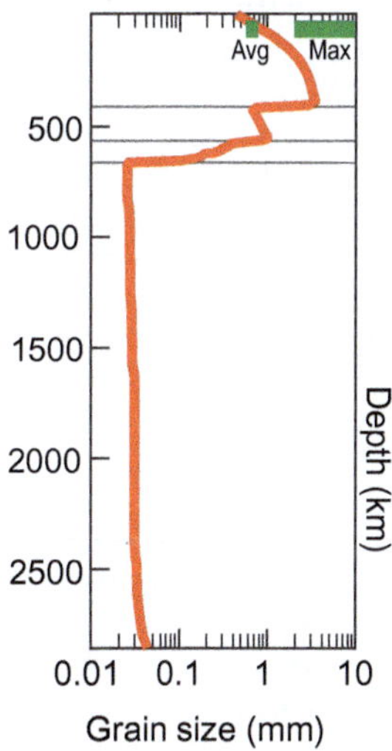

Fig. 3. Average (Avg) and maximum (Max) grain size (green bars) in Antarctic mantle xenoliths taken from references reported in Table 1 and plotted against depth. The grain-size axis is logarithmic. The average modelled grain size (orange curve) is shown for comparison (Dannberg *et al.* 2017). Horizontal lines mark phase transitions between olivine, wadsleyite, ringwoodite and bridgmanite and periclase (after Dannberg *et al.* 2017, their table S1).

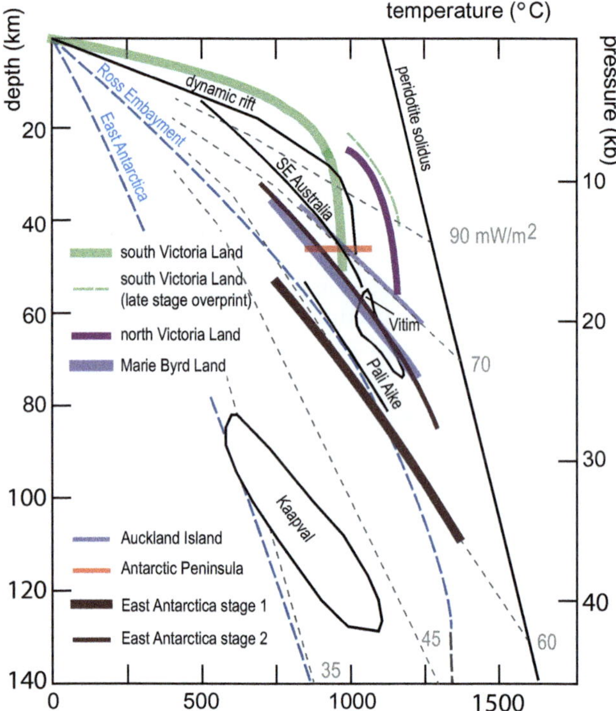

Fig. 4. Petrological geotherm estimates based on data for Antarctica xenoliths compared with other published geotherms. South Victoria Land (Berg *et al.* 1989; Martin *et al.* 2015); north Victoria Land (Perinelli *et al.* 2006) (Coltorti *et al.* 2021, this volume); Marie Byrd Land (Chatzaras *et al.* 2016); Antarctic Peninsula (Leat *et al.* 2021, this volume); East Antarctica stage 1 (early- pre-rift) and 2 (syn-rift; Foley *et al.* 2006); Auckland Island (Delpech *et al.* 2021, this volume); Pali Aike (Kempton *et al.* 1999; Gibson *et al.* 2020); Vitim (Glaser *et al.* 1999); Kaapvaal (Bell *et al.* 2003); dynamic rift (Chapman 1986); present-day Ross Embayment and East Antarctica (ten Brink *et al.* 1997); SE Australia (O'Relly and Griffin 1985); continental geotherms (grey dashed lines; Hasterok and Chapman 2011).

seismic structure (Shen *et al.* 2020) or comparable to other geophysical (± heat flow measurements) methods of determining heat flow (Shapiro and Ritzwoller 2004; Maule *et al.* 2005; An *et al.* 2015; Martos *et al.* 2017; Burton-Johnson *et al.* 2020*a*, *b*; Stål *et al.* 2020).

In Marie Byrd Land, a geotherm of *c.* 65 mW m^{-2} has been calculated (Fig. 4) at 1.5 Ma (Chatzaras *et al.* 2016) using mantle xenoliths. This is lower than the minimum average geothermal flux of 114 ± 10 mW m^{-2} calculated for the Thwaites Glacier catchment (Fig. 2) (Schroeder *et al.* 2014), but does overlap with the lower estimate of geothermal heat flux made by Dziadek *et al.* (2017) of between 68 and 110 mW m^{-2} in the region. This is consistent with interpretations of spatially complex heat flow in Marie Byrd Land (Shen *et al.* 2020; Burton-Johnson *et al.* 2020*a*). In the Antarctic Peninsula a geotherm has not been calculated from mantle xenoliths, but temperatures between 1050 and 850°C have been calculated for mantle xenoliths entrained in host rocks erupted between *c.* 8 and 5 Ma (Leat *et al.* 2021, this volume). This is in the range of mean heat flow estimates of between 67 and 81 mW m^{-2} calculated for the Antarctic Peninsula (Burton-Johnson *et al.* 2017).

In East Antarctica, the mantle xenolith data represent a fossil geotherm of *c.* 65 mW m^{-2} (Fig. 4; stage 2) at *c.* 140 Ma (Foley *et al.* 2006). The East Antarctica geotherm of ten Brink *et al.* (1997) is modelled assuming a 250 km-thick cooling plate model and suggests heat flow of 35 mW m^{-2}, comparable to the African Kaapvaal Craton geotherm (Fig. 4). Empirically constrained seismic structure estimates of heat flow show an average of *c.* 52 mW m^{-2} for East Antarctica (Shen *et al.* 2020), much closer to the *c.* 60 mW m^{-2} geotherm inferred for the pre-rift mantle heat flow from mantle xenoliths in East Antarctica (Fig. 4; stage 1).

In summary the temperature–pressure profiles in Victoria Land appear more similar to a dynamic rift geotherm (Fig. 4). Heat flow from mantle xenoliths in Marie Byrd Land is *c.* 65 mW m^{-2}, but with anomalously high heat flow around the Thwaites Glacier area calculated in other studies. In the Antarctic Peninsula, mantle temperatures are consistent with independently calculated geotherms. The difference in heat flow between East and West Antarctica indicated in mantle xenolith data supports similar inferences made through geophysical and heat flow measurement studies (e.g. Pappa and Ebbing 2021, this volume; Wiens *et al.* 2021, this volume).

Water contents in the Antarctic lithospheric mantle

High water contents in mantle rocks enhance creep, lower the melting temperature and can reduce seismic velocity by increasing anelasticity (Karato 2004; Ohtani 2020). The water content in the shallow upper mantle (above the 410 km discontinuity) varies depending upon setting (Fig. 5). Depleted mantle (DM on Fig. 5) sources have water contents of between 100 and 200 ppm (Hirschmann 2006; Marty 2012), with peridotite xenoliths from continental margin and off-craton settings having up to 220 ppm bulk rock H_2O. Pyroxenite xenolith bulk rock H_2O values can be up to 400 ppm from oceanic settings (Gibson *et al.* 2020).

In Antarctica, mantle water content has been estimated by secondary-ion mass spectrometry or Fourier transform infrared spectroscopy. These methods have been applied to either mantle xenolith minerals or igneous olivine in relatively

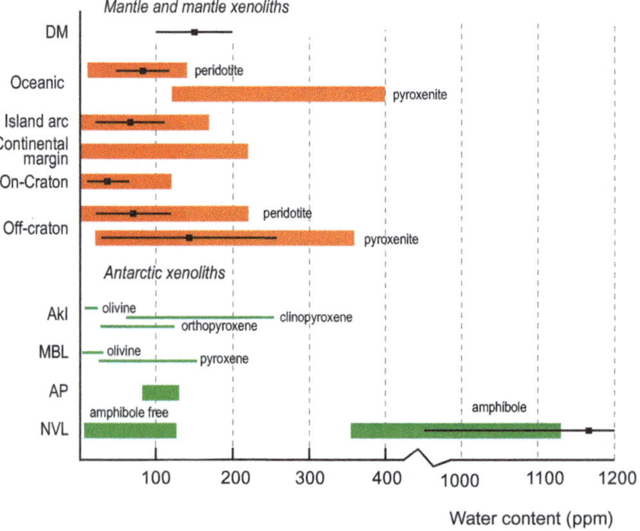

Fig. 5. Estimates of water contents in the global mantle from depleted mantle (DM; Marty 2012) and mantle xenoliths (orange; Gibson *et al.* 2020) from various geological settings, compared with values reported in Antarctic mantle xenoliths. The following references are used for Antarctica: Auckland Island (AkI; Delpech *et al.* 2021, this volume); Marie Byrd Land (MBL; Chatzaras *et al.* 2016) (Chatzaras and Kruckenberg 2021, this volume); Antarctic Peninsula (AP; Gibson *et al.* 2020) (Leat *et al.* 2021, this volume); north Victoria Land (NVL; green boxes; Coltorti *et al.* 2021, this volume) and values (black square) with errors (black bar) from Giacomoni *et al.* (2020).

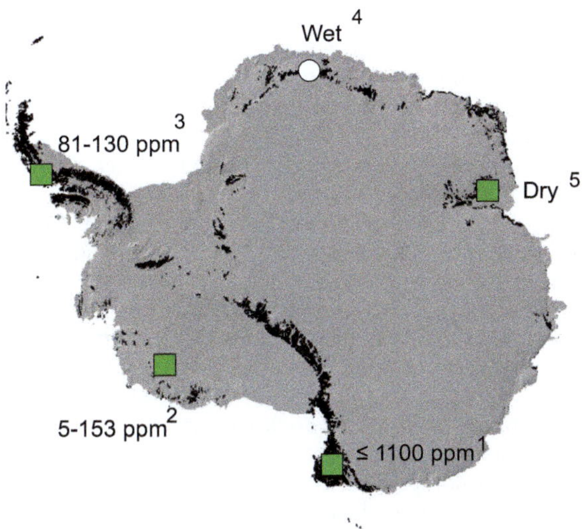

Fig. 6. A diagram showing the location and values of mantle water contents from xenoliths (green squares) and other mantle rocks (white circle). The 'wet' and 'dry' nomenclature follows van der Wal *et al.* (2015). Black indicates rock outcrop, grey indicates snow or ice. References are as follows: 1. (Giacomoni *et al.* 2020; Coltorti *et al.* 2021, this volume); 2. (Chatzaras *et al.* 2016); 3. (Leat *et al.* 2021, this volume); 4. (Markl *et al.* 2003); 5. (Foley *et al.* 2006).

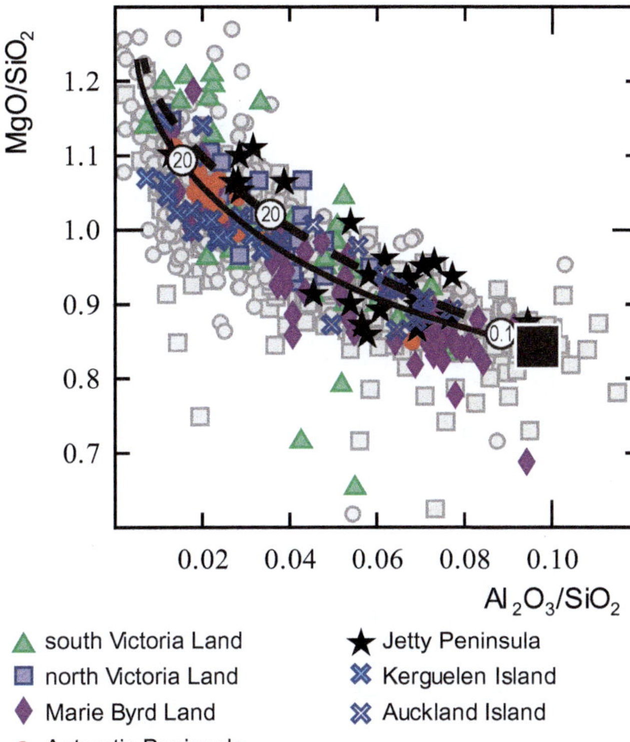

Fig. 7. Bulk rock Al_2O_3/SiO_2 v. MgO/SiO_2 plot for Antarctic mantle xenoliths. Antarctic data are from Table 1 and data for comparison from Pearson *et al.* (2003) are on-craton (grey circles) and off-craton (grey squares) spinel peridotite xenoliths from volcanic rocks. Melt extraction curves of Herzberg (2004) are also shown for 2 GPa melt depletion (dashed line) and 1 GPa (solid line) with indications of expected residual composition after 20% melt extraction. Black square is primitive mantle (McDonough and Sun 1995).

undifferentiated rocks (see references in Table 1 for more discussion). There are various limitations in comparing data collected in this way, and care should be taken not to overinterpret the results. A striking observation is, however, that the upper mantle beneath Marie Byrd Land and the Antarctic Peninsula, as well as the Auckland Islands is relatively dry (<300 ppm H_2O) compared to that of Victoria Land where mantle water contents are up to and above 1000 ppm (Figs 5 & 6). In detail (Fig. 5), the relatively dry mantle (also seen in amphibole-free peridotite in north Victoria Land) is equivalent to water content recorded in off-craton peridotite xenolith samples seen world-wide (Gibson *et al.* 2020) and in the source for depleted mantle (Hirschmann 2006; Marty 2012).

In amphibole-bearing mantle sources in north Victoria Land, the water content is up to 1160 ± 436 ppm (Giacomoni *et al.* 2020; Coltorti *et al.* 2021, this volume; Fig. 6). In the extreme case of amphibole-dominated veins in north Victoria Land, water contents up to 1.42 wt% have been reported (Bonadiman *et al.* 2014). A component of slab dewatering is required to account for the high mantle water contents in Victoria Land (Giacomoni *et al.* 2020), consistent with the tectonic setting and history of the region (Rocchi and Smellie 2021; Martin *et al.* 2021*b*).

Chemistry

Whole rock

Mantle xenolith bulk rock chemistry is fundamental to understanding lithospheric mantle thermal conductivity and other physical conditions, the mantle composition and formation of mafic igneous rocks (e.g. Pearson *et al.* 2003; Gradmann *et al.* 2013). Here, perhaps for the first time, the bulk rock chemistry of Antarctic mantle xenoliths is compared across the continent, and to mantle xenolith compositions globally.

The full range of Al_2O_3/SiO_2 v. MgO/SiO_2 compositions from a global compilation of mantle xenolith data (Pearson *et al.* 2003), is overlapped to a very large degree by the Antarctic mantle xenolith data (Fig. 7), reflecting the diverse geological and tectonic settings of the Antarctic continent. The data include both on-craton and off-craton settings (Pearson *et al.* 2003) (Fig. 7). In general, all Antarctic mantle xenolith data plot along the terrestrial array (Jagoutz *et al.* 1979).

At *c.* 0.02 Al_2O_3/SiO_2 (Fig. 7), differences in melt extraction depth are inferred between Kerguelen Island (maximum depth), the Antarctic Peninsula (moderate depth) and Victoria Land (shallowest depth, both north and south). This is consistent with pressure–temperature estimates (Fig. 4). In general, the Marie Byrd Land mantle xenolith data have higher Al_2O_3/SiO_2 and lower MgO/SiO_2 relative to mantle xenoliths from Victoria Land or the Antarctic Peninsula, which is consistent with lower modelled melt extraction values (Fig. 7; Herzberg 2004). Mantle xenoliths from the Auckland Islands have compositions reflecting less melt extraction at shallow depths than was the case for those from Kerguelen Island, which is consistent with observations made in Delpech *et al.* (2021, this volume).

In general, lithospheric mantle has become less depleted in bulk rock Al and Ca with time allowing a rule-of-thumb link between mantle age and composition (Poudjom Djomani *et al.* 2001). On a bulk rock wt% Al_2O_3 v. CaO plot, Kerguelen Island mantle xenolith data overlap with the Archean mantle age field of O'Reilly *et al.* (2001) and plot separately from Auckland Island mantle xenolith data that overlap with younger mantle fields (Fig. 8a). There is a spread in bulk rock wt% Al_2O_3 v. CaO values for Marie Byrd Land and Victoria Land samples (Fig. 8a). In general, however, the median Marie Byrd Land value (Fig. 8a – star symbol) overlaps with the Phanerozoic mantle field and the median Victoria Land value overlaps with the Proterozoic field. Both are separate from the

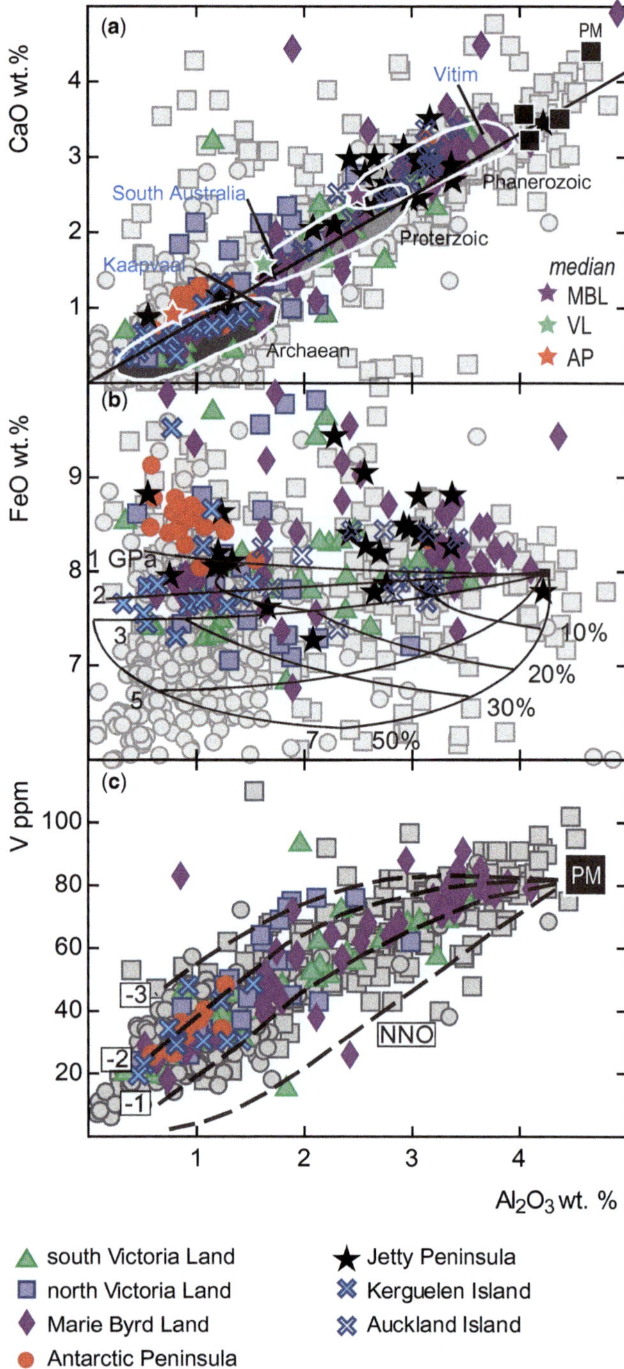

Fig. 8. Bulk rock Antarctica mantle peridotite xenolith plots of weight percent (wt%) Al_2O_3 v. **(a)** CaO, **(b)** FeO and **(c)** parts per million (ppm) V. Antarctic data are from Table 1 and these are compared with those from Pearson *et al.* (2003). The latter include on-craton (grey circles) and off-craton (grey squares) spinel peridotite xenoliths hosted in volcanic rocks. In panel a. comparisons are made with various estimates for primitive upper mantle (PM; black squares) and the ratio in chondritic meteorites (black line), taken from Pearson *et al.* (2003). The mantle age fields and data for comparison (Kaapvaal, South Australia, Vitim) are from O'Reilly *et al.* (2001). MBL, Marie Byrd Land; VL, Victoria Land; AP, Antarctic Peninsula. The melt grid in panel (b) is a function of pressure (GPa) and melt extraction (%) and is from Walter (2003). In panel (c), oxygen fugacity at 1.5 GPa is compared with the nickel–nickel–oxide (NNO) buffer (Canil 2002). Primitive mantle (PM) is from Palme and O'Neill (2003).

Antarctica Peninsula median value, which overlaps with the Archean field. There is a strong inference that the mantle lithosphere beneath Victoria Land and Marie Byrd Land and the Antarctic Peninsula consist of separate blocks (Fig. 8a and see Panter and Martin 2021, this volume). The Antarctic mantle xenolith data plot about the reference ratio for chondritic meteorites and overlap with values from a global dataset of mantle xenoliths (Pearson *et al.* 2003; Fig. 8a). In general, off-craton mantle xenoliths globally have higher wt% bulk rock Al_2O_3 and CaO relative to on-craton data, as is seen in, for example, the Auckland Island data and median Marie Byrd Land median value (Fig. 8a). It should be noted that refertilization (or *stealth metasomatism*: O'Reilly and Griffin 2013) can also increase CaO and Al_2O_3 and, as discussed below, refertilization and metasomatism are important components of the Antarctic lithospheric mantle.

In general, Antarctic bulk rock wt% Al_2O_3 v. FeO mantle xenolith data overlap with off-craton mantle xenolith data from a global compilation (Pearson *et al.* 2003) and do not overlap with on-craton data (Fig. 8b). As shown on Figure 7, estimates of depths of melt extraction are deeper for Kerguelen Island samples than for Antarctic Peninsula or Auckland Island data. These differences are also seen when the data are compared to the melt grid of Herzberg (2004; Fig. 8b).

Plotted for comparison on a bulk rock wt% Al_2O_3 v. parts per million (ppm) V diagram (Fig. 8c) are residual trends as a function of oxygen fugacity relative to the nickel–nickel oxide (NNO) buffer. Typically, the Antarctic mantle xenolith data plot between −2 and −1 of the NNO oxygen fugacity regardless of setting, which is generally consistent with oxygen fugacity values calculated independently for the Victoria Land mantle (Perinelli *et al.* 2012; Bonadiman *et al.* 2014; Martin *et al.* 2015) and Jetty Peninsula (Foley *et al.* 2006). Coltorti *et al.* (2021, this volume) have newly calculated oxygen fugacity values for mantle xenoliths for north Victoria Land, and they identify a group of $\Delta\log fO_2$ values closer to −3 relative to the fayalite–magnetite–quartz (FMQ) buffer. Only four Antarctic mantle xenolith values plot above −1 NNO (Fig. 8c).

Clinopyroxene

Select, clinopyroxene rare earth element (REE) data normalized to chondritic values of McDonough and Sun (1995) have been plotted in Figure 9. Two groups of clinopyroxene are observed in Victoria Land and Marie Byrd Land, a light (L)REE depleted group and a LREE enriched group (Fig. 9a). This suggests multiple generations of clinopyroxene, as has been suggested from petrography, geochemistry and heat flow calculations by various authors in this Memoir (Table 1). Similarly, at least two clinopyroxene patterns have been defined in Auckland Island mantle xenolith samples (Fig. 9b), and – though only one clinopyroxene REE pattern is shown for Kerguelen Island – petrography suggests multiple episodes of clinopyroxene formation (Delpech *et al.* 2021, this volume). Three groups have been identified from Jetty Peninsula samples (see Foley *et al.* 2021, this volume, for a discussion) and at least three distinct patterns are apparent for Antarctic Peninsula mantle xenolith samples (Fig. 9c). The principal conclusion from the clinopyroxene REE patterns is that they demonstrate that a complex history of multiple melt depletion and (re-)enrichment (refertilization and/or metasomatism) events has affected every single mantle xenolith locality in Antarctica.

Sr and Nd isotopes

Although limited, Sr and Nd isotope data are available for mantle xenolith samples from Victoria Land, Marie Byrd

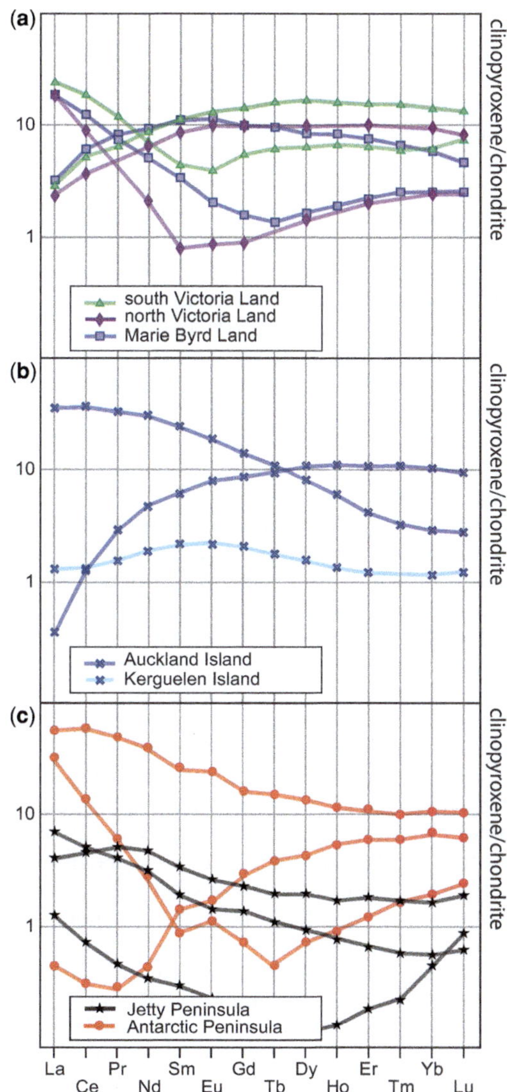

Fig. 9. Representative clinopyroxene rare earth element, chondrite-normalized (McDonough and Sun 1995) patterns for mantle peridotite xenoliths from Antarctica. South Victoria Land (Martin *et al.* 2021*a*, this volume); north Victoria Land (Coltorti *et al.* 2021, this volume); subantarctic islands Kerguelen (Delpech *et al.* 2021, this volume) and Auckland (Scott *et al.* 2014); East Antarctica (Foley *et al.* 2021, this volume) and the Antarctic Peninsula (Gibson *et al.* 2020).

Land, Kerguelen Island and Auckland Island (Fig. 10). There are no such data for xenoliths from the Antarctic Peninsula. In East Antarctica, only Sm–Nd isotope data exist on the mantle xenoliths, with Sr–Nd isotope data available only from igneous rocks. Antarctic mantle xenoliths have Sr–Nd isotope compositions that overlap to a large degree with the field defined by mantle xenolith samples from Zealandia. These trend towards high-μ (HIMU). Exceptions are samples with higher $^{143}Nd/^{144}Nd$ and lower $^{87}Sr/^{86}$ Sr isotopic values that plot towards the depleted mantle (DM) field (Fig. 10). Three Victoria Land samples and Kerguelen Island data have Sr–Nd isotope compositions similar to those of mantle xenoliths from SE Australia, plotting around bulk silicate earth (BSE; Fig. 10). A much more complete isotopic dataset is available for relatively undifferentiated igneous rocks from Antarctica and the sub-Antarctic islands, and the reader is referred to Panter and Martin (2021, this volume) and chapters in Smellie *et al.* (2021) for thorough reviews.

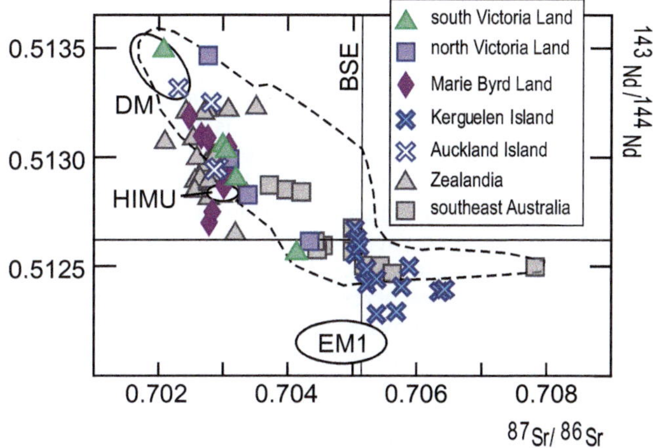

Fig. 10. A $^{87}Sr/^{86}Sr$ v. $^{143}Nd/^{144}Nd$ plot for mantle xenoliths from Victoria Land, Marie Byrd land and Auckland Island (Table 1) compared to mantle xenolith data from SE Australia (Stolz and Davies 1988) and southern New Zealand (Scott *et al.* 2014). Mantle fields for depleted mantle (DM), high-μ (HIMU) enriched mantle (EM 1) and bulk silicate earth (BSE; grey lines) are from Zindler and Hart (1986). The range of values from a global compilation of mantle xenoliths is shown for comparison (dashed line; Pearson *et al.* 2003).

Implications and future work

Implications for geophysics

The properties of mantle xenoliths discussed in this chapter, namely grain size, heat flow, water content and whole rock and mineral chemistry, provide key constraints for geophysical modelling and interpretation of geophysical data. Of particular importance are the following observations:

1. Regionally, there can be variations in bulk rock chemistry of >1 wt%.
2. Heat flow varies by up to 10s of mW m^{-2} by region.
3. Water content is higher (>350 ppm in amphibole-bearing specimens) in the Victoria Land mantle relative to other parts of West Antarctica or Auckland Island (<*c.* 200 ppm).
4. Depletion and (re-)enrichment are likely to be a ubiquitous feature. This observation was first made by Foley *et al.* (2021, this volume).
5. Average grain size in mantle xenoliths is 0.6 mm, however, maximum grain size (up to 10 mm) is larger than currently used in many models.

The following comments derive from the observations made by Foley *et al.* (2021, this volume) and apply to all of the Antarctic lithospheric mantle. Reference Earth models are commonly based on a presumption that the mantle is homogenous peridotite, but evidence presented in this Memoir (Table 1; Fig. 9) suggests this is not the case for Antarctic mantle xenoliths. Foley *et al.* (2021, this volume) contend that the entire Antarctic upper mantle is heterogeneous. This occurs at the local scale, for example within the same volcanic centre, implying vertical heterogeneity, or within the same volcanic field, implying lateral heterogeneity. Heterogeneity also occurs at regional (e.g. Figs 5 & 6) and continental scales (Fig. 7). Additionally, modal mineralogy (e.g. variation in modal plagioclase, spinel, garnet, secondary minerals), rock type (peridotite v. pyroxenite), and the variable (but ubiquitous) presence of melts or fluids adds to the complexity of mantle heterogeneity. Furthermore, there are few garnet-bearing xenoliths and relatively more plagioclase-bearing mantle xenoliths, emphasizing the samples we have of the Antarctic mantle are unusually shallow relative to other continents which also affects constraints on water content, viscosity, etc. This leads to the important observation that choosing appropriate parameters to use in geophysical modelling is challenging.

Generally, current geophysical models for the Antarctic mantle use either a single reference Earth model (e.g. Whitehouse *et al.* 2012; Ivins *et al.* 2013; Lloyd *et al.* 2019) or a 3D reference Earth model highlighting the divide in Antarctica between East and West (Geruo *et al.* 2013; van der Wal *et al.* 2015) but still implicitly or explicitly assuming a homogeneous mantle rock type across Antarctica. The small grain sizes (0.6 mm) and high water content seen in Victoria Land would lead to very low viscosities if directly entered in an olivine flow law. This questions whether flow laws for pyroxene may also be required to best model the shallow upper mantle of Victoria Land, cf. a pure olivine flow law.

The boundary constraints recorded in mantle xenoliths from East Antarctica are derived from only three localities (Fig. 2). The most important of these, the Jetty Peninsula site, provides information about the mantle at the time the xenolith host was erupted at 140 Ma, and there is uncertainty about how data from this site might constrain boundary conditions for modern geophysical studies. Certainly, East Antarctic xenoliths provide useful grain size information, but chemistry and heat flow are likely to have changed due to subsequent depletion and re-enrichment events (Foley *et al.* 2021, this volume), to be more heterogeneous and colder since 140 Ma (based on a modern-day geotherm of *c.* 35 mW m^{-2}, Fig. 4), respectively. Additionally, the nature of the mantle beneath most of East Antarctica is unknown from a mantle xenolith perspective. Of the suggested three tectonic domains of mantle beneath East Antarctica (Fig. 1), mantle xenoliths are known only from a single (Indian) domain, but useful correlations might be made by examining mantle xenoliths from adjacent continental blocks in Australia, India and Africa (Foley *et al.* 2021, this volume), or sub-Antarctic Islands (Delpech *et al.* 2021, this volume).

There is apparently more mantle variability in West Antarctica relative to East Antarctica. This is almost certainly an artefact caused by more intensive sampling, which provides a clearer picture of mantle heterogeneity across the entire continent. The host rocks to mantle xenoliths in West Antarctica tend to have erupted within the past 1 Ma. This makes the boundary conditions they record for the upper mantle relevant to modern geophysical studies. The Antarctic Peninsula and Marie Byrd Land share some commonality in grain size and mantle water content that would be useful to include in geophysical models. However, there are strong compositional contrasts (e.g. Fig. 7; Panter and Martin 2021, this volume). Victoria Land has higher water contents relative to the rest of West Antarctica and heat flow varies widely across all of Antarctica.

Thus, the variation recorded in mantle xenoliths is currently greater than the variation that can (or has) been incorporated into Antarctica-wide geophysical models. This has the following implications. When a mismatch between geophysical model outputs v. other observations occurs, the heterogeneity observed in the Antarctic upper mantle should be recalled. The degree to which regional-scale variation will impact on geophysical models should be quantified, as discussed below as future work. If the real-world mantle heterogeneity causes errors in geophysical models that are still smaller than what is required to answer a particular scientific question, then incorporating this variability becomes academic.

Implications for studies of the mantle

The occurrence, petrology and geochemistry of mantle xenoliths from Victoria Land were reviewed by Kyle *et al.* (1987) and limited East Antarctic examples were described by Nixon (1987*a*), both in the landmark work by Nixon (1987*b*). Now, probably for the first time, mantle xenoliths in igneous rocks from across Antarctica (Fig. 2) can be considered. This data set (from the references in Table 1) can now be included in global mantle xenolith compilations (e.g. Jagoutz *et al.* 1979; Nixon 1987*b*; Pearson *et al.* 2003).

Important boundary conditions for the Antarctic lithospheric mantle are set, for example, upper mantle oxygen fugacity (Fig. 8c) and grain size (Fig. 3). Regionally, there is important overlap between: Marie Byrd Land–Victoria Land and New Zealand data (e.g. Fig. 10); the Antarctic Peninsula and southern South America (e.g. Leat *et al.* 2021, this volume); and East Antarctica and adjacent continental blocks (Foley *et al.* 2021, this volume). There appears to be an unusual amount of plagioclase in the Antarctic lithospheric mantle relative to what is reported globally (Martin *et al.* 2014*b*), that implies the presence of melt and/or an atypically shallow Moho that can affect seismic properties.

None of the Antarctic mantle xenolith data are globally unique in their bulk rock composition, for example by comparison with the compilation of Pearson *et al.* (2003) shown in Figure 7. This is not unexpected given the geological affinity with adjacent crustal-blocks (Fig. 1) that have well-studied and documented mantle xenoliths in igneous rocks. There is the opportunity, however, to use the Antarctica occurrences of mantle xenoliths to inform about the formation of adjacent crustal-blocks and about general mantle petrogenesis. The reason for this is that despite more than 100 years of collection of mantle xenoliths (Martin *et al.* 2021*a*, this volume – Introduction), the study of Antarctic specimens lags behind those of other continents, especially in terms of the availability of data obtained using modern, ultra-low method detection limit techniques and analysis for novel elements, isotopes, gasses and volatiles. Exceptions to this, of course, occur. For example, He isotope (Melchiorre *et al.* 2011; Day *et al.* 2019), Re–Os isotope (Handler *et al.* 2003; Melchiorre *et al.* 2011; Doherty *et al.* 2013; Day *et al.* 2019; Scott *et al.* 2019), volatile (Gibson *et al.* 2020) and other noble gas studies (Buikin *et al.* 2014; Broadley *et al.* 2016; Correale *et al.* 2019). This discussion leads us into future research directions.

Future work

The following is a list of research tasks that should be prioritized and research questions that could be answered by further study of mantle xenoliths and related rocks in Antarctica. These should be undertaken because there will always be a dearth of direct information from mantle xenoliths in Antarctica with respect to other continents. The questions are:

Related to geophysical studies

- Quantifying which parameters (chemistry, grain size, water content or temperature) best constrain geophysical models. The relative impact of these parameters should be ranked in terms of the geophysical approach being taken. Incorporation of boundary conditions derived from mantle xenoliths will help geophysical models address pressing research questions (e.g. glacial isostatic adjustment, large-scale mantle structures, mantle plumes).
- Quantifying how a coarser grain size (up to 10 mm as described in this Memoir) in the upper mantle would affect geophysical models.
- How can the regional-scale variability indicated by mantle xenoliths be incorporated into geophysical models?
- How is the (near-) ubiquitous heterogeneity in the Antarctic lithospheric mantle affecting interpretation of geophysical signals? Can this be quantified and accounted for?

- Is there any advantage in updating the reference Earth model used in Antarctica, based on boundary conditions set by the study of Antarctic mantle xenoliths?

Related to mantle xenolith studies

- Collection of a consistent dataset of bulk rock and mineral chemistry from mantle xenoliths across Antarctica, including a full suite of elements and common isotopes.
- Determination of the timing of depletion and (re-)enrichment events recorded in mantle xenoliths to help constrain petrogenesis of the Antarctic lithospheric mantle.
- More explicit linkage of mantle xenolith studies from continental blocks adjacent to the Antarctic mantle, and vice versa.
- Quantification of variability (with well-constrained errors) in water chemistry across the Antarctic mantle.
- Characterization of the Antarctic upper mantle budget for other volatile elements and noble gasses to better understand their distribution and what affect they might have on geophysical studies.
- Extension of the microstructure studies commenced in Marie Byrd Land to other regions of Antarctica to better understand how seismic properties (velocities, anisotropy) are affected.
- Investigation of the applicability of chemical tomography (e.g. O'Reilly and Griffin 2006) for better interpreting geophysical models and to understand mantle petrogenesis.

Related to other rocks with mantle affinity

- Better investigation of the oceanic mantle around Antarctica either from tectonically emplaced, onshore slices or offshore abyssal peridotite samples.
- Investigation of mantle source variation using relatively undifferentiated volcanic rocks. This will involve pairing of the investigation of mantle sources using relatively undifferentiated volcanic rocks with studies of mantle xenoliths to maximize the information obtained about the mantle; samples can be linked in time and space but will reflect differences in depth.
- Study of the lithospheric–asthenospheric boundary depth through paired investigation of mantle xenoliths and relatively undifferentiated volcanic rocks.

Conclusions

This chapter provides an overview of detailed studies of various aspects of Antarctic mantle xenoliths that are presented in this Memoir. These include descriptions and interpretations of the petrology, grain size, water content, pressure–temperature conditions of formation and chemistry of Antarctic mantle xenoliths. The potential for this work to yield unique and novel discoveries is high. There are many significant research questions yet to be addressed by detailed fundamental studies and cutting-edge techniques for both chemistry and geophysics. Special attention should be given to how information from mantle xenoliths can be integrated with geophysical studies. This is because geophysical studies are vital in understanding the mantle beneath an ice-covered continent with only 2% rock exposure and a history of continuous human occupation that is only a few decades long. Study of the Antarctic mantle can contribute to significant outcomes including: quantifying the Antarctica contribution to sea-level rise; mapping mantle topography variations caused by density; and understanding how the Antarctic mantle integrates with the global mantle model. It is anticipated that this review chapter will encourage readers to investigate the other chapters in this volume and even to begin, or continue, their own studies into the fascinating topic that is the Antarctic mantle.

Acknowledgements I thank all the authors who contributed to this Memoir making this chapter possible. I especially thank those whose data were instrumental in constructing the plots. I thank R. Price for comments on an early draft, W. van der Wal for editorial handling and reviewers S. Foley and M. Coltorti and the editor for their helpful comments.

Author contributions **APM**: conceptualization (lead), data curation (lead), writing – original draft (lead).

Funding This contribution was in part supported by contract ANT1801.

Data availability All data generated or analysed during this study are included in this published Memoir 56 (and its supplementary information files) and in select references summarized in Table 1.

References

An, M., Wiens, D.A. *et al.* 2015. Temperature, lithosphere–asthenosphere boundary, and heat flux beneath the Antarctic Plate inferred from seismic velocities. *Journal of Geophysical Research: Solid Earth*, **120**, 8720–8742, https://doi.org/10.1002/2015JB011917

Andronikov, A.V., Mikhalsky, E.V. and Belyatsky, B.V. 1994. Abyssal xenoliths from the lamprophyres of the Vestfold Hills, East Antarctica. *Petrology*, **2**, 288–296.

Barletta, V.R., Nield, G.A., van der Wal, W. and van Calcar, C.J. 2022. Glacial isostatic adjustment and postseismic deformation in Antarctica. *Geological Society, London, Memoirs*, **56**, in press.

Bell, D.R., Schmitz, M.D. and Janney, P.E. 2003. Mesozoic thermal evolution of the southern African mantle lithosphere. *Lithos*, **71**, 273–287, https://doi.org/10.1016/S0024-4937(03)00117-8

Berg, J.H., Moscati, R.J. and Herz, D.L. 1989. A petrologic geotherm from a continental rift in Antarctica. *Earth and Planetary Science Letters*, **93**, 98–108, https://doi.org/10.1016/0012-821X(89)90187-8

Blackman, D.K., Von Herzen, R.P. and Lawver, L.A. 1987. Heat flow and tectonics in the western Ross Sea. *In*: Cooper, A.K. and Davey, F.J. (eds) *The Antarctic Continental Margin, Geology and Geophysics of the Western Ross Sea.* Circum-Pacific Council for Energy and Mineral Resources, Houston, Texas, 179–190.

Bonadiman, C., Nazzareni, S., Coltorti, M., Comodi, P., Giuli, G. and Faccini, B. 2014. Crystal chemistry of amphiboles: implications for oxygen fugacity and water activity in lithospheric mantle beneath Victoria Land, Antarctica. *Contributions to Mineralogy and Petrology*, **167**, 1–17, https://doi.org/10.1007/s00410-014-0984-8

Bredow, E., Steinberger, B., Gassmöller, R. and Dannberg, J. 2021. Mantle convection and possible mantle plumes beneath Antarctica – insights from geodynamic models and implications for topography. *Geological Society, London, Memoirs*, **56**, https://doi.org/10.1144/M56-2020-2

Broadley, M.W., Ballentine, C.J., Chavrit, D., Dallai, L. and Burgess, R ., 2016. Sedimentary halogens and noble gases within Western Antarctic xenoliths: implications of extensive volatile recycling to the sub continental lithospheric mantle. *Geochimica et Cosmochimica Acta*, **176**, 139–156, https://doi.org/10.1016/j.gca.2015.12.013

Buikin, A.I., Solovova, I.P., Verchovsky, A.B., Kogarko, L.N. and Averin, A.A. 2014. PVT parameters of fluid inclusions and the C, O, N, and Ar isotopic composition in a garnet lherzolite xenolith from the Oasis Jetty, East Antarctica. *Geochemistry International*, **52**, 805–821, https://doi.org/10.1134/S0016702914100036

Burton-Johnson, A., Halpin, J.A., Whittaker, J.M., Graham, F.S. and Watson, S.J. 2017. A new heat flux model for the Antarctic Peninsula incorporating spatially variable upper crustal radiogenic heat production. *Geophysical Research Letters*, **44**, 5436–5446, https://doi.org/10.1002/2017GL073596

Burton-Johnson, A., Dziadek, R. and Martin, C. 2020*a*. Geothermal heat flow in Antarctica: current and future directions. *The Cryosphere*, **14**, 3843–3873, https://doi.org/10.5194/tc-14-3843-2020

Burton-Johnson, A., Dziadek, R. *et al.* 2020*b*. *Antarctic Geothermal Heat Flow: Future Research Directions*. Scientific Committee on Antarctic Research (SCAR) White Paper, Miscellaneous Publication.

Canil, D. 2002. Vanadium in peridotites, mantle redox and tectonic environments: Archean to present. *Earth and Planetary Science Letters*, **195**, 75–90, https://doi.org/10.1016/S0012-821X(01)00582-9

Chapman, D.S. 1986. Thermal gradients in the continental crust. *Geological Society, London, Special Publications*, **24**, 63–70, https://doi.org/10.1144/GSL.SP.1986.024.01.07

Chatzaras, V. and Kruckenberg, S.C. 2021. Effects of melt-percolation, refertilization and deformation on upper mantle seismic anisotropy: constraints from peridotite xenoliths, Marie Byrd Land, West Antarctica. *Geological Society, London, Memoirs*, **56**, https://doi.org/10.1144/M56-2020-16

Chatzaras, V., Kruckenberg, S.C., Cohen, S.M., Medaris, L.G., Jr, Withers, A.C. and Bagley, B. 2016. Axial-type olivine crystallographic preferred orientations: the effect of strain geometry on mantle texture. *Journal of Geophysical Research: Solid Earth*, **121**, 4895–4922, https://doi.org/10.1002/2015JB012628

Coltorti, M., Beccaluva, L., Bonadiman, C., Faccini, B., Ntaflos, T. and Siena, F. 2004. Amphibole genesis via metasomatic reaction with clinopyroxene in mantle xenoliths from Victoria Land, Antarctica. *Lithos*, **75**, 115–139, https://doi.org/10.1016/j.lithos.2003.12.021

Coltorti, M., Bonadiman, C., Casetta, F., Faccini, B., Giacomoni, P.P., Pelorosso, B. and Perinelli, C. 2021. Nature and evolution of the northern Victoria Land lithospheric mantle (Antarctica) as revealed by ultramafic xenoliths. *Geological Society, London, Memoirs*, **56**, https://doi.org/10.1144/M56-2020-11

Cooper, A.F., Adam, L.J., Coulter, R.F., Eby, G.N. and McIntosh, W.C. 2007. Geology, geochronology and geochemistry of a basanitic volcano, White Island, Ross Sea, Antarctica. *Journal of Volcanology and Geothermal Research*, **165**, 189–216, https://doi.org/10.1016/j.jvolgeores.2007.06.003

Correale, A., Pelorosso, B., Rizzo, A.L., Coltorti, M., Italiano, F., Bonadiman, C. and Giacomoni, P.P. 2019. The nature of the West Antarctic Rift System as revealed by noble gases in mantle minerals. *Chemical Geology*, **524**, 104–118, https://doi.org/10.1016/j.chemgeo.2019.06.020

Dannberg, J., Eilon, Z., Faul, U., Gassmöller, R., Moulik, P. and Myhill, R. 2017. The importance of grain size to mantle dynamics and seismological observations. *Geochemistry, Geophysics, Geosystems*, **18**, 3034–3061, https://doi.org/10.1002/2017GC006944

Day, J.M.D., Harvey, R.P. and Hilton, D.R. 2019. Melt-modified lithosphere beneath Ross Island and its role in the tectono-magmatic evolution of the West Antarctic Rift System. *Chemical Geology*, **518**, 45–54, https://doi.org/10.1016/j.chemgeo.2019.04.012

Delpech, G., Grégoire, M., O'Reilly, S.Y., Cottin, J.Y., Moine, B., Michon, G. and Giret, A. 2004. Feldspar from carbonate-rich silicate metasomatism in the shallow oceanic mantle under Kerguelen Islands (South Indian Ocean). *Lithos*, **75**, 209–237, https://doi.org/10.1016/j.lithos.2003.12.018

Delpech, G., Scott, J.M. *et al.* 2021. The subantarctic lithospheric mantle. *Geological Society, London, Memoirs*, **56**, https://doi.org/10.1144/M56-2020-13

Doherty, C., Class, C. *et al.* 2013. Re–Os systematics of the lithospheric mantle beneath the Western Ross Sea area, Antarctica: depletion ages and dynamic response during rifting. AGU, 9–13 December, San Francisco, T13A-2516.

Dziadek, R., Gohl, K., Diehl, A. and Kaul, N. 2017. Geothermal heat flux in the Amundsen Sea sector of West Antarctica: new insights from temperature measurements, depth to the bottom of the magnetic source estimation, and thermal modeling. *Geochemistry, Geophysics, Geosystems*, **18**, 2657–2672, https://doi.org/10.1002/2016GC006755

Foley, S.F., Andronikov, A.V., Jacob, D.E. and Melzer, S. 2006. Evidence from Antarctic mantle peridotite xenoliths for changes in mineralogy, geochemistry and geothermal gradients beneath a developing rift. *Geochimica et Cosmochimica Acta*, **70**, 3096–3120, https://doi.org/10.1016/j.gca.2006.03.010

Foley, S.F., Andronikov, A.V., Halpin, J.A., Daczko, N.R. and Jacob, D.E. 2021. Mantle rocks in East Antarctica. *Geological Society, London, Memoirs*, **56**, https://doi.org/10.1144/M56-2020-8

Gaete, A.V.V. 2017. *Petrografía y geoquímica de xenolitos mantélicos de las Isla James Ross, Antártica*. PhD thesis, Universidad Andres Bello, Santiago, Chile.

Gamble, J.A. and Kyle, P.R. 1987. The origins of glass and amphibole in spinel-wehrlite xenoliths from Foster Crater, McMurdo Volcanic Group, Antarctica. *Journal of Petrology*, **28**, 755–779, https://doi.org/10.1093/petrology/28.5.755

Gamble, J.A., McGibbon, F., Kyle, P.R., Menzies, M. and Kirsch, I. 1988. Metasomatised xenoliths from Foster Crater, Antarctica: implications for lithospheric structure and process beneath the Transantarctic Mountain Front. *Journal of Petrology*, **1**, 109–138. Special Volume, https://doi.org/10.1093/petrology/Special_Volume.1.109

Geruo, A., Wahr, J. and Zhong, S. 2013. Computations of the viscoelastic response of a 3-D compressible earth to surface loading: an application to glacial isostatic adjustment in Antarctica and Canada. *Geophysical Journal International*, **192**, 557–572, https://doi.org/10.1093/gji/ggs030

Giacomoni, P.P., Bonadiman, C. *et al.* 2020. Long-term storage of subduction-related volatiles in northern Victoria Land lithospheric mantle: insight from olivine-hosted melt inclusions from McMurdo basic lavas (Antarctica). *Lithos*, **378–379**, 105826, https://doi.org/10.1016/j.lithos.2020.105826

Gibson, S.A., Rooks, E.E., Day, J.A., Petrone, C.M. and Leat, P.T. 2020. The role of sub-continental mantle as both 'sink' and 'source' in deep earth volatile cycles. *Geochimica et Cosmochimica Acta*, **275**, 140–162, https://doi.org/10.1016/j.gca.2020.02.018

Glaser, S.M., Foley, S.F. and Günther, D. 1999. Trace element compositions of minerals in garnet and spinel peridotite xenoliths from the Vitim volcanic field, Transbaikalia, eastern Siberia. *In*: Hilst, R.D.V.D. and McDonough, W.F. (eds) *Developments in Geotectonics*. Elsevier, 263–285.

Gradmann, S., Ebbing, J. and Fullea, J. 2013. Integrated geophysical modelling of a lateral transition zone in the lithospheric mantle under Norway and Sweden. *Geophysical Journal International*, **194**, 1358–1373, https://doi.org/10.1093/gji/ggt213

Grégoire, M., Moine, B.N., O'Reilly, S.Y., Cottin, J.Y. and Giret, A. 2000. Trace element residence and partitioning in mantle xenoliths metasomatized by highly alkaline, silicate- and carbonate-rich melts (Kerguelen Islands, Indian Ocean). *Journal of Petrology*, **41**, 477–509, https://doi.org/10.1093/petrology/41.4.477

Handler, M.R., Wysoczanski, R.J. and Gamble, J.A. 2003. Proterozoic lithosphere in Marie Byrd Land, West Antarctica: Re–Os systematics of spinel peridotite xenoliths. *Chemical Geology*, **196**, 131–145, https://doi.org/10.1016/S0009-2541(02)00410-2

Handler, M.R., Wysoczanski, R.J. and Gamble, J.A. 2021. Marie Byrd Land lithospheric mantle: a review of the xenolith record. *Geological Society, London, Memoirs*, **56**, https://doi.org/10.1144/M56-2020-17

Harte, B. 1977. Rock nomenclature with particular relation to deformation and recrystallization textures in olivine-bearing xenoliths. *Journal of Geology*, **85**, 279–288, https://doi.org/10.1086/628299

Hasterok, D. and Chapman, D.S. 2011. Heat production and geotherms for the continental lithosphere. *Earth and Planetary*

Science Letters, **307**, 59–70, https://doi.org/10.1016/j.epsl.2011.04.034

Herzberg, C. 2004. Geodynamic information in peridotite petrology. *Journal of Petrology*, **45**, 2507–2530, https://doi.org/10.1093/petrology/egh039

Hirschmann, M.M. 2006. Water, melting, and the deep earth H_2O cycle. *Annual Review of Earth and Planetary Sciences*, **34**, 629–653, https://doi.org/10.1146/annurev.earth.34.031405.125211

Hirth, G. and Kohlstedt, D. 2003. Rheology of the upper mantle and the mantle wedge: a view from the experimentalists. *AGU Geophysical Monographs*, **138**, 83–105, https://doi.org/10.1029/138GM06

Ivins, E.R., James, T.S., Wahr, J.O., Schrama, E.J., Landerer, F.W. and Simon, K.M. 2013. Antarctic contribution to sea level rise observed by GRACE with improved GIA correction. *Journal of Geophysical Research: Solid Earth*, **118**, 3126–3141, https://doi.org/10.1002/jgrb.50208

Jagoutz, E., Palme, H. *et al.* 1979. The abundances of major, minor and trace elements in the earth's mantle as derived from primitive ultramafic nodules. *Lunar and Planetary Science Conference Proceedings*, Houston, TX, 19–23 March, 2031–2050.

Karato, S.-I. 2004. Mapping water content in the upper mantle, *AGU Geophysical Monographs*, **138**, 135–152, https://doi.org/10.1029/138GM08

Kempton, P.D., Lopez-Escobar, L., Hawkesworth, C.J., Pearson, G., Wright, D.W. and Ware, A.J. 1999. Spinel ± garnet lherzolite xenoliths from Pali-Aike, part 1: petrography, mineral chemistry and geothermobarometry. *In*: Dawson, J.B., Gurney, J.J., Pascoe, M.D. and Richardson, S.H. (eds) *Proceedings of the Seventh International Kimberlite Conference*. Red Roof, Cape Town, 403–414.

Kyle, P.R., Wright, A.C. and Kirsch, I. 1987. Ultramafic xenoliths in the late Cenozoic McMurdo Volcanic Group, Western Ross Sea embayment, Antarctica. *In*: Nixon, P.H. (ed.) *Mantle Xenoliths*. John Wiley & Sons, 287–293.

Leat, P.T., Ross, A.J. and Gibson, S.A. 2021. Ultramafic mantle xenoliths in the late Cenozoic volcanic rocks of the Antarctic peninsula and Jones Mountains, west Antarctica. *Geological Society, London, Memoirs*, **56**, https://doi.org/10.1144/M56-2019-44

Lloyd, A., Wiens, D. *et al.* 2019. Seismic structure of the Antarctic upper mantle imaged with adjoint tomography. *Journal of Geophysical Research: Solid Earth*, **125**, https://doi.org/10.1029/2019JB017823

Markl, G., Abart, R., Vennemann, T. and Sommer, H. 2003. Mid-crustal metasomatic reaction veins in a spinel peridotite. *Journal of Petrology*, **44**, 1097–1120, https://doi.org/10.1093/petrology/44.6.1097

Martin, A.P., Cooper, A.F. and Price, R.C. 2014*a*. Increased mantle heat flow with on-going rifting of the West Antarctic Rift System inferred from characterisation of plagioclase peridotite in the shallow Antarctic mantle. *Lithos*, **190–191**, 173–190, https://doi.org/10.1016/j.lithos.2013.12.012

Martin, A.P., Price, R.C. and Cooper, A.F. 2014*b*. Constraints on the composition, source and petrogenesis of plagioclase-bearing mantle peridotite. *Earth-Science Reviews*, **138**, 89–101, https://doi.org/10.1016/j.earscirev.2014.08.006

Martin, A.P., Price, R.C., Cooper, A.F. and McCammon, C.A. 2015. Petrogenesis of the rifted southern Victoria Land lithospheric mantle, Antarctica, inferred from petrography, geochemistry, thermobarometry and oxybarometry of peridotite and pyroxenite xenoliths from the Mount Morning eruptive centre. *Journal of Petrology*, **56**, 193–226, https://doi.org/10.1093/petrology/egu075

Martin, A.P., Cooper, A.F., Price, R.C., Doherty, C.L. and Gamble, J.A. 2021*a*. A review of mantle xenoliths in volcanic rocks from southern Victoria Land, Antarctica. *Geological Society, London, Memoirs*, **56**, https://doi.org/10.1144/M56-2019-42

Martin, A.P., Cooper, A.F., Price, R.C., Kyle, P.R. and Gamble, J.A. 2021*b*. Erebus Volcanic Province: petrology. *Geological Society, London, Memoirs*, **55**, 447–489, https://doi.org/10.1144/M55-2018-80

Martin, A.P. and van der Wal, W. 2022. Introduction to the geochemistry and geophysics of the Antarctic mantle. *Geological Society, London, Memoirs*, **56**, in press.

Martos, Y.M., Catalán, M., Jordan, T.A., Golynsky, A., Golynsky, D., Eagles, G. and Vaughan, D.G. 2017. Heat flux distribution of Antarctica unveiled. *Geophysical Research Letters*, **44**, 11,417–11,426, https://doi.org/10.1002/2017GL075609

Marty, B. 2012. The origins and concentrations of water, carbon, nitrogen and noble gases on earth. *Earth and Planetary Science Letters*, **313–314**, 56–66, https://doi.org/10.1016/j.epsl.2011.10.040

Maule, C.F., Purucker, M.E., Olsen, N. and Mosegaard, K. 2005. Heat flux anomalies in Antarctica revealed by satellite magnetic data. *Science (New York, NY)*, **309**, 464–467, https://doi.org/10.1126/science.1106888

McDonough, W.F. and Sun, S.-S. 1995. The composition of the earth. *Chemical Geology*, **120**, 223–253, https://doi.org/10.1016/0009-2541(94)00140-4

Melchiorre, M., Coltorti, M., Bonadiman, C., Faccini, B., O'Reilly, S.Y. and Pearson, N.J. 2011. The role of eclogite in the rift-related metasomatism and Cenozoic magmatism of northern Victoria Land, Antarctica. *Lithos*, **124**, 319–330, https://doi.org/10.1016/j.lithos.2010.11.012

Nixon, P.H. 1987*a*. Indian–Australian and Antartic plates – introduction. *In*: Nixon, P.H. (ed.) *Mantle Xenoliths*. Wiley, New York, 241–248.

Nixon, P.H. 1987*b*. *Mantle Xenoliths*. John Wiley & Sons, Great Britain.

O'Relly, S.Y. and Griffin, W.L. 1985. A xenolith-derived geotherm for southeastern Australia and its geophysical implications. *Tectonophysics*, **111**, 41–63, https://doi.org/10.1016/0040-1951(85)90065-4

O'Reilly, S.Y. and Griffin, W.L. 2006. Imaging global chemical and thermal heterogeneity in the subcontinental lithospheric mantle with garnets and xenoliths: geophysical implications. *Tectonophysics*, **416**, 289–309, https://doi.org/10.1016/j.tecto.2005.11.014

O'Reilly, S.Y. and Griffin, W.L. 2013. Mantle metasomatism. *In*: *Metasomatism and the Chemical Transformation of Rock*. Lecture Notes in Earth System Sciences. Springer, Berlin, https://doi.org/10.1007/978-3-642-28394-9_12

O'Reilly, S.Y., Griffin, W.L., Poudjom, Y.H. and Morgan, P. 2001. Are lithosphere forever? tracking changes in subcontinental lithospheric mantle through time. *GSA Today*, **11**, 4–10, https://doi.org/10.1130/1052-5173(2001)011<0004:ALFTCI>2.0.CO;2

Ohtani, E. 2020. The role of water in earth's mantle. *National Science Review*, **7**, 224–232, https://doi.org/10.1093/nsr/nwz071

Palme, H. and O'Neill, H. St. C. 2003. Cosmochemical estimates of mantle composition. *In*: Holland, H.D. and Turrekian, K.K. (eds) *Treatise on Geochemistry*, Elsevier, 1–38, https://doi.org/10.1016/B0-08-043751-6/02177-0

Panter, K.S. and Martin, A.P. 2021. West Antarctic mantle deduced from mafic magmatism. *Geological Society, London, Memoirs*, **56**, https://doi.org/10.1144/M56-2021-10

Pappa, F. and Ebbing, J. 2021. Gravity, magnetics and geothermal heat flow of the Antarctic lithospheric crust and mantle. *Geological Society, London, Memoirs*, **56**, https://doi.org/10.1144/M56-2020-5

Pearson, D., Canil, D. and Shirey, S. 2003. Mantle samples included in volcanic rocks: xenoliths and diamonds. *Treatise on Geochemistry*, **2**, 568, https://doi.org/10.1016/B0-08-043751-6/02005-3

Perinelli, C., Armienti, P. and Dallai, L. 2006. Geochemical and O-isotope constraints on the evolution of lithospheric mantle in the Ross Sea rift area (Antarctica). *Contributions to Mineralogy and Petrology*, **151**, 245–266, https://doi.org/10.1007/s00410-006-0065-8

Perinelli, C., Andreozzi, G., Conte, A., Oberti, R. and Armienti, P. 2012. Redox state of subcontinental lithospheric mantle and relationships with metasomatism: insights from spinel peridotites from northern Victoria Land (Antarctica). *Contributions to Mineralogy and Petrology*, **164**, 1053–1067, https://doi.org/10.1007/s00410-012-0788-7

Poudjom Djomani, Y.H., O'Reilly, S.Y., Griffin, W.L. and Morgan, P. 2001. The density structure of subcontinental lithosphere through time. *Earth and Planetary Science Letters*, **184**, 605–621, https://doi.org/10.1016/S0012-821X(00)00362-9

Ranalli, G. 1995. *Rheology of the Earth*, 2nd edn. Chapman & Hall, Great Britain.

Rocchi, S. and Smellie, J.L. 2021. Northern Victoria Land: petrology. *Geological Society, London, Memoirs*, **55**, 383–413, https://doi.org/10.1144/M55-2019-19

Schroeder, D.M., Blankenship, D.D., Young, D.A. and Quartini, E., 2014. Evidence for elevated and spatially variable geothermal flux beneath the West Antarctic Ice Sheet. *Proceedings of the National Academy of Sciences*, **111**, 9070, https://doi.org/10.1073/pnas.1405184111

Scott, J.M., Waight, T.E., van der Meer, Q.H.A., Palin, J.M., Cooper, A.F. and Münker, C. 2014. Metasomatized ancient lithospheric mantle beneath the young Zealandia microcontinent and its role in HIMU-like intraplate magmatism. *Geochemistry, Geophysics, Geosystems*, **15**, https://doi.org/10.1002/2014GC005300

Scott, J.M., Liu, J. *et al.* 2019. Continent stabilisation by lateral accretion of subduction zone-processed depleted mantle residues; insights from Zealandia. *Earth and Planetary Science Letters*, **507**, 175–186, https://doi.org/10.1016/j.epsl.2018.11.039

Shapiro, N.M. and Ritzwoller, M.H. 2004. Inferring surface heat flux distributions guided by a global seismic model: particular application to Antarctica. *Earth and Planetary Science Letters*, **223**, 213–224, https://doi.org/10.1016/j.epsl.2004.04.011

Shen, W., Wiens, D.A., Lloyd, A.J. and Nyblade, A.A. 2020. A geothermal heat flux map of Antarctica empirically constrained by seismic structure. *Geophysical Research Letters*, **47**, e2020GL086955, https://doi.org/10.1029/2020GL086955

Smellie, J.L., Panter, K.S. and Geyer, A. (eds) 2021. *Volcanism in Antarctica: 200 Million Years of Subduction, Rifting and Continental Break-up*. Geological Society, London, Memoirs, **55**, https://doi.org/10.1144/M55

Stål, T., Reading, A.M., Halpin, J.A., Phipps, S.J. and Whittaker, J.M. 2020. The Antarctic crust and upper mantle: a flexible 3D model and software framework for interdisciplinary research. *Frontiers in Earth Science*, **8**, https://doi.org/10.3389/feart.2020.577502

Stolz, A.J. and Davies, G.R. 1988. Chemical and isotopic evidence from spinel lherzolite xenoliths for episodic metasomatism of the upper mantle beneath southeast Australia. *Journal of Petrology*, Special Volume, 303–330, https://doi.org/10.1093/petrology/Special_Volume.1.303

Storey, B. and Kyle, P. 1997. An active mantle mechanism for Gondwana breakup. *South African Journal of Geology*, **100**, 283–290.

ten Brink, U.S., Hackney, R.I., Bannister, S., Stern, T.A. and Makovsky, Y. 1997. Uplift of the Transantarctic Mountains and the bedrock beneath the East Antarctic Ice Sheet. *Journal of Geophysical Research: Solid Earth*, **102**, 27603–27621, https://doi.org/10.1029/97JB02483

van der Wal, W., Whitehouse, P.L. and Schrama, E.J.O. 2015. Effect of GIA models with 3D composite mantle viscosity on GRACE mass balance estimates for Antarctica. *Earth and Planetary Science Letters*, **414**, 134–143, https://doi.org/10.1016/j.epsl.2015.01.001

Walter, M.J. 2003. 2.08 – melt extraction and compositional variability in mantle lithosphere. *In*: Holland, H.D. and Turekian, K.K. (eds) *Treatise on Geochemistry*. Pergamon, Oxford, 363–394.

Whitehouse, P.L., Bentley, M.J., Milne, G.A., King, M.A. and Thomas, I.D. 2012. A new glacial isostatic adjustment model for Antarctica: calibrated and tested using observations of relative sea-level change and present-day uplift rates. *Geophysical Journal International*, **190**, 1464–1482, https://doi.org/10.1111/j.1365-246X.2012.05557.x

Wiens, D.A., Shen, W. and Lloyd, A.J. 2021. The seismic structure of the Antarctic upper mantle. *Geological Society, London, Memoirs*, **56**, https://doi.org/10.1144/M56-2020-18

Wörner, G. and Zipfel, J. 1996. A mantle P–T path for the Ross Sea rift margin (Antarctica) derived from Ca-in-olivine zonation patterns in peridotites xenoliths of the Plio-Pleistocene Mt. Melbourne Volcanic Field. *Geologisches Jahrbuch*, **B89**, 157–167.

Zindler, A. and Hart, S. 1986. Chemical geodynamics. *Annual Review of Earth and Planetary Sciences*, **14**, 493–571, https://doi.org/10.1146/annurev.ea.14.050186.002425

Zipfel, J. and Wörner, G. 1992. Four- and five-phase peridotites from a continental rift system: evidence for upper mantle uplift and cooling at the Ross Sea margin (Antarctica). *Contributions to Mineralogy and Petrology*, **111**, 24–36, https://doi.org/10.1007/BF00296575

Index

Page numbers in *italic* refer to Figures. Page numbers in **bold** refer to Tables.